Mathematical Thought
from Ancient to Modern Times

Mathematical Thought
from Ancient to Modern Times

MORRIS KLINE

Professor of Mathematics
Courant Institute of Mathematical Sciences
New York University

New York   OXFORD UNIVERSITY PRESS   1972

Copyright © 1972 by Morris Kline
Library of Congress Catalogue Card Number 77–170263
Printed in the United States of America

To my wife, Helen Mann Kline

Preface

If we wish to foresee the future of mathematics our proper
course is to study the history and present condition of the
science.   HENRI POINCARÉ

This book treats the major mathematical creations and developments
from ancient times through the first few decades of the twentieth century.
It aims to present the central ideas, with particular emphasis on those
currents of activity that have loomed largest in the main periods of the life
of mathematics and have been influential in promoting and shaping subsequent mathematical activity. The very concept of mathematics, the
changes in that concept in different periods, and the mathematicians' own
understanding of what they were achieving have also been vital concerns.

This work must be regarded as a survey of the history. When one
considers that Euler's works fill some seventy volumes, Cauchy's twenty-six
volumes, and Gauss's twelve volumes, one can readily appreciate that a
one-volume work cannot present a full account. Some chapters of this work
present only samples of what has been created in the areas involved, though
I trust that these samples are the most representative ones. Moreover, in
citing theorems or results, I have often omitted minor conditions required for
strict correctness in order to keep the main ideas in focus. Restricted as this
work may be, I believe that some perspective on the entire history has been
presented.

The book's organization emphasizes the leading mathematical themes
rather than the men. Every branch of mathematics bears the stamp of its
founders, and great men have played decisive roles in determining the course
of mathematics. But it is their ideas that have been featured; biography is
entirely subordinate. In this respect, I have followed the advice of Pascal:
"When we cite authors we cite their demonstrations, not their names."

To achieve coherence, particularly in the period after 1700, I have
treated each development at that stage where it became mature, prominent,
and influential in the mathematical realm. Thus non-Euclidean geometry is
presented in the nineteenth century even though the history of the efforts to

replace or prove the Euclidean parallel axiom date from Euclid's time onward. Of course, many topics recur at various periods.

To keep the material within bounds I have ignored several civilizations such as the Chinese,[1] Japanese, and Mayan because their work had no material impact on the main line of mathematical thought. Also some developments in mathematics, such as the theory of probability and the calculus of finite differences, which are important today, did not play major roles during the period covered and have accordingly received very little attention. The vast expansion of the last few decades has obliged me to include only those creations of the twentieth century that became significant in that period. To continue into the twentieth century the extensions of such subjects as ordinary differential equations or the calculus of variations would call for highly specialized material of interest only to research men in those fields and would have added inordinately to the size of the work. Beyond these considerations, the importance of many of the more recent developments cannot be evaluated objectively at this time. The history of mathematics teaches us that many subjects which aroused tremendous enthusiasm and engaged the attention of the best mathematicians ultimately faded into oblivion. One has but to recall Cayley's dictum that projective geometry is all geometry, and Sylvester's assertion that the theory of algebraic invariants summed up all that is valuable in mathematics. Indeed one of the interesting questions that the history answers is what survives in mathematics. History makes its own and sounder evaluations.

Readers of even a basic account of the dozens of major developments cannot be expected to know the substance of all these developments. Hence except for some very elementary areas the contents of the subjects whose history is being treated are also described, thus fusing exposition with history. These explanations of the various creations may not clarify them completely but should give some idea of their nature. Consequently this book may serve to some extent as a historical introduction to mathematics. This approach is certainly one of the best ways to acquire understanding and appreciation.

I hope that this work will be helpful to professional and prospective mathematicians. The professional man is obliged today to devote so much of his time and energy to his specialty that he has little opportunity to familiarize himself with the history of his subject. Yet this background is important. The roots of the present lie deep in the past and almost nothing in that past is irrelevant to the man who seeks to understand how the present came to be what it is. Moreover, mathematics, despite the proliferation into hundreds of branches, is a unity and has its major problems and goals. Unless the various specialties contribute to the heart of mathematics they are likely to be

1. A fine account of the history of Chinese mathematics is available in Joseph Needham's *Science and Civilization in China*, Cambridge University Press, 1959, Vol. 3, pp. 1–168.

sterile. Perhaps the surest way to combat the dangers which beset our fragmented subject is to acquire some knowledge of the past achievements, traditions, and objectives of mathematics so that one can direct his research into fruitful channels. As Hilbert put it, "Mathematics is an organism for whose vital strength the indissoluble union of the parts is a necessary condition."

For students of mathematics this work may have other values. The usual courses present segments of mathematics that seem to have little relationship to each other. The history may give perspective on the entire subject and relate the subject matter of the courses not only to each other but also to the main body of mathematical thought.

The usual courses in mathematics are also deceptive in a basic respect. They give an organized logical presentation which leaves the impression that mathematicians go from theorem to theorem almost naturally, that mathematicians can master any difficulty, and that the subjects are completely thrashed out and settled. The succession of theorems overwhelms the student, especially if he is just learning the subject.

The history, by contrast, teaches us that the development of a subject is made bit by bit with results coming from various directions. We learn, too, that often decades and even hundreds of years of effort were required before significant steps could be made. In place of the impression that the subjects are completely thrashed out one finds that what is attained is often but a start, that many gaps have to be filled, or that the really important extensions remain to be created.

The polished presentations in the courses fail to show the struggles of the creative process, the frustrations, and the long arduous road mathematicians must travel to attain a sizable structure. Once aware of this, the student will not only gain insight but derive courage to pursue tenaciously his own problems and not be dismayed by the incompleteness or deficiencies in his own work. Indeed the account of how mathematicians stumbled, groped their way through obscurities, and arrived piecemeal at their results should give heart to any tyro in research.

To cover the large area which this work comprises I have tried to select the most reliable sources. In the pre-calculus period these sources, such as T. L. Heath's *A History of Greek Mathematics*, are admittedly secondary, though I have not relied on just one such source. For the subsequent development it has usually been possible to go directly to the original papers, which fortunately can be found in the journals or in the collected works of the prominent mathematicians. I have also been aided by numerous accounts and surveys of research, some in fact to be found in the collected works. I have tried to give references for all of the major results; but to do so for all assertions would have meant a mass of references and the consumption of space that is better devoted to the account itself.

The sources have been indicated in the bibliographies of the various chapters. The interested reader can obtain much more information from these sources than I have extracted. These bibliographies also contain many references which should not and did not serve as sources. However, they have been included either because they offer additional information, because the level of presentation may be helpful to some readers, or because they may be more accessible than the original sources.

I wish to express thanks to my colleagues Martin Burrow, Bruce Chandler, Martin Davis, Donald Ludwig, Wilhelm Magnus, Carlos Moreno, Harold N. Shapiro, and Marvin Tretkoff, who answered numerous questions, read many chapters, and gave valuable criticisms. I am especially indebted to my wife Helen for her critical editing of the manuscript, extensive checking of names, dates, and sources, and most careful reading of the galleys and page proofs. Mrs. Eleanore M. Gross, who did the bulk of the typing, was enormously helpful. To the staff of Oxford University Press, I wish to express my gratitude for their scrupulous production of this work.

New York   M. K.
May 1972

Contents

15. Coordinate Geometry, 302

16. The Mathematization of Science, 325

17. The Creation of the Calculus, 342

18. Mathematics as of 1700, 391

19. Calculus in the Eighteenth Century, 400

20. Infinite Series, 436

21. Ordinary Differential Equations in the Eighteenth Century, 468

22. Partial Differential Equations in the Eighteenth Century, 502

Mathematical Thought
from Ancient to Modern Times

I
Mathematics in Mesopotamia

1. *Where Did Mathematics Begin?*

Mathematics as an organized, independent, and reasoned discipline did not exist before the classical Greeks of the period from 600 to 300 B.C. entered upon the scene. There were, however, prior civilizations in which the beginnings or rudiments of mathematics were created. Many of these primitive civilizations did not get beyond distinguishing among one, two, and many; others possessed, and were able to operate with, large whole numbers. Still others achieved the recognition of numbers as abstract concepts, the adoption of special words for the individual numbers, the introduction of symbols for numbers, and even the use of a base such as ten, twenty, or five to denote a larger unit of quantity. One also finds the four operations of arithmetic, but confined to small numbers, and the concept of a fraction, limited, however, to 1/2, 1/3, and the like, and expressed in words. In addition, the simplest geometric notions, line, circle, and angle, were recognized. It is perhaps of interest that the concept of angle must have arisen from observation of the angle formed by man's thigh and lower leg or his forearm and upper arm because in most languages the word for the side of an angle is either the word for leg or the word for arm. In English, for example, we speak of the arms of a right triangle. The uses of mathematics in these primitive civilizations were limited to simple trading, the crude calculation of areas of fields, geometric decoration on pottery, patterns woven into cloth, and the recording of time.

Until we reach the mathematics of the Babylonians and the Egyptians of about 3000 B.C. we do not find more advanced steps in mathematics. Since primitive peoples settled down in one area, built homes, and relied upon agriculture and animal husbandry as far back as 10,000 B.C., we see how slowly the most elementary mathematics made its first steps; moreover, the existence of vast numbers of civilizations with no mathematics to speak of shows how sparsely this science was cultivated.

3

2. *Political History in Mesopotamia*

The Babylonians were the first of these two early civilizations to contribute to the main course of mathematics. Since our knowledge of the ancient civilizations of the Near East and of Babylonia in particular is largely the product of archaeological research of the last hundred years, this knowledge is incomplete and subject to correction as new discoveries are made. The term "Babylonian" covers a series of peoples, who concurrently or successively occupied the area around and between the Tigris and Euphrates rivers, a region known as Mesopotamia and now part of modern Iraq. These peoples lived in independent cities such as Babylon, Ur, Nippur, Susa, Aššur, Uruk, Lagash, Kish, and others. About 4000 B.C. the Sumerians, racially distinct from the Semitic and Indo-Germanic peoples, settled down in part of Mesopotamia. Their capital city was Ur, and the land area they controlled was called Sumer. Though their culture reached its height about 2250 B.C., even before this, about 2500 B.C., the Sumerians came under the political control of the Akkadians, a Semitic people whose major city was Akkad and who were led at that time by the ruler Sargon. Sumerian civilization was submerged in the Akkadian. A period of high culture occurred during the reign of King Hammurabi (about 1700 B.C.), who is known as the formulator of a famous code of law.

Around 1000 B.C. migrations and the introduction of iron resulted in further changes. Later, in the eighth century B.C., the area was ruled by the Assyrians, who settled primarily in the upper Tigris region. To our knowledge, they did not add anything new to the culture. A century later the Assyrian empire was shared by the Chaldeans and the Medes, the latter being close racially to the Persians further east. This period in Mesopotamian history (7th cent. B.C.) is often referred to as Chaldean. The Near East was conquered by the Persians under Cyrus about 540 B.C. Persian mathematicians such as Nabu-rimanni (*c.* 490 B.C.) and Kidinu (*c.* 480 B.C.) became known to the Greeks.

In 330 B.C., Alexander the Great, the Greek military leader, conquered Mesopotamia. The period from 300 B.C. to the birth of Christ is called Seleucid after the Greek general who first took control of the region following the death of Alexander in 323 B.C. However, the flowering of Greek mathematics had already taken place, and from the time of Alexander until the seventh century A.D., when the Arabs arrived on the scene, the Greek influence predominated in the Near East. Most of what the Babylonians contributed to mathematics predates the Seleucid period.

Despite the numerous changes in the rulers of Mesopotamia, there was, in mathematics, a continuity of knowledge, tradition, and practice from ancient times at least to the time of Alexander.

3. *The Number Symbols*

Our main information about the civilization and mathematics of Babylonia, ancient and more recent, comes from texts in the form of clay tablets. These tablets were inscribed when the clay was still soft and then baked. Hence those that survived destruction are well preserved. They date mainly from two periods, some from about 2000 B.C., and a larger number from the period 600 B.C.–A.D. 300. The earlier tablets are the more important in the history of mathematics.

The language and script of the older tablets is Akkadian, which was superimposed on the older Sumerian language and script. The words of the Akkadian language consisted of one or more syllables; each syllable was represented by a collection of what are essentially line segments. The Akkadians used a stylus with a triangular cross-section held at an angle to the clay, which produced wedge-shaped impressions that could be oriented in different ways. From *cuneus*, the Latin word for "wedge," the script became known as cuneiform.

The most highly developed arithmetic of the Babylonian civilization is the Akkadian. Whole numbers were written as follows:

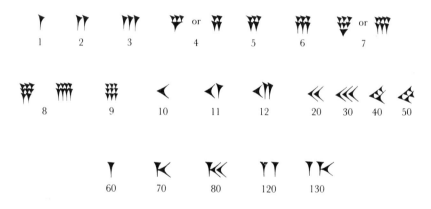

The striking features of the Babylonian number system are the base 60 and positional notation.

At first the Babylonians had no symbol to indicate the absence of a number in any one position, and consequently their numbers were ambiguous. Thus 𝕂𝕂 could mean 80 or 3620, depending upon whether the first symbol meant 60 or 3600. Spacing was often used to indicate no quantity in a given position, but of course this might be misinterpreted. In the

Seleucid period a special separation symbol was introduced to indicate the absence of a number. Thus 𒑖𒑚 $= 1{\cdot}60^2 + 0{\cdot}60 + 4 = 3604$. However, even in this period no symbol was used to indicate the absence of a quantity at the right-hand end as in our 20. In both periods the content had to be relied upon to indicate the precise value of the entire number.

The Babylonians used positional notation to represent fractions also. Thus 𒌋𒌋 intended as a fraction meant 20/60 and 𒌋𒌋𒁹 , intended as a fraction, could be 21/60 or 20/60 + 1/60². The ambiguity in their number system is therefore even greater than previously indicated.

A few fractions had special symbols. Thus one finds:

$$\underset{\frac{1}{2}}{\text{𒈦}} \qquad \underset{\frac{1}{3}}{\text{𒑚}} \qquad \underset{\frac{2}{3}}{\text{𒑚}}$$

These special fractions, 1/2, 1/3, and 2/3, were to the Babylonians "wholes," in the sense of measures of quantities, and not a division of unity into parts, though they arose as measures of quantity having these respective relations to another quantity. Thus we might write 10 cents as 1/10 in relation to the dollar but think of the 1/10 as a unit in itself.

Actually the Babylonians did not use base 60 exclusively. Sometimes the years were written 2 *me* 25, where *me* stood for hundred—in our symbols, 225. Also *limu* was used for 1000, generally in nonmathematical texts, though it occurs even in mathematical texts of the Seleucid period. Sometimes 10 and 60 were mixed, as in 2 *me* 1, 10, which means $2 \times 100 + 1 \times 60 + 10$ or 270. Mixed systems involving units of 60, 24, 12, 10, 6, and 2 were used for dates, areas, measures of weight, and coinage, just as we use 12 for hours, 60 for minutes and seconds, 12 for inches, and 10 for ordinary counting. The Babylonian systems, like ours, were a composite of many historical and regional customs. However, in the mathematical and astronomical texts they did use base 60 quite consistently.

We do not know definitely how base 60 came to be used. The suggestion may have come from the systems of weight measures. Suppose we have a system of weight measures with values in it having the ratios

$$1/2,\ 1/3,\ 2/3,\ 1,\ 10.$$

Now suppose there is another system with a different unit but the same ratios, and political or social forces compel the fusing of the two systems. (We have meters and yards, for example.) If the larger unit were 60 times the smaller, then 1/2, 1/3, and 2/3 of the larger unit would be integral multiples of the smaller one. Thus the larger unit might have been adopted because it was so convenient.

As to the origin of positional notation, there are two likely explanations. In an older scheme of writing numbers 1 multiplied by 60 was written with a larger 𒁹 than the same symbol for 1. When the writing was simplified the larger 𒁹 was reduced in size but kept in the usual place for 60. Hence position became the denotation of a multiple of 60. Another possible explanation comes from the coinage system. One talent and 10 mana may have been written $\blacktriangleleft$, where the $\mathbf{I}$ meant 1 talent, which is 60 mana. We do the same when we write $1.20 and mean 100 cents by the 1. The scheme for writing amounts of money may then have been taken over to arithmetic generally.

4. Arithmetic Operations

In the Babylonian system the symbols for 1 and 10 were basic. Numbers from 1 to 59 were formed by combining fewer or more of these symbols. Hence the processes of addition and subtraction were merely a matter of adding or taking away symbols. To indicate addition the Babylonians joined numbers together as in $\blacktriangleleft$, which indicates 16. Subtraction was often indicated by the symbol $\mathbf{I}$. Thus $\blacktriangleleft$ is 40 − 3. In astronomical texts of a later period the word *tab* appears, signifying addition.

Multiplication of integers was also performed. To multiply by 37, say, would mean multiplying by 30, then by 7, and adding the results. The symbol for multiplication was $\mathbf{I}$, pronounced *a-rá*. It meant "to go."

The Babylonians divided one whole number by another. Since to divide by an integer a is to multiply by the reciprocal $1/a$, to this extent fractions were involved. The Babylonians converted the reciprocals to sexagesimal "decimals" and, except for the few mentioned above, did not use special symbols for fractions. They had tables showing how numbers of the form $1/a$, where $a = 2^\alpha 3^\beta 5^\gamma$, could be written as terminating sexagesimal numbers. Some tables gave approximate values for $1/7$, $1/11$, $1/13$, etc., because these fractions led to infinite repeating sexagesimals. Where fractions involving denominators other than 2, 3, or 5 occurred in the older problems, the same troublesome factors occurred in the numerator and were cancelled.

The Babylonians relied entirely upon the tables of reciprocals. Their tables show, for example,

igi 2 *gál-bi* 30	*igi* 8 *gál-bi* 7, 30
igi 3 *gál-bi* 20	*igi* 9 *gál-bi* 6, 40
igi 4 *gál-bi* 15	. . . .
igi 6 *gál-bi* 10	*igi* 27 *gál-bi* 2, 13, 20

obviously meaning $1/2 = 30/60$, $1/3 = 20/60$, etc. The precise meanings of *igi* and *gál-bi* are not known. Sexagesimal fractions, that is, numbers less than 1, expressed in inverse powers of 60, 60^2, etc., but with denominators merely understood, continued to be used by the Greeks Hipparchus and Ptolemy and in Renaissance Europe up to the sixteenth century, when they were replaced by decimals in base 10.

The Babylonians also had tables expressing squares, square roots, cubes, and cube roots. When the root was a whole number it was given exactly. For other roots, the corresponding sexagesimal numbers were only approximate. Of course, irrationals cannot be expressed in a finite number of decimals or sexagesimals. However, there is no evidence that the Babylonians were aware of this. They might well have believed that the irrationals could be converted exactly to sexagesimal numbers if more places were used. An excellent Babylonian approximation to $\sqrt{2}$ gives $\sqrt{2} = 1.414213\ldots$ instead of $1.414214\ldots$.

Roots occur in their calculations of the diagonal d of a rectangle of height h and width w. One problem asks for the diagonal of a rectangular gate with given height and width. The answer is given without explanation and amounts to using the approximate formula for the diagonal d, namely,

$$d \approx h + \frac{w^2}{2h}.$$

The formula is a good approximation to d for $h > w$. Thus for $h > w$, as is the case in one problem, we can see that the answer is reasonable by noting that

$$d = \sqrt{h^2 + w^2} = h\sqrt{1 + \frac{w^2}{h^2}} = h\left(1 + \frac{w^2}{h^2}\right)^{1/2}.$$

If we now expand the binomial and retain only the first two terms we get the formula. Other approximate answers to square-root problems were given, obtained presumably by using numbers in Babylonian tables.

5. *Babylonian Algebra*

Distinct from the table texts, which yield much information on the Babylonian number system and operations with numbers, are texts that deal with algebraic and geometric problems. A fundamental problem of the older Babylonian algebra asks for a number which, added to its reciprocal, yields a given number. In modern notation, the Babylonians sought x and $\bar{x}$ such that

$$x\bar{x} = 1, \qquad x + \bar{x} = b.$$

These two equations yield a quadratic equation in x, namely, $x^2 - bx + 1 = 0$. They formed $\left(\frac{b}{2}\right)^2$; then $\sqrt{\left(\frac{b}{2}\right)^2 - 1}$; and then

$$\frac{b}{2} + \sqrt{\left(\frac{b}{2}\right)^2 - 1} \quad \text{and} \quad \frac{b}{2} - \sqrt{\left(\frac{b}{2}\right)^2 - 1},$$

which yield the answers. In effect, the Babylonians had the quadratic formula. Other problems, such as finding two numbers whose sum and product are given, were reduced to the above problem. Since the Babylonians had no negative numbers, negative roots of quadratic equations were neglected. Though only concrete examples were given, many were intended to illustrate a general procedure for quadratics. More complicated algebraic problems were reduced by transformations to simpler ones.

The Babylonians were able to solve special problems involving five unknowns in five equations. One problem, which arose in connection with the adjustment of astronomical observations, involved ten equations and ten unknowns, mostly linear. The solution used a special method of combining the equations until the unknowns were evaluated.

The algebraic problems were stated and solved verbally. The words *uš* (length), *sag* (breadth), and *aša* (area) were often used for the unknowns, not because the unknowns necessarily represented these geometric quantities, but probably because many algebraic problems came from geometric situations and the geometric terminology became standard. An illustration of the way in which these terms were employed for the unknowns and of the way in which problems were stated may be gathered from the following: "I have multiplied Length and Breadth and the Area is 10. I have multiplied the Length by itself and have obtained an Area. The excess of Length over Breadth I have multiplied by itself and this result by 9. And this Area is Area obtained by multiplying the Length by itself. What are the Length and Breadth?" It is obvious here that the words *Length, Breadth,* and *Area* are merely convenient terms for two unknowns and their product, respectively.[1]

Today we would write this problem as

$$xy = 10$$
$$9(x - y)^2 = x^2.$$

The solution, incidentally, leads to a fourth-degree equation in x, with the x and x^3 terms missing so that it can be and was solved as a quadratic in x^2.

Problems leading to a cube root also occurred. The modern formulation of one such problem would be

$$12x = z, \qquad y = x, \qquad xyz = V,$$

where V is some known volume. To find x here we must extract a cube root. The Babylonians calculated this root from the cube-root tables we mentioned earlier. They also did compound-interest problems that called for finding the value of an unknown exponent.

The Babylonians sometimes employed symbols for unknowns, but the

1. Many examples of algebraic problems can be found in van der Waerden, pp. 65–73. See the bibliography at the end of the chapter.

symbolism came about inadvertently. In some problems two special Sumerian words (somewhat modified by Akkadian endings) were used to represent two unknowns that were reciprocals of each other. Moreover, the ancient Sumerian pictorial symbols were used for these words, and since these pictorial symbols were no longer in the current language, the effect was to use special symbols for unknowns. These symbols were used repeatedly and one could identify them without even knowing how they were pronounced in Akkadian.

The algebraic problems were solved by describing only the steps required to execute the solution. For example, square 10 obtaining 100; subtract 100 from 1000; this gives 900; and so forth. Since no reasons for steps were given, we can only infer how they knew what to do.

They summed arithmetic progressions and geometric progressions in concrete problems; for the latter we find, in our notation,

$$1 + 2 + 4 + \cdots + 2^9 = 2^9 + (2^9 - 1) = 2^{10} - 1.$$

Also, the sum of the squares of the integers from 1 to 10 was given as though they had applied the formula

$$1^2 + 2^2 + \cdots + n^2 = \left(1 \times \frac{1}{3} + n \times \frac{2}{3}\right)(1 + 2 + 3 + \cdots + n).$$

No derivation accompanied the special cases treated in their texts.

Babylonian algebra included a bit of the theory of numbers. Many sets of Pythagorean triples were found, probably by the correct rule; that is, if $x = p^2 - q^2$, $y = 2pq$, $z = p^2 + q^2$, then $x^2 + y^2 = z^2$. They also solved $x^2 + y^2 = 2z^2$ in integers.

6. *Babylonian Geometry*

The role of geometry in Babylonia was insignificant. Geometry was not a separate mathematical discipline. Problems involving division of a field or the sizes of bricks needed for some construction were readily converted into algebraic problems. Some computations of area and volume were given in accordance with rules or formulas. However, the figures illustrating geometric problems were roughly drawn and the formulas may have been incorrect. In the Babylonians' calculations of areas, for example, we cannot tell whether their triangles were right triangles or whether their quadrilaterals were squares and, hence, whether their formulas were the right ones for the figures concerned. However, the Pythagorean relationship, the similarity of triangles, and the proportionality of corresponding sides in similar triangles were known. The area of a circle was obtained, apparently by following the rule that $A = \frac{c^2}{12}$ where c is the circumference. This rule amounts to using 3 for π. However, another of their results giving the relation between the circumference of a regular hexagon and its circumscribed circle implies a value of 3 1/8 for π. A few volumes were computed, some correctly and some incorrectly, in the course of solving particular physical problems.

Apart from a few special facts, such as the calculation of the radius of a circle circumscribing a known isosceles triangle, the substance of Babylonian geometry was a collection of rules for the areas of simple plane figures, including regular polygons, and the volumes of simple solids. The geometry was not studied in and for itself but always in connection with practical problems.

7. *The Uses of Mathematics in Babylonia*

Despite the limited extent of the Babylonians' mathematics, it entered into many phases of their lives. Babylonia was in the path of trade routes and commerce was extensive. The Babylonians used their knowledge of arithmetic and simple algebra to express lengths and weights, to exchange money and merchandise, to compute simple and compound interest, to calculate taxes, and to apportion shares of a harvest to the farmer, church, and state. The division of fields and inheritances led to algebraic problems. The majority of the cuneiform texts involving mathematics (exclusive of the tables and problem texts) were concerned with economic problems. There is no doubt as to the influence of economics on the development of arithmetic in the older period.

Canals, dams, and other irrigation projects required calculations. The use of bricks raised numerous numerical and geometrical problems. Volumes of granaries and buildings and the areas of fields had to be determined. The close relationship between Babylonian mathematics and practical problems is typified by the following: A canal whose cross-section was a trapezoid and whose dimensions were known was to be dug. The amount of digging one man could do in a day was known, as was the sum of the number of men employed and the days they worked. The problem was to calculate the number of men and the number of days of work.

Because the connection between mathematics and astronomy became vital from Greek times on, we shall note what the Babylonians knew and did in astronomy. Nothing is known about Sumerian astronomy, and the astronomy of the Akkadian period was crude and qualitative; the development of mathematics preceded the development of any significant astronomy. In the Assyrian period (about 700 B.C.), astronomy did begin to include mathematical description of phenomena and a systematic compilation of observational data. The use of mathematics increased in the last three centuries B.C. and was directed especially to the study of lunar and planetary motion. Most astronomical texts are from this Seleucid period. They fall into two groups, procedural texts and ephemerides, tables of positions of the heavenly bodies at various times. The procedural texts show how to compute the ephemerides.

The arithmetic behind the lunar and solar observations shows that the Babylonians calculated first and second differences of successive data, observed the constancy of the first or second differences, and extrapolated or interpolated data. Their procedure was equivalent to using the fact that the data can be fit by polynomial functions and enabled them to predict the daily positions of the planets. They knew the periods of the planets with some accuracy, and also used eclipses as a basis for calculation. There was, however, no geometrical scheme of planetary or lunar motion in Babylonian astronomy.

The Babylonians of the Seleucid period did have extensive tables on the motions of the sun and moon which gave variable velocities and positions. Also special conjunctions and eclipses of sun and moon were either in the data or readily obtained from them. Astronomers could predict the new moon and eclipses to within a few minutes. Their data indicate that they knew the length of the solar or tropical year (the year of the seasons) to be $12 + 22/60 + 8/60^2$ months (from new moon to new moon) and the length of the sidereal year (the time for the sun to regain the same position relative to the stars) to within 4 1/2 minutes.

The constellations that lent their names to the twelve signs of the zodiac were known earlier but the zodiac itself first appears in a text in 419 B.C. Each sector of the zodiac was a 30° arc. Positions of planets in the sky were fixed by reference to the stars and also by position in the zodiac.

Astronomy served many purposes. For one thing, it was needed to keep a calendar, which is determined by the positions of the sun, moon, and stars. The year, the month, and the day are astronomical quantities, which had to be obtained accurately to know planting times and religious holidays. In Babylonia, partly because of the connection of the calendar with religious holidays and ceremonies and partly because the heavenly bodies were believed to be gods, the priests kept the calendar.

The calendar was lunar. The month began when the crescent first appeared after the moon was fully dark (our new moon). The day began in the evening of the first appearance of the crescent and lasted from sunset to sunset. The lunar calendar is difficult to maintain because, although it is convenient to have the month contain an integral number of days, lunar months, reckoned as the time between successive conjunctions of sun and moon (that is, from new moon to new moon), vary from 29 to 30 days. Hence a problem arises in deciding which months are to have 29 and which 30. A more important problem is making the lunar calendar agree with the seasons. The answer is quite complicated because it depends upon the paths and velocities of the moon and sun. The lunar calendar contained extra months intercalated so that 7 such intercalations in each 19 years just *about* kept the lunar calendar in time with the solar year. Thus 235 lunar months were equal to 19 solar years. The summer solstice was systematically computed, and the winter solstice and the equinoxes were placed at equal intervals. This calendar was used by the Jews and Greeks and by the Romans up to 45 B.C., the year the Julian calendar was adopted.

The division of the circle into 360 units originated in the Babylonian astronomy of the last centuries before the Christian era. It had nothing to do with the earlier use of base 60; however, base 60 was used to divide the degree and the minute into 60 parts. The astronomer Ptolemy (2nd cent. A.D.) followed the Babylonians in this practice.

Closely connected with astronomy was astrology. In Babylonia, as in many ancient civilizations, the heavenly bodies were thought to be gods and so were presumed to have influence and even control over the affairs of man. When one takes into account the importance of the sun for light, heat, and the growth of plants, the dread inspired by its eclipses, and such seasonal phenomena as the mating of animals, one can well understand the belief that the heavenly bodies do affect even the daily events in man's life.

Pseudoscientific schemes of prediction in ancient civilizations did not always involve astronomy. Numbers themselves had mystic properties and could be used to make predictions. One finds some Babylonian usages in the Book of Daniel and in the writings of the Old and New Testament prophets. The Hebrew "science" of *gematria* (a form of cabbalistic mysticism) was based on the fact that each letter of the alphabet had a number value because the Hebrews used letters to represent numbers. If the sum of the numerical values of the letters in two words was the same, an important connection between the two ideas or people or events represented by the words was inferred. In the prophecy of Isaiah (21:8), the lion proclaims the fall of Babylon because the letters in the Hebrew word for lion and those in the word for Babylon add up to the same sum.

8. *Evaluation of Babylonian Mathematics*

The Babylonians' use of special terms and symbols for unknowns, their employment of a few operational symbols, and their solution of a few types of equations involving one or more unknowns, especially quadratic equations, constituted a start in algebra. Their development of a systematic way of writing whole numbers and fractions enabled them to carry arithmetic to a fairly advanced stage and to employ it in many practical situations, especially in astronomy. They possessed some numerical and what we would call algebraic skill in the solution of special equations of high degree, but on the whole their arithmetic and algebra were very elementary. Though they worked with concrete numbers and problems, they evidenced a partial grasp of abstract mathematics in their recognition that some procedures were typical of certain classes of equations.

The question arises as to what extent the Babylonians employed mathematical proof. They did solve by correct systematic procedures rather complicated equations involving unknowns. However, they gave verbal instructions only on the steps to be made and offered no justification of the steps. Almost surely, the arithmetic and algebraic processes and the geometrical rules were the end result of physical evidence, trial and error, and insight. That the methods worked was sufficient justification to the Babylonians for their continued use. The concept of proof, the notion of a logical structure based on principles warranting acceptance on one ground or another, and the consideration of such questions as under what conditions solutions to problems can exist, are not found in Babylonian mathematics.

Bibliography

Bell, E. T.: *The Development of Mathematics*, 2nd ed., McGraw-Hill, Chaps. 1–2.

Boyer, Carl B.: *A History of Mathematics*, John Wiley and Sons, 1968, Chap. 3.

Cantor, Moritz: *Vorlesungen über Geschichte der Mathematik*, 2nd ed., B. G. Teubner, 1894, Vol. 1, Chap. 1.

Chiera, E.: *They Wrote on Clay*, Chicago University Press, 1938.

Childe, V. Gordon: *Man Makes Himself*, New American Library, 1951, Chaps. 6–8.

Dantzig, Tobias: *Number: The Language of Science*, 4th ed., Macmillan, 1954, Chaps. 1–2.

Karpinski, Louis C.: *The History of Arithmetic*, Rand McNally, 1925.

Menninger, K.: *Number Words and Number Symbols: A Cultural History of Numbers*, Massachusetts Institute of Technology Press, 1969.

Neugebauer, Otto: *The Exact Sciences in Antiquity*, Princeton University Press, 1952, Chaps. 1–3 and 5.

———: *Vorgriechische Mathematik*, Julius Springer, 1934, Chaps. 1–3 and 5.

Sarton, George: *A History of Science*, Harvard University Press, 1952, Vol. 1, Chap. 3.

———: *The Study of the History of Mathematics and the History of Science*, Dover (reprint), 1954.

Smith, David Eugene: *History of Mathematics*, Dover (reprint), 1958, Vol. 1, Chap. 1.

Struik, Dirk J.: *A Concise History of Mathematics*, 3rd ed., Dover, 1967, Chaps. 1–2.

van der Waerden, B. L.: *Science Awakening*, P. Noordhoff, 1954, Chaps. 2–3.

2
Egyptian Mathematics

> All science, logic and mathematics included, is a function of the epoch—all science, in its ideals as well as in its achievements.
>
> E. H. MOORE

1. *Background*

While Mesopotamia experienced many changes in its ruling peoples, with resulting new cultural influences, the Egyptian civilization developed unaffected by foreign influences. The origins of the civilization are unknown but it surely existed even before 4000 B.C. Egypt, as the Greek historian Herodotus says, is a gift of the Nile. Once a year this river, drawing its water from the south, floods the territory all along its banks and leaves behind rich soil. Most of the people did and still do make their living by tilling this soil. The rest of the country is desert.

There were two kingdoms, one in the north and one in the south of what is present-day Egypt. Some time between 3500 and 3000 B.C., the ruler, Mena, or Menes, unified upper and lower Egypt. From this time on the major periods of Egyptian history are referred to in terms of the ruling dynasties, Menes having been the founder of the first dynasty. The height of Egyptian culture occurred during the third dynasty (about 2500 B.C.), during which period the rulers built the pyramids. The civilization went its own way until Alexander the Great conquered it in 332 B.C. Thereafter, until about A.D. 600, its history and mathematics belong to the Greek civilization. Thus, apart from one minor invasion by the Hyksos (1700–1600 B.C.) and slight contact with the Babylonian civilization (inferred from the discovery in the Nile valley of the cuneiform Tell al-Amarna tablets of about 1500 B.C.), Egyptian civilization was the product of its native people.

The ancient Egyptians developed their own systems of writing. One system, hieroglyphics, was pictorial, that is, each symbol was a picture of some object. Hieroglyphics were used on monuments until about the time of Christ. From about 2500 B.C. the Egyptians used for daily purposes what is called hieratic writing. This system employed conventional symbols, which at first were merely simplifications of the hieroglyphics. Hieratic writing is syllabic; each syllable is represented by an ideogram and an entire word is a collection of ideograms. The meaning of the word is not tied to the separate ideograms.

The writing was done with ink on papyrus, sheets made by pressing the pith of a plant and then slicing it. Since papyrus dries up and crumbles, very few documents of ancient Egypt have survived, apart from the hieroglyphic inscriptions on stone.

The main surviving mathematical documents are two sizable papyri: the Moscow papyrus, which is in Moscow, and the Rhind papyrus, discovered in 1858 by a Britisher, Henry Rhind, and now in the British Museum. The Rhind papyrus is also known as the Ahmes papyrus after its author, who opens with the words "Directions for Obtaining the Knowledge of All Dark Things." Both papyri date from about 1700 B.C. There are also fragments of other papyri written at this time and later. The mathematical papyri were written by scribes who were workers in the Egyptian state and church administrations.

The papyri contain problems and their solutions—85 in the Rhind papyrus and 25 in the Moscow papyrus. Presumably such problems occurred in the work of the scribes and they were expected to know how to solve them. It is most likely that the problems in the two major papyri were intended as examples of typical problems and solutions. Though these papyri date from about 1700 B.C., the mathematics in them was known to the Egyptians as far back as 3500 B.C., and little was added from that time to the Greek conquest.

2. *The Arithmetic*

The hieroglyphic number symbols used by the Egyptians were Ⅰ for 1, ∩ for 10, Ⓒ and ⊙ for 100, ⚶ for 1000, Ⅻ for 10,000, and other symbols for larger units. These symbols were combined to form intermediate numbers. The direction of the writing was from right to left, so that ⅠⅠⅠⅠ∩∩ represented 24. This system of writing numbers uses the base 10 but is not positional.

Egyptian hieratic whole numbers are illustrated by the following symbols:

| 1 | 2 | 3 | 4 | 5 | 6 | 7 | 8 | 9 | 10 |

The arithmetic was essentially additive. For ordinary additions and subtractions they could simply combine or take away symbols to reach the proper result. Multiplication and division were also reduced to additive processes. To calculate 12 times 12, say, the Egyptians did the following:

1	12
2	24
4	48
8	96.

Each line was derived from the preceding one by doubling. Now since $4 \cdot 12 = 48$ and $8 \cdot 12 = 96$, adding 48 and 96 gave the value of $12 \cdot 12$. This process was, of course, quite different from multiplying by 10 and by 2 and adding. Multiplication by 10 was also performed and consisted of replacing unit symbols by the symbol for 10 and replacing the 10 symbol by the symbol for 100.

Division of one whole number by another as carried out by the Egyptians is equally interesting. For example, 19 was divided by 8 as follows:

1	8
2	16
1/2	4
1/4	2
1/8	1

Therefore, the answer was $2 + 1/4 + 1/8$. The idea was simply to take the number of eights and parts of eight that totaled 19.

The denotation of fractions in the Egyptian number system was much more complicated than our own. The symbol $\bigcirc$, pronounced *ro*, which originally indicated 1/320 of a bushel, came to denote a fraction. In hieratic writing the oval was replaced by a dot. The $\bigcirc$ or dot was generally placed above the whole number to indicate the fraction. Thus, in hieroglyphic writing,

$$\overset{\bigcirc}{\underset{\substack{\vert\vert\vert \\ \vert\vert}}{}} = \frac{1}{5}, \qquad \overset{\bigcirc}{\cap} = \frac{1}{10}, \qquad \overset{\bigcirc}{\underset{\cap\ \vert\vert}{}}\ ^{\vert\vert\vert} = \frac{1}{15}.$$

A few fractions were denoted by special symbols. Thus the hieroglyph $\subset\!\!=$ denoted 1/2; π, 2/3; and $\times$, 1/4.

Aside from a few special ones, all fractions were decomposed into what are called unit fractions. Thus Ahmes writes 2/5 as $1/3 + 1/15$. The plus sign did not appear but was understood. The Rhind papyrus contains a table for expressing fractions with numerator 2 and odd denominators from 5 to 101 as sums of fractions with numerator 1. By means of this table a fraction such as our 7/29, which to Ahmes is the integer 7 divided by the integer 29, could also be expressed as a sum of unit fractions. Inasmuch as $7 = 2 + 2 + 2 + 1$, he proceeds by converting each 2/29 to a sum of fractions with numerator 1. By combining these results and by further conversion he ends up with a sum of unit fractions, each with a different denominator. The final expression for 7/29 is

$$\frac{1}{6} + \frac{1}{24} + \frac{1}{58} + \frac{1}{87} + \frac{1}{232}.$$

It so happens that 7/29 can also be expressed as $1/5 + 1/29 + 1/145$, but because Ahmes' $2/n$ table leads to the former expression, this is the one used. The expression of our a/b as a sum of unit fractions was done systematically according to age-old procedures. Using unit fractions, the Egyptians could carry out the four arithmetic operations with fractions. The extensive and complicated computations with fractions were one reason the Egyptians never developed arithmetic or algebra to an advanced state.

The nature of irrational numbers was not recognized in Egyptian arithmetic any more than it was in the Babylonian. The simple square roots that occurred in algebraic problems could be and were expressed in terms of whole numbers and fractions.

3. *Algebra and Geometry*

The papyri contain solutions of problems involving an unknown that are in the main comparable to our linear equations in one unknown. However, the processes were purely *arithmetical* and did not, in Egyptian minds, amount to a distinct subject, the solution of equations. The problems were stated verbally with bare directions for obtaining the solutions and without explanation of why the methods were used or why they worked. For example, problem 31 of the Ahmes papyrus, translated literally, reads: "A quantity, its 2/3, its 1/2, its 1/7, its whole, amount to 33." This means, for us:

$$\frac{2}{3}x + \frac{x}{2} + \frac{x}{7} + x = 33.$$

Simple arithmetic of the Egyptian variety gives the solution in this case.

Problem 63 of the papyrus runs as follows: "Directions for dividing 700 breads among four people, 2/3 for one, 1/2 for the second, 1/3 for the third, 1/4 for the fourth." For us this means

$$\frac{2}{3}x + \frac{1}{2}x + \frac{1}{3}x + \frac{1}{4}x = 700.$$

The solution, as given by Ahmes, is: "Add $\frac{2}{3}, \frac{1}{2}, \frac{1}{3}, \frac{1}{4}$. This gives $1\frac{1}{2}\frac{1}{4}$. Divide 1 by $1\frac{1}{2}\frac{1}{4}$. This gives $\frac{1}{2}\frac{1}{14}$. Now find $\frac{1}{2}\frac{1}{14}$ of 700. This is 400."

In some solutions Ahmes uses the "rule of false position." Thus to determine five numbers in arithmetic progression subject to a further condition and such that the sum is 100, he first chooses d, the common difference, to be 5 1/2 times the smallest number. He then picks 1 as the smallest and gets the progression: 1, 6 1/2, 12, 17 1/2, 23. But these numbers add up to 60 whereas they should add up to 100. He then multiplies each term by 5/3.

Only the simplest types of quadratic equations, such as $ax^2 = b$, are considered. Even when two unknowns occur, the type is

$$x^2 + y^2 = 100, \qquad y = \frac{3}{4}x,$$

so that after eliminating y, the equation in x reduces to the first type. Some concrete problems involving arithmetic and geometric progressions also can be found in the papyri. To infer general rules from all these problems and solutions is not very difficult.

The limited Egyptian algebra employed practically no symbolism. In the Ahmes papyrus, addition and subtraction are represented respectively by the legs of a man coming and going, $\diagup\!\!\!\rfloor$ and $\diagdown\!\!\!\llcorner$, and the symbol $\lceil$ is used to denote square root.

What of Egyptian geometry? The Egyptians did not separate arithmetic and geometry. We find problems from both fields in the papyri. Like the Babylonians, the Egyptians regarded geometry as a practical tool. One merely applied arithmetic and algebra to the problems involving areas, volumes, and other geometrical situations. Egyptian geometry is said by Herodotus to have originated in the need created by the annual overflow of the Nile to redetermine the boundaries of the lands owned by the farmers. However, Babylonia did as much in geometry without such a need. The Egyptians had prescriptions for the areas of rectangles, triangles, and trapezoids. In the case of the area of a triangle, though they multiplied one number by half another, we cannot be sure that the method is correct because we are not sure from the words used whether the lengths multiplied stood for base and altitude or just for two sides. Also the figures were so poorly drawn that one cannot be sure of just what area or volume were being found. Their calculation of the area of a circle, surprisingly good, followed the formula $A = (8d/9)^2$ where d is the diameter. This amounts to using 3.1605 for π.

An example may illustrate the "accuracy" of Egyptian formulas for area. On the walls of à temple in Edfu is a list of fields that were gifts to the temple. These fields generally have four sides, which we shall denote by a, b, c, d, where a and b and c and d are pairs of opposite sides. The inscriptions give the area of these various fields as $\frac{(a + b)}{2} \cdot \frac{(c + d)}{2}$. But some fields are triangles. In this case, d is said to be nothing and the calculation is changed to $\frac{(a + b)}{2} \cdot \frac{c}{2}$. Even for quadrangles the rule is just a crude approximation.

The Egyptians also had rules for the volume of a cube, box, cylinder, and other figures. Some of the rules were correct and others only approximations. The papyri give as the volume of a truncated conical clepsydra (water clock), in our notation,

$$V = \frac{h}{12}\left(\frac{3}{2}(D+d)\right)^2,$$

where h is the height and $(D+d)/2$ is the mean circumference. This formula amounts to using 3 for π.

The most striking rule of Egyptian geometry is the one for the volume of a truncated pyramid of square base, which in modern notation is

$$V = \frac{h}{3}(a^2 + ab + b^2),$$

where h is the height and a and b are sides of top and bottom. The formula is surprising because it is correct and because it is symmetrically expressed (but of course not in our notation). It is given only for concrete numbers. However, we do not know whether the pyramid was square-based or not because the figure in the papyrus is not carefully drawn.

Neither do we know whether the Egyptians recognized the Pythagorean theorem. We know there were rope-stretchers, that is, surveyors, but the story that they used a rope knotted at points to divide the total length into parts of ratios 3 to 4 to 5, which could then be used to form a right triangle, is not confirmed in any document.

The rules were not expressed in symbols. The Egyptians stated the problems verbally; and their procedure in solving them was essentially what we do when we calculate according to a formula. Thus an almost literal translation of the geometrical problem of finding the volume of the frustum of a pyramid reads: "If you are told: a truncated pyramid of 6 for the vertical height by 4 on the base, by 2 on the top. You are to square this 4, result 16. You are to double, result 8. You are to square 2, result 4. You are to add the 16, and 8, and the 4, result 28. You are to take one-third of 6, result 2. You are to take 28 twice, result 56. See, it is 56. You will find it right."

Did the Egyptians know proofs or justifications of their procedures and formulas? One belief is that the Ahmes papyrus was written in the style of a textbook for students of that day and hence, even though no general rules or principles for solving types of equations were formulated by Ahmes, it is very likely that he knew them but wanted the student to formulate them himself or have a teacher formulate them for him. Under this view the Ahmes papyrus is a rather advanced arithmetic text. Others say it is the notebook of a pupil. In either case, the papyri almost surely recorded the types of problems that had to be solved by business and administrative clerks, and the methods of solution were just practical rules known by experience to work. No one believes that the Egyptians had a deductive structure based on sound axioms that established the correctness of their rules.

4. *Egyptian Uses of Mathematics*

The Egyptians used mathematics in the administration of the affairs of the state and church, to determine wages paid to laborers, to find the volumes of granaries and the areas of fields, to collect taxes assessed according to the land area, to convert from one system of measures to another, and to calculate the number of bricks needed for the construction of buildings and ramps. The papyri also contain problems dealing with the amounts of corn needed to make given quantities of beer and the amount of corn of one quality needed to give the same result as corn of another quality whose strength relative to the first is known.

As in Babylonia a major use of mathematics was in astronomy, which dates from the first dynasty. Astronomical knowledge was essential. To the Egyptian the Nile was his life's blood. He made his living by tilling the soil which the Nile covered with rich silt in its annual overflow. However, he had to be well prepared for the dangerous aspects of the flood; his home, equipment, and cattle had to be temporarily removed from the area and arrangements made for sowing immediately afterwards. Hence the coming of the flood had to be predicted, which was done by learning what heavenly events preceded it.

Astronomy also made the calendar possible. Beyond the need for a calendar in commerce was the need to predict religious holidays. It was believed essential, to ensure the goodwill of the gods, that holidays be celebrated at the proper time. As in Babylonia, keeping the calendar was largely the task of the priests.

The Egyptians arrived at their estimate of the length of the solar year by observing the star Sirius. On one day in the summer this star became visible on the horizon just before sunrise. On succeeding days it was visible for a longer time before the sun's growing light blotted it out. The first day on which it was visible just before sunrise was known as the heliacal rising of Sirius, and the interval between two such days was about 365 1/4 days; so the Egyptians adopted, supposedly in 4241 B.C., a civil calendar of 365 days for the year. The concentration on Sirius is undoubtedly accounted for by the fact that the waters of the Nile began to rise on that day, which was chosen as the first day of the year.

The 365-day year was divided into 12 months of 30 days, plus 5 extra days at the end. Because the Egyptians did not intercalate the additional day every four years, the civil calendar lost all relation to the seasons. It takes 1460 years for the calendar to set itself right again; this interval is known as the Sothic cycle, from the Egyptian name for Sirius. Whether the Egyptians knew of the Sothic cycle is open to question. Their calendar was adopted by Julius Caesar in 45 B.C., but changed to a 365 1/4-day year on the advice of the Alexandrian Greek Sosigenes. Though the Egyptian determination of the year and the calendar were valuable contributions, they did not result from a well-developed astronomy, which in fact was crude and far inferior to Babylonian astronomy.

The Egyptians combined their knowledge of astronomy and geometry to construct their temples in such a manner that on certain days of the year the sun would strike them in a particular way. Thus some were built so that on the longest day of the year the sun would shine directly into the temple and illuminate the god at the altar. This orientation of temples is also found to some extent among the Babylonians and Greeks. The pyramids too were oriented to special directions of the heavens, and the Sphinx faces east. While the details of the construction of these works are unimportant for us, it is worth noting that the pyramids represent another application of Egyptian geometry. They are the tombs of kings; and because the Egyptians believed in immortality they believed that the proper construction of a tomb was material for the dead person's afterlife. In fact, an entire apartment for the future residence of king and queen was installed in each pyramid. They took great care to make the bases of the pyramids of the correct shape; the relative dimensions of base and height were also highly significant. However, one should not overemphasize the complexity or depth of the ideas involved. Egyptian mathematics was simple and crude and no deep principles were involved, contrary to what is often asserted.

5. *Summary*

Let us review the status of mathematics before the Greeks enter the picture. We find in the Babylonian and Egyptian civilizations an arithmetic of integers and fractions, including positional notation, the beginnings of algebra, and some empirical formulas in geometry. There was almost no symbolism, hardly any conscious thought about abstractions, no formulation of general methodology, and no concept of proof or even of plausible arguments that might convince one of the correctness of a procedure or formula. There was, in fact, no conception of any kind of theoretical science.

Apart from a few incidental results in Babylonia, mathematics in the two civilizations was not a distinct discipline, nor was it pursued for its own sake. It was a tool in the form of disconnected, simple rules which answered questions arising in the daily life of the people. Certainly nothing was done in mathematics that altered or affected the way of life. Although Babylonian mathematics was more advanced than the Egyptian, about the best one can say for both is that they showed some vigor, if not rigor, and more perseverance than brilliance.

All evaluation implies some sort of standard. It may be unfair but it is natural to compare the two civilizations with the Greek, which succeeded them. By this standard the Egyptians and Babylonians were crude carpenters, whereas the Greeks were magnificent architects. One does find more favorable, even laudatory, descriptions of the Babylonian and Egyptian achievements. But these are made by specialists who become, perhaps unconsciously, overimpressed by their own field of interest.

Bibliography

Boyer, Carl B.: *A History of Mathematics*, John Wiley and Sons, 1968, Chap. 2.

Cantor, Moritz: *Vorlesungen über Geschichte der Mathematik*, 2nd ed., B. G. Teubner, 1894, Vol. 1, Chap. 3.

Chace, A. B., *et al.*, eds.: *The Rhind Mathematical Papyrus*, 2 vols., Mathematical Association of America, 1927–29.

Childe, V. Gordon: *Man Makes Himself*, New American Library, 1951.

Karpinski, Louis C.: *The History of Arithmetic*, Rand McNally, 1925.

Neugebauer, O.: *The Exact Sciences in Antiquity*, Princeton University Press, 1952, Chap. 4.

——: *Vorgriechische Mathematik*, Julius Springer, 1934.

Sarton, George: *A History of Science*, Harvard University Press, 1952, Vol. 1, Chap. 2.

Smith, David Eugene: *History of Mathematics*, Dover (reprint), 1958, Vol. 1, Chap. 2; Vol. 2, Chaps. 2 and 4.

van der Waerden, B. L.: *Science Awakening*, P. Noordhoff, 1954, Chap. 1.

3
The Creation of Classical
Greek Mathematics

> This, therefore, is mathematics: she reminds you of the invisible form of the soul; she gives life to her own discoveries; she awakens the mind and purifies the intellect; she brings light to our intrinsic ideas; she abolishes oblivion and ignorance which are ours by birth.　　　PROCLUS

1. *Background*

In the history of civilization the Greeks are preeminent, and in the history of mathematics the Greeks are the supreme event. Though they did borrow from the surrounding civilizations, the Greeks built a civilization and culture of their own which is the most impressive of all civilizations, the most influential in the development of modern Western culture, and decisive in founding mathematics as we understand the subject today. One of the great problems of the history of civilization is how to account for the brilliance and creativity of the ancient Greeks.

Though our knowledge of their early history is subject to correction and amplification as more archeological research is carried on, we now have reason to believe, on the basis of the *Iliad* and the *Odyssey* of Homer, the decipherment of ancient languages and scripts, and archeological investigations, that the Greek civilization dates back to 2800 B.C. The Greeks settled in Asia Minor, which may have been their original home, on the mainland of Europe in the area of modern Greece, and in southern Italy, Sicily, Crete, Rhodes, Delos, and North Africa. About 775 B.C. the Greeks replaced various hieroglyphic systems of writing with the Phoenician alphabet (which was also used by the Hebrews). With the adoption of an alphabet the Greeks became more literate, more capable of recording their history and ideas.

As the Greeks became established they visited and traded with the Egyptians and Babylonians. There are many references in classical Greek writings to the knowledge of the Egyptians, whom some Greeks erroneously considered the founders of science, particularly surveying, astronomy, and arithmetic. Many Greeks went to Egypt to travel and study. Others visited Babylonia and learned mathematics and science there.

The influence of the Egyptians and Babylonians was almost surely felt in Miletus, a city of Ionia in Asia Minor and the birthplace of Greek philosophy, mathematics, and science. Miletus was a great and wealthy trading city on the Mediterranean. Ships from the Greek mainland, Phoenicia, and Egypt came to its harbors; Babylonia was connected by caravan routes leading eastward. Ionia fell to Persia about 540 B.C., though Miletus was allowed some independence. After an Ionian revolt against Persia in 494 B.C. was crushed, Ionia declined in importance. It became Greek again in 479 B.C. when Greece defeated Persia, but by then cultural activity had shifted to the mainland of Greece with Athens as its center.

Though the ancient Greek civilization lasted until about A.D. 600, from the standpoint of the history of mathematics it is desirable to distinguish two periods, the classical, which lasted from 600 to 300 B.C., and the Alexandrian or Hellenistic, from 300 B.C. to A.D. 600. The adoption of the alphabet, already mentioned, and the fact that papyrus became available in Greece during the seventh century B.C. may account for the blossoming of cultural activity about 600 B.C. The availability of this writing paper undoubtedly helped the spread of ideas.

2. The General Sources

The sources of our knowledge of Greek mathematics are, peculiarly, less authentic and less reliable than our sources for the much older Babylonian and Egyptian mathematics, because no original manuscripts of the important Greek mathematicians are extant. One reason is that papyrus is perishable; though the Egyptians also used papyrus, by luck a few of their mathematical documents did survive. Some of the voluminous Greek writings might still be available to us if their great libraries had not been destroyed.

Our chief sources for the Greek mathematical works are Byzantine Greek codices (manuscript books) written from 500 to 1500 years after the Greek works were originally composed. These codices are not literal reproductions but critical editions, so that we cannot be sure what changes may have been made by the editors. We also have Arabic translations of the Greek works and Latin versions derived from Arabic works. Here again we do not know what changes the translators may have made or how well they understood the original texts. Moreover, even the Greek texts used by the Arabic and Byzantine authors were questionable. For example, though we do not have the Alexandrian Greek Heron's manuscript, we know that he made a number of changes in Euclid's *Elements*. He gave different proofs and added new cases of the theorems and converses. Likewise Theon of Alexandria (end of 4th cent. A.D.) tells us that he altered sections of the *Elements* in his edition.

The Greek and Arabic versions we have may come from such versions of the originals. However, in one or another of these forms we do have the works of Euclid, Apollonius, Archimedes, Ptolemy, Diophantus, and other Greek authors. Many Greek texts written during the classical and Alexandrian periods did not come down to us because even in Greek times they were superseded by the writings of these men.

The Greeks wrote some histories of mathematics and science. Eudemus (4th cent. B.C.), a member of Aristotle's school, wrote a history of arithmetic, a history of geometry, and a history of astronomy. Except for fragments quoted by later writers, these histories are lost. The history of geometry dealt with the period preceding Euclid's and would be invaluable were it available. Theophrastus (c. 372–c. 287 B.C.), another disciple of Aristotle, wrote a history of physics, and this, too, except for a few fragments, is lost.

In addition to the above, we have two important commentaries. Pappus (end of 3rd cent. A.D.) wrote the *Synagoge* or *Mathematical Collection*; almost the whole of it is extant in a twelfth-century copy. This is an account of much of the work of the classical and Alexandrian Greeks from Euclid to Ptolemy, supplemented by a number of lemmas and theorems that Pappus added as an aid to understanding. Pappus had also written the *Treasury of Analysis*, a collection of the Greek works themselves. This book is lost, but in Book VII of his *Mathematical Collection* he tells us what his *Treasury* contained.

The second important commentator is Proclus (A.D. 410–485), a prolific writer. Proclus drew material from the texts of the Greek mathematicians and from prior commentaries. Of his surviving works, the *Commentary*, which treats Book I of Euclid's *Elements*, is the most valuable. Proclus apparently intended to discuss more of the *Elements*, but there is no evidence that he ever did so. The *Commentary* contains one of the three quotations traditionally credited to Eudemus' history of geometry (see sec. 10) but probably taken from a later modification. This particular extract, the longest of the three, is referred to as the Eudemian summary. Proclus also tells us something about Pappus' work. Thus, besides the later editions and versions of some of the Greek classics themselves, Pappus' *Mathematical Collection* and Proclus' *Commentary* are the two main sources of the history of Greek mathematics.

Of original wordings (though not the manuscripts) we have only a fragment concerning the lunes of Hippocrates, quoted by Simplicius (first half of 6th cent. A.D.) and taken from Eudemus' lost *History of Geometry*, and a fragment of Archytas on the duplication of the cube. And of original manuscripts we have some papyri written in Alexandrian Greek times. Related sources on Greek mathematics are also immensely valuable. For example, the Greek philosophers, especially Plato and Aristotle, had much to say about mathematics and their writings have survived somewhat in the same way as have the mathematical works.

The reconstruction of the history of Greek mathematics, based on sources such as we have described, has been an enormous and complicated task. Despite the extensive efforts of scholars, there are gaps in our knowledge and some conclusions are arguable. Nevertheless the basic facts are clear.

3. The Major Schools of the Classical Period

The cream of the classical period's contributions are Euclid's *Elements* and Apollonius' *Conic Sections*. Appreciation of these works requires some knowledge of the great changes made in the very nature of mathematics and of the problems the Greeks faced and solved. Moreover, these polished works give little indication of the three hundred years of creative activity preceding them or of the issues which became vital in the subsequent history.

Classical Greek mathematics developed in several centers that succeeded one another, each building on the work of its predecessors. At each center an informal group of scholars carried on its activities under one or more great leaders. This kind of organization is common in modern times also and its reason for being is understandable. Today, when one great man locates at a particular place—generally a university—other scholars follow, to learn from the master.

The first of the schools, the Ionian, was founded by Thales (*c*. 640–*c*. 546 B.C.) in Miletus. We do not know the full extent to which Thales may have educated others, but we do know that the philosophers Anaximander (*c*. 610–*c*. 547 B.C.) and Anaximenes (*c*. 550–480 B.C.) were his pupils. Anaxagoras (*c*. 500–*c*. 428 B.C.) belonged to this school, and Pythagoras (*c*. 585–*c*. 500 B.C.) is supposed to have learned mathematics from Thales. Pythagoras then formed his own large school in southern Italy. Toward the end of the sixth century, Xenophanes of Colophon in Ionia migrated to Sicily and founded a center to which the philosophers Parmenides (5th cent. B.C.) and Zeno (5th cent. B.C.) belonged. The latter two resided in Elea in southern Italy, to which the school had moved, and so the group became known as the Eleatic school. The Sophists, active from the latter half of the fifth century onward, were concentrated mainly in Athens. The most celebrated school is the Academy of Plato in Athens, where Aristotle was a student. The Academy had unparalleled importance for Greek thought. Its pupils and associates were the greatest philosophers, mathematicians, and astronomers of their age; the school retained its preeminence in philosophy even after the leadership in mathematics passed to Alexandria. Eudoxus, who learned mathematics chiefly from Archytas of Tarentum (Sicily), founded his own school in Cyzicus, a city of northern Asia Minor. When Aristotle left Plato's Academy he founded another school, the Lyceum, in Athens. The Lyceum is commonly referred to as the Peripatetic school. Not all of the great mathematicians of the classical period can be identified with a school, but for the sake of coherence we shall occasionally discuss the work of a man in connection with a particular school even though his association with it was not close.

4. *The Ionian School*

The leader and founder of this school was Thales. Though there is no sure knowledge about Thales' life and work, he probably was born and lived in Miletus. He traveled extensively and for a while resided in Egypt, where he carried on business activities and reportedly learned much about Egyptian mathematics. He is, incidentally, supposed to have been a shrewd business-man. During a good season for olive growing, he cornered all the olive presses in Miletus and Chios and rented them out at a high fee. Thales is said to have predicted an eclipse of the sun in 585 B.C., but this is disputed on the ground that astronomical knowledge was not adequate at that time.

He is reputed to have calculated the heights of pyramids by comparing their shadows with the shadow cast by a stick of known height at the same time. By some such use of similar triangles he is supposed to have calculated the distance of a ship from shore. He is also credited with having made mathematics abstract and with having given deductive proofs for some theorems. These last two claims, however, are dubious. Discovery of the attractive power of magnets and of static electricity is also attributed to Thales.

The Ionian school warrants only brief mention so far as contributions to mathematics proper are concerned, but its importance for philosophy and the philosophy of science in particular is unparalleled (see Chap. 7, sec. 2). The school declined in importance when the Persians conquered the area.

5. *The Pythagoreans*

The torch was picked up by Pythagoras who, supposedly having learned from Thales, founded his own school in Croton, a Greek settlement in southern Italy. There are no written works by the Pythagoreans; we know about them through the writings of others, including Plato and Herodotus. In particular we are hazy about the personal life of Pythagoras and his followers; nor can we be sure of what is to be credited to him personally or to his followers. Hence when one speaks of the work of Pythagoras one really refers to the work done by the group between 585 B.C., the reputed date of his birth, and roughly 400 B.C. Philolaus (5th cent. B.C.) and Archytas (428–347 B.C.) were prominent members of this school.

Pythagoras was born on the island of Samos, just off the coast of Asia Minor. After spending some time with Thales in Miletus, he traveled to other places, including Egypt and Babylon, where he may have picked up some mathematics and mystical doctrines. He then settled in Croton. There he founded a religious, scientific, and philosophical brotherhood. It was a formal school, in that membership was limited and members learned from leaders. The teachings of the group were kept secret by the members, though the secrecy as to mathematics and physics is denied by some historians. The Pythagoreans were supposed to have mixed in politics; they allied themselves with the aristocratic faction and were driven out by the popular or democratic party. Pythagoras fled to nearby Metapontum and was murdered there about 497 B.C. His followers spread to other Greek centers and continued his teachings.

One of the great Greek contributions to the very concept of mathematics was the conscious recognition and emphasis of the fact that mathematical entities, numbers, and geometrical figures are abstractions, ideas entertained by the mind and sharply distinguished from physical objects or pictures. It is true that even some primitive civilizations and certainly the Egyptians and Babylonians had learned to think about numbers as divorced from physical objects. Yet there is some question as to how much they were consciously aware of the abstract nature of such thinking. Moreover, geometrical thinking in all pre-Greek civilizations was definitely tied to matter. To the Egyptians, for example, a line was no more than either a stretched rope or the edge of a field and a rectangle was the boundary of a field.

The recognition that mathematics deals with abstractions may with some confidence be attributed to the Pythagoreans. However, this may not have been true at the outset of their work. Aristotle declared that the Pythagoreans regarded numbers as the ultimate components of real, material objects.[1] Numbers did not have a detached existence apart from objects of sense. When the early Pythagoreans said that all objects were composed of (whole) numbers or that numbers were the essence of the universe, they meant it literally, because numbers to them were like atoms are to us. It is also believed that the sixth- and fifth-century Pythagoreans did not really distinguish numbers from geometrical dots. Geometrically, then, a number was an extended point or a very small sphere. However, Eudemus, as reported by Proclus, says that Pythagoras rose to higher principles (than had the Egyptians and Babylonians) and considered abstract problems for the pure intelligence. Eudemus adds that Pythagoras was the creator of pure mathematics, which he made into a liberal art.

The Pythagoreans usually depicted numbers as dots in sand or as pebbles. They classified the numbers according to the shapes made by the arrangements of the dots or pebbles. Thus the numbers 1, 3, 6, and 10 were called triangular because the corresponding dots could be arranged as triangles (Fig. 3.1). The fourth triangular number, 10, especially fascinated the Pythagoreans because it was a prized number for them, and had 4 dots on

1. *Metaphys.* I, v, 986a and 986a 21, Loeb Classical Library ed.

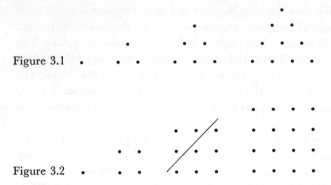

Figure 3.1

Figure 3.2

each side, 4 being another favorite number. They realized that the sums 1, 1 + 2, 1 + 2 + 3, and so forth gave the triangular numbers and that 1 + 2 + ⋯ + n = (n/2)(n + 1).

The numbers 1, 4, 9, 16, ... were called square numbers because as dots they could be arranged as squares (Fig. 3.2). Composite (nonprime) numbers which were not perfect squares were called oblong.

From the geometrical arrangements certain properties of the whole numbers became evident. Introducing the slash, as in the third illustration of Figure 3.2, shows that the sum of two consecutive triangular numbers is a square number. This is true generally, for as we can see, in modern notation,

$$\frac{n}{2}(n + 1) + \frac{n + 1}{2}(n + 2) = (n + 1)^2.$$

That the Pythagoreans could prove this general conclusion, however, is doubtful.

To pass from one square number to the next one, the Pythagoreans had the scheme shown in Figure 3.3. The dots to the right of and below the lines in the figure formed what they called a gnomon. Symbolically, what they saw here was that $n^2 + (2n + 1) = (n + 1)^2$. Further, if we start with 1 and

Figure 3.3

Figure 3.4. The shaded area is the gnomon.

add the gnomon 3 and then the gnomon 5, and so forth, what we have in our symbolism is

$$1 + 3 + 5 + \cdots + (2n - 1) = n^2.$$

As to the word "gnomon," originally in Babylonia it probably meant an upright stick whose shadow was used to tell time. In Pythagoras' time it meant a carpenter's square, and this is the shape of the above gnomon. It also meant what was left over from a square when a smaller square was cut out of one corner. Later, with Euclid, it meant what was left from a parallelogram when a smaller one was cut out of one corner provided that the parallelogram in the lower right-hand corner was similar to the one cut out (Fig. 3.4).

The Pythagoreans also worked with polygonal numbers such as pentagonal, hexagonal, and higher ones. As we can see from Figure 3.5, where each dot represents a unit, the first pentagonal number is 1, the second, whose dots form the vertices of a pentagon, is 5; the third is $1 + 4 + 7$, or 12, and so forth. The nth pentagonal number, in our notation, is $(3n^2 - n)/2$. Likewise the hexagonal numbers (Fig. 3.6) are 1, 6, 15, 28, ... and generally $2n^2 - n$.

A number that equaled the sum of its divisors including 1 but not the number itself was called perfect; for example, 6, 28, and 496. Those exceeding the sum of the divisors were called excessive and those which were less were called defective. Two numbers were called amicable if each was the sum of the divisors of the other, for example, 284 and 220.

The Pythagoreans devised a rule for finding triples of integers which could be the sides of a right triangle. This rule implies knowledge of the Pythagorean theorem, about which we shall say more later. They found that when m is odd, then m, $(m^2 - 1)/2$, and $(m^2 + 1)/2$ are such a triple. However,

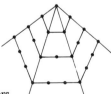

Figure 3.5. Pentagonal numbers

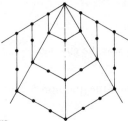

Figure 3.6. Hexagonal numbers

this rule gives only some sets of such triples. Any set of three integers which can be the sides of a right triangle is now called a Pythagorean triple.

The Pythagoreans studied prime numbers, progressions, and those ratios and proportions they regarded as beautiful. Thus if p and q are two numbers, the arithmetic mean A is $(p + q)/2$, the geometric mean G is $\sqrt{pq}$, and the harmonic mean H, which is the reciprocal of the arithmetic mean of $1/p$ and $1/q$, is $2pq/(p + q)$. Now G is seen to be the geometric mean of A and H. The proportion $A/G = G/H$ was called the perfect proportion and the proportion $p:(p + q)/2 = 2pq/(p + q):q$ was called the musical proportion.

Numbers to the Pythagoreans meant whole numbers only. A ratio of two whole numbers was not a fraction and therefore another kind of number, as it is in modern times. Actual fractions, expressing parts of a monetary unit or a measure, were employed in commerce, but such commercial uses of arithmetic were outside the pale of Greek mathematics proper. Hence the Pythagoreans were startled and disturbed by the discovery that some ratios— for example, the ratio of the hypotenuse of an isosceles right triangle to an arm or the ratio of a diagonal to a side of a square—cannot be expressed by whole numbers. Since the Pythagoreans had concerned themselves with whole-number triples that could be the sides of a right triangle, it is most likely that they discovered these new ratios in this work. They called ratios expressed by whole numbers commensurable ratios, which means that the two quantities are measured by a common unit, and they called ratios not so expressible, incommensurable ratios. Thus what we express as $\sqrt{2}/2$ is an incommensurable ratio. The ratio of incommensurable magnitudes was called αλογος (alogos, inexpressible). The term αρρητος (arratos, not having a ratio) was also used. The discovery of incommensurable ratios is attributed to Hippasus of Metapontum (5th cent. B.C.). The Pythagoreans were supposed to have been at sea at the time and to have thrown Hippasus overboard for having produced an element in the universe which denied the Pythagorean doctrine that all phenomena in the universe can be reduced to whole numbers or their ratios.

The proof that $\sqrt{2}$ is incommensurable with 1 was given by the Pythagoreans. According to Aristotle, their method was a *reductio ad absurdum*—that is, the indirect method. The proof showed that if the hypotenuse were commensurable with an arm then the same number would be both odd and even. It runs as follows: Let the ratio of hypotenuse to arm of an isosceles right triangle be $\alpha:\beta$ and let this ratio be expressed in the smallest numbers. Then $\alpha^2 = 2\beta^2$ by the Pythagorean theorem. Since α^2 is even, α must be even, for the square of any odd number is odd.[2] Now the ratio $\alpha:\beta$ is in its lowest terms. Hence β must be odd. Since α is even, let $\alpha = 2\gamma$. Then $\alpha^2 = 4\gamma^2 = 2\beta^2$. Hence $\beta^2 = 2\gamma^2$ and so β^2 is even. Then β is even. But β is also odd and so there is a contradiction.

This proof, which is of course the same as the modern one that $\sqrt{2}$ is irrational, was included in older editions of Euclid's *Elements* as Proposition 117 of Book X. However, it was most likely not in Euclid's original text and so is omitted in modern editions.

Incommensurable ratios are expressed in modern mathematics by irrational numbers. But the Pythagoreans would not accept such numbers. The Babylonians did work with such numbers by approximating them, though they probably did not know that their sexagesimal fractional approximations could never be made exact. Nor did the Egyptians recognize the distinctive nature of irrationals. The Pythagoreans did at least recognize that incommensurable ratios are entirely different in character from commensurable ones.

This discovery posed a problem that was central in Greek mathematics. The Pythagoreans had, up to this point, identified number with geometry. But the existence of incommensurable ratios shattered this identification. They did not cease to consider all kinds of lengths, areas, and ratios in geometry, but they restricted the consideration of numerical ratios to commensurable ones. The theory of proportions for incommensurable ratios and all kinds of magnitudes was provided by Eudoxus, whose work we shall consider shortly.

Some geometrical results are also credited to the Pythagoreans. The most famous is the Pythagorean theorem itself, a key theorem of Euclidean geometry. The Pythagoreans are also supposed to have discovered what we learn as theorems about triangles, parallel lines, polygons, circles, spheres, and the regular polyhedra. They knew in particular that the sum of the angles of a triangle is 180°. A limited theory of similar figures and the fact that a plane can be filled out with equilateral triangles, squares, and regular hexagons are included among their results.

The Pythagoreans started work on a class of problems known as

---

2. Any odd whole number can be expressed as $2n + 1$ for some n. Then $(2n + 1)^2 = 4n^2 + 4n + 1$, and this is necessarily odd.

application of areas. The simplest of these was to construct a polygon equal in area to a given polygon and similar to another given one. Another was to construct a specified figure with an area exceeding or falling short of another by a given area. The most important form of the problem of application of areas is: Given a line segment, construct on part of it or on the line segment extended a parallelogram equal to a given rectilinear figure in area and falling short (in the first case) or exceeding (in the second case) by a parallelogram similar to a given parallelogram. We shall discuss application of areas when we study Euclid's work.

The most vital contribution of the Greeks to mathematics is the insistence that all mathematical results be established deductively on the basis of explicit axioms. Hence the question arises as to whether the Pythagoreans proved their geometric results. No unequivocal answer can be given, but it is very doubtful that deductive proof on any kind of axiomatic basis, explicit or implicit, was a requirement in the early or middle period of Pythagorean mathematics. Proclus does affirm that they proved the angle sum theorem; this may have been done by the late Pythagoreans. The question of whether they proved the Pythagorean theorem has been extensively pursued, and the answer is that they probably did not. It is relatively easy to prove it by using facts about similar triangles, but the Pythagoreans did not have a complete theory of similar figures. The proof given in Proposition 47 of Book I of Euclid's *Elements* (Chap. 4, sec. 4) is a difficult one because it does not use the theory of similar figures, and this proof was credited by Proclus to Euclid himself. The most likely conclusion about proof in Pythagorean geometry is that during most of the life of the school the members affirmed results on the basis of special cases, much as they did in their arithmetic. However, by the time of the late Pythagoreans, that is, about 400 B.C., the status of proof had changed because of other developments; so these latter-day members of the brotherhood may have given legitimate proofs.

6. *The Eleatic School*

The Pythagorean discovery of incommensurable ratios brought to the fore a difficulty that preoccupied all the Greeks, namely, the relation of the discrete to the continuous. Whole numbers represent discrete objects, and a commensurable ratio represents a relation between two collections of discrete objects, or two lengths that have a common unit measure so that each length is a discrete collection of units. However, lengths in general are not discrete collections of units; this is why ratios of incommensurable lengths appear. Lengths, areas, volumes, time, and other quantities are, in other words, continuous. We would say that line segments, for example, can have irrational as well as rational lengths in terms of some unit. But the Greeks had not attained this view.

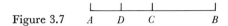

Figure 3.7 A D C B

The problem of the relation of the discrete to the continuous was brought into the limelight by Zeno, who lived in the southern Italian city of Elea. Born some time between 495 and 480 B.C., Zeno was a philosopher rather than a mathematician, and like his master Parmenides was said to have been a Pythagorean originally. He proposed a number of paradoxes, of which four deal with motion. His purpose in posing these paradoxes is not clear because not enough of the history of Greek philosophy is known. He was said to be defending Parmenides, who had argued that motion or change is impossible. He was also attacking the Pythagoreans, who believed in extended but indivisible units, the points of geometry. We do not know precisely what Zeno said but must rely upon quotations from Aristotle, who cites Zeno in order to criticize him, and from Simplicius, who lived in the sixth century A.D. and based his statements on Aristotle's writings.

The four paradoxes on motion are distinct, but the import of all four taken together was probably intended to be the significant argument. Two opposing views of space and time were held in Zeno's day: one, that space and time are infinitely divisible, in which case motion is continuous and smooth; and the other, that space and time are made up of indivisible small intervals (like a movie), in which case motion is a succession of minute jerks. Zeno's arguments are directed against both theories, the first two paradoxes being against the first theory and the latter two against the second theory. The first paradox of each pair considers the motion of a single body and the second considers the relative motion of bodies.

Aristotle in his *Physics* states the first paradox, called the Dichotomy, as follows: "The first asserts the nonexistence of motion on the ground that that which is in motion must arrive at the half-way stage before it arrives at the goal." This means that to traverse AB (Fig. 3.7), one must first arrive at C; to arrive at C one must first arrive at D; and so forth. In other words, on the assumption that space is infinitely divisible and therefore that a finite length contains an infinite number of points, it is impossible to cover even a finite length in a finite time.

Aristotle, refuting Zeno, says there are two senses in which a thing may be infinite: in divisibility or in extent. In a finite time one can come into contact with things infinite in respect to divisibility, for in this sense time is also infinite; and so a finite extent of time can suffice to cover a finite length. Zeno's argument has been construed by others to mean that to go a finite length one must cover an infinite number of points and so must get to the end of something that has no end.

The second paradox is called Achilles and the Tortoise. According to

Figure 3.8

Aristotle: "It says that the slowest moving object cannot be overtaken by the fastest since the pursuer must first arrive at the point from which the pursued started so that necessarily the slower one is always ahead. The argument is similar to that of the Dichotomy, but the difference is that we are not dividing in halves the distances which have to be passed over." Aristotle then says that if the slowly moving object covers a finite distance, it can be overtaken for the same reason he gives in answering the first paradox.

The next two paradoxes are directed against "cinematographic" motion. The third paradox, called the Arrow, is given by Aristotle as follows: "The third paradox he [Zeno] spoke about, is that a moving arrow is at a standstill. This he concludes from the assumption that time is made up of instants. If it would not be for this supposition, there would be no such conclusion." According to Aristotle, Zeno means that at any instant during its motion the arrow occupies a definite position and so is at rest. Hence it cannot be in motion. Aristotle says that this paradox fails if we do not grant indivisible units of time.

The fourth paradox, called the Stadium or the Moving Rows, is put by Aristotle in these words: "The fourth is the argument about a set of bodies moving on a race-course and passing another set of bodies equal in number and moving in the opposite direction, the one starting from the end, the other from the middle and both moving at equal speed; he [Zeno] concluded that it follows that half the time is equal to double the time. The mistake is to assume that two bodies moving at equal speeds take equal times in passing, the one a body which is in motion, and the other a body of equal size which is at rest, an assumption which is false."

The probable point of Zeno's fourth paradox can be stated as follows: Suppose that there are three rows of soldiers, A, B, and C (Fig. 3.8), and that in the smallest unit of time B moves one position to the left, while in that time C moves one position to the right. Then relative to B, C has moved two positions. Hence there must have been a smaller unit of time in which C was one position to the right of B or else half the unit of time equals the unit of time.

It is possible that Zeno merely intended to point out that speed is relative. C's speed relative to B is not C's speed relative to A. Or he may have meant there is no absolute space to which to refer speeds. Aristotle says that Zeno's fallacy consists in supposing that things that move with the same speed past a moving object and past a fixed object take the same time. Neither Zeno's argument nor Aristotle's answer is clear. But if we think of

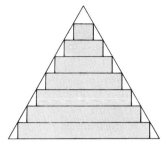

Figure 3.9

this paradox as attacking indivisible smallest intervals of time and indivisible smallest segments of space, which Zeno was attacking, then his argument makes sense.

We may include with the Eleatics Democritus (*c.* 460–*c.* 370 B.C.) of Abdera in Thrace. He is reputed to have been a man of great wisdom who worked in many fields, including astronomy. Since Democritus belonged to the school of Leucippus and the latter was a pupil of Zeno, many of the mathematical questions Democritus considered must have been suggested by Zeno's ideas. He wrote works on geometry, on number, and on continuous lines and solids. The works on geometry could very well have been significant predecessors of Euclid's *Elements*.

Archimedes says Democritus discovered that the volumes of a cone and a pyramid are 1/3 of the volumes of the cylinder and prism having the same base and height, but that the proofs were made by Eudoxus. Democritus regarded the cone as a series of thin indivisible layers (Fig. 3.9), but was troubled by the fact that if the layers were equal they should yield a cylinder and if unequal the cone could not be smooth.

7. *The Sophist School*

After the final defeat of the Persians at Mycale in 479 B.C., Athens became the major city in a league of Greek cities and a commercial center. The wealth acquired through trading, which made Athens the richest city of its time, was used by the famous leader Pericles to build up and adorn the city. Ionians, Pythagoreans, and intellectuals generally were attracted to Athens. Here emphasis was given to abstract reasoning and the goal of extending the domain of reason over the whole of nature and man was set.

The first Athenian school, the Sophist, embraced learned teachers of grammar, rhetoric, dialectics, eloquence, morals, and—what is of interest to us—geometry, astronomy, and philosophy. One of their chief pursuits was the use of mathematics to understand the functioning of the universe.

Many of the mathematical results obtained were by-products of efforts to solve the three famous construction problems: to construct a square equal in area to a given circle; to construct the side of a cube whose volume is double that of a cube of given edge; and to trisect any angle—all to be performed with straightedge and compass only.

The origin of these famous construction problems is accounted for in various ways. For example, one version of the origin of the problem of doubling the cube, found in a work of Eratosthenes (c. 284–192 B.C.), relates that the Delians, suffering from a pestilence, consulted the oracle, who advised constructing an altar double the size of the existing one. The Delians realized that doubling the side would not double the volume and turned to Plato, who told them that the god of the oracle had not so answered because he wanted or needed a doubled altar, but in order to censure the Greeks for their indifference to mathematics and their lack of respect for geometry. Plutarch also gives this story.

Actually, these construction problems are extensions of problems already solved by the Greeks. Since any angle could be bisected, it was natural to consider trisection. Since the diagonal of a square is the side of a square double in area to that of the original square, the corresponding problem for the cube becomes relevant. The problem of squaring the circle is typical of many Greek problems of constructing a figure of prescribed shape equal in area to a given figure. Another problem not quite so famous was to construct regular polygons of seven and more sides. Here, too, the construction of the square, regular pentagon, and regular hexagon suggested the next step.

Various explanations of the restriction to straightedge and compass have been given. The straight line and the circle were, in the Greek view, the basic figures, and the straightedge and compass are their physical analogues. Hence constructions with these tools were preferable. The reason is also given that Plato objected to other mechanical instruments because they involved too much of the world of the senses rather than the world of ideas, which he regarded as primary. It is very likely, however, that in the fifth century the restriction to straightedge and compass was not rigid. But, as we shall see, constructions played a vital role in Greek geometry and Euclid's axioms did limit constructions to those made with straightedge and compass. Hence from his time on, this restriction may have been taken more seriously. Pappus, for example, says that if a construction can be carried out with straightedge and compass, a solution using other means is not satisfactory.

The earliest known attempt to solve any of the three famous problems was made by the Ionian Anaxagoras, who is supposed to have worked on squaring the circle while in prison. We do not know any more about this work. One of the most famous attempts is due to Hippias of Elis, a city in the Peloponnesus. Hippias, a leading Sophist, was born about 460 B.C. and was a contemporary of Socrates.

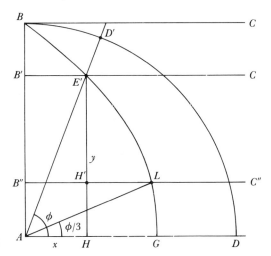

Figure 3.10

In his attempts to trisect an angle, Hippias invented a new curve, which, unfortunately, is not itself constructible with straightedge and compass. His curve is called the quadratrix and is generated as follows: Let AB (Fig. 3.10) rotate clockwise about A at a constant speed to the position AD. In the *same time* let BC move downward parallel to itself at a uniform speed to AD. Suppose AB reaches AD' as BC reaches $B'C'$. Let E' be the intersection of AD' and $B'C'$. Then E' is a typical point on the quadratrix $BE'G$. G is the final point on the quadratrix.[3]

The equation of the quadratrix in terms of rectangular Cartesian coordinates can be obtained as follows: Let AD' reach AD in some fraction t/T of the total time T that AB takes to reach AD. Since AD' and $B'C'$ move at constant speeds, $B'C'$ covers that part $E'H$ of BA in the same fraction of the total time. Hence

$$\frac{\phi}{\pi/2} = \frac{E'H}{BA}.$$

If we denote $E'H$ by y and BA by a, then

(1)
$$\frac{\phi}{\pi/2} = \frac{y}{a}$$

3. The point G cannot be obtained directly from the definition of the curve because AB reaches AD at the same instant as BC reaches AD and so there is no point of intersection of the rotating line and the horizontal line. G can be obtained only as the limit of preceding points of the quadratrix. By using the calculus we can show that $AG = 2a/\pi$ where $a = AB$.

or

$$y = a \cdot \phi \cdot \frac{2}{\pi}.$$

But if $AH = x$, then

$$\phi = \arctan \frac{y}{x}.$$

Hence

$$y = \frac{2a}{\pi} \arctan \frac{y}{x}$$

or

$$y = x \tan \frac{\pi y}{2a}.$$

The curve, if constructible, could be used to trisect any acute angle. Let ϕ be such an angle. Then trisect y so that $E'H' = 2H'H$. Draw $B''C''$ through H' and let it cut the quadratrix in L. Draw AL. Then $\angle LAD = \phi/3$, for, by the argument which led to (1),

$$\frac{\angle LAD}{\pi/2} = \frac{H'H}{a}$$

or

$$\frac{\angle LAD}{\pi/2} = \frac{y/3}{a}.$$

But by (1)

$$\frac{\phi}{\pi/2} = \frac{y}{a}.$$

Hence

$$\angle LAD = \frac{\phi}{3}.$$

Another famous discovery that resulted from the work on the construction problems was made by Hippocrates of Chios (5th cent. B.C.), the most famous mathematician of his century, who is to be distinguished from his contemporary Hippocrates of Cos, the father of Greek medicine. The mathematician Hippocrates flourished in Athens during the second half of the century; he was not a Sophist, but most likely a Pythagorean. He is credited with the idea of arranging theorems so that later ones can be proven on the

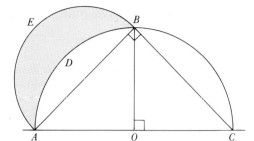

Figure 3.11

basis of earlier ones, in the manner familiar to us from the study of Euclid. He is also credited with introducing the indirect method of proof into mathe-. matics. His text on geometry, called the *Elements*, is lost.

Hippocrates did not, of course, solve the problem of squaring the circle, but he did solve a related one. Let *ABC* be an isosceles right triangle (Fig. 3.11) and let it be inscribed in the semicircle with center at *O*. Let *AEB* be the semicircle with *AB* as diamter. Then

$$\frac{\text{Area semicircle } ABC}{\text{Area semicircle } AEB} = \frac{AC^2}{AB^2} = \frac{2}{1}.$$

Hence the area *OADB* equals the area of the semicircle *AEB*. Now subtract the area *ADB* common to both. Then the area of the lune (shaded area) equals the area of triangle *AOB*. Thus the area of the lune, an area bounded by arcs, equals the area of a rectilinear figure; or, a curvilinear figure is reduced to a rectilinear one. This result is called a quadrature; that is, a curvilinear area has been computed in effect because it is equal to an area bounded by straight lines and the latter can be computed.

In this proof Hippocrates had to use the fact that the areas of two circles are to each other as the squares of their diameters. It is doubtful that Hippocrates really had a proof of this fact because the proof depends upon the method of exhaustion invented later by Eudoxus.

Hippocrates also squared three other lunes. This work on the lunes is known to us through the writings of Simplicius and is the only sizable fragment of *classical* Greek mathematics that we have as originally written.

Hippocrates also showed that the problem of doubling the cube can be reduced to finding two mean proportionals between the given side and one twice as long. In our algebraic notation, let *x* and *y* be such that

$$\frac{a}{x} = \frac{x}{y} = \frac{y}{2a}.$$

Then

$$x^2 = ay \quad \text{and} \quad y^2 = 2ax.$$

Since $y = x^2/a$ we obtain from the second equation that

$$x^3 = 2a^3.$$

This x is the desired answer. It cannot be constructed by straightedge and compass. Of course Hippocrates must have reasoned geometrically, in a manner that will be clearer when we examine Apollonius' *Conic Sections*.

One more very important idea was hit upon by the Sophists Antiphon (5th cent. B.C.) and Bryson (*c.* 450 B.C.). While trying to square the circle, Antiphon conceived the idea of approaching the area of a circle by inscribing polygons of more and more sides. Bryson added to this the idea of using circumscribed polygons. Antiphon further suggested that the circle be considered a polygon of an infinite number of sides. We shall see how these ideas were taken up by Eudoxus in the method of exhaustion (Chap. 4, sec. 9).

8. *The Platonic School*

The Platonic school succeeded the Sophists in the leadership of mathematical activity. Its forerunners, Theodorus of Cyrene in North Africa (born *c.* 470 B.C.) and Archytas of Tarentum in southern Italy (428–347 B.C.), were Pythagoreans and both taught Plato. Their teachings may have produced the strong Pythagorean influence in the entire Platonic school.

Theodorus is noted for having proved that the ratios that we represent as $\sqrt{3}, \sqrt{5}, \sqrt{7}, \ldots, \sqrt{17}$ are incommensurable with a unit. Archytas introduced the idea of regarding a curve as generated by a moving point and a surface as generated by a moving curve. He solved the duplication of the cube problem by finding two mean proportionals between two given quantities. These mean proportionals were constructed geometrically by finding the intersection of three surfaces: a circle rotated about a tangent, a cone, and a cylinder. The construction is quite detailed and does not warrant space here. Archytas also wrote on mathematical mechanics, designed machines, studied sound, and contributed inventions and some theory on musical scales.

The Platonic school was headed by Plato and included Menaechmus (4th cent. B.C.), his brother Dinostratus (4th cent. B.C.), and Theaetetus (*c.* 415–*c.* 369 B.C.). Many other members are known to us only by name.

Plato (427–347 B.C.) was born of a distinguished family and early in life had political ambitions. But the fate of Socrates convinced him there was no place in politics for a man of conscience. He traveled in Egypt and among the Pythagoreans in lower Italy; the Pythagorean influence may have been generated by these contacts. About 387 B.C. Plato founded in Athens his Academy, which in most respects was like a modern university. The Academy had grounds, buildings, students, and formal courses taught by Plato and his aides. During the classical period the study of mathematics and philosophy was favored there. Though the main center for mathematics shifted to Alexandria about 300 B.C., the Academy remained preeminent in philosophy throughout the Alexandrian period. It lasted nine hundred years until it was closed by the Christian emperor Justinian in A.D. 529 because it taught " pagan and perverse learning."

Plato, one of the most informed men of his day, was not a mathematician; but his enthusiasm for the subject and his belief in its importance for philosophy and for the understanding of the universe encouraged mathematicians to pursue it. It is noteworthy that almost all of the important mathematical work of the fourth century was done by the friends and pupils of Plato. Plato himself seems to have been more concerned to improve and perfect what was known.

Though we may not be sure to what extent the concepts of mathematics were treated as abstractions prior to Plato's time, there is no question that Plato and his successors did so regard them. Plato says that numbers and geometrical concepts have nothing material in them and are distinct from physical things. The concepts of mathematics are independent of experience and have a reality of their own. They are discovered, not invented or fashioned. This distinction between abstractions and material objects may have come from Socrates.

A quotation from Plato's *Republic* may serve to illustrate the contemporary view of the mathematical concepts. Socrates addresses Glaucon:

> And all arithmetic and calculation have to do with number.
> Yes....
> Then this is knowledge of the kind for which we are seeking, having a double use, military and philosophical; for the man of war must learn the art of number or he will not know how to array his troops, and the philosopher also, because he has to rise out of the sea of change and lay hold of true being, and therefore he must be an arithmetician.... Then this is a kind of knowledge which legislation may fitly prescribe; and we must endeavour to persuade those who are to be the principal men of our State to go and learn arithmetic, not as amateurs, but they must carry on the study until they see the nature of numbers with the mind only; nor again, like merchants or retail-traders, with a view to buying or selling, but for the sake of their military use, and of the soul herself; and because this will be the easiest way for her to pass from becoming to truth and being.... I mean, as I was saying, that arithmetic has a very great and elevating effect, compelling the soul to reason about abstract number, and rebelling against the introduction of visible or tangible objects into the argument....[4]

4. Book VII, sec. 525; in B. Jowett, *The Dialogues of Plato*, Clarendon Press, 1953, Vol. 2.

Another quotation[5] discusses the concepts of geometry. Speaking about mathematicians, Plato says: "And do you not know also that although they further make use of the visible forms and reason about them, they are thinking not of these, but of the ideals which they resemble . . . but they are really seeking to behold the things themselves, which can be seen only with the eye of the mind."

It is clear from these quotations that Plato and other Greeks for whom he speaks valued abstract ideas and preferred mathematical ideas as a preparation for philosophy. The abstract ideas with which mathematics deals are akin to others such as goodness and justice, the understanding of which is the goal of Plato's philosophy. Mathematics is the preparation for knowledge about the ideal universe.

Why did the Greeks prefer and stress the abstract concepts of mathematics? We cannot answer the question, but we should note that the early Greek mathematicians were philosophers and philosophers generally exerted a formative influence in the development of Greek mathematics. Philosophers are interested in ideas and typically show their propensity for abstractions in many domains. Thus the Greek philosophers thought about truth, goodness, charity, and intelligence. They speculated about the ideal society and the perfect state. The later Pythagoreans and the Platonists distinguished sharply between the world of ideas and the world of things. Relationships in the material world were subject to change and hence did not represent ultimate truth, but relationships in the ideal world were unchanging and absolute truths; these were the proper concern of the philosopher.

Plato in particular believed that the perfect ideals of physical objects are the reality. The world of ideals and relationships among them is permanent, ageless, incorruptible, and universal. The physical world is an imperfect realization of the ideal world and is subject also to decay. Hence the ideal world alone is worthy of study. Infallible knowledge can be obtained only about pure intelligible forms. About the physical world we can have only opinions; and physical science is sunk in the dregs of a sensuous world.

We are not sure whether the Platonists contributed the deductive structure of mathematics. They were concerned with proof and with the methodology of reasoning. Proclus and Diogenes Laertius (3rd cent. A.D.) credit the Platonists with two types of methodology. The first is the method of analysis, where what is to be established is regarded as known and the consequences deduced until a known truth or a contradiction is reached. If a contradiction is reached then the desired conclusion is false. If a known truth is reached then the steps are reversed, if possible, and the proof is made. The second is the method of *reductio ad absurdum* or the indirect method. The first method was probably not new with Plato, but perhaps he emphasized

5. *Republic*, Book VI, sec. 510.

the necessity for the subsequent synthesis. The indirect method, as already noted, is also attributed to Hippocrates.

The status of deductive structure with Plato is best indicated by a passage in *The Republic*.[6] He says

> You are aware that students of geometry, arithmetic and the kindred sciences assume the odd and the even and the figures and three kinds of angles and the like in their several branches of science; these are their hypotheses, which they and everybody are supposed to know, and therefore they do not deign to give any account of them either to themselves or others; but they begin with them, and go on until they arrive at last, and in a consistent manner, at their conclusion.

If this passage is indeed descriptive of the mathematics of the time, then proofs were certainly made, but the axiomatic basis was implicit or may have varied somewhat from one mathematician to another.

Plato did affirm the desirability of a deductive organization of knowledge. The task of science was to discover the structure of (ideal) nature and to give it an articulation in a deductive system. Plato was the first to systematize the rules of rigorous demonstration, and his followers are supposed to have arranged theorems in logical order. Also, we know that in Plato's Academy the question arose whether a given problem can be solved at all on the basis of the known truths and the hypotheses given in the problem. Whether or not mathematics was actually deductively organized on the basis of explicit axioms by the Platonists, there is no question that deductive proof from some accepted principles was required from at least Plato's time onward. By insisting on this form of proof the Greeks were discarding all rules, procedures, and facts that had been accepted in the body of mathematics for thousands of years preceding the Greek period.

Why did the Greeks insist on deductive proof? Since induction, observation, and experimentation were and still are vital sources of knowledge heavily and advantageously employed by the sciences, why did the Greeks prefer in mathematics deductive reasoning to the exclusion of all other methods? We know that the Greeks, the philosophical geometers as they were called, liked reasoning and speculation, as evidenced by their great contributions to philosophy, logic, and theoretical science. Moreover, philosophers are interested in obtaining truths. Whereas induction, experimentation, and generalizations based on experience yield only probable knowledge, deduction gives absolutely certain results if the premises are correct. Mathematics in the classical Greek world was part of the body of truths philosophers sought and accordingly had to be deductive.

Still another reason for the Greek preference for deduction may be found in the contempt shown by the educated class of the classical Greek

6. Book VI, sec. 510.

period toward practical affairs. Though Athens was a commercial center, business as well as such professions as medicine were carried on by the slave class. Plato contended that the engagement of freemen in trade should be punished as a crime, and Aristotle said that in the perfect state no citizen (as opposed to slaves) should practice any mechanical art. To thinkers in such a society, experimentation and observation would be alien. Hence no results, scientific or mathematical, would be derived from such sources.

There is, incidentally, evidence that in the sixth and fifth centuries B.C. the Greek attitude toward work, trade, and technical skills had been quite different and that mathematics had been applied to the practical arts. Thales used his mathematics to improve navigation. Solon, a ruler of the sixth century, invested the crafts with honor and inventors were esteemed. *Sophia*, the Greek word usually taken to mean wisdom and abstract thought, at that time meant technical skill. It was the Pythagoreans, Proclus says, who "transformed mathematics into a free education," that is, an education for free men rather than a skill for slaves.

Plutarch in his life of Marcellus substantiates the change in attitude toward such devices as mechanical instruments:

> Eudoxus and Archytas had been the first originators of this far-famed and highly-prized art of mechanics, which they employed as an elegant illustration of geometrical truths, and as means of sustaining experimentally, to the satisfaction of the senses, conclusions too intricate for proof by words and diagrams. As, for example, to solve the problem, so often required in constructing geometrical figures, given the two extremes, to find the two mean lines of a proportion, both of these mathematicians had recourse to the aid of instruments, adapting to their purpose certain curves and sections of lines. But what with Plato's indignation at it, and his invectives against it as the mere corruption and annihilation of the one good of geometry, which was thus shamefully turning its back upon the unembodied objects of pure intelligence to recur to sensation, and to ask help (not to be obtained without base supervisions and deprivation) from matter, so it was that mechanics came to be separated from geometry, and, repudiated and neglected by philosophers, took its place as a military art.

This accounts for the poor development of experimental science and the science of mechanics in the classical Greek period.

Whether or not historical research has isolated the relevant factors to explain the Greek preference for deductive reasoning, we do know that they were the first to insist on deductive reasoning as the sole method of *proof* in mathematics. This requirement has been characteristic of mathematics ever since and has distinguished mathematics from all other fields of knowledge or investigation. However, we have yet to see to what extent later mathematicians remained true to this principle.

Figure 3.12

As far as the content of mathematics is concerned, Plato and his school improved the definitions and are also supposed to have proved new theorems of plane geometry. Further, they gave impetus to solid geometry. In Book VII, section 528 of *The Republic*, Plato says that before one can consider astronomy, which treats solids in motion, one needs a science of such solids. But this science, he says, has been neglected. He complains that the investigators of solid figures have not received due support from the state. Plato and his associates proceeded to study solid geometry and are supposed to have proved new theorems. They studied the properties of the prism, pyramid, cylinder, and cone; and they knew that there can be at most five regular polyhedra. The Pythagoreans undoubtedly knew that one can form three of these solids with 4, 8, and 20 equilateral triangles, the cube with squares, and the dodecahedron with 12 pentagons, but the proof that there can be no more than five is probably due to Theaetetus.

The Platonic school's most significant discovery was the conic sections. The discovery is attributed by the Alexandrian Eratosthenes to Menaechmus, a geometer and astronomer, who was a pupil of Eudoxus but a member of Plato's Academy. While it is not known for certain what led to the discovery of the conic sections, a common belief is that it resulted from the work on the famous construction problems. We know that Hippocrates of Chios solved the problem of doubling the cube by finding an x and y such that

$$a:x = x:y = y:2a.$$

But these equations say that

$$x^2 = ay, \qquad y^2 = 2ax, \quad \text{and} \quad xy = 2a^2.$$

Hence we can see through coordinate geometry that x and y are the coordinates of the point of intersection of two parabolas or a parabola and a hyperbola. Menaechmus worked on the problem and saw both ways of solving it through pure geometry. According to the mathematical historian Otto Neugebauer (1899–), the conic sections might have originated in work on the construction of sundials.

Menaechmus introduced the conic sections by using three types of cones (Fig. 3.12), right-angled, acute-angled, and obtuse-angled, and cutting each by a plane perpendicular to an element. Only one branch of the hyperbola was recognized at this time.

Among other mathematical studies made by the Platonists was Theaetetus' work on incommensurables. Previously Theodorus of Cyrene had proved that (in our notation and language) $\sqrt{3}$, $\sqrt{5}$, $\sqrt{7}$, and other square roots are irrational. Theaetetus investigated other and higher types of irrationals and classified them. We shall note these types when we study Book X of Euclid's *Elements*. In this work of Theaetetus we see how the number system was being extended to more irrationals, but only those incommensurable ratios were studied which arose from geometrical thinking and could be constructed geometrically as lengths. Dinostratus, another Platonist, showed how to use the quadratrix of Hippias to square the circle. Aristaeus the Elder (*c*. 320 B.C.) is said by Pappus to have written a work in five books called *Elements of Conic Sections*.

9. *The School of Eudoxus*

The greatest of the classical Greek mathematicians and second only to Archimedes in all antiquity was Eudoxus. Eratosthenes called him "godlike." He was born in Cnidos in Asia Minor about 408 B.C., studied under Archytas in Tarentum, traveled in Egypt, where he learned some astronomy, and then founded a school at Cyzicus in northern Asia Minor. About 368 B.C. he and his followers joined Plato. Some years later he returned to Cnidos and died there about 355 B.C. An astronomer, physician, geometer, legislator, and geographer, he is most noted for the creation of the first astronomical theory of the heavenly motions (Chap. 7).

His first great contribution to mathematics was a new theory of proportion. The discovery of more and more irrationals (incommensurable ratios) made it necessary for the Greeks to face these numbers. Were they in fact numbers? They occurred in geometrical arguments, whereas the whole numbers and ratios of whole numbers occurred both in geometry and in the general study of quantity. Moreover, how could proofs of geometry which had been made for commensurable lengths, areas, and volumes be extended to incommensurable ones?

Eudoxus introduced the notion of a magnitude (Chap. 4, sec. 5). It was not a number but stood for entities such as line segments, angles, areas, volumes, and time which could vary, as we would say, continuously. Magnitudes were opposed to numbers, which jumped from one value to another, as from 4 to 5. No quantitative values were assigned to magnitudes. Eudoxus then defined a ratio of magnitudes and a proportion, that is, an equality of two ratios, to cover commensurable and incommensurable ratios. However, again, no numbers were used to express such ratios. The concepts of ratio and proportion were tied to geometry, as we shall see when we study Book V of Euclid.

What Eudoxus accomplished was to avoid irrational numbers as numbers. In effect, he avoided giving numerical values to lengths of line segments, sizes of angles, and other magnitudes, and to ratios of magnitudes. While Eudoxus' theory enabled the Greek mathematicians to make tremendous progress in geometry by supplying the necessary logical foundation for incommensurable ratios, it had several unfortunate consequences.

For one thing, it forced a sharp separation between number and geometry, for only geometry could handle incommensurable ratios. It also drove mathematicians into the ranks of the geometers, and geometry became the basis for almost all rigorous mathematics for the next two thousand years. We still speak of x^2 as x square and x^3 as x cube instead of x second or x third, say, because the magnitudes x^2 and x^3 had only geometric meaning to the Greeks.

The Eudoxian solution to the problem of treating incommensurable lengths or the irrational number actually reversed the emphasis of previous Greek mathematics. The early Pythagoreans had certainly emphasized number as the fundamental concept, and Archytas of Tarentum, Eudoxus' teacher, stated that arithmetic alone, not geometry, could supply satisfactory proofs. However, in turning to geometry to handle irrational numbers, the classical Greeks abandoned algebra and irrational numbers as such. What did they do about solving quadratic equations, where the solutions can indeed be irrational numbers? And what did they do about the simple problem of finding the area of a rectangle whose sides are incommensurable? The answer is that they converted most of algebra to geometry, in a manner we shall examine in the next chapter. The geometric representation of irrationals and of operations with irrationals was, of course, not practical. It might be logically satisfactory to think of $\sqrt{2} \cdot \sqrt{3}$ as an area of a rectangle, but if one needed to know the product in order to buy floor covering, he would not have it.

Though the Greeks devoted their deepest efforts in mathematics to geometry, we must keep in mind that whole numbers and ratios of whole numbers were still acceptable concepts. This area of mathematics, as we shall see, was built up deductively in Books VII, VIII, and IX of Euclid's *Elements*. The material is essentially what we call the theory of numbers or the properties of integers.

The question also arises: What did the classical Greeks do about the need for numbers in scientific work and in commerce and other practical affairs? Classical Greek science, as we shall see, was qualitative. As for the practical uses of numbers, we mentioned earlier that the intellectuals of that period confined themselves to philosophical and scientific activities and took no hand in commerce or the trades; educated people did not concern themselves with practical problems. But one could think about all rectangles in geometry without concerning himself in the least with the actual dimensions of even one rectangle. Mathematical thought was thus separated from practical needs, and there was no compulsion for the mathematicians to improve arithmetical and algebraic techniques. When the barrier between the cultured and slave classes was breached in the Alexandrian period (300 B.C. to about A.D. 600) and educated men interested themselves in practical affairs, the emphasis shifted to quantitative knowledge and the development of arithmetic and algebra.

To return to the contributions of Eudoxus, the powerful Greek method of establishing the areas and volumes of curved figures, now called the method of exhaustion, is also due to him. We shall examine the method and its use, as given by Euclid, later. It is really the first step in the calculus but does not use an explicit theory of limits. With it Eudoxus proved, for example, that the areas of two circles are to each other as the squares of their radii, the volumes of two spheres are to each other as the cubes of their radii, the volume of a pyramid is one-third the volume of a prism of the same base and altitude, and the volume of a cone is one-third the volume of the corresponding cylinder.

Some authority can be found to credit every school from Thales' onward with having introduced the deductive organization of mathematics. There is no question, however, that the work of Eudoxus established the deductive organization on *the basis of explicit axioms*. The necessity for understanding and operating with incommensurable ratios is undoubtedly the reason for this step. Since Eudoxus undertook to provide the precise logical basis for these ratios, he most likely saw the need to formulate axioms and deduce consequences one by one so that no mistakes would be made with these unfamiliar and troublesome magnitudes. This need to work with incommensurable ratios also undoubtedly reinforced the earlier decision to rely only on deductive reasoning for proof.

Because the Greeks sought truths and had decided on deductive proof, they had to obtain axioms that were themselves truths. They did find statements whose truth was self-evident to them, though the justifications given for accepting the axioms as indisputable truths varied. Almost all Greeks believed that the mind was capable of recognizing truths. Plato applied his theory of anamnesis, that we have had direct experience of truths in a period of existence as souls in another world before coming to earth, and we have but to recall this experience to know that these truths included the axioms of geometry. No experience on earth is necessary. Some historians read into statements by Plato and Proclus the idea that there can be some arbitrariness in the axioms, provided only that they are clear and true in the mind of the individual. The important thing is to reason deductively on the basis of the ones chosen. Aristotle had a good deal to say about axioms and we shall note his views in a moment.

10. *Aristotle and His School*

Aristotle (384–322 B.C.) was born in Stageira, a city in Macedonia. For twenty years he was a pupil and colleague of Plato, and for three years, from 343 to 340 B.C., he was the tutor of Alexander the Great. In 335 B.C. he founded his own school, the Lyceum. It had a garden, a lecture room, and an altar to the Muses.

Aristotle wrote on mechanics, physics, mathematics, logic, meteorology, botany, psychology, zoology, ethics, literature, metaphysics, economics, and many other fields. There is no one book on mathematics but discussions of the subject occur in a variety of places, and he uses it to illustrate a number of points.

He viewed the sciences as falling into three types—theoretical, productive, and practical. The theoretical ones, which seek truth, are mathematics, physics (optics, harmonics, and astronomy), and metaphysics; of these mathematics is the most exact. The productive sciences are the arts; and the practical ones, for example ethics and politics, seek to regulate human actions. In the theoretical sciences, logic is preliminary to the several subjects included there, and the metaphysician discusses and explains what the mathematician and natural philosopher (scientist) take for granted, for example, the being or reality of the subject matter and the nature of axioms.

Though Aristotle did not contribute significant new mathematical results (a few theorems in Euclid are his), his views on the nature of mathematics and its relation to the physical world were highly influential. Whereas Plato believed that there was an independent, eternally existing world of ideas which constituted the reality of the universe and that mathematical concepts were part of this world, Aristotle favored concrete matter or substance. However, he too arrived at an *emphasis* on ideas, namely, the universal essences of physical objects, such as hardness, softness, heaviness, lightness, sphericity, coldness, and warmth. Numbers and geometrical forms, too, were properties of real objects; they were recognized by abstraction but belonged to the objects. Thus mathematics deals with abstract concepts, which are derived from properties of physical bodies.

Aristotle discusses definition. His notion of definition is modern; he calls it a name for a collection of words. He also points out that definition must be in terms of something prior to the thing defined. Thus he criticizes the definition, "a point is that which has no part," because the words "that which" do not say what they refer to, except possibly "point" and so the definition is not proper. He grants the need for undefined terms, since there must be a starting point for the series of definitions, but later mathematicians lost sight of this need until the end of the nineteenth century.

He also notes (as Plato, according to Plutarch, did earlier) that a definition tells us what a thing is but not that the thing exists. The existence of defined things has to be proved except in the case of a few primary things such as point and line, whose existence is assumed along with the first principles or axioms. Thus one can define a square, but such a figure may not exist; that is, the properties demanded in the definition may be incompatible. Leibniz gave the example of a regular polyhedron with ten faces; one can define such a figure but it does not exist. If one did not realize that this figure did not exist, and proceeded to prove theorems about it, his results would be nonsensical. The method of proving existence that Aristotle and Euclid adopted was construction. The first three axioms in Euclid's *Elements* grant the construction of straight lines and circles; all other mathematical concepts must be constructed to establish their existence. Thus angle trisectors, though definable, are not constructible with straight lines and circles and so could not be considered in Greek geometry.

Aristotle also treats the basic principles of mathematics. He distinguishes between the axioms or common notions, which are truths common to all sciences, and the postulates, which are acceptable first principles for any one science. Among axioms he includes logical principles, such as the law of contradiction, the law of excluded middle, the axiom that if equals are added or subtracted from equals the results are equal, and other such principles. The postulates need not be self-evident but their truth must then be attested to by the consequences derived from them. The collection of axioms and postulates should be of the fewest possible, provided they enable all the results to be proved. Though, as we shall see, Euclid uses Aristotle's distinction between common notions and postulates, all mathematicians up to the late nineteenth century overlooked this distinction and treated axioms and postulates as equally self-evident. According to Aristotle, the axioms are obtained from the observation of physical objects. They are immediately apprehended generalizations. He and his followers gave many definitions and axioms or improved earlier ones. Some of the Aristotelian versions were taken up by Euclid.

Aristotle discusses the fundamental problem of how points and lines can be related. A point, he says, is indivisible and has position. But then no accumulation of points, however far it may be carried, can give us anything divisible, whereas of course a line is a divisible magnitude. Hence points cannot make up anything continuous like a line, for point cannot be continuous with point. A point, he says, is like the now in time; now is indivisible and not a part of time. A point may be an extremity, beginning, or divider of a line but is not a part of it or of magnitude. It is only by *motion* that a point can generate a line and thus be the origin of magnitude. He also argues that a point has no length and so if a line were composed of points, it would have no length. Similarly, if time were composed of instants there would be no interval of time. His definition of continuity, which a line possesses, is: A thing is continuous when the limits at which any two successive parts touch are one and the same and are, as the word continuous implies, held together. Actually Aristotle makes many statements about continuous magnitudes which are not in agreement with each other. The substance of his doctrine, nevertheless, is that points and numbers are discrete quantities and must be distinguished from the continuous magnitudes of geometry. There is no continuum in arithmetic. As to the relation of the two fields, he considers arithmetic—that is, the theory of numbers—more accurate, because numbers lend themselves to abstraction more readily than the geometric concepts. He also considers arithmetic to be prior to geometry because the number three is needed to consider a triangle.

In discussing infinity he makes a distinction, which is important today, between the potentially infinite and the actually infinite. The age of the earth, if it had a sudden beginning, is potentially infinite but is never at any time actually infinite. According to Aristotle, only the potentially infinite exists. The positive integers, he grants, are potentially infinite in that we can always add 1 to any number and get a new one, but the infinite set as such does not exist. Further, most magnitudes cannot be even potentially infinite, because if they were continually added to they could exceed the bounds of the universe. Space, however, is potentially infinite, in that it can be repeatedly subdivided, and time is potentially infinite in both ways.

A major achievement of Aristotle was the founding of the science of logic. In producing correct laws of mathematical reasoning the Greeks had laid the groundwork for logic, but it took Aristotle to codify and systematize these laws into a separate discipline. Aristotle's writings make it abundantly clear that he derived logic from mathematics. His basic principles of logic— the law of contradiction, which asserts that a proposition cannot be both true and false, and the law of excluded middle, which maintains that a proposition must be either true or false—are the heart of the indirect method of mathematical proof. Further, Aristotle used mathematical examples taken from contemporary texts to illustrate his principles of reasoning. Aristotelian logic remained unchallenged until the nineteenth century.

Though the science of logic was derived from mathematics, logic eventually came to be considered independent of and prior to mathematics and applicable to all reasoning. Even Aristotle, as already noted, regarded logic as preliminary to science and philosophy. In mathematics he emphasized deductive proof as the sole basis for establishing facts. For Plato, who believed that mathematical truths preexist or exist in a world independent of man, reasoning was not the guarantee of the correctness of theorems; the logical powers played only a secondary role. They made explicit, so to speak, what was already known to be true.

One member of Aristotle's school especially worthy of note is Eudemus of Rhodes, who lived in the last part of the fourth century B.C. and was the author of the Eudemian summary quoted by Proclus and Simplicius. As we noted earlier, Eudemus wrote histories of arithmetic, geometry, and astronomy. He is the first historian of science on record. But what is more significant is that the knowledge already existing in his time should have been sufficiently extensive to warrant histories.

The last of the men of the classical period we shall mention here is Autolycus of Pitane, an astronomer and geometer, who flourished about 310 B.C. He was not a member of Plato's or Aristotle's schools, though he did teach one of the leaders who succeeded Plato. Of three books he wrote, two have come down to us; they are the earliest Greek books we have intact, though only through manuscripts that presumably are copies of Autolycus' work. These books, *On the Moving Sphere* and *On Risings and Settings*, were eventually included in a collection called the *Little Astronomy* (as distinguished from Ptolemy's later *Great Collection*, or the *Almagest*). *On the Moving Sphere* treats meridian circles, great circles generally, and what we would call parallels of latitude, as well as the visible and invisible areas produced by a distant light source shining on a rotating sphere, as the sun does on the earth. The book presupposes theorems of spherical geometry which must, therefore, have been known to the Greeks of that time. Autolycus' second book, on the rising and setting of stars, belongs to observational astronomy.

The form of the book on moving spheres is significant. Letters denote points on diagrams. The propositions are logically ordered. Each proposition is first stated generally, then repeated, but with explicit reference to the figure; finally, the proof is given. This is the style Euclid uses.

Bibliography

Apostle, H. G.: *Aristotle's Philosophy of Mathematics*, University of Chicago Press, 1952.

Ball, W. W. Rouse: *A Short Account of the History of Mathematics*, Dover (reprint), 1960, Chaps. 2–3.

Boyer, Carl B.: *A History of Mathematics*, John Wiley and Sons, 1968, Chaps. 4–6.

Brumbaugh, Robert S.: *Plato's Mathematical Imagination*, Indiana University Press, 1954.

Gomperz, Theodor: *Greek Thinkers*, 4 vols., John Murray, 1920.

Guthrie, W. K. C.: *A History of Greek Philosophy*, Cambridge University Press, 1962 and 1965, Vols. 1 and 2.

Hamilton, Edith: *The Greek Way to Western Civilization*, New American Library, 1948.

Heath, Thomas L.: *A History of Greek Mathematics*, Oxford University Press, 1921, Vol. 1.

———: *A Manual of Greek Mathematics*, Dover (reprint), 1963, Chaps. 4–9.

———: *Mathematics in Aristotle*, Oxford University Press, 1949.

Jaeger, Werner: *Paideia*, 3 vols., Oxford University Press, 1939–44.

Lasserre, François: *The Birth of Mathematics in the Age of Plato*, American Research Council, 1964.

Maziarz, Edward A., and Thomas Greenwood: *Greek Mathematical Philosophy*,
    F. Unger, 1968.
Sarton, George: *A History of Science*, Harvard University Press, 1952, Vol. 1, Chaps.
    7, 11, 16, 17, and 20.
Scott, J. F.: *A History of Mathematics*, Taylor and Francis, 1958, Chap. 2.
Smith, David Eugene: *History of Mathematics*, Dover (reprint), 1958, Vol. 1,
    Chap. 3.
van der Waerden, B. L.: *Science Awakening*, P. Noordhoff, 1954, Chaps. 4–6.
Wedberg, Anders: *Plato's Philosophy of Mathematics*, Almqvist and Wiksell, 1955.

4
Euclid and Apollonius

> We have learned from the very pioneers of this science not to
> have any regard to mere plausible imaginings when it is a
> question of the reasonings to be included in our geometrical
> doctrine.   PROCLUS

1. *Introduction*

The cream of the mathematical work created by the men of the classical
period has fortunately come down to us in the writings of two men, Euclid
and Apollonius. Chronologically, both belong to the second great period of
Greek history, the Hellenistic or Alexandrian. It is quite certain that Euclid
lived in Alexandria about 300 B.C. and trained students there, though his
own education was probably acquired in Plato's Academy. This information, incidentally, is about all we have on Euclid's personal life and even this
comes from a one-paragraph passage in Proclus' *Commentary*. Apollonius died
in 190 B.C., so his life too falls within the Alexandrian period. It is customary,
however, to identify Euclid's work with the classical period, because his
books are accounts of what was developed in that age. Euclid's work is
actually an organization of the separate discoveries of the classical Greeks;
this is clear from a comparison of its contents with what is known of the
earlier work. The *Elements* in particular is as much a mathematical history
of the age just brought to a close as it is the logical development of a subject.
Apollonius' work is generally classed with that of the Alexandrian period but
the content and spirit of his major work, *Conic Sections*, is of the classical
period. In fact Apollonius acknowledged that the first four of the work's
eight books were a revision of Euclid's lost work on the same subject. Pappus
mentions that Apollonius spent a long time with the pupils of Euclid at
Alexandria, which readily explains the kinship with Euclid. The justification
for identifying Apollonius with the work of the classical period will be more
apparent when we have learned the characteristics of the work of the
Alexandrian period.

2. *The Background of Euclid's* Elements

Euclid's most famous work is the *Elements*. The major sources of the material in this work can generally be identified despite our slim knowledge of the classical period. Euclid undoubtedly owes much of his material to the Platonists with whom he studied. Moreover, Proclus says that Euclid put into his *Elements* many of Eudoxus' theorems, perfected theorems of Theaetetus, and made irrefragable demonstrations of results loosely proved by his predecessors.

The particular choice of axioms, the arrangement of the theorems, and some proofs are his, as are the polish and the rigor of the demonstrations. The form of presentation of proof has, however, already been noted in Autolycus and was pretty surely used by others who preceded Euclid. Despite all he may have taken from earlier texts and other sources, Euclid was unquestionably a great mathematician. His other writings support this judgment despite any question as to how much of the *Elements* is original with Euclid. Proclus makes it clear that the *Elements* was highly valued in Greece and refers by way of substantiation to the numerous commentaries on it. Among the most important must have been those by Heron (*c.* 100 B.C.– *c.* A.D. 100), Porphyry (3rd cent. A.D.), and Pappus (end of 3rd cent. A.D.). Presumably Euclid's book was so good that it superseded the ones supposed to have been written by Hippocrates of Chios and by the Platonists Leon and Theudius.

There are no extant manuscripts written by Euclid himself. Hence his writings have had to be reconstructed from numerous recensions, commentaries, and remarks by other writers. All of the English and Latin editions of Euclid's *Elements* stem originally from Greek manuscripts. These were Theon of Alexandria's recension of Euclid's *Elements* (end of the 4th cent. A.D.), copies of Theon's recension, written versions of lectures by Theon, and one Greek manuscript found by François Peyrard (1760–1822) in the Vatican Library. This tenth-century manuscript is a copy of an edition of Euclid that precedes Theon's. Hence the historians J. L. Heiberg and Thomas L. Heath have used principally this manuscript for their study of Euclid, comparing it of course with the other available manuscripts and commentaries. There are also Arabic translations of Greek works and Arabic commentaries, presumably based on Greek manuscripts no longer available. These, too, have been used to decide what was in Euclid's *Elements*. But the Arabic translations and revisions are on the whole inferior to the Greek manuscripts. Of course, the reconstruction, since it is based on so many sources, leaves some matters in doubt. The purpose of Euclid's *Elements* is in question. It is considered by some as a treatise for learned mathematicians and by others as a text for students. Proclus gives some weight to the latter belief.

In view of the length and incomparable historical importance of this work, we shall devote several sections of this chapter to a review of and comment on the contents. Since we still learn Euclidean geometry, we may be somewhat surprised by the contents of the *Elements*. The high school versions most widely used during our century are patterned on Legendre's modification of Euclid's work. Some algebra used by Legendre is not in the *Elements*, though, as we shall see, the equivalent geometrical material is.

3. *The Definitions and Axioms of the* Elements

The *Elements* contains thirteen books. Two others, surely by later writers, were included in some editions. Book I begins with the definitions of the concepts to be used in the first part of the work. We shall note only the most important ones; these are numbered as in Heath's edition.[1]

DEFINITIONS

1. A point is that which has no part.

2. A line is breadthless length.
   The word line means curve.

3. The extremities of a line are points.
   This definition makes clear that a line or curve is always finite in length. A curve extending to infinity does not occur in the *Elements*.

4. A straight line is a line which lies evenly with the points on itself.
   In keeping with definition 3, the straight line of Euclid is our line segment. The definition is believed to be suggested by the mason's level or an eye looking along a line.

5. A surface is that which has length and breadth only.

6. The extremities of a surface are lines.
   Hence a surface, too, is a bounded figure.

7. A plane surface is a surface which lies evenly with the straight lines on itself.

15. A circle is a plane figure contained by one line such that all the straight lines falling upon it from one point among those lying within the figure are equal to one another.

16. And the point is called the center of the circle.

17. A diameter of the circle is any straight line drawn through the center and terminated in both directions by the circumference [not defined explicitly] of the circle, and such a straight line also bisects the circle.

1. T. L. Heath: *The Thirteen Books of Euclid's* Elements, Dover (reprint), 1956, 3 vols.

23. Parallel straight lines are straight lines which, being in the same plane and being produced indefinitely in both directions, do not meet one another in either direction.

The opening definitions are framed in terms of concepts that are not defined, and hence serve no logical purpose. Euclid might not have realized that the initial concepts must be undefined and was naïvely explaining their meaning in terms of physical concepts. Some commentators say he appreciated that the definitions were not logically helpful but wanted to explain what his terms represented intuitively so that his readers would be convinced that the axioms and postulates were applicable to these concepts.

Euclid next lays down five postulates and five common notions (Proclus uses the term "axiom" for "common notion"). He adopts the distinction already made by Aristotle, namely, that the common notions are truths applicable to all sciences whereas the postulates apply only to geometry. As we have noted, Aristotle said that the postulates need not be known to be true but that their truth would be tested by whether the results deduced from them agreed with reality. Proclus even speaks of all of mathematics as hypothetical; that is, it merely deduces what must follow from the assumptions, whether or not the latter are true. Presumably Euclid accepted Aristotle's views concerning the truth of the postulates. However, in the subsequent history of mathematics, both the postulates and the common notions were accepted as unquestionable truths, at least until the advent of non-Euclidean geometry.

Euclid postulates the following:

POSTULATES

1. [It is possible] to draw a straight line from any point to any point.

2. [It is possible] to extend a finite straight line continuously in a straight line.

3. [It is possible] to describe a circle with any center and distance [radius].

4. That all right angles are equal to one another.

5. That if a straight line falling on two straight lines makes the interior angles on the same side less than two right angles, the two straight lines, if produced indefinitely, meet on that side on which the angles are less than the two right angles.

COMMON NOTIONS

1. Things which are equal to the same thing are also equal to one another.

2. If equals be added to equals, the wholes are equal.

3. If equals be subtracted from equals, the remainders are equal.

4. Things which coincide with one another are equal to one another.

5. The whole is greater than the part.

Euclid does not naïvely assume that the defined concepts exist or are consistent; as Aristotle had pointed out, one might define something that had incompatible properties. The first three postulates, since they declare the possibility of constructing lines and circles, are existence assertions for these two entities. In the development of Book I, Euclid proves the existence of the other entities by constructing them. An exception is the plane.

Euclid presupposes that the line in Postulate 1 is unique; this assumption is implicit in Book I, Proposition 4. It would have been better, however, to make it explicit. Likewise, in Postulate 2 Euclid assumes the extension is unique. He uses the uniqueness explicitly in Book XI, Proposition 1, but has actually already used it unconsciously at the very beginning of Book I.

Postulate 5 is Euclid's own; it is a mark of his genius that he recognized its necessity. Many Greeks objected to this postulate because it was not clearly self-evident and hence lacked the appeal of the others. The attempts to prove it from the other axioms and postulates—which, according to Proclus, commenced even in Euclid's own time—all failed. The full history of these efforts will be related in the discussion of non-Euclidean geometry.

As to the common notions, there are differences of opinion over which ones were in Euclid's original version. Common Notion 4, which is the basis for proof by superposition, is geometrical in character and should be a postulate. Euclid uses superposition in Book I, Propositions 4 and 8, though apparently he was unhappy about the method; he could have used it to prove Proposition 26 ($a.s.a. = a.s.a.$ and $s.a.a. = s.a.a.$), but instead uses a longer proof. He probably found the method in works of older geometers and did not know how to avoid it. More axioms were added to Euclid's by Pappus and others who found Euclid's set inadequate.

4. *Books I to IV of the* Elements

Books I to IV treat the basic properties of rectilinear figures and circles. Book I contains familiar theorems on congruence, parallel lines, the Pythagorean theorem, elementary constructions, equivalent figures (figures with equal area), and parallelograms. All figures are rectilinear, that is, composed of line segments. Of special note are the following theorems (the wording is not verbatim):

Proposition 1. On a given straight line to construct an equilateral triangle.
    The proof is simple. With A as center (Fig. 4.1) and AB as radius con-

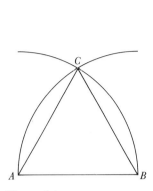

Figure 4.1

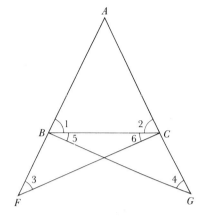

Figure 4.2

struct a circle. With *B* as center and *BA* as radius construct a circle. Let *C* be a point of intersection. Then *ABC* is the required triangle.

Proposition 2. To place at a given point (as an extremity) a straight line equal to a given straight line.

One would think this can be done immediately by using Postulate 3. But to do so means that the compass must keep its spread of the given length while being transferred to the point at which the equal length is to be constructed. Euclid, however, assumes a collapsible compass and gives a more complicated construction. Of course, he does assume that the compass remains fixed while describing a circle with given center and radius, that is, as long as the compass remains on the paper.

Proposition 4. Two triangles are congruent if two sides and the included angle of one triangle are equal to corresponding parts of the other.

The proof is made by placing one triangle on the other and showing that they must coincide.

Proposition 5. The base angles of an isosceles triangle are equal.

The proof is better than in many current elementary books, which assume at this stage the existence of the bisector of angle *A*. But the proof of its existence depends upon Proposition 5. Euclid extends *AB* to *F* (Fig. 4.2) and *AC* to *G* so that $BF = CG$. Then $\triangle AFC \cong \triangle AGB$. Hence $FC = GB$, $\angle ACF = \angle ABG$ and $\angle 3 = \angle 4$. Now $\triangle CBF \cong \triangle BCG$. Hence $\angle 5 = \angle 6$. Therefore $\angle 1 = \angle 2$. Pappus proves the theorem by regarding the given triangle as $\triangle ABC$ and $\triangle ACB$. Then Proposition 4 applies and the base angles are equal.

Proposition 16. An exterior angle of a triangle is greater than either remote interior angle.

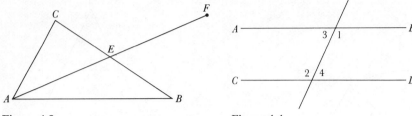

Figure 4.3   Figure 4.4

The proof (Fig. 4.3) requires an infinitely extensible straight line, because in it AE is extended its own length to F, and it must be possible to do this.

Proposition 20. The sum of any two sides of a triangle is greater than the third side.

This theorem is as close as one comes in Euclidean geometry to the fact that the straight line is the shortest distance between two points.

Proposition 27. If a straight line falling on two straight lines makes the alternate interior angles equal to one another, the straight lines will be parallel to one another.

The proof is made by contradiction using the proposition on the exterior angle of a triangle. The theorem establishes the existence of at least one parallel line to a given line and through a given point.

Proposition 29. A straight line falling on parallel straight lines makes the alternate interior angles equal to one another, the exterior angle equal to the interior and opposite angle [corresponding angles are equal], and the interior angles on the same side equal to two right angles.

The proof (Fig. 4.4) supposes ∢1 ≠ ∢2. If ∢2 is greater, add ∢4 to each. Then ∢2 + ∢4 > ∢1 + ∢4. This implies that ∢1 + ∢4 is less than two right angles. By the parallel postulate, which is used for the first time, the two given lines AB and CD would have to meet whereas they are parallel by hypothesis.

Proposition 44. To a given straight line to apply, in a given rectilinear angle, a parallelogram equal to a given triangle.

This proposition (Fig. 4.5) says that we are given a triangle C, an angle D, and a line segment AB. We are to construct on AB as one side a parallelogram equal in area to C and containing angle D as one angle. We shall not give Euclid's proof, which depends upon preceding propositions. The major point of note is that this is the first of the problems included under the theory of application of areas, the theory ascribed to the Pythagoreans by Eudemus (as reported by Proclus). In this case we apply (exactly) an area to AB.

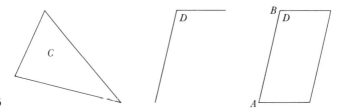

Figure 4.5

Secondly, this is an example of transformation from one area to another. Thirdly, in the special case when D is a right angle the parallelogram must be a rectangle. Then the area of the given triangle and AB may be regarded as given quantities. The other side of the rectangle is then the quotient of the given area C and AB. Hence we have division performed geometrically; this theorem is an example of geometrical algebra.

Proposition 47. In right-angled triangles the square on the side subtending the right angle is equal to the sum of the squares on the sides containing the right angle.

This is of course the Pythagorean theorem. The proof is made by means of areas, as in many high school texts. One shows (Fig. 4.6) that $\triangle ABD \cong \triangle FBC$, that rectangle $BL = 2\triangle ABD$, and rectangle $GB = 2\triangle FBC$. Then rectangle $BL =$ square GB. Likewise rectangle $CL =$ square AK.

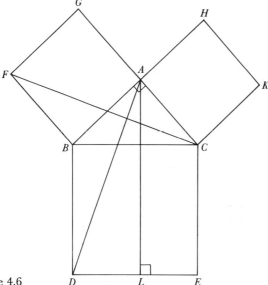

Figure 4.6

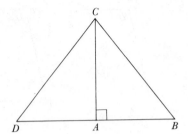

Figure 4.7

The theorem also shows how to obtain a given square whose area is a sum of two given squares, that is, how to find x such that $x^2 = a^2 + b^2$. Hence it is another example of geometrical algebra.

Proposition 48. If in a triangle the square on one side be equal to the sum of the squares on the remaining two sides of the triangle, the angle contained by the remaining two sides is a right angle.

This proposition is the converse of the Pythagorean theorem. The proof in Euclid (Fig. 4.7) constructs AD perpendicular to AC and equal to AB. We have by hypothesis that

$$AB^2 + AC^2 = BC^2$$

and we have from the right triangle ADC that

$$AD^2 + AC^2 = DC^2.$$

Since $AB = AD$ then $BC^2 = DC^2$ and so $DC = BC$. Hence the two triangles are congruent by *s.s.s.*, so that angle CAB must be a right angle.

The outstanding material in Book II is the contribution to geometrical algebra. We have already pointed out that the Greeks did not recognize the existence of irrational numbers and so could not handle all lengths, areas, angles, and volumes numerically. In Book II all quantities are represented geometrically, and thereby the problem of assigning numerical values is avoided. Thus numbers are replaced by line segments. The product of two numbers becomes the area of a rectangle with sides whose lengths are the two numbers. The product of three numbers is a volume. Addition of two numbers is translated into extending one line by an amount equal to the length of the other and subtraction into cutting off from one line the length of a second. Division of two numbers, which are treated as lengths, is merely indicated by a statement that expresses a ratio of the two lines; this is in accord with the principles introduced later in Books V and VI.

Division of a product (an area) by a third number is performed by finding a rectangle with the third number (length) as one side and equal in

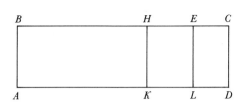

Figure 4.8

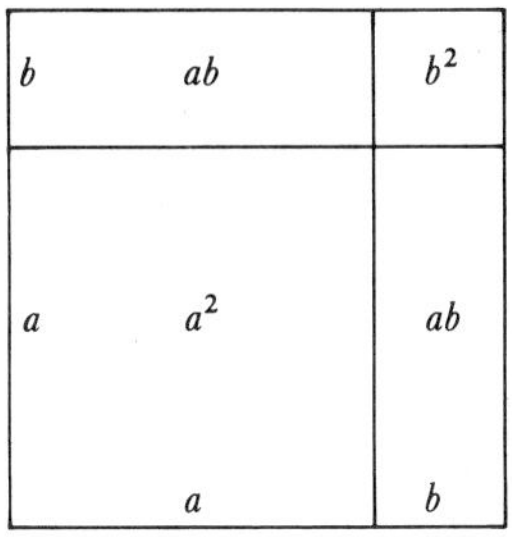

Figure 4.9

area to the given product. The other side of the rectangle is, of course, the quotient. The construction uses the theory of application of areas already touched upon in Proposition 44 of Book I. The addition and subtraction of products are the addition and subtraction of rectangles. The sum or difference is transformed into a single rectangle by means of the method of application of areas. The extraction of a square root is, in this geometrical algebra, the finding of a square equal in area to a rectangle whose area is the given quantity; this is done in Proposition 14 (see below).

The first ten propositions of Book II deal geometrically with the following equivalent algebraic propositions. Stated in our notation some of these are:

1. $a(b + c + d + \cdots) = ab + ac + ad + \cdots$;
2. $(a + b)a + (a + b)b = (a + b)^2$;
3. $(a + b)a = ab + a^2$;
4. $(a + b)^2 = a^2 + 2ab + b^2$;
5. $ab + \left\{\frac{1}{2}(a + b) - b\right\}^2 = \left\{\frac{1}{2}(a + b)\right\}^2$;
6. $(2a + b)b + a^2 = (a + b)^2$.

The geometric statement of (1) is contained in:

Proposition 1. If there be two straight lines (Fig. 4.8) and one of them be cut into any number of segments whatever, the rectangle contained by the two straight lines is equal to the rectangles contained by the uncut straight line and each of the segments.

Propositions 2 and 3 are really special cases of Proposition 1 but are separately stated and proved by Euclid. The geometric form of (4) above is well known. Euclid's statement is:

Proposition 4. If a straight line be cut at random (Fig. 4.9), the square on the whole is equal to the squares on the segments and twice the rectangle contained by the segments.

The proof gives the obvious geometrical facts shown in the figure.

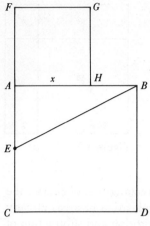

Figure 4.10

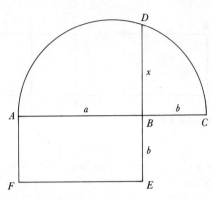

Figure 4.11

Proposition 11. To cut a given straight line so that the rectangle contained by the whole and one of the segments is equal to the square on the remaining segment.

This requires that we divide AB (Fig. 4.10) at some point H so that $AB \cdot BH = AH \cdot AH$. Euclid's construction is as follows: Let AB be the given line. Construct the square $ABDC$. Let E be the midpoint of AC. Draw BE. Let F on CA produced be such that $EF = EB$. Construct the square $AFGH$. Then H is the desired point on AB, that is,

$$AB \cdot BH = AH \cdot AH.$$

The proof is made by means of areas, using preceding theorems, including the Pythagorean theorem—the crucial theorem being Proposition 6.

The importance of the theorem is that AB of length a is divided into two segments of lengths x and $a - x$ so that

$$(a - x)a = x^2$$

or

$$x^2 + ax = a^2.$$

Hence we have a geometric way of solving this quadratic equation. Also, AB is divided in extreme and mean ratio, that is, from $AB \cdot BH = AH \cdot AH$ we have $AB:AH = AH:BH$. Other propositions in Book II amount to solving the quadratics $ax - x^2 = b^2$ and $ax + x^2 = b^2$.

Proposition 14. To construct a square equal to a given rectilinear figure.

The given rectilinear figure can be any polygon, but if the given figure is a *rectangle*, $ABEF$ (Fig. 4.11), then Euclid's method amounts to the follow-

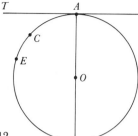

Figure 4.12

ing: Extend AB to C so that $BC = BE$. Construct the circle on AC as diameter. Erect the perpendicular DB at B. The desired square is the square on DB. Euclid gives a proof in terms of areas. This theorem solves $x^2 = ab$ or yields the square root of ab. As we shall see, in Book VI more complicated quadratic equations are solved geometrically.

Book III, which contains 37 propositions, begins with some definitions appropriate to the geometry of circles and then proceeds to discuss properties of chords, tangents, secants, central and inscribed angles, and so on. These theorems are in the main familiar to us from high school geometry. The following theorems are worthy of special note.

Proposition 16. The straight line drawn at right angles to the diameter of a circle from its extremity will fall outside the circle, and into the space between the straight line and the circumference another straight line cannot be interposed; further the angle of the semicircle is greater, and the remaining angle less, than any acute rectilinear angle.

The novelty in the theorem is that Euclid considers the space (Fig. 4.12) between the tangent TA and the arc ACE; he not only says that no line can be drawn in this space and through A, which falls entirely outside the circle, but also considers the angle formed by TA and the arc ACE. Whether this angle, which the Greeks called the hornlike angle, had a definite magnitude was a subject of controversy. Proposition 16 says this angle is smaller than any acute angle formed by straight lines but it does not say the angle is of zero magnitude.

Proclus speaks of horn angles as true angles. The subject of the size of horn angles was debated also in the late Middle Ages and in the Renaissance by Cardan, Peletier, Vieta, Galileo, Wallis, and others. What made the horn angle especially troublesome to later commentators on Euclid is that one can draw circles of smaller and smaller diameters passing through A and tangent to TA and it seems intuitively clear that the horn angle should increase in size but, according to the above proposition, it does not. On the other hand, if any two horn angles are of zero size and therefore equal, they should be superposable. But they are not. Some commentators concluded that a horn angle was not an angle.[2]

Book IV, in its 16 propositions, deals with figures inscribed in and circumscribed about circles, for example, triangles, squares, regular pentagons, and regular hexagons. The last proposition, which shows how to inscribe a 15-sided regular polygon in a given circle, is said to have been used in astronomy; the angle of the ecliptic (the angle made by the plane of the earth's equator and the plane of the earth's orbit around the sun) up to the time of Eratosthenes was taken to be 24°, or 1/15 of 360°.

5. *Book V: The Theory of Proportion*

Book V, based on Eudoxus' work, is considered to be the greatest achievement of Euclidean geometry; its contents have been more extensively discussed and its meaning more heavily debated than those of any other portion of the *Elements*. The Pythagoreans are supposed to have had a theory of proportion, that is, the equality of two ratios, for commensurable magnitudes, or magnitudes whose ratio could be expressed by a ratio of whole numbers. Though we do not know the details of this theory, it is presumed to be what we shall encounter in Book VII and is supposed to have been applied to statements about similar triangles. The mathematicians before Eudoxus who used proportions in general did not have a secure foundation for incommensurable magnitudes. Book V extends the theory of proportion to incommensurable ratios but avoids irrational numbers.

The notion of magnitude is intended to cover quantities or entities that are commensurable or incommensurable with each other. Thus lengths, areas, volumes, angles, weights, and time are magnitudes. The magnitudes length and area have already appeared, for example, in Book II. But thus far Euclid has had no occasion to treat other kinds of magnitudes or to treat ratios of magnitudes and proportions. Hence he does not introduce the general notion of magnitude until this point. His particular emphasis now is on proportions for all kinds of magnitudes.

Despite the fact that the definitions are important in this book, there is no definition of magnitude as such. Euclid starts with:

Definition 1. A magnitude is part of a magnitude, the less of the greater, when it measures the greater.

Part is used here to mean submultiple, as 2 is of 6, whereas 4 is not a submultiple of 6.

---

2. By the usual current definition of an angle between two curves, the horn angle is of zero size.

Definition 2. The greater is a multiple of the less when it is measured by the less.

Thus multiple means integral multiple.

Definition 3. A ratio is a sort of relation in respect to size between two magnitudes of the same kind.

It is difficult to separate the significance of this third definition from that of the next.

Definition 4. Magnitudes are said to have a ratio to one another which are capable, when multiplied, of exceeding one another.

The meaning of this definition is that the magnitudes a and b have a ratio if some integral multiple n (including 1) of a exceeds b and some integral multiple (including 1) of b exceeds a. The definition excludes the concept that appeared later, namely, the infinitely small quantity which is not 0, called the infinitesimal. Euclid's definition does not allow a ratio between two magnitudes if one is so small that some finite multiple of it does not exceed the other. The definition also excludes infinitely large magnitudes because then no finite multiple of the smaller one will exceed the larger. The next definition is the key one.

Definition 5. Magnitudes are said to be in the same ratio, the first to the second and the third to the fourth, when, if any equimultiples whatever be taken of the first and third, and any equimultiples whatever of the second and fourth, the former equimultiples alike exceed, are alike equal to, or alike fall short of, the latter equimultiples respectively taken in corresponding order.

The definition says that

$$\frac{a}{b} = \frac{c}{d}$$

if when we multiply a and c by *any* whole number m, say, and b and d by *any* whole number n, then for all such choices of m and n,

$$ma < nb \quad \text{implies} \quad mc < nd,$$
$$ma = nb \quad \text{implies} \quad mc = nd,$$

and

$$ma > nb \quad \text{implies} \quad mc > nd.$$

Just to see the meaning of this definition, let us use modern numbers. To test whether

$$\frac{\sqrt{2}}{1} = \frac{\sqrt{6}}{\sqrt{3}}$$

we should, at least theoretically, know that if m is *any* whole number and n *any* other, then

$$m\sqrt{2} < n \cdot 1 \quad \text{implies} \quad m\sqrt{6} < n\sqrt{3}$$

and

$$m\sqrt{2} = n \cdot 1 \quad \text{implies} \quad m\sqrt{6} = n\sqrt{3}$$

and

$$m\sqrt{2} > n \cdot 1 \quad \text{implies} \quad m\sqrt{6} > n\sqrt{3}.$$

Of course, in this present example the equality $m\sqrt{2} = n \cdot 1$ will not occur because m and n are whole numbers whereas $\sqrt{2}$ is irrational, but this means that $m\sqrt{6} = n\sqrt{3}$ need not occur. The definition merely says that if the left side of any one of the three possibilities occurs then the right side must occur. An alternative statement of Definition 5 is that the integral m and n for which $ma < nb$ are the same as those m' and n' for which $m'c < n'd$.

It may be desirable to point out at once what Euclid does with the above definitions. When we wish to prove that if $a/b = c/d$, then $(a + b)/b = (c + d)/d$, we regard the ratios and the proportion as numbers even if the ratios are incommensurable, and we use algebra to prove the result. We know that we can operate with irrationals by the laws of algebra. Euclid cannot and does not. The Greeks had not thus far justified operations with ratios of incommensurable magnitudes; hence Euclid proves this theorem by using the definitions he has given and, in particular, Definition 5. In effect, he is laying the basis for an algebra of magnitudes.

Definition 6. Let magnitudes which have the same ratio be called proportional.

Definition 7. When, of the equimultiples, the multiple of the first magnitude exceeds the multiple of the second, but the multiple of the third does not exceed the multiple of the fourth, then the first is said to have a greater ratio to the second than the third has to the fourth.

The definition states that if for even one m and one n, $ma > nb$ but mc is not greater than nd, then $a/b > c/d$. Hence, given an incommensurable ratio a/b, it is possible to place it among all other such ratios, namely those less and those greater.

Definition 8. A proportion in three terms is the least possible.

In this case $a/b = b/c$.

Definition 9. When three magnitudes are proportional, the first is said to have to the third the duplicate ratio of that which it has to the second.

Thus if $A/B = B/C$, then A has the duplicate ratio to C that it has to B. This means that $A/C = A^2/B^2$, for $A = B^2/C$ so that $A/C = B^2/C^2 = A^2/B^2$.

Definition 10. When four magnitudes are continuously proportional the first is said to have to the fourth the triplicate ratio of that which it has to the second, and so on continually, whatever the proportion.

Thus if $A/B = B/C = C/D$, then A has the triplicate ratio to D that it has to B. That is, $A/D = A^3/B^3$, for $A = B^2/C$ so that $A/D = B^2/CD = (B^2/C^2)(C/D) = A^3/B^3$.

Definitions 11 to 18 define corresponding magnitudes, alternation, inversion, composition, separation, conversion, etc. These refer to forming $(a + b)/b$, $(a - b)/b$, and other ratios from a/b.

Book V now proceeds to prove twenty-five theorems about magnitudes and ratios of magnitudes. The proofs are verbal and depend only on the definitions and on the common notions or axioms, such as that equals subtracted from equals give equals. The postulates are not used. Euclid uses line segments as illustrations of magnitudes to help his readers perceive the meaning of the theorems and proofs, but the theorems apply to all kinds of magnitudes.

We shall state some of the propositions of Book V in modern algebraic language, using m, n, and p for integers and a, b, and c for magnitudes. However, to illustrate Euclid's language, let us note his wording of the first proposition.

Proposition 1. If there be any number of magnitudes whatever which are, respectively, equimultiples of any magnitudes equal in multitude, then, whatever multiple one of the magnitudes is of one, that multiple also will be of all.

In algebraic language this means that $ma + mb + mc + \cdots = m(a + b + c + \cdots)$.

Proposition 4. If $a/b = c/d$, then $ma/nb = mc/nd$.

Proposition 11. If $a/b = c/d$ and $c/d = e/f$ then $a/b = e/f$.

Note the equality of ratios depends on the definition of proportion and Euclid is careful to prove that equality is transitive.

Proposition 12. If $a/b = c/d = e/f$ then $a/b = (a + c + e)/(b + d + f)$.

Proposition 17. If $a/b = c/d$ then $(a - b)/b = (c - d)/d$.

Proposition 18. If $a/b = c/d$ then $(a + b)/b = (c + d)/d$.

Some of the propositions seem to duplicate propositions in Book II. However, the propositions in the latter book are asserted and proved only for line segments, whereas Book V provides the theory for all kinds of magnitudes.

Book V was crucial for the subsequent history of mathematics. The classical Greeks did not introduce irrational numbers and sought to avoid them in part by working geometrically, as we have already noted in our review of Books I to IV. However, this use of geometry did not take care of ratios and proportions of incommensurable magnitudes of all sorts, and this lack was filled by Book V, which started anew with a general theory of magnitude. It thereby placed all of Greek geometry that dealt with magnitudes on a sound basis. The critical question, however, has been whether the theory of magnitudes provided a logical basis for a theory of real numbers, including, of course, the irrational numbers.

There is no question about how succeeding generations of mathematicians interpreted Euclid's theory of magnitudes. They regarded it as applicable only to geometry and therefore took the attitude that only geometry was rigorous. Hence, when in the Renaissance and in the following centuries irrational numbers were reintroduced and used, many mathematicians objected because these numbers had no logical foundation.

A critical examination of Book V seems to establish that they were right. It is true that the definitions and proofs as presented by Euclid in Book V make no use of geometry. As already noted his use of line segments in his presentation of the propositions and proofs is pedagogical only. However, if Euclid had really offered a theory of irrationals in his theory of magnitudes, it would have had to come from either one of two possible interpretations. The first is that magnitudes themselves could be taken to be the irrational numbers, and the second is that the ratios of two magnitudes could be the irrational numbers.

Let us suppose the magnitudes themselves could be the irrational numbers. Then, even if we leave aside any criticisms of Euclid's rigor when judged by modern standards, the following difficulties enter. Euclid never defines what he means by a magnitude, or the equality or equivalence of magnitudes. Moreover, Euclid works not with the magnitudes themselves but with proportions. A product of two magnitudes a and b occurs only when a and b are lengths, thereby enabling Euclid to consider the product as an area. The product ab could not then be a number because the product has no general meaning in Euclid. Further, Euclid proves in Book V a number of theorems on proportion which in themselves can readily be restated, as in fact we did above, as algebraic theorems. However, to prove Proposition 18 of Book V he needs to establish the fourth proportional to three given magnitudes, which he is able to do only for magnitudes that are line segments (Book VI, Proposition 12). Hence, not only is his theory of general magnitudes incomplete (even for proofs he himself makes in Book XII), but what he does establish for lengths is dependent on geometry. Moreover, Euclid insists in Definition 3 that a ratio can be formed only of magnitudes of the same kind. Clearly if magnitudes were numbers this limitation would be meaningless. His concept of magnitude as used later adheres to the definition and so is tied to geometry. Another difficulty is that there is no system of rational numbers to which a theory of irrationals could be added. Ratios of whole numbers occur, but only as members of a proportion, and even these ratios are not regarded as fractions. Finally, there is no product of a/b and c/d even when all four quantities are whole numbers, let alone magnitudes.

Now let us consider interpreting Euclid's ratios of magnitudes as numbers, so that the incommensurable ratios would be the irrational numbers and the commensurable ratios, the rational numbers. If these ratios are numbers it should be possible at least to add and multiply them. But nowhere in Euclid does one find what $(a/b) + (c/d)$ means when a, b, c, and d are magnitudes. In Euclid's usage the ratios occur only as elements of a proportion and hence have no general significance. Finally, as noted above, Euclid does not have the concept of a rational number on which to build a theory of irrationals.

Thus the course that the history of mathematics actually took until about 1800, namely, treating continuous quantities rigorously solely on a geometric basis, was necessary. As far as Euclid's *Elements* is concerned, there was no foundation for irrational numbers.

6. *Book VI: Similar Figures*

Book VI, which treats similar figures and uses the theory of proportion of Book V, opens with some definitions. We note just a few:

Definition 1. Similar rectilinear figures are those having corresponding angles equal and the sides about the equal angles proportional.

Definition 3. A straight line is cut in extreme and mean ratio when the whole line is to the greater segment as the greater to the less.

Definition 4. The height of any figure is the perpendicular drawn from the vertex to the base.

This definition is certainly vague but Euclid does employ it.

In the proofs of the theorems in this book, Euclid, using his theory of proportion, does not have to treat separately the commensurable and incommensurable cases, a separation introduced by Legendre, who used an algebraic definition of proportion limited to commensurable quantities and so had to treat the incommensurable cases by another argument such as a *reductio ad absurdum*.

We shall note only some of the thirty-three theorems. Again we shall find some basic results of modern algebra treated geometrically.

Proposition 1. Triangles and parallelograms [i.e. the areas] which are under the same height [have the same altitudes] are to one another as their bases.

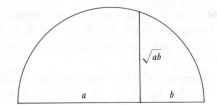

Figure 4.13

Here Euclid uses a proportion among four magnitudes two of which are areas.

Proposition 4. In equiangular triangles the sides about the equal angles are proportional, and those are the corresponding sides which subtend the equal angles.

Proposition 5. If two triangles have their sides proportional, the triangles will be equiangular and will have those angles equal which the corresponding sides subtend.

Proposition 12. To three given straight lines to find a fourth proportional.

Proposition 13. To two given straight lines to find a mean proportional.
    The method is the familiar one (Fig. 4.13). It means from an algebraic standpoint that, given a and b, we can find $\sqrt{ab}$.

Proposition 19. [The areas of] similar triangles are to one another in the duplicate ratio of the corresponding sides.
    We express this theorem today by the statement that the ratio of the areas of two similar triangles equals the ratio of the squares of two corresponding sides.

Proposition 27. Of all the parallelograms applied to the same straight line [constructed on part of the straight line] and deficient [from the parallelogram on the entire straight line] by parallelogramic figures similar and similarly situated to that [given parallelogram] described on the half of the straight line, [the area of] that parallelogram is greatest which is applied to the half of the straight line and is similar to the defect.
    The meaning of this proposition is as follows: We start (Fig. 4.14) with a given parallelogram AD constructed on AC, which is one-half of a given line segment AB. Now we consider a parallelogram AF on AK which is part of AB subject to the condition that the defect of AF, which is FB, is a parallelogram similar to AD. We can of course obtain many parallelograms meeting the conditions that AF does. Euclid's theorem states that of all such parallelograms, the one constructed on AC, which is half of AB, has the largest area.

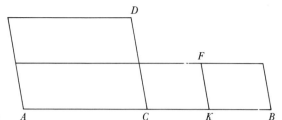

Figure 4.14

The proposition has a vital algebraic meaning. Suppose the given parallelogram AD is a rectangle (Fig. 4.15) and that the ratio of its sides is c to b, where b is AC. Now consider the rectangle AF which is required to meet the condition that its defect, rectangle FB, is similar to AD. If we denote FK by x, then KB is bx/c. Let the length of AB be a; then $AK = a - (bx/c)$. Hence the areas S of AF is

$$(1) \qquad\qquad S = x\left(a - \frac{bx}{c}\right).$$

Proposition 27 says that S is a maximum when AF is AD. But $AC = a/2$ and $CD = ac/2b$. Hence

$$S \leq \frac{a^2 c}{4b}.$$

On the other hand the condition that equation (1) regarded as a quadratic in x have a real root is that its discriminant be greater than or equal to 0. That is,

$$a^2 - 4\frac{b}{c} S \geq 0$$

or

$$S \leq \frac{a^2 c}{4b}.$$

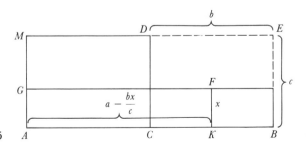

Figure 4.15

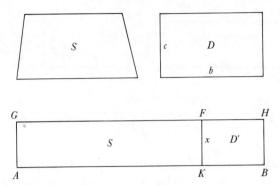

Figure 4.16   A

Thus the proposition tells us not only what the largest possible value of S is but that for all possible values there is an x which satisfies (1) and which geometrically furnishes one side, KF, of the rectangle AF. This result will be used in the next proposition.

Before considering it let us note an interesting special case of Proposition 27. Suppose the given parallelogram AD (Fig. 4.15) is a square. Then of all rectangles on AB deficient by a square similar to AD, the square on AC is the greatest. But the rectangle AF on (part of) AB has area $AK \cdot KF$ and since $KF = KB$, the perimeter of this rectangle is the same as that of square DB or square AD. But AD has larger area than AF. Thus of all rectangles with the same perimeter the square has the greatest area.

Proposition 28. To a given straight line to apply [on part of the line as side] a parallelogram equal to a given rectilinear figure [S] and deficient [from the parallelogram on the entire line] by a parallelogramic figure similar to a given one [D]. Thus [by Proposition 27] the given rectilinear figure [S] must not be greater than the parallelogram described on half of the straight line and similar to the defect.

This theorem is the geometrical equivalent of the solution of the quadratic equation $ax - (b/c)x^2 = S$, where S is the area of the given rectilinear figure and is subject to the condition for a real solution, namely, S is not greater than $a^2c/4b$. To see this, suppose (for convenience) that the parallelograms are rectangles (Fig. 4.16), S is the given rectilinear figure, D is the other given rectangular figure with sides c and b, a is AB, and x is one side of the desired rectangle. What Euclid constructs is the rectangle $AKFG$ of area S and such that the defect D' is similar to D. But $AKFG = ABHG - D'$. Since D' is similar to D its area is bx^2/c. Hence

(2) $$S = ax - \frac{b}{c} x^2.$$

Thus to construct $AKFG$ is to find AK and x such that x satisfies equation (2).

Proposition 29. To a given straight line to apply a parallelogram equal to a given rectilinear figure [S] and exceeding by a parallelogramic figure similar to a given one [D].

In algebraic terms this theorem solves

$$ax + \frac{b}{c} x^2 = S$$

where a, b, c, and S are given. S is not limited because for all positive S the equation has a real solution. With Propositions 28 and 29, Euclid has shown how, in modern language, one can solve any quadratic equation when one or both roots are positive. His constructions furnish the roots as lengths.

In Proposition 28 the parallelogram constructed falls short of the parallelogram on the entire line AB and in Proposition 29 the parallelogram constructed exceeds the one on the given line AB. The respective parallelograms were called in Greek *elleipsis* and *hyperbolè*. A parallelogram of specified area constructed on the entire given line as base, as in Book I, Proposition 44, was called *parabolè*. These terms were carried over to the conic sections for a reason which will be obvious when we study Apollonius' work.

Proposition 31. In right-angled triangles, the figure on the side subtending the right angle is equal to the similar and similarly described figures on the sides containing the right angle.

This is a generalization of the Pythagorean theorem.

7. *Books VII, VIII, and IX: The Theory of Numbers*

Books VII, VIII, and IX treat the theory of numbers, that is, the properties of whole numbers and ratios of whole numbers. These three books are the only ones in the *Elements* that treat arithmetic as such. In them Euclid represents numbers as line segments and the product of two numbers as a rectangle, but the arguments do not depend on the geometry. The statements and proofs are verbal as opposed to the modern symbolic form.

Many of the definitions and theorems, particularly those on proportion, duplicate what was done in Book V. Hence historians have considered the question of why Euclid proves all over again propositions for numbers instead of referring to propositions already proven in Book V.

The answers vary. Aristotle did include number as one kind of magnitude, but he also emphasized the cleavage between the discrete and the continuous, and we do not know whether Euclid was influenced by either of Aristotle's views on this matter. Nor can one decide on the basis of the vague definitions in Book V whether he meant his notion of magnitude to include whole number. If one judged by the fact that he treated number independently one would conclude that his magnitudes do not include numbers.

Another explanation of this independent treatment of numbers is that the theory of numbers and of commensurable ratios existed before Eudoxus' work and Euclid followed tradition in presenting what were two independent developments, the pre-Eudoxian and largely Pythagorean theory and the Eudoxian theory. He may also have believed that since the theory of numbers can be built up on simpler foundations than the theory of magnitudes, it was wise to do so. One finds alternative approaches in modern contributions to mathematics too, and for the same reason—that they are simpler. Although Euclid separates number and magnitude, he does have a few theorems that relate them. For example, Proposition 5 of Book X states that commensurable magnitudes have to one another the ratio which a number has to a number.

In the three books under discussion, as in other books, Euclid assumes facts that he does not state explicitly. Thus he assumes without mention that if A divides (evenly into) B and B divides C, then A divides C. Also, if A divides B and divides C, it divides $B + C$ and $B - C$.

Book VII begins with some definitions:

Definition 3. A number is a part of a number, the less of the greater, when it measures the greater. [The number which is part of another divides evenly into the latter.]

Definition 5. The greater number is a multiple of the less when it is measured by the less.

Definition 11. A prime number is that which is measured by a unit alone.

Definition 12. Numbers prime to one another [relatively prime] are those which are measured by unit alone as a common measure.

Definition 13. A composite number is that which is measured by some number [other than 1].

Definition 16. And when two numbers having multiplied one another make some number, the number so produced is called plane, and its sides are the numbers which have multiplied one another.

Definition 17. And when three numbers having multiplied one another make some number, the number so produced is solid, and its sides are the numbers which have multiplied one another.

Definition 20. Numbers are proportional when the first is the same multiple, or the same part, or the same parts of the second that the third is of the fourth.

Definition 22. A perfect number is that which is equal to [the sum of] its own parts.

Propositions 1 and 2 give the process of finding the greatest common measure (divisor) of two numbers. Euclid describes the process by saying that if A and B are the numbers and $B < A$, then subtract B from A enough times until a number C less than B is left. Then subtract C from B enough times until a number less than C is left. And so on. If A and B are relatively prime we arrive at 1 as the last remainder. Then 1 is the greatest common divisor. If A and B are not relatively prime we arrive at some stage where the last number measures the one before it. This last number is the greatest common divisor of A and B. This process is referred to today as the Euclidean algorithm.

Simple theorems about numbers follow. For example, if $a = b/n$ and $c = d/n$, then $a \pm c = (b \pm d)/n$. Some are just the theorems on proportion previously proved for magnitudes and now proved all over again for numbers. Thus if $a/b = c/d$, then $(a - c)/(b - d) = a/b$. Also, in Definition 15 $a \cdot b$ is defined as b added to itself a times. Hence Euclid proves that $ab = ba$.

Proposition 30. If two numbers by multiplying one another make some number, and any prime number measures the product, it will also measure one of the original numbers.

This result is fundamental in the modern theory of numbers. We say that if a prime p divides a product of two whole numbers it must divide at least one of the factors.

Proposition 31. Any composite number is measured by some prime number.

Euclid's proof says that if A is composite it is, by definition, measured by some number B. If B is not prime and hence composite, B is measured by C. Then C measures A. If C is not prime, etc. Then he says, "If the investigation be continued in this way, some prime number will be found which will measure the number before it, which will also measure A. For if it is not found, an infinite series of numbers will measure the number A, each of which is less than the other: which is impossible in numbers." He assumes here that any set of whole numbers has a least number.

Book VIII continues with the theory of numbers; no new definitions are needed. In essence the book treats geometrical progressions. A geometrical progression is to Euclid a set of numbers in continued proportion, that is, $a/b = b/c = c/d = d/e = \cdots$. This continued proportion satisfies our definition of geometric progression, for if $a, b, c, d, e, \ldots$ are in geometrical progression the ratio of any term to the next one is a constant.

Book IX concludes the work on the theory of numbers. There are theorems on square and cube numbers, plane and solid numbers, and more theorems on continued proportions. Of note are the following:

Proposition 14. If a number be the least that is measured by prime numbers, it will not be measured by any other prime number except those originally measuring it.

This means that if a is the product of the primes $p, q, \ldots$, then this decomposition of a into primes is unique.

Proposition 20. Prime numbers are more than any assigned multitude of prime numbers.

In other words, the number of primes is infinite. Euclid's proof of this proposition is a classic. He supposes that there is just a finite number of primes, $p_1, p_2, \ldots, p_n$. He then forms $p_1 \cdot p_2 \cdot \ \cdots \ \cdot p_n + 1$ and argues that if this new number is a prime, we have a contradiction, because this prime is larger than any of the n primes and so we would have more than n primes. On the other hand, if this new number is composite it must be divisible (exactly) by a prime. But this prime divisor is not $p_1, p_2, \ldots$, or p_n because these leave a remainder of 1. Hence there must be some other prime; and again we have a contradiction of the assumption that there are just the n primes $p_1, p_2, \ldots, p_n$.

Proposition 35 of Book IX gives an elegant proof for the sum of a geometric progression. Proposition 36 gives a famous theorem on perfect numbers, namely, if the sum of the terms (starting with 1) of the geometric progression

$$1 + 2 + 2^2 + \cdots + 2^{n-1}$$

is prime, the product of that sum and the last term, that is,

$$(1 + 2 + \cdots + 2^{n-1})2^{n-1} \quad \text{or} \quad (2^n - 1)2^{n-1},$$

is a perfect number. The first four perfect numbers, 6, 28, 496, and 8128, and perhaps the fifth, were known to the Greeks.

8. *Book X: The Classification of Incommensurables*

Book X of the *Elements* undertakes to classify types of irrationals, that is, magnitudes incommensurable with given magnitudes. Augustus De Morgan describes the general contents of this book by saying, "Euclid investigates every possible variety of line which can be represented [in modern algebra] by $\sqrt{\sqrt{a} \pm \sqrt{b}}$, a and b representing two commensurable lines." Of course not all irrationals are so representable, and Euclid covers only those that arise in his geometrical algebra.

The first proposition in Book X is important for developments in later books of the *Elements*.

Proposition 1. Two unequal magnitudes being set out, if from the greater there be subtracted a magnitude greater than its half, and from that which is left a magnitude greater than its half, and if this process be repeated continually, then there will be left some magnitude which will be less than the lesser magnitude set out.

At the conclusion of the proof Euclid says the theorem can be proven if the parts subtracted be halves. One step in the proof utilizes an axiom, not consciously recognized as such by Euclid, to the effect that of two unequal magnitudes the smaller can be added to itself a finite number of times, so as to have the sum exceed the larger. Euclid bases the questionable step on the definition of a ratio between two magnitudes (Definition 4 of Book V). But this definition does not justify the step. It says that two magnitudes have a ratio when either can be added to itself enough times to have the sum exceed the other; hence Euclid should prove that this can be done for the magnitudes he deals with. Instead he assumes that his magnitudes have a ratio and uses the fact that the smaller can be added to itself enough times to exceed the greater. According to Archimedes, the axiom in question (strictly an equivalent statement) was used by Eudoxus, who had established it as a lemma. Archimedes uses this lemma without proof and so, in effect, he too uses it as an axiom. It is called today the axiom of Archimedes–Eudoxus.

There are 115 propositions in Book X, though Propositions 116 and 117 are found in some editions of Euclid. The latter gives the proof already described in Chapter 3 of the irrationality of $\sqrt{2}$.

9. Books XI, XII, and XIII: Solid Geometry and the Method of Exhaustion

Book XI begins the treatment of solid geometry, though some important theorems on plane geometry are yet to come. The book opens with definitions.

Definition 1. A solid is that which has length, breadth, and depth.

Definition 2. An extremity of a solid is a surface.

Definition 3. A straight line is at right angles to a plane when it makes right angles with all the straight lines which meet it and are in the plane.

Definition 4. A plane is at right angles to a plane when the straight lines drawn, in one of the planes, at right angles to the common section of the planes, are at right angles to the remaining plane.

Definition 6. The inclination of a plane to a plane is the acute angle contained by the straight lines drawn at right angles to the common section, at the same point, one in each of the planes.

We call this acute angle the plane angle of the dihedral angle.

Also defined are parallel planes, similar solid figures, solid angle, pyramid, prism, sphere, cone, cylinder, cube, the regular octahedron, the

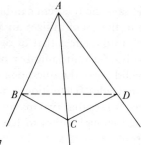

Figure 4.17

regular icosahedron, and other figures. The sphere is defined as the figure comprehended by a semicircle rotated around a diameter. The cone is defined as the figure comprehended by rotating a right-angled triangle about one of the arms. Then if the fixed arm or axis is less, equal to, or greater than the other arm, the cone is obtuse-angled, right-angled, or acute-angled, respectively. The cylinder is the figure comprehended by rotating a rectangle around one side. The significance of these last three definitions is that solid figures, except for the regular polyhedra, arise from rotating plane figures about an axis.

The definitions are loose, unclear, and often assume theorems. For example, in Definition 6 it is assumed that the acute angle is the same no matter where it is formed on the common section of the planes. Also Euclid intends to consider convex solids only, but does not specify this in his definitions of the regular polyhedra.

The book considers only figures formed by plane elements. The first 19 of the 39 theorems of this book treat properties of lines and planes, for example, theorems on lines perpendicular and parallel to planes. The proofs of the early theorems in this book are not adequate, and many general theorems about polyhedra are proved only for special cases.

Proposition 20. If a solid angle be contained by three plane angles, any two, taken together in any manner, are greater than the remaining one.

That is, of the three plane angles (Fig. 4.17) *CAB*, *CAD*, and *BAD*, the sum of any two is greater than the third one.

Proposition 21. Any solid angle is contained by plane angles [the sum of which is] less than four right angles.

Proposition 31. Parallelepipedal solids which are on equal bases and of the same height are equal [equivalent] to one another.

Proposition 32. Parallelepipedal solids which are of the same height are to one another as their bases.

Book XII contains 18 theorems on areas and volumes, particularly of curvilinear figures and figures bounded by surfaces. The dominant idea of the book is the method of exhaustion, which comes from Eudoxus. To prove, for example, that the areas of two circles are to each other as the squares on their diameters, the method approximates the areas of the two circles more and more closely by inscribed regular polygons and, since the theorem in question is true for the polygons, it is proved to be true for the circles. The term exhaustion comes from the fact that the successive inscribed polygons "exhaust" the circle. This term was not used by the Greeks; it was introduced in the seventeenth century. The term, as well as this loose description, may suggest that the method is approximate and just a step in the direction of the rigorous limit concept. But, as we shall see, the method is rigorous. There is no explicit limiting process in it; it rests on the indirect method of proof and in this way avoids the use of a limit. Actually Euclid's work on areas and volumes is sounder than that of Newton and Leibniz, who tried to build on algebra and the number system and sought to use the limit concept.

For a better understanding of the method of exhaustion, let us consider one example in some detail. (In the next chapter we shall consider some examples in Archimedes' work.) Book XII opens with:

Proposition 1. Similar polygons inscribed in circles are to one another as the squares on the diameters of the circles.

We shall not give the proof because no special feature is involved. We come now to the crucial proposition.

Proposition 2. Circles are to one another as the squares on the diameters.

The following describes the essence of Euclid's proof: He first proves that the circle can be "exhausted" by polygons. Inscribe a square in the circle (Fig. 4.18). The area of the square is more than 1/2 of the area of the circle because the former area is 1/2 of that of the circumscribed square and this area is larger than the area of the circle. Now let AB be any side of the inscribed square. Bisect arc AB at the point C and join AC and CB. Draw the tangent at C and then draw AD and BE perpendicular to the tangent. $\sphericalangle 1 = \sphericalangle 2$ because each is 1/2 of arc CB. It follows that DE is parallel to AB, and so $ABED$ is a rectangle whose area is greater than that of segment $ABFCG$. Hence triangle ABC, which is half the rectangle, is greater than 1/2 of segment $ABFCG$. By repeating the process at each side of the square, we obtain a regular octagon, which encloses not only the square but more than half of the difference between the area of the circle and the area of the square. On each side of the octagon we may construct a triangle, just as triangle ACB was constructed on AB. We then obtain a sixteen-sided regular polygon, which encloses the octagon and more than half of the difference between the area of the circle and the area of the octagon. The process may

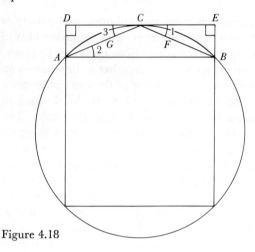

Figure 4.18

be repeated as often as one wishes. Euclid next employs Proposition 1 of Book X to affirm that the difference between the area of the circle and the area of some regular polygon of a sufficiently large number of sides can be made less than any given magnitude.

Now let S and S' be the areas of two circles (Fig. 4.19) and let d and d' be their diameters. Euclid wishes to prove that

(3) $S:S' = d^2:d'^2.$

Suppose that this equality does not hold but that

(4) $S:S'' = d^2:d'^2,$

where S'' is some area greater or less than S'. (The existence of the fourth proportional as an area is assumed here and elsewhere in Book XII.) Suppose $S'' < S'$. We construct regular polygons of more and more sides in

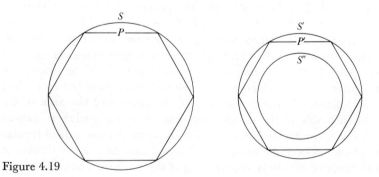

Figure 4.19

S' until we arrive at one, P' say, which is such that its area differs from S' by less than $S' - S''$. This polygon can be constructed because we proved above that the difference between the circle S' and inscribed regular polygons can be made less than any given magnitude and so less than $S' - S''$. Then

(5) $$S' > P' > S''.$$

Inscribe in S a polygon P similar to P'. Then, by Proposition 1,

$$P:P' = d^2:d'^2.$$

By reason of (4) we also have

$$P:P' = S:S''$$

or

$$P:S = P':S''.$$

However, since $P < S$ then

$$P' < S''.$$

But by (5) this is a contradiction.

Similarly, one can show that S'' cannot be greater than S'. Hence $S'' = S'$, and in view of (4), the proportion (3) is established.

This method is used to prove such critical and difficult theorems as:

Proposition 5. Pyramids which are of the same height and have triangular bases are to one another as the bases.

Proposition 10. Any cone is a third part of the cylinder which has the same base with it and equal height.

Proposition 11. Cones and cylinders which are of the same height are to one another as their bases.

Proposition 12. Similar cones and cylinders are to one another in the triplicate ratio [ratio of the cubes] of the diameters in their bases.

Proposition 18. Spheres are to one other in the triplicate ratio of their respective diameters.

Book XIII considers properties of regular polygons as such and when inscribed in a circle, and the problem of how to inscribe the five regular solids in a sphere. It also proves that no more than the five types of (convex) regular solids (polyhedra) exist. This last result is a corollary to Proposition 18, the last in the book.

The proof that no more than five regular solids can exist depends upon an earlier theorem, Book XI, Proposition 21, that the faces of a solid angle must contain less than 360°. Hence if we put together equilateral triangles we can have three meet at each vertex of the regular solid to form a tetrahedron; we can use four at a time to form an octahedron; and we can use five to form the icosahedron. Six equilateral triangles at one vertex would add up to 360° and so cannot be used. We can use three squares at any one vertex and thus form the cube. Then we can use three regular pentagons at each vertex to form the dodecahedron. No other regular polygons can be used, for even three coming together at one point will form an angle of 360° or more. Note that Euclid assumes convex regular solids. There are other, nonconvex regular solids.

The thirteen books of the *Elements* contain 467 propositions. In some of the old editions there are two more books, both of which contain more results on regular solids, though Book XV is unclear and inaccurate. However, both postdate Euclid. Book XIV is due to Hypsicles (*c.* 150 B.C.), and parts of Book XV were probably written as late as the sixth century A.D.

10. *The Merits and Defects of the* Elements

Since the *Elements* is the first substantial source of mathematical knowledge and one that was used by all succeeding generations, it influenced the course of mathematics as no other book has. The very concept of mathematics, the notion of proof, and the logical ordering of theorems were learned by studying it, and its contents determined the course of subsequent thinking. Hence we should note the characteristics that influenced so strongly the future of mathematics.

Though, as mentioned before, the form of presentation of the individual propositions is not original with Euclid, the form of presentation of the entire work—the statement of all the axioms at the outset, the explicit statement of all definitions, and the orderly chain of theorems—is his own. Moreover, the theorems are arranged to go from the simple to more and more complex ones.

Euclid also selected the theorems that he regarded as prior in importance. Thus he does not give, for example, the theorem that the altitudes of a triangle meet in a point. There are theorems in other works by Euclid, which we shall discuss shortly, that he did not deem worth including in the *Elements*.

Though the requirement that the existence of figures must be demonstrated before the figures can be received into the logical structure antedates Euclid, he carries out this prerequisite with skill and sophistication. In accordance with Postulates 1, 2, and 3, the constructions permitted involve only the drawing of straight lines and circles. This means, in effect, straightedge and compass constructions. It is because Euclid could not establish the existence of angle trisectors that he proved no theorems involving them.

Despite some omissions and errors of proof that we shall point out shortly, Euclid's choice of axioms is remarkable. From a small set he was able to prove hundreds of theorems, many of them deep ones. Moreover, his choice was sophisticated. His handling of the parallel axiom is especially clever. Euclid undoubtedly knew that any such axiom states explicitly or implicitly what must happen in the infinite reaches of space and that any pronouncement on what must be true of infinite space is physically dubious because man's experiences are limited. Nevertheless, he also realized that some such axiom is indispensable. He therefore chose a version that states conditions under which two lines will meet at a finitely distant point. Moreover, he proved all the theorems he could before calling upon this axiom.

Though Euclid used superposition of figures to establish congruence, a method which rests on Common Notion 4, he was evidently concerned about the soundness of the method. There are two objections to it: First, the concept of motion is utilized and there is no logical basis for this concept; and second, the method of superposition assumes that a figure retains all its properties when moved from one position to another. The displaced figure may indeed be proved congruent to a second one but the first figure in its original position may not be congruent to the second one. To assume that moving a figure does not change its properties is to make a strong assumption about physical space. Indeed the very purpose of Euclidean geometry is to compare figures in different positions. The evidence for Euclid's concern about the method's soundness is that he did not use it for proofs that he could make by other means, even though superposition would have permitted a simpler proof.

Though mathematicians generally did regard Euclid's work as a model of rigor until well into the nineteenth century, there are serious defects that a few mathematicians recognized and struggled with. The first is the use of superposition. The second is the vagueness of some of his definitions and the pointlessness of others. The initial definitions of point, line, and surface have no precise mathematical meanings and, as we now recognize, could not have been given any because any independent mathematical development must have undefined terms (see sec. 3). As to the vagueness of many definitions, we have but to refer back to those in Book V as an example. An additional objection to the definitions is that several, such as Definition 17 of Book I, presuppose an axiom.

A critical study of Euclid, with, of course, the advantage of present insights, shows that he uses dozens of assumptions that he never states and undoubtedly did not recognize. A few have been mentioned in our survey. What Euclid and hundreds of the best mathematicians of later generations did was to use facts either evident from the figures or intuitively so evident that they did not realize they were using them. In a few instances the unconscious assumptions could be obviated by proofs based on the explicit assumptions, but this is not true generally.

Among the assumptions made unconsciously are those concerning the continuity of lines and circles. The proof of Proposition 1 of Book I assumes that the two circles have a point in common. Each circle is a collection of points, and it could be that though the circles cross each other there is no common point on the two circles at the supposed point or points of intersection. The same criticism applies to the straight line. Two lines may cross each other and yet not have a common point as far as the logical basis in the *Elements* is concerned.

There are also defects in the proofs actually given. Some of these are mistakes made by Euclid that can be remedied, though new proofs would be needed in a few instances. Another kind of defect that runs throughout the *Elements* is the statement of a general theorem that is proved only for special cases or for special positions of the given data.

Though we have praised Euclid for the overall organization of the contents of the *Elements*, the thirteen books are not a unity, but are to an extent compilations of previous works. For example, we have already noted that Books VII, VIII, and IX repeat for whole numbers many results given for magnitudes. The first part of Book XIII repeats results of Books II and IV. Books X and XIII probably were a unit before Euclid and were due to Theaetetus.

Despite these defects, many of which were pointed out by later commentators (Chap. 42, sec. 1) and very likely also by immediate successors of Euclid, the *Elements* was so successful that it displaced all previous texts on geometry. In the third century B.C., when others were still extant, even Apollonius and Archimedes referred to the *Elements* for prior results.

11. *Other Mathematical Works by Euclid*

Euclid wrote a number of other mathematical and physical works, many significant in the development of mathematics. We shall reserve discussion of his chief physical works, the *Optics* and the *Catoptrica*, for a later chapter.

Euclid's *Data* was included by Pappus in his *Treasury of Analysis*. Pappus describes it as consisting of supplementary geometrical material concerned with "algebraic problems." When certain magnitudes are given or determined, the theorems determine other magnitudes. The material is not different in nature from what appears in the *Elements* but the specific theorems are different. The *Data* may have been intended as a set of exercises to review the *Elements*. It is known in full.

Of the works of Euclid, next to the *Elements*, his *Conics* played the most vital role in the history of mathematics. According to Pappus, the contents of this lost work of four books became substantially the first three books of Apollonius' *Conic Sections.* Euclid treated the conics as sections of the three different types of cones (right-angled, acute-angled, and obtuse-angled). The ellipse was also obtained as a section of any cone and of a circular cylinder. As we shall see, Apollonius changed the approach to the conic sections.

The *Pseudaria* of Euclid contained correct and false geometric proofs and was intended for the training of students. The work is lost.

On Divisions [of figures], mentioned by Proclus, treats the subdivision of a given figure into other figures, as a triangle into smaller triangles or a triangle into triangles and quadrilaterals. A Latin translation, probably due to Gerard of Cremona (1114–87), of an incorrect and incomplete Arabic version exists. In 1851 Franz Woepcke found and translated another Arabic version that seems to be correct. There is an English translation by R. C. Archibald.

Another lost work is the *Porisms.* The contents and even the nature of the work are largely unknown. Pappus in his *Mathematical Collection* says that the *Porisms* consisted of three books. It is believed from the remarks of Pappus and Proclus that the *Porisms* dealt essentially with constructions of geometric objects whose existence was already assured. Thus these problems were intermediate between pure theorems and constructions establishing existence. To find the center of a circle under some given conditions would be typical of the problems in the *Porisms.*

The work *Surface-Loci,* composed of two books, is mentioned by Pappus in his *Collection.* This work, which is not extant, probably dealt with loci that are surfaces.

Euclid's *Phaenomena,* though a text on astronomy, contains 18 propositions on spherical geometry and others on uniformly rotating spheres. The earth is treated as a sphere. Some versions are extant.

12. *The Mathematical Work of Apollonius*

The other great Greek who belongs to the classical period, in the two senses of summarizing and adding to the kind of mathematics the classical period produced, is Apollonius (*c.* 262–190 B.C.). Apollonius was born in Perga, a city in the northwestern part of Asia Minor, which was under the rule of Pergamum during his lifetime. He came to Alexandria in his youth and learned mathematics from Euclid's successors. As far as we know, he remained in Alexandria and became an associate of the great mathematicians who worked there. His chief work was on the conic sections but he also wrote on other subjects. His mathematical powers were so extraordinary that he became known in his time and thereafter as "the Great Geometer." His reputation as an astronomer was almost as great.

The conic sections, as we know, were studied long before Apollonius' time. In particular, Aristaeus the Elder and Euclid had written works on them. Also Archimedes' work, which we shall study later, contains some results on this subject. Apollonius, however, stripped the knowledge of all irrelevancies and fashioned it systematically. Besides being comprehensive, his *Conic Sections* contains highly original material and is ingenious, extremely adroit, and excellently organized. As an achievement it is so monumental that it practically closed the subject to later thinkers, at least from the purely geometrical standpoint. It may truly be regarded as the culmination of classical Greek geometry.

The *Conic Sections* was written in eight books and contained 487 propositions. Of these books we have the first four reproduced in Greek manuscripts of the twelfth and thirteenth centuries and the next three in an Arabic translation written in A.D. 1290. The eighth book is lost, though a restoration based on indications by Pappus was made by Halley in the seventeenth century.

Euclid's predecessors, Euclid himself, and Archimedes treated the conic sections as arising from the three kinds of right circular cones—as they had been introduced by the Platonist Menaechmus. Euclid and Archimedes were aware that the ellipse can also be obtained as a section of the two other types of right circular cones, and Archimedes knew in addition that sections of *oblique* circular cones made by planes cutting all the generators are ellipses. He probably realized that the other conic sections can be obtained from oblique circular cones.

Apollonius, however, was the first to base the theory of all three conics on sections of one circular cone, right or oblique. He was also the first to recognize both branches of the hyperbola. One presumed reason that Menaechmus and the other predecessors of Apollonius used sections perpendicular to one of the elements of the three types of right circular cones is not that they did not see that other sections can be made of these cones, but rather that they wanted to treat the converse problem. Given curves whose geometric properties are those of the conic sections, the proof that these curves are obtainable as sections of a cone is more readily made when the plane of the section is perpendicular to an element of the cone.

We consider first the definitions and basic properties of the conics, which are in Book I. Given a circle BC and any point A (Fig. 4.20) outside the plane of the circle, a double cone is generated by a line through A and moving around the circumference of the circle. The circle is called the base of the cone. The axis of the cone is the line from A to the center of the circle (not shown in the figure). If this line is perpendicular to the base, the cone is right circular; otherwise it is scalene or oblique. A section of the cone by a plane cuts the plane of the base in a line DE. Take the diameter BC of the base circle perpendicular to DE. Then ABC is a triangle in whose interior

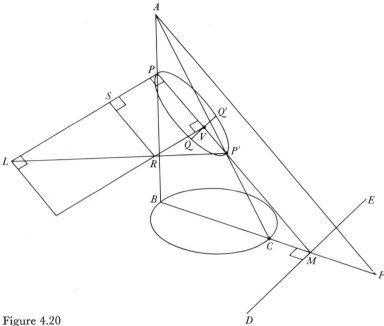

Figure 4.20

the axis of the cone lies; *ABC* is called an axial triangle. Let this triangle cut the conic in *PP'*. (*PP'* need not be an axis of the conic section.) *PP'M* is the line determined by the intersection of the cutting plane and the axial triangle.[3] In the conic section let *Q'Q* be any chord which is parallel to *DE*. Hence *QQ'* need not be perpendicular to *PP'*. Apollonius then proves that *Q'Q* is bisected by *PP'* so that *VQ* is half of *Q'Q*.

Now draw *AF* parallel to *PM* to meet *BM* in *F*, say. Next draw *PL* perpendicular to *PM* and in the plane of the section. For the ellipse and hyperbola *L* is chosen to satisfy the condition

$$\frac{PL}{PP'} = \frac{BF \cdot FC}{AF^2},$$

and for the parabola *L* is chosen to satisfy

$$\frac{PL}{PA} = \frac{BC^2}{BA \cdot AC}.$$

---

3. Apollonius points out that if the cone is scalene then *PM* is not necessarily perpendicular to *DE*. Perpendicularity holds only for right circular cones or when the plane of *ABC* is perpendicular to the base of a scalene cone.

In the cases of ellipse and hyperbola we now draw $P'L$. From V draw VR parallel to PL to meet $P'L$ in R. (In the case of the hyperbola the location of P' is on the other branch and $P'L$ has to be extended to locate R.)

After some subordinate constructions which we do not give Apollonius proves that for the ellipse and hyperbola

(6) $$QV^2 = PV \cdot VR.$$

Now Apollonius refers to QV as an ordinate and the result (6) says that the square of the ordinate equals a rectangle applied to PL, namely $PV \cdot VR$. Moreover, he proves that in the case of the ellipse this rectangle falls short of the entire rectangle $PV \cdot PL$ by the rectangle LR which is similar to the entire rectangle formed by PL and PP'. Hence the term "ellipse" (sec. 6).

In the case of the hyperbola, (6) still holds but the construction would show that VR is longer than PL so that the rectangle $PV \cdot VR$ exceeds the rectangle applied to PL, that is, $PL \cdot PV$, by the rectangle LR which is similar to the rectangle formed by PL and PP'. Hence the term hyperbola. In the case of the parabola Apollonius shows that in place of (6),

(7) $$QV^2 = PV \cdot PL$$

so that the rectangle which equals QV^2 is exactly the rectangle applied to PL with width PV. Hence the term "parabola."

Apollonius introduced the terminology parabola, ellipse, and hyperbola for the conics in place of Menaechmus' sections of the right-angled, acute-angled, and obtuse-angled cone. Where the words parabola and ellipse occur in Archimedes, as in his *Quadrature of the Parabola* (Chap. 5, sec. 3) they were introduced by later transcribers.

Equations (6) and (7) are the basic plane properties of the conic sections. Having derived them, Apollonius disregards the cone and derives further properties from these equations. In effect, where we now use abscissa, ordinate, and the equation of a conic to derive properties, Apollonius uses PV, the ordinate or semichord QV, and a geometric equality, namely (6) or (7). Of course, no algebra appears in Apollonius' treatment.

We can readily transcribe Apollonius' basic properties into modern coordinate geometry. If we denote PL, which Apollonius calls the latus rectum or the parameter of the ordinates, by $2p$ and denote the length of the diameter PP' by d, if x is the distance PV measured from P, and if y is the distance QV (which means we are using oblique coordinates), then one sees immediately from (7) that the equation of the parabola is

$$y^2 = 2px.$$

For the ellipse, we note that we first get from the defining equation (6) that

$$y^2 = PV \cdot VR.$$

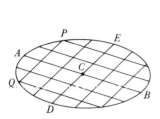

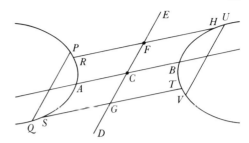

Figure 4.21   Figure 4.22

But $PV \cdot VR = x(2p - LS)$. Also, because the rectangle LR is similar to the rectangle determined by PL and PP',

$$\frac{LS}{PL} = \frac{x}{d}.$$

Hence $LS = 2px/d$. Then

$$y^2 = x\left(2p - \frac{2px}{d}\right) = 2px - \frac{2px^2}{d}.$$

For the hyperbola we get

$$y^2 = 2px + \frac{2px^2}{d}.$$

In the Apollonian construction, d is infinite for the parabola and so we see how the parabola appears as a limiting case of ellipse or hyperbola.

To pursue further Apollonius' treatment of the conics we need some definitions of concepts that are still important in modern geometry. Consider a set of parallel chords in an ellipse, say the set parallel to PQ in Figure 4.21. Apollonius proves that the centers of these chords lie on one line AB, which is called a diameter of the conic. (The line PP' in the basic Figure 4.20 is a diameter.) He then proves that if through C, the midpoint of AB, a line DE be drawn parallel to the original family of chords, this line will bisect all the chords parallel to AB. The line DE is called the conjugate diameter to AB. In the case of the hyperbola (Fig. 4.22), the chords may be inside the branches, e.g. PQ, and the length of the diameter is that cut off (if it is cut off) between the two branches, AB in the figure. The chords parallel to AB, for example RH, then lie between the branches. The conjugate diameter to AB, namely DE, which is defined to be the mean proportional between AB and the latus rectum of the hyperbola, does not cut the curve. In the parabola any diameter, that, is a line passing through the midpoints of a family of parallel chords, is always parallel to the axis of symmetry, but there is no diameter conjugate to a given diameter because each chord of the family of chords parallel to the given diameter is infinite in length. The axes of an

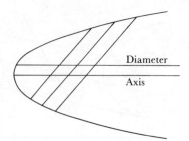

Figure 4.23

ellipse or a hyperbola are two diameters that are perpendicular to each other. For the parabola (Fig. 4.23), the axis is the diameter whose corresponding chords are perpendicular to the diameter.

After introducing the basic properties of the conic sections, Apollonius proves simple facts about conjugate diameters. Book I also treats tangents to conics. Apollonius conceives of a tangent as a line which has just one point in common with a conic section but which everywhere else lies outside of it. He then shows that a straight line drawn through an extremity of a diameter (point P in the basic Figure 4.20) and parallel to the corresponding chords of that diameter (parallel to QQ' in that figure) will fall outside the conic and no other straight line can fall between the said straight line and the conic (see *Elements*, Book III, Proposition 16). Therefore the said straight line touches the conic, that is, is the tangent at P.

Another theorem on tangents asserts the following: Suppose PP' (Fig. 4.24) is a diameter of a parabola and QV is one of the chords corresponding to that diameter. Then if a point T be taken on the diameter but outside the curve and such that $TP = PV$, where V is the foot of the ordinate (chord) from Q to the diameter PP', then the line TQ will touch the parabola at Q. There are analogous theorems for the ellipse and hyperbola.

Apollonius proves next that if any diameter of the conic other than PP' in the basic figure (4.20) be taken, the definitive property of the conic, equations (6) and (7), remains the same; of course QV then refers to the

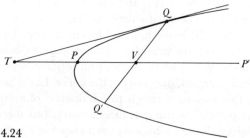

Figure 4.24

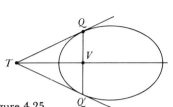

Figure 4.25

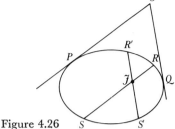

Figure 4.26

chords of that diameter. What he has done amounts in our language to transformation from one oblique coordinate system to another. In this connection, he also proves that from any diameter and the ordinates to it, one can transform to a diameter (axis) to which the ordinates are perpendicular. Then, in our language, the coordinate system is rectangular. Apollonius also shows how to construct conics given certain data—for example, a diameter, the latus rectum, and the inclination of the ordinates to the diameter. He does this by first constructing the cone of which the desired conic is then a section.

Book II starts with the construction and properties of the asymptotes to a hyperbola. He shows, for example, not only the existence of the asymptotes but also that the distance between a point on the curve and the asymptote becomes smaller than any given length by going far enough out on the curve. Then the conjugate hyperbola to a given one is introduced. It is shown to have the same asymptotes.

Additional theorems of Book II show how to find a diameter of a conic, the center of a central conic, the axis of a parabola, and the axes of a central conic. For example, if T (Fig. 4.25) is external to a conic, TQ and TQ' are tangents at the points Q and Q' of the conic, and V is the midpoint of the chord QQ', then TV is a diameter. Another method of finding the diameter of a conic is to draw two parallel chords; the line joining the midpoints of the chords is a diameter. The point of intersection of any two diameters is the center of a central conic. The book concludes with methods of constructing tangents to conics satisfying given conditions, as, for example, passing through a given point.

Book III begins with theorems about areas of figures formed by tangents and diameters. One of the chief results here (Fig. 4.26) is that if OP and OQ are tangents to a conic, if RS is any chord parallel to OP, and $R'S'$ any chord parallel to OQ, and if RS and $R'S'$ intersect in J (internally or externally), then

$$\frac{RJ \cdot JS}{R'J \cdot JS'} = \frac{OP^2}{OQ^2}.$$

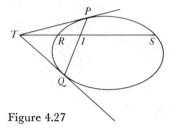

Figure 4.27

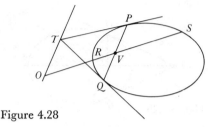

Figure 4.28

The theorem is a generalization of a well-known theorem in elementary geometry, namely, that if two chords intersect in a circle the product of the segments of one equals the product of the segments of the other, for in this case $OP^2/OQ^2 = 1$.

Book III then treats what we call the harmonic properties of pole and polar. Thus if TP and TQ are tangents to a conic (Fig. 4.27) and if TRS is any line meeting the conic in R and S and meeting PQ in I, then

$$\frac{TR}{TS} = \frac{IR}{IS}.$$

That is, T divides RS externally in the same ratio as I divides it internally. The line PQ is called the polar of the point T, and T, R, I, and S are said to form a harmonic set of points. Also if any line through V (Fig. 4.28), the midpoint of PQ, meets the conic in R and S and meets the parallel to PQ through T in O, then

$$\frac{OR}{OS} = \frac{VR}{VS}.$$

The line through T is the polar of V and O, R, V, and S are a harmonic set of points.

The book continues with the subject of the focal properties of central conics; the focus of a parabola is not mentioned here. The foci (the word is not used by Apollonius) are defined for the ellipse and the hyperbola (Fig. 4.29) as the points F and F' on the (major) axis AA' such that $AF \cdot FA' = AF' \cdot F'A' = 2p \cdot AA'/4$. Apollonius proves for the ellipse and the hyperbola that the lines PF and PF' from a point P on the conic make equal angles with the tangent at P and that the sum (for the ellipse) of the focal distances PF and PF' equals AA' and the difference of the focal distances (for the hyperbola) equals AA'.

No concept of directrix appears in this work, but the fact that a conic is a locus of points the ratio of whose distances from a fixed point (focus) and a fixed line (directrix) is constant was known to Euclid and is stated and proved by Pappus (Chap. 5, sec. 7).

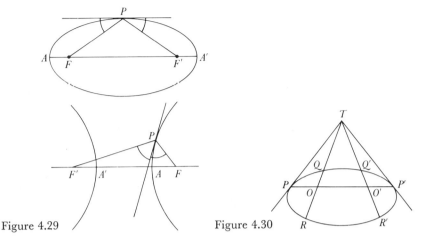

Figure 4.29   Figure 4.30

There is a famous problem, partly solved by Euclid, of determining the locus of points for each of which the distances p, q, r, and s to four given lines satisfy the condition $pq = \alpha rs$, where α is a given number. Apollonius says in his preface to the *Conic Sections* that this problem can be solved by the propositions of Book III. This can indeed be done; and Pappus too knew that this locus is a conic.

Book IV treats further properties of pole and polar. For example, one proposition gives a method of drawing two tangents to a conic from an external point T (Fig. 4.30). Draw TQR and $TQ'R'$. Let O be the fourth harmonic point to T on QR; that is, $TQ:TR = OQ:OR$, and let O' be the fourth harmonic point to T on $Q'R'$. Draw OO'. Then P and P' are the points of tangency.

The remainder of the book deals with the number of possible intersections of conics in various positions. Apollonius proves that two conics can intersect in at most four points.

Book V is the most remarkable for its novelty and originality. It deals with the maximum and minimum lengths that can be drawn from particular points to a conic. Apollonius starts with special points on the major axis of a central conic or on the axis of a parabola and finds the lines of maximum and minimum distances from such points to the curve. Then he takes points on the minor axis of an ellipse and does the same thing. He also proves that if O be any point within any conic and if OP is a maximum or minimum straight line from O to the conic, then the line perpendicular to OP at P is tangent at P; and if O' be any point on OP produced outside the conic, then $O'P$ is a minimum line from O' to the conic. The perpendicular to a tangent at a point of tangency we now call a normal, and so the maximum and minimum

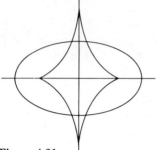

Figure 4.31

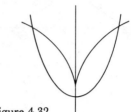

Figure 4.32

lines are normals. Apollonius next considers properties of normals to any conic. For example, in a parabola or an ellipse a normal at any point will meet the curve again. He then shows how from given points within or without a conic one can construct the normals to the conic.

In the course of his investigation of the (relative) maximum and minimum lines that can be drawn from a point to any conic, Apollonius determines the positions of points from which two, three, and four such lines can be drawn. For each of the conics he determines the locus of points such that from points on one side one number of normals can be drawn, and from points on the other, another number of normals can be drawn. The locus itself, which Apollonius does not discuss, is what we now call the evolute of the conic, or the locus of points of intersection of "nearby" normals to the conic, or the envelope of the family of normals to the conic. Thus from any point inside the evolute of the ellipse (Fig. 4.31) four normals to the ellipse can be drawn but from any point outside two normals can be drawn. (There are exceptional points.) In the case of a parabola, the evolute (Fig. 4.32) is the curve called a semicubical parabola (first studied by William Neile [1637–70]). From any point in the plane above the semicubical parabola three normals to the parabola can be drawn and from any point below, only one can be drawn. From a point on the semicubical parabola two can be drawn.

Book VI treats congruent and similar conics and segments of conics. Segments are regions cut off by a chord just as in the circle. Apollonius also shows how, given a right circular cone, one can construct on it a conic section equal to a given one.

Book VII has no outstanding propositions. It treats properties of conjugate diameters of a central conic. Apollonius compares these properties with the corresponding properties of the axes. Thus if a and b are the axes and a' and b' are two conjugate diameters of an ellipse or a hyperbola, $a + b < a' + b'$. Further, the sum of the squares on any two conjugate diameters of an ellipse equals the sum of the squares on the axes. For the hyperbola the corresponding proposition holds but with difference instead of sum. Also in an ellipse or a hyperbola the area of the parallelogram determined by any two conjugate diameters and the angle at which they intersect equals the area of the rectangle determined by the axes.

Book VIII is lost. It probably contained propositions on how to determine conjugate diameters of a (central) conic such that certain functions of their lengths have given values.

Pappus mentions six other mathematical works of Apollonius. One of these, *On Contacts*, whose contents were reconstructed by Vieta, contained the famous Apollonian problem: Given any three points, lines, or circles, or any combination of three of these, to construct a circle passing through the points and tangent to the given lines and circles. Many mathematicians, including Vieta and Newton, gave solutions to this problem.

The strict deductive mathematics of Euclid and Apollonius has given rise to the impression that mathematicians create by reasoning deductively. Our review of the three hundred years of activity preceding Euclid should show that conjectures preceded proofs and that analysis preceded synthesis. In fact, the Greeks did not think much of propositions obtained by simple deduction. Results that sprung readily from a theorem the Greeks called corollaries or porisms. Such results, obtained without additional labor, were regarded by Proclus as windfalls or bonuses.

We have not exhausted the contributions of the Greek genius to mathematics. What we have discussed thus far belongs to the classical Greek period; the significant epoch extending from about 300 B.C. to A.D. 600 still awaits us. Before turning the page let us recall that the classical period contributed more than content; it created mathematics in the sense in which we understand the word today. The insistence on deduction as a method of proof and the preference for the abstract as opposed to the concrete determined the character of mathematics, while the selection of a most fruitful and highly acceptable set of axioms and the divination and proof of hundreds of theorems sent the science well on its way.

Bibliography

Ball, W. W. Rouse: *A Short Account of the History of Mathematics*, Dover (reprint), 1960, Chaps. 2–3.

Boyer, Carl B.: *A History of Mathematics*, John Wiley and Sons, 1968, Chaps. 7 and 9.

Coolidge, Julian L.: *A History of Geometrical Methods*, Dover (reprint), 1963, Book 1, Chaps. 2–3.

Heath, Thomas L.: *A Manual of Greek Mathematics*, Dover (reprint), 1963, Chaps. 3–9 and 12.

———: *A History of Greek Mathematics*, Oxford University Press, 1921, Vol. 1, Chaps. 3–11; Vol. 2, Chap. 14.

———: *The Thirteen Books of Euclid's* Elements, 3 vols., Dover (reprint), 1956.

————: *Apollonius of Perga*, Barnes and Noble (reprint), 1961.

Neugebauer, Otto: *The Exact Sciences in Antiquity*, Princeton University Press, 1952, Chap. 6.

Proclus: *A Commentary on the First Book of Euclid's* Elements, Princeton University Press, 1970.

Sarton, George: *A History of Science*, Harvard University Press, 1952 and 1959, Vol. 1, Chaps. 8, 10, 11, 17, 20; Vol. 2, Chap. 3.

Scott, J. F.: *A History of Mathematics*, Taylor and Francis, 1958, Chap. 2.

Smith, David Eugene: *History of Mathematics*, Dover (reprint), 1958, Vol. 1, Chap. 3; Vol. 2, Chap. 5.

Struik, Dirk J.: *A Concise History of Mathematics*, 3rd ed., Dover, 1967, Chap. 3.

van der Waerden, B. L.: *Science Awakening*, P. Noordhoff, 1954, Chaps. 4–6.

5

The Alexandrian Greek Period: Geometry and Trigonometry

> Without the concepts, methods and results found and developed by previous generations right down to Greek antiquity one cannot understand either the aims or the achievements of mathematics in the last fifty years. HERMANN WEYL

1. *The Founding of Alexandria*

The course of mathematics has been very much dependent on the course of history. Conquests launched by the Macedonians, a Greek people living in the northern part of the mainland of Greece, led to the destruction of the classical Greek civilization and paved the way for another essentially Greek civilization of quite different character. The conquests were begun in 352 B.C. by Philip II of Macedonia. Athens was beaten in 338 B.C. In 336 B.C. Alexander the Great, son of Philip, took command and conquered Greece, Egypt, and the Near East as far east as India and as far south as the cataracts of the Nile. He constructed new cities everywhere, both as strongholds and as centers of commerce. The main one, Alexandria, centrally located in Alexander's empire and intended as his capital, was founded in Egypt in 332 B.C. Alexander chose the site and drew up the plans for the buildings and for colonizing the city, but the work was not completed for many years thereafter.

Alexander envisioned a cosmopolitan culture in his new empire. Because the only other leading civilization was Persian, Alexander deliberately sought to fuse the two. In 325 B.C. he himself married Statira, daughter of the Persian ruler Darius, and compelled a hundred of his generals and ten thousand of his soldiers to marry Persians. He incorporated twenty thousand Persians into his army and mixed them with Macedonians in the same phalanxes. He also brought colonists of all nations to the various cities he founded. After his death, written orders were found to transport large groups of Europeans to Asia and vice versa.

Alexander died in 323 B.C. before he could complete his capital and while still engaged in conquests. After his death, his generals fought each other for power. Following several decades of political instability, the empire was split into three independent parts. The European portion became the Antigonid empire (from the Greek general Antigonus); the Asian part, the Seleucid empire (after the Greek general Seleucus); and Egypt, ruled by the Greek Ptolemy dynasty, became the third empire. Antigonid Greece and Macedonia gradually fell under Roman domination and became unimportant as far as the development of mathematics is concerned. The mathematics generated in the Seleucid empire was largely a continuation of Babylonian mathematics, though influenced by the developments we are about to consider. The major creations following the classical Greek period were made in the Ptolemaic empire, primarily in Alexandria.

That the Ptolemaic empire became the mathematical heir of classical Greece was not accidental. The kings of the empire were wise Greeks and pursued Alexander's plan to build a cultural center at Alexandria. Ptolemy Soter, who ruled from 323 to 285 B.C., his immediate successors, Ptolemy II, called Philadelphus, who ruled from 285 to 247 B.C., and Ptolemy Euergetes, who reigned from 247 to 222 B.C., were well aware of the cultural importance of the great Greek schools such as those of Pythagoras, Plato, and Aristotle. These rulers therefore brought to Alexandria scholars from all the existing centers of civilization and supported them with state funds. About 290 B.C Ptolemy Soter built a center in which the scholars could study and teach. This building, dedicated to the muses, became known as the Museum, and it housed poets, philosophers, philologists, astronomers, geographers, physicians, historians, artists, and most of the famous mathematicians of the Alexandrian Greek civilization.

Adjacent to the Museum, Ptolemy built a library, not only for the preservation of important documents but for the use of the general public. This famous library was said at one time to contain 750,000 volumes, including the personal library of Aristotle and his successor Theophrastus. Books, incidentally, were more readily available in Alexandria than in classical Greece because Egyptian papyrus was at hand. In fact, Alexandria became the center of the book-copying trade of the ancient world.

The Ptolemies also pursued Alexander's plan of encouraging a mixture of peoples, so that Greeks, Persians, Jews, Ethiopians, Arabs, Romans, Indians, and Negroes came unhindered to Alexandria and mingled freely in the city. Aristocrat, citizen, and slave jostled each other and, in fact, the class distinctions of the older Greek civilization broke down. The civilization in Egypt was influenced further by the knowledge brought in by traders and by the special expeditions organized by the scholars to learn more about other parts of the world. Consequently, intellectual horizons were broadened. The long sea voyages of the Alexandrians called for far better knowledge of geography, methods of telling time, and navigational techniques, while commercial competition generated an interest in materials, in efficiency of production, and in improvement of skills. Arts that had been despised in the classical period were taken up with new zest and training schools were established. Pure science continued to be pursued but was also applied.

The mechanical devices created by the Alexandrians are astonishing even by modern standards. Pumps to bring up water from wells and cisterns, pulleys, wedges, tackles, systems of gears, and a mileage-measuring device no different from what may be found in the modern automobile were used commonly. Steam power was employed to drive a vehicle along the city streets in the annual religious parade. Water or air heated by fire in secret vessels of temple altars was used to make statues move. The awe-struck audience observed gods who raised their hands to bless the worshipers, gods shedding tears, and statues pouring out libations. Water power operated a musical organ and made figures on a fountain move automatically while compressed air was used to operate a gun. New mechanical instruments, including an improved sundial, were invented to refine astronomical measurements.

The Alexandrians had an advanced knowledge of such phenomena as sound and light. They knew the law of reflection of light and had an empirical grip on the law of refraction (Chap. 7, sec. 7), knowledge which they applied to the design of mirrors and lenses. In this period there appeared for the first time a work on metallurgy, which contained far more chemistry than the few empirical facts known to the earlier Egyptian and Greek scholars. Poisons were a specialty. Medicine flourished, partly because the dissection of human bodies, forbidden in classical Greece, was now permitted, and the art of healing reached its pinnacle in the work of Galen (129–c. 201), who, however, lived chiefly in Pergamum and Rome. Hydrostatics, the science of the equilibrium of bodies immersed in fluids, was investigated intensively and indeed founded in systematic form. The greatest of their scientific achievements was the first truly quantitative astronomical theory (Chap. 7, sec. 4).

2. The Character of Alexandrian Greek Mathematics

The work of the scholars at the Museum was divided into four departments— literature, mathematics, astronomy, and medicine. Since two of these were essentially mathematical and medicine, through astrology, involved some mathematics, we see that mathematics occupied a dominant place in the Alexandrian world. The character of the mathematics was very much affected by the new civilization and culture. No matter what mathematicians may say about the purity of their subject and their indifference to and elevation above their environment, the new Hellenistic civilization produced a mathematics entirely different in character from that of the classical period.

Of course Euclid and Apollonius were Alexandrians; but, as we have already noted, Euclid organized the work of the classical period, and Apollonius is exceptional in that he too organized and extended classical Greek mathematics—though in his astronomy and his work on irrational numbers (both of which will be presented in later chapters), he was somewhat affected by the Alexandrian culture. To be sure, the other great Alexandrian mathematicians, Archimedes, Eratosthenes, Hipparchus, Nicomedes, Heron, Menelaus, Ptolemy, Diophantus, and Pappus, continued to display the Greek genius for theoretical and abstract mathematics, but with striking differences. Alexandrian geometry was devoted in the main to results useful in the calculation of length, area, and volume. It is true that some such theorems are also in Euclid's *Elements*. For example, Proposition 10 of Book XII asserts that any cone is a third part of the cylinder which has the same base and height. Hence if one knows the volume of a cylinder he can compute the volume of a cone. However, such theorems are relatively scarce in Euclid, whereas they were the major concern of the Alexandrian geometers. Thus, while Euclid was content to prove that the areas of two circles are to each other as the squares on their diameters—which leaves us with the knowledge that the area $A = kd^2$ but without a value for k— Archimedes obtained a close approximation to π so that circular areas could be computed.

Further, the classical Greeks, because they would not entertain irrationals as numbers, had produced a purely qualitative geometry. The Alexandrians, following the practice of the Babylonians, did not hesitate to use irrationals and in fact applied numbers freely to lengths, areas, and volumes. The climax of this work was the development of trigonometry.

Even more significant is the fact that the Alexandrians revived and extended arithmetic and algebra, which became subjects in their own right. This development of the science of number was, of course, necessary if quantitative knowledge was to be obtained either from geometrical results or from the direct use of algebra.

The Alexandrian mathematicians took an active hand in the work on mechanics. They calculated centers of gravity of bodies of various shapes; they dealt with forces, inclined planes, pulleys, and gears; and they were often inventors. They were also the chief contributors of their time to the work on light, mathematical geography, and astronomy.

In the classical period mathematics had embraced arithmetic (of the whole numbers only), geometry, music, and astronomy. The scope of mathematics was broadened immeasurably in the Alexandrian period. Proclus, who drew material from Geminus of Rhodes (1st cent. B.C.), cites the latter on the divisions of mathematics (presumably in Geminus' time): arithmetic (our theory of numbers), geometry, mechanics, astronomy, optics, geodesy, canonic (science of musical harmony), and logistic (applied arithmetic). According to Proclus, Geminus says: "The entire mathematics was separated into two main divisions with the following distinction: one part concerned itself with the intellectual concepts and the other with material concepts." Arithmetic and geometry were intellectual. The other division was material. However, this distinction was gradually lost sight of, if it was still significant as late as the first century B.C. One can say, as a broad generalization, that the mathematicians of the Alexandrian period severed their relation with philosophy and allied themselves with engineering.

We shall treat first the Alexandrian work in geometry and trigonometry. In the next chapter we shall discuss the arithmetic and algebra.

3. Areas and Volumes in the Work of Archimedes

There is no one individual whose work epitomizes the character of the Alexandrian age so well as Archimedes (287–212 B.C.), the greatest mathematician in antiquity. The son of an astronomer, he was born in Syracuse, a Greek settlement in Sicily. As a youth he came to Alexandria, where he received his education. Though he returned to Syracuse and spent the rest of his life there, he kept contact with Alexandria. He was well known in the Greek world and greatly admired and respected by his contemporaries.

Archimedes was possessed of a lofty intellect, great breadth of interests—both practical and theoretical—and excellent mechanical skill. His work in mathematics included the finding of areas and volumes by the method of exhaustion, the calculation of π (in the course of which he approximated square roots of small and large numbers), and a new scheme for representing large numbers in verbal language. In mechanics, he found the centers of gravity of many plane and solid figures and gave theorems on the lever. The area of hydrostatics that deals with the equilibrium of bodies floating in water was founded by him. He is also reputed to have been a good astronomer.

His inventions so far excelled the technique of his times that endless stories and legends grew up about him. Indeed in popular esteem his inventions overshadowed his mathematics, though he is ranked with Newton and Gauss as one of the three greatest in that field. In his youth he constructed a planetarium, a contrivance operated by water power that reproduced the motions of the sun, moon, and planets. He invented a pump (Archimedean screw) for raising water from a river; he showed how to use the lever to move great weights; he used compound pulleys to launch a galley for King Hieron of Syracuse; and he invented military engines and catapults to protect Syracuse when it was under attack by the Romans. Taking advantage of the focusing properties of a paraboloidal mirror, he concentrated the sun's rays on the Roman ships besieging Syracuse and burned them.

Perhaps the most famous of the stories about Archimedes is his discovery of the method of testing the debasement of a crown of gold. The king of

Syracuse had ordered the crown. When it was delivered, he suspected that it was filled with baser metals and sent it to Archimedes to devise some method of testing the contents without, of course, destroying the workmanship. Archimedes pondered the problem; one day while bathing he observed that his body was partly buoyed up by the water and suddenly grasped the principle that enabled him to solve the problem. He was so excited by this discovery that he ran out into the street naked shouting "Eureka!" ("I have found it!") He had discovered that a body immersed in water is buoyed up by a force equal to the weight of the water displaced, and by means of this principle was able to determine the contents of the crown (see Chap. 7, sec. 6).

Though Archimedes was a remarkably ingenious and successful inventor, Plutarch says that these inventions were merely "the diversions of geometry at play." According to Plutarch, Archimedes "possessed so high a spirit, so profound a soul, and such treasures of scientific knowledge that, though these inventions had obtained for him the renown of more than human sagacity, he yet would not deign to leave behind him any written work on such subjects, but, regarding as ignoble and sordid the business of mechanics and every sort of art which is directed to use and profit, he placed his whole ambition in those speculations in whose beauty and subtlety there is no admixture of the common needs for life." However Plutarch's status as a storyteller is far higher than his status as a historian. Archimedes did write books on mechanics and we have one entitled *On Floating Bodies* and another, *On the Equilibrium of Planes*; two others, *On Levers* and *On Centers of Gravity*, are lost. He also wrote a work on optics that is lost, and did deign to write about his inventions; though the work has vanished, we know definitely that he wrote *On Sphere-making*, which describes an invention displaying the motions of the sun, the moon, and the five planets about the (fixed) earth.

The death of Archimedes portended what was to happen to the entire Greek world. In 216 B.C. Syracuse allied itself with Carthage in the second Punic war between that city and Rome. The Romans attacked Syracuse in 212 B.C. While drawing mathematical figures in the sand, Archimedes was challenged by one of the Roman soldiers who had just taken the city. Story has it that Archimedes was so lost in thought that he did not hear the challenge of the Roman soldier. The soldier thereupon killed him, despite the order of the Roman commander, Marcellus, that Archimedes be unharmed. Archimedes was then seventy-five and still in full possession of his powers. By way of "compensation," the Romans built an elaborate tomb upon which they inscribed a famous Archimedean theorem.

Archimedes' writings took the form of small tracts rather than large books. Our knowledge of these works comes from extant Greek manuscripts and from Latin manuscripts translated from the Greek from the thirteenth century onward. Some of the Latin versions were made from Greek

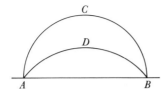

Figure 5.1

manuscripts available to those translators but not to us. In 1543 Tartaglia did a translation into Latin of some works of Archimedes.

Archimedes' geometrical work is the zenith of Alexandrian Greek mathematics. In his mathematical derivations Archimedes uses theorems of Euclid and Aristaeus, and still other results which he says are manifest, that is, can readily be proved from the known results. His proofs are therefore solidly established but not easy for us to follow because we are not familiar with many of the methods and results of the Greek geometers.

In his work *On the Sphere and Cylinder*, Archimedes starts with definitions and assumptions. The first assumption or axiom is that of all lines (curves) which have the same extremities the straight line is shortest. Other axioms involve the lengths of concave curves and surfaces. For example ADB (Fig. 5.1) is assumed to be less than ACB. These axioms enable Archimedes to compare perimeters of inscribed and circumscribed polygons with the perimeter of the circle.

After some preliminary propositions, he proves in Book I:

Proposition 13. The surface of any right circular cylinder excluding the bases is equal to [the area of] a circle whose radius is a mean proportional between the side [a generator] and the diameter of its base.

This is followed by many theorems about the volumes of cones. Of great interest are:

Proposition 33. The surface of any sphere is four times the [area of the] greatest circle on it.
Corollary to Prop. 34. Every cylinder whose base is the greatest circle in a sphere and whose height is equal to the diameter of the sphere is 3/2 of [the volume of] the sphere, and its surface together with its bases is 3/2 of the surface of the sphere.

That is, he compares the surface area and volume of a sphere and a cylinder circumscribed about the sphere. This is the famous theorem which in accordance with Archimedes' wishes, was inscribed on his tombstone.

He then proves in Propositions 42 and 43 that the surface of the segment $ALMNP$ of a sphere is the area of a circle whose radius is AL (Fig. 5.2). The segment can be less or more than a hemisphere.

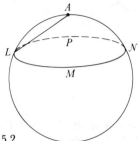

Figure 5.2

The theorems on surface area and volume are proved by the method of exhaustion. Archimedes uses inscribed and circumscribed rectilinear figures to "exhaust" the area or volume, and then, like Euclid, uses the indirect method of proof to complete the argument.

Some theorems in the second book of *On the Sphere and Cylinder*, which is concerned largely with segments of a sphere, are significant because they contain new geometrical algebra. For example, he gives:

Proposition 4. To cut a given sphere by a plane so that the volumes of the segments are to one another in a given ratio.

This problem amounts algebraically to the solution of the cubic equation

$$(a - x):c = b^2:x^2$$

and Archimedes solves it geometrically by finding the intersection of a parabola and a rectangular hyperbola.

The work *On Conoids and Spheroids* treats properties of figures of revolution generated by conics. Archimedes' right-angled conoid is a paraboloid of revolution. (In Archimedes' time the parabola was still regarded as a section of a right-angled cone.) The obtuse-angled conoid is one branch of a hyperboloid of revolution. Archimedes' spheroids are what we call oblate and prolate spheroids, which are figures of revolution generated by ellipses. The main object of the work is the determination of volumes of segments cut off from the three solids by planes. The book also contains some of Archimedes' work on the conic sections already alluded to in the discussion of Apollonius. As in other works, he presupposes theorems that he deems easily proved or that can be proved by methods he has previously used. Many of the proofs use the method of exhaustion. Some examples of the contents are furnished by the following propositions:

Proposition 5. If *AA′* and *BB′* be the major and minor axes of an ellipse and if *d* be the diameter of any circle, then the area of the ellipse is to the area of the circle as $AA' \cdot BB'$ is to d^2.

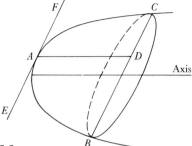

Figure 5.3

The theorem says that if $2a$ is the major axis and $2b$ is the minor axis and s and s' are the areas of the ellipse and circle, then $s/s' = 4ab/d^2$. Since $s' = (\pi/4)d^2$, $s = \pi ab$.

Proposition 7. Given an ellipse with center C and a line CO perpendicular to the plane of the ellipse, it is possible to find a circular cone with vertex O and such that the ellipse is a section of it.

Clearly Archimedes realized that some, at least, of the several conic sections can be obtained from the same cone, a fact that Apollonius utilized.

Proposition 11. If a paraboloid of revolution be cut by a plane through, or parallel to, the axis [of revolution] the section will be a parabola equal to the original parabola which generated the paraboloid . . . . If the paraboloid be cut by a plane at right angles to its axis, the section will be a circle whose center is on the axis.

There are similar results for the hyperboloid and spheroid.

Among principal results of the work are:

Proposition 21. Any segment [the volume] of a paraboloid of revolution is half as large again as the cone or segment of a cone which has the same base and the same axis.

The base is the area (Fig. 5.3) of the plane figure, ellipse or circle, which is cut out on the paraboloid by the plane determining the segment. The parabolic section BAC and BC on the base are cut out by a plane through the axis of the paraboloid and perpendicular to the original cutting plane. EF is the tangent to the parabola that is parallel to BC, and A is the point of tangency. AD, drawn parallel to the axis of the paraboloid, is the axis of the segment. It can be shown that D is the midpoint of CB. Also, if the base is an ellipse then CB is its major axis; if the base is a circle then CB is its diameter. The cone has the same base as the segment, and has vertex A and axis AD.

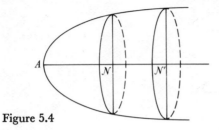

Figure 5.4

Proposition 24. If from a paraboloid of revolution two segments be cut off by planes drawn in *any* manner, the segments will be to one another as the squares on their axes.

To illustrate the theorem, suppose the planes are perpendicular to the axis of the paraboloid (Fig. 5.4); then the two volumes are to each other as AN^2 is to AN'^2. There are similar theorems for segments of hyperboloids and spheroids.

One of the very novel works of Archimedes is a short treatise known as *The Method*, in which he shows how he used ideas from mechanics to obtain correct mathematical theorems. This work was discovered as recently as 1906 in a library in Constantinople. The manuscript was written in the tenth century on a parchment that contains other works of Archimedes already known through other sources. Archimedes illustrates his method of discovery with the problem of finding the area of a parabolic segment CBA (Fig. 5.5). In this basically physical argument he uses theorems on centers of gravity established elsewhere by him.

ABC (Fig. 5.5) is any segment of a parabola bounded by the straight line AC and the arc ABC. Let CE be the tangent to the parabola at C; let D be the midpoint of CA; and let DBE be the diameter through D (line parallel to the axis of the parabola). Then Archimedes, referring to Euclid's *Conics*, states that

(1) $EB = BD,$

though Euclid's proof of this fact is not known. Now draw AF parallel to ED and let CB cut AF in K. Then, by (1) and the use of similar triangles, one proves that $FK = KA$. Produce CK to H so that $CK = KH$. Further, let $MNPO$ be any diameter of the parabola. Then $MN = NO$, because of (1) and the use of similar triangles.

Now Archimedes compares the area of the segment and the area of triangle CFA. He regards the first area as the sum of line segments such as PO and the area of the triangle as the sum of line segments such as MO. He then proves that

$$HK \cdot OP = KN \cdot MO.$$

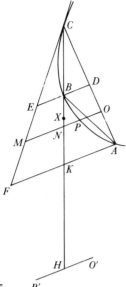

Figure 5.5

Physically this means that if we regard KH and KN as arms of a lever with a fulcrum at K, then OP regarded as a weight placed at H would balance the weight of MO placed at N. Consequently the sum of all the line segments such as PO placed at H will balance the sum of all the line segments such as MO each concentrated at its midpoint, which is the center of gravity of a line segment. But the collection of segments MO each placed at its center of gravity is "equivalent" to the triangle CAF placed at its center of gravity. In his book *On the Equilibrium of Planes*, Archimedes shows that this center is X on CK where $KX = (1/3)CK$. By the law of the lever, $KX \cdot$ the area of triangle $CFA = HK \cdot$ area of parabolic segment or

(2)
$$\frac{\triangle CFA}{\text{segment } CBA} = \frac{HK}{KX} = \frac{3}{1}.$$

Archimedes wishes to relate the area of the segment to triangle ABC. He points out that (the area of) this triangle is one half of triangle CKA because both have the same base CA and the altitude of one is readily shown to be half of the altitude of the other. Moreover, triangle CKA is one half of triangle CFA (because KA is half of FA). Hence triangle ABC is one fourth of triangle CFA and, from (2), he has that the area of segment ABC is to the area of triangle ABC as 4 is to 3.

In this mechanical method Archimedes regards the area of the parabolic segment and of triangle CFA as sums of an infinite number of line segments.

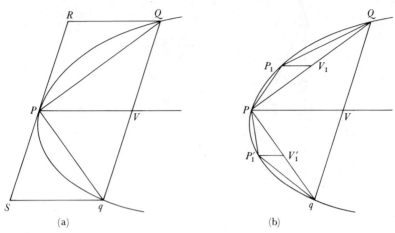

Figure 5.6, a and b

This method, he says, is one of discovery but not of rigorous geometrical proof. He shows in this treatise how effective the method is by using it to discover other theorems on segments of spheres, cylinders, spheroids, and paraboloids of revolution.

In his book *Quadrature of the Parabola* Archimedes gives two methods for finding the area of the parabolic segment. The first of these is similar to the mechanical argument just examined in that he again balances areas by means of the principle of the lever, but his choice of areas is different. His conclusion is of course the same as that in (2) above. It is given in Proposition 16. Now Archimedes knows the result he wishes to prove and he proceeds to do so by rigorous mathematics in a sequence of theorems (Propositions 18–24).

The first step is to prove that the parabolic segment can be "exhausted" by a series of triangles. Let QPq (Fig. 5.6a) be the parabolic segment and let PV be the diameter bisecting all chords parallel to the base Qq of the segment so that V is the midpoint of Qq. It is intuitively apparent and proved in Proposition 18 that the tangent at P is parallel to Qq. Next take QR and qS parallel to PV. Then triangle QPq is one half of parallelogram $QRSq$ and so triangle QPq is greater than one half of the parabolic segment.

As a corollary to this result, Archimedes shows that the parabolic segment can be approximated by a polygon as closely as one pleases, for by constructing a triangle in the segment cut off by PQ (Fig. 5.6b) wherein P_1V_1 is the diameter of that segment one can prove by simple geometry (Proposition 21) that (the area of) triangle $PP_1Q = (1/8)$ triangle PQq. Hence triangle PP_1Q and triangle $PP_1'q$, which is constructed on Pq and has the same properties as triangle PP_1Q, are together 1/4 of triangle PQq;

AREAS AND VOLUMES IN THE WORK OF ARCHIMEDES   113

moreover, by the result in the preceding paragraph, the two smaller triangles fill out more than half of the parabolic segments in which they lie. The process of constructing triangles on the new chords QP_1, P_1P, PP'_1, and P'_1q can be continued. This part of the proof is entirely analogous to the corresponding part of Euclid's theorem on the areas of two circles.

Hence we have the conditions sufficient to apply Proposition 1 of Book X of Euclid's *Elements*; that is, we can assert that the area of the polygonal figure obtained by adding triangles to the original triangle PQq, that is, the area

(3) $$\triangle PQq + (1/4)\triangle PQq + (1/16)\triangle PQq \cdots$$

to a *finite* number of terms can approximate the parabolic segment as closely as one pleases; that is, the difference between the area of the segment and the finite sum (3) can be made less than any preassigned quantity.

Now Archimedes applies the indirect method of proof that completes the proof by the method of exhaustion. He first proves that for n terms of a geometrical progression in which the common ratio is $1/4$,

(4) $$A_1 + A_2 + \cdots + A_n + (1/3)A_n = (4/3)A_1.$$

This is readily proven in many ways; we can do it by our formula for the sum of n terms of a geometrical progression. In the application of (4), A_1 is triangle PQq.

Then Archimedes shows that the area A of the parabolic segment cannot be greater or less than $(4/3)A_1$. His proof is simply that if the area A exceeded $(4/3)A_1$ then he could get a (finite) set of triangles whose sum S would differ from the area of the segment by less than any given magnitude, and hence the sum S would exceed $(4/3)A_1$. That is,

$$A > S > (4/3)A_1.$$

But by (4) if S contains m terms, say, then

$$S + (1/3)A_m = (4/3)A_1$$

or

$$S < (4/3)A_1.$$

Hence there is a contradiction.

Likewise, suppose one assumes that the area A of the parabolic segment is less than $(4/3)A_1$. Then $(4/3)A_1 - A$ is a definite number. Since the triangles Archimedes forms get smaller he can get a sequence of inscribed triangles such that

(5) $$(4/3)A_1 - A > A_m$$

where A_m is the mth term of the sequence and geometrically represents the sum of 2^{m-1} triangles. But since by (4)

$$(6) \qquad A_1 + A_2 + \cdots + A_m + \frac{1}{3} A_m = \frac{4}{3} A_1,$$

then

$$\frac{4}{3} A_1 - (A_1 + A_2 + \cdots + A_m) = \frac{1}{3} A_m,$$

or

$$(7) \qquad \frac{4}{3} A_1 - (A_1 + A_2 + \cdots + A_m) < A_m.$$

It follows from (5) and (7) that

$$(8) \qquad A_1 + A_2 + \cdots + A_m > A.$$

But a sum consisting of inscribed triangles is always less than the area of the segment. Hence (8) is impossible.

Of course, in effect Archimedes has summed an infinite geometric progression, because when n becomes infinite in (4), A_n approaches 0, and the sum of the infinite progression is $4A_1/3$.

The work of Archimedes on the mechanical and mathematical methods of obtaining the area of a parabolic segment shows how clearly he distinguishes physical from mathematical reasoning. His rigor is far superior to that which we shall find in the work of Newton and Leibniz.

In the work *On Spirals* Archimedes defines the spiral as follows. Suppose a line (ray) rotates at a constant rate about one end while remaining in one plane and a point starting from the fixed end moves out at a constant speed along the line; then the point will describe a spiral. In our polar coordinates the equation of the spiral is $\rho = a\theta$. As the curve is drawn in Figure 5.7, θ is a positive clockwise. The deepest result in the work is:

Proposition 24. The area bounded by the first turn of the spiral and the initial line [the shaded area in the figure] is equal to one third of the first circle.

The first circle is the circle with radius OA, which equals $2\pi a$, and so the shaded area is $\pi (2\pi a)^2/3$.

The proof is by the method of exhaustion. In preceding theorems, which prepare the ground, the area of a region bounded by an arc of a spiral, the arc $BPQRC$ in Figure 5.8, and by two radii vectors OB and OC, is enclosed between two sets of circular sectors. Thus $Bp', Pq', Qr', \ldots$ are arcs of circles with center at O and likewise $Pb, Qp, Rq, \ldots$ are arcs of circles. The circular sectors of the inscribed set are $OBp', OPq', OQr', \ldots$ and the circular sectors of the circumscribed set are $OPb, OQp, ORq, \ldots$. Thus circular

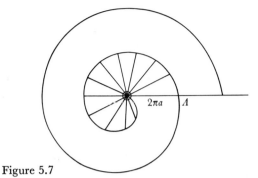

Figure 5.7

sectors replace inscribed and circumscribed polygons as the approximating figures in the method of exhaustion. (We use such approximating figures in the calculus when we determine areas in polar coordinates.) The novel feature in this application of the method of exhaustion is that Archimedes chooses smaller and smaller circular sectors so that the difference between the area under the arc of the spiral and the sum of the areas of the finite number of "inscribed" circular sectors (and the sum of the areas of the finite number of "circumscribed" circular sectors) can be made less than any given magnitude. This manner of approximating the desired area is not the same as "exhausting" it by adding more and more rectilinear figures. However, in the last part of the proof Archimedes uses the indirect method of proof as he does in the work on the parabola and as Euclid does in his proofs by the method of exhaustion. There is no explicit limit process.

Archimedes also gives the result for the area bounded by the arc of the spiral after the radius vector has rotated twice completely around O; and there are other related results on area. Incidentally, later mathematicians used the spiral to trisect an angle and in fact to divide an angle into any number of equal parts.

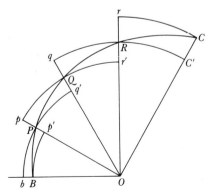

Figure 5.8

It is at once apparent from a study of the geometrical work of Archimedes that he is concerned to obtain useful results on area and volume. This work and his mathematical work in general are not spectacular as to conclusions nor especially new in method or subject matter, but he tackles very difficult and original problems. He often says that the suggestions for problems came from reading the works of his predecessors; for example, the work of Eudoxus on the pyramid, cone, and cylinder (given in Euclid's *Elements*) suggested to Archimedes his work on the sphere and cylinder, and the problem of squaring the circle suggested the quadrature of the parabolic segment. Archimedes' work on hydrostatics, however, is entirely novel; and his work on mechanics is new in that he gives mathematical demonstrations (Chap. 7, sec. 6). His writing is elegant, ordered, finished, and to the point.

4. *Areas and Volumes in the Work of Heron*

Heron, who lived sometime between 100 B.C. and A.D. 100, is of great interest not only from the standpoint of the history of mathematics but also in exhibiting the characteristics of the Alexandrian period. Proclus refers to Heron as *mechanicus*, which might mean a mechanical engineer today, and discusses him in connection with the inventor Ctesibius, his teacher. Heron was also a good surveyor.

The striking fact about Heron's work is his commingling of rigorous mathematics and the approximate procedures and formulas of the Egyptians. On the one hand, he wrote a commentary on Euclid, used the exact results of Archimedes (indeed he refers to him often), and in original works proved a number of new theorems of Euclidean geometry. On the other hand, he was concerned with applied geometry and mechanics and gave all sorts of approximate results without apology. He used Egyptian formulas freely and much of his geometry was also Egyptian in character.

In his *Metrica* and *Geometrica*, the latter known to us only through a book based on his work, Heron gives theorems and rules for plane areas, surface areas, and volumes of a great number of figures. The theorems in these books are not new. For figures with curved boundaries he uses Archimedes' results. In addition he wrote *Geodesy* and *Stereometry* (calculation of volumes of figures), both of which are concerned with the same subjects as the first two books. In all of these works he is primarily interested in numerical results.

In his *Dioptra* (theodolite), a treatise on geodesy, Heron shows how to find the distance between two points of which only one is accessible and between two points that are visible but not accessible. He also shows how to draw a perpendicular from a given point to a line that cannot be reached and how to find the area of a field without entering it. The formula for the area of a triangle, credited to him though due to Archimedes, namely,

$$\sqrt{s(s-a)(s-b)(s-c)}$$

wherein a, b, and c are the sides and s is half the perimeter, illustrates the last-mentioned idea. This formula appears in his *Geodesy*, and the formula and a proof are in both the *Dioptra* and the *Metrica*. In the *Dioptra* he shows how to dig a straight tunnel under a mountain by working simultaneously from both ends.

Though many of the formulas are proven, Heron gives many without proof and also gives many approximate ones. Thus he gives an inexact formula for the area of a triangle along with the above correct one. One reason that Heron gave many Egyptian formulas may be that the exact ones involved square roots or cube roots and the surveyors could not execute these operations. In fact, there was a distinction between pure geometry and geodesy or metrics. The calculation of areas and volumes belonged to geodesy and was not part of a liberal education. It was taught to surveyors, masons, carpenters, and other technicians. There is no doubt that Heron continued and enriched the Egyptian science of field measurements; his writings on geodesy were used for hundreds of years.

Heron applied many of his theorems and rules to the design of theaters, banquet halls, and baths. His applied works include *Mechanics*, *The Construction of Catapults*, *Measurements*, *The Design of Guns*, *Pneumatica* (the theory and use of air pressure), and *On the Art of Construction of Automata*. He gives designs for water clocks, measuring instruments, automatic machines, weight-lifting machines, and war engines.

5. *Some Exceptional Curves*

Though the classical Greeks did introduce and study some unusual curves, such as the quadratrix, the dictate that geometry was to be devoted to figures constructible with line and circle banished those curves to limbo. The Alexandrians, however, felt freer to ignore the restriction; thus Archimedes did not hesitate to introduce the spiral. A number of other curves were introduced in the Alexandrian period.

Nicomedes (*c.* 200 B.C.) is known for his definition of the conchoid. He starts with a point P and a line AB (Fig. 5.9). He then chooses a length a and lays this off on all rays from P that cross AB, the length a starting from the point of intersection of the ray and AB and in the direction away from P. The endpoints so determined are the points of the conchoid. Thus P_1, P_2, and P_3 of the figure are points of the conchoid.

If b is the perpendicular distance from P to AB and if the lengths a are measured along the rays through P and starting from AB but in the direction

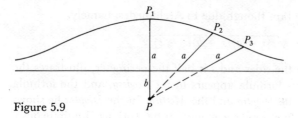

Figure 5.9

of P we get three other curves according as $a > b$, $a = b$, and $a < b$. Hence there are four types of conchoids, all due to Nicomedes. The modern polar equation is $r = a + b \sec \theta$. Nicomedes used the curve to trisect an angle and double the cube.[1]

Nicomedes is supposed to have invented a mechanism to construct the conchoids. The nature of the mechanism is of far less interest than the fact that mathematicians of the period were interested in devising them. The conchoids of Nicomedes, together with the line and circle, are the oldest mechanically constructible curves about which we possess satisfactory information.

Diocles (end of 2nd cent. B.C.), in his book *On Burning-glasses*, solved the problem of doubling the cube by introducing the curve called the cissoid. The curve is defined as follows: AB and CD are perpendicular diameters of a circle (Fig. 5.10) and EB and BZ are equal arcs. Draw ZH perpendicular to CD and then draw ED. The intersection of ZH and ED gives a point P on the cissoid. For Diocles the cissoid is the locus of all points P determined by all positions of E on arc BC and Z on arc BD with arc BE = arc BZ. One

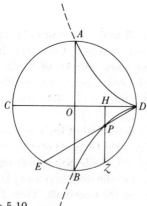

Figure 5.10

1. The method of trisection is given in T. L. Heath: *The Thirteen Books of Euclid's* Elements, Dover (reprint), 1956, Vol. 1, p. 266.

can prove that

$$CH:HZ = HZ:HD = HD:HP.$$

Thus HZ and HD are two mean proportionals between CH and HP. This solves the Delian problem. The equation of the cissoid in rectangular coordinates is $y^2(a + x) = (a - x)^3$, where O is the origin, a the radius of the circle, and OD and OA the coordinate axes. This equation includes the broken-line portions of the curve shown in the figure, which were not considered by Diocles.

6. The Creation of Trigonometry

Entirely new in the Alexandrian Greek quantitative geometry was trigonometry, a creation of Hipparchus, Menelaus, and Ptolemy. This work was motivated by the desire to build a quantitative astronomy that could be used to predict the paths and positions of the heavenly bodies and to aid the telling of time, calendar-reckoning, navigation, and geography.

The trigonometry of the Alexandrian Greeks is what we call spherical trigonometry though, as we shall see, the essentials of plane trigonometry were also involved. Spherical trigonometry presupposes spherical geometry, for example the properties of great circles and spherical triangles, much of which was already known; it had been investigated as soon as astronomy became mathematical, during the time of the later Pythagoreans. Euclid's *Phaenomena*, itself based on earlier work, contains some spherical geometry. Many of its theorems were intended to deal with the apparent motion of the stars. Theodosius (*c.* 20 B.C.) collected the then available knowledge of spherical geometry in his *Sphaericae*, but his work was not numerical and so could not be used to handle the fundamental problem of Greek astronomy, namely, to tell time at night by observation of the stars.

The founder of trigonometry is Hipparchus, who lived in Rhodes and Alexandria and died about 125 B.C. We know rather little about him. Most of what we do know comes from Ptolemy, who credits Hipparchus with a number of ideas in trigonometry and astronomy. We owe to him many astronomical observations and discoveries, the most influential astronomical theory of ancient times (Chap. 7, sec. 4), and works on geography. Only one work by Hipparchus, his *Commentary on the* Phaenomena *of Eudoxus and Aratus*, is preserved. Geminus of Rhodes wrote an introduction to astronomy that we do have, and it contains a description of Hipparchus' work on the sun.

Hipparchus' method of approaching trigonometry, as described and used by Ptolemy, is the following. The circumference of a circle is divided into 360°, as was first done by Hypsicles of Alexandria (*c.* 150 B.C.) in his book *On the Risings of the Stars* and by the Babylonians of the last centuries before

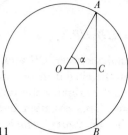

Figure 5.11

Christ, and a diameter is divided into 120 parts. Each part of the circumference and diameter is further divided into 60 parts and each of these into 60 more, and so on according to the Babylonian system of sexagesimal fractions. Then for a given arc AB of some number of degrees, Hipparchus—in a book, now lost, on chords in a circle—gives the number of units in the corresponding chord AB. Just how he calculated these will be described in the discussion of Ptolemy's work, which presents their combined thoughts and results.

The number of units in the chord corresponding to an arc of a given number of degrees is equivalent to the modern sine function. If 2α is the central angle of arc AB (Fig. 5.11), then for us $\sin \alpha = AC/OA$, whereas, instead of $\sin \alpha$, Hipparchus gives the number of units in $2 \cdot AC$ when the radius OA contains 60 units. For example, if the chord of 2α is 40 units, then for us $\sin \alpha = 20/60$, or, more generally,

$$(9) \qquad \sin \alpha = \frac{1}{60} \cdot \frac{1}{2} \text{ chord } 2\alpha = \frac{1}{120} \text{ chord } 2\alpha.$$

Greek trigonometry reached a high point with Menelaus (c. A.D. 98). His *Sphaerica* is his chief work, but apparently he also wrote *Chords in a Circle* in six books, and a treatise on the setting (or rising) of arcs of the zodiac. The Arabs attribute additional works to him.

The *Sphaerica*, extant in an Arab version, is in three books. In the first book, on spherical geometry, we find the concept of a spherical triangle, that is, the figure formed by three arcs of great circles on a sphere, each arc being less than a semicircle. The object of the book is to prove theorems for spherical triangles analogous to what Euclid proved for plane triangles. Thus the sum of two sides of a spherical triangle is greater than the third side and the sum of the angles of a triangle is greater than two right angles. Equal sides subtend equal angles. Then Menelaus proves the theorem that has no analogue in plane triangles, namely, that if the angles of one spherical triangle equal respectively the angles of another, the triangles are congruent. He also has other congruence theorems and theorems about isosceles triangles.

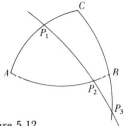

Figure 5.12

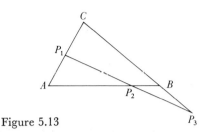

Figure 5.13

The second book of Menelaus' *Sphaerica* is chiefly about astronomy and only indirectly concerned with spherical geometry. The third book contains some spherical trigonometry and bases the development on the first theorem of the book, which supposes that we have a spherical triangle ABC (Fig. 5.12) and any great circle cutting the sides of the triangle (produced where necessary). To state the theorem we shall use our modern sine notion, but for Menelaus the sine of an arc such as AB (or the sine of the corresponding central angle at the center of the sphere) is replaced by the chord of double the arc AB. In terms of our sines, then, Menelaus' theorem says

$$\sin P_1 A \cdot \sin P_2 B \cdot \sin P_3 C = \sin P_1 C \cdot \sin P_2 A \cdot \sin P_3 B.$$

The proof of this theorem rests upon the corresponding theorem for plane triangles, which we still call Menelaus' theorem. For plane triangles the theorem states (Fig. 5.13):

$$P_1 A \cdot P_2 B \cdot P_3 C = P_1 C \cdot P_2 A \cdot P_3 B.$$

Menelaus does not prove the plane theorem. One may conclude that it was already known or perhaps proved by Menelaus in an earlier writing.

The second theorem of Book III, in the notation that arc a lies opposite angle A in triangle ABC, states that if ABC and $A'B'C'$ are two spherical triangles and if $A = A'$ and $C = C'$, or C is supplementary to C', then

$$\frac{\sin c}{\sin a} = \frac{\sin c'}{\sin a'}.$$

Theorem 5 of Book III uses a property of arcs that was presumably known by Menelaus' time, namely (Fig. 5.14), if four great circular arcs emanate from a point O and $ABCD$ and $A'B'C'D'$ are great circles cutting the four, then

$$\frac{\sin AD}{\sin DC} \cdot \frac{\sin BC}{\sin AB} = \frac{\sin A'D'}{\sin D'C'} \cdot \frac{\sin B'C'}{\sin A'B'}.$$

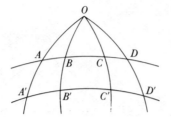

Figure 5.14

We shall find that an expression corresponding to the left or right member reappears under the concept of anharmonic ratio or cross ratio in the work of Pappus and in later work on projective geometry. Many more theorems on spherical trigonometry are due to Menelaus.

The development of Greek trigonometry and its application to astronomy culminated in the work of the Egyptian Claudius Ptolemy (d. A.D. 168), who was a member at least of the royal family of mathematicians though not of the royal house of Egypt. Ptolemy lived in Alexandria and worked at the Museum.

In his *Syntaxis Mathematica* or *Mathematical Collection* (the work was referred to by the Arabs as *Megale Syntaxis*, *Megiste*, and finally *Almagest*), Ptolemy presents the continuation and completion of the work of Hipparchus and Menelaus in trigonometry and astronomy. Astronomy and trigonometry are commingled in the thirteen books of the *Almagest*, though Book I is largely on spherical trigonometry and the others are devoted largely to astronomy, which will be discussed in Chapter 7.

Ptolemy's *Almagest* is thoroughly mathematical, except where he uses Aristotelian physics to refute the heliocentric hypothesis, which Aristarchus had suggested. He says that since only mathematical knowledge, approached inquiringly, will give its practitioners trustworthy knowledge, he was led to cultivate as far as lay in his power this theoretical discipline. Ptolemy also says he wishes to base his astronomy "on the incontrovertible ways of arithmetic and geometry."

In Chapter IX of Book I Ptolemy begins by calculating the chords of arcs of a circle, thereby extending the work of Hipparchus and Menelaus. As already noted, the circumference is divided into 360 parts or units (he does not use the word "degree") and the diameter into 120 units. Then he proposes, given an arc containing a given number of the 360 units, to find the length of the chord expressed in terms of the number of units which a full diameter contains, that is, 120 units.

He begins with the calculation of the chords of 36° and 72° arcs. In Figure 5.15 ADC is a diameter of a circle with center D and BD is perpendicular to ADC. E is the midpoint of DC and F is chosen so that $EF = BE$. Ptolemy proves geometrically that FD equals a side of the regular inscribed

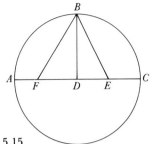

Figure 5.15

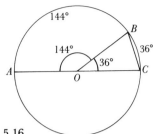

Figure 5.16

decagon and BF, a side of the regular inscribed pentagon. But ED contains 30 units and BD, 60 units. Since $EB^2 = ED^2 + BD^2$, $EB^2 = 4500$ and $EB = 67\ 4'55''$ (which means $67 + 4/60 + 55/60^2$ units). Now $EF = EB$ and so he knows EF. Then $FD = EF - DE = 67\ 4'55'' - 30 = 37\ 4'55''$. Since FD equals the side of the decagon, it is the chord of the $36°$ arc. Hence he knows the chord of this arc. By using FD and the right triangle FDB, he can calculate BF. It is $70\ 32'3''$. But BF is the side of the pentagon. Hence he has the chord of the $72°$ arc.

Of course for the side of a regular hexagon, since it equals the radius, he has at once that the chord of length 60 belongs to the arc of length 60. Also, since the side of the inscribed square is immediately calculable in terms of the radius, he has the chord of $90°$. It is $84\ 51'10''$. Further, since the side of the inscribed equilateral triangle is also immediately calculable in terms of the radius, he gets that the chord of $120°$ is $103\ 55'23''$.

By using the right triangle ABC (Fig. 5.16) on the diameter AC one can immediately get the chord of the supplementary arc AB if one knows the chord of the arc BC. Thus, since Ptolemy knows the chord of $36°$ he can find the chord of $144°$, which turns out to be $114\ 7'37''$.

The relationship established here is equivalent to $\sin^2 A + \cos^2 A = 1$ where A is any acute angle. This can be seen as follows. Ptolemy has proved that if S is any arc less than $180°$ then

$$(\text{chord } S)^2 + \text{chord } (180 - S)^2 = (120)^2.$$

But by (9) above

$$(\text{chord } S)^2 = (120)^2 \sin^2 \frac{S}{2}.$$

Hence

$$(120)^2 \sin^2 \frac{S}{2} + (120)^2 \sin^2 \left(\frac{180 - S}{2}\right) = 120^2$$

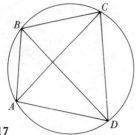

Figure 5.17

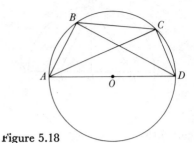

Figure 5.18

or

$$\sin^2 \frac{S}{2} + \sin^2 \left(90 - \frac{S}{2}\right) = 1;$$

that is,

$$\sin^2 \frac{S}{2} + \cos^2 \frac{S}{2} = 1.$$

Now Ptolemy proves what he calls a lemma but which is known today as Ptolemy's theorem. Given any quadrilateral inscribed in a circle (Fig. 5.17), he proves that $AC \cdot BD = AB \cdot DC + AD \cdot BC$. The proof is straightforward. He then takes the special quadrilateral $ABCD$ in which AD is a diameter (Fig. 5.18). Suppose we know AB and AC. Ptolemy now shows how to find BC. BD is the chord of the supplement of arc AB and CD is the chord of the supplement of arc AC. If one now applies the lemma, he sees that five of the six lengths involved in it are known and so the sixth length, which in this case is BC, can be calculated. But arc BC = arc AC − arc AB. Hence we can calculate the chord of the difference of two arcs if their chords are known. In modern terms this means that if we know $\sin A$ and $\sin B$ we can calculate $\sin (A - B)$. Ptolemy points out that, since he knows the chords of 72° and 60°, he can calculate the chord of 12°.

He shows next how, given any chord in a circle, he can find the chord of half the arc of the given chord. In modern terms this means finding $\sin A/2$ from $\sin A$. This result is powerful, as Ptolemy points out, because we can start with an arc whose chord is known and find the chords of successive halves of this arc. He also shows that if one knows the chords of arcs AB and BC, then one can find the chord of arc AC. In modern terms, this result is the formula for $\sin (A + B)$. As a special case, he points out that we can calculate, in modern terms, $\sin 2A$ from $\sin A$.

Since Ptolemy can get the chord of 3/4° from the chord of 12° by halving, he can add this arc of 3/4° or subtract it from any arc whose chord is known;

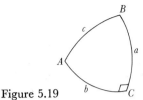

Figure 5.19

and by using the above theorems, he can calculate the chord of the sum or difference of the two arcs. Thus he is in a position to calculate the chords of all arcs in steps of $3/4°$. However, he wants to obtain the chords of arcs in steps of $1/2°$. Here he is stuck and resorts to clever reasoning with inequalities. The approximate result is that the chord of $1/2° = 0\ 31'25''$.

He is now in a position to make up a table of the chords of arcs for arcs differing by $1/2°$, from $0°$ to $180°$. This is the first trigonometric table.

Then Ptolemy proceeds (Chapter XI of Book I) to solve astronomical problems that call for finding arcs of great circles on a sphere. These arcs are sides of spherical triangles, some of whose parts are known through either observation or prior calculation. To determine the unknown arcs Ptolemy proves relationships that are theorems of spherical trigonometry some of which had already been proved in Book III of Menelaus' *Sphaerica*. Ptolemy's basic method is to use Menelaus' theorem for spherical triangles. Thus he proves, in our notation, that in the spherical triangle with right angle at C (Fig. 5.19) and with arc a denoting the side opposite angle A,

$$\sin a = \sin c \sin A$$
$$\tan a = \sin b \tan A$$
$$\cos c = \cos a \cos b$$
$$\tan b = \tan c \cos A.$$

Of course for Ptolemy the various trigonometric functions are chords of arcs. To treat oblique spherical triangles, he breaks them up into right-angled spherical triangles. There is no systematic presentation of spherical trigonometry; he proves just those theorems that he needs to solve specific astronomical problems.

The *Almagest* put trigonometry into the definitive form it retained for over a thousand years. We generally speak of this trigonometry as spherical, but the distinction between plane and spherical trigonometry really has little point in assessing Ptolemy's material. Ptolemy certainly works with spherical triangles, but in having calculated the chords of arcs, he has really given the theoretical basis for plane trigonometry. For by knowing $\sin A$, and, in effect, $\cos A$ for any A from $0°$ to $90°$, one can solve plane triangles.

We should note that trigonometry was created for use in astronomy; and, because spherical trigonometry was for this purpose the more useful tool, it was the first to be developed. The use of plane trigonometry in indirect measurement and in surveying is foreign to Greek mathematics. This may seem strange to us, but historically it is readily understandable, since astronomy was the major concern of the Greek mathematicians. Surveying did come to the fore in the Alexandrian period; but a mathematician such as Heron, who was interested in surveying and would have been capable of developing plane trigonometry, contented himself with applying Euclidean geometry. The uneducated surveyors were not in a position to create the necessary trigonometry.

7. Late Alexandrian Activity in Geometry

Mathematical activity in general, and in geometry in particular, declined in Alexandria from about the beginning of the Christian era. We shall look into possible reasons for the decline in Chapter 8. What we know about the geometrical work of the early Christian era comes from the major commentators Pappus, Theon of Alexandria (end of 4th cent. A.D.), and Proclus.

On the whole, very few original theorems were discovered in this period. The geometers seem to have occupied themselves primarily with the study and elucidation of the works of the great mathematicians who preceded them. They supplied proofs that the original authors had omitted, either because the proofs were regarded as easy enough to be left to the readers or because they were given in treatises that had been lost. These proofs, incidentally, were called lemmas, in an older use of the word.

Both Theon and Pappus report on the work of Zenodorus who lived sometime between 200 B.C. and A.D. 100. Apparently Zenodorus wrote a book on isoperimetric figures, that is, figures with equal perimeters, and in it proved the following theorems:

1. Among n-sided polygons of the same perimeter the regular one has most area.

2. Among regular polygons of equal perimeter, the one having more sides has greater area.

3. The circle has greater area than a regular polygon of the same perimeter.

4. Of all solids with the same surface the sphere has the greatest volume.

The subject of the theorems, which today we would describe as maxima and minima problems, was novel in Greek mathematics.

Late in the Alexandrian period, Pappus' additions to geometry came as a sort of anticlimax. The eight books of his *Mathematical Collection* contain some original material. Pappus' new work was not of the highest order but some of it is worthy of note.

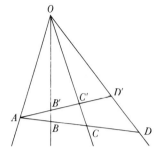

Figure 5.20

Book V gives the proofs, results, and extensions of Zenodorus' work on the areas bounded by curves of equal perimeter. Pappus added the theorem that of all segments of a circle having the same perimeter the semicircle has the greatest area. He also proves that the sphere has more volume than any cone, cylinder, or regular polyhedron with the same surface area.

Proposition 129 of Book VII is a special case of the theorem that the cross ratio (Fig. 5.20)

$$\frac{AB}{AD} \bigg/ \frac{BC}{CD}$$

is the same for every transversal cutting the four lines emanating from O. Pappus requires that all the transversals pass through A.

Proposition 130 states, in our language, that if five of the points in which the six sides of a complete quadrilateral (the four sides and the two diagonals) meet a straight line are fixed, then the sixth one is also fixed. Thus if $ABCD$ (Fig. 5.21) is the quadrilateral the six points in which its six sides meet an arbitrary straight line EK are E, F, G, H, J, and K. If five of these are fixed, the sixth is also. Pappus notes that these six points satisfy the condition

$$\frac{EK}{EH} \bigg/ \frac{JK}{JH} = \frac{EK}{EF} \bigg/ \frac{GK}{GF}.$$

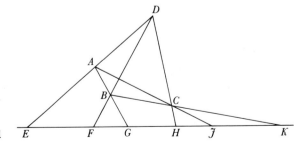

Figure 5.21

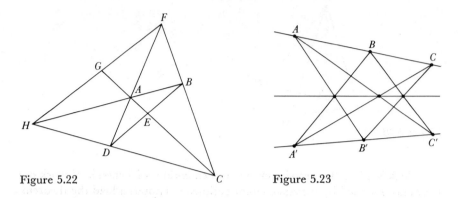

Figure 5.22   Figure 5.23

This condition states that the cross ratio determined by E, K, J, and H equals the cross ratio determined by E, K, G, and F. The condition is equivalent to one we shall find introduced by Desargues, who calls six such points "points of an involution."

Proposition 131 of Book VII amounts to the statement that in each quadrilateral a diagonal is cut harmonically by the other diagonal and by the line joining the points of intersection of the pairs of opposite sides. Thus $ABCD$ is a quadrilateral (Fig. 5.22); CA is a diagonal; CA is cut by the other diagonal BD and by FH which joins the intersection of AD and BC to the intersection of AB and CD. Then the points C, E, A, and G in the figure form a harmonic set; that is, E divides AC internally in the same ratio that G divides AC externally.

Proposition 139 of Book VII gives what is still called Pappus' theorem. If A, B, and C are three points on a line (Fig. 5.23) and A', B', and C' are three points on another, then AB' and $A'B$, BC' and $B'C$, and AC' and $A'C$ meet in three points that lie on one straight line.

One of the last lemmas, Proposition 238, establishes a fundamental property of all conic sections: The locus of all points whose distances from a fixed point (focus) and a fixed line (directrix) are in a constant ratio is a conic section. This basic property of the conics is not in Apollonius' *Conic Sections*, but, as noted in the preceding chapter, Euclid probably knew it.

In the introduction to Book VII Pappus restates Apollonius' assertion that his methods enable one to find the locus of points the product of whose distances from two lines equals a constant times the product of its distances from two other lines. Pappus knows but does not prove that the locus is a conic. He also points out that the problem can be generalized to include five, six, or more lines. We shall say more about this problem in connection with Descartes's work.

Book VIII is of special interest in that it is mainly on mechanics, which, in accordance with the Alexandrian outlook, is regarded as part of mathematics. In fact, Pappus prefaces the book with this contention. He cites Archimedes, Heron, and lesser known figures as the leaders in mathematical mechanics. The center of gravity of a body is defined as the point within the body (it need not be within) such that if the body is suspended from that point it will remain at rest in any position in which it is put. He then explains mathematical methods for determining the point. He also treats the subject of moving a body along an inclined plane and attempts to compare the force required to slide the body along a horizontal plane with that needed to slide it up the inclined plane.

Book VII also contains a famous theorem sometimes called Pappus' theorem and sometimes Guldin's theorem because Paul Guldin (1577–1643) rediscovered it independently. The theorem states that the volume generated by the complete revolution of a plane closed curve that lies wholly on one side of the axis of revolution equals the area bounded by the curve multiplied by the circumference of the circle traversed by the center of gravity. The result is a very general one and Pappus was aware of this. He does not give a proof of the theorem and it is very likely that the theorem and proof were known before his time.

As far as geometry is concerned, the Alexandrian period ends with the works of a number of commentators. Theon of Alexandria wrote a commentary on Ptolemy's *Almagest* and new editions of Euclid's *Elements* and *Optics*. His daughter Hypatia (d. 415), a learned mathematician, wrote commentaries on Diophantus and Apollonius.

Proclus Diadochus, whom we have often mentioned, wrote a commentary on Book I of Euclid's *Elements* (Chap. 3, sec. 2). This commentary is important because Proclus had access to works now lost, including the *History of Geometry* by Eudemus and the book of Geminus that was probably titled the *Doctrine* or the *Theory of Mathematics*.

Proclus received his education in Alexandria and then went to Athens, where he became head of Plato's Academy. He was a foremost neo-Platonist and wrote many books on Plato's work and on philosophy generally; his interests included poetry as well as mathematics. Like Plato he believed that mathematics is a handmaiden to philosophy. It is a propaedeutic because it clears the eye of the soul, removing the impediments that the senses place in the way of knowing universals.

There was another, nonmathematical side to Proclus, who accepted many myths and religious mysteries and was a devout worshiper of Greek and Oriental divinities. Ptolemaic theory he rejected because a Chaldean "whom it is not lawful to disbelieve" thought otherwise. It has been remarked that it was fortunate for Proclus that the Chaldean oracles did not contradict or deny Euclid.

Among many other commentators we shall mention just a few. Simplicius, a commentator on Aristotle, studied at Alexandria and at Plato's Academy, and went to Persia when Justinian closed down the Academy in A.D. 529. He repeated material from Eudemus' *History*, including a long extract on Antiphon's attempt to square the circle and on Hippocrates' quadrature of lunes. Isidorus of Miletus (6th cent. A.D.), who seems to have had a school in Constantinople (which had become the capital of the Eastern Roman Empire and the center of some mathematical activity), wrote commentaries and may have written part of the fifteenth book of Euclid's *Elements*. Eutocius (6th cent. A.D.), probably a pupil of Isidorus, wrote a commentary on Archimedes' work.

Bibliography

Aaboe, Asger: *Episodes from the Early History of Mathematics*, Random House, 1964, Chaps. 3–4.

Ball, W. W. R.: *A Short Account of the History of Mathematics*, Dover (reprint), 1960, Chaps. 4–6.

Cajori, Florian: *A History of Mathematics*, Macmillan, 1919, pp. 29–52.

Dijksterhuis, E. J.: *Archimedes* (English trans.), Ejnar Munksgaard, 1956.

Heath, Thomas L.: *A History of Greek Mathematics*, Oxford University Press, 1921, Vol. 2, Chaps. 13, 15, 17–19, 21.

————: *The Works of Archimedes*, Dover (reprint), 1953.

Pappus d'Alexandrie: *La Collection mathématique*, ed. Paul Ver Eecke, 2 vols., Albert Blanchard, 1933.

Parsons, Edward Alexander: *The Alexandrian Library*, The Elsevier Press, 1952.

Sarton, George: *A History of Science*, Harvard University Press, 1959, Vol. 2, Chaps. 1–3, 5, 18.

Scott, J. F.: *A History of Mathematics*, Taylor and Francis, 1958, Chaps. 3–4.

Smith, David Eugene: *History of Mathematics*, Dover (reprint), 1958, Vol. 1, Chap. 4; Vol. 2, Chaps. 8 and 10.

van der Waerden, B. L.: *Science Awakening*, P. Noordhoff, 1954, Chaps. 7–8.

6

The Alexandrian Period:
The Reemergence of
Arithmetic and Algebra

> Wherever there is number, there is beauty.
> PROCLUS

1. *The Symbols and Operations of Greek Arithmetic*

Let us go back for a moment to pick up the threads of arithmetic in the classical period. The classical Greeks called the art of calculation *logistica*; they reserved the word *arithmetica* for the theory of numbers. The classical mathematicians scorned *logistica* because it was concerned with the practical calculations needed in the trades and commerce. We, however, shall consider both *logistica* and *arithmetica* in order to see what the Alexandrian Greeks had at their disposal.

The classical Greek art of writing and working with numbers did not continue where the Babylonians left off. In *logistica* the Greeks seem to have fashioned their own beginnings. Archaic Greek numerals found in Crete antedate the classical period by about five hundred years. There are no noteworthy features in this scheme, just particular number symbols for 1, 2, 3, 4, 10, 200, 1000, and so forth. Very early in the classical period the Greeks introduced other special symbols for numbers, and used some form of an abacus for calculation. Later, about 500 B.C., they used the Attic system, of which the earliest known occurrence is an inscription of 450 B.C. The Attic system used strokes for the numbers from 1 to 4; Π, the first letter of *penta*, and later Γ were used for 5; Δ from *deka* was 10; H from *hekaton* represented 100; X from *chilioi* stood for 1000; and M from *myrioi* represented 10,000. Combinations of these special symbols produced the in-between numbers. Thus ΓI = 6; ΓΔ = 50; ΓH = 500; ΔΓIII = 18.

However, no one knows how early classical mathematicians—for example, the Pythagoreans—wrote numbers. They may have used pebbles to calculate, for the word "calculus" means "pebble." The original Greek

131

meaning of "abacus" was "sand," which suggests that before the introduction of the abacus, and probably afterward, they drew numbers as dots
in sand. In the three hundred years from Thales to Euclid the mathematicians paid no attention to computation, and this art made no progress.
It is significant that the books do not tell us about the practice of arithmetic.

For some unknown reason the classical Greeks changed their way of
writing numbers to the Ionic or Alexandrian system, which uses letters of
the alphabet. This alphabetic system was the most common one in Alexandrian Greek mathematics and is found, in particular, in Ptolemy's
Almagest. It was used also in ancient Syria and Israel.

The details of the Greek system are as follows:

α	β	γ	δ	ε	ς	ζ	η	θ
1	2	3	4	5	6	7	8	9

ι	κ	λ	μ	ν	ξ	ο	π	ϙ
10	20	30	40	50	60	70	80	90

ρ	σ	τ	υ	φ	χ	ψ	ω	ϡ
100	200	300	400	500	600	700	800	900

The intermediate numbers were written by combining the above symbols.
Thus $\iota\alpha = 11$, $\iota\beta = 12$, $\kappa\alpha = 21$, and $\rho\nu\gamma = 153$. The symbols for 6, 90,
and 900 and the symbol M from the Attic system were not in the then-
current Greek alphabet; the first three, now called stigma (or digamma),
koppa, and sampi, belonged to an older alphabet that the Greeks had taken
over from the Phoenicians (who did not, however, use letters for numbers).
From the fact that these older letters were used it is inferred that this system
of writing numbers dates back as far as 800 B.C. and probably came from
Miletus in Asia Minor.

For numbers larger than 1000 the alphabet was repeated, but a stroke
was placed before the letter to avoid confusion. Also, horizontal lines were
drawn over numbers to distinguish them from words. Thus,

$$,\overline{\alpha\tau\epsilon} = 1305.$$

Various Greek writers used minor variations of the above scheme and of
schemes given below.

Greek papyri of the first part of the Alexandrian period (first three
centuries B.C.) contain symbols for zero such as $0 \overline{} 0$, $\overline{0,}$ $\overline{0}$, and $\overline{0}$. The
zero of the Greek Alexandrian period was used, as was the zero of the Seleucid
Babylonian period, to indicate missing numbers. According to Byzantine
manuscripts, which are all we have of Ptolemy's work, he used 0 for zero
both in the middle and at the end of a number.

Archimedes' *Sand-Reckoner* presented a scheme for writing very large
numbers. He sought to show that he could write a number as large as the

number of grains of sand in the universe. He takes the largest number then expressed in Greek numerals, that is 10^8, a myriad myriads, and uses it as the starting point of a new series of numbers that goes up to $10^8 \times 10^8$ or 10^{16}. Then he uses 10^{16} as a new starting point for a series of numbers that goes from 10^{16} to 10^{24}, and so forth. He next estimates the number of grains of sand in the universe and shows it is less than the largest number he can express. What is important in this work of Archimedes is not a practical scheme for actually writing any large number, but the thought that one can construct indefinitely large numbers. Apollonius had a similar scheme.

The arithmetic operations, with the whole numbers written as described above, were like ours. Thus in adding, the Greeks wrote the numbers one below the other to form a unit column, tens column, and so forth, added the numbers in each column, and carried from one column to the other. These methods are a great step forward, compared with the Egyptian ones. The latter, however, were also taught by the Alexandrian Greeks.

As for fractions, there was the special symbol, L'' for $1/2$. Thus (sometimes with one accent), $\alpha L'' = 1\frac{1}{2}$, $\beta L'' = 2\frac{1}{2}$, and $\gamma L'' = 3\frac{1}{2}$. Small fractions were denoted by writing the numerator marked with an accent, then the denominator written once or twice, each time with two accents. Thus $\iota\gamma'\kappa\theta''\kappa\theta'' = 13/29$. Diophantus often wrote the denominator above the numerator.

The Egyptian scheme of writing fractions whose numerators are larger than 1 as a sum of unit fractions is also found. Thus Heron writes $\frac{163}{224}$ as $\frac{1}{2}\ \frac{1}{7}\ \frac{1}{14}\ \frac{1}{112}\ \frac{1}{224}$, but gives as well $\frac{1}{2}\ \frac{1}{8}\ \frac{1}{16}\ \frac{1}{32}\ \frac{1}{112}$ and other such expressions for the same fraction. He also uses the above Greek form. Ptolemy likewise writes *some* fractions like the Egyptians, for example, $\frac{23}{25}$ as $\frac{1}{2}\ \frac{1}{3}\ \frac{1}{15}\ \frac{1}{50}$. The plus sign was always understood, and of course letters of the Greek alphabet were used where we use numerals.

Common fractions written in the Greek or Egyptian system were too awkward for astronomical calculations. Hence the Alexandrian Greek astronomers adopted the Babylonian sexagesimal fractions. Just when this practice began is not known, but it is used in Ptolemy's *Almagest*. Thus when Ptolemy writes 31 25 he means $\frac{31}{60} + \frac{25}{60^2}$. Ptolemy says that he used sexagesimal fractions to avoid the difficulty of ordinary fractions. He wrote *whole* numbers in the decimal base but not in positional notation. However, large whole numbers occur so rarely in his astronomical calculations that one may say he used sexagesimal positional notation. The use of the sexagesimal system of place value for fractions and of the nonpositional alphabetic numerals for whole numbers seems peculiar and irrational. Yet we write 130°15′17″.5.

As the above discussion implies, the Alexandrians used fractions as numbers in their own right, whereas the mathematicians of the classical period spoke only of a ratio of integers, not of parts of a whole and the ratios were used only in proportions. However, even in the classical period genuine fractions, that is, fractions as entities in their own right, were used in commerce. In the Alexandrian period, Archimedes, Heron, Diophantus, and others used fractions freely and performed operations with them. Though, as far as the records show, they did not discuss the concept of fractions, apparently these were intuitively sufficiently clear to be accepted and used.

The square-root operation, though considered in classical Greece, was in effect bypassed there. There are indications in Plato's writings that the Pythagoreans approximated $\sqrt{2}$ by using 49/25 for 2, thus obtaining 7/5. Likewise, Theodorus probably approximated $\sqrt{3}$ by using 49/16 for 3, obtaining 7/4. The irrational number as such had no mathematical status in classical Greece.

The next information we have on the Greek handling of roots comes from Archimedes. In his *Measurement of a Circle* he seeks primarily to find a good approximation to π, that is, the ratio of the circumference to the diameter of a circle; in the course of the work he operates with large whole numbers and fractions. He also obtains an excellent approximation to $\sqrt{3}$, namely,

$$\frac{1351}{780} > \sqrt{3} > \frac{265}{153},$$

but does not explain how he got this result. Among the many conjectures in the historical literature concerning its derivation the following is very plausible. Given a number A, if one writes it as $a^2 \pm b$ where a^2 is the rational square nearest to A, larger or smaller, and b is the remainder, then

$$a \pm \frac{b}{2a} > \sqrt{a^2 \pm b} > a \pm \frac{b}{2a \pm 1}.$$

Several applications of this procedure do produce Archimedes' result. To obtain the approximation to π, Archimedes first proves, in Proposition 1, that the area of a circle is equal to the area of a right triangle whose base is as long as the circumference of the circle and whose altitude equals the radius. He must now find the circumference. This he approximates more and more closely by using inscribed and circumscribed regular polygons and calculates the perimeters of these polygons. His result for π is

$$3\frac{10}{71} < \pi < 3\frac{1}{7}.$$

Apollonius, too, wrote a book on the quadrature of the circle, called *Okytokion* (The Rapid Delivery), in which he is supposed to have improved on Archimedes' determination of π by using better arithmetic methods. This is the only book in which Apollonius departs from classical Greek mathematics.

Heron approximates square roots frequently by means of

$$\sqrt{A} = \sqrt{a^2 \pm b} \sim a \pm \frac{b}{2a},$$

where a and b have the meaning explained above. He gets this approximation by starting with the approximation $\alpha = (c + A/c)/2$, where c is any guess as to $\sqrt{A}$; if one writes A as $a^2 + b$ and chooses $c = a$, then $\alpha = a + b/2a$. Heron also improves on α by finding $\alpha_1 = (\alpha + A/\alpha)/2$. Clearly the closer α is to $\sqrt{A}$, the better the approximation α_1 will be. Heron's basic expression for α was also used by the Babylonians.

In later Alexandrian times the square root process uses, as ours does, the principle of $(a + b)^2 = a^2 + 2ab + b^2$. The successive approximations are obtained by trial, always keeping the square of the approximation obtained less than the number whose root is desired. Explaining Ptolemy's use of this method, Theon points out that a geometrical figure is used to aid the thinking; this is the figure used in Euclid's *Elements*, Book II, Proposition 4 and is the geometrical way of stating $(a + b)^2$. Thus Ptolemy gives

$$\frac{103}{60} \, (+) \, \frac{55}{60^2} \, (+) \, \frac{23}{60^3}$$

for $\sqrt{3}$, which amounts to 1.7320509 and is correct to six decimal places.

2. *Arithmetic and Algebra as an Independent Development*

We have been reviewing the methods of doing arithmetic employed by the Greeks in both periods but more especially in the Alexandrian period when the geometry and trigonometry became quantitative. But the major development with which this chapter is concerned is the rise of arithmetic and algebra as subjects *independent* of geometry. The arithmetical work of Archimedes, Apollonius, and Ptolemy was a step in this direction, but they used arithmetic to calculate geometric quantities. One might infer that the numbers were meaningful to them because they represented geometric magnitudes and the logic of the operations was guaranteed by the geometrical algebra. But there is no question that Heron, Nichomachus (*c.* A.D. 100), who was probably an Arabian from Gerasa in Judea, and Diophantus (*c.* A.D. 250), a Greek of Alexandria, did treat arithmetical and algebraic problems in and for themselves and did not depend upon geometry either for motivation or to bolster the logic.

More significant than Heron's arithmetical work of finding square and cube roots is the fact that he formulated and solved algebraic problems by purely arithmetic procedures. He did not use any special symbols; the account is verbal. For example, he treats the problem: Given a square such that the sum of its area and perimeter is 896 feet, find the side. The problem, in our notation, is to find x satisfying $x^2 + 4x = 896$. Heron completes the square by adding 4 to both sides and takes the square root. He does not prove but merely describes what operations to perform. There are many such problems in his work. Of course this is precisely the old Egyptian and Babylonian style of presentation, and there is no doubt that Heron took much material from the ancient Egyptian and Babylonian texts. There, we may remember, algebra was independent of geometry and, as for Heron, an extension of arithmetic.

In his *Geometrica*, Heron speaks of adding an area, a circumference, and a diameter. In using such words he means, of course, that he wants to add their numerical values. Likewise, when he says that he multiplies a square by a square, he means that he is finding the product of the numerical values. Heron also translated much of the Greek geometrical algebra into arithmetic and algebraic processes.

This work of Heron (as well as his use of approximate Egyptian area and volume formulas) is sometimes evaluated as the beginning of the decline of Greek geometry. It is more fitting to regard it as a Hellenized improvement on Babylonian and Egyptian mathematics. When Heron adds areas and line segments, he is not misapplying classical Greek geometry but merely continuing the practice of the Babylonians, for whom area and length were just words for certain arithmetic unknowns.

More remarkable from the standpoint of the reemergence of an independent arithmetic is the work of Nichomachus, who wrote the *Introductio Arithmetica* in two books. It was the first sizable book in which arithmetic (in the sense of the theory of numbers) was treated entirely independently of geometry. Historically its importance for arithmetic is comparable to Euclid's *Elements* for geometry. Not only was this book itself studied, referred to, and copied by dozens of later writers, but it is known to be typical of many books by other authors of the same period and so reflects the interests of the times. Numbers stood for quantities of objects and were no longer visualized as line segments as in Euclid. Nichomachus uses words throughout, whereas Euclid used a letter, such as A, or two letters such as BC—referring, in the second case, to a line segment—to speak about numbers. Hence Nichomachus' phrasing is clumsier. He treats only whole numbers and ratios of whole numbers

Nichomachus was a Pythagorean; and though the Pythagorean tradition was not dead, he reanimated it. Of the four subjects stressed by Plato— arithmetic, geometry, music, and astronomy—Nichomachus says arithmetic is the mother of the others. This, he maintains is

> not solely because we said that it existed before all the others in the mind
> of the creating God like some universal and exemplary plan, relying upon
> which as a design and archetypal example the creator of the universe sets
> in order his material creations and makes them attain to their proper ends;
> but also because it is naturally prior in birth....

Arithmetic, he continues, is essential to all the other sciences because they could not exist without it. However, if the other sciences were abolished, arithmetic would still exist.

The essence of the *Introductio* is the arithmetical work of the early Pythagoreans. Nichomachus treats even and odd, square, rectangular, and polygonal numbers. He treats also prime and composite numbers and parallelopipedal numbers (of the form $n^2(n + 1)$) and defines many other kinds. He gives the multiplication table for numbers from 1 to 9 precisely as we learn it.

Nichomachus repeats many Pythagorean statements, such as that the sum of two consecutive triangular numbers is a square number, and conversely. He goes beyond the Pythagoreans in seeing, though not proving, general relations. Thus he asserts that the $(n - 1)$st triangular number added to the nth k-gonal number gives the nth $(k + 1)$-gonal number. For example, the $(n - 1)$st triangular number added to the nth square number gives the nth pentagonal number. In our symbols,

$$\frac{n(n - 1)}{2} + n^2 = \frac{n}{2}(3n - 1).$$

Also the nth triangular number, the nth square number, the nth pentagonal number, and so on form an arithmetic progression with the $(n - 1)$st triangular number as the common difference.

He discovered the following proposition: If one writes down the odd numbers

$$1, 3, 5, 7, 9, 11, 13, 15, 17, \ldots$$

then the first is the cube of 1, the sum of the next two is the cube of 2, the sum of the next three is the cube of 3, and so on. There are other propositions on progressions.

Nichomachus gives four perfect numbers, 6, 28, 496, and 8128, and repeats Euclid's formula for perfect numbers. He classifies all sorts of ratios, including $m + 1:m$, $2m + n:m + n$, and $mn + 1:n$, and gives them names. These were important for music.

He also studies proportion, which, he says, is very necessary for "natural science, music, spherical trigonometry and planimetry and particularly for the study of the ancient mathematicians." He gives many types of proportion, among them the musical proportion

$$a:\frac{a+b}{2} = \frac{2ab}{a+b}:b.$$

The *Introductio* also gives the sieve of Eratosthenes (about whom we shall say more in Chapter 7); it is a method of obtaining prime numbers quickly. One writes down all the odd numbers from 3 on as far as one wishes, then crosses out all multiples of 3, that is to say every third number after 3. Next one crosses out all multiples of 5, or every fifth number after 5, counting any that may have been crossed out before. Then every seventh number after 7, and so forth. No crossed-out number can be the starting point of a new crossing-out process. The number 2 must now be included with those that are not crossed out. These numbers are the primes.

Nichomachus always uses specific numbers to discuss various categories and proportions. The examples illustrate and explain what he asserts but there is no support for any general assertions beyond the examples. He does not prove deductively.

The *Introductio* had value because it was a systematic, orderly, clear, and comprehensive presentation of the arithmetic of integers and ratios of integers freed of geometry. It was not original as far as ideas were concerned, but was a very useful compilation. It also incorporated speculative, aesthetic, mystical, and moral properties of numbers, but no practical applications. The *Introductio* was the standard text in arithmetic for a thousand years. At Alexandria, from the time of Nichomachus, arithmetic rather than geometry became the favorite study.

At this time, too, algebra came to the fore. Books of problems appeared that were solved by algebraic techniques. Some of these problems were exactly those that appeared in Babylonian texts of 2000 B.C. or in the Rhind papyrus. This Greek algebraic work was written in verbal form; no symbolism was used. Also, no proofs of the procedures were given. From the time of Nichomachus onward, problems leading to equations were a common form of puzzle. Between fifty and sixty of these are preserved in the Palatine Codex of Greek Epigrams (10th cent. A.D.). At least thirty of them are attributed to Metrodorus (*c.* A.D. 500), but are surely older. One is the Archimedes cattle problem, which calls for finding the number of bulls and cows of different colors subject to given information. Another is due to Euclid and involves a mule and a donkey carrying corn. Another calls for the time required for pipes to fill a cistern. There were also age problems such as appear in our algebra texts.

The highest point of Alexandrian Greek algebra is reached with Diophantus. We know almost nothing about his origins or life; he was probably

Greek. One of the algebraic problems found in a Greek collection gives the following facts about his life: His boyhood lasted 1/6 of his life; his beard grew after 1/12 more; he married after 1/7 more; and his son was born 5 years later. The son lived to half of his father's age and the father died 4 years after the son. The problem is to find how long Diophantus lived. The answer is easily found to be 84. His work towers above that of his contemporaries; unfortunately, it came too late to be highly influential in his time because a destructive tide was already engulfing the civilization.

Diophantus wrote several books that are lost in their entirety. Part of a tract *On Polygonal Numbers* is known, in which he states and proves theorems in the deductive manner of Books VII, VIII, and IX of the *Elements*; however, the theorems are not striking. His great work is the *Arithmetica* which, Diophantus says, comprises thirteen books. We have six, which come from a thirteenth-century manuscript that is a Greek copy of an older one and from later versions.

The *Arithmetica*, like the Rhind papyrus, is a collection of separate problems. The dedication says it was written as a series of exercises to help one of his students learn the subject. One of Diophantus' major steps is the introduction of symbolism in algebra. Since we do not have the manuscript written by him but only much later ones, we do not know the precise symbols. It is believed that the symbol he used for the unknown was ς, which served as our x. This ς may have been the same letter as the Greek σ when used at the end of a word, as in $\dot{\alpha}\rho\iota\theta\mu\acute{o}s$ (*arithmos*), and may have been chosen because it was not a number in the Greek system of using letters for numbers. Diophantus called the unknown "the number of the problem." For our x^2 Diophantus used Δ^Y, the Δ being the first letter of $\delta\acute{v}\nu\alpha\mu\iota s$ (*dýnamis*, "power"). x^3 is K^Y; the K is from $\kappa\acute{v}\beta os$ (*cubos*). x^4 is $\Delta^Y\Delta$; x^5 is ΔK^Y; x^6 is $K^Y K$. In this system K^Y is not clearly the cube of ς as our x^3 is of x. For Diophantus $\varsigma^x = 1/x$. He also uses names for these various powers, e.g. number for x, square for x^2, cube for x^3, square-square (dynamodynamis) for x^4, square-cube for x^5, and cubocube for x^6.[2]

The appearance of such symbolism is of course remarkable but the use of powers higher than three is even more extraordinary. The classical Greeks could not and would not consider a product of more than three factors because such a product had no geometrical significance. On a purely arithmetic basis, however, such products do have a meaning; and this is precisely the basis Diophantus adopts.

Addition is indicated in Diophantus by putting terms alongside one another. Thus

$$\Delta^Y \bar{\gamma} \overset{\circ}{\mathrm{M}} \iota \bar{\beta} \quad \text{means} \quad x^2 \cdot 3 + 12.$$

2. Some modern authors use δ, K, and $\bar{\nu}$ for Δ, K, and Y.

The $\overset{\circ}{M}$ is a symbol for unity and indicates that a pure number not involving the unknown follows. Again

$$\Delta^Y \bar{\alpha} s \beta \overset{\circ}{M} \bar{\gamma} \quad \text{means} \quad x^2 + x \cdot 2 + 3.$$

For subtraction he uses the symbol $\wedge$. Thus for $x^6 - 5x^4 + x^2 - 3x - 2$ he writes

$$K^Y \kappa \bar{\alpha} \Delta^Y \bar{\alpha} \ \wedge \ \Delta^Y \Delta \bar{\varepsilon} s \bar{\gamma} \overset{\circ}{M} \bar{\beta},$$

putting all the negative terms after the positive ones. There are no symbols for addition, multiplication, or division as operations. The symbol ι^σ is used (at least in the extant versions of the *Arithmetica*) to denote equality. The coefficients of the algebraic expressions are specific numbers; there are no symbols for general coefficients. Because he does use some symbolism, Diophantus' algebra has been called syncopated, whereas that of the Egyptians, the Babylonians, Heron, and Nichomachus is called rhetorical.

Diophantus writes out his solutions in a continuous text, as we write prose. His execution of the operations is entirely arithmetical; that is, there is no appeal to geometry to illustrate or substantiate his assertions. Thus $(x - 1)(x - 2)$ is carried out algebraically as we do it. He also applies algebraic identities such as

$$\left(\frac{p+q}{2}\right)^2 - \left(\frac{p-q}{2}\right)^2 = pq$$

to expressions such as $x + 2$ for p and $x + 3$ for q. That is, he makes steps that use the identities but the identities themselves do not appear.

The first book of the *Arithmetica* consists mainly of problems that lead to determinate equations of the first degree in one or more unknowns. The remaining five books treat mainly indeterminate equations of second degree. But this segregation is not sharply adhered to. In the case of the determinate equations (i.e. equations leading to unique solutions) with more than one unknown involved, he uses given information to eliminate all but one unknown and, at worst, ends with quadratics of the form $ax^2 = b$. Thus Problem 27 of Book I states: Find two numbers such that their sum is 20 and product is 96. Diophantus proceeds thus: Given sum 20, given product 96, $2x$ the difference of the required numbers. Therefore the numbers are $10 + x$, $10 - x$. Hence $100 - x^2 = 96$. Then $x = 2$ and the required numbers are 12, 8.

The most striking feature of Diophantus' algebra is his solution of indeterminate equations. Such equations had been considered before, as for example in the Pythagorean work on solutions of $x^2 + y^2 = z^2$, in the Archimedean cattle problem, which leads to seven equations in eight unknowns (plus two supplementary conditions), and in other odd writings. Diophantus, however, pursues indeterminate equations extensively and is the founder of this branch of algebra now called, in fact, Diophantine analysis

He solves linear equations in two unknowns, e.g.

$$x + y - 5 = 0.$$

In such equations he gives a value to one unknown and then solves for a positive rational value of the other. He recognizes that the value assigned to the first unknown is merely typical. (In modern Diophantine analysis one seeks integral solutions only.) Very little is done with this type of equation, and the work is hardly significant since positive rational solutions are obtainable at once.

He then solves quadratic equations in two unknowns, of which the most general type is (in our notation)

$$(1) \qquad y^2 = Ax^2 + Bx + C.$$

Diophantus does not write y^2 but says that the quadratic expression must equal a square number (square of a rational number). He considers (1) for special values of A, B, and C and treats these types in separate cases. For example, when C is absent he lets $y = mx/n$, where m and n are specific whole numbers, obtains

$$Ax^2 + Bx = \frac{m^2}{n^2} x^2,$$

and then cancels x and solves. When A and C do not vanish but $A = a^2$, he assumes $y = ax - m$. If $C = c^2$ he assumes $y = (mx - c)$. In all cases m is a specific number.

He also treats the case of simultaneous quadratics, namely,

$$(2) \qquad y^2 = Ax^2 + Bx + C$$

$$(3) \qquad z^2 = Dx^2 + Ex + F.$$

Here, too, he undertakes only particular cases, that is, where $A, B, \ldots, F$ are special numbers or satisfy special conditions, and his method is to assume expressions for y and z in terms of x and then solve for x.

In effect he is solving determinate equations in one unknown. He realizes, however, that in choosing expressions for y and z in (2) and (3) and for y in (1), he is giving merely typical solutions and that the values assigned to y and z are somewhat arbitrary.

He also has problems in which cubic and higher degree expressions in x must equal a square number, e.g.

$$Ax^3 + Bx^2 + Cx + d^2 = y^2.$$

Here he lets $y = mx + d$ and fixes m so that the coefficient of x vanishes. Since the d^2 terms cancel and he can then divide through by x^2, he obtains a first degree equation in x. There are also special cases of a quadratic in x equaling y^3.

All his quadratics in x reduce to the types

$$ax^2 = bx, \quad ax^2 = b, \quad ax^2 + bx = c, \quad ax^2 + c = bx, \quad ax^2 = bx + c,$$

and he solves each of these types. Only one cubic in x, of no significance, is solved.

The above equations show the *types* of problems Diophantus solves. The actual wording of the problems is illustrated by the following examples:

Book I, Problem 8. To divide a given square number into two squares.

Here he takes 16 as the given square number and obtains 256/25 and 144/25. This is the problem which Fermat generalized and which led to his assertion that $x^m + y^m = z^m$ is not solvable for $m > 2$.

Book II, Problem 9. To divide a given number which is a sum of two squares into two other squares.

He takes 13 or 4 + 9 as the given number and obtains 324/25 and 1/25.

Book III, Problem 6. To find three numbers such that their sum and the sum of any two is a square.

Diophantus gives 80, 320, and 41 as the three numbers.

Book IV, Problem 1. To divide a given number into two cubes such that the sum of their sides is a given number.

With the given number of 370 and the given sum of the sides as 10, he finds 343 and 27. The sides are the cube roots of the cubes.

Book IV, Problem 29. To express given number as the sum of four squares plus the sum of their sides.

Given the number 12 he finds 121/100, 49/100, 361/100, 169/100 as the four squares; their sides are the square root of each square.

In Book VI Diophantus solves a number of problems involving the (rational) sides of a right-angled triangle. The use of the geometrical language is incidental even where the term area occurs. He seeks rational numbers a, b, and c such that $a^2 + b^2 = c^2$ and subject to some other condition. Thus the first problem is to find a (rational) right-angled triangle such that the hypotenuse minus each of the sides gives a cube. In this case he happens to obtain the integral answers 40, 96, 104. However, in general he gets rational answers.

Diophantus shows great skill in reducing equations of various types to forms he can handle. We do not know how he arrived at his methods. Since he makes no appeal to geometry, it is not likely that he translated Euclid's methods for solving quadratics. Moreover, indeterminate problems are not in Euclid and as a class are new with Diophantus. Because we lack information on the continuity of thought in the later Alexandrian period, we cannot find traces of Diophantus' work in his predecessors. As far as we can tell, his work in pure algebra is remarkably different from past work.

He accepts only positive rational roots and ignores all others. Even when a quadratic equation has two positive roots he gives only one, the larger one. When an equation, as it is being solved, clearly leads to two negative or imaginary roots he rejects the equation and says it is not solvable. In the case of irrational roots, he retraces his steps and shows how by altering the equation he can get a new one that has rational roots. Here Diophantus differs from Heron and Archimedes. Heron was a surveyor and the geometrical quantities he sought could be irrational. Hence he accepted them, though of course he approximated them to obtain a useful value. Archimedes also sought exact answers, and when they were irrational he obtained inequalities to bound the irrational. Diophantus is a pure algebraist; and since algebra in his time did not recognize irrational, negative, and complex numbers, he rejected equations with such solutions. It is, however, worthy of note that fractions for Diophantus are numbers, rather than just a ratio of two whole numbers.

He has no *general methods*. Each of the 189 problems in the *Arithmetica* is solved by a different method. There are more than 50 different types of problems but no attempt is made to classify them by type. His methods are closer to the Babylonian ones than to those of his Greek predecessors, and there are indications of Babylonian influences. In fact, he does solve some problems just as the Babylonians did. But it has not been established that there was any direct connection between Diophantus' work and Babylonian algebra. His advance in algebra over the Babylonians consists in the use of symbolism and the solution of indeterminate equations. In determinate equations he went no further than they did, but his *Arithmetica* assimilated *logistica*, which Plato, among others, had banned from mathematics.

Diophantus' variety of methods for the separate problems dazzles rather than delights. He was a shrewd and clever virtuoso but apparently not deep enough to see the essence of his methods and thereby attain generality. (It is still true that Diophantine analysis is a maze of separate problems.) Unlike a speculative thinker who seeks general ideas, Diophantus sought only correct answers. There are a few results that might be called general, such as that no prime number of the form $4n + 3$ can be the sum of two squares. Euler did credit Diophantus with illustrating general methods that he could not display as such because he did not have literal coefficients. And there are others who credit Diophantus with recognizing that his material belonged to an abstract and basic science. But this view is not shared by all. His work as a whole, however, is a monument in algebra.

An element of mathematics that is of enormous importance today, which was missing in Greek algebra, is the use of letters to represent a class of numbers, as, for example, coefficients in equations. Aristotle did use letters of the Greek alphabet to indicate an arbitrary time or an arbitrary distance and in discussions of motion employed such phrases as "the half of B." Euclid, too, used letters for classes of numbers in Books VII to IX of the *Elements*, a practice that Pappus followed. However, there was no recognition of the enormous contribution that letters could make in increasing the effectiveness and generality of algebraic methodology.

Another feature of Alexandrian algebra is the absence of any explicit deductive structure. The various types of numbers—whole numbers, fractions, and irrationals—were certainly not defined. Nor was there any axiomatic basis on which a deductive structure could be erected. The work of Heron, Nichomachus, and Diophantus, and of Archimedes as far as his arithmetic is concerned, reads like the procedural texts of the Egyptians and Babylonians, which tell us how to do things. The deductive, orderly proof of Euclid and Apollonius, and of Archimedes' geometry is gone. The problems are inductive in spirit, in that they show methods for concrete problems that presumably apply to general classes whose extent is not specified. In view of the fact that as a consequence of the work of the classical Greeks mathematical results were supposed to be derived deductively from an explicit axiomatic basis, the emergence of an independent arithmetic and algebra with no logical structure of its own raised what became one of the great problems of the history of mathematics. This approach to arithmetic and algebra is the clearest indication of the Egyptian and Babylonian influences in the Alexandrian world. Though the Alexandrian Greek algebraists did not seem to be concerned about this deficiency, we shall find that it did trouble deeply the European mathematicians.

Bibliography

Ball, W. W. R.: *A Short Account of the History of Mathematics*, Dover (reprint), 1960, Chaps. 5 and 7.

Cajori, Florian: *A History of Mathematics*, Macmillan, 1919, pp. 52–62.

Heath, Thomas L.: *Diophantus of Alexandria*, Dover (reprint), 1964.

————: *The Works of Archimedes*, Dover (reprint), 1953, Chaps. 4 and 6 of the Introduction, pp. 91–98, 319–326.

————: *A History of Greek Mathematics*, Oxford University Press, 1921, Vol. 1, Chaps. 1–3; Vol. 2, Chap. 20.

————: *A Manual of Greek Mathematics*, Dover (reprint), 1963, Chaps. 2–3, and 17.

D'Ooge, Martin Luther: *Nichomachus of Gerasa*, University of Michigan Press, 1938.

van der Waerden, B. L.: *Science Awakening*, P. Noordhoff, 1954, pp. 278–86.

7
The Greek Rationalization
of Nature

> Mathematics is the gate and key of the sciences.
>
> ROGER BACON

1. *The Inspiration for Greek Mathematics*

Unfortunately, except for occasional hints, the Greek classics, such as Euclid's *Elements*, Apollonius' *Conic Sections*, and the geometrical works of Archimedes, give no indication of why these authors investigated their subjects. They give only the formal, polished deductive mathematics. In this respect, the Greek texts are no different from modern mathematics textbooks and treatises. Such books seek only to organize and present the mathematical results that have been attained and so omit the motivations for the mathematics, the clues and suggestions for the theorems, and the uses to which the mathematical knowledge is put.

To understand why the Greeks created so much vital mathematics, one must investigate their objectives. It was the urgent and irrepressible desire of the Greeks to understand the physical world that impelled them to create and value mathematics. Mathematics was part and parcel of the investigation of nature and the key to comprehension of the universe, for mathematical laws are the essence of its design.

What evidence do we have that this was the role of mathematics? It is difficult to demonstrate that any one theorem or body of theorems was created for a specific purpose because we do not have enough information about the Greek mathematicians. Ptolemy's direct statement that he created trigonometry for astronomy is an exception. However when one finds that Eudoxus was primarily an astronomer and that Euclid wrote not just the *Elements* but the *Phaenomena* (a work on the geometry of the sphere as applied to the motion of the sphere of stars), the *Optics* and *Catoptrica*, the *Elements of Music*, and small works on mechanics, all of which were mathematical, one cannot escape the conclusion that mathematics was more than an isolated discipline. Knowing how the human mind works and knowing in great detail how men such as Euler and Gauss worked, we may be fairly certain that the investigations in astronomy, optics, and music must have suggested mathematical problems, and it is most likely that the motivation for the mathematics was its application to these other areas. It is also relevant that the geometry of the sphere, known in Greek times as "sphaeric," was studied just as soon as astronomy became mathematical, which happened even before Eudoxus' time. The word "sphaeric" meant "astronomy" to the Pythagoreans.

Fortunately the inferences we may draw from the works of the mathematicians, though reasonable enough, are established beyond doubt by the overwhelming evidence in the writings of the Greek philosophers, many of whom were also prominent mathematicians, and of the Greek scientists. The bounds of mathematics were not mathematics proper. In the classical period mathematics comprised arithmetic, geometry, astronomy, and music; and in the Alexandrian period, as we have already noted in Chapter 5, the divisions of the mathematical sciences were arithmetic (theory of numbers), geometry, mechanics, astronomy, optics, geodesy, canonic (musical harmony), and logistics (applied arithmetic).

2. *The Beginnings of a Rational View of Nature*

The civilizations that preceded the Greek or were contemporary with it regarded nature as chaotic, mysterious, capricious, and terrifying. The happenings in nature were manipulated by gods. Prayers and magic might induce the gods to be kind and even to perform miracles but the life and fate of man were entirely subject to their will.

From the time our knowledge of Greek civilization and culture begins to be reasonably definite and specific, that is, from about 600 B.C., we find among the intellectuals a totally new attitude toward nature: rational, critical, and secular. Mythology was discarded, as was the belief that the gods manipulate man and the physical world according to their whims. The new doctrine holds that nature is orderly and functions invariably according to a plan. Moreover, the conviction is manifest that the human mind is powerful and even supreme; not only can the ways of nature be learned by man, but he can even predict the occurrences.

It is true that the rational approach was entertained only by the intellectuals, a small group in both the classical and Alexandrian periods. Whereas these men opposed the attribution of events to gods and demons and defied the mysteries and terrors of nature, people in general were deeply religious and believed that the gods controlled all events. They accepted mystical doctrines and superstitions as credulously as did the Egyptians and the Babylonians. In fact Greek mythology was vast and highly developed.

The Ionians began the task of determining the nature of reality. We shall not describe the qualitative theories of Thales, Anaxagoras, and their colleagues, each of whom fixed on a single substance persisting through all apparent change. The underlying identity of this prime substance is conserved but all forms of matter can be explained in terms of it. This natural philosophy of the Ionians was a series of bold speculations, shrewd guesses, and brilliant intuitions rather than the outcome of extensive and careful scientific investigations. They were perhaps a little too eager to see the whole picture and so naïvely jumped to broad conclusions. But they did substitute material and objective explanations of the structure and design of the universe for the older mythical stories. They offered a reasoned approach in place of the fanciful and uncritical accounts of the poets and they defended their contentions by reason. At least these men dared to tackle the universe with their minds and refused to rely upon gods, spirits, ghosts, devils, angels, and other mythical agents.

3. The Development of the Belief in Mathematical Design

The decisive step in removing the mystery, mysticism, and arbitrariness from the workings of nature and in reducing the seeming chaos to an understandable ordered pattern was the application of mathematics. The first major group to offer a rational and mathematical philosophy of nature were the Pythagoreans. They did draw some inspiration from the mystical side of Greek religion; their religious doctrines centered about the purification of the soul and its redemption from the taint and prison of the body. The members lived simply and devoted themselves to the study of philosophy, science, and mathematics. New members were pledged to secrecy at least as to religious beliefs and required to join up for life. Membership in the community was open to men and women.

The Pythagoreans' religious thinking was undoubtedly mystical, but their natural philosophy was decidedly rational. They were struck by the fact that phenomena that are most diverse from a qualitative point of view exhibit identical mathematical properties. Hence, mathematical properties must be the essence of these phenomena. More specifically, the Pythagoreans found this essence in number and in numerical relationships. Number was their first principle in the explanation of nature. All objects were made up of points or "units of existence" in combinations corresponding to the various geometrical figures. Since they thought of numbers both as points and as elementary particles of matter, number was the matter and form of the universe and the cause of every phenomenon. Hence the Pythagorean doctrine "All things are numbers." Says Philolaus, a famous fifth-century Pythagorean, "Were it not for number and its nature, nothing that exists would be clear to anybody either in itself or in its relation to other things. . . . You can observe the power of number exercising itself not only in the affairs of demons and gods but in all the acts and the thoughts of men, in all handicrafts and music."

The reduction of music, for example, to simple relationships among numbers became possible for the Pythagoreans when they discovered two facts: first, that the sound caused by a plucked string depends upon the length of the string; and second, that harmonious sounds are given off by equally taut strings whose lengths are to each other as the ratios of whole numbers. For example, a harmonious sound is produced by plucking two equally taut strings, one twice as long as the other. In our language, the interval between the two notes is an octave. Another harmonious combination is formed by two strings whose lengths are in the ratio 3 to 2; in this case the shorter one gives forth a note called the fifth above that given off by the first string. In fact, the relative lengths in every harmonious combination of plucked strings can be expressed as ratios of whole numbers. The Pythagoreans also developed a famous Greek musical scale. Though we shall not devote space to the music of the Greek period, we note that many Greek mathematicians, including Euclid and Ptolemy, wrote on the subject, especially on harmonious combinations of sounds and the construction of scales.

The Pythagoreans reduced the motions of the planets to number relations. They believed that bodies moving in space produce sounds; perhaps this was suggested by the swishing of an object whirled on the end of a string. They believed, further, that a rapidly moving body gives forth a higher note than one that moves slowly. Now according to their astronomy the greater the distance of a planet from the earth the more rapidly it moved. Hence the sounds produced by the planets, which we do not hear because we are accustomed to them from birth, varied with their distances from the earth and all harmonized. But since this "music of the spheres," like all harmony, reduced to no more than number relationships, so did the motions of the planets.

The Pythagoreans and probably Pythagoras himself wanted not just to observe and describe the heavenly motions but to find regularity in them. The idea of uniform circular motion, seemingly obvious in the case of the moon and sun, suggested that all the planetary motions were explainable in terms of uniform circular motions. The later Pythagoreans made a more striking break with tradition; they were the first to believe that the earth was spherical. Moreover, because 10 was their ideal number, they decided that the moving bodies in the heavens must be 10 in number. First, there was a central fire around which the heavenly bodies, *including the earth,* moved. They knew five planets in addition to the earth. These six bodies, the sun, the moon, and the sphere to which the stars were attached made only 9 moving bodies. Hence they asserted the existence of a tenth one, called the counter-earth, which also revolved around the central fire. We cannot see this tenth one because it moves at exactly the same speed as the earth on the opposite side of the central fire and also because the inhabited part of earth faces away from the central fire. Here we have the first theory to put the earth in motion. However, the Pythagoreans did not assert the rotation of the earth; rather, the sphere of fixed stars revolves about the center of the universe.

The belief that the celestial bodies are eternal, divine, perfect, and unchangeable and that the sublunar bodies, that is the earth and (according to the Greeks) the comets, are subject to change, decomposition, decay, and death may also have come from the Pythagoreans. The doctrine of uniform circular motion and the distinction between celestial and sublunar bodies became embedded in Greek thought.

Other features of nature also "reduced" to number. The numbers 1, 2, 3, and 4, the *tetractys*, were especially valued because they added up to 10. In fact the Pythagorean oath is reported to have been: "I swear in the name of the Tetractys which has been bestowed on our soul. The source and roots of the everflowing nature are contained in it." The Pythagoreans asserted that nature was composed of fournesses; for example, point, line, surface, and solid, and the four elements, earth, air, fire, and water. The four elements were also central in Plato's natural philosophy. Because 10 was ideal, 10 represented the universe. The ideality of 10 required that the whole universe be describable in terms of 10 categories of opposites: odd and even, bounded and unbounded, good and evil, right and left, one and many, male and female, straight and curved, square and oblong, light and darkness, and rest and motion.

Clearly, Pythagorean philosophy mingled serious thoughts with what we would consider fanciful, useless, and unscientific doctrines. Their obsession with the importance of numbers resulted in a natural philosophy that certainly had little correspondence with nature. But they did stress the understanding of nature, not, like the Ionians, through a single substance, but through the formal structure of number relationships. Moreover they and the Ionians both saw that underlying mere sense data there must be a harmonious account of nature.

We can now see why the discovery of incommensurable lengths was so disastrous to Pythagorean philosophy: a ratio of incommensurable lengths could not be expressed as a ratio of whole numbers. In addition, they had believed that a line is made up of a finite number of points (which they identified with physical particles); but this could not be the case for a length such as $\sqrt{2}$. Their philosophy, based on the primariness of the whole numbers, would have been shattered if they had accepted irrationals as numbers.

Because the Pythagoreans "reduced" astronomy and music to number, these subjects came to be linked to arithmetic and geometry; these four were regarded as the mathematical subjects. They became and remained part of the school curriculum even into medieval times, when they were called, collectively, "the quadrivium." As we have noted, the Pythagorean interest in arithmetic (i.e. the theory of numbers) was due not to the purely aesthetic value of that subject but to a search for the meaning of natural phenomena in numerical terms; and this value caused the emphasis on special proportions and on triangular, square, pentagonal, and higher forms into which numbers could be arranged. Further, it was the Pythagorean natural philosophy centering about number that gave the subject importance with such men as Nichomachus. In fact, modern science adheres to the Pythagorean emphasis on number—though, as we shall see, in a much more sophisticated form— while the purely aesthetic modern theory of numbers derives from Pythagorean arithmetic per se.

The philosophers who came chronologically between the Pythagoreans and Plato were equally concerned with the nature of reality but did not involve mathematics directly. The arguments and views of men such as Parmenides (5th cent. B.C.), Zeno (5th cent. B.C.), Empedocles (*c.* 484–*c.* 424 B.C.), Leucippus (*c.* 440 B.C.), and Democritus (*c.* 460–*c.* 370 B.C.) were, like those of their Ionian predecessors, qualitative. They made broad assertions about reality that were, at best, barely suggested by observation. Nevertheless, each affirmed that nature is intelligible and that reality can be grasped by thought. Each was a link in the chain that led to the mathematical investigation of nature. Leucippus and Democritus are notable because they were the most explicit in affirming the doctrine of atomism. Their common philosophy was that the world is composed of an infinite number of simple, eternal atoms. These differ in shape, size, order, and position, but every object is some combination of these atoms. Though geometrical magnitudes are infinitely divisible, the atoms are ultimate indivisible particles. (The word *atom* in Greek means indivisible.) Hardness, shape, and size are physically real properties of the atoms. All other properties, such as taste, heat, and color are not in the atoms but in the perceiver; thus sensuous knowledge is unreliable because it varies with the perceiver. Like the Pythagoreans, the atomists asserted that the reality underlying the constantly changing diversity of the physical world was expressible in terms of mathematics and, moreover, that the happenings in this world were strictly determined by mathematical laws.

Plato, the foremost Pythagorean next to Pythagoras, was the most influential propagator of the doctrine that the reality and intelligibility of the physical world can be comprehended only through mathematics. For him there was no question that the world was mathematically designed, for "God eternally geometrizes." The world perceived by the senses is confused and deceptive and in any case imperfect and impermanent. Physical knowledge is unimportant, because material objects change and decay; thus the direct study of nature and purely physical investigations are worthless. The physical world is but an imperfect copy of the ideal world, the one that mathematicians and philosophers should study. Mathematical laws, eternal and unchanging, are the essence of reality.

Plato went further than the Pythagoreans in wishing not merely to understand nature through mathematics but to substitute mathematics for nature itself. He believed that a few penetrating glances at the physical world would supply some basic truths with which reason could then carry on unaided. From that point there would be no nature, just mathematics, which would substitute for physical investigations as it does in geometry.

Plato's attitude toward astronomy illustrates his position on the knowledge to be sought. This science is not concerned with the movements of the visible heavenly bodies. The arrangement of the stars in the heavens and their apparent movements are indeed wonderful and beautiful to behold, but mere observation and explanation of the motions fall far short of true astronomy. Before we can attain to the latter we "must leave the heavens alone," for true astronomy deals with the laws of motion of true stars in a mathematical heaven of which the visible heaven is but an imperfect expression. Plato encourages devotion to a theoretical astronomy, whose problems please the mind, not the eye, and whose objects are apprehended by the mind, not visually. The varied figures that the sky presents to the eye are to be used only as diagrams to assist the search for the higher truths. The uses of astronomy in navigation, calendar-reckoning, and the measurement of time were alien to Plato.

Plato's views on the role of mathematics in astronomy are an integral part of his philosophy, which held that there is an objective, universally valid reality consisting of forms or ideas. These realities were independent of human beings and were immutable, eternal, and timeless. We become aware of these ideas through recollection or anamnesis; although they are present in the soul, it must be stimulated to recall them or fetch them up from its depths. These ideas are the only reality. Included among them but occupying a lesser rank are mathematical ideas, which are regarded as intermediate between the sensible world and such higher ideas as goodness, truth, justice, and beauty. In this comprehensive philosophy, mathematical ideas played a double role; not only were they part of reality themselves but, as we have already pointed out in Chapter 3, they helped train the mind to view eternal ideas. As Plato put it in Book VII of *The Republic*, the study of geometry made easier the vision of the idea of goodness: "Geometry will draw the soul toward truth, and create the spirit of philosophy. . . ."

Aristotle, while deriving many ideas from his teacher Plato, had a quite different concept of the study of the real world and of the relation of mathematics to reality. He criticized Plato's otherworldliness and his reduction of science to mathematics. Aristotle was a physicist; he believed in material things as the primary substance and source of reality. Physics and science generally must study the physical world to obtain truths; genuine knowledge is obtained from sense experience by intuition and abstraction. Then reason can be applied to the knowledge so obtained.

Matter alone is not significant. As such it is indeterminate, simply the potentiality of form; matter becomes significant when it is organized into various forms. Form and the changes in matter that give rise to new forms are the interesting features of reality and the real concern of science.

According to Aristotle matter is not, as some earlier Greeks believed, composed of one primitive substance. The matter we see and touch is composed of four basic elements: earth, water, fire, and air. Also, each element has its own characteristic qualities. Earth is cold and dry; water is cold and moist; air is hot and moist; and fire is hot and dry. Hence the qualities of any given object depend upon the proportions of the elements that enter into it; and thereby solidity, hardness, coarseness, and other qualities are determined.

The four elements have other qualities. Earth and water have gravity; air and fire have levity. Gravity causes an element to seek to be at rest at the center of the earth; levity causes it to seek the heavens. Thus by knowing the proportions of the elements that enter into a given object, one can also determine its motion.

Aristotle regarded solids, fluids, and gases as three different types of matter, distinguished by the possession of different substantial qualities. The transition from solid to fluid, for example, meant the loss of one quality and the substitution of another. Thus changing mercury into rigid gold involved taking from mercury the substance that possessed fluidity and substituting some other substance.

Science also had to consider the causes of change. For Aristotle there were four types of causes. The first was the material or immanent cause; for a statue made of bronze, bronze is the immanent cause. The second was the formal cause; for the statue it is the design or shape. The formal cause of harmony is the pattern of 2 to 1 in the octave. The third cause was the effective cause, the agent or doer; the artist and his chisel are the effective cause of the statue. The fourth was the final cause, or the purpose that the phenomenon served; the statue serves to please people, to offer beauty. Final cause was the most important of the four because it gave the ultimate reasons for events or phenomena. Everything had a final cause.

Where was mathematics in this scheme of things? The physical sciences were fundamental to the study of nature, and mathematics helped by describing formal properties such as shape and quantity. It also provided explanations of facts observed in material phenomena. Thus geometry provided the reasons for facts provided by optics and astronomy, and arithmetical proportions could give the reasons for harmony. But mathematics was definitely an abstraction from the real world, since mathematical objects are not independent of or prior to experience. They exist in human minds as a class of ideas mediating between the sensible objects themselves and the essence of objects. Because they are abstracted from the physical world, they are applicable to it; but they have no reality apart from visible and tangible things. Mathematics alone can never provide an adequate definition of substance. Qualitative differences, as among colors, cannot be reduced to differences in geometry. Hence in the study of causes, mathematics can provide at best some knowledge of the formal cause—that is, a description. It can describe what happens in the physical world, can correlate concomitant variations, but can say nothing about the efficient and final causes of movement or change. Thus Aristotle distinguished sharply between mathematics and physics and assigned a minor role to mathematics. He was not interested in prediction.

From this survey we may see that all the philosophers who forged and molded the Greek intellectual world stressed the study of nature for comprehension and appreciation of its underlying reality. From the time of the Pythagoreans, practically all asserted that nature was designed mathematically. During the classical period, the doctrine of the mathematical design of nature was established and the search for the mathematical laws instituted. Though this doctrine did not motivate all of the mathematics subsequently created, once established it was accepted and consciously pursued by most of the great mathematicians. During the time that this doctrine held sway, which was until the latter part of the nineteenth century, the search for the mathematical design was identified with the search for truth. Though a few Greeks—for example, Ptolemy—realized that mathematical theories were merely human attempts to provide a coherent account, the belief that mathematical laws were the truth about nature attracted some of the deepest and noblest thinkers to mathematics.

We should also note, in order to appreciate more readily what happened in the seventeenth century, the Greek emphasis on the power of the mind. Because the Greek philosophers believed that the mind was the most powerful agent in comprehending nature, they adopted first principles that appealed to the mind. Thus the belief that circular motion was the basic type, defended by Aristotle on the ground that the circle is complete whereas a rectilinear figure, because it is bounded by many curves (line segments), is incomplete and therefore secondary in importance, appealed to the mind on aesthetic grounds. That the heavenly bodies should move with only constant or uniform velocity, was a conception which appealed to the mind perhaps because it was simpler than nonuniform motion. The combination of uniform and circular motion seemed to befit heavenly bodies. That the sublunar bodies should be different from the planets, sun, and stars seemed reasonable also, because the heavenly bodies preserved a constant appearance whereas change on earth was evident. Even Aristotle, who stressed abstractions only insofar as they helped to understand the observable world, said that we must start from principles that are known and manifest to the mind and then proceed to analyze things found in nature. We proceed, he said, from universals to particulars, from man to men, just as children call all men father and then learn to distinguish. Thus even the abstractions made from concrete objects presuppose some general principles emanating from the mind. This doctrine, the power of the mind to yield first principles, was overthrown in the seventeenth century.

4. *Greek Mathematical Astronomy*

Let us examine now what the Greeks produced in the mathematical description of natural phenomena. It is from the time of Plato that the several sciences created by the Greeks take on significant content and direction. Though we intend to review Greek astronomy, let us note in passing one aspect of Euclidean geometry. We have already observed that spherical geometry was developed for astronomy. Geometry was in fact part of the larger study of cosmology. Geometric principles were, to the Greeks, embodied in the entire structure of the universe, of which space was the primary component. Hence the study of space itself and of figures in space was of importance to the larger goal. Geometry, in other words, was in itself a science, the science of physical space.

It was Plato who, though fully aware of the impressive number of astronomical observations made by the Babylonians and Egyptians, emphasized the lack of an underlying or unifying theory or explanation of the seemingly irregular motions of the planets. Eudoxus, who was for a while a student at the Academy, took up Plato's problem of "saving the appearances." His answer is the first reasonably complete astronomical theory. He wrote four books on astronomy, *Mirror, Phenomena, Eight-Year Period*, and *On Speeds*, only fragments of which are known. From these fragments and accounts by other writers, we know the essence of Eudoxus' theory.

The motions of the sun and moon, as viewed from the earth, can be crudely described as circular with constant speed. However, their deviations from circular orbits are great enough to have been observed and to require explanation. The motions of the planets as seen from the earth are even more complex, for during any one revolution they reverse their course, go backward for a while, and then go forward again. Moreover, their speeds on these paths are variable.

To show that the actual, rather complicated, and apparently lawless motions could be understood in terms of simple circular geometrical motions, Eudoxus proposed the following scheme: For any one heavenly body there was a set of three or four spheres, all concentric with the earth as the center, each rotating about an axis. The innermost sphere carried the body, which moved along what can be called the equator of that sphere; that is, the axis of rotation was perpendicular to the circular path of the body. However, while rotating on its axis, this sphere was being carried along by the rotation of the next of the concentric spheres by the following device. Imagine that the axis of rotation of the first sphere is extended at each end to reach the second sphere and that its endpoints are fixed on that second sphere. If, now, the second sphere rotates about an axis of its own, then this sphere will carry the axis of the first sphere around while the latter rotates about that axis. The axis of the second sphere is in turn carried around by the rotation of a third sphere on its own axis. Eudoxus found that, for the sun and moon, a combination of three spheres sufficed to reproduce the actual motions as viewed from the earth. For each planet a fourth sphere was required, to which the third was related in the manner just described. The outermost sphere of each combination rotated on an axis through the celestial poles once in each 24 hours. In all, Eudoxus used 27 spheres. Their axes of rotation, speeds of rotation, and radii were chosen to make the theory fit as well as possible the observations available in his time.

Eudoxus' scheme was mathematically elegant and remarkable in many ways. The very idea of using combinations of spheres was ingenious; and the task of choosing the axes, speeds, and radii to make the resultant motion of the heavenly body fit the actual observations called for tremendous mathematical skill in working with surfaces and curves (i.e. the paths of the planets) in space.

It is especially worthy of note that Eudoxus' theory is purely mathematical. His spheres, except for the "sphere" of fixed stars, were not material bodies but mathematical constructions. Nor did he try to account for the forces that would make the spheres rotate as he said they did. His theory is thoroughly modern in spirit, for today mathematical description and not physical explanation is the goal in science.

The Eudoxian system had serious shortcomings. It did not include the variable velocity of the sun and was slightly in error about its actual path. His theory did not fit at all well the actual motion of Mars and was unsatisfactory for Venus. That Eudoxus tolerated such inadequacies may be accounted for by the fact that he might not have had at his command a sufficient number of observations. He had probably learned in Egypt only the main facts about the stationary points, retrogressions, and periods of revolution of the outer planets (Mars, Jupiter, and Saturn). Perhaps also for this reason his values for the sizes and distances of the celestial bodies were very crude. Aristarchus says that Eudoxus believed the diameter of the sun to be nine times that of the moon.

Aristotle did not appreciate a purely mathematical scheme and hence was not satisfied with Eudoxus' solution. To devise a real mechanism that made one sphere force another to rotate, he added 29 spheres. These were

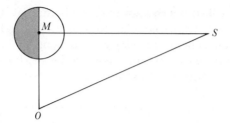

Figure 7.1

inserted among Eudoxus' so as to make the motion of one sphere drive another through actual contact and so that all derived their motive power from the outermost one. In some Aristotelian writings the sphere of stars, itself moving, is the prime mover of the other spheres. In others there is an unmoved mover behind the sphere of stars. His 56 spheres so complicated the system that it was discredited by scientists, though it was popular among educated laymen in medieval Europe. Aristotle, too, believed that the earth is spherical, for reasons of symmetry and equilibrium and because the shadow of the earth on the moon, seen in lunar eclipses, is circular.

Writings on astronomy continue after Aristotle in an almost unbroken sequence. After the works of Autolycus (Chap. 3, sec. 10) and the *Phaenomena* of Euclid (Chap. 4, sec. 11) the next major astronomical works are Alexandrian. Chaldean observations, observations made by the Babylonians of the Seleucid period, and measurements made by the Alexandrians themselves increased enormously the number and accuracy of the available data.

The first great Alexandrian astronomer is Aristarchus (*c.* 310–230 B.C.), who was learned in geometry, astronomy, music, and other branches of science. His book *On the Sizes and Distances of the Sun and Moon,* manuscripts of which are extant in Greek and Arabic, is the first major attempt to measure the distances of the sun and moon from the earth and the relative sizes of these bodies. These calculations are, incidentally, another example of the quantitative interests of the Alexandrians. Aristarchus did not have trigonometry at his disposal, nor did he have a good value for π (Archimedes' work on this came later), but he used Euclidean geometry very capably.

He knew that the moon's light is reflected light. When exactly half of the moon is illuminated, the angle at *M* (Fig. 7.1) is a right angle. The observer at *O* can measure the angle there and then the *relative* distances *OM* and *OS* can at least be estimated. Aristarchus' measurement of the angle was 87°; it is, to the nearest minute, 89°52′. Hence he estimated that the sun is more than 18 times and less than 20 times more distant than the moon. The correct ratio is 346 times more distant.

Knowing the relative distances, Aristarchus calculated the relative sizes by measuring the sizes of the discs the sun and moon present to the earth. He concluded that the volume of the sun was 7000 times larger than that of the moon. He was far from the truth here; the correct number is 2,296,000. He also found the ratio of the diameter of the sun to the diameter of the earth to be between 19/3 and 43/6; but the correct ratio is about 107.

Aristarchus is equally famous for having been the first to propose the heliocentric hypothesis—that the earth and the planets all revolve in circles around the fixed sun. The stars too are fixed and their apparent motion is due to the rotation of the earth on its axis. The moon revolves about the earth. Though, as we know today, Aristarchus had the right idea, it was not accepted for many reasons. For one thing, Greek mechanics (see below), already well developed by Aristotle, could not account for objects staying on a moving earth. According to Aristotle, heavy objects seek the center of the universe. This principle accounted for the fall of objects to earth as long as it was the center of the universe; but if it moved, the objects would stay behind. That argument was used by Ptolemy against Aristarchus and, in fact, was used later against Copernicus because the prevailing system of mechanics was still Aristotle's. Ptolemy also said that the clouds would lag behind a moving earth. Further, Aristotle's mechanics required a force to keep earthly objects in motion and there was no apparent force. We do not know how Aristarchus answered these arguments.

Another argument advanced against Aristarchus was that if the earth were in motion its distance from the fixed stars would vary, but it apparently does not. To this Aristarchus gave the correct rebuttal; he said that the radius of the sphere of fixed stars is so large that the earth's orbit is too small to matter. Aristarchus' heliocentric idea was rejected by many because it was impious to identify the corruptible matter of the earth with the incorruptible matter of the heavenly bodies. The hypothesis that the planets move about the sun in circles is, of course, unsatisfactory, because the motion is actually more complicated. But the heliocentric idea could have been refined, as Copernicus did later. It was, however, too radical for Greek thought.

The founder of quantitative mathematical astronomy is Apollonius. He was called Epsilon because the symbol ϵ was used to denote the moon, and much of his astronomy was devoted to the motion of that body. Before considering his work and that of Hipparchus and Ptolemy, to which it is closely related, we shall examine the basic scheme that had entered Greek astronomy between the times of Eudoxus and Apollonius, the scheme of epicycle and deferent. In this scheme a planet P moves at a constant speed on a circle (Fig. 7.2) with center S, while S itself moves with constant speed on a circle with center at the earth E. The circle on which S moves is called the deferent; the circle on which P moves is called the epicycle. The point S for some planets is the sun but in other cases it is just a mathematical point. The direction of the motion of P may agree with or be opposite to the direction of motion of S. The latter is the case for the sun and moon.

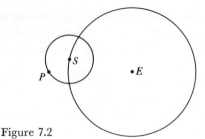

Figure 7.2

Apollonius is supposed to have known thoroughly the scheme of epi-cyclic motion and the details which made possible the representation of the motions of the planets, moon, and sun. Ptolemy credits Apollonius specifically with determining the points at which a planet appears to be stationary and reverses its direction of motion.

The climax of Greek astronomy is the work of Hipparchus and Ptolemy. The scheme of deferent and epicycle was taken over by Hipparchus (died *c.* 125 B.C.) and applied to the motion of the five planets then known, the moon, the sun, and the stars. We know Hipparchus' work through Ptolemy's *Almagest,* though it is difficult to distinguish what is due to Hipparchus and what to Ptolemy. After making observations at Rhodes for thirty-five years and utilizing Babylonian observations, Hipparchus worked out the details for the epicyclic theory of motion. By properly selecting the radii of the epicycle and deferent and the speeds of a body on its epicycle and the center of the epicycle on the deferent, he was able to get an improved description of the motions. He was quite successful for the sun and moon but only partially so for the planets. From the time of Hipparchus, an eclipse of the moon could be predicted to within an hour or two, though eclipses of the sun were predicted less accurately. This theory also accounted for the seasons.

Hipparchus' most original contribution is his discovery of the precession of the equinoxes. To understand this phenomenon we suppose that the earth's axis of rotation extends up to the stars. The point in which it hits the sphere of stars moves in a circle and takes 2600 years to traverse the circle. In other words the earth's axis continually changes its direction relative to the stars and the motion is periodic. The star to which it points at any one time is called the polar star. The angle subtended by a diameter of the above-mentioned circle at the earth is 45°.

Hipparchus made many other contributions to astronomy, such as the construction of instruments for observation, the determination of the angle of the ecliptic, measurements of irregularities in the motion of the moon, improvement of the determination of the length of the solar year (which he estimated to be 365 days, 5 hours, 55 minutes, and 12 seconds—or about

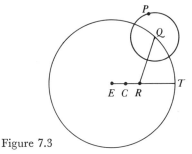

Figure 7.3

6 1/2 minutes too long), and a catalogue of about a thousand stars. He found that the ratio of the distance to the moon to the radius of the earth was 67.74; the modern figure is 60.4. He calculated the moon's radius to be 1/3 the earth's radius; the present figure is 27/100.

Ptolemy extended the work of Hipparchus by further improving the mathematical descriptions of the motions of all the heavenly bodies. The extended theory, presented in the *Almagest*, offers a complete exposition of the geocentric theory of deferent and epicycle, which has come to be known as the Ptolemaic theory.

To make the geometrical account fit observations, Ptolemy also introduced a variation on epicyclic motion known as uniform equant motion. In this scheme (Fig. 7.3) the planet moves on an epicycle with center Q while Q moves on a circle with center C which is not the earth but somewhat offset. He fixed the speed of Q by introducing the point R such that $EC = CR$ and required that $\sphericalangle QRT$ increase uniformly. Thus Q moves with uniform angular velocity, but not with uniform linear velocity.

The approach and understanding that the Greek astronomers attained are thoroughly modern. The Alexandrian Greek astronomers, notably Hipparchus and Ptolemy, made observations of their own; in fact, Hipparchus did not trust many of the older Egyptian and Chaldean observations and repeated them. The classical and Alexandrian astronomers not only constructed theories but fully realized that these theories were not the true design but just descriptions that fit the observations. Ptolemy says in the *Almagest*[1] that in astronomy one ought to seek as simple as possible a mathematical model. These men, unlike other Greeks, did not look for some physical explanation of the motions. On this point Ptolemy says,[2] "After all, generally speaking, the cause of the first principles is either nothing or hard to interpret in its nature." But his own mathematical model was later taken as the literal truth by the Christian world.

1. Book XIII, Chap. 2, last paragraph.
2. *Almagest*, Book IX.

Ptolemaic theory offered the first reasonably complete evidence of the uniformity and invariability of nature and was the final Greek answer to Plato's problem of rationalizing the apparent motions of the heavenly bodies. No other product of the entire Greek era rivaled the *Almagest* for its profound influence on conceptions of the universe and none, except Euclid's *Elements*, achieved such unquestioned authority.

This brief account of Greek astronomy has not revealed the full depth and extent of the work done even by the men discussed here, and omits many other contributions. Almost every Greek mathematician, including Archimedes, devoted himself to the subject. Greek astronomy was masterful and comprehensive and it employed a vast amount of mathematics.

5. *Geography*

Another science that received its foundation in Greek times is geography. Though a few classical Greeks such as Anaximander and Hecataeus of Miletus (died *c.* 475 B.C.) made maps of the earth as it was known at that time, it was the Alexandrians who made the great strides in geography. They measured or calculated distances along the earth, the heights of mountains, the depths of valleys, and the extent of the seas. The Alexandrians were especially stimulated to study geography because the Greek world had widened.

The first great Alexandrian geographer was Eratosthenes of Cyrene (*c.* 284–*c.* 192 B.C.), director of the library at Alexandria, a mathematician, poet, philosopher, historian, philologist, chronologist, and by reputation one of the most learned men of antiquity. He studied in Plato's school in Athens and was invited to Alexandria by Ptolemy Euergetes. Eratosthenes worked at Alexandria until blindness overtook him in his old age; because of this affliction he starved himself to death.

Eratosthenes collected the available geographical knowledge and made numerous calculations of distances on the earth between significant places (such as cities). His most famous calculation is the length of the circumference of the earth. At noon on the summer solstice, the sun was observed to be practically overhead at Syene, the city that today is called Aswan (Fig. 7.4). (This was confirmed by observing that the sun shone directly down a well there.) At the same time, in Alexandria, which is (within 1°) on the same meridian as Syene but north of it, the angle between the overhead direction for that location (*OB* in the figure) and the direction of the sun (*AD* in the figure) was observed to be 1/50 of 360°. The sun is so far from the earth that *SE* and *AD* may be considered parallel. Hence $\sphericalangle SOA$ is 1/50 × 360°. This means that arc *SA* is 1/50 of the circumference of the earth. Eratosthenes estimated the distance from Alexandria to Syene by using the fact that

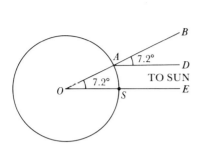

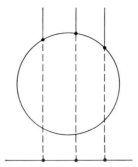

Figure 7.4   Figure 7.5

camel trains, which usually traveled 100 stadia a day, took 50 days to reach Syene. Hence this distance is 5000 stadia and the circumference of the earth is 250,000 stadia. It is believed that a stadium was 157 meters so that Eratosthenes' result is 24,662 miles. This result was far more accurate than all previous estimates.

Eratosthenes wrote the *Geography*, in which he incorporated the methods and results of his measurements and calculations. It includes explanations of the nature and causes of the changes that had taken place on the earth's surface. He also made a map of the world.

Scientific map-making became part of the work of geography. Hipparchus is generally credited with introducing latitude and longitude, though the scheme was known earlier. The use of latitude and longitude, of course, permitted accurate description of locations on the earth. Hipparchus did invent orthographic projection, in which "light rays" from infinity project the earth on a plane (Fig. 7.5). Our view of the moon, for example, is practically orthographic. This method enabled him to map a portion of the earth onto a flat surface.

Ptolemy in his *Planisphaerium*, and probably Hipparchus before him, used the method of stereographic projection. A line from *O* (Fig. 7.6) through *P* on the earth's surface is continued until it hits the equatorial plane or a tangent plane at the opposite pole. Hipparchus supposedy used

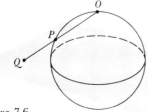

Figure 7.6

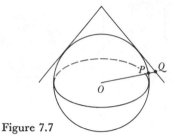

Figure 7.7

the latter and Ptolemy the equatorial plane. Thus points on the sphere are transferred to a plane. In this scheme all points on the map have true direction from the center of the map. Also angles are preserved locally (conformal mapping), though Ptolemy does not mention this. The meridians and the parallels of latitude are therefore at right angles. Circles on the sphere become circles on the plane, but area is not preserved. Ptolemy himself invented conical projection, that is, the projection of a region on the earth from the center of the sphere onto a tangent cone (Fig. 7.7).

In his *Geographia*, a work in eight books, Ptolemy teaches methods of map-making. Chapter 24 of Book I is the oldest work extant that is devoted by title and contents to mapping a sphere onto a plane. The entire *Geographia* is the first atlas and gazeteer. It gives the latitude and longitude of 8000 places on the earth and was the standard reference for hundreds of years.

6. *Mechanics*

The Greeks initiated the science of mechanics. In his *Physics*, Aristotle put together a theory of motion that is the high point of Greek mechanics. Like all of his physics, his mechanics is based on rational, seemingly self-evident principles only slightly checked by or drawn from observation and experiment.

There are, according to Aristotle, two kinds of motion, natural and violent or man-made. The heavenly spheres possess only natural motion, which is circular. As for earthly objects, he taught that a natural motion (as opposed to violent motions caused by throwing or pulling a body from one place to another) arises from the fact that each body has a natural place in the universe where it is in equilibrium or at rest. Heavy bodies have their natural place at the center of the universe, which is the center of the earth. Light bodies, such as gases, have their natural place in the sky. Natural motion results when a body seeks its natural place. In natural motions earthly objects move up or down in straight lines. If an earthly object were not in its natural place it would seek that place as soon as possible. Violent motions, that is, man-made ones, are composed of circular and rectilinear parts. Thus a stone thrown out and up follows a straight line path up and a straight line path down.

Any body in motion is subject to a force and a resistance. In the case of natural motion, the force is the weight of the body, and the resistance comes from the medium in which the body moves. In violent motion the force is applied by the hand or by some mechanism and the resistance comes from the body's weight. Without force there could be no motion; without resistance the motion would be accomplished in an instant. The velocity of any motion, then, depends upon the force and the resistance. These principles can be summed up in modern form by the formula $V \propto F/R$; that is, the velocity depends directly upon the force and inversely upon the resistance.

Since in violent motion the resistance is furnished by the weight, for lighter bodies the resistance, R, is smaller. By the above formula, the velocity V must then be larger; that is, lighter bodies move faster under the same force. In the case of natural motion, the force is the weight, and so heavier bodies fall faster. Since in natural motion the resistance is furnished by the medium, in a vacuum the velocity would be infinite. Hence a vacuum is impossible.

Aristotle had difficulty accounting for some phenomena. To explain the increasing velocities of falling bodies he said that the speed increases as the body comes closer to its natural place because the body moves more jubilantly; but this is not consistent with the speed being dependent on the fixed weight. In the case of an arrow shot from a bow, Aristotle said the arrow continued to move, although no longer in contact with the bow, because the hand or bowstring communicated a power of movement to the air nearby and this air to the next layer of air, and so on. Alternatively the air in front of the arrow is compressed and rushes around to the rear of the arrow to prevent a void and so the arrow is pushed forward. He did not explain the decay of the motive power.

The greatest mathematical physicist of Greek times is Archimedes. He, more than any other Greek writer, tied geometry to mechanics and used geometrical arguments ingeniously to make his proofs. In mechanics he wrote *On the Equilibrium of Planes or The Centers of Gravity of Planes*, a work in two books. By the center of gravity of a body or of a collection of bodies rigidly attached to each other he means, as we do, the point at which the body or collection of bodies can be supported so as to be in equilibrium under the pull of gravity. He starts with postulates about the lever and center of gravity. For example (the numbering follows Archimedes):

1. Equal weights at equal distances are in equilibrium and equal weights at unequal distances are not in equilibrium but incline towards the weight which is at greater distance.

2. If, when weights at certain distances are in equilibrium, something be added to one of the weights, they are not in equilibrium but incline towards the weight to which the addition was made.

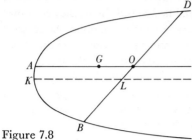

Figure 7.8

5. In figures which are unequal but similar the centers of gravity will be similarly situated. . . .

7. In any figure whose perimeter is concave in the same direction the center of gravity must be within the figure.

He follows these postulates with a number of propositions, some of whose proofs depend upon results in a lost treatise, *On Levers*:

Proposition 4. If two equal weights have not the same center of gravity, the center of gravity of both taken together is at the middle point of the lines joining their centers of gravity.

Propositions 6 and 7. Two magnitudes whether commensurable or incommensurable balance at distances reciprocally proportional to the magnitudes.

Proposition 10. The center of gravity of any parallelogram is the point of intersection of its diagonals.

Proposition 14. The center of gravity of any triangle is at the intersection of the lines drawn from any two vertices to the middle points of the opposite sides respectively.

Book II deals with the center of gravity of a parabolic segment. Among the principal theorems are:

Proposition 4. The center of gravity of any parabolic segment cut off by a straight line lies on the diameter of the segment.

The diameter is AO (Fig. 7.8), where O is the midpoint of BD and AO is parallel to the axis of the parabola. The proof uses results obtained in his *Quadrature of the Parabola*.

Proposition 8. If AO be the diameter of a parabolic segment and G its center of gravity, then $AG = (3/2)GO$.

Work on centers of gravity is found in many books of the Alexandrian Greek period. Examples are Heron's *Mechanica* and Book VIII of Pappus' *Mathematical Collection* (Chap. 5, sec. 7).

The science of hydrostatics—the study of fluid pressure when the fluid is at rest—was founded by Archimedes. In his book *On Floating Bodies* he is interested in the pressure exerted by water on objects placed in it. He gives two postulates. The first is to the effect that the pressure exerted by any part of a fluid on the fluid is downward. The second postulate states that the pressure of the fluid on a body placed in it is exerted upward along the perpendicular through the center of gravity of the body. Some of the theorems he proves in Book I are:

Proposition 2. The surface of any fluid at rest is the surface of a sphere whose center is the same as that of the earth.

Proposition 3. Of solids those which, size for size, are of equal weight with a fluid will, if let down into the fluid, be immersed so that they do not project above the surface but do not sink lower.

Proposition 5. Any solid lighter than a fluid will, if placed in the fluid, be so far immersed that the weight of the solid [in air] will be equal to the weight of the fluid displaced.

Proposition 7. A solid heavier than a fluid will, if placed in it, descend to the bottom of the fluid, and the solid will, when weighed in the fluid, be lighter than its true weight by the weight of the fluid displaced.

This last proposition is believed to be the one Archimedes used to determine the contents of the famous crown (Chap. 5, sec. 3). He must have argued as follows: Let W be the weight of the crown. Take a crown of pure gold and of weight W and weigh it in the fluid. It will weigh less by some amount, F_1, which is the weight of the displaced water. Likewise a weight W of pure silver will displace water of weight F_2 which can be measured by weighing the silver in the water. Then if the original crown contains weight w_1 of gold and weight w_2 of silver, the original crown will displace a weight of water equal to

$$\frac{w_1}{W} F_1 + \frac{w_2}{W} F_2.$$

Let F be the actual weight of the water displaced by the crown. Then

$$\frac{w_1}{W} F_1 + \frac{w_2}{W} F_2 = F$$

or

$$w_1 F_1 + w_2 F_2 = (w_1 + w_2) F$$

or

$$\frac{w_1}{w_2} = \frac{F_2 - F}{F - F_1}.$$

Thus Archimedes was able to determine the ratio of the gold to the silver in the crown without destroying the crown. Vitruvius' account of this story is that Archimedes used *volumes* of displaced water instead of weights. In this case, the F, F_1, and F_2 above are, respectively, the volumes of water displaced by the crown, a weight W of pure gold, and a weight W of pure silver. The same algebra follows but Proposition 7 is not involved. Archimedes did find that the gold had been debased with silver.

To gain some appreciation of the mathematical and physical complexity of the problems treated by Archimedes in this work, we shall cite one of the simple propositions of Book II.

Proposition 2. If a right segment of a paraboloid of revolution whose axis is not greater than $3p/4$ [p is the principal parameter or latus rectum of the generating parabola], and whose specific gravity is less than that of a fluid, be placed in the fluid with its axis inclined to the vertical at any angle, but so that the base of the segment does not touch the surface of the fluid (Fig. 7.9), the segment of the paraboloid will not remain in that position but will return to the position in which its axis is vertical.

The subject Archimedes is treating is the stability of bodies placed in water. He shows under what conditions a body will, when placed in water, either turn to or remain in a position of equilibrium. The problems are clearly idealizations of how ships would behave when obliged to assume different tilts in water.

7. *Optics*

Next to astronomy, optics has been the most constantly pursued and most successful of the mathematical sciences. It was founded by the Greeks. Almost all of the Greek philosophers, beginning with the Pythagoreans, speculated on the nature of light, vision, and color. Our concern, however, is with the mathematical accomplishments. The first of these is the assertion on a priori grounds by Empedocles of Agrigentum in Sicily (*c.* 490 B.C.) that light travels with finite velocity.

The first systematic treatments that we have are Euclid's *Optics* and *Catoptrica*. The *Optics* is concerned with the problem of vision and with the use of vision to determine sizes of objects. Euclid begins with definitions (which are really postulates), the first of which states (as had Plato) that vision is possible because rays of light emitted by the eye travel along straight lines and impinge on the object seen. Definition 2 states that the figure formed by

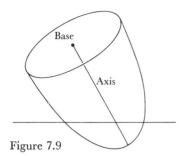

Figure 7.9

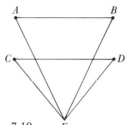

Figure 7.10   *E*

the visual rays is a cone that has its vertex in the eye and its base at the extremities of the object seen. Definition 4 says that of two objects the one that determines a greater vertex angle of the cone appears greater. Then in Proposition 8 Euclid proves that the apparent sizes of two equal and parallel objects (*AB* and *CD* in Fig. 7.10) are not proportional to their distances from the eye. Propositions 23 to 27 prove that the eye looking at a sphere really sees less than half of it, and that the contour of what is seen is a circle. Propositions 32 to 37 point out that the eye looking at a circle will see a circle only if the eye is located on the perpendicular to the plane of the circle and the perpendicular strikes the center. Euclid also shows how to calculate the sizes of objects that are seen as images in a plane mirror. There are 58 propositions in the book.

The *Catoptrica* (theory of mirrors) describes the behavior of light rays reflected from plane, concave, and convex mirrors and the effect of this behavior on what we see. Like the *Optics*, it starts with definitions that are really postulates. Theorem 1, the law of reflection, is now fundamental in what is called geometrical optics. It says that the angle *A*, which the incident ray makes with the mirror (Fig. 7.11), equals the angle *B*, which the reflected ray makes with the mirror. It is more customary today to say that $\angle C = \angle D$ and to speak of $\angle C$ as the angle of incidence and $\angle D$ as the angle of reflection. Euclid also proves the law for a ray striking a convex or a concave mirror by substituting a tangent for the mirror at the point where the ray strikes.

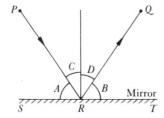

Figure 7.11

From the law of reflection Heron drew an important consequence. If P and Q (Fig. 7.11) are any two points on one side of the line ST, then of all the paths one could follow in going from point P to the line and then to point Q the shortest path is by way of the point R such that the two line segments PR and QR make equal angles with the line—which is exactly the path a light ray takes. Hence, the light ray takes the shortest path in going from P to the mirror to Q. Apparently nature is well acquainted with geometry and employs it to full advantage. This proposition appears in Heron's *Catoptrica*, which also treats concave and convex mirrors and combinations of mirrors.

Any number of works were written on the reflection of light by mirrors of various shapes. Among these are the now lost *Catoptrica* of Archimedes and the two works, both called *On Burning-Mirrors*, by Diocles and Apollonius. Burning-mirrors were undoubtedly concave mirrors shaped like spheres, paraboloids of revolution, and ellipsoids, the last formed by revolving an ellipse about its major axis. Apollonius undoubtedly knew that a paraboloidal mirror will reflect light emanating from the focus into a beam parallel to the axis of the mirror. Conversely, rays coming in parallel to the axis will, after reflection, be concentrated at the focus. The sun's rays thus concentrated produce great heat at the focus; hence the term burning-mirror. This is the property of the paraboloidal mirror that Archimedes is supposed to have used to concentrate the sun's rays on the Roman ships and set them afire. Apollonius also knew the reflection properties of the other conic sections—for example, that all rays emanating from one focus of the ellipsoidal mirror will be reflected to the other focus. He gives the relevant geometrical properties of the ellipse and hyperbola in Book III of his *Conic Sections* (Chap. 4, sec. 12). Later Greeks, Pappus in particular, certainly knew the focusing property of the paraboloid.

The phenomenon of refraction of light—that is, the bending of light rays as they pass through a medium whose properties change from point to point, or the sudden change in the direction of a light ray as it passes from one medium into another, say from air to water, was studied by the Alexandrian Greeks. Ptolemy noted the effect of refraction by the atmosphere on rays coming from the sun and stars and attempted, unsuccessfully, to find the correct law of refraction when light passes from air to water or air to glass. His *Optics*, which deals with both mirrors and refraction, is extant.

8. *Astrology*

Though astrology is not accepted as a science today, in earlier civilizations it did have that standing. The astrology developed by the Alexandrian Greeks of about the second century B.C. was different from the Babylonian astrology of the Assyrian period. The latter merely drew from observations of the positions of the planets some conclusions about the king and affairs of state. There were no computations, and the appearance of the sky at the moment of birth played no role. However, Hellenistic or Alexandrian astrology was personal; it predicted the future and fate of specific individuals on the basis of the computed positions of the sun, moon, and the five planets in the zodiac at the moment of birth. To evaluate these data an enormous body of doctrines was built up.

Certainly the science was taken seriously by the Alexandrian Greeks. Ptolemy wrote a well-known work on the subject, the *Quadripartite* or *Tetrabiblos*, or *Four Books Concerning the Influence of the Stars*, in which he gave rules for astrological predictions that were used for a thousand years.

The importance of astrology in the history of science lies in the motivation it supplied for the study of astronomy, not only in Greece but in India, Arabia, and medieval Europe. Astrology nourished astronomy much as alchemy nourished chemistry. Curiously, errors in astrological predictions were ascribed to errors in astronomy, not to the unreliability of the astrological doctrines.

Alexandrian Greece witnessed the beginnings of the application of mathematics to medicine, peculiarly enough through the medium of astrology. Doctors, called iatromathematicians, employed astrological signs to decide courses of treatment. Galen, the great physician of Greek times, was a firm believer in astrology, perhaps excusably so since Ptolemy, the most renowned astronomer, was also. This connection between mathematics and medicine became stronger in the Middle Ages.

Our account of Greek science, in view of our concern with mathematics, has dealt with the mathematical sciences. The Greeks carried on other investigations, in areas where mathematics, at least at that time, played no role. Moreover, they performed experiments and made observations, the latter particularly in astronomy. Nevertheless their vital achievement is that they established the value of mathematics in the scientific enterprise. The Platonic dialogue *Philebus* first gave expression to the thought that each science is a science only so far as it contains mathematics; this doctrine gained immense support from the Greek accomplishments. Moreover, the Greeks produced ample evidence that nature is mathematically designed. It was their vision of nature and their initiation of the mathematical investigation of nature that inspired the creation of mathematics, in Greek times and in all succeeding centuries.

Bibliography

Apostle, H. G.: *Aristotle's Philosophy of Mathematics*, University of Chicago Press, 1952.

Berry, Arthur: *A Short History of Astronomy*, Dover (reprint), 1961, Chaps. 1–2.

Clagett, Marshall: *Greek Science in Antiquity*, Abelard-Schuman, 1955.

Dreyer, J. L. E.: *A History of Astronomy from Thales to Kepler*, Dover (reprint), 1953, Chaps. 1–9.

Farrington, Benjamin: *Greek Science*, 2 vols., Penguin Books, 1944 and 1949.

Gomperz, Theodor: *Greek Thinkers*, 4 vols., John Murray, 1920.

Heath, Thomas L.: *Greek Astronomy*, J. M. Dent and Sons, 1932.

———: *Aristarchus of Samos*, Oxford University Press, 1913.

———: *The Works of Archimedes*, Dover (reprint), 1953, pp. 189–220 and 253–300.

Jaeger, Werner: *Paideia*, 3 vols., Oxford University Press, 1939–44.

Pannekoek, A.: *A History of Astronomy*, John Wiley and Sons, 1961.

Sambursky, S.: *The Physical World of the Greeks*, Routledge and Kegan Paul, 1956.

Santillana, G. de: *The Origins of Scientific Thought from Anaximander to Proclus, 600 B.C. to 300 A.D.*, University of Chicago Press, 1961.

Sarton, George: *A History of Science*, Harvard University Press, 1952 and 1959, Vols. 1 and 2.

Ver Eecke, Paul: *Euclide, L'Optique et la catoptrique*, Albert Blanchard, 1959.

Wedberg, Anders: *Plato's Philosophy of Mathematics*, Almqvist and Wiksell, 1955.

8

The Demise
of the Greek World

> He who understands Archimedes and Apollonius will admire
> less the achievements of the foremost men of later times.
>
> G. W. LEIBNIZ

1. *A Review of the Greek Achievements*

Though the Alexandrian Greek civilization lasted until A.D. 640, when it was finally destroyed by the Mohammedans, it is apparent, from its decreasing productiveness, that the civilization was already declining in the early centuries of the Christian era. Before investigating the reasons for the decline, we shall summarize the achievements and deficiencies of Greek mathematics and note the problems it left for subsequent generations. The Greeks accomplished so much, and the pursuit of mathematics when taken over by the Europeans, after minor interpolations by the Hindus and Arabs, was so completely determined by what the Greeks bequeathed, that it is important to be clear as to where mathematics stood.

The Greeks are to be credited with making mathematics abstract. This major contribution is of immeasurable significance and value, for the fact that the same abstract triangle or algebraic equation may apply to hundreds of different physical situations has proved to be the secret of the power of mathematics.

The Greeks insisted on deductive proof. This was indeed an extraordinary step. Of the hundreds of civilizations that have existed a number did develop some crude arithmetic and geometry. However, no civilization but the Greeks conceived the idea of establishing conclusions exclusively by deductive reasoning. The decision to require deductive proof is entirely at odds with the methods mankind has utilized in all other fields; it is, in fact, almost irrational, since so much highly reliable knowledge is acquired by experience, induction, reasoning by analogy, and experimentation. But the Greeks wanted truths, and saw that they could obtain them only by the unquestionable methods of deductive reasoning. They also realized that to secure truths they had to start from truths, and be sure not to assume any unwarranted facts. Hence they stated all their axioms explicitly and in addition adopted the practice of stating them at the very outset of their work so that they could be examined critically at once.

Beyond conceiving this highly remarkable plan to establish secure knowledge, the Greeks showed a sophistication one could hardly expect from innovators. Their recognition that the concepts must be free of contradiction and that one cannot build a consistent structure by working with nonexistent figures (such as a ten-faced regular polyhedron) shows an almost superhuman and certainly unprecedented keenness of thought. As we now know, their method of establishing the existence of concepts, so they could deal with them, was to demonstrate their constructibility with straight-edge and compass.

The power of the Greeks to divine theorems and proofs is attested to by the fact that Euclid's *Elements* contains 467 propositions and Apollonius' *Conic Sections* contains 487, all derived from the 10 axioms in the *Elements*. No doubt secondary in importance, and perhaps secondary in intent, is the coherence that the deductive structures provided. Were the same results obtainable from numerous different—though equally reliable—sets of axioms, they might have presented far less manageable and assimilable knowledge.

The Greek contribution to the content of mathematics—plane and solid geometry, plane and spherical trigonometry, the beginnings of the theory of numbers, the extension of Babylonian and Egyptian arithmetic and algebra —is enormous, especially in view of the small number of people involved and the few centuries of extensive activity. To these contributions we must add the geometrical algebra, which awaited only the recognition of irrational numbers and translation into symbolic language to become the basis of considerable elementary algebra. The treatment of curvilinear figures by the method of exhaustion, though part of their geometry, also warrants special mention because it is the beginning of the calculus.

An equally vital contribution and inspiration to later generations was the Greek conception of nature. The Greeks identified mathematics with the reality of the physical world and saw in mathematics the ultimate truth about the structure and design of the universe. They founded the alliance between mathematics and the disinterested study of nature, which has since become the very basis of modern science. Moreover, they went far enough in rationalizing nature to establish firmly the conviction that the universe is indeed mathematically designed, that it is controlled, lawful, and intelligible to man.

The aesthetic appeal of mathematics was by no means overlooked. In Greek times the subject was also valued as an art; beauty, harmony, simplicity, clarity, and order were recognized in it. Arithmetic, geometry, and astronomy were considered the art of the mind and music for the soul. Plato delighted in geometry; Aristotle would not divorce mathematics from aesthetics, for order and symmetry were to him important elements of beauty, and these he found in mathematics. Indeed, rational and aesthetic, as well as moral, interests can hardly be separated in Greek thought. Repeatedly one reads that the sphere has the most beautiful shape of all bodies and is therefore divine and good. The circle shared aesthetic appeal with the sphere; it therefore seemed obvious that the circle should be the path of those bodies that represented the changeless, eternal order of the heavens, whereas straight-line motion prevailed on the imperfect earth. There is no doubt that it was the aesthetic appeal of the subject that caused Greek mathematicians to carry the exploration of particular topics beyond their use in understanding the physical world.

2. The Limitations of Greek Mathematics

Despite its marvelous achievements, Greek mathematics was flawed. Its limitations indicate the avenues of progress that were yet to be opened up.

The first limitation was the inability to grasp the concept of the irrational number. This meant not only a restriction of arithmetic and algebra but a turn to and an emphasis on geometry, because geometrical thinking avoided explicit confrontation of the irrational as a number. Had the Greeks faced the irrational number, they might have furthered the development of arithmetic and algebra; and even if they themselves had not done so, they would not have hindered later generations, which were induced to think that only geometry offered a secure foundation for the treatment of any magnitude whose values might include irrationals. Archimedes, Heron, and Ptolemy started to work with irrationals as numbers, but did not alter the tenor of Greek mathematics or the subsequent impress of Greek thought. The Greek concentration on geometry blurred the vision of later generations by masking the intimate correspondence between geometric and arithmetic concepts and operations. The failure to define, accept, and conceptualize the irrational as a number forced a distinction between number and magnitude. Consequently algebra and geometry were regarded as unrelated disciplines.

Had the Greeks been less concerned to be logical and precise they might have casually accepted and operated with irrational numbers as had the Babylonians, and as civilizations succeeding the Greeks did. But the intuitive basis of the idealization was not clear, and the logical construction was not quite within their power. The Greek virtue of insisting on exact concepts and proofs was a defect so far as creative mathematics was concerned.

The restriction of rigorous mathematics (apart from the theory of numbers) to geometry imposed another serious disadvantage. The use of geometric methods led to more and more complicated proofs as mathematics was extended, particularly in the area of solid geometry. Moreover, even in the simpler proofs there is a lack of general methods, which is clear to us now that we have analytic geometry and the calculus. When one considers how hard Archimedes worked to find the area of a parabolic segment or the area under an arc of his spiral and compares this with the modern calculus treatment, one appreciates the effectiveness of the calculus.

Not only did the Greeks restrict mathematics largely to geometry; they even limited that subject to figures that could be obtained from the line and circle. Accordingly, the only surfaces admitted were those obtained by revolving lines and circles about an axis, for example the cylinder, cone, and sphere, formed by the revolution of a rectangle, triangle, and circle, respectively, about a line. A few exceptions were made: the plane, which is the analogue of a line; the prism, which is a special cylinder; and the pyramid, which results from decomposition of a prism. The conic sections were introduced by cutting cones by a plane. Curves such as the quadratrix of Hippias, the conchoid of Nicomedes, and the cissoid of Diocles were kept on the fringe of geometry; they were called mechanical, rather than geometrical.

Pappus' classification of curves shows an attempt to keep within fixed bounds. The Greeks, according to Pappus, distinguished curves as follows: Plane loci or plane curves were those constructed from straight lines and circles; the conics were called solid loci because they originated from the cone; the linear curves, such as quadratrices, conchoids, cissoids, and spirals constituted the third class. Similarly, they distinguished among plane, solid, and linear problems. Plane problems were solved by means of straight lines and circles; solid problems were solved by one or more of the conic sections. Problems that could not be solved by means of straight lines, circles, or conics were called linear, because they used lines (curves) having a more complicated or less natural origin than the others. Pappus stressed the importance of solving problems by means of plane or solid loci because then the criterion for the possibility of a real solution can be given.

Why did the Greeks limit their geometry to line and circle and figures readily derived from them? One reason was that this solved the problem of establishing the existence of geometrical figures. As we saw, Aristotle, in particular, pointed out that we must be sure that the concepts introduced are not self-contradictory; that is, they must be shown to exist. To settle this point, the Greeks, in principle at least, admitted only those concepts that were constructible. Line and circle were accepted as constructible in the postulates but all other figures had to be constructed with line and circle.

However, the use of constructions to establish existence was not carried over to figures in three dimensions. Here the Greeks apparently accepted what was intuitively clear, for example the existence of figures of revolution such as sphere, cylinder, and cone. Sections of these figures by planes produced curves such as the conic sections; thus even figures in the plane whose existence was not established were accepted—but reluctantly. Descartes notes this point near the beginning of Book II of *La Géométrie*:

"It is true that the conic sections were never freely received into ancient geometry. . . ."

Another reason for the limitation to line, circle, and figures derived from them stems from Plato, according to whom ideas had to be clear to be acceptable. While the whole number seemed to be acceptable as a clear idea in itself, even though never explicitly defined by the Greeks, the geometric figures had to be made precise. Lines and circles and figures derived from them were clear, whereas curves introduced by mechanical instruments (other than straightedge and compass) were not, and so were not admissible. The restriction to clear figures produced a simple, well-arranged, harmonious, and beautiful geometry.

By insisting on a unity, a completeness, and a simplicity for their geometry, and by separating speculative thought from utility, classical Greek geometry became a limited accomplishment. It narrowed people's vision and closed their minds to new thoughts and methods. It carried within itself the seeds of its own death. The narrowness of its field of action, the exclusiveness of its point of view, and the aesthetic demands on it might have arrested its development, had not the influences of the Alexandrian civilization broadened the outlook of Greek mathematicians.

The philosophical doctrines of the Greeks limited mathematics in another way. Throughout the classical period they believed that man does not create the mathematical facts: they preexist. He is limited to ascertaining and recording them. In the *Theaetetus* Plato compares the search for knowledge to a bird-hunt pursued in an aviary. The birds, already captive, need only to be seized. This belief about the nature of mathematics did not prevail.

The Greeks failed to comprehend the infinitely large, the infinitely small, and infinite processes. They "shrank before the silence of infinite spaces." The Pythagoreans associated good and evil with the limited and unlimited, respectively. Aristotle says the infinite is imperfect, unfinished, and therefore unthinkable; it is formless and confused. Only as objects are delimited and distinct do they have a nature.

To avoid any assertion about the infinitude of the straight line, Euclid says a line segment (he uses the word "line" in this sense) can be extended as far as necessary. Unwillingness to involve the infinitely large is seen also in Euclid's statement of the parallel axiom. Instead of considering two lines that extend to infinity and giving a direct condition or assumption under which parallel lines might exist, his parallel axiom gives a condition under which two lines will meet at some finite point.

The concept of the infinitely small is involved in the relation of points to a line or the relation of the discrete to the continuous, and Zeno's paradoxes may have caused the Greeks to shy away from this subject. The relationship of point to line bothered the Greeks and led Aristotle to separate the two. Though he admits points are on lines, he says that a line is not made up of points and that the continuous cannot be made up of the discrete (Chap. 3, sec. 10). This distinction contributed also to the presumed need for separating number from geometry, since to the Greeks numbers were discrete and geometry dealt with continuous magnitudes.

Because they feared infinite processes they missed the limit process. In approximating a circle by a polygon they were content to make the difference smaller than any given quantity, but something positive was always left over. Thus the process remained clear to the intuition; the limit process, on the other hand, would have involved the infinitely small.

3. The Problems Bequeathed by the Greeks

The limitations of Greek mathematical thought almost automatically imply the problems the Greeks left to later generations. The failure to accept the irrational as a number certainly left open the question of whether number could be assigned to incommensurable ratios so that these could be treated arithmetically. With the irrational number, algebra could also be extended. Instead of turning to geometry to solve quadratic and other equations that might have irrational roots, these problems could be treated in terms of number, and algebra could develop from the stage where the Egyptians and Babylonians or where Diophantus, who refused to consider irrationals, left it.

Even for whole numbers and ratios of whole numbers, the Greeks gave no logical foundation; they supplied only some rather vague definitions, which Euclid states in Books VII to IX of the *Elements*. The need for a logical foundation of the number system was aggravated by the Alexandrians' free use of numbers, including irrationals; in this respect they merely continued the empirical traditions of the Egyptians and Babylonians. Thus the Greeks bequeathed two sharply different, unequally developed branches of mathematics. On the one hand, there was the rigorous, deductive, systematic geometry and on the other, the heuristic, empirical arithmetic and its extension to algebra.

The failure to build a deductive algebra meant that rigorous mathematics was confined to geometry; indeed, this continued to be the case as late as the seventeenth and eighteenth centuries, when algebra and the calculus had already become extensive. Even then rigorous mathematics still meant geometry.

The limitation of Euclidean geometry to concepts constructible with straightedge and compass left mathematics with two tasks. The first was specific, to show that one could square the circle, trisect any angle, and double the cube with straightedge and compass. These three problems exerted an enormous fascination and even today intrigue people although, as we shall see, they were disposed of in the nineteenth century.

The second task was to broaden the criteria for existence. Constructibility as a means of proving existence turned out to be too restrictive of the concepts with which mathematics should (and later did) deal. Moreover, since some lengths are not constructible, the Euclidean line is incomplete; that is, strictly it does not contain those lengths that are not constructible. To be internally complete and more useful in the study of the physical world, mathematics had to free itself of a narrow technique for establishing existence.

As we saw, the attempt to avoid a direct affirmation about infinite parallel straight lines caused Euclid to phrase the parallel axiom in a rather complicated way. He realized that, so worded, this axiom lacked the self-evidency of the other nine axioms, and there is good reason to believe that he avoided using it until he had to. Many Greeks tried to find substitute axioms for the parallel axiom or to prove it on the basis of the other nine. Ptolemy wrote a tract on this subject; Proclus, in his commentary on Euclid, gives Ptolemy's attempt to prove the parallel postulate and also tries to prove it himself. Simplicius cites others who worked on the problem and says further that people "in ancient times" objected to the use of the parallel postulate.

Closely related to the problem of the parallel postulate is the problem of whether physical space is infinite. Euclid assumes in Postulate 2 that a straight-line segment can be extended as far as necessary; he uses this fact, but only to obtain a larger finite length—for example in Book I, Propositions 11, 16, and 20. For these theorems Heron gave new proofs that avoid extending the lines, in order to meet the objection of anyone who would deny that the space was available for the extension. Aristotle had considered the question of whether space is infinite and gave six nonmathematical arguments to prove that it is finite; he foresaw that this question would be troublesome.

Another major problem left to posterity was the calculation of areas bounded by curves and volumes bounded by surfaces. The Greeks, notably Eudoxus and Archimedes, had not only tackled such problems but had, as we have seen, made significant progress by using the method of exhaustion. But the method was deficient in at least two respects. First, each problem called for some ingenious scheme of approximating the area or volume in question; but human inventiveness simply could not have handled in this manner the areas and volumes that had to be calculated later. Second, the Greek result was usually the equivalence of the desired area or volume to the area or volume of some simpler figure whose size still was not known quantitatively. But it is quantitative knowledge that applications require.

4. *The Demise of the Greek Civilization*

Beginning approximately with the start of the Christian era, the vitality of Greek mathematical activity declined steadily. The only important contributions in the new era were those of Ptolemy and Diophantus. The great commentators Pappus and Proclus warrant attention, but they merely close the record. The decline of this civilization, which for five or six centuries made contributions far surpassing in extent and brilliance those of any other, calls for explanation.

Unfortunately, mathematicians are as subject to the forces of history as the lowliest peasant. One has only to familiarize himself with the most superficial facts about the political history of the Alexandrian Greeks of the Christian era to see that not only mathematics but every cultural activity was destined to suffer. As long as the Alexandrian Greek civilization was ruled by the dynasty of the Ptolemies, it flourished. The first disaster was the advent of the Romans, whose entire role in the history of mathematics was that of an agent of destruction.

Before discussing their impact on the Alexandrian Greek civilization, let us note a few facts about Roman mathematics and the nature of the Roman civilization. Roman mathematics hardly warrants mention. The period during which the Romans figured in history extends from about 750 B.C. to A.D. 476, roughly the same period during which the Greek civilization flourished. Moreover, as we shall see, from at least 200 B.C. onward, the Romans were in close contact with the Greeks. Yet in all of the eleven hundred years there was not one Roman mathematician; apart from a few details this fact in itself tells virtually the whole story of Roman mathematics.

The Romans did have a crude arithmetic and some approximate geometric formulas that were later supplemented by borrowings from the Alexandrian Greeks. Their symbols for the whole numbers are familiar to us. To calculate with whole numbers, they used various forms of the abacus. Calculations were also done with the fingers and with the aid of specially prepared tables.

Roman fractions were in base 12. Special symbols and words were used for $1/12, 2/12, \ldots, 11/12, 1/24, 1/36, 1/48, 1/96, \ldots$. The origin of the base 12 may be the relation of the lunar month to the year. The unit of weight, incidentally, was the *as*; one twelfth of this was the *uncia*, from which we get our ounce and inch.

The principal use of arithmetic and geometry by the Romans was in surveying, to fix city boundaries and plots of land for homes and temples. Surveyors could calculate most of the quantities they sought using only simple instruments and congruent triangles.

We do owe to the Romans an improvement in the calender. Up to the time of Julius Caesar (100–44 B.C.), the basic Roman year had 12 months, totalling 355 days. An intercalary month of 22 or 23 days was used every other year so that the average year consisted of 366 1/4 days. To improve this calendar, Caesar called in Sosigenes, an Alexandrian, who advised a year of 365 days and a leap year every 4 years. The Julian calendar was adopted in 45 B.C.

From about 50 B.C. on, the Romans wrote their own technical books; all of the basic material, however, was taken from Greek sources. The most famous of these technical works is Vitruvius' ten books on architecture, which date from 14 B.C. Here, too, the material is Greek. Peculiarly, Vitruvius says that the three greatest mathematical discoveries are the 3, 4, 5 right triangle, the irrationality of the diagonal of the unit square, and the Archimedean solution of the crown problem. He does give other facts involving mathematics, such as the proportions of the parts of the ideal human body, some harmonious arithmetical relationships, and arithmetic facts about the capacities of catapults.

Among the Romans the term "mathematics" came into disrepute because the astrologers were called *mathematicii*, and astrology was condemned by the Roman emperors. The emperor Diocletian (A.D. 245–316) distinguished between geometry and mathematics. The former was to be learned and applied in the public service; but the "art of mathematics"—that is, astrology—was damnable and forbidden in its entirety. The "code of mathematics and evil deeds," the Roman law forbidding astrology, was also applied in Europe during the Middle Ages. The Roman and Christian emperors nonetheless employed astrologers at their courts on the chance that there might be some truth in their prophecies. The distinction between the terms "mathematician" and "geometer" lasted until well past the Renaissance. Even in the seventeenth and eighteenth centuries, "geometer" meant what we mean by "mathematician."

The Romans were practical people and they boasted of their practicality. They undertook and completed vast engineering projects—viaducts, magnificent roads that survive even today, bridges, public buildings, and land surveys—but they refused to consider any ideas beyond the particular concrete applications they were making at the moment. The Roman attitude toward mathematics is stated by Cicero: "The Greeks held the geometer in the highest honor; accordingly, nothing made more brilliant progress among them than mathematics. But we have established as the limit of this art its usefulness in measuring and counting."

The Roman emperors did not support mathematics as did the Ptolemys of Egypt. Nor did the Romans understand pure science. Their failure to develop mathematics is striking, because they ruled a worldwide empire and because they did seek to solve practical problems. The lesson one can learn from the history of the Romans is that people who scorn the highly theoretical work of mathematicians and scientists and decry its usefulness are ignorant of the manner in which important practical developments have arisen.

Let us turn to the role the Romans played in the political and military history of Greece. After having secured control of central and northern Italy, they conquered the Greek cities in southern Italy and Sicily. (Archimedes, we recall, contributed to the defense of Syracuse when the Romans attacked the city and was killed by a Roman soldier.) The Romans conquered Greece proper in 146 B.C., and Mesopotamia in 64 B.C. By intervening in the internal strife in Egypt between Cleopatra, the last of the Ptolemy dynasty, and her brother, Caesar managed to secure a hold on the country. In 47 B.C., Caesar set fire to the Egyptian fleet riding at anchor in the harbor of Alexandria; the fire spread to the city and burned the library. Two and a half centuries of book-collecting and half a million manuscripts, which represented the flower of ancient culture, were wiped out. Fortunately, an overflow of books that could no longer be accommodated in the overcrowded library was by this time stored in the temple of Serapis and these were not burned. Also, Attalus III of Pergamum, who died in 133 B.C., had left his great collection of books to Rome. Mark Anthony gave this collection to Cleopatra and it was added to the books in the temple. The total collection was once more enormous.

The Romans returned at the death of Cleopatra in 31 B.C., and from that time on controlled Egypt. Their interest in extending their political power did not include spreading their culture. The subjugated areas became colonies, from which great wealth was extracted by expropriation and taxation. Since most of the Roman emperors were self-seekers, they ruined every country they controlled. When uprisings occurred, as they did, for example, in Alexandria, the Romans did not hesitate to starve and, when finally victorious, to kill off thousands of inhabitants.

The late history of the Roman Empire is also relevant. The Emperor Theodosius (ruled 379–95) divided the extensive empire between his two sons, Honorius, who was to rule Italy and western Europe, and Arcadius, who was to rule Greece, Egypt, and the Near East. The western part was conquered by the Goths in the fifth century A.D. and its subsequent history belongs to that of medieval Europe. The eastern part, which included Egypt (for a while), Greece, and what is now Turkey, preserved its independence until it was conquered by the Turks in 1453. Since the Eastern Roman Empire, known also as the Byzantine Empire, included Greece proper, Greek culture and works were to some extent preserved.

From the standpoint of the history of mathematics, the rise of Christianity had unfortunate consequences. Though the Christian leaders adopted many Greek and Oriental myths and customs with the intent of making Christianity more acceptable to converts, they opposed pagan learning and ridiculed mathematics, astronomy, and physical science; Christians were forbidden to contaminate themselves with Greek learning. Despite cruel persecution by the Romans, Christianity spread and became so powerful that the emperor Constantine (272–337) was obliged to adopt it as the official religion of the Roman Empire. The Christians were now able to effect even greater destruction of Greek culture. The emperor Theodosius proscribed the pagan religions and, in 392, ordered that the Greek temples be destroyed. Many of these were converted to churches, though often still adorned with Greek sculpture. Pagans were attacked and murdered throughout the empire. The fate of Hypatia, an Alexandrian mathematician of note and the daughter of Theon of Alexandria, symbolizes the end of the era. Because she refused to abandon the Greek religion, Christian fanatics seized her in the streets of Alexandria and tore her to pieces.

Greek books were burned by the thousands. In the year that Theodosius banned the pagan religions, the Christians destroyed the temple of Serapis, which still housed the only extensive collection of Greek works. It is estimated that 300,000 manuscripts were destroyed. Many other works written on parchment were expunged by the Christians so that they could use the parchment for their own writings. In 529 the Eastern Roman emperor Justinian closed all the Greek schools of philosophy, including Plato's Academy. Many Greek scholars left the country and some—for example, Simplicius—settled in Persia.

The final blow to Alexandria was the conquest of Egypt by the up-surging Moslems in A.D. 640. The remaining books were destroyed on the ground given by Omar, the Arab conqueror: "Either the books contain what is in the Koran, in which case we do not have to read them, or they contain the opposite of what is in the Koran, in which case we must not read them." And so for six months the baths of Alexandria were heated by burning rolls of parchment.

After the capture of Alexandria by the Mohammedans, the majority of the scholars migrated to Constantinople, which had become the capital of the Eastern Roman Empire. Though no activity along the lines of Greek thought could flourish in the unfriendly Christian atmosphere of Byzantium, this flux of scholars and their works to comparative safety increased the treasury of knowledge that was to reach Europe eight hundred years later.

It is perhaps pointless to contemplate what might have been. But one cannot help observe that the Alexandrian Greek civilization ended its active scientific life on the threshold of the modern age. It had the unusual combination of theoretical and practical interests that proved so fertile a thousand years later. Until the last few centuries of its existence, it enjoyed freedom of thought, which is also essential to a flourishing culture. And it tackled and made major advances in several fields that were to become all-important in the Renaissance: quantitative plane and solid geometry; trigonometry; algebra; calculus; and astronomy.

It has often been said that man proposes and God disposes. It is more accurate to say of the Greeks that God proposed them and man disposed of them. The Greek mathematicians were wiped out. But the fruits of their work did reach Europe in a way we have yet to relate.

Bibliography

Cajori, Florian: *A History of Mathematics*, Macmillan, 1919, pp. 63–68.

Gibbon, Edward: *The Decline and Fall of the Roman Empire* (many editions), Chaps. 20, 21, 28, 29, 32, 34.

Parsons, Edward Alexander: *The Alexandrian Library*, The Elsevier Press, 1952.

9

The Mathematics
of the Hindus and Arabs

> As the sun eclipses the stars by its brilliancy, so the man of
> knowledge will eclipse the fame of others in assemblies of the
> people if he proposes algebraic problems, and still more if he
> solves them.   BRAHMAGUPTA

1. *Early Hindu Mathematics*

The successors of the Greeks in the history of mathematics were the Hindus
of India. Though Hindu mathematics became significant only after it was
influenced by Greek achievements, there were earlier, indigenous developments that are worth noting.

The Hindu civilization dates back to at least 2000 B.C. but as far as we
know there was no mathematics prior to 800 B.C. During the Śulvasūtra
period, from 800 B.C. to A.D. 200, the Hindus did produce some primitive
mathematics. There were no separate mathematical documents but a few
facts can be gleaned from other writings and from coins and inscriptions.

From about the third century B.C., number symbols appear, which
varied considerably from one century to another. Typical are the Brahmi
symbols:

$$- = \equiv \; \Upsilon \; \Gamma \; \varphi \; \text{?} \; \text{5} \; \text{?} \; \alpha \; o \; \text{?} \; \text{?} \; \text{J} \; \dashv$$

$$1 \quad 2 \quad 3 \quad 4 \quad 5 \quad 6 \quad 7 \quad 8 \quad 9 \quad 10 \quad 20 \quad 30 \quad 40 \quad 50 \quad 60.$$

What is significant in this set is a separate, single symbol for each number
from 1 to 9. There was no zero and no positional notation as yet. The wisdom
of using separate symbols was undoubtedly not foreseen by this mathematically illiterate people; the practice may have arisen from using the first
letters of the words for these numbers.

Among the religious writings was a class called Śulvasūtras (rules of the
cord), containing instructions for the construction of altars. In one of the
Śulvasūtras of the fourth or fifth century B.C. an approximation to $\sqrt{2}$ is

given, but there is no indication that it is just an approximation. Almost nothing else is known about the arithmetic of the period.

The geometry of this ancient Hindu period is somewhat better known. The rules contained in the Śulvasūtras prescribed conditions for the shapes and sizes of altars. The three shapes most commonly used were square, circle, and semicircle; and no matter which of these shapes was used, the areas had to be the same. Hence the Hindus had to construct circles equal in area to the squares, or twice as large so that the semicircle could be used. Another shape used was an isosceles trapezoid; here it was permissible to use a similar shape. Hence additional geometrical problems were involved in constructing the similar figure.

In designing the permissible altars the Hindus got to know a few basic geometrical facts, such as the Pythagorean theorem, in the form: "The diagonal of an oblong [rectangle] produces by itself both the areas which the two sides of the oblong produce separately." In general, the geometry of this period consisted of a disconnected set of approximate, verbal rules for areas and volumes. Āpastamba (4th or 5th cent. B.C.) gave a construction for a circle equal in area to a square which, in effect, used the value of 3.09 for π; but he thought the construction was exact. In all of the geometry of this early period there were no proofs; the rules were empirical.

2. Hindu Arithmetic and Algebra of the Period A.D. 200–1200

The second period of Hindu mathematics, the high period, may be roughly dated from about A.D. 200 to 1200. During the first part of this period, the civilization at Alexandria definitely influenced the Hindus. Varāhamihira (c. 500), an astronomer, says, "The Greeks, though impure [anyone having a different faith is impure], must be honored, since they were trained in the sciences and therein excelled others. What, then, are we to say of a Brahman if he combines with his purity the height of science?" The geometry of the Hindus was certainly Greek, but they did have a special gift for arithmetic. As to algebra, they may have borrowed from Alexandria and possibly directly from Babylonia; but here, too, they went far on their own. India was also somewhat indebted to China.

The most important mathematicians of the second period are Āryabhata (b. 476), Brahmagupta (b. 598), Mahāvīra (9th cent.), and Bhāskara (b. 1114). Most of their work and that of Hindu mathematicians generally was motivated by astronomy and astrology. In fact, there were no separate mathematical texts; the mathematical material is presented in chapters of works on astronomy.

The Hindu methods of writing numbers up to A.D. 600 were numerous and even included using words or syllables for number symbols. By 600 they reverted to the older Brahmi symbols, though the precise form of these symbols changed throughout the period. Positional notation in base 10, which had been in limited use for about a hundred years, now became standard. Also, the zero, which the Alexandrian Greeks had earlier used only to denote the absence of a number, was treated as a complete number. Mahāvīra says that multiplication of a number by 0 gives 0 and that subtracting 0 does not diminish a number. He also says, however, that a number divided by 0 remains unchanged. Bhāskara, in talking about a fraction whose denominator is 0, says that this fraction remains the same though much be added and subtracted, just as no change takes place in the immutable Deity when worlds are created and destroyed. A number divided by zero, he adds, is termed an infinite quantity.

For fractions in astronomy, the Hindus used sexagesimal positional notation. For other purposes they used a ratio of integers but without the bar, thus: $\frac{3}{4}$.

The arithmetic operations of the Hindus were much like ours. For example, Mahāvīra gives our rule for division by a fraction: Invert and multiply.

The Hindus introduced negative numbers to represent debts; in such situations, positive numbers represented assets. The first known use is by Brahmagupta about 628; he also states the rules for the four operations with negative numbers. Bhāskara points out that the square root of a positive number is twofold, positive and negative. He brings up the matter of the square root of a negative number but says that there is no square root because a negative number is not a square. No definitions, axioms, or theorems are given.

The Hindus did not unreservedly accept negative numbers. Even Bhāskara, while giving 50 and -5 as two solutions of a problem, says, "The second value is in this case not to be taken, for it is inadequate; people do not approve of negative solutions." However, negative numbers gained acceptance slowly.

The Hindus took another great step in arithmetic by facing up to the problem of irrational numbers; that is, they started to operate with these numbers by correct procedures, which, though not proven generally by them, at least permitted useful conclusions to be drawn. For example, Bhāskara says, "Term the sum of two irrationals the greater surd; and twice their product the lesser one. The sum and difference of them reckoned like integers are so." He then shows how to add them, as follows: Given the irrationals $\sqrt{3}$ and $\sqrt{12}$,

$$\sqrt{3} + \sqrt{12} = \sqrt{(3 + 12) + 2\sqrt{3 \cdot 12}} = \sqrt{27} = 3\sqrt{3}.$$

The general principle, in our notation, is

$$(1) \qquad \sqrt{a} + \sqrt{b} = \sqrt{(a + b) + 2\sqrt{ab}}.$$

We should note the phrase "reckoned like integers" in the above quotation. Irrationals were treated as though they possessed the same properties as integers. Thus if we had integers c and d we would certainly write

$$(2) \qquad c + d = \sqrt{(c + d)^2} = \sqrt{c^2 + d^2 + 2cd}.$$

Now if $c = \sqrt{a}$ and $d = \sqrt{b}$, then (2) is exactly (1).

Bhāskara also gives the following rule for the sum of two irrationals: "The root of the quotient of the greater irrational divided by the lesser one being increased by one; the sum being squared and multiplied by the smaller irrational quantity is the sum of the two surd roots." This means, for example,

$$\sqrt{3} + \sqrt{12} = \sqrt{\left(\sqrt{\frac{12}{3}} + 1\right)^2 \cdot 3},$$

which yields $3\sqrt{3}$. He also gives rules for the multiplication, division, and square root of irrational expressions.

The Hindus were less sophisticated than the Greeks in that they failed to see the logical difficulties involved in the concept of irrational numbers. Their interest in calculation caused them to overlook philosophic distinctions, or distinctions based on principles that in Greek thought were fundamental. But in blithely applying to irrationals procedures like those used for rationals, they helped mathematics progress. Moreover, their entire arithmetic was completely independent of geometry.

The Hindus also made some progress in algebra. They used abbreviations of words and a few symbols to describe operations. As in Diophantus, there was no symbol for addition; a dot over the subtrahend indicated subtraction; other operations were called for by key words or abbreviations; thus *ka* from the word *karana* called for the square root of what followed. For the unknowns, when more than one was involved, they had words that denoted colors. The first one was called the unknown and the remaining ones black, blue, yellow, and so forth. The initial letter of each word was also used as a symbol. This symbolism, though not extensive, was enough to classify Hindu algebra as almost symbolic and certainly more so than Diophantus' syncopated algebra. Problems and solutions were written in this quasi-symbolic style. Only the steps were given; no reasons or proofs accompanied them.

The Hindus recognized that quadratic equations have two roots and included negative roots as well as irrational roots. The three types of quadratics $ax^2 + bx = c$, $ax^2 = bx + c$, $ax^2 + c = bx$, a, b, c positive, separately treated by Diophantus, were treated as one case $px^2 + qx + r = 0$, because the Hindus allowed some coefficients to be negative. They used the method of completing the square, which of course was not new with them. Since they did not recognize square roots for negative numbers, they could not solve all quadratics. Mahāvīra also solves $x/4 + 2\sqrt{x} + 15 = x$, which arises from a verbal problem.

In indeterminate equations the Hindus advanced beyond Diophantus. These equations arose in problems of astronomy; the solutions showed when certain constellations would appear in the heavens. The Hindus sought all integral solutions, whereas Diophantus sought one rational solution. The method of obtaining integral solutions of $ax \pm by = c$, where a, b, and c are positive integers, was introduced by Āryabhata and improved by his successors. It is the same as the modern one. Let us consider $ax + by = c$. If a and b have a common factor m that does not divide c, then no integral solutions are possible because the left side is divisible by m whereas the right side is not. If a, b, and c have a common factor it is removed, and, in the light of the preceding remark, one need consider only the case where a and b are relatively prime. Now Euclid's algorithm for finding the greatest common divisor of two integers a and b, with $a > b$, calls first for dividing b into a so that $a = a_1 b + r$, where a_1 is the quotient and r is the remainder. Hence $a/b = a_1 + r/b$. This can be expressed as

(3) $a/b = a_1 + 1/(b/r)$.

The second step in Euclid's algorithm is to divide r into b. Then $b = a_2 r + r_1$ or $b/r = a_2 + r_1/r$. If we insert this value of b/r in (3), we can write

(4)
$$\frac{a}{b} = a_1 + \cfrac{1}{a_2 + \cfrac{1}{r/r_1}}.$$

The continuation of Euclid's algorithm leads to what is called a continued fraction

$$\frac{a}{b} = a_1 + \cfrac{1}{a_2 + \cfrac{1}{a_3 + \cdots}},$$

which is also written as

$$\frac{a}{b} = a_1 + \frac{1}{a_2 +} \ \frac{1}{a_3 +} \cdots.$$

The process applies also when $a < b$. In this case a_1 is zero, and then the process is continued as before. For the case of a and b integral, the continued fraction terminates.

The fractions obtained by stopping with the first, second, third, and generally nth quotient are called the first, second, third, and nth convergent, respectively. Since in the case of a and b integral the continued fraction terminates, there is a convergent that just precedes the exact expression for a/b. If p/q is the value of this convergent, then one can show that

$$aq - bp = \pm 1.$$

Let us consider the positive value. We can now return to our original indeterminate equation, and since $aq - bp = 1$ we can write

$$ax + by = c(aq - bp)$$

and by rearranging terms obtain

$$\frac{cq - x}{b} = \frac{y + cp}{a}.$$

If we let t represent each of these fractions we have

(5) $x = cq - bt$ and $y = at - cp$.

We may now assign integral values to t, and since all the other quantities are integers, we thereby obtain integral values of x and y. The minor modifications to take care of the cases where $aq - bp = -1$, or when the original equation is $ax - by = c$, are readily devised. Brahmagupta gave solution (5), though not of course in terms of general letters a, b, p, and q.

The Hindus also worked with indeterminate quadratic equations. They solved the type

$$y^2 = ax^2 + 1, \qquad a \text{ not a perfect square,}$$

and recognized that this type was fundamental in treating

$$cy^2 = ax^2 + b.$$

The methods involved are too specialized to warrant consideration here.

It is noteworthy that they found pleasure in many mathematical problems and stated them in fanciful or verse form, or in some historical context, to please and attract people. The original reason for doing so may have been to aid the memory, because the old Brahman practice was to trust to memory and avoid writing things down.

Algebra was applied to the usual commercial problems: the calculation of interest, discount, division of the profits of a partnership, and the allocation of shares in an estate; but astronomy was the main application.

3. *Hindu Geometry and Trigonometry of the Period* A.D. 200–1200

Geometry during this period showed no notable advances; it consisted of formulas (correct and incorrect) for areas and volumes. Many of these, such as Heron's formula for the area of a triangle and Ptolemy's theorem, came from the Alexandrian Greeks. Sometimes the Hindus were aware that a formula was only approximately correct and sometimes they were not. Their values of π were generally inaccurate; $\sqrt{10}$ was commonly used, though the better value of 3.1416 appears at times. For the area of any quadrilateral they gave the formula $\sqrt{(s - a)(s - b)(s - c)(s - d)}$, where s is half the perimeter and a, b, c, and d are the sides, a formula which is correct only for quadrilaterals inscribed in circles. They offered no geometric proofs; on the whole they cared little for geometry.

In trigonometry the Hindus made a few minor advances. Ptolemy had used the chords of arcs, calculated on the basis of 120 units in the diameter of a circle. Varāhamihira used 120 units for the radius. Hence Ptolemy's table of chords became for him a table of half chords, but still associated with the full arc. Āryabhata then made two changes. First he associated the half chord with half of the arc of the full chord; this Hindu notion of sine was used by all later Hindu mathematicians. Secondly, he introduced a radius of 3438 units. This number comes from assigning $360 \cdot 60$ units (the number of minutes) to the circumference of a circle and using $C = 2\pi r$, with π approximated by 3.14. Thus in Āryabhata's scheme the sine of an arc of 30°, that is, the length of the half chord corresponding to an arc of 30°, was 1719. While the Hindus used the equivalent of our cosine, they more often used the sine of the complementary arc. They also used the notion of versed sine, or $1 - \text{cosine}$.

Since the radius of a circle now contained 3438 units, Ptolemy's values for the chords were no longer suitable, and the Hindus recomputed a table of half chords, starting from the fact that the half chord corresponding to an arc of 90° is 3438 and the half chord corresponding to an arc of 30° is 1719. Then, by using trigonometric identities such as Ptolemy had established, they were able to calculate the half chords of arcs at intervals of 3°45′. This angle resulted from dividing each quadrant of 90° into 24 parts. It is noteworthy that they used the identities in algebraic form, unlike the geometrical arguments of Ptolemy, and made arithmetic calculations on the basis of the algebraic relations. Their practice was, in principle, like ours.

The motivation for the trigonometry was astronomy, of which practically all of Hindu trigonometry was a by-product. Standard astronomical works included the *Sûrya Siddhânta* (System of the Sun, 4th cent.) and the *Āryabhatiya* of Āryabhata (6th cent.). The major work was the *Siddhānta Siromani* (Diadem of an Astronomical System) written by Bhāskara in 1150. Two chapters of this work are titled *Lilāvatī* (The Beautiful) and *Vija-ganita* (Root Extraction); these are devoted to arithmetic and algebra.

Though astronomy was a primary interest in the period after A.D. 200, the Hindus made no significant progress. They took over a minor Hellenistic activity in arithmetical astronomy (of Babylonian origin), which predicts planetary and lunar positions by extrapolation from observational data. Even the Hindu words for center, minute, and other terms were just the Greek words transliterated. The Hindus were only slightly concerned with the geometrical theory of deferent and epicycle, though they did teach the sphericity of the earth.

About the year 1200 scientific activity in India declined and progress in mathematics ceased. After the British conquered India in the eighteenth century, a few Indian scholars went to England to study and on their return did initiate some research. However, this modern activity is part of European mathematics.

As our survey indicates, the Hindus were interested in and contributed to the arithmetical and computational activities of mathematics rather than to the deductive patterns. Their name for mathematics was *ganita*, which means "the science of calculation." There is much good procedure and technical facility, but no evidence that they considered proof at all. They had rules, but apparently no logical scruples. Moreover, no general methods or new viewpoints were arrived at in any area of mathematics.

It is fairly certain that the Hindus did not appreciate the significance of their own contributions. The few good ideas they had, such as separate symbols for the numbers from 1 to 9, the conversion to base 10, and negative numbers, were introduced casually with no realization that they were valuable innovations. They were not sensitive to mathematical values. Along with the ideas they themselves advanced, they accepted and incorporated the crudest ideas of the Egyptians and Babylonians. The Persian historian al-Bîrûnî (973–1048) says of the Hindus, "I can only compare their mathematical and astronomical literature . . . to a mixture of pearl shells and sour dates, or of pearl and dung, or of costly crystals and common pebbles. Both kinds of things are equal in their eyes, since they cannot raise themselves to the methods of a strictly scientific deduction."

4. *The Arabs*

Thus far the Arab role in the history of mathematics has been to deliver the final blow to the Alexandrian civilization. Before starting their conquests, they had been nomads occupying the region of modern Arabia. They were stirred to activity and unity by Mohammed, and less than a century after his death in 632, had conquered lands from India to Spain, including North Africa and southern Italy. In 755 the Arab empire split into two independent kingdoms, the eastern part having its capital at Bagdad and the western one at Cordova in Spain.

Their conquests completed, the former nomads settled down to build a civilization and a culture. Rather quickly the Arabs became interested in the arts and sciences. Both centers attracted scientists and supported their work, though Bagdad proved to be the greater; an academy, a library, and an astronomical observatory were established there.

The cultural resources available to the Arabs were considerable. They invited Hindu scientists to settle in Bagdad. When Justinian closed Plato's Academy in A.D. 529, many of its Greek scholars went to Persia, and the Greek learning that flourished there became, a century later, part of the Arab world. The Arabs also established contacts with the Greeks of the independent Byzantine Empire; in fact the Arab caliphs bought Greek manuscripts from the Byzantines. Egypt, the center of Greek learning in the Alexandrian period, had been conquered by the Arabs, so the learning that survived there contributed to the activity in the Arab empire. The Syrian schools of Antioch, Emesa, and Damascus and the school of Nestorian Christians at Edessa, which had become the chief repositories in the Near East of Greek works after the destruction of Alexandria in 640, and Christian monasteries in the Near East, which also possessed these works, all came under Arab rule. Thus the Arabs had control over, or access to, the men and culture of the Byzantine Empire, Egypt, Syria, Persia, and the lands farther east, including India.

One speaks of Arabic mathematics, but it was Arabic in language primarily. Most of the scholars were Greeks, Christians, Persians, and Jews. However, it is to the credit of the Arabs that after the period of conquest, which was marked by religious fanaticism, they were liberal to other peoples and sects and the infidels were able to function freely.

Fundamentally, what the Arabs possessed was Greek knowledge obtained directly from Greek manuscripts or from Syrian and Hebrew versions. All the major works became accessible to them. They obtained a copy of Euclid's *Elements* from the Byzantines about 800 and translated it into Arabic. Ptolemy's *Mathematical Syntaxis* was translated into Arabic in 827 and to the Arabs became a preeminent, almost divine book; it became known as the *Almagest*, meaning the greatest work. They also translated Ptolemy's *Tetrabiblos* and this work on astrology was popular among them. In the course of time the works of Aristotle, Apollonius, Archimedes, Heron, Diophantus, and of the Hindus became accessible in Arabic. The Arabs then improved the translations and made commentaries. It is these translations, some still extant, that became available to Europe later, the Greek originals having been lost. Until 1300 the Arab civilization was dynamic and its learning became widespread.

5. *Arabic Arithmetic and Algebra*

When the Arabs were still nomads, they had words for numbers but no symbols. They took over and improved the Hindu number symbols and the idea of positional notation. They used these number symbols for whole numbers and common fractions (adding a bar to the Hindu scheme) in their mathematical texts and Arabic alphabetic numerals, on the Greek pattern, for astronomical texts. For astronomy they used the sexagesimal fractions in imitation of Ptolemy.

Like the Hindus, the Arabs worked freely with irrationals. In fact, Omar Khayyam (1048?–1122) and Nasîr-Eddin (1201–74) clearly state that every ratio of magnitudes, whether commensurable or incommensurable, may be called a number, a statement Newton felt obliged to reaffirm in his *Universal Arithmetic* of 1707. The Arabs took over the operations with irrational numbers that the Hindus had introduced, and such transformations as $\sqrt{a^2 b} = a\sqrt{b}$ and $\sqrt{ab} = \sqrt{a}\sqrt{b}$ became common.

In arithmetic the Arabs took one step backward. Though they were familiar with negative numbers and the rules for operating with them through the work of the Hindus, they rejected negative numbers.

To algebra the Arabs contributed first of all the name. The word "algebra" comes from a book written in 830 by the astronomer Mohammed ibn Musa al-Khowârizmî (c. 825), titled *Al-jabr w'al muqâbala*. The word *al-jabr* meant "restoring," in this context, restoring the balance in an equation by placing on one side of an equation a term that has been removed from the other; thus if -7 is removed from $x^2 - 7 = 3$, the balance is restored by writing $x^2 = 7 + 3$. *Al' muqâbala* meant "simplification," as by combining $3x$ and $4x$ into $7x$ or by subtracting equal terms from both sides of an equation. *Al-jabr* also came to mean "bonesetter," that is, a restorer of broken bones. When the Moors brought the word into Spain, it became *algebrista* and meant "bonesetter." At one time it was not uncommon in Spain to see a sign "Algebrista y Sangrador" (bonesetter and blood-letter) over the entrance to a barbershop because barbers at that time and even centuries later administered the simpler medical treatments. In sixteenth-century Italy, algebra meant the art of bonesetting. When al-Khowârizmî's book was first translated into Latin in the twelfth century, the title was rendered as *Ludus algebrae et almucgrabalaeque*, though other titles were also used. The name of the subject was finally shortened to algebra.

The algebra of al-Khowârizmî is founded on Brahmagupta's work but also shows Babylonian and Greek influences. Al-Khowârizmî performs some operations just as Diophantus does. For example, in equations involving several unknowns, he reduces to one unknown and then solves. Diophantus refers to his unknown s as the side when s^2 also occurs; so does al-Khowârizmî. Al-Khowârizmî calls the square of the unknown "power," which is Diophantus' word. He also uses, as did Diophantus, special names for the powers of the unknown. The unknown he refers to as the "thing" or the "root" (of a plant), whence our term root. Al-Karkhî of Bagdad (died c. 1029), who wrote a superior Arabic algebra text in the early part of the eleventh century, certainly follows the Greeks and especially Diophantus. However, the Arabs did not use symbolism. Their algebra is entirely rhetorical and, in this respect, a step backward, as compared with the Hindus and even Diophantus.

In his algebra al-Khowârizmî gives the product of $(x \pm a)$ and $(y \pm b)$. He shows how to add and subtract terms from an expression of the form $ax^2 + bx + c$. He solves linear and quadratic equations, but keeps the six separate forms, such as $ax^2 = bx$, $ax^2 = c$, $ax^2 + c = bx$, $ax^2 + bx = c$, and

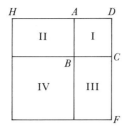

Figure 9.1

$ax^2 = bx + c$, so that a, b, and c are always positive. This avoids negative numbers standing alone and the subtraction of quantities that may be larger than the minuend. In this practice of using the separate forms, al-Khowârizmî follows Diophantus. Al-Khowârizmî recognizes that there can be two roots of quadratics, but gives only the real positive roots, which can be irrational. Some writers give both positive and negative roots.

One example of a quadratic treated by al-Khowârizmî reads as follows: "A square and ten of its roots are equal to nine and thirty dirhems, that is you add ten roots to one square, the sum is equal to nine and thirty." He gives the solution thus: "Take half the number of roots, that is, in this case five, then multiply this by itself and the result is five and twenty. Add this to the nine and thirty, which gives sixty-four; take the square root, or eight, and subtract from it half the number of roots, namely five, and there remains three. This is the root." The solution is exactly what the process of completing the square calls for.

Though the Arabs gave algebraic solutions of quadratic equations, they explained or justified their processes geometrically. Undoubtedly they were influenced by the Greek reliance upon geometrical algebra; while they arithmetized the processes, they must have believed that the proof had to be made geometrically. Thus to solve the equation, which is $x^2 + 10x = 39$, al-Khowârizmî gives the following geometrical method. Let AB (Fig. 9.1) represent the value of the unknown x. Construct the square $ABCD$. Produce DA to H and DC to F so that $AH = CF = 5$, which is one-half of the coefficient of x. Complete the square on DH and DF. Then the areas I, II, and III are x^2, $5x$, and $5x$, respectively. The sum of these is the left side of the equation. To both sides we now add area IV, which is 25. Hence the entire square is $39 + 25$ or 64 and its side must be 8. Then AB or AD is $8 - 5$ or 3. This is the value of x. The geometric argument rests on Proposition 4 of Book II of the *Elements*.

The Arabs solved some cubics algebraically and gave a geometrical explanation in the manner just illustrated for quadratics. This was done, for example, by Tâbit ibn Qorra (836–901), a pagan of Bagdad, who was also a physician, philosopher, and astronomer, and by the Egyptian al-Hasan ibn al-Haitham, known generally as Alhazen (*c.* 965–1039). As for the general

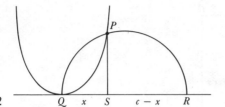

Figure 9.2

cubic, Omar Khayyam believed this could be solved only geometrically, by using conic sections. We shall illustrate the method he used in his *Algebra* (*c*. 1079) to solve some types of cubics by considering one of the simpler cases he treats, namely, $x^3 + Bx = C$, where B and C are positive.

Khayyam writes the equation as $x^3 + b^2x = b^2c$ wherein $b^2 = B$ and $b^2c = C$. He then constructs a parabola (Fig. 9.2), whose latus rectum is b. This quantity does indeed fix the parabola, and though the curve itself cannot be constructed with straightedge and compass, as many points as desired can be so constructed. He constructs next the semicircle on the diameter QR which has length c. Then the intersection P of the parabola and semicircle determines the perpendicular PS, and QS is the solution of the cubic equation.

Khayyam's proof is purely synthetic. From the geometric property of the parabola as given in Apollonius (or as we can see from the equation $x^2 = by$)

(6) $$x^2 = b \cdot PS$$

or

(7) $$\frac{b}{x} = \frac{x}{PS}.$$

Now consider the right triangle QPR. The altitude PS is the mean proportional between QS and SR. Hence

(8) $$\frac{x}{PS} = \frac{PS}{c - x}.$$

From (7) and (8) we have

(9) $$\frac{b}{x} = \frac{PS}{c - x}.$$

But from (7)

$$PS = \frac{x^2}{b}.$$

If we substitute this value of PS in (9) we see that x satisfies the equation $x^3 + b^2x = b^2c$.

Khayyam also solved the type $x^3 + ax^2 = c^3$, whose roots are determined by the intersection of a hyperbola and a parabola, and the type $x^3 \pm ax^2 + b^2x = b^2c$, whose roots are determined by the intersection of an ellipse and a hyperbola. He also solved one quartic, $(100 - x^2)(10 - x)^2 = 8100$, the roots of which are determined by the intersection of a hyperbola and a circle. He gives only positive roots.

The solution of cubic equations by using intersecting conics is the greatest Arab step in algebra. The mathematics is of precisely the same sort as the Greek geometrical algebra, though it uses conic sections. The goal should be an arithmetic answer, but the Arabs could obtain this only by measuring the final length representing x. In this work the influence of Greek geometry is clear.

The Arabs also solved indeterminate equations of the second and third degree. A couple of writers stated and tried to prove that $x^3 + y^3 = z^3$ cannot be solved integrally. They also gave the sums of the first, second, third, and fourth powers of the first n natural numbers.

6. Arabic Geometry and Trigonometry

Arabic geometry was influenced mainly by Euclid, Archimedes, and Heron. The Arabs did make critical commentaries of Euclid's *Elements,* which is surprising because it shows appreciation of rigor despite their usual indifference to it in algebra. These commentaries included work on the parallel axiom, which we shall consider later (Chap. 36). They are valuable less for what they have to offer in the way of new results or new proofs than for the information they supply about the Greek manuscripts the Arabs had at their disposal that have since been lost. One new problem, which became popular in Renaissance Europe, was explored by the Persian Abû'l-Wefâ or Albuzjani (940–98): constructions with a straightedge and a fixed circle (i.e., fixed compass opening).

The Arabs made a little progress in trigonometry. Theirs, like the Hindus', is arithmetical, rather than geometric as in Hipparchus and Ptolemy. Thus, to calculate some cosine values from sine values they used an identity such as $\sin^2 A + \cos^2 A = 1$ and algebraic steps. Like the Hindus, they used sines of arcs rather than chords of double the arcs, though (as in the Hindu work) the number of units in the sine or half chord depends upon the number of units taken for the radius. Tâbit ibn Qorra and the astronomer al-Battânî (*c.* 858–929) introduced this use of sines among the Arabs.

Arab astronomers introduced what we call the tangent and cotangent ratios, but as lines containing a number of units, just as the sine of an arc was a length containing a number of units. These two ratios are to be found in the work of al-Battânî. Abû'l-Wefâ introduced the secant and cosecant as lengths in a work on astronomy. He also computed tables of sines and tangents for every 10 minutes of angles. Al-Bîrûnî gave the law of sines for plane triangles and a proof.

The systematization of plane and spherical trigonometry in a work independent of astronomy was achieved by Nasîr-Eddin in his *Treatise on the Quadrilateral*. This work contains six fundamental formulas for the solution of right spherical triangles and shows how to solve more general triangles by what we now call the polar triangle. Unfortunately the Europeans did not know Nasîr's work until about 1450; until then trigonometry remained, as it had been from its inception, an appendage of astronomy, in the texts as well as in application.

The Arab scientific effort, though not original, was extensive; however, we cannot do more here than note the continuation of lines inaugurated by the Greeks. Unlike the Hindus, the Arabs did take over Ptolemy's astronomy. Astronomy was emphasized so that the hours for prayers could be accurately known and so that Arabs throughout the vast empire could face Mecca during prayers. Astronomical tables were enlarged; instruments were improved; and observatories were built and used. As in India, practically all mathematicians were primarily astronomers. Astrology also played a big role in stimulating work in astronomy and, therefore, mathematics.

Another science pursued by the Arabs was optics. Alhazen, who was a physicist as well as a mathematician, wrote the great treatise *Kitab al-manazer* or *Treasury of Optics*, which exercised an enormous influence. In it he stated the full law of reflection, including the fact that the incident ray, the reflected ray, and the normal to the reflecting surface all lie in one plane. But, like Ptolemy, he did not succeed in finding the law for the angle of refraction, despite much effort and experimentation. He discussed spherical and parabolic mirrors, lenses, the camera obscura, and vision. Optics was a favorite subject with the Arabs because it lent itself to occult and mystic thoughts. However, they made no important original contributions.

The Arab uses of mathematics were of the sort we have encountered before. Astronomy, astrology, optics, and medicine (through astrology) required it, though some of the algebra, as one of the Arab mathematicians put it, was "most needed in problems of distribution, inheritance, partnership, land measurement. . . ." Mathematics was studied by the Arabs mainly to further the few sciences they pursued, not for its own sake. Nor did they study science for its own sake. They were not interested in the Greek goal of understanding the mathematical design of nature or in comprehending God's ways, as were the medieval Europeans. The Arab objective, new in the history of science, was power over nature. They thought they could achieve that power through alchemy, magic, and astrology, which were an integral part of their scientific effort. This objective was taken up later by more critical minds, who could distinguish science from pseudoscience and were more profound in their approach.

The Arabs made no significant advance in mathematics. What they did was absorb Greek and Hindu mathematics, preserve it, and, ultimately, through events we have yet to look into, transmit it to Europe. Arab activity reached its peak about the year 1000. Between 1100 and 1300 the Christian attacks in the Crusades weakened the eastern Arabs. Subsequently, their territory was overrun and conquered by the Mongols; after 1258 the caliphate at Bagdad ceased to exist. Further destruction by the Tartars under Tamerlane just about wiped out this Arab civilization, though a bit of mathematical work was done there after the Tartar invasion. In Spain the Arabs were constantly attacked and finally conquered in 1492 by the Christians; this ended the mathematical and scientific activity in that region.

7. *Mathematics* circa 1300

Though the mathematical work of the Hindus and Arabs was not brilliant, it did bring about some changes in the content and character of mathematics that were material for the future of the subject. Positional notation in base 10 (using special number symbols for the quantities from 1 to 9 and zero as a number), the introduction of negative numbers, and the free use of irrationals as numbers not only extended arithmetic vastly but paved the way for a more significant algebra, an algebra in which letters and operations could apply to a far broader class of numbers.

Both peoples worked with equations, determinate and indeterminate, on an arithmetic rather than a geometric basis. Though algebra, as initiated by the Egyptians and Babylonians, was arithmetically grounded, the Greeks had subverted it by requiring a geometric basis. Hence the Hindu and Arabic work not only restored algebra to its proper foundation but even advanced the art in several ways. The Hindus used more symbolism and made progress in indeterminate equations, while the Arabs ventured further in the subject of third degree equations, though Khayyam's work still relied upon geometry.

Euclidean geometry was not furthered but trigonometry was. The introduction of the sine or half chord did prove to be a technical advantage. The arithmetical or algebraic technique for handling identities and for calculation in trigonometry was an expeditious step; and the segregation of trigonometric knowledge from astronomy revealed a more broadly applicable science.

Two developments occurred that were significant for the forthcoming recognition that algebra is coextensive with geometry. The acceptance of irrational numbers made it possible to assign numerical values to all line segments and two- and three-dimensional figures, that is, to express lengths, areas, and volumes by numbers. Moreover, the Arab practice of solving equations algebraically while justifying the processes by means of a geometric representation exhibited the parallelism of the two subjects. The fuller development of that parallelism was to lead to analytic geometry.

Perhaps most interesting is the Hindus' and Arabs' self-contradictory concept of mathematics. Both worked freely in arithmetic and algebra and yet did not concern themselves at all with the notion of proof. That the Egyptians and Babylonians were content to accept their few arithmetic and geometric rules on an empirical basis is not surprising; this is a natural basis for almost all human knowledge. But the Hindus and Arabs were aware of the totally new concept of mathematical proof promulgated by the Greeks. The Hindu behavior can be somewhat rationalized; though they did indeed possess some knowledge of the classic Greek works, they paid them little attention and followed primarily the Alexandrian Greek treatment of arithmetic and algebra. Even this preference for one kind of mathematics over another raises a question. But the Arabs were fully aware of Greek geometry and even made critical studies of Euclid and other Greek authors. Moreover, conditions for the pursuit of pure science were favorable over a period of several centuries, so that the pressure to produce practically useful results need not have caused mathematicians to sacrifice proof for immediate utility. How could both peoples have treated the two areas of mathematics so differently from the Greeks?

There are numerous possible answers. Both civilizations were on the whole uncritical, despite the Arabic commentaries on Euclid. Hence they may have been content to take mathematics as they found it; that is, geometry was to be deductive but arithmetic and algebra could be empirical or heuristic. A second possibility is that both of these peoples—more likely, the Arabs—recognized the widely different standards for geometry as opposed to arithmetic and algebra but did not see how to supply a logical foundation for arithmetic. One fact that seems to support such an explanation is that the Arabs did at least explain their solution of quadratics by giving the geometric basis.

There are other possible explanations. Both the Hindus and Arabs favored arithmetic, algebra, and the algebraic formulation of and operation with trigonometric relationships. This predisposition may bespeak a different mentality or it may reflect a response to the needs of the civilizations. Both of these civilizations were practically oriented and, as we have had occasion to note in connection with the Alexandrian Greeks, practical needs do call for quantitative results, which are supplied by arithmetic and algebra. One bit of evidence favoring the thesis that different mentalities were involved is the reaction of the Europeans to the very same mathematical heritage which the Hindus and Arabs received. As we shall see, the Europeans were far more troubled by the disparate states of arithmetic and geometry.

In the absence of any exhaustive and definitive study we must take the position that the Hindus and Arabs were conscious of the precarious basis for arithmetic and algebra but had the audacity, reinforced by practical needs, to develop these branches. Though they undoubtedly did not appreciate what they were accomplishing, they took the only course mathematical innovation can pursue. New ideas can come about only by the free and bold pursuit of heuristic and intuitive insights. The logical justification and corrective measures, should the latter be needed, can be brought into play only when there is something to logicize. Hindu and Arab venturesomeness brought arithmetic and algebra to the fore once again and placed it almost on a par with geometry.

Two independent traditions or concepts of mathematics had now become established: on the one hand, the logical deductive body of knowledge that the Greeks established, which served the larger purpose of understanding nature; and on the other, the empirically grounded, practically oriented mathematics founded by the Egyptians and Babylonians, resuscitated by some of the Alexandrian Greeks, and extended by the Hindus and Arabs. The one favored geometry and the other arithmetic and algebra. Both traditions and both goals were to continue to operate.

Bibliography

Ball, W. W. R.: *A Short Account of the History of Mathematics*, Dover (reprint), 1960, Chap. 9.

Berry, Arthur: *A Short History of Astronomy*, Dover (reprint), 1961, pp. 76–83.

Boyer, Carl B.: *A History of Mathematics*, John Wiley and Sons, 1968, Chaps. 12–13.

Cajori, Florian: *A History of Mathematics*, Macmillan, 1919, pp. 83–112.

Cantor, Moritz: *Vorlesungen über Geschichte der Mathematik*, 2nd ed., B. G. Teubner, 1894, Johnson Reprint Corp., 1965, Vol. 1, Chaps. 28–30, 32–37.

Coolidge, Julian L.: *The Mathematics of Great Amateurs*, Dover (reprint), 1963, Chap. 2.

Datta, B., and A. N. Singh: *History of Hindu Mathematics*, 2 vols., Asia Publishing House (reprint), 1962.

Dreyer, J. L. E.: *A History of Astronomy from Thales to Kepler*, Dover (reprint), 1953, Chap. 11.

Karpinski, L. C.: *Robert of Chester's Latin Translation of the Algebra of al-Khowarizmi*, Macmillan, 1915. English version also.

Kasir, D. S.: "The Algebra of Omar Khayyam," Columbia University Teachers College thesis, 1931.

O'Leary, De Lacy: *How Greek Science Passed to the Arabs*, Routledge and Kegan Paul, 1949.

Pannekoek, A.: *A History of Astronomy*, John Wiley and Sons, 1961, Chap. 15.

Scott, J. F.: *A History of Mathematics*, Taylor and Francis, 1958, Chap. 5.

Smith, David Eugene: *History of Mathematics*, Dover (reprint), 1958, Vol. 1, pp. 138–47, 152–92, 283–90.

Struik, D. J.: "Omar Khayyam, Mathematician," *The Mathematics Teacher*, 51, 1958, 280–85.

10

The Medieval Period in Europe

> In most sciences one generation tears down what another has
> built and what one has established another undoes. In mathe-
> matics alone each generation adds a new story to the old
> structure. HERMANN HANKEL

1. *The Beginnings of a European Civilization*

Western and central Europe entered into the development of mathematics
as the Arab civilization began to decline. However, to familiarize ourselves
with the state of affairs in medieval Europe, to learn how the European
civilization got its start, and to understand the directions it took, we must
go back, briefly at least, to its beginnings.

During the ages when the Babylonians, Egyptians, Greeks, and Romans
flourished, the area now called Europe (except for Italy and Greece)
possessed a primitive civilization. The Germanic tribes who lived there
mastered neither writing nor learning. The Roman historian Tacitus (1st
cent. A.D.) describes these tribes of about the time of Christ as honest,
hospitable, hard-drinking, hating peace, and proud of the loyalty of their
wives. Cattle-raising, hunting, and the cultivation of grain were the main
occupations. Beginning in the fourth century of the Christian era the Huns
drove the Goths and the Germanic tribes occupying central Europe farther
west. In the fifth century the Goths took over the western Roman Empire
itself.

Though parts of France and England had previously acquired some
learning while under the domination of the Roman Empire, by A.D. 500
new civilizing influences began to operate in Europe. Even before the empire
collapsed the Catholic Church was organized and powerful. The Church
gradually converted the Germanic and Gothic barbarians to Christianity
and began to found schools; these were attached to already existing monasteries that possessed fragments of Greek and Roman learning and had been
teaching people how to read the church services and the sacred books. A
little later, the need to train men for ecclesiastical posts motivated the
development of higher schools.

In the latter half of the eighth century some secular rulers founded additional schools. In Charlemagne's empire schools were organized by Alcuin of York (730–804), an Englishman who came to Europe at the invitation of Charlemagne himself. These schools too were attached to cathedrals or monasteries and emphasized Christian theology and music. Ultimately the universities of Europe grew out of the church schools, with teachers supplied by church orders such as the Franciscans and Dominicans. Bologna, the first university, was founded in 1088. The universities of Paris, Salerno, Oxford, and Cambridge were established about the year 1200. Of course at the outset these were hardly universities in the modern sense. Also, though formally independent, they were essentially devoted to the interests of the Church.

2. The Materials Available for Learning

As the Church extended its influence it imposed the culture it favored. Latin was the official language of the Church and so Latin became the international language of Europe and the language of mathematics and science. It was also the language of instruction in European schools until well into the eighteenth century. It became inevitable that Europeans would seek their knowledge largely from Latin—that is, Roman—books. Since Roman mathematics was insignificant, all the Europeans learned was a very primitive number system and a few facts of arithmetic. They also acquired a bit of Greek mathematics through a few translators.

The principal translator, whose works were widely used until the twelfth century, was Anicius Manlius Severinus Boethius (c. 480–524), a descendant of one of the oldest Roman families. Using Greek sources, he compiled in Latin selections from elementary treatises on arithmetic, geometry, and astronomy. Of Euclid's *Elements* he may have translated as many as five books or as few as two, and these constituted part of his *Geometry*. In this subject he gave definitions and theorems but no proofs. He also included in this work some material on the geometry of mensuration. Some results are incorrect and others only approximations. Peculiarly, the *Geometry* also contained material on the abacus and on fractions, the latter preliminary to material on astronomy (which we do not have). Boethius also wrote *Institutis arithmetica*, a translation of Nichomachus' *Introductio arithmetica*, though he omitted some of Nichomachus' results. This book was the source of all arithmetic taught in the schools for almost a thousand years. Finally, Boethius translated some works of Aristotle and wrote an astronomy based on Ptolemy and a book on music based on Euclid's, Ptolemy's, and Nichomachus' works. It is very likely that Boethius did not understand all he translated. It was he who introduced the word "quadrivium" for arithmetic, geometry, music, and astronomy. His best-known work, still read today, is the *Consolations of Philosophy*, which he wrote while imprisoned for alleged treason (for which he was ultimately beheaded).

Other translators were the Roman Aurelius Cassiodorus (*c.* 475–570), who rendered a few parts of the Greek works on mathematics and astronomy in his own poor version; Isidore of Seville (*c.* 560–636), who wrote the *Etymologies*, a work in twenty books on material ranging from mathematics to medicine; and the Englishman, the Venerable Bede (674–735). These men were the main links between Greek mathematics and the early medieval world.

In all of the problems appearing in books written by early medieval mathematicians, only the four operations with integers were involved. Since in practice calculation was done on various forms of the abacus, the rules of operation were specially adapted to it. Fractions were rarely employed, and where they were, the Roman fractions with names rather than special symbolism were used; for example, *uncia* for 1/12, *quincunx* for 5/12, *dodrans* for 9/12. Irrational numbers did not appear at all. Good calculators were known in the Middle Ages as practitioners of the "Black Art," magic.

In the tenth century the study of mathematics was improved somewhat by Gerbert (d. 1003), a native of Auvergne, later Pope Sylvester II. His writings, however, were confined to elementary arithmetic and elementary geometry.

3. *The Role of Mathematics in Early Medieval Europe*

Though the mathematical material available for instruction was scanty, mathematics was relatively important in the curriculum of even the early medieval schools. The curriculum was divided into the quadrivium and the trivium. The quadrivium included arithmetic, considered as the science of pure numbers; music, regarded as an application of numbers; geometry, or the study of magnitudes such as length, area, and volumes at rest; and astronomy, the study of magnitudes in motion. The trivium covered rhetoric, dialectic, and grammar.

Learning even the little mathematics we have described served several purposes. After the time of Gerbert it was applied to finding heights and distances, the astrolabe and mirror being the field instruments. The clergy was expected to defend the theology and rebut arguments by reasoning, and mathematics was regarded as good training for theological reasoning, just as Plato had regarded it as good training for philosophy. The Church advocated teaching mathematics for its application in keeping the calendar and predicting holidays. Each monastery had at least one man who could do the necessary calculations, and various improvements in arithmetic and in the method of keeping the calendar were devised in the course of this work.

Another motivation for the study of some mathematics was astrology.

This pseudoscience, which had had some vogue in Babylonia, in Hellenistic Greece, and among the Arabs, was almost universally accepted in medieval Europe. The basic doctrine of astrology was, of course, that the heavenly bodies influenced and controlled human bodies and fortunes. To understand the influences of the heavenly bodies and to predict what special heavenly events, such as conjunctions and eclipses, portended, some astronomical knowledge was needed; and thus some mathematics was indispensable.

Astrology was especially important in the late medieval centuries. Every court had astrologers and the universities had professors of astrology and courses in the subject. Astrologers advised princes and kings on political decisions, military campaigns, and personal matters. The curious thing is that even rulers who had become learned in and attached to Greek thought relied upon astrologers. In the late medieval period and in the Renaissance astrology not only became a major activity but was regarded as a branch of mathematics.

Through astrology mathematics was linked to medicine (Chap. 7, sec. 8). Although the Church dismissed the physical body as relatively unimportant, the physicians could not subscribe to this belief. Since the heavenly bodies presumably influenced health, physicians sought the relations between heavenly events and particular constellations, on the one hand, and the health of individuals, on the other. Records were kept of the constellations that appeared at the births, marriages, sicknesses, and deaths of thousands of people and used to predict the success of medical treatment. Such a wide knowledge of mathematics was required for this purpose that physicians had to become learned in the field. In fact they were astrologers and mathematicians far more than students of the human body.

The application of mathematics to medicine through astrology became more widespread in the latter part of the medieval period. Bologna had a school of medicine and mathematics in the twelfth century. When the astronomer Tycho Brahe attended the University of Rostock in 1566, there were no astronomers there but there were astrologers, alchemists, mathematicians, and physicians. In many universities, professors of astrology were more common than professors of medicine and astronomy proper. Galileo did lecture to medical students on astronomy, but for the sake of astrology.

4. *The Stagnation in Mathematics*

The early medieval period extends from about 400 to about 1100: seven hundred years during which the European civilization might have developed some mathematics. It could have derived immense help from the Greek works had it pursued the few available leads on the vast knowledge embodied in them. Yet during this period mathematics made no progress, nor were there any serious attempts to build mathematics. The reasons are of interest to those who seek to understand under what circumstances mathematics can flourish.

The primary reason for the low level of mathematics was lack of interest in the physical world. Christianity, which dominated Europe, prescribed its own goals, values, and way of life. The important concerns were spiritual, so much so that inquiries into nature stimulated by curiosity or practical ends were regarded as frivolous or unworthy. Christianity and even the later Greek philosophers, the Stoics, Epicureans, and neo-Platonists, emphasized lifting mind above flesh and matter and preparing the soul for an after-life in heaven. The ultimate reality was the everlasting life of the soul; and the soul's health was strengthened by learning moral and spiritual truths. The doctrines of sin, fear of hell, salvation, and aspiration to heaven were dominant. Since the study of nature did not help achieve such goals or prepare for the afterlife, it was opposed as worthless and even heretical.

Where, then, did the Europeans secure any knowledge about the nature and design of the universe and of man? The answer is that all knowledge was derived from the study of the Scriptures. The creeds and dogmas of the Church Fathers, which were amplifications and interpretations of the Scriptures, were taken as the supreme authority. St. Augustine (354–430), a very learned man and most influential in spreading neo-Platonism, said, "Whatever knowledge man has acquired outside of Holy Writ, if it be harmful it is there condemned; if it be wholesome it is there contained." This quotation, though not representative of Augustine, is representative of the early medieval attitude toward nature.

This brief sketch of the early medieval civilization, rather one-sided as it is, because we have been concerned largely with its relation to mathematics, may nevertheless give some idea of what was indigenous to Europe and what Europe, building on a meager legacy from Rome, produced under the leadership of the Church. Until 1100 the medieval period did not produce any great culture in intellectual spheres. The characteristics of its intellectual state were an indistinctness of ideas, dogmatism, mysticism, and a reliance upon authorities, who were constantly consulted, analyzed, and commented on. The mystical leanings caused people to elevate vague ideas into realities and even accept them as religious truths. What little theoretical science existed was static. Theology embraced all knowledge and the Fathers of the Church authored systems of universal knowledge. But they did not conceive or search for principles other than those contained in Christian doctrines.

The Roman civilization was unproductive in mathematics because it was too much concerned with practical and immediately applicable results. The civilization of medieval Europe was unproductive in mathematics for exactly the opposite reason. It was not at all concerned with the physical world. Mundane matters and problems were unimportant. Christianity put its emphasis on life after death and on preparation for that life.

Apparently mathematics cannot flourish in either an earthbound or a heavenbound civilization. We shall see that it has been most successful in a free intellectual atmosphere which couples an interest in the problems presented by the physical world with a willingness to think about ideas suggested by these problems in an abstract form that makes no promise of immediate or practical return. Nature is the matrix from which ideas are born. The ideas must then be studied for themselves. Then, paradoxically, a new insight into nature, a richer, broader, more powerful understanding, is achieved, which in turn generates deeper mathematical activities.

5. The First Revival of the Greek Works

By 1100 the civilization of Europe was somewhat stabilized. Though the society was largely feudal, there were already numerous independent merchants, the beginnings of industry, arts and crafts carried on by free people, large-scale farming, manufacturing, mining, banking, and cattle-raising. Foreign trade, notably with the Arabs and the Near East, had been established. Finally, the wealth necessary to support scholarship and the arts was acquired by princes, church officials, and merchants.

Though there was a stable society there is little indication that the Europeans, if left to pursue their own way of life, would ever have abandoned the outlook and emphasis already sketched and turned to a serious study of mathematics. Western Europe was the successor of Christianized Rome and neither Rome nor Christianity was inclined toward mathematics. But about 1100, new influences began to affect the intellectual atmosphere. Through trade and travel the Europeans had come into contact with the Arabs of the Mediterranean area and the Near East and with the Byzantines of the Eastern Roman Empire. The Crusades (c. 1100–c. 1300), military campaigns to conquer territory, brought Europeans into Arab lands. The Crusaders were men of action rather than learning; it may be that the importance of the contact through the Crusades has been overestimated. At any rate the Europeans began to learn about the Greek works from the Arabs and Byzantine Greeks.

Awareness of the Greek learning created great excitement; Europeans energetically sought out copies of the Greek works, their Arabic versions, and texts written by Arabs. Princes and church leaders supported scholars in the hunt for these treasures. The scholars went to Arab centers in Africa, Spain, southern France, Sicily, and the Near East to study the works and bring back what they could purchase. Adelard of Bath (c. 1090–c. 1150) went to Syria, which was under Arab control, to Cordova, disguised as a Moham-medan student, and to southern Italy. Leonardo of Pisa learned arithmetic in North Africa. The republics of northern Italy and the papacy sent missions and ambassadors to the Byzantine Empire and to Sicily, which was the original home of famous Greek centers and which up to 878 had been under Byzantine rule. In 1085 Toledo was captured by the Christians and thus a major center for the Arabic works was opened up to European scholars. Sicily was taken from the Arabs by the Christians in 1091 and the works there became freely accessible. A search in Rome, which possessed Greek works from the days of the Empire, unearthed more manuscripts.

As they secured these works the Europeans undertook more and more to translate them into Latin. The twelfth-century translations from Greek were, on the whole, not good because Greek was not well known. They were *de verbo ad verbum*; but they were better than the translations of Greek works that had passed through Arabic, a language quite dissimilar to Greek. Hence until well into the seventeenth century there was a steady output of new, improved translations.

Thus Europe got to know the works of Euclid and Ptolemy, al-Khow-ârizmî's *Arithmetic* and *Algebra*, the *Sphaerica* of Theodosius, many works of Aristotle and Heron, and a couple of Archimedes' works, particularly his *Measurement of a Circle*. (The rest of his work was translated into Latin in 1544 by Hervagius of Basle.) Neither Apollonius nor Diophantus was translated during the twelfth and thirteenth centuries. Works on philosophy, medicine, science, theology, and astrology were also translated. Since the Arabs did have almost all the Greek works, the Europeans acquired a tremendous literature. They admired these works so much and were so fascinated by the novel ideas that they became disciples of Greek thought. They valued these works far more than their own creations.

6. *The Revival of Rationalism and Interest in Nature*

A rational approach to natural phenomena and explanation in terms of natural causes, as opposed to moral or purposive explanations, began to show signs of life almost immediately after the first translations of the Greek and Arabic works reached Europe. A group at Chartres in France, Gilbert de la Porée (*c*. 1076–1154), Thierry of Chartres (d. *c*. 1155), and Bernard Sylvester (*c*. 1150), had begun to seek rational explanations even of biblical passages and spoke, at least, of the need to use mathematics in the study of nature. Their doctrines followed Plato's *Timaeus* but were more rational than that dialogue. However, their pronouncements on physical phenomena, though noteworthy in medieval thought, were neither significant nor influential enough to warrant attention here.

With the influx of Greek works, the trend to rational explanations, study of the physical world, and interest in the enjoyment of the real world through food, a physical life, and the pleasures of nature became marked. Some men even began to pit their own reason against the authority of the Church. Thus Adelard of Bath said he would not listen to those who are "led in a halter; . . . Wherefore if you want to hear anything from me, give and take reason."

Peculiarly enough, the introduction of some of the Greek works retarded the awakening of Europe for a couple of centuries. By 1200 or so the extensive writings of Aristotle became reasonably well known. The European intellectuals were pleased and impressed by his vast store of facts, his acute distinctions, his cogent arguments, and his logical arrangement of knowledge. The defect in Aristotle's doctrines was that he accepted those that appealed to the mind almost without regard for their correspondence with experience. He offered concepts, theories, and explanations, such as the doctrine of basic substances, the distinction between earthly and heavenly bodies (Chap. 7, sec. 3), and the emphasis on final cause, which had little basis in reality, or were not fruitful. Since all of these doctrines were accepted uncritically, new ideas were either not entertained or failed to gain a hearing, and progress was delayed. It was perhaps also a hindrance that Aristotle assigned a minor role to mathematics, certainly a role subordinate to physical explanation, which, for Aristotle, was qualitative.

The scientific work of the period from about 1100 to 1450 was done by the Scholastics, who espoused doctrines based on the authority of the Christian Fathers and of Aristotle; the work suffered accordingly. Some of the Scholastics revolted against the prevailing dogmatism and against the insistence on the absolute correctness of Aristotle's science. One who felt the need for obtaining general principles from experimentation and for deductions in which mathematics would play a part and which could then be tested against facts was the natural philosopher Robert Grosseteste (c. 1168–1253), Bishop of Lincoln.

The most eloquent protester against authority, and one who had genuine ideas to offer, was Roger Bacon (1214–94), the *Doctor Mirabilis*. He declared, "If I had the power over the works of Aristotle, I would have them all burned; for it is only a loss of time to study in them and a cause of error, and a multiplication of ignorance beyond expression." Bacon's enormous knowledge covered the sciences of his time and many languages, including Arabic. Long before they became widely known he was informed on the latest inventions and scientific advances: gunpowder, the action of lenses, mechanical clocks, the construction of the calendar, and the formation of the rainbow. He even discussed ideas on submarines, airplanes, and automobiles. His writings on mathematics, mechanics, optics, vision, astronomy, geography, chronology, chemistry, perspective, music, medicine, grammar, logic, metaphysics, ethics, and theology were sound.

What is especially striking about Bacon is that he understood how reliable knowledge is obtained. He inquired into the causes that produce or prevent the advance of science and speculated on the reform of the methods of inquiry. Though he did recommend study of the Scriptures, he emphasized mathematics and experimentation and foresaw great prospects that could be realized through science.

Mathematical ideas, he affirms, are innate in us and identical with things as they are in nature, for nature is written in the language of geometry. Hence mathematics offers truth. It is prior to the other sciences because it takes cognizance of quantity, which is apprehended by the intuition. He "proves" in one chapter of his *Opus Majus* that all science requires mathematics, and his arguments show a just appreciation of the office of mathematics in science. Though he stresses mathematics, he also has a full appreciation of the role and importance of experimentation as a means of discovery and as a test of results obtained theoretically or in any other way. "Argument concludes a question; but it does not make us feel certain, or acquiesce in the contemplation of truth, except the truth also be found to be so by experience."

Bacon's *Opus Majus* has much on the usefulness of mathematics for geography, chronology, music, the explanation of the rainbow, calendar-reckoning, and the certification of faith. He also treats the role of mathematics in state administration, meteorology, hydrography, astrology, perspective, optics, and vision.

However, even Bacon was a product of his times. He believed in magic and astrology and maintained that theology is the goal of all learning. He was also a victim of his times: he ended up in prison, as did many other intellectual leaders who had begun to assert the priority and independence of human reason and the importance of observation and experimentation. His influence on his age was not great.

William of Ockham (*c.* 1300–49) continued the weighty attacks on Aristotle, criticizing Aristotle's views on causation. Final cause, he said, is pure metaphor. All causes are immediate, and the total cause is the aggregate of all the antecedents that suffice to bring about an event. This knowledge of connections has a universal validity because of the uniformity of nature. The primary function of science is to establish sequences of observations. As for substance, Ockham said, we know only properties, not a fundamental substantial form.

He also attacked contemporary physics and metaphysics, saying that knowledge gained from experience is real, whereas rational constructs are not; they are merely invented to explain the observed facts. His famous principle is "Ockham's razor" (already stated by Grosseteste and Duns Scotus [1266–1308]): It is futile to work with more entities when fewer suffice. He divorced theology from natural philosophy (science) on the ground that theology derives knowledge from revelations whereas natural philosophy should derive it from experience.

These dissenters did not suggest new scientific ideas. But they did press for freedom of speculation, thought, and inquiry and urged experience as the source of scientific knowledge.

7. Progress in Mathematics Proper

Despite the rigid bounds to thought in the period from 1100 to about 1450, some mathematical activity did take place, the main centers being the universities of Oxford, Paris, Vienna (founded in 1365), and Erfurt (founded in 1392). The initial work was a direct response to the Greek and Arabic literature.

The first European worthy of mention is Leonardo of Pisa (c. 1170–1250), also called Fibonacci. He was educated in Africa, traveled extensively in Europe and Asia Minor, and was famous for his sovereign possession of the entire mathematical knowledge of his own and preceding generations. He resided at Pisa and was well known to Frederick II of Sicily and the philosophers of the court, to whom most of his extant works are dedicated.

In 1202 Leonardo wrote the epoch-making and long-used *Liber Abaci*, a free rendition of Arabic and Greek materials into Latin. Arabic notation for numbers and Hindu methods of calculation were already known to some extent in Europe, but only in the monasteries. People in general used Roman numerals and avoided zero because they did not understand it. Leonardo's book exerted great influence and changed the picture; it taught the Hindu methods of calculation with integers and fractions, square roots, and cube roots. These methods were subsequently improved by the Florentine merchants.

In both the *Liber Abaci* and a later work, the *Liber Quadratorum* (1225), Leonardo treated algebra. He followed the Arabs in using words rather than symbols and in basing the algebra on arithmetical methods. He presented the solution of determinate and indeterminate equations of the first and second degree and some cubic equations. Like Khayyam, he believed that general cubic equations could not be solved algebraically.

On the geometric side, Leonardo in his *Practica Geometriae* (1220) reproduced much of Euclid's *Elements* and Greek trigonometry. His teaching of surveying by trigonometric methods rather than by the Roman geometric methods represented a slight advance.

The most significant new feature of Leonardo's work is the observation that Euclid's classification of irrationals in Book X of the *Elements* did not include all irrationals. Leonardo showed that the roots of $x^3 + 2x^2 + 10x = 20$ are not constructible with straightedge and compass. This was the first indication that the number system contained more than the Greek criterion of existence by construction allowed. Leonardo also introduced the notion of what are still called Fibonacci sequences, wherein each term is the sum of the preceding two.

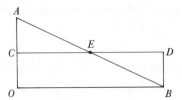

Figure 10.1

Beyond the observation on irrationals made by Leonardo, there were some germinal ideas in the work of Nicole Oresme (*c.* 1323–82), Bishop of Lisieux and a teacher in the Parisian College of Navarre. In his unpublished *Algorismus Proportionum* (*c.* 1360), he introduced both a notation and some computation with fractional exponents. His thinking was that since (in our notation) $4^3 = 64$ and $(4^3)^{1/2} = 8$, then $4^{3/2} = 8$. The notion of fractional exponents reappeared in the work of several sixteenth-century writers, but was not widely used until the seventeenth century.

Another contribution of Oresme's involves the study of change. We may recall that Aristotle distinguished sharply between quality and quantity. Intensity of heat was a quality to Aristotle. To change intensity some substance, a species of heat, had to be lost and another added. Oresme affirmed that there were not different kinds of heat but only more or less of the same kind. Several fourteenth-century Scholastics at Oxford and Paris, including Oresme, began to think about change and rate of change quantitatively. These men studied uniform motion (motion with constant velocity), difform motion (motion with varying velocity), and uniformly difform motion (motion with constant acceleration).

This line of thought culminated in that period with Oresme's doctrine of the latitude of forms. On this subject he wrote *De Uniformitate et Difformitate Intensionum* (*c.* 1350) and *Tractatus de Latitudinibus Formarum* (n.d.). To study change and rate of change, Oresme followed the Greek tradition in asserting that measurable quantities other than numbers can be represented by points, lines, and surfaces. Thus, to represent velocity changing with time, he represents time along a horizontal line, which he called the longitude, and the velocities at various times by vertical lines, which he called the latitudes. To represent a velocity decreasing uniformly from the value *OA* (Fig. 10.1) at *O* to zero at *B*, he gives a triangular figure. He also points out that the rectangle *OBDC*, determined by *E*, the midpoint of *AB*, has the same area as the triangle *OAB* and represents a uniform motion over the same time interval. Oresme associated physical change with the entire geometrical figure. The whole area represented the variation in question; numerical values were not involved.

It is often said that Oresme contributed to the formation of the function concept, to the functional representation of physical laws, and to the classification of functions. He is also credited with the creation of coordinate geometry and the graphical representation of functions. Actually the latitude of forms is a dim idea, at best a kind of graph. Though Oresme's representation of intensity under the name *latitudines formarum* was a major technique in the Scholastic attempt to study physical change, was taught in the universities, and was applied in efforts to revise Aristotle's theory of motion, its influence on subsequent thought was minor. Galileo did use this figure, but with far more clarity and pointedness. Since Descartes carefully avoided any reference to his predecessors, we do not know if he was influenced by Oresme's ideas.

8. *Progress in Physical Science*

Because the progress of mathematics depended vitally on the renewal of an interest in science, we shall note briefly the efforts made in this area by the medieval men.

In mechanics, they took over the highly acceptable Greek works on the lever and centers of gravity and Archimedes' work on hydrostatics. They did little more than comprehend the theory of the lever, though slight additions were made by Jordanus Nemorarius (d. 1237). It was the theory of motion that received most attention.

Because Aristotle's science had acquired ascendancy, it was his theory that became the starting point for investigations of motion. As we pointed out in Chapter 7, there were several apparent difficulties in Aristotle's theory. These the earlier medieval scientists sought to resolve within the basic Aristotelian framework. Thus, to account for the acceleration of falling bodies, several thirteenth-century men interpreted Aristotle's vague notions about gravity to mean that the weight of a body increases as it nears the center of the earth. Then, since the force increases, so does the velocity. Others questioned the correctness of Aristotle's fundamental law that velocity is force divided by resistance.

The school at Chartres was succeeded in the fourteenth century by a school at Paris whose leaders were Oresme and Jean Buridan (*c.* 1300– *c.* 1360). Here at the university Aristotelian views had been dominant. To explain the continuing motion of objects launched by a force, Buridan put forth a new theory, the theory of impetus. Following the sixth-century Christian scholar Philoponus, Buridan said that the motive power impressed on an arrow or projectile was impressed on the object itself and not on the air. This impetus, rather than the propelling action of the air, would maintain the object's uniform velocity indefinitely, were it not for the action of external forces. In the case of falling bodies, impetus was gradually increased by natural gravity acting to add successive increments of impetus

to that which the object already had. In rising bodies—for example, projectiles—the impetus given to the object was gradually decreased by the air
resistance and natural gravity. Impetus was what God gave to the celestial
spheres and these needed no heavenly agents to keep them running. Buridan
defined impetus as the quantity of matter multiplied by the velocity; hence,
in modern terms, it is momentum.

This new theory was remarkable for several reasons. By applying it to
heavenly and earthly motions, Buridan linked the two in one theory.
Furthermore, the theory implied that, contrary to Aristotle's law, a force
would alter motion and not just sustain it. Third, the concept of impetus
itself was a great advance; it transferred the motive power from the medium
to the moving object and also made possible the consideration of a void.
Buridan is one of the founders of modern dynamics. His theory gained wide
acceptance in his own century and for two centuries thereafter.

Perhaps projectile motion received this attention because by the thirteenth century improved weaponry, such as catapults, crossbows, and long-
bows, could throw projectiles over long and highly arched paths; a century
later cannonballs were in use. Aristotle had said that a body can move only
under one type of force at a time; if two were acting, one would destroy the
other. Hence a body thrown up and out would move along a straight line
until the "violent" motion was spent and then the body would drop straight
to earth under a natural motion. Of the various revisions of this theory, the
idea of Jordanus Nemorarius proved most helpful; he showed that the force
under which a body thrown straight out moved at any instant could be
resolved into two components, natural gravity acting downward and a
"violent" horizontal force of projection. This idea was taken up by da Vinci,
Stevin, Galileo, and Descartes.

The Parisian school of Buridan and Oresme went on to consider not
only uniform but difform and uniformly difform motion and proved to its
own satisfaction that the effective velocity in uniformly difform motion was
the average of the initial and final velocities. Perhaps most significant about
the efforts in thirteenth- and fourteenth-century mechanics was the attempt
to introduce quantitative considerations and to replace qualitative by
quantitative arguments.

The major interest of the medieval scientists lay in the field of optics.
One reason for this is that the Greeks had laid a firmer foundation in what
we now call geometrical optics than in other physical fields; and by the late
medieval period their numerous works in optics were known in Europe. In
addition, the Arabs had made some advances over the Greeks. By 1200 some
of the basic laws of light were well known, including the straight-line motion
of light in a uniform medium; the law of reflection; and Ptolemy's incorrect
law of refraction (he believed the angle of refraction was proportional to the
angle of incidence). Also the knowledge of spherical and parabolic mirrors,

spherical aberration, the pinhole camera, the uses of lenses, the functioning of the eye, atmospheric refraction, and the possibility of magnification had passed on to Europe from the Greeks and the Arabs.

The scientists Grosseteste, Roger Bacon, Vitello (13th cent.), John Peckham (d. 1292), and Theodoric of Freiberg (d. *c.* 1311) made advances in light. Having learned of the refraction of light by lenses, they determined the focal lengths of some lenses, studied combinations of lenses, suggested magnification by combinations of lenses, and made improvements in the theory of the rainbow. Glass mirrors were perfected during the thirteenth century; spectacles date from 1299. Vitello observed the dispersion of light under refraction; that is, he produced colors from white light passed through a hexagonal crystal. He also directed light through a bowl of water to study the rainbow, since he had previously noted that when sunlight passes through a bowl of water the colors of the rainbow appear in the emerging light. Optics continued to be a major science; we shall find Kepler, Galileo, Descartes, Fermat, Huygens, and Newton active in it.

9. *Summary*

In science, as in other fields, the Middle Ages devoted itself to time-tested and authoritative works. The schools produced diligent excerpting from older manuscripts, summarizing, and commentaries. The spirit of the times forced minds to follow trusted, prescribed, and rigid ways. The search for a universal philosophy covering all phenomena of man, nature, and God is characteristic of the late medieval period. But the contributions suffered from an indistinctness of ideas, mysticism, dogmatism, and a commentatorial spirit directed toward the analysis of authorities.

Nevertheless, as the world gradually changed, awareness of discrepancy and conflict between beliefs and overt facts grew stronger, and the need for a revision of learning and beliefs grew clearer. Before Galileo demonstrated the value of experience, before Descartes taught people to look within themselves, and before Pascal formulated the idea of progress, it was the unconventional thinkers, largely the dissident Scholastics, who attempted to advance along new lines, to challenge established outlooks, and to place stronger reliance upon observation of nature than had the Greeks.

Experimentation, motivated in part by the search for magical powers, and the use of induction to obtain general principles or scientific laws began to be important sources of knowledge, despite the fact that the major scientific method of the late Middle Ages was rational explanation, presented in a formal or geometrical demonstration based on a priori principles.

The value of mathematics in the investigation of nature also received some recognition. Though on the whole the medieval scientists followed

Aristotle in looking for material or physical explanations, these were hard to obtain and not too helpful; they found more and more that it was easier to correlate observations and experimental results mathematically and then check the mathematical law. Thus in astronomy it was not Aristotle's physical modification of Eudoxus' theory that was used by the scientists concerned with astronomical theory proper, navigation, and the calendar, but Ptolemy's. As a consequence, mathematics began to play a larger role than Aristotle had assigned to it.

Despite the new trends and activities, it is doubtful that medieval Europe, if permitted to pursue an unchanging course, would have ever developed any real science and mathematics. Free inquiry was not permitted. The few universities already in existence by 1400 were controlled by the Church and the professors were not free to teach what they thought correct. If no scientific doctrines were condemned in the medieval period, it was only because no new major ones were promulgated. Any real dissent from Christian thought that appeared in any sphere was suppressed with dispatch, cruelty, and viciousness unequaled in history, largely through the Inquisitions initiated by Pope Innocent III in the thirteenth century.

Other factors, relatively minor, delayed changes in Europe. The revived Greek knowledge reached only the few scholars who had the leisure and opportunity to study it. Manuscripts were expensive; many who wanted to acquire them could not. Also, Europe in the period from 1100 to 1500 was broken up into a number of independent dukedoms, principalities, more or less democratic or oligarchic city-states, and the Papal States. The wars among all these political units were continuous and absorbed the energies of the people. The Crusades, which began about 1100, wasted a fantastic number of lives. The Black Death in the second half of the fourteenth century took about a third of the population of Europe and set back the entire civilization. Fortunately, forces of revolutionary strength did begin to exert their effects on the European intellectual, political, and social scene.

Bibliography

Ball, W. W. R.: *A Short Account of the History of Mathematics*, Dover (reprint), 1960, Chaps. 8, 10, 11.

Boyer, Carl B.: *A History of Mathematics*, John Wiley and Sons, 1968, Chap. 14.

Cajori, Florian: *A History of Mathematics*, Macmillan, 1919, pp. 113–29.

Clagett, Marshall: *The Science of Mechanics in the Middle Ages*, University of Wisconsin Press, 1959.

————: *Nicole Oresme and the Geometry of Qualities and Motions*, University of Wisconsin Press, 1968.

Crombie, A. C.: *Augustine to Galileo*, Falcon Press, 1952, Chaps. 1–5.

————: *Robert Grosseteste and the Origins of Experimental Science*, Oxford University Press, 1953.

Easton, Stewart: *Roger Bacon and His Search for a Universal Science*, Columbia University Press, 1952.

Hofmann, J. E.: *The History of Mathematics*, Philosophical Library, 1957, Chaps. 3–4.

Smith, David Eugene: *History of Mathematics*, Dover (reprint), 1958, Vol. 1, pp. 177–265.

I I

The Renaissance

It appears to me that if one wants to make progress in mathematics, one should study the masters and not the pupils.

N. H. ABEL

1. *Revolutionary Influences in Europe*

In the period from about 1400 to about 1600, which we shall adopt as the period of the Renaissance (though this term is used to describe different chronological periods by various writers), Europe was profoundly shaken by a number of events that ultimately altered drastically the intellectual outlook and stirred up mathematical activity of unprecedented scale and depth.

The revolutionary influences were widespread. Political changes resulted from almost incessant wars involving every city and state of Europe. Italy, the mother of the Renaissance, is a prime example. Though the history of the Italian states in the fifteenth and sixteenth centuries is marred by constant intrigues, mass murders, and destruction caused by wars, the very fluid nature of political conditions and the establishment of some democratic governments were favorable to the rise of the individual. Wars against the papacy, a leading political and military power in that era, not only freed people from the domination of the Church but encouraged intellectual opposition as well.

Italy acquired great wealth during the latter part of the Middle Ages. This was largely due to its geographical position. Its seaports were most favorably located to import goods from Africa and Asia for shipment to the rest of Europe. Great banking houses made Italy the financial center. This wealth was essential to the support of learning. And it was in Italy, the most bedeviled of all countries and seething with turmoil, that the thoughts that were to mold Western civilization were first conceived and expressed.

During the fifteenth century the Greek works reached Europe in enormously greater numbers. In the early part of the century the connections between Rome and the Byzantine Empire, which possessed the largest collection of Greek documents but had been isolated, became stronger. The Byzantines, at war with the Turks, sought the help of the Italian states.

Under the improved relationships teachers of Greek were brought to Italy and Italians went to the Byzantine Empire to learn Greek. When the Turks conquered Constantinople in 1453, the Greek scholars fled to Italy, bringing more manuscripts with them. Thus not only were many more Greek works made available to Europe, but many of the newly acquired manuscripts were far better than the ones previously acquired in the twelfth and thirteenth centuries. The subsequent translations, made into Latin directly from the Greek, were more reliable than those made from Arabic.

Johann Gutenberg's invention, about 1450, of printing with movable type speeded up the spread of knowledge. Linen and cotton paper, which the Europeans learned about from the Chinese through the Arabians, replaced parchment and papyrus from the twelfth century on. Commencing in 1474, mathematical, astronomical, and astrological works appeared in printed form. For example, the first printed edition of Euclid's *Elements* in a Latin translation by Johannes Campanus (13th cent.) appeared in Venice in 1482. During the next century the first four books of Apollonius' *Conic Sections*, Pappus' works, and Diophantus' *Arithmetica*, as well as other treatises, appeared in printed editions.

The introduction of the compass and gunpowder had significant effects. The compass made possible navigation out of sight of land. Gunpowder, introduced in the thirteenth century, changed the method of warfare and the design of fortifications and made the study of the motion of projectiles important.

A new economic era was initiated by a tremendous growth in manufacturing, mining, large-scale agriculture and a great variety of trades. Each of these enterprises encountered technical problems that were tackled more vigorously than in any previous civilization. In contrast to the slave societies of Egypt, Greece, and Rome, and the serfdom of medieval feudalism, the new society possessed an expanding class of free artisans and laborers. Independent mechanics and wage-paying employers had an incentive to search for labor-saving devices. The competition of a capitalist economy also stimulated direct studies of physical phenomena and causal connections to improve materials and methods of production. Since the Church had offered explanations of many of these phenomena, conflicts arose. We may be sure that whenever the physical explanation proved more useful than the theological, the latter was ignored.

The merchant class contributed to a new order in Europe by promoting the geographical explorations of the fifteenth and sixteenth centuries. Prompted by the need for better trade routes and for sources of commodities, the explorations brought to Europe much knowledge of strange lands, plants, animals, climates, ways of life, beliefs, and customs. This knowledge challenged medieval dogmas and stimulated imaginations.

Doubts as to the soundness of Church science and cosmology raised by direct observation or by information filtering into Europe from explorers and traders, objections to the Church's suppression of experimentation and thought on the problems created by the new social order, the moral degeneration of some Church leaders, corrupt Church practices such as the sale of salvation, and, finally, serious doctrinal differences, all culminated in the Protestant Reformation. The reformers were supported by the merchants and princes, who were anxious to break the power of the Church.

The Reformation as such did not liberalize thought or free men's minds. The Protestant leaders aimed only to set up their own brand of dogmatism. However, in raising questions concerning the nature of the sacraments, the authority of Church government, and the meaning of passages in the Scriptures, Luther, Calvin, and Zwingli unintentionally encouraged many people to think and do what they would not otherwise have dared. Thought was stimulated and argumentation provoked. Further, to win adherents the Protestants professed that individual judgment, rather than papal authority, was the basis for belief. Thus, variations in belief were sanctioned. Called upon to choose between Catholic and Protestant claims, many declared "a plague on both your houses" and turned from the two faiths to nature, observation, and experimentation as the sources of knowledge.

2. *The New Intellectual Outlook*

The Church rested on authority, revered Aristotle, and branded doubt as criminal. It also deprecated material satisfactions and emphasized salvation of the soul for an afterlife. These tenets contrasted sharply with values the Europeans learned from the Greeks, though they were not pronounced in Aristotle: the study of nature, the enjoyment of the physical world, the perfecting of mind and body, freedom of inquiry and expression, and confidence in human reason. Irked by Church authority, restrictions on the physical life, and reliance on Scripture as the source of all knowledge and the authority for all assertions, the intellectuals grasped eagerly at the new values. Instead of endless disputations and wrangling as to the meaning of biblical passages to determine truth, men turned to nature itself.

Almost as a corollary to the revival of Greek knowledge and values came the revival of interest in mathematics. From the works of Plato especially, which became known in the fifteenth century, the Europeans learned that nature is mathematically designed and that this design is harmonious, aesthetically pleasing, and the inner truth. Nature is rational, simple and orderly, and it acts in accordance with immutable laws. Platonic and Pythagorean works also emphasized number as the essence of reality, a doctrine that had already received some attention from the deviating Scholastics of the thirteenth and fourteenth centuries. The revival of Platonism clarified and crystallized the ideas and methods with which these men had been struggling. The Pythagorean-Platonic emphasis on quantitative relations as the essence of reality gradually became dominant. Copernicus, Kepler, Galileo, Descartes, Huygens, and Newton were in this respect essentially Pythagoreans and by their work established the principle that the goal of scientific activity should be quantitative mathematical laws.

To the intellectuals of the Renaissance, mathematics appealed for still another reason. The Renaissance was a period in which the medieval civilization and culture were discredited as new influences, information, and revolutionary movements swept Europe. These men sought new and sound bases for the erection of knowledge, and mathematics offered such a foundation. Mathematics remained the one accepted body of truths amid crumbling philosophical systems, disputed theological beliefs, and changing ethical values. Mathematical knowledge was certain and offered a secure foothold in a morass; the search for truth was redirected toward it.

Mathematicians and scientists did receive some inspiration from the theological bias of the Middle Ages, which had inculcated the view that all the phenomena of nature are not only interconnected but operate in accordance with an overall plan: all the actions of nature follow the plan laid down by a single first cause. How, then, was the theological view of God's universe to be reconciled with the search for the mathematical laws of nature? The answer was a new doctrine, namely, that God designed the universe mathematically. In other words, by crediting God with being a supreme mathematician, it became possible to regard the search for the mathematical laws of nature as a religious quest.

This doctrine inspired the work of the sixteenth-, seventeenth-, and even some eighteenth-century mathematicians. The search for the mathematical laws of nature was an act of devotion; it was the study of the ways and nature of God and of His plan of the universe. The Renaissance scientist was a theologian, with nature instead of the Bible for his subject. Copernicus, Brahe, Kepler, Galileo, Pascal, Descartes, Newton, and Leibniz speak repeatedly of the harmony that God imparted to the universe through His *mathematical* design. Mathematical knowledge, since it is in itself truth about the universe, is as sacrosanct as any line of Scripture, indeed superior, because it is clear, undisputed knowledge. Galileo said, "Nor does God less admirably discover Himself to us in Nature's actions than in the Scriptures' sacred dictions." To which Leibniz added, "*Cum Deus calculat, fit mundus*" (As God calculates, so the world is made). These men sought mathematical relations that would reveal the grandeur and glory of God's handiwork. Man could not hope to learn the divine plan as clearly as God himself understood it, but with humility and modesty, man could seek at least to approach the mind of God.

The scientists persisted in the search for mathematical laws underlying natural phenomena because they were convinced that God had incorporated these laws in His construction of the universe. Each discovery of a law of nature was hailed more as evidence of God's brilliance than of the brilliance of the investigator. Kepler in particular wrote paeans to God on the occasion of each discovery. The beliefs and attitudes of the mathematicians and scientists exemplify the larger cultural phenomenon that swept Renaissance Europe. The Greek works impinged on a deeply devout Christian world, and the intellectual leaders born in one and attracted by the other fused the doctrines of both.

3. *The Spread of Learning*

For several reasons the diffusion of the new values took place slowly. The Greek works were, at first, to be found only at the courts of secular and ecclesiastical princes and were not accessible to the people at large. Printing aided enormously in making books generally available, but the effect was gradual because even printed books were expensive. The problem of spreading knowledge was complicated by two additional factors. First, many of those who would gladly have put mathematics and science to use in industry, crafts, navigation, architecture, and other pursuits were uneducated; schooling was by no means common. The second factor was language. The learned—scholars, professors, and theologians—were at home with Latin and, to some extent, with Greek. But artists, artisans, and engineers knew only the vernacular—French, German, and the several Italian languages— and were not benefited by the Latin translations of the Greek works.

Beginning in the sixteenth century, many Greek classics were translated into the popular languages. Mathematicians themselves took a hand in this activity. For example, Tartaglia translated Euclid's *Elements* from Latin into Italian in 1543. The translation movement continued well into the seventeenth century; but it proceeded slowly because many scholars were hostile to the common people. The former were contemptuous of the latter; they preferred Latin because they believed that the weight of tradition attached to it would give authority to their pronouncements. To counter such scholars and to reach the public and obtain its support for his ideas, Galileo deliberately wrote in Italian. Descartes wrote in French for the same reason; he hoped that those who used their natural reason might be better judges of his teachings than those who slavishly respected the ancient writings.

Another measure to educate the public, adopted in a few cities of Italy, was the establishment of libraries. The Medici family financed libraries in Florence and some of the popes did the same in Rome. Partly to spread education and partly to provide a meeting place for communication among scholars, several liberal rulers founded academies. Most notable among these was the Florentine Academy of Design, founded by Cosimo I de' Medici (1519–74) in 1563, which became a center for mathematical studies, and the

Roman Accademia dei Lincei (Lynx-like) founded in 1603. Members of these academies translated Latin works into the popular languages, gave lectures to the public, and extended and deepened their own knowledge by communication with each other. These academies were the forerunners of the more famous ones founded later in England, France, Italy, and Germany that were so enormously helpful in spreading knowledge.

Regrettably the universities of the fifteenth and sixteenth centuries played almost no role in these developments. Theology ruled the universities and its study was the purpose of learning. Knowledge was considered complete and final. Hence experimentation was unnecessary, and the new discoveries of outsiders were ignored. The conservative university professors did their best to cling to medieval learning as formulated by the Scholastics since the thirteenth century. The universities did teach arithmetic, geometry, astronomy, and music, but the astronomy was based on Ptolemy and was not observational. Natural philosophy meant studying Aristotle's *Physics*.

4. *Humanistic Activity in Mathematics*

While the Scholastics adhered firmly to late medieval doctrines, a new group of humanists devoted themselves to collecting, organizing, and critically studying Greek and Roman learning. These men made assiduous studies, and with on the whole questionable sagacity strove to cleanse the texts of errors and to restore lost material. They slavishly accepted, repeated, and endlessly interpreted what they found in the ancient and medieval manuscripts, even undertaking philological studies to determine precise meanings. They also wrote many books that redid the ancient works in a Scholastic reinterpretation. Though this activity may have aroused an interest in learning, it nevertheless fostered the deception that learning consisted in deepening and confirming the accepted body of knowledge.

Typical of the humanists of the sixteenth century was the algebraist Jerome Cardan (Gerolamo Cardano), who was born in Pavia in 1501. His career as rogue and scholar is one of the most fascinating among the fantastic careers of Renaissance men. He gives an account of his life in his *De Vita Propria* (Book of My Life), written in his old age, in which he praises and abases himself. He says that his parents endowed him only with misery and scorn; he passed a wretched boyhood and was so poor for the first forty years of his life that he ceased to regard himself as poor because, as he put it, he had nothing left to lose. He was high-tempered, devoted to erotic pleasures, vindictive, quarrelsome, conceited, humorless, incapable of compunction, and purposely cruel in speech. Though he had no passion for gambling he played dice every day for twenty-five years and chess for forty years as an escape from poverty, chronic illnesses, calumnies, and injustices. In his *Liber de Ludo Aleae* (Book on Games of Chance), published posthumously in

1663, he says one should gamble for stakes to compensate for the time lost and he gives advice on how to cheat to insure that compensation.

After devoting his youth to mathematics, physics, and gambling, he graduated from the University of Pavia in medicine. He practiced this art, later taught it in Milan and Bologna, and became celebrated all over Europe as a physician. He also served as professor of mathematics in several Italian universities. In 1570 he was sent to prison for the heresy of casting the horoscope of Jesus. Astonishingly, the pope subsequently hired him as an astrologer. At seventy-five, shortly before his death in 1576, he boasted of fame, a grandson, wealth, learning, powerful friends, belief in God, and fourteen good teeth.

His writings embraced mathematics, astronomy, astrology, physics, medicine, and an enormous variety of other subjects, including moral aphorisms (to atone for his cheating at cards). Despite much training in the sciences, Cardan was a man of his times; he firmly believed in astrology, dreams, charms, palmistry, portents, and superstitions, and wrote many volumes on these subjects. He is the rational apologist for these occult arts, which, he contended, permit as much certitude as do navigation and medicine. He also wrote encyclopedic treatises on the inhabitants of the universe, that is, angels, demons, and various intelligences, including in these books material undoubtedly stolen from his father's distinguished friend, Leonardo da Vinci. The extant material fills about 7000 pages.

The syncretic tendency, exhibited by the natural philosophers in their writings, which united all of reality in gigantic works, is represented in mathematics by Cardan. With uncritical diligence he brought together from ancient, medieval, and contemporary sources the available mathematical knowledge in an encyclopedic mass, using theoretical and empirical sources equally. With his cherished magical and mystical number theory was associated his predilection for algebraic speculation, which he carried further than his contemporaries. In addition to being a celebrated doctor, Cardan is distinguished from the other learned natural philosophers of the sixteenth century by his deeper interest in mathematics. But mathematics was not method to him; it was a special magical talent and an emotionally charged form of speculation.

One of the lesser-known humanists, Ignazio Danti (1537–86), a professor of mathematics at Bologna, wrote a book on mathematics for laymen that reduced all pure and applied mathematics to a series of synoptic tables. *Le scienze matematiche ridotte in tavole* (Bologna, 1577) is characteristic of the classifying spirit of the times; it served to direct the course of mathematical instruction in the schools of the late sixteenth century. Danti was one of the few mathematicians and astronomers who advocated applied mathematics as a branch of learning (as did Galileo later). The subjects covered are significant because they show the range of mathematics at that time: arithmetic, geometry, music, astrology, goniometry (especially measurement of volumes), meteorology, dioptrics, geography, hydrography, mechanics, architecture, military architecture, painting, and sculpture. The first four topics represented pure mathematics; the remainder, applied mathematics.

Characteristic humanist efforts are also evident in the investigations into mechanics by such learned mathematicians as Guidobaldo del Monte (1545–1607), Bernadino Baldi (1553–1617), and Giovanni Battista Benedetti (1530–90). These men hardly grasped the theorems of Archimedes; the works of Pappus made more sense to them and had greater appeal because Pappus elaborated on the proofs given in the earlier Greek classics. They deviated little from the Scholastics in tackling the standard problems and limited themselves to the correction of individual statements and theorems. They accepted much that was false and, in addition, lacked the ability to separate the living, vital ideas from the dead ones. Their humanistic schooling inclined them to incorporate in Euclidean deductions all old and new knowledge, regardless of how this squared with experiments. Consequently their critical faculties were weakened and their own experience lost its value. Their experimentation was free of magical ingredients and their erudition was indeed primarily humanistic, but in principle and in essence they were the last of the medievalists rather than the founders of new methods of thought and research. The Italian mathematicians and physicists Francesco Maurolycus (1494–1575), Benedetti, Baldi, and del Monte, the men whom Galileo later generously called his teachers, and who in some respects prepared the way for him, did not, because of their dependence on ancient modes of thought, make ground-breaking contributions in formulating or solving mathematical or physical problems.

5. *The Clamor for the Reform of Science*

As in the previous centuries of its existence, mathematics was to derive its major inspirations and themes from physical science. However, for science to flourish it was essential that the Europeans break away from slavish adherence to authority. A number of men realized that the methodology of science must be changed; they initiated a real break with Scholasticism and the uncritical acceptance of Greek knowledge.

One of the earliest to call clearly for a new approach to knowledge was the famous Renaissance artist Leonardo da Vinci (1452–1519). Incredibly endowed both physically and mentally, he achieved greatness as a linguist, botanist, zoologist, anatomist, geologist, musician, sculptor, painter, architect, inventor, and engineer. Leonardo made quite a point of distrusting the knowledge that scholars professed so dogmatically. These men of book learning he described as strutting about puffed up and pompous, adorned not by their own labors but by the labors of others whose work they merely repeated; they were but the reciters and trumpeters of other people's learning. He also criticized the concepts, methods, and goals of the bookish scholars because they did not deal with the real world. He almost boasted that he was not a man of letters and could do bigger and better things by learning from experience. And indeed he learned for himself many facts of mathematics, some principles of mechanics, and the laws of equilibrium of the lever. He made remarkable observations on the flight of birds, the flow of water, the structure of rocks, and the structure of the human body. He studied light, color, plants, and animals. Famous are his words, "If you do not rest on the good foundations of nature you will labor with little honor and less profit." Experience, he says, is never deceptive though our judgment may be. "In the study of the sciences which depend upon mathematics, those who do not consult nature but authors, are not the children of nature but only her grandchildren."

Leonardo did believe in the combination of theory and practice. He says, "He who loves practice without theory is like the sailor who boards ship without a rudder and compass and never knows where he may be cast." On the other hand, he said, theory without practice cannot survive and dies as quickly as it lives. "Theory is the general; experiments are the soldiers." He wished to use theory to direct experiments.

Nevertheless, Leonardo did not fully grasp the true method of science. In fact, he had no methodology, nor any underlying philosophy. His work was that of a practical investigator of nature, motivated by aesthetic drives but otherwise undirected. He was interested in and sought quantitative relationships and in this respect was a forerunner of modern science. However, he was not as consciously quantitative as Galileo. While his writings on mechanics and science were used by such sixteenth-century men as Cardan, Baldi, Tartaglia, and Benedetti, they were not the stimulus for Galileo, Descartes, Stevin, and Roberval.

Leonardo's views on mathematics and his working knowledge and use of it were peculiar to his time and illustrate its spirit and approach. Reading Leonardo one finds many statements suggesting that he was a learned mathematician and profound philosopher who worked on the level of the professional mathematician. He says, for example, "The man who discredits the supreme certainty of mathematics is feeding on confusion, and can never silence the contradictions of sophistical sciences, which lead to eternal quackery . . . for no human inquiry can be called science unless it pursues its path through mathematical exposition and demonstration." To pass beyond observation and experience there was for him only one trustworthy road through deceptions and mirages—mathematics. Only by holding fast to mathematics could the mind safely penetrate the labyrinth of intangible and insubstantial thought. Nature works through mathematical laws and nature's forces and operations must be studied through quantitative investigations. These mathematical laws, which one must approach through experience, are the goal of the study of nature. On the basis of such pronouncements, no doubt, Leonardo is often credited with being a greater mathematician than he actually was. When one examines Leonardo's notebooks one realizes how little he knew of mathematics and that his approach was empirical and intuitive.

More influential in urging a reform of the methods of science was Francis Bacon (1561–1626). Bacon sought methods of obtaining truths in intellectual, moral, political, and physical spheres. Though changes were already occurring in the methods of physical science during the sixteenth century, the public at large and even many men of letters were not aware of this. Bacon's lofty eloquence, wide learning, comprehensive views, and bold pronouncements as to the future made men turn more than a passing gaze upon what was going on and note the "Great Instauration" he depicted. He formulated trenchant aphorisms that caught their attention. When people finally noted that science was beginning to make the advances Bacon advocated, they hailed him as the author and leader of the revolution he had merely perceived earlier. Actually he did understand the changes taking place better than his contemporaries.

The salient feature of his philosophy was the confident and emphatic announcement of a new era in the progress of science. In 1605 he published his treatise *Advancement of Learning*; it was followed by the *Novum Organum* (New Method) of 1620. In the latter book he is more explicit. He points out the feebleness and scanty results of prior efforts to study nature. Science, he says, has served only medicine and mathematics or been used to train immature youths. Progress lies in a change of method. All knowledge begins with observation. Then he makes his extraordinary contribution: an insistence on a "graduate and successive induction" instead of hasty generalization. Bacon says, "There are two ways, and can only be two, of seeking and finding truth. The one, from senses and particulars, takes a flight to the most general axioms, and from those principles and their truths, settled once for all, invents and judges of intermediate axioms. The other method collects axioms from senses and particulars, ascending continuously and by degrees, so that in the end it arrives at the more general axioms; this latter way is the true one, but hitherto untried." By "axioms" he means general propositions arrived at by induction and suited to be the starting point of deductive reasoning.

Bacon attacked the Scholastic approach to natural phenomena in these words: "It cannot be that the axioms discovered by argumentations should avail for the discovery of new works; since the subtlety of nature is greater many times over than the subtlety of arguments. . . . Radical errors in the first concoction of the mind are not to be cured by the excellence of functions and remedies subsequent. . . . We must lead men to the particulars themselves, while men on their side must force themselves for a while to lay their notions by and begin to familiarize themselves with the facts."

Bacon did not realize that science must measure so as to obtain quantitative laws. He did not see, in other words, what kinds of gradual inquiries were needed and the order in which they must be taken. Nor did he appreciate the inventive genius that all discovery requires. In fact, he says that "there is not much left to acuteness and strength of genius, but all degrees of genius and intellect are brought to the same level."

Though he did not create it, Bacon issued the manifesto for the experimental method. He attacked preconceived philosophical systems, brain creations, and idle displays of learning. Scientific work, he said, should not be entangled in a search for final causes, which belongs to philosophy. Logic and rhetoric are useful only in organizing what we already know. Let us close in on nature and come to grips with her. Let us not have desultory, haphazard experimentation; let it be systematic, thorough, and directed. Mathematics is to be a handmaiden to physics. In all, Bacon offered a fascinating program for future generations.

Another doctrine and program is associated with Francis Bacon, though it antedates him. The Greeks had, on the whole, been content to derive from their mathematics and science an understanding of nature's ways. The few early medieval scientists and the Scholastics studied nature largely to determine the final cause or purpose served by phenomena. However, the Arabs, a more practical people, studied nature to acquire power over it. Their astrologers, seers, and alchemists sought the elixir of life, the philosopher's stone, methods of converting less useful to more useful metals, and magic properties of plants and animals, in order to prolong man's life, heal his sicknesses, and enrich him materially. While these pseudosciences continued to flourish in medieval times, some of the more rational Scholastics—for example, Robert Grosseteste and Roger Bacon—began to envision the same goal, but by more proper scientific investigations. Thanks to Francis Bacon's exhortations, the mastery of nature became a positive doctrine and a pervasive motivation.

Bacon wished to put knowledge to use. He wanted to command nature for the service and welfare of man, not to please and delight scholars. As he put it, science was to ascend to axioms and then descend to works. In *The New Atlantis* Bacon describes a society of scholars provided with space and equipment for the acquisition of useful knowledge. He foresaw that science could provide man with "infinite commodities," "endow human life with inventions and riches, and minister to the conveniences and comforts of man." These, he says, are the true and lawful goals of science.

Descartes, in his *Discourse on Method*, echoed this thought:

> It is possible to attain knowledge which is very useful in life, and instead of that Speculative Philosophy which is taught in the Schools, we may

find a practical philosophy by means of which, knowing the force and the action of fire, water, air, the stars, the heavens, and all other bodies that environ us, as distinctly as we know the different crafts of our artisans, we can in the same way employ them in all those uses to which they are adapted, and thus render ourselves the masters and possessors of nature.

The chemist Robert Boyle said, "The good of mankind may be much increased by the naturalist's insight into the trades."

The challenge thrown out by Bacon and Descartes was quickly accepted, and scientists plunged optimistically into the task of mastering as well as understanding nature. These two motivations are still the major driving forces, and indeed the interconnections between science and engineering grew rapidly from the seventeenth century onward.

This program was taken up most seriously even by governments. The French Academy of Sciences, founded by Colbert in 1666, and the Royal Society of London, founded in 1662, were dedicated to the cultivation of "such knowledge as has a tendency to use" and to making science "useful as well as attractive."

6. *The Rise of Empiricism*

While reformers of science urged the return to nature and the need for experimental facts, practically oriented artisans, engineers, and painters were actually obtaining the hard facts of experience. Using the natural intuitive approach of ordinary men and seeking not ultimate meanings but merely helpful explanations of phenomena they encountered in their work, these technicians obtained knowledge that in effect mocked the sophistical distinctions, the long etymological derivations of meanings, the involved logical arguments, and the pompous citation of Greek and Roman authorities advanced by the learned scholars and even the humanists. Because the technical accomplishments of Renaissance Europe were superior to and more numerous than those achieved by any other civilization, the empirical knowledge acquired in their pursuit was immense.

The artisans, engineers, and artists had to grapple with real and work-able mechanical ideas and properties of materials. Nevertheless, the new physical insights gained in this way were impressive. It was the spectacle makers who, without discovering a single law of optics, invented the telescope and the microscope. The technicians arrived at laws by attending to phenomena. By measured, gradual steps, such as no speculative view of scientific method has suggested, they obtained truths as profound and comprehensive as any conjecture had dared to anticipate. Whereas the theoretical reformers were bold, self-confident, hasty, ambitious, and contemptuous of antiquity, the practical reformers were cautious, modest, slow, and receptive to all knowledge, whether derived from tradition or from observation. They worked rather than speculated, dealt with particulars rather than generalities, and added to science instead of defining it or proposing how to obtain it. In physics, the plastic arts, and technical fields generally, experience rather than theory and speculation became the new source of knowledge.

Coupled with the pure empiricism of the artisans and, indeed, in part suggested by the problems they presented, was the gradual rise of systematic observation and experimentation, carried out largely by a more learned group. Greeks such as Aristotle and Galen had observed a great deal and had discussed the inductions that might be made on the basis of observations; but one cannot say that the Greeks ever possessed an experimental science. The Renaissance activity, very modest in extent, marks the beginning of the now vast scientific enterprise. The most significant groups of Renaissance experimentalists were the physiologists led by Andreas Vesalius (1514–64), the zoologists led by Ulysses Aldrovandi (1522–1605), and the botanists led by Andrea Cesalpino (1519–1603).

In the area of the physical sciences, the experimental work of William Gilbert (1540–1603) on magnetism was by far the most outstanding. In his famous *De Magnete* (1600), he stated explicitly that we must start from experiments. Though he respected the ancients because a stream of wisdom came from them, he scorned those who quoted others as authorities and did not experiment or verify what they were told. His series of carefully conducted, detailed, and simple experiments is a classic in the experimental method. He notes, incidentally, that Cardan, in *De Rerum Varietate*, had described a perpetual-motion machine and comments, "May the gods damn all such sham, pilfered, distorted works, which do but muddle the minds of the students."

We have been pointing out the variegated practical interests that led to a vast expansion of the study of nature and the consequent impulse to systematic experimentation. Side by side with this practical work, largely independent but not oblivious of it, some men pursued the larger goal of science—the understanding of nature. The work of the later Scholastics on falling bodies, described in the preceding chapter, was continued in the sixteenth century. Their predominant goal was to secure the basic laws of motion. The work on projectile motion, often described as a response to practical needs, was motivated far more by broad scientific interests in mechanics. The work of Copernicus and Kepler in astronomy (Chap. 12, sec. 5) was certainly motivated by the desire to improve astronomical theory. Even the artists of the Renaissance sought to penetrate to the essence of reality.

Fortunately technicians and scientists began to recognize common interests and to appreciate the assistance each could derive from the other. The technicians of the fifteenth and sixteenth centuries, the early engineers who had relied on manual dexterity, mechanical ingenuity, and sheer inventiveness and cared little for principles, became aware of the aid to practice they could secure from theory. On their side scientists became aware that the artisans were obtaining facts of nature that correct theory had to comprehend, and that they could secure pregnant suggestions for investigation from the work of the artisans. In the opening paragraph of his *Dialogues Concerning Two New Sciences* (the sciences are strength of materials and the theory of motion), Galileo acknowledges this inspiration for his investigations. "The constant activity which you Venetians display in your famous arsenal suggests to the studious mind a large field for investigation, especially that part of the work which involves mechanics; for in this department all types of instruments and machines are constantly being constructed by many artisans, among whom there must be some who, partly by inherited experiences and partly by their own observations, have become highly expert and clever in explanation."

The practical and purely scientific interests were fused in the seventeenth century. When the larger principles and problems had emerged from the empirical needs, and the mathematical knowledge of the Greeks had become fully available to the scientists, the latter were able to proceed more effectively with pure science. Without losing sight of the goal of understanding the design of the universe, they also willingly sought to aid practice. The outcome was an expansion in scientific activity on an unprecedented scale, plus far-reaching and weighty technical improvements that culminated in the Industrial Revolution.

The great importance of the beginnings of modern science for us is, of course, that it paved the way for the major developments in mathematics. Its immediate effect was involvement with concrete problems. Since the Renaissance mathematicians worked for the republics and princes and collaborated with architects and handworkers—Maurolycus was an engineer for the city of Messina, Baldi was a mathematician for the Duke of Urbino, Benedetti was the chief engineer for the Duke of Savoy, and Galileo was court mathematician for the Grand Duke of Tuscany—they took up the observations and experiences of the practical people. Up to the time of Galileo, the impact of the technicians and architects can be seen largely in the work of Nicolò Tartaglia (1506–57), a genius who was self-taught in the science of his time. Tartaglia made the transition from the practical to the learned mathematician, singling out with discernment useful problems and observations from empirical knowledge. His uniqueness lies in this achievement and in his complete independence from the magical influences that characterize the work of his rival Cardan. Tartaglia's position is midway between Leonardo and Galileo—not merely chronologically, but because his work on the mathematics of dynamical problems raised that subject to a new science and influenced the forerunners of Galileo.

The long-range effect was that modern mathematics, guided by the

Platonic doctrine that it is the essence of reality, grew almost entirely out of the problems of science. Under the new directive to study nature and to obtain laws embracing observations and experimental results, mathematics broke away from philosophy and became tied to physical science. The consequence for mathematics was a burst of activity and original creation that was the most prolific in its history.

Bibliography

Ball, W. W. R.: *A Short Account of the History of Mathematics*, Dover (reprint), 1960, Chaps. 12–13.

Burtt, E. A.: *The Metaphysical Foundations of Modern Physical Science*, Routledge and Kegan Paul, 1932.

Butterfield, Herbert: *The Origins of Modern Science*, Macmillan, 1951, pp. 1–87.

Cajori, Florian: *A History of Mathematics*, 2nd ed., Macmillan, 1919, pp. 128–45.

Cardano, Gerolamo: *Opera Omnia*, Johnson Reprint Corp., 1964.

————: *The Book of My Life*, Dover (reprint), 1962.

————: *The Book on Games of Chance*, Holt, Rinehart and Winston, 1961.

Clagett, Marshall, ed.: *Critical Problems in the History of Science*, University of Wisconsin Press, 1959, pp. 3–196.

Crombie, A. C.: *Augustine to Galileo*, Falcon Press, 1952, Chaps. 5–6.

————: *Robert Grosseteste and the Origins of Experimental Science*, Oxford University Press, 1953, Chap. 11.

Dampier-Whetham, W. C. D.: *A History of Science*, Cambridge University Press, 1929, Chap. 3.

Farrington, B.: *Francis Bacon*, Henry Schuman, 1949.

Mason, S. F.: *A History of the Sciences*, Routledge and Kegan Paul, 1953, Chaps. 13, 16, 19, and 20.

Ore, O.: *Cardano: The Gambling Scholar*, Princeton University Press, 1953.

Randall, John H., Jr.: *The Making of the Modern Mind*, Houghton Mifflin, 1940, Chaps. 6–9.

Russell, Bertrand: *A History of Western Philosophy*, Simon and Schuster, 1945, pp. 491–557.

Smith, David Eugene: *History of Mathematics*, Dover (reprint), 1958, Vol. 1, pp. 242–65, and Chap. 8.

Smith, Preserved: *A History of Modern Culture*, Holt, Rinehart and Winston, 1940, Vol. 1, Chaps. 5–6.

Strong, Edward W.: *Procedures and Metaphysics*, University of California Press, 1936; reprinted by Georg Olms, 1966, pp. 1–134.

Taton, René, ed.: *The Beginnings of Modern Science*, Basic Books, 1964, pp. 3–51, pp. 82–177.

Vallentin, Antonina: *Leonardo da Vinci*, Viking Press, 1938.

White, Andrew D.: *A History of the Warfare of Science with Theology*, George Braziller (reprint), 1955.

12

Mathematical Contributions
in the Renaissance

> The chief aim of all investigations of the external world should
> be to discover the rational order and harmony which has
> been imposed on it by God and which He revealed to us in
> the language of mathematics.　　　JOHANNES KEPLER

1. *Perspective*

Though the Renaissance men grasped only dimly the outlooks, values, and
goals of the Greek works, they did take some original steps in mathematics;
and they made advances in other fields that paved the way for the tremendous seventeenth-century upsurge in our subject.

The artists were the first to manifest the renewal of interest in nature
and to apply seriously the Greek doctrine that mathematics is the essence
of nature's reality. The artists were self-taught and learned through practice.
Fragments of Greek knowledge filtered down to them, but on the whole
they sensed rather than grasped the Greek ideas and intellectual outlook.
To an extent this was an advantage because, lacking formal schooling, they
were free of indoctrination. Also, they enjoyed freedom of expression because
their work was deemed "harmless."

The Renaissance artists were by profession universal men—that is,
they were hired by princes to perform all sorts of tasks from the creation of
great paintings to the design of fortifications, canals, bridges, war machines,
palaces, public buildings, and churches. Hence they were obliged to learn
mathematics, physics, architecture, engineering, stonecutting, metalworking,
anatomy, woodworking, optics, statics, and hydraulics. They performed
manual work and yet tackled the most abstract problems. In the fifteenth
century, at least, they were the best mathematical physicists.

To appreciate their contribution to geometry we must note their new
goals in painting. In the medieval period the glorification of God and the
illustration of biblical themes were the purposes of painting. Gilt back-
grounds suggested that the people and objects portrayed existed in some

heavenly region. Also the figures were intended to be symbolic rather than realistic. The painters produced forms that were flat and unnatural and did not deviate from the pattern. In the Renaissance the depiction of the real world became the goal. Hence the artists undertook to study nature in order to reproduce it faithfully on their canvases and were confronted with the mathematical problem of representing the three-dimensional real world on a two-dimensional canvas.

Filippo Brunelleschi (1377–1446) was the first artist to study and employ mathematics intensively. Giorgio Vasari (1511–74), the Italian artist and biographer, says that Brunelleschi's interest in mathematics led him to study perspective and that he undertook painting just to apply geometry. He read Euclid, Hipparchus, and Vitello on mathematics and optics and learned mathematics from the Florentine mathematician Paolo del Pozzo Toscanelli (1397–1482). The painters Paolo Uccello (1397–1475) and Masaccio (1401–28) also sought mathematical principles for a system of realistic perspective.

The theoretical genius in mathematical perspective was Leone Battista Alberti (1404–72), who presented his ideas in *Della pittura* (1435), printed in 1511. This book, thoroughly mathematical in character, also includes some work on optics. His other important mathematical work is *Ludi mathematici* (1450), which contains applications to mechanics, surveying, time-reckoning, and artillery fire. Alberti conceived the principle that became the basis for the mathematical system of perspective adopted and perfected by his artist successors. He proposed to paint what one eye sees, though he was well aware that in normal vision both eyes see the same scene from slightly different positions and that only through the brain's reconciliation of the two images is depth perceived. His plan was to further the illusion of depth by such devices as light and shade and the diminution of color with distance. His basic principle can be explained in the following terms. Between the eye and the scene he interposed a glass screen standing upright. He then imagined lines of light running from the eye or station point to each point in the scene itself. These lines he called a pyramid of rays or a projection. Where these rays pierced the glass screen (the picture plane), he imagined points marked out; the collection of points he called a section. The significant fact about it is that it creates the same impression on the eye as the scene itself, because the same lines of light come from the section as from the original scene. Hence the problem of painting realistically is to get a true section onto the glass screen or, in practice, on a canvas. Of course the section depends upon the position of the eye and the position of the screen. This means merely that different paintings of the same scene can be made.

Since the painter does not look through his canvas to determine the section, he must have rules based on mathematical theorems, which tell him how to draw it. Alberti furnished some correct rules[1] in his *Della pittura*, but did not give all the details. He intended his book as a summary to be supplemented by discussions with his fellow painters and, in fact, apologizes for his brevity. He tried to make his material concrete, rather than formal and rigorous, and so gave theorems and constructions without proof.

Beyond introducing the concepts of projection and section, Alberti raised a very significant question. If two glass screens are interposed between the eye and the scene itself, the sections on them will be different. Further, if the eye looks at the same scene from two different positions and in each case a glass screen is interposed between the eye and the scene, again the sections will be different. Yet all these sections convey the original figure. Hence they must have some properties in common. The question is, What is the mathematical relation between any two of these sections, or what mathematical properties do they have in common? This question became the starting point of the development of projective geometry.

Though a number of artists wrote books on mathematical perspective and shared Alberti's philosophy of art, we can mention here only one or two leaders. Leonardo believed that painting must be an exact reproduction of reality and that mathematical perspective would permit this. It was "the rudder and guide rope of painting" and amounted to applied optics and geometry. Painting for him was a science because it reveals the reality in nature; for this reason it is superior to poetry, music, and architecture. Leonardo's writings on perspective are contained in his *Trattato della pittura* (1651), compiled by some unknown author who used the most valuable of Leonardo's notes on the subject.

The painter who set forth the mathematical principles of perspective in fairly complete form is Piero della Francesca (*c.* 1410–92). He, too, regarded perspective as the science of painting and sought to correct and extend empirical knowledge through mathematics. His main work, *De prospettiva pingendi* (1482–87), made advances on Alberti's idea of projection and section. In general, he gives procedures useful to artists and his directions employ strips of paper, wood, and the like. To help the artist he, like Alberti, gives intuitively understandable definitions. He then offers theorems which he "demonstrates" by constructions or by an arithmetical calculation of ratios. He was the painter-mathematician and the scientific artist par excellence, and his contemporaries so regarded him. He was also the best geometer of his time.

However, of all the Renaissance artists, the best mathematician was the German Albrecht Dürer (1471–1528). His *Underweysung der Messung mid dem*

---

1. For some of these rules, and paintings constructed in accordance with them, see the author's *Mathematics in Western Culture*, Oxford University Press, 1953.

Zyrkel und Rychtscheyd (Instruction in Measuring with Compass and Straight-edge, 1525), a work primarily on geometry, was intended to pass on to the Germans knowledge Dürer had acquired in Italy and in particular to help the artists with perspective. His book, more concerned with practice than theory, was very influential.

The theory of perspective was taught in painting schools from the sixteenth century onward according to the principles laid down by the masters we have been discussing. However, their treatises on perspective had on the whole been precept, rule, and ad hoc procedure; they lacked a solid mathematical basis. In the period from 1500 to 1600 artists and subse-quently mathematicians put the subject on a satisfactory deductive basis, and it passed from a quasi-empirical art to a true science. Definitive works on perspective were written much later by the eighteenth-century mathe-maticians Brook Taylor and J. H. Lambert.

2. Geometry Proper

The developments in geometry apart from perspective during the fifteenth and sixteenth centuries were not impressive. One of the geometric topics discussed by Dürer, Leonardo, and Luca Pacioli (*c.* 1445–*c.* 1514), an Italian monk who was a pupil of Piero della Francesca and a friend and teacher of Leonardo, was the inscription of regular polygons in circles. These men attempted such constructions with a straightedge and a compass of fixed opening, a limitation already considered by the Arab Abû'l-Wefâ, but they gave only approximate methods.

The construction of the regular pentagon was a problem of great in-terest because it arose in the design of fortifications. In the *Elements*, Book IV, Proposition 11, Euclid had given a construction not limited by a com-pass of fixed opening. The problem of giving an exact construction with this limitation was tackled by Tartaglia, Ferrari, Cardan, del Monte, Bene-detti, and many other sixteenth-century mathematicians. Benedetti then broadened the problem and sought to solve all Euclidean constructions with a straightedge and a compass of fixed opening. The general problem was solved by the Dane George Mohr (1640–97) in his *Compendium Euclidis Curiosi* (1673).

Mohr also showed, in his *Euclides Danicus* (1672), that the constructions that can be performed with straightedge and compass can be performed with only a compass. Of course without a straightedge, one cannot draw the straight line joining two points; but given the two points, one can con-struct the points of intersection of the line and the circle, and given two pairs of points, one can construct the point of intersection of the two lines deter-mined by the two pairs. The fact that a compass alone suffices to perform the Euclidean constructions was rediscovered by Lorenzo Mascheroni (1750–1800) and published in his *La geometria del compasso* (1797).

Another of the Greek interests, centers of gravity of bodies, was also taken up by the Renaissance geometers. Leonardo, for example, gave a correct and an incorrect method of finding the center of gravity of an isosceles trapezoid. He then gave, without proof, the location of the center of gravity of a tetrahedron, namely, the center is one-quarter of the way up the line joining the center of gravity of the triangular base to the opposite vertex.

Two novel geometric ideas appear in minor works by Dürer. The first of these is space curves. He starts with helical space curves and considers the projection of these curves on the plane. The projections are various types of spirals and Dürer shows how to construct them. He also introduces the epicycloid, which is the locus of a point on a circle that rolls on the outside of a fixed circle. The second idea is the orthogonal projection of curves and of human figures on two and three mutually perpendicular planes. This idea, which Dürer merely touched on, was developed in the late eighteenth century into the subject of descriptive geometry by Gaspard Monge.

The work of Leonardo, Piero, Pacioli, and Dürer in pure geometry is certainly not significant from the standpoint of new results. Its chief value was that it spread widely some knowledge of geometry, crude as that knowledge was by Greek standards. The fourth part of Dürer's *Underweysung*, together with Piero's *De Corporibus Regularibus* (1487) and Pacioli's *De Divina Proportione* (1509), renewed interest in stereometry (mensuration of solid figures), which flourished in Kepler's time.

Another geometrical activity, map-making, served to stimulate further geometrical investigations. The geographical explorations had revealed the inadequacies of the existing maps; at the same time new geographical knowledge was being uncovered. The making and printing of maps was begun in the second half of the fifteenth century at centers such as Antwerp and Amsterdam.

The problem of map-making arises from the fact that a sphere cannot be slit open and laid flat without distorting distance. In addition, directions (angles) or area, or both, may be distorted. The most significant new method of map-making is due to Gerhard Kremer, known also as Mercator (1512–94), who devoted his life to the science. In 1569 he put out a map using the famous Mercator projection. In this scheme the lines of latitude and longitude are straight. The longitude lines are equally spaced, but the spacing between latitude lines is increased. The purpose of this increase is to keep the ratio of a length of one minute of longitude to one minute of latitude correct. On the sphere a change of 1′ in latitude equals 6087 feet; but only on the equator is a change of 1′ in longitude equal to 6087 feet. For example, at latitude 20° a change of 1′ in longitude equals 5722 feet, producing the ratio

$$\frac{1'\text{ change in longitude}}{1'\text{ change in latitude}} = \frac{5722}{6087}.$$

For this true ratio to hold on Mercator's straight-line map, where longitude lines are equally spaced and each minute of change equals 6087 feet, he increases the spaces between latitude lines by the factor $1/\cos L$ as latitude L increases. At 20° latitude on his map a 1′ change in latitude equals a distance of 6087 $(1/\cos 20°)$, or 6450 feet. Thus at latitude 20°

$$\frac{1'\text{ change in longitude}}{1'\text{ change in latitude}} = \frac{6087}{6450},$$

and this ratio equals the true ratio of 5722/6087.

The Mercator map has several advantages. Only on this projection are two points on the map at correct compass course from one another. Then a course of constant compass bearing on the sphere, that is, a curve called a loxodrome or rhumb line, which cuts all meridians at the same angle, becomes a straight line on the map. Distance and area are not preserved; in fact, the map distorts badly around the poles. However, since direction is preserved, so is the angle between two directions at a point, and the map is said to be conformal.

Though no large new mathematical ideas emerged from the sixteenth-century work on map-making, the problem was taken over later by mathematicians and led to work in differential geometry.

3. Algebra

Up to the appearance of Cardan's *Ars Magna* (1545), which we shall treat in the next chapter, there were no Renaissance developments of any consequence in algebra. However, the work of Pacioli is worth noting. Like most others of his century he believed that mathematics is the broadest systematic learning and that it applies to the practical and spiritual life of all people. He also realized the advantages of theoretical knowledge for practical work. Theory must be master and guide, he tells the mathematicians and technicians. Like Cardan he belonged to the humanist circle. Pacioli's major publication is the *Summa de Arithmetica, Geometria, Proportione et Proportionalita* (1494). The *Summa* was a compendium of the available knowledge and was representative of the times because it linked mathematics with a great variety of practical applications.

The contents covered the Hindu-Arabic number symbols, which were already in use in Europe, business arithmetic, including bookkeeping, the algebra thus far created, a poor summary of Euclid's *Elements*, and some

trigonometry taken from Ptolemy. The application of the concept of proportion to reveal design in all phases of nature and in the universe itself was
a major theme. Pacioli called proportion "mother" and "queen" and
applied it to the sizes of the parts of the human body, to perspective, and
even to color mixing. His algebra is rhetorical; he follows Leonardo and the
Arabs in calling the unknown quantity the "thing." The square of the
unknown Pacioli called *census*, which he sometimes abbreviates as *ce* or *Z*;
the cube of the unknown, *cuba*, is sometimes represented by *cu* or *C*. Other
abbreviations for words, such as *p* for plus and *æ* for *æqualis*, also occur. In
writing equations, whose coefficients are always numerical, he puts terms
on the side that permits the use of positive coefficients. Though an occasional
subtraction of a term, for example, $-3x$, does appear, no purely negative
numbers are used; only the positive roots of equations are given. He used
algebra to calculate geometrical quantities much as we would use an
arithmetic proportion to relate the lengths of sides of similar triangles and
perhaps to find one unknown length, though Pacioli's uses are often more
involved. He closes his book with the remark that the solution of $x^3 + mx = n$ and of $x^3 + n = mx$ (we use modern notation) are as impossible as the
quadrature of the circle.

Though there was nothing original in the *Summa*, this book and his *De
Divina Proportione* were valuable because they contained much more than was
taught in the universities. Pacioli served as the intermediary between what
existed in the scholarly works and the knowledge acquired by artists and
technicians. He tried to help these men learn and use mathematics. Nevertheless it is a significant commentary on the mathematical development of
arithmetic and algebra between 1200 and 1500 that Pacioli's *Summa*, appearing in 1494, contained hardly anything more than Leonardo of Pisa's *Liber
Abaci* of 1202. In fact, the arithmetic and algebra in the *Summa* were based
on Leonardo's book.

4. Trigonometry

Until 1450, trigonometry was largely spherical trigonometry; surveying
continued to use the geometric methods of the Romans. About that date
plane trigonometry became important in surveying, though Leonardo of
Pisa in his *Practica Geometriae* (1220) had already initiated the method.

New work in trigonometry was done by Germans of the late fifteenth
and early sixteenth centuries, who usually studied in Italy and then returned
to their native cities. At the time Germany had become prosperous, some of
the wealth having been acquired by the Hanseatic League of North Germany, which controlled much trade; hence merchant patrons were able
to support the work of many of the men we shall mention. The trigonometric
work was motivated by navigation, calendar-reckoning, and astronomy,

interest in the last-mentioned field having been heightened by the creation of the heliocentric theory, about which we shall say more later.

George Peurbach (1423–61) of Vienna started to correct the Latin translations of the *Almagest*, which had been made from Arabic versions, but which he proposed to make from the original Greek. He also began to make more accurate trigonometric tables. However, Peurbach died young and his work was taken up by his pupil Johannes Müller (1436–76), known as Regiomontanus, who vitalized trigonometry in Europe. Having studied astronomy and trigonometry at Vienna under Peurbach Regiomontanus went to Rome, studied Greek under Cardinal Bessarion (*c.* 1400–72), and collected Greek manuscripts from the Greek scholars who had fled the Turks. In 1471 he settled in Nuremberg under the patronage of Bernard Walther. Regiomontanus made translations of several Greek works—the *Conic Sections* of Apollonius and parts of Archimedes and Heron—and founded his own press to print them.

Following Peurbach, he adopted the Hindu sine, that is, the half chord of the half arc, and then constructed a table of sines based on a radius of 600,000 units and another based on a radius of 10,000,000 units. He also calculated a table of tangents. In the *Tabulae Directionum* (written in 1464–67), he gave five-place tangent tables and decimal subdivision of the angles, a very unusual procedure for those times.

Many men of the fifteenth and sixteenth centuries constructed tables, among them George Joachim Rhaeticus (1514–76), Copernicus, François Vieta (1540–1603), and Bartholomäus Pitiscus (1561–1613). Characteristic of this work was the use of a larger and larger number of units in the radius so that the values of the trigonometric quantities could be obtained more accurately without the use of fractions or decimals. For example, Rhaeticus calculated a table of sines based on a radius of 10^{10} units and another based on 10^{15} units, and gave values for every 10 seconds of arc. Pitiscus in his *Thesaurus* (1613) corrected and published the second table of Rhaeticus. The word "trigonometry" is his.

More fundamental was the work on the solution of plane and spherical triangles. Until about 1450 spherical trigonometry consisted of loose rules based on Greek, Hindu, and Arabic versions, the last of which came from Spain. The works of the Eastern Arabs Abû'l-Wefâ and Nâsir-Eddin were not known to Europe until this time. Regiomontanus was able to take advantage of Nâsir-Eddin's work, and in *De Triangulis*, written in 1462–63, put together in more effective fashion the available knowledge of plane trigonometry, spherical geometry, and spherical trigonometry. He gave the law of sines for spherical triangles, namely

$$\frac{\sin a}{\sin A} = \frac{\sin b}{\sin B} = \frac{\sin c}{\sin C}$$

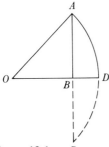

Figure 12.1   C

and the law of cosines involving the sides, that is,

$$\cos a = \cos b \cos c + \sin b \sin c \cos A.$$

The *De Triangulis* was not published until 1533; in the meantime, Johann Werner (1468–1528) improved and published Regiomontanus's ideas in *De Triangulis Sphaericis* (1514).

For many years after the work of Regiomontanus spherical trigonometry continued to be troubled by the need for a multitude of formulas, partly because Regiomontanus in his *De Triangulis*, and even Copernicus a century later, used only the sine and cosine functions. Also the negative values of the cosine and tangent functions for obtuse angles were not recognized as numbers.

Rhaeticus, who was a pupil of Copernicus, changed the meaning of sine. Instead of speaking of *AB* (Fig. 12.1) as the sine of *AD*, he spoke of *AB* as the sine of angle *AOB*. However, the length of *AB* was still expressed in a number of units dependent on the number of units chosen for the length of the radius. As a consequence of Rhaeticus's change, the triangle *OAB* became the basic structure and the circle with radius *OA* incidental. Rhaeticus used all six functions.

Plane and spherical trigonometry were further systematized and extended slightly by François Vieta, a lawyer by profession but recognized far more as the foremost mathematician of the sixteenth century. His *Canon Mathematicus* (1579) was the first of his many works on trigonometry. Here he gathered together the formulas for the solution of right and oblique plane triangles, including his own contribution, the law of tangents:

$$\frac{a - b}{a + b} = \frac{\tan\left(\dfrac{A - B}{2}\right)}{\tan\left(\dfrac{A + B}{2}\right)}.$$

For spherical right triangles he gave the complete set of formulas needed to calculate any one part in terms of two other known parts, and the rule for remembering this collection of formulas, which we now call Napier's rule. He also contributed the law of cosines involving the angles of an oblique spherical triangle:

$$\cos A = -\cos B \cos C + \sin B \sin C \cos a.$$

Many trigonometric identities had been established by Ptolemy; Vieta added to these. For example, he gave the identity

$$\sin A - \sin B = 2 \cos \frac{A + B}{2} \sin \frac{A - B}{2}$$

and identities for $\sin n\theta$ and $\cos n\theta$ in terms of $\sin \theta$ and $\cos \theta$. The latter identities are contained in his *Sectiones Angulares*, published posthumously in 1615.[2] Vieta expressed and worked with the identities in algebraic form, though the notation was by no means modern.

He used the formula for $\sin n\theta$ to solve the problem posed by the Belgian mathematician Adrianus Romanus (1561–1615) in his book *Ideae Mathematicae* (1593) as a challenge to all Frenchmen. The problem was to solve an equation of the forty-fifth degree in x. Henry IV of France sent for Vieta, who recognized that the problem amounted to this: Given the chord of an arc, to find the chord of the forty-fifth part of that arc. This is equivalent to expressing $\sin 45A$ in terms of $\sin A$ and finding $\sin A$. If $x = \sin A$ then the algebraic equation is of the forty-fifth degree in x. Vieta knew that problem could be solved by breaking this equation up into a fifth-degree equation and two third-degree equations; these he solved quickly. He gave the 23 positive roots but ignored the negative ones. In his *Responsum* (1595)[3] he explained his method of solution.

In the sixteenth century, trigonometry began to break away from astronomy and acquire status as a branch of mathematics. The application to astronomy continued to be extensive, but other applications—as, for example, surveying—warranted the study of the subject from a more detached viewpoint.

5. *The Major Scientific Progress in the Renaissance*

The Renaissance mathematicians prepared the ground for the upsurge of mathematical study in Europe by translating the Greek and Arabic works and by compiling encyclopedic works on the existing knowledge. But the motivations and directions of the subsequent mathematical creations of the Europeans stemmed primarily from scientific and technological problems,

2. *Opera*, Leyden, 1646, 287–304.
3. *Opera*, 305–24.

though there were some exceptions. The rise of algebra was, at least at the outset, a continuation of Arab lines of activity; and some new work in geometry was suggested by problems raised by artists.

By far the most significant Renaissance development in motivating the mathematics of the next two centuries was the revolution in astronomy led by Copernicus and Kepler. When the Greek works became available, after about 1200, both Aristotle's astronomical theory (a modification of Eudoxus') and Ptolemy's theory became widely known and were pitted against each other. Strictly speaking, various additions to both schemes had been introduced by the Arabs and the late medieval astronomers to improve the accuracy of both systems or to accommodate Aristotle's scheme to Christian theology. Ptolemy's scheme, reasonably accurate for the time, was purely mathematical and hence regarded only as a hypothesis, not a description of real structures. Aristotle's theory was the one most men accepted, though Ptolemy's was more useful for astronomical predictions, navigation, and calendar-reckoning.

A few Arab, late medieval, and Renaissance figures, including al-Bîrûnî (973–1048), Oresme, and Cardinal Nicholas of Cusa (1401–64), perhaps responding to Greek ideas, seriously considered that the earth might be rotating and that it might be equally possible to build an astronomical theory on the basis of the earth moving around the sun, but none worked out a new theory.

Among astronomers, Nicholas Copernicus suddenly appeared as a colossus. Born in Thorn, Poland, in 1473, Copernicus studied mathematics and science at the University of Cracow. At the age of twenty-three he went to Bologna to further his studies and there became familiar with Pythagorean and other Greek doctrines, including astronomical theory. He also studied medicine and church law. In 1512 he returned to Poland to become canon (an administrator) at the Cathedral of Frauenberg where he remained until his death in 1543. While performing his duties he devoted himself to intensive studies and observations that culminated in a revolutionary astronomical theory. This achievement in the domain of thought outclasses in significance, boldness, and grandeur the conquest of the seas.

It is difficult to determine what caused Copernicus to overthrow the fourteen-hundred-year-old Ptolemaic theory. The indications in the preface of his classic work, *De Revolutionibus Orbium Coelestium* (On the Revolutions of the Heavenly Spheres, 1543) are incomplete and somewhat enigmatic. Copernicus states that he was aroused by divergent views on the accuracy of the Ptolemaic system, by the view that the Ptolemaic theory was just a convenient hypothesis, and by the conflict between adherents of the Aristotelian and Ptolemaic theories.

Copernicus retained some principles of Ptolemaic astronomy. He used the circle as the basic curve on which his explanation of the motions of the heavenly bodies was to be constructed. Moreover, like Ptolemy, he used the fact that the motion of the planets must be built up through a sequence of motions with constant speed. His reason was that a change in speed could be caused only by a change in motive power, and since God, the cause of motion, was constant, the effect had to be, as well. He also took over the Greek scheme of epicyclic motion on a deferent. But Copernicus objected to the uniform equant motion utilized by Ptolemy because this motion did not call for uniform linear velocity.

By using Aristarchus' idea of putting the sun rather than the earth at the center of each deferent, Copernicus was able to replace the complicated diagrams formerly required to portray the motion of each heavenly body by much simpler ones. Instead of 77 circles he needed only 34 to explain the motion of the moon and the six known planets. Later he refined this scheme somewhat by putting the sun near but not quite at the center of the system.

Copernicus's theory was not in better accord with observations than the current modifications of Ptolemy's theory. The merit of his system was rather that it made the motion of the earth around the sun account for the major irregularities in the motions of the planets without employing as many epicycles. Moreover, his scheme treated all the planets in the same general way, whereas Ptolemy had used somewhat different schemes for the inner planets, Mercury and Venus, than for the outer planets, Mars, Jupiter, and Saturn. Finally, the calculation of the positions of the heavenly bodies was simpler in Copernicus's scheme, so much so that even in 1542 astronomers, using his theory, began preparation of new tables of celestial positions.

Copernicus's theory met both sound and prejudiced opposition. The discrepancies between Copernican theory and observations caused Tycho Brahe (1546–1601) to abandon the theory and seek a compromise. Vieta, for the same reason, rejected it altogether and turned to improving Ptolemaic theory. Most intellectuals rejected the theory either because they failed to understand it or because they would not entertain revolutionary ideas. The mathematics certainly was difficult to understand; as Copernicus himself said in his preface, the book was addressed to mathematicians. The observation of a new star by Brahe and German astronomers in 1572 did help; the sudden appearance and disappearance of stars contradicted the Aristotelian and Scholastic dogma of the invariability of the heavens.

The fate of the heliocentric theory would have been very uncertain were it not for the work of Johannes Kepler (1571–1630). He was born in Weil, a city in the Duchy of Württemberg. His father was a drunkard who turned from being a soldier of fortune to keeping a tavern. While attending elementary school, Kepler had to help in the tavern. He was then withdrawn from school and sent to work as a laborer in the fields. When still a boy he contracted smallpox, which left him with crippled hands and impaired eyesight. However, he managed to get a bachelor's degree at the College of Maulbronn in 1588; then, headed for the ministry, he studied at the University of Tübingen. There a friendly professor of mathematics and astronomy, Michael Mästlin (Möstlin, 1550–1631) taught him Copernican theory privately. Kepler's superiors at the university questioned his devoutness and in 1594 offered him a professorship of mathematics and morals at Grätz in Austria, which Kepler accepted. To fulfill his duties he was required to know astrology; this turned him further toward astronomy.

Kepler was expelled from Grätz when the city came under Catholic control, and he became an assistant to Tycho Brahe in the latter's observatory at Prague. When Brahe died, Kepler was appointed to his place. Part of his job was to cast horoscopes for his employer, Emperor Rudolph II. Kepler consoled himself with the thought that astrology enabled astronomers to make a living.

During his entire life Kepler was harassed by all sorts of difficulties. His first wife and several children died. As a Protestant, he suffered in various ways from persecution by Catholics. He was often in desperate financial state. His mother was accused of witchcraft and Kepler had to defend her. Yet throughout all his misfortunes he pursued his scientific work with perseverance, extraordinary labor, and fertile imagination.

In his approach to scientific problems Kepler is a transitional figure. Like Copernicus and the medieval thinkers he was attracted by a pretty and rational theory. He accepted the Platonic doctrine that the universe is ordered according to a preestablished mathematical plan. But unlike his predecessors, he had an immense regard for facts. His more mature work was based entirely on facts and advanced from facts to laws. In the search for laws he displayed inventiveness in hypotheses, a love of truth, and a lively fancy playing with but not obstructing reason. Though he devised a great number of hypotheses he did not hesitate to discard them when they did not fit the facts.

Moved by the beauty and harmony of Copernicus's system, he decided to devote himself to the search for whatever additional geometrical harmonies the much more accurate observations supplied by Tycho Brahe might permit. His search for the mathematical relationships of whose existence he was convinced led him to spend years following up false trails. In the preface to his *Mysterium Cosmographicum* (1596), we find him saying: "I undertake to prove that God, in creating the universe and regulating the order of the cosmos, had in view the five regular bodies of geometry as known since the days of Pythagoras and Plato, and that he has fixed according to those dimensions, the number of heavens, their proportions, and the relations of their movements."

And so he postulated that the radii of the orbits of the six planets were the radii of spheres related to the five regular solids in the following way. The

Figure 12.2

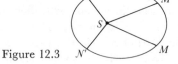

Figure 12.3

largest radius was that of the orbit of Saturn. In a sphere of this radius he supposed a cube to be inscribed. In this cube a sphere was inscribed whose radius should be that of the orbit of Jupiter. In this sphere he supposed a tetrahedron to be inscribed and in this, in turn, another sphere, whose radius was to be that of the orbit of Mars, and so on through the five regular solids. This allowed for six spheres, just enough for the number of planets then known. However, the deductions from this hypothesis were not in accord with observations and he abandoned the idea, but not before he had made extraordinary efforts to apply it in modified form.

Although the attempt to use the five regular solids to ferret out nature's secrets did not succeed, Kepler was eminently successful in later efforts to find harmonious mathematical relations. His most famous and important results are known today as Kepler's three laws of planetary motion. The first two were published in a book of 1609 bearing a long title, which is sometimes shortened to *Astronomia Nova* and sometimes to *Commentaries on the Motions of Mars*.

The first law states that the path of each planet is not the resultant of a combination of moving circles but is an ellipse with the sun at one focus (Fig. 12.2). Kepler's second law is best understood in terms of a diagram (Fig. 12.3). The Greeks, we saw, believed that the motion of a planet must be explained in terms of constant linear speeds. Kepler, like Copernicus, at first held firmly to the doctrine of constant speeds. But his observations compelled him to abandon this cherished belief too. His joy was great when he was able to replace it by something equally attractive, for his conviction that nature follows mathematical laws was reaffirmed. If MM' and NN' are distances traversed by a planet in equal intervals of time, then, according to the principle of constant speed, MM' and NN' would have to be equal distances. However, according to Kepler's second law, MM' and NN' are generally not equal but the *areas* SMM' and SNN' are equal. Thus Kepler replaced equal distances by equal areas. To wrest such a secret from the planets was indeed a triumph, for the relationship described is by no means as easily discernible as it may appear to be on paper.

Kepler made even more extraordinary efforts to obtain the third law of motion. This law says the square of the period of revolution of any planet is equal to the cube of its average distance from the sun, provided the period of the earth's revolution and its distance from the sun are the units of time and distance.[4] Kepler published this result in *The Harmony of the World* (1619).

Kepler's work is far more revolutionary than Copernicus's; equally daring in adopting heliocentrism, Kepler broke radically from authority and tradition by utilizing the ellipse (as opposed to a composition of circular motions) and nonuniform velocities. He hewed firmly to the position that scientific investigations are independent of all philosophical and theological doctrines, that mathematical considerations alone should determine the wisdom of any hypothesis, and that the hypotheses and deductions from them must stand the test of empirical confirmation.

The work of Copernicus and Kepler is remarkable on many accounts, but we must confine ourselves to its relevance to the history of mathematics. In view of the many serious counter-arguments against a heliocentric theory, their work demonstrates how strongly the Greek view that the truths of nature lie in mathematical laws had already taken hold in Europe.

There were weighty scientific objections, many of which had already been advanced by Ptolemy against Aristarchus' suggestion. How could such a heavy body as the earth be set into and kept in motion? The other planets were in motion, even according to Ptolemaic theory, but the Greeks and medieval thinkers had maintained that these were composed of some special light substance. There were other objections. Why, if the earth rotates from west to east, does an object thrown up into the air not fall back to the west of its original position? Why doesn't the earth fly apart in its rotation? Copernicus's very weak answer to the latter question was that the sphere is a natural shape and moves naturally, and hence the earth would not destroy itself. Further, it was asked, why do objects on the earth and the air itself stay with an earth rotating at about 3/10 of a mile per second and revolving around the sun at the rate of about 18 miles per second? If, as Ptolemy and Copernicus believed, the speed of a body in natural motion is proportional to its weight, the earth should leave behind it objects of lesser weight. Copernicus replied that the air possessed "earthiness" and would therefore stay with the earth.

There were additional scientific objections from astronomers. If the earth moved, why did the direction of the "fixed" stars not change? An angle of parallax of 2' required that the distance of the stars be at least four million times the radius of the earth; such a distance was inconceivable at that time. Not detecting any parallax of the stars (which implied that they had to be even farther away), Copernicus declared that "the heavens are immense by comparison with the Earth and appear to be of infinite size.... The bounds of the universe are unknown and unknowable." Then, realizing the inadequacy of this answer, he assigned the problem to the philosophers

4. Though Kepler stated it this way, the correct statement calls for replacing average distance by the semimajor axis.

and thereby evaded it. Not until 1838 did the mathematician Bessel measure the parallax of one of the nearest stars and find it to be 0.31″.

If Copernicus and Kepler had been "sensible" men, they would never have defied their senses. We do not feel either the rotation or the revolution of the earth despite the high speeds involved. On the other hand, we do see the motion of the sun.

Copernicus and Kepler were highly religious; yet both denied one of the central doctrines of Christianity, that man, the chief concern of God, was at the center of the universe, and everything in the universe revolved around him. In contrast, the heliocentric theory, by putting the sun at the center of the universe, undermined this comforting dogma of the Church, because it made man appear to be just one of a possible host of wanderers drifting through a cold sky. It seemed less likely that he was born to live gloriously and to attain paradise upon his death. Less likely, too, was it that he was the object of God's ministrations. Thus, by displacing the earth, Copernicus and Kepler removed a cornerstone of Catholic theology and imperiled its structure. Copernicus pointed out that the universe is so immense, compared to the earth, that to speak of a center is meaningless. However, this argument put him all the more into opposition with religion.

Against all the objections Copernicus and Kepler had only one retort, but a weighty one. Each had achieved mathematical simplification and, indeed, overwhelmingly harmonious and aesthetically superior theory. If mathematical relationships were to be the goal of scientific work, and if a better mathematical account could be given, then this fact, reinforced by the belief that God had designed the world and would clearly have used the superior theory, was sufficient to outweigh all objections. Each felt, and clearly stated, that his work revealed the harmony, symmetry, and design of the divine workshop and overpowering evidence of God's presence. Copernicus could not restrain his gratification: "We find, therefore, under this orderly arrangement, a wonderful symmetry in the universe, and a definite relation of harmony in the motion and magnitude of the orbs, of a kind that it is not possible to obtain in any other way." The very title of Kepler's work of 1619, *The Harmony of the World*, and endless paeans to God, expressing satisfaction with the grandeur of God's mathematical design, attest to his beliefs.

It is not surprising that at first only mathematicians supported the heliocentric theory. Only a mathematician, and one convinced that the universe was mathematically designed, would have had the mental fortitude to disregard the prevailing philosophical, religious, and physical beliefs. Not until Galileo focused his telescope on the heavens did astronomical evidence favor the mathematical argument. Galileo's observations, made in the early 1600s, revealed four moons circling around Jupiter, showing that planets could have moons. It followed that the earth, too, need not be more than a planet just because it had a moon. Galileo also observed that the moon had a rough surface and mountains and valleys like the earth. Hence the earth was also likely to be just one more heavenly body and not necessarily the center of the universe.

The heliocentric theory finally won acceptance because it was simpler for calculations, because of its superior mathematics, and because observations supported it. This meant that the science of motion had to be recast in the light of a rotating and revolving earth. In brief, a new science of mechanics was needed.

Investigations into light and optics continued in an unbroken line from what we have already noted in the medieval period. In the sixteenth century the astronomers became more interested in these subjects because the refractive effect of air on light changes the direction of light rays as they come from the planets and stars and so gives misleading information as to the directions of these bodies. Toward the end of the sixteenth century, the telescope and microscope were invented. The scientific uses of these instruments are obvious; they opened up new worlds, and interest in optics, already extensive, surged still higher. Almost all of the seventeenth-century mathematicians worked on light and lenses.

6. Remarks on the Renaissance

The Renaissance did not produce any brilliant new results in mathematics. The minor progress in this area contrasts with the achievements in literature, painting, and architecture, where masterpieces that still form part of our culture were created, and in science, where the heliocentric theory eclipsed the best of Greek astronomy and dwarfed any Arabic or medieval contributions. For mathematics the period was primarily one of absorption of the Greek works. It was not so much a rebirth as a recovery of an older culture.

Equally important for the health and growth of mathematics was that it had once more, as in Alexandrian times, reestablished its intimate connections with science and technology. In science the realization that mathematical laws are the goal, the be-all and the end-all, and in technology, the appreciation that the mathematical formulation of the results of investigations was the soundest and most useful form of knowledge and the surest guide to design and construction, guaranteed that mathematics was to be a major force in modern times and gave promise of new developments.

Bibliography

Armitage, Angus: *Copernicus*, W. W. Norton, 1938.
————: *John Kepler*, Faber and Faber, 1966.
————: *Sun, Stand Then Still*, Henry Schuman, 1947; in paperback as *The World of Copernicus*, New American Library, 1951.

Ball, W. W. Rouse: *A Short Account of the History of Mathematics*, Dover (reprint), 1960, Chap. 12.

Baumgardt, Carola: *Johannes Kepler, Life and Letters*, Victor Gollancz, 1952.

Berry, Arthur: *A Short History of Astronomy*, Dover (reprint), 1961, pp. 86–197.

Braunmühl, A. von: *Vorlesungen über die Geschichte der Trigonometrie*, 2 vols., B. G. Teubner, 1900 and 1903, reprinted by M. Sändig, 1970.

Burtt, E. A.: *The Metaphysical Foundations of Modern Physical Science*, 2nd ed., Routledge and Kegan Paul, 1932, Chaps. 1–2.

Butterfield, Herbert: *The Origins of Modern Science*, Macmillan, 1951, Chaps. 1–7.

Cantor, Moritz: *Vorlesungen über Geschichte der Mathematik*, 2nd ed., B. G. Teubner, 1900, Vol. 2, pp. 1–344.

Caspar, Max: *Johannes Kepler*, trans. Doris Hellman, Abelard-Schuman, 1960.

Cohen, I. Bernard: *The Birth of a New Physics*, Doubleday, 1960.

Coolidge, Julian L.: *The Mathematics of Great Amateurs*, Dover (reprint), 1963, Chaps. 3–5.

Copernicus, Nicolaus: *De Revolutionibus Orbium Coelestium* (1543), Johnson Reprint Corp., 1965.

Crombie, A. C.: *Augustine to Galileo*, Falcon Press, 1952, Chap. 4.

Da Vinci, Leonardo: *Philosophical Diary*, Philosophical Library, 1959.

————: *Treatise on Painting*, Princeton University Press, 1956.

Dampier-Whetham, William C. D.: *A History of Science*, Cambridge University Press, 1929, Chap. 3.

Dijksterhuis, E. J.: *The Mechanization of the World Picture*, Oxford University Press, 1961, Parts 3 and 4.

Drake, Stillman and I. E. Drabkin: *Mechanics in Sixteenth-Century Italy*, University of Wisconsin Press, 1969.

Dreyer, J. L. E.: *A History of Astronomy from Thales to Kepler*, Dover (reprint), 1953, Chaps. 12–16.

————: *Tycho Brahe, A Picture of Scientific Life and Work in the Sixteenth Century*, Dover (reprint), 1963.

Gade, John A.: *The Life and Times of Tycho Brahe*, Princeton University Press, 1947.

Galilei, Galileo: *Dialogue on the Great World Systems* (1632), University of Chicago Press, 1953.

Hall, A. R.: *The Scientific Revolution*, Longmans Green, 1954, Chaps. 1–6.

Hallerberg, Arthur E.: "George Mohr and *Euclidis Curiosi*," *The Mathematics Teacher*, 53, 1960, 127–32.

————: "The Geometry of the Fixed-Compass," *The Mathematics Teacher*, 52, 1959, 230–44.

Hart, Ivor B.: *The World of Leonardo da Vinci*, Viking Press, 1962.

Hofmann, Joseph E.: *The History of Mathematics*, Philosophical Library, 1957.

Hughes, Barnabas: *Regiomontanus on Triangles*, University of Wisconsin Press, 1967. A translation of *De Triangulis*.

Ivins, W. M., Jr.: *Art and Geometry* (1946), Dover (reprint), 1965.

Kepler, Johannes: *Gesammelte Werke*, C. H. Beck'sche Verlagsbuchhandlung, 1938–59.

————: *Concerning the More Certain Foundations of Astrology*, Cancy Publications, 1942. Many of Kepler's books have been reprinted and a few translated.

Koyré, Alexandre: *From the Closed World to the Infinite Universe*, Johns Hopkins Press, 1957.

——: *La Révolution astronomique*, Hermann, 1961.

Kuhn, Thomas S.: *The Copernican Revolution*, Harvard University Press, 1957.

MacCurdy, Edward: *The Notebooks of Leonardo da Vinci*, George Braziller (reprint), 1954.

Pannekoek, A.: *A History of Astronomy*, John Wiley and Sons, 1961, Chaps. 16–25.

Panofsky, Erwin: *"Dürer as a Mathematician,"* in James R. Newman, *The World of Mathematics*, Simon and Schuster, 1956, pp. 603–21.

Santillana, G. de: *The Crime of Galileo*, University of Chicago Press, 1955.

Sarton, George: *The Appreciation of Ancient and Medieval Science During the Renaissance*, University of Pennsylvania Press, 1955.

——: *Six Wings: Men of Science in the Renaissance*, Indiana University Press, 1957.

Smith, David Eugene: *History of Mathematics*, Dover (reprint), 1958, Vol. 1, Chap. 8; Vol. 2, Chap. 8.

Smith, Preserved: *A History of Modern Culture*, Holt, Rinehart and Winston, 1930, Vol. 1, Chaps. 2–3.

Taylor, Henry Osborn: *Thought and Expression in the Sixteenth Century*, Crowell-Collier (reprint), 1962, Part V.

Taylor, R. Emmet: *No Royal Road: Luca Pacioli and His Times*, University of North Carolina Press, 1942.

Tropfke, Johannes: *Geschichte der Elementarmathematik*, 7 vols., 2nd. ed., W. De Gruyter, 1921–24.

Vasari, Giorgio: *Lives of the Most Eminent Painters, Sculptors, and Architects* (many editions).

Wolf, Abraham: *A History of Science, Technology and Philosophy in the Sixteenth and Seventeenth Centuries*, George Allen and Unwin, 1950, Chaps. 1–6.

Zeller, Sister Mary Claudia: *"*The Development of Trigonometry from Regiomontanus to Pitiscus*"*, Ph.D. dissertation, University of Michigan, 1944; Edwards Brothers, 1946.

13
Arithmetic and Algebra in the Sixteenth and Seventeenth Centuries

Algebra is the intellectual instrument for rendering clear the quantitative aspects of the world.

ALFRED NORTH WHITEHEAD

1. *Introduction*

The first major new European mathematical developments took place in arithmetic and algebra. The Hindu and Arabic work had put practical arithmetical calculations in the forefront of mathematics and had placed algebra on an arithmetic instead of a geometric basis. This work also attracted attention to the problem of solving equations.

In the first half of the sixteenth century there was hardly any change from the Arab attitude or spirit, but merely an increase in the kind of activity the Europeans had learned about from the Arab works. By the middle of the century the practical and scientific needs of European civilization were prompting further steps in arithmetic and algebra. Technological applications of scientific work and practical needs require, as we have pointed out, quantitative results. For example, the far-ranging geographical explorations required more accurate astronomical knowledge. At the same time the interest in correlating the new astronomical theory with increasingly accurate observations demanded better astronomical tables, which in turn meant more accurate trigonometric tables. In fact, a good deal of the sixteenth-century interest in algebra was motivated by the need to solve equations and work with identities in making trigonometric tables. Developing banking and commercial activities called for an improved arithmetic. The response to these interests is evident in the writings of Pacioli, Tartaglia, and Stevin, among others. Pacioli's *Summa* and Tartaglia's *General trattato de' numeri e misure* (1556) contain an immense number of problems in mercantile arithmetic. Finally, the technical work of the artisans, especially in architecture, cannon-making and projectile motion, called for quantitative thinking. Beyond these applications, a totally new use for algebra—the representation of curves—motivated an immense amount of work. Under the pressure of these needs, progress in algebra accelerated.

We shall consider the new developments under four headings: arithmetic, symbolism, the theory of equations, and the theory of numbers.

2. The Status of the Number System and Arithmetic

By 1500 or so, zero was accepted as a number and irrational numbers were used more freely. Pacioli, the German mathematician Michael Stifel (1486?–1567), the military engineer Simon Stevin (1548–1620), and Cardan used irrational numbers in the tradition of the Hindus and Arabs and introduced more and more types. Thus Stifel worked with irrationals of the form $\sqrt[m]{a + \sqrt[n]{b}}$. Cardan rationalized fractions with cube roots in them. The extent to which irrationals were used is exemplified by Vieta's expression for π.[1] By considering regular polygons of 4, 8, 16, . . . sides inscribed in a circle of unit radius, Vieta found that the value of π is given by

$$\frac{2}{\pi} = \cos\frac{90°}{2} \cos\frac{90°}{4} \cos\frac{90°}{8} \cdots$$

$$= \sqrt{\frac{1}{2}} \sqrt{\frac{1}{2} + \frac{1}{2}\sqrt{\frac{1}{2}}} \sqrt{\frac{1}{2} + \frac{1}{2}\sqrt{\frac{1}{2} + \frac{1}{2}\sqrt{\frac{1}{2}}}} \cdots.$$

Though calculations with irrationals were carried on freely, the problem of whether irrationals were really numbers still troubled people. In his major work, the *Arithmetica Integra* (1544), which deals with arithmetic, the irrationals in the tenth book of Euclid, and algebra, Stifel considers expressing irrationals in the decimal notation. On the one hand, he argues:

> Since, in proving geometrical figures, when rational numbers fail us irrational numbers take their place and prove exactly those things which rational numbers could not prove ... we are moved and compelled to assert that they truly are numbers, compelled that is, by the results which follow from their use—results which we perceive to be real, certain, and constant. On the other hand, other considerations compel us to deny that irrational numbers are numbers at all. To wit, when we seek to subject them to numeration [decimal representation] ... we find that they flee away perpetually, so that not one of them can be apprehended precisely in itself. ... Now that cannot be called a true number which is of such a nature that it lacks precision. ... Therefore, just as an infinite number is not a number, so an irrational number is not a true number, but lies hidden in a kind of cloud of infinity.

He then argues that real numbers are either whole numbers or fractions; obviously irrationals are neither, and so are not real numbers. A

1. The symbol π was first used by William Jones (1706).

century later, Pascal and Barrow said that a number such as $\sqrt{3}$ can be understood only as a geometric magnitude; irrational numbers are mere symbols that have no existence independent of continuous geometrical magnitude, and the logic of operations with irrationals must be justified by the Eudoxian theory of magnitudes. This was also the view of Newton in his *Arithmetica Universalis* (published in 1707 but based on lectures of thirty years earlier).

Others made positive assertions that the irrational numbers were independent entities. Stevin recognized irrationals as numbers and approximated them more and more closely by rationals; John Wallis in *Algebra* (1685), also accepted irrationals as numbers in the full sense. He regarded the fifth book of Euclid's *Elements* as essentially arithmetical in nature. Descartes, too, in *Rules for the Direction of the Mind* (*c.* 1628), admitted irrationals as abstract numbers that can represent continuous magnitudes.

As for negative numbers, though they had become known in Europe through the Arab texts, most mathematicians of the sixteenth and seventeenth centuries did not accept them as numbers, or if they did, would not accept them as roots of equations. In the fifteenth century Nicolas Chuquet (1445?–1500?) and, in the sixteenth, Stifel (1553) both spoke of negative numbers as absurd numbers. Cardan gave negative numbers as roots of equations but considered them impossible solutions, mere symbols; he called them fictitious, whereas he called positive roots real. Vieta discarded negative numbers entirely. Descartes accepted them, in part. He called negative roots of equations false, on the ground that they claim to represent numbers less than nothing. However, he had shown (see sec. 5) that, given an equation, one can obtain another whose roots are larger than the original one by any given quantity. Thus an equation with negative roots could be transformed into one with positive roots. Since we can turn false roots into real roots, Descartes was willing to accept negative numbers. Pascal regarded the subtraction of 4 from 0 as utter nonsense.

An interesting argument against negative numbers was given by Antoine Arnauld (1612–94), a theologian and mathematician who was a close friend of Pascal. Arnauld questioned that $-1:1 = 1:-1$ because, he said, -1 is less than $+1$; hence, How could a smaller be to a greater as a greater is to a smaller? The problem was discussed by many men. In 1712 Leibniz agreed[2] that there was a valid objection but argued that one can calculate with such proportions because their form is correct, just as one calculates with imaginary quantities.

One of the first algebraists to accept negative numbers was Thomas Harriot (1560–1621), who occasionally placed a negative number by itself on one side of an equation. But he did not accept negative roots. Raphael

---

2. *Acta Erud.*, 1712, 167–69 = *Math. Schriften*, 5, 387–89.

Bombelli (16th cent.) gave clear definitions for negative numbers. Stevin used positive and negative coefficients in equations and also accepted negative roots. In his *L'Invention nouvelle en l'algèbre* (1629), Albert Girard (1595–1632) placed negative numbers on a par with positive numbers and gave both roots of a quadratic equation, even when both were negative. Both Girard and Harriot used the minus sign for the operation of subtraction and for negative numbers.

On the whole not many sixteenth- and seventeenth-century mathematicians felt at ease with or accepted negative numbers as such, let alone recognizing them as true roots of equations. There were some curious beliefs about them. Though Wallis was advanced for his times and accepted negative numbers, he thought they were larger than infinity but not less than zero. In his *Arithmetica Infinitorum* (1655), he argued that since the ratio $a/0$, when a is positive, is infinite, then, when the denominator is changed to a negative number, as in a/b with b negative, the ratio must be greater than infinity.

Without having fully overcome their difficulties with irrational and negative numbers the Europeans added to their problems by blundering into what we now call complex numbers. They obtained these new numbers by extending the arithmetic operation of square root to whatever numbers appeared in solving quadratic equations by the usual method of completing the square. Thus, Cardan, in Chapter 37 of *Ars Magna* (1545), sets up and solves the problem of dividing 10 into two parts whose product is 40. The equation is $x(10 - x) = 40$. He obtains the roots $5 + \sqrt{-15}$ and $5 - \sqrt{-15}$ and then says, "Putting aside the mental tortures involved," multiply $5 + \sqrt{-15}$ and $5 - \sqrt{-15}$; the product is $25 - (-15)$ or 40. He then states, "So progresses arithmetic subtlety the end of which, as is said, is as refined as it is useless." As we shall soon see, Cardan became further involved with complex numbers in his solution of cubic equations (sec. 4). Bombelli too considered complex numbers in the solution of cubic equations and formulated in practically modern form the four operations with complex numbers; but he still regarded them as useless and "sophistic." Albert Girard did recognize complex numbers as at least formal solutions of equations. In *L'Invention nouvelle en l'algèbre* he says, "One could say: Of what use are these impossible solutions [complex roots]? I answer: For three things—for the certitude of the general rules, for their utility, and because there are no other solutions." However, Girard's advanced views were not influential.

Descartes also rejected complex roots and coined the term "imaginary." He says in *La Géométrie*, "Neither the true nor the false [negative] roots are always real; sometimes they are imaginary." He argued that, whereas negative roots can at least be made "real" by transforming the equation in which they occur into another equation whose roots are positive, this cannot be done for complex roots. These, therefore, are not real but imaginary; they are not numbers. Descartes did make a clearer distinction than his predecessors between real and imaginary roots of equations.

Even Newton did not regard complex roots as significant, most likely because in his day they lacked physical meaning. In fact he says, in *Universal Arithmetic*,[3] "But it is just that the Roots of Equations should be often impossible [complex], lest they should exhibit the cases of Problems that are impossible as if they were possible." That is, problems that do not have a physically or geometrically real solution should have complex roots.

The lack of clarity about complex numbers is illustrated by the oft-quoted statement by Leibniz, "The Divine Spirit found a sublime outlet in that wonder of analysis, that portent of the ideal world, that amphibian between being and not-being, which we call the imaginary root of negative unity."[4] Though Leibniz worked formally with complex numbers, he had no understanding of their nature.

During the sixteenth and seventeenth centuries, the operational procedures with real numbers were improved and extended. In Belgium (then part of the Netherlands), we find Stevin advocating in *La Disme* (Decimal Arithmetic, 1585) the use of decimals, as opposed to the sexagesimal system, for writing and operating with fractions. Others—Christoff Rudolff (*c*. 1500–*c*. 1545), Vieta, and the Arab al-Kashî (d. *c*. 1436)—had used them previously. Stevin advocated a decimal system of weights and measures; he was concerned to save the time and labor of bookkeepers (he himself had started his career as a clerk). He writes 5.912 as 5 ⓪ 9 ① 1 ② 2 ③, or as 5, 9′ 1″ 2‴. Vieta improved and extended the methods of performing square and cube roots.

The use of continued fractions in arithmetic is another development of the period. We may recall that the Hindus—Āryabhata in particular—had used continued fractions to solve linear indeterminate equations. Bombelli, in his *Algebra* (1572), was the first to use them in approximating square roots. To approximate $\sqrt{2}$, he writes

$$(1) \qquad\qquad \sqrt{2} = 1 + \frac{1}{y}.$$

From this one finds

$$(2) \qquad\qquad y = 1 + \sqrt{2}.$$

By adding 1 to both sides of (1) and using (2), it follows that

$$(3) \qquad\qquad y = 2 + \frac{1}{y}.$$

3. Second ed., 1728, p. 193.
4. *Acta Erud.*, 1702 = *Math. Schriften*, 5, 350–61.

Hence, again by (1) and (3),

$$\sqrt{2} = 1 + \cfrac{1}{2 + \cfrac{1}{y}}.$$

And, since y is given by (3),

$$\sqrt{2} = 1 + \cfrac{1}{2 + \cfrac{1}{2 + \cfrac{1}{y}}}.$$

By repeated substitution of the value of y, Bombelli obtained

$$\sqrt{2} = 1 + \cfrac{1}{2 + \cfrac{1}{2 + \cfrac{1}{2 + \cfrac{1}{2} + \cdots}}}.$$

The right-hand side is also written as

$$\sqrt{2} = 1 + \frac{1}{2+} \frac{1}{2+} \frac{1}{2+} \cdots.$$

This continued fraction is simple because the numerators are all 1; it is periodic because the denominators repeat. Bombelli gave other examples of how to obtain continued fractions. He did not, however, consider the question of whether the expansion converged to the number it was supposed to represent.

The English mathematician John Wallis, in his *Arithmetica Infinitorum* (1655), represented $\frac{4}{\pi}$ as the infinite product $\frac{3 \cdot 3 \cdot 5 \cdot 5 \cdot 7 \cdot 7 \cdots}{2 \cdot 4 \cdot 4 \cdot 6 \cdot 6 \cdot 8 \cdots}$. In this book he also stated that Lord William Brouncker (1620–84), the first president of the Royal Society, had transformed this product into the continued fraction

$$\frac{4}{\pi} = 1 + \frac{1}{2+} \frac{9}{2+} \frac{25}{2+} \frac{49}{2+} \cdots.$$

Brouncker made no further use of this form. Wallis, however, took up the work. In his *Opera Mathematica*, I (1695), wherein he introduced the term "continued fraction," he gave the general rule for calculating the convergents of a continued fraction. That is, if p_n/q_n is the nth convergent of the continued fraction

$$\frac{b_1}{a_1+} \frac{b_2}{a_2+} \frac{b_3}{a_3+} \cdots, \quad \text{then} \quad \frac{p_n}{q_n} = \frac{a_n p_{n-1} + b_n p_{n-2}}{a_n q_{n-1} + b_n q_{n-2}}.$$

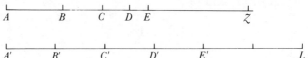

Figure 13.1

No definitive result on the convergence of p_n/q_n to the number that the continued fraction represents was obtained at this time.

The biggest improvement in arithmetic during the sixteenth and seventeenth centuries was the invention of logarithms. The basic idea was noted by Stifel. In *Arithmetica Integra* he observed that the terms of the geometric progression

$$1, r, r^2, r^3, \ldots$$

correspond to the terms in the arithmetic progression formed by the exponents

$$0, 1, 2, 3, \ldots.$$

Multiplication of two terms in the geometric progression yields a term whose exponent is the sum of the corresponding terms in the arithmetic progression. Division of two terms in the geometric progression yields a term whose exponent is the difference of the corresponding terms in the arithmetic progression. This observation had also been made by Chuquet in *Le Triparty en la science des nombres* (1484). Stifel extended this connection between the two progressions to negative and fractional exponents. Thus the division of r^2 by r^3 yields r^{-1}, which corresponds to the term -1 in the extended arithmetic progression. Stifel, however, did not make use of this connection between the two progressions to introduce logarithms.

John Napier (1550–1617), the Scotsman who did develop logarithms about 1594, was guided by this correspondence between the terms of a geometric progression and those of the corresponding arithmetic progression. Napier was interested in facilitating calculations in spherical trigonometry that were being made on behalf of astronomical problems. In fact, he sent his preliminary results to Tycho Brahe for approval.

Napier explained his ideas in *Mirifici Logarithmorum Canonis Descriptio* (1614) and in the posthumously published *Mirifici Logarithmorum Canonis Constructio* (1619). Since he was concerned with spherical trigonometry, he dealt with the logarithms of sines. Following Regiomontanus, who used half chords and a circle whose radius contained 10^7 units, Napier started with 10^7 as the largest number to be considered. He let the sine values from 10^7 to 0 be represented on the line AZ (Fig. 13.1) and supposed that A moves toward Z with a velocity proportional to its distance from Z.

Strictly speaking, the velocity of the moving point varies continually with the distance from A and the proper expression of it calls for the calculus. However, if we consider any small interval of time t and let the lengths AB, BC, CD, . . . be traversed in this interval, and if we assume that the velocity during the interval t is constant and the one possessed by the moving point at the beginning of the interval, then the lengths AZ, BZ, CZ, . . . are in geometric progression. For, consider DZ, and let k be the proportionality constant relating the velocity and the distance of the moving point from Z. Now

$$DZ = CZ - CD.$$

Moreover, the velocity of the moving point at C is $k(CZ)$. Then the distance $CD = k(CZ)t$. Hence

$$DZ = CZ - k(CZ)t = CZ(1 - kt).$$

Thus any length in the sequence AZ, BZ, CZ, . . . is $1 - kt$ times the preceding length.

Next Napier supposed that another point starting at the same time as A moves at a *constant* velocity on the line $A'L$ (Fig. 13.1), which extends indefinitely to the right, so that this point reaches B', C', D', . . . as the first point reaches B, C, D, . . ., respectively. The distances $A'B'$, $A'C'$, $A'D'$, . . . are obviously in arithmetic progression. These distances $A'B'$, $A'C'$, $A'D'$, . . . were taken by Napier to be the logarithms of BZ, CZ, DZ, . . ., respectively. The logarithm of AZ or 10^7 was taken to be 0. Thus the logarithms increase in arithmetic progression while the numbers (sine values) decrease in geometric progression. The quantity here denoted by $1 - kt$ is $1 - \frac{1}{10^7}$ in Napier's original work. However, he changed this ratio as he proceeded in his calculations of logarithms.

Note that the smaller we take t the less the decrease in the sine values from AZ to BZ to CZ, etc. Hence the numbers of the logarithm table are closer together.

Napier took the distances $A'B'$, $A'C'$, . . . to be 1, 2, 3, . . ., though there was no need to do that. They could have been 1/2, 1, 1 1/2, 2, . . . and the scheme would have worked just as well. Moreover, the original numbers were meaningful just as quantities, independent of the fact that they were sine values; so Napier's scheme really gave the logarithms of numbers. Napier himself applied logarithms to computations of spherical trigonometry.

The word "logarithm," coined by Napier, means "number of the ratio." "Ratio" refers to the common ratio of the sequence of numbers AZ, BZ, CZ, ... He also referred to logarithms as "artificial numbers."

Henry Briggs (1561–1631), a professor of mathematics and astronomy, suggested to Napier in 1615 that 10 be used as a base and that the logarithm of a number be the exponent in the power of 10 equal to that number. Here, unlike Napier's scheme, a base is chosen first. Briggs calculated his logarithms by taking successive square roots of 10, that is, $\sqrt{10}$, $\sqrt{\sqrt{10}}$, ..., until he reached, after 54 such root extractions, a number slightly greater than 1. That is, he obtained a number $A = 10^{[(1/2)^{54}]}$. Then he took $\log_{10} A$ to be $(1/2)^{54}$. By using the fact that the logarithm of a product of two numbers is the sum of their logarithms, he built up a table of logarithms for numbers that were closely spaced. The tables of common logarithms in current use are derived from those of Briggs.

Joost Bürgi (1552–1632), a Swiss watch and instrument maker and an assistant to Kepler in Prague, was also interested in facilitating astronomical calculations; he invented logarithms independently of Napier about 1600 but did not publish his work, *Progress Tabulen*, until 1620. Bürgi too was stimulated by Stifel's remarks that multiplication and division of terms in a geometric progression can be performed by adding and subtracting the exponents. His arithmetical work was similar to Napier's.

Variations of Napier's idea were gradually introduced. Also, many different tables of logarithms were calculated by algebraic means. The calculation of logarithms by the use of infinite series was made later by James Gregory, Lord Brouncker, Nicholas Mercator (born Kaufman, 1620–87), Wallis, and Edmond Halley (see Chap. 20, sec. 2).

Though the definition of logarithms as exponents of the powers that represent the numbers in a fixed base, as in Briggs's scheme, became the common approach, they were not defined as exponents in the early seventeenth century because fractional and irrational exponents were not in use. By the end of the century a number of men recognized that logarithms could be so defined, but the first systematic exposition of this approach was not made until 1742, when William Jones (1675–1749) gave such a presentation in the introduction to William Gardiner's *Table of Logarithms*. Euler had already defined logarithms as exponents and in 1728, in an unpublished manuscript (*Opera Posthuma*, II, 800–804), introduced e for the base of the natural logarithms.

The next development in arithmetic (whose further realization has proved momentous in recent times) was the invention of mechanical devices and machines to speed up the execution of arithmetic processes. The slide rule comes from the work of Edmund Gunter (1581–1626), who utilized Napier's logarithms. William Oughtred (1574–1660) introduced circular slide rules.

In 1642 Pascal invented a computing machine that handled addition by carrying from units to tens, tens to hundreds, etc., automatically. Leibniz saw it in Paris and then invented a machine which could perform multiplication. He showed the Royal Society of London his idea in 1677; a description was published by the Berlin Academy in 1710. In the late seventeenth century Samuel Morland (1625–95), independently invented one machine for addition and subtraction and another for multiplication.

We shall not pursue the history of calculating machines because until at least 1940, they were simply mechanical devices that performed only arithmetic and had no influence on the course of mathematics. We shall, however, note that the most significant step between those already described and modern electronic computers was made by Charles Babbage (1792–1871), who introduced a machine intended for astronomical and navigational computations. His "analytic engine" was designed to perform a whole series of arithmetic operations on the basis of instructions fed into the machine at the start. It would then work automatically under steam power. With support from the British government, he built demonstration models. Unfortunately the machine made excessive demands on the engineering competence of his time.

3. Symbolism

The advance in algebra that proved far more significant for its development and for analysis than the technical progress of the sixteenth century was the introduction of better symbolism. Indeed, this step made possible a science of algebra. Prior to the sixteenth century, the only man who had consciously introduced symbolism to make algebraic thinking and writing more compact and more effective was Diophantus. All other changes in notation were essentially abbreviations of normal words, rather casually introduced. In the Renaissance the common style was still rhetorical, that is, the use of special words, abbreviations, and of course number symbols.

The pressure in the sixteenth century to introduce symbolism undoubtedly came from the rapidly expanding scientific demands on mathematicians, just as the improvements in methods of calculation were a response to the increasing uses of that art. However, improvement was intermittent. Many changes were made by accident; and it is clear that the men of the sixteenth century certainly did not have an appreciation of what symbolism could do for algebra. Even after a decided advance in symbolism was made, it was not taken up at once by mathematicians.

Perhaps the first abbreviations, used from the fifteenth century on, were p for plus and m for minus. However, in the Renaissance and especially during the sixteenth and seventeenth centuries, special symbols were introduced. The symbols $+$ and $-$ were introduced by Germans of the fifteenth century to denote excess and defective weights of chests and were taken over by mathematicians; they appear in manuscripts after 1481. The symbol $\times$ for "times" is due to Oughtred; but Leibniz rightly objected to it because it can be confused with the letter x.

The sign $=$ was introduced in 1557 by Robert Recorde (1510–58) of Cambridge, who wrote the first English treatise on algebra, *The Whetstone of Witte* (1557). He said he knew no two things more nearly alike than parallel lines and so two such lines should denote equality. Vieta, who at first wrote out "aequalis," later used $\sim$ for equality. Descartes used $\propto$. The symbols $>$ and $<$ are due to Thomas Harriot. Parentheses appear in 1544. Square brackets and braces, introduced by Vieta, date from about 1593. The square root symbol, $\sqrt{}$, was used by Descartes; but for cube root he wrote $\sqrt{c.}$.

As examples of some of the writing, we might note the following. Using R for square root, p for plus, and m for minus, Cardan wrote

$$(5 + \sqrt{-15}) \cdot (5 - \sqrt{-15}) = 25 - (-15) = 40$$

as

5p: R m:15

5m: R m:15

25m:m:15 qd. est 40.

He also wrote

$$\sqrt{7 + \sqrt{14}} \quad \text{as} \quad \text{R.V. } 7.\text{p: R} 14.$$

The V indicated that all that followed was under the radical sign.

The use of symbols for the unknown and powers of the unknown had a rather surprisingly slow rise, in view of the simplicity and yet extraordinary value of the practice. (Of course Diophantus had used such symbols.) Early sixteenth-century authors, like Pacioli, referred to the unknown as *radix* (Latin for "root") or *res* (Latin for "thing"), *cosa* ("thing" in Italian), and *coss* ("thing" in German), for which reason algebra became known as the "cossic" art. In his *Ars Magna* Cardan referred to the unknown as *rem ignotam*. He wrote $x^2 = 4x + 32$ as *qdratu aeqtur 4 rebus p*:32.[5] The constant term, 32, was called the *numero*. The terms and notation varied a great deal; many symbols were derived from abbreviations. For example, one symbol for the unknown was the letter R, an abbreviation of *res*. The second power, represented by Z (from *zensus*), was called the *quadratum* or *censo*. C, taken from *cubus*, denoted x^3.

Gradually exponents were introduced to denote the powers of x. Exponents attached to numbers, we may recall, were used by Oresme in the fourteenth century. In 1484 Chuquet in *Triparty* wrote 12^3, 10^5, and 120^8 for $12x^3$, $10x^5$, and $120x^8$. He also wrote 12^0 for $12x^0$ and 7^{1m} for $7x^{-1}$. Thus 8^3, 7^{1m} equals 56^2 stood for $8x^3 \cdot 7x^{-1} = 56x^2$.

In his *Algebra* Bombelli used the word *tanto*, instead of *cosa*. For x, x^2, and x^3, he wrote $\underset{\smile}{1}$, $\underset{\smile}{2}$, and $\underset{\smile}{3}$. Thus $1 + 3x + 6x^2 + x^3$ is $1\,p.\ 3\overset{1}{\smile}p.\ 6\overset{2}{\smile}p.\ 1\overset{3}{\smile}$. In 1585 Stevin wrote for this expression $1\,⓪ + 3\,① + 6\ ② + ③$. Stevin also used fractional exponents, $\left(\dfrac{1}{2}\right)$ for square root, $\left(\dfrac{1}{3}\right)$ for cube root, and so on.

---

5. *Rem* and *rebus* are forms in the declension of *res*.

Claude Bachet de Méziriac (1581–1638) preferred to write $x^3 + 13x^2 + 5x + 2$ as $1C + 13Q + 5N + 2$. Vieta used the same notation for equations with numerical coefficients.

Descartes made rather systematic use of positive integral exponents. He expressed

$$1 + 3x + 6x^2 + x^3 \quad \text{as} \quad 1 + 3x + 6xx + x^3.$$

Occasionally he and others used x^2 also. For higher powers he used x^4, $x^5, \ldots$ but not x^n. Newton used positive, negative, integral, and fractional exponents, as in $x^{5/3}$ and x^{-3}. When, in 1801, Gauss adopted x^2 for xx, the former became standard.

The most significant change in the character of algebra was introduced in connection with symbolism by François Vieta. Trained as a lawyer, he served in that capacity for the parliament in Brittany; he was later privy councillor to Henry of Navarre. When he was out of office from 1584 to 1589, as a result of political opposition, he devoted himself entirely to mathematics. Generally, he pursued mathematics as a hobby and printed and circulated his work at his own expense—a guarantee, as one writer put it, of oblivion.

Vieta was a humanist in spirit and intention; he wished to be the preserver, rediscoverer, and continuator of ancient mathematics. Innovation was for him renovation. He describes his *In Artem Analyticam Isagoge* [6] as the "work of the restored mathematical analysis." For this book he drew on the seventh book of Pappus' *Mathematical Collection* and on Diophantus' *Arithmetica*. He believed that the ancients had used a general algebraic type of calculation, which he reintroduced in his algebra, thus merely reactivating an art known and approved of in antiquity.

During the hiatus in his political career Vieta studied the works of Cardan, Tartaglia, Bombelli, Stevin, and Diophantus. From them and particularly from Diophantus he got the idea of using letters. Though a number of men, including Euclid and Aristotle, had used letters in place of specific numbers, these uses were infrequent, sporadic, and incidental. Vieta was the first to use letters purposefully and systematically, not just to represent an unknown or powers of the unknown but as general coefficients. Usually he used consonants for the known quantities and vowels for the unknown quantities. He called his symbolic algebra *logistica speciosa*, as opposed to *logistica numerosa*. Vieta was fully aware that when he studied the general quadratic equation $ax^2 + bx + c = 0$ (in our notation), he was studying an entire class of expressions. In making the distinction between *logistica numerosa* and *logistica speciosa* in his *Isagoge*, Vieta drew the line between arithmetic and algebra. Algebra, the *logistica speciosa*, he said, was a method of operating on species or forms of things. Arithmetic, the *numerosa*,

6. Introduction to the Analytic Art, 1591 = *Opera*, 1–12.

dealt with numbers. Thus in this one step, algebra became a study of general types of forms and equations, since what is done for the general case covers an infinity of special cases. Vieta used literal coefficients for positive numbers only.

Vieta sought to establish the algebraic identities concealed in geometrical form in the old Greek works, but to his mind clearly recognizable in Diophantus. Indeed, as we noted in Chapter 6, the latter had made many transformations of algebraic expressions by using identities that he did not cite explicitly. In his *Zeteticorum Libri Quinque*[7] Vieta sought to recapture these identities. He completed the square of a general quadratic expression and expressed general identities such as

$$a^3 + 3a^2b + 3ab^2 + b^3 = (a + b)^3,$$

though he wrote

a cubus + b in a quadr. 3 + a in b quad. 3 + b cubo aequalia

$$\overline{a + b} \text{ cubo.}$$

It would seem that Vieta's successors would immediately have been struck by the idea of general coefficients. But as far as one can judge, the introduction of letters for classes of numbers was accepted as a minor move in the development of symbolism. The idea of literal coefficients slipped almost casually into mathematics. However Vieta's ideas on symbolism were appreciated and made more flexible by Harriot, Girard, and Oughtred.

Improvements in Vieta's use of letters are due to Descartes. He used first letters of the alphabet for known quantities and last letters for unknowns, which is the modern practice. However, like Vieta, Descartes used letters for positive numbers only, though he did not hesitate to subtract terms with literal coefficients. Not until John Hudde (1633–1704) did so in 1657 was a letter used for positive and negative numbers. Newton did so freely.

Leibniz must be mentioned in the history of symbolism, though he postdates the significant steps in algebra. He made prolonged studies of various notations, experimented with symbols, asked the opinions of his contemporaries, and then chose the best. We shall encounter some of his symbolism in our survey of the calculus. He certainly appreciated the great saving of thought that good symbols make possible.

Thus, by the end of the seventeenth century, the deliberate use of symbolism—as opposed to incidental and accidental use—and the awareness of the power and generality it confers entered mathematics. Unfortunately, far too many symbols introduced in a hit-or-miss and thoughtless

7. Five Books of Analysis, 1593 = *Opera*, 41–81.

manner by men who did not appreciate the importance of this device became standard. In noting this, the historian Florian Cajori was impelled to say that "our symbols today are a mosaic of individual signs of rejected systems."

4. The Solution of Third and Fourth Degree Equations

The solution of quadratic equations by the method of completing the square had been known since Babylonian times, and about the only progress in that subject until 1500 was made by the Hindus, who treated quadratics such as $x^2 + 3x + 2 = 0$ and $x^2 - 3x - 2 = 0$ as one type, whereas their predecessors and even most of their Renaissance successors preferred to treat the latter equation in the form $x^2 = 3x + 2$. Cardan, as we noted, did solve one quadratic having complex roots but dismissed the solutions as useless. The cubic equation, except for isolated cases, had thus far defied the mathematicians; as late as 1494 Pacioli had asserted that the solution of general cubic equations was impossible.

Scipione dal Ferro (1465–1526), a professor of mathematics at Bologna, solved cubics of the type $x^3 + mx = n$ about 1500. He did not publish his method because in the sixteenth and seventeenth centuries discoveries were often kept secret and rivals were challenged to solve the same problem. However, about 1510, he did confide his method to Antonio Maria Fior (first half of 16th cent.) and to his son-in-law and successor Annibale della Nave (1500?–58).

Nothing more happened until Niccolò Fontana of Brescia (1499?–1557) entered the scene. As a boy he had received a saber cut in the face from a French soldier, which caused him to stammer; as a consequence he was called Tartaglia, "Stammerer." Brought up in poverty, he taught himself Latin, Greek, and mathematics. He earned his living by teaching science in various Italian cities. In 1535 Fior challenged Tartaglia to solve thirty cubic equations. Tartaglia, who said he had already solved cubics of the form $x^3 + mx^2 = n$, m and n positive, solved all thirty, including the type $x^3 + mx = n$.

Pressed by Cardan to reveal his method, Tartaglia gave it to him in an obscure verse form after a pledge from Cardan to keep it secret. This was in 1539. In 1542 Cardan and his pupil Lodovico Ferrari (1522–65), on the occasion of a visit by della Nave, determined that dal Ferro's method was the same as Tartaglia's. Despite his pledge, Cardan published his version of the method in his *Ars Magna*. In Chapter 11 he says that, "Scipio Ferro of Bologna well-nigh thirty years ago discovered this rule and handed it on to Antonio Maria Fior of Venice whose contest with Niccolò Tartaglia of Brescia gave Niccolò occasion to discover it. He gave it to me in response to my entreaties, though withholding the demonstration. Armed with this assistance, I sought out its demonstration in [various] forms. This was very difficult. My version of it follows."

Tartaglia protested the breach of promise, and in *Quesiti ed invenzioni diverse* (1546) presented his own case. However, neither in this book nor in his *General trattato de' numeri e misure* (1556), which is a good presentation of the arithmetical and geometrical knowledge of the times, did he give more on the cubic equation itself. The dispute as to who first solved the cubic led to an open conflict between Tartaglia and Ferrari that exhausted itself in wild quarreling in which Cardan took no part. Tartaglia himself was not above reproach; he published a translation of some of Archimedes' work which he actually took over from William of Moerbecke (d. *c.* 1281), and he claimed to have discovered the law of motion of an object on an inclined plane— which really came from Jordanus Nemorarius.

The method Cardan published he first illustrates with the equation $x^3 + 6x = 20$. However, to see the generality of the method we shall consider

$$(4) \qquad\qquad x^3 + mx = n$$

with m and n positive. Cardan introduces two quantities t and u and lets

$$(5) \qquad\qquad t - u = n$$

and

$$(6) \qquad\qquad (tu) = \left(\frac{m}{3}\right)^3.$$

He then asserts that

$$(7) \qquad\qquad x = \sqrt[3]{t} - \sqrt[3]{u}.$$

By elimination, using (5) and (6), and by solving the resulting quadratic equation, he obtains

$$(8) \qquad t = \sqrt{\left(\frac{n}{2}\right)^2 + \left(\frac{m}{3}\right)^3} + \frac{n}{2}, \qquad u = \sqrt{\left(\frac{n}{2}\right)^2 + \left(\frac{m}{3}\right)^3} - \frac{n}{2}.$$

Here we have taken the positive radical as Cardan does. Having obtained t and u Cardan takes the positive cube root of each and by (7) obtains one value of x. Presumably this is the same root that Tartaglia got.

The above is Cardan's method. However, he had to prove that (7) gives a correct value for x. His proof is geometrical; for Cardan, t and u were volumes of cubes whose sides were $\sqrt[3]{t}$ and $\sqrt[3]{u}$, and the product $\sqrt[3]{t} \cdot \sqrt[3]{u}$ was a rectangle formed by the two sides whose area was $m/3$. Also, where we said $t - u = n$, Cardan says the difference in the volumes is n. Then he says that the solution x is the difference in the edges of the two cubes, i.e. $x = \sqrt[3]{t} - \sqrt[3]{u}$. To prove that this value of x is correct, he states and proves a

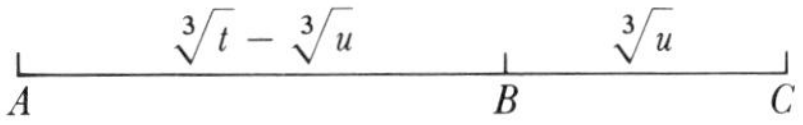

Figure 13.2

$$\underset{A}{\mathrel{\rule{0pt}{1ex}}}\overset{\sqrt[3]{t} - \sqrt[3]{u}}{\underset{B}{\rule{0pt}{1ex}}}\overset{\sqrt[3]{u}}{\underset{C}{\rule{0pt}{1ex}}}$$

geometrical lemma to the effect that, if from a line segment AC (Fig. 13.2) a segment BC is cut off, then the cube on AB will equal the cube on AC minus the cube on BC minus three times the right parallelepiped whose edges are AC, AB, and BC. This geometrical lemma is of course no more than that

(9) $$(\sqrt[3]{t} - \sqrt[3]{u})^3 = t - u - 3(\sqrt[3]{t} - \sqrt[3]{u})\sqrt[3]{t}\,\sqrt[3]{u}.$$

Granted this lemma (which, by using the binomial theorem, we see must be correct, but which Cardan establishes by citing theorems of Euclid), Cardan has but to observe that if he lets $x = \sqrt[3]{t} - \sqrt[3]{u}$, $t - u = n$, and $\sqrt[3]{t}\,\sqrt[3]{u} = \dfrac{m}{3}$, then the lemma says that $x^3 = n - mx$. Hence if he chooses t and u to satisfy the conditions (5) and (6), the value of x given by (7) in terms of t and u will satisfy the cubic. He then gives a purely verbal arithmetical rule for the method, which tells us to form $\sqrt[3]{t} - \sqrt[3]{u}$ where t and u are given by (8) in terms of m and n.

Cardan also solves (as did Tartaglia) particular equations of the types

$$x^3 = mx + n, \qquad x^3 + mx + n = 0, \qquad x^3 + n = mx.$$

He has to treat each of these cases separately and all three separately from equation (4) because, first, up to this time the Europeans wrote equations so that only terms with positive numbers appeared in them, and second, because he had to give a separate geometrical justification for the rule in each case.

Cardan also shows how to solve equations such as $x^3 + 6x^2 = 100$. He knew how to eliminate the x^2 term; that is, since the coefficient was 6, he replaced x by $y - 2$ and obtained $y^3 = 12y + 84$. He also recognized that one can treat an equation such as $x^6 + 6x^4 = 100$ as a cubic by letting $x^2 = y$. Throughout the book he gives positive and negative roots, despite the fact that he calls negative numbers fictitious. However he ignored complex roots. In fact, in Chapter 37 he calls problems leading to neither true nor false (positive or negative) roots false problems. The book is detailed— even boring, to a modern reader—because Cardan treats separately the many cases not only of the cubic equation but also of the auxiliary quadratic equations he must solve to find t and u. In each case, he writes the equation so that the coefficients of the terms are positive.

There is a difficulty with Cardan's solution of the cubic, which he observed but did not resolve. When the roots of the cubic are all real and distinct, it can be shown that t and u will be complex because the radicand in (8) is negative; yet we need $\sqrt[3]{t}$ and $\sqrt[3]{u}$ to obtain x. This means that real numbers can be expressed in terms of the cube roots of complex numbers. However, these three real roots cannot be obtained by algebraic means, that is, by radicals. This case was called irreducible by Tartaglia. One would think that the fact that real numbers can be expressed as combinations of complex numbers would have caused Cardan to take complex numbers seriously, but it did not.

Vieta, in *De Aequationum Recognitione et Emendatione*, written in 1591 and published in 1615,[8] was able to solve the irreducible case of cubics by using a trigonometric identity, and so avoided the use of Cardan's formula. This method is also used today. He started with the identity

$$(10) \qquad \cos 3A \equiv 4 \cos^3 A - 3 \cos A.$$

By letting $z = \cos A$ the identity becomes

$$(11) \qquad z^3 - \frac{3}{4} z - \frac{1}{4} \cos 3A \equiv 0.$$

Suppose the given cubic is (Vieta worked with $x^3 - 3a^2x = a^2b$ and $a > b/2$)

$$(12) \qquad y^3 + py + q = 0.$$

By introducing $y = nz$, where n is at our disposal, we can make the coefficients of (12) the same as those of (11). Substituting $y = nz$ in (12) gives

$$(13) \qquad z^3 + \frac{p}{n^2} z + \frac{q}{n^3} = 0.$$

Now we require of n that $p/n^2 = -3/4$ so that

$$(14) \qquad n = \sqrt{-4p/3}.$$

With this value of n chosen, we select a value A so that

$$(15) \qquad \frac{q}{n^3} = -\frac{1}{4} \cos 3A$$

or so that

$$(16) \qquad \cos 3A = -\frac{4q}{n^3} = \frac{-q/2}{\sqrt{-p^3/27}}.$$

One can show that if the three roots are real then p is negative so that n is real. Moreover, one can show that $|\cos 3A| < 1$. Hence one can find $3A$ from a table.

Whatever the value of A, $\cos A$ satisfies (11) because (11) is an identity. Now A was selected so that (13) is a particular case of (11). For this value of A, $\cos A$ satisfies (13). However, the value of A is determined by (16), and

8. On the Review and Correction of Equations, *Opera*, 82–162.

this fixes $3A$. But for any given A that satisfies (16), so do $A + 120°$ and $A + 240°$. Since $z = \cos A$, there are, then, three values satisfying (13):

$$\cos A, \qquad \cos (A + 120°), \quad \text{and} \quad \cos (A + 240°).$$

The three values satisfying (12) are n times these values of z, where n is given by (14). Vieta obtained just one root.

Of course the cubic equation has three roots. It was Leonhard Euler who, in 1732, gave the first complete discussion of Cardan's solution of the cubic, in which he emphasized that there are always three roots and pointed out how these are obtained.[9] If ω and ω^2 are the complex roots of $x^3 - 1 = 0$, that is, the roots of $x^2 + x + 1 = 0$, then the three cube roots of t and u in (8) are

$$\sqrt[3]{t}, \ \omega\sqrt[3]{t}, \ \omega^2\sqrt[3]{t} \quad \text{and} \quad \sqrt[3]{u}, \ \omega\sqrt[3]{u}, \ \omega^2\sqrt[3]{u}.$$

We must now choose a member of the first set and a member of the second so that the product is the real number $m/3$. (See equation (6) in Cardan's solution.) Since ω and ω^2 are roots of unity, $\omega \cdot \omega^2 = \omega^3 = 1$; hence the proper choices for x, in view of (7), are

(17) $\qquad x_1 = \sqrt[3]{t} - \sqrt[3]{u}, \qquad x_2 = \omega\sqrt[3]{t} - \omega^2\sqrt[3]{u}, \qquad x_3 = \omega^2\sqrt[3]{t} - \omega\sqrt[3]{u}.$

Success in solving the cubic equation was followed almost immediately by success in solving the quartic. The method is due to Lodovico Ferrari and was published in Cardan's *Ars Magna*; we describe it in modern notation and with literal coefficients to show the generality. The equation is

(18) $\qquad\qquad\qquad x^4 + bx^3 + cx^2 + dx + e = 0.$

By transposition we get

(19) $\qquad\qquad\qquad x^4 + bx^3 = -cx^2 - dx - e.$

Now complete the square on the left side by adding $\left(\dfrac{1}{2} bx\right)^2$. This gives

(20) $\qquad\qquad \left(x^2 + \dfrac{1}{2} bx\right)^2 = \left(\dfrac{1}{4} b^2 - c\right)x^2 - dx - e.$

Now add $\left(x^2 + \dfrac{1}{2} bx\right)y + \dfrac{1}{4} y^2$ to each side. Thus

(21) $\qquad \left(x^2 + \dfrac{1}{2} bx\right)^2 + \left(x^2 + \dfrac{1}{2} bx\right)y + \dfrac{1}{4} y^2$

$$= \left(\frac{1}{4} b^2 - c + y\right)x^2 + \left(\frac{1}{2} by - d\right)x + \frac{1}{4} y^2 - e.$$

9. *Comm. Acad. Sci. Petrop.*, 6, 1732/33, 217–31, pub. 1738 = *Opera*, (1), 6, 1–19.

By making the discriminant of the quadratic expression in x on the right side zero, we can make this side a perfect square of a first degree expression in x. Hence we set

$$(22) \qquad \left(\frac{1}{2}by - d\right)^2 - 4\left(\frac{1}{4}b^2 - c + y\right)\left(\frac{1}{4}y^2 - e\right) = 0.$$

This is a *cubic* equation in y. Choose any root of this cubic and substitute it for y in (21). Using the fact that the left side is also a perfect square, and taking the square root, we get a quadratic in x equaling either of two linear functions of x, one the negative of the other. Solving these two quadratics gives 4 roots for x. Choosing another root from (22) would give a different equation in (21) but the same four roots.

In presenting Ferrari's method Cardan, in Chapter 39 of *Ars Magna*, solves a multitude of special cases, each with numerical coefficients. Thus he solves equations of the type

$$x^4 = bx^2 + ax + n, \qquad x^4 = bx^2 + cx^3 + n,$$
$$x^4 = cx^3 + n, \qquad x^4 = ax + n.$$

As in the case of the cubic equation, he gives a geometrical proof of the basic algebraic steps and then gives the rule for solution in words.

Cardan, Tartaglia, and Ferrari showed, by solving numerous instances of cubic and quartic equations, that they had sought and obtained methods that would work for all cases of the respective degrees. The interest in generality is a new feature. Their work preceded Vieta's introduction of literal coefficients, so they could not take advantage of this device. Vieta, who had already made possible a generality in proof by introducing literal coefficients, now sought another kind of generality. He noted that the methods of solving the second, third, and fourth degree equations were quite different. He therefore sought a method that would work for equations of each degree. His first idea was to get rid of the term next to the highest in degree by a substitution. Tartaglia had done this for the cubic but did not try it for all equations.

In the *Isagoge* Vieta does the following. To solve the quadratic equation

$$x^2 + 2bx = c$$

he lets

$$x + b = y.$$

Then

$$y^2 = x^2 + 2bx + b^2.$$

In view of the original equation,

$$y = \sqrt{c + b^2}.$$

Then

$$x = y - b = \sqrt{c + b^2} - b.$$

In the case of the third degree equation

$$x^3 + bx^2 + cx + d = 0,$$

Vieta starts by letting $x = y - b/3$. This substitution gives the reduced cubic

(23) $$y^3 + py + q = 0.$$

Next he introduces a further transformation, indeed the one we learn today. He lets

(24) $$y = z - \frac{p}{3z}$$

and obtains

$$z^3 - \frac{p^3}{27z^3} + q = 0.$$

He then solves the quadratic in z^3 and obtains

$$z^3 = -\frac{q}{2} \pm \sqrt{R} \quad \text{where} \quad R = \left(\frac{p}{3}\right)^3 + \left(\frac{q}{2}\right)^2.$$

Here, as in Cardan's method, there are two values for z^3. Though Vieta used only the positive cube root of z^3, one can use all six (complex) roots. The use of (24) would show that only three distinct values of y result from the six values of z.

To solve the general fourth degree equation

$$x^4 + bx^3 + cx^2 + dx + e = 0,$$

Vieta lets $x = y - b/4$ and reduces the equation to

$$x^4 + px^2 + qx + r = 0.$$

He then transposes the last three terms and adds $2x^2y^2 + y^4$ to both sides. This makes the left side a perfect square and, as in Ferrari's method, by properly choosing y he makes the right side a perfect square of the form $(Ax + B)^2$. To choose y properly he applies the discriminant condition for a quadratic equation and is led to a sixth degree equation in y, which fortunately is a cubic in y^2. This step and the rest of the work is precisely the same as in Ferrari's method.

Another general method explored by Vieta was to factor the polynomial into first degree factors just as we might factor $x^2 + 5x + 6$ into $(x + 2)(x + 3)$. He was unsuccessful partly because he rejected all but positive roots and partly because he did not have enough theory, such as the factor theorem, on which to base a general method. Thomas Harriot had the same idea and failed for the same reasons.

The search for general algebraic methods turned next to solving equations of degree higher than four. James Gregory, who had given his own methods of solving cubic and quartic equations, tried to use them to solve the quintic equation. He and Ehrenfried Walter von Tschirnhausen (1651–1708) tried to use transformations to reduce higher-degree equations to two terms, a power of x and a constant. These attempts to solve equations beyond the quartic failed. In later work on integration Gregory surmised that one could not solve algebraically the general nth-degree equation for $n > 4$.

5. *The Theory of Equations*

The work on methods of solving equations produced a number of related theorems and observations that are studied nowadays in the elementary theory of equations. The number of roots an equation can have was considered. Cardan had introduced complex roots and it seemed to him for a while that an equation might have any number of roots. But he soon recognized that a third degree equation has 3 roots, a fourth degree 4, and so on. In *L'Invention nouvelle*, Albert Girard inferred and stated that an nth-degree polynomial equation has n roots if one counts the impossible roots, that is, the complex roots, and if one takes into account the repeated roots. But Girard gave no proof. Descartes, in the third book of *La Géométrie*, said that an equation can have as many distinct roots as the number of dimensions (degree) of the unknown. He said "can have" because he considered negative roots as false roots. Later, by including imaginary and negative roots for the purpose of counting roots, he concluded that there are as many as the degree.

The next significant question was how to predict the number of positive, negative, and complex roots. Cardan observed that the complex roots of an equation (with real coefficients) occur in pairs. Newton proved this in his *Arithmetica Universalis*. Descartes in *La Géométrie* stated without proof the rule of signs, known as Descartes's rule, which states that the maximum number of positive roots of $f(x) = 0$, where f is a polynomial, is the number of alterations in sign of the coefficients and that the maximum number of negative roots is the number of times two + signs or two − signs occur in succession. As now given, the latter part of the rule states that the maximum number of negative roots is the number of alterations in $f(-x) = 0$. The rule was proved by several eighteenth-century mathematicians. The proof usually presented today is due to Abbé Jean-Paul de Gua de Malves (1712–85), who showed also that the absence of $2m$ successive terms indicates $2m + 2$ or $2m$ complex roots, according as the two terms between which the deficiency occurs have like or unlike signs.

In his *Arithmetica Universalis* Newton described but did not prove another method for determining the maximum number of positive and negative real roots and hence the least possible number of complex roots. This method is more complicated to apply but gives better results than Descartes's rule of signs. It was finally proved as a special case of a more general theorem by Sylvester.[10] Somewhat earlier Gauss showed that if the number of positive roots falls short of the number of variations of sign, it falls short by an even number.

Another class of results concerns the relationships between the roots and coefficients of an equation. Cardan discovered that the sum of the roots is the negative of the coefficient of x^{n-1}, that the sum of the products two at a time is the coefficient of x^{n-2}, and so forth. Both Cardan and Vieta (in *De Aequationum Recognitione et Emendatione*) used the first relationship between roots and coefficients of low degree equations to eliminate the x^{n-1} term in polynomial equations in manners we have described earlier. Newton stated the relationship between roots and coefficients in his *Arithmetica Universalis*; so did James Gregory in a letter to John Collins (1625–83), the secretary of the Royal Society. However, no proofs were given by any of these men.

Vieta and Descartes constructed equations whose roots were more or less than the roots of a given equation. The process is merely to replace x by $y + m$. Both men also used the transformation $y = mx$ to obtain an equation whose roots are m times the roots of a given equation. For Descartes the former process had the significance we mentioned earlier, namely, that false (negative) roots can be made true (positive) roots, and conversely.

Descartes also proved that if a cubic equation with rational coefficients has a rational root, then the polynomial can be expressed as the product of factors with rational coefficients.

Another major result is now known as the factor theorem. In the third book of *La Géométrie* Descartes asserted that $f(x)$ is divisible by $x - a$, a positive, if and only if a is a root of $f(x) = 0$, and by $x + a$ if a is a false root. With this fact and others he had asserted, Descartes established the modern method of finding the rational roots of a polynomial equation. After making the highest coefficient 1, he made all the coefficients integral by multiplying the roots of the given equation by the necessary factor. This is done by using the rule he had given of replacing x by y/m in the equation. The rational roots of the original equation must now be integral factors of the constant term in the new equation. If by trial one finds that a, say, is a root, then by the factor theorem, $y - a$ is a factor of the new polynomial in y. Descartes points out that by eliminating this factor one reduces the degree of the equation and can then work with the reduced equation.

Newton in *Arithmetica Universalis* and others earlier gave theorems on the upper bound for the roots of equations. One of these theorems involves

10. *Proc. London Math. Soc.*, 1, 1865, 1–16 = *Math. Papers*, 2, 498–513.

the calculus and will be stated in Chapter 17 (sec. 7). Newton discovered the relation between the roots and discriminant of an equation, namely that, for example, $ax^2 + bx + c = 0$ has equal, real, or nonreal roots according as $b^2 - 4ac$ equals 0, is greater than 0 or is less than 0.

In *La Géométrie* Descartes introduced the principle of undetermined coefficients. The principle can be illustrated thus: To factor $x^2 - 1$ into two linear factors, one supposes that

$$x^2 - 1 = (x + b)(x + d).$$

By multiplying out the right-hand side and equating coefficients of like powers of x, one finds that

$$b + d = 0$$
$$bd = -1.$$

One can now solve for b and d. Descartes stressed the usefulness of this method.

One other method, the method of mathematical induction, entered algebra explicitly in the late sixteenth century. Of course the method is implicit even in Euclid's proof of the infinitude of the number of primes. As he proves the theorem, he shows that if there are n primes, there must be $n + 1$ primes; and since there is a first prime, the number of primes must be infinite. The method was recognized explicitly by Maurolycus in his *Arithmetica* of 1575 and was used by him to prove, for example, that $1 + 3 + 5 + \cdots + (2n - 1) = (n + 1)^2$. Pascal in one of his letters acknowledged Maurolycus's introduction of the method and used it himself in his *Traité du triangle arithmétique* (1665), wherein he presents what we now call the Pascal triangle (sec. 6).

6. *The Binomial Theorem and Allied Topics*

The binomial theorem for positive integral exponents, that is, the expansion of $(a + b)^n$ for n positive integral, was known by the Arabs of the thirteenth century. About 1544 Stifel introduced the term "binomial coefficient" and showed how to calculate $(1 + a)^n$ from $(1 + a)^{n-1}$. The arrangement of numbers

$$
\begin{array}{ccccccc}
 & & & 1 & & & \\
 & & 1 & & 1 & & \\
 & & 1 & 2 & 1 & & \\
 & 1 & 3 & 3 & 1 & & \\
 1 & & 4 & 6 & 4 & & 1 \\
1 & 5 & 10 & 10 & 5 & 1 \\
\end{array}
$$

. . . .

in which each number is the sum of the two immediately above it, already known to Tartaglia, Stifel, and Stevin, was used by Pascal (1654) to obtain the coefficients of the binomial expansion. Thus the numbers in the fourth row are the coefficients in the expansion of $(a + b)^3$. Despite the fact that this arrangement was known to many predecessors, it has been called Pascal's triangle.

Newton in 1665 showed that we may compute $(1 + a)^n$ directly without reference to $(1 + a)^{n-1}$. Then he became convinced that the expansion held for fractional and negative n (it is an infinite series in this case) and so stated, but never proved, this generalization. He did verify that the series of $(1 + x)^{1/2}$ times itself gave $1 + x$, but neither he nor James Gregory (who arrived at the theorem independently) thought a proof necessary. In two letters, of June 6 and October 4, 1676, to Henry Oldenburg ($c.$ 1615–77), secretary of the Royal Society, Newton stated the more general result, which he knew before 1669, namely, the expansion of $(P + PQ)^{m/n}$. He thought of it as a useful method of extracting roots, because if Q is less than 1 (the P being factored out), the successive terms, being powers of Q, are smaller and smaller in value.

Independently of the work on the binomial theorem, the formulas for the number of permutations and the number of combinations of n things taken r at a time appeared in the works of a number of mathematicians, for example, Bhāskara and the Frenchman Levi ben Gerson (1321). Pascal observed that the formula for combinations, often denoted by $_nC_r$ or $\binom{n}{r}$, also gives the binomial coefficients. That is, for n fixed and r running from 0 to n, the formula yields the successive coefficients. James Bernoulli, in his *Ars Conjectandi* (1713), extended the theory of combinations and then proved the binomial theorem for the case of n positive integral by using the formula for combinations.

The work on permutations and combinations is connected with another development, the theory of probability, which was to assume major importance in the late nineteenth century but which barely warrants mention in the sixteenth and seventeenth centuries. The problem of the probability of throwing a particular number on a throw of two dice had been raised even in medieval times. Another problem, how to divide the stake between two players when the stake is to go to the player who first wins n points, but the play is interrupted after the first player has made p points and the second q points, appears in Pacioli's *Summa* and in books by Cardan, Tartaglia, and others. This problem acquired some importance when, after it was proposed to Pascal by Antoine Gombaud, Chevalier de Méré (1610–85), Pascal and Fermat corresponded about it. The problem and their solutions are unimportant, but their work on it does mark the beginning of the theory of probability. Both applied the theory of combinations.

The first significant book on probability was Bernoulli's *Ars Conjectandi*. The most important new result, still called Bernoulli's theorem, states that if p is the probability of a single event happening and q the probability of its failing to happen, then the probability of the event happening at least m times in n trials is the sum of the terms in the expansion of $(p + q)^n$ from p^n to the term involving $p^m q^{n-m}$.

7. The Theory of Numbers

While practical interests stimulated the improvements in calculations, symbolism, and the theory of equations, interest in purely mathematical problems led to renewed activity in the theory of numbers. This subject had, of course, been initiated by the classical Greeks; and Diophantus had added the topic of indeterminate equations. The Hindus and Arabs had, at least, kept the subject from falling into oblivion. Though almost all the Renaissance algebraists we have mentioned made conjectures and observations, the European who first made extensive and impressive contributions to the theory of numbers and gave the subject enormous impetus was Pierre de Fermat (1601–65).

Born to a family of tradespeople, he was trained as a lawyer in the French city of Toulouse and made his living in this profession. For a time he was a councillor of the parliament of Toulouse. Though mathematics was but a hobby for Fermat and he could devote only spare time to it, he contributed first-class results to the theory of numbers and the calculus, was one of the two creators of coordinate geometry, and, together with Pascal, as we have seen, initiated work on probability. Like all mathematicians of his century, he worked on problems of science and made a lasting contribution to optics: Fermat's Principle of Least Time (Chap. 24, sec. 3). Most of Fermat's results are known through letters he wrote to friends. He published only a few papers, but some of his books and papers were published after his death.

Fermat believed that the theory of numbers had been neglected. He complained on one occasion that hardly anyone propounded or understood arithmetical questions and asked, "Is it due to the fact that up to now arithmetic has been treated geometrically rather than arithmetically?" Even Diophantus, he observed, was somewhat tied to geometry. Arithmetic, he believed, had a special domain of its own, the theory of integral numbers.

Fermat's work in the theory of numbers determined the direction of the work in this area until Gauss made his contributions. Fermat's point of departure was Diophantus. Many translations of the latter's *Arithmetica* had been made by Renaissance mathematicians. In 1621 Bachet de Méziriac published the Greek text and a Latin translation. This was the edition that Fermat owned; he noted most of his results in the margins of the book, though a few were communicated in letters to friends. The copy with Fermat's marginal notes was published in 1670 by his son.

Fermat stated many theorems on the theory of numbers but only in one case did he give a proof, and this was a sketch. The best mathematicians of the eighteenth century worked hard to prove his results, (Chap. 25, sec. 4). These all turned out to be correct except for one error (which we shall note later) and one still-unproved famous "theorem," for which the indications are all favorable. There is no doubt that he had great intuition, but it is unlikely that he had proofs for all of his affirmations.

A document discovered in 1879 among the manuscripts of Huygens gives a famous method, called the method of infinite descent, which was introduced and used by Fermat. To understand the method let us consider the theorem asserted by Fermat in a letter to Marin Mersenne (1588–1648) of December 25, 1640, which states that a prime of the form $4n + 1$ can be expressed in one and only one way as the sum of two squares. Thus $17 = 16 + 1$ and $29 = 25 + 4$. The method proceeds by showing that if there is one prime of the form $4n + 1$ that does *not* possess the required property, then there will be a smaller prime of the form $4n + 1$ not possessing it. Then, since n is arbitrary, there must be a still smaller one. By descending through the positive integral values of n one must reach $n = 1$ and thus the prime $4 \cdot 1 + 1$ or 5. Then 5 cannot possess the required property. But, since 5 is expressible as a sum of two squares and in only one way, so is every prime of the form $4n + 1$. This sketch of his method Fermat sent to his friend Pierre de Carcavi (d. 1684) in 1659. Fermat said that he used the method to prove the theorem just described, but his proof was never found. He also said that he proved other theorems by this method.

The method of infinite descent differs from mathematical induction. First of all, the method does not require that one exhibit even one case in which the proposed theorem is satisfied because one can conclude the argument by the fact that the case $n = 1$ merely leads to a contradiction of some other known fact. Moreover, after the appropriate hypothesis is made for one value of n, the method shows that there is a smaller, but not necessarily the next, value of n for which the hypothesis is true. Finally the method disproves certain assertions and is in fact more useful for this purpose.

Fermat also stated that no prime of the form $4n + 3$ can be expressed as a sum of two squares. In a note in his copy of Diophantus and in the letter to Mersenne, Fermat generalized on the well-known 3, 4, 5 right-triangle relationship by asserting the following theorems: A prime of the form $4n + 1$ is the hypotenuse of one and only one right triangle with integral arms. The square of $(4n + 1)$ is the hypotenuse of two and only two such right triangles; its cube, of three; its biquadrate, of 4; and so on, ad infinitum. As an example, consider the case of $n = 1$. Then $4n + 1 = 5$ and 3, 4, 5 are the sides of the one and only right triangle with 5 as hypotenuse. However, 5^2 is the hypotenuse of the two, and only two, right triangles 15, 20, 25 and 7, 24, 25. Also 5^3 is the hypotenuse of the three, and only three, right triangles 75, 100, 125; 35, 120, 125; and 44, 117, 125.

In the letter to Mersenne, Fermat declared that the same prime number $4n + 1$ and its square are each the sum of two squares in one way only; its cube and biquadrate, each in two ways; its fifth and sixth powers, each in three ways, and so on ad infinitum. Thus for $n = 1$, $5 = 4 + 1$ and $5^2 = 9 + 16$; $5^3 = 4 + 121 = 25 + 100$; and so forth. The letter continues: If a prime number that is a sum of two squares be multiplied into another prime that is also the sum of two squares, the product will be the sum of two squares in two ways. If the first prime be multiplied into the square of the second one, the product will be the sum of two squares in three ways; if multiplied into the cube of the second one, then the product will be the sum of two squares in four ways; and so on ad infinitum.

Fermat stated many theorems on the representation of prime numbers in the form $x^2 + 2y^2$, $x^2 + 3y^2$, $x^2 + 5y^2$, $x^2 - 2y^2$, and other such forms, which are extensions of the representation as a sum of squares. Thus every prime of the form $6n + 1$ can be represented as $x^2 + 3y^2$; every prime of the form $8n + 1$ and $8n + 3$ can be represented as $x^2 + 2y^2$. An odd prime number (every prime but 2) can be expressed as the difference of two squares in one and only one way.

Two theorems asserted by Fermat have since been referred to as the minor and major theorems, the latter also known as the last theorem. The minor one, communicated by Fermat in a letter of October 18, 1640, to his friend Bernard Frénicle de Bessy (1605-75), states that if p is a prime and if a is prime to p, then $a^p - a$ is divisible by p.

The major Fermat "theorem," which he believed he had proved, states that for $n > 2$ no integral solutions of $x^n + y^n = z^n$ are possible. This theorem was stated by Fermat in a marginal note in his copy of Diophantus alongside Diophantus' problem: To divide a given square number into (a sum of) two squares. Fermat added, "On the other hand it is impossible to separate a cube into two cubes, or a biquadrate into two biquadrates, or generally any power except a square into two powers with the same exponent. I have discovered a truly marvelous proof of this, which however the margin is not large enough to contain." Unfortunately, Fermat's proof, if he had one, was never found and hundreds of the best mathematicians have not been able to prove it. Fermat stated in a letter to Carcavi that he had used the method of infinite descent to prove the case of $n = 4$ but did not give full details. Frénicle, using Fermat's few indications, did give a proof for that case in 1676, in his posthumously published *Traité des triangles rectangles en nombres* (Treatise on Numerical Properties of Right Triangles).[11]

If we may anticipate, Euler proved the theorem for $n = 3$ (Chap. 25, sec. 4). Since the theorem is true for $n = 3$, it is true for any multiple of 3;

11. *Mém. de l'Acad. des Sci., Paris*, 5, 1729, 83-166.

for if it were not true for $n = 6$, say, then there would be integers x, y, and z such that

$$x^6 + y^6 = z^6.$$

But then

$$(x^2)^3 + (y^2)^3 = (z^2)^3,$$

and the theorem would be false for $n = 3$. Hence we know that Fermat's theorem is true for an infinite number of values of n, but we still do not know that it is true for all values of n. It is actually necessary to prove the theorem only for $n = 4$ and for n an odd prime. For suppose first that n is not divisible by an odd prime; it must then be a power of 2, and since it is larger than 2 it must be 4 or divisible by 4. Let $n = 4m$. Then the equation $x^n + y^n = z^n$ becomes

$$(x^m)^4 + (y^m)^4 = (z^m)^4.$$

If the theorem were not true for n, it would therefore not be true for $n = 4$. Hence if it is true for $n = 4$, it is true for all n not divisible by an odd prime. If $n = pm$ where p is an odd prime, then if the theorem were not true for n it would not be true for the exponent p. Hence if true for $n = p$ it is true for any n divisible by an odd prime.

Fermat did make some mistakes. He believed that he had found a solution to the long-standing problem of producing a formula that would yield primes for values of the variable n. Now it is not hard to show that $2^m + 1$ cannot be a prime unless m is a power of 2. In many letters dating from 1640 on[12] Fermat asserted the converse—namely, that $(2)^{2^n} + 1$ represents a series of primes—though he admitted that he could not prove this assertion. Later he doubted its correctness. Thus far only the five primes 3, 5, 17, 257, and 65,537 yielded by the formula are known. (See Chap. 25, sec. 4.)

Fermat stated and sketched[13] the proof by infinite descent of the theorem: The area of a right-angled triangle the sides of which are rational numbers cannot be a square number. This sketch is the only detailed one ever given by him, and it follows as a corollary that the solution of $x^4 + y^4 = z^4$ in integers is impossible.

On polygonal numbers Fermat stated in his copy of Diophantus the important theorem that every positive integer is itself triangular or the sum of 2 or 3 triangular numbers; every positive integer is itself square or a sum of 2, 3, or 4 squares; every positive integer is either pentagonal or a sum of 2, 3, 4, or 5 pentagonal numbers; and so on for higher polygonal numbers. Much work was required to prove these results, which are correct only if 0

12. *Œuvres*, 2, 206.
13. *Œuvres*, 1, 340; 3, 271.

and 1 are included as polygonal numbers. Fermat asserts that he proved them by the method of infinite descent.

Perfect numbers, as we know, were studied by the Greeks, and Euclid gave the basic result that $2^{n-1}(2^n - 1)$ is perfect if $2^n - 1$ is prime. For $n = 2, 3, 5$, and 7, the values of $2^n - 1$ are prime so that $6, 28, 496$, and 8128 are perfect numbers (as noted by Nichomachus). A manuscript of 1456 correctly gave 33,550,336 as the fifth perfect number; it corresponds to $n = 13$. In his *Epitome* (1536) Hudalrich Regius also gave this fifth perfect number. Pietro Antonio Cataldi (1552–1626) noted in 1607 that $2^n - 1$ is composite if n is composite, and verified that $2^n - 1$ is prime for $n = 13, 17$, and 19. Marin Mersenne in 1644 gave other values. Fermat worked on the subject of perfect numbers also. He considered when $2^n - 1$ is prime and in a letter to Mersenne of June 1640 stated these theorems: (a) If n is not a prime, $2^n - 1$ is not a prime. (b) If n is a prime, $2^n - 1$ is divisible only by primes of the form $2kn + 1$ if divisible at all. About twenty perfect numbers are now known. Whether any odd ones exist is an open question.

By rediscovering a rule that was first stated by Tâbit ibn Qorra, Fermat in 1636 gave a second pair of amicable numbers, 17,926 and 18,416 (the first, 220 and 284, was given by Pythagoras), and Descartes in a letter to Mersenne gave a third pair, 9,363,548 and 9,437,506.

Fermat rediscovered the problem of solving $x^2 - Ay^2 = 1$, wherein A is integral and not a square. The problem has a long history among the Greeks and Hindus. In a letter of February 1657 to Frénicle, Fermat stated the theorem that $x^2 - Ay^2 = 1$ has an unlimited number of solutions when A is positive and not a perfect square.[14] Euler erroneously called the equation Pell's equation; it is now so known. In the same letter[15] Fermat challenged all mathematicians to find an infinity of integral solutions. Lord Brouncker gave solutions, though he did not prove that there was an infinity of them. Wallis did solve the full problem and gave his solutions in letters of 1657 and 1658[16] and in Chapter 98 of his *Algebra*. Fermat also asserted that he could show when $x^2 - Ay^2 = B$, for given A and B, is solvable and could solve it. We do not know how Fermat solved either equation though he said in a letter of 1658 that he used the method of descent for the former.

8. *The Relationship of Algebra to Geometry*

Algebra, as we can see, expanded enormously during the sixteenth and seventeenth centuries. Because it had been tied to geometry, prior to 1500 equations of higher degree than the third were considered unreal. When the study of higher-degree equations was forced upon mathematicians (as for

14. *Œuvres*, 2, 333–35.
15. *Œuvres*, 2, 333–35; 3, 312–13.
16. Fermat, *Œuvres*, 3, 457–80, 490–503.

example by the use of trigonometric identities to aid in the calculation of tables) or suggested as the natural extension of third degree equations, the idea struck many mathematicians as absurd. Thus Stifel in his edition of Rudolff's *Coss* (Algebra) says, "Going beyond the cube just as if there were more than three dimensions . . . is against nature."

Nevertheless, algebra did rise above the limitations imposed by geometric thinking. But the relationship of algebra to geometry remained complicated. The major problem was how to justify algebraic reasoning; and the answer, during the sixteenth century and a good deal of the seventeenth, was to fall back upon the equivalent geometrical meaning of the algebra. Pacioli, Cardan, Tartaglia, Ferrari, and others gave geometrical proofs of algebraic rules. Vieta too was largely tied to geometry. Thus he writes $A^3 + 3B^2A = Z^3$, where A is the unknown and B and Z are constants, in order that each term be of the third degree and so represent a volume. However, as we shall see, Vieta's position on algebra was transitional. Barrow and Pascal actually objected to algebra, and later to analytic methods in coordinate geometry and the calculus, because the algebra lacked justification.

The dependence of algebra on geometry began to be reversed somewhat when Vieta, and later Descartes, used algebra to help solve geometric construction problems. The motivation for much of the algebra that appears in Vieta's *In Artem Analyticam Isagoge*, is solving geometric problems and systematizing geometrical constructions. Typical of the application of algebra to geometry by Vieta is the following problem from his *Zeteticorum Libri Quinque*: Given the area of a rectangle and the ratio of its sides, to find the sides of the rectangle. He takes the area as *B planum* and the ratio of the larger side to the smaller as S to R. Let A be the larger side. Then RA/S is the smaller side. Hence *B planum* equals $(R/S)(A$ squared$)$. Multiplying by S gives the final equation, $BS = RA^2$. Vieta then shows how from this equation A can be constructed by ruler and compass starting from the known quantities B and R/S. The idea here is that if one finds that a desired length x satisfies the equation $ax^2 + bx + c = 0$, one knows that

$$x = \frac{-b + \sqrt{b^2 - 4ac}}{2a}$$

and one can construct x by performing on a, b, and c the geometrical constructions called for by the algebraic expression on the right.

Algebra for Vieta meant a special procedure for discovery; it was analysis in the sense of Plato, who opposed it to synthesis. Theon of Alexandria, who introduced the term "analysis," defined it as the process that begins with the assumption of what is sought and by deduction arrives at a known truth. This is why Vieta called his algebra the analytic art. It performed the process of analysis, particularly for geometric problems. In fact this was the starting point of Descartes's thinking on coordinate geometry and his work on the theory of equations was motivated by the desire to further their use in solving geometric constructions.

The interdependence of algebra and geometry can be seen also in the work of Marino Ghetaldi (1566–1627), a pupil of Vieta. He made a systematic study of the algebraic solution of determinate geometric problems in one book of his *Apollonius Redivivus* (Apollonius Modernized, 1607). Conversely, he gave geometric proofs of algebraic rules. He also constructed geometrically the roots of algebraic equations. A full work on this subject is his *De resolutione et compositione mathematica* (1630), published posthumously.

We also find in the sixteenth and seventeenth centuries the recognition that algebra had to be developed to replace the geometrical methods introduced by the Greeks. Vieta saw the possibility of using algebra to deal with equality and proportion of magnitudes, no matter whether these magnitudes arose in geometrical, physical, or commercial problems. Hence he did not hesitate to consider higher-degree equations and algebraic methodology; he envisioned a deductive science of magnitudes employing symbolism. While algebra was to Vieta largely a royal road to geometry, his vision was great enough to see that algebra had a life and meaning of its own. Bombelli gave algebraic proofs acceptable for his time, without the use of geometry. Stevin asserted that what could be done in geometry could be done in arithmetic and algebra. Harriot's book *Artis Analyticae Praxis* (1631) extended, systematized, and brought out some of the implications of Vieta's work. The book is much like a modern text on algebra; it is more analytical than any algebra preceding it and presents a great advance in symbolism. It was widely used.

Descartes too began to see great potentialities in algebra. He says he begins where Vieta left off. He does not regard algebra as a science in the sense of giving knowledge of the physical world. Indeed such knowledge, he says, consists of geometry and mechanics; he sees in algebra a powerful method wherewith to carry on reasoning, particularly about abstract and unknown quantities. In his view algebra mechanizes mathematics so that thinking and processes become simple and do not require a great effort of the mind. Mathematical creation might become almost automatic.

Algebra, for Descartes, precedes the other branches of mathematics. It is an extension of logic useful for handling quantity, and in this sense more fundamental even than geometry; that is, it is logically prior to geometry. He therefore sought an independent and systematic algebra instead of an unplanned and unfounded collection of symbols and procedures tied to geometry. There is a sketch of a treatise on algebra, known as *Le Calcul* (1638), written either by Descartes himself or under his direction, that treats algebra as a distinct science. His algebra is devoid of meaning. It is a technique of calculation, or a method, and is part of his general search for method.

Descartes's view of algebra as an extension of logic in treating quantity suggested to him that a broader science of algebra might be created, which would embrace other concepts than quantity and be used to approach all problems. Even the logical principles and methods might be expressed symbolically, and the whole system employed to mechanize all reasoning. Descartes called this idea a "universal mathematics." The idea is vague in his works and was not pursued far by him. Nevertheless, he was the first to assign to algebra a fundamental place in the system of knowledge.

This view of algebra, first envisioned fully by Leibniz and ultimately developed into symbolic logic (Chap. 51, sec. 4), was also entertained by Isaac Barrow, though in more limited scope. Barrow, Newton's friend, teacher, and predecessor in the Lucasian chair of mathematics at Cambridge, did not regard algebra as part of mathematics proper but rather as a formalization of logic. To him only geometry was mathematical, and arithmetic and algebra dealt with geometrical magnitudes expressed in symbols.

No matter what philosophy of algebra may have been entertained by Descartes and Barrow, and no matter what potentiality they may have seen in it as a universal science of reasoning, the practical effect of the expanding use of arithmetic and algebraic techniques was to set up algebra as a branch of mathematics independent of geometry. For the time, it was a significant step that Descartes used a^2 to represent a length as well as an area, whereas Vieta had insisted that a second power must represent an area. Descartes called attention to his use of a^2 as a number, noting explicitly that he had departed from his predecessors. He says x^2 is a quantity such that $x^2 : x = x : 1$. Likewise, he says in *La Géométrie* that a product of lines can be a line; he was thinking of the quantities involved and not geometrically, as had the Greeks. It was clear to him that algebraic calculation was independent of geometry.

John Wallis, influenced by Vieta, Descartes, Fermat, and Harriot, went far beyond these men in freeing arithmetic and algebra from geometric representation. In his *Algebra* (1685) he derived all the results of Book V of Euclid algebraically. He too abandoned the limitation to homogeneous algebraic equations in x and y, a concept that had been maintained because such equations were derived from geometrical problems. He saw in algebra brevity and perspicuity.

Though Newton loved geometry, in his *Arithmetica Universalis* we find for the first time an affirmation of the basic importance of arithmetic and algebra as opposed to geometry; Descartes and Barrow had still favored geometry as the fundamental branch of mathematics. Newton needed and used the algebraic language for the development of the calculus, which could be handled best algebraically. And as far as the supremacy of algebra over geometry was concerned, the needs of the differential and integral calculus were to be decisive.

By 1700, then, algebra had reached the point where it could stand on its own feet. The only difficulty was that there was no ground on which to place them. From Egyptian and Babylonian times, intuition and trial and error had supplied some working rules; the reinterpretation of Greek geometrical algebra had supplied others; and independent algebraic work in the sixteenth and seventeenth centuries, partly guided by geometric interpretation, led to many new results. But logical foundations of algebra analogous to those Euclid had provided for geometry were nonexistent. The general lack of concern, apart from the objections of Pascal and Barrow, is surprising, in view of the fact that the Europeans were now fully aware of what rigorous deductive mathematics called for.

How did the mathematicians know what was correct? The properties of the positive integers and fractions are so readily derived from experience with collections of objects that they seem self-evident. Even Euclid failed to supply a logical basis for those books in the *Elements* that dealt with the theory of numbers. As new types of numbers were added to the number system, the rules of operation already accepted for the positive integers and fractions were applied to the new elements, with geometrical thinking as a handy guide. Letters, when introduced, were just representations of numbers and so could be treated as such. The more complicated algebraic techniques seemed justified either by geometrical arguments like those Cardan used or by sheer induction on specific cases. But none of these procedures was logically satisfactory. Geometry, even where called upon, did not supply the logic for negative, irrational, and complex numbers, nor did it justify arguing, for example, that if a polynomial is negative for $x = a$ and positive for $x = b$, that it must be zero between a and b.

Nevertheless, the mathematicians proceeded blithely and confidently to employ the new algebra. Wallis, in fact, affirmed that the procedures of algebra were not less legitimate than those of geometry. Without realizing it the mathematicians were about to enter a new era in which induction, intuition, trial and error, and physical arguments were to be the bases for proof. The problem of building a logical foundation for the number system and algebra was a difficult one, far more difficult than any seventeenth-century mathematician could possibly have appreciated. And it is fortunate that the mathematicians were so credulous and even naive, rather than logically scrupulous. For free creation must precede formalization and logical foundations, and the greatest period of mathematical creativity was already under way.

Bibliography

Ball, W. W. R.: *A Short Account of the History of Mathematics*, Dover (reprint), 1960, Chap. 12.

Boyer, Carl B.: *History of Analytic Geometry*, Scripta Mathematica, 1956, Chap. 4.

————: *A History of Mathematics*, John Wiley and Sons, 1968, Chaps. 15–16.

Cajori, Florian: *Oughtred, A Great Seventeenth Century Teacher of Mathematics*, Open Court, 1916.

————: *A History of Mathematics*, 2nd ed., Macmillan, 1919, pp. 130–59.

————: *A History of Mathematical Notations*, Open Court, 1928, Vol. 1.

Cantor, Moritz: *Vorlesungen über Geschichte der Mathematik*, 2nd ed., B. G. Teubner, 1900, Johnson Reprint Corp., 1965, Vol. 2, pp. 369 806.

Cardan, G.: *Opera Omnia*, 10 vols., 1663, Johnson Reprint Corp., 1964.

Cardano, Girolamo: *The Great Art*, trans. T. R. Witmer, Massachusetts Institute of Technology Press, 1968.

Coolidge, Julian L.: *The Mathematics of Great Amateurs*, Dover (reprint), 1963, Chaps. 6–7.

David, F. N.: *Games, Gods and Gambling: The Origins and History of Probability*, Hafner, 1962.

Descartes, René: *The Geometry*, Dover (reprint), 1954, Book 3.

Dickson, Leonard E.: *History of the Theory of Numbers*, Carnegie Institution, 1919–23, Chelsea (reprint), 1951, Vol. 2, Chap. 26.

Fermat, Pierre de: *Œuvres*, 4 vols. and Supplement, Gauthier-Villars, 1891–1912, 1922.

Heath, Sir Thomas L.: *Diophantus of Alexandria*, 2nd ed., Cambridge University Press, 1910; Dover (reprint), 1964, Supplement, Secs. 1–5.

Hobson, E. W.: *John Napier and the Invention of Logarithms*, Cambridge University Press, 1914.

Klein, Jacob: *Greek Mathematical Thought and the Origin of Algebra*, Massachusetts Institute of Technology Press, 1968. Contains a translation of Vieta's *Isagoge*.

Knott, C. G.: *Napier Tercentenary Memorial Volume*, Longmans Green, 1915.

Montucla, J. F.: *Histoire des mathématiques*, Albert Blanchard (reprint), 1960, Vol. 1, Part 3, Book 3; Vol. 2, Part 4, Book 1.

Mordell, J. L.: *Three Lectures on Fermat's Last Theorem*, Cambridge University Press, 1921.

Morley, Henry: *Life of Cardan*, 2 vols., Chapman and Hall, 1854.

Newton, Sir Isaac: *Mathematical Works*, Vol. 2, ed. D. T. Whiteside, Johnson Reprint Corp., 1967. This volume contains a translation of *Arithmetica Universalis*.

Ore, Oystein: *Cardano: The Gambling Scholar*, Princeton University Press, 1953.

————: "Pascal and the Invention of Probability Theory," *Amer. Math. Monthly*, 67, 1960, 409–19.

Pascal, Blaise: *Œuvres complètes*, Hachette, 1909.

Sarton, George: *Six Wings: Men of Science in the Renaissance*, Indiana University Press, 1957, "Second Wing."

Schneider, I.: "Der Mathematiker Abraham de Moivre (1667–1754)," Archive for History of Exact Sciences, 5, 1968, 177–317.

Scott, J. F.: *A History of Mathematics*, Taylor and Francis, 1958, Chaps. 6 and 9.

————: *The Scientific Work of René Descartes*, Taylor and Francis, 1952, Chap. 9.

————: *The Mathematical Work of John Wallis*, Oxford University Press, 1938.

Smith, David Eugene: *History of Mathematics*, Dover (reprint), 1958, Vol. 1, Chaps. 8–9; Vol. 2, Chaps. 4 and 6.

————: *A Source Book in Mathematics*, 2 vols., Dover (reprint), 1959.

Struik, D. J.: *A Source Book in Mathematics, 1200–1800*, Harvard University Press, 1969, Chaps. 1–2.

Todhunter, I.: *A History of the Mathematical Theory of Probability*, Chelsea (reprint), 1949, Chaps. 1–7.

Turnbull, H. W.: *James Gregory Tercentenary Memorial Volume*, Bell and Sons, 1939.

————: *The Mathematical Discoveries of Newton*, Blackie and Son, 1945.

Vieta, F.: *Opera mathematica*, (1646), Georg Olms (reprint), 1970.

14
The Beginnings of Projective Geometry

> I freely confess that I never had a taste for study or research either in physics or geometry except in so far as they could serve as a means of arriving at some sort of knowledge of the proximate causes ... for the good and convenience of life, in maintaining health, in the practice of some art ... having observed that a good part of the arts is based on geometry, among others the cutting of stone in architecture, that of sundials, that of perspective in particular.
>
> GIRARD DESARGUES

1. *The Rebirth of Geometry*

The resurgence of significant creative activity in geometry lagged behind that in algebra. Apart from the creation of the mathematical system of perspective and the incidental geometrical work of the Renaissance artists (Chap. 12, sec. 2), very little of consequence was done in geometry from the time of Pappus to about 1600. Some interest was created by the appearance of numerous printed editions of Apollonius' *Conic Sections*, especially the notable Latin translation by Federigo Commandino (1509–75) of Books I to IV in 1566. Books V to VII were made available by other translators, and a number of men, including Vieta, Willebrord Snell (1580–1626), and Ghetaldi, undertook to reconstruct the lost eighth book.

What was needed and did arise to direct the minds of mathematicians into new channels was new problems. One problem had already been raised by Alberti: What geometrical properties do two sections of the same projection of an actual figure have in common? A large number of problems came from science and practical needs. Kepler's use of the conic sections in his work of 1609 gave enormous impetus to the reexamination of these curves and to the search for properties useful in astronomy. Optics, an interest of mathematicians since Greek times, received greatly increased attention after the invention of the telescope and microscope in the beginning of the seventeenth century. The design of lenses for these instruments became a major problem; this meant an interest in the shapes of the surfaces or, since these surfaces are figures of revolution, the shapes of the generating curves. The geographical explorations had created a need for maps and for studying the paths of ships as represented on the sphere and on the map. The introduction of the notion of a moving earth called for new principles of mechanics to account for the paths of moving objects; this, too, meant a study of curves. Among moving objects, projectiles became more important because cannons could now fire balls over hundreds of yards; prediction of path and range was vital. The practical problem of finding areas and volumes began to attract more attention. Kepler's *Nova Stereometria Doliorum Vinariorum* (The New Science of Measuring Volumes of Wine Casks, 1615) initiated a new burst of activity in this area.

Another kind of problem came to the fore as a consequence of assimilation of the Greek works. Mathematicians began to realize that the Greek methods of proof lacked generality. A special method had to be devised for nearly every theorem. This point had been made by Agrippa von Nettesheim (1486–1532), as far back as 1527, and by Maurolycus, who had translated Greek works and had written books of his own on the conic sections and other mathematical subjects.

Much of the response to the new problems resulted in minor variations on old themes. The approach to the conic sections was altered. The curves were defined at the outset as loci in the plane instead of sections of a cone as in Apollonius. Guidobaldo del Monte, for example, in 1579 defined the ellipse as the locus of points the sum of whose distances from the foci is a constant. Not only the conics but older Greek curves such as the conchoid of Nicomedes, the cissoid of Diocles, the spiral of Archimedes, and the quadratrix of Hippias were restudied. New curves were introduced, notably the cycloid (see Chap. 17, sec. 2). While all of this work was helpful in disseminating the Greek contributions, none of it offered new theorems or new methods of proof. The first innovation of consequence came in answer to the problems raised by the painters.

2. The Problems Raised by the Work on Perspective

The basic idea in the system of focused perspective created by the painters is the principle of projection and section (Chap. 12, sec. 1). A real scene is viewed by the eye regarded as a point. The lines of light from various points of the scene to the eye are said to constitute a projection. According to the system, the painting itself must contain a section of that projection, the section being mathematically what a plane passing through the projection would contain.

Now suppose the eye at O (Fig. 14.1) looks at the horizontal rectangle

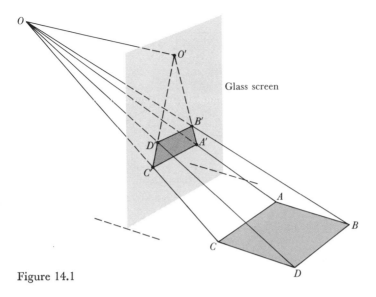

Figure 14.1

ABCD. The lines from *O* to the points on the four sides of this rectangle constitute a projection of which *OA*, *OB*, *OC*, and *OD* are typical lines. If a plane is now interposed between the eye and the rectangle, the lines of the projection will cut through the plane and mark out on it the quadrangle *A'B'C'D'*. Since the section, namely *A'B'C'D'*, creates the same impression on the eye as does the original rectangle, it is reasonable to ask, as Alberti did, what geometrical properties do the section and the original rectangle have in common? It is intuitively apparent that the original figure and the section will be neither congruent nor similar; nor will they contain the same area. In fact the section need not be a rectangle.

There is an extension of this problem: Suppose two different sections of this same projection are made by two different planes that cut the projection at any angle. What properties would the two sections have in common?

The problem may be further extended. Suppose a rectangle *ABCD* is viewed from two different locations *O'* and *O''* (Fig. 14.2). Then there are two projections, one determined by *O'* and the rectangle and the second determined by *O''* and the rectangle. If a section is made of each projection, then, in view of the fact that each section should have some geometrical properties in common with the rectangle, the two sections should have some common geometrical properties.

Some of the seventeenth-century geometers undertook to answer these questions. They viewed the methods and results they obtained as part of Euclidean geometry. However, these methods and results, while indeed contributing much to that subject, proved to be the beginning of a new

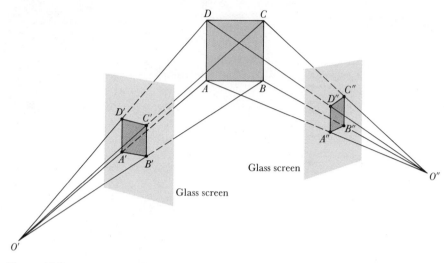

Figure 14.2

branch of geometry, which in the nineteenth century became known as projective geometry. We shall refer to the work by that name, but the distinction between Euclidean and projective geometry was not made in the seventeenth century.

3. *The Work of Desargues*

The man who first took up directly the problems just sketched was the self-educated Girard Desargues (1591–1661), who became an army officer and later an engineer and architect. Desargues knew the work of Apollonius and thought that he could introduce new methods for proving theorems about conics. He did and, indeed, was fully aware of the power of these methods. Desargues was also concerned to improve the education and technique of artists, engineers, and stonecutters; he had little use for theory for its own sake. He began by organizing numerous useful theorems and at first distributed his results through letters and handbills. He also gave lectures gratis in Paris. Afterwards he wrote several books, one of which taught children how to sing. Another applied geometry to masonry and stonecutting.

His major text, which was preceded in 1636, by a pamphlet on perspective, was *Brouillon project d'une atteinte aux événemens des rencontres du cône avec un plan* (Proposed Draft of an Attempt to Treat the Results of a Cone Intersecting a Plane, 1639).[1] This book deals with what we would now call

1. *Œuvres,* 1, 103–230.

projective methods in geometry. Desargues printed about fifty copies and circulated them among his friends. Not long afterwards all copies were lost. A manuscript copy made by La Hire was found accidentally in 1845 by Michel Chasles and reproduced by N. G. Poudra, who edited Desargues's work in 1864. However, an original edition of the 1639 work was discovered about 1950 by Pierre Moisy in the Bibliothèque nationale (Paris) and has been reproduced. This newly discovered copy also contains an appendix and errata by Desargues that are important. Desargues's main theorem on triangles and other theorems of his were published in 1648 in an appendix to a book on perspective by his friend Abraham Bosse (1602–76). In this book, *Manière universelle de M. Desargues, pour pratiquer la perspective*, Bosse tried to popularize Desargues's practical methods.

Desargues used curious terminology, some of which had appeared in Alberti's work. A straight line he called a "palm." When points were marked out on the line he called it a "trunk." However, a line with three pairs of points in involution (see below) was a "tree." Desargues's intent in introducing this new terminology was to gain clarity by avoiding the ambiguities in more customary terms. But the language and the strange ideas made his book difficult reading. His contemporaries, except for his friends Mersenne, Descartes, Pascal, and Fermat, called him crazy. Even Descartes, when he heard that Desargues was introducing a new method to treat conics, wrote to Mersenne that he could not see how anyone could do anything new on conics except with the aid of algebra. However, after Descartes learned the details of what Desargues was doing, he respected it highly. Fermat regarded Desargues as the real founder of the theory of conic sections and found his book, which Fermat apparently owned, rich in ideas. But the general lack of appreciation disgusted Desargues and he retired to his estate.

Before we note some of the theorems Desargues established, we must introduce one new convention on parallel lines. Alberti had pointed out that two lines that are parallel in some actual scene being painted (unless they are parallel to the glass screen or canvas) must be drawn on the canvas so as to meet in a point. Thus the lines $A'B'$ and $C'D'$ in Figure 14.1 above, which correspond to the parallel lines AB and CD, must, according to the principle of projection and section, meet at some point O'. As a matter of fact, O and AB determine a plane and so do O and CD. These two planes cut the glass screen in $A'B'$ and $C'D'$ but, since they meet at O, these two planes must have a line in common; this line cuts the glass screen at some point O', which is also the intersection of $A'B'$ and $C'D'$. The point O' does not correspond to any ordinary point on AB or CD. In fact, the line OO' is horizontal, and so parallel to AB and CD. The point O' is called a vanishing point because it does not have a corresponding point on the lines AB or CD, whereas any other point on the lines $A'B'$ or $C'D'$ corresponds to some definite point on AB or CD, respectively.

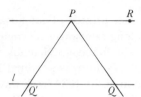

Figure 14.3

To complete the correspondence between points on the lines $A'B'$ and AB as well as between those on $C'D'$ and CD, Desargues introduced a new point on AB and on CD. Called the point at infinity, it is additional to the usual points on the two lines, and is to be regarded as the common point of the two lines. Moreover any other line parallel to AB or CD is to have this same point on it and meet AB and CD at this point. Any set of parallel lines having a different direction from that of AB or CD is likewise to have a common point at infinity. Since each set of parallel lines has a point in common and there is an infinite number of different sets of such lines, Desargues's convention introduces an infinity of new points in the Euclidean plane. He made the further assumption that all these new points lie on one line, which corresponds to the horizon line or vanishing line on the section. Thus a new line is added to the already existing lines in the Euclidean plane. A set of parallel planes is assumed to have the line at infinity on each in common; that is, all parallel planes meet on one line.

The addition of a new point on each line does not contradict any axiom or theorem of Euclidean geometry, though it does call for a change in wording. Non-parallel lines continue to meet in ordinary points while parallel lines meet in the "point at infinity" on each of the lines. The agreement as to the points at infinity is actually a convenience in Euclidean geometry, in that it avoids special cases. For example, one could now say that any two lines meet in exactly one point. We shall soon see more fully the advantages of this convention.

Kepler, too, decided (1604) to add the point at infinity to parallel lines but for a different reason. To each line through P (Fig. 14.3) and cutting l there is one point, Q, on l. However, no point on l corresponds to PR, the parallel to l through P. By adding a point at infinity common to PR and l, Kepler could affirm that every line through P cuts l. Moreover, after Q has moved to "infinity" on the right and PQ has become PR, the point of intersection of PR and l can be thought of as being at infinity to the left of P; and as PR continues to rotate around P, the point of intersection Q' of PR and l will move in from the left. Thus continuity of the intersection of PR and l is maintained. In other words, Kepler (and Desargues) regarded the two "ends" of the line as meeting at "infinity" so that the line has the structure of a circle. In fact, Kepler actually thought of a line as a circle with its center at infinity.

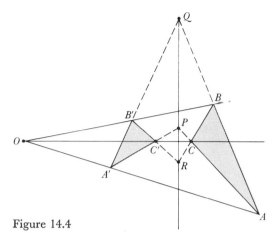

Figure 14.4

Having introduced his points and lines at infinity, Desargues was able to state a basic theorem, still called Desargues's theorem. Consider the point O (Fig. 14.4) and triangle ABC. The lines from O to the various points on the sides of the triangle constitute, as we know, a projection. A section of this projection will then contain a triangle $A'B'C'$, where A' corresponds to A, B' to B, and C' to C. The two triangles ABC and $A'B'C'$ are said to be perspective from the point O. Desargues's theorem then states: The pairs of corresponding sides AB and $A'B'$, BC and $B'C'$, and AC and $A'C'$ (or their prolongations) of two triangles perspective from a point meet in three points which lie on one straight line. Conversely, if the three pairs of corresponding sides of the two triangles meet in three points which lie on one straight line, then the lines joining corresponding vertices meet in one point. With specific reference to the figure, the theorem proper says that because AA', BB', and CC' meet in a point O, the sides AC and $A'C'$ meet in a point P; AB and $A'B'$ meet in a point Q; and BC and $B'C'$ meet in a point R; and P, Q, and R lie on a straight line.

Though the theorem is true whether the triangles ABC and $A'B'C'$ lie in the same or different planes, its proof is simple only in the latter case. Desargues proved the theorem and its converse for both the two- and three-dimensional cases.

In the appendix to Bosse's 1648 work, there appears another fundamental result due to Desargues, the invariance of cross ratio under projection. The cross ratio of the line segments formed by four points A, B, C, and D on one line (Fig. 14.5) is by definition $\dfrac{BA}{BC} \Big/ \dfrac{DA}{DC}$. Pappus had already introduced this ratio (Chap. 5, sec. 7) and proved that it is the same on the two lines AD and $A'D'$. Menelaus also had a similar theorem about arcs of great circles on

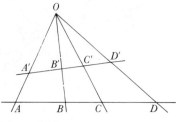

Figure 14.5

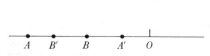

Figure 14.6

a sphere (Chap. 5, sec. 6). But neither of these men thought in terms of projection and section. Desargues did, and proved that the cross ratio is the same for every section of the projection.

In his major work (1639), he treated the concept of involution, which Pappus had also introduced but which Desargues named. Two pairs of points A, B and A', B' are said to be an involution if there is a special point O, called the center of the involution, on the line containing the four such that $OA \cdot OB = OA' \cdot OB'$ (Fig. 14.6). Likewise, three pairs of points, A, B, A', B', and A'', B'' are said to be an involution if $OA \cdot OB = OA' \cdot OB' = OA'' \cdot OB''$. The points A and B, A' and B', and A'' and B'' are said to be conjugate points. If there is a point E such that $OA \cdot OB = \overline{OE}^2$, then E is called a double point. In this case there is a second double point F, and O is the midpoint of EF. The conjugate of O is the point at infinity. Desargues used Menelaus' theorem on a line cutting the sides of a triangle to prove that if the four points A, B, A', and B' are an involution (Fig. 14.7), and if they are projected from P onto the points A_1, B_1, A'_1, and B'_1 of another line, then the second set of four points is also an involution.

In the subject of involutions, Desargues proved a major theorem. To get at it, let us first consider the concept of a complete quadrilateral, a notion already partly treated by Pappus. Let B, C, D, and E be any four points in a plane (Fig. 14.8), no three collinear. Then they determine six lines that are the sides of the complete quadrilateral. Opposite sides are two sides that do not have one of the four points in common. Thus BC and DE are opposite,

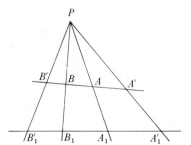

Figure 14.7

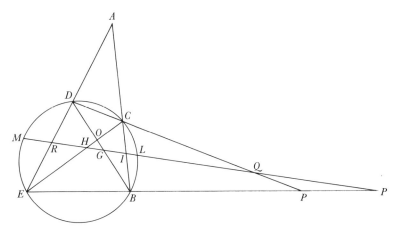

Figure 14.8

as are *CD* and *BE* and *BD* and *CE*. The intersections *O*, *F*, and *A* of the three pairs of opposite sides are the diagonal points of the quadrilateral. Now take the four vertices *B*, *C*, *D*, *E* on a circle. Suppose a line *PM* meets the pairs of sides in *P*, *Q*; *I*, *K*; and *G*, *H*; and the circle in the pair *L*, *M*. These four pairs of points are four pairs of an involution.

Further, suppose the entire figure is projected from some point outside the plane of the figure and a section is made of this projection. The circle will give rise to a conic in the section, and each line in the original figure will give rise to some line in the section. In particular, the quadrilateral in the circle will give rise to a quadrilateral in the conic. Since an involution projects into an involution, there follows the important and general result: If a quadrilateral is inscribed in a conic, any straight line not passing through a vertex intersects the conic and the pairs of opposite sides of the complete quadrilateral in four pairs of points of an involution.

Desargues next introduces the notion of a harmonic set of points. The points *A*, *B*, *E*, and *F* form a harmonic set if *A* and *B* are a pair of conjugate points with respect to the double points *E* and *F* of an involution. (The current definition that the cross ratio of a harmonic set is −1 is a later approach.)[2] Since an involution projects into an involution, so does a harmonic set project into a harmonic set. Then Desargues shows that if one member of a harmonic set of points is the point at infinity (on the line of the four), the other point of that pair bisects the line segment joining the other two points of the set. Further if *A*, *B*, *A'*, and *B'* are a harmonic set (Fig. 14.9), and if the projection from *O* is formed, then if *OA'* is perpendicular to *OB'*, *OA'* bisects angle *AOB*, and *OB'* bisects the supplementary angle.

With the notion of harmonic set available, Desargues proceeds to pole

2. It is due to Möbius: *Barycentrische Calcul* (1827), p. 269.

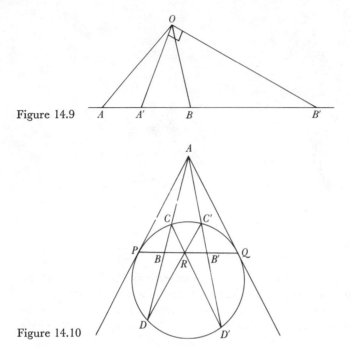

Figure 14.9

Figure 14.10

and polar theory, a subject already introduced by Apollonius. Desargues starts with a circle and the point A outside (Fig. 14.10). Then on any line from A cutting the circle in C and D, say, there is a fourth harmonic point B. For all such lines from A the fourth harmonic points lie on one line, the polar of the point A. Thus the line BB' is the polar of A. Moreover, suppose we introduce any complete quadrilateral for which A is one diagonal point and whose four vertices are on the circle. Thus we could choose C, D, D', and C' as the vertices of such a complete quadrilateral. Then the polar of A will go through the other two diagonal points of this complete quadrilateral. (In the figure, R is one of the two diagonal points.) The same assertions hold when the point A is inside the circle. If the point A is outside the circle, the polar of A joins the points of contact of the tangents from A to the circle (the points P and Q in the figure).

Having proved the above assertions for the circle, Desargues again uses projection from a point outside the plane of the figure and a section of the projection to prove that the assertions hold for any conic.

Desargues treats a diameter of a conic as the polar of a point at infinity. We shall see in a moment that this definition agrees with Apollonius'. Consider a family of parallel lines that cut the conic (Fig. 14.11). These lines have a point at infinity in common. If $A'B'$ is one such line, the har-

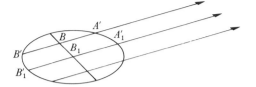

Figure 14.11

monic conjugate with respect to A' and B' of the point at infinity on $A'B'$ is the midpoint B of the chord $A'B'$. Likewise B_1, the harmonic conjugate of the point at infinity with respect to A'_1 and B'_1, is the midpoint of the chord $A'_1 B'_1$. These midpoints of a family of parallel chords lie on one straight line, which is a diameter in Apollonius' definition also. Desargues then proves a number of facts about diameters, conjugate diameters, and asymptotes to hyperbolas.

As we can now see, Desargues not only introduced new concepts, notably the elements at infinity, and many new theorems, but above all he introduced projection and section as a new method of proof and unified the approach to the several types of conics through projection and section, whereas Apollonius had treated each type of conic separately. Desargues was one of the most original mathematicians in a century rich in genius.

4. The Work of Pascal and La Hire

The second major contributor to projective geometry was Blaise Pascal (1623–62). Born in Clermont, France, he was a sickly child and in poor health during his short life. His father, Etienne Pascal, intended to keep his son from mathematics until he was fifteen or sixteen because he believed that a subject should not be introduced to a child until he was old enough to absorb it. But at the age of twelve Blaise insisted on knowing what geometry was and, on being told, set to work on it himself.

The family had moved to Paris when Pascal was eight. Even as a child he went with his father to the weekly meetings of the "Académie Mersenne," which later became the Académie Libre and, in 1666, the Académie des Sciences. Among the members were Father Mersenne, Desargues, Roberval (professor of mathematics at the Collège de France), Claude Mydorge (1585–1647), and Fermat.

Pascal devoted considerable time and energy to projective geometry. He was one of the founders of the calculus and in this subject influenced Leibniz. As we have seen, he also took a hand in the start of the work on probability. At the age of nineteen he invented the first calculating machine to help his father in the latter's work as a tax assessor. He also contributed to physics some experimental work on an original device for the creation of vacuums, the fact that the weight of the air decreases as the altitude increases, and a clarification of the concept of pressure in liquids. The originality of the work in physics has been questioned; in fact, some historians of science have described it as popularization, or plagiarism, depending upon whether or not they wished to be charitable.

Pascal was great in many other fields. He became a master of French prose; his *Pensées* and *Lettres provinciales* are literary classics. He also became famous as a polemicist in theology. From childhood on, he attempted to reconcile religious faith with the rationalism of mathematics and science, and throughout his life the two interests competed for his energy and time. Pascal, like Descartes, believed that truths of science must either appeal clearly and distinctly to the senses or the reason or be logical consequences of such truths. He saw no room for mystery-mongering in matters of science and mathematics. "Nothing that has to do with faith can be the concern of reason." In matters of science, in which only our natural thinking is involved, authority is useless; reason alone has grounds for such knowledge. However, the mysteries of faith are hidden from sense and reason and must be accepted on the authority of the Bible. He condemns those who use authority in science or reason in theology. However, the level of faith was above the level of reason.

Religion dominated his thoughts after the age of twenty-four though he continued to do mathematical and scientific work. He believed that the pursuit of science for mere enjoyment was wrong. To make enjoyment the chief end of research was to corrupt the research, for then one acquired "a greed or lust for learning, a profligate appetite for knowledge. . . . Such a study of science sprang from a prior concern for self as the center of things rather than a concern for seeking out, amid all surrounding natural phenomena, the presence of God and His glory."

In mathematical work he was largely intuitive; he anticipated great results, made superb guesses, and saw shortcuts. Later in life, he favored intuition as a source of all truths. Several of his declarations bearing on this have become famous. "The heart has its own reasons, which reason does not know." "Reason is the slow and tortuous method by which those who do not know the truth discover it." "Humble thyself, impotent reason."

If one may judge from a letter Pascal wrote to Fermat on August 10, 1660, toward the end of his life, Pascal seems to have turned somewhat against mathematics. He wrote: "To speak freely of mathematics, I find it the highest exercise of the spirit; but at the same time I know that it is so useless that I make little distinction between a man who is only a mathematician and a common artisan. Also, I call it the most beautiful occupation in the world; but it is only an occupation; and I have often said that it is good to make the attempt [to study mathematics], but not to use our forces: so that I would not take two steps for mathematics, and I am confident that

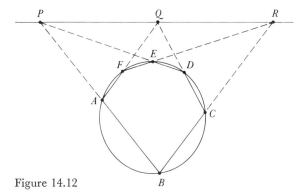

Figure 14.12

you are strongly of my opinion." Pascal was a man of manifold but contradictory qualities.

Desargues urged Pascal to work on the method of projection and section and suggested in particular the goal of reducing the many properties of the conic sections to a small number of basic propositions. Pascal took up these recommendations. In 1639, at the age of sixteen, he wrote a work on conics that used projective methods, that is projection and section. This work is now lost, but Leibniz did see it in Paris in 1676 and described it to Pascal's nephew. An eight-page *Essay on Conics* (1640), known to a few of Pascal's contemporaries, was also lost until 1779, when it was recovered.[3] Descartes, who saw the 1640 essay, regarded it as so brilliant that he could not believe it was written by so young a man.

Pascal's most famous result in projective geometry, which appeared in both of the works just mentioned, is a theorem now named after him. The theorem in modern language asserts the following: If a hexagon is inscribed in a conic, the three points of intersection of the pairs of opposite sides lie on one line. Thus (Fig. 14.12) *P*, *Q*, and *R* lie on one straight line. If the opposite sides of the hexagon are parallel, *P*, *Q*, and *R* will lie on the line at infinity.

We have only indications of how Pascal proved this theorem. He says that since it is true of the circle, it must by projection and section be true of all conics. And it is clear that if one forms a projection of the above figure from a point outside the plane and then a section of this projection, the section will contain a conic and a hexagon inscribed in it. Moreover, the opposite sides of this hexagon will meet in three points of a straight line, the points and straight line that correspond to *P*, *Q*, *R*, and the line *PQR* of the original figure. Incidentally Pappus' theorem (Chap. 5, sec. 7), which refers to three points on each of two lines, is a special case of the above

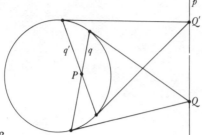

Figure 14.13

theorem. When the conic degenerates into two straight lines, as when a hyperbola degenerates into its asymptotes, then the situation described by Pappus results.

The converse of Pascal's theorem, namely, if a hexagon is such that the points of intersection of its three pairs of opposite sides lie on one straight line, then the vertices of the hexagon lie on a conic, is also correct but was not considered by Pascal. There are other results in Pascal's 1640 work but they do not warrant attention here.

The method of projection and section was taken up by Philippe de La Hire (1640–1718), who was a painter in his youth and then turned to mathematics and astronomy. Like Pascal, La Hire was influenced by Desargues and did a considerable amount of work on the conic sections. Some of it, in publications of 1673 and 1679, employed the synthetic manner of the Greeks but with new approaches, such as the focus-distance definition of the ellipse and hyperbola, and some of it used the analytic geometry of Descartes and Fermat. His greatest work, however, is the *Sectiones Conicae* (1685) and this is devoted to projective geometry.

Like Desargues and Pascal, La Hire first proved properties of the circle, chiefly involving harmonic sets, and then carried these properties over to the other conic sections by projection and section. Thus he could carry the properties of the circle over to any type of conic section in one method of proof. Though there were a few omissions, such as Desargues's involution theorem and Pascal's theorem, in this 1685 work of La Hire we find practically all the now familiar properties of conic sections synthetically proved and systematically established. In fact, La Hire proves almost all of Apollonius' 364 theorems on the conics. He also has the harmonic properties of quadrilaterals. In all, La Hire proved about 300 theorems. He tried to show that projective methods were superior to those of Apollonius and to the new analytic methods of Descartes and Fermat (Chap. 15) which had already been created.

On the whole, La Hire's results do not go beyond Desargues's and Pascal's. However, in pole and polar theory he has one major new result.

He proves that if a point traces a straight line, then the polar of the point will rotate around the pole of that straight line. Thus if Q (Fig. 14.13) moves along the line p, then the polar of Q rotates around the pole P of the line p.

5. *The Emergence of New Principles*

Over and above the specific theorems created by men such as Desargues, Pascal, and La Hire, several new ideas and outlooks were beginning to appear. The first is the idea of *continuous change of a mathematical entity* from one state to another; in the present instance, the entity is a geometrical figure. It was Kepler, in his *Astronomiae pars Optica*[4] of 1604, who first seemed to grasp the fact that parabola, ellipse, hyperbola, circle, and the degenerate conic consisting of a pair of lines are continuously derivable from each other. To effect this process of continuous change from one figure to another, say from ellipse to parabola and then to hyperbola, Kepler imagined that one focus is held fixed and the other is allowed to move along the straight line joining them. By letting the moving focus go off to infinity (while letting the eccentricity approach 1), the ellipse becomes the parabola; then, by having the moving focus reappear on the line but on the other side of the stationary focus, the parabola becomes a hyperbola. When the two foci of an ellipse move together, the ellipse becomes a circle; and when the two foci of a hyperbola move together, the hyperbola degenerates into two straight lines. So that one focus could go off to infinity in one direction and then reappear on the other side, Kepler assumed, as already noted, that the straight line extended in either direction meets itself at infinity at a single point, so that the line has the property of a circle. Though it is not intuitively satisfying to view the line in this manner, the idea is logically sound, and indeed became fundamental in nineteenth-century projective geometry. Kepler also pointed out that the various sections of the cone can be obtained by continuously varying the inclination of the plane that makes the section.

The notion of continuous change in a figure was also employed by Pascal. He allowed two consecutive vertices of his hexagon to approach each other so that the figure became a pentagon. He then argued from properties of the hexagon to properties of the pentagon by considering what happened to these properties under the continuous change. In the same manner he passed from pentagons to quadrilaterals.

The second idea to emerge clearly from the work of the projective geometers is that of *transformation and invariance*. To project a figure from some point and then take a section of that projection is to transform the figure to a new one. The properties of the original figure that are of interest are those that remain invariant under the transformation. Other geometers of the

4. *Ad Vitellionem Paralipomena, quibus Astronomiae pars Optica Traditur* (Supplement to Vitello, Giving the Optical Part of Astronomy).

seventeenth century, for example, Gregory of St. Vincent (1584–1667) and Newton,[5] introduced transformations other than projection and section.

The projective geometers, too, undertook the search for general methods that had been initiated by the algebraists, particularly Vieta. In Greek times the methods of proof had been limited in power. Each theorem required another plan of attack. Neither Euclid nor Apollonius seemed to have been concerned with general methods. But Desargues emphasized projection and section because he saw in it a general procedure for proving theorems about all conics once they were proved for the circle. And in the notions of involution and harmonic sets, he saw more general concepts than those of Euclidean geometry. Indeed a harmonic set of four points, when one of the four is at infinity, reduces to three points, one of which is the midpoint of the line segment joining the other two. Hence, the notion of harmonic set and theorems about harmonic sets are more general than the notion of a point bisecting a line segment. Desargues and Pascal sought to derive the largest possible number of results from a single theorem. Bosse says that Desargues derived sixty theorems of Apollonius from his involution theorem and that Pascal praised him for it. Pascal, by seeking the relationships of different figures, such as the hexagon and pentagon, also sought some common approach to these figures. In fact, he was supposed to have deduced some 400 corollaries from his theorem on the hexagon, by examining the consequences of the theorem for related figures. However, there are no extant works of his in which this is done. The interest in method is clear in the 1685 work of La Hire, for its major point was to show that the method of projection and section was superior to Apollonius' methods and even to Descartes's algebraic methods. This search for generality in results and methods became a powerful force in subsequent mathematics.

Though the geometers were emphasizing generality of method, they were unconsciously unearthing another kind of generality. Many of the theorems, such as Desargues's triangle theorem, deal with the intersection of points and lines rather than with the sizes of line segments, angles, and areas, as is true of Euclidean geometry. The fact that lines intersect is logically prior to any consideration of size because the very fact of intersection determines the formation of a figure. A new and fundamental branch of geometry, in which location and intersection properties, rather than size or metric properties, mattered, was being born. However, the seventeenth-century workers in projective geometry used Euclidean geometry as a base, in particular the concepts of distance and size of angle. Moreover, these geometers, far from thinking in terms of a new geometry, were in fact trying to improve the methods of Euclidean geometry. The realization that a new branch of geometry was implicit in their work did not come about until the nineteenth century.

5. *Principia*, 3rd ed., Book 1, Lemma 22 and Prop. 25.

Although at the outset the motivation for the work in projective geometry was a desire to help the painters, its goal became diverted to and merged with the rising interest in the conic sections. But pure mathematics was not congenial to the temper of the seventeenth century, whose mathematicians were far more concerned with understanding and mastering nature—in short, with scientific problems. Algebraic methods of working with mathematical problems proved in general to be more effective, and in particular to yield the quantitative knowledge that science and technology needed. The qualitative results the projective geometers produced by their synthetic methods were not nearly so helpful. Hence projective geometry was abandoned in favor of algebra, analytic geometry, and the calculus, which themselves blossomed into still other subjects of central importance in modern mathematics. The results of Desargues, Pascal, and La Hire were forgotten and had to be rediscovered later, chiefly in the nineteenth century, by which time intervening creations and new points of view enabled mathematicians to fructify the large ideas still dormant in projective geometry.

Bibliography

Chasles, Michel: *Aperçu historique des méthodes en géométrie*, 3rd ed., Gauthier-Villars et Fils, 1889, pp. 68–95, 118–37. (Same as first edition of 1837.)

Coolidge, Julian L.: *A History of Geometrical Methods*, Dover (reprint), 1963, pp. 88–92.

———: *A History of the Conic Sections and Quadric Surfaces*, Dover (reprint), 1968, Chap. 3.

———: "The Rise and Fall of Projective Geometry," *Amer. Math. Monthly*, 41, 1934, 217–28.

Desargues, Girard: *Œuvres*, 2 vols., Leiber, 1864.

Ivins, W. M., Jr.: "A Note on Girard Desargues," *Scripta Math.*, 9, 1943, 33–48.

———: "A Note on Desargues's Theorem," *Scripta Math.*, 13, 1947, 203–10.

Mortimer, Ernest: *Blaise Pascal: The Life and Work of a Realist*, Harper and Bros., 1959.

Pascal, Blaise: *Œuvres*, Hachette, 1914–21.

Smith, David Eugene: *A Source Book in Mathematics*, Dover (reprint), 1959, Vol. 2, pp. 307–14, 326–30.

Struik, D. J.: *A Source Book in Mathematics, 1200–1800*, Harvard University Press, 1969, pp. 157–68.

Taton, René: *L'Œuvre mathématique de G. Desargues*, Presses Universitaires de France, 1951.

15
Coordinate Geometry

> ... I have resolved to quit only abstract geometry, that is to
> say, the consideration of questions which *serve only to exercise
> the mind*, and this, in order to study another kind of geometry,
> which has for its object the explanation of the phenomena of
> nature.   RENÉ DESCARTES

1. *The Motivation for Coordinate Geometry*

Fermat and Descartes, the two men primarily responsible for the next major
creation in mathematics, were, like Desargues and his followers, concerned
with general methods for studying curves. But Fermat and Descartes were
very much involved in scientific work, keenly aware of the need for quantitative methods, and impressed with the power of algebra to supply that method.
And so Fermat and Descartes turned to the application of algebra to the
study of geometry. The subject they created is called coordinate, or analytic,
geometry; its central idea is the association of algebraic equations with curves
and surfaces. This creation ranks as one of the richest and most fruitful veins
of thought ever struck in mathematics.

That the needs of science and an interest in methodology motivated
both Fermat and Descartes is beyond doubt. Fermat's contributions to the
calculus such as the construction of tangents to curves and the calculation of
maxima and minima, were, as we shall see more clearly in connection with
the history of the calculus, designed to answer scientific problems; he was
also a first-rate contributor to optics. His interest in methodology is attested
to by an explicit statement in his brief book, *Ad Locos Planos et Solidos Isagoge*
(Introduction to Plane and Solid Loci[1]), written in 1629 but published in
1679.[2] He says there that he sought a universal approach to problems involving curves. As for Descartes, he was one of the greatest seventeenth-
century scientists, and he made methodology a prime objective in all of his
work.

---

1. Fermat uses these terms in the sense explained by Pappus. See Chap. 8, sec. 2.
2. *Œuvres*, 1, 91–103.

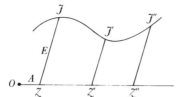

Figure 15.1

2. *The Coordinate Geometry of Fermat*

In his work on the theory of numbers, Fermat started with Diophantus. His work on curves began with his study of the Greek geometers, notably Apollonius, whose lost book, *On Plane Loci*, he, among others, had reconstructed. Having contributed to algebra, he was prepared to apply it to the study of curves, which he did in *Ad Locos*. He says that he proposed to open up a general study of loci, which the Greeks had failed to do. Just how Fermat's ideas on coordinate geometry evolved is not known. He was familiar with Vieta's use of algebra to solve geometric problems, but it is more likely that he translated Apollonius' results directly into algebraic form.

He considers any curve and a typical point J on it (Fig. 15.1). The position of J is fixed by a length A, measured from a point O on a base line to a point Z, and the length E from Z to J. Thus Fermat uses what we call oblique coordinates, though no y-axis appears explicitly and no negative coordinates are used. His A and E are our x and y.

Fermat had stated earlier his general principle: "Whenever in a final equation two unknown quantities are found we have a locus, the extremity of one of these describing a line straight or curved." Thus the extremities $J, J', J'', \ldots$ of E in its various positions describe the "line." His unknown quantities, A and E, are really variables or, one can say, the equation in A and E is indeterminate. Here Fermat makes use of Vieta's idea of having a letter stand for a class of numbers. Fermat then gives various algebraic equations in A and E and states what curves they describe. Thus he writes "*D in A aequetur B in E*" (in our notation, $Dx = By$) and states that this represents a straight line. He also gives (in our notation) the more general equation $d(a - x) = by$ and affirms that this too represents a straight line. The equation "*B quad. – A quad. aequetur E quad.*" (in our notation, $B^2 - x^2 = y^2$) represents a circle. Similarly (in our notation), $a^2 - x^2 = ky^2$ represents an ellipse; $a^2 + x^2 = ky^2$ and $xy = a$ represent hyperbolas; and $x^2 = ay$ represents a parabola. Since Fermat did not use negative coordinates, his equations could not represent the full curve that he said they described. He did appreciate that one can translate and rotate axes, because he gives more complicated second-degree equations and states the simpler forms to which they can be reduced. In fact, he affirms that an equation of the first degree in A and E has a straight-line locus and all second degree equations in A and E have conics as their loci. In his *Methodus ad Disquirendam Maximam et Minimam* (Method of Finding Maxima and Minima, 1637),[3] he introduced the curves of $y = x^n$ and $y = x^{-n}$.

3. René Descartes

Descartes was the first great modern philosopher, a founder of modern biology, a first-rate physicist, and only incidentally a mathematician. However, when a man of his power of intellect devotes even part of his time to a subject, his work cannot but be significant.

He was born in La Haye in Touraine on March 31, 1596. His father, a moderately wealthy lawyer, sent him at the age of eight to the Jesuit school of La Flèche in Anjou. Because he was of delicate health, he was allowed to spend the mornings in bed, during which time he worked. He followed this custom throughout his life. At sixteen he left La Flèche and at twenty he was graduated from the University of Poitiers as a lawyer and went to Paris. There he met Mydorge and Father Marin Mersenne and spent a year with them in the study of mathematics. However, Descartes became restless and entered the army of Prince Maurice of Orange in 1617. During the next nine years he alternated between service in several armies and carousing in Paris, but throughout this period continued to study mathematics. His ability to solve a problem that had been posted on a billboard in Breda in the Netherlands as a challenge convinced him that he had mathematical ability and he began to think seriously in this subject. He returned to Paris and, having become excited by the power of the telescope, secluded himself to study the theory and construction of optical instruments. In 1628 he moved to Holland to secure a quieter and freer intellectual atmosphere. There he lived for twenty years and wrote his famous works. In 1649 he was invited to instruct Queen Christina of Sweden. Tempted by the honor and the glamor of royalty, he accepted. He died there of pneumonia in 1650.

His first work, *Regulae ad Directionem Ingenii* (Rules for the Direction of the Mind),[4] was written in 1628 but published posthumously. His next major work was *Le Monde* (System of the World, 1634), which contains a cosmological theory of vortices to explain how the planets are kept in motion and in their paths around the sun. However, he did not publish it for fear of persecution by the Church. In 1637 he published his *Discours de la méthode pour bien conduire sa raison, et chercher la vérité dans les sciences*.[5] This book, a

---

3. *Œuvres*, 1, 133–79; 3, 121–56.
4. Published in Dutch in 1692; *Œuvres*, 10, 359–469.
5. *Œuvres*, 6, 1–78.

classic of literature and philosophy, contains three famous appendices, *La Géométrie*, *La Dioptrique*, and *Les Météores*. *La Géométrie*, which is the only book Descartes wrote on mathematics, contains his ideas on coordinate geometry and algebra, though he did communicate many other ideas on mathematics in numerous letters. The *Discours* brought him great fame immediately. As time passed, both he and his public became more impressed with his work. In 1644 he published *Principia Philosophiae*, which is devoted to physical science and especially to the laws of motion and the theory of vortices. It contains material from his *System*, which he believed he had now made more acceptable to the Church. In 1650 he published *Musicae Compendium*.

Descartes's scientific ideas came to dominate the seventeenth century. His teachings and writings became popular even among non-scientists because he presented them so clearly and attractively. Only the Church rejected him. Actually Descartes was devout, and happy to have (as he believed) established the existence of God. But he had taught that the Bible was not the source of scientific knowledge, that reason alone sufficed to establish the existence of God, and that man should accept only what he could understand. The Church reacted to these teachings by putting his books on the *Index of Prohibited Books* shortly after his death and by preventing a funeral oration on the occasion of his interment in Paris.

Descartes approached mathematics through three avenues, as a philosopher, as a student of nature, and as a man concerned with the uses of science. It is difficult and perhaps artificial to try to separate these three lines of thought. He lived when the Protestant-Catholic controversy was at its height and when science was beginning to reveal laws of nature that challenged major religious doctrines. Hence Descartes began to doubt all the knowledge he had acquired at school. As early as the conclusion of his course of study at La Flèche, he decided that his education had advanced only his perplexity. He found himself so beset with doubts that he was convinced he had progressed no further than to recognize his ignorance. And yet, because he had been in one of the most celebrated schools in Europe, and because he believed he had not been an inferior student, he felt justified in doubting whether there was any sure body of knowledge anywhere. He then pondered the question: How do we know anything?

He soon decided that logic in itself was barren: "As for Logic, its syllogisms and the majority of its other precepts are of avail rather in the communication of what we already know, or ... even in speaking without judgment of things of which we are ignorant, than in the investigation of the unknown." Logic, then, did not supply the fundamental truths.

But where were these to be found? He rejected the current philosophy, largely Scholastic, which, though appealing, seemed to have no clear-cut foundations and employed reasoning that was not always irreproachable.

Philosophy, he decided, afforded merely "the means of discoursing with an appearance of truth on all matters." Theology pointed out the path to heaven and he aspired to go there as much as any man, but was the path correct?

The method of establishing truths in all fields came to him, he says, in a dream, on November 10, 1619, when he was on one of his military campaigns; it was the method of mathematics. Mathematics appealed to him because the proofs based on its axioms were unimpeachable and because authority counted for naught. Mathematics provided the method of achieving certainties and effectively demonstrating them. Moreover, he saw clearly that the method of mathematics transcended its subject matter. He says, "It is a more powerful instrument of knowledge than any other that has been bequeathed to us by human agency, as being the source of all others." In the same vein he continues:

> ... All the sciences which have for their end investigations concerning order and measure are related to mathematics, it being of small importance whether this measure be sought in numbers, forms, stars, sounds, or any other object; that accordingly, there ought to exist a general science which should explain all that can be known about order and measure, considered independently of any application to a particular subject, and that, indeed, this science has its own proper name, consecrated by long usage, to wit, mathematics. And a proof that it far surpasses in facility and importance the sciences which depend upon it is that it embraces at once all the objects to which these are devoted and a great many others besides. ...

And so he concluded that "The long chains of simple and easy reasonings by means of which geometers are accustomed to reach the conclusions of their most difficult demonstrations had led me to imagine that all things to the knowledge of which man is competent are mutually connected in the same way."

From his study of mathematical method he isolated in his *Rules for the Direction of the Mind* the following principles for securing exact knowledge in any field. He would accept nothing as true that was not so clear and distinct in his own mind as to exclude all doubt; he would divide difficulties into smaller ones; he would proceed from the simple to the complex; and, lastly, he would enumerate and review the steps of his reasoning so thoroughly that nothing could be omitted.

With these essentials of method, which he distilled from the practice of mathematicians, Descartes hoped to solve problems in philosophy, physics, anatomy, astronomy, mathematics, and other fields. Although he did not succeed in this ambitious program, he did make remarkable contributions to philosophy, science, and mathematics. The mind's immediate apprehension of basic, clear, and distinct truths, this intuitive power, and the deduction of consequences are the essence of his philosophy of knowledge. Purported knowledge otherwise obtained should be rejected as suspect of error and dangerous. The three appendices to his *Discours* were intended to show that his method is effective; he believed that he had shown this.

Descartes inaugurated modern philosophy. We cannot pursue his system except to note a few points relevant to mathematics. In philosophy he sought as axioms truths so clear to him that he could accept them readily. He finally decided on four: (a) *cogito, ergo sum* (I think, therefore I am); (b) each phenomenon must have a cause; (c) an effect cannot be greater than its cause; (d) the mind has innate in it the ideas of perfection, space, time, and motion. The idea of perfection, of a perfect being, could not be derived from or created by the imperfect mind of man in view of axiom (c). It could be obtained only from a perfect being. Hence God exists. Since God would not deceive us, we can be sure that the axioms of mathematics, which are clear to our intuitions, and the deductions we make from them by purely mental processes, really apply to the physical world and so are truths. It follows, then, that God must have established nature according to mathematical laws.

As for mathematics itself, he believed that he had distinct and clear mathematical ideas, such as that of a triangle. These ideas did exist and were eternal and immutable. They did not depend on his thinking them or not. Thus mathematics had an external, objective existence.

Descartes's second major interest, shared by most thinkers of his age, was the understanding of nature. He devoted many years to scientific problems and even experimented extensively in mechanics, hydrostatics, optics, and biology. His theory of vortices was the dominant cosmological theory of the seventeenth century. He is the founder of the philosophy of mechanism—that all natural phenomena, including the functioning of the human body, reduce to motions obeying the laws of mechanics—though Descartes exempted the soul. Optics, and the design of lenses in particular, was of special interest to him; part of *La Géométrie* is devoted to optics, as is *La Dioptrique*. Descartes shares with Willebrord Snell the honor of discovering the correct law of refraction of light. As in philosophy, his work in science was basic and revolutionary.

Also important in Descartes's scientific work is his emphasis on putting the fruits of science to use (Chap. 11, sec. 5). In this attitude he breaks clearly and openly with the Greeks. To master nature for the good of man, he pursued many scientific problems. And, being impressed with the power of mathematics, he naturally sought to use that subject; for him it was not contemplative discipline but a constructive and useful science. Unlike Fermat, he cared little for its beauty and harmony; he did not value pure mathematics. He says that mathematical method applied only to mathematics is without value because it is not a study of nature. Those who cultivate mathematics for its own sake are idle searchers given to a vain play of the spirit.

4. *Descartes's Work in Coordinate Geometry*

Having decided that method was important and that mathematics could be effectively employed in scientific work, Descartes turned to the application of method to geometry. Here his general interest in method and his particular knowledge of algebra joined forces. He was disturbed by the fact that every proof in Euclidean geometry called for some new, often ingenious, approach. He explicitly criticized the geometry of the ancients as being too abstract, and so much tied to figures "that it can exercise the understanding only on condition of greatly fatiguing the imagination." The algebra that he found prevalent he also criticized because it was so completely subject to rules and formulas "that there results an art full of confusion and obscurity calculated to hamper instead of a science fitted to improve the mind." Descartes proposed, therefore, to take all that was best in geometry and algebra and correct the defects of one with the help of the other.

Actually it was the use of algebra in geometry that he undertook to exploit. He saw fully the power of algebra and its superiority over the Greek geometrical methods in providing a broad methodology. He also stressed the generality of algebra and its value in mechanizing the reasoning processes and minimizing the work in solving problems. He saw its potential as a universal science of method. The product of his application of algebra to geometry was *La Géométrie*.

Though in this book Descartes used the improvements in algebraic notation already noted in Chapter 13, the essay is not easy reading. Much of the obscurity was deliberate; Descartes boasted that few mathematicians in Europe would understand his work. He indicated the constructions and demonstrations, leaving it to others to fill in the details. In one of his letters he compares his writing to that of an architect who lays the plans and prescribes what should be done but leaves the manual work to the carpenters and bricklayers. He says also, "I have omitted nothing inadvertently but I have foreseen that certain persons who boast that they know everything would not miss the opportunity of saying that I have written nothing that they did not already know, were I to make myself sufficiently intelligible for them to understand me." He gave other reasons in *La Géométrie*, such as not wishing to deprive his readers of the pleasure of working things out for themselves. Many explanatory commentaries were written to make Descartes's book clear.

His ideas must be inferred from a number of examples worked out in the book. He says that he omits the demonstration of most of his general statements because if one takes the trouble to examine systematically these

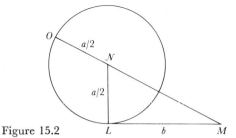

Figure 15.2

examples, the demonstrations of the general results will become apparent, and it is of more value to learn them in that way.

He begins in *La Géométrie* with the use of algebra to solve geometrical construction problems in the manner of Vieta; only gradually does the idea of the equation of a curve emerge. He points out first that geometrical constructions call for adding, subtracting, multiplying, and dividing lines and taking the square root of particular lines. Since all of these operations also exist in algebra, they can be expressed in algebraic terms.

In tackling a given problem, Descartes says we must suppose the solution of the problem already known and represent with letters all the lines, known and unknown, that seem necessary for the required construction. Then, making no distinction between known and unknown lines, we must "unravel" the difficulty by showing in what way the lines are related to each other, aiming at expressing one and the same quantity in two ways. This gives an equation. We must find as many equations as there are unknown lines. If several equations remain, we must combine them until there remains a single unknown line expressed in terms of known lines. Descartes then shows how to construct the unknown line by utilizing the fact that it satisfies the algebraic equation.

Thus, suppose a geometric problem leads to finding an unknown length x, and after algebraic formulation x is found to satisfy the equation $x^2 = ax + b^2$ where a and b are known lengths. Then we know by algebra that

(1)
$$x = \frac{a}{2} + \sqrt{\frac{a^2}{4} + b^2}.$$

(Descartes ignored the second root, which is negative.) Descartes now gives a construction for x. He constructs the right triangle NLM (Fig. 15.2) with $LM = b$ and $NL = a/2$, and prolongs MN to O so that $NO = NL = a/2$. Then the solution x is the length OM. The proof that OM is the correct length is not given by Descartes but it is immediately apparent for

$$OM = ON + MN = \frac{a}{2} + \sqrt{\frac{a^2}{4} + b^2}.$$

Thus the expression (1) for x, which was obtained by solving an algebraic equation, indicates the proper construction for x.

In the first half of Book I, Descartes solves only classical geometric construction problems with the aid of algebra. This is an application of algebra to geometry, but not analytic geometry in our present sense. The problems thus far are what one might call determinate construction problems because they lead to a unique length. He considers next indeterminate construction problems, that is, problems in which there are many possible lengths that serve as answers. The endpoints of the many lengths fill out a curve; and here Descartes says, "It is also required to discover and trace the curve containing all such points." This curve is described by the final indeterminate equation expressing the unknown lengths y in terms of the arbitrary lengths x. Moreover, Descartes stresses that for each x, y satisfies a determinate equation and so can be constructed. If the equation is of the first or second degree, y can be constructed by the methods of Book I, using only lines and circles. For higher-degree equations, he says he will show in Book III how y can be constructed.

Descartes uses the problem of Pappus (Chap. 5, sec. 7) to illustrate what happens when a problem leads to one equation in two unknowns. This problem, which had not been solved in full generality, is as follows: Given the position of three lines in a plane, find the position of all points (the locus) from which we can construct lines, one to each of the given lines and making a known angle with each of these given lines (the angle may be different from line to line), such that the rectangle contained by two of the constructed lines has a given ratio to the square on the third constructed line; if there are four given lines, then the constructed lines, making given angles with the given lines, must be such that the rectangle contained by two must have a given ratio to the rectangle contained by the other two; if there are five given lines, then the five constructed lines, each making a given angle with one of the given lines, must be such that the product of three of them has a given ratio to the product of the remaining two. The condition on the locus when there are more than five given lines is an obvious extension of the above.

Pappus had declared that when three or four lines are given, the locus is a conic section. In Book II Descartes treats the Pappus problem for the case of four lines. The given lines (Fig. 15.3) are AG, GH, EF, and AD. Consider a point C and the four lines from C to each of the four given lines and making a specified angle with each of the four given lines. The angle can be different from one line to another. Let us denote the four lines by CP, CQ, CR, and CS. It is required to find the locus of C satisfying the condition $CP \cdot CR = CS \cdot CQ$.

Descartes denotes AP by x and PC by y. By simple geometric considerations, he obtains the values of CR, CQ, and CS in terms of known

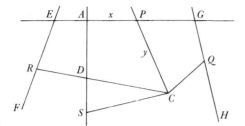

Figure 15.3

quantities. He uses these values to form $CP \cdot CR = CS \cdot CQ$ and obtains a second degree equation in x and y of the form

(2) $$y^2 = Ay + Bxy + Cx + Dx^2$$

where A, B, C, and D are simple algebraic expressions in terms of the known quantities. Now Descartes points out that if we select any value of x we have a quadratic equation for y that can be solved for y; and then y can be constructed by straightedge and compass as he has shown in Book I. Hence if one takes an infinite number of values for x, one obtains an infinite number of values for y and hence an infinite number of points C. The locus of all these points C is a curve whose equation is (2).

What Descartes has done is to set up one line (AG in the above figure) as a base line with an origin at the point A. The x-values are then lengths along this line, and the y-values are lengths that start at this base line and make a fixed angle with it. This coordinate system is what we now call an oblique system. Descartes's x and y stand for positive numbers only; yet his equations cover portions of curves in other than what we would call the first quadrant. He simply assumes that the locus lies primarily in the first quadrant and makes passing reference to what might happen elsewhere. That there is a length for each positive real number is assumed unconsciously.

Having arrived at the idea of the equation of a curve, Descartes now develops it. It is easily demonstrated, he asserts, that the degree of a curve is independent of the choice of the reference axis; he advises choosing this axis so that the resulting equation is as simple as possible. In another great stride, he considers two different curves, expresses their equations with respect to the same reference axis, and finds the points of intersection by solving the equations simultaneously.

Also in Book II, Descartes considers critically the Greek distinctions among plane, solid, and linear curves. The Greeks had said plane curves were those constructible by straightedge and compass; the solid curves were the conic sections; and the linear curves were all the others, such as the conchoid, spiral, quadratrix, and cissoid. The linear curves were also called mechanical by the Greeks because some special mechanism was required to construct them. But, Descartes says, even the straight line and circle require some instrument. Nor can the accuracy of the mechanical construction matter, because in mathematics only the reasoning counts. Possibly, he continues, the ancients objected to linear curves because they were insecurely defined. On these grounds, Descartes rejects the idea that only the curves constructible with straightedge and compass[6] are legitimate and even proposes some new curves generated by mechanical constructions. He concludes with the highly significant statement that geometric curves are those that can be expressed by a unique algebraic equation (of finite degree) in x and y. Thus Descartes accepts the conchoid and cissoid. All other curves, such as the spiral and the quadratrix, he calls mechanical.

Descartes's insistence that an acceptable curve is one that has an algebraic equation is the beginning of the elimination of constructibility as a criterion of existence. Leibniz went farther than Descartes. Using the words "algebraic" and "transcendental" for Descartes's terms "geometrical" and "mechanical," he protested the requirement that a curve must have an algebraic equation.[7] Actually Descartes and his contemporaries ignored the requirement and worked just as enthusiastically with the cycloid, the logarithmic curve, the logarithmic spiral ($\log \rho = a\theta$), and other non-algebraic curves.

In broadening the concept of admissible curves, Descartes made a major step. He not only admitted curves formerly rejected but opened up the whole field of curves, because, given any algebraic equation in x and y, one can find its curve and so obtain totally new curves. In *Arithmetica Universalis* Newton says (1707), "But the Moderns advancing yet much further [than the plane, solid and linear loci of the Greeks] have received into Geometry all Lines that can be expressed by Equations."

Descartes next considers the classes of geometric curves. Curves of the first and second degree in x and y are in the first and simplest class. Descartes says, in this connection, that the equations of the conic sections are of the second degree, but does not prove this. Curves whose equations are of the third and fourth degree constitute the second class. Curves whose equations are of the fifth and sixth degree are of the third class and so on. His reason for grouping third and fourth, as well as fifth and sixth degree curves, is that he believed the higher one in each class could be reduced to the lower, as the solution of quartic equations could be effected by the solution of cubics. This belief was of course incorrect.

The third book of *La Géométrie* returns to the theme of Book I. Its objective is the solution of geometric construction problems, which, when formulated algebraically, lead to determinate equations of third and higher degree and which, in accordance with the algebra, call for the conic sections

6. Compare the discussion in Chap. 8, sec. 2.
7. *Acta Erud.*, 1684, pp. 470, 587; 1686, p. 292 = *Math. Schriften*, 5, 127, 223, 226.

and higher-degree curves. Thus Descartes considers the construction problem of finding the two mean proportionals between two given quantities a and q. The special case when $q = 2a$ was attempted many times by the classical Greeks and was important because it is a way to solve the problem of doubling the cube. Descartes proceeds as follows: Let z be one of these mean proportionals; then z^2/a must be the second, for we must have

$$\frac{a}{z} = \frac{z}{z^2/a} = \frac{z^2/a}{z^3/a^2}.$$

Then, if we take z^3/a^2 to be q, we have the equation z must satisfy. Hence, given q and a, we must find z such that

(3) $$z^3 = a^2q,$$

or, we must solve a cubic equation. Descartes now shows that such quantities z and z^2/a can be obtained by a geometrical construction that utilizes a parabola and a circle.

As the construction is described by Descartes, seemingly no coordinate geometry is involved. However, the parabola is not constructible with straightedge and compass, except point by point, and so one must use the equation to plot the curve accurately.

Descartes does *not* obtain z by writing the equations in x and y of circle and parabola and finding the coordinates of the point of intersection by solving equations simultaneously. In other words, he is not solving equations graphically in our sense. Rather he uses purely geometric constructions (except for supposing that a parabola can be drawn), the knowledge of the fact that z satisfies an equation, and the geometric properties of the circle and parabola (which can be more readily seen through their equations). Descartes does here just what he did in Book I, except that he is now solving geometric construction problems in which the unknown length satisfies a third or higher-degree equation instead of a first or second degree equation. His solution of the purely algebraic aspect of the problem and the subsequent construction is practically the same one the Arabs gave, except that he was able to use the equations of the conic sections to deduce facts about the curves and to draw them.

Descartes not only wished to show how some solid problems could be solved with the aid of algebra and the conic sections but was interested in classifying problems so that one would know what they involved and how to go about solving them. His classification is based on the degree of the algebraic equation to which one is led when the construction problem is formulated algebraically. If that degree is one or two, then the construction can be performed with straight line and circle. If the degree is three or four,

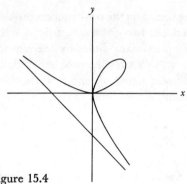

Figure 15.4

the conic sections must be employed. He does affirm, incidentally, that all cubic problems can be reduced to trisecting the angle and doubling the cube and that no cubic problems can be solved without the use of a curve more complex than the circle. If the degree of the equation is higher than four, curves more complicated than the conic sections may be required to perform the construction.

Descartes also emphasized the degree of the equation of a curve as the measure of its simplicity. One should use the simplest curve, that is, the lowest degree possible, to solve a construction problem. The emphasis on the degree of a curve became so strong that a complicated curve such as the folium of Descartes (Fig. 15.4), whose equation is $x^3 + y^3 - 3axy = 0$, was considered simpler than $y = x^4$.

What is far more significant than Descartes's insight into construction problems and their classification is the importance he assigned to algebra. This key makes it possible to recognize the typical problems of geometry and to bring together problems that in geometrical form would not appear to be related at all. Algebra brings to geometry the most natural principles of classification and the most natural hierarchy of method. Not only can questions of solvability and geometrical constructibility be decided elegantly, quickly, and fully from the parallel algebra, but without it they cannot be decided at all. Thus, system and structure were transferred from geometry to algebra.

Part of Book II of *La Géométrie* as well as *La Dioptrique* Descartes devoted to optics, using coordinate geometry as an aid. He was very much concerned with the design of lenses for the telescope, microscope, and other optical instruments because he appreciated the importance of these instruments for astronomy and biology. His *Dioptrique* takes up the phenomenon of refraction. Kepler and Alhazen before him had noted that the belief that the angle of refraction is proportional to the angle of incidence, the proportionality constant being dependent on the medium doing the refracting, was

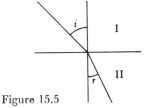

Figure 15.5

incorrect for large angles, but they did not discover the true law. Before 1626 Willebrord Snell discovered but did not publish the correct relationship,

$$\frac{\sin i}{\sin r} = \frac{v_1}{v_2},$$

where v_1 is the velocity of light in the first medium (Fig. 15.5) and v_2 the velocity in the medium into which the light passes. Descartes gave this same law in 1637 in the *Dioptrique*. There is some question as to whether he discovered it independently. His argument was wrong, and Fermat immediately attacked both the law and the proof. A controversy arose between them which lasted ten years. Fermat was not satisfied that the law was correct, until he derived it from his Principle of Least Time (Chap. 24, sec. 3).

In *La Dioptrique*, after describing the operation of the eye, Descartes considers the problem of designing properly focusing lenses for telescopes, microscopes, and spectacles. It was well known even in antiquity that a spherical lens will not cause parallel rays or rays diverging from a source S to focus on one point. Hence the question was open as to what shape would so focus the incoming rays. Kepler had suggested that some conic section would serve. Descartes sought to design a lens that would focus the rays perfectly.

He proceeded to solve the general problem of what surface should separate two media such that light rays starting from one point in the first medium would strike the surface, refract into the second medium, and there converge to one point. He discovered that the curve generating the desired surface of revolution is an oval, now known as the oval of Descartes. This curve and its refracting properties are discussed in *La Dioptrique*, and the discussion is supplemented in Book II of *La Géométrie*.

The modern definition is that the curve is the locus of points M satisfying the condition

$$FM \pm nF'M = 2a$$

where F and F' are fixed points, $2a$ is any real number larger than FF', and n is any real number. If $n = 1$ the curve becomes an ellipse. In the general case, the equation of the oval is of the fourth degree in x and y, and the curve

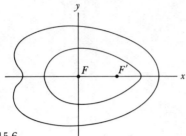

Figure 15.6

consists of two closed, distinct portions (Fig. 15.6) without common point and one inside the other. The inner curve is convex like an ellipse and the outer one can be convex or may have points of inflection, as in the figure.

As we can now see, Descartes's approach to coordinate geometry differs profoundly from Fermat's. Descartes criticized and proposed to break with the Greek tradition, whereas Fermat believed in continuity with Greek thought and regarded his work in coordinate geometry only as a reformulation of the work of Apollonius. The real discovery—the power of algebraic methods—is Descartes's; and he realized he was supplanting the ancient methods. Though the idea of equations for curves is clearer with Fermat than with Descartes, Fermat's work is primarily a technical achievement that completes the work of Apollonius and uses Vieta's idea of letters to represent classes of numbers. Descartes's methodology is universally applicable and potentially applies to the transcendental curves, too.

Despite these significant differences in approach to coordinate geometry and in goals, Descartes and Fermat became embroiled in controversy as to priority of discovery. Fermat's work was not published until 1679; however, his discovery of the basic ideas of coordinate geometry in 1629 predates Descartes's publication of *La Géométrie* in 1637. Descartes was by this time fully aware of many of Fermat's discoveries, but he denied having learned his ideas from Fermat. Descartes's ideas on coordinate geometry, according to the Dutch mathematician Isaac Beeckman (1588–1637), went back to 1619; and furthermore, there is no question about the originality of many of his basic ideas in coordinate geometry.

When *La Géométrie* was published, Fermat criticized it because it omitted ideas such as maxima and minima, tangents to curves, and the construction of solid loci, which, he had decided, merited the attention of all geometers. Descartes in turn said Fermat had done little, in fact no more than could be easily arrived at without industry or previous knowledge, whereas he himself had used a full knowledge of the nature of equations, which he had expounded in the third book of *La Géométrie*. Descartes referred sarcastically to Fermat as *vostre Conseiller De Maximis et Minimis* and said Fermat was indebted to him. Roberval, Pascal, and others sided with Fermat, and Mydorge and Desargues sided with Descartes. Fermat's friends wrote bitter letters against Descartes. Later the attitudes of the two men toward each other softened, and in a work of 1660, Fermat, while calling attention to an error in *La Géométrie*, declared that he admired that genius so much that even when he made mistakes Descartes's work was worth more than that of others who did correct things. Descartes had not been so generous.

The emphasis placed by posterity on *La Géométrie* was not what Descartes had intended. While the salient idea for the future of mathematics was the association of equation and curve, for Descartes this idea was just a means to an end—the solution of geometric construction problems. Fermat's emphasis on the equations of loci is, from the modern standpoint, more to the point. The geometric construction problems that Descartes stressed in Books I and III have dwindled in importance, largely because construction is no longer used, as it was by the Greeks, to establish existence.

One portion of Book III has also found a permanent place in mathematics. Since Descartes solved geometric construction problems by first formulating them algebraically, solving the algebraic equations, and then constructing what the solutions called for, he gathered together work of his own and of others on the theory of equations that might expedite their solution. Because algebraic equations continued to arise in hundreds of different contexts having nothing to do with geometrical construction problems, this theory of equations has become a basic part of elementary algebra.

5. *Seventeenth-Century Extensions of Coordinate Geometry*

The main idea of coordinate geometry—the use of algebraic equations to represent and study curves—was not eagerly seized upon by mathematicians for many reasons. Fermat's book, the *Ad Locos*, though circulated among friends, was not published until 1679. Descartes's emphasis on the solution of geometric construction problems obscured the main idea of equation and curve. In fact, many of his contemporaries thought of coordinate geometry primarily as a tool for solving the construction problems. Even Leibniz spoke of Descartes's work as a regression to the ancients. Descartes himself did realize that he had contributed far more than a new method of solving construction problems. In the introduction to *La Géométrie* he says, "Moreover, what I have given in the second book on the nature and properties of curved lines, and the method of examining them, is, it seems to me, as far beyond the treatment of ordinary geometry as the rhetoric of Cicero is beyond the a, b, c of children." Nevertheless, the uses he made of the equations of the curves, such as solving the Pappus problem, finding normals to curves, and obtaining properties of the ovals, were far overshadowed by the attention given to the construction problems. Another reason for the slow spread of analytic geometry was Descartes's insistence on making his presentation difficult to follow.

In addition many mathematicians objected to confounding algebra and geometry, or arithmetic and geometry. This objection had been voiced even in the sixteenth century, when algebra was on the rise. For example, Tartaglia insisted on the distinction between the Greek operations with geometrical objects and operations with numbers. He reproached a translator of Euclid for using interchangeably *multiplicare* and *ducere*. The first belongs to numbers, he says, and the second to magnitude. Vieta, too, considered the sciences of number and of geometric magnitudes as parallel but distinct. Even Newton, in his *Arithmetica Universalis*, objected to confounding algebra and geometry, though he contributed to coordinate geometry and used it in the calculus. He says,[8]

> Equations are expressions of arithmetical computation and properly have no place in geometry except insofar as truly geometrical quantities (that is, lines, surfaces, solids and proportions) are thereby shown equal, some to others. Multiplications, divisions and computations of that kind have been recently introduced into geometry, unadvisedly and against the first principles of this science. . . . Therefore these two sciences ought not to be confounded, and recent generations by confounding them have lost that simplicity in which all geometrical elegance consists.

A reasonable interpretation of Newton's position is that he wanted to keep algebra out of elementary geometry but did find it useful to treat the conics and higher-degree curves.

Still another reason for the slowness with which coordinate geometry was accepted was the objection to the lack of rigor in algebra. We have already mentioned Barrow's unwillingness to accept irrational numbers as more than symbols for continuous geometrical magnitudes (Chap. 13, sec. 2). Arithmetic and algebra found their logical justification in geometry; hence algebra could not replace geometry or exist as its equal. The philosopher Thomas Hobbes (1588–1679), though only a minor figure in mathematics, nevertheless spoke for many mathematicians when he objected to the "whole herd of them who apply their algebra to geometry." Hobbes said that these algebraists mistook the symbols for geometry and characterized John Wallis's book on the conics as scurvy and as a "scab of symbols."

Despite the hindrances to appreciation of what Descartes and Fermat had contributed, a number of men gradually took up and expanded coordinate geometry. The first task was to explain Descartes's idea. A Latin translation of *La Géométrie* by Frans van Schooten (1615–60), first published in 1649 and republished several times, not only made the book available in

8. *Arithmetica Universalis*, 1707, p. 282.

the language all scholars could read but contained a commentary which expanded Descartes's compact presentation. In the edition of 1659–61, van Schooten actually gave the algebraic form of a transformation of co-ordinates from one base line (*x*-axis) to another. He was so impressed with the power of Descartes's method that he claimed the Greek geometers had used it to derive their results. Having the algebraic work, the Greeks, according to van Schooten, saw how to obtain the results synthetically—he showed how this could be done—and then published their synthetic methods, which are less perspicuous than the algebraic, to amaze the world. Van Schooten may have been misled by the word "analysis," which to the Greeks meant analyzing a problem, and the term "analytic geometry," which specifically described Descartes's use of algebra as a method.

John Wallis, in *De Sectionibus Conicis* (1655), first derived the equations of the conics by translating Apollonius' geometric conditions into algebraic form (much as we did in Chap. 4, sec. 12) in order to elucidate Apollonius' results. He then defined the conics as curves corresponding to second degree equations in *x* and *y* and proved that these curves were indeed the conic sections as known geometrically. He was probably the first to use equations to prove properties of the conics. His book helped immensely to spread the idea of coordinate geometry and to popularize treatment of the conics as curves in the plane instead of as sections of a cone, though the latter ap-proach persisted. Moreover, Wallis emphasized the validity of the algebraic reasoning whereas Descartes, at least in his *Géométrie*, really rested on the geometry, regarding algebra as just a tool. Wallis was also the first to con-sciously introduce negative abscissas and ordinates. Newton, who did this later, may have gotten the idea from Wallis. We can contrast van Schooten's remark on method with one by Wallis, who said that Archimedes and nearly all the ancients so hid from posterity their method of discovery and analysis that the moderns found it easier to invent a new analysis than to seek out the old.

Newton's *The Method of Fluxions and Infinite Series*, written about 1671 but first published in an English translation by John Colson (d. 1760) under the above title in 1736, contains many uses of coordinate geometry, such as sketching curves from equations. One of the original ideas it offers is the use of new coordinate systems. The seventeenth- and even many of the eighteenth-century men generally used one axis, with the *y*-values drawn at an oblique or right angle to that axis. Among the new coordinate systems introduced by Newton is the location of points by reference to a fixed point and a fixed line through that point. The scheme is essentially our polar coordinate system. The book contains many variations on the polar coordinate idea. Newton also introduced bipolar coordinates. In this scheme a point is located by its distance from two fixed points (Fig. 15.7). Because this work of Newton did not become known until 1736, credit for the discovery of polar coordinates

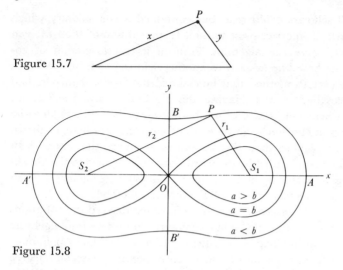

Figure 15.7

Figure 15.8

is usually given to James (Jakob) Bernoulli who published a paper on what was essentially this scheme in the *Acta Eruditorum* of 1691.

Many new curves and their equations were introduced. In 1694 Bernoulli introduced the lemniscate,[9] which played a major role in eighteenth-century analysis. This curve is a special case of a class of curves called the Cassinian ovals (general lemniscates) introduced by Jean-Dominique Cassini (1625–1712), though they did not appear in print until his son Jacques (1677–1756) published the *Eléments d'astronomie* in 1749. The Cassinian ovals (Fig. 15.8) are defined by the condition that the product $r_1 r_2$ of the distances of any point on the curve from two fixed points S_1 and S_2 equals b^2 where b is a constant. Let the distance $S_1 S_2$ be $2a$. Then if $b > a$ we get the non-self-intersecting oval. If $b = a$ we get the lemniscate introduced by James Bernoulli. And if $b < a$ we get the two separate ovals. The rectangular coordinate equation of the Cassinian ovals is of the fourth degree. Descartes himself introduced the logarithmic spiral,[10] which in polar coordinates has the equation $\rho = a^\theta$, and discovered many of its properties. Still other curves, among them the catenary and cycloid, will be noted in other connections.

The beginning of an extension of coordinate geometry to three dimensions was made in the seventeenth century. In Book II of his *Géométrie* Descartes remarks that his ideas can easily be made to apply to all those curves that can be conceived of as generated by the regular movements of a point in three-dimensional space. To represent such curves algebraically his plan is to drop perpendiculars from each point of the curve upon two planes

9. *Acta Erud.*, Sept. 1694 = *Opera*, 2, 608–12.
10. Letter to Mersenne of Sept. 12, 1638 = *Œuvres*, 2, 360.

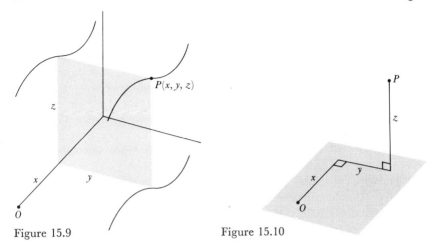

Figure 15.9   Figure 15.10

intersecting at right angles (Fig. 15.9). The ends of these perpendiculars will each describe a curve in the respective plane. These plane curves can then be treated by the method already given. Earlier in Book II Descartes observes that one equation in three unknowns for the determination of the typical point C of a locus represents a plane, a sphere, or a more complex surface. Clearly he appreciated thas his method could be extended to curves and surfaces in three-dimensional space, but he did not himself go further with the extension.

Fermat, in a letter of 1643, gave a brief sketch of his ideas on analytic geometry of three dimensions. He speaks of cyclindrical surfaces, elliptic paraboloids, hyperboloids of two sheets, and ellipsoids. He then says that, to crown the introduction of plane curves, one should study curves on surfaces. "This theory is susceptible of being treated by a general method which if I have leisure I will explain." In a work of half a page, *Novus Secundarum*,[11] he says that an equation in three unknowns gives a surface.

La Hire, in his *Nouveaux élémens des sections coniques* (1679), was a little more specific about three-dimensional coordinate geometry. To represent a surface, he first represented a point P in space by the three coordinates indicated in Figure 15.10 and actually wrote the equation of a surface. However, the development of three-dimensional coordinate geometry is the work of the eighteenth century and will be discussed later.

6. *The Importance of Coordinate Geometry*

In light of the fact that algebra had made considerable progress before Fermat and Descartes entered the mathematical scene, coordinate geometry

11. *Œuvres*, 1, 186–87; 3, 161–62.

was not a great technical achievement. For Fermat it was an algebraic rephrasing of Apollonius. With Descartes it arose as an almost accidental discovery as he continued the work of Vieta and others in expediting the solution of determinate construction problems by the introduction of algebra. But coordinate geometry changed the face of mathematics.

By arguing that a curve is any locus that has an algebraic equation, Descartes broadened in one swoop the domain of mathematics. When one considers the variety of curves that have come to be accepted and used in mathematics and compares this assemblage with what the Greeks had accepted, one sees how important it was that the Greek barriers be stormed.

Through coordinate geometry Descartes sought to introduce method in geometry. He achieved far more than he envisioned. It is commonplace today to recognize not only how readily one can prove, with the aid of the algebra, any number of facts about curves, but also that the method of approaching the problems is almost automatic. The methodology is even more powerful. When letters began to be used by Wallis and Newton to stand for positive and negative numbers and later even for complex numbers, it became possible to subsume under one algebraic treatment many cases that pure geometry would have had to treat separately. For example, in synthetic geometry, to prove that the altitudes of a triangle meet in a point, intersections inside and outside the triangle are considered separately. In coordinate geometry they are considered together.

Coordinate geometry made mathematics a double-edged tool. Geometric concepts could be formulated algebraically and geometric goals attained through the algebra. Conversely, by interpreting algebraic statements geometrically one could gain an intuitive grasp of their meanings as well as suggestions for the deduction of new conclusions. These virtues were cited by Lagrange in his *Leçons élémentaires sur les mathématiques*:[12] "As long as algebra and geometry travelled separate paths their advance was slow and their applications limited. But when these two sciences joined company, they drew from each other fresh vitality and thenceforward marched on at a rapid pace towards perfection." Indeed the enormous power mathematics developed from the seventeenth century on must be attributed, to a very large extent, to coordinate geometry.

The most significant virtue of coordinate geometry was that it provided science with just that mathematical facility it had always sorely needed and which, in the seventeenth century, was being openly demanded—quantitative tools. The study of the physical world does seem to call primarily for geometry. Objects are basically geometrical figures, and the paths of moving bodies are curves. Indeed Descartes himself thought that all of physics could be reduced to geometry. But, as we have pointed out, the uses of science in geodesy, navigation, calendar-reckoning, astronomical predictions, projectile motion,

12. *Œuvres*, 7, 183–287, p. 271 in part.

and even the design of lenses, which Descartes himself undertook, call for quantitative knowledge. Coordinate geometry made possible the expression of shapes and paths in algebraic form, from which quantitative knowledge could be derived.

Thus algebra, which Descartes had thought was just a tool, an extension of logic rather than part of mathematics proper, became more vital than geometry. In fact, coordinate geometry paved the way for a complete reversal of the roles of algebra and geometry. Whereas from Greek times until about 1600 geometry dominated mathematics and algebra was subordinate, after 1600 algebra became the basic mathematical subject; in this transposition of roles the calculus was to be the decisive factor. The ascendancy of algebra aggravated the difficulty to which we have already called attention, namely, that there was no logical foundation for arithmetic and algebra; but nothing was done about it until the late nineteenth century.

The fact that algebra was built up on an empirical basis has led to confusion in mathematical terminology. The subject created by Fermat and Descartes is usually referred to as analytic geometry. The word "analytic" is inappropriate; coordinate geometry or algebraic geometry (which now has another meaning) would be more suitable. The word "analysis" had been used since Plato's time to mean the process of analyzing by working backward from what is to be proved until one arrives at something known. In this sense it was opposed to "synthesis," which describes the deductive presentation. About 1590 Vieta rejected the word "algebra" as having no meaning in the European language and proposed the term "analysis" (Chap. 13, sec. 8); the suggestion was not adopted. However, for him and for Descartes, the word "analysis" was still somewhat appropriate to describe the application of algebra to geometry because the algebra served to analyze the geometric construction problem. One assumed the desired geometric length was known, found an equation that this length satisfied, manipulated the equation, and then saw how to construct the required length. Thus Jacques Ozanam (1640–1717) said in his *Dictionary* (1690) that moderns did their analysis by algebra. In the famous eighteenth-century *Encyclopédie*, d'Alembert used "algebra" and "analysis" as synonyms. Gradually, "analysis" came to mean the algebraic method, though the new coordinate geometry, up to about the end of the eighteenth century, was most often formally described as the application of algebra to geometry. By the end of the century the term "analytic geometry" became standard and was frequently used in titles of books.

However, as algebra became the dominant subject, mathematicians came to regard it as having a much greater function than the analysis of a problem in the Greek sense. In the eighteenth century the view that algebra as applied to geometry was more than a tool—that algebra itself was a basic method of introducing and studying curves and surfaces (the supposed view of Fermat as opposed to Descartes)—won out, as a result of the work of Euler, Lagrange, and Monge. Hence the term "analytic geometry" implied proof as well as the use of the algebraic method. Consequently we now speak of analytic geometry as opposed to synthetic geometry, and we no longer mean that one is a method of invention and the other of proof. Both are deductive.

In the meantime the calculus and extensions such as infinite series entered mathematics. Both Newton and Leibniz regarded the calculus as an extension of algebra; it was the algebra of the infinite, or the algebra that dealt with an infinite number of terms, as in the case of infinite series. As late as 1797, Lagrange, in *Théorie des fonctions analytiques*, said that the calculus and its developments were only a generalization of elementary algebra. Since algebra and analysis had been synonyms, the calculus was referred to as analysis. In a famous calculus text of 1748 Euler used the term "infinitesimal analysis" to describe the calculus. This term was used until the late nineteenth century, when the word "analysis" was adopted to describe the calculus and those branches of mathematics built on it. Thus we are left with a confusing situation in which the term "analysis" embraces all the developments based on limits, but "analytic geometry" involves no limit processes.

Bibliography

Boyer, Carl B.: *History of Analytic Geometry*, Scripta Mathematica, 1956.

Cantor, Moritz: *Vorlesungen über Geschichte der Mathematik*, 2nd ed., B. G. Teubner, 1900, Johnson Reprint Corp., 1965, Vol. 2, pp. 806–76.

Chasles, Michel: *Aperçu historique sur l'origine et le développement des méthodes en géométrie*, 3rd ed., Gauthier-Villars et Fils, 1889, Chaps. 2–3 and relevant notes.

Coolidge, Julian L.: *A History of Geometrical Methods*, Dover (reprint), 1963, pp. 117–31.

Descartes, René: *La Géométrie* (French and English), Dover (reprint), 1954.

————: *Œuvres*, 12 vols., Cerf, 1897–1913.

Fermat, Pierre de: *Œuvres*, 4 vols. and Supplement, Gauthier-Villars, 1891–1912; Supplement, 1922.

Montucla, J. F.: *Histoire des mathématiques*, Albert Blanchard (reprint), 1960, Vol. 2, pp. 102–77.

Scott, J. F.: *The Scientific Work of René Descartes*, Taylor and Francis, 1952.

Smith, David E.: *A Source Book in Mathematics*, Dover (reprint), 1959, Vol. 2, pp. 389–402.

Struik, D. J.: *A Source Book in Mathematics, 1200–1800*, Harvard University Press, 1969, pp. 87–93, 143–57.

Wallis, John: *Opera*, 3 vols., (1693–99), Georg Olms (reprint), 1968.

Vrooman, Jack R.: *René Descartes: A Biography*, G. P. Putnam's Sons, 1970.

Vuillemin, Jules: *Mathématiques et métaphysique chez Descartes*, Presses Universitaires de France, 1960.

16

The Mathematization of Science

So that we may say the door is now opened, for the first time, to a new method fraught with numerous and wonderful results which in future years will command the attention of other minds.   GALILEO GALILEI

1. *Introduction*

By 1600 the European scientists were unquestionably impressed with the importance of mathematics for the study of nature. The strongest evidence of this conviction was the willingness of Copernicus and Kepler to overturn the accepted laws of astronomy and mechanics and religious doctrines for the sake of a theory which in their time had only mathematical advantages. However, the astonishing successes of modern science and the enormous impetus to creative work that mathematics derived from that source probably would not have come about if science had continued in the footsteps of the past. But in the seventeenth century two men, Descartes and Galileo, revolutionized the very nature of scientific activity. They selected the concepts science should employ, redefined the goals of scientific activity, and altered the methodology of science. Their reformulation not only imparted unexpected and unprecedented power to science but bound it indissolubly to mathematics. In fact, their plan practically reduced theoretical science to mathematics. To understand the spirit that animated mathematics from the seventeenth through the nineteenth centuries, we must first examine the ideas of Descartes and Galileo.

2. *Descartes's Concept of Science*

Descartes proclaimed explicitly that the essence of science was mathematics. He says that he "neither admits nor hopes for any principles in Physics other than those which are in Geometry or in abstract mathematics, because thus all phenomena of nature are explained and some demonstrations of them can be given." The objective world is space solidified, or geometry incarnate. Its properties should therefore be deducible from the first principle of geometry.

325

Descartes elaborated on why the world must be accessible and reducible to mathematics. He insisted that the most fundamental and reliable properties of matter are shape, extension, and motion in space and time. Since shape is just extension, Descartes asserted, "Give me extension and motion and I shall construct the universe." Motion itself resulted from the action of forces on molecules. Descartes was convinced that these forces obeyed invariable mathematical laws; and, since extension and motion were mathematically expressible, all phenomena were mathematically describable.

Descartes's mechanistic philosophy extended even to the functioning of the human body. He believed that laws of mechanics would explain life in man and animals, and in his work in physiology he used heat, hydraulics, tubes, valves, and the mechanical actions of levers to explain the actions of the body. However, God and the soul were exempt from mechanism.

If Descartes regarded the external world as consisting only of matter in motion, how did he account for tastes, smells, colors, and the qualities of sounds? Here he adopted the old Greek doctrine of primary and secondary qualities which, as stated by Democritus, maintained that "sweet and bitter, cold and warm, as well as the colors, all these things exist but in opinion and not in reality; what really exist are unchangeable particles, atoms, and their motions in empty space." The primary qualities, matter and motion, exist in the physical world; the secondary qualities are only effects the primary qualities induce in the sense organs of human beings by the impact of external atoms on these organs.

Thus to Descartes there are two worlds; one, a huge, harmoniously designed mathematical machine existing in space and time, and the other, the world of thinking minds. The effect of elements in the first world on the second produces the nonmathematical or secondary qualities of matter. Descartes affirmed further that the laws of nature are invariable, since they are but part of a predetermined mathematical pattern, and that God could not alter invariable nature. Here Descartes denied the prevailing belief that God continually intervened in the functioning of the universe.

Though Descartes's philosophical and scientific doctrines subverted Aristotelianism and Scholasticism, he was a Scholastic in one fundamental respect: he drew from his own mind propositions about the nature of being and reality. He believed that there are a priori truths and that the intellect by its own power may arrive at a perfect knowledge of all things; he stated laws of motion, for example, on the basis of a priori reasoning. (Actually in his biological work he did experiment, and he drew vital conclusions from the experiments.) However, apart from his reliance upon a priori principles, he did promulgate a general and systematic philosophy that shattered the hold of Scholasticism and opened up fresh channels of thought. His attempt to sweep away all preconceptions and prejudices was a clear declaration of revolt from the past. By reducing natural phenomena to purely physical happenings he did much to rid science of mysticism and occult forces. Descartes's writings were highly influential; his deductive and systematic philosophy pervaded the seventeenth century and impressed Newton, especially, with the importance of motion. Daintily bound expositions of his philosophy even adorned ladies' dressing tables.

3. Galileo's Approach to Science

Though Galileo Galilei's philosophy of science agreed in large part with Descartes's, it was Galileo who formulated the more radical, more effective, and more concrete procedures for modern science and who by his own work demonstrated their effectiveness.

Galileo (1564–1642), born in Pisa to a cloth merchant, entered the University of Pisa to study medicine. The courses there were still at about the level of the medieval curriculum; Galileo learned his mathematics privately from a practical engineer, and at the age of seventeen switched from medicine to mathematics. After about eight years of study he applied for a teaching position at the University of Bologna but was refused as not sufficiently distinguished. He did secure a professorship of mathematics at Pisa. While there, he began to attack Aristotelian science; and he did not hesitate to express his views even though his criticisms alienated his colleagues. He had also begun to write important mathematical papers that aroused jealousy in the less competent. Galileo was made to feel uncomfortable and left in 1592 to accept the position of professor of mathematics at the University of Padua. There he wrote a short book, *Le mecaniche* (1604). After eighteen years at Padua he was invited to Florence by the Grand Duke Cosimo II de' Medici, who appointed him Chief Mathematician of his court, gave him a home and handsome salary, and protected him from the Jesuits, who dominated the papacy and had already threatened Galileo because he championed the Copernican theory. To express his gratitude, Galileo named the satellites of Jupiter, which he discovered in the first year of his service under Cosimo, the Medicean stars. In Florence Galileo had the leisure to pursue his studies and to write.

His advocacy of the Copernican theory irked the Roman Inquisition, and in 1616 he was called to Rome. His teachings on the heliocentric theory were condemned by the Inquisition; he had to promise not to publish any more on this subject. In 1630 Pope Urban VIII did give him permission to publish if he would make his book mathematical and not doctrinal. Thereupon, in 1632, he published his classic *Dialogo dei massimi sistemi* (Dialogue on the Great World Systems). The Roman Inquisition summoned him again in 1633 and under the threat of torture impelled him to recant his advocacy of the heliocentric theory. He was again forbidden to publish and required to live practically under house arrest. But he undertook to write up his years of thought and work on the phenomena of motion and on the strength of materials. The manuscript, entitled *Discorsi e dimostrazioni matematiche intorno à due nuove scienze* (Discourses and Mathematical Demonstrations Concerning Two New Sciences, also referred to as Dialogues Concerning Two New Sciences), was secretly transported to Holland and published there in 1638. This is the classic in which Galileo presented his new scientific method. He defended his actions with the words that he had never "declined in piety and reverence for the Church and my own conscience."

Galileo was an extraordinary man in many fields. He was a keen astronomical observer. He is often called the father of modern invention; though he did not invent the telescope or "perplexive glasses," as Ben Jonson called them, he was immediately able to construct one when he heard of the idea. He was an independent inventor of the microscope, and he designed the first pendulum clock. He also designed and made a compass with scales that automatically yielded the results of numerical computations so the user could read the scales and avoid having to do the calculations. This device was so much in demand that he produced many for sale.

Galileo was the first important modern student of sound. He suggested a wave theory of sound and began work on pitch, harmonics, and the vibrations of strings. This work was continued by Mersenne and Newton and became a major inspiration for mathematical work in the eighteenth century.

Galileo's major writings, though concerned with scientific subjects, are still regarded as literary masterpieces. His *Sidereus Nuncius* (Sidereal Messenger) of 1610, in which he announced his astronomical observations and declared himself in support of Copernican theory, was an immediate success, and he was elected to the prestigious Academy of the Lynx-like in Rome. His two greatest classics, the *Dialogue on the Great World Systems* and *Dialogues Concerning Two New Sciences*, are clear, direct, witty, yet profound. In both, Galileo has one character present the current views, against which another argues cleverly and tenaciously to show the fallacies and weaknesses of these views and the strengths of the new ones.

In his philosophy of science Galileo broke sharply from the speculative and mystical in favor of a mechanical and mathematical view of nature. He also believed that scientific problems should not become enmeshed in and beclouded by theological arguments. Indeed, one of his achievements in science, though somewhat apart from the method we are about to examine, is that he recognized clearly the domain of science and severed it sharply from religious doctrines.

Galileo, like Descartes, was certain that nature is mathematically designed. His statement of 1610 is famous:

> Philosophy [nature] is written in that great book which ever lies before our eyes—I mean the universe—but we cannot understand it if we do

not first learn the language and grasp the symbols in which it is written. The book is written in the mathematical language, and the symbols are triangles, circles and other geometrical figures, without whose help it is impossible to comprehend a single word of it; without which one wanders in vain through a dark labyrinth.[1]

Nature is simple and orderly and its behavior is regular and necessary. It acts in accordance with perfect and immutable mathematical laws. Divine reason is the source of the rational in nature; God put into the world that rigorous mathematical necessity that men reach only laboriously. Mathematical knowledge is therefore not only absolute truth, but as sacrosanct as any line of Scripture. In fact it is superior, for there is much disagreement about the Scriptures, but there can be none about mathematical truths.

Another doctrine, the atomism of the Greek Democritus, is clearer in Galileo than in Descartes. Atomism presupposed empty space (which Descartes did not accept) and individual, indestructible atoms. Change consisted in the combination and separation of atoms. All qualitative varieties in bodies were due to quantitative variety in number, size, shape, and spatial arrangement of the atoms. The atom's chief properties were impenetrability and indestructibility; these properties served to explain chemical and physical phenomena. Galileo's espousal of atomism placed it in the forefront of scientific doctrines.

Atomism led Galileo to the doctrine of primary and secondary qualities. He says, "If ears, tongues, and noses were removed, I am of the opinion that shape, quantity [size] and motion would remain, but there would be an end of smells, tastes, and sounds, which abstracted from the living creature, I take to be mere words." Thus in one swoop Galileo, like Descartes, stripped away a thousand phenomena and qualities to concentrate on matter and motion, properties that are mathematically describable. It is perhaps not too surprising that in the century in which problems of motion were the most prominent and serious, scientists should find motion to be a fundamental physical phenomenon.

The concentration on matter and motion was only the first step in Galileo's new approach to nature. His next thought, also voiced by Descartes, was that any branch of science should be patterned on the model of mathematics. This implies two essential steps. Mathematics starts with axioms—clear, self-evident truths—and from these proceeds by deductive reasoning to establish new truths. Any branch of science, then, should start with axioms or principles and then proceed deductively. Moreover, one should extract from the axioms as many consequences as possible. This thought, of course, goes back to Aristotle, who also aimed at deductive structure in science with the mathematical model in mind.

However, Galileo departed radically from the Greeks, the medieval

1. *Opere*, 4, 171.

scientists, and even Descartes in his method of obtaining first principles. The pre-Galileans and Descartes had believed that the mind supplied the basic principles; it had but to think about any class of phenomena and it would immediately recognize fundamental truths. This power of the mind was clearly evidenced in mathematics. Axioms such as "equals added to equals give equals" and "two points determine a line" suggested themselves immediately in thinking about number or geometrical figures, and were indubitable truths. So too had the Greeks found some physical principles equally appealing. That all objects in the universe should have a natural place was no more than fitting. The state of rest seemed clearly more natural than the state of motion. It seemed indubitable, too, that force must be applied to put and keep bodies in motion. To believe that the mind supplies fundamental principles did not deny that observations might play a role in obtaining these principles. But the observations merely evoked the correct principles, just as the sight of a familiar face might call to mind facts about that person.

The Greek and medieval scientists were so convinced that there were a priori fundamental principles that when occasional observations did not fit they invented special explanations to preserve the principles but still account for the anomalies. These men, as Galileo put it, first decided how the world should function and then fitted what they saw into their preconceived principles.

Galileo decided that in physics, as opposed to mathematics, first principles must come from experience and experimentation. The way to obtain correct and basic principles is to pay attention to what nature says rather than what the mind prefers. Nature, he argued, did not first make men's brains and then arrange the world to be acceptable to human intellects. To the medieval thinkers who kept repeating Aristotle and debating what he meant, Galileo addressed the criticism that knowledge comes from observation and not from books, and that it was useless to debate about Aristotle. He says, "When we have the decrees of nature, authority goes for nothing...." Of course some Renaissance thinkers and Galileo's contemporary Francis Bacon had also arrived at the conclusion that experimentation was necessary; in this particular aspect of his new method, Galileo was not ahead of all others. Yet the modernist Descartes did not grant the wisdom of Galileo's reliance upon experimentation. The facts of the senses, Descartes said, can only lead to delusion, but reason penetrates such delusions. From the innate general principles supplied by the mind, we can deduce particular phenomena of nature and understand them. Galileo did appreciate that one may glean an incorrect principle from experimentation and that as a consequence the deductions from it could be incorrect. Hence he proposed the use of experiments to check the conclusions of his reasonings as well as to acquire basic principles.

Galileo was actually a transitional figure as far as experimentation is concerned. He, and Isaac Newton fifty years later, believed that a few key or critical experiments would yield correct fundamental principles. Moreover, many of Galileo's so-called experiments were really thought-experiments; that is, he relied upon common experience to imagine what would happen if an experiment were performed. He then drew a conclusion as confidently as if he had actually performed the experiment. When in the *Dialogue on the Great World Systems* he describes the motion of a ball dropped from the mast of a moving ship, he is asked by Simplicio, one of the characters, whether he had made an experiment. Galileo replies, "No, and I do not need it, as without any experience I can confirm that it is so, because it cannot be otherwise." He says in fact that he experimented rarely, and then primarily to refute those who did not follow the mathematics. Though Newton performed some famous and ingenious experiments, he too says that he used experiments to make his *results* physically intelligible and to convince the common people.

The truth of the matter is that Galileo had some preconceptions about nature, which made him confident that a few experiments would suffice. He believed, for example, that nature was simple. Hence when he considered freely falling bodies, which fall with increasing velocity, he supposed that the increase in velocity is the same for each second of fall. This was the simplest "truth." He believed also that nature is mathematically designed, and hence any mathematical law that seemed to fit even on the basis of rather limited experimentation appeared to him to be correct.

For Galileo, as well as for Huygens and Newton, the deductive, mathematical part of the scientific enterprise played a greater part than the experimental. Galileo was no less proud of the abundance of theorems that flow from a single principle than of the discovery of the principle itself. The men who fashioned modern science—Descartes, Galileo, Huygens, and Newton (we can also include Copernicus and Kepler)—approached the study of nature as mathematicians, in their general method and in their concrete investigations. They were primarily speculative thinkers who expected to apprehend broad, deep (but also simple), clear, and immutable mathematical principles either through intuition or through crucial observations and experiments, and then to deduce new laws from these fundamental truths, entirely in the manner in which mathematics proper had constructed its geometry. The bulk of the activity was to be the deductive portion; whole systems of thought were to be so derived.

What the great thinkers of the seventeenth century envisaged as the proper procedure for science did indeed prove to be the profitable course. The rational search for laws of nature produced, by Newton's time, extremely valuable results on the basis of the slimmest observational and experimental knowledge. The great scientific advances of the sixteenth and seventeenth centuries were in astronomy, where observation offered little that was new, and in mechanics, where the experimental results were hardly startling and certainly not decisive, whereas the mathematical theory attained comprehensiveness and perfection. And for the next two centuries scientists produced deep and sweeping laws of nature on the basis of very few, almost trivial, observations and experiments.

The expectation of Galileo, Huygens, and Newton that just a few experiments would suffice can be readily understood. Because these men were convinced that nature is mathematically designed, they saw no reason why they could not proceed in scientific matters much as mathematicians had proceeded in their domain. As John Herman Randall says in *Making of the Modern Mind,* "Science was born of a faith in the mathematical interpretation of nature...."

Galileo did, however, obtain a few principles from experience; and in this work also his approach was a radical departure from that of his predecessors. He decided that one must penetrate to what is fundamental in phenomena and start there. In *Two New Sciences* he says that it is not possible to treat the infinite variety of weights, shapes, and velocities. He had observed that the speeds with which dissimilar objects fall differ less in air than in water. Hence the thinner the medium, the less difference in speed of fall among bodies. "Having observed this I came to the conclusion that in a medium totally devoid of resistance all bodies would fall with the same speed." What Galileo was doing here was to strip away the incidental or minor effects in an effort to get at the major one.

Of course, actual bodies do fall in resisting media. What could Galileo say about such motions? His answer was "... hence, in order to handle this matter in a scientific way, it is necessary to cut loose from these difficulties [air resistance, friction, etc.] and having discovered and demonstrated the theorems in the case of no resistance, to use them and apply them with such limitations as experience will teach."

Having stripped away air resistance and friction, Galileo sought basic laws for motion in a vacuum. Thus he not only contradicted Aristotle and even Descartes by thinking of bodies moving in empty space, but did just what the mathematician does in studying real figures. The mathematician strips away molecular structure, color, and thickness of lines to get at some basic properties and concentrates on these. So did Galileo penetrate to basic physical factors. The mathematical method of abstraction is indeed a step away from reality but, paradoxically, it leads back to reality with greater power than if all the factors actually present are taken into account at once.

Thus far Galileo had formulated a number of methodological principles, many of which were suggested by the approach mathematics had employed in geometry. His next principle was to use mathematics itself, but in a special way. Unlike the Aristotelians and the late medieval scientists, who had fastened upon qualities they regarded as fundamental and studied the acquisition and loss of qualities or debated the meaning of the qualities, Galileo proposed to seek *quantitative* axioms. This change is most important; we shall see the full significance of it later, but an elementary example may be useful now. The Aristotelians said that a ball falls because it has weight, and that it falls to the earth because every object seeks its natural place and the natural place of heavy bodies is the center of the earth. These principles are qualitative. Even Kepler's first law of motion, that the path of each planet is an ellipse, is a qualitative statement. By contrast, let us consider the statement that the speed (in feet per second) with which a ball falls is 32 times the number of seconds it has been falling, or in symbols, $v = 32t$. This is a quantitative statement about how a ball falls. Galileo intended to seek such quantitative statements as his axioms, and he expected to deduce new ones by mathematical means. These deductions would also give quantitative knowledge. Moreover, as we have seen, mathematics was to be his essential medium.

The decision to seek quantitative knowledge expressed in formulas carried with it another radical decision, though first contact with it hardly reveals its full significance. The Aristotelians believed that one of the tasks of science was to explain why things happened; explanation meant unearthing the causes of a phenomenon. The statement that a body falls because it has weight gives the effective cause of the fall and the statement that it seeks its natural place gives the final cause. But the quantitative statement $v = 32t$, for whatever it may be worth, gives no explanation of why a ball falls; it tells only how the speed changes with the time. In other words, formulas do not explain; they describe. The knowledge of nature Galileo sought was descriptive. He says in *Two New Sciences*, "The cause of the acceleration of the motion of falling bodies is not a necessary part of the investigation." More generally, he points out that he will investigate and demonstrate some of the properties of motion without regard to what the causes might be. Positive scientific inquiries were to be separated from questions of ultimate causation, and speculation as to physical causes was to be abandoned.

First reactions to this principle of Galileo are likely to be negative. Description of phenomena in terms of formulas hardly seems to be more than a first step. It would seem that the true function of science had really been grasped by the Aristotelians, namely, to explain why phenomena happened. Even Descartes protested Galileo's decision to seek descriptive formulas. He said, "Everything that Galileo says about bodies falling in empty space is built without foundation: he ought first to have determined the nature of weight." Further, said Descartes, Galileo should reflect on ultimate reasons. But we shall see clearly after a few chapters that Galileo's decision to aim for description was the deepest and most fruitful idea that anyone has had about scientific methodology.

Whereas the Aristotelians had talked in terms of qualities such as fluidity, rigidity, essences, natural places, natural and violent motion, and potentiality, Galileo chose an entirely new set of concepts, which, moreover, were measurable, so that their measures could be related by formulas. Some of them are: distance, time, speed, acceleration, force, mass, and weight. These concepts are too familiar to surprise us. But in Galileo's time they were radical choices, at least as fundamental concepts; and these are the ones that proved most instrumental in the task of understanding and mastering nature.

We have described the essential features of Galileo's program. Some of the ideas in it had been espoused by others; some were entirely original with him. But what establishes Galileo's greatness is that he saw so clearly what was wrong or deficient in the current scientific efforts, shed completely the older ways, and formulated the new procedures so clearly. Moreover, in applying them to problems of motion he not only exemplified the method but succeeded in obtaining brilliant results—in other words, he showed that it worked. The unity of his work, the clarity of his thoughts and expressions, and the force of his argumentation influenced almost all of his contemporaries and successors. More than any other man, Galileo is the founder of the methodology of modern science. He was fully conscious of what he had accomplished (see the chapter legend); so were others. The philosopher Hobbes said of Galileo, "He has been the first to open to us the door to the whole realm of physics."

We cannot pursue the history of the methodology of science. However, since mathematics became so important in this methodology and profited so much from its adoption, we should note how completely Galileo's program was accepted by giants such as Newton. He asserts that experiments are needed to furnish basic laws. Newton is also clear that the function of science, after having obtained some basic principles, is to deduce new facts from these principles. In the preface to his *Principia,* he says:

> Since the ancients (as we are told by Pappus) esteemed the science of mechanics of greatest importance in the investigation of natural things, and the moderns, rejecting substantial forms and occult qualities, have endeavored to subject the phenomena of nature to the laws of mathematics, I have in this treatise cultivated mathematics as far as it relates to philosophy [science] ... and therefore I offer this work as the mathematical principles of philosophy, for the whole burden in philosophy seems to consist in this—from the phenomena of motions to investigate the forces of nature, and then from these forces to demonstrate the other phenomena....

Of course, mathematical principles, to Newton as to Galileo, were quantitative principles. He says in the *Principia* that his purpose is to discover and set forth the exact manner in which "all things had been ordered in measure, number and weight." Newton had good reason to emphasize quantitative mathematical laws, as opposed to physical explanation, because the central physical concept in his celestial mechanics was the force of gravitation, whose action could not be explained at all in physical terms. In lieu of explanation Newton had a quantitative formulation of how gravity acted that was significant and usable. And this is why he says, at the beginning of the *Principia*, "For I here design only to give a mathematical notion of these forces, without considering their physical causes and seats." Toward the end of the book he repeats this thought:

> But our purpose is only to trace out the quantity and properties of this force from the phenomena, and to apply what we discover in some simple cases as principles, by which, in a mathematical way, we may estimate the effects thereof in more involved cases ... We said, in *a mathematical way* [italics Newton's], to avoid all questions about the nature or quality of this force, which we would not be understood to determine by any hypothesis..."

The abandonment of physical mechanism in favor of mathematical description shocked even great scientists. Huygens regarded the idea of gravitation as "absurd," because its action through empty space precluded any mechanism. He expressed surprise that Newton should have taken the trouble to make such a number of laborious calculations with no foundation but the mathematical principle of gravitation. Leibniz attacked gravitation as an incorporeal and inexplicable power; John Bernoulli (James's brother) denounced it as "revolting to minds accustomed to receiving no principle in physics save those which are incontestable and evident." But this reliance on mathematical description even where physical understanding was completely lacking made possible Newton's amazing contributions, to say nothing of subsequent developments.

Because science became heavily dependent upon—almost subordinate to—mathematics, it was the scientists who extended the domain and techniques of mathematics; and the multiplicity of problems provided by science gave mathematicians numerous and weighty directions for creative work.

4. The Function Concept

The first mathematical gain from scientific investigations conducted in accordance with Galileo's program came from the study of motion. This problem engrossed the scientists and mathematicians of the seventeenth century. It is easy to see why. Though Kepler's astronomy was accepted early in the seventeenth century, especially after Galileo's observations supplied additional evidence for a heliocentric theory, Kepler's law of elliptical motion is only approximately correct, though it would be exact if there were just the sun and one planet in the heavens. The ideas that the other planets disturb the elliptical motion of any one planet and that the sun disturbs the elliptical motion of the moon around the earth were already being considered; in fact, the notion of a gravitational force acting between any two bodies was suggested by Kepler, among others. Hence the problem of improving the calculation of the planets' positions was open. Moreover, Kepler had obtained his laws essentially by fitting curves to astronomical data, with no explanation in terms of fundamental laws of motion of why the planets moved in elliptical paths. The basic problem of deriving Kepler's laws from principles of motion posed a clear challenge.

The improvement of astronomical theory also had a practical objective. In their search for raw materials and trade, the Europeans had undertaken large-scale navigation that involved sailing long distances out of sight of land. Mariners therefore needed accurate methods of determining latitude and longitude. The determination of latitude can be made by direct observation of the sun or the stars, but determination of longitude is far more difficult. In the sixteenth century the methods of doing it were so inaccurate that navigators were often in error as much as 500 miles. After about 1514, the direction of the moon relative to the stars was used to determine longitude. These directions, as seen from some standard place at various times, were tabulated. A navigator would determine the direction of the moon, which was not affected much by his being in a different location, and determine his local time by using, for example, the directions of the stars. Directly from the tables or by interpolation he could find the time at the standard location when the moon had the measured direction and so compute the difference in time between his position and the standard one. Each hour of difference means a 15-degree difference in longitude. This method, however, was not accurate. Because the ships of those times were constantly heaving, it was difficult to obtain the moon's direction accurately; but, because the moon does not move much relative to the stars in a few hours, the direction of the moon had to be rather precisely determined. A mistake of one minute of angle means an error of half a degree of longitude; but even a measure accurate to within one minute was far beyond the capabilities of those times. Though other methods of determining longitude were suggested and tried, better knowledge of the moon's path to extend and improve the tables seemed indispensable and many scientists, including Newton, worked on the problem. Even in Newton's time the knowledge of the moon's position was so inaccurate that use of the tables led to errors of as much as 100 miles in determining position at sea.

The governments of Europe were very much concerned, because

shipping losses were considerable. In 1675 King Charles II of England set up the Royal Observatory at Greenwich to obtain better observations on the moon's motion and to serve as a fixed station for longitude. In 1712 the British government established a Commission for the Discovery of Longitude and offered rewards of up to £20,000 for ideas on how to measure longitude.

The problem of explaining terrestrial motions also faced seventeenth-century scientists. Under the heliocentric theory the earth was both rotating and revolving around the sun. Why then should objects stay with the earth? Why should dropped objects fall to earth if it was no longer the center of the universe? Moreover, all motions, projectile motion for example, seemed to take place as though the earth were at rest. These questions engaged the attention of many men, including Cardan, Tartaglia, Galileo, and Newton. The paths of projectiles, their ranges, the heights they could reach, and the effect of muzzle velocity on height and range were basic questions and the princes then, like nations now, spent great sums on the solutions. New principles of motion were needed to account for these terrestrial phenomena; and it occurred to the scientists that, since the universe was believed to be constructed according to one master plan, the same principles that explained terrestrial motions would also account for heavenly motions.

From the study of the various problems of motion there emerged the specific problem of designing more accurate methods of measuring time. Mechanical clocks, which had been in use since 1348, were not very accurate. The Flemish cartographer Gemma Frisius (1508–55) had suggested the use of a clock to determine longitude. A ship could carry a clock set to the time of a place of known longitude; since the determination of local time by the sun's position, for example, was relatively simple, the navigator need merely note the difference in time and translate this at once into the difference in longitude. But no durable, accurate, seaworthy clocks were available even by 1600.

The motion of a pendulum seemed to provide the basic mechanism for measuring time. Galileo had observed that the time for one complete oscillation of a pendulum was constant and ostensibly independent of the amplitude of the swing. He prepared the design of a pendulum clock and had his son construct one; but it was Robert Hooke and Huygens who did the basic work on the pendulum. Though the pendulum clock was unsuitable for a ship (an accuracy of two or three seconds a day was needed for the purpose of longitude-reckoning, and pendulums were too much affected by ship's motion), it proved immensely valuable in scientific work, as well as for timekeeping in homes and business. A clock appropriate for navigation was finally designed by John Harrison (1693–1776) in 1761 and began to be used by the end of the eighteenth century. Because a proper clock was not available earlier, accurate determination of the motion of the moon was still the chief scientific problem in that century.

From the study of motion mathematics derived a fundamental concept that was central to practically all of the work for the next two hundred years —the concept of a function or a relation between variables. One finds this notion almost throughout Galileo's *Two New Sciences*, the book in which he founded modern mechanics. Galileo expresses his functional relationships in words and in the language of proportion. Thus in his work on the strength of materials, he has occasion to state, "The areas of two cylinders of equal volumes, neglecting the bases, bear to each other a ratio which is the square root of the ratio of their lengths." Again, "The volumes of right cylinders having equal curved surfaces are inversely proportional to their altitudes." In his work on motion he states, for example, "The spaces described by a body falling from rest with a uniformly accelerated motion are to each other as the squares of the time intervals employed in traversing these distances." "The times of descent along inclined planes of the same height, but of different slopes, are to each other as the lengths of these planes." The language shows clearly that he is dealing with variables and functions; it was but a short step to write these statements in symbolic form. Since the symbolism of algebra was being extended at this time, Galileo's statement on the spaces described by a falling body soon was written as $s = kt^2$ and his statement on times of descent as $t = kl$.

Most of the functions introduced during the seventeenth century were first studied as curves, before the function concept was fully recognized. This was true, for example, of the elementary transcendental functions such as $\log x$, $\sin x$, and a^x. Thus Evangelista Torricelli (1608–47), a pupil of Galileo, in a letter of 1644 described his research on the curve we would represent by $y = ae^{-cx}$ with $x \geq 0$ (the manuscript in which he wrote up this research was not edited until 1900). The curve was suggested to Torricelli by the current work on logarithms. Descartes encountered the same curve in 1639 but did not speak of its connection with logarithms. The sine curve entered mathematics as the companion curve to the cycloid in Roberval's work on the cycloid (Chap. 17, sec. 2) and appears graphed for two periods in Wallis's *Mechanica* (1670). Of course the tabular values of the trigonometric and logarithmic functions were, by this time, known with great precision.

It is also relevant that old and new curves were introduced by means of motions. In Greek times, a few curves, such as the quadratrix and the Archimedean spiral, were defined in terms of motion, but in that period such curves were outside the pale of legitimate mathematics. The attitude was quite different in the seventeenth century. Mersenne in 1615 defined the cycloid (which had been known earlier) as the locus of a point on a wheel that rolls along the ground. Galileo, who had shown that the path of a projectile shot up into the air at an angle to the ground is a parabola, regarded the curve as the locus of a moving point.

With Roberval, Barrow, and Newton the concept of a curve as the path of a moving point attains explicit recognition and acceptance. Newton says in *Quadrature of Curves* (written in 1676), "I consider mathematical quantities in this place not as consisting of very small parts, but as described by a continued motion. Lines [curves] are described, and thereby generated, not by the apposition of parts but by the continued motion of points.... These geneses really take place in the nature of things, and are daily seen in the motion of bodies."

Gradually the terms and symbolism for the various types of functions represented by these curves were introduced. There were many subtle difficulties that were hardly recognized. For example, the use of functions of the form a^x, with x taking on positive and negative integral and fractional values, became common in the seventeenth century. It was assumed (until the nineteenth century, when irrational numbers were first defined) that the function was also defined for irrational values of x, so that no one questioned an expression of the form $2^{\sqrt{2}}$. The implicit understanding was that such a value was intermediate between that obtained for any two rational exponents above and below $\sqrt{2}$.

Descartes's distinction between geometric and mechanical curves (Chap. 15, sec. 4) gave rise to the distinction between algebraic and transcendental functions. Fortunately his contemporaries ignored his banishment of what he called mechanical curves. Through quadratures, the summation of series, and other operations that entered with the calculus, many types of transcendental functions arose and were studied. The distinction between algebraic and transcendental functions was clearly made by James Gregory in 1667, when he sought to show that the area of a circular sector could not be an algebriac function of the radius and the chord. Leibniz showed that $\sin x$ cannot be an algebraic function of x and incidentally proved the result sought by Gregory.[2] The full understanding and use of the transcendental functions came gradually.

The most explicit definition of the function concept in the seventeenth century was given by James Gregory in his *Vera Circuli et Hyperbolae Quadratura* (1667). He defined a function as a quantity obtained from other quantities by a succession of algebraic operations or by any other operation imaginable. By the last phrase he meant, as he explains, that it is necessary to add to the five operations of algebra a sixth operation, which he defines as passage to the limit. (Gregory, as we shall see in Chapter 17, was concerned with quadrature problems.) Gregory's concept of function was lost sight of; but in any case, it would soon have proved too narrow, because the series representation of functions became widely used.

From the very beginning of his work on the calculus, that is from 1665

2. *Math. Schriften*, 5, 97–98.

on, Newton used the term "fluent" to represent any relationship between variables. In a manuscript of 1673 Leibniz used the word "function" to mean any quantity varying from point to point of a curve—for example, the length of the tangent, the normal, the subtangent, and the ordinate. The curve itself was said to be given by an equation. Leibniz also introduced the words "constant," "variable," and "parameter," the latter used in connection with a family of curves.[3] In working with functions John Bernoulli spoke from 1697 on of a quantity formed, in any manner whatever, of variables and of constants;[4] by "any manner" he meant to cover algebraic and transcendental expressions. He adopted Leibniz's phrase "function of x" for this quantity in 1698. In his *Historia* (1714), Leibniz used the word "function" to mean quantities that depend on a variable.

As to notation, John Bernoulli wrote X or ξ for a general function of x, though in 1718 he changed to ϕx. Leibniz approved of this, but proposed also x^1 and x^2 for functions of x, the superscript to be used when several functions were involved. The notation $f(x)$ was introduced by Euler in 1734.[5] The function concept immediately became central in the work on the calculus. We shall see later how the concept was extended.

Bibliography

Bell, A. E.: *Christian Huygens and the Development of Science in the Seventeenth Century*, Edward Arnold, 1947.

Burtt, A. E.: *The Metaphysical Foundations of Modern Physical Science*, 2nd ed., Routledge and Kegan Paul, 1932, Chaps. 1–7.

Butterfield, Herbert: *The Origins of Modern Science*, Macmillan, 1951, Chaps. 4–7.

Cohen, I. Bernard: *The Birth of a New Physics*, Doubleday, 1960.

Coolidge, J. L.: *The Mathematics of Great Amateurs*, Dover (reprint), 1963, pp. 119–27.

Crombie, A. C.: *Augustine to Galileo*, Falcon Press, 1952, Chap. 6.

Dampier-Whetham, W. C. D.: *A History of Science and Its Relations with Philosophy and Religion*, Cambridge University Press, 1929, Chap. 3.

Dijksterhuis, E. J.: *The Mechanization of the World Picture*, Oxford University Press, 1961.

Drabkin, I. E., and Stillman Drake: *Galileo Galilei: On Motion and Mechanics*, University of Wisconsin Press, 1960.

Drake, Stillman: *Discoveries and Opinions of Galileo*, Doubleday, 1957.

Galilei, Galileo: *Opere*, 20 vols., 1890–1909, reprinted by G. Barbera, 1964–66.

———: *Dialogues Concerning Two New Sciences*, Dover (reprint), 1952.

Hall, A. R.: *The Scientific Revolution*, Longmans Green, 1954, Chaps. 1–8.

———: *From Galileo to Newton*, Collins, 1963, Chaps. 1–5.

3. *Math. Schriften*, 5, 266–69.
4. *Mém de l'Acad des Sci.*, Paris, 1718, 100 ff. = *Opera*, 2, 235–69, p. 241 in particular.
5. *Comm. Acad. Sci. Petrop.*, 7, 1734/35, 184–200, pub. 1740 = *Opera*, (1), 22, 57–75.

Huygens, C.: *Œuvres complètes*, 22 vols., M. Nyhoff, 1888–1950.

Newton, I.: *Mathematical Principles of Natural Philosophy*, University of California Press, 1946.

Randall, John H., Jr.: *Making of the Modern Mind*, rev. ed., Houghton Mifflin, 1940, Chap. 10.

Scott, J. F.: *The Scientific Work of René Descartes*, Taylor and Francis, 1952, Chaps. 10–12.

Smith, Preserved: *A History of Modern Culture*, Henry Holt, 1930, Vol. 1, Chaps. 3, 6, and 7.

Strong, Edward W.: *Procedures and Metaphysics*, University of California Press, 1936, Chaps. 5–8.

Whitehead, Alfred North: *Science and the Modern World*, Cambridge University Press, 1926, Chap. 3.

Wolf, Abraham: *A History of Science, Technology and Philosophy in the 16th and 17th Centuries*, 2nd ed., George Allen and Unwin, 1950, Chap. 3.

17
The Creation of the Calculus

Who, by a vigor of mind almost divine, the motions and figures of the planets, the paths of comets, and the tides of the seas first demonstrated.   NEWTON'S EPITAPH

1. *The Motivation for the Calculus*

Following hard on the adoption of the function concept came the calculus, which, next to Euclidean geometry, is the greatest creation in all of mathematics. Though it was to some extent the answer to problems already tackled by the Greeks, the calculus was created primarily to treat the major scientific problems of the seventeenth century.

There were four major types of problems. The first was: Given the formula for the distance a body covers as a function of the time, to find the velocity and acceleration at any instant; and, conversely, given the formula describing the acceleration of a body as a function of the time, to find the velocity and the distance traveled. This problem arose directly in the study of motion and the difficulty it posed was that the velocities and the acceleration of concern to the seventeenth century varied from instant to instant. In calculating an instantaneous velocity, for example, one cannot, as one can in the case of average velocity, divide the distance traveled by the time of travel, because at a given instant both the distance traveled and time are zero, and 0/0 is meaningless. Nevertheless, it was clear on physical grounds that moving objects do have a velocity at each instant of their travel. The inverse problem of finding the distance covered, knowing the formula for velocity, involves the corresponding difficulty; one cannot multiply the velocity at any one instant by the time of travel to obtain the distance traveled because the velocity varies from instant to instant.

The second type of problem was to find the tangent to a curve. Interest in this problem stemmed from more than one source; it was a problem of pure geometry, and it was of great importance for scientific applications. Optics, as we know, was one of the major scientific pursuits of the seventeenth century; the design of lenses was of direct interest to Fermat, Descartes, Huygens, and Newton. To study the passage of light through a lens, one

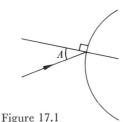

Figure 17.1

must know the angle at which the ray strikes the lens in order to apply the law of refraction. The significant angle is that between the ray and the normal to the curve (Fig. 17.1), the normal being the perpendicular to the tangent. Hence the problem was to find either the normal or the tangent. Another scientific problem involving the tangent to a curve arose in the study of motion. The direction of motion of a moving body at any point of its path is the direction of the tangent to the path.

Actually, even the very meaning of "tangent" was open. For the conic sections the definition of a tangent as a line touching a curve at only one point and lying on one side of the curve sufficed; this definition was used by the Greeks. But it was inadequate for the more complicated curves already in use in the seventeenth century.

The third problem was that of finding the maximum or minimum value of a function. When a cannonball is shot from a cannon, the distance it will travel horizontally—the range—depends on the angle at which the cannon is inclined to the ground. One "practical" problem was to find the angle that would maximize the range. Early in the seventeenth century, Galileo determined that (in a vacuum) the maximum range is obtained for an angle of fire of 45°; he also obtained the maximum heights reached by projectiles fired at various angles to the ground. The study of the motion of the planets also involved maxima and minima problems, such as finding the greatest and least distances of a planet from the sun.

The fourth problem was finding the lengths of curves, for example, the distance covered by a planet in a given period of time; the areas bounded by curves; volumes bounded by surfaces; centers of gravity of bodies; and the gravitational attraction that an *extended* body, a planet for example, exerts on another body. The Greeks had used the method of exhaustion to find some areas and volumes. Despite the fact that they used it for relatively simple areas and volumes, they had to apply much ingenuity, because the method lacked generality. Nor did they often come up with numerical answers. Interest in finding lengths, areas, volumes, and centers of gravity was revived when the work of Archimedes became known in Europe. The method of exhaustion was first modified gradually, and then radically by the invention of the calculus.

2. Early Seventeenth-Century Work on the Calculus

The problems of the calculus were tackled by at least a dozen of the greatest mathematicians of the seventeenth century and by several dozen minor ones. All of their contributions were crowned by the achievements of Newton and Leibniz. Here we shall be able to note only the principal contributions of the precursors of these two masters.

The problem of calculating the instantaneous velocity from a knowledge of the distance traveled as a function of the time, and its converse, were soon seen to be special cases of calculating the instantaneous rate of change of one variable with respect to another and its converse. The first significant treatment of general rate problems is due to Newton; we shall examine it later.

Several methods were advanced to find the tangent to a curve. In his *Traité des indivisibles*, which dates from 1634 (though not published until 1693), Gilles Persone de Roberval (1602–75) generalized a method Archimedes had used to find the tangent at any point on his spiral. Like Archimedes, Roberval thought of a curve as the locus of a point moving under the action of two velocities. Thus a projectile shot from a cannon is acted on by a horizontal velocity, PQ in Figure 17.2, and a vertical velocity, PR. The resultant of these two velocities is the diagonal of the rectangle formed on PQ and PR. Roberval took the line of this diagonal to be the tangent at P. As Torricelli pointed out, Roberval's method used a principle already asserted by Galileo, namely, that the horizontal and vertical velocities acted independently of each other. Torricelli himself applied Roberval's method to obtain tangents to curves whose equations we now write as $y = x^n$.

While the notion of a tangent as a line having the direction of the resultant velocity was more complicated than the Greek definition of a line touching a curve, this newer concept applied to many curves for which the older one failed. It was also valuable because it linked pure geometry and dynamics, which before Galileo's work had been regarded as essentially distinct. On the other hand, this notion of a tangent was objectionable on mathematical grounds, because it based the definition of tangent on physical concepts. Many curves arose in situations having nothing to do with motion and the definition of tangent was accordingly inapplicable. Hence other methods of finding tangents gained favor.

Fermat's method, which he had devised by 1629 and which is found in his 1637 manuscript *Methodus ad Disquirendam Maximam et Minimam* (Method of Finding Maxima and Minima),[1] is in substance the present method. Let PT be the desired tangent at P on a curve (Fig. 17.3). The length TQ is called the subtangent. Fermat's plan is to find the length of TQ, from which one knows the position of T and can then draw TP.

1. *Œuvres*, 1, 133–79; 3, 121–56.

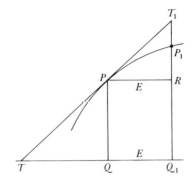

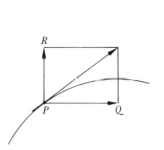

Figure 17.2   Figure 17.3

Let QQ_1 be an increment in TQ of amount E. Since triangle TQP is similar to triangle PRT_1,

$$TQ:PQ = E:T_1R.$$

But, Fermat says, T_1R is almost P_1R; therefore,

$$TQ:PQ = E:(P_1Q_1 - QP).$$

Calling $PQ, f(x)$ in our modern notation, we have

$$TQ:f(x) = E:[f(x + E) - f(x)].$$

Hence

$$TQ = \frac{E \cdot f(x)}{f(x + E) - f(x)}.$$

For the $f(x)$ Fermat treated, it was immediately possible to divide numerator and denominator of the above fraction by E. He then set $E = 0$ (he says, remove the E term) and so obtained TQ.

Fermat applied his method of tangents to many difficult problems. The method has the *form* of the now-standard method of the differential calculus, though it begs entirely the difficult theory of limits.

To Descartes the problem of finding a tangent to a curve was important because it enables one to obtain properties of curves—for example, the angle of intersection of two curves. He says, "This is the most useful, and the most general problem, not only that I know, but even that I have any desire to know in geometry." He gave his method in the second book of *La Géométrie*. It was purely algebraic and did not involve any concept of limit, whereas Fermat's did, if rigorously formulated. However, Descartes's method was useful only for curves whose equations were of the form $y = f(x)$, where $f(x)$ was a simple polynomial. Though Fermat's method was general, Descartes

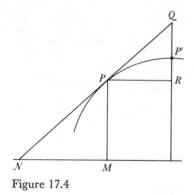

Figure 17.4

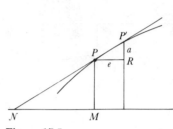

Figure 17.5

thought his own method was better; he criticized Fermat's, which admittedly was not clear as presented then, and tried to interpret it in terms of his own ideas. Fermat in turn claimed his method was superior and saw advantages in his use of the little increments E.

Isaac Barrow (1630–77) also gave a method of finding tangents to curves. Barrow was a professor of mathematics at Cambridge University. Well versed in Greek and Arabic, he was able to translate some of Euclid's works and to improve a number of other translations of the writings of Euclid, Apollonius, Archimedes, and Theodosius. His chief work, the *Lectiones Geometricae* (1669), is one of the great contributions to the calculus. In it he used geometrical methods, "freed," as he put it, "from the loathsome burdens of calculation." In 1669 Barrow resigned his professorship in favor of Newton and turned to theological studies.

Barrow's geometrical method is quite involved and makes use of auxiliary curves. However, one feature is worth noting because it illustrates the thinking of the time; it is the use of what is called the differential, or characteristic, triangle. He starts with the triangle PRQ (Fig. 17.4), which results from the increment PR, and uses the fact that this triangle is similar to triangle PMN to assert that the slope QR/PR of the tangent is equal to PM/MN. However, Barrow says, when the arc PP' is sufficiently small we may safely identify it with the segment PQ of the tangent at P. The triangle PRP' (Fig. 17.5), in which PP' is regarded both as an arc of the curve and as part of the tangent, is the characteristic triangle. It had been used much earlier by Pascal, in connection with finding areas, and by others before him.

In Lecture 10 of the *Lectiones*, Barrow does resort to calculation to find the tangent to a curve. Here the method is essentially the same as Fermat's. He uses the equation of the curve, say $y^2 = px$, and replaces x by $x + e$ and y by $y + a$. Then

$$y^2 + 2ay + a^2 = px + pe.$$

He subtracts $y^2 = px$ and obtains

$$2ay + a^2 = pe.$$

Then he discards higher powers of a and e (where present), which amounts to replacing PRP' of Figure 17.4 by PRP' of Figure 17.5, and concludes that

$$\frac{a}{e} = \frac{p}{2y}.$$

Now he argues that $a/e = PM/NM$, so that

$$\frac{PM}{NM} = \frac{p}{2y}.$$

Since PM is y, he has calculated NM, the subtangent, and knows the position of N.

The work on the third class of problems, finding the maxima and minima of functions, may be said to begin with an observation by Kepler. He was interested in the shape of casks for wine; in his *Stereometria Doliorum* (1615) he showed that, of all right parallelepipeds inscribed in a sphere and having square bases, the cube is the largest. His method was to calculate the volumes for particular choices of dimensions. This in itself was not significant; but he noted that as the maximum volume was approached, the *change* in volume for a fixed change in dimensions grew smaller and smaller.

Fermat in his *Methodus ad Disquirendam* gave his method, which he illustrated with the following example: Given a straight line (segment), it is required to find a point on it such that the rectangle contained by the two segments of the line is a maximum. He calls the whole segment B and lets one part of it be A. Then the rectangle is $AB - A^2$. He now replaces A by $A + E$. The other part is then $B - (A + E)$, and the rectangle becomes $(A + E)(B - A - E)$. He equates the two areas because, he argues, at a maximum the two function values—that is, the two areas—should be equal. Thus

$$AB + EB - A^2 - 2AE - E^2 = AB - A^2.$$

By subtracting common terms from the two sides and dividing by E, he gets

$$B = 2A + E.$$

He then sets $E = 0$ (he says, discard the E term) and gets $B = 2A$. Thus the rectangle is a square.

The method, Fermat says, is quite general; he describes it thus: If A is the independent variable, and if A is increased to $A + E$, then when E becomes indefinitely small and when the function is passing through a maximum or minimum, the two values of the function will be equal. These two values are equated; the equation is divided by E; and E is now made to vanish, so that from the resulting equation the value of A that makes the function a maximum or minimum can be determined. The method is essentially the one he used to find the tangent to a curve. However, the basic fact there is a similarity of two triangles; here it is the equality of two function values. Fermat did not see the need to justify introducing a non-zero E and then, after dividing by E, setting $E = 0$.[2]

The seventeenth-century work on finding areas, volumes, centers of gravity, and lengths of curves begins with Kepler, who is said to have been attracted to the volume problem because he noted the inaccuracy of methods used by wine dealers to find the volumes of kegs. This work (in *Stereometria Doliorum*) is crude by modern standards. For example, the area of a circle is to him the area of an infinite number of triangles, each with a vertex at the center and a base on the circumference. Then from the formula for the area of a regular inscribed polygon, 1/2 the perimeter times the apothem, he obtained the area of the circle. In an analogous manner he regarded the volume of a sphere as the sum of the volumes of small cones with vertices at the center of the sphere and bases on its surface. He then proceeded to show that the volume of the sphere is 1/3 the radius times the surface. The cone he regarded as a sum of very thin circular discs and was able thereby to compute its volume. Stimulated by Archimedes' *Spheroids and Conoids*, he generated new figures by rotation of areas and calculated the volumes. Thus he rotated the segment of a circle cut out by a chord around the chord and found the volume.

The identification of curvilinear areas and volumes with the sum of an infinite number of infinitesimal elements of the same dimension is the essence of Kepler's method. That the circle could be regarded as the sum of an infinite number of triangles was in his mind justified by the principle of continuity (Chap. 14, sec. 5). He saw no difference in kind between the two figures. For the same reason a line and an infinitesimal area were really the same; and he did, in some problems, regard an area as a sum of lines.

In *Two New Sciences* Galileo conceives of areas in a manner similar to Kepler's; in treating the problem of uniformly accelerated motion, he gives an argument to show that the area under the time-velocity curve is the distance. Suppose an object moves with varying velocity $v = 32t$, represented by the straight line in Figure 17.6; then the distance covered in time OA is the area OAB. Galileo arrived at this conclusion by regarding $A'B'$, say, as a typical velocity at some instant and also as the infinitesimal distance covered (as it would be if multiplied by a very small element of time), then arguing that the area OAB, which is made up of lines $A'B'$, must therefore be

2. For the equations that precede his setting $E = 0$, Fermat used the term *adaequalitas*, which Carl B. Boyer in *The Concepts of the Calculus*, p. 156, has aptly translated as "pseudo-equality."

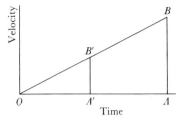

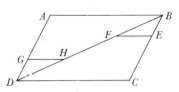

Figure 17.6   Figure 17.7

the total distance. Since AB is $32t$ and OA is t, the area of OAB is $16t^2$. The reasoning is of course unclear. It was supported in Galileo's mind by philosophical considerations that amount to regarding the area OAB as made up of an infinite number of indivisible units such as $A'B'$. He spent much time on the problem of the structure of continuous magnitudes such as line segments and areas but did not resolve it.

Bonaventura Cavalieri (1598–1647), a pupil of Galileo and professor in a lyceum in Bologna, was influenced by Kepler and Galileo and urged by the latter to look into problems of the calculus. Cavalieri developed the thoughts of Galileo and others on indivisibles into a geometrical method and published a work on the subject, *Geometria Indivisibilibus Continuorum Nova quadam Ratione Promota* (Geometry Advanced by a thus far Unknown Method, Indivisibles of Continua, 1635). He regards an area as made up of an indefinite number of equidistant parallel line segments and a volume as composed of an indefinite number of parallel plane areas; these elements he calls the indivisibles of area and volume, respectively. Cavalieri recognizes that the number of indivisibles making up an area or volume must be indefinitely large but does not try to elaborate on this. Roughly speaking, the indivisibilitists held, as Cavalieri put it in his *Exercitationes Geometricae Sex* (1647), that a line is made up of points as a string is of beads; a plane is made up of lines as a cloth is of threads; and a solid is made up of plane areas as a book is made up of pages. However, they allowed for an infinite number of the constituent elements.

Cavalieri's method or principle is illustrated by the following proposition, which of course can be proved in other ways. To show that the parallelogram $ABCD$ (Fig. 17.7) has twice the area of either triangle ABD or BCD, he argued that when $GD = BE$, then $GH = FE$. Hence triangles ABD and BCD are made up of an equal number of equal lines, such as GH and EF, and therefore must have equal areas.

The same principle is incorporated in the proposition now taught in solid geometry books and known as Cavalieri's Theorem. The principle says that if two solids have equal altitudes and if sections made by planes parallel

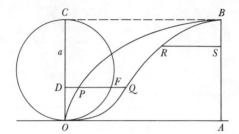

Figure 17.8

to the bases and at equal distances from them always have a given ratio, the volumes of the two solids have this given ratio to each other. Using essentially this principle, Cavalieri proved that the volume of a cone is 1/3 that of the circumscribed cylinder. Likewise he treated the area under two curves, say $y = f(x)$ and $y = g(x)$ in our notation, and over the same range of x-values; considering the areas as the sums of ordinates, if the ordinates of one are in a constant ratio to those of the other, then, says Cavalieri, the areas are in the same ratio. He showed by his methods in *Centuria di varii problemi* (1639) that, in our notation,

$$\int_0^a x^n \, dx = \frac{a^{n+1}}{n + 1}$$

for positive integral values of n up to 9. However, his method was entirely geometrical. He was successful in obtaining correct results because he applied his principle to calculate ratios of areas and volumes where the ratio of the indivisibles making up the respective areas and volumes was constant.

Cavalieri's indivisibles were criticized by contemporaries, and Cavalieri attempted to answer them; but he had no rigorous justification. At times he claimed his method was just a pragmatic device to avoid the method of exhaustion. Despite criticism of the method, it was intensively employed by many mathematicians. Others, such as Fermat, Pascal, and Roberval, used the method and even the language, sum of ordinates, but thought of area as a sum of infinitely small rectangles rather than as a sum of lines.

In 1634 Roberval, who says he studied the "divine Archimedes," used essentially the method of indivisibles to find the area under one arch of the cycloid, a problem Mersenne called to his attention in 1629. Roberval is sometimes credited with independent discovery of the method of indivisibles, but actually he believed in the infinite divisibility of lines, surfaces, and volumes, so that there are no ultimate parts. He called his method the "method of infinities," though he used as the title of his work *Traité des indivisibles*.

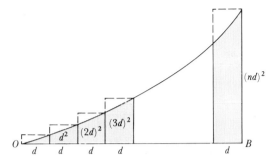

Figure 17.9

Roberval's method of obtaining the area under the cycloid is instructive. Let $OABP$ (Fig. 17.8) be the area under half of an arch of a cycloid. OC is the diameter of the generating circle and P is any point on the arch. Take $PQ = DF$. The locus of Q is called the companion curve to the cycloid. (The curve OQB is, in our notation, $y = a \sin x/a$ where a is the radius of the generating circle, provided the origin is at the midpoint of OQB and the x-axis is parallel to OA.) Roberval affirms that the curve OQB divides the rectangle $OABC$ into two equal parts because, basically, to each line DQ in $OQBC$ there corresponds an equal line RS in $OABQ$. Thus Cavalieri's principle is employed. The rectangle $OABC$ has its base and altitude equal, respectively, to the semicircumference and diameter of the generating circle; hence its area is twice that of the circle. Then $OABQ$ has the same area as the generating circle. Also, the area between OPB and OQB equals the area of the semicircle OFC, since by the very definition of Q, $DF = PQ$, so that these two areas are everywhere of the same width. Hence the area under the half-arch is 1 1/2 times the area of the generating circle. Roberval also found the area under one arch of the sine curve, the volume generated by revolving the arch about its base, other volumes connected with the cycloid, and the centroid of its area.

The most important new method of calculating areas, volumes, and other quantities started with modifications of the Greek method of exhaustion. Let us consider a typical example. Suppose one seeks to calculate the area under the parabola $y = x^2$ from $x = O$ to $x = B$ (Fig. 17.9). Whereas the method of exhaustion used different kinds of rectilinear approximating figures, depending on the curvilinear area in question, some seventeenth-century men adopted a systematic procedure using rectangles as shown. As the width d of these rectangles becomes smaller, the sum of the areas of the rectangles approaches the area under the curve. This sum, if the bases are all d in width, and if one uses the characteristic property of the parabola that the ordinate is the square of the abscissa, is

(1) $$d \cdot d^2 + d(2d)^2 + d(3d)^2 + \cdots + d(nd)^2$$

or
$$d^3(1 + 2^2 + 3^2 + \cdots + n^2).$$

Now the sum of the mth powers of the first n natural numbers had been obtained by Pascal and Fermat for use in just such problems; so the mathematicians could readily replace the last expression by

(2) $$d^3\left(\frac{2n^3 + 3n^2 + n}{6}\right).$$

But d is the fixed length OB divided by n. Hence (2) becomes

(3) $$OB^3\left(\frac{1}{3} + \frac{1}{2n} + \frac{1}{6n^2}\right).$$

Now if one argues, as these men did, that the last two terms can be neglected when n is infinite, the correct result is obtained. The limit process had not yet been introduced—or was only crudely perceived—and so the neglect of terms such as the last two was not justified.

We see that the method calls for approximating the curvilinear figure by rectilinear ones, as in the method of exhaustion. However, there is a vital shift in the final step: in place of the indirect proof used in the older method, here the number of rectangles becomes infinite and one takes the limit of (3) as n becomes infinite—though the thinking in terms of limit was at this stage by no means explicit. This new approach, used as early as 1586 by Stevin in his *Statics*, was pursued by many men, including Fermat.[3]

If the curve involved was not the parabola, then one had to replace the characteristic property of the parabola by that of the curve in question and so obtain some other series in place of (1) above. Summing the analogue of (1) to obtain the analogue of (2) did call for ingenuity. Hence the results on areas, volumes, and centers of gravity were limited. Of course the powerful method of evaluating the limit of such sums by reversing differentiation was not yet envisaged.

Using essentially the kind of summation technique we have just illustrated, Fermat knew before 1636 that (in our notation)

$$\int_0^a x^n \, dx = \frac{a^{n+1}}{n+1}$$

for all rational n except -1.[4] This result was also obtained independently by Roberval, Torricelli, and Cavalieri, though in some cases only in geometrical form and for more limited n.

Among those who used summation in geometrical form was Pascal. In 1658 he took up problems of the cycloid.[5] He calculated the area of any

3. *Œuvres*, 1, 255–59; 3, 216–19.
4. *Œuvres*, 1, 255–59; 3, 216–19.
5. *Traité des sinus du quart de cercle*, 1659 = *Œuvres*, 9, 60–76.

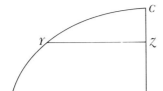

Figure 17.10

segment of the curve cut off by a line parallel to the base, the centroid of the segment, and the volumes of solids generated by such segments when revolved around their bases (YZ in Fig. 17.10) or a vertical line (the axis of symmetry). In this work, as well as in earlier work on areas under the curves of the family $y = x^n$, he summed small rectangles in the manner described in connection with (1) above, though his work and results were stated geometrically. Under the pseudonym of Dettonville, he proposed the problems he had solved as a challenge to other mathematicians, then published his own superior solutions (*Lettres de Dettonville*, 1659).

Before Newton and Leibniz, the man who did most to introduce analytical methods in the calculus was John Wallis (1616–1703). Though he did not begin to learn mathematics until he was about twenty—his university education at Cambridge was devoted to theology—he became professor of geometry at Oxford and the ablest British mathematician of the century, next to Newton. In his *Arithmetica Infinitorum* (1655), he applied analysis and the method of indivisibles to effect many quadratures and obtain broad and useful results.

One of Wallis's notable results, obtained in his efforts to calculate the area of the circle analytically, was a new expression for π. He calculated the area bounded by the axes, the ordinate at x, and the curve for the functions

$$y = (1 - x^2)^0, y = (1 - x^2)^1, y = (1 - x^2)^2, y = (1 - x^2)^3, \cdots$$

and obtained the areas

$$x, x - \frac{1}{3}x^3, x - \frac{2}{3}x^3 + \frac{1}{5}x^5, x - \frac{3}{3}x^3 + \frac{3}{5}x^5 - \frac{1}{7}x^7, \cdots$$

respectively. When $x = 1$, these areas are

(4) $$1, \frac{2}{3}, \frac{8}{15}, \frac{48}{105}, \cdots.$$

Now the circle is given by $y = (1 - x^2)^{1/2}$. Using induction and interpolation, Wallis calculated its area, and by further complicated reasoning arrived at

$$\frac{\pi}{2} = \frac{2 \cdot 2 \cdot 4 \cdot 4 \cdot 6 \cdot 6 \cdot 8 \cdot 8 \cdots}{1 \cdot 3 \cdot 3 \cdot 5 \cdot 5 \cdot 7 \cdot 7 \cdot 9 \cdots}.$$

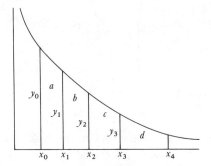

Figure 17.11

Gregory of St. Vincent in his *Opus Geometricum* (1647), gave the basis for the important connection between the rectangular hyperbola and the logarithm function. He showed, using the method of exhaustion, that if for the curve of $y = 1/x$ (Fig. 17.11) the x_i are chosen so that the areas a, b, c, $d, \ldots$ are equal, then the y_i are in geometric progression. This means that the sum of the areas from x_0 to x_i, which sums form an arithmetical progression, is proportional to the logarithm of the y_i values or, in our notation,

$$\int_{x_0}^{x} \frac{dx}{x} = k \log y.$$

This agrees with our familiar calculus result, because $y = 1/x$. The observation that the areas can be interpreted as logarithms is actually due to Gregory's pupil, the Belgian Jesuit Alfons A. de Sarasa (1618–67), in his *Solutio Problematis a Mersenno Propositi* (1649). About 1665 Newton also noted the connection between the area under the hyperbola and logarithms and included this relation in his *Method of Fluxions*. He expanded $1/(1 + x)$ by the binomial theorem and integrated term by term to obtain

$$\log_e (1 + x) = x - \frac{x^2}{2} + \frac{x^3}{3} - \cdots.$$

Nicholas Mercator, using Gregory's results, gave the same series independently (though he did not state it explicitly) in his *Logarithmotechnia* of 1668. Other men soon found series which, as we would put it, converged more rapidly. The work on the quadrature of the hyperbola and its relation to the logarithm function was done by many men, and much of it was communicated in letters, so that it is hard to trace the order of discovery and to assign credit.

Up to about 1650 no one believed that the length of a curve could equal exactly the length of a line. In fact, in the second book of *La Géométrie*, Descartes says the relation between curved lines and straight lines is not nor ever can be known. But Roberval found the length of an arch of the cycloid. The architect Christopher Wren (1632–1723) rectified the cycloid by showing

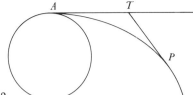

Figure 17.12

(Fig. 17.12) that arc $PA = 2PT$.[6] William Neile (1637–70) also obtained (1659) the length of an arch and, using a suggestion of Wallis, rectified the semicubical parabola ($y^3 = ax^2$).[7] Fermat, too, calculated some lengths of curves. These men usually used an inscribed polygon to approximate the curve, found the sum of the segments, then let the number of segments become infinite as each got smaller. James Gregory (1638–75), a professor at St. Andrews and Edinburgh (whose work was known slightly to his contemporaries but not known generally until a memorial volume, edited by H. W. Turnbull, appeared in 1939), gave in his *Geometriae Pars Universalis* (Universal Part of Geometry, 1668) a method of rectifying curves.

Further results on rectification were obtained by Christian Huygens (1629–95). In particular, he gave the length of arc of the cissoid. He also contributed to the work on areas and volumes and was the first to give results on the areas of surfaces beyond that of the sphere. Thus he obtained the areas of portions of the surfaces of the paraboloid and hyperboloid. Huygens obtained all these results by purely geometric methods, though he did use arithmetic, as Archimedes did occasionally, to obtain quantitative answers.

The rectification of the ellipse defied the mathematicians. In fact, James Gregory asserted that the rectification of the ellipse and the hyperbola could not be achieved in terms of known functions. For a while mathematicians were discouraged from further work on this problem and no new results were obtained until the next century.

We have been discussing the chief contributions of the predecessors of Newton and Leibniz to the four major problems that motivated the work on the calculus. The four problems were regarded as distinct; yet relationships among them had been noted and even utilized. For example, Fermat had used the very same method for finding tangents as for finding the maximum value of a function. Also, the problem of the rate of change of a function with respect to the independent variable and the tangent problem were readily seen to be the same. In fact, Fermat's and Barrow's method of finding tangents is merely the geometrical counterpart of finding the rate of change. But the major feature of the calculus, next to the very concepts of the

---

6. The method was published by Wallis in *Tractatus Duo* (1659 = *Opera*, 1, 550–69). Wren gave only the result.

7. Neile's work was published by Wallis in the reference in footnote 6.

derivative and of the integral as a limit of a sum, is the fact that the integral can be found by reversing the differentiation process or, as we say, by finding the antiderivative. Much evidence of this relationship had been encountered, but its significance was not appreciated. Torricelli saw in special cases that the rate problem was essentially the inverse of the area problem. It was, in fact, involved in Galileo's use of the fact that the area under a velocity-time graph gives distance. Since the rate of change of distance must be velocity, the rate of change of area, regarded as a "sum," must be the derivative of the area function. But Torricelli did not see the general point. Fermat, too, knew the relationship between area and derivative in special cases but did not appreciate its generality or importance. James Gregory, in his *Geometriae* of 1668, proved that the tangent and area problems are inverse problems but his book went unnoticed. In *Geometrical Lectures*, Barrow had the relation-ship between finding the tangent to a curve and the area problem, but it was in geometrical form, and he himself did not recognize its significance.

Actually an immense amount of knowledge of the calculus had accumu-lated before Newton and Leibniz made their impact. A survey of even the one book by Barrow shows a method of finding tangents, theorems on the differentiation of the product and quotient of two functions, the differentia-tion of powers of x, the rectification of curves, change of variable in a definite integral, and even the differentiation of implicit functions. Though in Barrow's case the geometric formulation made the discernment of the general ideas difficult, in Wallis's *Arithmetica Infinitorum* comparable results were in algebraic form.

One wonders then what remained to be achieved in the way of major new results. The answer is greater generality of method and the recognition of the generality of what had already been established in particular problems. The work on the calculus during the first two thirds of the century lost itself in details. Also, in their efforts to attain rigor through geometry, many men failed to utilize or explore the implications of the new algebra and coordinate geometry, and exhausted themselves in abortive subtle reasonings. What ultimately fostered the necessary insight and the attainment of generality was the arithmetical work of Fermat, Gregory of St. Vincent, and Wallis, the men whom Hobbes criticized for substituting symbols for geometry. James Gregory stated in the preface to *Geometriae* that the true division of mathe-matics was not into geometry and arithmetic but into the universal and the particular. The universal was supplied by the two all-embracing minds, Newton and Leibniz.

3. The Work of Newton

Great advances in mathematics and science are almost always built on the work of many men who contribute bit by bit over hundreds of years; eventually one man sharp enough to distinguish the valuable ideas of his predecessors from the welter of suggestions and pronouncements, imaginative enough to fit the bits into a new account, and audacious enough to build a master plan takes the culminating and definitive step. In the case of the calculus, this was Isaac Newton.

Newton (1642–1727) was born in the hamlet of Woolsthorpe, England, where his mother managed the farm left by her husband, who died two months before Isaac was born. He was educated at local schools of low educational standards and as a youth showed no special flair, except for an interest in mechanical devices. Having passed entrance examinations with a deficiency in Euclidean geometry, he entered Trinity College of Cambridge University in 1661 and studied quietly and unobstrusively. At one time he almost changed his course from natural philosophy (science) to law. Apparently receiving very little stimulation from his teachers, except possibly Barrow, he experimented by himself and studied Descartes's *Géométrie*, as well as the works of Copernicus, Kepler, Galileo, Wallis, and Barrow.

Just after Newton finished his undergraduate work the university was closed down because the plague was widespread in the London area. He left Cambridge and spent the years 1665 and 1666 in the quiet of the family home at Woolsthorpe. There he initiated his great work in mechanics, mathematics, and optics. At this time he realized that the inverse square law of gravitation, a concept advanced by others, including Kepler, as far back as 1612, was the key to an embracing science of mechanics; he obtained a general method for treating the problems of the calculus; and through experiments with light he made the epochal discovery that white light, such as sunlight, is really composed of all colors from violet to red. "All this," Newton said later in life, "was in the two plague years of 1665 and 1666, for in those days I was in the prime of my age for invention, and minded mathematics and philosophy [science] more than at any other time since."

Newton said nothing about these discoveries. He returned to Cambridge in 1667 to secure a master's degree and was elected a fellow of Trinity College. In 1669 Isaac Barrow resigned his professorship and Newton was appointed in Barrow's place as Lucasian professor of mathematics. Apparently he was not a successful teacher, for few students attended his lectures; nor was the originality of the material he presented noticed by his colleagues. Only Barrow and, somewhat later, the astronomer Edmond Halley (1656–1742) recognized his greatness and encouraged him.

At first Newton did not publish his discoveries. He is said to have had an abnormal fear of criticism; De Morgan says that "a morbid fear of opposition from others ruled his whole life." When in 1672 he did publish his work on light, accompanied by his philosophy of science, he was severely criticized by most of his contemporaries, including Robert Hooke and Huygens, who had different ideas on the nature of light. Newton was so taken aback that he decided not to publish in the future. However, in 1675 he did publish another paper on light, which contained his idea that light was a stream of particles—the corpuscular theory of light. Again he was met by a storm of criticism and even claims by others that they had already discovered these ideas. This time Newton resolved that his results would be published after his death. Nonetheless he did publish subsequent papers and several famous books, the *Principia*, the *Opticks* (English edition 1704, Latin edition 1706), and the *Arithmetica Universalis* (1707).

From 1665 on he applied the law of gravitation to planetary motion; in this area the works of Hooke and Huygens influenced him considerably. In 1684 his friend Halley urged him to publish his results, but aside from his reluctance to publish Newton lacked a proof that the gravitational attraction exerted by a solid sphere acts as though the sphere's mass were concentrated at the center. He says, in a letter to Halley of June 20, 1686, that until 1685 he suspected that it was false. In that year he showed that a sphere whose density varies only with distance to the center does in fact attract an external particle as though the sphere's mass were concentrated at its center, and agreed to write up his work.

Halley then assisted Newton editorially and paid for the publication. In 1687 the first edition of the *Philosophiae Naturalis Principia Mathematica* (*The Mathematical Principles of Natural Philosophy*) appeared. There were two subsequent editions, in 1713 and 1726, the second edition containing improvements. Though the book brought Newton great fame, it was very difficult to understand. He told a friend that he had purposely made it difficult "to avoid being bated by little smatterers in mathematics." He no doubt hoped in this way to avoid the criticism that his earlier papers on light had received.

Newton was also a major chemist. Though there are no great discoveries associated with his work in this area, one must take into account that chemistry was then in its infancy. He had the correct idea of trying to explain chemical phenomena in terms of ultimate particles, and he had a profound knowledge of experimental chemistry. In this subject he wrote one major paper, "De natura acidorum" (written in 1692 and published in 1710). In the Philosophical Transactions of the Royal Society of 1701, he published a paper on heat that contains his famous law on cooling. Though he read the works of alchemists, he did not accept their cloudy and mystical views. The chemical and physical properties of bodies could, he believed, be accounted for in terms of the size, shape, and motion of the ultimate particles; he rejected the alchemists' occult forces, such as sympathy, antipathy, congruity, and attraction.

In addition to his work on celestial mechanics, light, and chemistry, Newton worked in hydrostatics and hydrodynamics. Beyond his superb experimental work on light, he experimented on the damping of pendulum motion by various media, the fall of spheres in air and water, and the flow of water from jets. Like most men of the time Newton constructed his own equipment. He built two reflecting telescopes, even making the alloy for the frames, molding the frames, making the mountings, and polishing the lenses.

After serving as a professor for thirty-five years Newton became depressed and suffered a nervous breakdown. He decided to give up research and in 1695 accepted an appointment as warden of the British Mint in London. During his twenty-seven years at the mint, except for work on an occasional problem, he did no research. He became president of the Royal Society in 1703, an office he held until his death; he was knighted in 1705.

It is evident that Newton was far more engrossed in science than in mathematics and was an active participant in the problems of his time. He considered the chief value of his scientific work to be its support of revealed religion and was, in fact, a learned theologian, though he never took orders. He thought scientific research hard and dreary but stuck to it because it gave evidence of God's handiwork. Like his predecessor Barrow, Newton turned to religious studies later in life. In *The Chronology of Ancient Kingdoms Amended*, he tried to date accurately events described in the Bible and other religious documents by relating them to astronomical events. His major religious work was the *Observations Upon the Prophecies of Daniel and the Apocalypse of St. John*. Biblical exegesis was a phase of the rational approach to religion that was popular in the Age of Reason; Leibniz, too, took a hand in it.

So far as the calculus is concerned, Newton generalized the ideas already advanced by many men, established full-fledged methods, and showed the interrelationships of several of the major problems described above. Though he learned much as a student of Barrow, in algebra and the calculus he was more influenced by the works of Wallis. He said that he was led to his discoveries in analysis by the *Arithmetica Infinitorum*; certainly in his own work on the calculus he made progress by thinking analytically. However, even Newton thought the geometry was necessary for a rigorous proof.

In 1669 Newton circulated among his friends a monograph entitled *De Analysi per Aequationes Numero Terminorum Infinitas* (On Analysis by Means of Equations with an Infinite Number of Terms); it was not published until 1711. He supposes that he has a curve and that the area z (Fig. 17.13) under this curve is given by

$$z = ax^m,$$ (5)

where m is integral or fractional. He calls an infinitesimal increase in x, the moment of x, and denotes it by o, a notation used by James Gregory and the equivalent of Fermat's E. The area bounded by the curve, the x-axis, the y-axis, and the ordinate at $x + o$ he denotes by $z + oy$, oy being the moment of area. Then

$$z + oy = a(x + o)^m.$$ (6)

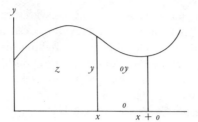

Figure 17.13

He applies the binomial theorem to the right side, obtaining an infinite series when m is fractional, subtracts (5) from (6), divides through by o, neglects those terms that still contain o, and obtains

$$y = max^{m-1}.$$

Thus, in our language, the rate of change of area at any x is the y-value of the curve at that value of x. Conversely, if the curve is $y = max^{m-1}$, the area under it is $z = ax^m$.

In this process Newton not only gave a general method for finding the instantaneous rate of change of one variable with respect to another (z with respect to x in the above example), but showed that area can be obtained by reversing the process of finding a rate of change. Since areas had also been expressed and obtained by the summation of infinitesimal areas, Newton also showed that such sums can be obtained by reversing the process of finding a rate of change. This fact, that summations (more properly, limits of sums) can be obtained by reversing differentiation, is what we now call the fundamental theorem of the calculus. Though it was known in special cases and dimly foreseen by Newton's predecessors, he saw it as general. He applied the method to obtain the area under many curves and to solve other problems that can be formulated as summations.

After showing that the derivative of the area is the y-value and asserting that the converse is true, Newton gave the rule that, if the y-value be a sum of terms, then the area is the sum of the areas that result from each of the terms. In modern terms, the indefinite integral of a sum of functions is the sum of the integrals of the separate functions.

His next contribution in the monograph carried further his use of infinite series. To integrate $y = a^2/(b + x)$, he divided a^2 by $b + x$ and obtained

$$y = \frac{a^2}{b} - \frac{a^2x}{b^2} + \frac{a^2x^2}{b^3} - \frac{a^2x^3}{b^4} + \cdots.$$

Having obtained this infinite series, he finds the integral by integrating term by term so that the area is

$$\frac{a^2x}{b} - \frac{a^2x^2}{2b^2} + \frac{a^2x^3}{3b^3} - \frac{a^2x^4}{4b^4} + \cdots.$$

He says of this infinite series that a few of the initial terms are exact enough for any use, provided that b be equal to x repeated some few times.

Likewise, to integrate $y = 1/(1 + x^2)$ he uses the binomial expansion to write

$$y = 1 - x^2 + x^4 - x^6 + x^8 - \cdots$$

and integrates term by term. He notes that if, instead, y is taken to be $1/(x^2 + 1)$, then by binomial expansion one would obtain

$$y = x^{-2} - x^{-4} + x^{-6} - x^{-8} + \cdots$$

and now one can integrate term by term. He then remarks that when x is small enough the first expansion is to be used; but when x is large, the second is to be used. Thus he was somewhat aware that what we call convergence is important, but had no precise notion about it.

Newton realized that he had extended term by term integration to infinite series but says in the *De Analysi*:

> And whatever the common Analysis performs by Means of Equations of a finite Number of Terms (provided that can be done) this can always perform the same by Means of infinite Equations so that I have not made any question of giving this the name of Analysis likewise. For the reasonings in this are no less certain than in the other; nor the equations less exact; albeit we Mortals whose reasoning powers are confined within narrow limits, can neither express, nor so conceive all the Terms of these Equations, as to know exactly from thence the quantities we want.

Thus far in his approach to the calculus Newton used what may be described as the method of infinitesimals. Moments are infinitely small quantities, indivisibles or infinitesimals. Of course the logic of what Newton did is not clear. He says in this work that his method is "shortly explained rather than accurately demonstrated."

Newton gave a second, more extensive, and more definitive exposition of his ideas in the book *Methodus Fluxionum et Serierum Infinitarum*, written in 1671 but not published until 1736. In this work he says he regards his variables as generated by the continuous motion of points, lines, and planes, rather than as static aggregates of infinitesimal elements, as in the earlier paper. A variable quantity he now called a fluent and its rate of change, the fluxion. His notation is $\dot{x}$ and $\dot{y}$ for fluxions of the fluents x and y. The fluxion of $\dot{x}$ is $\ddot{x}$, etc. The fluent of which x is the fluxion is $\dot{x}$, and the fluent of the latter is $\ddot{x}$.

In this second work Newton states somewhat more clearly the fundamental problem of the calculus: Given a relation between two fluents, find the relation between their fluxions, and conversely. The two variables whose relation is given can represent any quantities. However, Newton thinks of them as changing with time because it is a useful way of thinking, though, he points out, not necessary. Hence if o is an "infinitely small interval of time," then $\dot{x}o$ and $\dot{y}o$ are the indefinitely small increments in x and y or the moments of x and y. To find the relation between $\dot{y}$ and $\dot{x}$, suppose, for example, the fluent is $y = x^n$. Newton first forms

$$y + \dot{y}o = (x + \dot{x}o)^n,$$

and then proceeds as in the earlier paper. He expands the right side by using the binomial theorem, subtracts $y = x^n$, divides through by o, neglects all terms still containing o, and obtains

$$\dot{y} = nx^{n-1}\dot{x}.$$

In modern notation this result can be written

$$\frac{dy}{dt} = nx^{n-1}\frac{dx}{dt},$$

and since $dy/dx = (dy/dt)/(dx/dt)$, Newton, in finding the ratio of dy/dt to dx/dt or $\dot{y}$ to $\dot{x}$, has found dy/dx.

The method of fluxions is not essentially different from the one used in the *De Analysi*, nor is the rigor any better; Newton drops terms such as $\dot{x}\dot{x}o$ and $\dot{x}\dot{x}o\dot{x}o$ (he writes $\dot{x}^3 oo$) on the ground that they are infinitely small compared to the one retained. However, his point of view in the *Method of Fluxions* is somewhat different. The moments $\dot{x}o$ and $\dot{y}o$ change with time o, whereas in the first paper the moments are ultimate fixed bits of x and z. This newer view follows the more dynamic thinking of Galileo; the older used the static indivisible of Cavalieri. The change served, as Newton put it, only to remove the harshness from the doctrine of indivisibles; however, the moments $\dot{x}o$ and $\dot{y}o$ are still some sort of infinitely small quantities. Moreover, $\dot{x}$ and $\dot{y}$, which are the fluxions or derivatives with respect to time of x and y, are never really defined; this central problem is evaded.

Given a relation between $\dot{x}$ and $\dot{y}$, finding the relation between x and y is more difficult than merely integrating a function of x. Newton treats several types: (1) when $\dot{x}$, $\dot{y}$, and x or y are present; (2) when $\dot{x}$, $\dot{y}$, x, and y are present; (3) when $\dot{x}$, $\dot{y}$, $\dot{z}$, and the fluents are present. The first type is the easiest and, in modern notation, calls for solving $dy/dx = f(x)$. Of the second type, Newton treats $\dot{y}/\dot{x} = 1 - 3x + y + x^2 + xy$ and solves it by a successive approximation process. He starts with $\dot{y}/\dot{x} = 1 - 3x + x^2$ as a first approximation, obtains y as a function of x, introduces this value of y on the right side of the original equation, and continues the process. Newton describes what he does but does not justify it. Of the third type, he treats $2\dot{x} - \dot{z} + \dot{y}x = 0$. He assumes a relation between x and y, say $x = y^2$, so that $\dot{x} = 2\dot{y}y$. Then the equation becomes $4\dot{y}y - \dot{z} - \dot{y}y^2 = 0$, from which he gets $2y^2 + (y^3/3) = z$. Thus, if the third type is regarded as a partial differential equation, Newton obtains only a particular integral.

Newton realized that in this paper he had presented a general method. In a letter to John Collins, dated December 10, 1672, wherein he gives the facts of his method and one example, he says,

> This is one particular, or rather corollary, of a general method, which extends itself, without any troublesome calculations, not only to the drawing of tangents to any curved lines, whether geometrical or mechanical... but also to resolving other abstruser kinds of problems about the crookedness, areas, lengths, centres of gravity of curves, etc.; nor is it... limited to equations which are free from surd quantities. This method I have interwoven with that other of working in equations, by reducing them to infinite series.

Newton emphasized the use of infinite series because thereby he could treat functions such as $(1 + x)^{3/2}$, whereas his predecessors had been limited on the whole to rational algebraic functions.

In his *Tractatus de Quadratura Curvarum* (Quadrature of Curves), a third paper on the calculus, written in 1676 but published in 1704, Newton says he has abandoned the infinitesimal or infinitely small quantity. He now criticizes the dropping of terms involving o for, he says,

> in mathematics the minutest errors are not to be neglected.... I consider mathematical quantities in this place not as consisting of very small parts, but as described by a continual motion. Lines are described, and thereby generated, not by the apposition of parts, but by the continued motion of points; superficies by the motion of lines; solids by the motions of superficies; angles by the rotation of the sides; portions of time by continued flux....
>
> Fluxions are, as near as we please, as the increments of fluents generated in times, equal and as small as possible, and to speak accurately, they are in the prime ratio of nascent increments; yet they can be expressed by any lines whatever, which are proportional to them.

Newton's new concept, the method of prime and ultimate ratio, amounts to this. He considers the function $y = x^n$. To find the fluxion of y or x^n, let x "by flowing" become $x + o$. Then x^n becomes

$$(x + o)^n = x^n + nox^{n-1} + \frac{n^2 - n}{2} o^2 x^{n-2} + \cdots.$$

The increases of x and y, namely, o and $nox^{n-1} + \dfrac{n^2 - n}{2} o^2 x^{n-2} + \cdots$ are to each other as (dividing both by o)

$$1 \text{ to } nx^{n-1} + \frac{n^2 - n}{2} ox^{n-2} + \cdots.$$

"Let now the increments vanish and their last proportion will be"

$$1 \text{ to } nx^{n-1}.$$

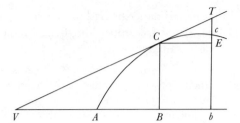

Figure 17.14

Then the fluxion of x is to the fluxion of x^n as 1 to nx^{n-1} or, as we would say today, the rate of change of y with respect to x is nx^{n-1}. This is the prime ratio of the nascent increments. Of course the logic of this version is no better than that of the preceding two; nevertheless Newton says this method is in harmony with the geometry of the ancients and that it is not necessary to introduce infinitely small quantities.

Newton also gave a geometrical interpretation. Given the data in Figure 17.14, suppose bc moves to BC so that c coincides with C. Then the curvilinear triangle CEc is "in the last form" similar to triangle CET, and its "evanescent" sides will be proportional to CE, ET, and CT. Hence the fluxions of the quantities AB, BC, and AC are, in the last ratio of their evanescent increments, proportional to the sides of the triangle CET or triangle VBC.

In the *Method of Fluxions* Newton made a number of applications of fluxions to differentiating implicit functions and to finding tangents of curves, maxima and minima of functions, curvature of curves, and points of inflection of curves. He also obtained areas and lengths of curves. In connection with curvature, he gave the correct formula for the radius of curvature, namely,

$$r = \frac{(1 + \dot{y}^2)^{3/2}}{\ddot{y}}$$

where $\dot{x}$ is taken as 1. He also gave this same quantity in polar coordinates. Finally, he included a brief table of integrals.

Newton did not publish his basic papers in the calculus until long after he had written them. The earliest printed account of his theory of fluxions appeared in Wallis's *Algebra* (2nd ed. in Latin, 1693), of which Newton wrote pages 390 to 396. Had he published at once he might have avoided the controversy with Leibniz on the priority of discovery.

Newton's first publication involving his calculus is the great *Mathematical Principles of Natural Philosophy*.[8] So far as the basic notion of the

8. The third edition was translated into English by Andrew Motte in 1729. This edition, revised and edited by Florian Cajori, was published by the University of California Press in 1946.

calculus, the fluxion, or, as we say, the derivative, is concerned, Newton makes several statements. He rejects infinitesimals or ultimate indivisible quantities in favor of "evanescent divisible quantities," quantities which can be diminished without end. In the first and third editions of the *Principia* Newton says, "Ultimate ratios in which quantities vanish are not, strictly speaking, ratios of ultimate quantities, but limits to which the ratios of these quantities, decreasing without limit, approach, and which, though they can come nearer than any given difference whatever, they can neither pass over nor attain before the quantities have diminished indefinitely."[9] This is the clearest statement he ever gave as to the meaning of his ultimate ratio. Apropos of the preceding quotation, he also says, "By the ultimate velocity is meant that with which the body is moved, neither before it arrives at its last place, when the motion ceases, nor after; but at the very instant when it arrives. . . . And, in like manner, by the ultimate ratio of evanescent quantities is to be understood the ratio of quantities, not before they vanish, nor after, but that with which they vanish."

In the *Principia* Newton used geometrical methods of proof. However, in what are called the Portsmouth Papers, containing unpublished work, he used analytical methods to find some of the theorems. These papers show that he also obtained analytically results beyond those he was able to translate into geometry. One reason he resorted to geometry is believed to be that the proofs would be more understandable to his contemporaries. Another is that he admired Huygens's geometrical work immensely and hoped to equal it. In these geometrical proofs Newton uses the basic limit processes of the calculus. Thus the area under a curve is considered essentially as the limit of the sum of the approximating rectangles, just as in the calculus today. However, instead of calculating such areas, he uses this concept to compare areas under different curves.

He proves that, when AR and BR (Fig. 17.15) are the perpendiculars to the tangents at A and B of the arc ACB, the ultimate ratio, when B approaches and coincides with A, of any two of the quantities chord AB, arc ACB, and AD, is 1. Hence he says in Corollary 3 to Lemma 2 of Book I, "And therefore in all our reasoning about ultimate ratios, we may freely use any one of these lines for any other." He then proves that when B approaches and coincides with A, the ratio of any two triangles (areas) RAB, $RACB$, and RAD will be 1. "And hence in all reasonings about ultimate ratios, we may use any one of these triangles for any other." Also, (Fig. 17.16) let BD and CE be perpendicular to AE (which is not necessarily tangent to arc ABC at A). When B and C approach and coincide with A, the ultimate ratio of the areas ACE and ABD will equal the ultimate ratio of AE^2 to AD^2.

The *Principia* contains a wealth of results, some of which we shall note.

9. Third edition, p. 39.

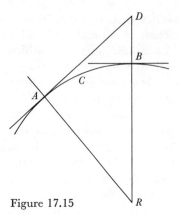

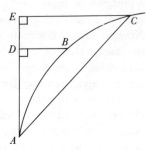

Figure 17.15   Figure 17.16

Though the book is devoted to celestial mechanics, it has enormous importance for the history of mathematics, not only because Newton's own work on
the calculus was motivated in large part by his overriding interest in the
problems treated therein, but because the *Principia* presented new topics and
approaches to problems that were explored during the next hundred years
in the course of which an enormous amount of analysis was created.

The *Principia* is divided into three books.[10] In a prefatory section Newton
defines concepts of mechanics such as inertia, momentum, and force, then
states the three famous axioms or laws of motion. In his words, they are:

Law I. Every body continues in its state of rest, or of uniform motion in a
right line, unless it is compelled to change that state by forces impressed
upon it.

Law II. The change [in the quantity] of motion is proportional to the motive
power impressed; and is made in the direction of the right line in which that
force is impressed.

By quantity of motion Newton means, as he has explained earlier, the
mass times the velocity. Hence the change in motion, if the mass is constant,
is the change in velocity, that is, the acceleration. This second law is now
often written as $F = ma$, when the force F is in poundals, the mass m is in
pounds, and the acceleration a is in feet per second per second. Newton's
second law is really a vector statement; that is, if the force has components in,
say, three mutually perpendicular directions, then each component causes an
acceleration in its own direction. Newton did use the vector character of
force in particular problems, but the full significance of the vector nature of

10. All references are to the edition mentioned in note 8.

the law was first fully recognized by Euler. This law incorporates the key change from the mechanics of Aristotle, which affirmed that force causes velocity. Aristotle had also affirmed that a force is needed to maintain velocity. Law I denies this.

Law III. To every action there is always opposed an equal reaction. . . .

We shall not digress into the history of mechanics except to note that the first two laws are more explicit and somewhat generalized statements of the principles of motion previously discovered and advanced by Galileo and Descartes. The distinction between mass, that is, the resistance a body offers to a change in its motion, and weight, the force gravity exerts on the mass of any object, is also due to these men; and the vector character of force generalizes Galileo's principle that the vertical and horizontal motions of a projectile, for example, can be treated independently.

Book I of the *Principia* begins with some theorems of the calculus, including the ones involving ultimate ratios cited above. It then discusses motion under central forces, that is, forces that always attract the moving object to one (fixed) point (the sun in practice), and proves in Proposition 1 that equal areas are swept out in equal time (which encompasses Kepler's law of areas). Newton considers next the motion of a body along a conic section and proves (Props. 11, 12, and 13) that the force must vary with the inverse square of the distance from some fixed point. He also proves the converse, which contains Kepler's first law. After some treatment of centripetal force, he deduces Kepler's third law (Prop. 15). There follow two sections devoted to properties of the conic sections. The principal problem is the construction of conics that satisfy five given conditions; in practice these are usually observational data. Then, given the time an object has been in motion along a conic section, he determines its velocity and position. He takes up the motion of the apse lines, that is, the lines joining the center of attraction (at one focus) to the maximum or minimum distance of a body moving along a conic that is itself rotating at some rate about the focus. Section 10 is devoted to the motion of bodies along surfaces with special reference to pendulum motion. Here Newton gives due acknowledgment to Huygens. In connection with the accelerating effect of gravity on motions, he investigates geometrical properties of cycloids, epicycloids, and hypocycloids and gives the length of the epicycloid (Prop. 49).

In Section 11 Newton deduces from the laws of motion and the law of gravitation the motion of two bodies, each attracting the other in accordance with the gravitational force. Their motion is reduced to the motion of one around the fixed second body. The moving body traverses an ellipse.

He then considers the attraction exerted by spheres and spheroids of uniform and varying density on a particle. He gives (Sec. 12, Prop. 70) a geometrical proof that a thin homogeneous spherical shell exerts no force on a particle in its interior. Since this result holds for a thin shell, it holds for a sum of such shells, that is, for a shell of finite thickness. (He proves later [Prop. 91, Cor. 3] that the same result holds for a homogeneous ellipsoidal shell, that is, a shell contained between two similar ellipsoidal surfaces, similarly placed.) Proposition 71 shows that the attraction of a thin homogeneous spherical shell on an *external* particle is equivalent to the attraction that would be exerted if the mass of the shell were concentrated at the center, so that the shell attracts the external particle toward the center and with a force varying inversely as the square of the distance from the center. Proposition 73 shows that a solid homogeneous sphere attracts a particle inside with a force proportional to the particle's distance from the center. As for the attraction that a solid homogeneous sphere exerts on an external point, Proposition 74 shows that it is the same as if the mass of the sphere were concentrated at its center. Then if two spheres attract each other, the first attracts every particle of the second as if the mass of the first were concentrated at its center. Thus the first sphere becomes a particle attracted by the distributed mass of the second; hence the second sphere can also be treated as a particle with its mass concentrated at its center. Thus both spheres can be treated as particles with their masses concentrated at their respective centers. All these results, original with Newton, are extended to spheres whose densities are spherically symmetric and to other laws of attraction in addition to the inverse square law.

Newton next takes up the motion of three bodies, each attracting the other two, and obtains some approximate results. The problem of the motion of three bodies has been a major one since Newton's time and has not been solved as yet.

The second book of the *Principia* is devoted to the motion of bodies in resisting media such as air and liquids. It is the beginning of the subject of hydrodynamics. Newton assumes in some problems that the resistance of the medium is proportional to the velocity and in others to the square of the velocity of the moving body. He considers what shape a body must have to encounter least resistance (see Chap. 24, sec. 1). He also considers the motion of pendulums and projectiles in air and in fluids. A section is devoted to the theory of waves in air (e.g., sound waves) and he obtains a formula for the velocity of sound in air. He also treats the motion of waves in water. Newton continues with a description of experiments he made to determine the resistance fluids offer to bodies moving in them. One major conclusion is that the planets move in a vacuum. In this book Newton broke entirely new ground; however, the definitive work on fluid motion was yet to be done

Book III, entitled *On the System of the World*, contains the application of the general theory developed in Book I to the solar system. It shows how the sun's mass can be calculated in terms of the earth's mass, and that the mass of any planet having a satellite can be found in the same way. He calculates the average density of the earth and finds it to be between 5 and 6 times that of water (today's figure is about 5.5).

He shows that the earth is not a true sphere but an oblate spheroid and calculates the flattening; his result is that the ellipticity of the oblate spheroid is 1/230 (the figure today is 1/297). From the observed oblateness of any planet, the length of its day is then calculated. Using the amount of flattening and the notion of centripetal force, Newton computes the variation of the earth's gravitational attraction over the surface and thus the variation in the weight of an object. He proves that the attractive force of a spheroid is not the same as if the spheroid's mass were concentrated at its center.

He then accounts for the precession of the equinoxes. The explanation is based on the fact that the earth is not spherical but bulges out along the equator. Consequently the gravitational attraction of the moon on the earth does not effectively act on the center of the earth but forces a periodic change in the direction of the earth's axis of rotation. The period of this change was calculated by Newton and found to be 26,000 years, the value obtained by Hipparchus by inference from observations available to him.

Newton explained the main features of the tides (Book I, Prop. 66, Book III, Props. 36, 37). The moon is the main cause; the sun, the second. Using the sun's mass he calculated the height of the solar tides. From the observed heights of the spring and neap tides (sun and moon in full conjunction or full opposition) he determined the lunar tide and made an estimate of the mass of the moon. Newton also managed to give some approximate treatment of the effect of the sun on the moon's motion around the earth. He determined the motion of the moon in latitude and longitude; the motion of the apse line (the line from the center of the earth to the maximum distance of the moon); the motion of the nodes (the points in which the moon's path cuts the plane of the earth's orbit; these points regress, that is, move slowly in a direction opposite to the motion of the moon itself); the evection (a periodic change in the eccentricity of the moon's orbit); the annual equation (the effect on the moon's motion of the daily change in distance between the earth and the sun); and the periodic change in the inclination of the plane of the moon's orbit to the plane of the earth's orbit. There were seven known irregularities in the motion of the moon and Newton discovered two more, the inequalities of the apogee (apse line) and of the nodes. His approximation gave only half of the motion of the apse line. Clairaut in 1752 improved the calculation and obtained the full 3° of rotation of the apse line; however, much later John Couch Adams found the correct calculation in Newton's papers. Finally Newton showed that the comets must be moving under the gravitational attraction of the sun because their paths, determined on the basis of observations, are conic sections. Newton devoted a great deal of time to the problem of the moon's motion because, as we noted in the preceding chapter, the knowledge was needed to improve the method of determining longitude. He worked so hard on this problem that he complained it made his head ache.

4. The Work of Leibniz

Though his contributions were quite different, the man who ranks with Newton in building the calculus is Gottfried Wilhelm Leibniz (1646–1716). He studied law and, after defending a thesis on logic, received a Bachelor of Philosophy degree. In 1666, he wrote the thesis *De Arte Combinatoria* (On the Art of Combinations),[11] a work on a universal method of reasoning; this completed his work for a doctorate in philosophy at the University of Altdorf and qualified him for a professorship. During the years 1670 and 1671 Leibniz wrote his first papers on mechanics, and, by 1671, had produced his calculating machine. He secured a job as an ambassador for the Elector of Mainz and in March of 1672 went to Paris on a political mission. This visit brought him into contact with mathematicians and scientists, notably Huygens, and stirred up his interest in mathematics. Though he had done a little reading in the subject and had written the paper of 1666, he says he knew almost no mathematics up to 1672. In 1673 he went to London and met other scientists and mathematicians, including Henry Oldenburg, at that time secretary of the Royal Society of London. While making his living as a diplomat, he delved further into mathematics and read Descartes and Pascal. In 1676 Leibniz was appointed librarian and councillor to the Elector of Hanover. Twenty-four years later the Elector of Brandenburg invited Leibniz to work for him in Berlin. While involved in all sorts of political maneuvers, including the succession of George Ludwig of Hanover to the English throne, Leibniz worked in many fields and his side activities covered an enormous range. He died neglected in 1716.

In addition to being a diplomat, Leibniz was a philosopher, lawyer, historian, philologist, and pioneer geologist. He did important work in logic, mechanics, optics, mathematics, hydrostatics, pneumatics, nautical science, and calculating machines. Though his profession was jurisprudence, his work in mathematics and philosophy is among the best the world has produced. He kept contact by letter with people as far away as China and Ceylon. He tried endlessly to reconcile the Catholic and Protestant faiths. It was he who proposed, in 1669, that a German Academy of Science be founded; finally the Berlin Academy was organized in 1700. His original recommendation had been for a society to make inventions in mechanics and discoveries in chemistry and physiology that would be useful to mankind; Leibniz wanted knowledge to be applied. He called the universities "monkish" and charged that they possessed learning but no judgment and were absorbed in trifles.

11. Published 1690 = *Die philosophische Schriften*, 4, 27–102.

Instead he urged the pursuit of real knowledge—mathematics, physics, geography, chemistry, anatomy, botany, zoology, and history. To Leibniz the skills of the artisan and the practical man were more valuable than the learned subtleties of the professional scholars. He favored the German language over Latin because Latin was allied to the older, useless thought. Men mask their ignorance, he said, by using the Latin language to impress people. German, on the other hand, was understood by the common people and could be developed to help clarity of thought and acuteness of reasoning.

Leibniz published papers on the calculus from 1684 on, and we shall say more about them later. However, many of his results, as well as the development of his ideas, are contained in hundreds of pages of notes made from 1673 on but never published by him. These notes, as one might expect, jump from one topic to another and contain changing notation as Leibniz's thinking developed. Some are simply ideas that occurred to him while reading books or articles by Gregory of St. Vincent, Fermat, Pascal, Descartes, and Barrow or trying to cast their thoughts into his own way of approaching the calculus. In 1714 Leibniz wrote *Historia et Origo Calculi Differentialis*, in which he gives an account of the development of his own thinking. However, this was written many years after he had done his work and, in view of the weaknesses of human memory and the greater insight he had acquired by that time, his history may not be accurate. Since his purpose was to defend himself against an accusation of plagiarism, he might have distorted unconsciously his account of the origins of his ideas.

Despite the confused state of Leibniz's notes we shall examine a few, because they reveal how one of the greatest intellects struggled to understand and create. By 1673 he was aware of the important direct and inverse problem of finding tangents to curves; he was also quite sure that the inverse method was equivalent to finding areas and volumes by summations. The somewhat systematic development of his ideas begins with notes of 1675. However, it seems helpful, in order to understand his thinking, to note that in his *De Arte Combinatoria* he had considered sequences of numbers, first differences, second differences, and higher-order differences. Thus for the sequence of squares

$$0, 1, 4, 9, 16, 25, 36,$$

the first differences are

$$1, 3, 5, 7, 9, 11$$

and the second differences are

$$2, 2, 2, 2, 2, 2.$$

Leibniz noted the vanishing of the second differences for the sequence of natural numbers, the third differences for the sequence of squares, and so on.

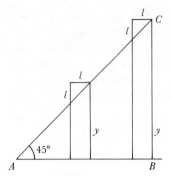

Figure 17.17

He also observed, of course, that if the original sequence starts from 0, the sum of the first differences is the last term of the sequence.

To relate these facts to the calculus he had to think of the sequence of numbers as the y-values of a function and the difference of any two as the difference of two nearby y-values. Initially he thought of x as representing the order of the term in the sequence and y as representing the value of that term.

The quantity dx, which he often writes as a, is then 1 because it is the difference of the orders of two successive terms, and dy is the actual difference in the values of two successive terms. Then using omn. as an abbreviation for the Latin *omnia*, to mean sum, and using l for dy, Leibniz concludes that omn. $l = y$, because omn. l is the sum of the first differences of a sequence whose terms begin with 0 and so gives the last term. However, omn. yl presents a new problem. Leibniz obtains the result that omn. yl is $y^2/2$ by thinking in terms of the function $y = x$. Thus, as Figure 17.17 shows, the area of triangle ABC is the sum of the yl (for "small" l) and it is also $y^2/2$. Leibniz says, "Straight lines which increase from nothing each multiplied by its corresponding element of increase form a triangle." These few facts already appear, among more complicated ones, in papers of 1673.

In the next stage he struggled with several difficulties. He had to make the transition from a discrete series of values to the case where dy and dx are increments of an arbitrary function y of x. Since he was still tied to sequences, wherein x is the order of the term, his a or dx was 1; so he inserted and omitted a freely. When he made the transition to the dy and dx of any function, this a was no longer 1. However, while still struggling with the notion of summation he ignored this fact.

Thus in a manuscript of October 29, 1675, Leibniz starts with

(7) $$\text{omn. } yl = \overline{\text{omn.omn. } l}\,\frac{l}{a},$$

which holds because y itself is omn. l. Here he divides l by a to preserve dimensions. Leibniz says that (7) holds, whatever l may be. But, as we saw in connection with Figure 17.17,

(8) $$\text{omn. } yl = \frac{y^2}{2}.$$

Hence from (7) and (8)

(9) $$\frac{y^2}{2} = \overline{\text{omn.omn. } l\frac{l}{a}}.$$

In our notation, he has shown that

$$\frac{y^2}{2} = \int \left\{ \int dy \right\} \frac{dy}{dx} = \int y \frac{dy}{dx}.$$

Leibniz says that this result is admirable.

Another theorem of the same kind, which Leibniz derived from a geometrical argument, is

(10) $$\text{omn. } xl = x \text{ omn. } l - \text{omn.omn. } l,$$

where l is the difference in values of two successive terms of a sequence and x is the number of the term. For us this equation is

$$\int x \, dy = xy - \int y \, dx.$$

Now Leibniz lets l itself in (10) be x, and obtains

$$\text{omn. } x^2 = x \text{ omn. } x - \text{omn.omn. } x.$$

But omn. x, he says, is $x^2/2$ (he has shown that omn. yl is $y^2/2$). Hence

$$\text{omn. } x^2 = x \frac{x^2}{2} - \text{omn. } \frac{x^2}{2}.$$

By transposing the last term he gets

$$\text{omn. } x^2 = \frac{x^3}{3}.$$

In this manuscript of October 29, 1675, Leibniz decided to write $\int$ for omn., so that

$$\int l = \text{omn. } l \quad \text{and} \quad \int x = \frac{x^2}{2}.$$

The symbol $\int$ is an elongated S for "sum."

Leibniz realized rather early, probably from studying the work of Barrow, that differentiation and integration as a summation must be inverse processes; so area, when differentiated, must give a length. Thus, in the same manuscript of October 29, Leibniz says, "Given l and its relation to x, to find $\int l$." Then, he says, "Suppose that $\int l = ya$. Let $l = ya/d$. [Here he puts d in the denominator. It would mean more to us if he wrote $l = d(ya)$.] Then just as $\int$ will increase, so d will diminish the dimensions. But $\int$ means a sum, and d, a difference. From the given y we can always find y/d or l, that is, the difference of the y's. Hence one equation may be transformed into the other; just as from the equation

$$\overline{\int c \int \overline{l^2}} = \frac{c \int \overline{l^3}}{3a^3},$$

we can obtain the equation

$$c \int \overline{l^2} = \frac{c \int \overline{l^3}}{3a^3 d}."$$

In this early paper Leibniz seems to be exploring the *operations* of $\int$ and d and sees that they are inverses. He finally realizes that $\int$ does not raise dimension nor d lower it, because $\int$ is really a summation of rectangles, and so a sum of areas. Thus he recognizes that, to get back to dy from y, he must form the difference of y's or take the differential of y. Then he says, "But $\int$ means a sum and d a difference." This may have been a later insertion. Hence a couple of weeks afterwards, in order to get from y to dy, he changes from dividing by d to taking the differential of y, and writes dy.

Up to this point Leibniz had been thinking of the y-values as values of terms of a sequence and of x usually as the order of these terms, but now, in this paper, says, "All these theorems are true for series in which the differences of the terms bear to the terms themselves a ratio that is less than any assignable quantity." That is, dy/y may be less than any assignable quantity.

In a manuscript dated November 11, 1675, entitled "Examples of the inverse method of tangents," Leibniz uses $\int$ for the sum and x/d for difference. He then says x/d is dx, the difference of two consecutive x-values, but apparently here dx is a constant and equal to unity.

From barely intelligible arguments such as the above, Leibniz asserted the fact that *integration as a summation process is the inverse of differentiation*. This idea is in the work of Barrow and Newton, who obtained areas by antidifferentiation, but it is first expressed as a relation between summation and differentiation by Leibniz. Despite this outright assertion, he was by no means clear as to how to obtain an area from what one might loosely write as $\sum y \, dx$—that is, how to obtain an area under a curve from a set of rectangles. Of course this difficulty beset all the seventeenth-century workers. Not possessing a clear concept of a limit, or even clear notions about area, Leibniz thought of the latter sometimes as a sum of rectangles so small and so

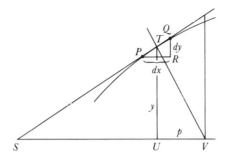

Figure 17.18

numerous that the difference between this sum and the true area under the curve could be neglected, and at other times as a sum of the ordinates or y-values. This latter concept of area was common, especially among the indivisibilists, who thought that the ultimate unit of area and the y-value were the same.

With respect to differentiation, even after recognizing that dy and dx can be arbitrarily small quantities, Leibniz had yet to overcome the fundamental difficulty that the ratio dy/dx is not quite the derivative in our sense. He based his argument on the characteristic triangle, which Pascal and Barrow had also used. This triangle (Fig. 17.18) consists of dy, dx, and the chord PQ, which Leibniz also thought of as *the curve between* P *and* Q *and part of the tangent at* T. Though he speaks of this triangle as indefinitely small, he maintains nevertheless that it is similar to a definite triangle, namely, the triangle STU formed by the subtangent SU, the ordinate at T, and the length of tangent ST. Hence dy and dx are ultimate elements, and their ratio has a definite meaning. In fact, he uses the argument that, from the similar triangles PRQ and SUT, $dy/dx = TU/SU$.

In the manuscript of November 11, 1675, Leibniz shows how he can solve a definite problem. He seeks the curve whose subnormal is inversely proportional to the ordinate. In Figure 17.18, the normal is TV and the subnormal p is UV. From the similarity of triangles PRQ and TUV, he has

$$\frac{dy}{dx} = \frac{p}{y}$$

or

$$p\,dx = y\,dy.$$

But the curve has the given property

$$p = \frac{b}{y},$$

where b is the proportionality constant. Hence

$$dx = \frac{y^2}{b} \, dy.$$

Then

$$\int dx = \int \frac{y^2}{b} \, dy$$

or

$$x = \frac{y^3}{3b}.$$

Leibniz also solved other inverse tangent problems.

In a paper of June 26, 1676, he realizes that the best method of finding tangents is to find dy/dx, where dy and dx are differences and dy/dx is the quotient. He ignores $dx \cdot dx$ and higher powers of dx.

By November of 1676, he is able to give the general rules $dx^n = nx^{n-1} \, dx$ for integral and fractional n and $\int x^n = x^{n+1}/n + 1$, and says, "The reasoning is general, and it does not depend upon what the progressions of the x's may be." Here x still means the order of the terms of a sequence. In this manuscript he also says that to differentiate $\sqrt{a + bz + cz^2}$, let $a + bz + cz^2 = x$, differentiate $\sqrt{x}$, and multiply by dx/dz. This is the chain rule.

By July 11, 1677, Leibniz could give the correct rules for the differential of sum, difference, product, and quotient of two functions and for powers and roots, but no proofs. In the manuscript of November 11, 1675, he had struggled with $d(uv)$ and $d(u/v)$, and thought that $d(uv) = du \, dv$.

In 1680, dx has become the difference of abscissas and dy the differences in the ordinates. He says, "... now these dx and dy are taken to be infinitely small, or the two points on the curve are understood to be a distance apart that is less than any given length. ..." He calls dy the "momentaneous increment" in y as the ordinate moves along the x-axis. But PQ in Figure 17.18 is still considered part of a straight line. It is "an element of the curve or a side of the infinite-angled polygon that stands for the curve. ..." He continues to use the usual differential form. Thus, if $y = a^2/x$, then

$$dy = -\frac{a^2}{x^2} \, dx.$$

He also says that differences are the opposite to sums. Then, to get the area under a curve (Fig. 17.19), he takes the sum of the rectangles and says one can neglect the remaining "triangles, since they are infinitely small com-

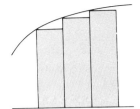

Figure 17.19

pared to the rectangles ... thus I represent in my calculus the area of the figure by $\int y\,dx.\ldots$" He also gives, for the element of arc,

$$ds = \sqrt{dx^2 + dy^2};$$

and, for the volume of a solid of revolution obtained by revolving a curve around the x-axis,

$$V = \pi \int y^2\,dx.$$

Despite prior statements that dx and dy are small differences, he still talks about sequences. He says, "Differences and sums are the inverses of one another, that is to say, the sum of the differences of a series [sequence] is a term of the series, and the difference of the sums of a series is a term of the series, and I enumerate the former thus, $\int dx = x$, and the latter thus, $d\int x = dx$." In fact, in a manuscript written after 1684, Leibniz says his method of infinitesimals has become widely known as the calculus of differences.

Leibniz's first publication on the calculus is in the *Acta Eruditorum* of 1684.[12] In this paper the meaning of dy and dx is still not clear. He says in one place, let dx be any arbitrary quantity, and dy is defined by (see Fig. 17.18)

$$dy:dx = y:\text{subtangent}.$$

This definition of dy presumes some expression for the subtangent; hence the definition is not complete. Moreover, Leibniz's definition of tangent as a line joining two infinitely near points is not satisfactory.

He also gives in this paper the rules he had obtained in 1677 for the differential of the sum, product, and quotient of two functions and the rule for finding $d(x^n)$. In this last case he sketches the proof for positive integral n but says the rule is true for all n; for the other rules he gives no proofs. He makes applications to finding tangents, maxima and minima, and points of inflection. This paper, six pages long, is so unclear that the Bernoulli brothers called it "an enigma rather than an explication."[13]

12. *Acta Erud.*, 3, 1684, 467–73 = *Math. Schriften*, 5, 220–26.
13. Leibniz: *Math. Schriften*, 3, Part 1, 5.

In a paper of 1686[14] Leibniz gives

$$y = \sqrt{2x - x^2} + \int \frac{dx}{\sqrt{2x - x^2}}$$

as the equation of the cycloid. His point here is to show that by his methods and notation some curves can be expressed as equations not obtainable in other ways. He reaffirms this in his *Historia* where he says that his *dx, ddx* (second difference), and the sums that are the inverses of these differences can be applied to all functions of x, not excepting the mechanical curves of Vieta and Descartes, which Descartes had said have no equations. Leibniz also says that he can include curves that Newton could not handle even with his method of series.

In the 1686 paper as well as in subsequent papers,[15] Leibniz gave the differentials of the logarithmic and exponential functions and recognized exponential functions as a class. He also treated curvature, the osculating circle, and the theory of envelopes (see Chap. 23). In a letter to John Bernoulli of 1697, he differentiated under the integral sign with respect to a parameter. He also had the idea that many indefinite integrals could be evaluated by reducing them to known forms and speaks of preparing tables for such reductions—in other words, a table of integrals. He tried to define the higher-order differentials such as *ddy* (d^2y) and *dddy* (d^3y), but the definitions were not satisfactory. Though he did not succeed, he also tried to find a meaning for $d^\alpha y$ where α is any real number.

With respect to notation, Leibniz worked painstakingly to achieve the best. His *dx, dy,* and *dy/dx* are, of course, still standard. He introduced the notation log x, d^n for the nth differential, and even d^{-1} and d^{-n} for $\int$ and the nth iteration of summation, respectively.

In general Leibniz's work, though richly suggestive and profound, was so incomplete and fragmentary that it was barely intelligible. Fortunately, the Bernoulli brothers, James and John, who were immensely impressed and stirred by Leibniz's ideas, elaborated his sketchy papers and contributed an immense number of new developments we shall discuss later. Leibniz agreed that the calculus was as much theirs as his.

5. *A Comparison of the Work of Newton and Leibniz*

Both Newton and Leibniz must be credited with seeing the calculus as a new and general method, applicable to many types of functions. After their work, the calculus was no longer an appendage and extension of Greek geometry, but an independent science capable of handling a vastly expanded range of problems.

14. *Acta Erud.*, 5, 1686, 292–300 = *Math. Schriften*, 5, 226–33.
15. *Acta Erud.*, 1692, 168–71 = *Math. Schriften*, 5, 266–69; *Acta Erud.*, 1694 = *Math. Schriften*, 5, 301–6.

Both also arithmetized the calculus; that is, they built on algebraic concepts. The algebraic notation and techniques used by Newton and Leibniz not only gave them a more effective tool than geometry, but also permitted many different geometric and physical problems to be treated by the same technique. A major change from the beginning to the end of the seventeenth century was the algebraicization of the calculus. This is comparable to what Vieta had done in the theory of equations and Descartes and Fermat in geometry.

The third vital contribution that Newton and Leibniz share is the reduction to antidifferentiation of area, volume, and other problems that were previously treated as summations. Thus the four main problems—rates, tangents, maxima and minima, and summation—were all reduced to differentiation and antidifferentiation.

The chief distinction between the work of the two men is that Newton used the infinitely small increments in x and y as a means of determining the fluxion or derivative. It was essentially the limit of the ratio of the increments as they became smaller and smaller. On the other hand, Leibniz dealt directly with the infinitely small increments in x and y, that is, with differentials, and determined the relationship between them. This difference reflects Newton's physical orientation, in which a concept such as velocity is central, and Leibniz's philosophical concern with ultimate particles of matter, which he called monads. As a consequence, Newton solved area and volume problems by thinking entirely in terms of rate of change. For him differentiation was basic; this process and its inverse solved all calculus problems, and in fact the use of summation to obtain an area, volume, or center of gravity rarely appears in his work. Leibniz, on the other hand, thought first in terms of summation, though of course these sums were evaluated by antidifferentiation.

A third distinction between the work of the two men lies in Newton's free use of series to represent functions; Leibniz preferred the closed form. In a letter to Leibniz of 1676, Newton stressed the use of series even to solve simple differential equations. Though Leibniz did use infinite series, he replied that the real goal should be to obtain results in finite terms, using the trigonometric and logarithmic functions where algebraic functions would not serve. He recalled to Newton James Gregory's assertion that the rectification of the ellipse and hyperbola could not be reduced to the circular and logarithmic functions and challenged Newton to determine by the use of series whether Gregory was correct. Newton replied that by the use of series he could decide whether some integrations could be achieved in finite terms, but gave no criteria. Again, in a letter of 1712 to John Bernoulli, Leibniz objected to the expansion of functions into series and stated that the calculus should be concerned with reducing its results to quadratures (integrations) and, where necessary, quadratures involving transcendental functions.

There are differences in their manner of working. Newton was empirical, concrete, and circumspect, whereas Leibniz was speculative, given to generalizations, and bold. Leibniz was more concerned with operational formulas to produce a calculus in the broad sense; for example, rules for the differential of a product or quotient of functions, his rule for $d^n(uv)$ (u and v being functions of x), and a table of integrals. It was Leibniz who set the canons of the calculus, the system of rules and formulas. Newton did not bother to formulate rules, even when he could easily have generalized his concrete results. He knew that if $z = uv$, then $\dot{z} = u\dot{v} + v\dot{u}$, but did not point out this general result. Though Newton initiated many methods, he did not stress them. His magnificent applications of the calculus not only demonstrated its value but, far more than Leibniz's work, stimulated and determined almost the entire direction of eighteenth-century analysis. Newton and Leibniz differed also in their concern for notation. Newton attached no importance to this matter, while Leibniz spent days choosing a suggestive notation.

6. *The Controversy over Priority*

Nothing of Newton's work on the calculus was published before 1687, though he had communicated results to friends during the years 1665 to 1687. In particular, he had sent his tract *De Analysi* in 1669 to Barrow, who had sent it to John Collins. Leibniz visited Paris in 1672 and London in 1673 and communicated with some of the people who knew Newton's work. However, he did not publish on the calculus until 1684. Hence the question of whether Leibniz had known the details of what Newton did was raised, and Leibniz was accused of plagiarism. However, investigations made long after the deaths of the two men show that Leibniz was an independent inventor of major ideas of the calculus, though Newton did much of his work before Leibniz did. Both owe much to Barrow, though Barrow used geometrical methods almost exclusively. The significance of the controversy lies not in the question of who was the victor but rather in the fact that the mathematicians took sides. The Continental mathematicians, the Bernoulli brothers in particular, sided with Leibniz, while the English mathematicians defended Newton. The two groups became unfriendly and even bitter toward each other; John Bernoulli went so far as to ridicule and inveigh against the English.

As a result, the English and Continental mathematicians ceased exchanging ideas. Because Newton's major work and first publication on the calculus, the *Principia*, used geometrical methods, the English continued to use mainly geometry for about a hundred years after his death. The Continentals took up Leibniz's analytical methods and extended and improved them. These proved to be far more effective; so not only did the English mathematicians fall behind, but mathematics was deprived of contributions that some of the ablest minds might have made.

7. Some Immediate Additions to the Calculus

The calculus is of course the beginning of that most weighty part of mathematics generally referred to as analysis. We shall be following the important developments of this field in succeeding chapters; we might note here, however, some additions that were made immediately after the basic work of Newton and Leibniz.

In his *Arithmetica Universalis* (1707) Newton established a theorem on the upper bound to the real roots of polynomial equations. The theorem says: A number a is an upper bound of the real roots of $f(x) = 0$ if, when a is substituted for x, it gives to $f(x)$ and to all its derivatives the same sign.

In his *De Analysi* and *Method of Fluxions*, he gave a general method of approximating the roots of $f(x) = 0$, which was published in Wallis's *Algebra* of 1685. In his tract *Analysis Aequationum Universalis* (1690), Joseph Raphson (1648–1715) improved on this method; though he applied it only to polynomials, it is much more broadly useful. It is this modification that is now known as Newton's method or the Newton-Raphson method. It consists in first choosing an approximation a. Then calculate $a - f(a)/f'(a)$. Call this b, and calculate $b - f(b)/f'(b)$. Call this last result c, and so forth. The numbers $a, b, c, \ldots$ are successive approximations to the root. (The notation is modern.) Actually the method does not necessarily give better and better approximations to the root. J. Raymond Mourraille showed in 1768 that a must be chosen so that the curve of $y = f(x)$ is convex toward the axis of x in the interval between a and the root. Much later Fourier discovered this fact independently.

In his *Démonstration d'une méthode pour résoudre les égalitéz de tous les dégrez* (16191), Michel Rolle (1652–1719) gave the famous theorem now named after him, namely, that if a function is 0 at two values of x, say, a and b, then the derivative is 0 at some value of x between a and b. Rolle stated the theorem but did not prove it.

After Newton and Leibniz the two most important founders of the calculus were the Bernoulli brothers, James and John. James (= Jakob = Jacques) Bernoulli (1655–1705) was self-taught in mathematics and so matured slowly in that subject. At the urging of his father he studied for the ministry, but eventually turned to mathematics, and in 1686 became a professor at the University of Basle. His chief interests thereafter were mathematics and astronomy. When, in the late 1670s, he began to work on mathematical problems, Newton's and Leibniz's work was still unknown to him. He too learned from Descartes's *La Géométrie*, Wallis's *Arithmetica Infinitorum* and Barrow's *Geometrical Lectures*. Though he took much from

Barrow, he put it into analytical form. He gradually became familiar with Leibniz's work, but because so little of the latter appeared in print, much of what James did overlapped Leibniz's results. Actually he, like the other mathematicians of the time, did not fully understand Leibniz's work.

James's activity is closely linked with that of his younger brother John (= Johann = Jean, 1667–1748). John was sent into business by his father but turned to medicine, while learning mathematics from his brother. He became a professor of mathematics at Groningen in Holland and then succeeded his brother at Basle.

Both James and John corresponded constantly with Leibniz, Huygens, other mathematicians, and each other. All these men worked on many common problems suggested in letters or posed as challenges. Since results, too, were in those days often communicated in letters with or without subsequent publication, the matter of priority is complicated. Sometimes credit was claimed for a result that was announced even though no proof was given at that time. The question is further complicated by the peculiar relationships that developed. John was extremely anxious to secure fame and began to compete with his brother; soon each was challenging the other on problems. John did not hesitate to use unscrupulous means to appear to be the discoverer of results he got from others, including his brother. James was very sensitive and reacted in kind. Each published papers that owed much to the other without acknowledging the origins of their ideas. John actually became a vitriolic critic of his brother, and Leibniz tried to mediate between the two. Though James had said earlier, while praising Barrow, that Leibniz's work should not be depreciated, he became more and more distrustful of Leibniz. Moreover, he resented Leibniz's superior insights and thought Leibniz was arrogant in pointing out that he had done things James thought were original with himself. He became convinced that Leibniz sought only to belittle his work and was favoring John in the disputes between the brothers. When Nicholas Fatio de Duillier (1664–1753) gave Newton credit for creating the calculus and became embroiled in controversy with Leibniz, James wrote letters to Fatio opposing Leibniz.

As to the Bernoullis' work in the calculus, they, too, tackled problems such as finding the curvature of curves, evolutes (envelopes of the normals to a curve), inflection points, the rectification of curves, and other basic calculus topics. The results of Newton and Leibniz were extended to spirals of various sorts, the catenary, and the tractrix, which was defined as the curve (Fig. 17.20) for which the ratio PT to OT is a constant. James also wrote five major papers on series (Chap. 20, sec. 4), which extended Newton's use of series to integrate complicated algebraic functions and transcendental functions. In 1691 both James and John gave the formula for the radius of curvature of a curve. James called it his "golden theorem" and wrote it as

$$z = dx\ ds : ddy = dy\ ds : ddx$$

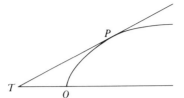

Figure 17.20

where z is the radius of curvature. If we divide numerator and denominator of each ratio by ds^2 we get

$$z = \frac{dx/ds}{d^2y/ds^2} = \frac{dy/ds}{d^2x/ds^2},$$

which are more familiar forms. James also gave the result in polar coordinates.

John produced a now-famous theorem for obtaining the limit approached by a fraction whose numerator and denominator approach 0. This theorem was incorporated by Guillaume F. A. l'Hospital (1661–1704), a pupil of John, in an influential book on the calculus, the *Analyse des infiniment petits* (1696), and is now known as L'Hospital's rule.

8. The Soundness of the Calculus

From the very introduction of the new methods of finding rates, tangents, maxima and minima, and so forth, the proofs were attacked as unsound. Cavalieri's use of indivisible ultimate elements and his arguments shocked those who still respected logical rigor. To their criticism Cavalieri responded that the contemporary geometers had been freer with logic than he—for example, Kepler, in his *Stereometria Doliorum*. These geometers, he continued, had been content in their calculation of areas to imitate Archimedes' method of summing lines, but had failed to give the complete proofs that the great Greek had used to make his work rigorous. They were satisfied with their calculations, provided only that the results were useful. Cavalieri felt justified in adopting the same point of view. He said that his procedures could lead to new inventions and that his method did not at all oblige one to consider a geometrical structure as composed of an infinite number of sections; it had no other object than to establish correct ratios between areas or volumes. But these ratios preserved their sense and value whatever opinion one might have about the composition of a continuum. In any case, said Cavalieri, "rigor is the concern of philosophy and not of geometry."

Fermat, Pascal, and Barrow recognized the looseness of their work on summation but believed that one could make precise proofs in the manner of Archimedes. Pascal, in *Letters of Dettonville* (1659), affirmed that the infinitesimal geometry and classical Greek geometry were in agreement. He concluded, "What is demonstrated by the true rules of indivisibles could be demonstrated also with the rigor and the manner of the ancients." Further, he said the method of indivisibles must be accepted by any mathematicians who pretend to rank among geometers. It differs only in language from the method of the ancients. Nevertheless, Pascal, too, had ambivalent feelings about rigor. At times he argued that the heart intervenes to assure us of the correctness of mathematical steps. The proper "finesse," rather than geometrical logic, is what is needed to do the correct work, just as the religious appreciation of grace is above reason. The paradoxes of geometry as used in the calculus are like the apparent absurdities of Christianity; and the indivisible in geometry has the same relation to the finite as man's justice has to God's.

The defenses Cavalieri and Pascal offered applied to the summation of infinitely small quantities. As to the derivative, early workers such as Fermat and Roberval thought they had a simple algebraic process that had a very clear geometric interpretation and so could be justified by geometrical arguments. Actually Fermat was careful not to assert general theorems when he advanced any idea he could not justify by the method of exhaustion. Barrow argued only geometrically and, despite his attacks on the algebraists for their lack of rigor, was less scrupulous about the soundness of his geometrical arguments.

Neither Newton nor Leibniz clearly understood nor rigorously defined his fundamental concepts. We have already observed that both vacillated in their definitions of the derivative and differentials. Newton did not really believe that he had departed from Greek geometry. Though he used algebra and coordinate geometry, which were not to his taste, he thought his underlying methods were but natural extensions of pure geometry. Leibniz, however, was a man of vision who thought in broad terms, like Descartes. He saw the long-term implications of the new ideas and did not hesitate to declare that a new science was coming to light. Hence he was not too concerned about the lack of rigor in the calculus.

In response to criticism of his ideas, Leibniz made various, unsatisfactory replies. In a letter to Wallis of March 30, 1690[16] he said:

> It is useful to consider quantities infinitely small such that when their ratio is sought, they may not be considered zero but which are rejected as often as they occur with quantities incomparably greater. Thus if we have $x + dx$, dx is rejected. But it is different if we seek the difference between $x + dx$ and x. Similarly we cannot have $x\, dx$ and $dx\, dx$ standing together. Hence if we are to differentiate xy we write $(x + dx)(y + dy) - xy = x\, dy + y\, dx + dx\, dy$. But here $dx\, dy$ is to be rejected as incomparably less than $x\, dy + y\, dx$. Thus in any particular case, the error is less than any finite quantity.

16. Leibniz: *Math. Schriften*, 4, 63.

As to the ultimate meanings of dy, dx and dy/dx, Leibniz remained vague. He spoke of dx as the difference in x values between two infinitely near points and of the tangent as the line joining such points. He dropped differentials of higher order with no justification, though he did distinguish among the various orders. The infinitely small dx and dy were sometimes described as vanishing or incipient quantities, as opposed to quantities already formed. These indefinitely small quantities were not zero, but were smaller than any finite quantity. Alternatively he appealed to geometry to say that a higher differential is to a lower one as a point is to a line[17] or that dx is to x as a point to the earth or as the radius of the earth to that of the heavens. The ratio of two infinitesimals he thought of as a quotient of inassignables or of indefinitely small quantities, but one which could nevertheless be expressed in terms of definite quantities such as the ratio of ordinate to subtangent.

A flurry of attacks and rebuttals was initiated in books of 1694 and 1695 by the Dutch physician and geometer Bernard Nieuwentijdt (1654–1718). Although he admitted that in general the new methods led to correct results, he criticized the obscurity and pointed out that sometimes the methods led to absurdities. He complained that he could not understand how the infinitely small quantities differed from zero and asked how a sum of infinitesimals could be finite. He also challenged the meaning and existence of differentials of higher order and the rejection of infinitely small quantities in portions of the arguments.

Leibniz, in a draft of a reply to Nieuwentijdt, probably written in 1695, and in an article in the *Acta Eruditorum* of 1695,[18] gives various answers. He speaks of "overprecise" critics and says that excessive scrupulousness should not cause us to reject the fruits of invention. He then says his method differs from Archimedes' only in the expressions used, but that his own are better adapted to the art of discovery. The words "infinite" and "infinitesimal" signify merely quantities that one can take as large or as small as one wishes in order to show that the error incurred is less than any number that can be assigned—in other words, that there is no error. One can use these ultimate things—that is, infinite and infinitely small quantities—as a tool, much as algebraists used imaginary roots with great profit.

Leibniz's argument thus far was that his calculus used only ordinary mathematical concepts. But since he could not satisfy his critics, he enunciated a philosophical principle known as the law of continuity, which was practically the same as one already stated by Kepler. In 1687, in a letter to Pierre Bayle,[19] Leibniz expressed this principle as follows: "In any supposed transition, ending in any terminus, it is permissible to institute a general reasoning, in which the final terminus may also be included." To support

17. *Math. Schriften*, 5, 322 ff.
18. *Acta Erud.*, 1695, 310–16 = *Math. Schriften*, 5, 320–28.
19. *Math. Schriften*, 5, 385.

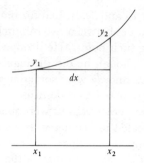

Figure 17.21

this principle he gives, in an unpublished manuscript of about 1695, the example of including under one argument ellipses and parabolas, though the parabola is a limiting case of the ellipse when one focus moves off to infinity. He then applies the principle to the calculation of $dy:dx$ for the parabola $y = x^2/a$. After obtaining

$$dy:dx = (2x + dx):a,$$

he says, "Now, since by our postulate it is permissible to include under the one general reasoning the case also in which [Fig. 17.21] the ordinate x_2y_2 is moved up nearer and nearer to the fixed ordinate x_1y_1 until it ultimately coincides with it, it is evident that in this case dx becomes equal to 0 and should be neglected...." Leibniz does not say what meaning should be given to the dx that appears at the left side of the equation.

Of course, he says, things that are absolutely equal have a difference that is absolutely nothing; therefore a parabola is not an ellipse.

> Yet a state of transition may be imagined, or one of evanescence, in which indeed there has not yet arisen exact equality or rest.... but in which it is passing into such a state that the difference is less than any assignable quantity; also that in this state there will still remain some difference, some velocity, some angle, but in each case one that is infinitely small....
>
> For the present, whether such a state of instantaneous transition from inequality to equality ... can be sustained in a rigorous or metaphysical sense, or whether infinite extensions successively greater and greater, or infinitely small ones successively less and less, are legitimate considerations, is a matter that I own to be possibly open to question. ...
>
> It will be sufficient if, when we speak of infinitely great (or, more strictly, unlimited) or of infinitely small quantities (i.e., the very least of those within our knowledge), it is understood that we mean quantities that are indefinitely great or indefinitely small, i.e., as great as you please, or as small as you please, so that the error that any one may assign may be less than a certain assigned quantity.

On these suppositions, all the rules of our algorithm, as set out in the *Acta Eruditorum* for October 1684, can be proved without much trouble.

Leibniz then goes over these rules. He introduces the quantities $(d)y$ and $(d)x$ and carries out the usual processes of differentiation with them. These he calls assignable or definite nonvanishing quantities. After obtaining the final result, he says, we can replace $(d)y$ and $(d)x$ by the evanescent or unassignable quantities dy and dx, by making "the supposition that the ratio of the evanescent quantities dy and dx is equal to the ratio of $(d)y$ and $(d)x$, because this supposition can always be reduced to an undoubtable truth."

Leibniz's principle of continuity is certainly not a mathematical axiom today, but he emphasized it and it became important later. He gave many arguments that are in accordance with this principle. For example, in a letter to Wallis,[20] Leibniz defended his use of the characteristic triangle as a form without magnitude, the form remaining after the magnitudes had been reduced to zero, and challengingly asked, "Who does not admit a form without magnitude?" Likewise, in a letter to Guido Grandi,[21] he said the infinitely small is not a simple and absolute zero but a relative zero, that is, an evanescent quantity which yet retains the character of that which is disappearing. However, Leibniz also said, at other times, that he did not believe in magnitudes truly infinite or truly infinitesimal.

Leibniz, less concerned with the ultimate justification of his procedures than Newton, felt that it lay in their effectiveness. He stressed the procedural or algorithmic value of what he had created. Somehow he had confidence that if he formulated clearly the rules of operation and these were properly applied, reasonable and correct results would be obtained, however doubtful might be the meanings of the symbols involved.

It is apparent that neither Newton nor Leibniz succeeded in making clear, let alone precise, the basic concepts of the calculus: the derivative and the integral. Not being able to grasp these properly, they relied upon the coherence of the results and the fecundity of the methods to push ahead without rigor.

Several examples may illustrate the lack of clarity even among the great immediate successors of Newton and Leibniz. John Bernoulli wrote the first text on the calculus in 1691 and 1692. The portion on the integral calculus was published in 1742;[22] the part on the differential calculus, *Die Differential-rechnung*, was not published until 1924. However, the Marquis de l'Hospital did publish a slightly altered French version (already referred to) under his own name in 1696. Bernoulli begins the *Differentialrechnung* with three postulates. The first reads: "A quantity which is diminished or increased by an

20. *Math. Schriften*, 4, 54.
21. *Math. Schriften*, 4, 218.
22. *Opera Omnia*, 3, 385–558.

infinitely small quantity is neither increased nor decreased." His second postulate is: "Each curved line consists of infinitely many straight lines, these themselves being infinitely small." In his reasoning he followed Leibniz and used infinitesimals. Thus to obtain dy from $y = x^2$, he uses e for dx and gets $(x + e)^2 - x^2$, or $2xe + e^2$, and then just drops e^2. Like Leibniz, he used vague analogies to explain what differentials were. Thus, he says, the infinitely large quantities are like astronomical distances and the infinitely small are like animalcules revealed by the microscope. In 1698 he argued that infinitesimals must exist.[23] One has only to consider the infinite series 1, 1/2, 1/4, ... If one takes 10 terms, then 1/10 exists; if one takes 100 terms, then 1/100 exists. Corresponding to the infinite number of terms there is the infinitesimal.

A few men, Wallis and John Bernoulli among them, tried to define the infinitesimal as the reciprocal of ∞, the latter being a definite number to them. Still others acted as though what was incomprehensible needed no further explanation. For most of the seventeenth-century men, rigor was not a matter of concern. What they often said could be rigorized by the method of Archimedes could actually not have been rigorized by an Archimedes; this is particularly true of the work on differentiation, which had no parallel in Greek mathematics.

Actually the new calculus was introducing concepts and methods that inaugurated a radical departure from earlier work. With the work of Newton and Leibniz, the calculus became a totally new discipline that required foundations of its own. Though they were not fully aware of it, the mathematicians had turned their backs on the past.

Germs of the correct new concepts can be found even in the seventeenth-century literature. Wallis, in the *Arithmetica Infinitorum*, advanced the arithmetical concept of the limit of a function as a number approached by the function so that the difference between this number and the function could be made less than any assignable quantity and would vanish ultimately when the process was continued to infinity. His wording is loose but contains the right idea.

James Gregory in his *Vera Circuli et Hyperbolae Quadratura* (1667) explicitly pointed out that the methods used to obtain areas, volumes, and lengths of curves involved a new process, the limit process. Moreover, he added, this operation was distinct from the five algebraic operations of addition, subtraction, multiplication, division, and extraction of roots. He put the method of exhaustion into algebraic form and recognized that the successive approximations obtained by using rectilinear figures circumscribed about a given area or volume and those obtained by using inscribed rectilinear figures both converged to the same "last term." He also noted that

23. Leibniz: *Math. Schriften*, 3, Part 2, 563 ff.

this limit process yields irrationals not obtainable as roots of rationals. But these insights of Wallis and Gregory were ignored in their century.

The foundations of the calculus remained unclear. Adding to the confusion was the fact that the proponents of Newton's work continued to speak of prime and ultimate ratios, while the followers of Leibniz used the infinitely small non-zero quantities. Many of the English mathematicians, perhaps because they were in the main still tied to the rigor of Greek geometry, distrusted all the work on the calculus. Thus the century ended with the calculus in a muddled state.

Bibliography

Armitage, A.: *Edmond Halley*, Thomas Nelson and Sons, 1966.

Auger, L.: *Un Savant méconnu: Gilles Persone de Roberval (1602–1675)*, A. Blanchard, 1962.

Ball, W. W. R.: *A Short Account of the History of Mathematics*, Dover (reprint), 1960, pp. 309–70.

Baron, Margaret E.: *The Origins of the Infinitesimal Calculus*, Pergamon Press, 1969.

Bell, Arthur E.: *Newtonian Science*, Edward Arnold, 1961.

Boyer, Carl B.: *The Concepts of the Calculus*, Dover (reprint), 1949.

———: *A History of Mathematics*, John Wiley and Sons, 1968, Chaps. 18–19.

Brewster, David: *Memoirs of the Life, Writings and Discoveries of Sir Isaac Newton*, 2 vols., 1855, Johnson Reprint Corp., 1965.

Cajori, Florian: *A History of the Conceptions of Limits and Fluxions in Great Britain from Newton to Woodhouse*, Open Court, 1919.

———: *A History of Mathematics*, Macmillan, 1919, 2nd ed., pp. 181–220.

Cantor, Moritz: *Vorlesungen über Geschichte der Mathematik*, 2nd ed., B. G. Teubner, 1900 and 1898, Vol. 2, pp. 821–922; Vol. 3, pp. 150–316.

Child, J. M.: *The Geometrical Lectures of Isaac Barrow*, Open Court, 1916.

———: *The Early Mathematical Manuscripts of Leibniz*, Open Court, 1920.

Cohen, I. B.: *Isaac Newton's Papers and Letters on Natural Philosophy*, Harvard University Press, 1958.

Coolidge, Julian L.: *The Mathematics of Great Amateurs*, Dover (reprint), 1963, Chaps. 7, 11, and 12.

De Morgan, Augustus: *Essays on the Life and Work of Newton*, Open Court, 1914.

Fermat, Pierre de: *Œuvres*, Gauthier-Villars, 1891–1912, Vol. 1, pp. 133–79, Vol. 3, pp. 121–56.

Gibson, G. A.: "James Gregory's Mathematical Work," *Proc. Edinburgh Math. Soc.*, 41, 1922/23, 2–25.

Huygens, C.: *Œuvres complètes*, 22 vols., Société Hollandaise des Sciences, Nyhoff, 1888–1950.

Leibniz, G. W.: *Œuvres*, Firmin-Didot, 1859–75.

———: *Mathematische Schriften*, ed. C. I. Gerhardt, 7 vols., Ascher-Schmidt, 1849–63. Reprinted by Georg Olms, 1962.

More, Louis T.: *Isaac Newton*, Dover (reprint), 1962.

Montucla, J. F.: *Histoire des mathématiques*, Albert Blanchard (reprint), 1960, Vol. 2, pp. 102–77, 348–403; Vol. 3, pp. 102–38.

Newton, Sir Isaac: *The Mathematical Works*, ed. D. T. Whiteside, 2 vols., Johnson Reprint Corp., 1964–67. Vol. 1 contains translations of the three basic papers on the calculus.

————: *Mathematical Papers*, ed. D. T. Whiteside, 4 vols., Cambridge University Press, 1967–71.

————: *Mathematical Principles of Natural Philosophy*, ed. Florian Cajori, 3rd ed., University of California Press, 1946.

————: *Opticks*, Dover (reprint), 1952.

Pascal, B.: *Œuvres*, Hachette, 1914–21.

Scott, Joseph F.: *The Mathematical Work of John Wallis*, Oxford University Press, 1938.

————: *A History of Mathematics*, Taylor and Francis, 1958, Chaps. 10–11.

Smith, D. E.: *A Source Book in Mathematics*, Dover (reprint), 1959, pp. 605–26.

Struik, D. J.: *A Source Book in Mathematics, 1200–1800*, Harvard University Press, 1969, pp. 188–316, 324–28.

Thayer, H. S.: *Newton's Philosophy of Nature*, Hafner, 1953.

Turnbull, H. W.: *The Mathematical Discoveries of Newton*, Blackie and Son, 1945.

————: *James Gregory Tercentenary Memorial Volume*, Royal Society of Edinburgh, 1939.

Turnbull, H. W. and J. F. Scott: *The Correspondence of Isaac Newton*, 4 vols., Cambridge University Press, 1959–1967.

Walker, Evelyn: *A Study of the* Traité des indivisibles *of Gilles Persone de Roberval*, Columbia University Press, 1932.

Wallis, John: *Opera Mathematica*, 3 vols., 1693–99, Georg Olms (reprint), 1968.

Whiteside, Derek T.: "Patterns of Mathematical Thought in the Seventeenth Century," *Archive for History of Exact Sciences*, 1, 1961, pp. 179–388.

Wolf, Abraham: *A History of Science, Technology and Philosophy in the 16th and 17th Centuries*, 2nd ed., George Allen and Unwin, 1950, Chaps. 7–14.

18

Mathematics as of 1700

> Those few things having been considered, the whole matter is
> reduced to pure geometry, which is the one aim of physics
> and mechanics.   G. W. LEIBNIZ

1. *The Transformation of Mathematics*

At the opening of the seventeenth century Galileo still found it necessary to
argue with the past. By the end of the century, mathematics had undergone
such extensive and radical changes that no one could fail to recognize the
arrival of a new era.

The European mathematicians produced far more between about 1550
and 1700 than the Greeks had in roughly ten centuries. This is readily
explained by the fact that, whereas mathematics in Greece was pursued by
only a handful of men, in Europe the spread of education, though by no
means universal, promoted the development of mathematicians in England,
France, Germany, Holland, and Italy. The invention of printing gave wide
access not only to the Greek works but to the results of the Europeans themselves, which, now readily available, served to stimulate new thoughts.

But the genius of the century is not evidenced solely by the expansion
of activity. The variety of new fields opened up in this brief period is impressive. The rise of algebra as a science (because the use of literal coefficients
permitted a measure of proof) as well as the vast expansion of its methods
and theory, the beginnings of projective geometry and the theory of probability, analytic geometry, the function concept, and above all the calculus
were major innovations, each destined to dwarf the one extensive accomplishment of the Greeks—Euclidean geometry.

Beyond the quantitative expansion and the new avenues of exploration
was the complete reversal of the roles of algebra and geometry. The Greeks
had favored geometry because it was the only way they could achieve rigor;
and even in the seventeenth century, mathematicians felt obliged to justify
algebraic methods with geometrical proofs. One could say that up to 1600
the body of mathematics was geometrical, with some algebraic and trigonometric appendages. After the work of Descartes, Fermat, and Wallis, algebra

became not only an effective methodology for its own ends but also the superior approach to the solution of geometric problems. The greater effectiveness of analytical methods in the calculus decided the competition, and algebra became the dominant substance of mathematics.

It was Wallis and Newton who saw clearly that algebra provided the superior methodology. Unlike Descartes, who regarded algebra as just technique, Wallis and Newton realized that it was vital subject matter. The work of Desargues, Pascal, and La Hire was depreciated and forgotten, and the geometric methods of Cavalieri, Gregory of Saint Vincent, Huygens, and Barrow were superseded. Pure geometry was eclipsed for about a hundred years, becoming at best an interpretation of algebra and a guide to algebraic thinking through coordinate geometry. It is true that excessive reverence for Newton's geometrical work in the *Principia*, reinforced by the enmity against the Continental mathematicians engendered by the dispute between Newton and Leibniz, caused the English mathematicians to persist in the geometrical development of the calculus. But their contributions were trivial compared to what the Continentals were able to achieve using the analytical approach. What was evident by 1700 was explicitly stated by no less an authority than Euler, who, in his *Introductio in Analysin Infinitorum* (1748), praises algebra as far superior to the synthetic methods of the Greeks.

It was with great reluctance that mathematicians abandoned the geometric approach. According to Henry Pemberton (1694–1771), who edited the third edition of Newton's *Principia*, Newton not only constantly expressed great admiration for the geometers of Greece but censured himself for not following them more closely than he did. In a letter to David Gregory (1661–1708), a nephew of James Gregory, Newton remarked that "algebra is the analysis of the bunglers in mathematics." But his own *Arithmetica Universalis* of 1707 did as much as any single work to establish the supremacy of algebra. Here he set up arithmetic and algebra as the basic science, allowing geometry only where it made demonstrations easier. Leibniz, too, noted the growing dominance of algebra and felt obliged to say, in an unpublished essay,[1] "Often the geometers could demonstrate in a few words what is very lengthy in the calculus . . . the view of algebra is assured, but it is not better."

Another, more subtle, change in the nature of mathematics had been unconsciously accepted by the masters. Up to 1550 the concepts of mathematics were immediate idealizations of or abstractions from experience. By that time negative and irrational numbers had made their appearance and were gradually gaining acceptance. When, in addition, complex numbers, an extensive algebra employing literal coefficients, and the notions of derivative and integral entered mathematics, the subject became dominated by

1. Couturat, L.: *Opuscules et fragments inédits de Leibniz*, 1903, reprinted by Georg Olms, 1961, p. 181.

concepts derived from the recesses of human minds. The notion of an instantaneous rate of change, in particular, though of course having some intuitive base in the physical phenomenon of velocity, is nevertheless far more of an intellectual construct and is also an entirely different contribution qualitatively than the mathematical triangle. Beyond these ideas, infinitely large quantities, which the Greeks had studiously avoided, and infinitely small ones, which the Greeks had skillfully circumvented, had to be contended with.

In other words, mathematicians were contributing concepts, rather than abstracting ideas from the real world. Nevertheless, these concepts were useful in physical investigations because (with the exception of complex numbers, which had yet to prove their worth) they had some ties to physical reality. Of course the Europeans were uneasy about the new types of numbers and the calculus notions without really discerning the cause of their concern. Yet as these concepts proved more and more useful in applications, they were at first grudgingly and later passively accepted. Familiarity bred not contempt but acceptability and even naturalness. After 1700, more and more notions, further removed from nature and springing full-blown from human minds, were to enter mathematics and be accepted with fewer qualms. For the genesis of its ideas mathematics gradually turned from the sensory to the intellectual faculties.

The incorporation of the calculus into the body of mathematics effected another change, in the very concept of mathematics, that subverted the ideal fashioned by the classical Greeks. We have already noted that the rise of algebra and the calculus introduced the problem of the logical foundations of these portions of mathematics and that this problem was not resolved. Throughout the century some mathematicians were upset by the abandonment of proof in the deductive sense, but their protests were drowned in the expanding content and use of algebra and the calculus; by the end of the century mathematicians had virtually dropped the requirement of clearly defined concepts and deductive proof. Rigorous axiomatic construction gave way to induction from particular examples, intuitive insights, loose geometrical evidence, and physical arguments. Since deductive proof had been the distinguishing feature of mathematics, the mathematicians were thus abandoning the hallmark of their subject.

In retrospect it is easy to see why they were forced into this position. As long as mathematicians derived their concepts from immediate experience, it was feasible to define the concepts and select the necessary axioms—though, at that, the logical basis for the theory of the integers that Euclid presented in Books VII to IX of the *Elements* was woefully deficient. But as they introduced concepts that no longer idealized immediate experiences, such as the irrational, negative, and complex numbers and the derivative and integral, they failed to recognize that these concepts were different in character, and so failed to realize that a basis for the axiomatic development other than self-evident truths was needed. It is true that the new concepts were far more subtle than the old ones; and the proper axiomatic basis, as we now know, could not have been readily erected.

How could critical mathematicians, well-versed in Greek mathematics, have been content to operate on a heuristic basis? They were concerned with major and in some cases pressing problems of science, and the mathematics they employed handled these problems. Rather than seeking full comprehension of the new creations or trying to erect the requisite deductive structure, they salved their consciences with their successes. An occasional recourse to philosophical or mystical doctrines succeeded in cloaking some difficulties so that they were no longer visible.

One new goal in particular characterizes the mathematics of the seventeenth and succeeding centuries—generality of methods and results. We have already noted the value placed on generality of method by Vieta, in his introduction of literal coefficients; by the projective geometers; by Fermat and Descartes in the exploration of curves; and by Newton and Leibniz in the treatment of functions. As far as generality of results is concerned, the accomplishments were limited. Many were just affirmations, such as that every polynomial equation of the nth degree has n roots, or that every equation of the second degree in x and y is a conic. Mathematical methods and notation were still too limited to permit the establishment of general results, but this became a goal of the mathematical efforts.

2. *Mathematics and Science*

Since classical Greek times, mathematics had been valued primarily for its role in the investigation of nature. Astronomy and music were constantly linked to mathematics, and mechanics and optics were certainly mathematical. However, the relationship of mathematics to science was altered in several ways by the work of the seventeenth century. First of all, because science, which was expanding enormously, had been directed by Galileo to use quantitative axioms and mathematical deduction (Chap. 16, sec. 3), the mathematical activity that was directly inspired by science became dominant.

Further, Galileo's injunction to seek mathematical description rather than causal explanation led to the acceptance of such a concept as the force of gravitation. This force and the laws of motion were the entire basis for Newton's system of mechanics. Since the only sure knowledge of gravitation was mathematical, mathematics became the substance of scientific theories. The insurgent seventeenth century found a qualitative world whose study was aided by mathematical abstractions and bequeathed a mathematical, quantitative world that subsumed under its mathematical laws the concreteness of the physical world.

Third, while the Greeks had employed mathematics freely in their science, there was, as long as the Euclidean basis sufficed for mathematics, a sharp distinction between it and science. Both Plato and Aristotle distinguished the two (Chap. 3, sec. 10 and Chap. 7, sec. 3), albeit in different ways; and Archimedes is especially clear about what is established mathematically and what is known physically. However, as the province of mathematics expanded, and mathematicians not only relied upon physical meanings to understand their concepts but accepted mathematical arguments because they gave sound physical results, the boundary between mathematics and science became blurred. Paradoxically, as science began to rely more and more upon mathematics to produce its physical conclusions, mathematics began to rely more and more upon scientific results to justify its own procedures.

The upshot of this interdependence was a virtual fusion of mathematics and vast areas of science. The compass of mathematics, as understood in the seventeenth century, may be seen from the *Cursus seu Mundus Mathematicus* (The Course or the World of Mathematics) by Claude-François Milliet Deschales (1621–78), published in 1674 and in an enlarged edition in 1690. Aside from arithmetic, trigonometry, and logarithms, he treats practical geometry, mechanics, statics, geography, magnetism, civil engineering, carpentry, stonecutting, military construction, hydrostatics, fluid flow, hydraulics, ship construction, optics, perspective, music, the design of firearms and cannons, the astrolabe, sundials, astronomy, calendar-reckoning, and horoscopy. Finally, he includes algebra, the theory of indivisibles, the theory of conics, and special curves such as the quadratrix and the spiral. This work was popular and esteemed. Though in the inclusion of some topics it reflects Renaissance interests, on the whole it presents a reasonable picture of the seventeenth- and even the eighteenth-century world of mathematics.

One might expect that the mathematicians would have been concerned to preserve the identity of their subject. But beyond the fact that they were obliged to depend upon physical meanings and results to defend their arguments, the greatest of the seventeenth- (and eighteenth-) century contributors to mathematics were either primarily scientists or at least equally concerned with both fields. Descartes, Huygens, and Newton, for example, were far greater physicists than mathematicians. Pascal, Fermat, and Leibniz were active in physics. In fact, it would be difficult to name an outstanding mathematician of the century who did not take a keen interest in science. As a consequence these men did not wish or seek to make any distinctions between the two fields. Descartes says in his *Rules for the Direction of the Mind* that mathematics is the science of order and measure and includes, besides algebra and geometry, astronomy, music, optics, and mechanics. Newton says in the *Principia*: "In mathematics we are to investigate the quantities of forces with their proportion consequent upon any conditions supposed; then, when we enter upon physics, we compare those proportions with the phenomena of Nature. . . ." Here physics refers to experimentation and observation. Newton's mathematics would be regarded as mathematical physics today.

3. Communication Among Mathematicians

Up to about 1550, mathematics was created by individuals or small groups headed by one or two prominent leaders. The results were communicated orally and occasionally written up in texts—which, however, were manuscripts. Since copies had also to be made by hand, they were scarce. By the seventeenth century printed books had become somewhat common, though even this improvement did not spread knowledge as widely as might be thought. Because the market for advanced mathematics was small, publishers had to charge high prices. Good printers were scarce. Publication was often followed by attacks on the authors from none-too-scrupulous opponents; it was not hard for such critics to find grounds for attack, especially because algebra and the calculus were not at all well grounded logically. Books in any case were not usually the answer for new creations because significant results did not warrant a book-sized publication.

As a consequence many mathematicians confined themselves to writing letters to friends in which they related their discoveries. Fearing that the letters would reach men who might take advantage of such unofficial documents, the writers often put their results in ciphers or anagrams, which they could then decode when challenged.

As more men began to participate in mathematical creation, the desire for exchange of information and for the stimulus of meeting people with the same intellectual interests resulted in the formation of scientific societies or academies. In 1601 the Accademia dei Lincei (Academy of the Lynx-like) was founded in Rome by young noblemen; it lasted thirty years. Galileo became a member in 1611. Another Italian society, the Accademia del Cimento (Academy of Experiments) was founded in Florence in 1657 as a formal organization of men who had been meeting in a laboratory founded by two members of the Medici family about ten years earlier. This academy included Vincenzo Viviani (1622–1703) and Torricelli, both pupils of Galileo, among its members. Unfortunately, the society was disbanded in 1667. In France, Desargues, Descartes, Gassendi, Fermat, and Pascal, among others, met privately under the leadership of Mersenne from 1630 on. This informal group was chartered by Louis XIV as the Académie Royale des Sciences in 1666, and its members were supported by the king. Paralleling what happened in France, an English group centered about John Wallis began in 1645 to hold meetings in Gresham College, London. These men emphasized mathematics and astronomy. The group was given a formal charter by

Charles II in 1662 and adopted the name of the Royal Society of London for the Promotion of Natural Knowledge. This society was concerned with putting mathematics and science to use and regarded dyeing, coinage, gunnery, the refinement of metals, and population statistics as subjects of interest. Finally, the Berlin Academy of Sciences, which Liebniz had advocated for some years, was opened in 1700 with Leibniz as its first president. In Russia, Peter the Great founded the Academy of Sciences of St. Petersburg in 1724.

The academies were important, not only in making possible direct contact and exchange of ideas, but because they also supported journals. The first of the scientific journals, though not sponsored by an academy, was the *Journal de Sçavans* or *Journal des Savants*, which began publication in 1665. This journal and the *Philosophical Transactions of the Royal Society*, which began publication in the same year, were the first journals to include mathematical and scientific articles. The French Académie des Sciences initiated the publication *Histoire de l'Académie Royale des Sciences avec les Mémoires de Mathématique et de Physique*. It also published the *Mémoires de Mathématique et de Physique Présentés à l'Académie Royale des Sciences par Divers Sçavans et Lus dans ses Assemblées*, also known as the *Mémoires des Savants Etrangers*. Another of the early scientific journals, the *Acta Eruditorum*, was begun in 1682 and, because it was published in Latin, soon acquired international readership. The Berlin Academy of Sciences sponsored the *Histoire de l'Académie Royale des Sciences et Belles-lettres* (whose title for some years was the *Miscellanea Berolinensia*).

The academies and their journals opened new outlets for scientific communication; these and later journals became the accepted medium for publication of new research. The academies furthered research in that most of them supported scientists. For example, Euler was supported by the Berlin Academy from 1741 to 1766 and Lagrange from 1766 to 1787. The St. Petersburg Academy supported Daniel and Nicholas Bernoulli at various times, and Euler from 1727 to 1741 and again from 1766 to his death in 1783. The founding of academies by the European governments marks also the official entry of governments into the area of science and the support of science. The usefulness of science had received recognition.

The institutions that a modern person would expect to play the major role in the creation and dissemination of knowledge—the universities—were ineffective. They were conservative and dogmatic, controlled by the official religions of the respective countries and very slow to incorporate new knowledge. On the whole they taught just a modicum of arithmetic, algebra, and geometry. Though there were some mathematicians at Cambridge University in the sixteenth century, from 1600 to 1630 there were none. In fact, in England during the early seventeenth century, mathematics did not form part of the curriculum. It was regarded as devilry. Wallis, who was born in 1616, says of the education common during his childhood, "Mathematics at that time with us was scarce looked on as academical but rather mechanical—as the business of tradesmen." He did attend Cambridge University and study mathematics there, but learned far more from independent study. Though prepared to be a professor of mathematics, he left Cambridge "because that study had died out there and no career was open to a teacher of that subject."

Professorships in mathematics were founded first at Oxford in 1619 and later at Cambridge. Prior to that, there had been only lecturers of low status. The Lucasian professorship at Cambridge, which Barrow was the first to hold, was founded in 1663. Wallis himself became a professor at Oxford in 1649 and held the chair until 1702. One obstacle to the enlistment of able professors was that they had to take holy orders, though exceptions were made, as in the case of Newton. The British universities generally (including also London, Glasgow, and Edinburgh) had roughly the same history: from about 1650 to 1750, they were somewhat active and then declined in activity until about 1825.

The French universities of the seventeenth and eighteenth centuries were inactive in mathematics. Not until the end of the eighteenth century, when Napoleon founded first-class technical schools, did they make any contribution. At German universities too, the mathematical activity of those two centuries was at a low level. Leibniz was isolated and, as we noted earlier, he railed against the teachings of the universities. The University of Göttingen was founded in 1731, but rose only slowly to any position of importance until Gauss became a professor there. The university centers of Geneva and Basel in Switzerland were exceptions in the period we are surveying; they could boast of the Bernoullis, Hermann, and others. The Italian universities were of some importance in the seventeenth century but lost ground in the eighteenth. When one notes that Pascal, Fermat, Descartes, Huygens, and Leibniz never taught at any university and that though Kepler and Galileo did teach for a while, they were court mathematicians for most of their lives, one sees how relatively unimportant the universities were.

4. The Prospects for the Eighteenth Century

The enormous seventeenth-century advances in algebra, analytic geometry, and the calculus; the heavy involvement of mathematics in science, which provided deep and intriguing problems; the excitement generated by Newton's astonishing successes in celestial mechanics; and the improvement in communications provided by the academies and journals all pointed to additional major developments and served to create immense exuberance about the future of mathematics.

There were obstacles to be overcome. The doubts as to the soundness of the calculus, the estrangement of the English and Continental mathematicians, the low state of the existing educational institutions, and the uncertainties of the support for careers in mathematics gave pause to young or would-be mathematicians. Nevertheless, the enthusiasm of the mathematicians was almost unbounded. They had glimpses of a promised land and were eager to press forward. They were, moreover, able to work in an atmosphere far more suitable for creation than at any time since 300 B.C. Classical Greek geometry had not only imposed restrictions on the domain of mathematics but had impressed a level of rigor for acceptable mathematics that hampered creativity. The seventeenth-century men had broken both of these bonds. Progress in mathematics almost demands a complete disregard of logical scruples; and, fortunately, the mathematicians now dared to place their confidence in intuitions and physical insights.

Bibliography

Hahn, Roger: *The Anatomy of a Scientific Institution: The Paris Academy of Sciences, 1666–1803*, University of California Press, 1971.

Hall, A. Rupert: *The Scientific Revolution, 1500–1800*, Longmans, Green, 1954, Chap. 7.

Hall, A. Rupert, and Marie Boas: *The Correspondence of Henry Oldenburg*, 4 vols., University of Wisconsin Press, 1968.

Hartley, Sir Harold: *The Royal Society: Its Origins and Founders*, The Royal Society, 1960.

Ornstein, M.: *The Role of Scientific Societies in the Seventeenth Century*, University of Chicago Press, 1938.

Purver, Margery: *The Royal Society, Concept and Creation*, Massachusetts Institute of Technology Press, 1967.

Wolf, Abraham: *A History of Science, Technology and Philosophy in the 16th and 17th Centuries*, 2nd ed., George Allen and Unwin, 1950, Chap. 4.

19
Calculus in the Eighteenth Century

> And therefore, whether the mathematicians of the present age
> act like men of science, in taking so much more pains to apply
> their principles than to understand them.   BISHOP BERKELEY

1. *Introduction*

The greatest achievement of the seventeenth century was the calculus. From this source there stemmed major new branches of mathematics: differential equations, infinite series, differential geometry, the calculus of variations, functions of complex variables, and many others. Indeed, the beginnings of some of these subjects were already present in the works of Newton and Leibniz. The eighteenth century was devoted largely to the development of some of these branches of analysis. But before this could be accomplished, the calculus itself had to be extended. Newton and Leibniz had created basic methods, but much remained to be done. Many new functions of one variable and functions of two or more variables had either to be recognized explicitly or created; the techniques of differentiation and integration had to be extended to some of the existing functions and to others yet to be introduced; and the logical foundation of the calculus was still missing. The first goal was to expand the subject matter of the calculus and this is the subject of the present chapter and the next one.

The eighteenth-century men did extend the calculus and founded new branches of analysis, though encountering in the process all the pangs, errors, incompleteness, and confusion of the creative process. The mathematicians produced a purely formal treatment of calculus and the ensuing branches of analysis. Their technical skill was unsurpassed; it was guided, however, not by sharp mathematical thinking but by intuitive and physical insights. These formal efforts withstood the test of subsequent critical examination and produced great lines of thought. The conquest of new domains of mathematics proceeds somewhat as do military conquests. Bold dashes into enemy territory capture strongholds. These incursions must then be followed up and supported by broader, more thorough and more cautious operations to secure what has been only tentatively and insecurely grasped.

It will be helpful, in appreciating the work and arguments of the eighteenth-century thinkers, to keep in mind that they did not distinguish between algebra and analysis. Because they did not appreciate the need for the limit concept and because they failed to recognize the problems introduced by the use of infinite series, they naively regarded the calculus as an extension of algebra.

The key figure in eighteenth-century mathematics and the dominant theoretical physicist of the century, the man who should be ranked with Archimedes, Newton, and Gauss, is Leonhard Euler (1707–83). Born near Basel to a preacher, who wanted him to study theology, Leonhard entered the university at Basel and completed his work at the age of fifteen. While at Basel he learned mathematics from John Bernoulli. He decided to pursue the subject and began to publish papers at eighteen. At nineteen he won a prize from the French Académie des Sciences for a work on the masting of ships. Through the younger Bernoullis, Nicholas (1695–1726) and Daniel (1700–82), sons of John, Euler in 1733 secured an appointment at the St. Petersburg Academy in Russia. He started as an assistant to Daniel Bernoulli but soon succeeded him as a professor. Though Euler passed some painful years (1733–41) under the autocratic government, he did an amazing amount of research, the results of which appeared in papers published by the St. Petersburg Academy. He also helped the Russian government on many physical problems. In 1741, at the call of Frederick the Great, he went to Berlin, where he remained until 1766. During this period, he gave lessons to the princess of Anhalt-Dessau, niece of the king of Prussia. These lessons, on a variety of subjects—mathematics, astronomy, physics, philosophy, and religion—were later published as the *Letters to a German Princess* and are still read with pleasure. At the request of Frederick the Great, Euler worked on state problems of insurance and the design of canals and waterworks. Even during his twenty-five years in Berlin he sent hundreds of papers to the St. Petersburg Academy and advised it on its affairs.

In 1766, at the request of Catherine the Great, Euler returned to Russia, although fearing the effect on his weakened sight (he had lost the sight of one eye in 1735) of the rigors of the climate there. He became, in effect, blind shortly after returning to Russia, and during the last seventeen years of his life was totally blind. Nevertheless, these years were no less fruitful than the preceding ones. Euler had a prodigious memory and knew by heart the formulas of trigonometry and analysis and the first six powers of the first 100 prime numbers, to say nothing of innumerable poems and the entire *Æneid*. His memory was so phenomenal that he could carry out in his head numerical calculations that competent mathematicians had difficulty doing on paper.

Euler's mathematical productivity is incredible. His major mathematical fields were the calculus, differential equations, analytic and differential geometry of curves and surfaces, the theory of numbers, series, and the calculus of variations. This mathematics he applied to the entire domain of physics. He created analytical mechanics (as opposed to the older geometrical mechanics) and the subject of rigid body mechanics. He calculated the perturbative effect of celestial bodies on the orbit of a planet and the paths of projectiles in resisting media. His theory of the tides and work on the design and sailing of ships aided navigation. In this area his *Scientia Navalis* (1749) and *Théorie complète de la construction et de la manœuvre des vaisseaux* (1773) are outstanding. He investigated the bending of beams and calculated the safety load of a column. In acoustics he studied the propagation of sound and musical consonance and dissonance. His three volumes on optical instruments contributed to the design of telescopes and microscopes. He was the first to treat the vibrations of light analytically and to deduce the equation of motion taking into account the dependence on the elasticity and density of the ether, and he obtained many results on the refraction and dispersion of light. In the subject of light he was the only physicist of the eighteenth century who favored the wave as opposed to the particle theory. The fundamental differential equations for the motion of an ideal fluid are his; and he applied them to the flow of blood in the human body. In the theory of heat, he (and Daniel Bernoulli) regarded heat as an oscillation of molecules, and his *Essay on Fire* (1738) won a prize. Chemistry, geography, and cartography also interested him, and he made a map of Russia. The applications were said to be an excuse for his mathematical investigations; but there can be no doubt that he liked both.

Euler wrote texts on mechanics, algebra, mathematical analysis, analytic and differential geometry, and the calculus of variations that were standard works for a hundred years and more afterward. The ones that will concern us in this chapter are the two-volume *Introductio in Analysin Infinitorum* (1748), the first connected presentation of the calculus and elementary analysis; the more comprehensive *Institutiones Calculi Differentialis* (1755); and the three-volume *Institutiones Calculi Integralis* (1768–70); all are landmarks. All of Euler's books contained some highly original features. His mechanics, as noted, was based on analytical rather than geometrical methods. He gave the first significant treatment of the calculus of variations. Beyond texts he published original research papers of high quality at the rate of about eight hundred pages a year during most of the years of his life. The quality of these papers may be judged from the fact that he won so many prizes for them that these awards became an almost regular addition to his income. Some of the books and four hundred of his research papers were written after he became totally blind. A current edition of his collected works, when completed, will contain seventy-four volumes.

Unlike Descartes or Newton before him or Cauchy after him, Euler did not open up new branches of mathematics. But no one was so prolific or could so cleverly handle mathematics; no one could muster and utilize the resources of algebra, geometry, and analysis to produce so many admirable results. Euler was superbly inventive in methodology and a skilled technician. One finds his name in all branches of mathematics: there are formulas of Euler, polynomials of Euler, Euler constants, Eulerian integrals, and Euler lines.

One might suspect that such a volume of activity could be carried on only at the expense of all other interests. But Euler married and fathered thirteen children. Always attentive to his family and its welfare he instructed his children and his grandchildren, constructed scientific games for them, and spent evenings reading the Bible to them. He also loved to express himself on matters of philosophy; but here he exhibited his only weakness, for which he was often chided by Voltaire. One day he was forced to confess that he had never studied any philosophy and regretted having believed that one could understand that subject without studying it. But Euler's spirit for philosophic disputes remained undampened and he continued to engage in them. He even enjoyed the sharp criticism he provoked from Voltaire.

Surrounded by universal respect—well merited by the nobility of his character—he could at the end of his life consider as his pupils all the mathematicians of Europe. On September 7, 1783, after having discussed the topics of the day, the Montgolfiers,[1] and the discovery of Uranus, according to the oft-cited words of J. A. N. C. de Condorcet, "He ceased to calculate and to live."

2. *The Function Concept*

As we have seen, the concept of a function and the simpler algebraic and transcendental functions were introduced and used during the seventeenth century. As Leibniz, James and John Bernoulli, L'Hospital, Huygens, and Pierre Varignon (1654–1722) took up problems such as the motion of a pendulum, the shape of a rope suspended from two fixed points, motion along curved paths, motion with fixed compass bearing on a sphere (the loxodrome), evolutes and involutes of curves, caustic curves arising in the reflection and refraction of light, and the path of a point on one curve that rolls on another, they not only employed the functions already known but arrived at more complicated forms of elementary functions. As a consequence of these researches and general work on the calculus, the elementary functions were fully recognized and developed practically in the manner in which we now have them. For example, the logarithm function, which originated

1. The Montgolfiers were two brothers who in 1783 first successfully made an ascent in a balloon filled with heated air.

as the relationship between terms in a geometric and arithmetic progression and was treated in the seventeenth century as the series obtained by integrating $1/(1 + x)$, (Chap. 17, sec. 2), was introduced on a new basis. The study of the exponential function by Wallis, Newton, Leibniz, and John Bernoulli showed that the logarithm function is the inverse of the exponential function whose properties are relatively simple. In 1742 William Jones (1675–1749) gave a systematic introduction to the logarithm function in this manner (Chap. 13, sec. 2). Euler in his *Introductio* defines the two functions as

$$e^x = \lim_{n \to \infty} \left(1 + \frac{x}{n}\right)^n, \qquad \log x = \lim_{n \to \infty} n(x^{1/n} - 1).$$

The mathematics of the trigonometric functions was also systematized. Newton and Leibniz gave series expansions for these functions. The development of the formulas for the functions of the sum and difference of two angles, that is $\sin (x + y)$, $\sin (x - y)$, and so forth, is due to a number of men, among them John Bernoulli and Thomas Fantet de Lagny (1660–1734); the latter wrote a paper in the *Mémoires* of the Paris Academy for 1703 on this subject. Frédéric-Christian Mayer (dates unknown), one of the first members of the St. Petersburg Academy of Sciences, then derived the common identities of analytical trigonometry on the basis of the sum and difference formulas.[2] Finally Euler, in a prize paper of 1748 on the subject of the inequalities in the motions of Jupiter and Saturn, gave the full systematic treatment of the trigonometric functions.[3] The periodicity of the trigonometric functions is clear in Euler's *Introductio* of 1748 wherein he also introduced the radian measure of angles.[4]

Study of the hyperbolic functions began when it was noticed that the area under a circle was given by $\int \sqrt{a^2 - x^2}\, dx$ whereas the area under the hyperbola is given by $\int \sqrt{x^2 - a^2}\, dx$. Since the two differ by a sign, and the area under a circle can be expressed by the trigonometric functions (let $x = a \sin \theta$), whereas the area under the hyperbola is related to the logarithm function, there should be a relation involving imaginary numbers between the trigonometric functions and the logarithm function. This idea was developed by many men (see sec. 3). Finally, J. H. Lambert studied the hyperbolic functions comprehensively.[5]

The concept of function had been formulated by John Bernoulli. Euler, at the very outset of the *Introductio*, defines a function as any analytical expression formed in any manner from a variable quantity and constants.

2. *Comm. Acad. Sci. Petrop.*, 2, 1727.
3. *Opera*, (2), 25, 45–157.
4. *Opera*, (1), 9, 217–39, 305–7.
5. *Hist. de l'Acad. de Berlin*, 24, 1768, 327–54, pub. 1770 = *Opera Math.*, 2, 245–69.

He includes polynomials, power series, and logarithmic and trigonometric expressions. He also defines a function of several variables. There follows the notion of an algebraic function, in which the operations on the independent variable involve only algebraic operations, which in turn are divided into two classes: the rational, involving only the four usual operations, and the irrational, involving roots. He then introduces the transcendental functions, namely, the trigonometric, the logarithmic, the exponential, variables to irrational powers, and some integrals.

The principal difference among functions, writes Euler, consists in the combination of variables and constants that compose them. Thus, he adds, the transcendental functions are distinguished from the algebraic functions because the former repeat an infinite number of times the combinations of the latter; that is, the transcendental functions could be given by infinite series. Euler and his contemporaries did not regard it as necessary to consider the validity of the expressions obtained by the unending application of the four rational operations.

Euler distinguished between explicit and implicit functions and between single-valued and multiple-valued functions, the latter being roots of higher-degree equations in two variables, the coefficients of which are functions of one variable. Here, he says, if a function, such as $\sqrt[3]{P}$, where P is a one-valued function, has real values for real values of the argument, then most often it can be included among the single-valued functions. From these definitions (which are not free of contradictions), Euler turns to rational integral functions or polynomials. Such functions with real coefficients, he affirms, can be decomposed into first and second degree factors with real coefficients (see sec. 4 and Chap. 25, sec. 2).

By a continuous function, Euler, like Leibniz and the other eighteenth-century writers, meant a function specified by an analytic formula; his word "continuous" really means "analytic" for us, except for an occasional discontinuity as in $y = 1/x$.[6] Other functions were recognized; the curves representing them were called "mechanical" or "freely drawn."

Euler's *Introductio* was the first work in which the function concept was made primary and used as a basis for ordering the material of the two volumes. Something of the spirit of this book may be gathered from Euler's remarks on the expansion of functions in power series.[7] He asserts that any function can be so expanded but then states that "if anyone doubts that every function can be so expanded then the doubt will be set aside by actually expanding functions. However in order that the present investigation extend over the widest possible domain, in addition to the positive integral powers

6. In Vol. 2, Chap. 1 of his *Introductio*, Euler introduces "discontinuous" or mixed functions which require different analytic expressions in different domains of the independent variable. But the concept plays no role in the work.
7. *Opera*, (1), 8, Chap. 4, p. 74.

of z, terms with arbitrary exponents will be admitted. Then it is surely indisputable that every function can be expanded in the form $Az^\alpha + Bz^\beta + Cz^\gamma + Dz^\delta + \cdots$ in which the exponents $\alpha, \beta, \gamma, \delta, \cdots$ can be any numbers." For Euler the possibility of expanding all functions into series was confirmed by his own experience and the experience of all his contemporaries. And in fact, it was true in those days that all functions given by analytic expressions were developable in series.

Though a controversy about the notion of a function did arise in connection with the vibrating-string problem (see Chap. 22) and caused Euler to generalize his own notion of what a function was, the concept that dominated the eighteenth century was still the notion of a function given by a single analytic expression, finite or infinite. Thus Lagrange, in his *Théorie des fonctions analytiques* (1797), defined a function of one or several variables as any expression useful for calculation in which these variables enter in any manner whatsoever. In *Leçons sur le calcul des fonctions* (1806), he says that functions represent different operations that must be performed on known quantities to obtain the values of unknown quantities, and that the latter are properly only the last result of the calculation. In other words, a function is a combination of operations.

3. *The Technique of Integration and Complex Quantities*

The basic method for integrating even somewhat complicated algebraic functions and the transcendental functions—the technique introduced by Newton—was to represent the functions as series and integrate term by term. Little by little, the mathematicians developed the techniques of going from one closed form to another.

The eighteenth-century use of the integral concept was limited. Newton had utilized the derivative and the antiderivative or indefinite integral, whereas Leibniz had emphasized differentials and the summation of differentials. John Bernoulli, presumably following Leibniz, treated the integral as the inverse of the differential, so that if $dy = f'(x)\,dx$, then $y = f(x)$. That is, the Newtonian antiderivative was chosen as the integral, but differentials were used in place of Newton's derivative. According to Bernoulli, the object of the integral calculus was to find from a given relation among *differentials* of variables, the relation of the variables. Euler emphasized that the derivative is the ratio of the evanescent differentials and said that the integral calculus was concerned with finding the function itself. The summation concept was used by him only for the approximate evaluation of integrals. In fact, all of the eighteenth-century mathematicians treated the integral as the inverse of the derivative or of the differential dy. The existence of an integral was never questioned; it was, of course, found explicitly in most of the applications made in the eighteenth century, so that the question did not occur.

A few instances of the development of the technique of integration are worth noting. To evaluate

$$\int \frac{a^2 \, dx}{a^2 - x^2}$$

James Bernoulli[8] had used the change of variable

$$x = a \frac{b^2 - t^2}{b^2 + t^2};$$

this converts the integral to the form

$$\int \frac{dt}{2at},$$

which is readily integrable as a logarithm function. John Bernoulli noticed in 1702 and published in the *Mémoires* of the Academy of Sciences of that year[9] the observation that

$$\frac{a^2}{a^2 - x^2} = \frac{a}{2} \left(\frac{1}{a + x} + \frac{1}{a - x} \right),$$

so that the integration can be performed at once. Thus the method of partial fractions was introduced. This method was also noted independently by Leibniz in the *Acta Eruditorum* of 1702.[10]

In correspondence between John Bernoulli and Leibniz, the method was applied to

$$\int \frac{dx}{ax^2 + bx + c}.$$

However, since the linear factors of $ax^2 + bx + c$ could be complex, the method of partial fractions led to integrals of the form

$$\int \frac{dx}{cx + d}$$

in which d at least was a complex number. Both Leibniz and John Bernoulli nevertheless integrated by using the logarithm rule and so involved the logarithms of complex numbers. Despite the confusion about complex numbers, neither hesitated to integrate in this manner. Leibniz said the presence of complex numbers did no harm.

John Bernoulli employed them repeatedly. In a paper published in 1702[11] he pointed out that, just as $adz/(b^2 - z^2)$ goes over by means of the

8. *Acta Erud.*, 1699 = *Opera*, 2, 868–70.
9. *Opera*, 1, 393–400.
10. *Math. Schriften*, 5, 350–66.
11. *Mém. de l'Acad. des Sci.*, Paris, 1702, 289 ff. = *Opera*, 1, 393–400.

substitution $z = b(t - 1)/(t + 1)$ into $adt/2bt$, so the differential

$$\frac{dz}{b^2 + z^2}$$

goes over by the substitution $z = \sqrt{-1}\, b(t - 1)/(t + 1)$ to

$$\frac{-dt}{\sqrt{-1}\, 2bt},$$

and the latter is the differential of the logarithm of an imaginary number. Since the original integral also leads to the arc tan function, Bernoulli had thus established a relation between the trigonometric and logarithmic functions.

However, these results soon raised lively discussions about the nature of the logarithms of negative and complex numbers. In his article of 1712[12] and in an exchange of letters with John Bernoulli during the years 1712–13, Leibniz affirmed that the logarithms of negative numbers are nonexistent (he said imaginary), while Bernoulli sought to prove that they must be real. Leibniz's argument was that positive logarithms are used for numbers greater than 1 and negative logarithms for numbers between 0 and 1. Hence there could be no logarithm for negative numbers. Moreover, if there were a logarithm for -1, then the logarithm of $\sqrt{-1}$ would be half of it; but surely there was no logarithm for $\sqrt{-1}$. That Leibniz could argue in this manner after having introduced the logarithms of complex numbers in integration is inexplicable. Bernoulli argued that since

(1) $$\frac{d(-x)}{-x} = \frac{dx}{x},$$

then $\log(-x) = \log x$; and since $\log 1 = 0$, so is $\log(-1)$. Leibniz countered that $d(\log x) = dx/x$ holds only for positive x. A second round of correspondence and disagreement took place between Euler and John Bernoulli during the years 1727–31. Bernoulli maintained his position, while Euler disagreed with it, though, at the time, he had no consistent position of his own.

The final clarification of what the logarithm of a complex number is became possible by virtue of related developments that are themselves significant and that led to the relationship between the exponential and the trigonometric functions. In 1714 Roger Cotes (1682–1716) published[13] a theorem on complex numbers, which, in modern notation, states that

(2) $$\sqrt{-1}\,\phi = \log_e(\cos\phi + \sqrt{-1}\sin\phi).$$

12. *Acta Erud.*, 1712, 167–69 = *Math. Schriften*, 5, 387–89.
13. *Phil. Trans.*, 29, 1714, 5–45.

In a letter to John Bernoulli of October 18, 1740, Euler stated that $y = 2 \cos x$ and $y = e^{\sqrt{-1}\,x} + e^{-\sqrt{-1}\,x}$ were both solutions of the same differential equation (which he recognized through series solutions), and so must be equal. He published this observation in 1743,[14] namely,

$$(3) \qquad \cos s = \frac{e^{\sqrt{-1}\,s} + e^{-\sqrt{-1}\,s}}{2}, \qquad \sin s = \frac{e^{\sqrt{-1}\,s} - e^{-\sqrt{-1}\,s}}{2\sqrt{-1}}.$$

In 1748 he rediscovered the result (2) of Cotes, which would also follow from (3).

While this development was taking place, Abraham de Moivre (1667–1754), who left France and settled in London when the Edict of Nantes protecting Huguenots was revoked, obtained, at least implicitly, the formula now named after him. In a note of 1722, which utilizes a result already published in 1707,[15] he says that one can obtain a relation between x and t, which represent the versines of two arcs (vers $\alpha = 1 - \cos \alpha$) that are in the ratio of 1 to n, by eliminating z from the two equations

$$1 - 2z^n + z^{2n} = -2z^n t \quad \text{and} \quad 1 - 2z + z^2 = -2zx.$$

In this result the de Moivre formula is implicit, for if one sets $x = 1 - \cos \phi$ and $t = 1 - \cos n\phi$, one can derive

$$(4) \qquad (\cos \phi \pm \sqrt{-1} \sin \phi)^n = \cos n\phi \pm \sqrt{-1} \sin n\phi.$$

For de Moivre n was an integer > 0. Actually, he never wrote the last result explicitly; the final formulation is due to Euler[16] and was generalized by him to all real n.

By 1747 Euler had enough experience with the relationship between exponentials, logarithms, and trigonometric functions to obtain the correct facts about the logarithms of complex numbers. In an article of 1749, entitled "De la controverse entre Mrs. [Messrs] Leibnitz et Bernoulli sur les logarithmes négatifs et imaginaires,"[17] Euler disagrees with Leibniz's counterargument that $d(\log x) = dx/x$ for positive x only. He says that Leibniz's objection, if correct, shatters the foundation of all analysis, namely, that the rules and operations apply no matter what the nature of the objects to which they are applied. He affirms that $d(\log x) = dx/x$ is correct for positive and negative x but adds that Bernoulli forgets that all one can conclude from (1) above is that $\log (-x)$ and $\log x$ differ by a constant. This constant must be $\log (-1)$ because $\log (-x) = \log (-1 \cdot x) = \log (-1) + \log x$. Hence, says Euler, Bernoulli has assumed, in effect, that $\log (-1) = 0$, but this must be proved. Bernoulli had given other arguments, which

14. *Miscellanea Berolinensia*, 7, 1743, 172–92 = *Opera*, (1), 14, 138–55.
15. *Phil. Trans.*, 25, 1707, 2368–71.
16. *Introductio*, Chap. 8.
17. *Hist. de l'Acad. de Berlin*, 5, 1749, 139–79, pub. 1751 = *Opera*, (1), 17, 195–232.

Euler also answers. For example, Bernoulli had argued that since $(-a)^2 = a^2$, then $\log (-a)^2 = \log a^2$ and so $2 \log (-a) = 2 \log a$ or $\log (-a) = \log a$. Euler counters that, since $(a\sqrt{-1})^4 = a^4$, then $\log a = \log (a\sqrt{-1}) = \log a + \log \sqrt{-1}$ and so in this case, presumably $\log \sqrt{-1}$ would have to be 0. But, Euler says, Bernoulli himself has proved in another connection that $\log \sqrt{-1} = \sqrt{-1}\, \pi/2$.

Leibniz had argued that since

$$(5) \qquad \log (1 + x) = x - \frac{1}{2} x^2 + \frac{1}{3} x^3 - \frac{1}{4} x^4 + \cdots$$

then for $x = -2$

$$\log (-1) = -2 - \frac{4}{2} - \frac{8}{3} \cdots,$$

from which one sees at least that $\log (-1)$ is not 0 (in fact, Leibniz had said that $\log (-1)$ is nonexistent). Euler's answer to this argument is that from

$$\frac{1}{1 + x} = 1 - x + x^2 - x^3 + x^4 \cdots,$$

for $x = -3$ one obtains

$$-\frac{1}{2} = 1 + 3 + 9 + 27 + \cdots$$

and for $x = 1$ one obtains

$$\frac{1}{2} = 1 - 1 + 1 - 1 + \cdots,$$

so that by adding the left and right sides

$$0 = 2 + 2 + 10 + 26 + \cdots.$$

Hence, says Euler, the argument from series proves nothing.

After refuting Leibniz and Bernoulli, Euler gives what is, by present standards, an incorrect argument. He writes

$$x = e^y = \left(1 + \frac{y}{i}\right)^i,$$

wherein i is an infinitely large number.[18] Then

$$x^{1/i} = 1 + \frac{y}{i}.$$

18. In his earlier work Euler used i (the first letter of *infinitus*) for an infinitely large quantity. After 1777 he used i for $\sqrt{-1}$.

and so

$$y = i(x^{1/i} - 1).$$

Since $x^{1/i}$, "the root with infinitely large exponent i," has infinitely many complex values, y has such values, and since $y = \log x$, then so does $\log x$. In fact Euler now writes [19]

$$x = a + b\sqrt{-1} = c(\cos \phi + \sqrt{-1} \sin \phi).$$

Letting c be e^C he has

$$x = e^C(\cos \phi + i \sin \phi) = e^C e^{\sqrt{-1}(\phi \pm 2\lambda\pi)},$$

and so

(6) $$y = \log x = C + (\phi \pm 2\lambda\pi)\sqrt{-1}$$

where λ is a positive integer or zero. Thus, Euler affirms, for positive real numbers only one value of the logarithm is real and the others are all imaginary; but for negative real numbers and for imaginary numbers all values of the logarithm are imaginary. Despite this successful resolution of the problem, Euler's work was not accepted. D'Alembert advanced meta-physical, analytical, and geometrical arguments to show that $\log (-1) = 0$.

4. Elliptic Integrals

John Bernoulli, having succeeded in integrating some rational functions by the method of partial fractions, asserted in the *Acta Eruditorum* of 1702 that the integral of any rational function need not involve any other tran-scendental functions than trigonometric or logarithmic functions. Since the denominator of a rational function can be an nth degree polynomial in x, the correctness of the assertion depended on whether any polynomial with real coefficients can be expressed as a product of first and second degree factors with real coefficients. Leibniz in his paper in the *Acta* of 1702 thought this was not possible and gave the example of $x^4 + a^4$. He pointed out that

$$
\begin{aligned}
x^4 + a^4 &= (x^2 - a^2\sqrt{-1})(x^2 + a^2\sqrt{-1}) \\
&= (x + a\sqrt{\sqrt{-1}})(x - a\sqrt{\sqrt{-1}}) \\
&\quad \times (x + a\sqrt{-\sqrt{-1}})(x - a\sqrt{-\sqrt{-1}})
\end{aligned}
$$

and, he said, no two of these four factors multiplied together give a quadratic factor with real coefficients. Had he been able to express the square root of

19. $a + b\sqrt{-1} = \sqrt{a^2 + b^2}\left(\dfrac{a}{\sqrt{a^2 + b^2}} + \sqrt{-1}\,\dfrac{b}{\sqrt{a^2 + b^2}}\right) = c(\cos \phi + i \sin \phi).$

$\sqrt{-1}$ and $-\sqrt{-1}$ as ordinary complex numbers he would have seen his error. Nicholas Bernoulli (1687–1759), a nephew of James and John, pointed out in the *Acta Eruditorum* of 1719 that $x^4 + a^4 = (a^2 + x^2)^2 - 2a^2x^2 = (a^2 + x^2 + ax\sqrt{2})(a^2 + x^2 - ax\sqrt{2})$; thus the function $1/(x^4 + a^4)$ could be integrated in terms of trigonometric and logarithmic functions.

The integration of irrational functions was also considered. James Bernoulli and Leibniz corresponded on this subject, because such integrands were being encountered frequently. James in 1694[20] was concerned with the elastica, the shape assumed by a thin rod when forces are applied to it—as, for example, at its ends. For one set of end-conditions he found that the equation of the curve is given by

$$dy = \frac{(x^2 + ab)\, dx}{\sqrt{a^4 - (x^2 + ab)^2}};$$

he could not integrate it in terms of the elementary functions. In connection with this work he introduced the lemniscate, whose rectangular coordinate equation is $(x^2 + y^2)^2 = a^2(x^2 - y^2)$ and whose polar coordinate equation is $r^2 = a^2 \cos 2\theta$. James tried to find the arc length, which from the vertex to an arbitrary point on the curve is given by

$$s = \int_0^r \frac{a^2}{\sqrt{a^4 - r^4}}\, dr,$$

and surmised that this integral, too, could not be integrated in terms of the elementary functions. Seventeenth-century attempts to rectify the ellipse, whose arc length is important for astronomy, led to the problem of evaluating

$$s = a \int_0^t \frac{(1 - k^2t^2)\, dt}{\sqrt{(1 - t^2)(1 - k^2t^2)}},$$

when the equation of the ellipse is taken as

$$\frac{x^2}{a^2} + \frac{y^2}{b^2} = 1$$

and in the integrand $k = (a^2 - b^2)/a^2$ and $t = x/a$. The problem of finding the period of a simple pendulum led to the integral

$$T = 4\sqrt{\frac{l}{g}} \int_0^{\pi/2} \frac{d\phi}{\sqrt{1 - k^2 \sin^2 \phi}}.$$

Such irrational integrands also occurred in finding the length of arc of a hyperbola, the trigonometric functions, and other curves. These integrals were already known by 1700; and others involving such integrands kept occurring throughout the eighteenth century. Thus Euler, in a definitive

20. *Acta Erud.*, 1694, 262–76 = *Opera*, 2, 576–600.

treatment of the elastica in the appendix to his 1744 book on the calculus of variations, obtained

$$dy = \frac{(\alpha + \beta x + \gamma x^2)\, dx}{\sqrt{a^4 - (\alpha + \beta x + \gamma x^2)^2}}$$

where the constants are for us immaterial. Like his predecessors, he resorted to series in order to obtain physical results.

The class of integrals comprised by the above examples is known as elliptic, the name coming from the problem of finding the length of arc of an ellipse. The eighteenth-century men did not know it, but these integrals cannot be evaluated in terms of the algebraic, circular, logarithmic, or exponential functions.[21]

The first investigations of elliptic integrals were directed not so much toward attempts to evaluate the integrals as toward the reduction of the more complicated ones to those arising in the rectification of ellipse and hyperbola. The reason for this approach is that from the geometrical point of view, which dominated at that time, the integrals for the elliptic and hyperbolic arcs seemed to be the simplest ones. A new point of view was opened up by the observation that the differential equation

(7) $$f(x)\, dx = \pm f(y)\, dy,$$

where $\int f(x)\, dx$ is a logarithmic function or an inverse trigonometric function, has as an integral an algebraic function of x and y; that is, in spite of the fact that it is impossible to find an algebraic integral of $f(x)\, dx$ itself, one can find an algebraic integral of the sum or difference of two such differentials. John Bernoulli then asked whether this property might not hold for other integrals than those of logarithms or inverse trigonometric functions.[22] He had discovered in 1698 that the difference of two arcs of the cubical parabola $(y = x^3)$ is integrable, a result he had obtained accidentally and regarded as most elegant. He then posed the more general problem of finding arcs of (higher) parabolas, ellipses, and hyperbolas whose sum or difference is equal to a rectilinear quantity, and affirmed that parabolic curves of the form $a^m y^p = b^n x^q$, $m + p = n + q$, are such that arcs of such curves, added or subtracted, equal a straight line. He gave no proof.

Count Giulio Carlo de' Toschi di Fagnano (1682–1766), an amateur mathematician, began in 1714 to take up these problems.[23] He considered the curves $y = (2/m + 2)x^{(m+2)/2}/a^{m/2}$ with m rational. For such curves it is rather straightforward to show (Fig. 19.1) that

$$\frac{m}{m+2} \int_{x_0}^{x_1} \frac{dx}{\sqrt{1 + (x/a)^m}} = \text{arc } PP_1 - (P_1 R_1 - PR)$$

21. This was proved by Liouville (*Jour. de l'Ecole Poly.*, 14, 1833, 124–93).
22. *Acta Erud.*, Oct. 1698, 462 ff. = *Opera*, 1, 249–53.
23. *Giornale dei Letterati d'Italia*, Vols. 19 ff.

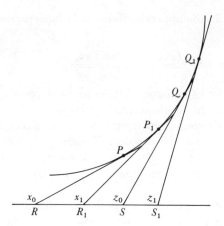

Figure 19.1

where x_0 and x_1 are the abscissas of R and R_1, and PR and P_1R_1 are the tangents at P and P_1 respectively. Likewise

$$\frac{m}{m+2} \int_{z_0}^{z_1} \frac{dz}{\sqrt{1+(z/a)^m}} = \text{arc } QQ_1 - (Q_1S_1 - QS).$$

If, therefore, for some relation of x to z we have

$$(8) \qquad \frac{dx}{\sqrt{1+(x/a)^m}} + \frac{dz}{\sqrt{1+(z/a)^m}} = 0,$$

then the sum of the two definite integrals would be 0, and we would have

$$(9) \qquad \text{arc } QQ_1 - \text{arc } PP_1 = (Q_1S_1 - QS) - (P_1R_1 - PR).$$

A solution of (8) for $m = 4$ is

$$(10) \qquad \frac{x}{a} \cdot \frac{z}{a} = 1.$$

Then for $m = 4$, on the curve $y = x^3/3a^2$, the difference of two arcs whose end-values x and z are related by (10) is expressible as a straight line segment. Fagnano also obtained integrals of (8) for $m = 6$ and $m = 3$.

Fagnano showed further that on the ellipse, as well as on the hyperbola, one can find infinitely many arcs whose difference can be expressed algebraically, even though the individual arcs cannot be rectified. Thus in 1716 he showed that the difference of any two elliptic arcs is algebraic. Analytically he had

$$(11) \qquad \frac{\sqrt{hx^2+l}}{\sqrt{fx^2+g}}\,dx + \frac{\sqrt{hz^2+l}}{\sqrt{fz^2+g}}\,dz = 0$$

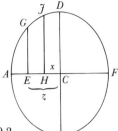

Figure 19.2

or, for brevity,

$$X\,dx + Z\,dz = 0$$

where h, l, f, g, x, and z satisfy the condition

(12) $$fhx^2z^2 + flx^2 + flz^2 + gl = 0.$$

Fagnano showed that

(13) $$\int X\,dx + \int Z\,dz = -\frac{hxz}{\sqrt{-fl}}.$$

What this means geometrically is that if $2a$ is the minor axis FA of an ellipse (Fig. 19.2), $CH = x$, $CE = z$, JH is the ordinate at H, and GE, the ordinate at E, then

(14) $$\text{arc } JD + \text{arc } DG = \frac{-hxz}{2a^2} + C.$$

(To identify this with the integrals, let p be the parameter [latus rectum] of the ellipse, let $p - 2a = h$, $l = 2a^3$, $f = -2a$, $g = 2a^3$. Then z is $a\sqrt{2a^3 - 2ax^2}/\sqrt{2a^3 + hx^2}$.) When $x = 0$, arc JD vanishes and the algebraic term in (14) vanishes. By (12), $z = a$, and so arc DG becomes arc DA. This is the value of C. Then one can say

$$\text{arc } JD + \text{arc } GD = \frac{-hxz}{2a^2} + \text{arc } DA$$

or

$$\text{arc } JD - \text{arc } GA = \frac{-hxz}{2a^2}.$$

One result from this work,[24] still called Fagnano's theorem and obtained in 1716, states the following: Let

$$\frac{x^2}{a^2} + \frac{y^2}{b^2} = 1$$

24. *Opere*, 2, 287–92.

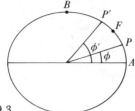

Figure 19.3

be the ellipse of eccentricity e and let $P(x, y)$ and $P'(x', y')$ be two points (Fig. 19.3) whose eccentric angles ϕ and ϕ' satisfy the condition

$$(15) \qquad\qquad \tan \phi \tan \phi' = \frac{b}{a}.$$

Then the theorem states that

$$(16) \qquad\qquad \text{arc } BP + \text{arc } BP' - \text{arc } BA = e^2 xx'/a.$$

The points P and P' may coincide (while satisfying [15]). For this common position F, called Fagnano's point, he showed that

$$(17) \qquad\qquad \text{arc } BF - \text{arc } AF = a - b.$$

From 1714 on Fagnano also concerned himself with the rectification of the lemniscate by means of elliptic and hyperbolic arcs.

In 1717 and 1720 Fagnano integrated other combinations of differentials. Thus he showed that

$$(18) \qquad\qquad \frac{dx}{\sqrt{1 - x^4}} = \frac{dy}{\sqrt{1 - y^4}}$$

has the integral

$$(19) \qquad\qquad x = -\sqrt{\frac{1 - y^2}{1 + y^2}}$$

or

$$(20) \qquad\qquad x^2 + y^2 + x^2 y^2 = 1.$$

One way of stating the result is: Between two integrals that express arcs of a lemniscate (for which $a = 1$), an algebraic relation exists, even though each integral separately is a transcendental function of a new kind.

Fagnano then proceeded to establish a number of similar relations in order to obtain special results about the lemniscate.[25] For example, he

25. *Giornale dei Letterati d'Italia*, 30, 1718, 87 ff. = *Opere*, 2, 304–13.

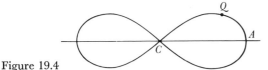

Figure 19.4

showed that if

(21)
$$\frac{dx}{\sqrt{1 - x^4}} = \frac{2\,dy}{\sqrt{1 - y^4}}$$

then

(22)
$$\frac{\sqrt{1 - y^4}}{y\sqrt{2}} = \frac{\sqrt{1 - x}}{\sqrt{1 + x}},$$

or by solving for x,

(23)
$$x = \frac{-1 + 2y^2 + y^4}{1 + 2y - y^4}.$$

From a variety of such results Fagnano showed how to find the points on the lemniscate (the values of r in $r^2 = a^2 \cos 2\theta$) that divide the quadrant, that is, the arc CQA in Figure 19.4, into n equal parts for certain values of n. He also showed how, given an arc CS, one can find the point I on this arc that divides CS into two equal parts. Further, he found the points on arc CQA which, when joined to C, divide the area under CQA and the horizontal axis into two, three, and five parts; and given the chords that divide this area into n equal parts, he found the chords bisecting each of these parts.

Thus Fagnano had done more than answer Bernoulli's question; he had shown that the same remarkable algebraic property that characterized the integrals representing logarithmic and inverse trigonometric functions held for at least certain classes of elliptic integrals.

About 1750 Euler noted Fagnano's work on the ellipse, hyperbola, and lemniscate and began a series of investigations of his own. In a paper, "Observationes de Comparatione Arcuum Curvarum Irrectificabilium,"[26] Euler, after repeating some of Fagnano's work, showed how to divide the area of a quadrant of the lemniscate into $n + 1$ parts if it has already been divided into n parts. He then points out that his and Fagnano's work has furnished some useful results on integration, for the equation (18) has, besides the obvious integral $x = y$, the additional particular integral

$$x = -\sqrt{(1 - y^2)/(1 + y^2)}.$$

26. *Novi Comm. Acad. Sci. Petrop.*, 6, 1756/7, 58–84, pub. 1761 = *Opera*, (1), 20, 80–107.

In his paper "De Integratione Æquationis Differentialis $m\,dx/\sqrt{1-x^4}$ $= n\,dy/\sqrt{1-y^4}$,"[27] Euler takes Fagnano's results as his point of departure. The integrals Fagnano had obtained for most of the differential equations he had considered were particular integrals. These were algebraic; but the complete integral might well be transcendental. Euler decided to look for complete integrals in algebraic form. He started with (18), but hoped to obtain the complete integral of

$$(24) \qquad\qquad \frac{m\,dx}{\sqrt{1-x^4}} = \frac{n\,dy}{\sqrt{1-y^4}}.$$

Here m/n is rational and the equation expresses the problem of finding two arcs of a lemniscate that have this ratio to each other. Euler says that he has been led by trial to believe that (24) has a complete integral that can be expressed algebraically when m/n is rational.

From the investigations of Fagnano it followed that equation (18) is satisfied by the particular integral (19) or (20). The integral of each side of (18) is an arc of a lemniscate with half-axis 1 and abscissa x, and the integration of the ordinary differential equation (18) amounts to finding two arcs that are of equal length. Euler had pointed out that $x = y$ is another particular integral of (18). The complete integral should be such that it reduces to each of these particular integrals for special values of the arbitrary constant. Guided by these facts, Euler found that the complete integral of (18) is

$$(25) \qquad\qquad x^2 + y^2 + c^2 y^2 x^2 = c^2 + 2xy\sqrt{1-c^4}$$

or

$$(26) \qquad\qquad x = \frac{y\sqrt{1-c^4} \pm c\sqrt{1-y^4}}{1+c^2 y^2}$$

where c is an arbitrary constant. Of course, given (25) we can readily verify that it is the complete integral of (18).

Implicit in the result (25) is what is often called Euler's addition theorem for these simple elliptic integrals. It is clear by straight differentiation that

$$(27) \qquad\qquad \int_0^x \frac{dx}{\sqrt{1-x^4}} = \int_0^y \frac{dx}{\sqrt{1-x^4}} + \int_0^c \frac{dx}{\sqrt{1-x^4}},$$

where c is a constant, is also a complete integral of (18). Hence x, y, and c must be related by (25). Thus the addition theorem says that if (27) holds for the elliptic integrals therein, then the upper limit x is an algebraic symmetric function, namely (26), of the arbitrarily chosen upper limits y and c of the other two integrals. The addition theorem applies to more general integrals, as we shall see.

27. *Novi Comm. Acad. Sci. Petrop.*, 6, 1756/7, 37–57, pub. 1761 = *Opera*, (1), 20, 58–79.

By utilizing these results (25) and (27), it is rather straightforward to show that if

$$(28) \qquad \int_0^y \frac{dx}{\sqrt{1 - x^4}} = n \int_0^x \frac{dx}{\sqrt{1 - x^4}},$$

then y is an algebraic function of x. This result is called Euler's multiplication theorem for the elliptic integral $\int_0^x dx/\sqrt{1 - x^4}$. From this result the complete integral of equation (28) follows; the important point is that it is an algebraic equation in x, y, and an arbitrary constant c. Euler shows how the complete integral might be obtained but does not give it explicitly.

In the same paper of 1756–57 and in Volume 7 of the same journal,[28] Euler tackled more general elliptic integrals. He gives the following result, which he says he obtained by trial and error. If one differentiates

$$(29) \qquad \alpha + 2\beta(x + y) + \gamma(x^2 + y^2) + 2xy + 2\varepsilon xy(x + y) + \zeta x^2 y^2 = 0,$$

then the differential equation can be put in the form

$$(30) \qquad \frac{dx}{\sqrt{X}} + \frac{dy}{\sqrt{Y}} = 0$$

where X and Y are polynomials of the fourth degree of which four coefficients (the same in X and Y) can be expressed in terms of the five in (29) with the help of an arbitrary constant. Then (29) is the complete integral of (30), and when (30) is specialized to (18), then (29) becomes (25). Euler remarks that it is wonderful that, even though an integral of $dx/\sqrt{X}$ cannot be obtained in terms of circular or logarithmic functions, the equation (30) is satisfied by an algebraic relation. He then generalizes the results to

$$(31) \qquad \frac{m\,dx}{\sqrt{X}} = \frac{n\,dy}{\sqrt{Y}}, \qquad m/n \text{ rational},$$

where X and Y are fourth degree polynomials with the same coefficients. The result is also in Euler's *Institutiones Calculi Integralis*,[29] where he shows what the results mean geometrically in terms of ellipse, hyperbola, and lemniscate.

From these results Euler was able to proceed with what is now called the addition theorem for elliptic integrals of the first kind. Consider the elliptic integral

$$(32) \qquad \int \frac{dx}{\sqrt{R(x)}}$$

28. *Novi Comm. Acad. Sci. Petrop.*, 7, 1758/9, 3–48, pub. 1761 = *Opera*, (1), 20, 153–200.
29. Vol. 1, Sec. 2, Chap. 6 = *Opera*, (1), 11, 391–423.

where $R(x) = Ax^4 + Bx^3 + Cx^2 + Dx + E$. Then the addition theorem states that the equation

$$(33) \qquad \frac{dx}{\sqrt{R(x)}} = \frac{dy}{\sqrt{R(y)}}$$

is satisfied by a certain definite algebraic equation in x and y such that y is rationally expressible in terms of x, the corresponding value of $\sqrt{R(x)}$, the arbitrary constants x_0 and y_0, and the corresponding values of $\sqrt{R(x_0)}$ and $\sqrt{R(y_0)}$. Also y takes on the arbitrarily preassigned value y_0 when x takes on the arbitrary value x_0.

This result leads readily to another theorem, which may be more enlightening. If the sum or the difference of two elliptic integrals of the form

$$(34) \qquad \int \frac{dx}{\sqrt{R(x)}}$$

is equated to a third integral of the same form, and if further the lower limit of integration and the coefficients under the radical are the same for all three integrals, then the upper limit of integration of the third integral is an algebraic function of the two other upper limits, the common lower limit, and the corresponding values of $\sqrt{R(x)}$ at the two upper limits and the lower limit.

Euler went further. Just as Fagnano's treatment of the difference of two lemniscate arcs led him to the general elliptic integral of the first kind, so Fagnano's treatment of the difference of two elliptic arcs (see [11]) led Euler to an addition theorem for integrals of a second kind.[30] He expressed regret that his methods were not extensible to higher roots than the square root or to radicands of higher than fourth degree. He also saw a great defect in his work in that he had not obtained his complete algebraic integrals by a general method of analysis; consequently his results were not naturally related to other parts of the calculus.

The definitive work on elliptic integrals was done by Adrien-Marie Legendre (1752–1833), a professor at the Ecole Militaire who served on several governmental committees; he later became an examiner of students at the Ecole Polytechnique. Up to his death in 1833 he never ceased to work with passion and regularity. His name lives in a great number of theorems, very varied, because he tackled most diverse questions. But he had neither the originality nor the profundity of Lagrange, Laplace, and Monge. Legendre's work gave birth to very important theories, but only after it was taken over by more powerful minds. He ranks just after the three contemporaries just mentioned.

30. *Novi Comm. Acad. Sci. Petrop.*, 7, 1758/9, 3–48, pub. 1761 = *Opera*, (1), 20, 153–200 and *Inst. Cal. Integ.*, 1, ¶645 = *Opera*, (1), 11, ¶645.

Euler's addition theorems were the main results of the theory of elliptic integrals when Legendre took up the subject in 1786. For four decades he was the only man who added investigations concerning these integrals to the literature. He devoted two basic papers to the subject,[31] then wrote the *Exercices de calcul intégral* (3 vols., 1811, 1817, 1826), the *Traité des fonctions elliptiques*[32] (2 vols., 1825–26), and three supplements giving accounts of the work of Abel and Jacobi in 1829 and 1832. Euler's results, like Fagnano's, were tied to geometrical considerations; Legendre concentrated on the analysis proper.

Legendre's chief result, which is in his *Traité*, was to show that the general elliptic integral

$$(35) \qquad \int \frac{P(x)}{\sqrt{R(x)}} \, dx$$

where $P(x)$ is any rational function of x and $R(x)$ is the usual general fourth degree polynomial, can be reduced to three types:

$$(36) \qquad \int \frac{dx}{\sqrt{1 - x^2}\sqrt{1 - l^2 x^2}},$$

$$(37) \qquad \int \frac{x^2 \, dx}{\sqrt{1 - x^2}\sqrt{1 - l^2 x^2}},$$

$$(38) \qquad \int \frac{dx}{(x - a)\sqrt{1 - x^2}\sqrt{1 - l^2 x^2}}.$$

Legendre designated these three types as elliptic integrals of the first, second, and third kinds, respectively.

He also showed that by further transformations these three integrals can be reduced to the three forms

$$(39) \qquad F(k, \phi) = \int_0^\phi \frac{d\phi}{\sqrt{1 - k^2 \sin^2 \phi}}, \qquad\qquad 0 < k < 1$$

$$(40) \qquad E(k, \phi) = \int_0^\phi \sqrt{1 - k^2 \sin^2 \phi} \, d\phi, \qquad\qquad 0 < k < 1$$

$$(41) \qquad \pi(n, k, \phi) = \int_0^\phi \frac{d\phi}{(1 + n \sin^2 \phi)\sqrt{1 - k^2 \sin^2 \phi}}, \qquad 0 < k < 1$$

where n is any constant. In these forms one sees that the values of the integrals from $\phi = 0$ to $\phi = \pi/2$ are repeated but in reverse order from $\phi = \pi/2$ to

31. *Hist. de l'Acad. des Sci., Paris*, 1786, 616–43 and 644–83.
32. The use of the word "function" in this text is misleading. He studied elliptic integrals and at times those with variable upper limits. These are, of course, functions of the upper limits. But the term "elliptic functions" refers today to the functions introduced later by Abel and Jacobi.

$\phi = \pi$. The notation $\Delta(k, \phi)$ for $\sqrt{1 - k^2 \sin^2 \phi}$ was also introduced by Legendre.

These forms can be converted, by the change of variable $x = \sin \phi$, to the Jacobi forms

$$(42) \qquad F(k, x) = \int_0^x \frac{dx}{\sqrt{1 - x^2}\sqrt{1 - k^2 x^2}}$$

$$(43) \qquad E(k, x) = \int_0^x \sqrt{\frac{1 - k^2 x^2}{1 - x^2}}\, dx$$

$$(44) \qquad \pi(n, k, x) = \int_0^x \frac{dx}{(1 + nx^2)\sqrt{(1 - x^2)(1 - k^2 x^2)}}.$$

The quantity k is called the modulus of each elliptic integral. If the limits of integration are $\phi = \pi/2$ or $x = 1$, then the integrals are called complete; otherwise, incomplete.

There was much merit in Legendre's work on elliptic integrals. He drew many inferences, previously unstated, from the work of his predecessors and organized the mathematical subject; but he did not add any basic ideas, nor did he attain the new insight of Abel and Jacobi (Chap. 27, sec. 6) who inverted these integrals and so conceived of the elliptic *functions*. Legendre did get to know the work of Abel and Jacobi and praised them, with much humility and, probably, some bitterness. In devoting the supplements to his 1825 work to their new ideas, he understood very well that this material threw into the shade all he had done on the subject. He had overlooked one of the great discoveries of his epoch.

5. *Further Special Functions*

The elliptic indefinite integrals are new transcendental functions. As the analytical work of the eighteenth century developed, more transcendental functions were obtained. Of these the most important is the gamma function, which arose from work on two problems, interpolation theory and anti-differentiation. The problem of interpolation had been considered by James Stirling (1692–1770), Daniel Bernoulli, and Christian Goldbach (1690–1764). It was posed to Euler and he announced his solution in a letter of October 13, 1729, to Goldbach.[33] A second letter, of January 8, 1730, brought in the integration problem.[34] In 1731 Euler published results on both in a paper, "De Progressionibus. . . ."[35]

33. Fuss, *Correspondance*, 1, 3–7.
34. Fuss, *Correspondance*, 1, 11–18.
35. *Comm. Acad. Sci. Petrop.*, 5, 1730/1, 36–57, pub. 1738 = *Opera*, (1), 14, 1–24.

The interpolation problem was to give meaning to $n!$ for nonintegral values of n. Euler noted that

$$(45) \qquad n! = \left[\left(\tfrac{2}{1}\right)^n \frac{1}{n+1}\right]\left[\left(\tfrac{3}{2}\right)^n \frac{2}{n+2}\right]\left[\left(\tfrac{4}{3}\right)^n \frac{3}{n+3}\right] \cdots$$

$$= \prod_{k=1}^{\infty} \left(\frac{k+1}{k}\right)^n \frac{k}{k+n}.$$

The equation is seen to be formally correct if one cancels common factors in the infinite product. However, this analytic expression for $n!$, unlike the basic definition $n(n-1)\cdots 2\cdot 1$, has a meaning for all n except negative integral values. Euler noticed that for $n = 1/2$ the right side yields, after a bit of manipulation, the infinite product of Wallis

$$(46) \qquad \frac{\pi}{2} = \left(\frac{2\cdot 2}{1\cdot 3}\right)\left(\frac{4\cdot 4}{3\cdot 5}\right)\left(\frac{6\cdot 6}{5\cdot 7}\right)\left(\frac{8\cdot 8}{7\cdot 9}\right)\cdots.$$

In the notation $\Gamma(n+1) = n!$, introduced later by Legendre, Euler also showed that $\Gamma(n+1) = n\Gamma(n)$ and so obtained $\Gamma(3/2)$, $\Gamma(5/2)$, and so forth.

Euler could have used (45) as his generalization of the factorial concept. In fact, it is often introduced today in the equivalent form, which Euler also gave, namely,

$$(47) \qquad \lim_{m\to\infty} \frac{m!\,(m+1)^n}{(n+1)(n+2)\cdots(n+m)}.$$

But the connection with Wallis's result led Euler to take up an integral already considered by Wallis, namely,

$$(48) \qquad \int_0^1 x^e (1-x)^n \, dx$$

wherein for Euler e and n are now arbitrary. Euler evaluated this integral by expanding $(1-x)^n$ by the binomial theorem and obtained

$$(49) \quad \int_0^1 x^e(1-x)^n \, dx$$

$$= \frac{1}{e+1} - \frac{n}{1(e+2)} + \frac{n(n-1)}{1\cdot 2(e+3)} - \frac{n(n-1)(n-2)}{1\cdot 2\cdot 3(e+4)} + \cdots.$$

For $n = 0, 1, 2, 3, \ldots$ the sums on the right side are respectively

$$(50) \quad \frac{1}{e+1}, \frac{1}{(e+1)(e+2)}, \frac{1\cdot 2}{(e+1)(e+2)(e+3)},$$

$$\frac{1\cdot 2\cdot 3}{(e+1)(e+2)(e+3)(e+4)}, \ldots.$$

Hence for positive integral n Euler found that

(51) $$\int_0^1 x^e(1 - x)^n \, dx = \frac{n!}{(e + 1)(e + 2) \cdots (e + n + 1)}.$$

Euler now sought an expression for $n!$ with n arbitrary. By a series of transformations that would not be wholly acceptable to us today, Euler arrived at

(52) $$n! = \int_0^1 (-\log x)^n \, dx.$$

This integral has meaning for almost all arbitrary n. It is called the second Eulerian integral, or, as Legendre later called it, the gamma function, and is denoted by $\Gamma(n + 1)$. [Gauss let $\pi(n) = \Gamma(n + 1)$]. Later, in 1781 (pub. 1794), Euler gave the modern form, which is obtained from (52) by letting $t = -\log x$,

(53) $$\Gamma(n + 1) = \int_0^\infty x^n e^{-x} \, dx.$$

The integral (48), which Legendre called the first Eulerian integral, became standardized as the beta function,

(54) $$B(m, n) = \int_0^1 x^{m-1}(1 - x)^{n-1} \, dx.$$

Euler discovered the relation between the two integrals,[36] namely,

$$B(m, n) = \frac{\Gamma(m)\Gamma(n)}{\Gamma(m + n)}.$$

In his *Exercices de calcul intégral*, Legendre made a profound study of the Eulerian integrals and arrived at the duplication formula

(55) $$\Gamma(2x) = (2\pi)^{-1/2} 2^{2x - (1/2)} \Gamma(x) \Gamma\left(x + \frac{1}{2}\right).$$

Gauss studied the gamma function in his work on the hypergeometric function[37] and extended Legendre's result to what is called the multiplication formula:

(56) $$\Gamma(nx) = (2\pi)^{(1-n)/2} n^{nx - (1/2)} \Gamma(x) \Gamma\left(x + \frac{1}{n}\right) \Gamma\left(x + \frac{2}{n}\right)$$

$$\cdots \Gamma\left(x + \frac{n - 1}{n}\right).$$

36. *Novi Comm. Acad. Sci. Petrop.*, 16, 1771, 91–139, pub. 1772 = *Opera*, (1), 17, 316–57.
37. *Comm. Soc. Gott.*, II, 1813 = *Werke*, 3, 123–162, p. 149 in part.

6. *The Calculus of Functions of Several Variables*

The development of the calculus of functions of two and three variables took place early in the century. We shall note just a few of the details.

Though Newton derived from polynomial equations in x and y, that is $f(x, y) = 0$, expressions which we now obtain by partial differentiation of f with respect to x and y, this work was not published. James Bernoulli also used partial derivatives in his work on isoperimetric problems, as did Nicholas Bernoulli (1687–1759) in a paper in the *Acta Eruditorum* of 1720 on orthogonal trajectories. However, it was Alexis Fontaine des Bertins (1705–71), Euler, Clairaut, and d'Alembert who created the theory of partial derivatives.

The difference between an ordinary and a partial derivative was at first not explicitly recognized, and the same symbol d was used for both. Physical meaning dictated that in the case of functions of several independent variables the desired derivative was to take account of the change in one variable only.

The condition that $dz = p\,dx + q\,dy$, where p and q are functions of x and y, be an exact differential, that is, be obtainable from $z = f(x, y)$ by forming the differential $dz = (\partial f/\partial x)\,dx + (\partial f/\partial y)\,dy$ was obtained by Clairaut.[38] His result is that $p\,dx + q\,dy$ is an exact differential, that is, there is an f such that $\partial f/\partial x = p$ and $\partial f/\partial y = q$, if and only if $\partial p/\partial y = \partial q/\partial x$.

The major impetus to work with derivatives of functions of two or more variables came from the early work on partial differential equations. Thus a calculus of partial derivatives was supplied by Euler in a series of papers devoted to problems of hydrodynamics. In a paper of 1734[39] he shows that if $z = f(x, y)$ then

$$\frac{\partial^2 z}{\partial x\,\partial y} = \frac{\partial^2 z}{\partial y\,\partial x}.$$

In other papers written from 1748 to 1766, he treats change of variable, inversion of partial derivatives, and functional determinants. In works of 1744 and 1745 on dynamics, d'Alembert extended the calculus of partial derivatives.

Multiple integrals were really involved in Newton's work in the *Principia* on the gravitational attraction exerted by spheres and spherical shells on particles. However, Newton used geometrical arguments. In the eighteenth century Newton's work was cast in analytical form and extended. Multiple integrals appear in the first half of the century and were used to denote the solution of $\partial^2 z/\partial x\,\partial y = f(x, y)$. They were also used, for example, to determine the gravitational attraction exerted by a lamina on particles.

38. *Mém. de l'Acad. des Sci.*, Paris, 1739, 425–36, and 1740, 293–323.
39. *Comm. Acad. Sci. Petrop.*, 7, 1734/5, 174–93, pub. 1740 = *Opera*, (1), 22, 36–56.

Thus the attraction of an elliptical lamina of thickness δc on a point directly over the center and distance c units is a constant times the integral

$$\delta c \int \int \frac{c \, dx \, dy}{(c^2 + x^2 + y^2)^{3/2}}$$

taken over the ellipse $(x^2/a^2) + (y^2/b^2) = 1$. This was evaluated by Euler in 1738 by repeated integration.[40] He integrated with respect to y and used infinite series to expand the new integrand as a function of x.

By 1770 Euler did have a clear conception of the definite double integral over a bounded domain enclosed by arcs, and he gave the procedure for evaluating such integrals by repeated integration.[41] Lagrange, in his work on the attraction by ellipsoids of revolution,[42] expressed the attraction as a triple integral. Finding it difficult to evaluate in rectangular coordinates, he transformed to spherical coordinates. He introduced

$$x = a + r \sin \phi \cos \theta$$
$$y = b + r \sin \phi \sin \theta$$
$$z = c + r \cos \phi,$$

where a, b, and c are the coordinates of the new origin, θ is the longitude angle, ϕ the colatitude, and $0 \leq \phi \leq \pi$, $0 \leq \theta \leq 2\pi$. The essence of the transformation of the integral is to replace $dx \, dy \, dz$ by $r^2 \sin \theta \, d\theta \, d\phi \, dr$. Thus Lagrange began the subject of the transformation of multiple integrals. In fact, he gave the general method, though not very clearly. Laplace also gave the spherical coordinate transformation almost simultaneously.[43]

7. The Attempts to Supply Rigor in the Calculus

Accompanying the expansion of the concepts and techniques of the calculus were efforts to supply the missing foundations. The books on the calculus that appeared after Newton's and Leibniz's unsuccessful attempts to explain the concepts and justify the procedures tried to clear up the confusion but actually added to it.

Newton's approach to the calculus was potentially easier to rigorize than Leibniz's, though the latter's methodology was more fluid and more convenient for application. The English thought they could secure the rigor for both approaches by trying to tie them to Euclid's geometry. But they confused Newton's moments (his indivisible increments) with his fluxions, which dealt with continuous variables. The Continentals, following Leibniz, worked with differentials and tried to rigorize this concept. Differentials

40. *Comm. Acad. Sci. Petrop.*, 10, 1738, 102–15, pub. 1747 = *Opera*, (2), 6, 175–88.
41. *Novi Comm. Acad. Sci. Petrop.*, 14, 1769, 72–103, pub. 1770 = *Opera*, (1), 17, 289–315.
42. *Nouv. Mém. de l'Acad. de Berlin*, 1773, 121–48, pub. 1775 = *Œuvres*, 3, 619–58.
43. *Mém. des sav. étrangers*, 1772, 536–44, pub. 1776 = *Œuvres*, 8, 369–477.

were treated either as infinitesimals, that is, quantities not zero but not of any finite size, or sometimes, as zero.

Brook Taylor (1685–1731), who was secretary of the Royal Society from 1714 to 1718, in his *Methodus Incrementorum Directa et Inversa* (1715), sought to clarify the ideas of the calculus but limited himself to algebraic functions and algebraic differential equations. He thought he could always deal with finite increments but was vague on their transition to fluxions. Taylor's exposition, based on what we would call finite differences, failed to obtain many backers because it was arithmetical in nature when the British were trying to tie the calculus to geometry or to the physical notion of velocity.

Some idea of the obscurity of the eighteenth-century efforts and their lack of success may also be gained from Thomas Simpson's (1710–61) *A New Treatise on Fluxions* (1737). After some preliminary definitions, he defines a fluxion thus: "The magnitude by which any flowing quantity would be uniformly increased in a given portion of time with the generating celerity at any proposed position or instant (was [were] it from thence to continue invariable) is the fluxion of the said quantity at that position or instant." In our language Simpson is defining the derivative by saying that it is (dy/dt) Δt. Some authors gave up. The French mathematician Michel Rolle at one point taught that the calculus was a collection of ingenious fallacies.

The eighteenth century also witnessed new attacks on the calculus. The strongest was made by Bishop George Berkeley (1685–1753), who feared the growing threat to religion posed by mechanism and determinism. In 1734 he published *The Analyst, Or A Discourse Addressed to an Infidel Mathematician. Wherein It is examined whether the Object, Principles, and Inferences of the modern Analysis are more distinctly conceived, or more evidently deduced, than Religious Mysteries and Points of Faith. "First cast out the beam out of thine own Eye; and then shalt thou see clearly to cast out the mote out of thy brother's Eye."* (The "infidel" was Edmond Halley.)[44]

Berkeley rightly pointed out that the mathematicians were proceeding inductively rather than deductively. Nor did they give the logic or reasons for their steps. He criticized many of Newton's arguments; for example, in the latter's *De Quadratura*, where he said he avoided the infinitely small, he gave x the increment denoted by o, expanded $(x + o)^n$, subtracted the x^n, divided by o to find the ratio of the increments of x^n and x, and then dropped terms involving o, thus obtaining the fluxion of x^n. Berkeley said that Newton first did give x an increment but then let it be zero. This, he said, is a defiance of the law of contradiction and the fluxion obtained was really $0/0$. Berkeley attacked also the method of differentials, as presented by l'Hospital and others on the Continent. The ratio of the differentials, he

44. George Berkeley: *The Works*, G. Bell and Sons, 1898, Vol. 3, 1–51.

said, should determine the secant and not the tangent; one undoes this error by neglecting higher differentials. Thus, "by virtue of a twofold mistake you arrive, though not at a science, yet at the truth," because errors were compensating for each other. He also picked on the second differential $d(dx)$ because it is the difference of a quantity dx that is itself the least discernible quantity. He says, "In every other science men prove their conclusions by their principles, and not their principles by the conclusions."

As for the derivative regarded as the ratio of the evanescent increments in y and x or dy and dx, these were "neither finite quantities, nor quantities infinitely small, nor yet nothing." These rates of change were but "the ghosts of departed quantities. Certainly . . . he who can digest a second or third fluxion . . . need not, methinks, be squeamish about any point in Divinity." He concluded that the principles of fluxions were no clearer than those of Christianity and denied that the object, principles, and inferences of the modern analysis were more distinctly conceived or more soundly deduced than religious mysteries and points of faith.

A reply to the *Analyst* was made by James Jurin (1684–1750). In 1734 he published *Geometry, No Friend to Infidelity*, in which he maintained that fluxions are clear to those versed in geometry. He then tried ineffectually to explain Newton's moments and fluxions. For example, Jurin defined the limit of a variable quantity as "some determinate quantity, to which the variable quantity is supposed continually to approach" and to come nearer than any given difference. However, then he added, "but never to go beyond it." This definition he applied to a variable ratio (the difference quotient). Berkeley's crushing answer, entitled *A Defense of Freethinking in Mathematics* (1735),[45] indicated that Jurin was trying to defend what he did not understand. Jurin replied, but did not clarify matters.

Benjamin Robins (1707–51) then entered the fray with articles and a book, *A Discourse Concerning the Nature and Certainty of Sir Isaac Newton's Method of Fluxions and of Prime and Ultimate Ratios* (1735). Robins neglected the moments of Newton's first paper but emphasized fluxions and prime and ultimate ratios. He defined a limit thus: "We define an ultimate magnitude to be a limit, to which a varying magnitude can approach within any degree of nearness whatever, though it can never be made absolutely equal to it." Fluxions he considered to be the right idea, and prime and ultimate ratios as only an explanation. He added that the method of fluxions is established without recourse to limits despite the fact that he gave explanations in terms of a variable approaching a limit. Infinitesimals he disavowed.

To answer Berkeley, Colin Maclaurin (1698–1746), in his *Treatise of Fluxions* (1742), attempted to establish the rigor of the calculus. It was a commendable effort but not correct. Like Newton, Maclaurin loved geom-

---

45. George Berkeley: *The Works*, G. Bell and Sons, 1898, Vol. 3, 53–89.

etry, and therefore tried to found the doctrine of fluxions on the geometry of the Greeks and the method of exhaustion, particularly as used by Archimedes. He hoped thereby to avoid the limit concept. His accomplishment was to use geometry so skillfully that he persuaded others to use it and neglect analysis.

The Continental mathematicians relied more upon the formal manipulation of algebraic expressions than on geometry. The most important representative of this approach is Euler, who rejected geometry as a basis for the calculus and tried to work purely formally with functions, that is, to argue from their algebraic (analytic) representation.

He denied the concept of an infinitesimal, a quantity less than any assignable magnitude and yet not 0. In his *Institutiones* of 1755 he argued,[46]

> There is no doubt that every quantity can be diminished to such an extent that it vanishes completely and disappears. But an infinitely small quantity is nothing other than a vanishing quantity and therefore the thing itself equals 0. It is in harmony also with that definition of infinitely small things, by which the things are said to be less than any assignable quantity; it certainly would have to be nothing; for unless it is equal to 0, an equal quantity can be assigned to it, which is contrary to the hypothesis.

Since Euler banished differentials he had to explain how dy/dx, which was $0/0$ for him, could equal a definite number. He does this as follows: Since for any number n, $n \cdot 0 = 0$, then $n = 0/0$. The derivative is just a convenient way of determining $0/0$. To justify dropping $(dx)^2$ in the presence of a dx Euler says $(dx)^2$ vanishes before dx does, so that, for example, the ratio of $dx + (dx)^2$ to dx is 1. He did accept ∞ as a number, for example, as the sum $1 + 2 + 3 + \cdots$. He also distinguished orders of ∞. Thus $a/0 = \infty$, but $a/(dx)^2$ is infinite to the second order, and so on.

Euler then proceeds to obtain the derivative of $y = x^2$. He gives x the increment ω; the increment of y is $\eta = 2x\omega + \omega^2$ and the ratio of η/ω is $2x + \omega$. He then says that this ratio approaches $2x$ the smaller ω is taken. However, he emphasizes that these differentials η and ω are absolutely zero and that nothing can be inferred from them other than their mutual ratio, which is in the end reduced to a finite quantity. Thus Euler accepts unqualifiedly that there exist quantities that are absolutely zero but whose ratios are finite numbers. There is more "reasoning" of this nature in Chapter 3 of the *Institutiones*, where he encourages the reader by remarking that the derivative does not hide so great a mystery as is commonly thought and which in the mind of many renders the calculus suspect.

As another example of Euler's reasoning, let us take his derivation of the

46. *Opera*, (1), 10, 69.

differential of $y = \log x$, in section 180 of his *Institutiones* (1755). Replacing x by $x + dx$ gives

$$dy = \log (x + dx) - \log x = \log \left(1 + \frac{dx}{x}\right).$$

Now he calls upon a result in Chapter 7 of Volume 1 of his *Introductio* (1748),

$$(57) \qquad \log_e (1 + z) = z - \frac{z^2}{2} + \frac{z^3}{3} - \frac{z^4}{4} + \cdots.$$

Replacing z by dx/x gives

$$dy = \frac{dx}{x} - \frac{dx^2}{2x^2} + \frac{dx^3}{3x^3} - \cdots.$$

Since all the terms beyond the first one are evanescent, we have

$$d(\log x) = \frac{dx}{x}.$$

We should keep in mind that Euler's texts were the standard of his day. What Euler did contribute in his formalistic approach was to free the calculus from geometry and base it on arithmetic and algebra. This step at least prepared the way for the ultimate justification of the calculus on the basis of the real number system.

Lagrange, in a paper of 1772 [47] and in his *Théorie des fonctions analytiques* [48] made the most ambitious attempt to rebuild the foundations of the calculus. The subtitle of his book reveals his folly. It reads: "Containing the principal theorems of the differential calculus without the use of the infinitely small, or vanishing quantities, or limits and fluxions, and reduced to the art of algebraic analysis of finite quantities."

Lagrange criticizes Newton's approach by pointing out that, regarding the limiting ratio of arc to chord, Newton considers chord and arc equal not before or after vanishing, but *when* they vanish. As Lagrange correctly points out, "That method has the great inconvenience of considering quantities in the state in which they cease, so to speak, to be quantities; for though we can always properly conceive the ratios of two quantities as long as they remain finite, that ratio offers to the mind no clear and precise idea, as soon as its terms both become nothing at the same time." Maclaurin's *Treatise of Fluxions*, he says, shows how difficult it is to demonstrate the method of fluxions. He is equally dissatisfied with the little zeros (infinitesimals) of Leibniz and Bernoulli and with the absolute zeros of Euler, all of which, "although correct in reality are not sufficiently clear to serve as foundation of a science whose certitude should rest on its own evidence."

47. *Nouv. Mém. de l'Acad. de Berlin*, 1772, pub. 1774 = *Œuvres*, 3, 441–76.
48. 1797; 2nd ed., 1813 = *Œuvres*, 9.

Lagrange wished to give the calculus all the rigor of the demonstrations of the ancients and proposed to do this by reducing the calculus to algebra, which, as we noted above, included infinite series as extensions of polynomials. In fact, for Lagrange, function theory is the part of algebra concerned with the derivatives of functions. Specifically Lagrange proposed to use power series. With becoming modesty he remarks that it is strange that this method did not occur to Newton.

He now wants to use the fact that any function $f(x)$ can be expressed thus:

$$(58) \qquad f(x + h) = f(x) + ph + qh^2 + rh^3 + sh^4 + \cdots$$

wherein the coefficients, $p, q, r, \ldots$ involve x but are independent of h. However, he wishes to be sure before proceeding that such a power series expansion is always possible. Of course, he says, this is known through any number of familiar examples, but he does agree that there are exceptional cases. As exceptional cases Lagrange has in mind those in which some derivative of $f(x)$ becomes infinite and those in which the function and its derivatives become infinite. These exceptions happen only at isolated points; hence, with Lagrange they do not count. In a similar cavalier fashion he deals with a second difficulty. Lagrange and Euler accepted without question that a series expansion containing integral and fractional powers of h was surely possible, but Lagrange wished to eliminate the need for fractional powers. Fractional powers, he believed, could arise only if $f(x)$ contained radicals. But these, too, he dismisses as exceptional cases. Hence he is ready to proceed with (58).

By a somewhat involved but purely formal argument Lagrange concludes that we get $2q$ from p in the same way that we get p from $f(x)$, and a similar conclusion holds for the other coefficients $r, s, \ldots$ in (58). Hence if we let p be denoted by $f'(x)$ and designate by $f''(x)$ a function derived from $f'(x)$ as $f'(x)$ is derived from $f(x)$, then

$$p = f'(x), \qquad q = \frac{1}{2!} f''(x), \qquad r = \frac{1}{3!} f'''(x), \ldots,$$

and so (58) gives

$$f(x + h) = f(x) + hf'(x) + \frac{h^2}{2!} f''(x) + \cdots.$$

Lagrange now concludes that the final "expression has the advantage of showing how the terms of the series depend on each other, and especially how when one knows how to form the first derivative function, one can form all the derivative functions which enter the series." A little later he adds, "For one who knows the rudiments of the differential calculus it is clear that these derivative functions coincide with $dy/dx, d^2y/dx^2, \cdots$."

Lagrange has yet to show how one derives p or $f'(x)$ from $f(x)$. Here he uses (58) and neglects all terms after the second. Then $f(x + h) - f(x) = ph$. He divides by h and concludes that $p = f'(x)$.

Actually Lagrange's assumption that a function can be expanded in a power series is one weak point in the scheme. The criteria now known for such an expansion involve the existence of derivatives, and this is what Lagrange sought to avoid. His arguments to justify the power series only added to the confusion about which functions can be so expanded. Even if such an expansion is possible, Lagrange shows how to calculate the coefficients only if we can get the first one, that is, $f'(x)$; and here he does the same crude thing his predecessors did. Finally, the question of the convergence of the series (58) is really not discussed. He does show that for h small enough the last term kept is greater than what is neglected. He also gives in this book the Lagrange form of the remainder in a Taylor expansion (Chap. 20, sec. 7), but it plays no role in the above development. Despite all these weaknesses, Lagrange's approach to the calculus met great favor for quite some time. Later it was abandoned.

Lagrange believed he had dispensed with the limit concept. He did agree[49] that the calculus could be founded on a theory of limits but said the kind of metaphysics one must employ is foreign to the spirit of analysis. Despite the inadequacies of his approach, he did contribute, as did Euler, to divorcing the foundations of analysis from geometry and mechanics; in this his influence was decisive. While this separation is not pedagogically desirable, since it bars intuitive understanding, it did make clear that logically analysis must stand on its own feet.

Toward the end of the century the mathematician, soldier, and administrator Lazare N. M. Carnot (1753–1823) wrote a popular, widely sold book, *Réflexions sur la métaphysique du calcul infinitésimal* (1797), in which he sought to make the calculus precise. He tried to show that the logic rested on the method of exhaustion and that all the ways of treating the calculus were but simplifications or shortcuts whose logic could be supplied by founding them on that method. After much thought he, like Berkeley, ended up concluding that errors in the usual arguments of the calculus were compensating for each other.

Among the multitude of efforts to rigorize the calculus, a few were on the right track. The most notable of these were d'Alembert's and, earlier, Wallis's. D'Alembert believed that Newton had the right idea and that he himself was merely explaining Newton's meaning. In his article "Différentiel" in the famous *Encyclopédie ou Dictionnaire Raisonné des Sciences, des Arts, et des Métiers* (1751–80), he says, "Newton has never regarded the differential calculus as a calculus of infinitesimals, but as a method of prime and ulti-

49. *Œuvres*, 1, 325.

mate ratios, that is to say, a method of finding the limit of these ratios." But d'Alembert defined a differential as "an infinitely small quantity or at least smaller than any assignable magnitude." He did believe the calculus of Leibniz could be built up on three rules of differentials; however, he favored the derivative as a limit. In his pursuit of the use of limits he, like Euler, argued that $0/0$ may be equal to any quantity one wishes.

In another article, "Limite," he says: "The theory of limits is the true metaphysics of the calculus. . . . It is never a question of infinitesimal quantities in the differential calculus: it is uniquely a question of limits of finite quantities. Thus the metaphysics of the infinite and infinitely small quantities, larger or smaller than one another, is totally useless to the differential calculus." Infinitesimals were merely a manner of speaking that avoided the lengthier description in terms of limits. In fact, d'Alembert gave a good approximation to the correct definition of limit in terms of a variable quantity approaching a fixed quantity more closely than any given quantity, though here too he talks about the variable never reaching the limit. However, he did not give a formal exposition of the calculus that incorporated and utilized his basically correct views.

He, too, was vague on a number of points; for example, he defined the tangent to a curve as the limit of the secant when the two points of intersection become one. This vagueness, especially in his statement of the notion of limit, caused many to debate the question of whether a variable can reach its limit. Since there was no explicit, correct presentation, d'Alembert advised students of the calculus, "Persist and faith will come to you."

Sylvestre-François Lacroix (1765–1843), in the second edition (1810–1819) of his *Traité du calcul différentiel et du calcul intégral* had the idea more explicitly that the ratio of two quantities, each of which approaches 0, can approach a definite number as a limit. He gives the ratio $ax/(ax + x^2)$ and notes that this ratio is the same as $a/(a + x)$, and the latter approaches 1 as x approaches 0. Moreover, he points out that 1 is the limit even when x approaches 0 through negative values. However, he also speaks of the ratio of the limits when these are 0 and even uses the symbol $0/0$. He does introduce the differential dy of a function $y = f(x)$ in terms of the derivative; that is, $dy = f'(x)\,dx$. Hence if $y = ax^0$, $dy = 3ax^2\,dx$. He first used the term "differential coefficient" for the derivative; thus $3ax^2$ is the differential coefficient.

Almost every mathematician of the eighteenth century made some effort or at least a pronouncement on the logic of the calculus, and though one or two were on the right track, all the efforts were abortive. The distinction between a very large number and an infinite "number" was hardly made. It seemed clear that a theorem that held for any n must hold for n infinite. Likewise a difference quotient was replaced by a derivative, and a sum of a finite number of terms and an integral were hardly distinguished.

Mathematicians passed from one to the other freely. In 1755, in his *Institutiones*, Euler distinguished between the increment in a function and the differential of that function and between a summation and the integral, but these distinctions were not immediately taken up. All the efforts could be summed up in Voltaire's description of the calculus as "the art of numbering and measuring exactly a Thing whose Existence cannot be conceived."

In view of the almost complete absence of any foundations, how could the mathematicians proceed with the manipulation of the variety of functions? In addition to their great reliance upon physical and intuitive meanings, they did have a model in mind—the simpler algebraic functions, such as polynomials and rational functions. They carried over to all functions the properties they found in these explicit, concrete functions: continuity, the existence of isolated infinities and discontinuities, expansion in power series, and the existence of derivatives and integrals. But when they were obliged, largely through the work on the vibrating string, to broaden the concept of function, as Euler put it, to any freely drawn curves (Euler's mixed or irregular or discontinuous functions), they could no longer use the simpler functions as a guide. And when the logarithmic function had to be extended to negative and complex numbers, they really proceeded without any reliable basis at all; this is why arguments on such matters were common. The rigorization of the calculus was not achieved until the nineteenth century.

Bibliography

Bernoulli, James: *Opera*, 2 vols., 1744, reprinted by Birkhaüser, 1968.

Bernoulli, John: *Opera Omnia*, 4 vols., 1742, reprinted by Georg Olms, 1968.

Boyer, Carl B.: *The Concepts of the Calculus*, Dover (reprint), 1949, Chap. 4.

Brill, A. and M. Nöther: "Die Entwicklung der Theorie der algebraischen Funktionen in älterer and neuerer Zeit," *Jahres. der Deut. Math.-Verein.*, 3, 1892/3, 107–566.

Cajori, Florian: "History of the Exponential and Logarithmic Concepts," *Amer. Math. Monthly*, 20, 1913, 5–14, 35–47, 75–84, 107–17, 148–51, 173–82, 205–10.

————: *A History of the Conceptions of Limits and Fluxions in Great Britain from Newton to Woodhouse*, Open Court, 1919.

————: "The History of Notations of the Calculus," *Annals of Math.*, (2), 25, 1923, 1–46.

Cantor, Moritz: *Vorlesungen über Geschichte der Mathematik*, B. G. Teubner, 1898 and 1924, Vols. 3 and 4, relevant sections.

Davis, Philip J.: "Leonhard Euler's Integral: A Historical Profile of the Gamma Function," *Amer. Math. Monthly*, 66, 1959, 849–69.

Euler, Leonhard: *Opera Omnia*, B. G. Teubner and Orell Füssli, 1911–; see references to specific volumes in the chapter.

Fagnano, Giulio Carlo: *Opere matematiche*, 3 vols., Albrighi Segati, 1911.

Fuss, Paul H. von: *Correspondance mathématique et physique de quelques célèbres géomètres du XVIIIème siècle*, 2 vols., 1843, Johnson Reprint Corp., 1967.

Hofmann, Joseph E.: "Über Jakob Bernoullis Beiträge zur Infinitesimal-mathematik," *L'Enseignement Mathématique*, (2), 2, 61–171, 1956; also published separately by Institut de Mathématiques, Genève, 1957.

Mittag-Leffler, G.: "An Introduction to the Theory of Elliptic Functions," *Annals of Math.*, (2), 24, 1922–23, 271–351.

Montucla, J. F.: *Histoire des mathématiques*, A. Blanchard (reprint), 1960, Vol. 3, pp. 110–380.

Pierpont, James: "Mathematical Rigor, Past and Present," *Amer. Math. Soc. Bulletin*, 34, 1928, 23–53.

Struik, D. J.: *A Source Book in Mathematics, 1200–1800*, Harvard University Press, 1969, pp. 333–38, 341–51, 374–91.

20
Infinite Series

1. *Introduction*

Infinite series were in the eighteenth century and are still today considered an essential part of the calculus. Indeed, Newton considered series inseparable from his method of fluxions because the only way he could handle even slightly complicated algebraic functions and the transcendental functions was to expand them into infinite series and differentiate or integrate term by term. Leibniz in his first published papers of 1684 and 1686 also emphasized "general or indefinite equations." The Bernoullis, Euler, and their contemporaries relied heavily on the use of series. Only gradually, as we pointed out in the preceding chapter, did the mathematicians learn to work with the elementary functions in closed form, that is, as simple analytical expressions. Nevertheless, series were still the only representation for some functions and the most effective means of calculating the elementary transcendental functions.

The successes obtained by using infinite series became more numerous as the mathematicians gradually extended their discipline. The difficulties in the new concept were not recognized, at least for a while. Series were just infinite polynomials and appeared to be treatable as such. Moreover, it seemed clear, as Euler and Lagrange believed, that every function could be expressed as a series.

2. *Initial Work on Infinite Series*

Infinite series, usually in the form of infinite geometric progressions with common ratio less than 1, appear very early in mathematics. Aristotle[1] even recognized that such series have a sum. They appear sporadically among the later medieval mathematicians, who considered infinite series to calculate

1. *Physica*, Book III, Chap. 6, 206b, 3–33.

the distance traveled by moving bodies when the velocity changes from one period of time to another. Oresme, who had considered a few such series, even proved in a tract, *Quæstiones Super Geometriam Euclidis* (*c.* 1360), that the harmonic series

$$1 + \frac{1}{2} + \frac{1}{3} + \frac{1}{4} + \frac{1}{5} + \cdots$$

is divergent by the method used today, namely, to replace the series by the series of lesser terms

$$\frac{1}{2} + \frac{1}{2} + \left(\frac{1}{4} + \frac{1}{4}\right) + \left(\frac{1}{8} + \frac{1}{8} + \frac{1}{8} + \frac{1}{8}\right) + \cdots$$

and to note that the latter series diverges because we can obtain as many groups of terms each of magnitude 1/2 as we please. However, one must not conclude that Oresme or mathematicians in general began to distinguish convergent and divergent series.

In his *Varia Responsa* (1593, *Opera*, 347–435) Vieta gave the formula for the sum of an *infinite* geometric progression. He took from Euclid's *Elements* that the sum of *n* terms of $a_1 + a_2 + \cdots + a_n$ is given by

$$\frac{s_n - a_n}{s_n - a_1} = \frac{a_1}{a_2}.$$

Then if $a_1/a_2 > 1$, a_n approaches 0 as *n* becomes infinite, so that

$$s_\infty = \frac{a_1^2}{a_1 - a_2}.$$

In the middle of the seventeenth century Gregory of Saint Vincent, in his *Opus Geometricum* (1647), showed that the Achilles and the Tortoise paradox could be resolved by summing an infinite geometric series. The finiteness of the sum showed that Achilles would overtake the tortoise at a definite time and place. Gregory gave the first explicit statement that an infinite series represents a magnitude, namely, the sum of the series, which he called the limit of the series. He says the "terminus of a progression is the end of the series to which the progression does not attain, even if continued to infinity, but to which it can approach more closely than by any given interval." He made many other statements that are less accurate and less clear, but he did contribute to the subject and influenced many pupils.

Mercator and Newton (Chap. 17, sec. 2) found the series

$$\log (1 + x) = x - \frac{1}{2} x^2 + \frac{1}{3} x^3 + \cdots.$$

The observation was made that the series has an infinite value for $x = 2$, whereas, according to the left side, it should yield log 3. Wallis noted this

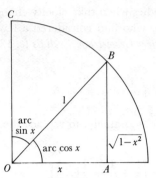

Figure 20.1

difficulty but could not explain it. Newton obtained many other series for algebraic and transcendental functions. Thus, to obtain the series for arc sin x, in 1666, he used the fact (Fig. 20.1) that the area $OBC = (1/2)$ arc sin x, so arc sin $x = \int_0^x \sqrt{1 - x^2}\, dx - x\sqrt{1 - x^2}/2$. He got the result by expanding the right side into series, integrating term by term, and combining the two series. He also obtained the series for arc tan x. In his *De Analysi* of 1669 he gave the series for sin x, cos x, arc sin x, and e^x. Some of these he got from others by inverting a series, that is, solving it for the independent variable in terms of the dependent variable. His method of doing this is crude and inductive. Nevertheless, Newton was immensely pleased with his derivation of so many series.

Collins received Newton's *De Analysi* in 1669 and communicated the results on series to James Gregory on December 24, 1670. Gregory answered (Turnbull, *Correspondence*, 1, 52–58 and 61–64) on February 15, 1671 that he had obtained other series, among them

$$\tan x = x + \frac{x^3}{3} + \frac{2}{15}x^5 + \frac{17}{315}x^7 + \cdots,$$

$$\sec x = 1 + \frac{x^2}{2} + \frac{5}{24}x^4 + \frac{61}{720}x^6 + \cdots.$$

His derivations of these series are not known. Leibniz, too, obtained series for sin x, cos x, and arc tan x, presumably independently, in 1673. The series for the transcendental functions were the most fertile method available in the early stages of the calculus for handling these functions and are a significant part of Newton's and Leibniz's work on the calculus.

These men and others who used the binomial theorem for fractional and negative exponents to obtain many of the series not only ignored the questions that arose from the use of series but had no proof of the binomial theorem. They also accepted unquestioningly that the series was equal to the function that was being expanded.

James Bernoulli in 1702[2] derived the series for sin x and cos x by using expressions he had derived for sin $n\alpha$ in terms of sin α and then letting α approach 0 while n becomes infinite, so that $n\alpha$ approaches x while n sin α, which equals $n\alpha$ sin α/α, also approaches x. Wallis had mentioned in the Latin edition of his *Algebra* (1693) that Newton had given these series again in 1676; Bernoulli noted this remark but failed to acknowledge Newton's priority. Moreover, de Moivre gave a proof of Newton's results in the *Philosophical Transactions* of 1698;[3] though Bernoulli used and referred to this journal in other work, he gave no indication that he was aware through this source of Newton's work.

One of the major uses of series beyond their service in differentiation and integration is to calculate special quantities, such as π and e, and the logarithmic and trigonometric functions. Newton, Leibniz, James Gregory, Cotes, Euler, and many others were interested in series for this purpose. However, some series converge so slowly that they are almost useless for calculation. Thus Leibniz in 1674[4] obtained the famous result

$$\frac{\pi}{4} = 1 - \frac{1}{3} + \frac{1}{5} - \frac{1}{7} + \cdots.$$

However, it would require about 100,000 terms to compute π, even to the accuracy obtained by Archimedes. Likewise, the series for log $(1 + x)$ converges very slowly, so that many terms have to be taken into account to achieve an accuracy of a few decimal places. This series was transformed in various ways to produce more rapidly converging series. Thus James Gregory (*Exercitationes Geometricae*, 1668) obtained

$$\frac{1}{2}\frac{\log (1 + z)}{\log (1 - z)} = z + \frac{1}{3}z^3 + \frac{1}{5}z^5 + \cdots,$$

which proved to be more useful for the calculation of logarithms. The problem of transforming a series into another that converges more rapidly was pursued by many men throughout the eighteenth century. One such transformation, due to Euler, is given in Section 4.

Still another use of series was initiated by Newton. Given the implicit function $f(x, y) = 0$, to work with y as a function of x one wants the explicit function. There may be several such explicit functions, as is evident in the trivial case of $x^2 + y^2 - 1 = 0$, which has two solutions $y = \pm\sqrt{1 - x^2}$, both emanating from the point $(1, 0)$. In this simple case the two solutions can be expressed in terms of closed analytic expressions. But generally each expression for y must be expressed as an infinite series in x. However, these series are not necessarily power series, particularly if the points at which the

2. *Opera*, 2, 921–29.
3. Vol. 20, 190–93.
4. *Math. Schriften*, 5, 88–92; also *Acta Erud.*, 1682 = *Math. Schriften*, 5, 118–22.

expansions are sought are singular points ($f_x = f_y = 0$). In his *Method of Fluxions* Newton published a scheme for determining the forms of the several series, one for each explicit solution. His method, which uses what is known as Newton's parallelogram, shows how to determine the first few exponents in a series of the form

$$y = a_1 x^m + a_2 x^{m+n} + a_3 x^{m+2n} + \cdots.$$

The coefficients of the series can then be determined by the method of undetermined coefficients. Actually Newton gave only specific examples, from which one must infer the method.

The problem of determining the exponents in each series is troublesome. Taylor, James Stirling, and Maclaurin gave rules; Maclaurin tried to extend and prove them but made no progress. A proof of Newton's method was given independently by Gabriel Cramer and Abraham G. Kästner (1719–1800).

3. *The Expansion of Functions*

One of the problems faced by mathematicians in the late seventeenth and eighteenth centuries was interpolation of table values. Greater accuracy of the interpolated values of the trigonometric, logarithmic, and nautical tables was necessary to keep pace with progress in navigation, astronomy, and geography. The common method of interpolation (the word is Wallis's) is called linear interpolation because it assumes that the function is a linear function of the independent variable in the interval between two known values. However, the functions in question are not linear; and the mathematicians realized that a better method of interpolation was needed.

The method we are about to describe was initiated by Briggs in his *Arithmetica Logarithmica* (1624), though the key formula was given by James Gregory in a letter to Collins (Turnbull, *Correspondence*, 1, 45–48) of November 23, 1670, and independently by Newton. Newton's work appears in Lemma 5 of Book III of the *Principia* and in the *Methodus Differentialis*, which, though published in 1711, was written by 1676. The method uses what are called finite differences and is the first major result in the calculus of finite differences.

Suppose $f(x)$ is a function whose values are known at $a, a + c, a + 2c, a + 3c, \ldots, a + nc$. Let

$$\begin{aligned}
\Delta f(a) &= f(a + c) - f(a), \\
\Delta f(a + c) &= f(a + 2c) - f(a + c), \\
\Delta f(a + 2c) &= f(a + 3c) - f(a + 2c), \\
&\cdots\cdots\cdots\cdots\cdots\cdots\cdots\cdots\cdots\cdots
\end{aligned}$$

Further, let

$$\Delta^2 f(a) = \Delta f(a + c) - \Delta f(a)$$
$$\Delta^3 f(a) = \Delta^2 f(a + c) - \Delta^2 f(a),$$
$$\dots\dots\dots\dots\dots\dots\dots\dots\dots$$

Then the Gregory-Newton formula states that

(1) $$f(a + h) = f(a) + \frac{h}{c} \Delta f(a) + \frac{\frac{h}{c}\left(\frac{h}{c} - 1\right)}{1 \cdot 2} \Delta^2 f(a) + \cdots.$$

Newton sketched a proof but Gregory did not.

To calculate a value of $f(x)$ at any value x between the known values, one simply gives h the value $x - a$. This calculated value is not necessarily the true value of the function; what the formula yields is the value of a polynomial in h that agrees with the true function at the special values a, $a + c, a + 2c, \ldots$.

The Gregory-Newton formula was also used to carry out approximate integration. Given a function, say $g(x)$, to be integrated, perhaps in order to find the area under the corresponding curve, one uses the values of $g(x)$ to obtain $g(a), g(a + c), g(a + 2c), \ldots$ and their differences and higher-order differences; these values are substituted in (1). Then (1) gives a polynomial approximation to $g(x)$, and, as Newton points out, since polynomials are readily integrated, one gets an approximation to the desired integral of $g(x)$.

Gregory also applied (1) to the function $(1 + d)^x$. He knew the value of this function at $x = 0, 1, 2, 3, \ldots$. Then $f(0) = 1$, $\Delta f(0) = d$, $\Delta^2 f(0) = d^2$, and so on. Thus by letting $a = 0$, $c = 1$, and $h = x - 0$ in (1), and using the values of $f(0), \Delta f(0), \ldots$ he got

(2) $$(1 + d)^x = 1 + dx + \frac{x(x - 1)}{1 \cdot 2} d^2 + \frac{x(x - 1)(x - 2)}{1 \cdot 2 \cdot 3} d^3 + \cdots.$$

Thus Gregory obtained the binomial expansion for general x.

The Gregory-Newton interpolation formula was used by Brook Taylor to develop the most powerful single method for expanding a function into an infinite series. The binomial theorem, division of the denominator of a rational function into the numerator, and the method of undetermined co-efficients are limited devices. In his *Methodus Incrementorum Directa et Inversa* (1715), the first publication in which he treated the calculus of finite differences, Taylor derived the theorem that still bears his name and which he had stated in 1712. Incidentally, he praises Newton but makes no mention of Leibniz's work of 1673 on finite differences, though Taylor knew this work. Taylor's theorem was known to James Gregory in 1670 and was discovered independently somewhat later by Leibniz; however, these two men did not publish it. John Bernoulli did publish practically the same result in the *Acta*

Eruditorum of 1694; and though Taylor knew this result he did not refer to it. His own "proof" was different. What he did amounts to letting c be Δx in the Gregory-Newton formula. Then, for example, the third term on the right side of (1) becomes

(3)
$$\frac{h(h - \Delta x)}{1 \cdot 2} \frac{\Delta^2 f(a)}{\Delta x^2}.$$

Taylor concluded that when $\Delta x = 0$, this term becomes $h^2 f''(a)/2!$, and so the entire Gregory-Newton formula becomes

(4) $$f(a + h) = f(a) + f'(a)h + f''(a)\frac{h^2}{2!} + f'''(a)\frac{h^3}{3!} + \cdots.$$

Of course Taylor's method was not rigorous, nor did he consider the question of convergence.

Taylor's theorem for $a = 0$ is now called Maclaurin's theorem. Colin Maclaurin, who succeeded James Gregory as professor at Edinburgh, gave this special case in his *Treatise of Fluxions* (1742) and stated that it was but a special case of Taylor's result. However, historically it has been credited to Maclaurin as a separate theorem. Incidentally, Stirling gave this special case for algebraic functions in 1717 and for general functions in his *Methodus Differentialis* of 1730.

Maclaurin's proof of his result is by the method of undetermined coefficients. He proceeds as follows. Let

(5) $$f(z) = A + Bz + Cz^2 + Dz^3 + \cdots.$$

Then

$$f'(z) = B + 2Cz + 3Dz^2 + \cdots$$
$$f''(z) = 2C + 6Dz + \cdots$$
$$\cdots\cdots\cdots\cdots\cdots\cdots\cdots\cdots$$

Let $z = 0$ in each equation and determine $A, B, C, \cdots$. He did not worry about convergence and proceeded to use the result.

4. *The Manipulation of Series*

James and John Bernoulli did a great deal of work with series. James wrote five papers between 1689 and 1704 that were published by his nephew Nicholas (1695–1726) (John's son) as a supplement to James's *Ars Conjectandi* (1713). Most of the work in these papers is devoted to the use of series representations of functions for the purposes of differentiating and integrating the functions and obtaining areas under curves and lengths of curves. While these applications were a substantial contribution to the calculus, there were no especially novel features. However, some of the methods he used to sum series are worth noting because they illustrate the nature of mathematical thought in the eighteenth century.

In the first paper (1689),[5] he starts with the series

$$(6) \qquad N = \frac{a}{c} + \frac{a}{2c} + \frac{a}{3c} + \cdots,$$

from which

$$(7) \qquad N - \frac{a}{c} = \frac{a}{2c} + \frac{a}{3c} + \frac{a}{4c} + \cdots.$$

He now subtracts (7) from (6); in this process each term on the right side of (7) is subtracted from the term above it. This yields

$$(8) \qquad \frac{a}{c} = \frac{a}{1 \cdot 2c} + \frac{a}{2 \cdot 3c} + \frac{a}{3 \cdot 4c} + \cdots.$$

This is a correct result but incorrectly derived, because the original series is divergent. James says that the procedure is questionable and should not be used without some circumspection.

He then considers the ordinary harmonic series and shows that its sum is infinite.[6] He considers the terms

$$(9) \qquad \frac{1}{n + 1} + \frac{1}{n + 2} + \cdots + \frac{1}{n^2}$$

and says this sum is larger than $(n^2 - n) \cdot (1/n^2)$ because there are $n^2 - n$ terms and each is at least as large as the last. But

$$(n^2 - n) \cdot \left(\frac{1}{n^2}\right) = 1 - \frac{1}{n}.$$

Hence if we add $1/n$ to (9)

$$\frac{1}{n} + \frac{1}{n + 1} + \frac{1}{n + 2} + \cdots + \frac{1}{n^2} > 1.$$

Thus, he says, we can go from one group of terms to another, each group having a sum greater than 1. Hence we can obtain a finite number of terms whose sum is as large as we please; and therefore the sum of the whole series must be infinite. Consequently, he also points out, the sum of an infinite series whose "last" term vanishes can be *infinite*; this is contrary to his earlier belief and the belief of many eighteenth-century mathematicians, including Lagrange.

5. *Opera*, 1, 375–402.
6. *Opera*, 1, 392.

John Bernoulli had previously given a different "proof" of the infinite sum of the harmonic series. It runs thus:

$$(10) \quad \frac{1}{2} + \frac{1}{3} + \frac{1}{4} + \frac{1}{5} + \cdots = \frac{1}{1 \cdot 2} + \frac{2}{2 \cdot 3} + \frac{3}{3 \cdot 4} + \frac{4}{4 \cdot 5} + \cdots$$

$$= \left(\frac{1}{1 \cdot 2} + \frac{1}{2 \cdot 3} + \frac{1}{3 \cdot 4} + \frac{1}{4 \cdot 5} + \cdots \right) + \left(\frac{1}{2 \cdot 3} + \frac{1}{3 \cdot 4} + \frac{1}{4 \cdot 5} + \cdots \right)$$

$$+ \left(\frac{1}{3 \cdot 4} + \frac{1}{4 \cdot 5} + \cdots \right) + \left(\frac{1}{4 \cdot 5} + \cdots \right) + \cdots.$$

Now, using (8), wherein we let a and c be 1, we get from (10) that

$$\frac{1}{2} + \frac{1}{3} + \frac{1}{4} + \cdots =$$

$$1 + \left(1 - \frac{1}{2} \right) + \left(1 - \frac{1}{2} - \frac{1}{6} \right) + \left(1 - \frac{1}{2} - \frac{1}{6} - \frac{1}{12} \right) + \cdots$$

$$= 1 + \frac{1}{2} + \frac{1}{3} + \frac{1}{4} + \cdots.$$

If we let $A = 1/2 + 1/3 + 1/4 + 1/5 + \cdots$, we have proved that $A = 1 + A$. If A were finite, this result would be impossible.

In the succeeding four tracts on series James Bernoulli does many things so loosely that it is difficult to believe he had ever recognized the need for caution with infinite series. For example, in the second tract (1692),[7] he argues as follows: From the formula for a geometrical progression we have $1 + 1/2 + 1/4 + 1/8 + \cdots = 2$. Then by taking 1/3 of both sides $1/3 + 1/6 + 1/12 + \cdots = 2/3$, and by taking 1/5 of both sides of the original series $1/5 + 1/10 + 1/20 + \cdots = 2/5$, and so on. Hence the sum of the left sides, which is the entire harmonic series, equals the sum of the right sides, or

$$1 + \frac{1}{2} + \frac{1}{3} + \frac{1}{4} + \cdots = 2 + \frac{2}{3} + \frac{2}{5} + \cdots = 2\left(1 + \frac{1}{3} + \frac{1}{5} + \cdots \right).$$

Hence the sum of the odd terms is half the sum of the harmonic series. Then $1/2 + 1/4 + 1/6 + 1/8 + \cdots$ is also 1/2 the harmonic series, so that $1 + 1/3 + 1/5 + \cdots = 1/2 + 1/4 + 1/6 + \cdots$.

In the third tract (1696)[8] he writes

$$\frac{l}{m+n} = \frac{l}{m}\left(1 + \frac{n}{m} \right)^{-1} = \frac{l}{m} - \frac{ln}{m^2} + \frac{ln^2}{m^3} - \cdots;$$

7. *Opera*, 1, 517–42.
8. *Opera*, 2, 745–64.

and when $n = m$,

(11)
$$\frac{l}{2m} = \frac{l}{m} - \frac{l}{m} + \frac{l}{m} - \cdots,$$

which he describes as a not inelegant paradox.

In the second paper on series he replaced the general term by a sum or difference of two other terms, and then performed other operations that lead to specific results. This replacement is correct for absolutely convergent series but not for conditionally convergent ones. Hence he got wrong results which he also described as paradoxes.

One of James's very interesting results deals with the series of reciprocals of the nth powers of the natural numbers, that is, with $1 + 1/2^n + 1/3^n + 1/4^n + \cdots$. James proved that the sum of the odd-numbered terms is to the sum of the even-numbered terms as $2^n - 1$ is to 1. This is correct for $n \geq 2$. However, James did not hesitate to apply it for the case $n = 1$ and $n = 1/2$. This last result he found paradoxical.

Another result on series due to James says that the sum of the series $1 + 1/\sqrt{2} + 1/\sqrt{3} + \cdots$ is infinite because each term is greater than the corresponding term of the harmonic series. Here the comparison test was used effectively.

The series that provoked the greatest discussion and controversy, which occurs when l and m in (11) are 1, is

(12)
$$1 - 1 + 1 - 1 + \cdots.$$

It seemed clear that by writing the series as

(13)
$$(1 - 1) + (1 - 1) + (1 - 1) + \cdots$$

the sum should be 0. It also seemed clear that by writing the series as

$$1 - (1 - 1) - (1 - 1) - (1 - 1) - \cdots$$

the sum should be 1. However, if one denotes the sum of (12) by S, then $S = 1 - S$, so that $S = 1/2$; and this is in fact Bernoulli's result in (11). Guido Grandi (1671–1742), a professor of mathematics at the University of Pisa, in his little book *Quadratura Circuli et Hyperbolae* (The Quadrature of Circles and Hyperbolas, 1703), obtained the third result by another method. He set $x = 1$ in the expansion

(14)
$$\frac{1}{1 + x} = 1 - x + x^2 - x^3 + \cdots$$

and obtained

$$\frac{1}{2} = 1 - 1 + 1 - 1 + \cdots.$$

Grandi therefore maintained that 1/2 was the sum of the series (12). He also argued that since the sum of (12) in the form (13) was 0, he had proved that the world could be created out of nothing.

In a letter to Christian Wolf (1678–1754), published in the *Acta*,[9] Leibniz also treated the series (12). He agreed with Grandi's result but thought it should be possible to obtain it without resorting to his argument. Instead, Leibniz argued that if one takes the first term, the sum of the first two, the sum of the first three, and so forth, one obtains 1, 0, 1, 0, 1, $\cdots$. Thus 1 and 0 are equally probable; one should therefore take the arithmetic mean, which is also the most probable value, as the sum. This solution was accepted by James and John Bernoulli, Daniel Bernoulli, and, as we shall see, Lagrange. Leibniz conceded that his argument was more metaphysical than mathematical but went on to say that there was more metaphysical truth in mathematics than was generally recognized. However, he was probably much more influenced by Grandi's argument than he himself realized. For when, in later correspondence, Wolf wished to conclude that

$$1 - 2 + 4 - 8 + 16 \cdots = \frac{1}{3}$$

$$1 - 3 + 9 - 27 + 81 \cdots = \frac{1}{4}$$

by using an extension of Leibniz's own probability argument, Leibniz objected. He pointed out that series that have sums have decreasing terms, and (12) is at least a limit of series with decreasing terms, as is evident from (14) by letting x approach 1 from below.

Really extensive work on series began about 1730 with Euler, who aroused tremendous interest in the subject. But there was much confusion in his thinking. To obtain the sum of

$$1 - 1 + 1 - 1 + 1 - \cdots$$

Euler argued that since

(15) $$\frac{1}{1 - x} = 1 + x + x^2 + x^3,$$

then when $x = -1$,

(16) $$\frac{1}{2} = 1 - 1 + 1 - 1 + \cdots$$

so that the sum is 1/2.

Also, when $x = -2$, (15) shows that

(17) $$\frac{1}{3} = 1 - 2 + 2^2 - 2^3 + \cdots ;$$

9. *Acta Erud. Supplementum*, 5, 1713, 264–70 = *Math. Schriften*, 5, 382–87.

hence the sum of the right-hand series is 1/3. As a third example, since

(18) $$\frac{1}{(1 + x)^2} = (1 + x)^{-2} = 1 - 2x + 3x^2 - 4x^3 + \cdots,$$

then for $x = 1$ we have

$$\frac{1}{4} = 1 - 2 + 3 - 4 + \cdots.$$

Again, since

$$\frac{1 - x}{(1 + x)^2} = (1 - x)(1 + x)^{-2} = 1 - 3x + 5x^2 - 7x^3 + \cdots,$$

then for $x = 1$ we have

(19) $$0 = 1 - 3 + 5 - 7 + \cdots.$$

There are numerous examples of such arguments in his work.

One sees from (18), for $x = -1$, that

(20) $$\infty = 1 + 2 + 3 + 4 + 5 + \cdots.$$

This Euler accepted. Moreover one sees from (15), for $x = 2$, that

(21) $$-1 = 1 + 2 + 4 + 8 + \cdots.$$

Since the right-hand side of (21) should exceed the right-hand side of (20), the sum $1 + 2 + 4 + 8 + \cdots$ should exceed ∞. According to (21), it yields -1. Euler concluded that ∞ must be a sort of limit between the positive and negative numbers and in this respect resembles 0.

Apropos of (19), Nicholas Bernoulli (1687–1759) said, in a letter to Euler of 1743, that the sum of this series $1 - 3 + 5 - 7 + \cdots$ is $-\infty(-1)^\infty$. Euler's result of 0 he called an unsolvable contradiction. Bernoulli also noted that from (15) one gets, for $x = 2$,

$$-1 = 1 + 2 + 4 + 8 + \cdots;$$

and from

(22) $$\frac{1}{1 - x - x^2} = 1 + x + 2x^2 + 3x^3 + \cdots,$$

for $x = 1$ one gets

$$-1 = 1 + 1 + 2 + 3 + \cdots.$$

The fact that two different series give -1 is also an unsolvable contradiction, for otherwise one could equate the two series.

In one paper Euler did point out that series can be used only for values of x for which they converge. Nevertheless, in the very same paper[10] he concluded that

$$(23) \qquad \cdots + \frac{1}{x^2} + \frac{1}{x} + 1 + x + x^2 + x^3 + \cdots = 0.$$

His argument was that

$$\frac{x}{1-x} = x + x^2 + x^3 + \cdots$$

and

$$\frac{x}{x-1} = \frac{1}{1 - \frac{1}{x}} = 1 + \frac{1}{x} + \frac{1}{x^2} + \frac{1}{x^3} + \cdots.$$

But the two left sides add up to 0, while the two right sides add up to the original series.

In an earlier paper,[11] Euler started with the series

$$(24) \qquad y = \sin x = x - \frac{x^3}{3!} + \frac{x^5}{5!} - \cdots$$

or

$$(25) \qquad 1 - \frac{x}{y} + \frac{x^3}{3!y} - \frac{x^5}{5!y} + \cdots = 0.$$

By using algebraic considerations applied to (25) as a *polynomial of infinite degree,* and by using the theorem on the relation between roots and coefficients of an algebraic equation, Euler proved that[12]

$$\frac{1}{1^2} + \frac{1}{3^2} + \frac{1}{5^2} + \cdots = \frac{\pi^2}{8}$$

$$\frac{1}{1^3} - \frac{1}{3^3} + \frac{1}{5^3} - \cdots = \frac{\pi^3}{32}$$

$$\frac{1}{1^4} + \frac{1}{3^4} + \frac{1}{5^4} + \cdots = \frac{\pi^4}{96}$$

$$\frac{1}{1^5} - \frac{1}{3^5} + \frac{1}{5^5} - \cdots = \frac{5\pi^5}{1536}$$

$$\frac{1}{1^6} + \frac{1}{3^6} + \frac{1}{5^6} + \cdots = \frac{\pi^6}{960}$$

$$\cdots\cdots\cdots\cdots\cdots\cdots\cdots.$$

10. *Comm. Acad. Sci. Petrop.,* 11, 1739, 116–27, pub. 1750 = *Opera,* (1), 14, 350–63.
11. *Comm. Acad. Sci. Petrop.,* 7, 1734/35, 123–34, pub. 1740 = *Opera,* (1), 14, 73–86.
12. He used the symbol p for π until 1739; π had been introduced by William Jones in 1706.

In the same paper, he first gave the product expansion

$$(26) \qquad \sin s = \left(1 - \frac{s^2}{\pi^2}\right)\left(1 - \frac{s^2}{4\pi^2}\right)\left(1 - \frac{s^2}{9\pi^2}\right)\cdots$$

His argument was simply that $\sin s$ has the zeros $\pm\pi$, $\pm 2\pi$, $\cdots$ (he discards the root 0), and so like every polynomial must have a linear factor corresponding to each of its roots. (In 1743[13] and in his *Introductio*,[14] he gave another derivation to meet criticism.) He treated the right side of (26) as a polynomial, which he set equal to zero, and again by using the relation between the roots and the coefficients he deduced that

$$\frac{1}{1^2} + \frac{1}{2^2} + \frac{1}{3^2} + \frac{1}{4^2} + \cdots = \frac{\pi^2}{6}$$

$$\frac{1}{1^4} + \frac{1}{2^4} + \frac{1}{3^4} + \frac{1}{4^4} + \cdots = \frac{\pi^4}{90},$$

and similar sums for higher even powers in the denominator.

In a later paper[15] Euler obtained one of his finest triumphs,

$$\sum_{\nu=1}^{\infty} \frac{1}{\nu^{2n}} = (-1)^{n-1} \frac{(2\pi)^{2n}}{2(2n)!} B_{2n},$$

where the B_{2n} are the Bernoulli numbers (see below). The connection with the Bernoulli numbers was actually established by Euler a little later in his *Institutiones* of 1755.[16] He also gave in the 1740 paper the sum $\sum_{\nu=1}^{\infty} (1/\nu^n)$ for the first few odd values of n but got no general expression for all odd n.

Euler also worked on harmonic series, that is, series such that the reciprocals of the terms are in arithmetic progression. In particular he showed[17] how one can sum a finite number of terms of the ordinary harmonic series by using the logarithm function. He starts with

$$(27) \qquad \log\left(1 + \frac{1}{x}\right) = \frac{1}{x} - \frac{1}{2x^2} + \frac{1}{3x^3} - \frac{1}{4x^4} + \cdots$$

Then

$$\frac{1}{x} = \log\left(\frac{x+1}{x}\right) + \frac{1}{2x^2} - \frac{1}{3x^3} + \frac{1}{4x^4} - \cdots.$$

13. *Opera*, (1), 14, 138–55.
14. *Opera*, (1), 8, 168.
15. *Comm. Acad. Sci. Petrop.*, 12, 1740, 53–96, pub. 1750 = *Opera*, (1), 14, 407–62.
16. Part II, Chap. 5, ¶124 = *Opera*, (1), 10, 327.
17. *Comm. Acad. Sci. Petrop.*, 7, 1734/35, 150–61, pub. 1740 = *Opera*, (1), 14, 87–100.

Now let $x = 1, 2, 3, \ldots, n$. These substitutions give

$$\frac{1}{1} = \log 2 + \frac{1}{2} - \frac{1}{3} + \frac{1}{4} - \frac{1}{5} + \cdots$$

$$\frac{1}{2} = \log \frac{3}{2} + \frac{1}{2\cdot4} - \frac{1}{3\cdot8} + \frac{1}{4\cdot16} - \frac{1}{5\cdot32} + \cdots$$

$$\frac{1}{3} = \log \frac{4}{3} + \frac{1}{2\cdot9} - \frac{1}{3\cdot27} + \frac{1}{4\cdot81} - \frac{1}{5\cdot243} + \cdots$$

$$\cdots\cdots\cdots\cdots\cdots\cdots\cdots\cdots\cdots\cdots\cdots\cdots\cdots$$

$$\frac{1}{n} = \log \frac{n+1}{n} + \frac{1}{2n^2} - \frac{1}{3n^3} + \frac{1}{4n^4} - \frac{1}{5n^5} + \cdots.$$

By adding and noting that each log term is a difference of two logarithms, one gets

$$\frac{1}{1} + \frac{1}{2} + \frac{1}{3} + \cdots + \frac{1}{n} = \log(n + 1) + \frac{1}{2}\left(1 + \frac{1}{4} + \frac{1}{9} + \cdots + \frac{1}{n^2}\right)$$

$$- \frac{1}{3}\left(1 + \frac{1}{8} + \frac{1}{27} + \cdots + \frac{1}{n^3}\right) + \frac{1}{4}\left(1 + \frac{1}{16} + \frac{1}{81} + \cdots + \frac{1}{n^4}\right) - \cdots,$$

or

$$(28) \qquad 1 + \frac{1}{2} + \frac{1}{3} + \cdots + \frac{1}{n} = \log(n + 1) + C,$$

where C represents the sum of the infinite set of finite arithmetic sums. The value of C was calculated approximately by Euler (it depends upon n, but for large n the value of n does not affect the result much) and he obtained 0.577218. This C is now known as Euler's constant and is denoted by γ. A more accurate representation of γ is obtained nowadays as follows. Subtract $\log n$ from both sides of (28). Now $\log(n + 1) - \log n = \log(1 + 1/n)$ and this approaches 0 as $n \to \infty$. Hence

$$(29) \qquad \gamma = \lim_{n \to \infty}\left(1 + \frac{1}{2} + \frac{1}{3} + \cdots + \frac{1}{n} - \log n\right).$$

Incidentally, no simpler form than (29) has been found for Euler's constant, whereas we do have various expressions for π and e. Moreover, we do not know today whether γ is rational or irrational.

In his "De Seriebus Divergentibus"[18] Euler investigated the divergent series

18. *Novi Comm. Acad. Sci. Petrop.*, 5, 1754/5, 205–37, pub. 1760 = *Opera*, (1), 14, 585–617.

(30) $$y = x - (1!)x^2 + (2!)x^3 - (3!)x^4 + \cdots.$$

Formally this series satisfies the differential equation

(31) $$x^2 y' + y = x.$$

But this differential equation has the integrating factor $x^2 e^{-1/x}$, so that

(32) $$y = e^{1/x} \int_0^x \frac{e^{-1/t}}{t}\, dt$$

is a solution that can be shown by L'Hospital's rule to vanish with x. Euler considered the series (30) to be the series expansion of the function in (32) and (32) as the sum of the series (30). In fact he lets $x = 1$ and obtains

$$1 - 1 + 2! - 3! + 4! - \cdots = e \int_0^1 \frac{e^{-1/t}}{t}\, dt.$$

The remarkable fact about the series (30) is that it can be used to obtain good numerical values for the function (32) because, given a value of x, if we neglect all terms beyond a certain one, the absolute value of the remainder can be shown to be smaller than the absolute value of the first of the neglected terms. Hence the series can be used to obtain good numerical approximations to the integral. Euler was using divergent series to advantage. The full significance of what these divergent series accomplished was not appreciated for another 150 years. (See Chap. 47.)

Another famous result of Euler's in the area of series should be noted. In *Ars Conjectandi* James Bernoulli, treating the subject of probability, introduced the now widely used Bernoulli numbers. He sought a formula for the sums of the positive integral powers of the integers and gave the following formula without demonstration:

(33) $$\sum_{k=1}^{n} k^c = \frac{1}{c+1} n^{c+1} + \frac{1}{2} n^c + \frac{c}{2} B_2 n^{c-1} + \frac{c(c-1)(c-2)}{2\cdot 3\cdot 4} B_4 n^{c-3}$$

$$+ \frac{c(c-1)(c-2)(c-3)(c-4)}{2\cdot 3\cdot 4\cdot 5\cdot 6} B_6 n^{c-5} + \cdots.$$

This series terminates at the last positive power of n. The $B_2, B_4, B_6, \ldots$ are the Bernoulli numbers

(34) $$B_2 = \frac{1}{6}, B_4 = -\frac{1}{30}, B_6 = \frac{1}{42}, B_8 = -\frac{1}{30}, B_{10} = \frac{5}{66}, \cdots.$$

Bernoulli also gave the recurrence relation, which permits one to calculate these coefficients.

Euler's result, the Euler-Maclaurin summation formula, is a generaliza-tion.[19] Let $f(x)$ be a real-valued function of the real variable x. Then (in modern notation) the formula reads

$$(35) \quad \sum_{i=0}^{n} f(i) = \int_0^n f(x)\, dx - \frac{1}{2}[f(n) - f(0)] + \frac{B_2}{2!}[f'(n) - f'(0)]$$

$$+ \frac{B_4}{4!}[f'''(n) - f'''(0)] + \cdots + \frac{B_{2k}}{(2k)!}[f^{(2k-1)}(n) - f^{(2k-1)}(0)] + R_k$$

where

$$(36) \qquad\qquad R_k = \int_0^n f^{(2k+1)}(x) P_{2k+1}(x)\, dx.$$

Here n and k are positive integers. $P_{2k+1}(x)$ is the $(2k+1)$th Bernoulli polynomial (which also appears in Bernoulli's *Ars Conjectandi*), which is given by

$$(37) \qquad P_k(x) = \frac{x^k}{k!} + \frac{B_1}{1!}\frac{x^{k-1}}{(k-1)!} + \frac{B_2}{2!}\frac{x^{k-2}}{(k-2)!} + \cdots + \frac{B_k}{k!},$$

wherein $B_1 = -1/2$ and $B_{2k+1} = 0$ for $k = 1, 2, \cdots$. The series

$$(38) \qquad\qquad \sum_{k=1}^{\infty} \frac{B_{2k}}{(2k)!}[f^{(2k-1)}(n) - f^{(2k-1)}(0)]$$

is divergent for almost all $f(x)$ that occur in applications. Nevertheless, the remainder R_k is less than the first term neglected and so the series in (35) gives a useful approximation to

$$\sum_{i=0}^{n} f(i).$$

The Bernoulli numbers B_i are often defined today by a relation given later by Euler,[20] namely,

$$(39) \qquad\qquad t(e^t - 1)^{-1} = \sum_{i=0}^{\infty} B_i \frac{t^i}{i!}.$$

Independently of Euler, Maclaurin[21] arrived at the same summation formula (35) but by a method a little surer and closer to that which we use today. The remainder was first added and seriously treated by Poisson.[22]

Euler also introduced[23] a transformation of series, still known and

19. *Comm. Acad. Sci. Petrop.*, 6, 1732/3, 68–97, pub. 1738 = *Opera*, (1), 14, 42–72; and *Comm. Acad. Sci. Petrop.*, 8, 1736, 147–58, pub. 1741 = *Opera*, (1), 14, 124–37.
20. *Opera*, (1), 14, 407–62.
21. *Treatise of Fluxions*, 1742, p. 672.
22. *Mém. de l'Acad. des Sci., Inst. France*, 6, 1823, 571–602, pub. 1827.
23. *Inst. Cal. Diff.*, 1755, p. 281.

used. Given a series $\sum_{n=0}^{\infty} b_n$, he wrote it as $\sum_{n=0}^{\infty} (-1)^n a_n$. Then by a number of formal algebraic steps he showed that

$$(40) \qquad \sum_{n=0}^{\infty} (-1)^n a_n = \sum_{n=0}^{\infty} (-1)^n \frac{\Delta^n a_0}{2^{n+1}},$$

wherein the Δ^n denotes the nth finite difference (sec. 3). The advantage of this transformation, in modern terms, is to convert a convergent series into a more rapidly converging one. However, for Euler, who did not usually distinguish convergent and divergent series, the transformation could also transform divergent series into convergent ones. If one applies (40) to

$$(41) \qquad 1 - 1 + 1 - 1 + \cdots,$$

then the right side of (40) yields $1/2$. Likewise for the series

$$(42) \qquad 1 - 2 + 2^2 - 2^3 + 2^4 \cdots$$

(40) gives

$$(43) \qquad \sum_{n=0}^{\infty} (-1)^n 2^n = \frac{1}{2}(1) + \frac{1}{4}(-1) + \frac{1}{8}(1) - \frac{1}{16}(-1) \cdots = \frac{1}{3}.$$

These results are, of course, the same as those Euler got above (see [16] and [17]) by taking the sum of the series to be the value of the function from which the series is derived.

The spirit of Euler's methods should be clear. He is the great manipulator and pointed the way to thousands of results later established rigorously.

One other famous series must be mentioned. In his *Methodus Differentialis*[24] James Stirling gave the series we now write as

$$(44) \quad \log n! = \left(n + \frac{1}{2}\right) \log n - n + \log \sqrt{2\pi} + \frac{B_2}{1 \cdot 2} \frac{1}{n} + \frac{B_4}{3 \cdot 4} \frac{1}{n^3} + \cdots$$

$$+ \frac{B_{2k}}{(2k-1)(2k)} \frac{1}{n^{2k-1}} + \cdots,$$

which is equivalent to

$$(45) \quad n! = \left(\frac{n}{e}\right)^n \sqrt{2\pi n} \exp\left[\frac{B_2}{1 \cdot 2} \frac{1}{n} + \cdots + \frac{B_{2k}}{(2k-1)2k} \frac{1}{n^{2k-1}} + \cdots\right].$$

Stirling gave the first five coefficients and a recurrence formula for determining the succeeding ones. Though the series for $\log n!$ is divergent, Stirling calculated $\log_{10}(1000!)$, which is 2567 plus a decimal, to ten decimal places by using only a few terms of his series. De Moivre in 1730 (*Miscellanea*

24. 1730, p. 135.

Analytica) gave a similar formula. For large n, $n! \sim (n/e)^n \sqrt{2\pi n}$; though given by de Moivre, it is known as Stirling's approximation.

5. *Trigonometric Series*

The eighteenth-century mathematicians also worked extensively with trigonometric series, especially in their astronomical theory. The usefulness of such series in astronomy is evident from the fact that they are periodic functions and astronomical phenomena are largely periodic. This work was the beginning of a vast subject whose full significance was not appreciated in the eighteenth century. The problem that launched the use of trigonometric series was interpolation, particularly to determine the positions of the planets between those obtained by observation. The same series were introduced in the early work on partial differential equations (see Chap. 22) but curiously the two lines of thought were kept separate even though the same men worked on both problems.

By a trigonometric series is meant any series of the form

(46)
$$\frac{1}{2} a_0 + \sum_{n=1}^{\infty} (a_n \cos nx + b_n \sin nx)$$

with a_n and b_n constant. If such a series represents a function $f(x)$, then

(47)
$$a_n = \frac{1}{\pi} \int_0^{2\pi} f(x) \cos nx \, dx, \qquad b_n = \frac{1}{\pi} \int_0^{2\pi} f(x) \sin nx \, dx$$

for $n = 0, 1, 2, \ldots$. The attainment of these formulas for the coefficients was one of the chief results of the theory, though we shall say nothing at present about the conditions under which these are necessarily the values of a_n and b_n.

As early as 1729 Euler had undertaken the problem of interpolation; that is, given a function $f(x)$ whose values for $x = n$, n positive and integral, are prescribed, to find $f(x)$ for other values of x. In 1747 he applied the method he had obtained to a function arising in the theory of planetary perturbations and secured a trigonometric series representation of the function. In 1753[25] he published the method he had found in 1729.

First he tackled the problem when the given conditions are $f(n) = 1$ for each n and sought a periodic solution that is 1 for integral x. His reasoning is interesting because it illustrates the analysis of the period. He lets $f(x) = y$, and by Taylor's theorem writes

(48)
$$f(x + 1) = y + y' + \frac{1}{2} y'' + \frac{1}{6} y''' + \cdots.$$

25. *Novi Comm. Acad. Sci. Petrop.*, 3, 1750/51, 36–85, pub. 1753 = *Opera*, (1), 14, 463–515.

Since $f(x + 1)$ is to equal $f(x)$, y must satisfy the linear differential equation of infinite order

(49) $$y' + \frac{1}{2}y'' + \frac{1}{6}y''' + \cdots = 0.$$

He now applied his method of solving linear ordinary differential equations of finite order published in 1743 (see Chap. 21). That is, he set up the auxiliary equation

(50) $$z + \frac{1}{2}z^2 + \frac{1}{6}z^3 + \cdots = 0.$$

This equation, in view of the series for e, is

$$e^z - 1 = 0.$$

Next he determines the roots of this last equation. He starts with the equation

$$\left(1 + \frac{z}{n}\right)^n = 1,$$

which is a polynomial of the nth degree. According to a theorem which Cotes (1722) and Euler independently in his *Introductio*[26] had proven, this polynomial has the linear factor z and the quadratic factors

$$\left(1 + \frac{z}{n}\right)^2 - 2\left(1 + \frac{z}{n}\right)\cos\frac{2k\pi}{n} + 1, \qquad k = 1, 2, \ldots, < \frac{n}{2}.$$

By virtue of the trigonometric identity for $\sin z$ in terms of $\cos 2z$ these factors are the same as

$$4\left(1 + \frac{z}{n}\right)\sin^2\frac{k\pi}{n} + \frac{z^2}{n^2}.$$

The roots of (50) are not affected if we divide each factor by $4\sin^2 k\pi/n$ (for the respective k), and so the quadratic factors are

$$1 + \frac{z}{n} + \frac{z^2}{4n^2\sin^2\dfrac{k\pi}{n}}.$$

For $n = \infty$, the term z/n is 0. The quantity $\sin k\pi/n$ is replaced by $k\pi/n$, and so the factors become

$$1 + \frac{z^2}{4k^2\pi^2}.$$

26. Vol. 1, Chap. 14.

To such a factor in the auxiliary equation (50) there correspond the roots $z = \pm i2k\pi$, and hence the integral

$$\alpha_k \sin 2k\pi x + A_k \cos 2k\pi x$$

of (49). The linear factor z mentioned above gives rise to a constant integral. Since $f(0) = 1$ is an initial condition, Euler finally obtains

$$y = 1 + \sum_{k=1}^{\infty} \{\alpha_k \sin 2k\pi x + A_k(\cos 2k\pi x - 1)\}.$$

The coefficients α_k and A_k are still subject to the condition that $f(n) = 1$ for each n.

This paper also contains a result which is formally identical with what came to be called the Fourier expansion of an arbitrary function, as well as the determination of the coefficients by integrals. Specifically Euler showed that the general solution of the functional equation

$$f(x) = f(x - 1) + X(x)$$

is

$$f(x) = \int_0^x X(\xi)\, d\xi + 2 \sum_{n=1}^{\infty} \cos 2n\pi x \int_0^x X(\xi) \cos 2n\pi\xi\, d\xi$$

$$+ 2 \sum_{n=1}^{\infty} \sin 2n\pi x \int_0^x X(\xi) \sin 2n\pi\xi\, d\xi.$$

Here we have a function expressed as a trigonometric series in the year 1750–51. Euler maintained that his was the most general solution of the interpolation problem. If so, it surely included the representation of polynomials by trigonometric series. But, as we shall see in Chapter 22, Euler denied this in the arguments on the vibrating string and related problems.

In 1754 d'Alembert [27] considered the problem of the expansion of the reciprocal of the distance between two planets in a series of cosines of the multiples of the angle between the rays from the origin to the planets, and here too the definite integral expressions for the coefficients in Fourier series can be found.

In another work Euler obtains trigonometric series representations of functions in a totally different fashion. [28] He starts with the geometric series

$$\sum_{n=0}^{\infty} a^n(\cos x + i \sin x)^n, \qquad i = \sqrt{-1}$$

27. *Recherches sur différens points importans du système du monde,* 1754, Vol. II, p. 66.
28. *Novi Comm. Acad. Sci. Petrop.,* 5, 1754/5, 164–204, pub. 1760 = *Opera,* (1), 14, 542–84; see also *Opera,* (1), 15, 435–97, for another method.

and by summing it obtains

$$\frac{1}{1 - a(\cos x + i \sin x)}.$$

He then uses standard formulas to replace powers of $\cos x$ and $\sin x$ by $\cos nx$ and $\sin nx$ (which amounts to de Moivre's theorem) and obtains

$$\frac{1}{1 - a(\cos x + i \sin x)} = \sum_{n=0}^{\infty} a^n (\cos nx + i \sin nx).$$

By multiplying numerator and denominator on the left by the complex conjugate of the denominator, separating the $n = 0$ term on the right and putting it on the left side, dividing through by a, and separating real and imaginary parts, he obtains

$$\frac{a \cos x - a^2}{1 - 2a \cos x + a^2} = \sum_{n=1}^{\infty} a^n \cos nx$$

$$\frac{a \sin x}{1 - 2a \cos x + a^2} = \sum_{n=1}^{\infty} a^n \sin nx.$$

So far his results are not surprising. He now lets $a = \pm 1$ and obtains, for example,

(51) $$\frac{1}{2} = 1 \pm \cos x + \cos 2x \pm \cos 3x + \cos 4x \pm \cdots.$$

(Actually the series are divergent.) He then integrates and obtains

(52) $$\frac{\pi - x}{2} = \sin x + \frac{1}{2} \sin 2x + \frac{1}{3} \sin 3x + \cdots,$$

(which holds for $0 < x < \pi$ and equals 0 for $x = 0$ and π) and

(53) $$\frac{x}{2} = \sin x - \frac{1}{2} \sin 2x + \frac{1}{3} \sin 3x - \frac{1}{4} \sin 4x + \cdots,$$

(which converges in $-\pi < x < \pi$). An integration of the latter and evaluation at $x = 0$ to determine the constant of integration gives

(54) $$\frac{x^2}{4} - \frac{\pi^2}{4} = -\cos x + \frac{1}{4} \cos 2x - \frac{1}{9} \cos 3x + \frac{1}{16} \cos 4x - \cdots.$$

Euler believed that the latter two series [which are convergent in $(-\pi < x < \pi)$] represent the respective functions for all values of x. Moreover, by successively differentiating (51), Euler deduced that

$$\sin x \pm 2 \sin 2x + 3 \sin 3x \pm \cdots = 0$$
$$\cos x \pm 4 \cos 2x + 9 \cos 3x \pm \cdots = 0$$

and other such equations. Daniel Bernoulli, who had also given expansions such as (52), (53), and (54), recognized that the series represent the functions only for certain ranges of x values.

In 1757, while studying perturbations caused by the sun, Clairaut[29] took a far bolder step. He says he will represent *any* function in the form

$$(55) \qquad f(x) = A_0 + 2 \sum_{n=1}^{\infty} A_n \cos nx.$$

He regards the problem as one of interpolation and so uses the function values at the x-values

$$\frac{2\pi}{k}, \quad \frac{4\pi}{k}, \quad \frac{6\pi}{k}, \dots$$

and after some manipulations obtains

$$A_0 = \frac{1}{k} \sum_{\mu} f\left(\frac{2\mu\pi}{k}\right)$$

$$A_n = \frac{1}{k} \sum_{\mu} f\left(\frac{2\mu\pi}{k}\right) \cos \frac{2\mu n\pi}{k}.$$

By letting k become infinite, Clairaut arrives at

$$(56) \qquad A_n = \frac{1}{2\pi} \int_0^{2\pi} f(x) \cos nx \, dx,$$

which is the correct formula for the A_n.

Lagrange, in his research on the propagation of sound,[30] obtained the series (51) and defended the fact that the sum is $1/2$. Yet neither Euler nor Lagrange commented on the remarkable fact that they had expressed non-periodic functions in the form of trigonometric series. However, somewhat later they did observe this fact in another connection. D'Alembert had often given the example of $x^{2/3}$ as a function that could not be expanded in a trigonometric series. Lagrange showed him in a letter[31] of August 15, 1768 that $x^{2/3}$ can indeed be expressed in the form

$$x^{2/3} = a + b \cos 2x + c \cos 4x + \cdots.$$

D'Alembert objected and gave counterarguments, such as that the derivatives of the two sides are not equal for $x = 0$. Also, by Lagrange's method one could express $\sin x$ as a cosine series; yet $\sin x$ is an odd function,

29. *Hist. de l'Acad. des Sci.*, Paris, 1754, 545 ff., pub. 1759.
30. *Misc. Taur.*, 1, 1759 = *Œuvres*, 1, 110.
31. *Lagrange, Œuvres*, 13, 116.

whereas the right side would be an even one. The problem was not resolved in the eighteenth century.

In 1777,[32] Euler, working on a problem in astronomy, actually obtained the coefficients of a trigonometric series by using the orthogonality of the trigonometric functions, the method we use today. That is, from

(57) $$f(x) = \frac{a_0}{2} + \sum_{k=1}^{\infty} a_k \cos \frac{k\pi x}{l}$$

he deduced that

$$a_k = \frac{2}{l} \int_0^l f(s) \cos \frac{k\pi s}{l} \, ds.$$

He had first obtained it, in the immediately preceding paper, in a somewhat complicated fashion, then realized that he could obtain it directly by multiplying both sides of (57) by $\cos(\nu\pi x/l)$, integrating term by term, and applying the relations

$$\int_0^l \cos \frac{\nu\pi x}{l} \cos \frac{k\pi x}{l} \, dx = \begin{cases} 0 & \text{if } \nu \neq k \\ l/2 & \text{if } \nu = k \neq 0 \\ l & \text{if } \nu = k = 0 \end{cases}.$$

Throughout all of the above work on trigonometric series ran the paradox that, although all sorts of functions were being represented by trigonometric series, Euler, d'Alembert, and Lagrange never abandoned the position that *arbitrary* functions could not be represented by such series. The paradox is partially explained by the fact that the trigonometric series were assumed to hold where other evidence, in some cases physical, seemed to assure this fact. They then felt free to assume the series and deduce the formulas for the coefficients. This issue of whether any function can be represented by a trigonometric series became central.

6. *Continued Fractions*

We have already noted (Chap. 13, sec. 2) the use of continued fractions to obtain approximations to irrational numbers. Euler took up this subject. In his first paper on it,[33] entitled "De Fractionibus Continuis," he derived a number of interesting results, such as that every rational number can be expressed as a finite continued fraction. He then gave the expansions

$$e - 1 = 1 + \frac{1}{1+} \frac{1}{2+} \frac{1}{1+} \frac{1}{1+} \frac{1}{4+} \frac{1}{1+} \frac{1}{1+} \frac{1}{6+} \cdots,$$

32. *Nova Acta Acad. Sci. Petrop.*, 11, 1793, 114–32, pub. 1798 = *Opera*, (1), 16, Part 1, 333–55.
33. *Comm. Acad. Sci. Petrop.*, 9, 1737, 98–137, pub. 1744 = *Opera*, (1), 14, 187–215.

which had already appeared in a paper by Cotes in the *Philosophical Transactions* of 1714, and

$$\frac{e+1}{e-1} = 2 + \frac{1}{6+} \frac{1}{10+} \frac{1}{14+} \cdots.$$

He showed substantially that e and e^2 are irrational.

The foundations of a theory of continued fractions were laid by Euler in his *Introductio* (Chap. 18). There he showed how to go from a series to a continued fraction representation of the series, and conversely.

Euler's work on continued fractions was used by Johann Heinrich Lambert (1728–77), a colleague of Euler and Lagrange at the Berlin Academy of Sciences, to prove[34] that if x is a rational number (not 0), then e^x and tan x cannot be rational. He thereby proved not only that e^x for positive integral x is irrational, but that all rational numbers have irrational natural (base e) logarithms. From the result on tan x, it follows, since $\tan(\pi/4) = 1$, that neither $\pi/4$ nor π can be rational. Lambert actually proved the convergence of the continued fraction expansion for tan x.

Lagrange[35] used continued fractions to find approximations to the irrational roots of equations, and, in another paper in the same journal,[36] he got approximate solutions of differential equations in the form of continued fractions. In the 1768 paper, Lagrange proved the converse of a theorem that Euler had proved in his 1744 paper. The converse states that a real root of a quadratic equation is a periodic continued fraction.

7. *The Problem of Convergence and Divergence*

We are aware today that the eighteenth-century work on series was largely formal, and that the question of convergence and divergence was certainly not taken too seriously; neither, however, was it entirely ignored.

Newton,[37] Leibniz, Euler, and even Lagrange regarded series as an extension of the algebra of polynomials and hardly realized that they were introducing new problems by extending sums to an infinite number of terms. Consequently, they were not quite prepared to face the problems that infinite series thrust upon them; but the apparent difficulties that did arise caused them at least occasionally to bring up these questions. What is especially interesting is that the correct resolution of the paradoxes and other difficulties was often voiced and just as often ignored.

Even some seventeenth-century men had observed the distinction between convergence and divergence. In 1668 Lord Brouncker, treating

34. *Hist. de l'Acad. de Berlin*, 1761, 265–322, pub. 1768 = *Opera*, 2, 112–59.
35. *Nouv. Mém. de l'Acad. de Berlin*, 23, 1767, 311–52, pub. 1769 = *Œuvres*, 2, 539–78, and 24, 1768, 111–80, pub. 1770 = *Œuvres*, 2, 581–652.
36. 1776 = *Œuvres*, 4, 301–34.
37. See the quotation from Newton in Chap. 17, sec. 3.

the relation between $\log x$ and the area under $y = 1/x$, demonstrated the convergence of the series for $\log 2$ and $\log 5/4$ by comparison with a geometric series. Newton and James Gregory, who made much use of numerical values of series to calculate logarithmic and other function tables and to evaluate integrals, were aware that the sums of series can be finite or infinite. The terms "convergent" and "divergent" were actually used by James Gregory in 1668, but he did not develop the ideas. Newton recognized the need to consider convergence but did no more than affirm that power series converge for small values of the variable at least as well as the geometric series. He also remarked that some series can be infinite for some values of x and so be useless, as, for example, the series for $y = \sqrt{ax - x^2}$ at $x = a$.

Leibniz, too, felt some concern about convergence and noted in a letter of October 25, 1713, to John Bernoulli what is now a theorem, that a series whose terms alternate in sign and decrease in absolute value monotonically to zero converges.[38]

Maclaurin, in his *Treatise of Fluxions* (1742), used series as a regular method for integration. He says, "When a fluent cannot be represented accurately in algebraic terms, it is then to be expressed by a converging series." That the terms of a convergent series must continually decrease and become less than any quantity howsoever small that can be assigned he also recognized. "In that case a few terms at the beginning of the series will be nearly equal to the value of the whole." In the *Treatise* Maclaurin gave the integral test (independently discovered by Cauchy) for the convergence of an infinite series: $\sum_n \phi(n)$ converges if and only if $\int_a^\infty \phi(x)$ is finite, provided that $\phi(x)$ is finite and of the same sign for $a \leq x \leq \infty$. Maclaurin gave it in geometrical form.

Some ideas about convergence were also expressed by Nicholas Bernoulli (1687–1759) in letters to Leibniz of 1712 and 1713. In a letter of April 7, 1713,[39] Bernoulli says the series

$$(1 + x)^n = 1 + nx + \frac{n(n - 1)}{2} x^2 + \cdots$$

has no sum when x is negative and numerically greater than 1 if n is fractional and has an even denominator. That is, the (arithmetical) divergence of a series is not the only reason for a series not to have a sum. Thus for $x > 1$ both series

$$(1 - x)^{-1/3} = 1 + \frac{1}{3} x + \frac{1 \cdot 4}{3 \cdot 6} x^2 + \frac{1 \cdot 4 \cdot 7}{3 \cdot 6 \cdot 9} x^3 + \cdots$$

$$(1 - x)^{-1/2} = 1 + \frac{1}{2} x + \frac{1 \cdot 3}{2 \cdot 4} x^2 + \frac{1 \cdot 3 \cdot 5}{2 \cdot 4 \cdot 6} x^3 + \cdots$$

38. *Math. Schriften*, 3, 922–23. Leibniz also gave an incorrect proof in a letter to John of January 10, 1714 = *Math. Schriften*, 3, 926.
39. Leibniz: *Math. Schriften*, 3, 980–84.

are divergent, but the first series has a possible value and the second an imaginary value. One cannot distinguish the two by examining the series because the *remainders are missing*. However, Nicholas did not set up a clear concept of convergence. In a reply of June 28, 1713, Leibniz [40] uses the term "advergent" for series that converge (roughly in our sense) and agrees that non-advergent series may be impossible or infinitely large.

There is no doubt that Euler saw some of the difficulties with divergent series, and in particular the difficulty in using them for computations, but he certainly was unclear about the concepts of convergence and divergence. He did recognize that the terms must become infinitely small for convergence. The letters described below tell us, indirectly, something of his views.

Nicholas Bernoulli (1687–1759), in correspondence with Euler during 1742–43, had challenged some of Euler's ideas and work. He pointed out that Euler's use in his paper of 1734/35 (see sec. 4) of

$$\sin s = s - \frac{s^3}{3!} + \frac{s^5}{5!} + \cdots = \left(1 - \frac{s^2}{\pi^2}\right)\left(1 - \frac{s^2}{4\pi^2}\right)\left(1 - \frac{s^2}{9\pi^2}\right)\cdots$$

does yield

$$\frac{\pi^2}{6} = 1 + \frac{1}{2^2} + \frac{1}{3^2} + \frac{1}{4^2} + \cdots,$$

but a proof of the convergence of the basic series in s is missing. In a letter of April 6, 1743,[41] he says he cannot imagine that Euler can believe a divergent series gives the exact value of some quantity or function. He points out that the remainder is lacking. Thus $1/(1 - x)$ cannot equal $1 + x + x^2 + \cdots$ because the remainder, namely, $x^{\infty + 1}/(1 - x)$, is missing.

In another letter of 1743, Bernoulli says Euler must distinguish between a finite sum and a sum of an infinite number of terms. There is no last term in the latter case. Hence one cannot use for infinite polynomials (as Euler did) the relation between the roots and coefficients of a polynomial of finite degree. For polynomials with an infinite number of terms one cannot speak of the sum of the roots.

Euler's answers to these letters of Bernoulli are not known. In writing to Goldbach on August 7, 1745,[42] Euler refers to Bernoulli's argument that divergent series such as

$$+1 - 2 + 6 - 24 + 120 - 720 - \cdots$$

have no sum but says that these series have a definite *value*. He notes that we should not use the term "sum" because this refers to actual addition. He then states the general principle which explains what he means by a

40. *Math. Schriften*, 3, 986.
41. Fuss: *Correspondance*, 2, 701 ff.
42. Fuss: *Correspondance*, 1, 324.

definite value. He points out that the divergent series come from finite algebraic expressions and then says that the value of the series is *the value of the algebraic expression from which the series comes*. In the paper of 1754/55 (sec. 4), he adds, "Whenever an infinite series is obtained as the development of some closed expression, it may be used in mathematical operations as the equivalent of that expression, even for values of the variable for which the series diverges." He repeats the first principle in his *Institutiones* of 1755:

> Let us say, therefore, that the sum of any infinite series is the finite expression, by the expansion of which the series is generated. In this sense the sum of the infinite series $1 - x + x^2 - x^3 + \cdots$ will be $1/(1 + x)$, because the series arises from the expansion of the fraction, whatever number is put in place of x. If this is agreed, the new definition of the word sum coincides with the ordinary meaning when a series converges; and since divergent series have no sum in the proper sense of the word, no inconvenience can arise from this terminology. Finally, by means of this definition, we can preserve the utility of divergent series and defend their use from all objections.[43]

It is fairly certain that Euler meant to limit the doctrine to power series.

In writing to Nicholas Bernoulli in 1743, Euler did say that he had had grave doubts as to the use of divergent series but that he had never been led into error by using his definition of sum.[44] To this Bernoulli replied that the same series might arise from the expansion of two different functions and, if so, the sum would not be unique.[45] Euler then wrote to Goldbach (in the letter of August 7, 1745): "Bernoulli gives no examples and I do not believe it possible that the same series could come from two truly different algebraic expressions. Hence it follows unquestionably that any series, divergent or convergent, has a definite sum or value."

There is an interesting sequel to this argument. Euler rested on his contention that the sum of series such as

$$(58) \qquad 1 - 1 + 1 - 1 + 1 \cdots$$

could be the value of the function from which the series comes. Thus the above series comes from $1/(1 + x)$ when $x = 1$, and so has the value $1/2$. However, Jean-Charles (François) Callet (1744–99), in an unpublished memorandum submitted to Lagrange (Lagrange approved it for publication in the *Mémoires* of the Academy of Sciences of Paris but it was never published), pointed out some forty years later that

$$(59) \qquad \frac{1 + x + \cdots + x^{m-1}}{1 + x + \cdots + x^{n-1}} = \frac{1 - x^m}{1 - x^n}$$
$$= 1 - x^m + x^n - x^{n+m} + x^{2n} - \cdots.$$

43. Paragraphs 108–11.
44. *Opera Posthuma*, 1, 536.
45. April 6, 1743; Fuss: *Correspondance*, 2, 701 ff.

Hence for $x = 1$ (and $m < n$), since the left side is m/n, the sum of the right side must also be m/n, where m and n are at our disposal.

Lagrange[46] considered Callet's objection and argued that it was incorrect. He used Leibniz's probability argument thus: Suppose $m = 3$ and $n = 5$. Then the *full* series on the right side of (59) is

$$1 + 0 + 0 - x^3 + 0 + x^5 + 0 + 0 - x^8 + 0 + x^{10} + 0 - \cdots.$$

Now if one takes for $x = 1$, the sum of the first term, the first two, the first three, ..., then in each five of these partial sums three are equal to 1 and two are equal to 0. Hence the most probable value (mean value) is $3/5$; and this is the value of the series in (59) for $m = 3$ and $n = 5$. Incidentally, Poisson, without mentioning Lagrange, repeats Lagrange's argument.[47]

Euler did say that great care should be exercised in the summation of divergent series. He also made a distinction between divergent series and semiconvergent series, such as (58), that oscillate in value as more and more terms are added but do not become infinite. Certainly he recognized the distinction between convergent series and divergent ones. In one case (1747), where he used infinite series to calculate the attraction that the earth, as an oblate spheroid, exerts on a particle at the pole, he says the series "converges vehemently."

Lagrange, too, showed some awareness of the distinction between convergence and divergence. In his earlier writings he was indeed lax on this matter. In one paper[48] he says that a series will represent a number if it converges to its extremity, that is, if its nth term approaches 0. Later, toward the end of the eighteenth century, when he worked with Taylor's series, he gave what we call Taylor's theorem,[49] namely,

$$f(x + h) = f(x) + f'(x)h + f''(x)\frac{h^2}{2!} + \cdots + f^{(n)}(x)\frac{h^n}{n!} + R_n$$

where

$$R_n = f^{(n+1)}(x + \theta h)\frac{h^{n+1}}{(n+1)!}$$

and θ is between 0 and 1 in value. This expression for R_n is still known as Lagrange's form of the remainder. Lagrange said that the Taylor (infinite)

46. *Mém. de l'Acad. des Sci.*, *Inst. France*, 3, 1796, 1–11, pub. 1799; this article does not appear in the *Œuvres*.

47. *Jour. de l'Ecole Poly.*, 12, 1823, 404–509. If one insists on using the full *power* series, then Lagrange's argument makes more sense. It can be rigorized by applying Frobenius's definition of summability (Chap. 47, sec. 4).

48. *Hist. de l'Acad. de Berlin*, 24, 1770 = *Œuvres*, 3, 5–73, p. 61 in particular.

49. *Théorie des fonctions*, 2nd ed., 1813, Chap. 6 = *Œuvres*, 9, 69–85. The mean value theorem of the differential calculus, $f(b) - f(a) = f'(c)(b - a)$, is due to Lagrange (1797). Later it was used to derive Taylor's theorem as in modern books.

series should not be used without consideration of the remainder. However, he did not investigate the idea of convergence or the relation of the value of the remainder to the convergence of the infinite series. He thought that one need consider only a finite number of terms of the series, enough to make the remainder small. Convergence was considered later by Cauchy, who stressed Taylor's theorem as primary, as well as the fact that to obtain a convergent series the remainder must approach 0.

D'Alembert, too, distinguished convergent from divergent series. In his article "Série" in the *Encyclopédie* he says, "When the progression or series approaches some finite quantity more and more, and, consequently, the terms of the series, or quantities of which it is composed, go on diminishing, one calls it a convergent series, and if one continues to infinity, it will finally become equal to this quantity. Thus $1/2 + 1/4 + 1/8 + 1/16 + \cdots$ form a series which always approaches 1 and which will become equal to it finally when the series is continued to infinity." In 1768 d'Alembert expressed doubts about the use of nonconvergent series. He said, "As for me, I avow that all the reasonings based on series that are not convergent . . . appear to me very suspect, even when the results are in accord with truths arrived at in other ways."[50] In view of the effective uses of series by John Bernoulli and Euler, the doubts such as d'Alembert expressed went unheeded in the eighteenth century. In this same volume d'Alembert gave a test for the absolute convergence of the series $u_1 + u_2 + u_3 + \cdots$; namely, if for all n greater than some fixed value r, the ratio $|u_{n+1}/u_n| < \rho$ where ρ is independent of n and less than 1, the series converges absolutely.[51]

Edward Waring (1734–98), Lucasian professor of mathematics at Cambridge University, held advanced views on convergence. He taught that

$$1 + \frac{1}{2^n} + \frac{1}{3^n} + \frac{1}{4^n} + \cdots$$

converges when $n > 1$ and diverges when $n < 1$. He also gave (1776) the well-known test for convergence and divergence, now known as the ratio test and attributed to Cauchy. The ratio of the $(n + 1)$st to the nth term is formed, and if the limit as $n \to \infty$ is less than 1, the series converges; if greater than 1, the series diverges. No conclusion may be drawn when the limit is 1.

Though Lacroix said several nonsensical things about series in the 1797 edition of his influential *Traité du calcul différentiel et du calcul intégral*, he was more cautious in his second edition. Speaking of

$$\frac{a}{a - x} = 1 + \frac{x}{a} + \frac{x^2}{a^2} + \frac{x^3}{a^3} + \cdots,$$

50. *Opuscules mathématiques*, 5, 1768, 183.
51. Pages 171–82.

he says that one should speak of the series as a *development* of the function because the series does not always have the *value* of the function to which it belongs.[52] The series, he says, gives the value of the function only for $|x| < |a|$. He continues with a thought already expressed by Euler, that the infinite series is nevertheless tied in with the function for all x. In any analytical work involving the series we would be right to conclude that we are dealing with the function. Thus if we discover some property of the series, we may be sure this property holds for the function. To perceive the truth of this assertion, it is sufficient to observe that the series verifies the equation that characterizes the function. For example, for $y = a/(a - x)$ we have

$$a - (a - x)y = 0.$$

But if one substitutes the series for y in this last equation, he will see that the series also satisfies it. One knows, Lacroix continues, that it would be the same for any other example; and he points to the great number presented in the text.

It is fair to say that in the eighteenth-century work on infinite series the formal view dominated. On the whole, the mathematicians even resented any limitations, such as the need to think about convergence. Their work produced useful results, and they were satisfied with this pragmatic sanction. They did exceed the bounds of what they could justify, but they were at least prudent in their use of divergent series. As we shall see, the insistence on restricting the use of series to convergent ones won out during most of the nineteenth century. But the eighteenth-century men were ultimately vindicated; two vital ideas that they glimpsed in infinite series were later to gain acceptance. The first was that divergent series can be useful for numerical approximations of functions; the second, that a series may represent a function in analytical operations, even though the series is divergent.

Bibliography

Bernoulli, James: *Ars Conjectandi*, 1713, reprinted by Culture et Civilisation, 1968.
———: *Opera*, 2 vols., 1744, reprinted by Birkhäuser, 1968.
Bernoulli, John: *Opera Omnia*, 4 vols., 1742, reprinted by Georg Olms, 1968.
Burkhardt, H.: "Trigonometrische Reihen und Integrale bis etwa 1850," *Encyk. der math. Wiss.*, B. G. Teubner, 1914–15, 2, Part 1, pp. 825–1354.
———: "Entwicklungen nach oscillirenden Functionen," *Jahres. der Deut. Math.-Verein.*, Vol. 10, 1908, pp. 1–1804.
———: "Über den Gebrauch divergenter Reihen in der Zeit 1750–1860," *Math. Ann.*, 70, 1911, 189–206.

52. 1810–19, 3 vols.; Vol. 1, p. 4.

Cantor, Moritz: *Vorlesungen über Geschichte der Mathematik*, B. G. Teubner, 1898, Vol. 3, Chaps. 85, 86, 97, 109, 110.

Dehn, M., and E. D. Hellinger: "Certain Mathematical Achievements of James Gregory," *Amer. Math. Monthly*, 50, 1943, 149–63.

Euler, Leonhard: *Opera Omnia*, (1), Vols. 10, 14, and 16 (2 parts), B. G. Teubner and Orell Füssli, 1913, 1924, 1933, and 1935.

Fuss, Paul Heinrich von: *Correspondance mathématique et physique de quelques célèbres géomètres du XVIIIème siècle*, 2 vols., 1843, Johnson Reprint Corp., 1967.

Hofmann, Joseph E.: "Über Jakob Bernoullis Beiträge zur Infinitesimal-mathematik," *L'Enseignement Mathématique*, (2), 2, 1956, 61–171; also published separately by Institut de Mathématiques, Genève, 1957.

Montucla, J. F.: *Histoire des mathématiques*, A. Blanchard (reprint), 1960, Vol. 3, pp. 206–43.

Reiff, R. A.: *Geschichte der unendlichen Reihen*, H. Lauppsche Buchhandlung, 1889; Martin Sändig (reprint), 1969.

Schneider, Ivo: "Der Mathematiker Abraham de Moivre (1667–1754)," *Archive for History of Exact Sciences*, 5, 1968, 177–317.

Smith, David Eugene: *A Source Book in Mathematics*, Dover (reprint), 1959, Vol. 1, pp. 85–90, 95–98.

Struik, D. J.: *A Source Book in Mathematics, 1200–1800*, Harvard University Press, 1969, pp. 111–15, 316–24, 328–33, 338–41, 369–74.

Turnbull, H. W.: *James Gregory Tercentenary Memorial Volume*, Royal Society of Edinburgh, 1939.

――――: *The Correspondence of Issac Newton*, Cambridge University Press, 1959, Vol. 1.

21

Ordinary Differential Equations in the Eighteenth Century

> A traveler who refuses to pass over a bridge until he has
> personally tested the soundness of every part of it is not likely
> to go far; something must be risked, even in mathematics.
>
> HORACE LAMB

1. *Motivations*

The mathematicians sought to use the calculus to solve more and more physical problems and soon found themselves obliged to handle a new class of problems. They wrought more than they had consciously sought. The simpler problems led to quadratures that could be evaluated in terms of the elementary functions. Somewhat more difficult ones led to quadratures which could not be so expressed, as was the case for elliptic integrals (Chap. 19, sec. 4). Both of these types fall within the purview of the calculus. However, solution of the still more complicated problems demanded specialized techniques; thus the subject of differential equations arose.

Several classes of physical problems motivated the investigations in differential equations. Problems in the area now generally known as the theory of elasticity were one class. A body is elastic if it deforms under the action of a force and recovers its original shape when the force is removed. The most practical problems are concerned with the shapes assumed by beams, vertical and horizontal, when loads are applied. These problems, treated empirically by the builders of the great medieval cathedrals, were approached mathematically in the seventeenth century by men such as Galileo, Edme Mariotte (1620?–84), Robert Hooke (1635–1703), and Wren. The behavior of beams is one of the two sciences Galileo treats in the *Dialogues Concerning Two New Sciences*. Hooke's investigation of springs led to his discovery of the law that states that the restoring force of a spring that is stretched or contracted is proportional to the stretch or contraction. The men of the eighteenth century, armed with more mathematics, began their work in elasticity by tackling such problems as the shape assumed by an inelastic but flexible rope suspended from two fixed points, the shape of an inelastic but flexible cord or chain suspended from one fixed point and set into vibration, the shape assumed by an elastic vibrating string held fixed at its ends, the shape of a rod when fixed at its ends and subject to a load, and the shape when the rod is set into vibration.

The pendulum continued to interest the mathematicians. The exact differential equation for the circular pendulum, $d^2\theta/dt^2 + mg \sin \theta = 0$, dcficd treatment, and even the approximate one obtained by replacing $\sin \theta$ by θ had yet to be treated analytically. Moreover, the period of a circular pendulum is not strictly independent of the amplitude of the motion, and the search was undertaken for a curve along which the bob of a pendulum must swing for the period to be strictly independent of the amplitude. Huygens had solved this geometrically by introducing the cycloid; but the analytical solution was yet to be fashioned.

The pendulum was closely connected with two other major investigations of the eighteenth century, the shape of the earth and the verification of the inverse square law of gravitational attraction. The approximate period of a pendulum, $T = 2\pi\sqrt{l/g}$, was used to measure the force of gravity at various points on the surface of the earth because the period depends on the acceleration g determined by that force. By measuring successive lengths along a meridian, each length corresponding to a change of one degree in latitude, one can, with the aid of some theory and the values of g, determine the shape of the earth. In fact by using the observed variation in the period at various places on the earth's surface, Newton deduced that the earth bulges at the equator.

After Newton had, by his theoretical argument, concluded that the equatorial radius was 1/230 longer than the polar radius (this value is 30 percent too large), the European scientists were eager to confirm it. One method would be to measure the length of a degree of latitude near the equator and near a pole. If the earth were flattened, one degree of latitude would be slightly longer at the poles than at the equator.

Jacques Cassini (1677–1756) and members of his family made such measurements and in 1720 gave an opposite result. They found that the pole-to-pole diameter was 1/95 longer than the equatorial diameter. To settle the question the French Academy of Sciences sent out two expeditions in the 1730s, one to Lapland under the mathematician Pierre L. M. de Maupertuis and the other to Peru. Maupertuis's party included a fellow mathematician Alexis-Claude Clairaut. Their measurements confirmed that the earth was flattened at the poles; Voltaire hailed Maupertuis as the "flattener of the poles and the Cassinis." Actually Maupertuis's value was 1/178, which was less accurate than Newton's. The question of the shape of the earth remained a major subject and for a long time it was open as to whether the shape was an oblate spheroid, a prolate spheroid, a general ellipsoid, or some other figure of revolution.

The related problem, verifying the law of gravitation, could be handled if the shape of the earth were known. Given the shape, one could determine the centripetal force needed to keep an object on or near the surface of the rotating earth. Then, knowing the acceleration g due to the force of gravity at the surface, one could test whether the full force of gravity, which supplies the centripetal acceleration and g, is indeed the inverse square law. Clairaut, one of the men who questioned the law, believed at one time that it should be of the form $F = A/r^2 + B/r^3$. The two problems of the law of attraction and the shape of the earth are further intertwined, because when the earth is treated as a rotating fluid in equilibrium, the conditions for equilibrium involve the attraction of the particles of the fluid on each other.

The physical field of interest that dominated the century was astronomy. Newton had solved what is called the two-body problem, that is, the motion of a single planet under the gravitational attraction of the sun, wherein each body is idealized to have a point mass. He had also made some steps toward treating the major three-body problem, the behavior of the moon under the attraction of the earth and sun (Chap. 17, sec. 3). However, this was just the beginning of the efforts to study the motions of the planets and their satellites under the gravitational attraction of the sun and the mutual attraction of all the other bodies. Even Newton's work in the *Principia*, though constituting in effect the solution of differential equations, had to be translated into analytical form, which was done gradually during the eighteenth century. It was begun, incidentally, by Pierre Varignon, a fine French mathematician and physicist, who sought to free dynamics from the encumbrance of geometry. Newton did solve some differential equations in analytical form, for example in his *Method of Fluxions* of 1671 (Chap. 17, sec. 3); and in his *Tractatus* of 1676, he observed that the solution of $d^n y/dx^n = f(x)$ is arbitrary to the extent of an $(n - 1)$st degree polynomial in x. In the third edition of the *Principia*, Proposition 34, Scholium, he confines himself to a statement of which shapes of surfaces of revolution offer least resistance to motion in a fluid; but in a letter to David Gregory of 1694 he explains how he got his results and in the explanation uses differential equations.

Among problems of astronomy, the motion of the moon received the greatest attention, because the common method of determining longitude of ships at sea (Chap. 16, sec. 4), as well as other methods recommended in the seventeenth century, depended on knowing at all times the direction of the moon from a standard position (which from late in the century was Greenwich, England). It was necessary to know this direction of the moon to within 15 seconds of angle to determine the time at Greenwich to within 1 minute; even such an error could lead to an error of 30 kilometers in the determination of a ship's position. But with the tables of the moon's position available in Newton's time such accuracy was far from attainable. Another reason for the interest in the theory of the motion of the moon is that it could be used to predict eclipses, which in turn were a check on the entire astronomical theory.

The subject of ordinary differential equations arose in the problems just sketched. As mathematics developed, the subject of partial differential equations led to further work in ordinary differential equations. So did the branches now known as differential geometry and the calculus of variations. In this chapter we shall consider the problems leading directly to the basic early work in ordinary differential equations, that is, equations involving derivatives with respect to only one independent variable.

2. First Order Ordinary Differential Equations

The earliest work in differential equations, like the late seventeenth- and early eighteenth-century work in the calculus, was first disclosed in letters from one mathematician to another, many of which are no longer available, or in publications that often repeated results established or claimed in letters. The announcement of a result by one man frequently provoked the claim by another that he had done precisely the same thing earlier, which, in view of the bitter rivalries that existed, might or might not have been true. Some proofs were merely sketched, and it is not clear that the authors had the complete story; likewise, purported general methods of solution were merely illustrated by particular examples. For these reasons, even if we ignore the entire question of rigor, it is difficult to credit the results obtained to the right man.

In the *Acta Eruditorum* of 1693[1] Huygens speaks explicitly of differential equations, and Leibniz in another article in the same journal and year[2] says differential equations are functions of pieces of the characteristic triangle. The point we usually learn first about ordinary differential equations, that they can arise by eliminating the arbitrary constants from a given function and its derivatives, was not made until about 1740 and is due to Alexis Fontaine des Bertins.

James Bernoulli was among the earliest to use the calculus in solving analytically problems of ordinary differential equations. In May of 1690[3] he published his solution of the problem of the isochrone, though an analytic solution had already been given by Leibniz. This problem is to find a curve along which a pendulum takes the same time to make a complete oscillation whether it swings through a wide or small arc. The differential equation, in Bernoulli's symbols, was

$$dy\sqrt{b^2y - a^3} = dx\sqrt{a^3}.$$

1. *Œuvres*, 10, 512–14.
2. *Math. Schriften*, 5, 306.
3. *Acta Erud.*, 1690, 217–19 = *Opera*, 1, 421–24.

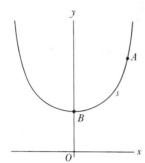

Figure 21.1

Bernoulli concluded from the equality of the differentials that the integrals (the word is used for the first time) must be equal and gave

$$\frac{2b^2y - 2a^3}{3b^2} \sqrt{b^2y - a^3} = x\sqrt{a^3}$$

as the solution. The curve is, of course, the cycloid.

In the same paper of 1690 James Bernoulli posed the problem of finding the curve assumed by a flexible inextensible cord hung freely from two fixed points, the curve Leibniz called the catenary. The problem had been considered as far back as the fifteenth century by Leonardo da Vinci. Galileo thought the curve was a parabola. Huygens affirmed that this was not correct and showed, largely by physical reasoning, that if the total load of cord and any weights suspended from it is uniform per horizontal foot, the curve is a parabola. For the true catenary the weight per foot *along the cable* is uniform.

In the *Acta* for June of 1691, Leibniz, Huygens, and John Bernoulli published independent solutions. Huygens's was geometrical and unclear. John Bernoulli[4] gave a solution by the method of the calculus. The full explanation is in his calculus text of 1691. It is the one now given in calculus and mechanics texts and is based on the equation

(1)
$$\frac{dy}{dx} = \frac{s}{c}$$

where s is arc length from B to some arbitrary point A (Fig. 21.1) and c depends upon the weight per unit length of the cord. This differential equation leads to what we now write as $y = c \cosh (x/c)$. Leibniz too obtained this result by calculus methods.

John Bernoulli was immensely proud that he had been able to solve the catenary problem and that his brother James, who had proposed it,

4. *Acta Erud.*, 1691, 274–76 = *Opera*, 1, 48–51.

had not. In a letter of September 29, 1718, to Pierre Rémond de Montmort (1678–1719) he boasts,[5]

> The efforts of my brother were without success; for my part, I was more fortunate, for I found the skill (I say it without boasting, why should I conceal the truth?) to solve it in full and to reduce it to the rectification of the parabola. It is true that it cost me study that robbed me of rest for an entire night. It was much for those days and for the slight age and practice I then had, but the next morning, filled with joy, I ran to my brother, who was still struggling miserably with this Gordian knot without getting anywhere, always thinking like Galileo that the catenary was a parabola. Stop! Stop! I say to him, don't torture yourself any more to try to prove the identity of the catenary with the parabola, since it is entirely false. The parabola indeed serves in the construction of the catenary, but the two curves are so different that one is algebraic, the other is transcendental. . . . But then you astonish me by concluding that my brother found a method of solving this problem. . . . I ask you, do you really think, if my brother had solved the problem in question, he would have been so obliging to me as not to appear among the solvers, just so as to cede me the glory of appearing alone on the stage in the quality of the first solver, along with Messrs. Huygens and Leibniz?

During the years 1691 and 1692 James and John also solved the problem of the shape assumed by a flexible inelastic hanging cord of variable density, an elastic cord of constant thickness, and a cord acted on at each point by a force directed to a fixed center. John also solved the converse problem: Given the equation of the curve assumed by an inelastic hanging cord, to find the law of variation of density of the cord with arc length. John's solutions are the ones often found in mechanics texts. James published in the *Acta* of 1691 the proof that of all shapes that a given cord hung from two fixed points can take, the catenary has the lowest center of gravity.

In the *Acta* of 1691 James Bernoulli derived the equation for the tractrix, the curve (Fig. 21.2) for which the ratio of PT to OT is constant for any point P on the curve. James first derived

$$\frac{dy}{ds} = \frac{y}{a},$$

where s is arc length. From this equation he deduced that

(2)
$$\int y \, dx = \int dy \sqrt{a^2 - y^2}$$

and

(3)
$$\int y^2 \, dx = -\frac{1}{3} \sqrt{(a^2 - y^2)^3},$$

5. Johann Bernoulli, *Der Briefwechsel von Johann Bernoulli*, Birkhäuser Verlag, 1955, 97–98.

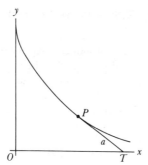

Figure 21.2

which he left as characteristic integrals for the curve. (Equation [2] can be integrated to yield $x + \sqrt{a^2 - y^2} = a \log [(a + \sqrt{a^2 - y^2})/y]$.)

Leibniz hit upon the technique of separating variables in ordinary differential equations and communicated it in a letter to Huygens of 1691. Thus he solved an equation of the form $y(dx/dy) = f(x)g(y)$, by writing $dx/f(x) = g(y) \, dy/y$, and then was able to integrate both sides. He did not formulate the general method. He also reduced (1691) the homogeneous differential equations of first order, $y' = f(y/x)$, to quadratures. He let $y = vx$ and substituted in the equation. The equation is then separable. Both these ideas, separation of variables and solution of homogeneous equations, were explained more fully by John Bernoulli in the *Acta Eruditorum* of 1694. Then Leibniz, in 1694, showed how to reduce the linear first order ordinary differential equation $y' + P(x)y = Q(x)$ to quadratures. His method utilized a change in the dependent variable. In general, Leibniz solved only first order ordinary differential equations.

James Bernoulli then proposed in the *Acta* of 1695[6] the problem of solving what is now called Bernoulli's equation:

$$(4) \qquad \frac{dy}{dx} = P(x)y + Q(x)y^n.$$

Leibniz in 1696[7] showed it can be reduced to a linear equation (first degree in y and y') by the change of variable $z = y^{1-n}$. John Bernoulli gave another method. In the *Acta* of 1696 James solved it essentially by separation of variables.

In 1694 Leibniz and John Bernoulli introduced the problem of finding the curve or family of curves that cut a given family of curves at a given angle. John Bernoulli called the cutting curves trajectories and pointed out, on the basis of Huygens's work on light, that this problem is important in finding the paths of rays of light traveling in a non-uniform medium because

6. Page 553.
7. *Acta Erud.*, 1696, 145.

these rays cut what are called the wave fronts of light orthogonally. The problem did not become public until 1697, when John posed it as a challenge to James, who solved a few special cases. John obtained the differential equation of the orthogonal trajectories of a particular family of curves and solved it in 1698.[8] Leibniz found the orthogonal trajectories of a family of curves thus: Consider $y^2 = 2bx$, where b is the parameter of the family (a term he introduced). From this equation $y \, dy/dx = b$. Leibniz then lets $b = -y \, dx/dy$, substitutes in $y^2 = 2bx$, and obtains $y^2 = -2xy \, dx/dy$ as the differential equation of the trajectories. The solution is $a^2 - x^2 = y^2/2$. Though he solved just special cases, he conceived the general problem and method.

The subject of orthogonal trajectories remained dormant until 1715, when Leibniz, aiming primarily at Newton, challenged the English mathematicians to discover the general method of finding the orthogonal trajectories of a given family of curves. Newton, tired from a day at the mint, solved the problem before going to sleep and the solution was published in the *Philosophical Transactions* of 1716.[9] Newton also showed how to find the curves cutting a given family at a constant angle or at an angle varying with each curve of the given family according to a given law. The method is not too different from the modern one, though Newton used ordinary differential equations of the second order.

Further work on this problem was done by Nicholas Bernoulli (1695–1726), in 1716. Jacob Hermann (1678–1733), a student of James Bernoulli, gave in the *Acta* of 1717 the rule that if $F(x, y, c) = 0$ is the given family of curves, then $y' = -F_x/F_y$, where F_x and F_y are the partial derivatives of F, and the orthogonal trajectory has F_y/F_x as its slope.[10] Hence, Hermann said, the ordinary differential equation of the orthogonal trajectories of $F(x, y, c) = 0$ is

(5) $$F_y \, dx = F_x \, dy.$$

He solved (5) for c, substituted this value in the original equation $F(x, y, c) = 0$, and solved the resulting differential equation. This method is really Leibniz's, but more explicitly stated. It is more customary today to find the true differential equation satisfied by $F = 0$; this equation does not contain parameter c. In it we replace y' by $-1/y'$ and so obtain the differential equation of the orthogonal trajectories.

John Bernoulli posed other trajectory problems for the English, his particular bête noire being Newton. Since the English and the Continentals were already at odds, the challenges were marked by bitterness and hostility.

8. *Opera*, 1, 266.
9. *Phil. Trans.*, 29, 1716, 399–400.
10. *Acta Erud.*, 1717, 349 ff. Also in John Bernoulli, *Opera*, 2, 275–79.

John Bernoulli then solved the problem of finding the motion of a projectile in a medium whose resistance is proportional to any power of the velocity. The differential equation here is

$$(6) \qquad\qquad m\frac{dv}{dt} - kv^n = mg.$$

Exact first order equations, that is, equations $M(x, y)\, dx + N(x, y) \cdot dy = 0$, for which $M\, dx + N\, dy$ is an exact differential of some function $z = f(x, y)$, were also recognized. Clairaut, who is famous for his work on the shape of the earth, had given the condition $\partial M/\partial y = \partial N/\partial x$ that the equation be exact in his papers of 1739 and 1740 (Chap. 19, sec. 6). The condition was also given independently by Euler in a paper written in 1734–35.[11] If the equation is exact then, as Clairaut and Euler pointed out, it can be integrated.

When a first order equation is not exact it is often possible to multiply the equation by a quantity, called an integrating factor, that makes it exact. Though integrating factors had been used in special problems of first order ordinary differential equations, it was Euler who realized (in the 1734/35 paper) that this concept furnishes a method; he set up classes of equations in which integrating factors will work. He also proved that if two integrating factors of any first order ordinary differential equations are known then their ratio is a solution of the equation. Clairaut independently introduced the idea of an integrating factor in his 1739 paper and in the 1740 paper added to the theory. All the elementary methods of solving first order equations were known by 1740.

3. Singular Solutions

Singular solutions are not obtainable from the general solution by giving a definite value to the constant of integration; that is, they are not particular solutions. This was observed by Brook Taylor in his *Methodus Incrementorum*[12] while solving a particular first order second degree equation. Leibniz in 1694 had already noted that an envelope of a family of solutions is also a solution. Singular solutions were more fully explored by Clairaut and Euler.

Clairaut's work of 1734[13] dealt with the equation that now bears his name,

$$(7) \qquad\qquad y = xy' + f(y').$$

Let y' be denoted by p. Then

$$(8) \qquad\qquad y = xp + f(p).$$

11. *Comm. Acad. Sci. Petrop.*, 7, 1734/35, 174–93, pub. 1740 = *Opera*, (1), 22, 36–56.
12. 1715, p. 26.
13. *Hist. de l'Acad. des Sci.*, Paris, 1734, 196–215.

By differentiating with respect to x Clairaut obtained

$$p = p + \{x + f'(p)\} \frac{dp}{dx}.$$

Then

(9) $$\frac{dp}{dx} = 0 \quad \text{and} \quad x + f'(p) = 0.$$

The equation $dp/dx = 0$ leads to $y' = c$, and then from the original equation we have

(10) $$y = cx + f(c).$$

This is the general solution and is a family of straight lines. The second factor, $x + f'(p) = 0$, may be used together with the original equation to eliminate p; this yields a new solution, which is the singular solution. To see that it is the envelope of the general solution we take (10) and differentiate with respect to c. Then

(11) $$x + f'(c) = 0.$$

The envelope is the curve that results from eliminating c between (10) and (11). But these two are exactly the same as the two equations that yield the singular solution. The fact that the singular solution was an envelope was not yet appreciated, but Clairaut was explicit that the singular solution was not included in the general solution.

Clairaut and Euler had given a method of finding the singular solution by working from the differential equation itself, that is, by eliminating y' from $f(x, y, y') = 0$ and $\partial f/\partial y' = 0$. This fact and the fact that singular solutions are not contained in the general solution puzzled Euler. In his *Institutiones* of 1768[14] he gave a criterion for distinguishing the singular solution from a particular integral, which could be used when the general solution was not known. D'Alembert[15] sharpened this criterion. Then Laplace[16] extended the notion of singular solutions (he called them particular integrals) to equations of higher order and to differential equations in three variables.

Lagrange[17] made a systematic study of singular solutions and their connection with the general solution. He gave the general method of obtaining the singular solution from the general solution by elimination of the constant in a clear and elegant way that surpasses Laplace's contribution. Given the general solution $V(x, y, \alpha) = 0$, Lagrange's method was to find $dy/d\alpha$,

14. Vol. 1, pp. 393 ff.
15. *Hist. de l'Acad. des Sci.*, Paris, 1769, 85 ff., pub. 1772.
16. *Hist. de l'Acad. des Sci.*, Paris, 1772, Part 1, 344 ff., pub. 1775 = *Œuvres* 8, 325–66.
17. *Nouv. Mém. de l'Acad. de Berlin*, 1774, pub. 1776 = *Œuvres*, 4, 5–108.

set it equal to 0, and eliminate α from this equation and $V = 0$. The same procedure can be used with $dx/d\alpha = 0$. He also gave further information on Clairaut's and Euler's method of obtaining the singular solution from the differential equation. Finally, Lagrange gave the geometrical interpretation of the singular solution as the envelope of the family of integral curves. There are a number of special difficulties in the theory of singular solutions which he did not recognize. For example, he did not realize that other singular curves, which may appear in the equation obtained by eliminating y' from $f(x, y, y') = 0$ and $\partial f/\partial y' = 0$ are not singular solutions, or that a singular solution may contain a branch that is a particular solution. The full theory of singular solutions was developed in the nineteenth century and given its present form by Cayley and Darboux in 1872.

4. Second Order Equations and the Riccati Equations

Second order ordinary differential equations arose in physical problems as early as 1691. James Bernoulli tackled the problem of the shape of a sail under the pressure of the wind, the *velaria* problem, and was led to a second order equation $d^2x/ds^2 = (dy/ds)^3$ where s is arc length. John Bernoulli treated this problem in his calculus text of 1691 and established that it is the same mathematically as the catenary problem. Second order equations appear next in the attack on the problem of determining the shape of a vibrating elastic string fixed at both ends—for example, the violin string. In tackling this problem Taylor was pursuing an old theme. The whole subject of mathematics and musical sounds, begun by the Pythagoreans, was continued by men of the medieval period and brought to the fore in the seventeenth century. Benedetti, Beeckman, Mersenne, Descartes, Huygens, and Galileo are prominent in this work, though no new mathematical results are worth noting here. The fact that a string can vibrate in many modes, that is, halves, thirds, etc., and that the tone produced by a string vibrating in k parts is the kth harmonic or $(k - 1)$st overtone (the fundamental is the first harmonic) was well known in England by 1700, largely through the experimental work of Joseph Sauveur (1653–1716).

Brook Taylor[18] derived the fundamental frequency of a stretched vibrating string. He solved the equation $a^2\ddot{x} = \dot{s}y\dot{y}$, where $\dot{s}$ equals $\sqrt{\dot{x}^2 + \dot{y}^2}$ and the differentiation is with respect to time, and gave $y = A \sin(x/a)$ as the form of the string at any time. Here $a = l/\pi$, where l is the length of the string. Taylor's result for the fundamental frequency (in modern notation) is

$$v = \frac{1}{2l} \sqrt{\frac{T}{\sigma}},$$

18. *Phil. Trans.*, 28, 1713, 26–32, pub. 1714; also in *Phil. Trans. Abridged*, 6, 1809, 7–12, 14–17.

where T is the tension in the string, $\sigma = m/g$, m is the mass per unit length, and g is the acceleration of gravity.

In his effort to treat the vibrating string, John Bernoulli, in a letter of 1727 to his son Daniel and in a paper,[19] considered the weightless elastic string loaded with n equal and equally spaced masses. He derived the fundamental frequency of the system when there are $1, 2, \ldots, 6$ masses. (There are other frequencies of oscillation of the system of masses.) John recognized that the force on each mass is $-K$ times its displacement, and solved $d^2x/dt^2 = -Kx$, thus integrating the equation of simple harmonic motion by analytic methods. He then passed to the continuous string which, like Taylor, he proved must have the shape of a sine curve (at any instant) and calculated the fundamental frequency. Here he solved $d^2y/dx^2 = -ky$. Neither Taylor nor John Bernoulli treated the higher modes of elastic vibrating bodies.

In 1728 Euler began to consider second order equations. His interest in these was aroused partly by his work in mechanics. He had worked, for example, on pendulum motion in resisting media, which leads to a second order differential equation. For the king of Prussia he worked on the effect of the resistance of air on projectiles. Here he took over the work of the Englishman Benjamin Robins, improved it, and wrote a German version (1745). This was translated into French and English and used by the artillery.

He also considered[20] a class of second order equations that he reduced to first order by a change of variables. For example, he considered the equation

$$(12) \qquad a x^m \, dx^p = y^n \, dy^{p-2} \, d^2y$$

or in derivative form

$$(13) \qquad \left(\frac{dy}{dx}\right)^{p-2} \frac{d^2y}{dx^2} = \frac{a x^m}{y^n}.$$

Euler introduced the new variables t and v by means of the equations

$$(14) \qquad y = e^v t(v), \qquad x = e^{\alpha v}$$

wherein α is a constant to be determined. The equations (14) may be regarded as parametric equations for x and y in terms of v, so that one can now calculate dy/dx and d^2y/dx^2 and by substitution in (13) obtain a second order equation in t as a function v. Euler then fixes α so as to eliminate the exponential factor, and v no longer appears explicitly. A further transformation, namely, $z = dv/dt$, reduces the second order equation to first order.

The details of this method are not worth pursuing because they apply to just one class of second order equations, but historically this piece of work

19. *Comm. Acad. Sci. Petrop.*, 3, 1728, 13–28, pub. 1732 = *Opera*, 3, 198–210.
20. *Comm. Acad. Sci. Petrop.*, 3, 1727, 124–37, pub. 1732 = *Opera*, (1), 22, 1–14.

Figure 21.3

is significant because it does initiate the systematic study of second order equations and because Euler here introduces the exponential function, which, as we shall see, plays an important role in solving second and higher order equations.

Before leaving St. Petersburg in 1733, Daniel Bernoulli completed a paper, "Theorems on Oscillations of Bodies Connected by a Flexible Thread and of a Vertically Suspended Chain." [21] He starts with the hanging chain suspended from the upper end, weightless but loaded with equally spaced weights. He finds, when the chain is set into vibration, that the system has different modes of (small) oscillation about a vertical line through the point of suspension. Each of these modes has its own characteristic frequency. [22] Then, for an oscillating, uniformly heavy hanging chain of length l, he gives that the displacement y at a distance x from the bottom (Fig. 21.3) satisfies the equation

$$(15) \qquad \alpha \frac{d}{dx}\left(x\frac{dy}{dx}\right) + y = 0,$$

and the solution is an infinite series, which (in modern notation) can be expressed as

$$(16) \qquad y = AJ_0(2\sqrt{x/\alpha}),$$

where J_0 is the zero-th order Bessel function (of the first kind). [23] Moreover, α is such that

$$(17) \qquad J_0(2\sqrt{l/\alpha}) = 0,$$

21. *Comm. Acad. Sci. Petrop.*, 6, 1732/33, 108–22, pub. 1738.
22. In the case of n masses each mass has its own motion, which is a sum of n sinusoidal terms, each having one of the characteristic frequencies. The entire system has n different principal modes, each with one of the characteristic frequencies. Which of these are present depends on the initial conditions.

23. $J_n(x) = \left(\frac{x}{2}\right)^n \sum_{k=0}^{\infty} \frac{(-1)^k (x/2)^{2k}}{k!\,(k+n)!}$ for n positive integral or 0.

where l is the length of the chain. He asserts that (17) has infinitely many roots, which become smaller and smaller and approach 0, and he gives the largest value for α. For each α there is a mode of oscillation and a characteristic frequency.

He now says, "Nor would it be difficult to derive from this theory a theory of musical strings agreeing with those given by Taylor and by my father . . . Experiment shows that in musical strings there are intersections [nodes] similar to those for vibrating chains." Actually, Bernoulli here goes beyond Taylor and his father in recognizing the higher modes or harmonics of a vibrating string.

His paper on the hanging chain also treats the oscillating chain of non-uniform thickness; here he introduces the differential equation

$$(18) \qquad \alpha \frac{d}{dx}\left(g(x)\frac{dy}{dx}\right) + y\frac{dg(x)}{dx} = 0,$$

where $g(x)$ is the distribution of the weight along the chain. For $g(x) = x^2/l^2$ he gives a series solution that can be expressed in modern notation as

$$(19) \qquad y = 2A\left(\frac{2x}{\alpha}\right)^{-1/2} J_1(2\sqrt{2x/\alpha})$$

with

$$J_1(2\sqrt{2l/\alpha}) = 0.$$

J_1 is the Bessel function of first order and first kind.

What is missing in Daniel Bernoulli's solutions is, first, any reference to the displacement as a function of the *time*, so that his work remains mathematically in the realm of ordinary differential equations; and second, any suggestion that the simple modes (the harmonics), which he explicitly recognizes as real motions, may be superposed to form more complicated ones.

After having entered upon the subject of musical sounds with a book, *Tentamen Novae Theoriae Musicae ex Certissimis Harmoniae Principiis Dilucide Expositae* (An Investigation into a New, Clearly Presented Theory of Music Based on Incontestable Principles of Harmony), written by 1731 and published in 1739,[24] Euler followed up Daniel Bernoulli's work in a paper, "On the Oscillations of a Flexible Thread Loaded with Arbitrarily Many Weights."[25] Euler's results are much the same as Bernoulli's, except that Euler's mathematics is clearer. For one form of the continuous chain, that is, the special case in which the weight is proportional to x^n, Euler has to solve

$$\frac{x}{n+1}\frac{d^2y}{dx^2} + \frac{dy}{dx} + \frac{y}{\alpha} = 0.$$

24. *Opera*, (3), 1, 197–427.
25. *Comm. Acad. Sci. Petrop.*, 8, 1736, 30–47, pub. 1741 = *Opera*, (2), 10, 35–49.

He derives the series solution, which in modern notation is[26]

$$y = Aq^{-n/2}I_n(2\sqrt{q}), \qquad q = -\frac{(n+1)x}{\alpha}.$$

The n here is general so that Euler is introducing Bessel functions of arbitrary real index. He also gives the integral solution

$$y = A\frac{\int_0^1 (1 - t^2)^{(2n-1)/2} \cosh\left(2t\sqrt{\frac{(n+1)x}{\alpha}}\right) dt}{\int_0^1 (1 - \tau^2)^{(2n-1)/2} d\tau}.$$

This is perhaps the earliest case of a solution of a second order differential equation expressed as an integral.

In a paper of 1739[27] Euler took up the differential equations of the harmonic oscillator $\ddot{x} + kx = 0$ and the forced oscillation of the harmonic oscillator

(20) $M\ddot{x} + Kx = F\sin \omega_\alpha t.$

He obtained the solutions by quadratures and discovered (really rediscovered, since others had found it earlier) the phenomenon of resonance; that is, if ω is the natural frequency $\sqrt{K/M}$ of the oscillator, which obtains when $F = 0$, then when ω_α/ω approaches 1 the forced oscillations have larger and larger amplitudes and become infinite.

In an attempt to set up a model for the transmission of sound in air, Euler considers in his paper "On the Propagation of Pulses Through an Elastic Medium"[28] n masses M connected by like (weightless) springs and lying in a horizontal line PQ. The motion considered is longitudinal, that is, along PQ. For the kth mass he obtains

$$M\ddot{x}_k = K(x_{k+1} - 2x_k + x_{k-1}), \qquad k = 1, 2, \ldots, n$$

where K is the spring constant and x_k is the displacement of the kth mass. He obtains the correct characteristic frequencies for each mass and the general solution

(21) $$x_k = \sum_{r=1}^{n} A_r \sin\frac{rk\pi}{n+1} \cos\left(2\sqrt{K/Mt}\,\frac{\sin r\cdot\pi/2}{n+1}\right),$$

26. For general ν (including complex values)

$$I_\nu(z) = \sum_{n=0}^{\infty} \frac{(z/2)^{\nu+2n}}{n!\,\Gamma(\nu+n+1)}.$$

The $I_\nu(z)$ are called modified Bessel functions.
27. *Comm. Acad. Sci. Petrop.*, 11, 1739, 128–49, pub. 1750 = *Opera*, (2), 10, 78–97.
28. *Novi Comm. Acad. Sci. Petrop.*, 1, 1747/48, 67–105, pub. 1750 = *Opera*, (2), 10, 98–131.

wherein $k = 1, 2, \ldots, n$. Thus he not only obtains the individual modes for each mass but the general motion of that mass as a sum of simple harmonic modes. The particular modes that may appear depend on the initial conditions, that is, on how the masses are set into motion. All these results are interpretable in terms of the transverse motion (perpendicular to PQ) of the loaded string.

Some of the equations already treated, for example, the Bernoulli equation, are nonlinear; that is, as an equation in the variables y, y', and y'' (if present), terms of the second or higher degree occur. Among such equations of first order, a few are of special interest because they are intimately involved with linear second order equations. In the early history of ordinary differential equations the nonlinear Riccati equation

$$(22) \qquad \frac{dy}{dx} = a_0(x) + a_1(x)y + a_2(x)y^2$$

commanded a great deal of attention.

The Riccati equation acquired importance when it was introduced by Jacopo Francesco, Count Riccati of Venice (1676–1754), who worked in acoustics, to help solve second order ordinary differential equations. He considered curves whose radii of curvature depend only on the ordinates and was led [29] to

$$x^m \frac{d^2 x}{dp^2} = \frac{d^2 y}{dp^2} + \left(\frac{dy}{dp}\right)^2$$

(Riccati wrote $x^m\, d^2 x = d^2 y + (dy)^2$), wherein we must understand that x and y depend upon p. By changes of variables Riccati obtained

$$x^m \frac{dq}{dx} = \frac{du}{dx} + \frac{u^2}{q},$$

which is of first order. He assumed next that q is a power of x, for example, x^n, and arrived at the form

$$(23) \qquad \frac{du}{dx} + \frac{u^2}{x^n} = nx^{m+n-1}.$$

He then showed how to solve (23) for special values of n by the method of separation of variables for ordinary differential equations. Later, several of the Bernoullis determined other values of n for which solution of (23) by separation of variables was possible.

Riccati's work is significant not only because he treated second order differential equations but because he had the idea of reducing second order equations to first order. This idea of reducing the order of an ordinary

29. *Acta Erud.*, 1724, 66–73.

differential equation by one device or another will be seen to be a major method in the treatment of higher order ordinary differential equations.

Euler in 1760[30] considered the Riccati equation

$$(24) \qquad \frac{dz}{dx} + z^2 = ax^n$$

and showed that if one knows a particular integral v, then the transformation

$$z = v + u^{-1}$$

produces a linear equation. Moreover, if one knows two particular integrals, one can reduce the problem of solving the original differential equation to quadratures.

D'Alembert[31] was the first to consider the general form (22) of the Riccati equation and to use the term "Riccati equation" for this form. He started with

$$(25) \qquad \frac{d^2S}{dx^2} = \frac{-\lambda^2 x \pi^2 S}{2aLe}$$

and let

$$(26) \qquad S = \exp\left[\int p\, dx\right], \qquad p = f(x),$$

from which he obtained the form (22) for an equation in p as a function of x.

5. *Higher Order Equations*

In December of 1734 Daniel Bernoulli wrote to Euler, who was in St. Petersburg, that he had solved the problem of the transverse displacement y of an elastic bar (a steel or wooden one-dimensional body) fixed at one end in a wall and free at the other. Bernoulli obtained the differential equation

$$(27) \qquad K^4 \frac{d^4y}{dx^4} = y,$$

where K is a constant, x is the distance from the free end of the bar, and y is the vertical displacement at that point from the unbent position of the bar. Euler, in a reply written before June 1735, said he too had discovered this equation and was unable to integrate it except by using series, and that he did obtain four separate series. These series represented circular and exponential functions, but Euler did not realize it at this time.

---

30. *Novi Comm. Acad. Sci. Petrop.*, 8, 1760/61, 3–63, pub. 1763 = *Opera*, (1), 22, 334–94, and 9, 1762/63, 154–69, pub. 1764 = *Opera*, (1), 22, 403–20.
31. *Hist. de l'Acad. de Berlin*, 19, 1763, 242 ff., pub. 1770.

Four years later, in a letter to John Bernoulli (September 15, 1739), Euler indicated that his solution can be represented as

$$(28) \qquad y = A\left[\left(\cos \frac{x}{K} + \cosh \frac{x}{K}\right) - \frac{1}{b}\left(\sin \frac{x}{K} + \sinh \frac{x}{K}\right)\right]$$

where b is determined by the condition that $y = 0$ when $x = l$ so that

$$b = \frac{\sin \dfrac{l}{K} + \sinh \dfrac{l}{K}}{\cos \dfrac{l}{K} + \cosh \dfrac{l}{K}}.$$

The problems of elasticity led Euler to consider the mathematical problem of solving general linear equations with constant coefficients; and in the letter of September 15, 1739, he wrote John Bernoulli that he had succeeded. Bernoulli wrote back that he had already considered such equations in 1700, even with variable coefficients. Actually he had considered only a special third order equation and had shown how to reduce it to one of second order.

In his publication of this work[32] Euler considers the equation

$$(29) \qquad 0 = Ay + B\frac{dy}{dx} + C\frac{d^2y}{dx^2} + D\frac{d^3y}{dx^3} + \cdots + L\frac{d^ny}{dx^n},$$

where the coefficients are constants. The equation is called homogeneous because the term independent of y and its derivatives is 0. He points out that the general solution must contain n arbitrary constants and that the solution will be a sum of n particular solutions, each multiplied by an arbitrary constant. Then he makes the substitution

$$y = \exp\left[\int r \, dx\right],$$

r constant, and obtains the equation in r,

$$A + Br + Cr^2 + \cdots + Lr^n = 0,$$

which is called the characteristic or indicial or auxiliary equation. When q is a simple real root of this equation, then

$$a \exp\left[\int q \, dx\right]$$

is a solution of the original differential equation. When the characteristic equation has a multiple root q, Euler lets $y = e^{qx}u(x)$ and substitutes in the differential equation. He finds that

$$(30) \qquad y = e^{qx}(\alpha + \beta x + \gamma x^2 + \cdots + \kappa x^{k-1})$$

32. *Misc. Berolin.*, 7, 1743, 193–242 = *Opera*, (1), 22, 108–49.

is a solution involving k arbitrary constants if the root q appears k times in the characteristic equation. He also treats the cases of conjugate complex roots and multiple complex roots. Thus Euler disposes completely of the homogeneous linear equation with constant coefficients.

Somewhat later[33] he treated the nonhomogeneous nth-order linear ordinary differential equation. His method was to multiply the differential by $e^{\alpha x} dx$, integrate both sides, and proceed to determine α so as to reduce the equation to one of lower order. Thus to treat

(31)
$$C\frac{d^2y}{dx^2} + B\frac{dy}{dx} + Ay = X(x),$$

he multiplies through by $e^{\alpha x} dx$ and obtains

$$\int \left[e^{\alpha x}C\frac{d^2y}{dx} + e^{\alpha x}B\frac{dy}{dx} + e^{\alpha x}Ay \right] dx = \int e^{\alpha x}X \, dx.$$

But for proper A', B', and α the left side must be

$$e^{\alpha x}\left(A'y + B'\frac{dy}{dx} \right).$$

By differentiating this quantity and comparing with the original equation he finds that

(32)
$$B' = C, \qquad A' = B - \alpha C, \qquad A' = \frac{A}{\alpha},$$

whence, from the second two equations,

(33)
$$A - B\alpha + C\alpha^2 = 0.$$

Thus α, A', and B' are found and the original equation is reduced to

(34)
$$A'y + B'\frac{dy}{dx} = e^{-\alpha x}\int e^{\alpha x}X \, dx.$$

An integrating factor for this equation is $e^{\beta x} dx$ where $\beta = A'/B'$, so that from (32) he has $\alpha\beta = A/C$ and $\alpha + \beta = B/C$ and therefore, by (33), α and β are the two roots of $A - B\alpha + C\alpha^2 = 0$.

The method applies to linear nth-order ordinary differential equations with constant coefficients. The order is reduced step by step as in the above example. Euler also took care of the cases of equal roots of the equation in α and of complex roots.

Following the work on linear ordinary differential equations with constant coefficients, Lagrange took the step to non-constant coefficients.[34]

33. *Novi Comm. Acad. Sci. Petrop.*, 3, 1750/51, 3–35, pub. 1753 = *Opera*, (1), 22, 181–213.
34. *Misc. Taur.*, 3, 1762/65, 179–86 = *Œuvres*, 1, 471–78.

This led, as we shall see, to the concept of the adjoint equation. Lagrange starts with

$$(35) \qquad Ly + M\frac{dy}{dt} + N\frac{d^2y}{dt^2} + \cdots = T$$

where L, M, N, and T are functions of t. For simplicity we shall stick to second order equations. Lagrange multiplies by $z\,dt$ where $z(t)$ is as yet undetermined, and integrates by parts, thus:

$$\int Mzy'\,dt = Mzy - \int (Mz)'y\,dt$$

$$\int Nzy''\,dt = Nzy' - (Nz)'y + \int (Nz)''y\,dt.$$

Then the original differential equation becomes

$$y[Mz - (Nz)'] + y'(Nz) + \int [Lz - (Mz)' + (Nz)'']y\,dt = \int Tz\,dt.$$

The bracket under the integral sign may be treated as an ordinary differential equation in z and set equal to 0. If it can be solved for $z(t)$, there remains an ordinary differential equation for y of lower order than the original. The new differential equation in z is called the adjoint of the original equation, a term introduced by Lazarus Fuchs in 1873. Lagrange used no special term for it.

To treat the equation for z (the adjoint equation), Lagrange proceeds in the same way to reduce the order. He multiplies by $w(t)\,dt$, proceeds as above, and arrives at an equation in w that will reduce the order of the equation in z. The equation in w turns out to be the original equation (35) except that the right side is 0. Hence Lagrange discovered the theorem that the adjoint of the adjoint of the original nonhomogeneous ordinary differential equation is the original but homogeneous equation. Euler did essentially the same thing in 1778. He had seen Lagrange's work but apparently forgot about it.

In further work on homogeneous linear ordinary differential equations with variable coefficients, Lagrange[35] extended to these equations some of the results Euler had obtained for linear ordinary differential equations with constant coefficients. Lagrange found that the general solution of the homogeneous equation is a sum of independent particular solutions, each multiplied by an arbitrary constant, and that knowing m particular integrals of the nth-order homogeneous equation, one can reduce the order by m.

35. *Misc. Taur.*, 3, 1762/65, 190–99 = *Œuvres*, 1, 481–90.

6. *The Method of Series*

We have already had occasion to note that some differential equations were solved by means of infinite series. The importance of this method even today warrants a few specific remarks on the subject. Series solutions have been used so widely since 1700 that we must restrict ourselves to a few examples.

We know that Newton used series to integrate somewhat complicated functions even where only quadrature was involved. He also used them to solve first order equations. Thus, to integrate

(36) $$\dot{y} = 2 + 3x - 2y + x^2 + x^2 y,$$

Newton assumes

(37) $$y = A_0 + A_1 x + A_2 x^2 + \cdots.$$

Then

(38) $$\dot{y} = A_1 + 2A_2 x + 3A_3 x^2 + \cdots.$$

Substitution of (37) and (38) in (36) and equating coefficients of like powers of x yields

$$A_1 = 2 - 2A_0, \qquad 2A_2 = 3 - 2A_1, \qquad 3A_3 = 1 + A_0 - 2A_2 + \cdots.$$

Thus we determine the A_i except for A_0. The fact that A_0 is undetermined and that therefore there is an infinite number of solutions was noted, but the significance of an arbitrary constant was not fully appreciated until about 1750. Leibniz solved some elementary differential equations by the use of infinite series [36] and also used the above method of undetermined coefficients.

Euler put the method of series in the fore from about 1750 on, to solve differential equations that could not be integrated in closed form. Though he worked with specific differential equations, and the details of what he did are often complicated, his method is what we use today. He assumes a solution of the form

$$y = x^\lambda (A + Bx + Cx^2 + \cdots),$$

substitutes for y and its derivatives in the differential equation, and determines λ and the coefficients $A, B, C, \ldots$ from the condition that each power of x in the resulting series must have a zero coefficient. Thus the ordinary differential equation which arose from his work on the oscillating membrane [37] (see Chap. 22, sec. 3), namely,

$$\frac{d^2 u}{dr^2} + \frac{1}{r}\frac{du}{dr} + \left(\alpha^2 - \frac{\beta^2}{r^2}\right) = 0,$$

36. *Acta Erud.*, 1693 = *Math. Schriften*, 5, 285–88.
37. *Novi Comm. Acad. Sci. Petrop.*, 10, 1764, 243–60, pub. 1766 = *Opera*, (2), 10, 344–59.

now called the Bessel equation, Euler solved by an infinite series. He gives
the solution

$$u(r) = r^\beta \left\{ 1 - \frac{1}{1 \cdot (\beta + 1)} \left(\frac{\alpha r}{2}\right)^2 + \frac{1}{1 \cdot 2(\beta + 1)(\beta + 2)} \left(\frac{\alpha r}{2}\right)^4 \right.$$
$$\left. - \frac{1}{1 \cdot 2 \cdot 3(\beta + 1)(\beta + 2)(\beta + 4)} \left(\frac{\alpha r}{2}\right)^6 + \cdots \right\},$$

which is, except for a factor depending only on β, what we now write as
$J_\beta(r)$. In further work on these functions he showed that for values of β
that are odd halves of an integer, the series reduce to elementary functions.
He noted further that $u(r)$, for real β, has an infinite number of zeros and
gave an integral representation for $u(r)$. Finally, for $\beta = 0$ and $\beta = 1$ he
gave the second linearly independent series solution of the differential
equation.

Euler in the *Institutiones Calculi Integralis* [38] treated the hypergeometric
differential equation

(39) $$x(1 - x)\frac{d^2 y}{dx^2} + [c - (a + b + 1)x]\frac{dy}{dx} - aby = 0$$

and gave the series solution

(40) $$y = 1 + \frac{a \cdot b}{1 \cdot c} x + \frac{a(a + 1)b(b + 1)}{1 \cdot 2 \cdot c(c + 1)} x^2$$
$$+ \frac{a(a + 1)(a + 2)b(b + 1)(b + 2)}{1 \cdot 2 \cdot 3 \cdot c(c + 1)(c + 2)} x^3 + \cdots.$$

The above form of (39) and of the solution (40) he gave again in his main
paper on this subject, written in 1778.[39] He had written other papers on
what he called the hypergeometric series, but there the term referred to
another series originally introduced by Wallis. The term "hypergeometric,"
to describe the differential equation (39) and the series (40), is due to
Johann Friedrich Pfaff (1765–1825), Gauss's friend and teacher. The series
(40) for y is now denoted by $F(a, b, c; z)$. In this notation Euler gave the
famous relations

$$F(-n, b, c; z) = (1 - z)^{c+n-b} F(c + n, c - b, c; z)$$

(41) $$F(-n, b, c; z) = \frac{n!}{c(c + 1) \cdots (c + n - 1)} \cdot$$
$$\int_0^1 t^{-n-1}(1 - t)^{c+n-1}(1 - tz)^{-b} dt.$$

38. Vol. 2, 1769, Chaps. 8–11.
39. *Nova Acta Acad. Sci. Petrop.*, 12, 1794, 58–70, pub. 1801 = *Opera*, (1), 16_2, 41–55.

7. Systems of Differential Equations

The differential equations involved thus far in the study of elasticity were rather simple, because the mathematicians were using crude physical principles and were still struggling to grasp better ones. In the area of astronomy, however, the physical principles, chiefly Newton's laws of motion and the law of gravitation, were clear and the mathematical problems far deeper. The basic mathematical problem in studying the motion of two or more bodies, each moving under the gravitational attraction of the others, is that of solving a system of ordinary differential equations, though the problem often reduces to solving a single equation.

Beyond isolated occurrences, the work on systems dealt primarily with problems of astronomy. The basis for writing the differential equations is Newton's second law of motion, $f = ma$, where f is the force of attraction. This is a vector law, which means that each component of f produces an acceleration in the direction of the component. Euler in a paper of 1750[40] gave the analytical form of Newton's second law as

$$(42) \qquad f_x = m\frac{d^2x}{dt^2}, \qquad f_y = m\frac{d^2y}{dt^2}, \qquad f_z = m\frac{d^2z}{dt^2}.$$

Here he assumes fixed rectangular axes. He also points out that for point bodies, that is, bodies that can be treated as though their masses were concentrated at one point, m is the total mass, and for distributed masses m is dM.

We shall consider briefly the formulation of the differential equations. Let us suppose that a fixed body of mass M is at the origin and that a moving body of mass m is at (x, y, z). Then the components of the gravitational force in the axial directions (Fig. 21.4) are

$$f_x = -\frac{GMmx}{r^3}, \qquad f_y = -\frac{GMmy}{r^3}, \qquad f_z = -\frac{GMmz}{r^3},$$

where G is the gravitational constant and $r = \sqrt{(x^2 + y^2 + z^2)}$. One can readily show that the moving body stays in one plane so that the system of equations (42) reduces to

$$(43) \qquad \frac{d^2x}{dt^2} = -\frac{kx}{r^3}, \qquad \frac{d^2y}{dt^2} = -\frac{ky}{r^3}$$

with $k = GM$. In polar coordinates these equations become

$$(44) \qquad \frac{d^2r}{dt^2} - r\left(\frac{d\theta}{dt}\right)^2 = -\frac{k}{r^2}$$

$$r\frac{d^2\theta}{dt^2} + 2\frac{dr}{dt}\frac{d\theta}{dt} = 0.$$

40. *Hist. de l'Acad. de Berlin*, 6, 1750, 185–217, pub. 1752 = *Opera*, (2), 5, 81–108.

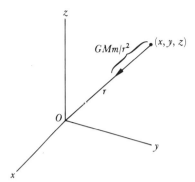

Figure 21.4

In this case of one body moving under the attracting force of another (fixed) body, the two differential equations can be combined into one involving x and y or r and θ because, for example, the second polar equation can be integrated to yield $r^2 \, d\theta/dt = C$, and the value of $d\theta/dt$ can be substituted in the first equation. It turns out that the moving body describes a conic section with the position of the first body as a focus.

If the two bodies move, each subject to the attraction of the other, then the differential equations are slightly different. Let m_1 and m_2 be the masses of two spherical bodies with spherically symmetric mass and with $m_1 + m_2 = M$. Choose a fixed coordinate system (usually the center is taken at the center of mass of the two bodies) and let (x_1, y_1, z_1) be the coordinates of one body and (x_2, y_2, z_2) the coordinates of the other; let r be the distance $\sqrt{(x_1 - x_2)^2 + (y_1 - y_2)^2 + (z_1 - z_2)^2}$. Then the system of equations that describes their motion is

$$m_1 \frac{d^2 x_1}{dt^2} = -k m_1 m_2 \frac{(x_1 - x_2)}{r^3}, \qquad m_1 \frac{d^2 y_1}{dt^2} = -k m_1 m_2 \frac{(y_1 - y_2)}{r^3}$$

$$m_1 \frac{d^2 z_1}{dt^2} = -k m_1 m_2 \frac{(z_1 - z_2)}{r^3}, \qquad m_2 \frac{d^2 x_2}{dt^2} = -k m_1 m_2 \frac{(x_2 - x_1)}{r^3}$$

$$m_2 \frac{d^2 y_2}{dt^2} = -k m_1 m_2 \frac{(y_2 - y_1)}{r^3}, \qquad m_2 \frac{d^2 z_2}{dt^2} = -k m_1 m_2 \frac{(z_2 - z_1)}{r^3}.$$

This is a system of six second order equations whose solution calls for twelve integrals, each with an arbitrary constant of integration. These are determined by the three coordinates of initial position and three components of initial velocity of each body. The equations can be solved, and the solution shows that each body moves in a conic section with respect to the common center of mass of the two bodies.

Actually this problem of the motion of two spheres under the mutual attractive force of gravitation was solved geometrically by Newton in the

Principia (Book I, Section 11). However, the analytical work was not taken up for some time. In mechanics the French followed Descartes's system until Voltaire, after visiting London in 1727, returned to champion Newton's. Even Cambridge, Newton's own university, continued to teach natural philosophy from the text of Jacques Rohault (1620–75), a Cartesian. In addition, the most eminent mathematicians of the late seventeenth century —Huygens, Leibniz, and John Bernoulli—were opposed to the concept of gravitation and hence to its application. Analytical methods of treating planetary motion were undertaken by Daniel Bernoulli, who received a prize from the French Academy of Sciences for a paper of 1734 on the two-body problem. It was handled completely by Euler in his book *Theoria Motuum Planetarum et Cometarum*.[41]

If we have n bodies, each spherical and with spherically symmetric mass distribution (density a function of radius), they will attract each other as though their masses were at their centers. Let $m_1, m_2, \ldots, m_n$ represent the masses and (x_i, y_i, z_i) the (variable) coordinates of the ith mass with respect to a fixed system of axes; let r_{ij} be the distance from m_i to m_j. Then the x-components of the forces acting on m_1 are

$$-\frac{k}{r_{12}^3} m_1 m_2 (x_1 - x_2), \quad -\frac{k}{r_{13}^3} m_1 m_3 (x_1 - x_3), \ldots, \quad -\frac{k}{r_{1n}^3} m_1 m_n (x_1 - x_n),$$

with similar expressions for the y- and z-components of the force. Each body has such components of force acting on it.

The differential equations of the motion of the ith body are then

$$m_i(d^2 x_i/dt^2) = -km_i \sum_{j=1}^{n} m_j[(x_i - x_j)/r_{ij}^3]$$

(45)
$$m_i(d^2 y_i/dt^2) = -km_i \sum_{j=1}^{n} m_j[(y_i - y_j)/r_{ij}^3]$$

$$m_i(d^2 z_i/dt^2) = -km_i \sum_{j=1}^{n} m_j[(z_i - z_j)/r_{ij}^3]$$

with $j \neq i$ and $i = 1, 2, \ldots, n$. There are $3n$ equations, each of second order. The origin can be chosen to be the center of mass of the n bodies or it can be chosen in one of the bodies, for example, the sun. There are $6n$ integrals, of which 10 can be found somewhat readily; these are the only ones known in the general problem.

The problem of n bodies, indeed even of three bodies, cannot be solved exactly. Hence the investigations of this problem have taken two general

41. 1744 = *Opera*, (2), 28, 105–251.

directions. The first is a search for whatever general theorems one can deduce, which may at least shed some light on the motions. The second is a search for approximate solutions that may be useful for a period of time subsequent to some instant at which data may be available; this is known as the method of perturbations.

The first type of investigation produced some theorems on the motion of the center of mass of n bodies, which were given by Newton in his *Principia*. For example, the center of mass of the n bodies moves with uniform speed in a straight line. The ten integrals mentioned above, which are consequences of what are called conservation laws of motion, also constitute theorems of the first type. These integrals were known to Euler. There are also some exact results for special cases of the problem of three bodies, and these are due to one of the masters of celestial mechanics, Joseph-Louis Lagrange.

Lagrange (1736–1813) was of French and Italian extraction. As a boy he was unimpressed with mathematics, but while still at school he read an essay by Halley on the merits of Newton's calculus and became excited about the subject. At the age of nineteen he became a professor of mathematics at the Royal Artillery School of Turin, the city of his birth. He soon contributed so much to mathematics that even at an early age he was recognized as one of the period's greatest mathematicians. Though Lagrange worked in many branches of mathematics—the theory of numbers, the theory of algebraic equations, the calculus, differential equations and the calculus of variations—and in many branches of physics, his chief interest was the application of the law of gravitation to planetary motion. He said, in 1775, "The arithmetical researches are those which have cost me most trouble and are perhaps the least valuable." Archimedes was Lagrange's idol.

Lagrange's most famous work, his *Mécanique analytique* (1788; second edition, 1811–15; posthumous edition, 1853), extended, formalized, and crowned Newton's work on mechanics. Lagrange had once complained that Newton was a most fortunate man, since there was but one universe and Newton had already discovered its mathematical laws. However, Lagrange had the honor of making apparent to the world the perfection of the Newtonian theory. Though the *Mécanique* is a classic of science and is significant also for the theory and use of ordinary differential equations, Lagrange had trouble finding a publisher.

The particular exact solutions obtained in the three-body problem were given by Lagrange in a prize paper of 1772, *Essai sur le problème des trois corps*.[42] One of these solutions states that it is possible to set these bodies in motion so that their orbits are similar ellipses all described in the same time and with the center of mass of the three bodies as a common focus.

42. *Hist. de l'Acad. des Sci.*, Paris, 9, 1772 = *Œuvres*, 6, 229–331.

Another solution assumes the three bodies are started from the three summits of an equilateral triangle. They will then move as though attached to the triangle which itself rotates about the center of mass of the bodies. The third solution assumes the three bodies are projected into motion from positions on a straight line. For appropriate initial conditions they will continue to be fixed on that line while the line rotates in a plane about the center of mass of the bodies. These three cases had no physical reality for Lagrange, but the equilateral triangle case was found in 1906 to apply to the sun, Jupiter, and an asteroid named Achilles.

The second type of problem involving n bodies deals, as already noted, with approximate solutions or the theory of perturbations. Two spherical bodies acted upon by their mutual gravitational attraction move along conic sections. This motion is said to be unperturbed. Any departure from such motions, whether in position or velocity, however caused, is perturbed motion. If there are two spheres but there is resistance from the medium in which they move, or if the two bodies are no longer spherical but, say, oblate spheroids, or if more than two bodies are involved, then the orbits of the bodies are no longer conic sections. Before the use of telescopes the perturbations were not striking. In the eighteenth century the calculation of perturbations became a major mathematical problem, and Clairaut, d'Alembert, Euler, Lagrange, and Laplace all made contributions. Laplace's work in this area was the most outstanding.

Pierre-Simon de Laplace (1749–1827) was born to reasonably well-to-do parents in the town of Beaumont, Normandy. It seemed likely that he would become a priest but at the University of Caen, which he entered at the age of sixteen, he took to mathematics. He spent five years at Caen and while there wrote a paper on the calculus of finite differences. After finishing his studies, Laplace went to Paris with letters of recommendation to d'Alembert, who ignored him. Then Laplace wrote d'Alembert a letter containing an exposition of the general principles of mechanics; this time d'Alembert took notice, sent for Laplace, and got him the position of professor of mathematics at the Ecole Militaire in Paris.

Even as a youth Laplace published prolifically. A statement made in the Paris Academy of Sciences shortly after his election in 1773 pointed out that no one so young had presented so many papers on such diverse and difficult subjects. In 1783 he replaced Bezout as an examiner in artillery and examined Napoleon. During the Revolution he was made a member of the Commission on Weights and Measures but was later expelled, along with Lavoisier and others, for not being a good republican. Laplace retired to Melun, a small city near Paris, where he worked on his celebrated and popular *Exposition du système du monde* (1796). After the Revolution he became a professor at the Ecole Normale, where Lagrange by this time was also teaching, and served on a number of government committees. Then he became successively Minister of the Interior, a member of the Senate, and chancellor of that body. Though honored by Napoleon with the title of count, Laplace voted against him in 1814 and rallied to Louis XVIII, who named Laplace marquis and peer of France.

During these years of political activity he continued to work in science. Between 1799 and 1825 there appeared the five volumes of his *Mécanique céleste*. In this work Laplace presented "complete" analytic solutions of the mechanical problems posed by the solar system. He took as little from observational data as possible. The *Mécanique céleste* embodies discoveries and results of Newton, Clairaut, d'Alembert, Euler, Lagrange, and Laplace himself. The masterpiece was so complete that his immediate successors could add little. Perhaps its only defect is that Laplace frequently neglected to acknowledge the sources of his results and left the impression that they were all his own.

In 1812 he published his *Théorie analytique des probabilités*. The introduction to the second edition (1814) consists of a popular essay known as the *Essai philosophique sur les probabilités*. It contains the famous passage to the effect that the future of the world is completely determined by the past and that one possessed of the mathematical knowledge of the state of the world at any given instant could predict the future.

Laplace made many important discoveries in mathematical physics, some of which we shall take up in later chapters. Indeed, he was interested in anything that helped to interpret nature. He worked on hydrodynamics, the wave propagation of sound, and the tides. In the field of chemistry, his work on the liquid state of matter is classic. His studies of the tension in the surface layer of water, which accounts for the rise of liquids inside a capillary tube, and of the cohesive forces in liquids, are fundamental. Laplace and Lavoisier designed an ice calorimeter (1784) to measure heat and measured the specific heat of numerous substances; heat to them was still a special kind of matter. Most of Laplace's life was, however, devoted to celestial mechanics. He died in 1827. His last words were reported to be, "What we know is very slight; what we don't know is immense"—though De Morgan states that they were "Man follows only phantoms."

Laplace is often linked with Lagrange, but they are not similar in their personal qualities or their work. Laplace's vanity kept him from giving sufficient credit to the works of those whom he considered rivals; in fact, he used many ideas of Lagrange without acknowledgment. When Laplace is mentioned in connection with Lagrange it is always to commend the personal qualities of the latter. Lagrange is the mathematician, very careful in his writings, very clear, and very elegant. Laplace created a number of new mathematical methods that were subsequently expanded into branches of mathematics, but he never cared for mathematics except as it helped him to study nature. When he encountered a mathematical problem in his physical research he solved it almost casually and merely stated, "It is easy to see that . . ." without bothering to show how he had worked out the result. He confessed, however, that it was not easy to reconstruct his own work. Nathaniel Bowditch (1773–1838), the American mathematician and astronomer, who translated four of the five volumes of the *Mécanique céleste* and added explanations, said that whenever he came across the phrase "It is easy to see that . . ." he knew he had hours of hard work ahead to fill the gaps. Actually, Laplace was impatient of the mathematics and wanted to get on with the applications. Pure mathematics did not interest him and his contributions to it were by-products of his great work in natural philosophy. Mathematics was a means, not an end, a tool one ought to perfect in order to solve the problems of science.

Laplace's work, so far as it is relevant to the subject of this chapter, deals with approximate solutions of problems of planetary motion. The possibility of obtaining useful solutions by any approximate method rests on the following factors. The solar system is dominated by the sun, which contains 99.87 percent of the entire matter in the system. This means that the orbits of the planets are nearly elliptical, because the perturbing forces of the planets on each other are small. Nevertheless, Jupiter has 70 percent of the planetary mass. Also, the earth's moon is quite close to the earth and so each is affected by the other. Hence perturbations do have to be considered.

The three-body problem, especially sun, earth, and moon, was studied most in the eighteenth century, partly because it is the next step after the two-body problem and partly because an accurate knowledge of the moon's motion was needed for navigation. In the case of the sun, earth, and moon, some advantage can be taken of the fact that the sun is far from the other two bodies and can be treated as exercising only a small influence on the relative motion of the earth and moon. In the case of the sun and two planets, one planet is usually regarded as perturbing the motion of the other around the sun. If one of the planets is small, its gravitational effect on the other planet can be neglected, but the effect of the larger one on the small one must be taken into account. These special cases of the three-body problem are called the restricted three-body problem.

The theory of perturbations for the three-body problem was first applied to the motion of the moon; this was done geometrically by Newton in the third book of the *Principia*. Euler and Clairaut tried to obtain exact solutions for the general three-body problem, complained about the difficulties, and of course resorted to approximate methods. Here Clairaut made the first real progress (1747) by using series solutions of the differential equations. He then had occasion to apply his result to the motion of Halley's comet, which had been observed in 1531, 1607, and 1682. It was expected to be at the perihelion of its path around the earth in 1759. Clairaut computed the perturbations due to the attraction of Jupiter and Saturn and predicted in a paper read to the Paris Academy on November 14, 1758, that the perihelion would occur on April 13, 1759. He remarked that the exact time was uncertain to the extent of a month because the masses of Jupiter and Saturn were not known precisely and because there were slight perturbations caused by other planets. The comet reached its perihelion on March 13.

To compute perturbations the method called variation of the elements or variation of parameters—or variation of the constants of integration—was created and is the most effective one. We are obliged to confine ourselves to its mathematical principles; we shall therefore examine it without taking into account the full physical background.

The mathematical method of variation of parameters for the problem of three bodies goes back to Newton's *Principia*. After treating the motion of the moon about the earth and obtaining the elliptical orbit, Newton took account of the effects of the sun on the moon's orbit by considering variations in the latter. The method was used in isolated instances to solve nonhomogeneous equations by John Bernoulli in the *Acta Eruditorum* of 1697[43] and by Euler in 1739, in treating the second order equation $y'' + k^2 y = X(x)$. It was first used to treat perturbations of planetary motions by Euler in his paper of 1748,[44] which treated the mutual perturbations of Jupiter and Saturn and won a prize from the French Academy. Laplace wrote many papers on the method.[45] It was fully developed by Lagrange in two papers.[46]

The method of variation of parameters for a single ordinary differential equation was applied by Lagrange to the nth-order equation

$$Py + Qy' + Ry'' + \cdots + Vy^{(n)} = X,$$

where X, P, Q, R, ... are functions of x. For simplicity we shall suppose that we have a second order equation.

In the case $X = 0$, Lagrange knew that the general solution is

(46) $$y = ap(x) + bq(x)$$

where a and b are integration constants and p and q are particular integrals of the homogeneous equation. Now, says Lagrange, let us regard a and b as functions of x. Then

(47) $$\frac{dy}{dx} = ap' + bq' + pa' + qb'.$$

43. Page 113.
44. *Opera*, (2), 25, 45–157.
45. See, for example, *Hist. de l'Acad. des Sci.*, Paris, 1772, Part 1, 651 ff., pub. 1775 = *Œuvres*, 8, 361–66, and *Hist. de l'Acad. des Sci.*, Paris, 1777, 373 ff., pub. 1780 = *Œuvres*, 9, 357–80.
46. *Nouv. Mém. de l'Acad. de Berlin*, 5, 1774, 201 ff., and 6, 1775, 190 ff. = *Œuvres*, 4, 5–108 and 151–251.

Lagrange now sets

(48) $$pa' + qb' = 0;$$

that is, the part of y' that results from the variability of a and b he sets equal to 0. From (47) we have, in view of (48),

(49) $$\frac{d^2y}{dx^2} = ap'' + bq'' + p'a' + q'b'.$$

If the equation were of higher order than the second, Lagrange would now set $p'a' + q'b' = 0$ and find d^3y/dx^3. Since, in our case, the equation is of second order, he would keep all the terms in (49).

He now substitutes the expressions for y, dy/dx and d^2y/dx^2 given by (46), (47), and (49) in the original equation. Since (46) is a solution of the homogeneous equation and (48) throws out some of the terms that result from the variability of a and b, there remains after the substitution

(50) $$p'a' + q'b' = \frac{X}{R}.$$

This equation and equation (48) give two algebraic equations in the two unknown functions a' and b'. These two simultaneous equations can be solved for a' and b' in terms of the known functions p, q, p', q', X, and R. Then a and b can each be obtained by an integration or are at least reduced to a quadrature. With these values of a and b, (46) gives a solution of the original nonhomogeneous equation. This solution together with the solution of the homogeneous equation is the complete solution of the nonhomogeneous equation.

The method of variation of parameters was treated in more general fashion by Lagrange[47] and he showed its applicability to many problems of physics. In the paper of 1808 he applied it to a system of three second order equations. The technique is of course more complicated, but again the basic idea is to treat the six constants of integration of the solution of the corresponding homogeneous system as variables and to determine them so that the expression will satisy the nonhomogeneous system.

During the period in which they were developing the method of variation of parameters and afterward, Lagrange and Laplace wrote a number of key papers on basic problems of the solar system. In his crowning work, the *Mécanique céleste*, Laplace summarized the scope of their results:

> We have given, in the first part of this work, the general principles of the equilibrium and motion of bodies. The application of these principles

47. *Mém. de l'Acad. des Sci., Inst. France*, 1808, 267 ff. = *Œuvres*, 6, 713–68.

to the motions of the heavenly bodies had conducted us, by geometrical [analytical] reasoning, without any hypothesis, to the law of universal attraction, the action of gravity and the motion of projectiles being particular cases of this law. We have then taken into consideration a system of bodies subjected to this great law of nature and have obtained, by a singular analysis, the general expressions of their motions, of their figures, and of the oscillations of the fluids which cover them. From these expressions we have deduced all the known phenomena of the flow and ebb of the tide; the variations of the degrees and the force of gravity at the surface of the earth; the precession of the equinoxes; the libration of the moon; and the figure and rotation of Saturn's rings. We have also pointed out the reason that these rings remain permanently in the plane of the equator of Saturn. Moreover, we have deduced, from the same theory of gravity, the principal equations of the motions of the planets, particularly those of Jupiter and Saturn, whose great inequalities have a period of above 900 years.[48]

Laplace concluded that nature ordered the celestial machine "for an eternal duration, upon the same principles which prevail so admirably upon the earth, for the preservation of individuals and for the perpetuity of the species."

As the mathematical methods for solving differential equations were improved and as new physical facts about the planets were acquired, efforts were made throughout the nineteenth and twentieth centuries to obtain better results on the various subjects Laplace mentions, in particular on the n-body problem and the stability of the solar system.

8. *Summary*

As we have seen, the attempt to solve physical problems, which at first involved no more than quadratures, led gradually to the realization that a new branch of mathematics was being created, namely, ordinary differential equations. By the middle of the eighteenth century the subject of differential equations became an independent discipline and the solution of such equations an end in itself.

The nature of what was regarded and sought as a solution gradually changed. At first, mathematicians looked for solutions in terms of elementary functions; soon they were content to express an answer as a quadrature that might not be effected. When the major attempts to find solutions in terms of elementary functions and quadratures failed, mathematicians became content to seek solutions in infinite series.

The problem of solution in closed form was not forgotten, but instead of attempting to solve in that manner the particular differential equations

48. Preface to Vol. 3.

that arose from physical problems, the mathematicians sought differential equations that permit solutions in terms of a finite number of elementary functions. A great number of differential equations integrable in this manner were found. D'Alembert (1767) worked on this problem and included elliptic integrals among acceptable answers. A typical approach to this problem, made by Euler (1769), among others, was to start with a differential equation whose integration in closed form could be effected and then derive other differential equations from the known one. Another approach was to look for conditions under which the series solution might contain only a finite number of terms.

An interesting but unfruitful piece of work by Marie-Jean-Antoine-Nicolas Caritat de Condorcet (1743–94) in *Du calcul intégral* (1765) was his attempt to bring order and method out of the many separate methods and tricks used to solve differential equations. He listed such operations as differentiation, elimination, and substitution and sought to reduce all methods to these canonical operations. The work led nowhere. In line with this plan Euler showed that where separation of variables is possible a multiplying (integrating) factor will work, but not conversely. Also he showed that separation of variables will not do for higher order differential equations. As for substitutions, he found no general principles for obtaining them.[49] Finding substitutions is as difficult as solving the differential equations directly. However, transformation can reduce the order of a differential equation. Euler used this idea to solve the nth-order nonhomogeneous linear ordinary differential equation, and even in the case of the homogeneous equation he thought of each $\exp\left[\int p\, dx\right]$ as giving, for the proper value of p, a first order factor of the ordinary differential equation. Reducing the order was also Riccati's plan. A number of other methods, including Lagrange's undetermined multipliers, were devised. At first this method was believed to be general but it did not prove to be so.

The search for general methods of integrating ordinary differential equations ended about 1775. Much new work was yet to be done with ordinary differential equations, particularly those resulting from the solution of partial differential equations. But no major new methods beyond those already surveyed here were discovered for a hundred years or so, until operator methods and the Laplace transform were introduced at the end of the nineteenth century. In fact interest in general methods of solution receded, because methods of one form or another adequate to those types required in applications were obtained. Broad, comprehensive principles for the solution of ordinary differential equations are still lacking. On the whole the subject has continued to be a series of separate techniques for the various types.

49. *Institutiones Calculi Integralis*, 1, 290.

Bibliography

Bernoulli, James: *Opera*, 2 vols., 1744, reprinted by Birkhaüser, 1968.

Bernoulli, John: *Opera Omnia*, 4 vols., 1742, reprinted by Georg Olms, 1968.

Berry, Arthur: *A Short History of Astronomy*, Dover (reprint), 1961, Chaps. 9–11.

Cantor, Moritz: *Vorlesungen über Geschichte der Mathematik*, B. G. Teubner, 1898 and 1924, Vol. 3, Chaps. 100 and 118, Vol. 4, Sec. 27.

Delambre, J. B. J.: *Histoire de l'astronomie moderne*, 2 vols., 1821, Johnson Reprint Corp., 1966.

Euler, Leonhard: *Opera Omnia*, Orell Füssli, Series 1, Vols. 22 and 23, 1936 and 1938; Series 2, Vols. 10 and 11, Part 1, 1947 and 1957.

Hofmann, J. E.: "Über Jakob Bernoullis Beiträge zur Infinitesimal-mathematik," *L'Enseignement Mathématique*, (2), 2, 61–171. Published separately by Institut de Mathématiques, Geneva, 1957.

Lagrange, Joseph-Louis: *Œuvres de Lagrange*, Gauthier-Villars, 1868–1873, relevant papers in Vols. 2, 3, 4, and 6.

————: *Mécanique analytique*, 1788; 4th ed., Gauthier-Villars, 1889. The fourth edition is an unchanged reproduction of the third edition of 1853.

Lalande, J. de: *Traité d'astronomie*, 3 vols., 1792, Johnson Reprint Corp., 1964.

Laplace, Pierre-Simon: *Œuvres complètes*, Gauthier-Villars, 1891–1904, relevant papers in Vols. 8, 11 and 13.

————: *Traité de mécanique céleste*, 5 vols., 1799–1825. Also in *Œuvres complètes*, Vols. 1–5, Gauthier-Villars, 1878–82. English trans. of Vols. 1–4 by Nathaniel Bowditch, 1829–39, Chelsea (reprint), 1966.

————: *Exposition du système du monde*, 1st ed., 1796, 6th ed. in *Œuvres complètes*, Gauthier-Villars, 1884, Vol. 6.

Montucla, J. F.: *Histoire des mathématiques*, 1802, Albert Blanchard (reprint), 1960, Vol. 3, 163–200; Vol. 4, 1–125.

Todhunter, I.: *A History of the Mathematical Theories of Attraction and the Figure of the Earth*, 1873, Dover (reprint), 1962.

Truesdell, Clifford E.: *Introduction to Leonhardi Euleri Opera Omnia, Vol. X et XI Seriei Secundae*, in Euler, *Opera Omnia*, (2), 11, Part 2, Orell Füssli, 1960.

22

Partial Differential Equations in the Eighteenth Century

Mathematical Analysis is as extensive as nature herself.
JOSEPH FOURIER

1. *Introduction*

As in the case of ordinary differential equations, the mathematicians did not consciously create the subject of partial differential equations. They continued to explore the same physical problems that had led to the former subject; and as they secured a better grasp of the physical principles underlying the phenomena, they formulated mathematical statements that are now comprised in partial differential equations. Thus, whereas the displacement of a vibrating string had been studied separately as a function of time and as a function of the distance of a point on the string from one end, the study of the displacement as a function of both variables and the attempt to comprehend all the possible motions led to a partial differential equation. The natural continuation of this study, namely, the investigation of the sounds created by the string as they propagate in air, introduced additional partial differential equations. After studying these sounds the mathematicians took up the sounds given off by horns of all shapes, organ pipes, bells, drums, and other instruments.

Air is one type of fluid, as the term is used in physics, and happens to be compressible. Liquids are (virtually) incompressible fluids. The laws of motion of such fluids and, in particular, the waves that can propagate in both became a broad field of investigation that now constitutes the subject of hydrodynamics. This field, too, gave rise to partial differential equations.

Throughout the eighteenth century, mathematicians continued to work on the problem of the gravitational attraction exerted by bodies of various shapes, notably the ellipsoid. While basically this is a problem of triple integration, it was converted by Laplace into a problem of partial differential equations in a manner we shall examine shortly.

2. The Wave Equation

Though specific partial differential equations appear as early as 1734 in the work of Euler[1] and in 1743 in d'Alembert's *Traité de dynamique*, nothing worth noting was done with them. The first real success with partial differential equations came in renewed attacks on the vibrating-string problem, typified by the violin string. The approximation that the vibrations are small was imposed to make the partial differential equation tractable. Jean Le Rond d'Alembert (1717–83), in his papers of 1746[2] entitled "Researches on the Curve Formed by a Stretched String Set into Vibrations," says he proposes to show that infinitely many curves other than the sine curve are modes of vibration.

We may recall from the preceding chapter that in the first approaches to the vibrating string, it was regarded as a "string of beads." That is, the string was considered to contain n discrete equal and equally spaced weights joined to each other by pieces of weightless, flexible, and elastic thread. To treat the continuous string, the number of weights was allowed to become infinite while the size and mass of each was decreased, so that the total mass of the increasing number of individual "beads" approached the mass of the continuous string. There were mathematical difficulties in passing to the limit, but these subtleties were ignored.

The case of a discrete number of masses had been treated by John Bernoulli in 1727 (Chap. 21, sec. 4). If the string is of length l and lies along $0 \le x \le l$, and if x_k is the abscissa of the kth mass, $k = 1, 2, \cdots, n$ (the nth mass at $x = l$ is motionless), then

$$x_k = k\frac{l}{n}, \qquad k = 1, 2, \cdots, n.$$

By analyzing the force on the kth mass, Bernoulli had shown that if y_k is the displacement of the kth mass, then

$$\frac{d^2 y_k}{dt^2} = \left(\frac{na}{l}\right)^2 (y_{k+1} - 2y_k + y_{k-1}), \qquad k = 1, 2, \cdots, n-1,$$

where $a^2 = lT/M$, T is the tension in the string (which is taken to be constant as the string vibrates), and M the total mass. D'Alembert replaced y_k by $y(t, x)$ and l/n by Δx. Then

$$\frac{\partial^2 y(t, x)}{\partial t^2} = a^2 \left[\frac{y(t, x + \Delta x) - 2y(t, x) + y(t, x - \Delta x)}{(\Delta x)^2}\right].$$

1. *Comm. Acad. Sci. Petrop.*, 7, 1734/35, 184–200, pub. 1740 = *Opera*, (1), 22, 57–75.
2. *Hist. de l'Acad. de Berlin*, 3, 1747, 214–19 and 220–49, pub. 1749.

He now observed that as n becomes infinite so that Δx approaches 0, the bracketed expression becomes $\partial^2 y/\partial x^2$. Hence

$$(1) \qquad \frac{\partial^2 y(t, x)}{\partial t^2} = a^2 \frac{\partial^2 y(t, x)}{\partial x^2},$$

where a^2 is now T/σ, σ being the mass per unit length. Thus what is now called the wave equation in one spatial dimension appears for the first time.

Since the string is fixed at the endpoints $x = 0$ and $x = l$, the solution must satisfy the boundary conditions

$$(2) \qquad y(t, 0) = 0, \qquad y(t, l) = 0.$$

At $t = 0$ the string is displaced into some shape $y = f(x)$ and then released, which means that each particle starts with zero initial velocity. These initial conditions are expressed mathematically as

$$(3) \qquad y(0, x) = f(x), \qquad \frac{\partial y(t, x)}{\partial t}\bigg|_{t=0} = 0$$

and they must also be satisfied by the solution.

This problem was solved by d'Alembert in so clever a manner that it is often reproduced in modern texts. We shall not take space for all the details. He proved first that

$$(4) \qquad y(t, x) = \frac{1}{2}\phi(at + x) + \frac{1}{2}\psi(at - x),$$

where ϕ and ψ are as yet unknown functions.

Thus far d'Alembert had deduced that *every* solution of the partial differential equation (1) is the sum of a function of $(at + x)$ and a function of $(at - x)$. The converse is easy to show by direct substitution of (4) into (1). Of course d'Alembert had yet to satisfy the boundary and initial conditions. The condition $y(t, 0) = 0$ applied to (4) gives, for all t,

$$(5) \qquad \frac{1}{2}\phi(at) + \frac{1}{2}\psi(at) = 0.$$

Since for any x, $ax + t = at'$ for some value of t', we may say that for any x and t

$$(6) \qquad \phi(x + at) = -\psi(x + at).$$

Then the condition $y(t, l) = 0$, becomes, in view of (4),

$$(7) \qquad \frac{1}{2}\phi(at + l) = \frac{1}{2}\phi(at - l);$$

and since this is an identity in t, it shows that ϕ must be periodic in $at + x$ with period $2l$.

The condition

$$(8) \qquad \frac{\partial y(t, x)}{\partial t}\bigg|_{t=0} = 0$$

yields, from (4) and the fact that $\phi = -\psi$,

$$(9) \qquad \phi'(x) = \phi'(\quad x).$$

On integration this becomes

$$(10) \qquad \phi(x) = -\phi(-x),$$

and thus ϕ is an odd function of x. If we now use the fact that $\phi = -\psi$ in (4), form $y(0, x)$ and use (10), we have that

$$(11) \qquad y(0, x) = \phi(x),$$

and since the initial condition is $y(0, x) = f(x)$ we have

$$(12) \qquad \phi(x) = f(x) \quad \text{for} \quad 0 \le x \le l.$$

To sum up,

$$(13) \qquad y(t, x) = \frac{1}{2} \phi(at + x) - \frac{1}{2} \phi(at - x),$$

where ϕ is subject to the above conditions of periodicity and oddness. Moreover, if the initial state is $y(0, x) = f(x)$, then (12) must hold between 0 and l. Thus there would be just one solution for a given $f(x)$. Now d'Alembert regarded functions as analytic expressions formed by the processes of algebra and the calculus. Hence if two such functions agree in one interval of x-values, they must agree for every value of x. Since $\phi(x) = f(x)$ in $0 \le x \le l$ and ϕ had to be odd and periodic, then $f(x)$ had to meet the same conditions. Finally, since $y(t, x)$ had to satisfy the differential equation, it had to be twice differentiable. But $y(0, x) = f(x)$, and so $f(x)$ had to be twice differentiable.

Within a few months of seeing d'Alembert's 1746 papers, Euler wrote his own paper, "On the Vibration of Strings," which was presented on May 16, 1748.[3] Though in method of solution he followed d'Alembert, Euler by this time had a totally different idea as to what functions could be admitted as initial curves and therefore as solutions of partial differential equations. Even before the debate on the vibrating-string problem, in fact in a work of 1734, he allowed functions formed from parts of different well-known curves and even formed by drawing curves freehand. Thus the curve (Fig. 22.1) formed by an arc of a parabola in the interval (a, c) and by an arc of a third degree curve in the interval (c, b) constituted one curve or one function under

3. *Nova Acta Erud.*, 1749, 512–27 = *Opera*, (2), 10, 50–62; also in French by Euler, *Hist. de l'Acad. de Berlin*, 4, 1748, 69–85 = *Opera*, (2), 10, 63–77.

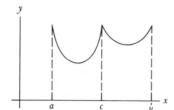

Figure 22.1

this concept. Euler called such curves discontinuous, though in modern terminology they are continuous with discontinuous derivatives. In his text, the *Introductio* of 1748, he stuck to the notion that was standard in the eighteenth century, that a function must be given by a single analytical expression. However, the physics of the vibrating-string problem seems to have been his compelling reason for bringing his new concept of function to the fore. He accepted any function defined by a formula $\phi(x)$ in $-l \le x \le l$, and regarded $\phi(x + 2l) = \phi(x)$ to be the definition of the curve outside $(-l, l)$. In a later paper[4] he goes further; he says that

(14) $y = \phi(ct + x) + \psi(ct - x),$

with arbitrary ϕ and ψ, is a solution of

(15) $$\frac{1}{c^2}\frac{\partial^2 y}{\partial t^2} = \frac{\partial^2 y}{\partial x^2}.$$

This follows by substitution in the differential equation. But the initial curve is equally satisfactory, whether it is expressed by some equation, or whether it is traced in any fashion not expressible by an equation. Of the initial curve, only the part in $0 \le x < l$ is relevant. The continuation of this part is not to be taken into consideration. The different parts of this curve are thus not joined to each other by any law of continuity (single analytic expression); it is only by the description that they are joined together. For this reason it may be impossible to comprise the entire curve in one equation, except when by chance the curve is some sine function.

In 1755 Euler gave as a new definition of function, "If some quantities depend on others in such a way as to undergo variation when the latter are varied, then the former are called functions of the latter." And in another paper[5] he says that the parts of a "discontinuous" function do not belong to one another and are not determined by one single equation for the whole extent of the function. Moreover, given the initial shape in $0 \le x \le l$, one repeats it in reverse order in $-l \le x \le 0$ (so as to make it odd) and conceives the continual repetition of this curve in each interval of length $2l$ to infinity.

4. *Hist. de l'Acad. de Berlin*, 9, 1753, 196–222, pub. 1755 = *Opera*, (2), 10, 232–54.
5. *Novi Comm. Acad. Sci. Petrop.*, 11, 1765, 67–102, pub. 1767 = *Opera*, (1), 23, 74–91.

Then, if this curve $[y = f(x)]$ is used to represent the initial function, after the time t the ordinate that will answer to the abscissa x of the string in vibration will be (cf. [13] and [12])

$$(16) \qquad y = \frac{1}{2}f(x + ct) + \frac{1}{2}f(x - ct).$$

In his basic 1749 paper Euler points out that all possible motions of the vibrating string are periodic in time whatever the shape of the string; that is, the period is (usually) the period of what we now call the fundamental. He also realized that individual modes whose periods are one half, one third, and so on of the basic (fundamental) period can occur as the vibrating figure. He gives such special solutions as

$$(17) \qquad y(t, x) = \sum A_n \sin \frac{n\pi x}{l} \cos \frac{n\pi ct}{l}$$

when the initial shape is

$$(18) \qquad y(0, x) = \sum A_n \sin \frac{n\pi x}{l},$$

but does not say whether the summation covers a finite or infinite number of terms. Nevertheless he has the idea of superposition of modes. Thus Euler's main point of disagreement with d'Alembert is that he would admit all kinds of initial curves, and therefore non-analytic solutions, whereas d'Alembert accepted only analytic initial curves and solutions.

In introducing his "discontinuous" functions, Euler appreciated that he had taken a big step forward. He wrote to d'Alembert on December 20, 1763, that "considering such functions as are subject to no law of continuity [analyticity] opens to us a wholly new range of analysis."[6]

The solution of the vibrating-string problem was given in entirely different form by Daniel Bernoulli; this work stirred up another ground for controversy about the allowable solutions. Daniel Bernoulli (1700–82), the son of John Bernoulli, was a professor of mathematics at St. Petersburg from 1725 to 1733 and then, successively, professor of medicine, metaphysics, and natural philosophy at Basle. His chief work was in hydrodynamics and elasticity. In the former area, he won a prize for a paper on the flow of the tides; he also contemplated the application of the theory of the flow of liquids to the flow of blood in human blood vessels. He was a skilled experimentalist and through experimental work discovered the law of attraction of static electric charges before 1760. This law is usually credited to Charles Coulomb. Bernoulli's *Hydrodynamica* (1738), which contains studies that appeared in a number of papers, is the first major text in its field. It has a chapter on the

6. *Opera* (2), 11, sec. 1, 2.

mechanical theory of heat (as opposed to heat as a substance) and gives many results on the theory of gases.

In his paper of 1732/33 cited in the preceding chapter, Bernoulli had expressly stated that the vibrating string could have higher modes of oscillations. In a later paper[7] on composite oscillation of weights on a loaded vertical flexible string, he made the following remark:

> Similarly, a taut musical string can produce its isochronous tremblings in many ways and even according to theory infinitely many, ... and moreover in each mode it emits a higher or lower note. The first and most natural mode occurs when the string in its oscillations produces a single arch; then it makes the slowest oscillations and gives out the deepest of all its possible tones, fundamental to all the rest. The next mode demands that the string produce two arches on the opposite sides [of the string's rest position] and then the oscillations are twice as fast, and now it gives out the octave of the fundamental sound.

Then he describes the higher modes. He does not give the mathematics but it seems evident that he had it.

In a paper on the vibrations of a bar and the sounds given off by the vibrating bar,[8] Bernoulli not only gives the separate modes in which the bar can vibrate but says distinctly that both sounds (the fundamental and a higher harmonic) can exist together. This is the first statement of the co-existence of small harmonic oscillations. Bernoulli based it on his physical understanding of how the bar and the sounds can act but gave no mathematical evidence that the sum of two modes is a solution.

When he read d'Alembert's first paper of 1746 and Euler's paper of 1749 on the vibrating string, he hastened to publish the ideas he had had for many years.[9] After indulging in sarcasm about the abstractness of d'Alembert's and Euler's works, he reasserts that many modes of a vibrating string can exist simultaneously (the string then responds to the sum or superposition of all the modes) and claims that this is all that Euler and d'Alembert have shown. Then comes a major point. He insists that *all* possible initial curves are representable as

$$(19) \qquad f(x) = \sum_{n=1}^{\infty} a_n \sin \frac{n\pi x}{l}$$

because there are enough constants a_n to make the series fit any curve. Hence, he asserts, *all* subsequent motions would be

$$(20) \qquad y(t, x) = \sum_{n=1}^{\infty} a_n \sin \frac{n\pi x}{l} \cos \frac{n\pi ct}{l}.$$

7. *Comm. Acad. Sci. Petrop.*, 12, 1740, 97–108, pub. 1750.
8. *Comm. Acad. Sci. Petrop.*, 13, 1741/43, 167–96, pub. 1751.
9. *Hist. de l'Acad. de Berlin*, 9, 1753, 147–72 and 173–95, pub. 1755.

Thus every motion corresponding to any initial curve is no more than a sum
of sinusoidal periodic modes, and the combination has the frequency of the
fundamental. However, he gives no mathematical arguments to back up his
contentions; he relies on the physics. In this paper of 1753 Bernoulli states:

> My conclusion is that all sonorous bodies include an infinity of sounds
> with a corresponding infinity of regular vibrations. . . . But it is not of
> this multitude of sounds that Messrs. d'Alembert and Euler claim to
> speak. . . . Each kind [each fundamental mode generated by some
> initial curve] multiplies an infinite number of times to accord to
> each interval an infinite number of curves, such that each point starts,
> and achieves at the same instant, these vibrations while, following the
> theory of Mr. Taylor, each interval between two nodes should assume
> the form of the companion of the cycloid [sine function] extremely
> elongated.
>
> We then remark that the chord AB cannot make vibrations only
> conforming to the first figure [fundamental] or second [second har-
> monic] or third and so forth to infinity but that it can make a combina-
> tion of these vibrations in all possible combinations, and that all new
> curves given by d'Alembert and Euler are only combinations of the
> Taylor vibrations.

In this last remark Bernoulli ascribes knowledge to Taylor that Taylor
never displayed. This apart, however, Bernoulli's contentions were enormously significant.

Euler objected at once to Bernoulli's last assertion. In fact Euler's 1753
paper presented to the Berlin Academy (already referred to above) was in
part a reply to Bernoulli's two papers. Euler emphasizes the importance of
the wave equation as the starting point for the treatment of the vibrating
string. He praises Bernoulli's recognition that many modes can exist simultaneously so that the string can emit many harmonics in one motion, but
denies, as did d'Alembert, that all possible motions can be expressed by (20).
He admits that an initial curve such as

$$(21) \qquad\qquad f(x) = \frac{c \sin(ax/l)}{1 - a \cos(ax/l)}, \qquad |a| < 1,$$

can be expressed by a series such as (19). Bernoulli would be borne out if
every function could be represented by an infinite trigonometric series, but
this Euler regards as impossible. A sum of sine functions is, Euler says, an
odd periodic function. But in his solution (see [16]),

$$(22) \qquad\qquad y(t, x) = \frac{1}{2}f(x + ct) + \frac{1}{2}f(x - ct),$$

f is arbitrary (discontinuous, in Euler's sense) and so cannot be expressed as a sum of sine functions. In fact, he says, f could be a combination of arcs spread out over the infinite x domain and be odd and periodic; yet because it is discontinuous (in Euler's sense), it cannot be expressed as a sum of sine curves. His own solution, he affirms, is not limited in any respect. The initial curve, in fact, need not be expressible by an equation (single analytic expression).

Euler also pointed, in this instance rightly, to the Maclaurin series and said this could not represent any arbitrary function; hence neither could an infinite sine series. All he would grant was that Bernoulli's trigonometric series represented special solutions; and indeed he (Euler) had obtained such solutions in his 1749 paper (see [17] and [18]).

D'Alembert, in his article "Fondamental" in Volume 7 (1757) of the *Encyclopédie*, also attacked Bernoulli. He did not believe that all odd and periodic functions could be represented by a series such as (19), because the series is twice differentiable but all odd and periodic functions need not be so. However, even when the initial curve is sufficiently differentiable—and d'Alembert did require that it be twice differentiable in his 1746 paper— it need not be representable in Bernoulli's form. On the same ground d'Alembert objected to Euler's discontinuous curves. Actually d'Alembert's requirement that the initial curve $y = f(x)$ must be twice differentiable was correct, because a solution derived from an $f(x)$ that does not have a second derivative at some value or values of x must satisfy special conditions at such singular points.

Bernoulli did not retreat from his position. In a letter of 1758[10] he repeats that he had in the a_n an infinite number of coefficients at his disposal, and so by choosing them properly could make the series in (19) agree with any function $f(x)$ at an infinite number of points. In any case he insisted that (20) was the most general solution. The argument between d'Alembert, Euler, and Bernoulli continued for a decade with no agreement reached. The essence of the problem was the extent of the class of functions that could be represented by the sine series, or, more generally, by Fourier series.

In 1759, Lagrange, then young and unknown, entered the controversy. In his paper, which dealt with the nature and propagation of sound,[11] he gave some results on that subject and then applied his method to the vibrating string. He proceeded as though he were tackling a new problem but repeated much that Euler and Daniel Bernoulli had done before. Lagrange, too, started with a string loaded with a finite number of equal and equally spaced masses and then passed to the limit of an infinite number of masses. Though he criticized Euler's method as restricting the results to continuous (analytic) curves, Lagrange said he would prove that Euler's

10. *Jour. des Sçavans*, March 1758, 157–66.
11. *Misc. Taur.*, 1_3, 1759, i-x, 1–112 = *Œuvres*, 1, 39–148.

conclusion, that any initial curve can serve, is correct. We shall pass at once to Lagrange's conclusion for the continuous string. He had obtained

$$(23) \quad y(x, t) = \frac{2}{l} \sum_{r=1}^{\infty} \sin \frac{r\pi x}{l} \sum_{q=1}^{\infty} \sin \frac{r\pi x}{l} \, dx \left[Y_q \cos \frac{c\pi rt}{l} + \frac{l}{r\pi c} V_q \sin \frac{c\pi rt}{l} \right].$$

Here Y_q and V_q are the initial displacement and initial velocity of the qth mass. He then replaced Y_q and V_q by $Y(x)$ and $V(x)$ respectively. Lagrange regarded the quantities

$$(24) \qquad \sum_{q=1}^{\infty} \sin \frac{r\pi x}{l} Y(x) \, dx \quad \text{and} \quad \sum_{q=1}^{\infty} \sin \frac{r\pi x}{l} V(x) \, dx$$

as integrals and he took the integration operation outside the summation $\sum_{r=1}^{\infty}$. From these moves there resulted

$$(25) \qquad y(x, t) = \left(\frac{2}{l} \int_0^l Y(x) \sum_{r=1}^{\infty} \sin \frac{r\pi x}{l} \, dx \right) \sin \frac{r\pi x}{l} \cos \frac{r\pi ct}{l}$$

$$+ \left(\frac{2}{\pi c} \int_0^l V(x) \sum_{r=1}^{\infty} \frac{1}{r} \sin \frac{r\pi x}{l} \, dx \right) \sin \frac{r\pi x}{l} \sin \frac{r\pi ct}{l}.$$

The interchange of summation and integration not only introduced divergent series but spoiled whatever chance Lagrange might have had to recognize

$$(26) \qquad \qquad \int_0^l Y(x) \sin \frac{r\pi x}{l} \, dx$$

as a Fourier coefficient. After other long, difficult, and dubious steps, Lagrange obtained Euler's and d'Alembert's result

$$(27) \qquad \qquad y = \phi(ct + x) + \psi(ct - x).$$

He concluded that the above derivation put the theory of this great geometer [Euler]

> beyond all doubt and established on direct and clear principles which rest in no way on the law of continuity [analyticity] which Mr. d'Alembert requires; this, moreover, is how it can happen that the same formula that has served to support and prove the theory of Mr. Bernoulli on the mixture of isochronous vibrations when the number of bodies is ... finite shows us its insufficiency ... when the number of these bodies becomes infinite. In fact, the change that this formula undergoes in passing from one case to the other is such that the simple motions which made up the absolute motions of the whole system annul each

other for the most part, and those which remain are so disfigured and altered as to become absolutely unrecognizable. It is truly annoying that so ingenious a theory ... is shown false in the principal case, to which all the small reciprocal motions occurring in nature may be related.

All of this is almost total nonsense.

Lagrange's main basis for contending that his solution did not require the initial curve $Y(x)$ and initial velocity $V(x)$ to be restricted is that he did not apply differentiation to them. But if one were to rigorize what he did do, restrictions would have to be made.

Euler and d'Alembert criticized Lagrange's work but actually did not hit at the main failings; they picked on details in his "prodigious calculations," as Euler put it. Lagrange tried to answer these criticisms. The replies and rebuttals on both sides are too extensive to relate here, though many are revealing of the thinking of the times. For example, Lagrange replaced $\sin \pi/m$ for $m = \infty$ by π/m and $\sin \nu\pi/2m$ by $\nu\pi/2m$ for $m = \infty$. D'Alembert allowed the first but not the second, because the values of ν involved were comparable with m. The objection that a series of the form

$$\cos x + \cos 2x + \cos 3x + \cdots$$

might be divergent was also raised by d'Alembert and answered by Lagrange with the argument, common at that time, that the value of the series is the value of the function from which the series comes.

Though Euler did criticize mathematical details, his overall response to Lagrange's paper, communicated in a letter of October 23, 1759,[12] was to commend Lagrange's mathematical skill and to state that it put the whole discussion beyond all quibbling, and that everyone must now recognize the use of irregular and discontinuous (in Euler's sense) functions in this class of problems.

On October 2, 1759, Euler wrote Lagrange: "I am delighted to learn that you approve my solution ... which d'Alembert has tried to undermine by various cavils, and that for the sole reason that he did not get it himself. He has threatened to publish a weighty refutation; whether he really will I do not know. He thinks he can deceive the semi-learned by his eloquence. I doubt whether he is serious, unless perhaps he is thoroughly blinded by self-love."[13]

In 1760/61 Lagrange, seeking to answer criticisms of d'Alembert and Bernoulli that had been communicated by letter, gave a different solution of the vibrating-string problem.[14] This time he starts directly with the wave

12. Lagrange, Œuvres, 14, 164–70.
13. Œuvres, 14, 162–64.
14. Misc. Taur., 2_2, 1760/61, 11–172, pub. 1762 = Œuvres, 1, 151–316.

equation (with $c = 1$), and by multiplying by an unknown function and further steps he reduces the partial differential equation to the solution of two ordinary differential equations. Then by still further steps, not all correct, Lagrange obtains the solution

$$y(t, x) = \frac{1}{2}f(x + t) + \frac{1}{2}f(x - t) - \frac{1}{2}\int_0^{x+t} g \, dx + \frac{1}{2}\int_0^{x-t} g \, dx,$$

wherein $f(x) = y(0, x)$ and $g(x) = \partial y/\partial t$ at $t = 0$ are the given initial data. As Lagrange shows, this agrees with d'Alembert's result. But then, without reference to his own work, he tries to convince his readers that he did not use any law of continuity (analyticity) for the initial curve. It is true that he did not use any direct operation of differentiation on the initial function. But, in this paper also, to justify rigorously his limit procedures, one cannot avoid assumptions about the continuity and differentiability of the initial functions.

The debate raged throughout the 1760s and 1770s. Even Laplace entered the fray in 1779,[15] and sided with d'Alembert. D'Alembert continued it in a series of booklets, entitled *Opuscules*, which began to appear in 1768. He argued against Euler, on the ground that Euler admitted too general initial curves, and against Daniel Bernoulli, on the ground that his (d'Alembert's) solutions could not be represented as a sum of sine curves, so that Bernoulli's solutions were not general enough. The idea that the infinite series of trigonometric functions $\sum_{n=1}^{\infty} a_n \sin nx$ might be made to fit any initial curve because there is an infinity of a_n's to be determined (Daniel Bernoulli had so contended) was rejected by Euler as impossible to execute. He also raised the question of how a trigonometric series could represent the initial curve when only a part of the string is disturbed initially. Euler, d'Alembert, and Lagrange continued throughout to deny that a trigonometric series could represent any analytic function, to say nothing of more arbitrary functions.

Many of the arguments each presented were grossly incorrect; and the results, in the eighteenth century, were inconclusive. One major issue, the representability of an arbitrary function by trigonometric series, was not settled until Fourier took it up. Euler, d'Alembert, and Lagrange were on the threshold of discovering the significance of Fourier series but did not appreciate what lay before them. Judging by the knowledge of the times, all three men and Bernoulli were correct in their main contentions. D'Alembert, following a tradition established since Leibniz's time, insisted that functions must be analytical, so that any problem not solvable in such terms was unsolvable. He was also correct in the argument he gave that $y(t, x)$

15. *Mém. de l'Acad. des Sci., Paris*, 1779, 207–309, pub. 1782 = *Œuvres*, 10, 1–89.

must be periodic in x. However, he failed to realize that, given any arbitrary function in, say, $0 \leq x \leq l$, this function could be repeated in every interval $[nl, (n + 1)l]$ for integral n and so be periodic. Of course, such a periodic function might not be representable by one (closed) formula. Euler and Lagrange were, at least in their time, justified in believing that not all "discontinuous" functions could be represented by Fourier series, yet equally right in believing (though they did not have proof) that the initial curve can be very general. It need not be analytic, nor need it be periodic. Bernoulli did adopt the correct position on physical grounds but could not back it up with the mathematics.

One of the very curious features of the debate on the trigonometric series representation of functions is that all the men involved knew that non-periodic functions can be represented (in an interval) by trigonometric series. Reference to Chapter 20 (sec. 5) will show that Clairaut, Euler, Daniel Bernoulli, and others had actually produced such representations; many of their papers also had the formulas for the coefficients of the trigonometric series. Practically all of this work was in print by 1759, the year in which Lagrange presented his basic paper on the vibrating string. He could then have inferred that any function has a trigonometric expansion and could have read off the formulas for the coefficients, but failed to do so. Only in 1773, when the heat of the controversy was past, did Daniel Bernoulli notice that the sum of a trigonometric series may represent different algebraic expressions in different intervals. Why did all these results have no influence on the controversy concerning the vibrating string? It may be explained in several ways. Many of the results on the representation of quite general functions by trigonometric series were in papers on astronomy, and Daniel Bernoulli may not have read these and so could not point to them in defense of his position. Euler and d'Alembert, who must have known Clairaut's work of 1757 (Chap. 20, sec. 5), were probably not inclined to study it, since it refuted their own arguments. Also, this astronomical work by Clairaut was soon superseded and forgotten. On the other hand, whereas Euler used trigonometric series, as in his work on interpolation theory, to represent polynomial expressions, he did not accept the general fact that quite arbitrary functions could be so represented; the existence of such a series representation, where he used it, was assured by other means.

Another issue, how a partial differential equation with analytic coefficients (e.g. constants) could have a non-analytic solution, was not really clarified. In the case of ordinary differential equations, if the coefficients are analytical, the solutions must be. However, this is not true for partial differential equations. Though Euler was correct in saying that solutions with corners are admissible (and he did insist on it), determination of the singularities that are admissible in the solution of partial differential equations was still far in the future.

3. *Extensions of the Wave Equation*

While the controversy over the vibrating string was being carried on, the interest in musical instruments prompted further work, not only on vibrations of physical structures but also on hydrodynamical questions which concern the propagation of sound in air. Mathematically, these involve extensions of the wave equation.

In 1762 Euler took up the problem of the vibrating string with variable thickness. He had been stimulated by one of the principal questions of musical aesthetics. Jean-Philippe Rameau (1683–1764) had explained (1726) that the consonance of a musical sound is due to the fact that the component tones of any one sound are harmonics of the fundamental tone; that is, their frequencies are integral multiples of the fundamental frequency. But Euler, in his *Tentamen Novae Theoriae Musicae* (1739)[16] maintained that only in proper musical instruments were the overtones harmonics of the fundamental tone. He therefore undertook to show that the string of variable thickness or nonuniform density $\sigma(x)$ and tension T gives off inharmonic overtones.

The partial differential equation becomes

$$(28) \qquad \frac{1}{c^2}\frac{\partial^2 y}{\partial t^2} = \frac{\partial^2 y}{\partial x^2},$$

with c now a function of x. The first substantial results were obtained by Euler in a paper, "On the Vibratory Motion of Non-Uniformly Thick Strings."[17] The general solution he declares to be beyond the power of analysis. He obtains a solution in the special case where the mass distribution σ is given by

$$\sigma = \frac{\sigma_0}{\left(1 + \dfrac{x}{\alpha}\right)^4},$$

wherein σ_0 and α are constants. Then

$$y = \left(1 + \frac{x}{\alpha}\right)\left[\phi\left(\frac{x}{1 + \dfrac{x}{\alpha}} + c_0 t\right) + \psi\left(\frac{x}{1 + \dfrac{x}{\alpha}} - c_0 t\right)\right],$$

where $c_0 = \sqrt{T/\sigma_0}$. The frequencies of the modes or harmonics are given by

$$\nu_k = \frac{k}{2l}\left(1 + \frac{l}{\alpha}\right)\sqrt{T/\sigma_0}, \qquad k = 1, 2, 3, \cdots.$$

16. *Opera*, (3), 1, 197–427.
17. *Novi Comm. Acad. Sci. Petrop.*, 9, 1762/63, 246–304, pub. 1764 = *Opera*, (2), 10, 293–343.

Thus the ratio of two successive frequencies is the same as for a string of uniform thickness, but the fundamental frequency is no longer inversely proportional to the length.

In this paper of 1762/63 Euler also considered the vibrations of a string composed of two lengths, a and b, of different thicknesses m and n. He derived the equation for the frequencies ω of the modes. These turn out to be solutions of

$$(29) \qquad m \tan \frac{\omega a}{m} + n \tan \frac{\omega b}{m} = 0,$$

and he solves for ω in special cases. The solutions of (29) are called the characteristic values or eigenvalues of the problem. These values are, as we shall see, of prime importance in the theory of partial differential equations. It is almost evident from (29) that the characteristic frequencies are not integral multiples of the fundamental one.

However, Euler took up this question again in another paper on the vibrating string of variable thickness,[18] and starting with (28) he shows that there are functions $c(x)$ for which the frequencies of the higher modes are not integral multiples of the fundamental.

D'Alembert, too, took up the string of variable thickness.[19] Here he used a significant method of solution that he had introduced earlier for the string of constant density. In this earlier attempt at the vibrating-string problem d'Alembert had introduced the idea of separation of variables, which is now a basic method of solution in partial differential equations.[20] To solve

$$\frac{\partial^2 y(t, x)}{\partial t^2} = a^2 \frac{\partial^2 y(t, x)}{\partial x^2}$$

d'Alembert sets

$$y = h(t)\, g(x),$$

substitutes this in the differential equation, and obtains

$$(30) \qquad \frac{1}{a^2} \frac{h''(t)}{h(t)} = \frac{g''(x)}{g(x)}.$$

He then argues, as we do now, that since g''/g does not vary when t does, it must be a constant, and by the like argument applied to h''/h, this expression too must be a constant. The two constants are equal and are denoted by A. Thus he gets the two separate ordinary differential equations

$$(31) \qquad \begin{aligned} h''(t) - a^2 A h(t) &= 0 \\ g''(x) - A g(x) &= 0. \end{aligned}$$

18. *Misc. Taur.*, 3, 1762/65, 25–59, pub. 1766 = *Opera*, (2), 10, 397–425.
19. *Hist. de l'Acad. de Berlin*, 19, 1763, 242 ff., pub. 1770.
20. *Hist. de l'Acad. de Berlin*, 6, 1750, 335–60, pub. 1752.

Since a and A are constants, each of these equations is readily solvable, and d'Alembert gets

$$y(t, x) = h(t)\, g(x) = [Me^{a\sqrt{A}\cdot t} + Ne^{-a\sqrt{A}\cdot t}][Pe^{\sqrt{A}\cdot x} + Qe^{-\sqrt{A}\cdot x}].$$

The end-conditions, $y(t, 0) = 0$ and $y(t, l) = 0$, led d'Alembert to assert that $g(x)$ must be of the form $k \sin Rx$ and that $h(t)$ must be of the same form because $y(t, x)$ must be periodic in t. He left the matter there. Daniel Bernoulli had used the idea of separation of variables in 1732 in his treatment of the vibrations of a chain suspended from one end, but d'Alembert was more explicit, despite the fact that he did not complete the solution.

In his 1763 paper d'Alembert wrote the wave equation as

$$\frac{\partial^2 y}{\partial t^2} = X(x)\, \frac{\partial^2 y}{\partial x^2}$$

and sought solutions of the form

$$u = \zeta(x)\, \cos \lambda \pi t.$$

He obtained for ζ the equation

(32)
$$\frac{d^2\zeta}{dx^2} = \frac{-\lambda^2\pi^2\zeta}{X(x)}.$$

D'Alembert now had to determine ζ so that it was 0 at both ends of the string. By a detailed analysis he showed that there are values of λ for which ζ meets this condition. He did not appreciate in this work that there are infinitely many values of λ. The significance of the investigation is that it is another step in the direction of boundary value or eigenvalue problems for ordinary differential equations.

The transverse oscillation of a *heavy* continuous horizontal cord was taken up by Euler. In the paper, "On the Modifying Effect of Their Own Weight on the Motion of Strings,"[21] Euler obtains the differential equation

$$\frac{1}{c^2} y_{tt} = \frac{g}{c^2} + y_{xx}.$$

For constant c and with fixed ends at $x = 0$ and $x = l$, Euler finds that

$$y = -\frac{(1/2)gx(x - l)}{c^2} + \phi(ct + x) + \psi(ct - x).$$

Thus the results are the same as for the "weightless" string (where the gravitational force is ignored), except that the oscillation takes place about the parabolic figure of equilibrium

$$y = -\frac{(1/2)gx(x - l)}{c^2}.$$

21. *Acta Acad. Sci Petrop.*, 1, 1781, 178–90, pub. 1784 = *Opera*, (2), 11, 324–34, but dating from 1774.

As we shall see in a moment, Euler had introduced all the Bessel functions of the first kind in a paper on the vibrating drum (see also Chap. 21, secs. 4 and 6) and in this 1781 paper he remarks that it is possible to express any motion by a series of Bessel functions (despite the fact that he had argued against Daniel Bernoulli's claim, in the vibrating-string problem, that any function can be represented as a series of trigonometric functions).

Papers on the vibrating string and the hanging chain, of which the above are just samples, were published by many other men up to the end of the century. The authors continued to disagree, correct each other, and make all sorts of errors in doing so, including contradicting what they themselves had previously said and even proven. They made assertions, contentions, and rebuttals on the basis of loose arguments and often just personal predilections and convictions. Their references to papers to prove their contentions did not prove what they claimed. They also resorted to sarcasm, irony, invective, and self-praise. Mingled with these attacks were seeming agreements expressed in order to curry favor, particularly with d'Alembert, who had considerable influence with Frederick II of Prussia and as director of the Berlin Academy of Sciences.

The second order partial differential equation problems described thus far involved only one space variable and time. The eighteenth century did not go much beyond this. In a paper of 1759[22] Euler took up the vibration of a rectangular drum, thus considering a two-dimensional body. He obtained for the vertical displacement z of the surface of the drum

$$(33) \qquad \frac{1}{c^2}\frac{\partial^2 z}{\partial t^2} = \frac{\partial^2 z}{\partial x^2} + \frac{\partial^2 z}{\partial y^2},$$

wherein x and y represent the coordinates of any point on the drum and c is determined by the mass and tension. Euler tried

$$z = v(x, y) \sin(\omega t + \alpha)$$

and found that

$$0 = \frac{\omega^2 v}{c^2} + \frac{\partial^2 v}{\partial x^2} + \frac{\partial^2 v}{\partial y^2}.$$

This equation has sinusoidal solutions of the form

$$v = \sin\left(\frac{\beta x}{a} + B\right) \sin\left(\frac{\gamma y}{b} + C\right)$$

where

$$\frac{\omega^2}{c^2} = \frac{\beta^2}{a^2} + \frac{\gamma^2}{b^2}.$$

22. *Novi Comm. Acad. Sci. Petrop.*, 10, 1764, 243–60, pub. 1766 = *Opera*, (2), 10, 344–59.

The dimensions of the drum are a and b, so that $0 \leq x \leq a$ and $0 \leq y \leq b$. When the initial velocity is 0, B and C may be taken to be 0. If the boundaries are fixed, then $\beta = m\pi$ and $\gamma = n\pi$ where m and n are integers. Then, since $\omega = 2\pi\nu$ where ν is the frequency per second, he obtains readily that the frequencies are

$$\nu = \frac{1}{2} c \sqrt{\frac{m^2}{a^2} + \frac{n^2}{b^2}}.$$

He then considers a circular drum and transforms (33) to polar coordinates (a highly original step), obtaining

$$(34) \qquad \frac{1}{c^2} \frac{\partial^2 z}{\partial t^2} = \frac{\partial^2 z}{\partial r^2} + \frac{1}{r} \frac{\partial z}{\partial r} + \frac{1}{r^2} \frac{\partial^2 z}{\partial \phi^2}.$$

He now tries solutions of the form

$$(35) \qquad z = u(r) \sin (\omega t + A) \sin (\beta \phi + B)$$

so that $u(r)$ satisfies

$$(36) \qquad u'' + \frac{1}{r} u' + \left(\frac{\omega^2}{c^2} - \frac{\beta^2}{r^2} \right) u = 0.$$

Here Bessel's equation appears in the current form (Cf. Chap. 21, sec. 6). Euler then calculates a power series solution

$$u\left(\frac{\omega}{c} r \right) = r^\beta \left\{ 1 - \frac{1}{1(\beta + 1)} \left(\frac{\omega}{c} \frac{r}{2} \right)^2 + \frac{1}{1 \cdot 2(\beta + 1)(\beta + 2)} \left(\frac{\omega}{c} \frac{r}{2} \right)^4 + \cdots \right\},$$

which we would write now as

$$u\left(\frac{\omega}{c} r \right) = \left(\frac{c}{\omega} \right)^\beta 2^\beta \Gamma(\beta + 1) J_\beta \left(\frac{\omega}{c} r \right).$$

Since the edge $r = a$ must remain fixed,

$$(37) \qquad J_\beta \left(\frac{\omega}{c} a \right) = 0.$$

It also follows from (35), since z must be of period 2π in ϕ, that β is an integer. Euler asserts that for a fixed β there are infinitely many roots ω so that infinitely many simple sounds result. However, he did not calculate these roots. He did attempt to find a second solution of (36) but failed to do so. The theory of the vibrating membrane was derived independently by Poisson[23] and is often credited solely to him.

Euler, Lagrange, and others worked on the propagation of sound in air. Euler wrote on the subject of sound frequently from the time he was twenty

23. *Mém. de l'Acad. des Sci., Paris*, (2), 8, 1829, 357–570.

years old (1727) and established this field as a branch of mathematical physics. His best work on the subject followed his major papers of the 1750s on hydrodynamics. Air is a compressible fluid, and the theory of the propagation of sound is part of fluid mechanics (and of elasticity, because air is also an elastic medium). However, to treat the propagation of sound he made reasonable simplifications of the general hydrodynamical equations.

Three fine and definitive papers were read to the Berlin Academy in 1759. In the first, "On the Propagation of Sound,"[24] Euler considers the propagation of sound in one space dimension. After some approximations, which amount to considering waves of small amplitude, he is led to the one-dimensional wave equation

$$\frac{\partial^2 y}{\partial t^2} = 2gh \frac{\partial^2 y}{\partial x^2},$$

where y is the amplitude of the wave at the point x and at time t, g is the acceleration of gravity and h is a constant relating the pressure and density. This equation, as Euler of course recognized, is the same as that for the vibrating string and he did nothing mathematically new in solving it.

In his second paper[25] Euler gives the two-dimensional equation of propagation in the form

(38) $\qquad \dfrac{\partial^2 x}{\partial t^2} = c^2 \dfrac{\partial^2 x}{\partial X^2} + c^2 \dfrac{\partial^2 y}{\partial X \partial Y}, \qquad \dfrac{\partial^2 y}{\partial t^2} = c^2 \dfrac{\partial^2 y}{\partial Y^2} + c^2 \dfrac{\partial^2 x}{\partial X \partial Y},$

where x and y are the wave amplitudes in the X-direction and Y-direction respectively, or the components of the displacement, and $c = \sqrt{2gh}$. He gives the plane wave solution

$$x = \alpha\phi(\alpha X + \beta Y + c\sqrt{\alpha^2 + \beta^2}\, t), y = \beta\phi(\alpha X + \beta Y + c\sqrt{\alpha^2 + \beta^2}\, t),$$

where ϕ is an arbitrary function and α and β are arbitrary constants. Then letting

(39) $\qquad\qquad\qquad\qquad v = \dfrac{\partial x}{\partial X} + \dfrac{\partial y}{\partial Y}$

(v is called the divergence of the displacement), he gets the two-dimensional wave equation

(40) $\qquad\qquad\qquad\qquad \dfrac{1}{c^2} \dfrac{\partial^2 v}{\partial t^2} = \dfrac{\partial^2 v}{\partial X^2} + \dfrac{\partial^2 v}{\partial Y^2}.$

He also states the need for the superposition of solutions to obtain the most general solution of the problem in order to meet some initial condition, that is, the value of v or of x and y at $t = 0$.

24. *Mém. de l'Acad. de Berlin*, 15, 1759, 185–209, pub. 1766, = *Opera*, (3), 1, 428–51.
25. *Mém. de l'Acad. de Berlin*, 15, 1759, 210–40, pub. 1766 = *Opera*, (3), 1, 452–83.

Euler then shows how he can get the differential equation whose solutions are called cylindrical waves because the wave spreads out like an expanding cylinder. He lets $Z = \sqrt{X^2 + Y^2}$ and introduces $v = f(Z, t)$ where f is arbitrary. By letting $x = vX$ and $y = vY$, he obtains from (40)

$$\frac{1}{c^2}\frac{\partial^2 v}{\partial t^2} = \frac{3}{Z}\frac{\partial v}{\partial Z} + \frac{\partial^2 v}{\partial Z^2}.$$

He also obtains in this paper in a similar manner the three-dimensional wave equation

(41) $$\frac{1}{c^2}\frac{\partial^2 v}{\partial t^2} = \frac{\partial^2 v}{\partial X^2} + \frac{\partial^2 v}{\partial Y^2} + \frac{\partial^2 v}{\partial Z^2},$$

where v is again the divergence of the displacement (x, y, z). Euler gives plane wave and spherical wave solutions using the kind of substitution just indicated for cylindrical waves. The basic equation for spherical waves is

$$\frac{1}{c^2}\frac{\partial^2 s}{\partial t^2} = \frac{4}{V}\frac{\partial s}{\partial V} + \frac{\partial^2 s}{\partial V^2}, \qquad V = \sqrt{X^2 + Y^2 + Z^2}.$$

Much of the above work on spherical and cylindrical waves was also done independently by Lagrange at the end of the year 1759. Each communicated his results to the other. Though there are many details in which Lagrange's work differs from Euler's, there are no major mathematical points worthy of being related here.

From the propagation of sound waves in air it was but a step to the study of the sounds given off by musical instruments that employ air motion. This study was initiated by Daniel Bernoulli in 1739. Bernoulli, Euler, and Lagrange wrote numerous papers on the tones given off by an almost incredible variety of such instruments. In a publication of 1762 Daniel Bernoulli showed that at the open end of a cylindrical tube (organ pipe) no condensation of air can take place.[26] At a closed end the air particles must be at rest. He concluded from this that a tube closed at both ends or open at both ends has the same fundamental mode as a tube of half the length but open at one end and closed at the other. He also discovered the theorem that for closed organ pipes the frequencies of the overtones are odd multiples of the frequency of the fundamental. In the same paper Bernoulli took up pipes of other than cylindrical form, in particular the conical pipe, for which he obtained expressions for the individual tones (modes) but recognized that these hold only for infinite cones and not for the truncated one. For the (infinite) conical pipe the overtones proved to be harmonic to the fundamental. Bernoulli confirmed many of his theoretical results by experiments.

26. *Mém. de l'Acad. des Sci., Paris*, 1762, 431–85, pub. 1764.

Euler, too, studied cylindrical pipes and non-cylindrical figures of revolution[27] and considered reflection at open and closed ends. The efforts of these men were directed toward understanding flutes; organ pipes; all sorts of horns of hyperboloidal, conical, and cylindrical shape; trumpets; bugles; and other wind instruments.

On the whole these efforts to solve partial differential equations in three and four variables were limited, mainly because the solutions were expressed in series involving several variables as opposed to simpler trigonometric series in x and t separately (cf. [20]). But the mathematicians knew too little about the functions that appeared in these more complicated series and about methods of determining the coefficients. Such methods were soon developed.

It is worth mentioning that Euler, in considering the sound of a bell and reconsidering some of the problems of the vibrations of rods, was led to fourth order partial differential equations. However, he was unable to do much with them and in fact, for the rest of the century no progress was made with them.

4. *Potential Theory*

The development of the subject of partial differential equations was furthered by another class of physical investigations. One of the major problems of the eighteenth century was the determination of the amount of gravitational attraction one mass exerts on another, the prime cases being the attraction of the sun on a planet, of the earth on a particle exterior or interior to it, and of the earth on another extended mass. When the two masses are very far apart compared to their sizes, it is possible to treat them as point masses; but in other cases, notably the earth attracting a particle, the extent of the earth must be taken into account. Clearly the shape of the earth must be known if one is to calculate the gravitational attraction its distributed mass exerts on a particle or another distributed mass. Although the precise shape remained a subject for investigation (Chap. 21, sec. 1), it was already clear by 1700 that it must be some form of ellipsoid, perhaps an oblate spheroid (an ellipsoid generated by revolving an ellipse around the minor axis). For the solid oblate spheroid the force of attraction both on an external and on an internal particle cannot be calculated as though the mass were concentrated at the center.

In a prize paper of 1740 on the tides, and in his *Treatise of Fluxions* (1742), Maclaurin proved that, for a fluid of uniform density under constant angular rotation, the oblate spheroid is an equilibrium shape. Then Maclaurin proved synthetically that, given two confocal homogeneous ellipsoids

27. *Novi Comm. Acad. Sci. Petrop.*, 16, 1771, 281–425, pub. 1772 = *Opera*, (2), 13, 262–369.

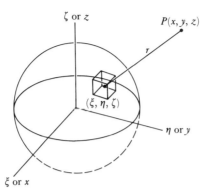

ζ or z

$P(x, y, z)$

r

(ξ, η, ζ)

η or y

Figure 22.2 ξ or x

of revolution, the attractions of the two bodies on the same particle external to both, provided the particle be on the prolongation of the axis of revolution or in the plane of the equator, will be proportional to the volumes. Some other limited results were also established geometrically in the nineteenth century by James Ivory (1765–1842) and Michel Chasles.

The geometrical approach to the problem of gravitational attraction used by Newton, Maclaurin, and others is good only for special bodies and special locations of the attracted masses. This approach soon gave way to analytical methods, which one finds first in papers by Clairaut before 1743 and especially in his famous book *Théorie de la figure de la terre* (1743), in which he considers both the shape of the earth and gravitational attraction.

Let us first note some facts about the analytical formulation. The gravitational force exerted by an extended body on a unit mass P regarded as a particle is the sum of the forces exerted by all the small masses that make up the body. If $d\xi \, d\eta \, d\zeta$ is a small volume of the body (Fig. 22.2), so small that it may be regarded as a particle centered at the point (ξ, η, ζ), and if P has the coordinates (x, y, z), the attraction exerted by the small mass of density ρ on the unit particle is a vector directed from P to the small mass and, in view of the Newtonian law of gravitation, the components of this vector are (Chap. 21, sec. 7)

$$-k\rho \, \frac{x-\xi}{r^3} \, d\xi \, d\eta \, d\zeta, \qquad -k\rho \, \frac{y-\eta}{r^3} \, d\xi \, d\eta \, d\zeta, \qquad -k\rho \, \frac{z-\zeta}{r^3} \, d\xi \, d\eta \, d\zeta,$$

where k is the constant in the Newtonian law and

$$r = \sqrt{(x-\xi)^2 + (y-\eta)^2 + (z-\zeta)^2}.$$

Of course ρ may be a function of ξ, η, and ζ or, in the case of a homogeneous body, a constant.

The force exerted by the entire body on the unit mass at P has the components

$$f_x = -k \iiint \rho \frac{x - \xi}{r^3} \, d\xi \, d\eta \, d\zeta$$

(42)
$$f_y = -k \iiint \rho \frac{y - \eta}{r^3} \, d\xi \, d\eta \, d\zeta$$

$$f_z = -k \iiint \rho \frac{z - \zeta}{r^3} \, d\xi \, d\eta \, d\zeta,$$

wherein the integral is extended over the entire attracting body. These integrals are finite and correct also when P is inside the attracting body.

Instead of treating each component of the force separately, it is possible to introduce one function $V(x, y, z)$ whose partial derivatives with respect to $x, y,$ and z respectively are the three components of the force. This function is

(43)
$$V(x, y, z) = \iiint \frac{\rho}{r} \, d\xi \, d\eta \, d\zeta.$$

By differentiating under the integral sign with respect to $x, y,$ and z which are involved in r, one obtains

$$\frac{\partial V}{\partial x} = \frac{1}{k} f_x, \qquad \frac{\partial V}{\partial y} = \frac{1}{k} f_y, \qquad \frac{\partial V}{\partial z} = \frac{1}{k} f_z,$$

and these equations also hold when P is inside the attracting body. The function V is called a potential function. When problems involving the three components $f_x, f_y,$ and f_z can be reduced to the problem of working with V, there is the advantage of working with one function instead of three.

If one knows the distribution of mass inside the body, which means knowing ρ as a function of $\xi, \eta,$ and ζ, and if one knows the precise shape of the body, one can sometimes calculate V by actually evaluating the integral. However, for most shapes of bodies this triple integral is not integrable in terms of simple functions. Moreover, we do not know the true distribution of mass inside the earth and other bodies. Hence V must be determined in other ways. The principal fact about V is that for points (x, y, z) *outside* the attracting body, it satisfies the partial differential equation

(44)
$$\frac{\partial^2 V}{\partial x^2} + \frac{\partial^2 V}{\partial y^2} + \frac{\partial^2 V}{\partial z^2} = 0,$$

in which we note that ρ does not appear. This differential equation is known as the potential equation and as Laplace's equation.

The idea that a force can be derived from a potential function, and even the term "potential function," were used by Daniel Bernoulli in

Hydrodynamica (1738). The potential equation itself appears for the first time in one of Euler's major papers composed in 1752, "Principles of the Motion of Fluids."[28] In dealing with the components u, v, and w of the velocity of any point in a fluid, Euler had shown that $u\,dx + v\,dy + w\,dz$ must be an exact differential. He introduces the function S such that $dS = u\,dx + v\,dy + w\,dz$. Then

$$u = \frac{\partial S}{\partial x}, \qquad v = \frac{\partial S}{\partial y}, \qquad w = \frac{\partial S}{\partial z}.$$

But the motion of incompressible fluids is subject to what is called the law of continuity, namely,

$$(45) \qquad \frac{\partial u}{\partial x} + \frac{\partial v}{\partial y} + \frac{\partial w}{\partial z} = 0,$$

which expresses mathematically the fact that no matter is destroyed or created during the motion. Then it follows that

$$\frac{\partial^2 S}{\partial x^2} + \frac{\partial^2 S}{\partial y^2} + \frac{\partial^2 S}{\partial z^2} = 0.$$

How to solve this equation generally, Euler says, is not known; so he considers just special cases where S is a polynomial in x, y, and z. The function S was later (1868) called by Helmholtz the velocity potential. In a paper published in 1762[29] Lagrange reproduced all of these quantities, which he took over from Euler without acknowledgment, though he did improve the order of the ideas and the expressions.

Before we can investigate the work done to solve the potential equation in behalf of gravitational attraction, we must review some efforts to evaluate this attraction directly by means of the integrals (42) or the equivalents in other coordinate systems.

In a paper written in 1782 but published in 1785, entitled "Recherches sur l'attraction des sphéroïdes,"[30] Legendre, interested in the attraction exerted by solids of revolution, proved the theorem: If the attraction of a solid of revolution is known for every external point on the prolongation of its axis, then it is known for every external point. He first expressed the component of the force of attraction in the direction of the radius vector r by means of

$$(46) \quad P(r, \theta, 0) = \iiint \frac{(r - r')\cos\gamma}{(r^2 - 2rr'\cos\gamma + r'^2)^{3/2}}\, r'^2 \sin\theta'\, d\theta'\, d\phi'\, dr',$$

28. *Novi Comm. Acad. Sci. Petrop.*, 6, 1756/57, 271–311, pub. 1761 = *Opera*, (2), 12, 133–68.
29. *Misc. Taur.*, 2_2 1760/61, 196–298, pub. 1762 = *Œuvres*, 1, 365–468.
30. *Mém. des sav. étrangers*, 10, 1785, 411–34.

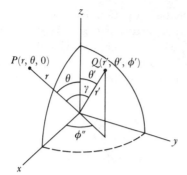

Figure 22.3

where (Fig. 22.3) r is the radius vector to the attracted point, r' is the radius vector to any point of the attracting body, and γ is the angle formed at the center of the body by the two radii vectors. The ϕ coordinate of the external point can be taken to be 0 because the solid is a figure of revolution around the z-axis. Then he expanded the integrand in powers of r'/r. This is done by writing the denominator as

$$r^3\left[1 - \left(2\frac{r'}{r}\cos\gamma - \frac{r'^2}{r^2}\right)\right]^{3/2}.$$

The quantity in the brackets can be put into the numerator and then expanded by the binomial theorem with the quantity in parentheses as the second term of the binomial. Legendre obtained for the integrand, apart from the volume element, the series

$$\frac{1}{r^2}\left\{1 + 3P_2(\cos\gamma)\frac{r'^2}{r^2} + 5P_4(\cos\gamma)\frac{r'^4}{r^4} + 7P_6(\cos\gamma)\frac{r'^6}{r^6} + \cdots\right\}.$$

The coefficients P_2, P_4, ... are rational integral functions of $\cos\gamma$. These functions are what we now call the Legendre polynomials or Laplace coefficients or zonal harmonics. Legendre gave the form of the functions so that the general P_n, namely,

(47) $$P_n(x) = \frac{(2n-1)(2n-3)\cdots 1}{n!}.$$

$$\left[x^n - \frac{n(n-1)}{2(2n-1)}x^{n-2} + \frac{n(n-1)(n-2)(n-3)}{2\cdot 4\cdot(2n-1)(2n-3)}x^{n-4} + \cdots\right],$$

could be inferred.

He could now integrate with respect to r', and he obtained

$$\frac{2}{r^2}\iint\left\{\frac{R^3}{3} + \frac{3}{5}P_2(\cos\gamma)\frac{R^5}{r^2} + \frac{5}{7}P_4(\cos\gamma)\frac{R^7}{r^4} + \cdots\right\}\sin\theta'\,d\theta'\,d\phi',$$

where $R = f(\theta')$ is the value of r' at a given θ' (it is independent of ϕ'). He then had to integrate with respect to ϕ'. For this he used[31]

$$\cos \gamma = \cos \theta \cos \theta' + \sin \theta \sin \theta' \cos \phi'.$$

Having established the subsidiary result

$$\frac{1}{\pi} \int_0^\pi P_{2n}(\cos \gamma) \, d\phi' = P_{2n}(\cos \theta) P_{2n}(\cos \theta'),$$

he obtained finally

$$P(r, \theta, 0) = \frac{3M}{r^2} \sum_{n=0}^\infty \frac{2n - 3}{2n - 1} P_{2n}(\cos \theta) \frac{\alpha_n}{r^{2n}},$$

where

$$\alpha_n = \frac{4\pi}{3M} \int_0^{\pi/2} R^{2n+3} P_{2n}(\cos \theta') \sin \theta' d\theta'.$$

The value of this integral depends upon the shape of the meridian curves $R = f(\theta')$.

From the above result, and on the basis of a communication from Laplace, Legendre then obtained the expression for the potential function for this problem, and from the potential derived the component of the force of attraction perpendicular to the radius vector.

In a second paper written in 1784,[32] Legendre derived some properties of the functions P_{2n}. Thus

(48) $$\int_0^1 f(x^2) P_{2n}(x) \, dx = 0$$

for each rational integral function of x^2 whose degree in x^2 is less than n. If n is any positive integer,

(49) $$\int_0^1 x^n P_{2m} \, dx = \frac{n(n - 2) \cdots (n - 2m + 2)}{(n + 1)(n + 3) \cdots (n + 2m + 1)}.$$

If m and n are positive integers,

(50) $$\int_0^1 P_{2n}(x) P_{2m}(x) \, dx = \begin{cases} 0 & \text{for } m \gtrless n, \\ \dfrac{1}{4m + 1} & \text{for } m = n. \end{cases}$$

---

31. This expression is derived as follows: In view of the equations of transformation from spherical to rectangular coordinates $x = r \sin \theta \cos \phi$, $y = r \sin \theta \sin \phi$, $z = r \cos \theta$, the rectangular coordinates of P are $(r \sin \theta \cos \phi, r \sin \theta \sin \phi, r \cos \theta)$ and the rectangular coordinates of Q are $(r' \sin \theta' \cos \phi', r' \sin \theta' \sin \phi', r' \cos \theta')$. Then using the distance formula we can express PQ. But by the law of cosines $PQ = r^2 + r'^2 + 2rr' \cos \gamma$. Equating the two expressions for PQ gives the above expression for $\cos \gamma$.

32. *Mém. de l'Acad. des Sci.*, Paris, 1784, 370–89, pub. 1787.

He also proved that the zeros of each of the P_{2n} are real, different from each other, symmetric with respect to 0, and in absolute value less than 1. Also for $0 < x < 1$, $P_{2n}(x) < 1$.

Then, with the help of the orthogonality condition (50), he proves (by integration of the series term by term) that a given function of x^2 can be expressed in only one way in a series of functions $P_{2n}(x)$.

Finally, using these and other properties of his polynomials, Legendre returns to the main problem of gravitational attraction and using the expression (43) for the potential and the condition for equilibrium of a rotating fluid mass, he obtains the equation for the meridian curve of such a mass in the form of a series of his polynomials. He believed that this equation included all possible equilibrium figures for a spheroid of revolution.

Now Laplace enters the picture. He had written several papers on the force of attraction exerted by volumes of revolution (1772, pub. 1776; 1773, pub. 1776; and 1775, pub. 1778), in which he worked with the components of the force but not the potential function. The article by Legendre of 1782, published in 1785, inspired a famous and remarkable fourth paper by Laplace, "Théorie des attractions des sphéroïdes et de la figure des planètes."[33] Without mentioning Legendre, Laplace took up the problem of the attraction exerted by an arbitrary spheroid as opposed to Legendre's figures of revolution. By a spheroid Laplace meant any surface given by one equation in r, θ, and ϕ.

He starts with the theorem that the potential V of the force an arbitrary body exerts on an external point, expressed in spherical coordinates r, θ, ϕ with $\mu = \cos \theta$, satisfies the potential equation

$$(51) \qquad \frac{\partial}{\partial \mu}\left((1 - \mu^2)\frac{\partial V}{\partial \mu}\right) + \frac{1}{1 - \mu^2}\frac{\partial^2 V}{\partial \phi^2} + r\frac{\partial^2(rV)}{\partial r^2} = 0.$$

Laplace does not say here how he obtained the equation. In a later paper[34] he gives the rectangular coordinate form (44). One may be fairly sure that he possessed the rectangular form first and derived the spherical coordinate form from it. In fact, both forms had already been given by Euler and Lagrange, but Laplace does not mention them. He may not have known their work, though this is doubtful.

In the 1782 paper Laplace sets

$$(52) \qquad V(r, \theta, \phi) = \frac{U_0}{r} + \frac{U_1}{r^2} + \frac{U_2}{r^3} + \cdots,$$

33. *Mém. de l'Acad. des Sci.*, Paris, 1782, 113–96, pub. 1785 = *Œuvres*, 10, 339–419.
34. *Mém. de l'Acad. des Sci.*, Paris, 1787, 249–67, pub. 1789 = *Œuvres*, 11, 275–92.

where $U_n = U_n(\theta, \phi)$, and substitutes this in (51). Then the individual U_n satisfy[35]

(53) $$\frac{\partial}{\partial \mu}\left[(1 - \mu^2)\frac{\partial U}{\partial \mu}\right] + \frac{1}{1 - \mu^2}\frac{\partial^2 U}{\partial \phi^2} + n(n + 1)U = 0.$$

With the help of Legendre's P_{2n} he is able to show that

(54) $$U_n(\theta, \phi) = \int\int\int r'^{n+2} P_{2n}(\cos \theta \cos \theta' + \sin \theta \sin \theta' \cos (\phi - \phi')) \cdot$$

$$\sin \theta' \, d\theta' \, d\phi' \, dr'.$$

Now Laplace uses this result and (52) to calculate the potential of a spheroid differing little from a sphere. He writes the equation of the surface of the spheroid as

(55) $$r = a(1 + \alpha y),$$

where α is small and y on the spheroid is a function of θ' and ϕ'. Laplace assumes that $y(\theta, \phi)$ can be expanded in a series of functions

(56) $$y = Y_0 + Y_1 + Y_2 + \cdots,$$

where the Y_n are functions of θ and ϕ and satisfy the differential equation (53). The result he obtains here is first that

(57) $$Y_n = \frac{2n + 1}{4\pi\alpha a^{n+1}} U_n.$$

These Y_n, then, may be used in (52). Also the expansion (56) may be recast as

(58) $$y(\mu, \phi) = \frac{1}{4\pi} \sum_{n=0}^{\infty} (2n + 1) \int_{-1}^{1}\int_{0}^{2\pi} Y_n(\mu', \phi')P_n(\mu, \phi, \mu', \phi') \, d\mu' \, d\phi',$$

where $\mu' = \cos \theta'$. With the value of y he now has an expression for r in (55) and with this and the U_n in (54) he obtains V in (52).

Laplace does not consider here the general problem of the development of *any* function of θ and ϕ into a series of the Y_n. In what he does do here and in later papers he presumes that such an expression is possible and is unique.

35. If we ignore the middle term (ϕ is absent) the resulting ordinary differential equation is what we now call Legendre's differential equation,

$$(1 - x^2)\frac{d^2z}{dx^2} - 2x\frac{dz}{dx} + n(n + 1)z = 0.$$

The $P_n(x)$ satisfy this equation. On the other hand the U_n (and the Y_n in [57]) regarded as functions of the two variables $\mu = \cos \theta$ and ϕ satisfy (53). The U_n and Y_n are called by the Germans spherical functions and by Lord Kelvin spherical harmonics or spherical surface harmonics.

He deals with rational integral functions of μ, $\sqrt{1 - \mu^2}\cos\phi$, and $\sqrt{1 - \mu^2}$ $\sin\phi$, and so his need for the general result is limited. He does prove the basic orthogonality property

$$(59) \qquad \int_{-1}^{1}\int_{0}^{2\pi} U_n(\mu, \phi) U_m(\mu, \phi) = 0, \qquad m \neq n.$$

However in his *Mécanique céleste*, Volume 2, he shows that an arbitrary function of θ and ϕ can be expanded in a series of the U_n (or the Y_n) and shows that (59) implies uniqueness of the expansion.

Laplace wrote several more papers on the attraction of spheroids and on the shape of the earth (e.g. 1783, pub. 1786; 1787, pub. 1789); and in these papers uses expansions in spherical functions. In the last paper, in which Laplace gave the rectangular coordinate form of the potential equation, he made one mistake of consequence. He assumed that this equation holds when the point mass being attracted by a body lies inside the body. This error was corrected by Poisson (Chap. 28, sec. 4).

During the 1780s Legendre continued his investigations. His fourth paper, written in 1790,[36] introduced the $P_n(x)$ for *odd* n. The expression for $P_n(x)$ given in (47) is the correct one for all n. Legendre proves that for any positive integral m and n

$$(60) \qquad \int_{-1}^{1} P_m(x)P_n(x)\,dx = \begin{cases} 0 \text{ for } m \neq n, \\ \dfrac{2}{2n+1} \text{ for } m = n. \end{cases}$$

Then he too introduces the spherical functions. That is, he lets Y_n be the coefficient of z^n in the expansion of $(1 - 2zt + z^2)^{-1/2}$ where $t = \cos\theta\cos\theta'$ $+ \sin\theta\sin\theta'\cos(\phi - \phi')$. Then, letting $\mu = \cos\theta$, $\mu' = \cos\theta'$ and $\psi = \phi - \phi'$, he shows that

$$Y_n(t) = P_n(\mu)P_n(\mu') + \frac{2}{n(n+1)}\frac{dP_n(\mu)}{d\mu}\frac{dP_n(\mu')}{d\mu'}\sin\theta\sin\theta'\cos\psi$$

$$+ \frac{2}{(n-1)(n)(n+1)(n+2)}\frac{d^2P_n(\mu)}{d\mu^2}\frac{d^2P_n(\mu')}{d\mu'^2}\sin^2\theta\sin^2\theta'\cos 2\psi + \cdots,$$

the higher terms containing higher derivatives of P_n. This equation is equivalent to

$$P_n(\cos\theta\cos\theta' + \sin\theta\sin\theta'\cos(\phi - \phi')) =$$

$$\sum_{m=1}^{n} P_n^m(\cos\theta)P_n^m(\cos\theta')\cos m(\phi - \phi'),$$

36. *Mém. de l'Acad. des Sci., Paris*, 1789, 372–454, pub. 1793.

wherein the m is a superscript and not an exponent. The $P_n^m(x)$ satisfy

(61) $$\frac{d}{dx}\left[(1 - x^2)\frac{dP_n^m}{dx}\right] + \left[n(n + 1) - \frac{\nu^2}{1 - x^2}\right]P_n^m = 0$$

and the $P_n^m(x)$ agree up to a constant factor with

$$(1 - x^2)^{m/2}\frac{d^m P_n(x)}{dx^m}.$$

The $P_n^m(x)$ thus introduced are now called the associated Legendre polynomials. Then Legendre proves that

(62) $$\int_{-1}^{1}\int_{0}^{2\pi} U_n(\mu', \phi')P_n(\mu, \phi, \mu', \phi')\, d\mu'\, d\phi' = \frac{4\pi}{2n + 1}\, U_n(\mu, \phi)$$

and that

(63) $$\int_{-1}^{1}\int_{0}^{2\pi} \{P_n(\mu)\}^2\, d\mu\, d\phi = \frac{4\pi}{2n + 1}.$$

The fact that the $P_n(x)$ satisfy Legendre's differential equation is used in this paper.

Many other special results involving the Legendre polynomials and the spherical harmonics were obtained by Legendre, Laplace, and others. A basic result is the formula of Olinde Rodrigues (1794–1851), given in 1816,[37]

(64) $$P_n(x) = \frac{1}{2^n n!}\frac{d^n(x^2 - 1)^n}{dx^n}.$$

Laplace's work on the solution of the potential equation for the attracting force of spheroids was the beginning of a vast amount of work on this subject. Equally important was his and Legendre's work on the Legendre polynomials $P_n(x)$, the associated Legendre polynomials $P_n^m(x)$, and the spherical (surface) harmonics $Y_n(\mu, \phi)$, because rather arbitrary functions can be expressed in terms of infinite series of the P_n, P_n^m and the Y_n. These series of functions are analogous to the trigonometric functions, which Daniel Bernoulli claimed could also be used to represent arbitrary functions. The choice of class of functions depends on the differential equation being solved and on the initial and boundary conditions. Of course far more had to be done and was to be done with these functions to render them more useful in the solution of partial differential equations.

5. First Order Partial Differential Equations

Up to the time of Lagrange there was very little systematic work on first order partial differential equations. The second order equations received

37. *Corresp. sur l'Ecole Poly.*, 3, 1816, 361–85.

primary attention because the physical problems led directly to them. A few special first order equations had been solved, but these were either readily integrated or integrated by tricks. There was one exception, the equation usually called today a total differential equation and having the form

(65) $$P\,dx + Q\,dy + R\,dz = 0,$$

where P, Q, and R are functions of x, y, and z. Such an equation, if integrable, defines z as a function of x and y. Clairaut encountered such equations in 1739 in his work on the shape of the earth.[38] If the expression on the left side of (65) is an exact differential, that is, if there is a function $u(x, y, z) = C$ such that

(66) $$du = P\,dx + Q\,dy + R\,dz,$$

then Clairaut points out that

(67) $$\frac{\partial P}{\partial y} = \frac{\partial Q}{\partial x}, \qquad \frac{\partial P}{\partial z} = \frac{\partial R}{\partial x}, \qquad \frac{\partial Q}{\partial z} = \frac{\partial R}{\partial y}.$$

Clairaut showed how to solve (65) by a method still used in modern texts. The interest in equation (65) stemmed from the fact that if P, Q, R are components of velocity in fluid motion, then (65) has to be an exact differential.

If (65) is not an exact differential, then Clairaut also showed that it may be possible to find an integrating factor, that is, a function $\mu(x, y, z)$ such that when multiplied into (65), it makes the new left side an exact differential. Clairaut[39] and later d'Alembert (*Traité de l'équilibre et du mouvement des fluides*, 1744) gave a necessary condition that it be integrable (with the aid of an integrating factor). This condition (which is also sufficient) is

(68) $$P\left(\frac{\partial Q}{\partial z} - \frac{\partial R}{\partial y}\right) + Q\left(\frac{\partial R}{\partial x} - \frac{\partial P}{\partial z}\right) + R\left(\frac{\partial P}{\partial y} - \frac{\partial Q}{\partial x}\right) = 0.$$

The *general* first order partial differential equation in two independent variables is of the form

(69) $$f(x, y, z, p, q) = 0,$$

where $p = \partial x/\partial z$ and $q = \partial z/\partial y$. If the equation is linear in p and q then it is a linear partial differential equation and if not, then nonlinear. The significant theory was contributed by Lagrange.

Lagrange's terminology, which is still current, must be noted first to understand his work. He classified solutions of nonlinear first order equations as follows. Any solution $V(x, y, z, a, b) = 0$ containing two arbitrary constants is the *complete solution* or *complete integral*. By letting $b = \phi(a)$ where ϕ

38. *Mém. de l'Acad. des Sci., Paris*, 1739, 425–36.
39. *Mém. de l'Acad. des Sci., Paris*, 1740, 293–323.

is arbitrary, we obtain a one-parameter family of solutions. When $\phi(a)$ is arbitrary, the envelope of this family is called a *general integral*. When a definite $\phi(a)$ is used, the envelope is a *particular case* of the general integral. The envelope of all the solutions in the complete integral is called the *singular integral*. We shall see later what these solutions are geometrically. The complete integral is not unique in that there can be many different ones, which are not obtainable from each other by a simple change in the arbitrary constants. But from any one we can get all the solutions given by another through the particular cases and the singular integral.

Between two important papers of 1772 and 1779, which we shall take up shortly, Lagrange wrote a paper in 1774 discussing the relationships among complete, general, and singular solutions of first order partial differential equations. The general integral is obtained by eliminating a from $V(x, y, z, a, \phi(a)) = 0$ and $\partial V/\partial a = 0$ where $\phi(a)$ is arbitrary.[40] (For a particular $\phi(a)$ we get a particular solution.) The singular solution is obtained by eliminating a and b from $V(x, y, z, a, b) = 0$, $\partial V/\partial a = 0$ and $\partial V/\partial b = 0$.

Lagrange gave first the general theory of *nonlinear* first order equations. In the paper of 1772[41] he considered a general first order equation with two independent variables x and y and the dependent variable z. Here he improved on and generalized what Euler did earlier. He regarded equation (69) as given in the form where q is a function of x, y, z, and p, namely,

$$(70) \qquad q - Q(x, y, z, p) = 0$$

and sought to determine p as a function P of x, y, and z so that the two equations

$$(71) \qquad q - Q(x, y, z, p) = 0 \quad \text{and} \quad p - P(x, y, z) = 0$$

have a single infinity of common integral surfaces, or, as Lagrange put it analytically, so that the expression

$$(72) \qquad dz - p\, dx - q\, dy$$

by multiplication by a suitable factor $M(x, y, z)$ becomes an exact differential dN of $N(x, y, z) = 0$. For this to be so he must have

$$\frac{\partial N}{\partial z} = M, \qquad \frac{\partial N}{\partial x} = -Mp, \qquad \frac{\partial N}{\partial y} = -Mq.$$

For these equations the integrability conditions (67) imply

$$\frac{\partial M}{\partial x} = -\frac{\partial (Mp)}{\partial x}, \qquad \frac{\partial M}{\partial y} = -\frac{\partial (Mq)}{\partial z}, \qquad \frac{\partial (Mp)}{\partial y} = \frac{\partial (Mq)}{\partial x}.$$

40. For arbitrary ϕ it is not generally possible to actually carry out the elimination of a. The general integral is a concept and amounts to a collection of particular solutions.
41. *Nouv. Mém. de l'Acad. de Berlin*, 1772 = *Œuvres*, 3, 549–75.

If one puts into the last of these three equations the values of $\partial M/\partial x$ and $\partial M/\partial y$ from the first two, one obtains

$$(73) \qquad \frac{\partial p}{\partial y} - \frac{\partial q}{\partial x} - p\frac{\partial q}{\partial z} + q\frac{\partial p}{\partial z} = 0.$$

This is the condition (68) for the integrability of (72), a condition known before, as Lagrange remarks. In (73) q can be taken as the given function Q of x, y, z, and p so that explicitly the equation becomes

$$(74) \qquad -Q_p\frac{\partial p}{\partial x} + \frac{\partial p}{\partial y} + (Q - pQ_p)\frac{\partial p}{\partial z} - Q_x - pQ_z = 0.$$

Lagrange's plan now is to find a solution $p = P$ of this first order equation, which is *linear* in the derivatives of p and whose solution contains an arbitrary constant α. Having obtained this, he integrates the two equations

$$(75) \qquad q - Q(x,y,z,p) = 0, \qquad p - P(x,y,z,\alpha) = 0,$$

which represent $\partial z/\partial x$ and $\partial z/\partial y$ as functions of x, y, z, and finds a family of ∞^2 integral surfaces of the original equation (70); that is, he finds the complete solution. Thus far, then, Lagrange has replaced the problem of solving the nonlinear equation (70) by the problem of solving the *linear* equation (74).

In 1779 Lagrange gave his method of solving linear first order partial differential equations.[42] He considers the equation still called Lagrange's linear equation

$$(76) \qquad Pp + Qq = R,$$

where P, Q, and R are functions of x, y, and z; the equation is called non-homogeneous because of the presence of the R term. This equation is intimately related to the homogeneous equation in three independent variables

$$(77) \qquad P\frac{\partial f}{\partial x} + Q\frac{\partial f}{\partial y} + R\frac{\partial f}{\partial z} = 0.$$

What Lagrange shows readily is that if $u(x, y, z) = c$ is a solution of (76), then $f = u(x, y, z)$ is a solution of (77) and conversely. Hence the problem of solving (76) is equivalent to that of solving (77). The equation (77) in turn is related to the system of ordinary differential equations

$$(78) \qquad \frac{dx}{P} = \frac{dy}{Q} = \frac{dz}{R} \quad \text{or} \quad \frac{dy}{dx} = \frac{Q}{P} \quad \text{and} \quad \frac{dz}{dx} = \frac{R}{P}.$$

In fact, if $f = u(x, y, z)$ and $f = v(x, y, z)$ are two independent solutions of (77), then $u = c_1$ and $v = c_2$ are a solution of (78) and conversely. Hence, if we can find the solutions $u = c_1$ and $v = c_2$ of (78), $f = u$ and $f = v$ will be solutions of (77) and $u = c$ and $v = c$ will be solutions of (76). Moreover,

42. *Nouv. Mém. de l'Acad. de Berlin*, 1779 = *Œuvres*, 4, 585–634.

one can show readily that $f = \phi(u, v)$ where ϕ is an arbitrary function of u and v also satisfies (77). Then $\phi(u, v) = c$ or, since ϕ is arbitrary, $\phi(u, v) = 0$ is a general solution of (76). Lagrange gave the above scheme in 1779 and gave a proof in 1785.[43] It is perhaps worth noting incidentally that Euler was aware that the solution of (77) can be reduced to the solution of (78).

If one takes this work on linear equations in connection with the 1772 work on nonlinear equations, one sees that Lagrange had succeeded in reducing an arbitrary first order equation in x, y, and z to a system of simultaneous ordinary differential equations. He does not state the result explicitly but it follows from the above work. Curiously, in 1785 he had to solve a particular first order partial differential equation and said it was impossible with present methods; he had forgotten his earlier (1772) work.

Then Paul Charpit (d. 1784) in 1784 presumably combined the methods for nonlinear and linear equations to reduce any $f(x, y, z, p, q) = 0$ to a system of ordinary differential equations. Lacroix said in 1798 that Charpit had submitted a paper in 1784 (which was not published) in which he reduced first order partial differential equations to systems of ordinary differential equations. Jacobi found Lacroix's statement striking and expressed the wish that Charpit's work be published. But this was never done and we do not know whether Lacroix's statement is correct. Actually Lagrange had done the full job and Charpit could have added nothing. The method given in modern texts, called the Lagrange, Lagrange-Charpit, or Charpit method, is the fusion of the ideas Lagrange presented in his 1772 and 1779 papers. It states that to solve the general first order partial differential equation $f(x, y, z, p, q) = 0$ one must solve the system of ordinary differential equations (the characteristic equations of $f = 0$),

$$\frac{dx}{dt} = f_p, \qquad \frac{dy}{dt} = f_q, \qquad \frac{dz}{dt} = pf_p + qf_q,$$

(79)

$$\frac{dp}{dt} = -f_x - f_z p, \qquad \frac{dq}{dt} = -f_y - f_z q.$$

The solution is effected by finding any one integral of (79), say $u(x, y, z, p, q) = A$. One solves this and $f = 0$ simultaneously for p and q and substitutes for p and q in $dz = pdx + qdy$ (see [72]). Then one integrates by the method used for (65).

Lagrange's method is often called Cauchy's method of characteristics because the generalization to n variables of the method of arriving at (79) used by Lagrange and Charpit for a differential equation in two independent variables presents difficulties which were surmounted by Cauchy in 1819.[44]

43. *Nouv. Mém. de l'Acad. de Berlin*, 1785 = *Œuvres*, 5, 543–62.
44. *Bull. de la Société Philomathique*, 1819, 10–21; see also *Exercices d'analyse et de phys. math.*, 2, 238–72 = *Œuvres*, (2), 12, 272–309.

6. *Monge and the Theory of Characteristics*

Lagrange worked purely analytically. Gaspard Monge (1746–1818) introduced the language of geometry. His work in partial differential equations
was not as great as that of Euler, Lagrange, and Legendre, but he started the
movement to interpret the analytic work geometrically and introduced
thereby many fruitful ideas. He saw that just as problems involving curves
led to ordinary differential equations, so problems involving surfaces lead to
partial differential equations. More generally, for Monge geometry and
analysis were one subject, whereas for the other mathematicians of the
century the two branches were distinct, with just points of contact. Monge
began his work in 1770 but did not publish until much later.

It was primarily in the subject of nonlinear first order equations that
Monge not only introduced the geometric interpretation but emphasized
a new concept, that of characteristic curves.[45] His ideas on characteristics
and on integrals as envelopes were not understood and were called a metaphysical principle by his contemporaries, but the theory of characteristics
was to become a very significant theme in later work. Monge developed his
ideas more fully in his lectures and in subsequent publications, notably the
Feuilles d'analyse appliquée à la géométrie (1795). The ideas are best illustrated
by his own example.

Consider the two-parameter family of spheres, all of constant radius R
and centers anywhere in the XY-plane. The equation of this family is

$$(80) \qquad (x - a)^2 + (y - b)^2 + z^2 = R^2.$$

This equation is the complete integral of the nonlinear first order partial
differential equation

$$(81) \qquad z^2(p^2 + q^2 + 1) = R^2,$$

because (80) contains two arbitrary constants a and b and clearly satisfies
(81). Any subfamily of spheres introduced by letting $b = \phi(a)$ is a family
of spheres with centers on a curve, the curve $y = \phi(x)$ in the XY-plane.
The envelope of this one-parameter family of spheres (a tubular surface) is
also a solution of (81). This particular solution is obtained by eliminating a
from

$$(82) \qquad (x - a)^2 + (y - \phi(a))^2 + z^2 = R^2$$

and the partial derivative of (82) with respect to a, namely,

$$(83) \qquad (x - a) + (y - \phi(a))\phi'(a) = 0.$$

For each particular choice of a, equations (82) and (83) represent two
particular surfaces, and therefore the two considered simultaneously repre-

45. *Hist. de l'Acad. des Sci., Paris*, 1784, 85–117, 118–92, pub. 1787.

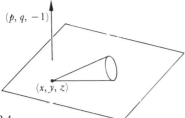

$(p, q, -1)$

(x, y, z)

Figure 22.4

sent a curve called a characteristic curve. This curve is also the curve of intersection of two "consecutive" members of the subfamily. The set of characteristic curves fills out the envelope; that is, the envelope touches each member of the subfamily along a characteristic. The general integral is the aggregate of surfaces (envelopes of one-parameter families), each generated by a set of characteristic curves. The envelope of all the solutions of (80), that is, the singular solution, obtained by eliminating a and b from (80) and the partial derivatives of (80) with respect to a and b respectively, is $z = \pm R$.

The characteristic curve appears in another way. Consider two sub-families of spheres whose envelopes are tangential along any one sphere. We might call such envelopes consecutive envelopes. The curve of intersection of these two consecutive envelopes is the same characteristic curve on the sphere as the one obtained by considering consecutive members of either subfamily. Any one sphere may belong to an infinity of different subfamilies whose envelopes are all different, and so there will be different characteristic curves on the same sphere. All are great circles in vertical planes.

Monge gave the analytical form of the differential equations of the characteristic curves, which amounts to the fact that equations (79) determine the characteristic curves of (69). (Monge used total differential equations to express the equations of the characteristic curves.)

Monge also introduced (1784) the notion of a characteristic cone. At any point (x, y, z) of space (Fig. 22.4) one may consider a plane whose normal has the direction numbers p, q, -1. For a fixed (x, y, z), the set of p and q which satisfy

(84) $$F(x, y, z, p, q) = 0$$

determines a one-parameter family of planes all passing through (x, y, z). This set of planes envelopes a cone with vertex at (x, y, z). This is the characteristic cone or Monge cone at (x, y, z). If we now consider a surface S whose equation is $z = g(x, y)$, then the surface has a tangent plane at each (x, y, z). A necessary and sufficient condition that such a surface be an integral surface of $F = 0$ is that at each point (x, y, z) the tangent plane of F

be a tangent plane of the Monge cone at (x, y, z). A curve C on an integral surface S is called a characteristic curve if at each point of C the tangent is a generator of the Monge cone at that point. These characteristic curves are the same as those Monge deduced from the complete integral illustrated by (80) and are the solutions of the simultaneous equations (79). He also points out in a paper of 1802 which he incorporated in his *Application de l'analyse à la géométrie*[46] (1807) that each integral surface is a locus of characteristic curves and only one characteristic curve passes through each point of the integral surface.

The significance of the characteristic curves lies in the following. If one chooses a space curve $x(t), y(t), z(t)$ (for some interval of t-values) that is not a characteristic curve, then there is just one integral surface of $F = 0$ that passes through this curve; that is, there is just one $z = g(x, y)$ such that $z(t) = g(x(t), y(t))$ (for the range of t-values). On the other hand, as Monge noted in lectures of 1806, through any characteristic curve one can pass an infinity of integral surfaces. Moreover, the infinity of integral surfaces that pass through the curve are tangent to each other on that curve.

7. Monge and Nonlinear Second Order Equations

In addition to the second order linear equations we have already reviewed, the eighteenth-century mathematicians had occasion to consider more general linear second order equations in two independent variables and even nonlinear ones. Thus they studied the linear equation

$$A \frac{\partial^2 z}{\partial x^2} + B \frac{\partial^2 z}{\partial x \partial y} + C \frac{\partial^2 z}{\partial y^2} + D \frac{\partial z}{\partial x} + E \frac{\partial z}{\partial y} + Fz + G = 0,$$

where $A, B, \cdots, G$ are functions of x and y. This equation is commonly written as

(85) $$Ar + Bs + Ct + Dp + Eq + Fz + G = 0,$$

where the letters $r, s, t, p,$ and q have the obvious meanings. Laplace showed in 1773[47] that equation (85) can, by a change of variables, be reduced to the form

(86) $$s + ap + bq + cz + g = 0,$$

where $a, b, c,$ and g are functions of x and y only, provided that $B^2 - 4AC \neq 0$. He then solved the equation in terms of an infinite series.

In his *Feuilles d'analyse*, Monge considered the nonlinear equation

(87) $$Rr + Ss + Tt = V$$

---

46. This is the title of the third edition of his *Feuilles d'analyse*.
47. *Hist. de l'Acad. des Sci., Paris*, 1773, pub. 1777 = *Œuvres*, 9, 5–68.

in which R, S, T, and V are functions of $x, y, z, p,$ and q, so that the equation is linear only in the second derivatives r, s, and t. This type of equation arose in Lagrange's work on minimal surfaces, that is, surfaces of least area bounded by given space curves, wherein the specific differential equation is $(1 + q^2)r - 2pqs + (1 + p^2)t = 0$. (See also Chap. 24, sec. 4). Though Monge had already done some work on equation (87), in the present work (1795) he was able to solve it elegantly by the method we shall sketch.

By using the immediate facts

(88) $$dz = p\,dx + q\,dy$$

(89) $$dp = r\,dx + s\,dy$$

(90) $$dq = s\,dx + t\,dy$$

and eliminating r and t from (87), (89), and (90) he obtained the equation

(91) $$s(R\,dy^2 - S\,dx\,dy + T\,dx^2) - (R\,dy\,dp + T\,dx\,dq - V\,dx\,dy) = 0.$$

His argument then was that whenever it is possible to solve simultaneously

(92) $$R\,dy^2 - S\,dx\,dy + T\,dx^2 = 0$$

and

(93) $$R\,dy\,dp + T\,dx\,dq - V\,dx\,dy = 0$$

then (91) will be satisfied and so will (87).

Equation (92) is equivalent to two first order equations

(94) $$dy - W_1(x, y, z, p, q)\,dx = 0 \quad \text{and} \quad dy - W_2(x, y, z, p, q) = 0.$$

Equations (88) and (93), together with either one of (94), constitute a system of three total differential equations in the five variables $x, y, z, p,$ and q. When these three equations can be solved it is possible to find two solutions

$$u_1(x, y, z, p, q) = C_1 \quad \text{and} \quad u_2(x, y, z, p, q) = C_2$$

and then

(95) $$u_1 = \phi(u_2),$$

where ϕ is arbitrary, is a first order partial differential equation. The equation (95) is called an intermediate integral. Its general solution is the solution of (87). If the other equation in (94) can be used together with (88) and (93), we get another function

(96) $$u_3 = \phi(u_4).$$

In this case (95) and (96) can be solved simultaneously for p and q and these values are substituted in (88); then this total differential equation can be solved. This at least is the general scheme, though there are details we shall not take time to cover. Monge's integration of the equation for minimal surfaces was one of his claims to glory.

For the equation (87) also Monge introduced the theory of characteristics. The total differential equation of the characteristics is (92), that is,

$$R \, dy^2 - S \, dx \, dy + T \, dx^2 = 0.$$

This equation, which appears in his work as early as 1784,[48] defines at each point of an integral surface two directions that are the characteristic directions at that point. Through each point on an integral surface there pass two characteristic curves, along each of which two consecutive integral surfaces touch each other.

8. *Systems of First Order Partial Differential Equations*

Systems of partial differential equations arose first in the eighteenth-century work on fluid dynamics or hydrodynamics. The work on incompressible fluids, for example, water, was motivated by such practical problems as designing the hulls of ships to reduce resistance to motion in water and calculating the tides, the flow of rivers, the flow of water from jets, and the pressure of water on the sides of a ship. The work on compressible fluids, air in particular, sought to analyze the action of air on sails of ships, the design of windmill vanes, and the propagation of sound. The work we studied earlier on the propagation of sound was historically an application of the work on hydrodynamics specialized to waves of small amplitude.

After having treated incompressible fluids in 1752 in a paper entitled "Principles of the Motion of Fluids,"[49] Euler generalized this work in a paper of 1755, entitled "General Principles of the Motion of Fluids."[50] Here he gave the still-famous equations of fluid flow for perfect (nonviscous) compressible and incompressible fluids. The fluid is regarded as a continuum and the particles are mathematical points. He considers the force acting on a small volume of the fluid subject to the pressure p, density ρ, and external forces with components P, Q, and R per unit mass.

In one of the two approaches to fluid dynamics that Euler created, known in the literature as the spatial description, the components u, v, and w of the fluid velocity are given at every point in the fluid by

$$(97) \qquad u = u(x, y, z, t), \qquad v = v(x, y, z, t), \qquad w = w(x, y, z, t).$$

Now

$$du = \frac{\partial u}{\partial x} \, dx + \frac{\partial u}{\partial y} \, dy + \frac{\partial u}{\partial z} \, dz + \frac{\partial u}{\partial t} \, dt.$$

48. *Hist. de l'Acad. des Sci.*, Paris, 1784, 118–92, pub. 1787.
49. *Novi Comm. Acad. Sci Petrop.*, 6, 1756/57, 271–311, pub. 1761 = *Opera*, (2), 12, 133–68.
50. *Hist. de l'Acad. de Berlin*, 11, 1755, 274–315, pub. 1757 = *Opera*, (2), 12, 54–91.

In time dt, the particle at (x, y, z) will travel a distance $u\ dt$ in the x-direction, $v\ dt$ in the y-direction, and $w\ dt$ in the z-direction. Then the actual changes dx, dy, and dz in the expression for du are given by these quantities, so that

$$du = \frac{\partial u}{\partial x} u\ dt + \frac{\partial u}{\partial y} v\ dt + \frac{\partial u}{\partial z} w\ dt + \frac{\partial u}{\partial t} dt$$

or

(98)
$$\frac{du}{dt} = u\frac{\partial u}{\partial x} + v\frac{\partial u}{\partial y} + w\frac{\partial u}{\partial z} + \frac{\partial u}{\partial t},$$

and there are the corresponding expressions for dv/dt and dw/dt. These quantities give what is called now the convective rate of change of the velocity at (x, y, z) or the convective acceleration. By calculating the forces acting on the particle at (x, y, z) and applying Newton's second law, Euler obtains the system of differential equations

(99)
$$P - \frac{1}{\rho}\frac{\partial p}{\partial x} = \frac{du}{dt}$$

$$Q - \frac{1}{\rho}\frac{\partial p}{\partial y} = \frac{dv}{dt}$$

$$R - \frac{1}{\rho}\frac{\partial p}{\partial z} = \frac{dw}{dt}.$$

Euler also generalized d'Alembert's differential equation of continuity (45) and obtained for compressible flow the equation

(100)
$$\frac{\partial \rho}{\partial t} + \frac{\partial(\rho u)}{\partial x} + \frac{\partial(\rho u)}{\partial y} + \frac{\partial(\rho w)}{\partial z} = 0.$$

There are four equations and five unknowns, but the pressure p as a function of the density, the equation of state, must be specified.

In the 1755 paper Euler says, "And if it is not permitted to us to penetrate to a complete knowledge concerning the motion of fluids, it is not to mechanics, or to the insufficiency of the known principles of motion, that we must attribute the cause. It is analysis itself which abandons us here, since all the theory of the motion of fluids has just been reduced to the solution of analytic formulas." Unfortunately analysis was still too weak to do much with these equations. He then proceeds to discuss some special solutions. He also wrote other papers on the subject and dealt with the resistance encountered by ships and ship propulsion. Euler's equations are not the final ones for hydrodynamics. He neglected viscosity, which was introduced seventy years later by Navier and Stokes (Chap. 28, sec. 7).

Lagrange, too, worked on fluid motion. In the first edition of his *Mécanique analytique*, which contains some of the work, he gave Euler's basic equations and generalized them. Here he gave credit to d'Alembert but none to Euler. He too says the equations of fluid motion are too difficult to be handled by analysis. Only the cases of infinitely small movements are susceptible of rigorous calculation.

In the area of systems of partial differential equations, the equations of hydrodynamics were, in the eighteenth century, the main inspiration for mathematical research on this subject. Actually the eighteenth century accomplished little in the solution of systems.

9. *The Rise of the Mathematical Subject*

Up to 1765 partial differential equations appeared only in the solution of physical problems. The first paper devoted to purely mathematical work on partial differential equations is by Euler: "Recherches sur l'intégration de l'équation $\left(\dfrac{d\,dz}{dt^2}\right) = aa\left(\dfrac{d\,dz}{dx^2}\right) + \dfrac{b}{x}\left(\dfrac{dz}{dx}\right)$."[51] Shortly thereafter Euler published a treatise on the subject in the third volume of his *Institutiones Calculi Integralis*.[52]

Before d'Alembert's work of 1747 on the vibrating string, partial differential equations were known as equations of condition and only special solutions were sought. After this work and d'Alembert's book on the general causes of winds (1746), the mathematicians realized the difference between special and general solutions. However, once aware of this distinction, they seemed to believe that general solutions would be more important. In the first volume of his *Mécanique céleste* (1799), Laplace still complained that the potential equation in spherical coordinates could not be integrated in general form. Appreciation of the fact that general solutions such as Euler and d'Alembert obtained for the vibrating string were not as useful as particular ones satisfying initial and boundary conditions was not attained in that century.

The mathematicians did realize that partial differential equations involved no new operational techniques but differed from ordinary differential equations in that arbitrary functions might appear in the solution. These they expected to determine by reducing partial differential equations to ordinary ones. Laplace (1773) and Lagrange (1784) say clearly that they regard a partial differential equation as integrated when it is reduced to a problem of ordinary differential equations. An alternative, such as Daniel Bernoulli used for the wave equation and Laplace for the potential equation, was to seek expansions in series of special functions.

The major achievement of the eighteenth-century work on partial differential equations was to reveal their importance for problems of elasticity,

51. *Misc. Taur.*, 3_2 1762/65, 60–91, pub. 1766 = *Opera*, (1), 23, 47–73.
52. 1770 = *Opera*, (1), 13.

hydrodynamics, and gravitational attraction. Except in the case of Lagrange's work on first order equations, no broad methods were developed, nor were the potentialities in the method of expansion in special functions appreciated. The efforts were directed to solving the special equations that arose in physical problems. The theory of the solution of partial differential equations remained to be fashioned and the subject as a whole was still in its infancy.

Bibliography

Burkhardt, H.: "Entwicklungen nach oscillirenden Funktionen und Integration der Differentialgleichungen der mathematischen Physik," *Jahres. der Deut. Math.-Verein.*, 10, 1908, 1–1804.

Burkhardt, H., and W. Franz Meyer: "Potentialtheorie," *Encyk. der Math. Wiss.*, B. G. Teubner, 1899–1916, 2, A7b, 464–503.

Cantor, Moritz: *Vorlesungen über Geschichte der Mathematik*, B. G. Teubner, 1898 and 1924; Johnson Reprint Corp., 1965, Vol. 3, 858–78, Vol. 4, 873–1047.

Euler, Leonard: *Opera Omnia*, Orell Füssli, (1), Vols. 13 (1914) and 23 (1938); (2), Vols. 10, 11, 12, and 13 (1947–55).

Lagrange, Joseph-Louis: *Œuvres*, Gauthier-Villars, 1868–70, relevant papers in Vols. 1, 3, 4, 5.

Laplace, Pierre-Simon: *Œuvres complètes*, Gauthier-Villars, 1893–94, relevant papers in Vols. 9 and 10.

Montucla, J. F.: *Histoire des mathématiques* (1802), Albert Blanchard (reprint), 1960, Vol. 3, pp. 342–52.

Langer, Rudolph E.: "Fourier Series: The Genesis and Evolution of a Theory," *Amer. Math. Monthly*, 54, No. 7, Part 2, 1947.

Taton, René: *L'Œuvre scientifique de Monge*, Presses Universitaires de France, 1951.

Todhunter, Isaac: *A History of the Mathematical Theories of Attraction and the Figure of the Earth* (1873), Dover (reprint), 1962.

Truesdell, Clifford E.: *Introduction to Leonhardi Euleri Opera Omnia Vol. X et XI Seriei Secundae*, in Euler, *Opera Omnia*, (2), 11, Part 2, Orell Füssli, 1960.

———: *Editor's Introduction* in Euler, *Opera Omnia*, (2), Vol. 12, Orell Füssli, 1954.

———: *Editor's Introduction* in Euler, *Opera Omnia*, (2), Vol. 13, Orell Füssli, 1956.

23
Analytic and Differential Geometry in the Eighteenth Century

> Geometry may sometimes appear to take the lead over analysis but in fact precedes it only as a servant goes before the master to clear the path and light him on his way.
>
> JAMES JOSEPH SYLVESTER

1. *Introduction*

The exploration of physical problems led inevitably to the search for greater knowledge of curves and surfaces, because the paths of moving objects are curves and the objects themselves are three-dimensional bodies bounded by surfaces. The mathematicians, already enthusiastic about the method of coordinate geometry and the power of the calculus, approached geometrical problems with these two major tools. The impressive results of the century were obtained in the already established area of coordinate geometry and the new field created by applying the calculus to geometrical problems, namely, differential geometry.

2. *Basic Analytic Geometry*

Two-dimensional analytic geometry was extensively explored in the eighteenth century. The improvements in elementary plane analytics are readily summarized. Whereas Newton and James Bernoulli had introduced and used what are essentially polar coordinates for special curves (Chap. 15, sec. 5), Jacob Hermann in 1729 not only proclaimed their general usefulness, but applied them freely to study curves. He also gave the transformation from rectangular to polar coordinates. Strictly Hermann used as variables p, $\cos \theta$, $\sin \theta$, which he designated by z, n, and m. Euler extended the use of polar coordinates and used trigonometric notation explicitly; with him the system is practically modern.

Though some seventeenth-century men—for example, Jan de Witt

544

(1625–72) in his *Elementa Curvarum Linearum* (1659)—did reduce some second degree equations in x and y to standard forms, James Stirling, in his *Lineae Tertii Ordinis Neutonianae* (1717), reduced the general second degree equation in x and y to the several standard forms.

In his *Introductio* (1748) Euler introduced the parametric representation of curves, wherein x and y are expressed in terms of a third variable. In this famous text Euler treated plane coordinate geometry systematically.

The suggestion of three-dimensional coordinate geometry can be found, as we know, in the work of Fermat, Descartes, and La Hire. The actual development was the work of the eighteenth century. Though some of the early work, Pitot's and Clairaut's for example, is tied up with the development of differential geometry, we shall consider only coordinate geometry proper at this point.

The first task was the improvement of La Hire's suggestion of a three-dimensional coordinate system. John Bernoulli, in a letter to Leibniz of 1715, introduced the three coordinate planes we use today. Through contributions too detailed to warrant space here, Antoine Parent (1666–1716), John Bernoulli, Clairaut, and Jacob Hermann clarified the notion that a surface can be represented by an equation in three coordinates. Clairaut, in his book *Recherche sur les courbes à double courbure* (Research on the Curves of Double Curvature, 1731), not only gave the equations of some surfaces but made clear that two such equations are needed to describe a curve in space. He also saw that certain combinations of the equations of two surfaces passing through a curve, the sum for example, give the equation of another surface passing through the curve. Using this fact, he explains how one can obtain the equations of the projections of these curves or, equivalently, the equations of the cylinders perpendicular to the planes of projection.

The quadric surfaces, e.g. sphere, cylinder, paraboloid, hyperboloid of two sheets, and ellipsoid, were of course known geometrically before 1700; in fact, some of them appear in Archimedes' work. Clairaut in his book of 1731 gave the equations of some of these surfaces. He also showed that an equation that is homogeneous in x, y, and z (all terms are of the same degree) represents a cone with vertex at the origin. To this result Jacob Hermann, in a paper of 1732,[1] added that the equation $x^2 + y^2 = f(z)$ is a surface of revolution about the z-axis. Both Clairaut and Hermann were primarily concerned with the shape of the earth, which by their time was believed to be some form of ellipsoid.

Though Euler had done some earlier work on the equations of surfaces, it is in Chapter 5 of the Appendix to the second volume of his *Introductio* (1748)[2] that he systematically takes up three-dimensional coordinate

1. *Comm. Acad. Sci. Petrop.*, 6, 1732/33, 36–67, pub. 1738.
2. *Opera*, (1), 9.

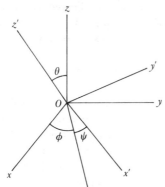

Figure 23.1   Line of nodes

geometry. He presents much of what had already been done and then studies the general second degree equation in three variables

(1) $\qquad ax^2 + by^2 + cz^2 + dxy + exz + fyz + gx + hy + kz = l.$

He now seeks to use change of axes to reduce this equation to the forms that result from having the principal axes of the quadric surfaces represented by (1) as the coordinate axes. He introduces the transformation from the xyz-system to the $x'y'z'$-system, whose equations are expressed (Fig. 23.1) in terms of the angles ϕ, ψ, and θ. The angle ϕ is measured in the xy-plane from the x-axis to the line of nodes, which is the line in which the $x'y'$-plane cuts the xy-plane. The angle ψ is measured in the $x'y'$-plane and locates x' with respect to the line of nodes. The angle shown is θ. Then the equations of transformation, including translation, are

$$
\begin{aligned}
x = x_0 &+ x'(\cos\psi\cos\phi - \cos\theta\sin\psi\sin\phi) \\
&- y'(\cos\psi\sin\phi + \cos\theta\sin\psi\sin\phi) + z'\sin\theta\sin\phi \\
(2)\quad y = y_0 &+ x'(\sin\psi\cos\phi + \cos\theta\cos\psi\sin\phi) \\
&- y'(\sin\psi\sin\phi - \cos\theta\cos\psi\sin\phi) - z'\sin\theta\sin\phi \\
z = z_0 &+ x'\sin\theta\sin\phi + y'\sin\theta\cos\phi + z'\cos\theta.
\end{aligned}
$$

Euler uses this transformation to reduce (1) to canonical forms and obtains six distinct cases: cone, cylinder, ellipsoid, hyperboloid of one and two sheets, hyperbolic paraboloid (which he discovered), and parabolic cylinder. Like Descartes, Euler maintained that classification by the degree of the equation was the correct principle; Euler's reason was that the degree is invariant under linear transformation.

After continuing work on this problem of change of axes, he wrote another paper,[3] in which he considers the transformation that will carry

3. *Novi Comm. Acad. Sci. Petrop.*, 15, 1770, 75–106, pub. 1771 = *Opera*, (1), 6, 287–315.

$x^2 + y^2 + z^2$ into $x'^2 + y'^2 + z'^2$. Here he—and Lagrange a little later, in a paper on the attraction of spheroids[4]—gave the symmetric form for the rotation of axes, the homogeneous linear orthogonal transformation

$$x = \lambda x' + \mu y' + \nu z'$$
$$y = \lambda' x' + \mu' y' + \nu' z'$$
$$z = \lambda'' x' + \mu'' y' + \nu'' z',$$

where

$$\lambda^2 + \lambda'^2 + \lambda''^2 = 1 \qquad \lambda\mu + \lambda'\mu' + \lambda''\mu'' = 0$$
$$\mu^2 + \mu'^2 + \mu''^2 = 1 \qquad \lambda\nu + \lambda'\nu' + \lambda''\nu'' = 0$$
$$\nu^2 + \nu'^2 + \nu''^2 = 1 \qquad \mu\nu + \mu'\nu' + \mu''\nu'' = 0.$$

The λ, μ, and ν, unprimed and primed, are of course direction cosines in today's terminology.

Gaspard Monge's writings contain a great deal of three-dimensional analytic geometry. His outstanding contribution to analytic geometry as such is to be found in the paper of 1802 written with his pupil Jean-Nicolas-Pierre Hachette (1769–1834), "Application de l'algèbre à la géométrie."[5] The authors show that every plane section of a second degree surface is a second degree curve, and that parallel planes cut out similar and similarly placed curves. These results parallel Archimedes' geometric theorems. The authors also show that the hyperboloid of one sheet and the hyperbolic paraboloid are ruled surfaces, that is, each can be generated in two different ways by the motion of a line or each surface is formed by two systems of lines. The result on the one-sheeted hyperboloid was known by 1669 to Christopher Wren, who said that this figure could be generated by revolving a line about another not in the same plane. With the work of Euler, Lagrange, and Monge, analytic geometry became an independent and full-fledged branch of mathematics.

3. Higher Plane Curves

The analytic geometry described thus far was devoted to curves and surfaces of the first and second degree. It was of course natural to investigate the curves of equations of higher degree. In fact, Descartes had already discussed such equations and their curves somewhat. The study of curves of degree higher than two became known as the theory of higher plane curves, though it is part of coordinate geometry. The curves studied in the eighteenth century were algebraic; that is, their equations are given by $f(x, y) = 0$ where f is a polynomial in x and y. The degree or order is the highest degree of the terms.

4. *Nouv. Mém. de l'Acad. de Berlin*, 1773, 85–120 = *Œuvres*, 3, 619–58.
5. *Jour. de l'Ecole Poly.*, 11 cahier, 1802, 143–69.

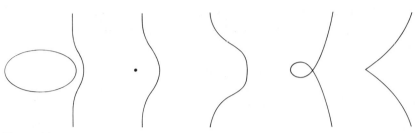

Figure 23.2

The first extensive study of higher plane curves was made by Newton. Impressed by Descartes's plan to classify curves according to the degree of their equations and then to study systematically each degree by methods suited to that degree, Newton undertook to study third degree curves. This work appeared in his *Enumeratio Linearum Tertii Ordinis*, which was published in 1704 as an appendix to the English edition of his *Opticks* but had been composed by 1676. Though the use of negative x- and y-values appears in works of La Hire and Wallis, Newton not only uses two axes and negative x- and y-values but plots in all four quadrants.

Newton showed how all curves comprised by the general third degree equation

$$(3) \quad ax^3 + bx^2y + cxy^2 + dy^3 + ex^2 + fxy + gy^2 + hx + jy + k = 0$$

can, by a change of axes, be reduced to one of the following four forms:

(a) $xy^2 + ey = ax^3 + bx^2 + cx + d$
(b) $\quad\quad xy = ax^3 + bx^2 + cx + d$
(c) $\quad\quad y^2 = ax^3 + bx^2 + cx + d$
(d) $\quad\quad y = ax^3 + bx^2 + cx + d.$

The third class, which Newton called diverging parabolas, contains five species of curves whose types are shown in Figure 23.2. The species are distinguished by the nature of the roots of the cubic right-hand member, as follows: all real and distinct; two roots complex; all real but two equal and the double root greater or less than the simple root; and all three equal. Newton affirmed that every cubic curve can be obtained by projection of one of these five types from a point and then by a section of the projection.

Newton gave no proofs of many of the assertions in his *Enumeratio*. In his *Lineae*, James Stirling proved or reproved in other ways most of the assertions but not the projection theorem, which Clairaut[6] and François Nicole (1683–1758)[7] proved. Also, whereas Newton recognized seventy-two

6. *Mém. de l'Acad. des Sci.*, Paris, 1731, 490–93, pub. 1733.
7. *Mém. de l'Acad. des Sci.*, Paris, 1731, 494–510, pub. 1733.

species of third degree curves, Stirling added four more and abbé Jean-Paul de Gua de Malves, in a little book of 1740 entitled *Usage de l'analyse de Descartes pour découvrir sans le secours du calcul différential* . . . , added two more.

Newton's work on third degree curves stimulated much other work on higher plane curves. The topic of classifying third and fourth degree curves in accordance with one or another principle continued to interest mathematicians of the eighteenth and nineteenth centuries. The number of classes found varied with the methods of classification.

As is evident from the figures of Newton's five species of cubic curves, the curves of higher-degree equations exhibit many peculiarities not found in first and second degree curves. The elementary peculiarities, called singular points, are inflection points and multiple points. Before proceeding, let us see what some of them look like.

Inflection points are familiar from the calculus. A point at which there are two or more tangents which may coincide, is called a multiple point. At such a point two or more branches of the curve intersect. If two branches intersect at the multiple point, it is called a double point. If three branches intersect then the point is called a triple point, and so on.

If we take the equation of an algebraic curve

$$f(x, y) = 0,$$

f being a polynomial in x and y, we can by a translation always remove the constant term. If this is done and if there are first degree terms in f, say $a_1 x + b_1 y$, then $a_1 x + b_1 y = 0$ gives the equation of the tangent to the curve at the origin. The origin is not in this case a multiple point. If there are no first degree terms, and if $a_2 x^2 + b_2 xy + c_2 y^2$ are the second degree terms, then several cases arise. The equation $a_2 x^2 + b_2 xy + c_2 y^2 = 0$ may represent two distinct lines. These lines are tangents at the origin (this can be proven), and since there are two distinct tangents the origin is a double point; it is called a node. Thus the equation of the lemniscate (Fig. 23.3) is

(4) $$a^2(y^2 - x^2) + (y^2 + x^2)^2 = 0$$

and the second degree terms yield $y^2 - x^2 = 0$. Then $y = x$ and $y = -x$ are the equations of the tangents. Likewise, the folium of Descartes (Fig. 23.4) has the equation

(5) $$x^3 + y^3 = 3axy$$

and the tangents at the origin, which is a node, are given by $x = 0$ and $y = 0$.

When the two tangent lines are coincident, the single line is considered as a double tangent and the two branches of the curve touch each other at the point of tangency, which is called a cusp. (Sometimes cusps are

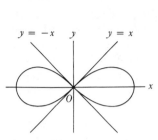

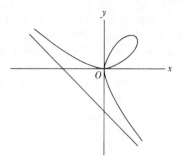

Figure 23.3. Lemniscate

Figure 23.4. Folium of Descartes

included among the double points.) Thus the semicubical parabola (Fig. 23.5)

$$(6) \qquad\qquad ay^2 = x^3$$

has a cusp at the origin, and the equation of the two coincident tangents is $y^2 = 0$. On the curve $(y - x^2)^2 = x^5$ (Fig. 23.6) the origin is a cusp. Here both branches lie on the same side of the double tangent, which is $y = 0$. De Gua in his *Usage* had tried to show that this type of cusp could not occur, but Euler[8] gave many examples. A cusp is also called a stationary point or point of retrogression because a point moving along the curve must come to rest before continuing its motion at a cusp.

When the two tangent lines are imaginary, the double point is called a conjugate point. The coordinates of the point satisfy the equation of the curve but the point is isolated from the rest of the curve. Thus the curve (Fig. 23.7) of $y^2 = x^2(2x - 1)$ has a conjugate point at the origin. The equation of the double tangent there is $y^2 = -x^2$ and the tangents are imaginary.

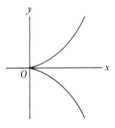

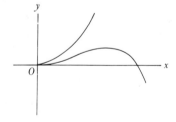

Figure 23.5. Semicubical parabola

Figure 23.6

8. *Mém. de l'Acad. de Berlin*, 5, 1749, 203–21, pub. 1751 = *Opera*, (1), 27, 236–52.

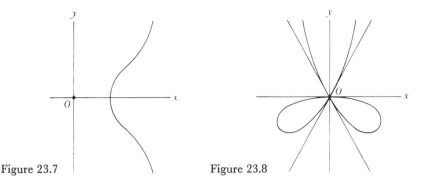

Figure 23.7   Figure 23.8

The curve of $ay^3 - 3ax^2y = x^4$ (Fig. 23.8) has a triple point at the origin. The equation of the three tangents is

$$ay^3 - 3ax^2y = 0$$

or $y = 0$ and $y = \pm x\sqrt{3}$.

The curve of $ay^4 - ax^2y^2 = x^5$ (Fig. 23.9) has a quadruple point at the origin. The origin is a combination of a node and a cusp. The tangents are $y = 0, y = 0, y = \pm x$.

Curves of the third degree (order) may have a double point (which may be a cusp) but no other multiple point. There are of course cubics with no double point.

To return to the history proper, many of these special or singular points on curves were studied by Leibniz and his successors. The analytical conditions for such points, such as that $\ddot{y} = 0$ at an inflection point and that $\dot{y}$ is indeterminate at a double point, were known even to the founders of the calculus.

Clairaut in the 1731 book referred to above assumed that a third degree curve cannot have more than three real inflection points and must have at least one. De Gua in *Usage* proved that if a third degree curve has three real inflection points, a line through two of them passes through the third

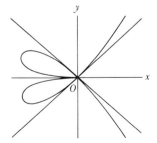

Figure 23.9

one. This theorem is often credited to Maclaurin. De Gua also investigated double points and gave the condition that if $f(x, y) = 0$ is the equation of the curve, then f_x and f_y must be 0 at a double point. A k-fold point is characterized by the vanishing of all partial derivatives through the $(k - 1)$st order. He showed that the singularities are compounded of cusps, ordinary points, and points of inflection. In addition, de Gua treated middle points of curves, the form of branches extending to infinity, and properties of such branches.

Maclaurin in his *Geometria Organica* (1720), written when he was nineteen, proved that the maximum number of double points of an irreducible curve of the nth degree is $(n - 1)(n - 2)/2$. For this purpose he counted a k-fold point as $k(k - 1)/2$ double points. He also gave upper bounds for the number of higher multiple points of each kind. He then introduced the notion of the deficiency (later called genus) of an algebraic curve as the maximum possible number of double points minus the actual number. Among curves those of deficiency 0 or possessing the maximum possible number of double points received a great deal of attention. Such curves are also called rational or unicursal. Geometrically a unicursal curve may be traversed by the continuous motion of a moving point (which may, however, pass through the point at infinity). Thus the conics, including the hyperbola, are unicursal curves.

In his *Method of Fluxions* Newton gave a method, commonly referred to as Newton's diagram or Newton's parallelogram, for determining series representations of the various branches of a curve at a multiple point (Chap. 20, sec. 2). De Gua in *Usage* replaced the Newton parallelogram by a *triangle algébrique*. Then if the origin is a singular point, for small x the equation of an algebraic curve breaks down into factors of the form $y^m - Ax^n$, where m is positive integral and n integral. The branches of the curve are given by those factors for which n is also positive. Euler noted (1749) that de Gua had neglected imaginary branches.

Gabriel Cramer (1704–52), in his *Introduction à l'analyse des lignes courbes algébriques* (1750), also tackled the expansion of y in terms of x when y and x are given in an implicit function, that is, $f(x, y) = 0$, in order to determine a series expression for each branch of the curve, particularly the branches that extend to infinity. He treated y as an ascending and as a descending series of powers of x. Like de Gua he used the triangle in place of Newton's parallelogram, and like others he neglected imaginary branches of the curves.

The conclusion that resulted from the work of obtaining series expansions for each branch of a curve issuing from a multiple point was drawn much later by Victor Puiseux (1820–83)[9] and is known as Puiseux's theorem:

9. *Jour. de Math.*, 13, 1850, 365–480.

The total neighborhood of a point (x_0, y_0) of an algebraic plane curve can be expressed by a finite number of developments

$$(7) \qquad y - y_0 = a_1(x - x_0)^{q_1/q_0} + a_2(x - x_0)^{q_2/q_0} + \cdots.$$

These developments converge in some interval about x_0 and all the q_i have no common factors. The points given by each development are called a branch of the algebraic curve.

The intersections of a curve and line and of two curves is another topic that received a great deal of attention. Stirling, in his *Lineae* of 1717, showed that an algebraic curve of the nth degree (in x and y) is determined by $n(n + 3)/2$ of its points because it has that number of essential coefficients. He also asserted that any two parallel lines cut a given curve in the same number of points, real or imaginary, and he showed that the number of branches of a curve that extend to infinity is even. Maclaurin's work, *Geometria Organica*, founded the theory of intersections of higher plane curves. He generalized on results for special cases and on this basis concluded that an equation of the mth degree and one of the nth degree intersect in mn points.

In 1748 Euler and Cramer sought to prove this result, but neither gave a correct proof. Euler[10] relied upon an argument by analogy; realizing that his argument was not complete, he said one should apply the method to particular examples. Cramer's "proof" in his book of 1750 relied entirely on examples and was certainly not acceptable. Both men took into account points of intersection with imaginary coordinates and infinitely distant common points and noted that the number mn will be attained only if both types of points are included and if any factor, such as $ax + by$, common to both curves is excluded. However, both failed to assign the proper multiplicity to several types of intersections. In 1764 Etienne Bezout (1730–83) gave a better proof of the theorem, but this was also incomplete in the count of the multiplicity assigned to points at infinity and multiple points. The proper count of the multiplicity was settled by Georges-Henri Halphen (1844–89) in 1873.[11]

In his book of 1750 Cramer took up a paradox noted by Maclaurin in his *Geometria* concerning the number of points common to two curves. A curve of degree n is determined by $n(n + 3)/2$ points. Two nth degree curves meet in n^2 points. Now if n is 3, the curve should, by the first statement, be determined by 9 points. But since two third degree curves meet in 9 points, these 9 points do not determine a unique third degree curve. A similar paradox arises when $n = 4$. Cramer's explanation of the paradox, now referred to as his, was that the n^2 equations that determine the n^2 points of intersection are not independent. All cubics that pass through 8 fixed

10. *Mém. de l'Acad. de Berlin*, 4, 1748, 234–48 = *Opera*, (1), 26, 46–59.
11. *Bull. Soc. Math. de France*, 1, 1873, 130–48; 2, 1873, 34–52; 3, 1875, 76–92 = *Œuvres*, 1, 98–157, 171–93, 337–57.

points on a given cubic must pass through the same ninth fixed point. That is, the ninth point is dependent on the first 8. Euler gave the same explanation in 1748.[12]

In 1756 Matthieu B. Goudin (1734–1817) and Achille-Pierre Dionis du Séjour (1734–94) wrote the *Traité des courbes algébriques*. Its new features are that a curve of order (degree) n cannot have more than $n(n - 1)$ tangents with a given direction, nor more than n asymptotes. As had Maclaurin, they pointed out that an asymptote cannot cut the curve in more than $n - 2$ points.

The two best eighteenth-century compendia of results on higher plane curves are the second volume of Euler's *Introductio* (1748) and Cramer's *Lignes courbes algébriques*. The latter book has a unity of viewpoint, is excellently set forth, and contains good examples. The work was often cited, even to the point of crediting Cramer with some results that were not original with him.

4. The Beginnings of Differential Geometry

Differential geometry was initiated while analytic geometry was being extended, and the two developments were often intertwined. Interest in the theory of algebraic curves waned during the latter part of the eighteenth century, and differential geometry became more important as far as geometry was concerned. This subject is the study of those properties of curves and surfaces that vary from point to point and therefore can be grasped only with the techniques of the calculus. The term "differential geometry" was first used by Luigi Bianchi (1856–1928) in 1894.

To a large extent, differential geometry was a natural outgrowth of problems of the calculus itself. Consideration of normals to curves, points of inflection, and curvature is actually the differential geometry of plane curves. However, many new problems of the late seventeenth and early eighteenth centuries, more knowledge about the curvature of plane and space curves, envelopes of families of curves, geodesics on surfaces, the study of rays of light and of wave surfaces of light, dynamical problems of motion along curves and constraints posed by surfaces, and, above all, map-making led to questions about curves and surfaces; it became evident that the calculus must be applied.

The eighteenth- and even early nineteenth-century workers in differential geometry used geometrical arguments along with analytic ones, although the latter dominated the picture. The analysis was still crude. An infinitesimal or differential of an independent variable was regarded as an extremely small constant. No real distinction was drawn between the increment of a

12. *Mém. de l'Acad. de Berlin*, 4, 1748, 219–33, pub. 1750 = *Opera*, (1), 26, 33–45.

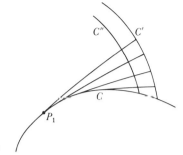

Figure 23.10

dependent variable and the differential. Differentials of higher orders were considered, but all were regarded as small and freely neglected. The mathematicians spoke of adjacent or next points on a curve as though there were no points between two adjacent points if the distance between two was sufficiently small; thus a tangent to a curve connected a point with the next one.

5. *Plane Curves*

The first applications of the calculus to curves dealt with plane curves. Some of the concepts subsequently treated by the calculus were introduced by Christian Huygens, who used purely geometrical methods. His work in this direction was motivated by his interest in light and in the design of pendulum clocks. In 1673, in the third chapter of his *Horologium Oscillatorium*, he introduced the involute of a plane curve C. Imagine a cord wrapped around C from P_1 to the right (Fig. 23.10). The end at P_1 on C is held fixed and the other unwound while the cord is kept taut. The locus C' of the free end is an involute of C. Huygens proved that at the free end the cord is perpendicular to the locus C'. Each point of the cord also describes an involute; thus C'' is also an involute, and Huygens proved that the involutes cannot touch one another. Since the cord is tangent to C at the point where it just leaves C, it follows that every orthogonal trajectory of the family of tangents to a curve is an involute of the curve.

Huygens then treated the evolute of a plane curve. Given a fixed normal at a point P on a curve, as an adjacent normal moves toward it the point of intersection of the two normals attains a limiting position on the fixed normal, which is called the center of curvature of the curve at P. The distance from the point on the curve along the fixed normal to the limiting position was shown by Huygens to be (in modern notation)

$$\frac{[1 + (dy/dx]^{3/2}}{d^2y/dx^2}.$$

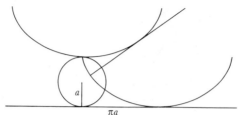

Figure 23.11

This length is the radius of the curvature of the curve at P. The locus of the centers of curvature, one on each normal, is called the evolute of the original curve. Thus the curve C above is the evolute of any one of its involutes. In this work Huygens proved that the evolute of a cycloid is a cycloid or, more precisely, the evolute of the left half of the lower cycloid in Figure 23.11 is the right half of the upper cycloid. This theorem was proved analytically by Euler in 1764.[13] The significance of the cycloid for Huygens's work on pendulum clocks is that a pendulum bob swinging along a cycloidal arc takes exactly the same time to complete swings of large and of small amplitude. For this reason the cycloid is called the tautochrone.

Newton, too, in his *Geometria Analytica* (published in 1736 though most of it was written by 1671) introduces the center of curvature as the limiting point of intersection of a normal at P with an adjacent normal. He then states that the circle with center at the center of curvature and radius equal to the radius of curvature is the circle of closest contact with the curve at P; that is, no other circle tangent to the curve at P can come between the curve and the circle of closest contact. This circle of closest contact is called the osculating circle, the term "osculating" having been used by Leibniz in a paper of 1686.[14] The curvature of this circle is the reciprocal of its radius and is the curvature of the curve at P. Newton also gave the formula for the curvature and calculated the curvature of several curves, including the cycloid. He noted that at a point of inflection a curve has zero curvature. These results duplicate those of Huygens, but probably Newton wished to show that he could use analytical methods to establish them.

In 1691 John Bernoulli took up the subject of plane curves and produced some new results on envelopes. The caustic of a family of light rays, that is, the envelope of the family, had been introduced by Tschirnhausen in 1682. In the *Acta Eruditorum* of 1692 Bernoulli obtained the equations of some caustics, for example, the caustic of rays reflected from a spherical mirror when a beam of parallel rays strikes it.[15] Then he tackled the problem posed

13. *Novi Comm. Acad. Sci. Petrop.*, 10, 1764, 179–98, pub. 1766 = *Opera*, (1), 27, 384–400.
14. *Acta Erud.*, 1686, 289–92 = *Math. Schriften*, 7, 326–29.
15. *Opera*, 1, 52–59.

to him by Fatio de Duillier, to find the envelope of the family of parabolas that are the paths of cannon balls fired from a cannon with the same initial velocity but at various angles of elevation. Bernoulli showed that the envelope is a parabola with focus at the gun. This result had already been established geometrically by Torricelli. In the *Acta Eruditorum* of 1692 and 1694[16] Leibniz gave the general method of finding the envelope of a family of curves. If the family is given by (in our notation) $f(x, y, \alpha) = 0$, where α is the parameter of the family, the method calls for eliminating α between $f = 0$ and $\partial f/\partial \alpha = 0$. L'Hospital's text, *L'Analyse des infiniment petits* (1696), helped to perfect and spread the theory of plane curves.

6. Space Curves

Clairaut launched the theory of space curves, the first major development in three-dimensional differential geometry. Alexis-Claude Clairaut (1713–65) was precocious. At the age of twelve he had already written a good work on curves. In 1731 he published *Recherche sur les courbes à double courbure*, which was written in 1729 when he was but sixteen. In this book he treated the analytics of surfaces and space curves (sec. 2). Another paper by Clairaut led to his election to the Paris Academy of Sciences at the unprecedented age of seventeen. In 1743 he produced his classic work on the shape of the earth. Here he treated in more complete form than Newton or Maclaurin the shape a rotating body such as the earth assumes under the mutual gravitational attraction of its parts. He also worked on the problem of three bodies, primarily to study the moon's motion (Chap. 21, sec. 7), and wrote several papers on it, one of which won a prize from the St. Petersburg Academy in 1750. In 1763 he published his *Théorie de la lune*. Clairaut had great personal charm and was a figure in Paris society.

In his 1731 work he treated analytically fundamental problems of curves in space. He called space curves "curves of double curvature" because, following Descartes, he considered their projections on two perpendicular planes. The space curve then partakes of the curvatures of the two curves on the planes. Geometrically he thought of a space curve as the intersection of two surfaces; analytically the equation of each surface was expressed as an equation in three variables (sec. 2). Clairaut then studied tangents to curves of double curvature. He saw that a space curve can have an infinity of normals located in a plane perpendicular to the tangent. The expressions for the arc length of a space curve and the quadrature of certain areas on surfaces are also due to him.

Though Clairaut had taken a few steps in the theory of space curves, very little had been done in this subject or in the theory of surfaces by 1750.

16. Page 311; see also *Math. Schriften*, 2, 166; 3, 967, 969.

This is reflected in Euler's *Introductio* of 1748, where he presented the differential geometry of planar and spatial figures. The first was rather complete, but the second was scanty.

The next major step in the differential geometry of space curves was taken by Euler. A great deal of his work in differential geometry was motivated by his use of curves and surfaces in mechanics. His *Mechanica* (1736),[17] written when he was twenty-nine, is a major contribution to the analytical foundation of mechanics. He gave another treatment of the subject in his *Theoria Motus Corporum Solidorum seu Rigidorum* (1765).[18] In this book he derived the currently used polar coordinate formulas for the radial and normal components of acceleration of a particle moving along a plane curve, namely,

$$a_r = \frac{d^2r}{dt^2} - r\left(\frac{d\theta}{dt}\right)^2, \qquad a_\theta = r\frac{d^2\theta}{dt^2} + 2\frac{dr}{dt}\frac{d\theta}{dt}.$$

He started to write on the theory of space curves in 1774. The particular problem that very likely motivated Euler to take up this theory was the study of the skew elastica, that is, the form assumed by an initially straight band when, under pressure at the ends, it is bent and twisted into the shape of a skew curve. To treat this problem he introduced some new concepts in 1774.[19] He then gave a full treatment of the theory of skew curves in a paper presented in 1775.[20]

Euler represented space curves by the parametric equations $x = x(s)$, $y = y(s)$, $z = z(s)$, where s is arc length, and like other writers of the eighteenth century he used spherical trigonometry to carry out the analysis. From the parametric equations he has

$$dx = p\, ds, \qquad dy = q\, ds, \qquad dz = r\, ds,$$

where p, q, and r are direction cosines, varying from point to point and, of course, with $p^2 + q^2 + r^2 = 1$. The quantity ds, the differential of the independent variable, he regarded as a constant.

To study the properties of the curve he introduced the spherical indicatrix. Around any point (x, y, z) of a curve Euler describes a sphere of radius 1. The spherical indicatrix may be defined as the locus on the unit sphere of the points whose position vectors emanating from the center O are equal to the unit tangent at (x, y, z) and the unit tangents at neighboring points. Thus the two radii in Figure 23.12 represent a unit tangent at (x, y, z) and at a neighboring point of the curve. Let ds' be the arc or the angle between

17. *Opera*, (2), 1 and 2.
18. *Opera*, (2), 3 and 4.
19. *Novi Comm. Acad. Sci. Petrop.*, 19, 1774, 340–70, pub. 1775 = *Opera*, (2), 11, 158–79.
20. *Acta Acad. Sci. Petrop.*, 1, 1782, 19–57, pub. 1786 = *Opera*, (1), 28, 348–81.

Figure 23.12

the two neighboring tangents of the two points that are ds apart along the curve. Euler's definition of the radius of curvature of the curve is

$$\frac{ds'}{ds}.$$

He then derives an analytical expression for the radius of curvature:

$$(8) \qquad \rho = \frac{ds^2}{\sqrt{(d^2x)^2 + (d^2y)^2 + (d^2z)^2}} \left[= \frac{1}{\sqrt{x''^2 + y''^2 + z''^2}} \right].$$

The plane through ds' and the center O is Euler's definition of the osculating plane at (x, y, z). John Bernoulli, who introduced the term, regarded the plane as determined by three "coincident" points. Its equation, as given by Euler, is

$$x(r\,dq - q\,dr) + y(p\,dr - r\,dp) + z(q\,dp - p\,dq) = t,$$

where t is determined by the point (x, y, z) on the curve through which the osculating plane passes. This equation is equivalent to the one we write today in vector notation as

$$(\mathbf{R} - \mathbf{r}) \cdot \mathbf{r}' \mathbf{\times} \mathbf{r}'' = 0,$$

where $\mathbf{r}(s)$ is the position vector with respect to some point in space of the point on the curve at which the osculating plane is determined, and $\mathbf{R}$ is the position vector of any point in the osculating plane. In vector form $\mathbf{r}$ is given by

$$x(s)\mathbf{i} + y(s)\mathbf{j} + z(s)\mathbf{k}$$

and $\mathbf{R}$ has the form $X\mathbf{i} + Y\mathbf{j} + Z\mathbf{k}$, where (X, Y, Z) arc the coordinates of $\mathbf{R}$.

Clairaut had introduced the idea that a space curve has two curvatures. One of these was standardized by Euler in the manner just described. The other, now called "torsion" and representing geometrically the rate at which a curve departs from a plane at a point (x, y, z), was formulated explicitly and analytically by Michel-Ange Lancret (1774–1807), an engineer and mathematician who was a student of Monge and worked in his spirit. He singled out[21] three principal directions at any point of a

21. *Mém. divers Savans*, 1, 1806, 416–54.

curve. The first is that of the tangent. "Successive" tangents lie in a plane, the osculating plane. The normal to the curve that lies in the osculating plane is the principal normal, and the perpendicular to the osculating plane, the binormal, is the third principal direction. Torsion is the rate of change of the direction of the binormal with respect to arc length; Lancret used the terminology, flexion of successive osculating planes or successive binormals.

Lancret represented a curve by

$$x = \phi(z), \qquad y = \psi(z)$$

and called $d\mu$ the angle between successive normal planes and $d\nu$ that between successive osculating ones. Then, in modern notation,

$$\frac{d\mu}{ds} = \frac{1}{\rho}, \qquad \frac{d\nu}{ds} = \frac{1}{\tau},$$

where ρ is the radius of curvature and τ is the radius of torsion.

Cauchy improved the formulation of the concepts and clarified much of the theory of space curves in his famous *Leçons sur les applications du calcul infinitésimal à la géométrie* (1826).[22] He discarded constant infinitesimals, the ds's, and straightened out the confusion between increments and differentials. He pointed out that when one writes

$$ds^2 = dx^2 + dy^2 + dz^2$$

one should mean

$$\left(\frac{ds}{dt}\right)^2 = \left(\frac{dx}{dt}\right)^2 + \left(\frac{dy}{dt}\right)^2 + \left(\frac{dz}{dt}\right)^2.$$

Cauchy preferred to write a surface as $w(x, y, z) = 0$ instead of the unsymmetric form $z = f(x, y)$, and he wrote the equation of a straight line through the point (ξ, η, ζ) as

$$\frac{\xi - x}{\cos \alpha} = \frac{\eta - y}{\cos \beta} = \frac{\zeta - z}{\cos \gamma},$$

where $\cos \alpha$, $\cos \beta$, and $\cos \gamma$ are the direction cosines of the line, though more often he used direction numbers instead of direction cosines.

Cauchy's development of the geometry of curves is practically modern. He got rid of the spherical trigonometry in the proofs, but he, too, took the arc length as the independent variable. He obtains for the direction cosines of the tangent at any point

$$\frac{dx}{ds}, \frac{dy}{ds}, \frac{dz}{ds}, \quad \text{or} \quad x'(s), y'(s), z'(s).$$

22. *Œuvres*, (2), 5.

The direction numbers of the principal normal are shown to be

$$\frac{d^2x}{ds^2}, \frac{d^2y}{ds^2}, \frac{d^2z}{ds^2}, \quad \text{or} \quad x''(s), y''(s), z''(s)$$

and the curvature k of the curve is

$$k = \frac{1}{\rho} = \sqrt{(x'')^2 + (y'')^2 + (z'')^2}.$$

Then he proves that, if the direction cosines of the tangent are $\cos \alpha$, $\cos \beta$, and $\cos \gamma$,

(9)
$$x'' = \frac{d(\cos \alpha)}{ds} = \frac{\cos \lambda}{\rho}, \qquad y'' = \frac{d(\cos \beta)}{ds} = \frac{\cos \mu}{\rho},$$

$$z'' = \frac{d(\cos \gamma)}{ds} = \frac{\cos \nu}{\rho},$$

where ρ is the radius of curvature already introduced and $\cos \lambda$, $\cos \mu$, and $\cos \nu$ are the direction cosines of a normal, which he takes to be the principal one. He shows next that

$$\frac{1}{\rho} = \frac{d\omega}{ds}$$

where ω is the angle between adjacent tangents.

He introduces the osculating plane as the plane of the tangent and principal normal. The normal to this plane is the binormal, and its direction cosines $\cos L$, $\cos M$, and $\cos N$ are given by the formulas

$$\frac{\cos L}{dy\, d^2z - dz\, d^2y} = \frac{\cos M}{dz\, d^2x - dx\, d^2z} = \frac{\cos N}{dx\, d^2y - dy\, d^2x}.$$

He can then prove that

(10)
$$\frac{d\cos L}{ds} = \frac{\cos \lambda}{\tau}, \qquad \frac{d\cos M}{ds} = \frac{\cos \mu}{\tau}, \qquad \frac{d\cos N}{ds} = \frac{\cos \nu}{\tau},$$

where $1/\tau$ is the torsion, and that the torsion equals $d\Omega/ds$, where Ω is the angle between osculating planes.

Formulas (9) and (10) are two of the three famous Serret-Frénet formulas, the third being

(11)
$$\frac{d\cos \lambda}{ds} = -\frac{\cos \alpha}{\rho} - \frac{\cos L}{\tau}, \qquad \frac{d\cos \mu}{ds} = -\frac{\cos \beta}{\rho} - \frac{\cos M}{\tau},$$

$$\frac{d\cos \nu}{ds} = -\frac{\cos \gamma}{\rho} - \frac{\cos N}{\tau},$$

where $1/\tau$ is the torsion and $1/\rho$ is the curvature. These formulas (9), (10), and (11), which give the derivatives of the direction cosines of the tangent, binormal, and normal respectively, were published by Joseph Alfred Serret (1819–85) in 1851[23] and Fréderic-Jean Frénet (1816–1900) in 1852.[24] The significance of curvature and torsion is that they are the two essential properties of space curves. Given the curvature and torsion as functions of arc length along the curve, the curve is completely determined except for position in space. This theorem is readily proven on the basis of the Serret-Frénet formulas.

7. The Theory of Surfaces

Like the theory of space curves, the theory of surfaces made a slow start. It began with the subject of geodesics on surfaces, with geodesics on the earth as the main concern. In the *Journal des Sçavans* of 1697, John Bernoulli posed the problem of finding the shortest arc between two points on a convex surface.[25] He wrote to Leibniz in 1698 to point out that the osculating plane (the plane of the osculating circle) at any point of a geodesic is perpendicular to the surface at that point. In 1698 James Bernoulli solved the geodesic problem on cylinders, cones, and surfaces of revolution. The method was a limited one, though in 1728 John Bernoulli[26] did have some success with the method and found geodesics on other kinds of surfaces.

In 1728 Euler[27] gave differential equations for geodesics on surfaces. Euler used the method he introduced in the calculus of variations (see Chap. 24, sec. 2). In 1732 Jacob Hermann[28] also found geodesics on particular surfaces.

Clairaut in 1733 and again in 1739[29] in his work on the shape of the earth treated more fully geodesics on surfaces of revolution. He proved in the 1733 paper that for any surface of revolution, the sine of the angle made by a geodesic curve and any meridian (any position of the generating curve) it crosses varies inversely as the length of the perpendicular from the point of intersection to the axis. In another paper[30] he also proved the nice theorem that if at any point M of a surface of revolution a plane be passed normal to the surface and to the plane of the meridian through M, then the curve cut out on the surface has a radius of curvature at M equal to the

23. *Jour. de Math.*, 16, 1851, 193–207.
24. *Jour. de Math.*, 17, 1852, 437–47.
25. *Opera*, 1, 204–5.
26. *Opera*, 4, 108–28.
27. *Comm. Acad. Sci. Petrop.*, 3, 1728, 110–24, pub. 1732 = *Opera*, (1), 25, 1–12.
28. *Comm. Acad. Sci. Petrop.*, 6, 1732/3, 36–67.
29. *Hist. le l'Acad. des Sci.*, Paris, 1733, 186–94, pub. 1735 and 1739, 83–96, pub. 1741.
30. *Mém. de l'Acad. des Sci.*, Paris, 1735, 117–22, pub. 1738.

length of the normal between M and the axis of revolution. Clairaut's methods were analytical, but like most of his predecessors he did not employ the ideas we now associate with the calculus of variations.

In 1760, in his *Recherches sur la courbure des surfaces*[31] Euler established the theory of surfaces. This work is Euler's most important contribution to differential geometry and a landmark in the subject. He represents a surface by $z = f(x, y)$ and introduces the now standard symbolism

$$p = \frac{\partial z}{\partial x}, \qquad q = \frac{\partial z}{\partial y}, \qquad r = \frac{\partial^2 z}{\partial x^2}, \qquad s = \frac{\partial^2 z}{\partial x \, \partial y}, \qquad t = \frac{\partial^2 z}{\partial y^2}.$$

He then says, "I begin by determining the radius of curvature of any plane section of a surface; then I apply this solution to sections which are perpendicular to the surface at any given point; and finally I compare the radii of curvature of these sections with respect to their mutual inclination, which puts us in a position to establish a proper idea of the curvature of surfaces."

He obtains first a rather complex expression for the radius of curvature of any curve made by cutting the surface with a plane. He then particularizes the result by applying it to normal sections (sections containing a normal to the surface). For normal sections the general expression for the radius of curvature simplifies a little. Next he defines the principal normal section as that normal section perpendicular to the xy-plane. (This use of "principal" is not followed today.) The radius of curvature of a normal section whose plane makes an angle ϕ with the plane of the principal normal section has the form

(12) $$\frac{1}{L + M \cos 2\phi + N \sin 2\phi},$$

where L, M, and N are functions of x and y. To obtain the greatest and least curvature of all normal sections through one point on the surface (or, when the form of the denominator in [12] is indefinite, to get the two greatest curvatures), he sets the derivative with respect to ϕ of the denominator equal to zero and (in both cases) obtains $\tan 2\phi = N/M$. There are two roots differing by $90°$ so that there are two mutually perpendicular normal planes. We call the corresponding curvatures the principal curvatures κ_1 and κ_2.

It follows from Euler's results that the curvature κ of any other normal section making an angle α with one of the sections with principal curvature is

(13) $$\kappa = \kappa_1 \cos^2 \alpha + \kappa_2 \sin^2 \alpha.$$

This result is called Euler's theorem.

The same results were obtained in 1776 in a more elegant manner by a student of Monge, Jean-Baptiste-Marie-Charles Meusnier de La Place (1754–93), who also worked in hydrostatics and in chemistry with Lavoisier.

31. *Mém. de l'Acad. de Berlin*, 16, 1760, 119–43, pub. 1767 = *Opera*, (1), 28, 1–22.

Meusnier[32] then took up the curvature of non-normal sections, for which Euler had obtained a very complex expression. Meusnier's result, still called Meusnier's theorem, is that the curvature of a plane section of a surface at a point P is the curvature of the normal section through the same tangent at P divided by the sine of the angle that the original plane makes with the tangent plane at P. There follows the beautiful result that if one considers the family of planes through the same tangent line MM' to a surface, the centers of curvature of the sections made by these planes lie on a circle in a plane perpendicular to MM' and whose diameter is the radius of curvature of the normal section. Meusnier then proves the theorem that the only surfaces for which the two principal curvatures are everywhere equal are planes or spheres. His paper was remarkably simple and fertile; it helped to make intuitive a number of results reached in the eighteenth century.

A major concern of the theory of surfaces, motivated by the needs of map-making, is the study of developable surfaces, that is, surfaces that can be flattened out on a plane without distortion. Since the sphere cannot be cut and then so flattened, the problem was to find surfaces with shapes close to that of a sphere that could be unrolled without distortion. Euler was the first to consider the subject. This work is contained in his "De Solidis Quorum Superficiem in Planum Explicare Licet."[33] Surfaces were regarded as boundaries of solids in the eighteenth century; this is why Euler speaks of solids whose surfaces may be unfolded on a plane. In this paper he introduces the parametric representation of surfaces, that is,

$$x = x(t, u), \qquad y = y(t, u), \qquad z = z(t, u),$$

and asks what conditions these functions must satisfy so that the surface may be developable on a plane. His approach is to represent t and u as rectangular coordinates in a plane and then to form a small right-angled triangle (t, u), $(t + dt, u)$, $(t, u + du)$. Because the surface is developable this triangle must be congruent to a small triangle on the surface. If we denote the partial derivatives of x, y, and z with respect to t by l, m, and n, and those with respect to u by λ, μ, ν, then the corresponding triangle on the surface is (x, y, z), $(x + l\,dt, y + m\,dt, z + n\,dt)$, and $(x + \lambda\,du, y + \mu\,du, z + \nu\,du)$. From the congruence of the two triangles Euler derives

$$(14) \quad l^2 + m^2 + n^2 = 1, \qquad \lambda^2 + \mu^2 + \nu^2 = 1, \qquad l\lambda + m\mu + n\nu = 0.$$

These are the analytic necessary and sufficient conditions for developability. The condition is equivalent to requiring that the line element on the surface equal the line element in the plane. Analytically the problem of determining whether a surface is developable is the problem of finding a parametric

32. *Mém. divers Savans*, 10, 1785, 477–85.
33. *Novi Comm. Acad. Sci. Petrop.*, 16, 1771, 3–34, pub. 1772 = *Opera*, (1), 28, 161–86.

representation $x(t, u)$, $y(t, u)$, and $z(t, u)$ such that the partial derivatives satisfy the conditions (14).

Euler then investigated the relationship between space curves and developable surfaces and showed that the family of tangents to any space curve fills out or constitutes a developable surface. He tried unsuccessfully to show that every developable surface is a ruled surface, that is, one generated by a moving straight line, and conversely. The converse is, in fact, not true.

The subject of developable surfaces was taken up independently by Gaspard Monge. With Monge geometry and analysis supported each other. He embraced simultaneously both aspects of the same problem and showed the usefulness of thinking both geometrically and analytically. Because analysis had dominated the eighteenth century, despite some analytic geometry and differential geometry by Euler and Clairaut, the effect of Monge's double view was to put geometry on at least an equal basis with analysis and then to inspire the revival of pure geometry. He is the first real innovator in synthetic geometry after Desargues.

His extensive work in descriptive geometry (which primarily serves architecture), analytic geometry, differential geometry, and ordinary and partial differential equations won the admiration and envy of Lagrange. The latter, after listening to a lecture by Monge, said to him, "You have just presented, my dear colleague, many elegant things. I wish I could have done them." Monge contributed much to physics, chemistry, metallurgy (problems of forges), and machinery. In chemistry he worked with Lavoisier. Monge saw the need for science in the development of industry and advocated industrialization as a path to the betterment of life. He was inspired by an active social concern, perhaps because he knew the hardships of humble origins. For this reason, too, he supported the French Revolution and served as Minister of the Navy and as a member of the Committee on Public Health in the succeeding governments. He designed armaments and instructed governmental staff in technical matters. His admiration for Bonaparte seduced him into following the latter in his counter-revolutionary measures.

Monge helped to organize the Ecole Polytechnique and as a professor there founded a school of geometers. He was a great teacher and a force in inspiring nineteenth-century mathematical activity. His vigorous and fertile lectures communicated enthusiasm to his students, among whom were at least a dozen of the most famous men of the early nineteenth century.

Monge created results in three-dimensional differential geometry which go far beyond Euler's. His paper of 1771, *Mémoire sur les développées, les rayons de courbure, et les différents genres d'inflexions des courbes à double courbure*, published much later,[34] was followed by his *Feuilles d'analyse appliquée à la géométrie* (1795, second ed. 1801). The *Feuilles* was as much differential geometry as

34. *Mém. divers Savans*, 10, 1785, 511–50.

analytic geometry and partial differential equations. Based on lecture notes, it offered a systematization and extension of old results, new results of some importance, and the translation of various properties of curves and surfaces into the language of partial differential equations. In pursuing the correspondence between the ideas of analysis and those of geometry, Monge recognized that a family of surfaces having a common geometric property or defined by the same method of generation should satisfy a partial differential equation (cf. Chap. 22, sec. 6).

Monge's first major work, the paper on the developable surfaces of curves of double curvature, takes up space curves and surfaces associated with them. At this time Monge did not know of Euler's work on developable surfaces. He treats space curves either as the intersections of two surfaces or by means of their projections on two perpendicular planes, given by $y = \phi(x)$ and $z = \psi(x)$. At any ordinary point there is an infinity of normals (perpendiculars to the tangent line) lying on one plane, the normal plane. In this plane there is a line he calls the (polar) axis, which is the limit of the intersection with a neighboring normal plane. As one moves along the curve the normal planes envelop a developable surface, called the polar developable. The surface is also swept out by the axes of the normal planes. The perpendicular from P to the axis in the normal plane at P is the principal normal and the foot Q of the perpendicular is the center of curvature.

To get the equation of the polar developable, he finds the equation of the normal plane and eliminates x between this equation and the partial derivative of the equation with respect to x. He also gives a rule for finding the envelope of any one-parameter family of planes that is the one we use today. The rule, which applies equally well to a one-parameter family of surfaces, is this: In our notation, he takes $F(x, y, z, \alpha) = 0$ as the family. To find where this meets "an infinitely near surface" he finds $\partial F/\partial \alpha = 0$. The curve of intersection of the two surfaces is called the characteristic curve. The equation of the envelope is found by eliminating α between $F = 0$ and $\partial F/\partial \alpha = 0$. He applies this method to the study of other developable surfaces, each of which he regards as the envelope of a one-parameter family of planes.

Monge also considered the edge of regression (*arête de rebroussement*) of a developable surface. The edge (curve) is formed by a set of lines generating the surface. The intersection of any two neighboring lines is a point and the locus of such points is the edge of regression Γ. Then the tangents to Γ are the generators, or Γ is the envelope of the family of generating lines. The edge of regression separates the developable surface into two nappes, just as a cusp separates a plane curve into two parts. Monge obtained the equations of the edge of regression. In the case of the polar developable surface of a space curve, the edge of regression is the locus of the centers of curvature of the original curve.

In 1775 Monge presented to the Académie des Sciences another paper on surfaces, particularly developable surfaces that occur in the theory of shadows and penumbras.[35] Using the definition that a developable surface can be flattened out on a plane without distortion, he argues intuitively that a developable surface is a ruled surface (but not conversely) on which two consecutive lines are concurrent or parallel and that any developable is equivalent to that formed by the tangents to a space curve. It is in this paper of 1775 that he gives a general representation of developable surfaces. Their equations are always of the form

$$z = x[F(q) - qF'(q)] + f(q) - qf'(q),$$

where $q = \partial z/\partial y$ and the partial differential equation which such surfaces, except for cylinders perpendicular to the xy-plane, always satisfy is

$$z_{xx}z_{yy} - z_{xy}^2 = 0.$$

Monge then studies ruled surfaces and gives a general representation for them. Also, he gives the third order partial differential equation which they satisfy and integrates this differential equation. He then proves that developable surfaces are a special case of ruled surfaces.

In the "Mémoire sur la théorie des déblais et des remblais" (Excavation and Fill) of 1776,[36] he considers the problem arising in building fortifications which involves the transport of earth and other materials from one place to another and seeks to do it in the most efficient way, that is, the material transported times the distance transported should be a minimum. Only part of this work is important for the differential geometry of surfaces; in fact, the results of the practical problem are not too realistic and, as Monge says, he published the paper for the sake of the geometrical results in it. Here he initiates the treatment of the subject of a family of lines depending upon two parameters or a congruence of lines. Then, following up on the work of Euler and Meusnier, he considers the family of normals to a surface S, which also forms a congruence of lines. Consider in particular the normals along a line of curvature, a line of curvature being a curve on the surface that has a principal curvature at each point on the curve. The surface normals at these points of a line of curvature form a developable surface, called a normal developable. Likewise, the surface normals along the line of curvature perpendicular to the first one form a developable surface. Since there are two families of lines of curvature on the surface, there are two families of developable surfaces, and the members of one intersect the members of the other at right angles. The intersection of any two in fact takes place along a surface normal. On each normal there are

35. *Mém. divers Savans*, 9, 1780, 382–440.
36. *Mém. de l'Acad. des Sci.*, Paris, 1781, 666–704, pub. 1784.

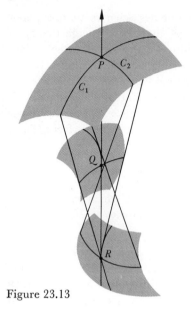

Figure 23.13

two points distant from the surface by the amount of the two principal curvatures. Each set of points determined by one of the principal curvatures, one on each of the normals to a line of curvature, lies on the edge of regression of the developable surface formed by that set of normals, so that the normals are tangent to that edge. The edges of regression of one family of developable surfaces all constitute a surface called a center surface. There are then two such center surfaces (Fig. 23.13). The envelope of each family of developable surfaces is called a focal surface.

The further details of Monge's work on families of surfaces, families which satisfy nonlinear and linear first, second, and even third order partial differential equations, are more significant for the subject of partial differential equations.

Monge tended to treat fully a number of concrete curves and surfaces through which he explained his ideas. The generalizations and the exploitation of his ideas were carried out by nineteenth-century men. Always practically oriented, Monge concluded his *Feuilles* with a projection of how his theory could be applied to architecture and in particular to the construction of large meeting halls.

Some additional contributions to the theory of surfaces were made by Charles Dupin (1784–1873), a pupil of Monge. Dupin was graduated from the Ecole Polytechnique as a naval engineer and, like Monge, constantly kept applications of geometry in mind. His text *Développements de géométrie*

(1813) was subtitled "With applications to the stability of ships, excavation and fill, fortifications, optics, and so forth," and in other writings, notably his *Applications de géométrie et de mécanique* (1822), he made many applications to geometry and to mechanics. The first few results we shall describe are in the 1813 book.

One of Dupin's contributions, which sums up and clarifies prior results of Euler and Meusnier, is called the Dupin indicatrix. Given the tangent plane of a surface at a point M, he laid off in each direction and starting from M a segment whose length is equal to the square root of the radius of curvature of the normal section of the surface in that direction. The locus of the endpoints of these segments is a conic, the indicatrix, which gives a first approximation to the form of the surface around M (because it is almost similar to a section of the surface made by a plane near M and parallel to the tangent plane at M). The lines of curvature of a surface at a point, that is, the curves on the surface through M having extreme (maximum or minimum) curvature, are the curves having as tangents at M the axes of the indicatrix.

In the three-dimensional rectangular coordinate system, the coordinate surfaces are the three families of surfaces x = const., y = const., and z = const. Suppose one has three families of surfaces, each given by an equation in x, y, and z. If each member of one family cuts the members of the other two families orthogonally, then the three families are called orthogonal. Dupin's foremost result in this subject, which is in his *Développements*, is the theorem that the three families of orthogonal surfaces cut each other along the lines of curvature (the curves of maximum or minimum normal curvature) on each surface.

Dupin also extended Monge's results on congruences of lines. If the congruence, the two-parameter family, is cut orthogonally by a family of surfaces, as is the case in optics where the lines are light rays and the surfaces are wave fronts, then the congruence is called normal. Apparently using Monge's results, though he does not refer to them, Etienne-Louis Malus (1775–1812), a French physicist, proved[37] that a normal congruence of lines emanating from one point (a homocentric set) remains such after reflection or refraction (according to the laws of optics) at a surface. In 1816[38] Dupin proved that this is true for any normal congruence after any number of reflections. Lambert A. J. Quetelet (1796–1874) then gave a proof that a normal congruence remains normal after any number of refractions.[39] The subject of congruences of lines and complexes of lines, families introduced by Malus and depending on three parameters, was pursued by many men in the nineteenth century.

37. *Jour. de l'Ecole Poly.*, Cahier 14, 1808, 1–44, 84–129.
38. *Annales de Chimie et de Physique*, 5, 1817, 85–88. Also in his *Applications* of 1822, 195–97.
39. *Correspondance mathématique et physique*, 1, 1825, 147–49.

8. *The Mapping Problem*

A great deal of the differential geometry of the eighteenth century was motivated by problems of geodesy and map-making. However, the map-making problem involves special considerations and difficulties that generated mathematical developments, notably conformal transformations or conformal mapping, running across many mathematical subjects. Map-making is of course much older than differential geometry, and even mathematical methods of mapping go far back. Of these, stereographic projection and other methods (Chap. 7, sec. 5) stem from Ptolemy, and Mercator's projection (Chap. 12, sec. 2) dates from the sixteenth century. It may have been intuitively clear prior to the eighteenth century that no true map of a sphere could be made on a plane, that is, one cannot map a sphere on a plane and still preserve lengths. If this were possible all the geometric properties would be preserved. Only developable surfaces can be so mapped and these, the eighteenth-century work revealed, are cylinders (not necessarily circular), cones, and any surface generated by the tangent lines of a space curve. Since a map of a sphere on a plane cannot preserve all the geometric properties, attention was directed toward maps that preserve angles.

In such maps, if two curves on one surface meet at an angle α and the corresponding curves on the other meet at the same angle, and if the direction of the angles is preserved, the map is said to be conformal. Stereographic projection and the Mercator projection are conformal. Conformality does not mean that two corresponding *finite* figures are similar, because the equality of the angles is a property holding at a point.

J. H. Lambert initiated a new epoch in theoretical cartography. He was the first to consider the conformal mapping of the sphere on the plane in full generality; and in a book of 1772, *Anmerkungen und Zusätze zur Entwerfung der Land-und Himmelscharten* (Notes and Additions on Designing Land Maps and Maps of the Heavens), he obtained the formulas for this mapping. Euler, too, made many contributions to the subject and actually made a map of Russia. In a paper presented to the St. Petersburg Academy in 1768,[40] Euler, by using complex functions, devised a method of representing conformal transformations from one plane to another. But he did not capitalize on it. Then, in two papers presented in 1775,[41] he showed that the sphere cannot be mapped congruently into a plane. Here again he used complex functions and treated rather general conformal representations. He also gave a full analysis of the Mercator and stereographic projections. Lagrange in 1779[42] obtained all conformal transformations of a portion of

40. *Novi Comm. Acad. Sci. Petrop.*, 14, 1769, 104–28, pub. 1770 = *Opera*, (1), 28, 99–119.
41. *Acta Acad. Sci. Petrop.*, 1, 1777, 107–32 and 133–42, pub. 1778 = *Opera*, (1), 28, 248–75 and 276–87.
42. *Nouv. Mém. de l'Acad. de Berlin*, 1779, 161–210, pub. 1781 = *Œuvres*, 4, 637–92.

the earth's surface onto a plane area that transform latitude and longitude circles into circular arcs.

Further progress in the mapping problem and in conformal mapping in particular awaited the extension of differential geometry and complex function theory.

Bibliography

Ball, W. W. Rouse: "On Newton's Classification of Cubic Curves," *Proceedings of the London Mathematical Society*, 22, 1890, 104–43.

Bernoulli, John: *Opera Omnia*, 4 vols., 1742, reprint by Georg Olms, 1968.

Berzolari, Luigi: "Allgemeine Theorie der höheren ebenen algebraischen Kurven," *Encyk. der Math. Wiss.*, B. G. Teubner, 1903–15, III C4, 313–455.

Boyer, Carl B.: *History of Analytic Geometry*, Scripta Mathematica, 1956, Chaps. 6–8.

————: A History of Mathematics, John Wiley and Sons, 1968, Chaps. 20 and 21.

Brill, A., and Max Noether: "Die Entwicklung der Theorie der algebraischen Funktionen in älterer und neuerer Zeit," *Jahres. der Deut. Math.-Verein.*, 3, 1892/3, 107–56.

Cantor, Moritz: *Vorlesungen über Geschichte der Mathematik*, B. G. Teubner, 1898 and 1924; Johnson Reprint Corp., 1965, Vol. 3, 18–35, 748–829; Vol. 4, 375–88.

Chasles, M.: *Aperçu historique sur l'origine et le développement des méthodes en géométrie* (1837), 3rd ed., Gauthier-Villars, 1889, pp. 142–252.

Coolidge, Julian L.: "The Beginnings of Analytic Geometry in Three Dimensions," *Amer. Math. Monthly*, 55, 1948, 76–86.

————: A History of Geometrical Methods, Dover (reprint), 1963, pp. 134–40, 318–46.

————: *The Mathematics of Great Amateurs*. Dover (reprint), 1963, Chap. 12.

————: *A History of Conic Sections and Quadric Surfaces*, Dover (reprint), 1968.

Euler, Leonhard: *Opera Omnia*, Orell Füssli, Series 1, Vol. 6 (1921), Vols. 26–29 (1953–56); Series 2, Vols. 3 and 4 (1948–50), Vol. 9 (1968).

Huygens, Christian: *Horologium Oscillatorium* (1673), reprint by Dawsons, 1966; also in Huygens, *Œuvres Complètes*, 18, 27–438.

Hofmann, Jos. E.: "Über Jakob Bernoullis Beiträge zur Infinitesimalmathematik," *L'Enseignement Mathématique*, (2), 2, 61–171, 1956; published separately by Institut de Mathématiques, Geneva, 1957.

Kötter, Ernst: "Die Entwickelung der synthetischen Geometrie von Monge bis auf Staudt," *Jahres. der Deut. Math.-Verein.*, 5, Part II, 1896, 1–486; also as a book, B. G. Teubner, 1901.

Lagrange, Joseph-Louis: *Œuvres*, Gauthier-Villars, 1867–69, Vol. 1, 3–20; Vol. 3, 619–92.

Lambert, J. H.: *Anmerkungen und Zusätze zur Entwerfung der Land-und Himmelscharten* (1772), Ostwald's Klassiker No. 54, Wilhelm Engelmann, Leipzig, 1896.

Loria, Gino: *Spezielle algebraische und tranzendente ebenen Kurven, Theorie und Geschichte*, 2 vols., 2nd ed., B. G. Teubner, 1910–11.

Montucla, J. F.: *Histoire des mathématiques* (1802), Albert Blanchard (reprint), 1960, Vol. 3, pp. 63–102.

Struik, D. J.: *A Source Book in Mathematics, 1200–1800*, Harvard University Press, 1969, pp. 168–78, 180–83, 263–69, 413–19.

Taton, René: *L'Œuvre scientifique de Monge*, Presses Universitaires de France, 1951, Chap. 4.

Whiteside, Derek T.: "Patterns of Mathematical Thought in the Later Seventeenth Century," *Archive for History of Exact Sciences*, 1, 1961, 179–388. See pp. 202–5 and 270–311.

————: *The Mathematical Works of Isaac Newton*, Johnson Reprint Corp., 1967, Vol. 2, 137–61. This contains Newton's *Enumeratio* in English.

24
The Calculus of Variations in the Eighteenth Century

> For since the fabric of the universe is most perfect and the
> work of a most wise Creator, nothing at all takes place in the
> universe in which some rule of maximum or minimum does
> not appear.   LEONHARD EULER

1. *The Initial Problems*

As in the areas of series and differential equations, the early work on the
calculus of variations could hardly be distinguished from the calculus proper.
But within a few years after Newton's death in 1727 it was clear that a
totally new branch of mathematics, with its own characteristic problems
and methodology, had come into being. This new subject, almost comparable
in importance with differential equations for mathematics and science,
supplied one of the grandest principles in all of mathematical physics.

To gain some preliminary notion of the nature of the calculus of variations, let us consider the problems that launched the mathematicians into
the subject. Historically the first significant problem was posed and solved
by Newton. In Book II of his *Principia*, he studied the motion of objects in
water; then, in the Scholium to Proposition 34 of the third edition, he
considered the shape that a surface of revolution moving at a constant velocity
in the direction of its axis must have if it is to offer the least resistance to the
motion. Newton assumed that the resistance of the fluid at any point on the
surface of the body is proportional to the component of the velocity normal
to the surface. In the *Principia* itself he gave only a geometrical characterization of the desired shape, but in a letter presumably written to David
Gregory in 1694 he gave his solution.

In modern form, Newton's problem is to find the minimum value of
the integral

$$J = \int_{x_1}^{x_2} \frac{y(x)[y'(x)]^3}{1 + [y'(x)]^2} \, dx$$

573

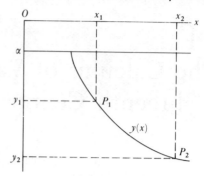

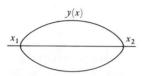

Figure 24.1   Figure 24.2

by choosing the proper function $y(x)$ for the shape of the curve that is to be rotated around the x-axis (Fig. 24.1). The peculiar feature of this problem (and of calculus of variations problems generally) is that it poses an integral whose value depends upon an unknown function $y(x)$ which appears in the integrand and which is to be determined so as to make the integral a minimum or a maximum.

Newton's solution, though it did use the idea of introducing a change in the shape of a part of the meridian arc $y(x)$, which is almost what the essential method of the calculus of variations involves, is not typical of the technique of the subject and so we shall not look into it. It may be of interest that the parametric equations of the proper $y(x)$ are

$$x = \frac{c}{p}(1 + p^2)^2, \qquad y = a + c\left(-\log p + p^2 + \frac{3}{4}p^4\right),$$

where p is the parameter. Of this work Newton says, "This proposition I conceive may be of use in the building of ships." Problems of this nature have become important in the design not only of ships but submarines and airplanes.

In the *Acta Eruditorum* of June 1696[1] John Bernoulli proposed as a challenge to other mathematicians the now famous brachistochrone problem. The problem is to determine the path down which a particle will slide from one given point to another not directly below in the shortest time. The initial velocity v_1 at P_1 (Fig. 24.2) is given; friction and air resistance are to be neglected. In modern form, this problem is to minimize the integral J which represents the time of descent where

$$J = \frac{1}{\sqrt{2g}} \int_{x_1}^{x_2} \sqrt{\frac{1 + [y'(x)]^2}{y(x) - \alpha}} \, dx.$$

1. Page 269 = *Opera*, 1, 161.

Here g is the gravitational acceleration and $\alpha = y_1 - v_1^2/2g$. Again the $y(x)$ in the integrand must be chosen so as to make J a minimum. The problem had been formulated and incorrectly solved by Galileo (1630 and 1638), who gave the arc of a circle as the answer. The correct answer is the arc of the unique cycloid joining P_1 and the second point P_2, which is concave upward; the line l on which the generating circle rolls must be at just the proper height, $y = \alpha$, above the given initial point of fall. Then there is one and only one cycloid through the two points.

Newton, Leibniz, L'Hospital, John Bernoulli, and his elder brother James found the correct solution. All these were published in the May issue of the *Acta Eruditorum* of 1697. The solutions of the two Bernoullis warrant further comment. John's method[2] was to see that the path of quickest descent is the same as the path of a ray of light in a medium with a suitably selected variable index of refraction, $n(x, y) = c/\sqrt{y - \alpha}$. The law of refraction at a sharp discontinuity (Snell's law) was known; so John broke up the medium into a finite number of layers with a sharp change in index from layer to layer and then let the number of layers go to infinity. James's method[3] was much more laborious and more geometrical. But it was also more general and was a bigger step in the direction of method for the calculus of variations.

The cycloid was well known through the work of Huygens and others on the pendulum problem (Chap. 23, sec. 5). When the Bernoulli brothers found that it was also the solution of the brachistochrone problem, they were amazed. John Bernoulli said,[4] "With justice we admire Huygens because he first discovered that a heavy particle traverses a cycloid in the same time, no matter what the starting point may be. But you will be struck with astonishment when I say that this very same cycloid, the tautochrone of Huygens, is the brachistochrone we are seeking."

Another important class of problems calls for geodesics, that is, paths of minimum length between two points on a surface. If the surface is a plane then the integral involved is

$$ J = \int_{x_1}^{x_2} \sqrt{1 + [y'(x)]^2} \, dx, $$

and the answer is of course a line segment. In the eighteenth century the geodesic problem of most interest concerned the shortest paths on the surface of the earth, whose precise shape was not known, though the mathematicians believed it was some form of ellipsoid and most likely a figure of revolution. The early work on geodesics already noted (Chap. 23, sec. 7) did not use the method of the calculus of variations but it was clear that special devices would not be powerful enough to treat the general geodesic problem.

2. *Acta Erud.*, 1697, 206–11 = *Opera*, 1, 187–93.
3. *Acta Erud.*, 1697, 211–17 = *Opera*, 2, 768–78.
4. *Opera*, 1, 187–93.

Analytically the problems thus far formulated are of the form

$$J = \int_{x_1}^{x_2} f(x, y, y') \, dx$$

and call for finding the $y(x)$ that extends from (x_1, y_1) to (x_2, y_2) and that minimizes or maximizes J. Another class of problems, called isoperimetrical problems, also entered the history of the calculus of variations at the end of the seventeenth century. The progenitor of this class of problems, of all closed plane curves with a given perimeter to find the one that bounds maximum area, may date back to pre-Greek times. There is a story that Princess Dido of the ancient Phoenician city of Tyre ran away from her home to settle on the Mediterranean coast of North Africa. There she bargained for some land and agreed to pay a fixed sum for as much land as could be encompassed by a bull's hide. The shrewd Dido cut the hide into very thin strips, tied the strips end to end and proceeded to enclose an area having the total length of these strips as its perimeter. Moreover, she chose land along the sea so that no hide would be needed along the shore. According to the legend Dido decided that the length of hide should form a semicircle—the correct shape to enclose maximum area.

Apart from the work of Zenodorus (Chap. 5, sec. 7), there was practically no work on isoperimetrical problems until the end of the seventeenth century. In a move to challenge and embarrass his brother, James Bernoulli posed a rather complicated isoperimetrical problem involving several cases in the *Acta Eruditorum* of May 1697.[5] James even offered John a prize of fifty ducats for a satisfactory solution. John gave several solutions, one of which was obtained in 1701,[6] but all were incorrect. James gave a correct solution.[7] The brothers quarreled about the correctness of each other's solutions. Actually James's method, as in the case of the brachistochrone problem, was a major step toward the general technique soon to be fashioned. In 1718 John[8] considerably improved his brother's solution.

Analytically the basic isoperimetric problem is formulated thus. The possible curves are represented parametrically by

$$x = x(t), \qquad y = y(t), \qquad t_1 \le t \le t_2$$

and because they are closed curves, $x(t_1) = x(t_2)$ and $y(t_1) = y(t_2)$. Moreover no curve must intersect itself. The problem then calls for determining the $x(t)$ and $y(t)$ such that the length

$$L = \int_{t_1}^{t_2} \sqrt{(x')^2 + (y')^2} \, dt$$

5. Page 214.
6. *Mém. de l'Acad. des. Sci., Paris*, 1706, 235 = *Opera*, 1, 424.
7. *Acta Erud.*, 1701, 213 ff. = *Opera*, 2, 897–920.
8. *Mém. de l'Acad. des Sci., Paris*, 1718, 100 ff. = *Opera*, 2, 235–69.

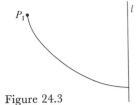

Figure 24.3

is a given constant and such that the area integral

$$J = \int_{t_1}^{t_2} (xy' - x'y)\, dt$$

is a maximum. There are two new features in this isoperimetrical problem. One, the use of the parametric representation, is incidental. The other is the presence of the auxiliary condition that L must be a constant.

Another problem that James posed in the May 1697 issue of the *Acta* was to determine the shape of the curve along which a particle slides from a given point P_1 with given initial velocity v_1 to any point of a line l (Fig. 24.3) so as to make the time of sliding from P_1 to l a minimum. This problem differs from the preceding ones in that the possible curves do not extend from one fixed point to another but from a fixed point to some line. The answer, given by James in the *Acta* of 1698 (though possessed by John in 1697 but not published by him), is an arc of a cycloid that cuts the line l at right angles. This problem was later generalized to the cases where l can be any given curve and where in place of P_1 another curve is given, so that the problem is to find the path requiring least time to slide from some point on one given curve to some point on another. This class of problems is described by the phrase "problems with variable endpoints."

2. The Early Work of Euler

In 1728 John Bernoulli proposed to Euler the problem of obtaining geodesics on surfaces by using the property that the osculating planes of geodesics cut the surface at right angles (Chap. 23, sec. 7). This problem started Euler off on the calculus of variations. He solved it in 1728.[9] In 1734 Euler generalized the brachistochrone problem to minimize quantities other than time and to take into account resisting media.[10]

Then Euler undertook to find a more general approach to problems in the subject. His method, which was a simplification of James Bernoulli's, was to replace the integral of a problem by a finite sum and to replace

9. *Comm. Acad. Sci. Petrop.*, 3, 1728, 110–24, pub. 1732 = *Opera*, (1), 25, 1–12.
10. *Comm. Acad. Sci. Petrop.*, 7, 1734/35, 135–49, pub. 1740 = *Opera*, (1), 25, 41–53.

derivatives in the integrand by difference quotients, thus making the integral a function of a finite number of ordinates of the arc $y(x)$. He then varied one or more arbitrarily selected ordinates and calculated the variation in the integral. By equating the variation of the integral to zero and by using a crude limiting process to transform the resulting difference equation, he obtained the differential equation which must be satisfied by the minimizing arc.

By the above described method applied to integrals of the form

$$(1) \qquad\qquad J = \int_{x_1}^{x_2} f(x, y, y') \, dx,$$

Euler succeeded in showing that the function $y(x)$ that minimizes or maximizes the value of J must satisfy the ordinary differential equation

$$(2) \qquad\qquad f_y - \frac{d}{dx}\,(f_{y'}) = 0.$$

This notation must be understood in the following sense. The integrand $f(x, y, y')$ is to be regarded as a function of the independent variables x, y, and y' insofar as f_y and $f_{y'}$ are concerned. However $df_{y'}/dx$ must be taken to be the derivative of $f_{y'}$ wherein $f_{y'}$ depends on x through x, y, and y'. That is, Euler's differential equation is equivalent to

$$(3) \qquad\qquad f_y - f_{y'x} - f_{y'y}y' - f_{y'y'}y'' = 0.$$

Since f is known, this equation is a second order, generally nonlinear, ordinary differential equation in $y(x)$. This famous equation, which Euler published[11] in 1736, is still the basic differential equation of the calculus of variations. It is, as we shall see more clearly later, a necessary condition that the minimizing or maximizing function $y(x)$ must satisfy.

Euler then tackled more difficult problems that involved special side conditions, as in the isoperimetric problems, but his procedure was still to solve the differential equation (3) in order to get first the possible minimizing or maximizing arcs and then to determine from the number of constants in the general solution of (2) or (3) what side conditions he could apply. One of the problems he tackled was called to his attention by Daniel Bernoulli in a letter of 1742. Bernoulli proposed to find the shape of an elastic rod subject to pressure at both ends by assuming that the square of the curvature along the curve taken by the bent rod, that is, $\int_0^L ds/R^2$, where s is arc length and R is the radius of curvature, is a minimum. This condition amounts to assuming that the potential energy stored up in the shape taken by the rod is a minimum.

11. *Comm. Acad. Sci. Petrop.*, 8, 1736, 159–90, pub. 1741 = *Opera*, (1), 25, 54–80.

The differential equation (3) is not the proper one when the integrands of the integrals to be minimized or maximized are more complicated than in (1). In the years from 1736 to 1744 Euler improved his methods and obtained the differential equations analogous to (3) for a large number of problems. These results he published in 1744 in a book, *Methodus Inveniendi Lineas Curvas Maximi Minimive Proprietate Gaudentes* (The Art of Finding Curved Lines Which Enjoy Some Maximum or Minimum Property).[12] Euler's work in his *Methodus* was cumbersome because he used geometric considerations, successive differences, and series and he changed derivatives to difference quotients and integrals to finite sums. He failed, in other words, to make most effective use of the calculus. But he ended with simple and elegant formulas applicable to a large variety of problems; and he treated a large number of examples to show the convenience and generality of his method. One example deals with minimal surfaces of revolution. Here the problem is to determine the plane curve $y = f(x)$ lying between (x_0, y_0) and (x_1, y_1) such that when revolved around the x-axis it generates the surface of least area. The integral to be minimized is

$$(4) \qquad\qquad A = \int_{x_0}^{x_1} 2\pi y \sqrt{1 + y'^2}\, dx.$$

Euler proved that the function $f(x)$ must be an arc of a catenary; the surface so generated is called a catenoid. In an appendix to his 1744 book Euler also gave a definitive solution of the elastic-rod problem referred to above. He not only deduced that the shape of the rod took the form of an elliptic integral, but also gave solutions for different kinds of end-conditions. This book brought him immediate fame and recognition as the greatest living mathematician.

With this work the calculus of variations came into existence as a new branch of mathematics. However, geometrical arguments were used extensively, and the combined analytical and geometrical arguments were not only complicated but hardly provided a systematic general method. Euler was fully aware of these limitations.

3. The Principle of Least Action

While progress in the solution of problems of the calculus of variations was being made, a new motivation for work in the subject came directly from physics. The contemporary development was the Principle of Least Action. To explain the basis for this principle we must go back a bit. Euclid had proved in his *Catoptrica* (Chap. 7, sec. 7) that light traveling from P to a mirror (Fig. 24.4) and then to Q takes the path for which $\angle 1 = \angle 2$. Then the Alexandrian Heron proved that the path PRQ, which light actually

12. *Opera*, (1), 24.

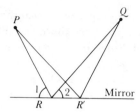

Figure 24.4

takes, is shorter than any other path, such as $PR'Q$, which it could conceivably take. Since the light takes the shortest path, if the medium on the upper side of the line RR' is homogeneous, then the light travels with constant velocity and so takes the path requiring least time. Heron applied this principle of shortest path and least time to problems of reflection from concave and convex spherical mirrors.

Basing their case on this phenomenon of reflection and on philosophic, theological, and aesthetic principles, philosophers and scientists after Greek times propounded the doctrine that nature acts in the shortest possible way or, as Olympiodorus (6th cent. A.D.) said in his *Catoptrica*, "Nature does nothing superfluous or any unnecessary work." Leonardo da Vinci said nature is economical and her economy is quantitative, and Robert Grosseteste believed that nature always acts in the mathematically shortest and best possible way. In medieval times it was commonly accepted that nature behaved in this manner.

The seventeenth-century scientists were at least receptive to this idea but, as scientists, tried to tie it to phenomena that supported it. Fermat knew that under reflection light takes the path requiring least time and, convinced that nature does indeed act simply and economically, affirmed in letters of 1657 and 1662[13] his Principle of Least Time, which states that light always takes the path requiring least time. He had doubted the correctness of the law of refraction of light (Chap. 15, sec. 4) but when he found in 1661[14] that he could deduce it from his Principle, he not only resolved his doubts about the law but felt all the more certain that his Principle was correct.

Fermat's Principle is stated mathematically in several equivalent forms. According to the law of refraction

$$\frac{\sin i}{\sin r} = \frac{v_1}{v_2},$$

where v_1 is the velocity of light in the first medium and v_2 in the second. The ratio of v_1 to v_2 is denoted by n and is called the index of refraction of the

13. *Œuvres*, 2, 354–59, 457–63.
14. *Œuvres*, 2, 457–63.

second medium relative to the first, or, if the first is a vacuum, n is called the absolute index of refraction of the nonvacuous medium. If c denotes the velocity of light in a vacuum, then the absolute index $n = c/v$ where v is the velocity of light in the medium. If the medium is variable in character from point to point, then n and v are functions of x, y, and z. Hence the time required for light to travel from a point P_1 to a point P_2 along a curve $x(\sigma)$, $y(\sigma)$, $z(\sigma)$ is given by

$$(5) \qquad J = \int_{\sigma_1}^{\sigma_2} \frac{ds}{v} = \int_{\sigma_1}^{\sigma_2} \frac{n}{c} \, ds = \frac{1}{c} \int_{\sigma_1}^{\sigma_2} n(x, y, z) \sqrt{\dot{x}^2 + \dot{y}^2 + \dot{z}^2} \, d\sigma,$$

where σ_1 is the value of σ at P_1 and σ_2 the value at P_2. Thus the Principle states that the path light actually takes in traveling from P_1 to P_2 is given by the curve which makes J a minimum.[15]

By the early eighteenth century the mathematicians had several impressive examples of the fact that nature does attempt to maximize or minimize some important quantities. Huygens, who had at first objected to Fermat's Principle, showed that it does hold for the propagation of light in media with variable indices of refraction. Even Newton's first law of motion, which states that the straight line or shortest distance is the natural motion of a body, showed nature's desire to economize. These examples suggested that there might be some more general principle. The search for such a principle was undertaken by Maupertuis.

Pierre-Louis Moreau de Maupertuis (1698–1759), while working with the theory of light in 1744, propounded his famous Principle of Least Action in a paper entitled "Accord des différentes lois de la nature qui avaient jusqu'ici paru incompatibles."[16] He started from Fermat's Principle, but in view of disagreements at that time as to whether the velocity of light was proportional to the index of refraction as Descartes and Newton believed, or inversely proportional as Fermat believed, Maupertuis abandoned least time. In fact he did not believe that it was always correct.

Action, Maupertuis said, is the integral of the product of mass, velocity, and distance traversed, and any changes in nature are such as to make the action least. Maupertuis was somewhat vague because he failed to specify the time interval over which the product of m, v, and s was to be taken and because he assigned a different meaning to action in each of the applications he made to optics and some problems of mechanics.

Though he had some physical examples to support his Principle, Maupertuis advocated it also for theological reasons. The laws of behavior of matter had to possess the perfection worthy of God's creation; and the

15. There are instances, as for example in the reflection of light from a concave mirror, where light takes the path requiring maximum time. This fact was known to Fermat and was explicitly stated by William R. Hamilton.
16. *Mém. de l'Acad. des Sci., Paris*, 1744.

least action principle seemed to satisfy this criterion because it showed that nature was economical. Maupertuis proclaimed his principle to be a universal law of nature and the first scientific proof of the existence of God. Euler, who had corresponded with Maupertuis on this subject between 1740 and 1744, agreed with Maupertuis that God must have constructed the universe in accordance with some such basic principle and that the existence of such a principle evidenced the hand of God.

In the second appendix to his 1744 book Euler formulated the Principle of Least Action as an exact dynamical theorem. He limited himself to the motion of a single particle moving along plane curves. Moreover, he supposed that the speed is dependent upon position or, in modern terms, that the force is derivable from a potential. Whereas Maupertuis wrote

$$mvs = \text{min.},$$

Euler wrote

$$\partial \int v \, ds = 0,$$

by which he meant that the rate of change of the integral for a change in the path must be zero. He also wrote that, since $ds = v \, dt$,

$$\partial \int v^2 \, dt = 0.$$

Just what Euler meant by the rate of change of the integral was still vague here even though he applied the principle correctly in specific problems by using his technique of the calculus of variations. At least he showed that Maupertuis's action was least for motions along plane curves.

Euler went further than Maupertuis in believing that all natural phenomena behave so as to maximize or minimize some function, so that the basic physical principles should be expressed to the effect that some function is maximized or minimized. In particular this should be true in dynamics, which studies the motions of bodies propelled by forces. Euler was not too far from the truth.

4. The Methodology of Lagrange

Euler's work attracted the attention of Lagrange, who began to concern himself with problems of the calculus of variations in 1750 when he was nineteen. He discarded the geometric-analytic arguments of the Bernoullis and Euler and introduced purely analytical methods. In 1755 he obtained a general procedure, systematic and uniform for a wide variety of problems, and worked on it for several years. His famous publication on the subject was the "Essai d'une nouvelle méthode pour déterminer les maxima et les minima des formules intégrales indéfinies." [17] In a letter to Euler of August

17. *Misc. Taur.*, 2, 1760/61, 173–95, pub. 1762 = *Œuvres*, 1, 333–62.

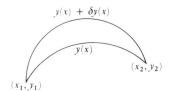

Figure 24.5 (x_1, y_1)

1755 he described the method, which he termed the method of variations, but which Euler in a paper presented to the Berlin Academy in 1756[18] named the calculus of variations.

Let us note Lagrange's method for the basic problem of the calculus of variations, namely, to minimize or maximize the integral

(6)
$$J = \int_{x_1}^{x_2} f(x, y, y') \, dx,$$

where $y(x)$ is to be determined. One of Lagrange's innovations was not to vary individual ordinates of the minimizing or maximizing curve $y(x)$ but to introduce new curves running between the endpoints (x_1, y_1) and (x_2, y_2). These new curves (Fig. 24.5) were represented by Lagrange in the form $y(x) + \delta y(x)$, the δ being a special symbol introduced by Lagrange to indicate a variation of the *entire curve* $y(x)$. The introduction of a new curve in the integrand of (6) of course changes the value of J. The increment in J, which we shall denote by ΔJ, is then

$$\Delta J = \int_{x_1}^{x_2} \{ f(x, y + \delta y, y' + \delta y') - f(x, y, y') \} \, dx.$$

Now Lagrange regards f as a function of three independent variables, but since x is not changed, the integrand can be expanded by means of Taylor's theorem applied to a function of two variables. The expansion gives first degree terms in δy and $\delta y'$, second degree terms in these increments, and so forth. Lagrange then writes

(7)
$$\Delta J = \delta J + \frac{1}{2} \delta^2 J + \frac{1}{3!} \delta^3 J + \cdots$$

where δJ indicates the integral of first degree terms in δy and $\delta y'$, $\delta^2 J$ indicates the integral of the second degree terms, and so forth. Thus

$$\delta J = \int_{x_1}^{x_2} (f_y \, \delta y + f_{y'} \, \delta y') \, dx$$

$$\delta^2 J = \int_{x_1}^{x_2} \{ f_{yy} (\delta y)^2 + 2 f_{yy'} (\delta y)(\delta y') + f_{y'y'} (\delta y')^2 \} \, dx.$$

18. "Elementa Calculi Variationum," *Novi Comm. Acad. Sci. Petrop.*, 10, 1764, 51–93, pub. 1766 = *Opera*, (1), 25, 141–76.

δJ is called the first variation of J; $\delta^2 J$, the second variation; and so forth.

Lagrange now argues that the value of δJ, since it contains the first order terms in the small variations δy and $\delta y'$, dominates the right side of (7), so that when δJ is positive or negative ΔJ will be positive or negative. But at a maximum or minimum of J, ΔJ must have the same sign, as in the case of ordinary maxima and minima of a function $f(x)$ of one variable, so that for $y(x)$ to be a maximizing function, δJ must be 0. Moreover, Lagrange says,

$$(8) \qquad \delta y' = \frac{d(\delta y)}{dx};$$

that is, the order of the operations d and δ can be interchanged. This is correct, though the reason was not clear to Lagrange's contemporaries and Euler clarified it later. [It is easily seen to be correct if we write $y + \delta y$ as $y + n(x)$, where $n(x)$ is the variation of $y(x)$. Then $\delta y = y + n(x) - y = n(x)$ and $\delta y' = y' + n'(x) - y' = n'(x)$. But $n'(x) = dn(x)/dx = d(\delta y)/dx$.] Using (8) Lagrange writes the first variation as

$$\delta J = \int_{x_1}^{x_2} \left[f_y\, \delta y + f_{y'} \frac{d}{dx}(\delta y) \right] dx.$$

Integrating the second term by parts and using the fact that δy must vanish at x_1 and at x_2 yields

$$(9) \qquad \delta J = \int_{x_1}^{x_2} \left(f_y\, \delta y - \left(\frac{d}{dx} f_{y'} \right) \delta y \right) dx.$$

Now δJ must be 0 for every variation δy. Hence Lagrange concludes that the coefficient of δy must be 0,[19] or that

$$(10) \qquad f_y - \frac{d}{dx}(f_{y'}) = 0.$$

Thus Lagrange arrived at the same ordinary differential equation for $y(x)$ that Euler had obtained. Lagrange's method of deriving (10) (except for his use of differentials) and even his notation are used today. Of course, (10) is a necessary condition on $y(x)$ but not sufficient.

In this paper of 1760/61 Lagrange also deduced for the first time end-conditions that must be satisfied by a minimizing curve for problems with variable endpoints. He found the transversality conditions that must hold at the intersections of the minimizing curve with the fixed curves or

19. The fact that the coefficient of δy must be 0 was intuitively accepted or incorrectly proven by every writer on the subject for one hundred years after Lagrange's work. Even Cauchy's proof was inadequate. The first correct proof was given by Pierre Frédéric Sarrus (1798–1861) (*Mém. divers Savans,* (2), 10, 1848, 1–128). The result is now known as the fundamental lemma of the calculus of variations.

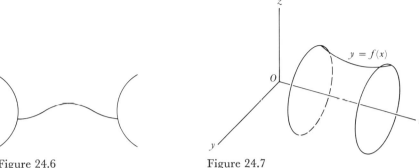

Figure 24.6   Figure 24.7

surfaces on which the endpoints of the comparison curves are allowed to vary (Fig. 24.6).

Though much more remains to be said about maximizing and minimizing integrals of the form (6), historically the next step, made by Lagrange in his 1760/61 paper and in a following one,[20] was to consider problems leading to multiple integrals. The integral to be maximized or minimized is of the form

$$(11) \qquad\qquad J = \iint f(x, y, z, p, q) \, dx \, dy$$

wherein z is a function of x and y, $p = \partial z/\partial x$, and $q = \partial z/\partial y$. The integration is over some area in the xy-plane. The problem, then, is to find the function $z(x, y)$ that maximizes or minimizes the value of J. One of the most important problems that comes under this class of double integrals is to find the surface of least area among all surfaces whose boundary is fixed in some way. Thus one might be given two closed non-intersecting curves in space and seek the surface of minimum area bounded by these two curves. As a special case of the minimal surface problem, the two curves can be circles parallel to the yz-plane (Fig. 24.7) and with centers on the x-axis. Then the possible minimal surfaces are necessarily surfaces of revolution bounded by the two curves and the problem is to find the surface of revolution of minimum area. This last problem, as we noted above, had already been solved by Euler in 1744. However, this special case of a surface of revolution can be treated by the theory applicable to the integral (11).

By a method similar to the one he had used for the simpler integral (6), Lagrange obtained the differential equation that the function $z(x, y)$ minimizing (11) must satisfy. If we use the common notation

$$\frac{\partial z}{\partial x} = p, \qquad \frac{\partial z}{\partial y} = q, \qquad \frac{\partial^2 z}{\partial x^2} = r, \qquad \frac{\partial^2 z}{\partial x \partial y} = s, \qquad \frac{\partial^2 z}{\partial y^2} = t,$$

20. *Misc. Taur.*, 4, 1766/69 = *Œuvres*, 2, 37–63.

then the equation is

(12) $$Rr + Ss + Tt = U$$

wherein R, S, T, and U are functions of x, y, z, p, and q. This nonlinear second order partial differential equation, called the equation of Monge, is not easy to solve; equations of this form had been the subject of research from the days of Euler onward (Chap. 22, sec. 7).

In the case of the minimal surface problem, the integral (11) becomes

(13) $$\iint (1 + p^2 + q^2)^{1/2} \, dx \, dy,$$

and for this special class of problems the partial differential equation (12) becomes

(14) $$(1 + q^2)r - 2pqs + (1 + p^2)t = 0.$$

This equation was given by Lagrange in his 1760/61 paper (though not quite in this form) and is a major analytical result in the theory of minimal surfaces. Geometrically, as Meusnier pointed out in a paper of 1785,[21] this partial differential equation expresses the fact that at any point on the minimizing surface the principal radii of curvature are equal and opposite or that the mean curvature, that is, the average of the principal curvatures, is zero.

In a later paper (1770)[22] Lagrange also considered single and multiple integrals in which higher derivatives than the first ones appear in the integrand. This topic has been well developed since Lagrange's time and is now standard material in the calculus of variations. However, since the principles are not basically different from the cases already considered, we shall not go into this extension of the subject. The content of Lagrange's papers on the calculus of variations is incorporated in his *Mécanique analytique*.

The calculus of variations was not well understood by the contemporaries of Lagrange and Euler. Euler explained Lagrange's method in numerous writings and used it to re-prove a number of old results. Though he realized that the calculus of variations was a new branch or technique, which he says is symbolized by the new operational symbol δ, he, like Lagrange, tried to base the logic of the calculus of variations on the ordinary calculus. Euler's idea[23] was to introduce a parameter t such that the curves of the family considered in a variations problem would vary with t, that is, for each t of some range there would be a curve $y_t(x)$. Then, says Euler, whereas $dy = (dy/dx) \, dx$, $\delta y = (dy/dt) \, dt$. Hence the variation δy is expressed by a partial differentiation with respect to t. He then formulated the technique

21. *Mém. divers Savans*, 10, 1785, 477–85.
22. *Nouv. Mém. de l'Acad. de Berlin*, 1770 = *Œuvres*, 3, 157–86.
23. *Novi Comm. Acad. Sci. Petrop.*, 16, 1771, 35–70, pub. 1772 = *Opera*, (1), 25, 208–35.

of the calculus of variations in terms of this new concept of differentiation with respect to t. His final results were, of course, the same as those already obtained.

Euler proceeded (1779)[24] to consider space curves with maximum or minimum properties and (1780) extensions of the brachistochrone problem when the applied force (which is gravity in the usual problem) operates in three dimensions or when a resisting medium is present.[25]

5. *Lagrange and Least Action*

Lagrange applied the calculus of variations to dynamics. He took over from Euler the Principle of Least Action and became the first to express the principle in concrete form, namely, that for a single particle the integral of the product of mass, velocity, and distance taken between two fixed points is a maximum or a minimum; that is, $\int mv \, ds$ must be a maximum or a minimum for the actual path taken by the particle. Alternatively, since $ds = v \, dt$, then $\int mv^2 \, dt$ must be a maximum or a minimum. The quantity mv^2 [today $(1/2)mv^2$] is called the kinetic energy; in Lagrange's day it was called living force. Lagrange also asserted that the principle is true for a collection of particles and even for extended masses, though he was not clear on the last case.

Using the Principle of Least Action and the method of the calculus of variations, Lagrange obtained his famous equations of motion. Let us consider the case where the kinetic energy is a function of x, y, and z. Then for a single particle the kinetic energy T is given by

$$(15) \qquad T = \frac{1}{2} m(\dot{x}^2 + \dot{y}^2 + \dot{z}^2).$$

Lagrange also supposed that the forces acting to cause the motion are derivable from a potential function V, which depends on x, y, and z. An additional condition, then, is that $T + V = $ const., that is, the total energy is constant. Lagrange's action is

$$(16) \qquad \int_{t_0}^{t_1} T \, dt$$

and his Principle of Least Action states that this action must be a minimum or a maximum, that is

$$(17) \qquad \delta \int_{t_0}^{t_1} T \, dt = 0.$$

24. *Mém. de l'Acad. des Sci. de St. Peters.*, 4, 1811, 18–42, pub. 1813 = *Opera*, (1), 25, 293–313.
25. *Mém. de l'Acad. des Sci. de St. Peters.*, 8, 1817/18, 17–45, pub. 1822 = *Opera*, (1), 25, 314–42.

In a minimizing or maximizing action, even though the motion takes place between two fixed points in space and the two fixed time values t_0 and t_1, the space and time variables must be varied.

By applying the method of the calculus of variations to the action integral, Lagrange derived equations analogous to Euler's equation (2), namely,

$$(18) \qquad \frac{d}{dt}\left(\frac{\partial T}{\partial x}\right) + \frac{\partial V}{\partial x} = 0$$

and the two corresponding equations with y and z. These equations are the equivalent of Newton's second law of motion.

Lagrange made the further step of introducing what are now called generalized coordinates. That is, in place of rectangular coordinates one may use polar coordinates or, in fact, any set of coordinates q_1, q_2, q_3, which are needed to fix the position of the particle (or extended mass). Then

$$x = x(q_1, q_2, q_3)$$
$$y = y(q_1, q_2, q_3)$$
$$z = z(q_1, q_2, q_3),$$

where the q_i are now functions of t. In terms of the new coordinates, T becomes a function of the q_i and $\dot{q}_i$ while V becomes a function of the q_i. Then the equations (18) become

$$(19) \qquad \frac{d}{dt}\left(\frac{\partial T}{\partial \dot{q}_i}\right) - \frac{\partial T}{\partial q_i} + \frac{\partial V}{\partial q_i} = 0, \qquad i = 1, 2, 3.$$

This is a set of 3 simultaneous second order ordinary differential equations in the q_i. They are the Euler (characteristic) equations for the action integral. If n coordinates are needed to fix the position of the moving object, for example, 2 particles require 6 coordinates, then equations (19) are replaced by n equations.[26]

These generalized coordinates need not have either geometrical or physical significance. One speaks of them today as coordinates in a configuration space and then the $q_i(t)$ are the equations of a path in configuration space. Thus Lagrange recognized that the variational principle, namely, that the action must be a minimum or maximum, can be used with any set

26. Lagrange is explicit that the number of variables in T and in V are the number required to determine the position of the mechanical system. Thus if there are N independent particles and each requires three coordinates (x_i, y_i, z_i) to describe its path in space, then $3N$ coordinates are required. In this case there will be $3N$ coordinates q_i, $3N$ equations relating the Cartesian coordinates to the q_i and $3N$ equations (19). The number of independent coordinates or the number of degrees of freedom, as physicists put it, depends on the system being treated and on the constraints in the motion.

of coordinates and the Lagrangian equations of motion (19) are invariant in form with respect to any coordinate transformation.

Lagrange's Principle, though it amounts to Newton's second law of motion, has several advantages over Newton's formulation. First of all, any convenient coordinate system is already, so to speak, built into the formulation. Secondly, it is easier to handle problems with constraints on the motion. Thirdly, instead of a series of separate differential equations, which may be numerous if many particles are involved, there is—to start with, at least—one principle from which the differential equations follow. Finally, though his principle supposes a knowledge of the kinetic and potential energies of a problem, it does not require a knowledge of the forces acting. With his principle Lagrange deduced major laws of mechanics and solved new problems, though it was not broad enough to include all the problems mechanics deals with. His work on the action principle is fully treated in his *Mécanique analytique*. He also started the movement to deduce the laws of other branches of physics from variational principles that would be the analogues of least action. He himself gave a variational principle for a broad class of hydrodynamical problems. This subject will be resumed in our study of the nineteenth-century work.

From the mathematical standpoint Lagrange's work on least action gave major importance to the calculus of variations. In particular Lagrange had derived the Euler equations for an integral whose integrand contains one independent variable but several dependent variables and their derivatives. This is an extension of the original calculus of variations problem, which contains only one dependent variable and its derivative. In this extended case the Euler equations are a system of second order ordinary differential equations in the q_i.

6. *The Second Variation*

The Euler differential equation, as Euler and Lagrange realized, is only a necessary condition for the solution to furnish a maximum or a minimum. They used the differential equation to find the solution and then decided on intuitive or physical grounds whether it furnished a maximum or a minimum. The role of the Euler equation is entirely analogous to the condition $f'(x) = 0$ in the ordinary calculus. A value of x that maximizes or minimizes $y = f(x)$ must satisfy $f'(x) = 0$, but the converse is not necessarily true.

The question of what additional conditions a solution of the Euler equation must satisfy to actually furnish a maximum or a minimum value of an integral depending on $y(x)$ was tackled unsuccessfully by Laplace in 1782. It was then taken up by Legendre in 1786.[27] Guided by the fact that

27. *Hist. de l'Acad. des Sci., Paris*, 1786, 7–37, pub. 1788.

in the ordinary calculus the sign of $f''(x)$ at a value of x for which $f'(x) = 0$ determines whether $f(x)$ has a maximum or a minimum, Legendre considered the second variation $\delta^2 J$, recast its form, and concluded that J is a maximum for the curve $y(x)$ which satisfies Euler's equation and passes through (x_0, y_0) and (x_1, y_1) provided that $f_{y'y'} \leq 0$ at each x along $y(x)$. Likewise, J is a minimum for a $y(x)$ satisfying the first two conditions provided that $f_{y'y'} \geq 0$ at each x along $y(x)$. Legendre then extended this result to more general integrals than (6). However, Legendre realized in 1787 that the condition on $f_{y'y'}$ was just a necessary condition on $y(x)$ in order that it be a maximizing or minimizing curve. The problem of finding sufficient conditions that a curve $y(x)$ maximize or minimize an integral such as (6) was not solved in the eighteenth century.

Bibliography

Bernoulli, James: *Opera*, 2 vols., 1744, reprint by Birkhaüser, 1968.

Bernoulli, John: *Opera Omnia*, 4 vols., 1742, Georg Olms (reprint), 1968.

Bliss, Gilbert A.: *The Calculus of Variations*, Open Court, 1925.

————: "The Evolution of Problems in the Calculus of Variations," *Amer. Math. Monthly*, 43, 1936, 598–609.

Cantor, Moritz: *Vorlesungen über Geschichte der Mathematik*, B. G. Teubner, 1898 and 1924, Vol. 3, Chap. 117, and Vol. 4, 1066–74.

Caratheodory, C.: Introduction to Series (1), Vol. 24 of Euler's *Opera Omnia*, viii–lxii, Orell Füssli, 1952. Also in C. Caratheodory: *Gesammelte mathematische Schriften*, C. H. Beck, 1957, Vol. 5, pp. 107–74.

Darboux, Gaston: *Leçons sur la théorie générale des surfaces*, 2nd ed., Gauthier-Villars, 1914, Vol. 1, Book III, Chaps. 1–2.

Euler, Leonhard: *Opera Omnia*, (1), Vols. 24–25, Orell Füssli, 1952.

Hofmann, Joseph E.: "Über Jakob Bernoullis Beiträge zur Infinitesimalmathematik," *L'Enseignement Mathématique*, (2), 2, 1956, 61–171; published separately by Institut de mathématiques, Geneva, 1957.

Huke, Aline: *An Historical and Critical Study of the Fundamental Lemma in the Calculus of Variations*, University of Chicago Contributions to the Calculus of Variations, University of Chicago Press, Vol, 1, 1930, pp. 45–160.

Lagrange, Joseph-Louis: *Œuvres de Lagrange*, Gauthier-Villars, 1867–69, relevant papers in Vols. 1–3.

————: *Mécanique analytique*, 2 vols., 4th ed., Gauthier-Villars, 1889.

Lecat, Maurice: *Bibliographie du calcul des variations depuis les origines jusqu'à 1850*, Gand, 1916.

Montucla, J. F.: *Histoire des mathématiques*, 1802, Albert Blanchard (reprint), 1960, Vol. 3, 643–58.

Porter, Thomas Isaac: "A History of the Classical Isoperimetric Problem," *University of Chicago Contributions to the Calculus of Variations*, University of Chicago Press, 1933, Vol. 2, pp. 475–517.

Smith, David E.: *A Source Book in Mathematics*, Dover (reprint), 1959, pp. 644–55.

Struik, D. J.: *A Source Book in Mathematics, 1200–1800*, Harvard University Press, 1969, pp. 391–413.

Todhunter, Isaac: *A History of the Calculus of Variations during the Nineteenth Century*, 1861, Chelsea (reprint), 1962.

Woodhouse, Robert: *A History of the Calculus of Variations in the Eighteenth Century*, 1810, Chelsea (reprint), 1964.

25
Algebra in the Eighteenth Century

I present higher analysis as it was in its childhood but you are bringing it to man's estate.

JOHN BERNOULLI, IN A LETTER TO EULER

1. *Status of the Number System*

Though algebra and analysis were hardly distinguished from each other in the eighteenth century because the significance of the limit concept was still obscure, it is desirable from our modern point of view to separate the two fields of activity. In the seventeenth century, algebra was a major center of interest; in the eighteenth century it became subordinate to analysis, and except in the theory of numbers the motivation for work in it came largely from analysis.

Since the basis of algebra is the number system, let us note the status of this subject. By 1700 all of the familiar members of the system—whole numbers, fractions, irrationals, and negative and complex numbers—were known. However, opposition to the newer types of numbers was expressed throughout the century. Typical are the objections of the English mathematician Baron Francis Masères (1731–1824), a Fellow of Clare College in Cambridge and a member of the Royal Society. Masères, who did write acceptable papers in mathematics and a substantial treatise on the theory of life insurance, published in 1759 his *Dissertation on the Use of the Negative Sign in Algebra.* He shows how to avoid negative numbers (except to indicate the subtraction of a larger quantity from a smaller one), and especially negative roots, by carefully segregating the types of quadratic equations so that those with negative roots are considered separately; and, of course, the negative roots are to be rejected. He does the same with cubics. Then he says of negative roots,

> ...they serve only, as far as I am able to judge, to puzzle the whole doctrine of equations, and to render obscure and mysterious things that are in their own nature exceeding plain and simple.... It were to be wished therefore that negative roots had never been admitted into

algebra or were again discarded from it: for if this were done, there is good reason to imagine, the objections which many learned and ingenious men now make to algebraic computations, as being obscure and perplexed with almost unintelligible notions, would be thereby removed; it being certain that Algebra, or universal arithmetic, is, in its own nature, a science no less simple, clear, and capable of demonstration, than geometry.

Certainly negative numbers were not really well understood until modern times. Euler, in the latter half of the eighteenth century, still believed that negative numbers were greater than ∞. He also argued that $(-1) \cdot (-1) = +1$ because the product must be $+1$ or -1 and since $1 \cdot (-1) = -1$, then $(-1) \cdot (-1) = +1$. Carnot, the noted French geometer, thought the use of negative numbers led to erroneous conclusions. As late as 1831 Augustus De Morgan (1806–71), professor of mathematics at University College, London, and a famous mathematical logician and contributor to algebra, in his *On the Study and Difficulties of Mathematics*, said,"The imaginary expression $\sqrt{-a}$ and the negative expression $-b$ have this resemblance, that either of them occurring as the solution of a problem indicates some inconsistency or absurdity. As far as real meaning is concerned, both are equally imaginary, since $0 - a$ is as inconceivable as $\sqrt{-a}$."

De Morgan illustrated this by means of a problem. A father is 56; his son is 29. When will the father be twice as old as the son? He solves $56 + x = 2(29 + x)$ and obtains $x = -2$. Thus the result, he says, is absurd. But, he continues, if we change x to $-x$ and solve $56 - x = 2(29 - x)$, we get $x = 2$. He concludes that we phrased the original problem wrongly and thus were led to the unacceptable negative answer. De Morgan insisted that it was absurd to consider numbers less than zero.

Though nothing was done in the eighteenth century to clarify the concept of irrational numbers, some progress was made in this subject. In 1737 Euler showed, substantially, that e and e^2 are irrational and Lambert showed that π is irrational (Chap. 20, sec. 6). The work on the irrationality of π was motivated largely by the desire to solve the problem of squaring the circle. Legendre's conjecture that π may not be a root of an algebraic equation with rational coefficients led to a distinction between types of irrationals. Any root, real or complex, of any algebraic (polynomial) equation with rational coefficients is called an algebraic number. Thus the roots of

$$a_0 x^n + a_1 x^{n-1} + \cdots + a_{n-1} x + a_n = 0,$$

where the a_i are rational numbers, are called algebraic numbers. Consequently, every rational number and some irrationals are algebraic numbers, for any rational number c is a root of $x - c = 0$, and $\sqrt{2}$ is a root of $x^2 - 2 = 0$. Those numbers that are not algebraic are called transcendental because, as Euler put it, "they transcend the power of algebraic methods."

This distinction between algebraic and transcendental numbers was recognized by Euler at least as early as 1744. He conjectured that the logarithm to a rational base of a rational number must be either rational or transcendental. However, no number was known to be transcendental in the eighteenth century and the problem of showing that there are transcendental numbers remained open.

Complex numbers were more of a bane to the eighteenth-century mathematicians. These numbers were practically ignored from their introduction by Cardan until about 1700. Then (Chap. 19, sec. 3) complex numbers were used to integrate by the method of partial fractions, which was followed by the lengthy controversy about complex numbers and the logarithms of negative and complex numbers. Despite his correct resolution of the problem of the logarithms of complex numbers, neither Euler nor the other mathematicians were clear about these numbers.

Euler tried to understand what complex numbers really are, and in his *Vollständige Anleitung zur Algebra* (Complete Introduction to Algebra), which first appeared in Russian in 1768–69 and in German in 1770 and is the best algebra text of the eighteenth century, says,

> Because all conceivable numbers are either greater than zero or less than 0 or equal to 0, then it is clear that the square roots of negative numbers cannot be included among the possible numbers [real numbers]. Consequently we must say that these are impossible numbers. And this circumstance leads us to the concept of such numbers, which by their nature are impossible, and ordinarily are called imaginary or fancied numbers, because they exist only in the imagination.

Euler made mistakes with complex numbers. In this *Algebra* he writes $\sqrt{-1} \cdot \sqrt{-4} = \sqrt{4} = 2$ because $\sqrt{a}\,\sqrt{b} = \sqrt{ab}$. He also gives $i^i = 0.2078795763$ but misses other values of this quantity. He gave this number originally in a letter to Goldbach of 1746 and also in his article of 1749 on the controversy between Leibniz and John Bernoulli (Chap. 19, sec. 3). Though he calls complex numbers impossible numbers, Euler says they are of use. The use he has in mind occurs when we tackle problems about which we do not know whether there is or is not an answer. Thus if we are asked to separate 12 into two parts whose product is 40, we would find that the parts are $6 + \sqrt{-4}$ and $6 - \sqrt{-4}$. Thereby, he says, we recognize that the problem cannot be solved.

Some positive steps beyond Euler's correct conclusion on the logarithms of complex numbers were made, but their influence on the eighteenth century was limited. In his *Algebra* (1685, Chaps. 66–69), John Wallis showed how to represent geometrically the complex roots of a quadratic equation with real coefficients. Wallis said, in effect, that complex numbers are no more absurd than negative numbers; and since the latter can be represented on a direct line, it should be possible to represent complex numbers in the plane. He started with an axis on which real numbers were marked out in relation to an origin; a distance from the origin along this axis represented the real part of the root, the distance being measured in the positive or negative direction of the axis according as this part of the root was positive or negative. From the point on the real axis thus determined, a line was drawn perpendicular to the axis of reals whose length represented the number which, multiplied by $\sqrt{-1}$, gave the imaginary part of the root, the line being drawn in one direction or the opposite according as the number was positive or negative. (He failed to introduce the y-axis itself as the axis of imaginaries.) Wallis then gave a geometrical construction for the roots of $ax^2 + bx + c = 0$ when the roots are real and when they are complex. His work was correct, but it was not a useful representation of $x + iy$ for other purposes. There were other attempts to represent complex numbers geometrically in the eighteenth century, but they were not broadly useful. Nor did geometrical representation at this time make complex numbers more acceptable.

Early in the century most mathematicians believed that different roots of complex numbers would introduce different types or orders of complex numbers and that there might be ideal roots whose nature they could not specify but which might somehow be calculated. But d'Alembert, in his prize work *Réflexions sur la cause générale des vents* (1747), affirmed that every expression built up from complex numbers by means of algebraic operations (in which he included raising to an arbitrary power), is a complex number of the form $A + B\sqrt{-1}$. The one difficulty he had in proving this assertion was the case of $(a + bi)^{g+hi}$. His demonstration of this fact had to be mended by Euler, Lagrange, and others. In the *Encyclopédie*, d'Alembert maintained a discrete silence on complex numbers.

Throughout the eighteenth century complex numbers were used effectively enough for mathematicians to acquire some confidence in them (Chap. 19, sec. 3; Chap. 27, sec. 2). Where used in intermediate stages of mathematical arguments, the results proved to be correct; this fact had telling effect. Yet there were doubts as to the validity of the arguments and often even of the results.

In 1799 Gauss gave his first proof of the fundamental theorem of algebra, and since this necessarily depended on the recognition of complex numbers, Gauss solidified the position of these numbers. The nineteenth century then plunged forward boldly with complex functions. But even long after this theory was developed and employed in hydrodynamics, the Cambridge University professors preserved "an invincible repulsion to the objectionable $\sqrt{-1}$, cumbrous devices being adopted to avoid its occurrence or use wherever possible."

The general attitude toward complex numbers even as late as 1831 may be learned from De Morgan's book *On the Study and Difficulties of Mathematics*. He says that this book contains nothing that could not be found in the best works then in use at Oxford and Cambridge. Turning to complex numbers, he states,

> We have shown the symbol $\sqrt{-a}$ to be void of meaning, or rather self-contradictory and absurd. Nevertheless, by means of such symbols, a part of algebra is established which is of great utility. It depends upon the fact, which must be verified by experience, that the common rules of algebra may be applied to these expressions [complex numbers] without leading to any false results. An appeal to experience of this nature appears to be contrary to the first principles laid down at the beginning of this work. We cannot deny that it is so in reality, but it must be recollected that this is but a small and isolated part of an immense subject, to all other branches of which these principles apply in their fullest extent.

The "principles" he refers to are that mathematical truths should be derived by deductive reasoning from axioms.

He then compares negative roots and complex roots.

> There is, then, this distinct difference between the negative and the imaginary result. When the answer to a problem is negative, by changing the sign of x in the equation which produced that result, we may either discover an error in the method of forming that equation or show that the question of the problem is too limited, and may be extended so as to admit of a satisfactory answer. When the answer to a problem is imaginary this is not the case.... We are not advocates for stopping the progress of the student by entering fully into all the arguments for and against such questions, as the use of negative quantities, etc., which he could not understand, and which are inconclusive on both sides; but he might be made aware that a difficulty does exist, the nature of which might be pointed out to him, and he might then, by the consideration of a sufficient number of examples, treated separately, acquire confidence in the results to which the rules lead.

By the time De Morgan wrote the above lines, the concepts of complex numbers and complex functions were well on the way to clarification. But the diffusion of the new knowledge was slow. Certainly throughout the eighteenth and the first half of the nineteenth centuries the meaning of complex numbers was debated hotly. All the arguments of John Bernoulli, d'Alembert, and Euler were continually rehashed. Even twentieth-century textbooks on trigonometry supplemented presentations that employed complex numbers by proofs not involving $\sqrt{-1}$.

We should note here another point whose significance is almost in inverse proportion to the conciseness with which it can be stated: In the eighteenth century no one worried about the logic of the real or complex number systems. What Euclid had done in Book V of the *Elements* to establish the properties of incommensurable magnitudes was disregarded. That this exposition had been tied to geometry, whereas by now arithmetic and algebra were independent of geometry, accounts in part for this disregard. Moreover this logical development, even if suitably modified to free it of geometry, could not establish the logical foundations of negative and complex numbers; and this fact too may have caused mathematicians to desist from any attempt to found the number system rigorously. Finally, the century was concerned primarily to use mathematics in science, and since the rules of operation were intuitively secure at least for real numbers no one really worried about the foundations. Typical is the statement of d'Alembert in his article on negative numbers in the *Encyclopédie*. The article is not at all clear and d'Alembert concludes that "the algebraic rules of operation with negative numbers are generally admitted by everyone and acknowledged as exact, whatever idea we may have about these quantities." The various types of numbers, never properly introduced to the world, nevertheless gained a firmer place in the eighteenth-century mathematical community.

2. *The Theory of Equations*

One of the investigations continued from the seventeenth century with hardly a break was the solution of polynomial equations. The subject is fundamental in mathematics and so the interest in obtaining better methods of solving equations of any degree, getting better methods of approximating roots of equations, and completing the theory—in particular by proving that every nth degree polynomial equation has n roots—was natural. In addition, the use of the method of partial fractions for integration raised the question of whether any polynomial with real coefficients can be decomposed into a product of linear factors or a product of linear and quadratic factors with real coefficients to avoid the use of complex numbers.

As we saw in Chapter 19 (sec. 4), Leibniz did not believe that every polynomial with real coefficients could be decomposed into linear and quadratic factors with real coefficients. Euler took the correct position. In a letter to Nicholas Bernoulli (1687–1759) of October 1, 1742, Euler affirmed without proof that a polynomial of arbitrary degree with real coefficients could be so expressed. Nicholas did not believe the assertion to be correct and gave the example of

$$x^4 - 4x^3 + 2x^2 + 4x + 4$$

with the zeroes $1 + \sqrt{2 + \sqrt{-3}}$, $1 - \sqrt{2 + \sqrt{-3}}$, $1 + \sqrt{2 - \sqrt{-3}}$, $1 - \sqrt{2 - \sqrt{-3}}$, which he said contradicts Euler's assertion. On December 15, 1742, Euler, writing to Goldbach (Fuss, Vol. 1, 169–71), pointed out that complex roots occur in conjugate pairs, so that the product of $x - (a + b\sqrt{-1})$ and $x - (a - b\sqrt{-1})$, wherein $a + b\sqrt{-1}$ and $a - b\sqrt{-1}$ are a conjugate pair, gives a quadratic expression with real coefficients. Euler then showed that this was true for Bernoulli's example. But Goldbach, too, rejected the idea that every polynomial with real coefficients can be factored into real factors and gave the example $x^4 + 72x - 20$. Euler then showed Goldbach that the latter had made a mistake and that he (Euler) had proved his theorem for polynomials up to the sixth degree. However, Goldbach was not convinced, because Euler did not succeed in giving a general proof of his assertion.

The kernel of the problem of factoring a real polynomial into linear and quadratic factors with real coefficients was to prove that every such polynomial had at least one real or complex root. The proof of this fact, called the fundamental theorem of algebra, became a major goal.

Proofs offered by d'Alembert and Euler were incomplete. In 1772[1] Lagrange, in a long and detailed argument, "completed" Euler's proof. But Lagrange, like Euler and his contemporaries, applied freely the ordinary properties of numbers to what were supposedly the roots without establishing that the roots must at worst be complex numbers. Since the nature of the roots was unknown, the proof was actually incomplete.

The first substantial proof of the fundamental theorem, though not rigorous by modern standards, was given by Gauss in his doctoral thesis of 1799 at Helmstädt.[2] He criticized the work of d'Alembert, Euler, and Lagrange and then gave his own proof. Gauss's method was not to calculate a root but to demonstrate its existence. He points out that the complex roots $a + ib$ of $P(x + iy) = 0$ correspond to points (a, b) of the plane, and if $P(x + iy) = u(x, y) + iv(x, y)$ then (a, b) must be an intersection of the curves $u = 0$ and $v = 0$. By a qualitative study of the curves he shows that a continuous arc of one joins the points of two distinct regions separated by the other. Then the curve $u = 0$ must cut the curve $v = 0$. The argument was highly original. However, he depended on the graphs of these curves, which were somewhat complicated, to show that they must cross. In this same paper Gauss proved that the nth degree polynomial can be expressed as a product of linear and quadratic factors with real coefficients.

Gauss gave three more proofs of the theorem. In the second proof[3] he dispensed with geometrical arguments. In this proof he also showed that the product of the differences of each two roots (which we, following Sylvester, call the discriminant) can be expressed as a linear combination of the polynomial and its derivative, so that a necessary and sufficient condition that the polynomial and its derivative have a common root is that the discriminant

1. *Nouv. Mém. de l'Acad. de Berlin*, 1772, 222 ff. = *Œuvres*, 3, 479–516.
2. *Werke*, 3, 1–30; also reproduced in Euler, *Opera*, (1), 6, 151–69.
3. *Comm. Soc. Gott.*, 3, 1814/15, 107–42 = *Werke*, 3, 33–56.

vanish. However, this second proof assumed that a polynomial cannot change signs at two different values of x without vanishing in between. The proof of this fact lay beyond the rigor of the time.

The third proof[4] used in effect what we now call the Cauchy integral theorem (Chap. 27, sec. 4).[5] The fourth proof[6] is a variation of the first as far as method is concerned. However, in this proof Gauss uses complex numbers more freely because, he says, they are now common knowledge. It is worth noting that the theorem was, in many proofs, not proved in all generality. Gauss's first three proofs and later proofs by Cauchy, Jacobi, and Abel assumed that the (literal) coefficients represented real numbers, whereas the full theorem includes the case of complex coefficients. Gauss's fourth proof did allow the coefficients of the polynomial to be complex numbers.

Gauss's approach to the fundamental theorem of algebra inaugurated a new approach to the entire question of mathematical existence. The Greeks had wisely recognized that the existence of mathematical entities must be established before theorems about them can be entertained. Their criterion of existence was constructibility. In the more explicit formal work of the succeeding centuries, existence was established by actually obtaining or exhibiting the quantity in question. For example, the existence of the solutions of a quadratic equation is established by exhibiting quantities that satisfy the equation. But in the case of equations of degree higher than four, this method is not available. Of course a proof of existence such as Gauss's may be of no help at all in computing the object whose existence is being established.

While the work that ultimately showed that every polynomial equation with real coefficients has at least one root was under way, the mathematicians also pushed hard to solve equations of degree higher than four by algebraic processes. Leibniz and his friend Tschirnhausen were the first to make serious efforts. Leibniz[7] reconsidered the irreducible case of the third degree equation and convinced himself that one could not avoid the use of complex numbers to solve this type. He then tackled the solution of the fifth degree equation but without success. Tschirnhausen[8] thought he had solved this problem by transforming the given equation into a new one by means of a transform $y = P(x)$, where $P(x)$ is a suitable fourth degree polynomial.

4. *Comm. Soc. Gott.*, 3, 1816 = *Werke*, 3, 59–64.

5. For a discussion of Gauss's third proof see M. Bocher, "Gauss's Third Proof of the Fundamental Theorem of Algebra," *Amer. Math. Soc., Bull.*, 1, 1895, 205–9. A translation of the third proof can be found in H. Meschkowski, *Ways of Thought of Great Mathematicians*, Holden-Day, 1964.

6. *Abhand. der Ges. der Wiss. zu Gött.*, 4, 1848/50, 3–34 = *Werke*, 3, 73–102.

7. *Der Briefwechsel von Gottfried Wilhelm Leibniz mit Mathematikern*, Georg Olms (reprint), 1961, Vol. 1, 547–64.

8. *Acta Erud.*, 2, 1683, 204–07.

This transformation eliminated all but the x^5 and constant terms of the equation. But Leibniz showed that to determine the coefficients of $P(x)$ one had to solve equations of degree higher than five, and so the method was useless.

For a while the problem of solving the nth degree equation centered on the special case $x^n - 1 = 0$, called the binomial equation. Cotes and De Moivre showed, through the use of complex numbers, that the solution of this problem amounts to the division of the circumference of a circle into n equal parts. To obtain the roots by radicals (trigonometric solutions are not necessarily algebraic) it is sufficient to consider the case of n an odd prime, because if $n = pm$, where p is a prime, then one can consider $(x^m)^p - 1 = 0$. If this equation can be solved for x^m, then $x^m - A$ can be considered where A is any of the roots of the preceding equation. Alexandre-Théophile Vandermonde (1735–96) affirmed in a paper of 1771[9] that every equation of the form $x^n - 1 = 0$, where n is a prime, is solvable by radicals. However, Vandermonde merely verified that this is so for values of n up to 11. The decisive work on binomial equations was done by Gauss (Chap. 31, sec. 2).

The major effort toward solving equations of degree higher than four concentrated on the general equation, and toward this end some subsidiary work on symmetric functions proved important. The expression $x_1 x_2 + x_2 x_3 + x_3 x_1$ is a symmetric function of x_1, x_2, and x_3 because replacement throughout of any x_i by an x_j and x_j by x_i leaves the entire expression unaltered. The interest in symmetric functions arose when the seventeenth-century algebraists noted and Newton proved that the various sums of the products of the roots of a polynomial equation can be expressed in terms of the coefficients. For example, for $n = 3$, the sum of the products taken two at a time,

$$a_1 a_2 + a_2 a_3 + a_3 a_1$$

is an elementary symmetric function; and if the equation is written as

$$x^3 - c_1 x^2 + c_2 x - c_3 = 0,$$

the sum equals c_2. The progress made by Vandermonde in the 1771 paper was to show that any symmetric function of the roots can be expressed in terms of the coefficients of the equation.

The outstanding work of the eighteenth century on the problem of the solution of equations by radicals, after efforts by many men including Euler,[10] was by Vandermonde in the 1771 paper and Lagrange, in his massive paper "Réflexions sur la résolution algébrique des équations."[11] Vandermonde's

9. *Hist. de l'Acad. des Sci., Paris*, 1771, 365–416, pub. 1774.

10. *Comm. Acad. Sci. Petrop.*, 6, 1732/3, 216–31, pub. 1738 = *Opera*, (1), 6, 1–19 and *Novi Comm. Acad. Sci. Petrop.*, 9, 1762/63, 70–98, pub. 1764 = *Opera*, (1), 6, 170–96.

11. *Nouv. Mém. de l'Acad. de Berlin*, 1770, 134–215, pub. 1772 and 1771, 138–254, pub. 1773 = *Œuvres*, 3, 205–421.

ideas are similar but not so extensive nor so clear. We shall therefore present Lagrange's version. Lagrange set himself the task of analyzing the methods of solving third and fourth degree equations, to see why these worked and to see what clue these methods might furnish for solving higher-degree equations.

For the third degree equation

(1) $$x^3 + nx + p = 0$$

Lagrange noted that if one introduces the transformation (Chap. 13, sec. 4)

(2) $$x = y - (n/3y),$$

one obtains the auxiliary equation

(3) $$y^6 + py^3 - n^3/27 = 0.$$

This equation is also known as the reduced equation because it is quadratic in y^3 and with $r = y^3$ becomes

(4) $$r^2 + pr - n^3/27 = 0.$$

Now we see that we can calculate the roots r_1 and r_2 of this equation in terms of the coefficients of the original one. But to go back to y from r we must introduce cube roots or solve

$$y^3 - r = 0.$$

Then if we let ω be the particular cube root of unity $(-1 + \sqrt{-3})/2$, the values of y are

$$\sqrt[3]{r_1}, \; \omega\sqrt[3]{r_1}, \; \omega^2\sqrt[3]{r_1}, \; \sqrt[3]{r_2}, \; \omega\sqrt[3]{r_2}, \; \omega^2\sqrt[3]{r_2}$$

and the distinct solutions of (1) are

$$x_1 = \sqrt[3]{r_1} + \sqrt[3]{r_2}, \qquad x_2 = \omega\sqrt[3]{r_1} + \omega^2\sqrt[3]{r_2}, \qquad x_3 = \omega^2\sqrt[3]{r_1} + \omega\sqrt[3]{r_2}.$$

Thus the solutions of the original equation are obtained in terms of those of the reduced equation.

Lagrange showed that the different procedures utilized by his predecessors all amount to the above method. Then he pointed out that we should turn our attention not to x as a function of the y-values but to y as a function of x because it is the reduced equation that permits us to solve at all, and the secret must lie in the relation that expresses the solutions of the reduced equation in terms of the solutions of the proposed equation.

Lagrange noted that each of the y-values can be written (since $1 + \omega + \omega^2 = 0$) in the form

(5) $$y = \frac{1}{3}(x_1 + \omega x_2 + \omega^2 x_3)$$

when x_1, x_2, and x_3 are taken in particular orders. Examination of this expression enables us to perceive two properties of the reduced equation in y. First, the roots x_1, x_2, and x_3 in the expression for y are not one fixed choice for x_1, x_2, and x_3, and the expression is, so to speak, ambiguous. Thus any of the three x-values can be x_1, any of the other two x_2, etc. But there are 3! permutations of the x's. Hence there are six values for y, and y should satisfy a sixth degree equation. Thus the degree of the reduced equation is determined by the number of permutations of the roots in the proposed equation.

In the second place, the relation (5) also shows why one can reduce the sixth degree reduced equation to the second degree equation. For among the six permutations, three (including the identity) come from interchanging all the x_i and three from interchanging just two and keeping one fixed. But then in view of the value of ω, the six values of y that result are related by

$$(6) \qquad\qquad y_1 = \omega^2 y_2 = \omega y_3; \qquad y_4 = \omega^2 y_5 = \omega y_6$$

and by cubing,

$$y_1^3 = y_2^3 = y_3^3, \qquad y_4^3 = y_5^3 = y_6^3.$$

Another way of putting this result is that the function

$$(x_1 + \omega x_2 + \omega^2 x_3)^3$$

can take on only two values under all six permutations of x_1, x_2, and x_3, and this is why the equation that y satisfies proves to be a quadratic in y^3. Moreover the coefficients of the sixth degree equation that y satisfies are rational functions of the coefficients of the original cubic.

In the case of the general fourth degree equation in x, Lagrange considers

$$y = x_1 x_2 + x_3 x_4.$$

This function of the four roots takes on only three distinct values under all 24 possible permutations of the four roots. Hence there should be a third degree equation that y satisfies and the coefficients of this equation should be rational functions of those of the original equation. These statements do apply to the fourth degree equation.

Lagrange then takes up the general nth degree equation

$$(7) \qquad\qquad x^n + a_1 x^{n-1} + \cdots + a_{n-1} x + a_n = 0.$$

The coefficients of this equation are assumed to be independent; that is, there must be no relation holding among the a_i. Then the roots must also be independent, because if there should be a relation among the roots one could show that this must be true of the coefficients (since, essentially, the coefficients are symmetric functions of the roots). Thus the n roots of the general equation must be considered as independent variables, and each function of them is a function of independent variables.

To understand Lagrange's plan of solution let us consider first

$$x^2 + bx + c = 0.$$

We know two functions of the roots, namely, $x_1 + x_2$ and $x_1 x_2$. These are symmetric functions, that is, they do not change when the roles of the roots are exchanged. When a function does not change under a permutation performed on its variables, one says that the function admits the permutation. Thus the function $x_1 + x_2$ admits the permutation of x_1 and x_2 but the function $x_1 - x_2$ does not.

Lagrange then demonstrated two important propositions. If a function $\phi(x_1, x_2, \ldots, x_n)$ of the roots of the general equation of degree n admits all the permutations of the x_i that another function $\psi(x_1, x_2, \ldots, x_n)$ admits (and possibly other permutations that ψ does not admit), the function ϕ can be expressed rationally in terms of ψ and the coefficients of the general equation (7). Thus the function x_1 of the quadratic admits all the permutations which $x_1 - x_2$ admits, (there is just one, the identity); then

$$x_1 = \frac{-b + (x_2 - x_1)}{2}.$$

Lagrange's proof of this proposition also shows how to express ϕ as a rational function of ψ.

Lagrange's second proposition states: If a function $\phi(x_1, x_2, \ldots, x_n)$ of the roots of the general equation does not admit all the permutations admitted by a function $\psi(x_1, x_2, \ldots, x_n)$ but takes on under the permutations that ψ admits r different values, then ϕ is a root of an equation of degree r whose coefficients are rational functions of ψ and of the given general equation of degree n. This equation of degree r can be constructed. Thus $x_1 - x_2$ does not admit all the permutations $x_1 + x_2$ admits but takes on the two values $x_1 - x_2$ and $x_2 - x_1$. Then $x_1 - x_2$ is a root of a second degree equation whose coefficients are rational functions of $x_1 + x_2$ and of b and c. As a matter of fact, $x_1 - x_2$ is a root of

$$t^2 - (b^2 - 4c) = 0$$

because $b^2 - 4ac = (x_1 - x_2)^2$. With the value of this root, namely $\sqrt{b^2 - 4c}$, we can find x_1 by means of the preceding equation for x_1.

Likewise for the equation $x^3 + px + q = 0$ the expression (the ϕ)

$$(x_1 + \omega x_2 + \omega^2 x_3)^3,$$

where $\omega = (-1 + i\sqrt{3})/2$, takes on two values under the six possible permutations of the roots, whereas $x_1 + x_2 + x_3$ (the ψ) admits all six permutations. If the two values are denoted by A and B, one can show that A and B are roots of a second degree equation whose coefficients are rational in p and q (since $x_1 + x_2 + x_3 = 0$). If we solve the second degree equation and the roots are A and B, we can then find x_1, x_2, and x_3 from

$$x_1 + x_2 + x_3 = 0$$
$$x_1 + \omega x_2 + \omega^2 x_3 = \sqrt[3]{A}$$
$$x_1 + \omega^2 x_2 + \omega x_3 = \sqrt[3]{B}.$$

For the fourth degree equation, Lagrange started with the function

(8) $x_1 x_2 + x_3 x_4,$

which takes on three different values under the 24 possible permutations of the roots, whereas $x_1 + x_2 + x_3 + x_4$ admits all 24. Hence (8) is a root of a third degree equation whose coefficients are rational functions of those of the original equation. And in fact the auxiliary or reduced equation for the general quartic is of the third degree.

For the nth degree equation with general coefficients, Lagrange's idea was to start with a symmetric function ϕ_0 of the roots which admits all $n!$ permutations of the roots. Such a function, he points out, could be $x_1 + x_2 + \cdots + x_n$. Then he would choose a function ϕ_1 which admits only some of the substitutions. Suppose ϕ_1 takes on under the $n!$ permutations, say, r different values. ϕ_1 will then be a root of an equation of degree r whose coefficients are rational functions of ϕ_0 and the coefficients of the given general equation. The equation of degree r can be constructed. Further, if ϕ_0 is taken to be one of the symmetric functions that relates roots and coefficients, then the coefficients of the equation of degree r are known entirely in terms of coefficients of the given general equation. If the equation of degree r can be solved algebraically, then ϕ_1 will be known in terms of the coefficients of the original equation. Then a function ϕ_2 is chosen that admits only some of the permutations ϕ_1 admits. ϕ_2 may take on, say, s different values under the substitutions that ϕ_1 admits. Then ϕ_2 will be a root of an equation of degree s whose coefficients are rational functions of ϕ_1 and of the coefficients of the given general equation. The coefficients of this sth degree equation will be known if the rth degree equation which has ϕ_1 as one of its roots can be solved. If the sth degree equation can be solved algebraically, then ϕ_2 will be known in terms of the coefficients of the original equation.

One continues thus to ϕ_3, ϕ_4, $\ldots$ until the final function which is chosen to be x_1. If then the equations of degree r, s, $\ldots$ can be solved algebraically, one will know x_1 in terms of the coefficients of the given general equation. The other roots x_2, x_3, $\ldots$, x_n come out of the same process. The equations of degree r, s, $\ldots$, are today called resolvent equations.[12]

---

12. Lagrange used the word "resolvent" for the special forms of the ϕ_i functions and not for the equations which the ϕ_i satisfy. Thus for the cubic equation, $x_1 + \omega x_2 + \omega^2 x_3$ is one form of Lagrange's resolvent.

Lagrange's method worked for the general second, third, and fourth degree equations. He tried to solve the fifth degree equation in this way but found the work so difficult that he abandoned it. Whereas for the cubic he had but to solve an equation of the second degree, for the quintic he had to solve a sixth degree equation. Lagrange sought in vain to find a resolving function (in his sense of the term) that would satisfy an equation of degree less than five. However his work did not give any criterion for picking ϕ_i's that would satisfy algebraically solvable equations. Also his method applied only to the general equation because his two basic propositions assume that the roots are independent.

Lagrange was drawn to the conclusion that the solution of the general higher-degree equation (for $n > 4$) by algebraic operations was likely to be impossible. (For special equations of higher degree he offered little.) He decided that either the problem was beyond human capacities or the nature of the expressions for the roots must be different from all those thus far known. Gauss, too, in his *Disquisitiones* of 1801, declared that the problem could not be solved.

Lagrange's method, despite its lack of success, does give insight into the reason for the successes when $n \leq 4$ and failures when $n > 4$; this insight was capitalized on by Abel and Galois (Chap. 31). Moreover, Lagrange's idea that one must consider the number of values that a rational function takes on when its variables are permuted led to the theory of permutation or substitution groups. In fact, he had in effect the theorem that the order of a subgroup (the number of elements) must be a divisor of the order of the group. In Lagrange's work, which preceded any work on group theory, this result takes the form that the number r of values that ϕ_1 takes on is a divisor of $n!$.

Influenced by Lagrange, Paolo Ruffini (1765–1822), a mathematician, doctor, politician, and an ardent disciple of Lagrange, made several attempts during the years 1799 to 1813 to prove that the general equation of degree higher than the fourth could not be solved algebraically. In his *Teoria generale delle equazioni*,[13] Ruffini succeeded in proving by Lagrange's own method that no resolving function (in Lagrange's sense) existed that would satisfy an equation of degree less than five. In fact he demonstrated that no rational function of n elements existed that took on 3 or 4 values under permutations of the n elements when $n > 4$. Then he boldly undertook to prove, in his *Riflessioni intorno alla soluzione delle equazioni algebraiche generali*,[14] that the algebraic solution of the general equation of degree $n > 4$ was impossible. This effort was not conclusive, though at first Ruffini believed it was correct. Ruffini used but did not prove the auxiliary theorem, now known as Abel's theorem, that if an equation is solvable by radicals, the

13. 1799 = *Opere Mat.*, 1, 1–324.
14. 1813 = *Opere Mat.*, 2, 155–268.

expressions for the roots can be given such a form that the radicals in them are rational functions with rational coefficients of the roots of the given equation and the roots of unity.

3. *Determinants and Elimination Theory*

The study of a system of linear equations which we write today as

$$(9) \qquad x_i = \sum_{j=1}^{n} a_{ij} y_j, \qquad i = 1, 2, \ldots, m$$

wherein the x_i are known and the y_j are the unknowns, was initiated before 1678 by Leibniz. In 1693 Leibniz[15] used a systematic set of indices for the coefficients of a system of three liner equations in two unknowns x and y. He eliminated the two unknowns from the system of three linear equations and obtained a determinant, now called the resultant of the system. The vanishing of this determinant expresses the fact that there is an x and y satisfying all three equations.

The solution of simultaneous linear equations in two, three, and four unknowns by the method of determinants was created by Maclaurin, probably in 1729, and published in his posthumous *Treatise of Algebra* (1748). Though not as good in notation, his rule is the one we use today and which Cramer published in his *Introduction à l'analyse des lignes courbes algébriques* (1750). Cramer gave the rule in connection with determining the coefficients of the general conic, $A + By + Cx + Dy^2 + Exy + x^2 = 0$, passing through five given points. His determinants were, as at present, the sum of the products formed by taking one and only one element from each row and column, with the sign of each product determined by the number of derangements of the elements from a standard order, the sign being positive if this number was even and negative if odd. In 1764, Bezout[16] systematized the process of determining the signs of the terms of a determinant. Given n homogeneous linear equations in n unknowns, Bezout showed that the vanishing of the determinant of the coefficients (the vanishing of the resultant) is the condition that non-zero solutions exist.

Vandermonde[17] was the first to give a connected and logical exposition of the theory of determinants as such,—that is, apart from the solution of linear equations,—though he, too, did apply them to the solution of linear equations. He also gave a rule for expanding a determinant by using second order minors and their complementary minors. In the sense that he concentrated on determinants, he is the founder of the theory.

15. *Math. Schriften*, 2, 229, 238–40, 245.
16. *Hist. de l'Acad. des Sci. Paris*, 1764, 288–388.
17. *Mém. de l'Acad. des Sci.*, Paris, 1772, 516–32, pub. 1776.

In a paper of 1772, "Recherches sur le calcul intégral et sur le système du monde,"[18] Laplace, who referred to Cramer's and Bezout's work, proved some of Vandermonde's rules and generalized his method of expanding determinants by using a set of minors of r rows and the complementary minors. This method is still known by his name.[19]

The condition that, say, a set of three linear nonhomogeneous equations in two unknowns have a common solution, the vanishing of the resultant, also expresses the result of eliminating x and y from the three equations. But the problem of elimination extends in other directions. Given two polynomials

$$f = a_0 x^m + \cdots + a_n$$
$$g = b_0 x^n + \cdots + b_n,$$

one can ask for the condition that $f = 0$ and $g = 0$ have a common root. Since the condition involves the fact that at least one value of x satisfying $f = 0$ also satisfies $g = 0$, the substitution of that value of x obtained from $f = 0$ in $g = 0$ should result in a condition on the a_i and b_i. This condition, or eliminant, or resultant, was first investigated by Newton. In his *Arithmetica Universalis* he gave rules for eliminating x from two equations, which could be of degrees two to four.

Euler, in Chapter 19 of the second volume of his *Introductio*, gives two methods of elimination. The second is the forerunner of Bezout's multiplier method and was better described by Euler in a paper of 1764.[20] Bezout's method proved to be the most widely accepted one, and we shall therefore examine it. In his *Cours de mathématique* (1764–69), he considers two equations of degree n,

$$\text{(10)} \quad \begin{aligned} f(x) &= a_n x^n + a_{n-1} x^{n-1} + \cdots + a_0 = 0 \\ \phi(x) &= b_n x^n + b_{n-1} x^{n-1} + \cdots + b_0 = 0. \end{aligned}$$

One multiplies f by b_n and ϕ by a_n and subtracts; next one multiplies f by $b_n x + b_{n-1}$ and ϕ by $a_n x + a_{n-1}$ and subtracts; then f is multiplied by $b_n x^2 + b_{n-1} x + b_{n-2}$ and ϕ by $a_n x^2 + a_{n-1} x + a_{n-2}$ and again one subtracts; etc. Each of the equations so obtained is of degree $n - 1$ in x. One can regard this set of equations as a system of n linear homogeneous equations in the unknowns x^{n-1}, x^{n-2}, ..., 1. The resultant of this system of linear equations, which is the determinant of the coefficients of the unknowns, is the resultant of the original two, $f = 0$ and $\phi = 0$. Bezout also gave a method of finding the resultant when the two equations are not of the same degree.[21]

18. *Mém. de l'Acad. des Sci.*, Paris, 1772, 267–376, pub. 1776 = *Œuvres*, 8, 365–406.

19. See M. Bocher, *Introduction to Higher Algebra*, Dover (reprint), 1964, p. 26.

20. *Mém. de l'Acad. de Berlin*, 20, 1764, 91–104, pub. 1766 = *Opera*, (1), 6, 197–211.

21. An exposition can be found in W. S. Burnside and A. W. Panton, *The Theory of Equations*, Dover (reprint), 1960, Vol. 2, p. 76.

Elimination theory was also applied to two equations, $f(x, y) = 0$ and $g(x, y) = 0$, of degree higher than 1. The motivation was to establish the number of common solutions of the two equations or, geometrically, to find the number of intersections of the curves corresponding to the equations. The outstanding method for eliminating one unknown from $f(x, y) = 0$ and $g(x, y) = 0$, sketched first by Bezout in a paper of 1764, was presented by him in his *Théorie générale des équations algébriques* (1779). Bezout's idea was that by multiplying $f(x, y)$ and $g(x, y)$ by suitable polynomials, $F(x)$ and $G(x)$ respectively, he could form

$$(11) \qquad R(y) = F(x)f(x, y) + G(x)g(x, y).$$

Moreover he sought an F and G such that the degree of $R(y)$ would be as small as possible.

The question of the degree of the eliminant was also answered by Bezout in his *Théorie* (and independently by Euler in the 1764 paper). Both gave the answer as mn, the product of the two degrees of f and g, and both proved this theorem by reducing the problem to one of elimination from an auxiliary set of linear equations. This product is the number of points of intersection of the two algebraic curves. Jacobi[22] and Minding[23] also gave Bezout's method of elimination for the two equations, but neither mentions Bezout. It could be that Bezout's work was not known to them.

4. *The Theory of Numbers*

The theory of numbers in the eighteenth century remained a series of disconnected results. The most important works in the subject were Euler's *Anleitung zur Algebra* (German edition, 1770) and Legendre's *Essai sur la théorie des nombres* (1798). The second edition appeared in 1808 under the title *Théorie des nombres*, and an expanded third edition in two volumes appeared in 1830. The problems and results to be described are a small sample of what was done.

In 1736 Euler demonstrated[24] the minor theorem of Fermat, namely, if p is a prime and a is prime to p, $a^p - a$ is divisible by p. Many proofs of this same theorem were given by other men in the eighteenth and nineteenth centuries. In 1760 Euler generalized this theorem[25] by introducing the ϕ function or totient of n; $\phi(n)$ is the number of integers $< n$ and prime to n so that if n is a prime, $\phi(n)$ is $n - 1$. [The notation $\phi(n)$ was introduced by Gauss.] Then Euler proved that if a is relatively prime to n,

$$a^{\phi(n)} - 1$$

is divisible by n.

22. *Jour. für Math.*, 15, 1836, 101–24 = *Gesam. Werke*, 3, 297–320.
23. *Jour. für Math.*, 22, 1841, 178–83.
24. *Comm. Acad. Sci. Petrop.*, 8, 1736, 141–46, pub. 1741 = *Opera*, (1), 2, 33–37.
25. *Novi Comm. Acad. Sci. Petrop.*, 8, 1760/1, 74–104, pub. 1763 = *Opera*, (1), 3, 531–55.

As for the famous Fermat conjecture on $x^n + y^n = z^n$, Euler proved[26] this was correct for $n = 3$ and $n = 4$; the case $n = 4$ had been proven by Frénicle de Bessy. This work of Euler had to be completed by Lagrange, Legendre, and Gauss. Legendre then proved[27] the conjecture for $n = 5$. As we shall see, the history of the efforts to prove Fermat's conjecture is extensive.

Fermat had also conjectured (Chap. 13, sec. 7) that the formula

$$2^{2^n} + 1$$

yielded primes for an unspecified set of values of n. This is true for $n = 0, 1, 2, 3,$ and 4. However, Euler showed in 1732[28] that for $n = 5$ this number is not prime, one of the factors being 641. In fact the formula is now known not to yield primes for many other values of n, whereas no other value beyond 4 has been found for which the formula does yield primes. However, the formula is of interest in that it reappears in the work of Gauss on the constructibility of regular polygons (Chap. 31, sec. 2).

A topic with many subdivisions concerns the decomposition of integers of various types into other classes of integers. Fermat had affirmed that each positive integer is the sum of at most four squares (repetition of a square as in $8 = 4 + 4$ is permitted if it is counted the number of times it appears). Over a period of forty years, Euler kept trying to prove this theorem and gave partial results.[29] Using some of Euler's work, Lagrange[30] proved the theorem. Neither Euler nor Lagrange obtained the number of representations.

Euler, in the 1754/5 paper just referred to and in another paper in the same journal[31] proved Fermat's assertion that every prime of the form $4n + 1$ is decomposable uniquely into a sum of two squares. However, Euler's proof did not follow the method of descent that Fermat had sketched for this theorem. In another paper[32] Euler also proved that every divisor of the sum of two relatively prime squares is the sum of two squares.

Edward Waring (1734–98) stated, in his *Meditationes Algebraicae* (1770), the theorem known now as "Waring's theorem," that every integer is either a cube or the sum of at most nine cubes; also every integer is either a fourth power or the sum of at most 19 fourth powers. He conjectured also that every positive integer can be expressed as the sum of at most r kth powers, the r depending upon k. These theorems were not proven by him.[33]

26. *Algebra*, Part II, Second Section, 509–16 = *Opera*, (1), 1, 484–89 (for $n = 3$); and *Comm. Acad. Sci. Petrop.*, 10, 1738, 125–46, pub. 1747 = *Opera*, (1), 2, 38–59 (for $n = 4$).
27. *Mém. de l'Acad. des Sci.*, Paris, 6, 1823, 1–60, pub. 1827.
28. *Comm. Acad. Sci Petrop.*, 6, 1732/3, 103–07 = *Opera*, (1), 2, 1–5.
29. *Novi Comm. Acad. Sci. Petrop.*, 5, 1754/5, 13–58, pub. 1760 = *Opera*, (1), 2, 338–72.
30. *Nouv. Mém. de l'Acad. de Berlin*, 1, 1770, 123–33, pub. 1772 = *Œuvres*, 3, 189–201.
31. *Novi*, 1754/5, 3–13, pub. 1760 = *Opera*, (1), 2, 328–37.
32. *Novi Comm. Acad. Sci. Petrop.*, 4, 1752/53, 3–40, pub. 1758 = *Opera*, (1) 2, 295–327.
33. The general theorem was proven by David Hilbert (*Math. Ann.*, 67, 1909, 281–300).

In a letter to Euler of June 7, 1742, Christian Goldbach, a Prussian envoy to Russia, stated without proof that every even integer is the sum of two primes and every odd integer is either a prime or a sum of three primes. The first part of the assertion is now known as Goldbach's hypothesis and is still an unsolved problem. The second assertion actually follows from the first, because if n is odd subtract any prime p from it. Then $n - p$ is even.

Among somewhat more specialized results dealing with the decomposition of numbers is Euler's proof that $x^4 - y^4$ and $x^4 + y^4$ cannot be squares.[34] Euler and Lagrange proved many of Fermat's assertions to the effect that certain primes can be expressed in particular ways. For example Euler proved[35] that a prime of the form $3n + 1$ can be expressed uniquely in the form $x^2 + 3y^2$.

Amicable and perfect numbers continued to interest the mathematicians. Euler[36] gave 62 pairs of amicable numbers, including the three pairs already known. Two of his pairs were incorrect. He also proved in a posthumously published paper[37] the converse of Euclid's theorem: Every even perfect number is of the form $2^{p-1}(2^p - 1)$ where the second factor is prime.

John Wilson (1741–93), who had been a prize student of mathematics at Cambridge but became a lawyer and judge, stated a theorem still named after him: For every prime p, the quantity $(p - 1)! + 1$ is divisible by p; in addition, if the quantity is divisible by q, then q is a prime. Waring published the statement in his *Meditationes Algebraicae* and Lagrange proved it in 1773.[38]

The problem of solving the equation $x^2 - Ay^2 = 1$ in integers has already been discussed (Chap. 13, sec. 7). Euler, in a paper of 1732/33, erroneously called it Pell's equation, and the name has stuck. He became interested in this equation because he needed its solutions to solve $ax^2 + bx + c = y^2$ in integers; he wrote several papers on this last theme. In 1759 Euler gave a method of solving the Pellian equation by expressing $\sqrt{A}$ as a continued fraction.[39] Euler's idea was that the values x and y which satisfy the equation are such that the ratios x/y are convergents (in the sense of continued fractions) to $\sqrt{A}$. He failed to prove that his method always gives solutions and that all its solutions are given by the continued fraction

34. *Comm. Acad. Sci. Petrop.*, 10, 1738, 125–46, pub. 1747 = *Opera*, (1), 2, 38–59; also in *Algebra* (1770), Part II, Ch. 13, arts. 202–8 = *Opera*, (1), 1, 436–43.
35. *Novi Comm. Acad. Sci. Petrop.*, 8, 1760/1, 105–28, pub. 1763 = *Opera*, (1), 2, 556–75.
36. "De numeris amicabilibus," *Opuscula varii argumenti*, 2, 1750, 23–107 = *Opera*, (1), 2, 86–162.
37. "De numeris amicabilibus," *Comm. Arith.*, 2, 1849, 627–36 = *Opera postuma*, 1, 1862, 85–100 = *Opera*, (1), 5, 353–65.
38. *Nouv. Mém. de l'Acad. de Berlin*, 2, 1771, 125 ff., pub. 1773 = *Œuvres*, 3, 425–38.
39. *Novi Comm. Acad. Sci. Petrop.*, 11, 1765, 28–66, pub. 1767 = *Opera*, (1), 3, 73–111.

development of $\sqrt{A}$. The existence of solutions of Pell's equation was shown in 1766 by Lagrange[40] and then more simply in later papers.[41]

Fermat had asserted that he could determine when the more general equation $x^2 - Ay^2 = B$ was solvable in integers and that he could solve it when solvable. The equation was solved by Lagrange in the two papers just mentioned.

The problem of giving all integral solutions of the general equation

$$ax^2 + 2bxy + cy^2 + 2dx + 2ey + f = 0,$$

where the coefficients are integers, was also tackled. Euler gave incomplete classes of solutions; then Lagrange[42] gave the complete solution. In the next volume of the *Mémoires*[43] he gave a simpler proof.

The most original and perhaps the most consequential discovery of the eighteenth century in the theory of numbers is the law of quadratic reciprocity. It uses the notion of quadratic remainders or residues. In the language introduced by Euler in the 1754/55 paper and adopted by Gauss, if there exists an x such that $x^2 - p$ is divisible by q, then p is said to be a quadratic residue of q; if there is no such x, p is said to be a quadratic nonresidue of q. Legendre (1808) invented a symbol that is now used to represent either state of affairs. The symbol is (p/q) and means the following: For any number p and any prime q,

$$(p/q) = \begin{cases} 1 & \text{if } p \text{ is a quadratic residue of } q \\ -1 & \text{if } p \text{ is a quadratic nonresidue of } q. \end{cases}$$

It is also understood that $(p/q) = 0$ if p divides evenly into q.

The law of quadratic reciprocity, in symbolic form, states that if p and q are distinct odd primes, then

$$(p/q)(q/p) = (-1)^{(p-1)(q-1)/4}.$$

This means that if the exponent of (-1) is even, p is a quadratic residue of q and q is a quadratic residue of p or neither is a quadratic residue of the other. When the exponent is odd, which occurs when p and q are of the form $4k + 3$, one prime will be a quadratic residue of the other but not the second of the first.

The history of this law is detailed. Euler in a paper of 1783[44] gave four theorems and a fifth summarizing theorem that states the law of quadratic reciprocity very clearly. However, he did not prove these theorems. The

40. *Misc. Taur.*, 4, 1766/69, 19 ff. = *Œuvres*, 1, 671–731.

41. *Mém. de l'Acad. de Berlin*, 23, 1767, 165–310, pub. 1769, and 24, 1768, 181–256, pub. 1770 = *Œuvres*, 2, 377–535 and 655–726; also in Lagrange's additions to his translation of Euler's *Algebra*; see the bibliography.

42. *Mém. de l'Acad. de Berlin*, 23, 1767, 165–310, pub. 1769 = *Œuvres*, 2, 377–535.

43. 24, 1768, 181–256, pub. 1770 = *Œuvres*, 2, 655–726.

44. *Opuscula Analytica*, 1, 1783, 64–84 = *Opera*, (1), 3, 497–512.

work on this paper dates from 1772 and incorporates even earlier work. Kronecker observed in 1875[45] that the statement of the law is actually contained in a much earlier paper by Euler.[46] However, Euler's "proof" was based on calculations. In 1785 Legendre announced the law independently, though he cites another paper by Euler, in the same volume of the *Opuscula*, in his own paper on the subject. His proof[47] was incomplete. In his *Théorie des nombres*[48] he again stated the law and gave another proof. However, this one, too, was incomplete, because he assumed that there are an infinite number of primes in certain arithmetic progressions. The desire to find what is behind the law and to derive its many implications has been a key theme in investigations since 1800 and has led to important discoveries, some of which we shall consider in a later chapter.

The eighteenth-century work on the theory of numbers closes with Legendre's classic *Théorie* of 1798. Though it contains a number of interesting results, in this domain as in others (such as the subject of elliptic integrals), Legendre made no great innovations. One could reproach him for presenting a collection of propositions from which general conceptions could have been extracted but were not. This was done by his successors.

Bibliography

Cajori, Florian: "Historical note on the Graphical Representation of Imaginaries Before the Time of Wessel," *Amer. Math. Monthly*, 19, 1912, 167–71.

Cantor, Moritz: *Vorlesungen über Geschichte der Mathematik*, B. G. Teubner, 1898 and 1924; Johnson Reprint Corp., 1965, Vol. 3, Chap. 107; Vol. 4, pp. 153–98.

Dickson, Leonard E.: *History of the Theory of Numbers*, 3 vols., Chelsea (reprint), 1951.

————: "Fermat's Last Theorem," *Annals of Math.*, 18, 1917, 161–87.

Euler, Leonhard: *Opera Omnia*, (1), Vols. 1–5, Orell Füssli, 1911–44.

————: *Vollständige Anleitung zur Algebra* (1770) = *Opera Omnia*, (1), 1.

Fuss, Paul H. von, ed.: *Correspondance mathématique et physique de quelques célèbres géomètres du XVIIIème siècle*, 2 vols. (1843), Johnson Reprint Corp., 1967.

Gauss, Carl Friedrich: *Werke*, Königliche Gesellschaft der Wissenschaften zu Göttingen, 1876, Vol. 3, pp. 3–121.

Gerhardt, C. I.: *Der Briefwechsel von Gottfried Wilhelm Leibniz mit Mathematikern*, Mayer und Müller, 1899; Georg Olms (reprint), 1962.

Heath, Thomas L.: *Diophantus of Alexandria*, 1910, Dover (reprint), 1964, pp. 267–380.

Jones, P. S.: "Complex Numbers: An Example of Recurring Themes in the

45. *Werke*, 2, 3–10.
46. *Comm. Acad. Sci. Petrop.*, 14, 1744/46, 151–81, pub. 1751 = *Opera*, (1), 2, 194–222.
47. *Hist. de l'Acad. des Sci.*, Paris, 1785, 465–559, pub. 1788.
48. 1798, 214–26; 2nd ed., 1808, 198–207.

Development of Mathematics," *The Mathematics Teacher*, 47, 1954, 106–14, 257–63, 340–45.

Lagrange, Joseph-Louis: *Œuvres*, Gauthier-Villars, 1867–69, Vols. 1–3, relevant papers.

————: "Additions aux éléments d'algèbre d'Euler," *Œuvres*, Gauthier-Villars, 1877, Vol. 7, pp. 5–179.

Legendre, Adrien-Marie, *Théorie des nombres*, 4th ed., 2 vols., A. Blanchard (reprint), 1955.

Muir, Thomas: *The Theory of Determinants in the Historical Order of Development*, 1906, Dover (reprint), 1960, Vol. 1, pp. 1–52.

Ore, Oystein: *Number Theory and its History*, McGraw-Hill, 1948.

Pierpont, James: "Lagrange's Place in the Theory of Substitutions," *Amer. Math. Soc. Bulletin*, 1, 1894/5, 196–204.

————: "Zur Geschichte der Gleichung des V. Grades (bis 1858)," *Monatshefte für Mathematik und Physik*, 6, 1895, 15–68.

Smith, H. J. S.: *Report on the Theory of Numbers*, 1867, Chelsea (reprint), 1965; also in Vol. 2 of the *Collected Mathematical Papers of H. J. S. Smith*, 1894, Chelsea (reprint), 1965.

Smith, David Eugene: *A Source Book in Mathematics*, 1929, Dover (reprint), 1959, Vol. 1, relevant selections. One of the selections is an English translation of Gauss's second proof of the fundamental theorem of algebra.

Struik, D. J.: *A Source Book in Mathematics, 1200–1800*, Harvard University Press, 1969, pp. 26–54, 99–122.

Vandiver, H. S.: "Fermat's Last Theorem," *Amer. Math. Monthly*, 53, 1946, 555–78.

Whiteside, Derek T.: *The Mathematical Works of Isaac Newton*, Johnson Reprint Corp., 1967, Vol. 2, pp. 3–134. This section contains Newton's *Universal Arithmetic* in English.

Wussing, H. L.: *Die Genesis der abstrakten Gruppenbegriffes*, VEB Deutscher Verlag der Wissenschaften, 1969.

26
Mathematics as of 1800

> When we cannot use the compass of mathematics or the torch
> of experience...it is certain that we cannot take a single step
> forward. VOLTAIRE

1. *The Rise of Analysis*

If the seventeenth century has correctly been called the century of genius,
then the eighteenth may be called the century of the ingenious. Though both
centuries were prolific, the eighteenth-century men, without introducing
any concept as original and as fundamental as the calculus, but by exercising
virtuosity in technique, exploited and advanced the power of the calculus
to produce what are now major branches: infinite series, ordinary and
partial differential equations, differential geometry, and the calculus of
variations. In extending the calculus to these several areas, they built what
is now the most extensive domain of mathematics, which we call analysis
(though the word now includes a couple of other branches upon which the
eighteenth-century men barely touched). The progress in coordinate geometry and algebra, on the other hand, was hardly more than a minor
extension of what the seventeenth century had initiated. Even the major
problem of algebra, the solution of the nth degree equation, received attention because it was needed for analysis, as for example, in integration by the
method of partial fractions.

During the first third or so of the century, geometrical methods were
used freely; but Euler and Lagrange, in particular, recognizing the greater
effectiveness of analytic methods, deliberately and gradually replaced
geometrical arguments by analytic ones. Euler's many texts showed how
analysis could be used. Toward the end of the century Monge did revive
pure geometry, though he used it largely to give intuitive meaning and
guidance to the work in analysis. Monge is often referred to as a geometer,
but this is because, working at a time when geometry had dried up, he put
new life into it by showing its importance, at least for the purposes just
described. In fact, in a paper published in 1786, he implicitly attached
greater importance to analysis, when he observed that geometry could make
progress because analysis could be used to study it. He, as well as the others,
did not, in the main, look for new geometrical ideas. The primary interest
and the end results were in analytical work.

The classic statement of the importance of analysis was made by Lagrange in his *Mécanique analytique*. He writes in his preface,

> We already have various treatises on Mechanics but the plan of this one is entirely new. I have set myself the problem of reducing this science [mechanics], and the art of solving the problems appertaining to it, to general formulas whose simple development gives all the equations necessary for the solutions of each problem.... No diagrams will be found in this work. The methods which I expound in it demand neither constructions nor geometrical or mechanical reasonings, but solely algebraic [analytic] operations subjected to a uniform and regular procedure. Those who like analysis will be pleased to see mechanics become a new branch of it, and will be obliged to me for having extended its domain.

Laplace, too, stressed the power of analysis. In his *Exposition du système du monde* he says,

> The algebraic analysis soon makes us forget the main object [of our researches] by focusing our attention on abstract combinations and it is only at the end that we return to the original objective. But in abandoning oneself to the operations of analysis, one is led by the generality of this method and the inestimable advantage of transforming the reasoning by mechanical procedures to results often inaccessible to geometry. Such is the fecundity of the analysis that it suffices to translate into this universal language particular truths in order to see emerge from their very expression a multitude of new and unexpected truths. No other language has the capacity for the elegance that arises from a long sequence of expressions linked one to the other and all stemming from one fundamental idea. Therefore the geometers [mathematicians] of this century convinced of its [analysis] superiority have applied themselves primarily to extending its domain and pushing back its bounds.[1]

A few features of the analysis warrant note. Newton's emphasis on the derivative and antidifferentiation was retained, so that the summation concept was rarely employed. On the other hand, Leibniz's *concept*, that is, the differential form of the derivative, and Leibniz's notation became standard, despite the fact that throughout the century the Leibnizian differentials had no precise meaning. The first differentials dy and dx of a function $y = f(x)$ were legitimized in the nineteenth century (Chap. 40, sec. 3), but the higher-order differentials, which the eighteenth-century men used freely, have not been put on a rigorous basis even today. The eighteenth century also continued the Leibnizian tradition of formal manipulation of analytical expressions and, in fact, accentuated this practice.

The importance assigned to analysis had implications that the eighteenth-century men did not appreciate. It furthered the separation of number

1. Book V, Chap. 5 = *Œuvres*, 6, 465–66.

from geometry and implicitly underscored the issue of the proper foundations for the number system, algebra, and analysis itself. This problem was to become critical in the nineteenth century. Peculiarly the eighteenth-century mathematicians still referred to themselves as geometers, the term in vogue during the preceding ages in which geometry had dominated mathematics.

2. *The Motivation for the Eighteenth-Century Work*

Far more than in any other century, the mathematical work of the eighteenth was directly inspired by physical problems. In fact, one can say that the goal of the work was not mathematics, but rather the solution of physical problems; mathematics was a means to physical ends. Laplace, though perhaps an extreme case, certainly regarded mathematics as just a tool for physics and was himself entirely concerned with its value for astronomy.

The major physical field was, of course, mechanics, and particularly celestial mechanics. Mechanics became the paradise for mathematics, much as Leonardo had predicted, because it suggested so many directions of research. So extensive was the concern of mathematics with the problems of mechanics that d'Alembert in the *Encyclopédie* and Denis Diderot (1713–84) in his *Pensées sur l'interprétation de la nature* (1754) wrote of a transition from the seventeenth-century age of mathematics to an age of mechanics. They believed, in fact, that the mathematical work of men such as Descartes, Pascal, and Newton was passé, and that mechanics was to be the major interest of mathematicians. Mathematics generally was useful, in their opinion, only so far as it served physics. The eighteenth century concentrated on the mechanics of discrete systems of masses and of continuous media. Optics was temporarily pushed into the background.

What was actually happening, however, was the opposite of what Diderot and d'Alembert maintained. The truer interpretation, certainly in light of future developments, was expressed by Lagrange when he said that those who liked analysis would be pleased to see mechanics become a new branch of it. Put more broadly, the program initiated by Galileo, and consciously pursued by Newton, of expressing the basic physical principles as quantitative mathematical statements and deducing new physical results by mathematical arguments, had been advanced immeasurably. Physics was becoming more and more mathematical, at least in those areas where the physical principles had become sufficiently well understood. The increasing incorporation of major branches of physics into mathematical frameworks established mathematical physics.

Mathematics not only began to embrace science, but, partly because there was no sharp separation between science and what we would call engineering, the mathematicians undertook technological problems as a matter of course. Euler, for example, worked on the design of ships, the action of sails, ballistics, cartography, and other practical problems. Monge took his work on excavation and fill and the design of windmill vanes as seriously as any problem of differential geometry or differential equations.

J. F. (Jean-Etienne) Montucla (1725–99), in his *Histoire des mathématiques* (2nd ed., 1799–1802) divided mathematics into two parts, the one "comprising those things that are pure and abstract, the other those that one calls compound, or more ordinarily physico-mathematics." His second part comprised fields that can be approached and treated mathematically, that is, mechanics, optics, astronomy, military and civil architecture, insurance, acoustics, and music. Under optics he included dioptrics and even optometry, catoptrics, and perspective. Mechanics included dynamics and statics, hydrodynamics and hydrostatics. Astronomy covered geography, theoretical astronomy, spherical astronomy, gnomonics (e.g. sundials), chronology, and navigation. Montucla also included astrology, the constructon of observatories, and the design of ships.

3. The Problem of Proof

That physical problems motivated most mathematical work was of course not peculiar to the eighteenth century, but in that century the merger of mathematics and physics was crucial. The major development, as we have noted, was analysis. However, the very basis of the calculus not only was not clear, but had been under attack almost from the beginning of the seventeenth-century work on the subject. Eighteenth-century thinking was certainly loose and intuitive. Any delicate questions of analysis, such as the convergence of series and integrals, the interchange of the order of differentiation and integration, the use of differentials of higher order, and questions of existence of integrals and solutions of differential equations, were all but ignored. That the mathematicians were able to proceed at all was due to the fact that the rules of operation were clear. Having formulated the physical problems mathematically, the virtuosos got to work, and new methodologies and conclusions emerged. Certainly the mathematics itself was purely formal. Euler was so fascinated by formulas that he could hardly refrain from manipulating them. How could the mathematicians have dared merely to apply rules and yet assert the reliability of their conclusions?

The physical meaning of the mathematics guided the mathematical steps and often supplied partial arguments to fill in nonmathematical steps. The reasoning was in essence no different from a proof of a theorem of geometry, wherein some facts entirely obvious in the figure are used even though no axiom or theorem supports them. Finally, the physical correctness of the conclusions gave assurance that the mathematics must be correct.

Another element in eighteenth-century mathematical thought supported the arguments. The men trusted the symbols far more than they did logic. Because infinite series had the same symbolic form for all values of x, the distinction between values of x for which the series converged and values for which it diverged did not seem to demand attention. And even though they recognized that some series, such as $1 + 2 + 3 + \ldots$, had an infinite sum, they preferred to try to give meaning to the sum rather than question the summation.

Likewise the somewhat free use of complex numbers rested on the confidence in symbols. Since second degree expressions, $ax^2 + bx + c$, could be expressed as a product of linear factors when the zeros were real, it was equally clear that there should be linear factors when the zeros were complex. The formal operations of the calculus, differentiation and anti-differentiation, were extended to new functions despite the fact that the mathematicians were conscious of their own lack of clarity about the ideas. This reliance upon formalism blinded them somewhat. Thus the difficulty they had in broadening their notion of function was caused by their adherence to the conviction that functions must be expressible by formulas.

The eighteenth-century men were fully aware of the mathematical requirement of proof. We have seen that Euler did try to justify his use of divergent series and Lagrange, among others, did offer a foundation for the calculus. But the few efforts to achieve rigor, significant because they show that standards of rigor vary with the times, did not logicize the work of the century; and the men almost willingly took the position that what cannot be cured must be endured. They were so intoxicated with their physical successes that most often they were indifferent to the missing rigor. Very striking is the extreme confidence in the conclusions while the theory was so ill-assured. Because the eighteenth-century mathematicians were willing to plunge ahead so boldly without logical support, this period has been called the heroic age in mathematics.

Perhaps because the few efforts to rigorize the calculus were unsuccessful and additional questions of rigor raised by subsequent analytical work were hopelessly beyond resolution, some mathematicians abandoned the effort to secure it and, like the fox with the grapes, consciously derided the rigor of the Greeks. Sylvestre-François Lacroix (1765–1843), in the second edition (1810–19) of his three-volume *Traité du calcul différentiel et du calcul intégral*, says, in the preface to Volume 1 (p. 11), "Such subtleties as the Greeks worried about we no longer need." The typical attitude of the century was: Why go to the trouble of proving by abstruse reasoning things which one never doubts in the first place, or of demonstrating what is more evident by means of what is less evident?

Even Euclidean geometry was criticized, on the ground that it offered proofs where none were deemed to be needed. Clairaut said in his *Eléments de géométrie* (1741),

> It is not surprising that Euclid goes to the trouble of demonstrating that two circles which cut one another do not have the same center,

that the sum of the sides of a triangle which is enclosed within another is smaller than the sum of the sides of the enclosing triangle. This geometer had to convince obstinate sophists who glory in rejecting the most evident truths; so that geometry must, like logic, rely on formal reasoning in order to rebut the quibblers.... But the tables have turned. All reasoning concerned with what common sense knows in advance, serves only to conceal the truth and to weary the reader and is today disregarded.

This attitude of the eighteenth century was also expressed by Josef Maria Hoene-Wronski (1778–1853), who was a great algorithmist but was not concerned with rigor. A paper of his was criticized by a Commission of the Paris Academy of Sciences as lacking in rigor; Wronski replied that this was "pedantry which prefers means to the end."

The men were, on the whole, aware of their own indifference to rigor. D'Alembert said in 1743, "Up to the present ... more concern has been given to enlarging the building than to illuminating the entrance, to raising it higher than to giving proper strength to the foundations." Accordingly, the eighteenth century broke new ground as has not been done since. The great creations, in view of the limited number of men involved, were more numerous than in any other century. The nineteenth- and twentieth-century men, inclined to look down on the crude, often brash, inductive work of the eighteenth, stressed its excesses and errors in order to belittle its triumphs.

4. The Metaphysical Basis

Though the mathematicians did recognize that their creations had not been reformulated in terms of the deductive model of Euclid, they were confident of the truth of the mathematics. This confidence rested in part, as noted, on the physical correctness of the conclusions, but also on philosophical and theological grounds. Truth was assured because mathematics was simply unearthing the mathematical design of the universe. The keynote of late seventeenth- and eighteenth-century philosophy, expressed especially by Thomas Hobbes, John Locke, and Leibniz, was the pre-established harmony between reason and nature. This doctrine had really been unchallenged since the time of the Greeks. Need one quibble, then, if the mathematical laws that so clearly applied to nature lacked the precision of purely mathematical proof? Though the eighteenth century uncovered only pieces, they were pieces of underlying truths. The remarkable accuracy of the mathematical deductions, especially in celestial mechanics, were glorious confirmation of the century's confidence in the mathematical design of the universe.

The eighteenth-century men were likewise convinced that certain mathematical principles must be true because the mathematical design must have incorporated them. Thus, since a perfect universe would not tolerate waste, its action was the least required to achieve its purposes. Hence the Principle of Least Action was, as Maupertuis affirmed and Euler seconded in his *Methodus Inveniendi*, unquestionable.

The conviction that the world was mathematically designed derived from the earlier linking of science and theology. We may recall that most of the leading figures of the sixteenth and seventeenth centuries were not only deeply religious but found in their theological views the inspiration and conviction vital to their scientific work. Copernicus and Kepler were certain that the heliocentric theory must be true because God must have preferred the mathematically simpler theory. Descartes's belief that our innate ideas, among them the axioms of mathematics, are true and that our reasoning is sound rested on the conviction that God would not deceive us, so that to deny the truth and clarity of mathematics would be to deny God. Newton, as already mentioned, regarded the chief value of his scientific efforts to be the study of God's work and the support of revealed religion. Many passages in his works are glorifications of God, and the General Scholium at the end of Book III of his *Principia* is largely a tribute to God reminiscent of Kepler. Leibniz's explanation of the concord between the real and the mathematical worlds, and his ultimate defense of the applicability of his calculus to the real world, was the unity of the world and God. The laws of reality could therefore not deviate from the ideal laws of mathematics. The universe was the most perfect conceivable, the best of all possible worlds, and rational thinking disclosed its laws.

Though it retained the belief in the mathematical design of nature, the eighteenth century ultimately discarded the philosophical and religious bases for this belief. The core of the entire philosophical position, the doctrine that the universe was designed by God, was gradually undermined by purely mathematico-physical explanations. The religious motivation for mathematical work had begun to lose ground even in the seventeenth century. Galileo had sounded the call for a break. He says in one of his letters, "Yet for my part any discussion of the Sacred Scriptures might have lain dormant forever; no astronomer or scientist who remained within proper bounds has ever got into such things." Then Descartes, by asserting the invariability of the laws of nature, had implicitly restricted God's role. Newton confined the actions of God to keeping the world functioning according to plan. He used the figure of speech of a watchmaker keeping a watch in repair. Thus the role attributed to God became more and more restricted; and, as universal laws embracing heavenly and earthly motions (which Newton himself revealed) began to dominate the intellectual scene and as the continued agreement between predictions and observations bespoke the perfection of these laws, God sank more and more into the background and the mathematical laws of the universe became the focus of attention. Leibniz saw that Newton's *Principia* implied a world functioning according to plan with or without God, and attacked the book as anti-Christian. But the further the development of mathematics proceeded in the eighteenth century the more the religious inspiration and motivation for mathematical work receded.

Unlike Maupertuis and Euler, Lagrange denied any metaphysical implication in the Principle of Least Action. The concern to obtain physically significant results replaced the concern for God's design. As far as mathematical physics was concerned, the complete rejection of God and any metaphysical principles resting on his existence was made by Laplace. There is a well-known story that when Laplace gave Napoleon a copy of his *Mécanique céleste*, Napoleon remarked, " M. Laplace, they tell me you have written this large book on the system of the universe and have never even mentioned its Creator." Laplace is said to have replied, " I have no need of this hypothesis." Toward the end of the century the label "metaphysical" applied to an argument became a term of reproach, though it was often used to condemn what mathematicians could not understand. Thus Monge's theory of characteristics, which was not understood by his contemporaries, was called metaphysical.

5. The Expansion of Mathematical Activity

In the eighteenth century the academies of science founded in the middle and at the end of the seventeenth century, rather than the universities, sponsored and supported mathematical research. The academies also supported the journals, which had become the regular outlet for new work. About the only change in the affairs of the academies is that in 1795 the Academy of Sciences of Paris was reconstituted as one of three divisions of the Institute of France.

In Germany, prior to 1800, the universities did no research. They offered a couple of required years of liberal arts followed by specialization in law, theology, or medicine. The great mathematicians were not in the universities but were attached to the Berlin Academy of Science. However, in 1810, Alexander von Humboldt (1769–1859) founded the University of Berlin and introduced the radical idea that professors should lecture on what they wished to and that students could take what they preferred. Hence for the first time professors could lecture on their research interests. Thus Jacobi lectured on his work on elliptic functions at Koenigsberg from 1826 on, though this was still unusual and other teachers had to cover his regular classes. In the nineteenth century many of the German kingdoms, duchies, and free cities founded universities that began to support research professors.

The French universities of the eighteenth century, at least until the Revolution, were no better than those in Germany. However the new government decided to found high-level universities for teaching and research. The organizational work was carried out by Nicolas de Condorcet, who had been active in mathematics. The Ecole Polytechnique, founded in 1794, had Monge and Lagrange as its first mathematics professors. Students competed for admission and were supposed to receive the training that would enable them to become engineers or army officers. Actually the mathematical level of the courses was very high, and the graduates were able to undertake mathematical research. Through this training and through the published lectures, the school exerted a wide and strong influence. In 1808 the French government founded the Ecole Normale Supérieure, which had been preceded by the Ecole Normale; the latter, founded in 1794, lasted only a few months. The new Ecole was devoted to the training of teachers and was divided into two sections, humanities and science. Here too students competed for admission; advanced courses were offered; facilities for study and research were good; and the better students were directed into research.

The nations of Europe differed considerably in mathematical productivity during the eighteenth century. The leading country was France, followed by Switzerland. Germany was relatively inactive, so far as German nationals were concerned, though Euler and Lagrange were supported by the Berlin Academy of Sciences. England too languished. Brook Taylor, Matthew Stewart (1717–85), and Colin Maclaurin were the only prominent mathematicians. England's poor performance in view of its great activity in the seventeenth century may be surprising, but the explanation is readily found. The English mathematicians had not only isolated themselves personally from the Continentals as a consequence of the controversy between Newton and Leibniz, but also suffered by following the geometrical methods of Newton. The English settled down to study Newton instead of nature. Even in their analytical work they used Newton's notation for fluxions and fluents and refused to read anything written in the notation of Leibniz. Moreover, at Oxford and Cambridge, no Jew or Dissenter from the Church of England could even be a student. By 1815 mathematics in England was at its last gasp and astronomy nearly so.

In the first quarter of the nineteenth century the British mathematicians began to take interest in the work on the calculus and its extensions, which had proceeded apace on the Continent. The Analytical Society was formed at Cambridge in 1813 to study this work. George Peacock (1791–1858), John Herschel (1792–1871), Charles Babbage and others undertook to study the principles of "d-ism"—that is, the Leibnizian notation in the calculus, as against those of "dot-age," or the Newtonian notation. Soon the quotient dy/dx replaced $\dot{y}$, and the Continental texts and papers became accessible to English students. Babbage, Peacock, and Herschel translated a one-volume edition of Lacroix's *Traité* and published it in 1816. By 1830 the English were able to join in the work of the Continentals. Analysis in England did prove to be largely mathematical physics, though some entirely new directions of work, algebraic invariant theory and symbolic logic, were also initiated in that country.

6. *A Glance Ahead*

As we know, by the end of the eighteenth century the mathematicians had created a number of new branches of mathematics. But the problems of these branches had become extremely complicated and, with few exceptions, no general methods had been devised to treat them. The mathematicians began to feel blocked. Lagrange wrote to d'Alembert on September 21, 1781, "It appears to me also that the mine [of mathematics] is already very deep and that unless one discovers new veins it will be necessary sooner or later to abandon it. Physics and chemistry now offer the most brilliant riches and easier exploitation; also our century's taste appears to be entirely in this direction and it is not impossible that the chairs of geometry in the Academy will one day become what the chairs of Arabic presently are in the universities."[2] Euler and d'Alembert agreed with Lagrange that mathematics had almost exhausted its ideas, and they saw no new great minds on the horizon. This fear was expressed even as early as 1754 by Diderot in *Thoughts on the Interpretation of Nature*: "I dare say that in less than a century we shall not have three great geometers [mathematicians] left in Europe. This science will very soon come to a standstill where the Bernoullis, Maupertuis, Clairauts, Fontaines, d'Alemberts and Lagranges will have left it. . . . We shall not go beyond this point."

Jean-Baptiste Delambre (1749–1822), who was permanent secretary of the mathematics and physics section of the Institut de France, in a report (*Rapport historique sur le progrès des sciences mathématiques depuis 1789 et sur leur état actuel*, Paris, 1810) said, "It would be difficult and rash to analyze the chances which the future offers to the advancement of mathematics; in almost all its branches one is blocked by insurmountable difficulties; perfection of detail seems to be the only thing which remains to be done. All these difficulties appear to announce that the power of our analysis is practically exhausted. . . ."

The wiser prediction was made in 1781 by Condorcet, who was impressed by Monge's work.

> . . . in spite of so many works often crowned by success, we are far from having exhausted all the applications of analysis to geometry, and instead of believing that we have approached the end where

2. Lagrange, *Œuvres*, 13, 368.

these sciences must stop because they have reached the limit of the forces of the human spirit, we ought to avow rather we are only at the first steps of an immense career. These new [practical] applications, independently of the utility which they may have in themselves, are necessary to the progress of analysis in general; they give birth to questions which one would not think to propose; they demand that one create new methods. Technical processes are the children of need; one can say the same for the methods of the most abstract sciences. But we owe the latter to needs of a more noble kind, the need to discover the new truths or to know better the laws of nature.

Thus one sees in the sciences many brilliant theories which have remained unapplied for a long time suddenly becoming the foundation of most important applications, and likewise applications very simple in appearance giving birth to ideas of the most abstract theories, for which no one would have felt the need, and directing the work of geometers [mathematicians] to these theories....

Condorcet was of course correct. In fact, mathematics expanded in the nineteenth century even more than in the eighteenth. In 1783, the year in which both Euler and d'Alembert died, Laplace was thirty-four years of age, Legendre thirty-one, Fourier fifteen, and Gauss six.

Just what the new mathematics was we shall see in succeeding chapters. But we shall note here some of the agencies for the propagation of results to which we shall refer later. There was, first of all, a vast expansion in the number of research journals. The *Journal de l'Ecole Polytechnique* was founded at the same time that the school was organized. In 1810 Joseph-Diez Gergonne (1771–1859) started the *Annales de Mathématiques Pures et Appliquées*, which lasted until 1831. This was the first purely mathematical journal. In the meantime August Leopold Crelle (1780–1855), a noteworthy figure because he was a good organizer and helped many young people to obtain university positions, started the *Journal für die reine und angewandte Mathematik* in 1826.[3] By this title Crelle intended to show that he wished to broaden the scope of mathematical interests. Despite Crelle's intentions, the journal was soon filled with specialized mathematical articles and was often referred to humorously as the *Journal für reine unangewandte Mathematik* (journal purely for unapplied mathematics). This journal is also referred to as *Crelle's Journal*, and, from 1855 to 1880, as *Borchardt's Journal*. In 1795 the *Mémoires de l'Académie des Sciences* became the *Mémoires de l'Académie des Sciences de l'Institut de France*. The Academy of Sciences of Paris also started, in 1835, the *Comptes Rendus*, to serve the purpose of giving in four pages or less brief notices of new results. There is an apocryphal story that the restriction to four pages was intended to limit Cauchy, who had been writing prolifically.

In 1836 Liouville founded the French analogue of *Crelle's Journal*, the

---

3. Hereafter referred to as *Jour. für Math.*

Journal de Mathématiques Pures et Appliquées, which is often referred to as *Liouville's Journal*. Louis Pasteur (1822–95) founded the *Annales Scientifiques de l'Ecole Normale Supérieure* in 1864, while Gaston Darboux (1842–1917) launched *Le Bulletin des Sciences Mathématiques* in 1870. Among the numerous other journals, we shall mention the *Mathematische Annalen* (1868), the *Acta Mathematica* (1882), and the *American Journal of Mathematics*, the first mathematical journal in the United States. This was founded in 1878 by J. J. Sylvester when he was a professor at Johns Hopkins University.

Another type of agency has promoted mathematical activity since the nineteenth century. The mathematicians of several countries formed professional societies such as the London Mathematical Society, the first of its kind and organized in 1865, the Société Mathématique de France (1872), the American Mathematical Society (1888), and the Deutsche Mathematiker-Vereinigung (1890). These societies hold regular meetings at which papers are presented; each sponsors one or more journals, in addition to those already mentioned, such as the *Bulletin de la Société Mathématique de France* and the *Proceedings of the London Mathematical Society*.

As the above sketch of newer organizations and publications implies, mathematics expanded enormously in the nineteenth century. In large part this expansion became possible because a small aristocracy of mathematicians was supplanted by a far broader group. The spread of learning made possible the entry of far more scholars from all economic levels. The trend had begun even in the eighteenth century. Euler was the son of a pastor; d'Alembert was an illegitimate child brought up by a poor family; Monge was the son of a peddler; and Laplace was born to a peasant family. The participation of universities in research, the writing of textbooks, and the systematic training of scientists initiated by Napoleon produced a far greater number of mathematicians.

Bibliography

Boutroux, Pierre: *L'Ideal scientifique des mathématiciens*, Libraire Felix Alcan, 1920.

Brunschvicg, Léon: *Les Etapes de la philosophie mathématique*, Presses Universitaires de France, 1947, Chaps. 10–12.

Hankins, Thomas L.: *Jean d'Alembert: Science and Enlightenment*, Oxford University Press, 1970.

Hille, Einar: "Mathematics and Mathematicians from Abel to Zermelo," *Mathematics Magazine*, 26, 1953, 127–46.

Montucla, J. F.: *Histoire des mathématiques*, 2nd ed., 4 vols., 1799–1802, Albert Blanchard (reprint), 1960.

27

Functions of a Complex Variable

The shortest path between two truths in the real domain passes through the complex domain.   JACQUES HADAMARD

1. *Introduction*

From the technical standpoint, the most original creation of the nineteenth century was the theory of functions of a complex variable. The subject is often referred to as the theory of functions, though the abbreviated description implies more than is intended. This new branch of mathematics dominated the nineteenth century almost as much as the direct extensions of the calculus dominated the eighteenth century. The theory of functions, a most fertile branch of mathematics, has been called the mathematical joy of the century. It has also been acclaimed as one of the most harmonious theories in the abstract sciences.

2. *The Beginnings of Complex Function Theory*

As we have seen, complex numbers and even complex functions entered mathematics effectively in connection with integration by partial fractions, the determination of the logarithms of negative and complex numbers, conformal mapping, and the decomposition of a polynomial with real coefficients. Actually the men of the eighteenth century did far more with complex numbers and functions.

In his essay on hydrodynamics, *Essay on a New Theory of the Resistance of Fluids* (1752), d'Alembert considers the motion of a body through a homogeneous, weightless, ideal fluid, and in connection with this study considers the following problem. He seeks to determine two functions p and q whose differentials are

$$(1) \qquad dq = M\,dx + N\,dy, \qquad dp = N\,dx - M\,dy.$$

Since the quantities N and M occur in both dp and dq, it follows that

$$(2) \qquad \frac{\partial p}{\partial x} = \frac{\partial q}{\partial y}, \qquad \frac{\partial p}{\partial y} = -\frac{\partial q}{\partial x}.$$

These equations are now called the Cauchy-Riemann equations. The equations (2) say (Chap. 19, sec. 6) that $q\,dx + p\,dy$ and $p\,dx - q\,dy$ are exact differentials of certain functions. Then the expressions (we shall use i for $\sqrt{-1}$, though this was done only occasionally by Euler and made common practice by Gauss)

$$q\,dx + p\,dy + i(p\,dx - q\,dy) = (q + ip)\left(dx + \frac{dy}{i}\right)$$

and

$$q\,dx + p\,dy - i(p\,dx - q\,dy) = (q - ip)\left(dx - \frac{dy}{i}\right)$$

are also complete differentials, and so $q + ip$ is a function of $x + y/i$ and $q - ip$ is a function of $x - y/i$. D'Alembert sets

$$(3) \qquad q + ip = \xi\left(x + \frac{y}{i}\right) + i\zeta\left(x + \frac{y}{i}\right)$$

$$(4) \qquad q - ip = \xi\left(x - \frac{y}{i}\right) - i\zeta\left(x - \frac{y}{i}\right)$$

where ξ and ζ are functions to be determined and which d'Alembert does determine in special cases. By adding and subtracting (3) and (4) he obtains p and q. The significance of this is that he has shown that p and q are the real and imaginary parts of a complex function.

Euler showed how to use complex functions to evaluate real integrals. He wrote a series of papers from 1776 to the time of his death in 1783 which were published from 1788 on. Two of these were published in 1793 and 1797.[1] Euler remarks that every function of z which for $z = x + iy$ takes the form $M + iN$, where M and N are real functions, also takes on, for $z = x - iy$, the form $M - iN$. This, he states, is the fundamental theorem of complex numbers. He uses this assertion to evaluate real integrals.

Suppose

$$(5) \qquad \int Z(z)\,dz = V$$

where z is real. He sets $z = x + iy$ so that V becomes $P + iQ$. Then

$$(6) \qquad P + iQ = \int (M + iN)(dx + i\,dy),$$

where $M + iN$ is now the complex form of $Z(z)$. By his basic assertion,

$$(7) \qquad P - iQ = \int (M - iN)(dx - i\,dy),$$

1. *Nova Acta Acad. Sci. Petrop.*, 7, 1789, 99–133, pub. 1793 = *Opera*, (1), 19, 1–44; *ibid.*, 10, 1792, 3–19, pub. 1797 = *Opera*, (1), 19, 268–86.

so that by separating real and imaginary parts,

$$(8) \qquad P = \int M \, dx + N \, dy, \qquad Q = \int N \, dx + M \, dy.$$

Then $M \, dx - N \, dy$ and $N \, dx + M \, dy$ are exact differentials of P and Q respectively, from which it follows that

$$(9) \qquad \frac{\partial M}{\partial y} = -\frac{\partial N}{\partial x}, \qquad \frac{\partial N}{\partial y} = \frac{\partial M}{\partial x}.$$

Thus by substituting $z = x + iy$ in $Z(z)$, "one obtains two functions M and N which possess the remarkable property that $\partial M/\partial y = -\partial N/\partial x$ and $\partial M/\partial x = \partial N/\partial y$; P and Q have similar properties." Here Euler stresses that M and N, the real and imaginary parts of a complex function, satisfy the Cauchy-Riemann equations. However, his main point is to use the integrals (8) to calculate (5), for P equals the original V. To reduce the integrals in (8) to integrals of functions of one variable, Euler replaces $z = x + iy$ in (5) by $z = r(\cos \theta + i \sin \theta)$ and keeps θ constant. This amounts to integrating along a ray through the origin of the complex plane. He then uses his method to evaluate some integrals.

Laplace too used complex functions to evaluate integrals. In a series of papers that start in 1782 and end with his famous *Théorie analytique des probabilités* (1812), he passes from real to complex integrals, much as Euler did, to evaluate the real integrals. Laplace claimed priority because Euler's papers were published later than his own. However, even the papers of 1793 and 1797 mentioned above were read to the St. Petersburg Academy in March of 1777. Incidentally, in this work Laplace introduced what we now call the Laplace transform method of solving differential equations.

The work of Euler, d'Alembert, and Laplace constituted significant progress in the theory of functions. However, there is an essential limitation in their work. They depended upon separating the real and imaginary parts of $f(x + iy)$ to carry out their analytical work. The complex function was not really the basic entity. It is clear that these men were still very uneasy about the use of complex functions. Laplace, in his 1812 book, remarks, "This transition from real to imaginary can be regarded as a heuristic method, which is like the method of induction long used by mathematicians. However if one uses the method with great care and restraint, one will always be able to prove the results obtained." He does emphasize that the results must be verified.

3. *The Geometrical Representation of Complex Numbers*

A vital step that made the erection of a theory of functions of a complex variable more intuitively reasonable was the geometrical representation of

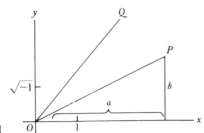

Figure 27.1

complex numbers and of the algebraic operations with these numbers. That many men—Cotes, De Moivre, Euler, and Vandermonde—really thought of complex numbers as points in the plane follows from the fact that all, in attempting to solve $x^n - 1 = 0$, thought of the solutions

$$\cos \frac{2k\pi}{n} + i \sin \frac{2k\pi}{n}$$

as the vertices of a regular polygon. Euler, for example, replaced x and y, which geometrically he visualized as a point in the coordinate plane, by $x + iy$ and then represented the latter by $r(\cos \theta + i \sin \theta)$, which in turn he plots as polar coordinates r and θ. Hence one could say that the plotting of complex numbers as coordinates of points in the plane was known by 1800. However, decisive identification of the two was not made, nor did the algebraic operations with complex numbers have as yet any geometrical meaning. Also missing was the idea of plotting the values of a complex function $u + iv$ of $x + iy$ as points in another plane.

In 1797 the self-taught Norwegian-born surveyor Caspar Wessel (1745–1818) wrote a paper entitled "On the Analytic Representation of Direction; an Attempt," which was published in the memoirs for 1799 of the Royal Academy of Denmark. Wessel sought to represent geometrically directed line segments (vectors) and operations with them. In this paper an axis of imaginaries with $\sqrt{-1}$ (he writes ϵ for $\sqrt{-1}$) as an associated unit is introduced along with the usual x-axis with real unit 1. In Wessel's geometric representation, the vector OP (Fig. 27.1) is the line segment OP drawn from the origin O in the plane of the units $+1$ and $\sqrt{-1}$, and the vector is represented by the complex number, $a + b\sqrt{-1}$. Similarly, the vector OQ is the line segment OQ and is represented by another number, say $c + d\sqrt{-1}$.

Wessel then defines the operations with vectors by defining the operations with complex numbers in geometrical terms. His definitions of the four operations are practically the ones we learn today. Thus the sum of $a + bi$ and $c + di$ is the diagonal of the parallelogram determined by the adjacent sides OP and OQ. The product of $a + bi$ and $c + di$ is a new vector OR, say,

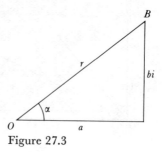

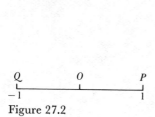

Figure 27.2   Figure 27.3

such that OR is to OQ as OP is to the real unit, and the angle made by OR and the x-axis is the sum of the angles made by OP and OQ. Clearly Wessel thought in terms of vectors rather than associating complex numbers with points in the plane. He applies his geometric representation of vectors to problems of geometry and trigonometry. Despite its great merit, Wessel's paper went unnoticed until 1897, when it was republished in a French translation.

A somewhat different geometric interpretation of complex numbers was given by a Swiss, Jean-Robert Argand (1768–1822). Argand, who was also self-taught and a bookkeeper, published a small book, *Essai sur une manière de représenter les quantités imaginaires dans les constructions géométriques* (1806).[2] He observes that negative numbers are an extension of positive numbers that results from combining direction with magnitude. He then asks, Can we extend the real number system by adding some new concept? Let us consider the sequence $1, x, -1$. Can we find an operation that turns 1 into x, and then repeated on x turns x into -1? If we rotate OP (Fig. 27.2) counterclockwise about O through 90° and then repeat this rotation, we do go from P to Q by repeating an operation twice. But, Argand notes, this is precisely what happens if we multiply 1 by $\sqrt{-1}$ and then multiply this product by $\sqrt{-1}$; that is, we obtain -1. Hence we can think of $\sqrt{-1}$ as a rotation through 90° counterclockwise, say, and $-\sqrt{-1}$ as a clockwise rotation through 90°.

To utilize this operational meaning of complex numbers, Argand decided that a typical line segment OB (Fig. 27.3) emanating from the origin, which he calls a directed line, should be represented as $r(\cos \alpha + i \sin \alpha)$, where r is the length. He also regarded the complex number $a + bi$ as symbolizing the geometrical combination OB of a and bi. Argand, like Wessel, showed how complex numbers can be added and multiplied geometrically and applied these geometrical ideas to prove theorems of trigo-

2. A number of papers on Argand's idea and ideas of other writers on the geometrical representation of complex numbers can be found in Volume 4 (1813–14) and Volume 5 (1814–15) of Gergonne's *Annales des Mathématiques*.

nometry, geometry, and algebra. Though Argand's book stirred up some controversy about the geometric interpretation of complex numbers, this was the only contribution to mathematics that he made and his work had little impact. We do, however, still speak of the Argand diagram.

Gauss was more effective in bringing about the acceptance of complex numbers. He used them in his several proofs of the fundamental theorem of algebra (Chap. 25, sec. 2). In the first three proofs (1799, 1815, and 1816) he presupposes the one-to-one correspondence of points of the Cartesian plane and complex numbers. There is no actual plotting of $x + iy$, but rather of x and y as coordinates of a point in the real plane. Moreover, the proofs do not really use complex function theory because he separates the real and imaginary parts of the functions involved. He is more explicit in a letter to Bessel of 1811,[3] wherein he says that $a + ib$ is represented by the point (a, b) and that one can go from one point to another of the complex plane by many paths. There is no doubt, if one judges from the thinking exhibited in the three proofs and in other unpublished works, some of which we shall discuss shortly, that by 1815 Gauss was in full possession of the geometrical theory of complex numbers and functions, though he did say in a letter of 1825 that "the true metaphysics of $\sqrt{-1}$ is elusive."

However, if Gauss still possessed any scruples, by 1831 he had overcome them, and he publicly described the geometric representation of complex numbers. In the second commentary to his paper "Theoria Residuorum Biquadraticorum"[4] and in the "Anzeige" (announcement and brief account) of this paper that he wrote himself for the *Göttingische gelehrte Anzeigen* of April 23, 1831,[5] Gauss is very explicit about the geometrical representation of complex numbers. In Article 38 of the paper he not only gives the representation of $a + bi$ as a point (not a vector, as with Wessel and Argand) in the complex plane, but describes the geometrical addition and multiplication of complex numbers. In the "Anzeige"[6] Gauss says that the transfer of the theory of biquadratic residues into the domain of complex numbers may disturb those who are not familiar with these numbers and may leave them with the impression that the theory of residues is left up in the air. He therefore repeats what he has said in the paper proper about the geometrical representation of complex numbers. He then points out that while fractions, negative numbers, and real numbers were now well understood, complex numbers had merely been tolerated, despite their great value. To many they have appeared to be just a play with symbols. But in this geometrical representation one finds the "*intuitive meaning of complex numbers completely established and more is not needed to admit these quantities*

3. *Werke*, 8, 90–92.
4. *Comm. Soc. Gott.*, 3, 1832 = *Werke*, 2, 95–148; the main content of this paper will be discussed in Chapter 34, sec. 2.
5. *Werke*, 2, 169–78.
6. *Werke*, 2, 174 ff.

into the domain of arithmetic [italics added]." He also says that if the units 1
−1, and $\sqrt{-1}$ had not been given the names positive, negative, and imaginary units but were called direct, inverse, and lateral, people would not
have gotten the impression that there was some dark mystery in these
numbers. The geometrical representation, he says, puts the true metaphysics of imaginary numbers in a new light. He introduced the term
"complex numbers" as opposed to imaginary numbers,[7] and used *i* for
$\sqrt{-1}$.

4. *The Foundation of Complex Function Theory*

Gauss also introduced some basic ideas about functions of a complex
variable. In the letter to Bessel of 1811,[8] apropos of a paper by Bessel on the
logarithmic integral, $\int dx/x$, Gauss points out the necessity of taking
imaginary (complex) limits into account. Then he asks, "What should one
mean by $\int \phi(x)\, dx$ [Gauss writes $\int \phi x \cdot dx$] when the upper limit is $a + bi$?
Manifestly, if one wants to have clear concepts, he must assume that x takes
on small increments from that value of x for which the integral is 0 to the
value $a + bi$ and then sum all the $\phi(x)\, dx$ .... But the continuous passage
from one value of x to another in the complex plane takes place over a
curve and is therefore possible over many paths. I affirm now that the
integral $\int \phi(x)\, dx$ has only one value even if taken over different paths provided $\phi(x)$ is single-valued and does not become infinite in the space
enclosed by the two paths. This is a very beautiful theorem whose proof is
not difficult and which I shall give on a convenient occasion." Gauss did
not give this proof. He also affirms that if $\phi(x)$ does become infinite, then
$\int \phi(x)\, dx$ can have many values, depending upon whether one chooses a
closed path that encloses the point at which $\phi(x)$ becomes infinite one, two,
or more times.

Then Gauss returns to the particular case of $\int dx/x$ and says that,
starting from $x = 1$ and going to some value $a + bi$, one obtains a unique
value for the integral if the path does not enclose $x = 0$; but if it does, one
must add $2\pi i$ or $-2\pi i$ to the value obtained by going from $x = 1$ to $x =
a + bi$ without enclosing $x = 0$. Thus there are many logarithms for a given
$a + bi$. Later in the letter, Gauss says that the investigation of the integrals
of functions of complex arguments should lead to most interesting results.
Thus, even before Gauss had published his second and third proofs of the
fundamental theorem—wherein, as in the first proof, he avoided the direct
use of complex numbers and functions except for an occasional writing of
$a + b\sqrt{-1}$—he had very definite ideas about complex functions and about
the integrals of such functions.

7. *Werke*, 2, p. 102.
8. *Werke*, 8, 90–92.

Poisson noted in 1815 and discussed in a paper of 1820[9] the use of integrals of complex functions taken over paths in the complex plane. As an example he gives

(10)
$$\int_{-1}^{1} \frac{dx}{x}.$$

Here he sets $x = e^{i\theta}$, where θ runs from $(2n + 1)\pi$ to 0, and obtains, by treating the integral as a limit of a sum, the value $-(2n + 1)\pi i$.

Then he notes that the value of an integral need not be the same when taken over an imaginary path as when over a real path. He gives the example

(11)
$$\int_{-\infty}^{\infty} \frac{\cos ax}{b^2 + x^2}\, dx,$$

where a and b are positive constants. He lets $x = t + ik$, where k is constant and positive, and obtains the values $\pi(e^{-ab} - e^{ab})/2\pi$ for $k > b$ and πe^{-ab} for $k < b$. The second value is also the correct one for $k = 0$. Thus for two different values of k, which means two different paths, one gets two different results. Poisson was the first man to carry out integrations along a path in the complex plane.

Though these observations of Gauss and Poisson are indeed significant, neither published a major paper on complex function theory. This theory was founded by Augustin-Louis Cauchy. Born in Paris in 1789, in 1805 he entered the Ecole Polytechnique, where he studied engineering. Because he was in poor health he was advised by Lagrange and Laplace to devote himself to mathematics. He held professorships at the Ecole Polytechnique, the Sorbonne, and the Collège de France. Politics had unexpected effects on his career. He was an ardent Royalist and supporter of the Bourbons. When in 1830 a distant branch of the Bourbons took control in France, he refused to swear allegiance to the new monarchy and resigned his professorship at the Ecole Polytechnique. He exiled himself to Turin and taught Latin and Italian for some years. In 1838 he returned to Paris, where he served as professor in several religious institutions, up to the time when the government that took over after the revolution of 1848 did away with oaths of allegiance. In 1848 Cauchy took the chair of mathematical astronomy in the Faculté des Sciences of the Sorbonne. Though Napoleon III restored the oath in 1852, he allowed Cauchy to forgo it. To the condescending gesture of the Emperor he responded by donating his salary to the poor of Sceaux, where he lived. Cauchy, an admirable professor and one of the greatest mathematicians, died in 1857.

Cauchy had universal interests. He knew the poetry of his time and was the author of a work on Hebrew prosody. In mathematics he wrote

9. *Jour. de l'Ecole Poly.*, 11, 1820, 295–341.

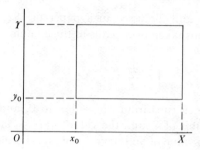

Figure 27.4

over seven hundred papers, second only to Euler in number. His works in a modern edition fill twenty-six volumes and embrace all branches of mathematics. In mechanics he wrote important works on the equilibrium of rods and of elastic membranes and on waves in elastic media. In the theory of light he occupied himself with the theory of waves that Fresnel had started and with the dispersion and polarization of light. He advanced the theory of determinants immensely and contributed basic theorems in ordinary and partial differential equations.

Cauchy's first significant paper in the direction of complex function theory is his "Mémoire sur la théorie des intégrales définies." This paper was read to the Paris Academy in 1814. However, it was not submitted for publication until 1825 and was published in 1827.[10] In the publication Cauchy added two notes that pretty surely reflect developments between 1814 and 1825 and the possible influence of Gauss's work during that period. Let us consider the paper proper for the present. Cauchy says in the preface that he was led to this work by trying to rigorize the passage from the real to the imaginary in processes used by Euler since 1759 and by Laplace since 1782 to evaluate definite integrals, and in fact Cauchy cites Laplace, who observed that the method needed rigorization. But the paper proper does not treat this problem. It treats the question of the interchange of the order of integration in double integrals that arose in hydrodynamical investigations. Euler had said in 1770[11] that this interchange was permissible when the limits for each of the variables under the integral sign were independent of each other, and Laplace apparently agreed because he used this fact repeatedly.

Specifically, Cauchy treats the relation

$$(12) \qquad \int_{x_0}^{X} \int_{y_0}^{Y} f(x, y) \, dy \, dx = \int_{y_0}^{Y} \int_{x_0}^{X} f(x, y) \, dx \, dy$$

wherein x_0, y_0, X, and Y are constants (Fig. 27.4). This interchange of the order of integration holds when $f(x, y)$ is continuous in and on the boundary

10. *Mém. des sav. étrangers*, (2), 1, 1827, 599–799 = *Œuvres*, (1), 1, 319–506.
11. *Novi Comm. Acad. Sci. Petrop.*, 14, 1769, 72–103, pub. 1770 = *Opera*, (1), 17, 289–315.

of the region. Then he introduces two functions $V(x, y)$ and $S(x, y)$ such that

(13)
$$\frac{\partial V}{\partial y} = \frac{\partial S}{\partial x} \quad \text{and} \quad \frac{\partial V}{\partial x} = -\frac{\partial S}{\partial y}.$$

Euler had already shown in 1777 how to obtain such functions (see [5], [8], and [9]). Now Cauchy considers an $f(x, y)$ which is given by $\partial V/\partial y = \partial S/\partial x$. In (12) he replaces f on the left side by $\partial V/\partial y$, and f on the right side by $\partial S/\partial x$, so that

(14)
$$\int_{x_0}^{X} \int_{y_0}^{Y} \frac{\partial V}{\partial y} \, dy \, dx = \int_{y_0}^{Y} \int_{x_0}^{X} \frac{\partial S}{\partial x} \, dx \, dy,$$

and by using the second equation in (13) he obtains

(15)
$$\int_{x_0}^{X} \int_{y_0}^{Y} \frac{\partial S}{\partial y} \, dy \, dx = -\int_{y_0}^{Y} \int_{x_0}^{X} \frac{\partial V}{\partial x} \, dx \, dy.$$

These equalities can be used to evaluate double integrals in either order of integration; however they do not involve complex functions. When Cauchy says in his Introduction[12] that he will "establish rigorously and directly the passage from the real to the imaginary (complex)," it is equations (13) that he has in mind. Cauchy states[13] that these two equations contain the whole theory of the passage from the real to the imaginary.

All of the above is in the body of the 1814 paper, and really gives no explicit indication of how complex function theory is involved. Moreover, though Cauchy used complex functions to evaluate definite real integrals in the same manner as had Euler and Laplace, this use did not involve complex functions as the basic entity. As late as 1821, in his *Cours d'analyse*,[14] he says

$$\cos a + \sqrt{-1} \sin a$$
$$\cos b + \sqrt{-1} \sin b$$
$$\cos (a + b) + \sqrt{-1} \sin (a + b)$$

"are three symbolic expressions which cannot be interpreted according to generally established conventions, and do not represent anything real." The fact that the product of the first and second expressions above equals the third, he says, does not make sense. To make sense of this equation one must equate the real parts and the coefficients of $\sqrt{-1}$. "Every imaginary equation is only the symbolic representation of two equations between real quantities." If we operate on complex expressions according to rules established for *real* quantities, we get exact results that are often important.

12. *Œuvres*, (1), 1, 330.
13. *Œuvres*, (1), 1, 338.
14. *Œuvres*, (2), 3, 154.

In this book he does treat complex numbers and complex variables $u + \sqrt{-1}v$, where u and v are functions of *one* real variable, but always in the sense that the two real components are their significant content. Complex-valued functions of a complex variable were not considered.

In the year 1822, Cauchy took some further steps. From the relations (14) and (15) he had

(16) $$\int_{x_0}^{X} [V(x, Y) - V(x, y_0)] \, dx = \int_{y_0}^{Y} [S(X, y) - S(x_0, y)] \, dy$$

and

(17) $$\int_{x_0}^{X} [S(x, Y) - S(x, y_0)] \, dx = -\int_{y_0}^{Y} [V(X, y) - V(x_0, y)] \, dy.$$

He now had the idea that he could combine these two equations and thus make a statement about $F(z) = F(x + iy) = S + iV$. Thus, by multiplying (16) through by i and adding the two equations, he obtains

$$\int_{x_0}^{X} F(x + iY) \, dx - \int_{x_0}^{X} F(x + iy_0) \, dx$$

$$= \int_{y_0}^{Y} F(X + iy)i \, dy - \int_{y_0}^{Y} F(x_0 + iy)i \, dy,$$

which, by rearranging terms, gives

(18) $$\int_{y_0}^{Y} F(x_0 + iy)i \, dy + \int_{x_0}^{X} F(x + iY) \, dx =$$

$$\int_{x_0}^{X} F(x + iy_0) \, dx + \int_{y_0}^{Y} F(X + iy)i \, dy.$$

This last result is the Cauchy integral theorem for the simple case of complex integration around the boundary of a rectangle (Fig. 27.4). One can express the result thus:

(19) $$\int_{ADC} F(z) \, dz = \int_{ABC} F(z) \, dz.$$

That is, the integral is independent of the path.

The above ideas Cauchy gives in a note of 1822, in his *Résumé des leçons sur le calcul infinitésimal*,[15] and in a footnote to the paper of 1814 which was published in 1827. From these later writings we see how Cauchy did go from real to complex functions.

In 1825 Cauchy wrote another paper, "Mémoire sur les intégrales définies prises entre des limites imaginaires," but this was not published

15. = *Œuvres* (2), 4, 13–256.

until 1874.[16] This paper is considered by many as his most important, and one of the most beautiful in the history of science, though for a long time Cauchy himself did not appreciate its worth.

In this paper he again considers the question of evaluating real integrals by the method of substituting complex values for constants and variables. He treats

$$(20) \qquad \int_{x_0 + iy_0}^{X + iY} f(z) \, dz,$$

where $z = x + iy$, and defines the integral carefully as the limit of the sum

$$\sum_{\nu = 0}^{n - 1} f(x_\nu + iy_\nu)[(x_{\nu+1} - x_\nu) + i(y_{\nu+1} - y_\nu)],$$

where $x_0, x_1, \ldots, X$, and $y_0, y_1, \ldots, Y$ are the points of subdivision along a path from (x_0, y_0) to (X, Y). Here $x + iy$ is definitely a point of the complex plane and the integral is over a complex path. He also shows that if we set $x = \phi(t)$, $y = \psi(t)$, where t is real, then the result is independent of the choice of ϕ and ψ, that is, independent of the path, provided that no discontinuity of $f(z)$ lies between two distinct paths. This result generalizes the result for rectangles.

Cauchy formulates his theorem thus: If $f(x + iy)$ is finite and continuous for $x_0 \leq x \leq X$ and $y_0 \leq y \leq Y$, then the value of the integral (20) is independent of the form of the functions $x = \phi(t)$ and $y = \psi(t)$. His proof of this theorem uses the method of the calculus of variations. He considers as an alternative path $\phi(t) + \varepsilon u(t)$, $\psi(t) + \varepsilon v(t)$, and shows that the first variation of the integral with respect to ε vanishes. This proof is not satisfactory. In it Cauchy not only uses the existence of the derivative of $f(z)$ but also the continuity of the derivative, though he does not assume either fact in the statement of the theorem. The explanation is that Cauchy believed a continuous function is always differentiable and that the derivative can be discontinuous only where the function itself is discontinuous. Cauchy's belief was reasonable in that in the earlier years of his work a function meant for him, as for others of the eighteenth and early nineteenth centuries, an analytic expression, and the derivative is given at once by the usual formal rules of differentiation.

In the 1825 paper Cauchy is clearer about a major idea he had already touched on in the paper proper of 1814 and in a footnote to that paper. He considers what happens when $f(z)$ is discontinuous inside or on the boundary

16. *Bull. des Sci. Math.*, 7, 1874, 265–304, and 8, 1875, 43–55, 148–59; this paper is not in Cauchy's *Œuvres*.

of the rectangle (Fig. 27.4). Then the value of the integral along two different paths may be different. If at $z_1 = a + ib$, $f(z)$ is infinite, but the limit

$$F = \lim_{z \to z_1} (z - z_1) f(z)$$

exists, that is, if f has a simple pole at z_1, then the difference in the integrals
is $\pm 2\pi \sqrt{-1} F$. Thus for the function $f(z) = 1/(1 + z^2)$, which is infinite
at $z = \sqrt{-1}$, so that $a = 0$ and $b = 1$,

$$(21) \quad F = \lim_{\substack{x \to 0 \\ y \to 1}} \frac{x + (y - 1)\sqrt{-1}}{[x + (y + 1)\sqrt{-1}][x + (y - 1)\sqrt{-1}]} = \frac{-\sqrt{-1}}{2}.$$

The quantity F itself is what Cauchy called the *résidu intégral* (integral
residue) in his *Exercices de mathématique*.[17] Also, when a function has several
poles in the region bounded by the two paths of integration, Cauchy points
out that one must take the sum of the residues to get the difference in the
integrals over the two paths. In this particular section on residues his two
paths still form a rectangle, but he takes a very large one and lets the sides
become infinite in length to include all the residues.

In the *Exercices*[18] Cauchy points out that the residue of $f(z)$ at z_1 is also
the coefficient of the term $(z - z_1)^{-1}$ in the development of $f(z)$ in powers
of $z - z_1$. Much later, in a paper of 1841,[19] Cauchy gives a new expression
for the residue at a pole, namely,

$$F(z_1) = E[f(z)]_{z_1} = \frac{1}{2\pi i} \int f(z) \, dz,$$

where the integral is taken around a small circle enclosing $z = z_1$. The
concept and development of residues is a major contribution by Cauchy.
His immediate use of all of his results thus far was to evaluate definite
integrals.

It is evident from what Cauchy added in the footnotes to his 1814
paper and from what he wrote in the masterful paper of 1825 that he must
have thought long and hard to realize that some relations between pairs of
real functions achieve their simplest form when complex quantities are
introduced. Just how much he may have learned from the work of Gauss and
Poisson is not known.

During the years 1830 to 1838, when he lived in Turin and in Prague,
his publications were disconnected. He refers to them and repeats most of
the work in his *Exercices d'analyse et de physique mathématique* (4 vols., 1840–47).

17. Four vols., 1826–30 = *Œuvres*, (2), 6–9.
18. Vol. 1, 1826, 23–37 = *Œuvres*, (2), 6, 23–37.
19. *Exercices d'analyse et de physique mathématique*, Vol. 2, 1841, 48–112 = *Œuvres*, (2), 12,
48–112.

In a paper of 1831, published later,[20] he obtained the following theorem: The function $f(z)$ can be expanded according to Maclaurin's formula in a power series that is convergent for all z whose absolute value is smaller than those for which the function or its derivative ceases to be finite or continuous. (The only singularities Cauchy knew at this time were what we now call poles.) He shows that this series is less term for term than a convergent geometric series whose sum is

$$\frac{Z}{Z - z}\overline{f(z)},$$

where Z is the first value for which $f(z)$ is discontinuous and $\overline{f(z)}$ is the largest value of $|f(z)|$ for all z whose absolute value equals $|Z|$. Thus Cauchy gives a powerful and easily applied criterion for the expansibility of a function in a Maclaurin series which uses a comparison series now called a majorant series.

In the proof of the theorem he first shows that

$$f(z) = \frac{1}{2\pi} \int_{-\pi}^{\pi} \frac{\bar{z} f(\bar{z})}{\bar{z} - z} \, d\phi,$$

where $\bar{z} = |Z| e^{i\phi}$. This result is practically what we now call the Cauchy integral formula. He then expands the fraction $\bar{z}/(\bar{z} - z)$ in a geometric series in powers of $z/\bar{z}$ and proves the theorem proper.

In this theorem too Cauchy assumes the existence and continuity of the derivative as necessarily following from the continuity of the function itself. By the time he reproduced this material in his *Exercices*, he had exchanged correspondence with Liouville and Sturm and added to the statement of the above theorem that the region of convergence ends at that value of z for which the function *and* its derivative cease to be finite or continuous. But he was not convinced that one must add conditions on the derivative and in later works he dropped them.

In another major paper on complex function theory, "Sur les intégrales qui s'étendent à tous les points d'une courbe fermée,"[21] Cauchy relates the integral of an [analytic] $f(z) = u + iv$ around a curve bounding a [simply connected] area to an integral over the area. Thus if u and v are functions of x and y,

$$(22) \quad \iint \left(\frac{\partial u}{\partial x} - \frac{\partial v}{\partial y} \right) dx \, dy = \int u \, dy + \int v \, dx$$

$$\iint \left(\frac{\partial u}{\partial y} + \frac{\partial v}{\partial x} \right) dx \, dy = \int -u \, dx + \int v \, dy,$$

20. *Comp. Rend.*, 4., 1837, 216–218 = *Œuvres*, (1), 4, 38–42; see also *Exercices d'analyse et de physique mathématique*, Vol. 2, 1841, 48–112 = *Œuvres*, (2), 12, 48–112.
21. *Comp. Rend.*, 23, 1846, 251–55 = *Œuvres*, (1), 10, 70–74.

where the double integrals are over the area and the single integrals over the bounding curve. Now, in view of the Cauchy-Riemann equations (cf. [13]), the left sides are 0 and the two right sides are the integrals that appear in

$$\int f(z)\, dz = \int (u + iv)(dx + i\, dy) = \int (u\, dx - v\, dy) + i \int (u\, dy + v\, dx).$$

Hence $\int f(z)\, dz = 0$, and Cauchy has a new proof of the basic theorem on independence of the path. He proves the theorem for a rectangle and then generalizes to a closed curve that does not cut itself. (The theorem was also obtained independently by Weierstrass in 1842). Whether Cauchy got this more fertile formulation of his earlier ideas by learning of Green's work in 1828 (Chap. 28, sec. 4) is not certain, but there are such indications, because Cauchy extended the above result to areas on curved surfaces.

In the above-mentioned paper of 1846 and another of the same year,[22] Cauchy changed his viewpoint toward complex functions as against his work of 1814, 1825, and 1826. Instead of being concerned with definite integrals and their evaluation, he turned to complex function theory proper and to building a base for this theory. In this second paper of 1846, he gives a new statement that involves $\int f(z)\, dz$ around an arbitrary closed curve: If the curve encloses poles, then the value of the integral is $2\pi i$ times the sum of the residues of the function at these poles; that is,

$$(23) \qquad\qquad \int f(z)\, dz = 2\pi i E[f(z)],$$

where $E[f(z)]$ is his notation for the sum of the residues.

He also took up the subject of the integrals of multiple-valued functions.[23] In the first part of the paper, where he treats integrals of single-valued functions, he does not state much more than what Gauss had pointed out in his letter to Bessel apropos of $\int dx/x$ or $\int dx/(1 + x^2)$. The integrals are indeed multiple-valued and their values depend on the path of integration.

But Cauchy goes further to consider multiple-valued functions under the integral sign. Here he says that if the integrand is an expression for the roots of an algebraic or transcendental equation, for example, $\int w^3\, dz$ where $w^3 = z$, and if one integrates over a closed path and returns to the starting point, then the integrand now represents another root. In these cases the value of the integral over the closed path is not independent of the starting

22. "Sur les intégrales dans lesquelles la fonction sous le signe ∫ change brusquement de valeur," *Comp. Rend.*, 23, 1846, 537 and 557–69 = *Œuvres*, (1), 10, 133–34 and 135–43.
23. "Considérations nouvelles sur les intégrales définies qui s'étendent à tous les points d'une courbe fermée," *Comp. Rend.*, 23, 1846, 689–702 = *Œuvres*, (1), 10, 153–68.

point; and continuation around the path produces different values of the integral. But if one goes around the path enough times so that w returns to its original value, then the values of the integral will repeat and the integral is a periodic function of z. The periodicity modules (*indices de périodicité*) of the integral are not any longer, as in the case of single-valued functions, representable by residues. Cauchy's ideas on the integrals of multiple-valued functions were still vague.

For about twenty-five years, from 1821 on, Cauchy singlehandedly developed complex function theory. In 1843 fellow countrymen began to take up threads of his work. Pierre-Alphonse Laurent (1813–54), who worked alone, published a major result obtained in 1843. He showed[24] that where a function is discontinuous at an isolated point, in place of a Taylor's expansion one must use an expansion in increasing and decreasing powers of the variable. If the function and its derivative are single-valued and continuous in a ring whose center is the isolated point a, then the integral of the function taken over the two circular boundaries of the ring but in opposite directions and properly expanded gives a convergent expansion within the ring itself and with increasing and decreasing powers of z. This Laurent expansion is

$$(24) \qquad f(z) = \sum_{-\infty}^{\infty} a_n(z - a)^n$$

and is an extension of the Taylor expansion. The result was known to Weierstrass in 1841 but he did not publish it.[25]

The subject of multiple-valued functions was taken up by Victor-Alexandre Puiseux. In 1850 Puiseux published a celebrated paper[26] on complex algebraic functions given by $f(u, z) = 0$, f a polynomial in u and z. He first made clear the distinction between poles and branch-points that Cauchy barely perceived and introduced the notion of an essential singular point (a pole of infinite order), to which Weierstrass independently had called attention. Such a point is exemplified by $e^{1/z}$ at $z = 0$. Though Cauchy in the 1846 paper did consider the variation of simple multiple-valued functions along paths that enclosed branch-points, Puiseux clarified this subject too. He shows that if u_1 is one solution of $f(u, z) = 0$ and z varies along some path, the final value of u_1 does not depend upon the path, provided that the path does not enclose any point at which u_1 is infinite or any point where u_1 becomes equal to some other solution (that is, a branch-point).

24. *Comp. Rend.*, 17, 1843, 348–49; the full paper was published in the *Jour. de l'Ecole Poly.*, 23, 1863, 75–204.
25. *Werke*, 1, 51–66.
26. *Jour. de Math.*, 15, 1850, 365–480.

Puiseux also showed that the development of a function of z about a branch-point $z = a$ must involve fractional powers of $z - a$. He then improved on Cauchy's theorem on the expansion of a function in a Maclaurin series. Puiseux obtains an expansion for a solution u of $f(u, z) = 0$ not in powers of z but in powers of $z - c$ and therefore valid in a circle with c as center and not containing any pole or branch-point. Then Puiseux lets c vary along a path so that the circles of convergence overlap and so that the development within one circle can be extended to another. Thus, starting with a value of u at one point, one can follow its variation along any path.

By his significant investigations of many-valued functions and their branch-points in the complex plane, and by his initial work on integrals of such functions, Puiseux brought Cauchy's pioneering work in function theory to the end of what might be called the first stage. The difficulties in the theory of multiple-valued functions and integrals of such functions were still to be overcome. Cauchy did write other papers on the integrals of multiple-valued functions,[27] in which he attempted to follow up on Puiseux's work; and though he introduced the notion of branch-cuts (*lignes d'arrêt*), he was still confused about the distinction between poles and branch-points. This subject of algebraic functions and their integrals was to be pursued by Riemann (sec. 8).

In several other papers in the *Comptes Rendus* for 1851,[28] Cauchy gave some more careful statements about properties of complex functions. In particular, Cauchy affirmed that the continuity of the derivatives, as well as the continuity of the complex function itself, is needed for expansion in a power series. He also pointed out that the derivative at $z = a$ of u as a function of z is independent of the direction in the $x + iy$ plane along which z approaches a and that u satisfies $\partial^2 u/\partial x^2 + \partial^2 u/\partial y^2 = 0$.

In these 1851 papers Cauchy introduced new terms. He used *monotypique* and also *monodrome* when the function was single-valued for each value of z in some domain. A function is *monogen* if for each z it has just one derivative (that is, the derivative is independent of the path). A monodrome, monogenic function that is never infinite he called *synectique*. Later Charles A. A. Briot (1817–82) and Jean-Claude Bouquet (1819–85) introduced "holomorphic" in place of *synectique* and "meromorphic" if the function possessed just poles in that domain.

5. *Weierstrass's Approach to Function Theory*

While Cauchy was developing function theory on the basis of the derivatives and integrals of functions represented by analytic expressions, Karl

27. *Comp. Rend.*, 32, 1851, 68–75 and 162–64 = *Œuvres*, (1), 11, 292–300 and 304–5.
28. *Œuvres*, (1), 11.

Weierstrass developed a new approach. Born in Westphalia in 1815, Weierstrass entered the University of Bonn to study law. After four years in this effort he turned to mathematics in 1838 but did not complete the doctoral work. Instead he secured a state license to become a *gymnasium* (high school) teacher and from 1841 to 1854 he taught youngsters such subjects as writing and gymnastics. During these years he had no contact with the mathematical world, though he worked hard in mathematical research. The few results he published during this period secured for him in 1856 a position teaching technical subject matter in the Industrial Institute in Berlin. In that same year he became a lecturer at the University of Berlin and then professor in 1864. He remained in this post until he died in 1897.

He was a methodical and painstaking man. Unlike Abel, Jacobi, and Riemann he did not have flashes of intuition. In fact he distrusted intuition and tried to put mathematical reasoning on a firm basis. Whereas Cauchy's theory rested on geometrical foundations, Weierstrass turned to constructing the theory of real numbers; when this was done (Chap. 41, sec. 3), about 1841, he built up the theory of analytic functions on the basis of power series, a technique he learned from his teacher Christof Gudermann (1798–1852), and the process of analytic continuation. This work was performed in the 1840s, though he did not publish it then. He contributed to many other topics in the theory of functions and worked on the n-body problem in astronomy and in the theory of light.

It is difficult to date Weierstrass's creations because he did not publish many when he first achieved them. Much of what he had done became known to the mathematical world through his lectures at the University of Berlin. When he published his *Werke* in the 1890s, he did not worry about priority because many of his results had been published by others in the meantime. He was more concerned to present his method of developing function theory.

Power series to represent complex functions already given in analytical form were, of course, known. However, the task, given a power series that defines a function in a restricted domain, to derive other power series that define the same function in other domains on the basis of theorems on power series, was tackled by Weierstrass. A power series in $z - a$ that is convergent in some circle C of radius r about a represents a function that is analytic at each value of z in the circle C. By choosing a point b in the circle and using the values of the function and its derivatives as given by the original series, one may be able to obtain a new power series in $z - b$ whose circle of convergence C' overlaps the first circle. At the points common to the two circles the two series give the same value of the function. However, at the points of C' which are exterior to C, the values of the second series are an analytic continuation of the function defined by the first series. By continuing as far as possible from C' successively to still other circles, one obtains the entire analytic continuation of $f(z)$. The complete $f(z)$ is the collection of values at all points z in all of the circles. Each series is called an element of the function.

It is possible that, during the extension of the domain of the function by the adjunction of more and more circles of convergence, one of the new circles may cover part of a circle not immediately preceding it in the chain and the values of the function in this common part of the new circle and an earlier one may not agree. Then the function is multiple-valued.

The singular points (poles or branch-points) that may arise in this process, which necessarily lie on the boundaries of the circles of convergence of the power series, are included in the function by Weierstrass if the order of a singular point is finite, because at such a point an expansion in powers of $(z - z_0)^{1/n}$ having only a finite number of terms with negative exponents is possible. To obtain expansions about $z = \infty$ Weierstrass uses series in $1/z$. If the function element converges in the entire plane, Weierstrass calls it an entire function, and if it is not a rational integral function, that is, not a polynomial, then it has an essential singularity at ∞ (e.g. $\sin z$).

Weierstrass also gave the first example of a power series whose circle of convergence is its natural boundary; that is, the circle is a curve of singular points, and an example of an analytic expression that can represent different analytic functions in different parts of the plane.

6. *Elliptic Functions*

Paralleling the development of the basic theorems of complex function theory during the first half of the century was a special development dealing with elliptic and later Abelian functions. There is no doubt that Gauss obtained a number of key results in the theory of elliptic functions, because many of these were found after his death in papers he had never published. However, the acknowledged founders of the theory of elliptic functions are Abel and Jacobi.

Niels Henrik Abel (1802–29) was the son of a poor pastor. As a student in Christiania (Oslo), Norway, he had the luck to have Berndt Michael Holmböe (1795–1850) as a teacher. The latter recognized Abel's genius and predicted when Abel was seventeen that he would become the greatest mathematician in the world. After studying at Christiania and at Copenhagen, Abel received a scholarship that permitted him to travel. In Paris he was presented to Legendre, Laplace, Cauchy, and Lacroix, but they ignored him. Having exhausted his funds, he departed for Berlin and spent the years 1825–1827 with Crelle. He returned to Christiania so exhausted that he found it necessary, he wrote, to hold on to the gate of a church. To earn money he gave lessons to young students. He began to receive attention through his published works, and Crelle thought he might be able to secure him a professorship at the University of Berlin. But Abel became ill with tuberculosis and died in 1829.

Abel knew the work of Euler, Lagrange, and Legendre on elliptic integrals and may have gotten suggestions for the work he undertook from remarks made by Gauss, especially in his *Disquisitiones Arithmeticae*. He himself started to write papers in 1825. He presented his major paper on integrals to the Academy of Sciences in Paris on October 30, 1826, for publication in its journal. This paper, "Mémoire sur une propriété générale d'une classe très-étendue de fonctions transcendantes," contained Abel's great theorem (sec. 7). Fourier, the secretary of the Academy at the time, read the introduction to the paper and then referred the paper to Legendre and Cauchy for evaluation, the latter being chiefly responsible. The paper was long and difficult, only because it contained many new ideas. Cauchy laid it aside to favor his own work. Legendre forgot about it. After Abel's death, when his fame was established, the Academy searched for the paper, found it, and published it in 1841.[29] Abel published other papers in *Crelle's Journal* and Gergonne's *Annales* on the theory of equations and elliptic functions. These appeared from 1827 on. Because Abel's main paper of 1826 was not published until 1841, other authors, learning the more limited theorems published in between these dates, obtained independently many of Abel's 1826 results.

The other discoverer of elliptic functions was Carl Gustav Jacob Jacobi (1804–51). Unlike Abel, he lived a quiet life. Born in Potsdam to a Jewish family, he studied at the University of Berlin and in 1827 became a professor at Königsberg. In 1842 he had to give up his post because of ill health. He was given a pension by the Prussian government and retired to Berlin, where he died in 1851. His fame was great even in his own lifetime, and his students spread his ideas to many centers.

Jacobi taught the subject of elliptic functions for many years. His approach to it became the model according to which the theory of functions itself was developed. He also worked in functional determinants (Jacobians), ordinary and partial differential equations, dynamics, celestial mechanics, fluid dynamics, and hyperelliptic integrals and functions. Jacobi has often been labeled a pure mathematician, but, like almost all mathematicians of his own and preceding centuries, he took most seriously the investigation of nature.

While Abel was working on elliptic functions, Jacobi, who had also read Legendre's work on elliptic integrals, started work in 1827 on the corresponding functions. He submitted a paper to the *Astronomische Nachrichten*[30] without proofs. Almost simultaneously, Abel published independently his

29. *Mém. des sav. étrangers*, 7, 1841, 176–264 = *Œuvres*, 145–211.
30. *Astron. Nach.*, 6, 1827, 33–38 = *Werke*, 1, 31–36.

"Recherches sur les fonctions elliptiques."[31] Both had arrived at the key idea of working with inverse functions of the elliptic integrals, an idea Abel had had since 1823. Jacobi next gave proofs of the results he had published in 1827 in several articles of *Crelle's Journal* for the years 1828 to 1830. Thereafter, both published on the subject of elliptic functions; but whereas Abel died in 1829, Jacobi lived until 1851 and was able to publish much more. In particular, Jacobi's *Fundamenta Nova Theoriae Functionum Ellipticarum* of 1829[32] became a key work on elliptic functions.

Through letters from Jacobi, Legendre became familiar with Jacobi's and Abel's work. He wrote to Jacobi on February 9, 1828, "It is a great satisfaction to me to see two young mathematicians so successfully cultivate a branch of analysis which has long been my favorite field, but not at all received as it deserves in my own country." Legendre then published three supplements (1829 and 1832) to his *Traité des fonctions elliptiques* (2 vols., 1825–26), in which he gave accounts of Jacobi's and Abel's work.

The general elliptic integral concerns

$$(25) \qquad u = \int R(x, \sqrt{P(x)}),$$

where $P(x)$ is a third or fourth degree polynomial with distinct roots and $R(x, y)$ is a rational function of x and y. The efforts to deduce some general facts about the function u of x had to fail because the very meaning of the integral was limited for Euler and Legendre. The coefficients of $P(x)$ were real, and the range of x was real and moreover did not contain a root of $P(x) = 0$. With more knowledge of the theory of complex functions some advance might have been made in learning something about u as a function of x, but this knowledge was not available. As it turned out, Abel and Jacobi had a better idea.

To be specific, Legendre had introduced (Chap. 19, sec. 4) the elliptic integrals $F(k, \phi)$, $E(k, \phi)$, and $\pi(n, k, \phi)$. It was Abel who, about 1826, observed that if, to consider $F(k, \phi)$ for example, one studied

$$(26) \qquad u = \int_0^x \frac{dx}{\sqrt{(1 - x^2)(1 - k^2 x^2)}} = \int_0^\phi \frac{d\phi}{\sqrt{1 - k^2 \sin^2 \phi}},$$

where $x = \sin \phi$, then one encountered the same difficulties as when one studied

$$u = \int_0^x \frac{dx}{\sqrt{1 - x^2}} = \text{arc sin } x.$$

The nicer relationships come from studying x as a function of u. Hence Abel proposed to study x as a function of u in the case of the elliptic integrals. Since $x = \sin \phi$, one can also study ϕ as a function of u.

31. *Jour. für Math.*, 2, 1827, 101–81, and 3, 1828, 160–90 = *Œuvres*, 263–388.
32. *Werke*, 1, 49–239.

Jacobi introduced[33] the notation

$$\phi = am\, u$$

for the function ϕ of u defined by (26). He also introduced

$$\cos\phi = \cos am\, u \quad \text{and} \quad \Delta\phi = \Delta am\, u = \sqrt{1 - k^2 \sin^2 \phi}.$$

This notation was abbreviated by Gudermann to

$$x = \sin\phi = \sin am\, u = sn\, u, \qquad \cos\phi = \cos am\, u = cn\, u,$$
$$\Delta\phi = \Delta am\, u = dn\, u.$$

We have at once that

$$sn^2\, u + cn^2\, u = 1, \qquad dn^2\, u + k^2\, sn^2\, u = 1.$$

If ϕ is changed to $-\phi$, then u changes sign. Hence

$$am(-u) = -am\, u, \qquad sn(-u) = -sn\, u, \qquad cn(-u) = cn\, u,$$
$$dn(-u) = dn\, u.$$

The role of π in the trigonometric functions is played here by the quantity K defined by

$$K = \int_0^1 \frac{dx}{\sqrt{(1 - x^2)(1 - k^2 x^2)}} = \int_0^{\pi/2} \frac{d\phi}{\sqrt{1 - k^2 \sin^2 \phi}} = F\left(k, \frac{\pi}{2}\right).$$

Associated with K is the transcendental quantity K', which is the same function of k' as K is of k and where k' is defined by $k^2 + k'^2 = 1$, $0 < k < 1$.

What is important (but not proved here) about K and K' is that

$$sn(u \pm 4K) = sn\, u, \qquad cn(u + 4K) = cn\, u, \qquad dn(u \pm 2K) = dn\, u.$$

Hence $4K$ is a period of the elliptic functions $sn\, u$ and $cn\, u$, and $2K$ is a period of $dn\, u$.

The functions $sn\, u$, $cn\, u$, and $dn\, u$ are defined thus far for real x and u. Abel had what he regarded as an element of each function because each was defined just for real values. His next thought was to define the elliptic functions in their totality by introducing complex values of u. As for the knowledge of complex functions, Abel in his visit to Paris had become acquainted with Cauchy's work. He had in fact studied the binomial theorem for complex values of the variable and the exponent. The extension first to pure imaginary values was achieved by what is called Jacobi's imaginary transformation. Abel introduced

$$\sin\theta = i \tan\phi, \qquad \cos\theta = \frac{1}{\cos\phi}, \qquad \Delta(\theta, k) = \frac{\Delta(\phi, k')}{\cos\phi},$$

33. *Fundamenta Nova*, 1829.

where $\theta = am\ i\,u$, so that

$$sn(iu, k) = i\,\frac{sn(u, k')}{cn(u, k')}, \qquad cn(iu, k) = \frac{1}{cn(u, k')}, \qquad dn(iu, k) = \frac{dn(u, k')}{cn(u, k')}.$$

In addition to allowing his variables to take on pure imaginary values, Abel also developed what is called the addition theorem for elliptic functions. In the case where

$$u = A(x) = \int_0^x \frac{1}{\sqrt{1 - x^2}}\,dx$$

we know that the integral is the multiple-valued function $A(x) = $ arc sin x, and it is true that

(27) $$A(x_1) + A(x_2) = A(x_1 y_2 + x_2 y_1),$$

wherein y_1 and y_2 are the corresponding cosine values; i.e. $y_1 = \sqrt{1 - x_1^2}$. But in this case a great simplification is obtained by introducing the inverse function $x = \sin u$, which is single-valued, and in place of (27) we have the familiar addition theorem for the sine function. Now in the case of

$$u = E(x) = \int_0^x \frac{dx}{\sqrt{R(x)}}$$

where $y^2 = R(x)$ is a polynomial of degree four, Euler had obtained the addition theorem (Chap. 19, sec. 4)

$$E(x_1) + E(x_2) = E(x_3)$$

where x_3 is a known rational function of x_1, x_2, y_1, and y_2, and $y = \sqrt{R(x)}$. Abel thought that for the inverse function $x = \phi(u)$ there might be a simple addition theorem, and this proved to be the case. This result too appeared in his 1827 paper. Thus for real u and v

(28) $$sn(u + v) = \frac{sn\ u\ cn\ v\ dn\ v + sn\ v\ cn\ u\ dn\ u}{1 - k^2 sn^2\ u\ sn^2\ v}$$

with analogous formulas for $cn(u + v)$ and $dn(u + v)$. These are the addition theorems for elliptic functions and the analogues of the addition theorems for elliptic integrals.

Having defined the elliptic functions for real and imaginary values of the arguments, Abel was able with the addition theorems to extend the definitions to complex values. For, if $z = u + iv$, $sn\ z = sn(u + iv)$ now has a meaning in view of the addition theorem, in terms of the sn, cn, and dn of u and of iv separately.

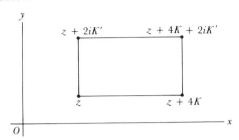

Figure 27.5

It also follows that

$$sn(iu + 2iK', k) = sn(iu, k)$$

(29)

$$cn(iu + 4iK', k) = cn(iu, k)$$

$$dn(iu + 4iK', k) = dn(iu, k).$$

Thus the periods (which are not unique) of $sn\ z$ are $4K$ and $2iK'$; those of $cn\ z$ are $4K$ and $2K + 2iK'$; and those of $dn\ z$ are $2K$ and $4iK'$. The important point about the periods is that there are two periods (whose ratio is not real), so that the elliptic functions are doubly periodic. This was one of Abel's great discoveries. The functions are single-valued. They need, therefore, to be studied only in one parallelogram (Fig. 27.5) of the complex plane, for they repeat their behavior in every congruent parallelogram. Elliptic functions besides being single-valued and doubly periodic, have only one essential singularity, at ∞. In fact these properties can be used to define elliptic functions. They do have poles within each period parallelogram.

Though Abel had taken from Legendre what could have been the cream of his life's work by introducing the idea of the inversion of elliptic integrals, which Legendre overlooked and which proved to be the key to exploring them, Legendre praised Abel by saying, "What a head this young Norwegian has." Charles Hermite said Abel left ideas on which mathematicians could work for 150 years.

Many of Abel's results were obtained independently by Jacobi, whose first publication in this field, as already noted, appeared in 1827. Jacobi was conscious of the fact that the basic method he used in his *Fundamenta Nova* was unsatisfactory and in that book to some extent, and in his subsequent lectures, used a different starting point. His lectures were never completely published, but their essential content is fairly well known through letters and notes made by his pupils. In his new approach he built his theory of elliptic functions on the basis of auxiliary functions called theta functions, which are illustrated by

(30)
$$\theta(z) = \sum_{n=-\infty}^{\infty} e^{-n^2 t + 2niz},$$

where z and t are complex and $Re(t) > 0$. The series converge absolutely and uniformly in any bounded region of the z-plane. Jacobi introduced four theta functions and then expressed $sn\ u$, $cn\ u$, and $dn\ u$ in terms of these functions. The theta functions are the simplest elements out of which the elliptic functions can be constructed. He also obtained various expressions for the theta functions in the form of infinite series and infinite products. Further study of ideas in Abel's work led Jacobi to relations between the theta functions and the theory of numbers. This connection was taken up later by Hermite, Kronecker, and others. The pursuit of relations among many different forms of theta functions was a major activity for mathematicians of the nineteenth century. It was one of the many fads that sweep through mathematics regularly.

In an important paper of 1835 [34] Jacobi showed that a single-valued function of a single variable which for every finite value of the argument has the character of a rational function (that is, is a meromorphic function) cannot have more than two periods, and the ratio of the periods is necessarily a nonreal number. This discovery opened up a new direction of work, namely, the problem of finding all doubly periodic functions. As early as 1844 [35] Liouville, in a communication to the French Academy of Sciences, showed how to develop a complete theory of doubly periodic elliptic functions starting from Jacobi's theorem. This theory was a major contribution to elliptic functions. In the double periodicity Liouville had discovered an essential property of the elliptic functions and a unifying point of view for their theory, though the doubly periodic functions are a more general class than those designated by Jacobi as elliptic. However, all the fundamental properties of elliptic functions do hold for the doubly periodic functions.

Weierstrass took up the subject of elliptic functions about 1860. He learned Jacobi's work from Gudermann, and Abel's work through his papers. These impressed him so much that he constantly urged his pupils to read Abel. For his teacher's certificate he took up a problem assigned to him by Gudermann, namely, to represent the elliptic functions as quotients of power series. This he did. In his lectures as a professor he constantly reworked his theory of elliptic functions.

Legendre had reduced the elliptic integrals to three standard forms involving the square root of a fourth degree polynomial. Weierstrass arrived at three different forms involving the square root of a third degree polynomial,[36] namely

$$\int \frac{dx}{\sqrt{4x^3 - g_2 x - g_3}}, \int \frac{x\ dx}{\sqrt{4x^3 - g_2 x - g_3}}, \int \frac{dx}{(x - a)\sqrt{4x^3 - g_2 x - g_3}},$$

34. *Jour. für Math.*, 13, 1835, 55–78 = *Werke*, 2, 23–50.
35. *Comp. Rend.*, 19, 1844, 1261–63, and 32, 1851, 450–52.
36. *Sitzungsber. Akad. Wiss. zu Berlin*, 1882, 443–51 = *Werke*, 2, 245–55; see also *Werke*, 5.

and he introduced as the fundamental elliptic function that which results from "inverting" the first integral. That is, if

$$u = \int_0^x \frac{dx}{\sqrt{4x^3 - g_2 x - g_3}},$$

then the elliptic function x of u is Weierstrass's

$$x = \wp(u) = \wp(u \,|\, g_2, g_3).$$

In order that $\wp(u)$ should not degenerate into an exponential or trigonometric function, it is necessary that the discriminant $g_2^3 - 27g_3^2 \neq 0$, or, in other words, the three roots of the cubic in x should be unequal. Weierstrass's doubly periodic $\wp(u)$ plays the role of $sn\, u$ in the theory of Jacobi and furnishes the simplest doubly periodic function. He showed that every elliptic function can be expressed very simply in terms of $\wp(u)$ and the derivative of this function. The "trigonometry" of elliptic functions is simpler in Weierstrass's approach, but the functions of Jacobi and the elliptic integrals of Legendre are better for numerical work.

Weierstrass actually started with an element of his $\wp(u)$, namely,

$$\wp(u) = \frac{1}{u^2} + \frac{g_2}{4 \cdot 5} u^2 + \frac{g_3}{4 \cdot 7} u^4 + \cdots, \qquad (g_2, g_3 \text{ complex}),$$

which he obtained by solving the differential equation for dx/du given by the above integral, and then used the addition theorem for $\wp(u)$, in a manner similar to Abel's, to obtain the full function. Weierstrass's work completed, remodeled, and filled with elegance the theory of elliptic functions.

Though we shall not enter into specific details, we cannot leave the subject of elliptic functions without mentioning the work of Charles Hermite (1822–1901), who was a professor at the Sorbonne and at the Ecole Polytechnique. From his student days on, Hermite constantly occupied himself with the subject of elliptic functions. In 1892 he wrote, "I cannot leave the elliptic domain. Where the goat is attached she must graze." He produced basic results in the theory proper and studied the connection with the theory of numbers. He applied elliptic functions to the solution of the fifth degree polynomial equation and treated problems of mechanics involving the functions. He is known also for his proof of the transcendence of e and his introduction of the Hermite polynomials.

7. Hyperelliptic Integrals and Abel's Theorem

The successes achieved in the study of the elliptic integrals (25) and the corresponding functions encouraged the mathematicians to tackle a more difficult type, hyperelliptic integrals.

The hyperelliptic integrals are of the form

$$(31) \qquad \int R(x, y) \, dx,$$

where $R(x, y)$ is a rational function of x and y, $y^2 = P(x)$, and the degree of $P(x)$ is at least five. When $P(x)$ is of degree five or six, the integrals were called ultraelliptic in the mid-nineteenth century. To emphasize complex values it is common to write

$$(32) \qquad \int R(u, z) \, dz;$$

and $P(z)$ is often written as

$$(33) \qquad u^2 \equiv P(z) = A(z - e_1) \cdots (z - e_n).$$

Of course u is a multiple-valued function of z.

Among the integrals of the form (32) there are some that are everywhere finite. These basic ones are

$$(34) \qquad u_1 = \int \frac{dz}{u}, u_2 = \int \frac{z \, dz}{u}, \ldots, u_p = \int \frac{z^{p-1} \, dz}{u},$$

where u is given by (33) and $p = (n - 2)/2$ or $(n - 1)/2$ according as n is even or odd. For $n = 6$ (and so $p = 2$), there are two such integrals. The general integrals (32) have at most poles and logarithmically singular points, that is, singular points that behave like $\log z$ at $z = 0$. Those integrals of the first class, that is, those which are everywhere finite and so have no singular points, can always be expressed in terms of the p integrals (34), which are linearly independent.

For the case $n = 6$ (and $p = 2$) integrals of the second class are exemplified by

$$(35) \qquad \int \frac{z^2 \, dz}{\sqrt{P(z)}}, \quad \int \frac{z^3 \, dz}{\sqrt{P(z)}},$$

where $P(z)$ is a polynomial of the sixth degree. The integrals of the first and second kind for $n = 6$ have four periods each.

The hyperelliptic integrals are functions of the upper limit z if the lower limit is fixed. Suppose we denote one such function by w. Then, as in the case of the elliptic integrals, one can raise the question as to what is the inverse function z of w. This problem was tackled by Abel, but he did not solve it. Jacobi then tackled it.[37] Let us consider with Jacobi the particular hyperelliptic integrals

$$(36) \qquad w = \int_0^z \frac{dz}{\sqrt{P(z)}}, \quad \text{and} \quad w = \int_0^z \frac{z \, dz}{\sqrt{P(z)}},$$

37. *Jour. für Math.*, 9, 1832, 394–403 = *Werke*, 2, 7–16, and *Jour. für Math.*, 13, 1835, 55–78 = *Werke*, 2, 25–50 and 516–21.

where $P(z)$ is a fifth or sixth degree polynomial. Here the determination of z as a single-valued function of w proved to be hopeless. Jacobi showed in fact that the mere inversion of such integrals with $P(z)$ of degree five did not lead to a monogenic function. The inverse functions appeared unreasonable to Jacobi because in each case z as a function of w is infinitely valued; such functions were not well understood at the time.

Jacobi decided to consider combinations of such integrals. Guided by Abel's theorem (see below), whose statement at least he knew because that much had been published by this time, Jacobi did the following. Consider the equations

$$(37) \qquad \int_0^{z_1} \frac{dz}{\sqrt{P(z)}} + \int_0^{z_2} \frac{dz}{\sqrt{P(z)}} = w_1$$

$$(38) \qquad \int_0^{z_1} \frac{z\,dz}{\sqrt{P(z)}} + \int_0^{z_2} \frac{z\,dz}{\sqrt{P(z)}} = w_2.$$

Jacobi succeeded in showing that the symmetric functions $z_1 + z_2$ and $z_1 z_2$ are each single-valued functions of w_1 and w_2 with a system of *four* periods. The functions z_1 and z_2 of the two variables w_1 and w_2 can then be obtained. He also gave an addition theorem for these functions. Jacobi left many points incomplete. "For Gaussian rigor," he said, "we have no time."

The study of a generalization of the elliptic and hyperelliptic integrals was begun by Galois, but the more significant initial steps were taken by Abel in his 1826 paper. He considered (32), that is,

$$(39) \qquad \int R(u, z)\, dz,$$

but in place of (33), where u and z are connected merely by a polynomial as in $u^2 = P(z)$, Abel considered a general algebraic equation in z and u,

$$(40) \qquad f(u, z) = 0.$$

Equations (39) and (40) define what is called an *Abelian integral*, which then includes as special cases the elliptic and hyperelliptic integrals.

Though Abel did not carry the study of Abelian integrals very far, he proved a key theorem in the subject. Abel's basic theorem is a very broad generalization of the addition theorem for elliptic integrals (Chap. 19, sec. 4). The theorem and proof is in his Paris paper of 1826 and the statement of it is in *Crelle's Journal* for 1829.[38] Let us consider the integral

$$(41) \qquad \int R(x, y)\, dx,$$

38. *Jour. für Math.*, 4, 212–15 = *Œuvres*, 515–17.

where x and y are related by $f(x, y) = 0$, f being a polynomial in x and y. Abel writes as though x and y are real variables, though occasionally complex numbers appear. Loosely stated, Abel's theorem is this: A sum of integrals of the form (41) can be expressed in terms of p such integrals plus algebraic and logarithmic terms. Moreover, the number p depends only on the equation $f(x, y) = 0$ and is in fact the genus of the equation.

To obtain a more precise statement, let y be the algebraic function of x defined by the equation

$$(42) \qquad f(x, y) = y^n + A_1 y^{n-1} + \cdots + A_n = 0,$$

where the A_i are polynomials in x and the polynomial (42) is irreducible into factors of the same form. Let $R(x, y)$ be any rational function of x and y. Then the sum of any number m of similar integrals

$$(43) \qquad \int^{(x_1, y_1)} R(x, y) \, dx + \cdots + \int^{(x_m, y_m)} R(x, y) \, dx$$

with fixed (but arbitrary) lower limits is expressible by rational functions of $x_1, y_1, \ldots, x_m,$ and y_m and logarithms of such rational functions, with the addition of a sum of a certain number p of integrals

$$(44) \qquad \int^{(z_1, s_1)} R(x, y) \, dx, \cdots, \int^{(z_p, s_p)} R(x, y) \, dx,$$

wherein $z_1, \cdots, z_p$ are values of x determinable from $x_1, y_1, \cdots, x_m,$ and y_m as the roots of an algebraic equation whose coefficients are rational functions of $x_1, y_1, \cdots, x_m,$ and $y_m,$ and $s_1, \cdots,$ and s_p are the corresponding values of y determined by (42) with any s_i determinable as a rational function of the z_i and the $x_1, y_1, \cdots, x_m,$ and y_m. The relations thus determining $(z_1, s_1), \cdots,$ (z_p, s_p) in terms of $(x_1, y_1), \cdots, (x_m, y_m)$ must be supposed to hold at all stages of the integration; in particular these relations determine the lower limits of the last p integrals in terms of the lower limits of the first m integrals. The number p does not depend on m nor on the form of the rational function $R(x, y)$, nor on the values $x_1, y_1, \cdots, x_m,$ and $y_m,$ but does depend on the fundamental equation (42) which relates y and x.

In the case of those hyperelliptic integrals where $f = y^2 - P(x)$ and $P(x)$ is a sixth degree polynomial and where p, which is $(n - 2)/2$, is 2, the main part of Abel's theorem says that

$$(45) \qquad \int_0^{x_1} R(x, y) \, dx + \cdots + \int_0^{x_m} R(x, y) \, dx$$

$$= \int_0^A R(x, y) \, dx + \int_0^B R(x, y) \, dx + R_1(x_1, y_1, \cdots, x_m, y_m, A, y(A), B, y(B))$$

$$+ \sum \text{const.} \log R_2(x_1, y_1, \cdots, x_m, y_m, A, y(A), B, y(B)),$$

where R_1 and R_2 are rational functions of their variables.

Abel actually calculated the number p for a few cases of general $f(x, y) = 0$. Though he did not see the full significance of his result, he certainly recognized the notion of genus before Riemann and founded the subject of Abelian integrals. His paper was very difficult to understand, partly because he tried to prove what we would today call an existence theorem by actually computing the result. Later proofs simplified Abel's considerably (see also Chap. 39, sec. 4). Abel did not consider the inversion problem. All of the work on the inversion of hyperelliptic and Abelian integrals up to the entry of Riemann on the scene was hampered by the limited methods of handling multiple-valued functions.

8. Riemann and Multiple-Valued Functions

About 1850 one period of achievement in the theory of functions came to an end. Rigorous methods, such as Weierstrass provided, sharp delineation of results, and unquestionable existence proofs denote in any mathematical discipline an important but also the last stage of a development. Further development must be preceded by a period of free, numerous, disconnected, often accidentally discovered, and perhaps disordered creations. Abel's theorem was one such step. A new period of discovery in the theory of algebraic functions, their integrals, and the inverse functions is due to Riemann. Riemann actually offered a much broader theory, namely, the treatment of multiple-valued functions, thus far only touched upon by Cauchy and Puiseux, and thereby paved the way for a number of different advances.

Georg Friedrich Bernhard Riemann (1826–66) was a student of Gauss and Wilhelm Weber. He came to Göttingen in 1846 to study theology but soon turned to mathematics. His doctoral thesis of 1851, written under Gauss, and entitled "Grundlagen für eine allgemeine Theorie der Funktionen einer veränderlichen complexen Grösse"[39] is a basic paper in complex function theory. Three years later he became a *Privatdozent* at Göttingen, that is, he was privileged to give lectures and charge students a fee. To qualify as *Privatdozent* he wrote the *Habilitationsschrift*, "Über die Darstellbarkeit einer Function durch eine trigonometrische Reihe," and gave a qualifying lecture, the *Habilitationsvortrag*, "Über die Hypothesen welche der Geometrie zu Grunde liegen." These were followed by a number of other famous papers. Riemann succeeded Dirichlet as a professor of mathematics at Göttingen in 1859. He died of tuberculosis.

Riemann is often described as a pure mathematician, but this is far from correct. Though he made numerous contributions to mathematics proper, he was deeply concerned with physics and the relationship of

39. *Werke*, 3–43.

mathematics to the physical world. He wrote papers on heat, light, the theory of gases, magnetism, fluid dynamics, and acoustics. He attempted to unify gravitation and light and investigated the mechanism of the human ear. His work on the foundations of geometry sought to ascertain what is absolutely reliable about our knowledge of physical space (Chap. 37). He himself states that his work on physical laws was his chief interest. As a mathematician, he used geometrical intuition and physical arguments freely. It seems very likely, on the basis of evidence given by Felix Klein, that Riemann's ideas on complex functions were suggested to him by his studies on the flow of electrical currents along a plane. The potential equation is central in that subject and became so in Riemann's approach to complex functions.

The key idea in Riemann's approach to multiple-valued functions is the notion of a Riemann surface. The function $w^2 = z$ is multiple-valued and in fact there are two values of w for each value of z. To work with this function and keep the two sets of values $\sqrt{z}$ and $-\sqrt{z}$ apart, that is, to separate the branches, Riemann introduced one plane of z-values for each branch. Incidentally he also introduced a point on each plane corresponding to $z = \infty$. The two planes are regarded as lying one above the other and joined, first of all, at those values of z where the branches give the same w-values. Thus for $w^2 = z$ the two planes, or sheets as they are called, are joined at $z = 0$ and $z = \infty$.

Now $w = +\sqrt{z}$ is represented by the z-values only on the upper sheet and $w = -\sqrt{z}$ by z-values on the lower sheet. As long as one considers z values on the upper sheet, one understands that one must compute $w_1 = +\sqrt{z}$. However, as z moves in a circle around the origin on that sheet (Fig. 27.6) so that the θ in $z = \rho(\cos \theta + i \sin \theta)$ varies from 0 to 2π, $\sqrt{z}$ covers one half of a circle in the complex plane in which the w-values are mapped. Now let z move into the second sheet as it crosses, say, the positive x-axis. As z moves on this second sheet, we take for w the values given by $w_2 = -\sqrt{z}$. As z makes another circuit of the origin, so that θ ranges from 2π to 4π while in the second sheet, we obtain the range of values of $w_2 = -\sqrt{z}$ for this path, and the polar angle of these w-values ranges from π to 2π. As z crosses the positive x-axis again, we regard it as traveling over the first sheet. Thus by two circuits of z-values around the origin, one on each sheet, we get the full range of w-values for the function $w^2 = z$. Moreover, and this is essential, w becomes a single-valued function of z if z ranges over the Riemann surface, which is the aggregate of the two sheets.

To distinguish paths on one sheet from those on the other we agree in the case of $w^2 = z$ to regard the positive x-axis as a branch-cut. This joins the points $z = 0$ and $z = \infty$. That is, whenever z crosses this cut, that branch of w must be taken which belongs to the sheet into which z passes.

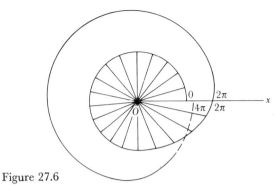

Figure 27.6

The branch-cut need not be the positive x-axis but it must, in the present case, join 0 and ∞. The points 0 and ∞ are called branch-points because the branches of $w^2 = z$ are interchanged when z describes a closed path about each.

The function $w^2 = z$ and, consequently, its associated Riemann surface, are especially simple. Let us consider the function $w^2 = z^3 - z$. This function also has two branches that become equal at $z = 0$, $z = 1$, $z = -1$, and $z = \infty$. Moreover, though we shall not present the full argument, all four of these points are branch-points, because if z makes a circuit around any one of them the value of w changes from that of one branch to that of another. The branch-cuts can be taken to be the line segments from 0 to 1, 0 to -1, 1 to ∞, and -1 to ∞. As z crosses any one of these cuts, the value of w changes from those it takes on one branch to those of the second.

For more complicated multiple-valued functions the Riemann surface is more complicated. An n-valued function requires an n-sheeted Riemann surface. There may be many branch-points, and one must introduce branch-cuts joining each two. Moreover the sheets that come together at one branch-point need not be the same as those that come together at another. If k sheets coincide at a branch-point, the order of the point is said to be $k - 1$. However, two sheets of a Riemann surface may touch at a point but the branches of the function may not change as z goes completely around the point. Then the point is not a branch-point.

It is not possible to represent Riemann surfaces accurately in three-dimensional space. For example, the two sheets for $w^2 = z$ must intersect along the positive x-axis if represented in three dimensions, so that one must be cut along the positive x-axis, whereas the mathematics requires smooth passage from the first sheet into the second and then, after a circuit around $z = 0$, back to the first sheet again.

Riemann surfaces are not merely a way of portraying multiple-valued functions but, in effect, make such functions single-valued on the surface as opposed to the z-plane. Then theorems about single-valued functions can be extended to multiple-valued functions. For example, Cauchy's theorem about integrals of *single-valued* functions being 0 over a curve bounding a domain (in which the function is analytic) was extended by Riemann to multiple-valued functions. The domain of analyticity must be simply connected (contractable to a point) on the *surface*.

Riemann thought of his surface as an n-sheeted duplication of the plane, each replica completed by a point at infinity. However, it is difficult to follow all of the arguments involving such a surface by visualizing in terms of the n interconnected planes. Hence mathematicians since Riemann's time have suggested equivalent models that are easier to contemplate. It is known that a plane may be transformed to a sphere by stereographic projection (Chap. 7, sec. 5). Hence we can construct a model of the Riemann surface by considering n concentric spheres of about the same radius. The sequence of spheres is to be the same as the sequence of plane sheets. The branch-points of the planes and the branch-cuts are likewise transformed to the spheres, so that these spheres wind into each other along branch-cuts. We now think of the set of spheres as the domain of z, and the multiple-valued function w of z is single-valued on this set of spherical sheets.

In our exposition thus far of Riemann's ideas, we have started with a function $f(w, z) = 0$, which is an irreducible polynomial in w and z, and have pointed out what its Riemann surface is. This was not Riemann's approach. He started with a Riemann surface and proposed to show that there is an equation $f(w, z) = 0$ that belongs to it and, further, that there are other single-valued and multiple-valued functions defined on this Riemann surface.

Riemann's definition of a single-valued analytic function $f(z) = u + iv$ is that the function is analytic at a point and in its neighborhood if it is continuous and differentiable and satisfies what we now call the Cauchy-Riemann equations

$$(46) \qquad \frac{\partial u}{\partial x} = \frac{\partial v}{\partial y} \quad \text{and} \quad \frac{\partial u}{\partial y} = -\frac{\partial v}{\partial x}.$$

These equations, as we know, appeared in the work of d'Alembert, Euler, and Cauchy. Incidentally, Riemann was the first to require that the existence of the derivative dw/dz mean that the limit of $\Delta w/\Delta z$ must be the same for *every* approach of $z + \Delta z$ to z. (This condition distinguishes complex functions, for in the case of a real function $u(x, y)$, the existence of the first derivatives of u for all directions of approach to some (x_0, y_0) does not guarantee analyticity.) He then sought what we might describe as the minimum conditions under which a function of $x + iy$ could be determined as a whole in whatever domain it existed. It is apparent from the Cauchy-

Riemann equations that u and v satisfy the two-dimensional potential equation

(47)
$$\frac{\partial^2 w}{\partial x^2} + \frac{\partial^2 w}{\partial y^2} = 0.$$

Riemann had the idea that the complex function could be determined at once in the totality of its domain of existence by using the fact that u satisfies the potential equation.

Specifically, Riemann supposes that a function w of position on the Riemann surface is determined, except as to an additive constant, by the real function $u(x, y)$ if u is subject to the conditions:

(1) It satisfies the potential equation at all points on the surface where its derivatives are not infinite.

(2) If u should be multiple-valued, its values at any point on the surface differ by linear combinations of integral multiples of real constants. (These real constants are the real parts of the periodicity moduli of w, which we shall discuss later.)

(3) u may have specified infinities (poles) of given form at assigned points on the surface. These infinities should belong to the real parts of the terms that give the infinities of w. He does assume further as a subsidiary condition that u may have finite values along a closed curve bounding a portion of the surface or that a relation between the boundary values of u and v may exist. Riemann is vague on how general such a relation may be.

These conditions should determine u. Once u is determined, then, in view of the Cauchy-Riemann equations,

(48)
$$v = \int \left(-\frac{\partial u}{\partial y}\, dx + \frac{\partial u}{\partial x}\, dy \right).$$

Thus v is also determined, and so is w. It is important to note that for Riemann the domain of u was any part of a Riemann surface, including possibly the whole surface. In his doctoral thesis he considered surfaces with boundaries and only later used closed surfaces, that is, surfaces without boundaries, such as a torus.

To determine u, Riemann's essential tool was what he called the Dirichlet principle, because he learned it from Dirichlet; but he extended it to domains on Riemann surfaces and, moreover, prescribed singularities for u in the domain and prescribed jumps (conditions 2 and 3 above). The Dirichlet principle says that the function u that minimizes the Dirichlet integral

$$\iint \left\{ \left(\frac{\partial u}{\partial x}\right)^2 + \left(\frac{\partial u}{\partial y}\right)^2 \right\} dx\, dy$$

satisfies the potential equation. The latter is in fact the necessary condition that the first variation of the Dirichlet integral vanish (see also Chap. 28, sec. 4). Since the integrand in the Dirichlet integral is positive and so has a lower bound which is greater than or at worst zero, Riemann concluded that there must be a function u that minimizes the integral and so satisfies the potential equation. Thus the existence of the function u and therefore by (48) of $f(z)$ which belongs to the Riemann surface and may even have prescribed singularities and complex jumps (periodicity modules) was assured as far as Riemann was concerned.

Once the existence of functions on a given Riemann surface as their domain has been established, one can show that there is a fundamental equation that can be associated with the given surface; that is, there is an $f(w, z) = 0$ that has the given surface as its surface. Just how the surface corresponds to the relationship between w and z Riemann does not state. Actually this $f(w, z) = 0$ is not unique. As a matter of fact, from every rational function w_1 of w and z on the surface one can obtain through $f(w, z) = 0$ another equation $f_1(w_1, z) = 0$ which, if irreducible, has the same Riemann surface. This is a feature of Riemann's method.

To investigate further the kinds of functions that can exist on a Riemann surface, it is necessary to become acquainted with Riemann's notion of the connectivity of a Riemann surface. A Riemann surface may have boundary curves or be closed like a sphere or a torus. If it is the Riemann surface of an algebraic function, that is, if $f(w, z) = 0$ defines w as a function of z and f is a polynomial in w and z, then the surface is closed. If f is irreducible, that is, cannot be expressed as a product of such polynomials, then the surface consists of one piece or is said to be connected.

A plane or a sphere is a surface such that any closed curve divides it into two parts so that it is not possible to pass continuously from a point in one part to a point in the second without crossing the closed curve. Such a surface is said to be *simply* connected. If, however, it is possible to draw some closed curve on a surface and the curve does not disconnect the surface, then the surface is not simply connected. For example one may draw two different closed curves on the torus (Fig. 27.7), and even the presence of both does not disconnect the surface.

Riemann wished to assign a number that indicated the connectivity of his surface. He regarded the poles and branch-points as part of the surface and because he had algebraic functions in mind, his surfaces were closed. By removing a small portion of one sheet the surface had a boundary curve C. He then thought of the surface as cut by a non-self-intersecting curve which runs from the boundary C to another point of the boundary C. Such a curve is called a cross-cut (*Querschnitt*). This cross-cut and C are regarded as a new boundary and a second cross-cut can be introduced that starts at one point of the (new) boundary and ends at another and does not cross the (new) boundary.

A sufficient number of these cross-cuts is introduced so as to cut up a

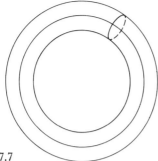

Figure 27.7

Riemann surface that may be multiply connected into one simply connected surface. Thus, if a surface is simply connected, no cross-cut is necessary and the surface has connectivity (*Grundzahl*) 1. A surface is called doubly connected if by one appropriate cross-cut it is changed into a single simply connected surface. Then the connectivity is 2. A plane ring and a spherical surface with two holes are examples. A surface is called triply connected when by two appropriate cross-cuts it is converted into a single simply connected surface. Then the connectivity is 3. An example is the (surface of a) torus with a hole in it. In general a surface will be said to be N-ply connected or has connectivity N if by $N - 1$ appropriate cross-cuts it can be changed into a single surface that is simply connected. A spherical surface with n holes in it has connectivity N.

It is now possible to relate the connectivity of a Riemann surface (with one boundary) and the number of branch-points. Each branch-point, r_i, say, is to be counted according to the multiplicity of the number of branches of the function which interchange at that point. If the number is $w_i, i = 1, 2, \ldots, r$, then the multiplicity of r_i is $w_i - 1$. Suppose the surface has q sheets. Then the connectivity N is given by

$$N = \sum_i w_i - 2q + 3.$$

One can show that the connectivity N of a closed surface with a single boundary is $2p + 1$. Hence

$$2p = \sum w_i - 2q + 2.$$

The integer p is called the genus of the Riemann surface and of the associated equation $f(w, z) = 0$. This relation was established by Riemann.

A special case of considerable importance is the surface of

$$w^2 = (z - a_1)(z - a_2) \cdots (z - a_n),$$

which has a two-leaved Riemann surface. There are n finite branch-points, and $z = \infty$ is a branch-point when n is odd. Then $\sum w_i = n$ or $n + 1$ and $2q = 4$. The genus p of the surface is given by

$$p = \begin{cases} \dfrac{n - 2}{2} & \text{when } n \text{ is even} \\[2em] \dfrac{n - 1}{2} & \text{when } n \text{ is odd.} \end{cases}$$

Given a Riemann surface determined by $f(w, z) = 0$, we know that w is a single-valued function of the points on the surface. Then every rational function of w and z is also a single-valued function of position on the surface (because we can replace w in the rational function by its value in terms of z). Also the branch-points of this rational function, though not its poles, are the same as those of f. Conversely, one can prove that every single-valued function of position on the surface having poles of finite order is a rational function of w and z.

Even in the case of simple single-valued functions defined on the ordinary plane, the integrals of such functions can be multiple-valued. Thus

$$\int_0^z \frac{dz}{1 + z^2} = w + n\pi$$

where w is the value of the integral along, say, a straight line path from 0 to z, and n depends upon how the path from 0 to z circles $\pm i$. Likewise, when considering functions single-valued on a Riemann surface, say a rational function of w and z on the surface, the integral of such a function may be multiple-valued. This does indeed occur. If one then introduces cross-cuts to make the surface simply connected, and if the integral takes a path from z_1 to z_2, each time the path crosses a cross-cut a constant value I is added to the basic value U of the integral for a path on a simply connected portion of the surface. If the path should cross the cross-cut m times in the same direction, then the value mI is added to U. The constant I is called a periodicity modulus. Each cross-cut introduces its own periodicity modulus, and if the connectivity of the surface is $N + 1$, there are N linearly independent periodicity moduli. Let these be $I_1, I_2, \ldots, I_n$. Then the value of the integral of the original single-valued function taken over its original path is

$$U + m_1 I_1 + m_2 I_2 + \cdots + m_n I_n,$$

where $m_1, m_2, \ldots, m_n$ are integers. The I_i are generally complex numbers.

9. *Abelian Integrals and Functions*

While four important papers Riemann published in the *Journal für Mathematik*[40] repeat many of the ideas in his dissertation, they are primarily devoted to Abelian integrals and functions. The fourth paper is the one that gave the subject its major development. All four were difficult to understand; "they were a book with seven seals." Fortunately many fine mathematicians later elaborated on and explained the material. Riemann pulled together the work of Abel and Jacobi, which stemmed largely from real functions, and Weierstrass's treatment, which used complex functions.

Since Riemann had clarified the concept of multiple-value functions, he could be clearer about Abelian integrals. Let $f(w, z) = 0$ be the equation of a Riemann surface and let $\int R(w, z) \, dz$ be an integral of a rational function of w and z on this Riemann surface. Riemann classified Abelian integrals as follows. Among the *integrals* of rational functions of w and z on a Riemann surface determined by an equation $f(w, z) = 0$, there are some which, though multiple-valued functions on the uncut surface, are everywhere finite. These are called integrals of the first kind. The number of such linearly independent integrals is equal to the genus p of the surface if the connectivity is $2p + 1$. If the $2p$ cross-cuts are introduced, each integral is a single-valued function for a path in a region bounded by cross-cuts. If the path should cross a cross-cut, then the periodicity moduli we discussed in the preceding section must be brought in and the value of the integral is this: If W is its value from a fixed point to z, then all possible values are

$$W + \sum_{r=1}^{2p} m_r \omega_r,$$

where the m_r are integers and the ω_r are periodicity moduli for this integral.

The integrals of the second kind have algebraic but not logarithmic infinities. An elementary integral of the second kind has an infinity of first order at one point on the Riemann surface. If $E(z)$ is a value of the integral at one point on the surface (the upper limit of the integral), then all the values of the integral are included in

$$E(z) + \sum_{r=1}^{2p} n_r \varepsilon_r,$$

where the n_r are integers and the ε_r are the periodicity moduli for this integral. Two elementary integrals with an infinity at a common point on the Riemann surface differ by an integral of the first kind. From this we can infer that there are $p + 1$ linearly independent elementary integrals of the second kind with an infinity at the same point on the Riemann surface.

40. Vol. 54, 1857, 115–55 = *Werke*, 88–144.

Integrals having logarithmic infinities are called integrals of the third kind. It turns out that each must have two logarithmic infinities. If such an integral has no algebraic infinities, that is, no algebraic terms in the expansion of the integral in the vicinity of either of the points at which it has logarithmic infinities, then the integral is called an *elementary* integral of the third kind. There are $p + 1$ linearly independent elementary integrals of the third kind having their logarithmic infinities at the same two points on the Riemann surface. Every Abelian integral is a sum of integrals of the three kinds.

The analysis of Abelian integrals sheds light on what kinds of functions can exist on a Riemann surface. Riemann treats two classes of functions; the first consists of single-valued functions on the surface whose singularities are poles. The second class consists of functions that are one-valued on the surface provided with cross-cuts but discontinuous along each cross-cut. Indeed, such a function differs by a complex constant h_v on one side of the vth cross-cut from its value on the other side. This second type of function may also have poles and logarithmic infinities. Riemann shows that the functions of the first class are algebraic and those of the second class are integrals of algebraic functions.

There are also everywhere finite functions on the surface. One can represent such a function by means of functions of the first of the above classes. One can also build up algebraic functions on a surface by combining integrals of the second and third kind. Thus Riemann showed that algebraic functions can be represented as a sum of transcendental functions. Also single-valued functions algebraically infinite in a given number of points can be represented by rational functions. A function that is one-valued on the entire surface is the integrand of an everywhere finite integral. The function can then be represented as a rational function in w and z and can have the form $\phi(w, z)/\partial f/\partial w$ where $f(w, z) = 0$ is the equation of the surface. The function ϕ, which enters here and into the construction of integrals of the first kind, is called the adjoint polynomial of $f(w, z) = 0$. Its degree is generally $n - 3$ when the degree of f is n.

The importance of rational functions on a Riemann surface derives from the fact, just mentioned, that every function single-valued on the surface and having no essential singularities is a rational function. Such a function has as many zeros as poles and takes on every value the same number of times. Moreover, once the equation $f(w, z) = 0$, which defines the surface, is fixed, all other functions of position on the surface are co-extensive in their totality with rational functions of w and z and integrals of such functions.

Weierstrass also worked on Abelian integrals during the 1860s. But he and the other successors of Riemann in this field set up transcendental functions from the algebraic functions, the reverse of Riemann's procedure.

They did so because they had reason to distrust the Dirichlet principle. Weierstrass, in a paper read in 1870,[41] pointed out that the existence of a function that minimizes the Dirichlet integral was not established. Riemann himself was of another mind. He recognized the problem of establishing the existence of a minimizing function for the Dirichlet integral before Weierstrass made his statement, but declared that the Dirichlet principle was just a convenient tool that happened to be available; the existence of the function u, he said, was nevertheless correct. Helmholtz's remark on this point is also interesting: ". . . for us physicists the [use of the] Dirichlet principle remains a proof."[42]

Another of the new investigations in complex function theory that Riemann launched is the inversion of Abelian integrals, that is, to determine the function z of u when

$$u = \int_0^z R(z, w) \, dz$$

and of course w and z are related by an algebraic equation. The function z of u is not only multiple-valued but is not clearly definable. As in the case of hyperelliptic integrals, Riemann took sums of p Abelian integrals and defined new *Abelian functions* of p variables that are one-valued and $2p$-tuply periodic. By a $2p$-tuply periodic function of p-variables is meant that there exist $2p$ sets of quantities $\omega_{1k}, \omega_{2k}, \ldots, \omega_{pk}, k = 1, 2, \ldots, 2p$, each set containing a period of each of the p-variables. Riemann proved that a single-valued function cannot have more than $2p$ sets of simultaneous periods. The Abelian functions, expressed in terms of theta functions in p-variables, are generalizations of the elliptic functions.

One of the noteworthy results on functions on a Riemann surface of genus p is now known as the Riemann-Roch theorem. The work on this result was begun by Riemann and completed by Gustav Roch (1839–66).[43] Essentially, the theorem determines the number of linearly independent meromorphic functions on the surface that have at most a specified finite set of poles. More specifically, suppose w is a single-valued function on the surface and has poles of first order at the points $c_1, \ldots, c_m$, but not elsewhere. The positions c_i need not be independent. If q linearly independent functions (adjoint functions) vanish on them, then w contains $m - p + q + 1$ arbitrary constants. It is a linear combination of arbitrary multiples of $m - p + q$ functions, each having $p - q + 1$ poles of the first order, $p - q$ of which are common to all the functions in the combination.

41. *Werke*, 2, 49–54.
42. For the subsequent history of the Dirichlet problem and Dirichlet principle, see Chap. 28, secs. 4 and 8.
43. *Jour. für Math.*, 64, 1864, 372–76.

10. *Conformal Mapping*

To complete the theory of his doctoral thesis, Riemann closes with some applications of the theory of functions to conformal mapping. The general problem of conformal mapping of a plane into a plane (which comes from map-making) was solved by Gauss in 1825. His result amounts to the fact that such a conformal map is set up by any analytic $f(z)$—though Gauss did not use complex function theory. Riemann knew that an analytic function establishes a conformal mapping from the z- to the w-plane, but he was concerned to extend this to Riemann surfaces. Thus a new chapter in conformal mapping was opened up.

At the close of his thesis Riemann gives the following theorem: Two given simply connected plane surfaces (he includes simply connected domains on Riemann surfaces) can be mapped one-to-one and conformally on each other, and one inner point and one boundary point on one surface can be assigned to arbitrarily chosen inner and boundary points on the other. Thereby the entire mapping is determined. This theorem contains as a special case the basic result that, given any simply connected domain D with a boundary that contains more than one point and given a point A of this domain and a direction T at this point, there exists a function $w = f(z)$ that is analytic in D and maps D conformally and biuniquely into a circle of radius 1 centered at the origin of the w-plane. Under this mapping A goes into the origin and T is sent into the direction of the positive real axis. This latter statement is usually described as the Riemann mapping theorem.

Riemann proved his theorem by using the Dirichlet principle, but since this principle was found faulty at the time, the mathematicians sought a sound proof. Carl Gottfried Neumann and Hermann Amandus Schwarz did prove (1870) that one could map a simply connected plane region onto a circle. However, they could not handle multileaved simply connected domains.

Incidentally, the emphasis on mapping a simply connected region conformally on a circle is explained by the fact that, to map one simply connected region on another conformally, it is sufficient to map each on a circle and then a product of two conformal mappings will do the trick.

While the proof of the Riemann mapping theorem remained open, a number of special results on conformal mapping were obtained. Of these a most useful one for the solution of partial differential equations was given by Schwarz[44] and Elwin Bruno Christoffel.[45] Their theorem shows how to map a polygon and its interior (Fig. 27.8) in the z-plane conformally into

44. *Jour. für Math.*, 70, 1869, 105–20 = *Ges. Abh.*, 2, 65–83.
45. *Annali di Mat.*, (2), 1, 1867, 95–103, and (2), 4, 1871, 1–9 = *Ges. Abh.*, 2, 56 ff.

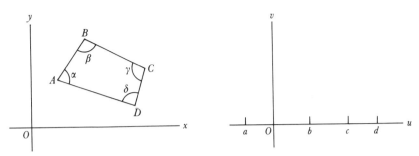

Figure 27.8

the upper half of the w-plane. The mapping is given by

$$z = c \int_a^z (w - a)^{(\alpha/\pi) - 1} (w - b)^{(\beta/\pi) - 1} \cdots dw + c',$$

where c and c' are determinable from the position of the polygon and where $a, b, c, \ldots$, correspond to $A, B, C, \ldots$. The mapping has proved to be very useful in solving the potential (Laplace) equation.

11. *The Representation of Functions and Exceptional Values*

The development of complex function theory proceeded apace in the latter part of the nineteenth century, and we shall have occasion in later chapters to consider some of these developments. However, a few of the many creations that bear primarily on complex functions themselves will be noted here.

Among single-valued complex functions, the entire functions, that is, those having no singularities in the finite part of the plane, which includes polynomials, e^z, $\sin z$, and $\cos z$, proved to be of considerable interest because they are roughly the analogues of the elementary real functions. For such functions Liouville's theorem states that every bounded entire function is a constant.[46] Weierstrass extended to entire functions the theorem on the decomposition of real polynomials into linear factors. The theorem Weierstrass established,[47] which probably dates from the 1840s, is called the factorization theorem, and states: If $G(z)$ is an entire function that does not vanish identically but has an infinite number of roots (that is, is not a polynomial), then $G(z)$ can be written as the infinite product

$$G(z) = \Gamma(z) z^m \prod_{n=1}^{\infty} \left(1 - \frac{z}{a_n}\right) e^{g_n(z)},$$

46. The theorem is due to Cauchy (*Comp. Rend.*, 19, 1844, 1377–81 = *Œuvres*, (1), 8, 378–85). C. W. Borchardt heard it in Liouville's lectures of 1847 and attributed the theorem to him.

47. *Abh. König. Akad. der Wiss.*, Berlin, 1876, 11–60 = *Werke*, 2, 77–124.

where

$$g_n(z) = \frac{z}{a_n} + \frac{1}{2}\left(\frac{z}{a_n}\right)^2 + \cdots + \frac{1}{m_n}\left(\frac{z}{a_n}\right)^{m_n}.$$

$\Gamma(z)$ is an entire function having no zeros; the a_n are the zeros of $G(z)$; and z^m represents the zero at $z = 0$ of multiplicity m if $G(z)$ has such a zero. The individual factors of the products are called the prime factors of $G(z)$.

Next to entire functions in order of complexity are the meromorphic functions, which can have only poles in the finite region of the complex plane. In his paper of 1876[48] Weierstrass showed that a meromorphic function can be expressed as the quotient of two entire functions. The theorem was extended by Gösta Mittag-Leffler (1846–1927) in a paper of 1877.[49] A function meromorphic in an arbitrary region can be expressed as the quotient of two functions, each analytic in the region. In both Weierstrass's and Mittag-Leffler's theorems the numerator and denominator do not vanish at the same point in the region.

Another topic that has engaged the attention of numerous mathematicians is the range of values that various types of complex functions can take on. A series of results was obtained by (Charles) Emile Picard (1856–1941), a professor of higher analysis at the Sorbonne and permanent secretary of the Paris Academy of Sciences. Picard in 1879[50] showed that an entire function can omit at most one finite value without reducing to a constant, and if there exist at least two values, each of which is taken on only a finite number of times, then the function is a polynomial. Otherwise the function takes on every value, other than the exceptional one, an infinite number of times. If the function is meromorphic, infinity being an admissible value, at most two values can be omitted without the function reducing to a constant.

In this same paper he extended a result of Julian W. Sochozki (1842–1927) and Weierstrass and proved that in any neighborhood of an isolated essential singular point a function takes on all values, with the possible exception of at most one (finite) value. The result is deep and has a multitude of consequences. Indeed a number of other results and alternative proofs were created, which carried the topic well into the twentieth century.

In the subject of complex functions, the nineteenth century ended with a return to fundamentals. The proofs of the Cauchy integral theorem in the nineteenth century used the fact that df/dz is continuous. Edouard Goursat (1858–1936) proved[51] Cauchy's theorem, $\int f(z)\,dz = 0$ around a closed

48. *Abh. König. Akad. der Wiss.*, Berlin, 1876, 11–60 = *Werke*, 2, 77–124.
49. *Öfversigt af Kongliga Vetenskops-Akademiens Förhandlingar*, 34, 1877, #1, 17–43; see also *Acta Math.*, 4, 1884, 1–79.
50. *Ann. de l'Ecole Norm. Sup.*, (2), 9, 1880, 145–66.
51. *Amer. Math. Soc., Trans.*, 1, 1900, 14–16.

curve C, without assuming the continuity of the derivative $f'(z)$ in the closed region bounded by the curve C. The existence of $f'(z)$ was sufficient. Goursat pointed out that the continuity of $f(z)$ and existence of the derivative are sufficient to characterize analyticity.

As our sketch of the rise of the theory of complex functions has shown, Cauchy, Riemann, and Weierstrass are the three major founders of the theory of functions. For a long time their respective ideas and methods were pursued independently by their followers. Then Cauchy's and Riemann's ideas were fused and Weierstrass's ideas were gradually deduced from the Cauchy-Riemann view, so that the idea of starting from the power series is no longer emphasized. Moreover, the rigor of the Cauchy-Riemann view was improved, so that from this standpoint also Weierstrass's approach is not essential. Full unification took place only at the beginning of the twentieth century.

Bibliography

Abel, N. H.: *Œuvres complètes*, 2 vols., 1881, Johnson Reprint Corp., 1964.
———: *Mémorial publié à l'occasion du centénaire de sa naissance*, Jacob Dybwad, Kristiania, 1902.
Brill, A., and M. Noether: "Die Entwicklung der Theorie der algebraischen Functionen in älterer und neuerer Zeit," *Jahres. der Deut. Math.-Verein.*, 3, 1892/3, 109–556, 155–86 in particular.
Brun, Viggo: "Niels Henrik Abel. Neue biographische Funde," *Jour. für Math.*, 193, 1954, 239–49.
Cauchy, A. L.: *Œuvres complètes*, 26 vols., Gauthier-Villars, 1882–1938, relevant papers.
Crowe, Michael J.: *A History of Vector Analysis*, University of Notre Dame Press, 1967, Chap. 1.
Enneper, A.: *Elliptische Funktionen: Theorie und Geschichte*, 2nd ed., L. Nebert, 1890.
Hadamard, Jacques: *Notice sur les travaux scientifiques de M. Jacques Hadamard*, Gauthier-Villars, 1901.
Jacobi, C. G. J.: *Gesammelte Werke*, 7 vols. and Supplement, G. Reimer, 1881–91; Chelsea reprint, 1968.
Jourdain, Philip E. B.: "The Theory of Functions with Cauchy and Gauss," *Bibliotecha Mathematica*, (3), 6, 1905, 190–207.
Klein, Felix: *Vorlesungen über die Entwicklung der Mathematik im 19. Jahrhundert*, 2 vols., Chelsea (reprint), 1950.
Lévy, Paul, et al.: "La Vie et l'œuvre de J. Hadamard," *L'Enseignement Mathématique*, (2), 13, 1967, 1–72.
Markuschewitsch, A. I.: *Skizzen zur Geschichte der analytischen Funktionen*, V E B Deutscher Verlag der Wissenschaften, 1955.
Mittag-Leffler, G.: "An Introduction to the Theory of Elliptic Functions," *Annals of Math.*, 24, 1922/3, 271–351.

————: "Die ersten 40 Jahre des Lebens von Weierstrass," *Acta Math.*, 39, 1923, 1–57.

Ore, O.: *Niels Henrik Abel, Mathematician Extraordinary*, University of Minnesota Press, 1957.

Osgood, W. F.: "Allgemeine Theorie der analytischen Funktionen," *Encyk. der Math. Wiss.*, B. G. Teubner, 1901–21, II B1, 1–114.

Reichardt, Hans, ed.: *Gauss: Leben und Werk*, Haude und Spenersche Verlagsbuchhandlung, 1960; B. G. Teubner, 1957, 151–82.

Riemann, Bernhard: *Gesammelte mathematische Werke*, 2nd ed., Dover (reprint), 1953.

Schlesinger, L.: "Über Gauss' Arbeiten zu Funktionenlehre," *Nachrichten König. Ges. der Wiss. zu Gött.*, 1912, Beiheft, 1–143; also in Gauss's *Werke*, 10_2, 77 ff.

Smith, David Eugene: *A Source Book in Mathematics*, 2 vols., Dover (reprint), 1959, pp. 55–66, 404–10.

Staeckel, Paul: "Integration durch imaginäres Gebiet," *Bibliotecha Mathematica*, (3), 1, 1900, 109–28.

Valson, C. A.: *La Vie et les travaux du baron Cauchy*, 2 vols., Gauthier-Villars, 1868.

Weierstrass, Karl: *Mathematische Werke*, 7 vols., Mayer und Müller, 1895–1924.

28

Partial Differential Equations in the Nineteenth Century

> The profound study of nature is the most fertile source of mathematical discoveries.　　　　JOSEPH FOURIER

1. *Introduction*

The subject of partial differential equations, which had its beginnings in the eighteenth century, burgeoned in the nineteenth. As physical science expanded, in both the variety and depth of the phenomena investigated, the number of new types of differential equations increased; and even the types already known, the wave equation and the potential equation, were applied to new areas of physics. Partial differential equations became and remain the heart of mathematics. Their importance for physical science is only one of the reasons for assigning them this central place. From the standpoint of mathematics itself, the solution of partial differential equations created the need for mathematical developments in the theory of functions, the calculus of variations, series expansions, ordinary differential equations, algebra, and differential geometry. The subject has become so extensive that in this chapter we can give only a few of the major results.

We are accustomed today to classifying partial differential equations according to types. At the beginning of the nineteenth century, so little was known about the subject that the idea of distinguishing the various types could not have occurred. The physical problems dictated which equations were to be pursued and the mathematicians passed freely from one type to another without recognizing some differences among them that we now consider fundamental. The physical world was and still is indifferent to the mathematicians' classification.

2. *The Heat Equation and Fourier Series*

The first big nineteenth-century step, and indeed one of enormous importance, was made by Joseph Fourier (1768–1830). Fourier did very well

671

as a young student of mathematics but had set his heart on becoming an army officer. Denied a commission because he was the son of a tailor, he turned to the priesthood. When he was offered a professorship at the military school he had attended he accepted and mathematics became his life interest.

Like other scientists of his time, Fourier took up the flow of heat. The flow was of interest as a practical problem in the handling of metals in industry and as a scientific problem in attempts to determine the temperature in the interior of the earth, the variation of that temperature with time, and other such questions. He submitted a basic paper on heat conduction to the Academy of Sciences of Paris in 1807.[1] The paper was judged by Lagrange, Laplace, and Legendre and was rejected. But the Academy did wish to encourage Fourier to develop his ideas, and so made the problem of the propagation of heat the subject of a grand prize to be awarded in 1812. Fourier submitted a revised paper in 1811, which was judged by the men already mentioned and others. It won the prize but was criticized for its lack of rigor and so not published at that time in the *Mémoires* of the Academy. Fourier resented the treatment he received. He continued to work on the subject of heat and, in 1822, published one of the classics of mathematics, *Théorie analytique de la chaleur.*[2] It incorporated the first part of his 1811 paper practically without change. This book is the main source for Fourier's ideas. Two years later he became secretary of the Academy and was able to have his 1811 paper published in its original form in the *Mémoires.*[3]

In the interior of a body that is gaining or losing heat, the temperature is generally distributed nonuniformly and changes at any one place with time. Thus the temperature T is a function of space and time. The precise form of the function depends upon the shape of the body, the density, the specific heat of the material, the initial distribution of T, that is, the distribution at time $t = 0$, and the conditions maintained at the surface of the body. The first major problem Fourier considered in his book was the determination of the temperature T in a homogeneous and isotropic body as a function of x, y, z, and t. He proved on the basis of physical principles that T must satisfy the partial differential equation, called the heat equation in three space dimensions,

$$(1) \qquad \left(\frac{\partial^2 T}{\partial x^2} + \frac{\partial^2 T}{\partial y^2} + \frac{\partial^2 T}{\partial z^2} \right) = k^2 \frac{\partial T}{\partial t},$$

where k^2 is a constant whose value depends on the material of the body.

1. The manuscript is in the library of the *Ecole des Ponts et Chaussées*.
2. *Œuvres*, 1.
3. *Mém. de l'Acad. des Sci.*, Paris, (2), 4, 1819/20, 185–555, pub. 1824, and 5, 1821/22, 153–246, pub. 1826; only the second part is reproduced in Fourier's *Œuvres*, 2, 3–94.

Fourier then solved specific heat conduction problems. We shall consider a case that is typical of his method, the problem of solving equation (1) for the cylindrical rod whose ends are kept at $0°$ temperature and whose lateral surface is insulated so that no heat flows through it. Since this rod involves only one space dimension, (1) becomes

$$(2) \qquad \frac{\partial^2 T}{\partial x^2} = k^2 \frac{\partial T}{\partial t}$$

subject to the boundary conditions

$$(3) \qquad T(0, t) = 0, \qquad T(l, t) = 0, \qquad \text{for } t > 0,$$

and the initial condition

$$(4) \qquad T(x, 0) = f(x) \qquad \text{for } 0 < x < l.$$

To solve this problem Fourier used the method of separation of variables. He let

$$(5) \qquad T(x, t) = \phi(x)\psi(t).$$

On substitution in the differential equation, he obtained

$$\frac{\phi''(x)}{k^2\phi(x)} = \frac{\psi'(t)}{\psi(t)}.$$

He then argued (cf. [30] of Chap. 22) that each of these ratios must be a constant, $-\lambda$ say, so that

$$(6) \qquad \phi''(x) + \lambda k^2 \phi(x) = 0$$

and

$$(7) \qquad \psi'(t) + \lambda\psi(t) = 0.$$

However the boundary conditions (3), in view of (5), imply that

$$(8) \qquad \phi(0) = 0 \quad \text{and} \quad \phi(l) = 0.$$

The general solution of (6) is

$$\phi(x) = b \sin (\sqrt{\lambda}kx + c).$$

The condition that $\phi(0) = 0$ implies that $c = 0$. The condition $\phi(l) = 0$ imposes a limitation on λ, namely that $\sqrt{\lambda}$ must be an integral multiple of π/kl. Hence there are an infinite number of admissible values λ_ν of λ or

$$(9) \qquad \lambda_\nu = \left(\frac{\nu\pi}{kl}\right)^2, \qquad \nu \text{ integral.}$$

These λ_ν are what we now call the eigenvalues or characteristic values.

Since the general solution of (7) is an exponential function but λ is now limited to the λ_ν, then, in view of (5), Fourier had, so far, that

$$T_\nu(x, t) = b_\nu e^{-(\nu^2\pi^2/k^2l^2)t} \sin \frac{\nu\pi x}{l},$$

where b_ν now denotes the constant in place of b and $\nu = 1, 2, 3, \ldots.$ However the equation (2) is linear, so that a sum of solutions is a solution. Hence one can assert that

$$(10) \qquad T(x, t) = \sum_{\nu=1}^{\infty} b_\nu e^{-(\nu^2\pi^2/k^2l^2)t} \sin \frac{\nu\pi x}{l}.$$

To satisfy the initial condition (4), one must have for $t = 0$

$$(11) \qquad f(x) = \sum_{\nu=1}^{\infty} b_\nu \sin \frac{\nu\pi x}{l}.$$

Fourier then faced the question, Can $f(x)$ be represented as a trigonometric series? In particular, can the b_ν be determined?

Fourier proceeded to answer these questions. Though by this time he was somewhat conscious of the problem of rigor, he proceeded formally in the eighteenth-century spirit. To follow Fourier's work we shall, for simplicity, let $l = \pi$. Thus we consider

$$(12) \qquad f(x) = \sum_{\nu=1}^{\infty} b_\nu \sin \nu x, \qquad \text{for } 0 < x < \pi.$$

Fourier takes each sine function and expands it by Maclaurin's theorem into a power series; that is, he uses

$$(13) \qquad \sin \nu x = \sum_{n=1}^{\infty} \frac{(-1)^{n-1}\nu^{2n-1}}{(2n-1)!} x^{2n-1}$$

to replace $\sin \nu x$ in (12). Then, by interchanging the order of the summations, an operation unquestioned at the time, he obtains

$$(14) \qquad f(x) = \sum_{n=1}^{\infty} \frac{(-1)^{n-1}}{(2n-1)!} \left(\sum_{\nu=1}^{\infty} \nu^{2n-1} b_\nu \right) x^{2n-1}.$$

Thus $f(x)$ is expressed as a power series in x, which implies a strong restriction on the admissible $f(x)$ that was not presupposed for the $f(x)$ Fourier treats. This power series must be the Maclaurin series for $f(x)$, so that

$$(15) \qquad f(x) = \sum_{k=0}^{\infty} \frac{1}{k!} f^{(k)}(0) x^k.$$

By equating coefficients of like powers of x in (14) and (15), Fourier finds that $f^{(k)}(0) = 0$ for even k and beyond that

$$\sum_{v=1}^{\infty} v^{2n-1} b_v = (-1)^{n-1} f^{(2n-1)}(0), \qquad n = 1, 2, 3, \cdots.$$

Now the derivatives of $f(x)$ are known, because $f(x)$ is a given initial condition. Hence the b_v are an infinite set of unknowns in an infinite system of linear algebraic equations.

In a previous problem, wherein he faced this same kind of system, Fourier took the first k terms and the right-hand constant of the first k equations, solved these, and by obtaining a general expression for $b_{v,k}$, which denotes the approximate value of b_v obtained from the first k equations, he boldly concluded that $b_v = \lim_{k \to \infty} b_{v,k}$. However, this time he had much difficulty in determining the b_v. He took several different $f(x)$'s and showed how to determine the b_v by very complicated procedures that involved divergent expressions. Using these special cases as a guide he obtained an expression for b_v involving infinite products and infinite sums. Fourier realized that this expression was rather useless, and by further bold and ingenious, though again often questionable, steps finally arrived at the formula

(16) $$b_v = \frac{2}{\pi} \int_0^{\pi} f(s) \sin vs \, ds.$$

The conclusion was to an extent not new. We have already related (Chap. 20, sec. 5) how Clairaut and Euler had expanded some functions in Fourier series and had obtained the formulas

(17) $\quad a_n = \frac{1}{\pi} \int_{-\pi}^{\pi} f(x) \cos nx \, dx, \qquad b_n = \frac{1}{\pi} \int_{-\pi}^{\pi} f(x) \sin nx \, dx, \qquad n \geq 1.$

Moreover, Fourier's results as derived thus far were limited, because he assumed his $f(x)$ had a Maclaurin expansion, which means an infinite number of derivatives. Finally, Fourier's method was certainly not rigorous and was more complicated than Euler's. Whereas Fourier had to use an infinite system of linear algebraic equations, Euler proceeded more simply by using the properties of trigonometric functions.

But Fourier now made some remarkable observations. He noted that each b_v can be interpreted as the area under the curve of $y = (2/\pi) f(x) \cdot \sin vx$ for x between 0 and π. Such an area makes sense even for very arbitrary functions. The functions need not be continuous or could be known only graphically. Hence Fourier concluded that *every* function $f(x)$ could be represented as

$$(18) \qquad f(x) = \sum_{\nu = 1}^{\infty} b_\nu \sin \nu x, \qquad \text{for } 0 < x < \pi.$$

This possibility had, of course, been rejected by the eighteenth-century masters except for Daniel Bernoulli.

How much Fourier knew of the work of his predecessors is not clear. In a paper of 1825 he says that Lacroix had informed him of Euler's work but he does not say when this happened. In any case Fourier was not deterred by the opinions of his predecessors. He took a great variety of functions $f(x)$, calculated the first few b_ν for each function, and plotted the sum of the first few terms of the sine series (18) for each one. From this graphical evidence he concluded that the series always represents $f(x)$ over $0 < x < \pi$, whether or not the representation holds outside this interval. He points out in his book (p. 198) that two functions may agree in a given interval but not necessarily outside that interval. The failure to see this explains why earlier mathematicians could not accept that an arbitrary function can be expanded in a trigonometric series. What the series does give is the function in the domain 0 to π, in the present case, and periodic repetitions of it outside.

Once Fourier obtained the above simple result for the b_ν, he, like Euler, realized that each b_ν can be obtained by multiplying the series (18) by $\sin \nu x$ and integrating from 0 to π. He also points out that this procedure is applicable to the representation

$$(19) \qquad f(x) = \frac{a_0}{2} + \sum_{\nu = 1}^{\infty} a_\nu \cos \nu x.$$

He considers next the representation of any $f(x)$ in the interval $(-\pi, \pi)$. The series (18) represents an odd function $[f(x) = -f(-x)]$ and the series (19) an even function $[f(x) = f(-x)]$. But any function can be represented as the sum of an odd function $f_0(x)$ and an even function $f_e(x)$ where

$$f_0(x) = \frac{1}{2} [f(x) - f(-x)], \qquad f_e = \frac{1}{2} [f(x) + f(-x)].$$

Then any $f(x)$ can be represented in the interval $(-\pi, \pi)$ by

$$(20) \qquad f(x) = \frac{a_0}{2} + \sum_{\nu = 1}^{\infty} (a_\nu \cos \nu x + b_\nu \sin \nu x)$$

and the coefficients can be determined by multiplying through by $\cos \nu x$ or $\sin \nu x$ and integrating from $-\pi$ to π, which yields (17).

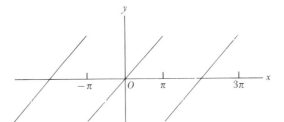

Figure 28.1

Fourier never gave any complete proof that an "arbitrary" function can be represented by a series such as (20). In the book he gives some loose arguments, and in his final discussion of this point (paragraphs 415, 416, and 423) he gives a sketch of a proof. But even there Fourier does not state the conditions that a function must satisfy to be expansible in a trigonometric series. Nevertheless Fourier's conviction that this was possible is expressed throughout the book. He also says[4] that his series are convergent no matter what $f(x)$ may be, whether or not one can assign an analytic expression to $f(x)$ and whether or not the function follows any regular law. Fourier's conviction that any function can be expanded in a Fourier series rested on the geometrical evidence described above. About this he says in his book (p. 206), "Nothing has appeared to us more suitable than geometrical constructions to demonstrate the truth of the new results and to render intelligible the forms which analysis employs for their expressions."

Fourier's work incorporated several major advances. Beyond furthering the theory of partial differential equations, he forced a revision in the notion of function. Suppose the function $y = x$ is represented by a Fourier series (20) in the interval $(-\pi, \pi)$. The series repeats its behavior in each interval of length 2π. Hence the function given by the series looks as shown in Figure 28.1. Such functions cannot be represented by a single (finite) analytic expression, whereas Fourier's predecessors had insisted that a function must be representable by a single expression. Since the entire function $y = x$ for all x is not represented by the series, they could not see how an arbitrary function, which is not periodic, could be represented by the series, though Euler and Lagrange had actually done so for particular nonperiodic functions. Fourier is explicit that his series can represent functions that also have different analytical expressions in different parts of the interval $(0, \pi)$ or $(-\pi, \pi)$, whether or not the expressions join one another continuously. He points out, finally, that his work settles the arguments on solutions of the vibrating-string problem in favor of Daniel Bernoulli. Fourier's work marked the break from analytic functions or functions developable in Taylor's series. It is also significant that a Fourier series

4. Page 196 = _Œuvres_, 1, 210.

represents a function over an entire interval, whereas a Taylor series represents a function only in a neighborhood of a point at which a function is analytic, though in special cases the radius of convergence may be infinite.

We have already noted that Fourier's paper of 1807, in which he had maintained that an arbitrary function can be expanded in a trigonometric series, was not well received by the Academy of Sciences of Paris. Lagrange in particular denied firmly the possibility of such expansions. Though he criticized only the lack of rigor in the paper, he was certainly disturbed by the generality of the functions that Fourier entertained, because Lagrange still believed that a function was determined by its values in an arbitrarily small interval, which is true of analytic functions. In fact Lagrange returned to the vibrating-string problem and, with no better insight than he had shown in earlier work, insisted on defending Euler's contention that an arbitrary function cannot be represented by a trigonometric series. Poisson did assert later that Lagrange had shown that an arbitrary function can be represented by a Fourier series but Poisson, who was envious of Fourier, said this to rob Fourier of the credit and give it to Lagrange.

Fourier's work made explicit another fact that was also implicit in the eighteenth-century work of Euler and Laplace. These men had expanded functions in series of Bessel functions and Legendre polynomials in order to solve specific problems. The general fact that a function might be expanded in a series of functions such as the trigonometric functions, Bessel functions, and Legendre polynomials was thrust into the light by Fourier's work. He showed, further, how the initial condition imposed on the solution of a partial differential equation could be met, and so advanced the technique of solving such equations. Fourier's paper of 1811, though not published until 1824–26, was accessible to other men in the meantime, and his ideas, at first grudgingly accepted, finally won favor.

Fourier's method was taken up immediately by Siméon-Denis Poisson (1781–1840), one of the greatest of nineteenth-century analysts and a first-class mathematical physicist. Though his father had wanted him to study medicine, he became a student and then professor at the fountainhead of nineteenth-century French mathematics, the Ecole Polytechnique. He worked in the theory of heat, was one of the founders of the mathematical theory of elasticity, and was one of the first to suggest that the theory of the gravitational potential be carried over to static electricity and magnetism.

Poisson was so much impressed with Fourier's evidence that arbitrary functions can be expanded in a series of functions that he believed all partial differential equations could be solved by series expansions; each term of the series would itself be a product of functions (cf. [10]), one for each independent variable. These expansions, he thought, embraced the most general solutions. He also believed that if an expansion diverged, this meant that one should seek an expansion in terms of other functions. Of course Poisson was far too optimistic.

From about 1815 on he himself solved a number of heat conduction problems and used expansions in trigonometric functions, Legendre polynomials, and Laplace surface harmonics. We shall encounter some of this work later. Much of Poisson's work on heat conduction was presented in his *Théorie mathématique de la chaleur* (1835).

3. *Closed Solutions; the Fourier Integral*

Despite the success and impact of Fourier's series solutions of partial differential equations, one of the major efforts during the nineteenth century was to find solutions in closed form, that is, in terms of elementary functions and integrals of such functions. Such solutions, at least of the kind known in the eighteenth and early nineteenth centuries, were more manageable, more perspicuous, and more readily used for calculation.

The most significant method of solving partial differential equations in closed form, which arose from work initiated by Laplace, was the Fourier integral. The idea is due to Fourier, Cauchy, and Poisson. It is impossible to assign priority for this important discovery, because all three presented papers orally to the Academy of Sciences that were not published until some time afterward. But each heard the others' papers, and one cannot tell from the publications what each may have taken from the verbal accounts.

In the last section of his prize paper of 1811, Fourier treated the propagation of heat in domains that extend to infinity in one direction. To obtain an answer for such problems, he starts with the general form of the solution of the heat equation for a bounded domain, namely (cf. [10]),

$$(21) \qquad u = \sum_{n=1}^{\infty} a_n e^{-kq_n^2 t} \cos q_n x,$$

where the q_n are determined by the boundary conditions and the a_n are determined by the initial conditions. Fourier now regards the q_n as abscissas of a curve and the a_n as the ordinates of that curve. Then $a_n = Q(q_n)$ where Q is some function of q. He then replaces (21) by

$$(22) \qquad u = \int_0^{\infty} Q(q) e^{-kq^2 t} \cos qx \, dq,$$

and seeks to determine Q. He goes back to the formula for the coefficients

$$a_n = \frac{2}{\pi} \int_0^{\pi} \phi(x) \cos nx \, dx,$$

where $\phi(x)$ would usually be the initial function. By a "limiting process" that replaces a_n by Q and n by q, he obtains

(23) $$Q = \frac{2}{\pi} \int_0^\infty F(x) \cos qx \, dx,$$

where $F(x)$, an even function, is the given initial temperature in the infinite domain. Then by using (23) in (22) and by an interchange of the order of integration, which Fourier does not bother to question, he has

$$u = \frac{2}{\pi} \int_0^\infty F(\alpha) \, d\alpha \int_0^\infty e^{-kq^2 t} \cos qx \cos q\alpha \, dq.$$

Fourier then does the analogous thing for an odd $F(x)$, and so finally obtains

(24) $$u = \frac{1}{\pi} \int_{-\infty}^\infty F(\alpha) \, d\alpha \int_0^\infty e^{-kq^2 t} \cos q(x - \alpha) \, dq.$$

Thus the solution is expressed in closed form. Now for $t = 0$, u is $F(x)$, which could be any given function. Hence Fourier asserts that, for an *arbitrary* $F(x)$,

(25) $$F(x) = \frac{1}{\pi} \int_{-\infty}^\infty F(\alpha) \, d\alpha \int_0^\infty \cos q(x - \alpha) \, dq,$$

and this is one form of the Fourier double-integral representation of an arbitrary function. In his book Fourier showed how to solve many types of differential equations with this integral. One use lies in the fact that if (24) is obtained by any process, then (25) shows that u satisfies the initial condition at $t = 0$. Another use is more evident if one writes the Fourier integral in exponential form, using the Euler relation, $e^{ix} = \cos x + i \sin x$. Then (25) becomes

$$F(x) = \frac{1}{2\pi} \int_{-\infty}^\infty e^{iqx} \, dq \int_{-\infty}^\infty F(\alpha) e^{-iq\alpha} \, d\alpha.$$

This form shows that $F(x)$ can be resolved into an infinite number of harmonic components with continuously varying frequency $q/2\pi$ and with amplitude $(1/2\pi) \int_{-\infty}^\infty F(\alpha) e^{-iq\alpha} \, d\alpha$, whereas the ordinary Fourier series resolves a given function into an infinite but discrete set of harmonic components.

Cauchy's derivation of the Fourier integral is somewhat similar. The paper in which it appeared, "Théorie de la propagation des ondes," received the prize of the Paris Academy for 1816.[5] This paper is the first large investigation of waves on the surface of a fluid, a subject initiated by

5. *Mém. divers savans*, 1, 1827, 3–312 = *Œuvres*, (1), 1, 5–318; see also Cauchy, *Nouv. Bull. de la Soc. Phil.*, 1817, 121–24 = *Œuvres*, (2), 2, 223–27.

Laplace in 1778. Though Cauchy sets up the general hydrodynamical equations he limits himself almost at once to special cases. In particular he considers the equation

$$\frac{\partial^2 q}{\partial x^2} + \frac{\partial^2 q}{\partial y^2} = 0$$

in which q is what was later called a velocity potential and x and y are spatial coordinates. He writes down without explanation the solution (cf. [22])

$$(26) \qquad q = \int_0^\infty \cos mx \, e^{-ym} f(m) \, dm,$$

wherein $f(m)$ is arbitrary thus far. Since $y = 0$ on the surface, q reduces to a given $F(x)$,

$$(27) \qquad F(x) = \int_0^\infty \cos mx \, f(m) \, dm.$$

Then Cauchy shows that

$$(28) \qquad f(m) = \frac{2}{\pi} \int_0^\infty \cos mu \, F(u) \, du.$$

With this value of $f(m)$

$$(29) \qquad F(x) = \frac{2}{\pi} \int_0^\infty \int_0^\infty \cos mx \cos mu \, F(u) \, du \, dm.$$

Cauchy thus obtains not only the Fourier double-integral representation of $F(x)$, but he also has the Fourier transform from $f(m)$ to $F(x)$ and the inverse transform. Given $F(x)$, $f(m)$ is determined by (28) and can be used in (26).

Shortly after Cauchy turned in his prize paper, Poisson, who could not compete for the prize because he was a member of the Academy, published a major work on water waves, "Mémoire sur la théorie des ondes."[6] In this work he derives the Fourier integral in about the same manner as Cauchy.

4. The Potential Equation and Green's Theorem

The next significant development centered about the potential equation, though the principal result, Green's theorem, has application to many other types of differential equations. The potential equation had figured in the eighteenth-century work on gravitation and had also appeared in the nineteenth-century work on heat conduction, for when the temperature

6. *Mém. de l'Acad. des Sci., Paris*, (2), 1, 1816, 71–186.

distribution in a body, though varying from point to point, remains the same as time varies, or is in the steady state, then T in (1) is independent of time and the heat equation reduces to the potential equation. The emphasis on the potential equation for the calculation of gravitational attraction continued in the early nineteenth century but was accentuated by a new class of applications to electrostatics and magnetostatics. Here too the attraction of ellipsoids was a key problem.

One correction in the theory of gravitational attraction as expressed by the potential equation was made by Poisson.[7] Laplace (Chap. 22, sec. 4) had assumed that the potential equation

$$(30) \qquad \frac{\partial^2 V}{\partial x^2} + \frac{\partial^2 V}{\partial y^2} + \frac{\partial^2 V}{\partial z^2} = 0,$$

wherein V is a function of x, y, and z, holds at any point (x, y, z) whether inside or outside of the body that exerts the gravitational attraction. Poisson showed that if (x, y, z) lies inside the attracting body, then V satisfies

$$(31) \qquad \frac{\partial^2 V}{\partial x^2} + \frac{\partial^2 V}{\partial y^2} + \frac{\partial^2 V}{\partial z^2} = -4\pi\rho,$$

where ρ is the density of the attracting body and is also a function of x, y, and z. Though (31) is still called Poisson's equation, his proof that it holds was not rigorous, as he himself recognized, even by the standards of that time.

In this same paper Poisson called attention to the utility of this function V in electrical investigations, remarking that its value over the surface of any conductor must be constant when electrical charge is allowed to distribute itself over the surface. In other papers he solved a number of problems calling for the distribution of charge on the surfaces of conducting bodies when the bodies are near each other. His basic principle was that the resultant electrostatic force in the interior of any one of the conductors must be zero.

Despite the work of Laplace, Poisson, Gauss, and others, almost nothing was known in the 1820s about the general properties of solutions of the potential equation. It was believed that the general integral must contain two arbitrary functions, of which one gives the value of the solution on the boundary and the other, the derivative of the solution on the boundary. Yet it was known in the case of steady-state heat conduction, in which the temperature satisfies the potential equation, that the temperature or heat distribution throughout the three-dimensional body is determined when the temperature alone is specified on the surface. Hence one of the

7. *Nouv. Bull. de la Soc. Philo.*, 3, 1813, 388–92.

arbitrary functions in the supposed general solution of the potential equation must somehow be fixed by some other condition.

At this point George Green (1793–1841), a self-taught English mathematician, undertook to treat static electricity and magnetism in a thoroughly mathematical fashion. In 1828 Green published a privately printed booklet, *An Essay on the Application of Mathematical Analysis to the Theories of Electricity and Magnetism.* This was neglected until Sir William Thomson (Lord Kelvin, 1824–1907) discovered it, recognized its great value, and had it published in the *Journal für Mathematik.*[8] Green, who learned much from Poisson's papers, also carried over the notion of the potential function to electricity and magnetism.

He started with (30) and proved the following theorems. Let U and V be any two continuous functions of x, y, and z whose derivatives are not infinite at any point of an arbitrary body. The major theorem asserts that (we shall used ΔV for the left hand side of [30], though it was not used by Green)

$$(32) \quad \iiint U \, \Delta V \, dv + \iint U \frac{\partial V}{\partial n} \, d\sigma$$
$$= \iiint V \, \Delta U \, dv + \iint V \frac{\partial U}{\partial n} \, d\sigma,$$

where n is the surface normal of the body directed inward and $d\sigma$ is a surface element. Theorem (32), incidentally, was also proved by Michel Ostrogradsky (1801–61), a Russian mathematician, who presented it to the St. Petersburg Academy of Sciences in 1828.[9]

Green then showed that the requirement that V and each of its first derivatives be continuous in the interior of the body can be imposed instead of a boundary condition on the derivatives of V. In light of this fact, Green represented V in the interior of the body in terms of its value $\overline{V}$ on the boundary (which function would be given) and in terms of another function U which has the properties: (a) U must be 0 on the surface; (b) at a fixed but undetermined point P in the interior, U becomes infinite as $1/r$ where r is the distance of any other point from P; (c) U must satisfy the potential equation (30) in the interior. If U is known, and it might be found more readily because it satisfies simpler conditions than V, then V can be represented at every interior point by

$$4\pi V = -\iint \overline{V} \frac{\partial U}{\partial n} \, d\sigma,$$

8. *Jour. für Math.*, 39, 1850, 73–89; 44, 1852, 356–74; and 47, 1854, 161–221 = *Green's Mathematical Papers*, 1871, 3–115.
9. *Mém. Acad. Sci. St. Peters.*, (6), 1, 1831, 39–53.

where the integral extends over the surface, and $\partial U/\partial n$ is the derivative of U in the direction perpendicular to the surface and into the body. It is understood that the coordinates of P are contained in $\partial U/\partial n$ and are the arguments at P. This function U, introduced by Green, which Riemann later called the Green's function, became a fundamental concept of partial differential equations. Green himself used the term "potential function" for this special function U as well as for V. His method of obtaining solutions of the potential equation, as opposed to the method of using series of special functions, is called the method of singularities. There is unfortunately no general expression for the function U, nor is there a general method for finding it. Green was content in this matter to give the physical meaning of U for the case of the potential created by electric charges.

Green applied his theorem and concepts to electrical and magnetic problems. He also took up in 1833[10] the problem of the gravitational potential of ellipsoids of variable densities. In this work Green showed that when V is given on the boundary of a body, there is just one function that satisfies $\Delta V = 0$ throughout the body, has no singularities, and has the given boundary values. To make his proof, Green assumed the existence of a function that minimizes

$$(33) \qquad \iiint \left[\left(\frac{\partial V}{\partial x} \right)^2 + \left(\frac{\partial V}{\partial y} \right)^2 + \left(\frac{\partial V}{\partial z} \right)^2 \right] dv.$$

This is the first use of the Dirichlet principle (cf. Chap. 27, sec. 8).

In this 1835 paper Green did much of the work in n dimensions instead of three and also gave important results on what are now called ultraspherical functions, which are a generalization to n variables of Laplace's spherical surface harmonics. Because Green's work did not become well known for some time, other men did some of this work independently.

Green is the first great English mathematician to take up the threads of the work done on the Continent after the introduction of analysis to England. His work inspired the great Cambridge school of mathematical physicists which included Sir William Thomson, Sir Gabriel Stokes, Lord Rayleigh, and Clerk Maxwell.

Green's achievements were followed by Gauss's masterful work of 1839,[11] "Allgemeine Lehrsätze in Beziehung auf die im verkehrten Verhältnisse des Quadrats der Entfernung wirkenden Anziehungs-und Abstossungs-kräfte" (General Theorems on Attractive and Repulsive Forces Which Act According to the Inverse Square of the Distance). Gauss proved rigorously Poisson's result, namely, that $\Delta V = -4\pi\rho$ at a point inside the acting mass, under the condition that ρ is continuous at that point and in a

---

10. *Trans. Camb. Phil. Soc.*, 5_3, 1835, 395–430 = *Mathematical Papers*, 187–222.
11. *Resultate aus den Beobachtungen des magnetischen Vereins*, Vol. 4, 1840 = *Werke*, 5, 197–242.

small domain around it. This condition is not fulfilled on the surface of the acting mass. On the surface the quantities $\partial^2 V/\partial x^2$, $\partial^2 V/\partial y^2$, and $\partial^2 V/\partial z^2$ have jumps.

The work thus far on the potential equation and on Poisson's equation assumed the existence of a solution. Green's proof of the existence of a Green's function rested entirely on a physical argument. From the existence standpoint the fundamental problem of potential theory was to show the existence of a potential function V, which William Thomson about 1850 called a harmonic function, whose values are given on the boundary of a region and which satisfies $\Delta V = 0$ in the region. One might establish this directly, or establish the existence of a Green's function U and from that obtain V. The problem of establishing the existence of the Green's function or of V itself is known as the Dirichlet problem or the first boundary-value problem of potential theory, the most basic and oldest existence problem of the subject. The problem of finding a V satisfying $\Delta V = 0$ in a region when the normal derivative of V is specified on the boundary is called the Neumann problem, after Carl G. Neumann (1832–1925), a professor at Leipzig. This problem is called the second fundamental problem of potential theory.

One approach to the problem of establishing the existence of a solution of $\Delta V = 0$, which Green had already used (see [33]), was brought into prominence by William Thomson. In 1847[12] Thomson announced the theorem or principle which in England is named after him and on the Continent is called the Dirichlet *principle* because Riemann so named it. Though Thomson stated it in a somewhat more general form, the essence of the principle may be put thus: Consider the class of all functions U that have continuous derivatives of the second order in the interior and exterior domains T and T' respectively separated by a surface S. The U's are to be continuous everywhere and assume on S the values of a continuous function f. The function V that minimizes the Dirichlet integral

$$(34) \qquad I = \iiint_T \left[\left(\frac{\partial U}{\partial x} \right)^2 + \left(\frac{\partial U}{\partial y} \right)^2 + \left(\frac{\partial U}{\partial z} \right)^2 \right] dv$$

is the one that satisfies $\Delta V = 0$ and takes on the value f on the boundary S. The connection between (34) and ΔV is that the first variation of I in the sense of the calculus of variations is ΔV, and this must be 0 for a minimizing V. Since for real U, I cannot be negative, it seemed clear that a minimizing function V must exist, and it is then not difficult to prove it is unique. The Dirichlet principle is then *one* approach to the Dirichlet problem of potential theory.

12. *Jour. de Math.*, 12, 1847, 493–96 = *Cambridge and Dublin Math. Jour.*, 3, 1848, 84–87 = *Math. and Physical Papers*, 1, 93–96.

Riemann's work on complex functions gave a new importance to the Dirichlet problem and the principle itself. Riemann's "proof" of the existence of V in his doctoral thesis used the two-dimensional case of the Dirichlet principle, but it was not rigorous, as he himself realized.

When Weierstrass in his paper of 1870[13] presented a critique of the Dirichlet principle, he showed that the *a priori* existence of a minimizing U was not supported by proper arguments. It was correct that for all continuous differentiable functions U that go continuously from the interior onto the prescribed boundary values the integral has a *lower bound*. But whether there is a function U_0 in the class of continuous, differentiable functions that furnishes the lower bound was not established.

Another technique for the solution of the potential equation employs complex function theory. Though d'Alembert in his work of 1752 (Chap. 27, sec. 2) and Euler in special problems had used this technique to solve the potential equation, it was not until the middle of the nineteenth century that complex function theory was vitally employed in potential theory. The relevance of function theory to potential theory rests on the fact that if $u + iv$ is an analytic function of z, then both u and v satisfy Laplace's equation. Moreover, if u satisfies Laplace's equation, then the conjugate function v such that $u + iv$ is analytic necessarily exists (Chap. 27, sec. 8).

Where the equation $\Delta u = 0$ is used in fluid flow, the function $u(x, y)$ is what Helmholtz called the velocity potential, and then $\partial u/\partial x$ and $\partial u/\partial y$ represent the components of the velocity of the fluid at any point (x, y). In the case of static electricity, u is the electrostatic potential and $\partial u/\partial x$ and $\partial u/\partial y$ are the components of electric force. In both cases the curves $u =$ const. are equipotential lines and the curves $v =$ const., which are orthogonal to $u =$ const., are the flow or stream lines (lines of force for electricity). The function $v(x, y)$ is called the stream function. The introduction of this function is clearly helpful because of its physical significance.

One advantage of the use of complex function theory in solving the potential equation derives from the fact that if $F(z) = F(x + iy)$ is an analytic function, so that its real and imaginary parts satisfy $\Delta V = 0$, then the transformation of x and y to ξ and η by

$$\xi = f(x, y), \qquad \eta = g(x, y)$$

where

$$\zeta = \xi + i\eta$$

produces another analytic function $G(\zeta) = G(\xi + i\eta)$, and its real and imaginary parts also satisfy $\Delta V(\xi, \eta) = 0$. Now if the original potential

13. Chap. 27, sec. 9.

problem $\Delta V = 0$ has to be solved in some domain D, then by proper choice of the transformation the domain D' in which the transformed $\Delta V = 0$ has to be solved can be much simpler. Here the use of conformal transformations, such as the Schwarz-Christoffel transformation, is of great service.

We shall not pursue the uses of complex function theory in potential theory because the details of its use go far beyond any basic methodology in the solution of partial differential equations. It is, however, again worth noting that many mathematicians resisted the use of complex functions because they were still not reconciled to complex numbers. At Cambridge University, even in 1850, cumbrous devices were used to avoid involving complex functions. Horace Lamb's *Treatise on the Mathematical Theory of the Motion of Fluids*, published in 1879 and still a classic (now known as *Hydrodynamics*), was the first book to acknowledge the acceptance of function theory at Cambridge.

5. *Curvilinear Coordinates*

Green introduced a number of major ideas whose significance extended far beyond the potential equation. Gabriel Lamé (1795–1870), a mathematician and engineer concerned primarily with the heat equation, introduced another major technique, the use of curvilinear coordinate systems, which could also be used for many types of equations. Lamé pointed out in 1833[14] that the heat equation had been solved only for conducting bodies whose surfaces are normal to the coordinate planes $x = $ const., $y = $ const., and $z = $ const. Lamé's idea was to introduce new systems of coordinates and the corresponding coordinate surfaces. To a very limited extent this had been done by Euler and Laplace, both of whom used spherical coordinates ρ, θ, and ϕ, in which case the coordinate surfaces $\rho = $ const., $\theta = $ const., and $\phi = $ const. are spheres, planes, and cones respectively. Knowing the equations that transform from rectangular to spherical coordinates, one can, as Euler and Laplace did, transform the potential equation from rectangular to spherical coordinates.

The value of the new coordinate systems and surfaces is twofold. First, a partial differential equation in rectangular coordinates might not be separable into ordinary differential equations in this system but might be separable in some other system. Secondly, the physical problem might call for a boundary condition on, say, an ellipsoid. Such a boundary is represented simply in a coordinate system wherein one family of surfaces consists of ellipsoids, whereas in the rectangular system a relatively complicated equation must be used. Moreover, after separation of variables in the proper coordinate system is employed, this boundary condition becomes applicable to just one of the resulting ordinary differential equations.

14. *Jour. de l'Ecole Poly.*, 14, 1833, 194–251.

Lamé introduced several new coordinate systems for the express purpose of solving the heat equation in these systems.[15] His chief system was the three families of surfaces given by the equations

$$\frac{x^2}{\lambda^2} + \frac{y^2}{\lambda^2 - b^2} + \frac{z^2}{\lambda^2 - c^2} - 1 = 0$$

$$\frac{x^2}{\mu^2} + \frac{y^2}{\mu^2 - b^2} + \frac{z^2}{\mu^2 - c^2} - 1 = 0$$

$$\frac{x^2}{\nu^2} + \frac{y^2}{\nu^2 - b^2} + \frac{z^2}{\nu^2 - c^2} - 1 = 0,$$

where $\lambda^2 > c^2 > \mu^2 > b^2 > \nu^2$. These three families are ellipsoids, hyperboloids of one sheet, and hyperboloids of two sheets, all of which possess the same foci. Any surface of one family cuts all the surfaces of the other two orthogonally, and in fact cuts them in lines of curvature (Chap. 23, sec. 7). Any point in space accordingly has coordinates (λ, μ, ν), namely the λ, μ, and ν of the surfaces, one from each family, which go through that point. This new coordinate system is called ellipsoidal, though Lamé called it elliptical, a term now used for another system.

Lamé transformed the heat equation for the steady-state case (temperature independent of time), that is, the potential equation, to these coordinates, and showed that he could use separation of variables to reduce the partial differential equation to three ordinary differential equations. Of course these equations must be solved subject to appropriate boundary conditions. In a paper of 1839[16] Lamé studied further the steady-state temperature distribution in a three-axis ellipsoid and gave a complete solution of the problem treated in his 1833 paper. In this 1839 paper he also introduced another curvilinear coordinate system, now called the spheroconal system, wherein the coordinate surfaces are a family of spheres and two families of cones. This system too Lamé used to solve heat conduction problems. Lamé wrote many other papers on heat conduction using ellipsoidal coordinates, including a second one of 1839 in the same volume of the *Journal de Mathématiques*, in which he treats special cases of the ellipsoid.[17]

The subject of mutually orthogonal families of surfaces had such obvious importance in the solution of partial differential equations that it became a subject of investigation in and for itself. In a paper of 1834[18] Lamé considered the general properties of any three families of mutually orthogonal surfaces and gave a procedure for expressing a partial differen-

15. *Annales de Chimie et Physique*, (2), 53, 1833, 190–204.
16. *Jour. de Math.*, 4, 1839, 126–63.
17. *Jour. de Math.*, 4, 1839, 351–85.
18. *Jour de l'Ecole Poly.*, 14, 1834, 191–288.

tial equation in any orthogonal coordinate system, a technique used continually ever since.

(Heinrich) Eduard Heine (1821–81) followed in Lamé's tracks. Heine in his doctoral dissertation of 1842[19] determined the potential (steady-state temperature) not merely for the interior of an ellipsoid of revolution when the value of the potential is given at the surface, but also for the exterior of such an ellipsoid and for the shell between confocal ellipsoids of revolution.

Lamé was so much impressed with what he and others accomplished by the use of triply orthogonal coordinate systems that he thought all partial differential equations could be solved by finding a suitable system. Later he realized that this was a mistake. In 1859 he published a book on the whole subject, *Leçons sur les coordonnées curvilignes*.

Though the use of mutually orthogonal families of surfaces as the coordinate surfaces did not solve all partial differential equations, it did open up a new technique that could be exploited to advantage in many problems. The use of curvilinear coordinates was carried over to other partial differential equations. Thus Emile-Léonard Mathieu (1835–1900), in a paper of 1868,[20] treated the vibrations of an elliptic membrane, which involves the wave equation, and here introduced elliptic cylinder coordinates and functions appropriate to these coordinates, now called Mathieu functions (Chap. 29, sec. 2). In the same year Heinrich Weber (1842–1913), working with the equation $\partial^2 u/\partial x^2 + \partial^2 u/\partial y^2 + k^2 u = 0$, solved it[21] for a domain bounded by a complete ellipse and also for the region bounded by two arcs of confocal ellipses and two arcs of hyperbolas confocal with the ellipses. The special case in which the ellipses and hyperbolas become confocal parabolas was also considered, and here Weber introduced functions appropriate to expansions in this coordinate system, now called Weber functions or parabolic cylinder functions. In his *Cours de physique mathématique* (1873), Mathieu took up new problems involving the ellipsoid and introduced still other new functions.

Our discussion of the idea initiated by Lamé, the use of curvilinear coordinates, describes just the beginning of this work. Many other coordinate systems have been introduced; corresponding special functions that result from solving the ordinary differential equations, which arise from separation of variables, have also been studied.[22] Most of this theory of special functions was created by physicists as they needed the functions and their properties in concrete problems (see also Chap. 29).

19. *Jour. für Math.*, 26, 1843, 185–216.
20. *Jour. de Math.*, (2), 13, 1868, 137–203.
21. *Math. Ann.*, 1, 1869, 1–36.
22. See William E. Byerly, *An Elementary Treatise on Fourier Series*, Dover (reprint), 1959, and E. W. Hobson, *The Theory of Spherical and Ellipsoidal Harmonics*, Chelsea (reprint), 1955.

6. The Wave Equation and the Reduced Wave Equation

Perhaps the most important type of partial differential equation is the wave equation. In three spatial dimensions the basic form is

$$(36) \qquad \frac{\partial^2 u}{\partial x^2} + \frac{\partial^2 u}{\partial y^2} + \frac{\partial^2 u}{\partial z^2} = a^2 \frac{\partial^2 u}{\partial t^2}.$$

As we know, this equation had already been introduced in the eighteenth century and had also been expressed in spherical coordinates. During the nineteenth century new uses of the wave equation were found, especially in the burgeoning field of elasticity. The vibrations of solid bodies of a variety of shapes with different initial and boundary conditions and the propagation of waves in elastic bodies produced a host of problems. Further work in the propagation of sound and light raised hundreds of additional problems.

Where separation of variables is possible, the technique of solving (37) is no different from what Fourier did with the heat equation or Lamé did after expressing the potential equation in some system of curvilinear coordinates. Mathieu's use of curvilinear coordinates to solve the wave equation by separation of variables is typical of hundreds of papers.

Quite another and important class of results dealing with the wave equation was obtained by treating the equation as an entirety. The first of such major results deals with initial-value problems and goes back to Poisson, who worked on this equation during the years 1808 to 1819. His principal achievement[23] was a formula for the propagation of a wave $u(x, y, z, t)$ whose initial state is described by the initial conditions

$$(37) \qquad u(x, y, z, 0) = \phi_0(x, y, z), \qquad u_t(x, y, z, 0) = \phi_1(x, y, z)$$

and which satisfies the partial differential equation

$$(38) \qquad \frac{\partial^2 u}{\partial x^2} + \frac{\partial^2 u}{\partial y^2} + \frac{\partial^2 u}{\partial z^2} = \frac{1}{a^2} \frac{\partial^2 u}{\partial t^2}$$

wherein a is a constant. The solution u is given by

$$(39) \quad u(x, y, z, t)$$

$$= \frac{1}{4\pi a} \int_0^\pi \int_0^{2\pi} \phi_1(x + at \sin \phi \cos \theta, y + at \sin \phi \sin \theta,$$

$$z + at \cos \phi) at \sin \phi \, d\theta \, d\phi$$

$$+ \frac{1}{4\pi a} \frac{\partial}{\partial t} \int_0^\pi \int_0^{2\pi} \phi_0(x + at \sin \phi \cos \theta, y + at \sin \phi \sin \theta,$$

$$z + at \cos \phi) at \sin \phi \, d\theta \, d\phi$$

wherein θ and ϕ are the usual spherical coordinates. The domain of integration is the surface of a sphere S_{at} with radius at about the point P with coordinates x, y, and z.

23. *Mém. de l'Acad. des Sci., Paris*, (2), 3, 1818, 121–76.

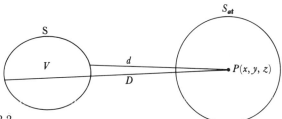

Figure 28.2

To obtain some indication of what Poisson's result means, let us consider a physical example. Suppose that the initial disturbance is set up by a body V (Fig. 28.2) with boundary S so that ϕ_0 and ϕ_1 are defined on V and are 0 outside of V. We say that the initial disturbance is localized in V. Physically a wave sets out from V and spreads out into space. Poisson's formula tells us what happens at any point $P(x, y, z)$ outside of V. Let d and D represent the minimum and maximum distances of P to the points of V. When $t < d/a$, the integrals in (39) are 0 because the domain of integration is the surface of the sphere S_{at} with radius at and center at P. Since ϕ_0 and ϕ_1 are 0 on S_{at}, then the function u is 0 at P. This means that the wave spreading out from S has not reached P. At $t = d/a$, the sphere S_{at} just touches S so the leading front of the wave emanating from S arrives at P. Between $t = d/a$ and $t = D/a$ the sphere S_{at} cuts V and so $u(P, t) \neq 0$. Finally for $t > D/a$, the sphere S_{at} will not cut S (the entire region V lies inside S_{at}); that is, the initial disturbance has passed through P. Hence again $u(P, t) = 0$. The instant $t = D/a$ corresponds to the passage of the trailing edge of the wave through P. At any given time t the leading edge of the wave takes the form of a surface which separates points not reached by the disturbance from those reached. This leading edge is the envelope of the family of spheres with centers on S and with radii at. The terminating edge of the wave at time t is a surface separating points at which the disturbance exists from those which the disturbance has passed. We see, then, that the disturbance which is localized in space gives rise at each point P to an effect that lasts only for a finite time. Moreover the wave (disturbance) has a leading and a terminating edge. This entire phenomenon is called Huygens's principle.

A quite different method of solving the initial-value problem for the wave equation was created by Riemann in the course of his work on the propagation of sound waves of finite amplitude.[24] Riemann considers a linear differential equation of second order that can be put in the form

$$(40) \qquad L(u) = \frac{\partial^2 u}{\partial x\, \partial y} + D \frac{\partial u}{\partial x} + E \frac{\partial u}{\partial y} + Fu = 0,$$

24. *Abh. der Ges. der Wiss. zu Gött.*, 8, 1858/59, 43–65 = *Werke*, 156–78.

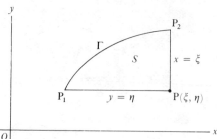

Figure 28.3

where D, E, and F are continuous and differentiable to second order functions of x and y. The problem calls for finding u at an arbitrary point P (Fig. 28.3) when one knows u and $\partial u/\partial n$ (which means knowing $\partial u/\partial x$ and $\partial u/\partial y$) along a curve Γ. His method depends on finding a function v (called a Riemann function or characteristic function)[25] that satisfies what is now called the adjoint equation

$$(41) \qquad M(v) = \frac{\partial^2 v}{\partial x\,\partial y} - \frac{\partial(Dv)}{\partial x} - \frac{\partial(Ev)}{\partial y} + Fv = 0$$

and other conditions we shall specify shortly.

Riemann introduced the segments PP_1 and PP_2 of the characteristics (he did not use the term) $x = \xi$ and $y = \eta$ through P. Now a generalized Green's theorem (in two dimensions) is applied to the differential expression $L(u)$. To express the theorem compactly, let us introduce

$$X = \frac{1}{2}\left(v\frac{\partial u}{\partial y} - u\frac{\partial v}{\partial y}\right) + Duv$$

$$Y = \frac{1}{2}\left(v\frac{\partial u}{\partial x} - u\frac{\partial v}{\partial x}\right) + Euv.$$

Then Green's theorem states

$$(42) \qquad \int_S [vL(u) - uM(v)]\,dS = \int_S \left(\frac{\partial X}{\partial x} + \frac{\partial Y}{\partial y}\right) dS$$

$$= \int_C \{X\cos(n, x) + Y\cos(n, y)\}\,ds,$$

where S is the area in the figure, C is the entire boundary of S, and $\cos(n, x)$ is the cosine of the angle between the normal to C and the x-axis.

25. v is not the same as a fundamental solution or a Green's function.

Beyond satisfying (41), Riemann requires of v that

(a) $v = 1$ at P,

(43)   (b) $\dfrac{\partial v}{\partial y} - Dv = 0$ on $x = \xi$,

(c) $\dfrac{\partial v}{\partial x} - Ev = 0$ on $y = \eta$.[26]

By using the condition that $M(v) = 0$ and the conditions (43), and by evaluation of the curvilinear integral over C, Riemann obtains

$$(44) \quad u(\xi, \eta) = \int_{\Gamma} \{X \cos (n, x) + Y \cos (n, y)\} \, ds + \frac{1}{2} \{(uv)_{P_1} + (uv)_{P_2}\}.$$

Thus the value of u at any arbitrary point P is given in terms of the values of u, $\partial u/\partial n$, v, and $\partial v/\partial n$ on Γ and the values of u and v at P_1 and P_2.

Now u is given at P_1 and P_2. The function v must itself be found by solving $M(v) = 0$ and meeting the conditions in (43). Hence what Riemann's method achieves is to change the original initial-value problem for u to another kind of initial-value problem, the one for v. The second problem is usually easier to solve. In Riemann's physical problem it was especially easy to find v. However the existence of such a v generally was not established by Riemann.

The Riemann method as just described is useful only for the type of equation exemplified by the wave equation (hyperbolic equations) in two independent variables and cannot be extended directly. The extension of the method to more than two independent variables meets with the difficulty that the Riemann function becomes singular on the boundary of the domain of integration and the integrals diverge. The method has been extended at the cost of increased complication.

Progress in the solution of the wave equation by other methods is intimately connected with what are called steady-state problems, which lead to the reduced wave equation. The wave equation, by its very form, involves the time variable. In many physical problems, where one is interested in simple harmonic waves, one assumes that $u = w(x, y, z)e^{ikt}$, and by substituting this into the wave equation one obtains

$$(45) \quad\quad \Delta w + k^2 w = \frac{\partial^2 w}{\partial x^2} + \frac{\partial^2 w}{\partial y^2} + \frac{\partial^2 w}{\partial z^2} + k^2 w = 0.$$

This is the reduced wave equation or the Helmholtz equation. The equation $\Delta w + k^2 w = 0$ represents all harmonic, acoustic, elastic, and electromagnetic waves. While the older authors were satisfied to find particular integrals, Hermann von Helmholtz (1821–94), in his work on the oscillations

26. For two-dimensional problems v is a function of four variables, ξ, η, x, and y. It satisfies $M(v) = 0$ as a function of x and y.

of air in a tube (organ pipe) with an open end, gave the first general investigation of its solutions.[27] He was concerned with the acoustical problem in which w is the velocity potential of a harmonically moving air mass, k is a constant determined by the elasticity of the air and the oscillation frequency, and λ, which equals $2\pi/k$, is the wavelength. By applying Green's theorem he showed that any solution of $\Delta w + k^2 w = 0$ that is continuous in a given domain can be represented as the effect of single and double layers of excitation points on the surface of the domain. Using $e^{-ikr}/4\pi r$ as one of the functions in Green's theorem, he obtained

$$(46) \qquad w(P) = -\frac{1}{4\pi} \int\int \frac{e^{-ikr}}{r} \frac{\partial w}{\partial n} \, dS + \frac{1}{4\pi} \int\int w \frac{\partial}{\partial n} \left(\frac{e^{-ikr}}{r} \right) dS$$

wherein r denotes the distance from P to a variable point on the boundary. Thus w at any point P within the domain in which the solution is sought is given in terms of the values of w and $\partial w/\partial n$ on the boundary S.

The work of Helmholtz was used by Gustav R. Kirchhoff (1824–87), one of the great German nineteenth-century mathematical physicists, to obtain another solution of the initial-value problem for the wave equation. Let us suppose that $\Delta w + k^2 w = 0$ comes from

$$\frac{\partial^2 u}{\partial t^2} = c^2 \Delta u$$

wherein we have let $u = we^{i\sigma t}$ so that $k = \sigma/c$. Then (46) may be written as

$$(47) \quad u(P, t) = -\frac{1}{4\pi} \int\int \frac{e^{i\sigma[t-(r/c)]}}{r} \frac{\partial u}{\partial n} \, dS + \frac{1}{4\pi} \int\int u \frac{\partial}{\partial n} \left(\frac{e^{i\sigma[t-(r/c)]}}{r} \right) dS.$$

This formula was generalized by Kirchhoff. If we let $\phi(t)$ be the value of u at any point (x, y, z) of the boundary at the instant τ and let $f(\tau)$ be the corresponding value of $\partial u/\partial n$, then Kirchhoff showed[28] that

$$(48) \quad u(P, t) = -\frac{1}{4\pi} \int\int \frac{f[t - (r/c)]}{r} \, dS + \frac{1}{4\pi} \int\int \frac{\partial}{\partial n} \left(\frac{\phi[t - (r/c)]}{r} \right) dS,$$

provided that in the last term the differentiation with respect to n applies to r only insofar as it appears explicitly in both numerator and denominator. Thus u is obtained at P in terms of values of u and $\partial u/\partial n$ at earlier times at points of the closed surface surrounding P. This result is called Huygens's principle of acoustics and it is a generalization of Poisson's formula.

We have noted that Riemann used a somewhat generalized Green's theorem. The full generalization of Green's theorem that employs the adjoint differential equation and which is also called Green's theorem, comes from a paper by Paul Du Bois-Reymond (1831–89)[29] and from

27. *Jour. für Math.*, 57, 1860, 1–72 = *Wissenschaftliche Abhandlungen*, 1, 303–82.
28. *Sitzungsber. Akad. Wiss. zu Berlin*, 1882, 641–69 = *Ges. Abh.*, 2, 22 ff.
29. *Jour. für Math.*, 104, 1889, 241–301.

Darboux in his *Théorie générale des surfaces*;[30] both cite Riemann's paper of 1858/59. If the given equation is

$$L(u) = A \frac{\partial^2 u}{\partial x^2} + 2B \frac{\partial^2 u}{\partial x \, \partial y} + C \frac{\partial^2 u}{\partial y^2} + D \frac{\partial u}{\partial x} + E \frac{\partial u}{\partial y} + Fu = 0,$$

wherein the coefficients are functions of x and y, one integrates the product $vL(u)$ over an arbitrary domain R of the xy-plane under the assumption that u, v, and their first and second partial derivatives are continuous. Then integration by parts yields the generalized Green's theorem, which states that

$$\iint uM(v) \, dx \, dy = -\iint vL(u) \, dx \, dy - \int (Q \, dy - P \, dx),$$

where the double integrals are over the interior of R and the single integrals over the boundary of R,

$$M(v) = \frac{\partial^2 (Av)}{\partial x^2} + 2 \frac{\partial^2 (Bv)}{\partial x \, \partial y} + \frac{\partial^2 (Cv)}{\partial y^2} - \frac{\partial (Dv)}{\partial x} - \frac{\partial (Ev)}{\partial y} + Fu$$

$$P = B\left(v \frac{\partial u}{\partial x} - u \frac{\partial v}{\partial x}\right) + C\left(v \frac{\partial u}{\partial y} - u \frac{\partial v}{\partial y}\right) + \left(E - \frac{\partial B}{\partial x} - \frac{\partial C}{\partial y}\right) uv$$

$$Q = A\left(v \frac{\partial u}{\partial x} - u \frac{\partial v}{\partial x}\right) + B\left(v \frac{\partial u}{\partial y} - u \frac{\partial v}{\partial y}\right) + \left(D - \frac{\partial A}{\partial x} - \frac{\partial B}{\partial y}\right) uv.$$

$M(v)$ is the adjoint expression of $L(u)$ and $M(v) = 0$ is the adjoint differential equation. Conversely $L(u)$ is the adjoint of $M(v)$.

The significance of Green's theorem is that it can be used to obtain solutions of some partial differential equations. Thus, since the elliptic equation can always be put in the form

$$L(u) = \Delta u + a \frac{\partial u}{\partial x} + b \frac{\partial u}{\partial y} + cu = 0,$$

then

$$M(v) = \Delta v - \frac{\partial (av)}{\partial x} - \frac{\partial (bv)}{\partial y} + cv = 0.$$

Let v be a solution of the adjoint equation that becomes logarithmically infinite at an arbitrary point (ξ, η); that is, it behaves like

$$v = U \log r + V,$$

where r is the distance from (ξ, η) to (x, y); U and V are continuous in the domain R being considered; and U is normalized so that $U(\xi, \eta) = 1$. Now exclude (ξ, η) from the domain of integration by enclosing it in a

30. Vol. 2, Book IV, Chap. 4, 2nd ed., 1915.

circle. Then the generalized Green's theorem gives, when the circle is contracted to (ξ, η):

$$(49) \quad 2\pi u(\xi, \eta) = -\iint vL(u) \, dx \, dy$$

$$+ \int \left[v \frac{\partial u}{\partial n} - u \frac{\partial v}{\partial n} + (a \cos (n, x) + b \cos (n, y)) \cdot uv \right] ds$$

where n is positive if directed to the outside of the domain; the single integral is taken counterclockwise over the boundary. Since u satisfies $L(u) = 0$, if we know v and if u and $\partial u/\partial n$ are given (both are not arbitrary) on the boundary, then we have u expressed as a single integral. The function v is called a Green's function, though often the condition that v vanish on the boundary of R is added in the definition of the Green's function. Various specializations and generalizations of this use of Green's theorem have been developed.

7. Systems of Partial Differential Equations

In the eighteenth century the differential equations of fluid motion presented the first important system of partial differential equations. In the nineteenth century three more fundamental systems, the fluid dynamical equations for viscous media, the equations of elastic media, and the equations of electromagnetic theory, were created.

The acquisition of the equations of fluid motion when viscosity is present (as it always is) took a tortuous path. Euler had given the equations of motion of a fluid that is nonviscous. Since the time of Lagrange the essential difference between the motion of a fluid when a velocity potential exists and when it does not had been recognized. Led by a formal analogy with the theory of elasticity and by the hypothesis of molecules animated by repulsive forces, Claude L. M. H. Navier (1785–1836), professor of mechanics at the Ecole Polytechnique and at the Ecole des Ponts et Chaussées, obtained the basic equations in 1821.[31] The Navier-Stokes equations, as they are now identified, are

$$\rho \frac{Du}{Dt} = \rho X - \frac{\partial p}{\partial x} + \frac{1}{3} \mu \frac{\partial \theta}{\partial x} + \mu \Delta u$$

$$(50) \qquad \rho \frac{Dv}{Dt} = \rho Y - \frac{\partial p}{\partial y} + \frac{1}{3} \mu \frac{\partial \theta}{\partial y} + \mu \Delta v$$

$$\rho \frac{Dw}{Dt} = \rho Z - \frac{\partial p}{\partial z} + \frac{1}{3} \mu \frac{\partial \theta}{\partial z} + \mu \Delta w$$

$$\theta = \frac{\partial u}{\partial x} + \frac{\partial v}{\partial y} + \frac{\partial w}{\partial z},$$

31. *Mém de l'Acad. des Sci.*, Paris, (2), 1827, 375–94.

where Δ has the usual meaning; ρ is the density of the fluid; p, the pressure; u, v, and w are the components of velocity of the fluid at any x, y, z, and t; X, Y, and Z are the components of an external force; the constant μ, which depends on the nature of the fluid, is called the coefficient of viscosity; and the derivative D/Dt has the meaning explained in Chapter 22, section 8. For an incompressible fluid, $\theta = 0$.

The equations were also obtained in 1829 by Poisson.[32] They were then rederived in 1845 on the basis of the mechanics of continua by George Gabriel Stokes (1819–1903), professor of mathematics at Cambridge University, in his essay "On the Theories of the Internal Friction of Fluids in Motion."[33] Stokes endeavored to account for the frictional action in all known liquids, which causes the motion to subside by converting kinetic energy into heat. Fluids, by virtue of their viscosity, stick to the surfaces of solids and thus exert tangential forces on them.

The subject of elasticity was founded by Galileo, Hooke, and Mariotte and was cultivated by the Bernoullis and Euler. But these men dealt with specific problems. To solve them they concocted *ad hoc* hypotheses on how beams, rods, and plates behaved under stresses, pressures, or loads. The theory proper is the creation of the nineteenth century. From the beginning of the nineteenth century on a number of great men worked persistently to obtain the equations that govern the behavior of elastic media, which includes the air. These men were primarily engineers and physicists. Cauchy and Poisson are the great mathematicians among them, though Cauchy was an engineer by training.

The problems of elasticity include the behavior of bodies under stress wherein one considers what equilibrium position they will assume, the vibrations of bodies when set in motion by an initial disturbance or by a continuously applied force, and, in the case of air and solid bodies, the propagation of waves through them. The interest in elasticity in the nineteenth century was heightened by the appearance about 1820 of a wave theory of light, initiated by the physician Thomas Young (1773–1829) and by Augustin-Jean Fresnel (1788–1827), an engineer. Light was regarded as a wave motion in ether, and ether was believed to be an elastic medium. Hence the propagation of light through ether became a basic problem. Another stimulus to a strong interest in elasticity in the early nineteenth century was Ernst F. F. Chladni's (1756–1827) experiments (1787) on the vibrations of glass and metal, in which he showed the nodal lines. These should be related to the sounds given off by, for example, a vibrating drumhead.

The work to obtain basic equations of elasticity was long and full of

32. *Jour. de l'Ecole Poly.*, 13, 1831, 1–74.
33. *Trans. Camb. Phil. Soc.*, 8, 1849, 287–319 = *Math. and Phys. Papers*, 1, 75–129.

pitfalls because little was known of the internal or molecular structure of matter; hence it was difficult to grasp any physical principles. The assumptions made as to solid bodies, the air, and ether varied from one writer to another and were disputed. In the case of ether, which presumably penetrated solid bodies because light passed through some and was absorbed by others, the relationship of the ether molecules to the molecules of the solid body also posed great difficulties. We do not intend to follow the physical theories of elastic bodies, nor is our understanding complete even today.

Navier[34] was the first (1821) to investigate the general equations of equilibrium and vibrations of elastic solids. The material was assumed to be isotropic, and the equations contained a single constant representing the nature of the solid. By 1822, stimulated by Fresnel's work, Cauchy had created another approach to the theory of elasticity.[35] Cauchy's equations contain two constants to represent the material of the body, and for an isotropic body are

$$(\lambda + \mu) \frac{\partial \theta}{\partial x} + \mu \Delta u = \rho \frac{\partial^2 u}{\partial t^2}$$

(51)

$$(\lambda + \mu) \frac{\partial \theta}{\partial y} + \mu \Delta v = \rho \frac{\partial^2 v}{\partial t^2}$$

$$(\lambda + \mu) \frac{\partial \theta}{\partial z} + \mu \Delta w = \rho \frac{\partial^2 w}{\partial t^2}$$

$$\theta = \frac{\partial u}{\partial x} + \frac{\partial v}{\partial y} + \frac{\partial w}{\partial z}.$$

Here u, v, and w are components of displacements, θ is called the dilatation, and λ and μ are constants of the body or medium. For general anisotropic media the equations are quite complicated, and it may be pointless to write them in all generality. The equations are given by Cauchy.[36]

The most spectacular triumph of the nineteenth century, with an enormous impact on science and technology, was Maxwell's derivation in 1864 of the laws of electromagnetism.[37] Maxwell, utilizing the electrical and magnetic researches of numerous predecessors, notably Faraday, introduced the notion of a displacement current—radio waves are one form of displacement current—and with this notion formulated the laws of electromagnetic wave propagation. His equations, which are most conveniently stated in the vector form adopted later by Oliver Heaviside, are four in number and involve the electric field intensity **E**, the magnetic field

34. *Mém. de l'Acad. des Sci., Paris*, (2), 7, 1827, 375–94.
35. *Exercices de math.*, 1828 = *Œuvres*, (2), 8, 195–226.
36. *Exercices de math.*, 1828 = *Œuvres*, (2), 8, 253–77.
37. *Phil. Trans.*, 155, 1865, 459–512 = *Scientific Papers*, 1, 526–97.

intensity **H**, the dielectric constant ε of the medium, the magnetic permeability μ of the medium, and the charge density ρ. The equations are

(52) $$\operatorname{curl} \mathbf{H} = \frac{1}{c} \frac{\partial(\varepsilon \mathbf{E})}{\partial t}, \quad \operatorname{curl} \mathbf{E} = -\frac{1}{c} \frac{\partial(\mu \mathbf{H})}{\partial t}$$

(53) $$\operatorname{div} \varepsilon \mathbf{E} = \rho \quad \text{and} \quad \operatorname{div} \mu \mathbf{H} = 0.^{38}$$

The first two equations are the primary ones and amount to six scalar (non-vectorial) partial differential equations. The displacement current is the term $\partial(\varepsilon \mathbf{E})/\partial t$.

By working with just these equations, Maxwell predicted that electromagnetic waves travel through space and at the speed of light. On the basis of the identity of the two speeds, he dared to assert that light is an electromagnetic phenomenon, a prediction that has been amply confirmed since his time.

No general methods for solving any of the above systems of equations are known. However, the nineteenth-century men gradually realized that in the case of partial differential equations, whether single equations or systems, general solutions are not nearly so useful as the solutions for specific problems where the initial and boundary conditions are given, and where experimental work might also aid one in making useful simplifying assumptions. The writings of Fourier, Cauchy, and Riemann furthered this realization. The work on the solution of the many initial- and boundary-value problems to which specializations of these systems gave rise is enormous, and almost all of the mathematicians of the century undertook such problems.

8. *Existence Theorems*

As the eighteenth- and nineteenth-century mathematicians created a vast number of types of differential equations, they found that methods of solving many of these equations were not available. Somewhat as in the case of polynomial equations, where efforts to solve equations of degree higher than four failed and Gauss turned to the proof of existence of a root (Chap. 25, sec. 2), so in the work on differential equations the failure to find explicit solutions, which of course *ipso facto* demonstrate existence, caused the mathematicians to turn to proof of the existence of solutions. Such proofs, even though they do not exhibit a solution or exhibit it in a useful form, serve several purposes. The differential equations were in nearly all cases the mathematical formulation of physical problems. No guarantees were available that the mathematical equations could be solved; hence a

---

38. For the meaning of curl and div see Chap. 32, sec. 5.

proof of the existence of a solution would at least insure that a search for a solution would not be attempting the impossible. The proof of existence would also answer the question: What must we know about a given physical situation, that is, what initial and boundary conditions insure a solution and preferably a unique one? Other objectives, perhaps not envisaged at the beginning of the work on existence theorems, were soon recognized. Does the solution change continuously with the initial conditions, or does some totally new phenomenon enter when the initial or boundary conditions are varied slightly? Thus a parabolic orbit that obtains for one value of the initial velocity of a planet may change to an elliptic orbit as a consequence of a slight change in the initial velocity. Such a difference in orbit is physically most significant. Further, some of the methodologies of solution, such as the use of the Dirichlet principle or Green's theorem, presupposed the existence of a particular solution. The existence of these particular solutions had not been established.

Before we give some brief indications of the work on existence theorems, it may be helpful to note a classification of partial differential equations that was actually made rather late in the century. Though some efforts toward classification by reducing these equations to normal or standard forms had been made by Laplace and Poisson, the classification introduced by Du Bois-Reymond has now become standard. In 1889[39] he classified the most general homogeneous linear equation of second order

$$(54) \qquad R\frac{\partial^2 u}{\partial x^2} + S\frac{\partial^2 u}{\partial x\,\partial y} + T\frac{\partial^2 u}{\partial y^2} + P\frac{\partial u}{\partial x} + Q\frac{\partial u}{\partial y} + Zu = 0,$$

where the coefficients are functions of x and y and they and their first and second derivatives are continuous, by means of the characteristics (Chap. 22, sec. 7). The projections of the characteristic curves onto the xy-plane (these projections are also called characteristics) satisfy

$$T\,dx^2 - S\,dx\,dy + R\,dy^2 = 0.$$

The characteristics are imaginary, real and distinct, or real and coincident according as

$$TR - S^2 > 0, \qquad TR - S^2 < 0, \qquad TR - S^2 = 0.$$

Du Bois-Reymond called these cases elliptic, hyperbolic, and parabolic, respectively. He then pointed out that by introducing new real independent variables

$$\xi = \phi(x, y), \qquad \eta = \psi(x, y),$$

39. *Jour. für Math.*, 104, 1889, 241–301.

the above equation can always be transformed into one of the three types of normal forms

(a) $R'\left(\dfrac{\partial^2 u}{\partial \xi^2} + \dfrac{\partial^2 u}{\partial \eta^2}\right) + P'\dfrac{\partial u}{\partial \xi} + Q'\dfrac{\partial u}{\partial \eta} + Zu = 0$

(b) $S'\dfrac{\partial^2 u}{\partial \xi \, \partial \eta} + P'\dfrac{\partial u}{\partial \xi} + Q'\dfrac{\partial u}{\partial \eta} + Zu = 0$

(c) $R'\dfrac{\partial^2 u}{\partial \xi^2} + P'\dfrac{\partial u}{\partial \xi} + Q'\dfrac{\partial u}{\partial \eta} + Zu = 0$

respectively. The two families $\phi(x, y) = $ const. and $\psi(x, y) = $ const. are the equations of two families of characteristics.

The supplementary conditions that can be imposed differ for the three types of equations. In the elliptic case (a) one considers a bounded domain of the xy-plane and specifies the value of u on the boundary (or an equivalent condition) and asks for the value of u in the domain. For the initial-value problem of the hyperbolic differential equation (b) one must specify u and $\partial u/\partial n$ on some initial curve. There may also be boundary conditions. The proper initial conditions for the parabolic case (c) were not specified at this time, though it is now known that one initial condition and boundary conditions can be imposed. This classification of partial differential equations was extended to equations in more independent variables, higher-order equations and to systems. Though the classification and the supplementary conditions were not known early in the century, the mathematicians gradually became aware of the distinctions and these figured in the existence theorems they were able to prove.

The work on existence theorems became a major activity with Cauchy, who emphasized that existence can often be established where an explicit solution is not available. In a series of papers[40] Cauchy noted that any partial differential equation of order greater than one can be reduced to a system of partial differential equations, and he treated the existence of a solution for the system. He called his method the *calcul des limites* but it is known today as the method of majorant functions. The essence of the method is to show that a power series in the independent variables with a definite domain of convergence does satisfy the system of equations. We shall illustrate the method in connection with Cauchy's work on ordinary differential equations (Chap. 29, sec. 4). His theorem covers only the case of analytic coefficients in the equations and analytic initial conditions.

To obtain some concrete idea about Cauchy's work, we shall consider what it implies for the second order equation in two independent variables

(55) $$r = f(z, x, y, p, q, s, t)$$

40. *Comp. Rend.*, 14, 1842, 1020–25 = *Œuvres*, (1), 6, 461–67, and *Comp. Rend.*, 15, 1842, 44–59, 85–101, 131–38 = *Œuvres*, (1), 7, 17–33, 33–49, 52–58.

where, as usual, $r = \partial^2 z/\partial x^2$ and where f is analytic in its variables. In this case one must specify on the initial line $x = 0$ that

$$z(0, y) = z_0(y), \qquad \frac{\partial z}{\partial x}(0, y) = z_1(y),$$

where z_0 and z_1 are analytic. (The initial line may be replaced by a curve, in which case $\partial z/\partial x$ must be replaced by $\partial z/\partial n$.) If the above conditions are fulfilled, then the solution $z = z(x, y)$ exists and is unique and analytic in some domain starting at the initial line.

Cauchy's work on systems was done independently and in somewhat improved form by Sophie Kowalewsky (1850–91),[41] who was a pupil of Weierstrass and who pursued his ideas. Kowalewsky is one of the few women mathematicians of distinction. In 1816 Sophie Germain (1776–1831) had won a prize awarded by the French Academy for a paper on elasticity. Kowalewsky, too, won the Paris Academy's prize, for a work of 1888 on the integration of the equations of motion for a solid body rotating around a fixed point; in 1889 she became a professor of mathematics in Stockholm. The proofs of Cauchy and Kowalewsky were later improved by Goursat.[42]

If, instead of (55), the given second order equation is in the form

(56) $$G(z, x, y, p, q, r, s, t) = 0,$$

it is necessary to solve for r before it can be put in the form (55). To consider a simple but vital case, if the equation is

$$G = A\frac{\partial^2 z}{\partial x^2} + 2B\frac{\partial^2 z}{\partial x\,\partial y} + C\frac{\partial^2 z}{\partial y^2} + D\frac{\partial z}{\partial x} + E\frac{\partial z}{\partial y} + Fz = 0,$$

where $A, B, \ldots, F$ are functions of x and y, then $\partial G/\partial r$ must not be 0 to solve for r. In case $\partial G/\partial r = 0$, the solution of the Cauchy problem need not exist, and when it does is not unique. In the case of three or more independent variables (let us consider three), and if the equation is written as

(57) $$\sum_{i,k} A_{ik}\frac{\partial^2 u}{\partial x_i\,\partial x_k} + \sum_i B_i\frac{\partial u}{\partial x_i} + Cu = f,$$

where the coefficients are functions of the independent variables x_1, x_2, and x_3, then the exceptional case occurs when the initial surface S satisfies the first order partial differential equation

(58) $$\sum A_{ik}\frac{\partial S}{\partial x_i}\frac{\partial S}{\partial x_k} = 0.$$

41. *Jour. für Math.*, 80, 1875, 1–32.
42. *Bull. Soc. Math. de France*, 26, 1898, 129–34.

Along such surfaces two solutions of (57) can be tangent and even have higher-order contact. This property is the same as for the characteristic curves of the first order equation $f(x, y, u, p, q) = 0$ (Chap. 22, sec. 5), and so these surfaces S are also called characteristics. Physically the surfaces S are wave fronts.

This theory of characteristics for the case of two independent variables was known to Monge and to André-Marie Ampère (1775–1836). Its extension to the case of second order equations in more than two independent variables was first made by Albert Victor Bäcklund (1845–1922),[43] but was not widely known until it was redone by Jules Beudon.[44]

In his *Leçons sur la propagation des ondes* (1903), Jacques Hadamard (1865–1963), the leading French mathematician of this century, generalized the theory of characteristics to partial differential equations of any order. As an example, let us consider a system of three partial differential equations of the second order in the dependent variables ξ, η, and ζ and the independent variables $x_1, x_2, \ldots, x_n$. The Cauchy problem for this system is: Given the values of ξ, η, ζ, and $\partial\xi/\xi x_n$, $\partial\eta/\xi x_n$, and $\partial\zeta/\partial x_n$ on a "surface" M_{n-1} of $n-1$ dimensions, to find the functions ξ, η, and ζ. The values of the second and higher derivatives of ξ, η, and ζ may then be computed unless M_{n-1} satisfies a first order partial differential equation of the sixth degree, say $H = 0$. All "surfaces" satisfying $H = 0$ are characteristic "surfaces." According to the theory of first order partial differential equations, the differential equation $H = 0$ has characteristic lines (curves) defined by

$$\frac{dx_1}{\partial H/\partial P_1} = \frac{dx_2}{\partial H/\partial P_2} = \cdots = \frac{dx_{n-1}}{\partial H/\partial P_{n-1}},$$

where $P_1, P_2, \ldots, P_{n-1}$ are the partial derivatives of x_n with respect to $x_1, x_2, \ldots, x_{n-1}$ taken along the "surface" M_{n-1}. These lines are called the bicharacteristics of the original second order system. In the theory of light they are the rays.

The characteristics now play a vital role in the theory of partial differential equations. For example, Darboux[45] has given a powerful method of integrating second order partial differential equations in two independent variables that rests on the theory of characteristics. It converts the problem to the integration of one or more ordinary differential equations and embraces the methods of Monge, Laplace, and others.

Another class of existence theorems dealt with the Dirichlet problem, that is, establishing the existence of a solution of $\Delta V = 0$ either directly or by means of the Dirichlet principle. The first existence proof of the Dirichlet

43. *Math. Ann.*, 13, 1878, 411–28.
44. *Bull. Soc. Math. de France*, 25, 1897, 108–20.
45. *Ann. de l'Ecole Norm. Sup.*, (1), 7, 1870, 175–80.

problem in two dimensions (but not of the Dirichlet principle of minimizing the Dirichlet integral) was given by Hermann Amandus Schwarz (1843–1921), a pupil of Weierstrass, whom he succeeded at Berlin in 1892 and who suggested the problem to him. Under general assumptions about the bounding curve, and by using a process called the alternating procedure,[46] he demonstrated the existence of a solution.[47]

In the same year, 1870, Carl G. Neumann gave another proof of the existence of a solution of the Dirichlet problem in three dimensions[48] by using the method of arithmetic means, though he too did not use the Dirichlet principle.[49] The principal exposition of his ideas is in his *Vorlesungen über Riemann's Theorie der Abel'schen Integrale*.[50]

Then Henri Poincaré[51] used the *methode de balayage*, the method of "sweeping out," which approaches the problem by building a succession of functions not harmonic in the domain *R* but taking on the correct boundary values, the functions becoming more and more harmonic.

Finally, David Hilbert reconstructed the calculus of variations method of Thomson and Dirichlet and established the Dirichlet *principle* as a method for proving the existence of a solution of the Dirichlet problem. In 1899[52] Hilbert showed that under proper conditions on the region, boundary values, and the admissible functions *U*, the Dirichlet principle does hold. He made the Dirichlet principle a powerful tool in function theory. In another publication, of work done in 1901,[53] Hilbert gave more general conditions.

The history of the Dirichlet principle is remarkable. Green, Dirichlet, Thomson, and others of their time regarded it as a completely sound method and used it freely. Then Riemann in his complex function theory showed it to be extraordinarily instrumental in leading to major results. All of these men were aware that the fundamental existence question was not settled, even before Weierstrass announced his critique in 1870, which discredited the method for several decades. The principle was then rescued by Hilbert and was used and extended in this century. Had the progress made with the use of the principle awaited Hilbert's work, a large segment

46. Schwarz's method is sketched in Felix Klein, *Vorlesungen über die Entwicklung der Mathematik im 19. Jahrhundert*, Chelsea (reprint), 1950, 1, p. 265, and given fully in A. R. Forsyth, *Theory of Functions*, Dover (reprint), 1965, 2, Chap. 17. Many references are given in the latter source.
47. *Monatsber. Berliner Akad.*, 1870, 767–95 = *Ges. Math. Abh.*, 2, 144–71.
48. *Königlich Sächsischen Ges. der Wiss. zu Leipzig*, 1870, 49–56, 264–321.
49. The method is described in O. D. Kellogg, *Foundations of Potential Theory*, Julius Springer, 1929, 281 ff.
50. Second ed., 1884, 238 ff.
51. *Amer. Jour. of Math.*, 12, 1890, 211–94 = *Œuvres*, 9, 28–113.
52. *Jahres. der Deut. Math.-Verein.*, 8, 1900, 184–88 = *Ges. Abh.*, 3, 10–14.
53. *Math. Ann.*, 59, 1904, 161–86 = *Ges. Abh.*, 3, 15–37.

of nineteenth-century work on potential theory and function theory would
have been lost.

The Laplace equation $\Delta V = 0$ is the basic form of elliptic differential
equations. Many more existence theorems were established for more
general elliptic differential equations, such as

$$(59) \qquad \frac{\partial^2 u}{\partial x^2} + \frac{\partial^2 u}{\partial y^2} + a \frac{\partial u}{\partial x} + b \frac{\partial u}{\partial y} + cu = 0.$$

The variety of such theorems is vast. We shall mention just one key result.
The existence and uniqueness of a solution of this equation (the value of the
solution is prescribed on the boundary) was demonstrated by Picard[54] for
domains of sufficiently small area. The result has been extended to more
variables, large domains, and in other respects by Picard and others. Picard
also established[55] that every equation of the above form (and even slightly
more general) whose coefficients are analytic functions possesses only analytic
solutions inside the domain in which the solution is sought, even though the
solution assumes non-analytic boundary values.

The theorems discussed thus far have generally dealt with analytic
differential equations and analytic initial or boundary data. However, such
conditions are too restrictive for applications because the given physical
data may not be analytic. Another major class of theorems deals with less
stringent conditions. We shall give just one example. Riemann's method,
which applies to the hyperbolic equation, relies upon the existence of his
characteristic function v and, as we pointed out, the existence of v was not
established by Riemann.

For this hyperbolic case (see [40]), Du Bois-Reymond in 1889 sought
the proper conditions and obtained results[56] which, expressed for the case
where $x = $ const. and $y = $ const. are the characteristics, read thus: Given
the continuous functions u and $\partial u/\partial n$ along a curve AB that is cut not more
than once by any characteristic line, then there exists one and only one
solution u of the differential equation which takes on the given values of u
and $\partial u/\partial n$ along AB. This solution is defined in the rectangle determined by
the characteristics through A and through B. If instead the values of a continuous function u on two segments of characteristics which abut one another
are given, then u is again uniquely determined in the rectangle determined
by the characteristics. In terms of $x, y,$ and u as spatial coordinates, the first
result states that the surface $u(x, y)$ goes through a given space curve and
with a given inclination. The second result means that the solution or
surface is enclosed in the space determined by two intersecting space curves.

54. *Comp. Rend.*, 107, 1888, 939–41, *Jour. de Math.*, (4), 6, 1890, 145–210, *Jour. de Math.*,
(5), 2, 1896, 295–304.
55. *Jour de l'Ecole Poly.*, 60, 1890, 89–105.
56. *Jour. für Math.*, 104, 1889, 241–301.

For the continuous initial conditions u and $\partial u/\partial n$ (and u in the second case), the solutions will be regular or satisfy the differential equation everywhere in the rectangles described above. Discontinuities in u and $\partial u/\partial n$ will propagate along the characteristics in the rectangle.

A great deal of work was done in the second half of the century on the existence of eigenvalues for $\Delta u + k^2 u = 0$ considered in a domain D. The main result is that for a given domain and under any one of the three boundary conditions $u = 0$, $\partial u/\partial n = 0$, $\partial u/\partial n + hu = 0$ ($h > 0$ when the positive normal is directed outside the domain), there are always an infinite number of discrete values of k^2, for each of which there is a solution. In two dimensions the vibrations of a membrane fixed along its boundary illustrate this theorem. The values of k are the frequencies of the infinitely many purely harmonic vibrations. The corresponding solutions give the deformation of the membrane in carrying out its characteristic oscillations.

The first major step was the proof by Schwarz[57] of the existence of the first eigenfunction of

$$\Delta u + \xi f(x, y)u = 0,$$

that is, the existence of a U_1 such that

$$\Delta U_1 + k_1^2 f(x, y) U_1 = 0$$

and $U_1 = 0$ on the boundary of the domain considered. His method gave a procedure for finding the solution and permitted the calculation of k_1^2. Picard[58] then established the existence of the second eigenvalue k_2^2.

Schwarz also showed in the 1885 paper that when the domain varies continuously, the value of k_1^2, the first characteristic number, also varies continuously; and as the domain becomes smaller, k_1^2 increases unboundedly. Thus a smaller membrane gives off a higher first harmonic.

In 1894 Poincaré[59] demonstrated the existence and the essential properties of all the eigenvalues of

(60) $$\Delta u + \lambda u = f,$$

λ complex, in a bounded, three-dimensional domain, with $u = 0$ on the boundary. The existence of u was demonstrated by a generalization of Schwarz's method. He proved next that $u(\lambda)$ is a meromorphic function of the complex variable λ and that the poles are real; these are just the eigenvalues λ_n. Then he obtained the characteristic solutions U_i, that is,

$$\Delta U_i + k_i^2 U_i = 0 \quad \text{(in the interior)}$$
$$U_i = 0 \quad \text{(on the boundary).}$$

57. *Acta Soc. Fennicae*, 15, 1885, 315–62 = *Ges. Math. Abh.*, 1, 223–69.
58. *Comp. Rend.*, 117, 1893, 502–7.
59. *Rendiconti del Circolo Matematico di Palermo*, 8, 1894, 57–155 = *Œuvres*, 9, 123–96.

The k_i^2 are the characteristic numbers (eigenvalues) and determine the frequencies of the respective characteristic solutions.

Physically Poincaré's result has the following significance. The function f in (60) can be thought of as an applied force. The free oscillations of a mechanical system are those at which the forced oscillations degenerate and become infinite. In fact, (60) is the equation of an oscillating system excited by a periodically varying force of amplitude f; and the characteristic solutions are the free oscillations of the system, which, once excited, continue indefinitely. The frequencies of the free oscillations, which are proportional to the k_i, are calculated by Poincaré's method as the values of $\sqrt{\lambda}$ for which the forced oscillation u becomes infinite.

At the end of the century the systematic theory of boundary- and initial-value problems for partial differential equations, which dates from Schwarz's fundamental 1885 paper, was still young. The work in this area expanded rapidly in the twentieth century.

Bibliography

Bacharach, Max: *Abriss der Geschichte der Potentialtheorie*, Vandenhoeck and Ruprecht, 1883.

Burkhardt, H.: "Entwicklungen nach oscillirenden Functionen und Integration der Differentialgleichungen der mathematischen Physik," *Jahres. der Deut. Math.-Verein.*, Vol. 10, 1908, 1–1804.

Burkhardt, H. and W. Franz Meyer: "Potentialtheorie," *Encyk. der Math. Wiss.*, B. G. Teubner, 1899–1916, II A7b, 464–503.

Cauchy, Augustin-Louis: *Œuvres complètes*, Gauthier-Villars, 1890, (2), 8.

Fourier, Joseph: *The Analytical Theory of Heat* (1822), Dover reprint of English translation, 1955.

———: *Œuvres*, 2 vols., Gauthier-Villars, 1888–90; Georg Olms (reprint), 1970.

Green, George: *Essay on the Application of Mathematical Analysis to the Theories of Electricity and Magnetism*, 1828, reprinted by Wezäta-Melins Aktiebolag, 1958; also in Ostwald's Klassiker #61 (in German), Wilhelm Engelmann, 1895.

———: *Mathematical Papers*, Macmillan, 1871; Chelsea (reprint), 1970.

Heine, Eduard: *Handbuch der Kugelfunktionen*, 2 vols., 1878–81, Physica Verlag (reprint), 1961.

Helmholtz, Hermann von: "Theorie der Luftschwingungen in Röhren mit offenen Enden," *Ostwald's Klassiker der exakten Wissenschaften*, Wilhelm Engelmann, 1896.

Klein, Felix: *Vorlesungen über die Entwicklung der Mathematik im 19. Jahrhundert*, 2 vols., Chelsea (reprint), 1950.

Langer, R. E.: "Fourier Series, the Genesis and Evolution of a Theory," *Amer. Math. Monthly*, 54, Part II, 1947, 1–86.

Pockels, Friedrich: *Über die partielle Differentialgleichung* $\Delta u + k^2 u = 0$, B. G. Teubner, 1891.

Poincaré, Henri: *Œuvres*, Gauthier-Villars, 1954, 9.

Rayleigh, Lord (Strutt, John William): *The Theory of Sound*, 2nd ed., 2 vols., Dover (reprint), 1945.

Riemann, Bernhard: *Gesammelte mathematische Werke*, 2nd ed., Dover (reprint), 1953, pp. 156–211.

Sommerfeld, Arnold: "Randwertaufgaben in der Theorie der partiellen Differentialgleichungen," *Encyk. der math. Wiss.*, B. G. Teubner, 1899–1916, II A7c, 504–70.

Todhunter, Isaac, and Karl Pearson: *A History of the Theory of Elasticity*, 2 vols., Dover (reprint), 1960.

Whittaker, Sir Edmund: *History of the Theories of Aether and Electricity*, rev. ed., Thomas Nelson and Sons, 1951, Vol. I.

29
Ordinary Differential Equations in the Nineteenth Century

> La Physique ne nous donne pas seulement l'occasion de résoudre des problèmes . . . elle nous fait préssentir la solution.
> HENRI POINCARÉ

1. *Introduction*

Ordinary differential equations arose in the eighteenth century in direct response to physical problems. By tackling more complicated physical phenomena, notably in the work on the vibrating string, the mathematicians arrived at partial differential equations. In the nineteenth century the roles of these two subjects were somewhat reversed. The efforts to solve partial differential equations by the method of separation of variables led to the problem of solving ordinary differential equations. Moreover, because the partial differential equations were expressed in various coordinate systems the ordinary differential equations that resulted were strange ones and not solvable in closed form. The mathematicians resorted to solutions in infinite series which are now known as special functions or higher transcendental functions as opposed to the elementary transcendental functions such as $\sin x$, e^x, and $\log x$.

After much work on the extended variety of ordinary differential equations, some deep theoretical studies were devoted to types of such equations. These theoretical investigations also differentiate the nineteenth-century work from that of the eighteenth century. The contributions of the new century were so vast that, as in the case of partial differential equations, we cannot hope to survey all of the major developments. Our topics are a sample of what was created during the century.

2. *Series Solutions and Special Functions*

As we have just observed, to solve the ordinary differential equations that resulted from the method of separation of variables applied to partial differential equations, the mathematicians, without worrying much about the existence and form the solutions should take, turned to the method of infinite series (Chap. 21, sec. 6). Of the ordinary differential equations that resulted from separation of variables the most important is the Bessel equation

(1) $$x^2 y'' + xy' + (x^2 - n^2)y = 0,$$

where n, a parameter, can be complex and x can also be complex. However, for Friedrich Wilhelm Bessel (1784–1846), a mathematician and director of the astronomical observatory in Königsberg, n and x were real. Special cases of this equation occurred as early as 1703, when Jacob Bernoulli mentioned it as a particular solution in a letter to Leibniz, and thereafter in more extensive work by Daniel Bernoulli and Euler (Chap. 21, secs. 4 and 6, and Chap. 22, sec. 3). Special cases also occurred in the writings of Fourier and Poisson. The first systematic study of solutions of this equation was made by Bessel[1] while working on the motion of the planets. The equation has two independent solutions for each n, denoted today by $J_n(x)$ and $Y_n(x)$ and called Bessel functions of the first and second kind, respectively. Bessel, whose work on the equation dates from 1816, gave first the integral relation (for integral n)

$$J_n(x) = \frac{1}{2\pi} \int_0^{2\pi} \cos(nu - x \sin u)\, du$$

(he wrote I_k^h and his k is our x). Bessel also obtained the series

(2)

$$J_n(x) = \frac{x^n}{2^n \Gamma(n+1)} \left\{ 1 - \frac{x^2}{2^2 \cdot 1\,!\,(n+1)} + \frac{x^4}{2^4 \cdot 2\,!\,(n+1)(n+2)} - \cdots \right\}.$$

In 1818 Bessel proved that $J_0(x)$ has an infinite number of real zeros. In the 1824 paper Bessel also gave the recursion formula (for integral n)

$$xJ_{n+1}(x) - 2nJ_n(x) + xJ_{n-1}(x) = 0,$$

and many other relations involving the Bessel function of the first kind. The generalization of the Bessel function $J_n(x)$ to complex n and x was made by several men[2] with (2) remaining as the correct form.

Since there should be two independent solutions for a second order equation, many mathematicians sought it. When n is not an integer this second solution is $J_{-n}(x)$. For integral n the second solution was given by

1. *Abh. Konig. Akad. der Wiss. Berlin*, 1824, 1–52, pub. 1826 = *Werke*, 1, 84–109.
2. Chiefly by Eugen C. J. Lommel (1837–99) in his *Studien über die Bessel'schen Functionen* (1868).

Carl G. Neumann.[3] However, the form adopted most commonly today was given by Hermann Hankel (1839–73),[4] namely,

$$Y_n(z) = \sum_{r=0}^{\infty} \frac{(-1)^r[(1/2)z]^{n+2r}}{r!\,(n+r)!} \left\{ 2\log\left(\frac{z}{2}\right) + 2\gamma - \sum_{m=1}^{n+r}\frac{1}{m} - \sum_{m=1}^{r}\frac{1}{m} \right\}$$
$$- \sum_{r=0}^{n-1} \frac{[(1/2)z]^{-n+2r}(n-r-1)!}{r!},$$

where γ is Euler's constant. Neumann[5] also gave the expansion of an analytic function $f(z)$, namely,

$$f(z) = \alpha_0 J_0(z) + \alpha_1 J_1(z) + \alpha_2 J_2(z) + \cdots,$$

where the α_i are constants and can be determined.

Many mathematicians, usually working in celestial mechanics, arrived independently at the Bessel functions and at hundreds of other relations and expressions for these functions. Some idea of the vast literature on these functions may be gained from perusing G. N. Watson's *A Treatise on the Theory of Bessel Functions*.[6]

The Legendre polynomials or spherical functions of one variable and the spherical surface functions, which are functions of two variables, had already been introduced by Legendre and Laplace (Chap. 22, sec. 4). The Legendre polynomials satisfy the Legendre differential equation

$$(1 - x^2)y'' - 2xy' + n(n+1)y = 0.$$

This equation, as we know, results from separation of variables applied to the potential equation expressed in spherical coordinates. In 1833, Robert Murphy (d. 1843), a fellow at Cambridge University, wrote a text, *Elementary Principles of the Theories of Electricity, Heat and Molecular Actions*. In it he put together some older results on Legendre polynomials and obtained some new ones. Since the major results were already known, we shall not present the details of Murphy's work except to point out that it was systematic and that he showed that "any" function $f(x)$ can be expanded in terms of the $P_n(x)$ by applying term-by-term integration and the orthogonality property (integral theorem).

Heine,[7] treating the problem of the potential for the exterior of an ellipsoid of revolution and for the shell between confocal ellipsoids of revolution (Chap. 28, sec. 5), introduced spherical harmonics of the second kind, usually denoted by $Q_n(x)$, which provide a second independent solution

3. *Theorie der Bessel'schen Funktionen*, 1867, 41.
4. *Math. Ann.*, 1, 1869, 467–501.
5. *Jour. für Math.*, 67, 1867, 310–14.
6. Cambridge University Press, 2nd ed., 1944.
7. *Jour. für Math.*, 26, 1843, 185–216.

of Legendre's equation. The Legendre functions, like the Bessel functions, have been extended to complex n and complex x, and a number of alternative representations and relationships among them have been obtained.[8]

The study of special functions that arise as series solutions of ordinary differential equations was furthered by Gauss in a famous paper of 1812 on the hypergeometric series.[9] In this paper Gauss made no use of the differential equation

$$x(1 - x)y'' + \{\gamma - (\alpha + \beta + 1)x\}y' - \alpha\beta y = 0,$$

but he did in unpublished material.[10] Of course the equation and the series solution

$$(3) \qquad F(\alpha, \beta, \gamma; x) = 1 + \frac{\alpha \cdot \beta}{1 \cdot \gamma} x + \frac{\alpha(\alpha + 1)\beta(\beta + 1)}{1 \cdot 2 \cdot \gamma(\gamma + 1)} x^2 + \cdots$$

were already known because they had been studied by Euler (Chap. 21, sec. 6). Gauss recognized that for special values of α, β, and γ the series included almost all the elementary functions then known and many higher transcendental functions such as the Bessel functions and spherical functions. In addition to proving any number of properties of the series, Gauss established the famous relation

$$F(\alpha, \beta, \gamma; 1) = \frac{\Gamma(\gamma)\Gamma(\gamma - \alpha - \beta)}{\Gamma(\gamma - \alpha)\Gamma(\gamma - \beta)}.$$

He also established the convergence of the series (cf. Chap. 40, sec. 5). The notation $F(\alpha, \beta, \gamma; x)$ is due to Gauss.

Another class of special functions was introduced by Lamé.[11] In Chapter 28, sec. 5, we pointed out that Lamé, working on a steady-state temperature distribution in an ellipsoid, separated Laplace's equation in ellipsoidal coordinates ρ, μ, and ν. This process gives the same ordinary differential equations in each of the three variables, namely,

$$(4) \quad (\rho^2 - h^2)(\rho^2 - k^2)\frac{d^2E(\rho)}{d\rho^2} + \rho(2\rho^2 - h^2 - k^2)\frac{dE(\rho)}{d\rho}$$

$$+ \{(h^2 + k^2)p - n(n + 1)\rho^2\}E(\rho) = 0,$$

with appropriate changes for μ and ν in place of ρ. Here h^2 and k^2 are the parameters in the equations of the families of the coordinate surfaces and p and n are constants. This equation is known as Lamé's differential equation.

8. See, for example, E. W. Hobson, *The Theory of Spherical and Ellipsoidal Harmonics*, 1931, Chelsea (reprint), 1955.
9. *Comm. Soc. Sci. Gott.*, 2, 1813 = *Werke*, 3, 123–62.
10. *Werke*, 3, 207–30.
11. *Jour. de Math.*, 2, 1837, 147–83; 4, 1839, 126–63.

The solutions $E(\rho)$ are called Lamé functions or ellipsoidal harmonics. For integral n these functions fall into four classes of the form

$$E_n^p(\rho) = a_0\rho^n + a_1\rho^{n-2} + \cdots$$

or such polynomials multiplied by $\sqrt{\rho^2 - h^2}$ or $\sqrt{\rho^2 - k^2}$ or both factors. For a given value of n, the number of such functions (to insure some properties of $E(\rho)$) is $2n + 1$.

The second solution of Lamé's equation (resulting from other conditions or properties of $E(\rho)$) is

$$F_n^p(\rho) = (2n + 1)E_n^p(\rho) \int_\rho^\infty \frac{d\rho}{\sqrt{(\rho^2 - h^2)(\rho^2 - k^2)}[E_n^p(\rho)]^2},$$

and such functions are called Lamé functions of the second kind. These were introduced by Liouville[12] and Heine.[13]

The differential equations

$$\frac{d^2\phi}{d\eta^2} + (a - 2k^2\cos 2\eta)\phi = 0 \qquad \frac{d^2\psi}{d\xi^2} - (a - 2k^2\cosh 2\xi)\psi = 0$$

arose in Mathieu's work on the vibrations of an elliptical membrane[14] and they also arise in problems on the potential of elliptical cylinders when separation of variables is applied to $\Delta u + k^2u = 0$ expressed in elliptical cylindrical coordinates in two dimensions. These elliptic coordinates, incidentally, are related to rectangular coordinates by the equations

$$x = h\cosh \xi \cos \eta, \qquad y = h\sinh \xi \sin \eta,$$

where $x = \pm h, y = 0$ are the foci of the confocal ellipses and hyperbolas of the family of ellipses and the family of hyperbolas in the elliptic coordinate system. The variety of forms in which Mathieu's equation is written and the many notations for the solutions by different authors present a confusing picture. The functions defined by either differential equation were called by Heine functions of the elliptic cylinder and are now called Mathieu functions. Mathieu and Heine first got series expressions for the solutions. Then they sought to fix the parameter a so that one class of solutions is periodic and of period 2π. The problem of finding periodic solutions, which are the most important ones for physical applications, was pursued throughout the century. In 1883[15] Gaston Floquet (1847–1920) published a complete discussion of the existence and properties of the periodic solutions of a linear differential equation of the nth order having periodic coefficients which have the same period ω. The general properties of the solutions having been

12. *Jour. de Math.*, 10, 1845, 222–28.
13. *Jour. für Math.*, 29, 1845, 185–208.
14. *Jour. de Math.*, (2), 13, 1868, 137–203.
15. *Ann. de l'Ecole Norm. Sup.*, (2), 12, 1883, 47–88.

determined, later writers devoted considerable attention to the problem of discovering practical methods of finding them. No general methods were found (cf. sec. 7).

A widely studied class of special functions was introduced by Heinrich Weber (1842–1913) in 1868.[16] Weber was interested in integrating $\Delta u + k^2 u = 0$ in a domain bounded by two parabolas. He therefore changed from rectangular coordinates to parabolic coordinates (which are a limiting case of elliptic coordinates) through the transformation

$$x = \xi^2 - \eta^2, \qquad y = 2\xi\eta.$$

For $\xi =$ const. and for $\eta =$ const., the two families of curves are families of parabolas, with each member of one family cutting the members of the other orthogonally. The ordinary differential equations which Weber derived from the reduced wave equation by separation of variables are

$$\frac{d^2 E}{d\xi^2} + (k^2\xi^2 + a)E = 0$$

$$\frac{d^2 H}{d\eta^2} + (k^2\eta^2 - a)H = 0.$$

Weber gave four particular solutions of the second equation in the form of definite integrals. The solutions are called parabolic cylinder functions. Weber also showed that the only case in which separation of variables can be applied to $\Delta u + k^2 u = 0$, among all orthogonal coordinate systems, is that of confocal surfaces of the second degree or specializations thereof.

The class of special functions is far more extensive than we can indicate here. The above-mentioned types and many others that were introduced serve to solve differential equations in some bounded domain and to represent arbitrary functions (usually the initial functions of a partial differential equations problem) in that domain. The limitation to bounded domains is imposed by the orthogonality property. In the basic case of the trigonometric functions this domain is $(-\pi, \pi)$ because, for example,

$$\int_{-\pi}^{\pi} \sin mx \sin nx \, dx = 0 \quad \text{if } m \neq n.$$

The problem of solving ordinary differential equations over infinite intervals or semi-infinite intervals and of obtaining expansions of arbitrary functions over such intervals was also tackled by many men during the second half of the century and such special functions as Hermite functions first introduced by Hermite in 1864[17] and Nikolai J. Sonine (1849–1915) in 1880[18] serve to solve this problem.

16. *Math. Ann.*, 1, 1869, 1–36.
17. *Comp. Rend.*, 58, 1864, 93–100 and 266–73 = *Œuvres*, 2, 293–308.
18. *Math. Ann.*, 16, 1880, 1–80.

To work with all these types of special functions one must know their properties as intimately as one knows the properties of the elementary functions. Also because these special functions are more complicated the properties are likewise so. The literature both in original papers and texts is almost incredibly vast. Whole treatises have been devoted to Bessel functions, spherical functions, ellipsoidal functions, Mathieu functions, and other types.

3. Sturm-Liouville Theory

The problems involving partial differential equations of mathematical physics usually contain boundary conditions such as the condition that the vibrating string must be fixed at the endpoints. When the method of separation of variables is applied to a partial differential equation this equation is resolved into two or more ordinary differential equations and the boundary conditions on the desired solution become boundary conditions on one ordinary differential equation. The ordinary equation generally contains a parameter, which in fact results from the separation of variables procedure, and solutions can usually be obtained for particular values of the parameter. These values are called eigenvalues or characteristic values and the solution for any one eigenvalue is called an eigenfunction. Moreover, to meet the initial condition or conditions of the original problem it is necessary to express a given function $f(x)$ in terms of the eigenfunctions (see, for example, [11] of Chap. 28).

These problems of determining the eigenvalues and eigenfunctions of an ordinary differential equation with boundary conditions and of expanding a given function in terms of an infinite series of the eigenfunctions, which date from about 1750, became more prominent as new coordinate systems were introduced and new classes of functions such as Bessel functions, Legendre polynomials, Lamé functions, and Mathieu functions arose as the eigenfunctions of ordinary differential equations. Two men, Charles Sturm (1803–55), professor of mechanics at the Sorbonne, and Joseph Liouville (1809–82), a friend of Sturm and professor of mathematics at the Collège de France, decided to tackle the general problem for any second order ordinary differential equation. Sturm had been working since 1833 on problems of partial differential equations, primarily on the flow of heat in a bar of variable density, and hence was fully aware of the eigenvalue and eigenfunction problem.

The mathematical ideas he applied to this problem[19] are closely related to his investigations of the reality and distribution of the roots of algebraic equations. His ideas on differential equations, he says, came from the study of difference equations and a passage to the limit.

19. *Jour. de Math.*, 1, 1836, 106–86 and 373–444.

Liouville, informed by Sturm of the problems he was working on, took up the same subject.[20] The results in the several papers of these two men are quite detailed and are most conveniently summarized in modern notation as follows. They considered the general second order equation

(5) $$Ly'' + My' + \lambda Ny = 0,$$

where L, M, and N are continuous functions of x, L is not zero, and λ is a parameter. Such an equation can be transformed, by multiplying through by $L^{-1} e^{\int ML^{-1} dx}$, into

$$\frac{d}{dx}\left[p(x)\frac{dy}{dx} \right] + \lambda\rho(x)y = 0, \qquad p(x) > 0, \rho(x) > 0.$$

The boundary conditions to be satisfied by the original or transformed equation can have the general form

$$\begin{aligned} y'(a) - h_1 y(a) &= 0, \\ y'(b) + h_2 y(b) &= 0 \end{aligned} \qquad h_1 \geq 0, h_2 \geq 0, a < b.$$

Sturm and Liouville demonstrated the following fundamental results:

(a) The problem has a non-zero solution only when λ takes on any one of a sequence of values λ_n of positive numbers which increase to ∞.

(b) For each λ_n the solutions are multiples of one function v_n, which one can normalize by the condition $\int_a^b \rho v_n^2 \, dx = 1$.

(c) The orthogonality property, $\int_a^b \rho v_m v_n \, dx = 0$ for $m \neq n$, holds.

(d) Each function f twice differentiable in (a, b) and satisfying the boundary conditions can be expanded in a uniformly convergent series

$$f(x) = \sum_{n=1}^{\infty} c_n v_n(x)$$

where

$$c_n = \int_a^b \rho f v_n(x) \, dx.$$

(e) The equality

$$\int_a^b \rho f^2 \, dx = \sum_{n=1}^{\infty} c_n^2$$

obtains. This last equality, called the Parseval equality, had already been demonstrated purely formally by Marc-Antoine Parseval (?–1836) in 1799[21]

20. *Jour. de Math.*, 1, 1836, 253–65; 2, 1837, 16–35 and 418–36.
21. *Mém. des sav. étrangers*, (2), 1, 1805, 639–48.

for the set of trigonometric functions. From it there follows the inequality demonstrated by Bessel in 1828 [22] also for trigonometric series, namely,

$$\sum_{n=1}^{\infty} c_n^2 \leq \int_a^b |f(x)|^2 \, dx.$$

Actually the Sturm-Liouville results were not satisfactorily established in all respects. The proof that $f(x)$ can be represented as an infinite sum of the eigenfunctions was inadequate. One difficulty was the matter of the completeness of the set of eigenfunctions, which for a continuous $f(x)$ on (a, b) is the condition (e) above and which loosely means that the set of eigenfunctions is large enough to represent "any" $f(x)$. Also the question of the sense in which the series $\sum c_n v_n(x)$ converges to $f(x)$, whether pointwise, uniformly, or in some more general sense, was not covered though Liouville did give convergence proofs in some cases, using theory developed by Cauchy and Dirichlet.

4. Existence Theorems

We have already noted under this same topic in the chapter on nineteenth-century partial differential equations that as mathematicians found the problem of obtaining solutions for specific differential equations more and more difficult they turned to the question, Given a differential equation does it have a solution for given initial conditions and boundary conditions? The same movement, of course to be expected, occurred in ordinary differential equations. That the question of existence was neglected so long is partly due to the fact that differential equations arose in physical and geometrical problems, and it was intuitively clear that these equations had solutions.

Cauchy was the first to consider the question of the existence of solutions of differential equations and succeeded in giving two methods. The first, applicable to

$$(6) \qquad\qquad y' = f(x, y),$$

was created sometime between 1820 and 1830 and summarized in his *Exercices d'analyse.*[23]

This method, the essence of which may be found in Euler,[24] utilizes the same idea as is involved in the integral as a limit of a sum. Cauchy wished to show that there is one and only one $y = f(x)$ that satisfies (6) and which meets the given initial condition that $y_0 = f(x_0)$. He divided (x_0, x) into n parts $\Delta x_0, \Delta x_1, \ldots, \Delta x_{n-1}$ and formed

$$y_{i+1} = y_i + f(x_i, y_i) \, \Delta x_i,$$

22. *Astronom. Nach.*, 6, 1828, 333–48.
23. Vol. 1, 1840, 327 ff. = *Œuvres*, (2), 11, 399–465.
24. *Inst. Cal. Int.*, 1, 1768, 493.

where x_i is any value of x in Δx_i. Then by definition

$$y_n = y_0 + \sum_{i=0}^{n-1} f(x_i, y_i) \, \Delta x_i.$$

Now Cauchy shows that as n becomes infinite y_n converges to a unique function

$$y = y_0 + \int_{x_0}^{x} f(x, y) \, dx$$

and that this function satisfies (6) and the initial conditions.

Cauchy assumed that $f(x, y)$ and f_y are continuous for all real values of x and y in the rectangle determined by the intervals (x_0, x) and (y_0, y). In 1876 Rudolph Lipschitz (1832–1903) weakened the hypotheses of the theorem.[25] His essential condition was that for all (x, y_1) and (x, y_2) in the rectangle $|x - x_0| \leq a$, $|y - y_0| \leq b$, that is, for any two points with the same abscissa, there is a constant K such that

$$|f(x, y_1) - f(x, y_2)| < K(y_1 - y_2).$$

This condition is known as the Lipschitz condition, and the existence theorem is called the Cauchy-Lipschitz theorem.

Cauchy's second method of establishing the existence of solutions of differential equations, the method of dominant or majorant functions, is more broadly applicable than his first one and was applied by Cauchy in the complex domain. The method was presented in a series of papers in the *Comptes Rendus* during the years 1839–42.[26] The method was called by Cauchy the *calcul des limites* because it provides lower limits within which the solution whose existence is established is sure to converge. The method was simplified by Briot and Bouquet and their version[27] has become standard.

To illustrate the method let us note how it applies to

$$y' = f(x, y),$$

where f is analytic in x and y. The theorem to be established reads thus: If for

(7) $$\frac{dy}{dx} = f(x, y)$$

the function $f(x, y)$ is analytic in the neighborhood of $P_0 = (x_0, y_0)$, the differential equation then has a unique solution $y(x)$ which is analytic in a

25. *Bull. des Sci. Math.*, (1), 10, 1876, 149–59.
26. *Œuvres*, (1), Vols. 4 to 7 and 10. The most important papers are in the *Comptes Rendus* for Aug. 5 and Nov. 21, 1839, June 29, Oct. 26, Nov. 2, and Nov. 9 of 1840, and June 20 and July 4 of 1842.
27. *Comp. Rend.*, 39, 1854, 368–71.

neighborhood of x_0 and which reduces to y_0 when $x = x_0$. The solution can be represented by the series

(8) $\qquad y = y_0 + y_0'(x - x_0) + \dfrac{y_0''}{2!}(x - x_0)^2 + \dfrac{y_0'''}{3!}(x - x_0)^3 + \cdots$

wherein $y_0' = dy/dx$ at (x_0, y_0) and similarly for $y_0'', y_0''', \ldots$, and where the derivatives are determined by successive differentiation of the original differential equation in which y is treated as a function of x.

The method of proof, which we merely sketch, uses first the fact that because $f(x, y)$ is analytic in the neighborhood of (x_0, y_0), which for convenience we take to be $(0, 0)$, there is a circle of radius a about $x_0 = 0$ and a circle of radius b about $y_0 = 0$ in which $f(x, y)$ is analytic. Then $f(x, y)$ has an upper bound M for all values of x and y in the respective circles. Now the very method of obtaining the series (8) guarantees that it formally satisfies (7). The problem is to show that the series converges.

Toward this end one sets up the majorant function

$$F(x, y) = \sum \frac{M}{a^p b^q} x^p y^q,$$

which is the expansion of

(9) $\qquad\qquad\qquad F(x, y) = \dfrac{M}{(1 - x/a)(1 - y/b)}.$

One shows next that the series solution of

(10) $\qquad\qquad\qquad \dfrac{dY}{dx} = F(x, Y)$

namely,

(11) $\qquad\qquad Y = Y_0'x + Y_0''\dfrac{x^2}{2!} + Y_0'''\dfrac{x^3}{3!} + \cdots,$

which is derived from (10) in the same way that (8) is derived from (7), dominates term for term the series (8). Hence if (11) converges then (8) does. To show that (11) converges one solves (10) explicitly using the value of F in (9) and shows that the series expansion of the solution, which must be (11), converges.

The method does not in itself determine the precise radius of convergence of the series for y. Numerous efforts were therefore devoted to showing that the radius can be extended. However, the papers do not give the full domain of convergence and are of little practical importance.

A third method of establishing the existence of solutions of ordinary differential equations, probably known to Cauchy, was first published by Liouville[28] for a second order equation. This is the method of successive

28. *Jour. de Math.*, (1), 3, 1838, 561–614.

approximation and is now credited to Emile Picard because he gave the method a general form.[29] For the equation in real x and y,

$$y' = f(x, y),$$

wherein $f(x, y)$ is analytic in x and y and whose solution $y = f(x)$ is to pass through (x_0, y_0), the method is to introduce the sequence of functions

$$y_1(x) = y_0 + \int_{x_0}^{x} f(t, y_0)\, dt$$

$$y_2(x) = y_0 + \int_{x_0}^{x} f(t, y_1(t))\, dt$$

$$\cdot \quad \cdot \quad \cdot \quad \cdot \quad \cdot \quad \cdot \quad \cdot \quad \cdot \quad \cdot \quad \cdot \quad \cdot \quad \cdot$$

$$y_n(x) = y_0 + \int_{x_0}^{x} f(t, y_{n-1}(t))\, dt.$$

Then one proves that $y_n(x)$ tends to a limit $y(x)$ which is the one and only continuous function of x satisfying the ordinary differential equation and such that $y(x_0) = y_0$. The method as usually presented today presupposes that $f(x, y)$ satisfies only the Lipschitz condition. The method was extended to second order equations by Picard in the 1893 paper and has also been extended to complex x and y.

The various methods described above were applied not only to higher order ordinary differential equations but to systems of differential equations for complex-valued variables. Thus Cauchy extended his second type of existence theorem to systems of first order ordinary differential equations in n dependent variables. He also extended this method of *calcul des limites* to systems in the complex domain.[30] Cauchy's result reads as follows: Given the system of equations

$$(12) \qquad \frac{dy_k}{dx} = f_k(x, y_0, \cdots, y_{n-1}), \qquad k = 0, 1, 2, \cdots, n - 1,$$

let $f_0, \ldots, f_{n-1}$ be monogenic (single-valued analytic) functions of their arguments and let them be developable in the neighborhood of the initial values

$$x = \xi, y_0 = \eta_0, \cdots, y_{n-1} = \eta_{n-1}$$

in positive integral powers of

$$x - \xi, y_0 - \eta_0, \cdots, y_{n-1} - \eta_{n-1}.$$

Then there are n power series in $x - \xi$ convergent in the neighborhood of $x = \xi$ which when substituted for $y_0, \ldots, y_{n-1}$ in (12) satisfy the equations.

29. *Jour. de Math.*, (4), 6, 1890, 145–210; and (4), 9, 1893, 217–71.
30. For different existence proofs on systems of first order differential equations, see the second reference to Painlevé in the bibliography at the end of this chapter.

These power series are unique. They give a regular solution of the system and take on the initial values. In this generality the result can be found in Cauchy's "Mémoire sur l'emploi du nouveau calcul, appelé calcul des limites, dans l'intégration d'un système d'équations différentielles."[31] Thus the idea was to content oneself with establishing the existence of and obtaining the solution in the neighborhood of one point in the complex plane. Weierstrass obtained the same result in the same year (1842) but did not publish it until his *Werke* came out in 1894.[32]

5. *The Theory of Singularities*

In the middle of the nineteenth century the study of ordinary differential equations took a new course. The existence theorems and Sturm-Liouville theory presuppose that the differential equations contain analytic functions or, at the very least, continuous functions in the domains in which solutions are considered. On the other hand some of the differential equations already considered, such as Bessel's, Legendre's, and the hypergeometric equation, when expressed so that the coefficient of the second derivative is unity, have coefficients that are singular, and the form of the series solutions in the neighborhood of the singular points, particularly that of the second solution, is peculiar. Hence mathematicians turned to the study of solutions in the neighborhood of singular points, that is, points at which one or more of the coefficients are singular. A point at which all the coefficients are at least continuous and usually analytic is called an ordinary point.

The solutions in the neighborhood of singular points are obtainable as series, and the knowledge of the proper form of the series must be at hand before calculating it. This knowledge can be obtained only from the differential equation. The new problem was described by Lazarus Fuchs (1833–1902) in a paper of 1866 (see below). "In the present condition of science the problem of the theory of differential equations is not so much to reduce a given differential equation to quadratures, as to deduce from the equation itself the behavior of its integrals at all points of the plane, that is, for all values of the complex variable." For this problem Gauss's work on the hypergeometric series pointed the way. The leaders were Riemann and Fuchs, the latter a student of Weierstrass and his successor at Berlin. The theory which resulted is called the Fuchsian theory of linear differential equations.

Attention in this new area concentrated on linear differential equations of the form

(13) $$y^{(n)} + p_1(z)y^{(n-1)} + \cdots + p_n(z)y = 0,$$

31. *Comp. Rend.*, 15, 1842, 14–25 = *Œuvres* (1), 7, 5–17.
32. *Math. Werke*, 1, 75–85.

where the $p_i(z)$ are single-valued analytic functions of complex z except at isolated singular points. This equation was emphasized because its solutions embrace all the elementary functions and even some higher functions, such as the modular and automorphic functions we shall encounter later.

Before considering solutions at and in the neighborhood of singular points, let us note a basic theorem that does follow from Cauchy's existence theorem on systems of ordinary differential equations but which was proven directly by Fuchs,[33] though he acknowledged his indebtedness to Weierstrass's lectures. If the coefficients $p_1, \ldots, p_n$ are analytic at a point a and in some neighborhood of that point, and if arbitrary initial conditions for y and its first $n - 1$ derivatives are given at $z = a$, then there is a unique power series solution for y in terms of z of the form

$$(14) \qquad y(z) = \sum_{r=0}^{\infty} \frac{1}{r!} y^{(r)}(a)(z - a)^r.$$

To Cauchy's result Fuchs added that the series is absolutely and uniformly convergent within any circle having a as a center and in which the $p_i(z)$ are analytic. It follows that the solutions can possess singularities only where the coefficients are singular.

The study of solutions in the neighborhood of singular points was initiated by Briot and Bouquet.[34] Since their results for first order linear equations were soon generalized we shall consider the more general treatments.

To get at the behavior of solutions in the neighborhood of singular points Riemann proposed an unusual approach. Though the $p_i(z)$ in (13) are assumed to be single-valued functions analytic except at isolated singular points, the solutions $y_i(z)$, analytic except possibly at the singular points, are not in general single-valued over the entire domain of z-values. Let us suppose that we have a fundamental set of solutions, $y_i(z)$, $i = 1, 2, \ldots, n$, that is, n independent solutions of the kind specified in the above theorem. Then the general solution is

$$y = c_1 y_1 + c_2 y_2 + \cdots + c_n y_n$$

wherein the c_i are constants.

If we now trace the behavior of an analytic y_i along a closed path enclosing a singular point, y_i will change its value to another branch of the same function though it remains a solution of the differential equation. Since any solution is a linear combination of n particular solutions, the altered y_i,

33. *Jour. für Math.*, 66, 1866, 121–60 = *Math. Werke*, 1, 159 ff.
34. *Jour. d'Ecole Poly.*, (1), 21, 1856, 85–132, 133–198, 199–254.

say y'_i, is still a linear combination of the y_i. Thus we obtain

(15)
$$
\begin{aligned}
y'_1 &= c_{11}y_1 + \cdots + c_{1n}y_n \\
y'_2 &= c_{21}y_1 + \cdots + c_{2n}y_n \\
&\ \cdot\quad\cdot\quad\cdot\quad\cdot\quad\cdot\quad\cdot\quad\cdot \\
y'_n &= c_{n1}y_1 + \cdots + c_{nn}y_n.
\end{aligned}
$$

That is, the $y_1, \ldots, y_n$ undergo a certain linear transformation when each is carried around a closed path enclosing a singular point. Such a transformation arises for any closed path around each of the singular points or combination of singular points. The set of transformations forms a group,[35] which is called the monodromy group of the differential equation, a term introduced by Hermite.[36]

Riemann's approach to obtaining the character of solutions in the neighborhood of singular points appeared in his paper of 1857 "Beiträge zur Theorie der durch die Gauss'sche Reihe $F(\alpha, \beta, \gamma, x)$ darstellbaren Functionen."[37] The hypergeometric differential equation, as Gauss knew, has three singular points, 0, 1, and ∞. Now Riemann showed that for complex x, to obtain conclusions about the behavior of the particular solutions around singular points of the second order equation, one does not need to know the differential equation itself but rather how two independent solutions behave as the independent variable traces closed paths around the three singular points. That is, we must know the transformations

$$
y'_1 = c_{11}y_1 + c_{12}y_2, \qquad y'_2 = c_{21}y + c_{22}y_2
$$

for each singular point.

Thus Riemann's idea in treating functions defined by differential equations was to derive the properties of the functions from a knowledge of the monodromy group. His 1857 paper dealt with the hypergeometric differential equation but his plan was to treat nth-order linear ordinary differential equations with algebraic coefficients. In a fragment written in 1857 but not published until his collected works appeared in 1876,[38] Riemann considered more general equations than the second order with three singular points. He accordingly assumes he has n functions uniform, finite, and continuous except at certain arbitrarily assigned points (the singular points) and undergoing an arbitrarily assigned linear substitution when z describes a closed circuit around such a point. He then shows that such a system of

35. By Riemann's time the algebraic notion of a group was known. It will be introduced in this book in Chap. 31. However, all that one needs to know here is that the application of two successive transformations is a transformation of the set and that the inverse of each transformation belongs to the set.
36. *Comp. Rend.*, 32, 1851, 458–61 = *Œuvres*, 1, 276–80.
37. *Werke*, 67–83.
38. *Werke*, 379–90.

functions will satisfy an nth-order linear differential equation. But he does not prove that the branch-points (singular points) and the substitutions may be chosen arbitrarily. His work here was incomplete and he left open a problem known as the Riemann problem: Given m points $a_1, \ldots, a_m$ in the complex plane with each of which is associated a linear transformation of the form (15), to prove on the basis of elementary assumptions about the behavior of the monodromy group associated with these singular points (so far as such behavior is not already determined) that a class of functions $y_1, \ldots, y_n$ is determined which satisfy a linear nth-order differential equation with the given a_i as singular (branch) points and such that when z traverses a closed path around a_i the y_i's undergo the linear transformation associated with a_i.

Guided by Riemann's 1857 paper on the hypergeometric equation the work on singularities was carried further by Fuchs. Beginning in 1865,[39] Fuchs and his students took up nth-order differential equations whereas Riemann had published only on Gauss's hypergeometric differential equation. Fuchs did not follow Riemann's approach but worked directly with the differential equation. Fuchs also brought not only linear differential equations but the entire theory of differential equations generally into the domain of complex function theory.

In the papers mentioned above Fuchs gave his major work on ordinary differential equations. He starts with the linear differential equation of the nth order whose coefficients are rational functions of x. By a careful examination of the convergence of the series which formally satisfy the equation he finds that the singular points of the equation are fixed, that is, independent of the constants of integration, and can be found before integrating because they are the poles of the coefficients of the differential equation.

He then shows that a fundamental system of solutions undergoes a linear transformation with constant coefficients when the independent variable z describes a circuit enclosing a singular point. From this behavior of the solutions he derives expressions for them valid in a circular region surrounding that point and extending to the next singular point. He thus establishes the existence of systems of n functions uniform, finite, and continuous except in the vicinity of certain points and undergoing linear substitutions with constant coefficients when the variable z describes closed circuits around these points.

Fuchs then considered what properties a differential equation of the form (13) must have in order that its solutions at a singular point $z = a$ have the form

$$(z - a)^s [\phi_0 + \phi_1 \log (z - a) + \cdots + \phi_\lambda \log^\lambda (z - a)],$$

where s is some number (which can be further specified) and the ϕ_i are single-valued functions in the neighborhood of $z = a$ which may have poles

39. *Jour. für Math.*, 66, 1866, 121–60; 68, 1868, 354–85.

of finite order. His answer was that a necessary and sufficient condition is that $p_r(z) = (z - a)^{-r}P(z)$, where $P(z)$ is analytic at and in the neighborhood of $z = a$. Thus $p_1(z)$ has a pole of order one and so on. Such a point a is called a regular singular point (Fuchs called it a point of determinateness).

Fuchs also studied a more specialized class of equations of the form (13). A homogeneous linear equation of this type is said to be of *Fuchsian type* when it has at worst regular singular points in the extended complex plane (including the point at ∞). In this case the $p_i(z)$ must be rational functions of z. For example, the hypergeometric equation has regular singular points at $z = 0$, 1, and ∞.

But the study of integrals of differential equations in the neighborhood of a given point does not necessarily furnish the integrals themselves. The study was taken as the point of departure for the investigation of the full integrals. Since the great researches of Fuchs, mathematicians have succeeded in extending the variety of linear ordinary differential equations that can be integrated explicitly. Previously only the nth-order linear equations with constant coefficients and Legendre's equation

$$(ax + b)^n \frac{d^ny}{dx^n} + A(ax + b)^{n-1} \frac{d^{n-1}y}{dx^{n-1}} + \cdots + L(ax + b) \frac{dy}{dx} + My = 0$$

could be integrated, the latter by the transformation $ax + b = e^t$. The new ones that can be integrated are those with integrals that are uniform (single-valued) functions of z. One recognizes that the integrals have this property by studying the singular points of the differential equation. The general integrals so obtained are usually new functions.

Beyond general results on the kinds of integrals which special classes of differential equations can have, there is the series approach to the solutions at a point $z = a$ where the equation has a regular singular point. If the origin is such a point then the equation must have the form

$$z^n \frac{d^nw}{dz^n} + z^{n-1}P_1(z) \frac{d^{n-1}w}{dz^{n-1}} + \cdots + zP_{n-1}(z) \frac{dw}{dz} + P_n(z)w = 0,$$

in which the $P_i(z)$ are analytic at and around $z = 0$. In this case one can obtain the n fundamental solutions in the form of series about $z = 0$ and show that the series converge for some range of z-values. The series are of the form

$$w = \sum_{v=0}^{\infty} c_v z^{\rho + v}$$

and the ρ and the c_v are determinable for each solution. The result is due to Georg Frobenius (1849–1917).[40]

40. *Jour. für Math.*, 76, 1874, 214–35 = *Ges. Abh.*, 1, 84–105.

The Riemann problem was also taken up during the latter part of the nineteenth century, but unsuccessfully until Hilbert in 1905 [41] and Oliver D. Kellogg (1878–1932),[42] with the help of the theory of integral equations, which was developed in the meantime, gave the first complete solution. They showed that the generating transformation of the monodromy group can be prescribed arbitrarily.

6. *Automorphic Functions*

The theory of solutions of linear differential equations was tackled next by Poincaré and Felix Klein. The subject they introduced is called automorphic functions, which, though important for various other applications, play a major role in the theory of differential equations.

Henri Poincaré (1854–1912) was a professor at the Sorbonne. His publications, almost as numerous as Euler's and Cauchy's, cover a wide range of mathematics and mathematical physics. His physical researches, which we shall not have occasion to discuss, included capillary attraction, elasticity, potential theory, hydrodynamics, the propagation of heat, electricity, optics, electromagnetic theory, relativity, and above all, celestial mechanics. Poincaré had penetrating insight, and in every problem he studied he brought out its essential character. He focused sharply on a problem and examined it minutely. He also believed in a qualitative study of all aspects of a problem.

Automorphic functions are generalizations of the circular, hyperbolic, elliptic, and other functions of elementary analysis. The function $\sin z$ is unchanged in value if z is replaced by $z + 2m\pi$ where m is any integer. One can also say that the function is unaltered in value if z is subjected to any transformation of the group $z' = z + 2m\pi$. The hyperbolic function $\sinh z$ is unchanged in value if z be subjected to any transformation of the group $z' = z + 2\pi mi$. An elliptic function remains invariant in value under transformations of the group $z' = z + m\omega + m'\omega'$ where ω and ω' are the periods of the function. All of these groups are discontinuous (a term introduced by Poincaré); that is, all the transforms of any point under the transformations of the group are finite in number in any closed bounded domain.

The term automorphic function is now used to cover functions that are invariant under the group of transformations

$$(16) \qquad z' = \frac{az + b}{cz + d},$$

41. *Proc. Third Internat. Math. Cong.*, 1905, 233–40; and *Nachrichten König. Ges. der Wiss. zu Gött.*, 1905, 307–88. Also in D. Hilbert, *Grundzüge einer allgemeinen Theorie der linearen Integralgleichungen*, 1912, Chelsea (reprint), 1953, 81–108.
42. *Math. Ann.*, 60, 1905, 424–33.

where a, b, c, and d may be real or complex numbers, and $ad - bc = 1$, or under some subgroup of this group. Moreover the group must be discontinuous in any finite part of the complex plane.

The earliest automorphic functions to be studied were the elliptic modular functions. These functions are invariant under the modular group, which is that subgroup of (16) wherein a, b, c, and d are real integers and $ad - bc = 1$, or under some subgroup of this group. These elliptic modular functions derive from the elliptic functions. We shall not pursue them here because they do not bear on the basic theory of differential equations.

More general automorphic functions were introduced to study linear differential equations of the second order

(17)
$$\frac{d^2\eta}{dz^2} + p_1 \frac{d\eta}{dz} + p_2\eta = 0,$$

where p_1 and p_2 were at first rational functions of z. A special case is the hypergeometric equation

(18)
$$\frac{d^2\eta}{dz^2} + \frac{\gamma - (\alpha + \beta + 1)}{z(1 - z)} \frac{d\eta}{dz} + \frac{\alpha\beta}{z(z - 1)} \eta = 0$$

with the three singular points 0, 1, and ∞.

Riemann, in his lectures of 1858–59 on the hypergeometric series and in a posthumous paper of 1867 on minimal surfaces, and Schwarz[43] independently established the following. Let η_1 and η_2 be any two particular solutions of the equation (17). All solutions can then be expressed as

$$\eta = m\eta_1 + n\eta_2.$$

When z traverses a closed path around a singular point, η_1 and η_2 go over into

$$\eta_1^1 = a\eta_1 + b\eta_2, \qquad \eta_2^2 = c\eta_1 + d\eta_2$$

and by letting z traverse closed paths around all the singular points one obtains a whole group of such linear transformations, which is the monodromy group of the differential equation.

Now let $\zeta(z) = \eta_1/\eta_2$. The quotient ζ as z traverses a closed path is transformed to

(19)
$$\zeta^1 = \frac{a\zeta + b}{c\zeta + d}.$$

From (17) we find that ζ satisfies the differential equation

(20)
$$\frac{\zeta'''}{\zeta'} - \frac{3}{2} \cdot \left(\frac{\zeta''}{\zeta'}\right) = 2p_2 - \frac{1}{2} p_1^2 - p_1'.$$

43. *Jour. für Math.*, 75, 1873, 292–335 = *Ges. Abh.*, 2, 211–59.

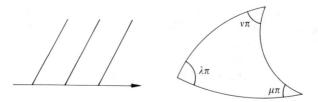

Figure 29.1

If we take for the p_1 and p_2 in (17) the particular functions in (18) we obtain

$$(21) \quad \frac{\zeta'''}{\zeta'} - \frac{3}{2}\left(\frac{\zeta''}{\zeta'}\right)^2 = \frac{1 - \lambda^2}{2z^2} + \frac{1 - \mu^2}{2(1 - z)^2} - \frac{\lambda^2 + \mu^2 - \nu^2 - 1}{2z(1 - z)},$$

where $\lambda^2 = 1 - \gamma^2$, $\mu^2 = (\gamma - \alpha - \beta)^2$, $\gamma^2 = (\alpha - \beta)^2$, and λ, μ, ν are taken positive (α, β, γ are real). The class of transformations (19) is the monodromy group of the differential equation (21).

Then Riemann and Schwarz showed that every particular solution $\zeta(z)$ of the equation (21), when λ, μ, and ν are real, is a conformal mapping of the upper half of the z-plane (Fig. 29.1) into a curvilinear triangle with circular arcs in the ζ-plane whose angles are $\lambda\pi$, $\mu\pi$, and $\nu\pi$.

In the case of a domain bounded by three arcs of circles, if the angles of the triangle satisfy certain conditions, the inverse function to $\zeta = \zeta(z)$ is an automorphic function $z = \phi(\zeta)$ whose entire domain of existence is a half-plane or a circle. This function remains invariant under transformation of ζ by elements of the group of linear transformations (19), which carry any curvilinear triangle of the form shown into another. The given "circular" triangle is the fundamental domain of the group. Under the group of transformations this domain is carried into analogous triangles whose union covers the half-plane or circle. The circular triangle is the analogue of the parallelogram in the case of elliptic functions.

The work of Poincaré and Klein carries on from this point. Klein did some basic work on automorphic functions before 1880. Then during the years 1881–82 he worked with Poincaré, who had also done previous work on the subject after his attention had been drawn to it by the above-described work of Fuchs. By 1884 Poincaré published five major papers on automorphic functions in the first five volumes of the *Acta Mathematica*. When the first of these was published in the first volume of the new *Acta Mathematica*, Kronecker warned the editor, Mittag-Leffler, that this immature and obscure article would kill the journal.

Guided by the theory of elliptic functions, Poincaré invented a new class of automorphic functions.[44] This class was obtained by considering the

44. *Acta Math.*, 1, 1882, 1–62 and 193–294 = *Œuvres*, 2, 108–68, 169–257.

inverse function of the ratio of two linearly independent solutions of the equation

$$\frac{d^2\eta}{dz^2} + P(w, z)\frac{d\eta}{dz} + Q(w, z)\eta = 0,$$

where w and z are connected by a polynomial equation $\phi(w, z) = 0$ and P and Q are rational functions. This is the class of Fuchsian automorphic functions and consists of uniform (single-valued) meromorphic functions within a circle (called the fundamental circle) which are invariant under the class of linear transformations of the form

(22) $$z' = \frac{az + b}{cz + d},$$

where a, b, c, and d are real and $ad - bc = 1$. These transformations, which leave the circle and its interior invariant, form a group called the Fuchsian group. Schwarz's function $\phi(\zeta)$ constitutes the simplest example of a Fuchsian function. Thus Poincaré demonstrated the existence of a class of auto-morphic functions more general than the elliptic modular functions.[45]

Poincaré's construction of automorphic functions (in the second paper of 1882) was based on his theta series. Let the transformations of the group (22) be

(23) $$z' = \frac{a_i z + b_i}{c_i z + d_i}, \qquad a_i d_i - b_i c_i = 1, \qquad i = 1, 2, \cdots.$$

Let $z_1, z_2, \ldots$ be the transforms of z under the various transformations of the group. Let $H(z)$ be a rational function (aside from other minor conditions). Then Poincaré's theta series is the function

(24) $$\theta(z) = \sum_{i=0}^{\infty} (c_i z + d_i)^{-2m} H(z_i), \qquad m > 1.$$

One can show that $\theta(z_j) = (c_j z + d_j)^{2m}\theta(z)$. Now let $\theta_1(z)$ and $\theta_2(z)$ be two theta series with the same m. These series are not only uniform functions but entire. Then

(25) $$F(z) = \frac{\theta_1(z)}{\theta_2(z)}$$

is an automorphic function of the group (23). Poincaré called the series (24) a theta-fuchsian series or a theta-kleinian series according as the group to which it belongs is Fuchsian or Kleinian (the latter will be described in a moment).

45. In this work on Fuchsian groups Poincaré used non-Euclidean geometry (Chap. 36) and showed that the study of Fuchsian groups reduces to that of the translation group of Lobatchevskian geometry.

Fuchsian functions are of two kinds, one existing in the entire plane, the other existing only in the interior of the fundamental circle. The inverse function of a Fuchsian function is, as we saw above, the ratio of two integrals of a second order linear differential equation with algebraic coefficients. Such an equation, which Poincaré called a Fuchsian equation, can be integrated by means of Fuchsian functions.

Then Poincaré[46] extended the group of transformations (22) to complex coefficients and considered several types of such groups, which he named Kleinian. We must be content here to note that a group is Kleinian if, essentially, it is not finite and not Fuchsian but, of course, is of the form (22) and discontinuous in any part of the complex plane. For these Kleinian groups Poincaré obtained new automorphic functions, that is, functions invariant under the Kleinian groups, which he called Kleinian functions. These functions have properties analogous to the Fuchsian ones; however, the fundamental region for the new functions is more complicated than a circle. Incidentally, Klein had considered Fuchsian functions while Lazarus Fuchs had not. Klein therefore protested to Poincaré. Poincaré responded by naming the next class of automorphic functions that he discovered Kleinian because, as someone wryly observed, they had never been considered by Klein.

Then Poincaré showed how to express the integrals of nth-order linear equations with *algebraic* coefficients having only regular singular points with the aid of the Kleinian functions. Thus this entire class of linear differential equations is solved by the use of these new transcendental functions of Poincaré.

7. *Hill's Work on Periodic Solutions of Linear Equations*

While the theory of automorphic functions was being created, the work in astronomy stimulated interest in a second order ordinary differential equation somewhat more general than Mathieu's equation. Since the n-body problem was not solvable explicitly and only complicated series solutions were at all available, the mathematicians turned to culling periodic solutions.

The importance of periodic solutions stems from the problem of the stability of a planetary or satellite orbit. If a planet is displaced slightly from its orbit and given a small velocity will it then oscillate about its orbit and perhaps return to the orbit after a time, or will it depart from the orbit? In the former case the orbit is stable, and in the latter unstable. Thus the question of whether the primary motion of the planets or any irregularities in the motions are periodic is vital.

As we know (Chap. 21, sec. 7), Lagrange had found special periodic solutions in the problem of three bodies. No new periodic solutions of the

46. *Acta Math.*, 3, 1883, 49–92; 4, 1884, 201–312 = *Œuvres*, 2, 258–99, 300–401.

three-body problem were found until George William Hill (1838–1914), the first great American mathematician, did his work on lunar theory. In 1877 Hill published privately a remarkably original paper on the motion of the moon's perigee.[47] He also published a very important paper on the motion of the moon in the *American Journal of Mathematics*.[48] His work founded the mathematical theory of homogeneous linear differential equations with periodic coefficients.

Hill's first fundamental idea (in his 1877 paper) was to determine a periodic solution of the differential equations for the motion of the moon that approximated the actual observed motion. He then wrote equations for variations from this periodic solution that led him to a fourth order system of linear differential equations with periodic coefficients. Knowing some integrals, he was able to reduce his fourth order system to a single linear differential equation of the second order

$$(26) \qquad \frac{d^2x}{dt^2} + \theta(t)x = 0,$$

with $\theta(t)$ periodic of period π and even. The form of Hill's equation can be put, by expanding $\theta(t)$ in a Fourier series, as

$$(27) \qquad \frac{d^2x}{dt^2} + x(q_0 + 2q_1 \cos 2t + 2q_2 \cos 4t + \cdots) = 0.$$

Hill put $\zeta = e^{it}$, $q_{-\alpha} = q_\alpha$ and wrote (27) as

$$(28) \qquad \frac{d^2x}{dt^2} + x \sum_{-\infty}^{\infty} q_\alpha \zeta^{2\alpha} = 0.$$

He then let

$$x = \sum_{j=-\infty}^{\infty} b_j \zeta^{\mu + 2j},$$

where μ and b_j were to be determined. By substituting this value of x in (28) and setting the coefficients of each power of ζ equal to 0 he obtained the doubly infinite system of linear equations

$$
\begin{aligned}
&\cdot \quad \cdot \quad \cdot \quad \cdot \quad \cdot \quad \cdot \quad \cdot \quad \cdot \quad \cdot \quad \cdot \quad \cdot \quad \cdot \\
\cdots &[-2]b_{-2} - q_1 b_{-1} - q_2 b_0 - q_3 b_1 - q_4 b_2 - \cdots = 0 \\
\cdots - &q_1 b_{-2} + [-1]b_{-1} - q_1 b_0 - q_2 b_1 - q_3 b_2 - \cdots = 0 \\
\cdots - &q_2 b_{-2} - q_1 b_{-1} + [0]b_0 - q_1 b_1 - q_2 b_2 - \cdots = 0 \\
\cdots - &q_3 b_{-2} - q_2 b_{-1} - q_1 b_0 + [1]b_1 - q_1 b_2 - \cdots = 0 \\
\cdots - &q_4 b_{-2} - q_3 b_{-1} - q_2 b_0 - q_1 b_1 + [2]b_2 - \cdots = 0 \\
&\cdot \quad \cdot \quad \cdot \quad \cdot \quad \cdot \quad \cdot \quad \cdot \quad \cdot \quad \cdot \quad \cdot \quad \cdot \quad \cdot
\end{aligned}
$$

47. This was reprinted in *Acta Math.*, 8, 1886, 1–36 = *Coll. Math. Works*, 1, 243–70.
48. *Amer. Jour. of Math.*, 1, 1878, 5–26, 129–47, 245–60 = *Coll. Math. Works*, 1, 284–335.

where

$$[j] = (\mu + 2j)^2 - q_0.$$

Hill set the determinant of the coefficients of the unknown b_j equal to 0. He first determined the properties of the infinitely many solutions for μ and gave explicit formulas for determining the μ. With these values of μ he then solved the system of an infinite number of linear homogeneous equations in the infinite number of b_j for the ratio of the b_j to b_0. Hill did show that the second order differential equation has a periodic solution and that the motion of the moon's perigee is periodic.

Hill's work was ridiculed until Poincaré[49] proved the convergence of the procedure and thereby put the theory of infinite determinants and infinite systems of linear equations on its feet. Poincaré's attention to and completion of Hill's efforts gave prominence to Hill and the subjects involved.

8. Nonlinear Differential Equations: The Qualitative Theory

A new approach to the search for periodic solutions of the differential equations governing planetary motion and the stability of planetary and satellite orbits was initiated by Poincaré under the stimulus of Hill's work. Because the relevant equations are nonlinear, Poincaré took up this class. Nonlinear ordinary differential equations had appeared practically from the beginnings of the subject as, for example, in the Riccati equation (Chap. 21, sec. 4), the pendulum equation, and the Euler equations of the calculus of variations (Chap. 24, sec. 2). No general methods of solving nonlinear equations had been developed.

In view of the fact that the equations for the motion of even three bodies cannot be solved explicitly in terms of known functions, the problem of stability cannot be solved by examining the solution. Poincaré therefore sought methods by which the problem could be answered by examining the differential equations themselves. The theory he initiated he called the qualitative theory of differential equations. It was presented in four papers all under substantially the same title, "Mémoire sur les courbes définies par une équation différentielle."[50] The questions he sought to answer were stated by him in these words: "Does the moving point describe a closed curve? Does it always remain in the interior of a certain portion of the plane? In other words, and speaking in the language of astronomy, we have inquired whether the orbit is stable or unstable."

49. *Bull. Soc. Math. de France*, 13, 1885, 19–27; 14, 1886, 77–90 = *Œuvres*, 5, 85–94, 95–107.
50. *Jour. de Math.*, (3), 7, 1881, 375–422; 8, 1882, 251–96; (4), 1, 1885, 167–244; 2, 1886, 151–217 = *Œuvres*, 1, 3–84, 90–161, 167–221.

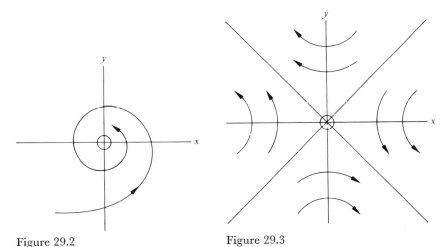

Figure 29.2   Figure 29.3

Poincaré started with nonlinear equations of the form

(29) $$\frac{dy}{dx} = \frac{P(x, y)}{Q(x, y)},$$

where P and Q are analytic in x and y. This form was chosen partly because some problems of planetary motion led to it and partly because it was the simplest mathematical type with which to commence the kind of investigation Poincaré had in mind. The solution of (29) is of the form $f(x, y) = 0$, and this equation is said to define a system of trajectories. In place of $f(x, y) = 0$ one can consider the parametric form $x = x(t), y = y(t)$.

In the analysis of the kinds of solutions equation (29) can have, Poincaré found that the singular points of the differential equation (the points at which P and Q both vanish) play a key role. These singular points are undetermined or irregular in Fuchs's sense. Here Poincaré took up earlier work by Briot and Bouquet (sec. 5) but limited himself to real values and to studying the behavior of the entire solution rather than just in the neighborhood of the singular points. He distinguished four types of singular points and described the behavior of solutions around such points.

The first type of singular point is the focus (*foyer*), the origin in Figure 29.2, and the solution spirals around and approaches the origin as t runs from $-\infty$ to ∞. This type of solution is considered stable. The second kind of singular point is the saddle point (*col*). It is the origin of Figure 29.3 and the trajectories approach this point and then depart from it. The lines AA' and BB' are the asymptotes of the trajectories. The motion is unstable. The third type of singular point, called a node (*nœud*), is a point where an infinity of solutions cross, and the fourth, called a center, is one around which closed trajectories exist, one enclosing the other and all enclosing the center.

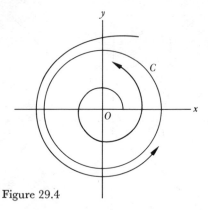

Figure 29.4

Among many results, Poincaré found that there can be closed curves which do not *touch* any of the curves satisfying the differential equation. He called these closed curves cycles without contact. A curve satisfying the differential equation cannot meet such a cycle in more than one point, and so if it crosses the cycle it cannot re-cross it. Such a curve, if it is an orbit of a planet, represents unstable motion.

Beyond cycles without contact there are closed curves that Poincaré called limit cycles. These are closed curves which satisfy the differential equation and which other solutions approach asymptotically, that is, without ever reaching the limit cycle. The approach can be from without or within the limit cycle C (Fig. 29.4). For some differential equations of the type (29) he determined the limit cycles and the regions in which they exist. In the case of limit cycles the trajectories approach a periodic curve, and the motion is again stable. If, however, the direction of the motions were away from the limit cycle the motion outside would be unstable and the motion inside would be a contracting spiral.

In the third of his papers on this subject, Poincaré studied first order equations of higher degree and of the form $F(x, y, y') = 0$, where F is a polynomial in x, y, and y'. To study these equations Poincaré regarded x, y, and y' as three Cartesian coordinates and considered the surface defined by the differential equation. If this surface has genus 0 (form of a sphere), then the integral curves have the same properties as in the case of first degree differential equations. For other genuses the results on integral curves can be quite different. Thus for a torus many new circumstances arise. Poincaré did not complete this study. In the fourth paper (1886) he studied second order equations and obtained some results analogous to those for first order equations.

While continuing his work on the types of solutions of the differential equation (29), Poincaré considered a more general theory directed to the three-body problem of astronomy. In a prize paper, "Sur le problème de trois corps et les équations de la dynamique,"[51] he considered the system of differential equations

(30) $$\frac{dx_i}{dt} = X_i(x_1, \cdots, x_n, \mu), \qquad i = 1, 2, \cdots, n.$$

He developed the X_i in powers of the small parameter μ and, supposing that the system had for $\mu = 0$ a known periodic solution

$$x_i = \phi_i(t), \qquad i = 1, 2, \cdots, n$$

of period T, he proposed to find the periodic solution of the system which for $\mu = 0$ reduces to the $\phi_i(t)$. The existence of periodic solutions for the three-body problem had been discovered by Hill, and Poincaré made use of this fact.

The details of Poincaré's work are too specialized to consider here. He first generalized earlier work of Cauchy on solutions of systems of ordinary differential equations wherein the latter had used his *calcul des limites*. Poincaré then demonstrated the existence of the periodic solutions he sought and applied what he learned to the study of periodic solutions of the three-body problem for the case where the masses of two of the bodies (but not that of the sun) are small. Thus one obtains such solutions by supposing that the two small masses move in concentric circles about the sun and are in the same plane. One can obtain others by supposing that for $\mu = 0$ the orbits are ellipses and that their periods are commensurable. With these solutions, and by using the theory he had developed for the system, he obtained other periodic solutions. In sum he showed that there is an infinity of initial positions and initial velocities such that the *mutual distances* of the three bodies are periodic functions of the time. (Such solutions are also called periodic.)

Poincaré in this paper of 1890 drew many other conclusions about periodic and almost periodic solutions of the system (30). Among them is the very notable discovery for such a system of a new class of solutions previously unknown. These he called asymptotic solutions. There are two kinds. In the first the solution approaches the periodic solution asymptotically as t approaches $-\infty$ or as t approaches $+\infty$. The second kind consists of doubly asymptotic solutions, that is, solutions that approach a periodic solution as t approaches $-\infty$ and $+\infty$. There is an infinity of such doubly asymptotic solutions. All of the results in this paper of 1890 and many others can be found also in Poincaré's *Les Méthodes nouvelles de la mécanique céleste*.[52]

Poincaré's work on the problem of stability of the solar system was only partially successful. The stability is still an open question. As a matter of

51. *Acta Math.*, 13, 1890, 1–270 = *Œuvres*, 7, 262–479.
52. Three volumes, 1892–99.

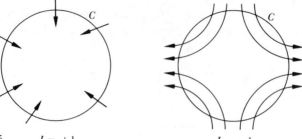

Figure 29.5 $I = +1$ $I = -1$

fact so is the question of whether the orbit of the moon is stable; most scientists today believe it is not.

The stability of the solutions of (29) can be analyzed by means of what is called the characteristic equation, namely,

$$(31) \qquad \begin{vmatrix} Q_x(x_0, y_0) - \lambda & P_x(x_0, y_0) \\ Q_y(x_0, y_0) & P_y(x_0, y_0) - \lambda \end{vmatrix} = 0,$$

where (x_0, y_0) is a singular point of (29). The stability in the neighborhood of (x_0, y_0), according to a theorem of the distinguished Russian mathematician Alexander Liapounoff (1857–1918) depends upon the roots of this characteristic equation.[53] The analysis of possible cases is detailed and includes many more types than those given in the discussion above of Poincaré's work. The basic result, according to Liapounoff, whose work on stability problems continued into the early years of this century, is that the solutions are stable in the neighborhood of a singular point when and only when the roots of the equation (31) in λ have negative real parts.

The qualitative study of nonlinear equations was advanced by Poincaré's introduction of topological arguments (in the first of the four papers in the *Journal de Mathématiques*). To describe the nature of a singular point he introduced the notion of an index. Consider a singular point P_0 and a simple closed curve C surrounding it. At each intersection of C with the solutions of

$$(32) \qquad \frac{dy}{dx} = \frac{P(x, y)}{Q(x, y)}$$

there is a direction angle of the trajectory, which we shall denote by ϕ and which can have any value from 0 to 2π radians. If a point now moves in a counterclockwise direction around C (Fig. 29.5), the angle ϕ will vary; and after completion of the circuit around C, ϕ will have the value $2\pi I$ where I is an integer or zero (since the direction angle of the trajectories has

53. *Ann. Fac. Sci. de Toulouse*, (2), 9, 1907, 203–474; originally published in Russian in 1892.

returned to its original value). The quantity I is the index of the curve. It can be proved that the index of a closed curve that contains several singularities is the algebraic sum of their indices. The index of a closed trajectory is $+1$ and conversely.

The nature of the trajectories can be determined by the characteristic equation, and so the index I of a curve should be determinable by knowing just the differential equation. One can prove that

$$I = \frac{1}{2\pi} \int_C d\left(\arctan \frac{P}{Q}\right) = \frac{1}{2\pi} \int_C \frac{P\, dQ - Q\, dP}{P^2 + Q^2},$$

where the path of integration is the closed curve C.

After Poincaré, the most significant work on solutions of equations of the form (32) is due to Ivar Bendixson (1861–1935). One of his major results[54] provides a criterion by means of which, in certain regions, one can show that no closed trajectory exists. Let D be a region in which $\partial Q/\partial x + \partial P/\partial y$ has the same sign. Then the equation (32) has no periodic solution in D.

The theorem now named after Poincaré and Bendixson, which is in the latter's 1901 paper, provides a positive criterion for the existence of a periodic solution of (32). If P and Q are defined and regular in $-\infty < x, y < \infty$ and if as t approaches ∞ a solution $x(t), y(t)$ remains within a bounded region of the (x, y)-plane without approaching singular points, then there exists at least one closed solution curve of the differential equation.

The study of nonlinear equations that Poincaré launched was broadened in various directions. One more topic begun in the nineteenth century will be mentioned. The linear differential equations studied by Fuchs have the property that the singular points are fixed and are, in fact, determined by the coefficients of the differential equation. In the case of nonlinear equations the singular points may vary with the initial conditions and are called movable singular points. Thus the equation $y' + y^2 = 0$ has the general solution $y = 1/(x - c)$ where c is arbitrary. The location of the singularity in the solution depends on the value of c. This phenomenon of movable singular points was discovered by Fuchs.[55] The study of movable singular points and of nonlinear second order equations with and without such singular points was taken up by many men, notably by Paul Painlevé (1863–1933). One interesting feature is that many of the types of second order equations of the form $y'' = f(x, y, y')$ require for their solution new types of transcendental functions now called Painlevé transcendents.[56]

The interest in nonlinear equations has become strong in the twentieth century. The applications have moved from astronomy to problems of

54. *Acta Math.*, 24, 1901, 1–88.
55. *Sitzungsber. Akad. Wiss. zu Berlin*, 1884, 699–710 = *Werke*, 2, 364 ff.
56. *Comp. Rend.*, 143, 1906, 1111–17.

communications, servomechanisms, automatic control systems, and electronics. The study has also moved from the qualitative stage to quantitative investigations.

Bibliography

Acta Mathematica, Vol. 38, 1921. This entire volume is devoted to articles on Poincaré's work by various leading mathematicians.

Bocher, M.: "Randwertaufgaben bei gewöhnlichen Differentialgleichungen," *Encyk. der Math. Wiss.*, B. G. Teubner, 1899–1916, II, A7a, 437–63.

——: "Boundary Problems in One Dimension," *Internat. Cong. of Math.*, Proc., Cambridge, 1912, 1, 163–95.

Burkhardt, H.: "Entwicklungen nach oscillirenden Functionen und Integration der Differentialgleichungen der mathematischen Physik," *Jahres. der Deut. Math.-Verein.*, 10, 1908, 1–1804.

Cauchy, A. L.: *Œuvres complètes*, (1), Vols. 4, 7, and 10, Gauthier-Villars, 1884, 1892, and 1897.

Craig, T.: "Some of the Developments in the Theory of Ordinary Differential Equations Between 1878 and 1893," *N.Y. Math. Soc. Bull.*, 2, 1893, 119–34.

Fuchs, Lazarus: *Gesammelte mathematische Werke*, 3 vols., 1904–09, Georg Olms (reprint), 1970.

Heine, Eduard: *Handbuch der Kugelfunktionen*, 2 vols., 1878–81, Physica Verlag (reprint), 1961.

Hilb, E.: "Lineare Differentialgleichungen im komplexen Gebiet," *Encyk. der Math. Wiss.*, B. G. Teubner, 1899–1916, II, B5.

——: "Nichtlineare Differentialgleichungen," *Encyk. der Math. Wiss.*, B. G. Teubner, 1899–1916, II, B6.

Hill, George W.: *Collected Mathematical Works*, 4 vols., 1905, Johnson Reprint Corp., 1965.

Klein, Felix: *Vorlesungen über die Entwicklung der Mathematik im 19. Jahrhundert*, Vol. I, Chelsea (reprint), 1950.

——: *Gesammelte mathematische Abhandlungen*, Julius Springer, 1923, Vol. 3.

Painlevé, P.: "Le Problème moderne de l'intégration des équations différentielles," *Third Internat. Math. Cong. in Heidelberg*, 1904, 86–99, B. G. Teubner, 1905.

——: "Gewöhnliche Differentialgleichungen, Existenz der Lösungen," *Encyk. der Math. Wiss.*, B. G. Teubner, 1899–1916, II, A4a.

Poincaré, Henri: *Œuvres*, 1, 2, and 5, Gauthier-Villars, 1928, 1916, and 1960.

Riemann, Bernhard: *Gesammelte mathematische Werke*, 2nd ed., 1892, Dover (reprint), 1953.

Schlesinger, L.: "Bericht über die Entwickelung der Theorie der linearen Differentialgleichungen seit 1865," *Jahres. der Deut. Math.-Verein.*, 18, 1909, 133–266.

Wangerin, A.: "Theorie der Kugelfunktionen und der verwandten Funktionen," *Encyk. der Math. Wiss.*, B. G. Teubner, 1899–1916, II, A10.

Wirtinger, W.: "Riemanns Vorlesungen über die hypergeometrische Reihe und ihre Bedeutung," *Third Internat. Math. Cong. in Heidelberg*, 1904, B. G. Teubner, 1905, 121–39.

30
The Calculus of Variations in the Nineteenth Century

> Although to penetrate into the intimate mysteries of nature and thence to learn the true causes of phenomena is not allowed to us, nevertheless it can happen that a certain fictive hypothesis may suffice for explaining many phenomena.
>
> LEONHARD EULER

1. *Introduction*

As we have seen, the calculus of variations was founded in the eighteenth century chiefly by Euler and Lagrange. Beyond mathematical and physical problems of various sorts there was one leading motivation for the study, namely, the Principle of Least Action, which in the hands of Maupertuis, Euler, and Lagrange became a leading principle of mathematical physics. The nineteenth-century men continued the work on least action and the greatest stimulus to the calculus of variations in the first half of the century came from this direction. Physically, the interest was in the science of mechanics and particularly in problems of astronomy.

2. *Mathematical Physics and the Calculus of Variations*

Lagrange's successful formulation of the laws of dynamics in terms of his Principle of Least Action suggested that the idea should be applicable to other branches of physics. Lagrange[1] gave a minimum principle for fluid dynamics (applicable to compressible and incompressible fluids) from which he derived Euler's equations for fluid dynamics (Chap. 22, sec. 8) and indeed he boasted that a minimum principle governed this field as it did the motion of particles and rigid bodies. Many problems of elasticity also were solved by the calculus of variations in the early part of the nineteenth century by Poisson, Sophie Germain, Cauchy, and others, and this work too helped to

---

1. *Misc. Taur.*, 2_2, 1760/61, 196–298, pub. 1762 = *Œuvres*, 1, 365–468.

keep the subject active, but no major new mathematical ideas of the calculus of variations are to be noted in this area or in Gauss's famous contribution to mechanics, The Principle of Least Constraint.[2]

The first new point worth noting is due to Poisson. Using Lagrange's generalized coordinates, he followed up immediately two papers by Lagrange and starts with[3] Lagrange's equations (Chap. 24, sec. 5)

$$(1) \qquad \frac{d}{dt}\left(\frac{\partial T}{\partial \dot{q}_i}\right) - \frac{\partial T}{\partial q_i} + \frac{\partial V}{\partial q_i} = 0, \qquad i = 1, 2, \cdots, n.$$

Here the kinetic energy T expressed in generalized coordinates is $2T = \sum_{i,j=1}^{n} a_{ij}\,\dot{q}_i\,\dot{q}_j$, V is the potential energy, and T and V are independent of t. He sets $L = T - V$. Where V depends only on the q_i and not on the $\dot{q}_i$ he can write

$$(2) \qquad \frac{\partial L}{\partial \dot{q}_i} = \frac{\partial T}{\partial \dot{q}_i},$$

so that the equations of motion read

$$(3) \qquad \frac{d}{dt}\left(\frac{\partial L}{\partial \dot{q}_i}\right) - \frac{\partial L}{\partial q_i} = 0, \qquad i = 1, 2, \cdots, n.$$

He also introduces

$$(4) \qquad p_i = \frac{\partial L}{\partial \dot{q}_i} = \frac{\partial T}{\partial \dot{q}_i},$$

and so from (3) he has

$$(5) \qquad \dot{p}_i = \frac{\partial L}{\partial q_i}, \qquad i = 1, 2, \cdots, n.$$

The p_i are momentum components when the q_i are rectangular coordinates of position. Equation (5) is a step in the direction we shall now look into.

The big change in the formulation of least action principles which is important for the calculus of variations and for ordinary and partial differential equations was made by William R. Hamilton. Hamilton came to dynamics through optics. His goal in optics was to fashion a deductive mathematical structure in the manner of Lagrange's treatment of mechanics.

Hamilton too started with a least action principle and was to deduce new ones. However, his attitude toward such principles was profoundly different from that of Maupertuis, Euler, and Lagrange. In a paper published in the *Dublin University Review*[4] he says, "But although the law of least action has thus attained a rank among the highest theorems of physics, yet its pretensions to a cosmological necessity, on the ground of economy in the

2. *Jour. für Math.*, 4, 1829, 232–35 = *Werke*, 5, 23–28.
3. *Jour. de l'Ecole Poly.*, 8, 1809, 266–344.
4. 1833, 795–826 = *Math. Papers*, 1, 311–32.

universe, are now generally rejected. And the rejection appears just, for this, among other reasons, that the quantity pretended to be economized is in fact often lavishly expended.'' Because in some phenomena of nature, even simple ones, the action is maximized, Hamilton preferred to speak of a principle of stationary action.

In a series of papers of the period 1824 to 1832 Hamilton developed his mathematical theory of optics and then carried over ideas he had introduced there to mechanics. He wrote two basic papers.[5] It is the second of these which is more pertinent. Here he introduces the action integral, namely, the time integral of the difference between kinetic and potential energies

$$(6) \qquad S = \int_{P_1, t_1}^{P_2, t_2} (T - V) \, dt.$$

The quantity $T - V$ is called the Lagrangian function though it was introduced by Poisson; P_1 stands for $q_1^1, q_2^1, \ldots, q_n^1$, and P_2 for $q_1^{(2)}, q_2^{(2)}, \ldots, q_n^{(2)}$. Now Hamilton generalizes the principle of Euler and Lagrange by allowing comparison paths that are not restricted except that the motion along them must start at P_1 at time t_1 and end at P_2 at time t_2. Also, the law of conservation of energy need not hold whereas in the Euler-Lagrange principle conservation of energy is presupposed, and as a consequence the time required by an object to traverse any one of the comparison paths differs from the time taken to traverse the actual path.

The Hamiltonian principle of least action asserts that the actual motion is the one that makes the action stationary. For conservative systems, that is, where the components of force are derivable from a potential that is a function of position only, $T + V = \text{const.}$ Hence $T - V = 2T - \text{const.}$ and so Hamilton's principle reduces to Lagrange's, but, as noted, Hamilton's principle also holds for nonconservative systems. Also the potential energy V can be a function of time and even of the velocities; that is, in generalized coordinates $V = V(q_1, \ldots, q_n, \dot{q}_1, \ldots, \dot{q}_n, t)$.

If, setting $T - V$ equal to L, we write the action integral (6) as

$$(7) \qquad S = \int_{t_1}^{t_2} L(q_1, \cdots, q_n, \dot{q}_1, \cdots, \dot{q}_n, t) \, dt,$$

with the condition that all comparison $q_i(t)$ must have the same given values at t_1 and t_2, then the problem is to determine the q_i as functions of t from the condition that the true q_i make the integral stationary. The Euler equations, which express the condition that the first variation of S is 0, become a system of simultaneous second order ordinary differential equations, namely,

$$(8) \qquad \frac{\partial L}{\partial q_k} - \frac{d}{dt}\left(\frac{\partial L}{\partial \dot{q}_k}\right) = 0, \qquad k = 1, 2, \cdots, n,$$

5. *Phil. Trans.*, 1834, Part II, 247–308; 1835, Part I, 95–144 = *Math. Papers*, 2, 103–211.

and the equations are to be solved in $t_1 \le t \le t_2$. These equations are still called the Lagrangian equations of motion even though L is now a different function. The choice of coordinate system is arbitrary and usually utilizes generalized coordinates. This is an essential advantage of variational principles.

Now introduce (see (4))

$$p_i = \frac{\partial L}{\partial \dot{q}_i}.$$

Then the equations (8) become

$$\dot{p}_i = \frac{\partial L}{\partial q_i}.$$

The introduction of the p_i as a new set of independent variables is credited to Hamilton though it was first done by Poisson. We now have the symmetrical system of differential equations

(9) $$p_i = \frac{\partial L}{\partial \dot{q}_i}, \qquad \dot{p}_i = \frac{\partial L}{\partial q_i}, \qquad i = 1, 2, \cdots, n.$$

This is a system of $2n$ first order ordinary differential equations in p_i and $\dot{p}_i$. However, the $\dot{p}_i$ are dp_i/dt.

In his second (1835) paper Hamilton simplifies this system of equations. He introduces a new function H which is defined by

(10) $$H(p_i, q_i, t) = -L + \sum_{i=1}^{n} p_i \dot{q}_i.$$

This function is physically the total energy, for the summation can be shown to be equal to $2T$. The transformation from L to H is called a Legendre transformation because it was used by Legendre in his work on ordinary differential equations. That H is a function of the p_i, q_i, and t, whereas $L = T - V$ is a function of the q_i, $\dot{q}_i$, and t, results from the fact that since $p_i = \partial L/\partial \dot{q}_i$ we can solve for the $\dot{q}_i$ and substitute in L.

With (10) the differential equations of motion (9) can be shown to take the form

(11) $$\dot{q}_i = \frac{\partial H}{\partial p_i}, \qquad \dot{p}_i = -\frac{\partial H}{\partial q_i}, \qquad i = 1, 2, \cdots, n.$$

The function H is assumed to be known in the application to physical problems. These equations are a system of $2n$ first order ordinary differential equations in the $2n$ dependent variables p_i and q_i as functions of t, whereas Lagrange's equations (1) are a system of n second order equations in the $q_i(t)$. Later Jacobi called Hamilton's equations the canonical differential equations. They are the variational equations (Euler equations) for the integral

$$S = \int_{P_1,t_1}^{P_2,t_2} (T - V) \, dt = \int L \, dt = \int \left\{ \sum_{i=1}^{n} p_i \dot{q}_i - H(p_i, q_i, t) \right\} dt.$$

This set of equations appears in one of Lagrange's papers of 1809 which deals with the perturbation theory of mechanical systems. However, whereas Lagrange did not recognize the basic connection of these equations with the equations of motion, Cauchy in an unpublished paper of 1831 did. Hamilton in 1835 made these equations the basis of his mechanical investigations.

To use Hamilton's equations of motion it is often possible to express H in an appropriate p and q coordinate system so that the system of equations (11) is solvable for the p_i and q_i as functions of the time. In particular, if we can choose coordinates so that H depends only on the p_i, the system is solvable.

In a paper of 1837[6] and in lectures on dynamics of the years 1842 and 1843, which were published in 1866 in the classic *Vorlesungen über Dynamik*, Jacobi showed that one can reverse Hamilton's process. In Hamilton's theory if one knows the action S or the Hamiltonian H one can form the $2n$ canonical differential equations and attempt to solve the system. Jacobi's thought was to try to find coordinates P_j and Q_j so that H is as simple as possible, and then the differential equations (11) would be easily integrated. Specifically, he sought a transformation

(12)
$$\begin{aligned} Q_j &= Q_j(p_i, q_i, t) \\ P_j &= P_j(p_i, q_i, t) \end{aligned}$$

such that

$$\delta \int_{t_1}^{t_2} \left(\sum_{i=1}^{n} p_i \dot{q}_i - H(p_i, q_i, t) \right) dt = 0$$

goes over by the transformation (12) into

$$\delta \int_{t_1}^{t_2} \left(\sum_{i=1}^{n} P_i \dot{Q}_i - K(P_i, Q_i, t) \right) dt = 0,$$

and so that the Hamiltonian differential equations become

(13)
$$\dot{Q}_i = \frac{\partial K}{\partial P_i}, \qquad \dot{P}_i = -\frac{\partial K}{\partial Q_i},$$

where $K(P_i, Q_i, t)$ is the new Hamiltonian. This path leads to

$$K = H(p_i, q_i, t) + \frac{\partial \Omega}{\partial t}(Q_i, q_i, t),$$

6. *Jour. für Math.*, 17, 1837, 97–162 = *Ges. Werke*, 4, 57–127.

where Ω is a new function, called the generating function of the transformation. Jacobi chose $K = 0$ so that by (13)

$$\dot{Q}_i = 0, \qquad \dot{P}_i = 0,$$

or Q_i and P_i are constants. Moreover,

(14) $$H + \frac{\partial \Omega}{\partial t} = 0,$$

and it can be shown that

$$p_i = \frac{\partial \Omega}{\partial q_i}.$$

Hence by (14), in view of the variables in H,

(15) $$H\left(\frac{\partial \Omega}{\partial q_i}, q_i, t\right) + \frac{\partial \Omega}{\partial t}(Q_i, q_i, t) = 0.$$

Since $\dot{Q}_i = 0$, $Q_i = \alpha_i$, and so the equation is of first order in Ω with the independent variables q_i and t. With this change equation (15) is the Hamilton-Jacobi partial differential equation for the function Ω. If this equation can be solved for a complete Ω, that is, one containing n arbitrary constants, the solution would have the form

$$\Omega(\alpha_1, \alpha_2, \cdots, \alpha_n, q_1, q_2, \cdots, q_n, t).$$

Now it is a fact of the Jacobi transformation theory that

$$P_i = -\frac{\partial \Omega}{\partial \alpha_i}$$

and that $P_i = \beta_i$, a constant, because $\dot{P}_i = 0$. Hence one solves the algebraic equations

$$\frac{\partial \Omega}{\partial \alpha_i} = -\beta_i, \qquad i = 1, 2, \cdots, n$$

for the q_i. These solutions

$$q_i = f_i(\alpha_1, \cdots, \alpha_n, \beta_1, \cdots, \beta_n, t), \qquad i = 1, 2, \cdots, n,$$

are the solutions of Hamilton's canonical equations. Thus Jacobi had shown that one can solve the system of ordinary equations (11) by solving the partial differential equation (15). Jacobi himself found the proper Ω for many problems of mechanics.

Hamilton's work was the culmination of a series of efforts to provide a broad principle from which the laws of motion of various problems of mechanics could be derived. It inspired efforts to obtain similar variational principles in other branches of mathematical physics such as elasticity, electromagnetic theory, relativity, and quantum theory. The principles that have been derived, and even Hamilton's principle, are not necessarily more practical approaches to the solution of particular problems. Rather the attractiveness of such broad formulations lies in philosophic and aesthetic interests though scientists no longer infer that the existence of a maximum-minimum principle is evidence of God's wisdom and efficiency.

From the standpoint of the history of mathematics the work of Hamilton and Jacobi is significant because it prompted further research not only in the calculus of variations but also on systems of ordinary differential equations and first order partial differential equations.

3. *Mathematical Extensions of the Calculus of Variations Proper*

We may recall that even in the simplest case, maximizing or minimizing the integral

(16) $$J = \int_a^b f(x, y, y') \, dx,$$

the results of Euler and Legendre provided only necessary conditions (Chap. 24). For about fifty years after the work of Legendre, mathematicians explored further the first and second variations but no decisive results were obtained. In 1837[7] Jacobi found out how to sharpen Legendre's condition so that it might yield a sufficient condition. His chief discovery in this connection was the concept of conjugate point. Let us note first what this is.

Consider the curves which satisfy Euler's (characteristic) equation; such curves are called extremals. For the basic problem of the calculus of variations, there is a one-parameter family of extremals passing through a given point A. Suppose now that A is one of the two endpoints between which we seek a maximizing or minimizing curve. Given any one extremal, the limiting point of intersection of other extremals as they come closer to that extremal is the conjugate point to A on that extremal. Another way of putting it is that we have a family of curves, and this family may have an envelope. The point of contact of any one extremal and the envelope of the family is the point conjugate to A on that extremal. Then Jacobi's condition is that if $y(x)$ is an extremal between the endpoints A and B of the original problem, no conjugate point must lie on that extremal between A and B or even be B itself.

Just what this means in a concrete case may be seen from an example. It can be shown that the parabolic paths of all trajectories (Fig. 30.1)

7. *Jour. für. Math.*, 17, 1837, 68–82 = *Ges. Werke*, 4, 39–55.

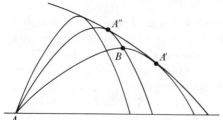

Figure 30.1

emanating from A with constant velocity v but with varying angles of fire are extremals of the problem of minimizing or maximizing the action integral

$$\frac{m}{2} \int_A^B v \, ds.$$

The problem of minimizing the action between two points A and B does in general have two solutions, the parabola $AA''B$ and the parabola ABA'. It is also the case that the family of parabolas through A has an envelope that touches the two parabolas at A'' and A'. The conjugate point on $AA''B$ is A'' and the one on ABA' is A'. According to Jacobi's condition, the extremal $AA''B$ could not furnish a maximum or a minimum but the extremal ABA' could.

Jacobi reconsidered the second variation $\delta^2 J$ (Chap. 24, sec. 4). If we write $y + \epsilon t(x)$ in place of Lagrange's $y + \delta y$ and if a and b are the abscissas of A and B then

(17) $$\delta^2 J = \frac{\epsilon^2}{2} \int_a^b (t^2 f_{yy} + 2tt' f_{yy'} + t'^2 f_{y'y'}) \, dx.$$

Jacobi showed that

$$\delta^2 J = \frac{\epsilon^2}{2} \int_a^b f_{y'y'} \left(t' - t \frac{u'}{u} \right)^2 dx,$$

where u is a solution of Jacobi's accessory equation

(18) $$\left\{ f_{yy} - \frac{d}{dx} f_{yy'} \right\} u - \frac{d}{dx} (f_{y'y'} u') = 0$$

and where the partial derivatives are evaluated along an extremal joining the two endpoints A and B. Now $u(x)$ is required to pass through A. Then all other points on the extremal $y(x)$ through A and B at which $u(x)$ vanishes are the conjugate points of A on that extremal. If $u = \beta_1 u_1 + \beta_2 u_2$ is the general solution of the accessory equation (18), then one can show that

$$\frac{u_1(x)}{u_2(x)} = \frac{u_1(a)}{u_2(a)},$$

where a is the abscissa of the point A, is the equation for the abscissas of all points conjugate to A.

Jacobi also showed that one need not solve the accessory equation. Since one must solve the Euler equation in any case, let $y = y(x, c_1, c_2)$ be the general solution of that equation, that is, the family of extremals. Then u_1 can be taken to be $\partial y/\partial c_1$, and u_2 to be $\partial y/\partial c_2$.

From his work on conjugate points Jacobi drew two conclusions. The first was that if along the extremal from A to B a conjugate point to A occurs then a maximum or a minimum is impossible. In this conclusion Jacobi was essentially correct.

On the basis of his considerations of conjugate points Jacobi also concluded that an extremal (a solution of Euler's equation) taken between A and B for which $f_{y'y'} > 0$ along the curve and for which no conjugate point exists between A and B (or at B) furnishes a minimum for the original integral. The corresponding statement with $f_{y'y'} < 0$, he asserted, holds for a maximum. Actually, these sufficient conditions were not correct, as we shall see in a few moments. In this 1837 paper Jacobi stated results and gave brief indications of proofs. The full proofs of the correct statements were supplied by later workers.

Aside from the specific value of Jacobi's results for the existence of a maximizing or minimizing function, his work made clear that progress in the calculus of variations could not be guided by the theory of maxima and minima of the ordinary calculus.

For thirty-five years both of Jacobi's conclusions were accepted as correct. During this period the papers on the subject were imprecise in statement and dubious in proof; problems were not sharply formulated and all sorts of errors were made. Then Weierstrass undertook work on the calculus of variations. He presented his material in his lectures at Berlin in 1872 but did not publish it himself. His ideas aroused a new interest, stimulated further activity in the subject, and sharpened the thinking, as did Weierstrass's work in other domains.

Weierstrass's first point was that the criteria for a minimum or a maximum hitherto established—Euler's, Legendre's, and Jacobi's—were limited because the supposed minimizing or maximizing curve $y(x)$ was compared with other curves $y(x) + \varepsilon t(x)$, wherein it was actually supposed that $\varepsilon t(x)$ and $\varepsilon t'(x)$, or what Lagrange called δy and $\delta y'$, were both small along the x-range from A to B. That is, $y(x)$ was being compared with a limited class of other curves, and by satisfying the three criteria it did better than any other one of *these* comparison curves. Such variations $\varepsilon t(x)$ were called by Adolf Kneser (1862–1930) weak variations. However, to find the curve that really maximizes or minimizes the integral J one must compare it with *all* other curves joining A and B, including those whose derivatives may not approach the derivatives of the maximizing (or minimizing) curve as

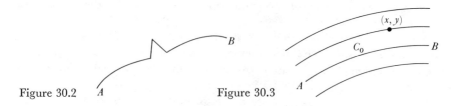

Figure 30.2   A   Figure 30.3

these comparison curves come closer in position to the maximizing curve. Thus a comparison curve may have a sharp corner (Fig. 30.2) at one or several places along the x-range from A to B. The comparison curves envisioned by Weierstrass are what Kneser called strong variations.

Weierstrass did prove in 1879 that for weak variations the three conditions, that a curve be an extremal (a solution of Euler's equation), that $f_{y'y'} > 0$ along the extremal, and that any conjugate point to A must lie beyond B, are indeed sufficient conditions that the extremal furnish a minimum of the integral J (and $f_{y'y'} < 0$ for a maximum).

Then Weierstrass considered strong variations. For these variations Weierstrass first introduced a fourth necessary condition. He introduced a new function called the E-function, or Excess function, defined by

$$(19) \qquad E(x, y, y', \tilde{p}) = f(x, y, \tilde{p}) - f(x, y, y') - (\tilde{p} - y')f_{y'}(x, y, y'),$$

and his result was: The fourth necessary condition that $y(x)$ furnish a minimum is that $E(x, y, y', \tilde{p}) \geq 0$ along the extremal $y(x)$ for every finite value of $\tilde{p}$. For a maximum $E \leq 0$.

Weierstrass then (1879) turned his attention to sufficient conditions for a maximum (or a minimum) when strong variations are permitted. To formulate his sufficient conditions it is necessary to introduce Weierstrass's concept of a field. Consider any one-parameter family (Fig. 30.3) of extremals $y = \Phi(x, \gamma)$ in which the particular extremal joining A and B is included, say for $\gamma = \gamma_0$. Aside from some details on the continuity and differentiability of $\Phi(x, \gamma)$, the essential fact about this family of extremals is that in a region about the extremal through A and B there passes through any point (x, y) of the region one and only one extremal of the family. A family of extremals satisfying this essential condition is called a field.

Given a field surrounding the extremal C_0 joining A and B (Fig. 30.4) then if at every point (x, y) lying between $x = a$ and $x = b$ and in the region covered by the field, $E(x, y, p(x, y), \tilde{p}) \geq 0$, where $p(x, y)$ denotes the slope at (x, y) of the extremal passing through (x, y) and $\tilde{p}$ is any finite value, then C_0 minimizes the integral J with respect to any other C lying within the field and joining A and B. (For a maximum, $E \leq 0$.)

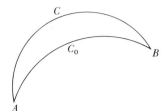

Figure 30.4   A

In 1900 Hilbert[8] introduced his invariant integral theory which greatly simplified the sufficiency proof. Hilbert asked the question: Is it possible to determine the function $p(x, y)$ so that the integral

(20) $$I = \int_{x_1}^{x_2} \left\{ f(x, y, p) - (y' - p) f_{y'}(x, y, p) \right\} dx$$

is independent of the path in a region of (x, y) values? He found that if $p(x, y)$ is so determined then the solutions of the differential equation

$$\frac{dy}{dx} = p(x, y)$$

are extremals of a field. Conversely, if $p(x, y)$ is the slope function of a field F then I is independent of the path in F. From this theorem Hilbert derived Weierstrass's sufficiency condition for strong variations.

4. Related Problems in the Calculus of Variations

Our exposition of the history of the calculus of variations has been concentrated largely on the integral

$$J = \int_{x_1}^{x_2} f(x, y, y') \, dx.$$

Some mention has been made of other problems, the isoperimetric problems, the problems of several functions of one variable such as arise in the Principle of Least Action, and the case of multiple integrals, which Lagrange first treated and which occurs in the minimal surface problem (Chap. 24, sec. 4). There are numerous types of related problems, such as those in which the minimizing or maximizing curve is treated in a parameteric representation $x = x(t)$ and $y = y(t)$—this problem was thoroughly discussed by Weierstrass—and problems largely in dynamics in which the variables that appear in the integrand are restricted by auxiliary or subsidiary equations,

8. *Nachrichten König. Ges. der Wiss. zu Gött.*, 1900, 291–96 = *Ges. Abh.*, 3, 323–29. There is an English translation by Mary Winston Newsom in the *Amer. Math. Soc. Bull.*, 8, 1902, 472–78. This material is part of Hilbert's famous paper of 1900, "Mathematical Problems."

called constraints. The last type of problem is somewhat related to the isoperimetric problem because there too a subsidiary condition, namely, the length of the curve bounding the maximum area, is specified—though in that problem the subsidiary condition is in the form of an integral that expresses the length of the curve, whereas in the case of dynamical constraints the subsidiary condition or conditions are in the form of equations involving the independent and dependent variables or even the differentials of the dependent variables. There is also the major problem called the Dirichlet problem, which has already been discussed (Chap. 28, secs. 4 and 8).

We shall not trace the detailed history of these problems because no major feature of the development of mathematics emerged from it, although the problems are significant and considerable work has been done on them down to the present time. It is perhaps worth noting that the subject of minimal surfaces, which calls for solving the equation

$$(1 + q^2)r - 2pqs + (1 + p^2)t = 0,$$

had been dormant after a paper by Ampère of 1817 until the Belgian physicist Joseph Plateau (1801–83) in a book of 1873, *Statique expérimentale et théorique des liquides soumis aux seules formes moléculaires*, showed that if one dips wires having the shapes of closed curves into a glycerine solution (or soapy water) and then withdraws them, a soap film which has the shape of the surface of least area will span the wire boundary. The mathematicians thus received a new stimulus to consider minimal surfaces bounded by a closed space curve. Since the boundary curve or curves can be quite complicated, the actual analytical explicit solution for the minimal surface may be impossible to obtain. This problem, now known as Plateau's problem, led to work on proving at least the existence of solutions, from which some properties of the solutions can be deduced.

Bibliography

Dresden, Arnold: "Some Recent Work in the Calculus of Variations," *Amer. Math. Soc. Bull.*, 32, 1926, 475–521.

Duren, W. L., Jr.: "The Development of Sufficient Conditions in the Calculus of Variations," *University of Chicago Contributions to the Calculus of Variations*, I, 1930, 245–349, University of Chicago Press, 1931.

Hamilton, W. R.: *The Mathematical Papers*, 3 vols., Cambridge University Press, 1931, 1940, and 1967.

Jacobi, C. G. J.: *Gesammelte Werke*, G. Reimer, 1886 and 1891, Chelsea (reprint), 1968, Vols. 4 and 7.

———: *Vorlesungen über Dynamik* (1866), Chelsea (reprint), 1968. Also in Vol. 8 of Jacobi's *Gesammelte Werke*.

McShane, E. J.: "Recent Developments in the Calculus of Variations," *Amer. Math. Soc. Semicentennial Publications*, II, 1938, 69–97.

Porter, Thomas Isaac: "A History of the Classical Isoperimetric Problem," *University of Chicago Contributions to the Calculus of Variations*, II, 475–517, University of Chicago Press, 1933.

Prange, Georg: "Die allgemeinen Integrationsmethoden der analytischen Mechanik," *Encyk. der Math. Wiss.*, B. G. Teubner, 1904–35, IV, 2, 509–804.

Todhunter, Isaac: *A History of the Calculus of Variations in the Nineteenth Century*, Chelsea (reprint), 1962.

Weierstrass, Karl: *Werke*, Akademische Verlagsgesellschaft, 1927, Vol. 7.

31

Galois Theory

Unfortunately what is little recognized is that the most worthwhile scientific books are those in which the author clearly indicates what he does not know; for an author most hurts his readers by concealing difficulties. EVARISTE GALOIS

1. Introduction

The basic problem of algebra, the solution of polynomial equations, continued to occupy the center of the stage in the algebra of the early nineteenth century. In this period, the broad question of which equations are solvable by algebraic operations was definitely and comprehensively answered by Galois. Moreover, he not only created the first significant coherent body of algebraic theory but he introduced new notions which were to be developed into still other broadly applicable theories of algebra. In particular the concepts of a group and a field emerged from his work and Abel's.

2. Binomial Equations

We have already discussed (Chap. 25, sec. 2) the fruitless efforts of Euler, Vandermonde, Lagrange, and Ruffini to solve algebraically equations of degree greater than 4 and the binomial equation $x^n - 1 = 0$. A major success was achieved by Gauss. In the last section of his *Disquisitiones Arithmeticae*[1] Gauss considered the equation

$$(1) \qquad\qquad x^p - 1 = 0,$$

where p is a prime.[2] This equation is often called the cyclotomic equation or the equation for the division of a circle. The latter term refers to the fact that the roots of this equation are, by de Moivre's theorem,

$$(2) \qquad x_j = \cos\frac{k2\pi\theta}{p} + i\sin\frac{k2\pi\theta}{p}, \qquad k = 1, 2, \ldots, p,$$

---

1. 1801, *Werke*, I.
2. The case of p prime takes care of $x^n - 1 = 0$ for if $n = pq$, let $y = x^q$. But $y^p - 1 = 0$ is solvable. Hence $x^q = $ const. can be solved if q is a prime and if not q can be decomposed in the same manner that n is.

and the complex numbers x_j when plotted geometrically are the vertices of a p-sided regular polygon that lie on the unit circle.

Gauss showed that the roots of this equation may be rationally expressed in terms of the roots of a sequence of equations

(3) $$Z_1 = 0, \, Z_2 = 0, \cdots,$$

whose coefficients are rational in the roots of the preceding equations of the sequence. The degrees of the equations (3) are precisely the prime factors of $p - 1$. There is a Z_i for each factor even if repeated. Moreover, each of the $Z_i = 0$ can be solved by radicals and so equation (1) can also be so solved.

This result is of course of great significance for the problem of solving the general nth-degree equation algebraically. It shows that some equations of high degree can be solved by radicals, for example a fifth degree equation if 5 is a factor of $p - 1$ or a seventh degree equation if 7 is a factor.

The result is of major importance also for the geometric problem of constructing regular polygons of p sides. If $p - 1$ contains no factors other than 2, then the polygon is constructible with straightedge and compass because the degrees of the equations (3) are each 2 and each of its roots is constructible in terms of its coefficients. Thus we are able to construct all polygons of a prime number of sides p if $p - 1$ is a power of 2. Such primes are 3, 5, 17, 257, 65537, ... Alternatively, a regular polygon can be constructed if p is a prime of the form $2^{2^h} + 1$.[3] Gauss remarks (Art. 365) that although the geometric construction of regular polygons of 3, 5, and 15 sides and those immediately derivable from them—for example, 2^n, $2^n \cdot 3$, $2^n \cdot 5$, $2^n \cdot 15$, wherein n is a positive integer—was known in Euclid's time, in an interval of 2000 years no new constructible polygons had been discovered and geometers had been unanimous in declaring that no others could be constructed.

Gauss thought that his result might lead to all sorts of attempts to find new constructible polygons of a prime number of sides. He warns then: "As often as $p - 1$ contains other prime factors besides 2, we arrive at higher equations, namely, to one or more cubic equations if 3 enters once or oftener as a factor of $p - 1$, to equations of the fifth degree if $p - 1$ is divisible by 5, etc. And we can prove with all rigor that these higher equations cannot be avoided or made to depend upon equations of lower degree; and although the limits of this work do not permit us to give the demonstration here, we still thought it necessary to note this fact in order that one should not seek to construct other polygons [of prime number of sides] than those given by our theory, as for example, polygons of 7, 11, 13, 19 sides, and so employ his time in vain."

Gauss then considers (Art. 366) polygons of any number of sides n and asserts that a regular polygon of n sides is constructible if and only if

---

3. In a prime of the form $2^\mu + 1$, μ is necessarily of the form 2^h, but $2^{2^h} + 1$ is not necessarily prime.

$n = 2^l p_1 \cdot p_2 \cdots p_n$, where $p_1, p_2, \ldots, p_n$ are distinct primes of the form $2^{2^h} + 1$ and where l is any positive integer or 0. The sufficiency of this condition does follow readily from Gauss's work on polygons of a prime number of sides but the necessity is not at all obvious and was not proven by Gauss.[4]

The construction of regular polygons had interested Gauss since 1796 when he conceived the first proof that the 17-sided polygon is constructible. There is a story about this discovery which may bear repeating. This construction problem was already famous. One day Gauss approached his professor A. G. Kästner at the University of Göttingen with the proof that this polygon is constructible. Kästner was incredulous and sought to dismiss Gauss, much as university teachers today dismiss angle-trisectors. Rather than take the time to examine Gauss's proof and find the supposed error in it, Kästner told Gauss the construction was unimportant because practical constructions were known. Of course Kästner knew that the existence of practical or approximate constructions was irrelevant for the theoretical problem. To interest Kästner in his proof Gauss pointed out that he had solved a seventeenth degree algebraic equation. Kästner replied that the solution was impossible. But Gauss rejoined that he had reduced the problem to solving an equation of lower degree. "Oh well," scoffed Kästner, "I have already done this." Later Gauss repaid Kästner, who also prided himself on his poetry, by lauding Kästner as the best poet among mathematicians and the best mathematician among poets.

3. Abel's Work on the Solution of Equations by Radicals

Abel read Lagrange's and Gauss's work on the theory of equations and while still a student in high school tackled the problem of the solvability of higher degree equations by following Gauss's treatment of the binomial equation. At first Abel thought he had solved the general fifth degree equation by radicals. But soon convinced of his error he tried to prove that such a solution was not possible (1824–26). First he succeeded in proving the theorem: The roots of an equation solvable by radicals can be given such a form that each of the radicals occurring in the expressions for the roots is expressible as a rational function of the roots of the equation and certain roots of unity. Abel then used this theorem to prove[5] the impossibility of solving by radicals the general equation of degree greater than four.

Abel's proof, done in ignorance of Ruffini's work (Chap. 25, sec. 2), is

---

4. See James Pierpont, "On an Undemonstrated Theorem of the *Disquisitiones Arith-meticae*," *Amer. Math. Soc. Bull.*, 2, 1895–96, 77–83. This article gives the proof. The fact that the Gauss condition is necessary was first proved by Pierre L. Wantzel (1814–48), *Jour. de Math.*, 2, 1837, 366–72.

5. *Jour. für Math.*, 1, 1826, 65–84 = *Œuvres*, 1, 66–94.

roundabout and unnecessarily complicated. His paper also contained an error in a classification of functions, which fortunately was not essential to the argument. He later published two more elaborate proofs. A simple, direct, and rigorous proof based on Abel's idea was given by Kronecker in 1879[6].

Thus the question of the solution of general equations of degree higher than four was settled by Abel. He also considered some special equations. He took up[7] the problem of the division of the lemniscate (solving $x^n - 1 = 0$ is the equivalent of the problem of the division of the circle into n equal arcs) and arrived at a class of algebraic equations, now called Abelian equations, that are solvable by radicals. The cyclotomic equation (1) is an example of an Abelian equation. More generally an equation is called Abelian if all its roots are rational functions of one of them, that is, if the roots are $x_1, \theta_1(x_1), \theta_2(x_1), \ldots, \theta_{n-1}(x_1)$ where the θ_i are rational functions. There is also the condition that $\theta_\alpha(\theta_\beta(x_1)) = \theta_\beta(\theta_\alpha(x_1))$ for all values of α and β from 1 to $n - 1$.

In this last work he introduced two notions (though not the terminology), field and polynomial irreducible in a given field. By a field of numbers he, like Galois later, meant a collection of numbers such that the sum, difference, product, and quotient of any two numbers in the collection (except division by 0) are also in the collection. Thus the rational numbers, real numbers, and complex numbers form a field. A polynomial is said to be reducible in a field (usually the field to which its coefficients belong) if it can be expressed as the product of two polynomials of lower degrees and with coefficients in the field. If the polynomial cannot be so expressed it is said to be irreducible.

Abel then tackled the problem of characterizing all equations which are solvable by radicals and had communicated some results to Crelle and to Legendre just before death overtook him in 1829.

4. Galois's Theory of Solvability

After Abel's work the situation was as follows: Although the general equation of degree higher than four was not solvable by radicals, there were many special equations, such as the binomial equations $x^p = a$, p a prime, and Abelian equations that were solvable by radicals. The task now became to determine which equations are solvable by radicals. This task, just begun by Abel, was taken up by Evariste Galois (1811–32). Born to well-to-do and educated parents he attended one of the celebrated lycées of Paris and started to study mathematics at the age of fifteen. This subject became his passion and he studied carefully the works of Lagrange, Gauss, Cauchy, and Abel.

6. *Monatsber. Berliner Akad.*, 1879, 205–29 = *Werke*, 4, 73–96. Kronecker's proof is explained by James Pierpoint: "On the Ruffini-Abelian Theorem, *Amer. Math. Soc. Bull.*, 2, 1895–96, 200–21.

7. *Jour. für Math.*, 4, 1829, 131–56 = *Œuvres*, 1, 478–507.

The other subjects he neglected. Galois sought to enter the Ecole Polytechnique but possibly because he failed to explain in sufficient detail the
questions he had to answer orally at the entrance examination or because the
examining professors did not understand him he was rejected in two tries.
He therefore entered the Ecole Préparatoire (the name then for the Ecole
Normale and a much inferior school at that time.) During the 1830 Revolution, which drove Charles X from the throne and installed Louis Philippe,
Galois publicly criticized the director of his school for failing to support
the Revolution and was expelled. He was twice arrested for political offenses,
spent most of the last year and a half of his life in prison, and was killed in a
duel on May 31, 1832.

During his first year at the Ecole Galois published four papers. In 1829
he submitted two papers on the solution of equations to the Academy of
Sciences. These were entrusted to Cauchy, who lost them. In January of
1830 he presented to the Academy of Sciences another carefully written
paper on his research. This was sent to Fourier, who died soon after, and
this paper too was lost. At the suggestion of Poisson Galois wrote (1831) a
new paper on his research. This article, "Sur les conditions de résolubilité
des équations par radicaux,"[8] the only finished article on his theory of
the solution of equations, was returned by Poisson as unintelligible, with the
recommendation that a fuller exposition be written. The night before his
death Galois drew up a hastily written account of his researches which he
entrusted to his friend, August Chevalier. This account has been preserved.

In 1846 Liouville edited and published in the *Journal de Mathématiques*[9]
part of Galois's papers including a revision of the 1831 paper. Then Serret's
Cours d'algèbre supérieure (3rd ed.) of 1866 gave an exposition of Galois's ideas.
The first full and clear presentation of Galois theory was given in 1870 by
Camille Jordan in his book *Traité des substitutions et des équations algébriques.*

Galois approached the problem of characterizing equations solvable by
radicals by improving on Lagrange's ideas, though he also derived some
suggestions from Legendre's, Gauss's, and Abel's work. He proposed to
consider the general equation, which is, of course,

$$(4) \qquad x^n + a_1 x^{n-1} + \cdots + a_{n-1} x + a_n = 0$$

wherein, as in the work of Lagrange, the coefficients must be independent or
completely arbitrary, and particular equations, such as

$$(5) \qquad x^4 + px^2 + q = 0,$$

wherein only two coefficients are independent. Galois's main thought was to
bypass the construction of the Lagrange resolvents (Chap. 25, sec. 2) of the

8. *Œuvres*, 1897, 33–50.
9. *Jour. de Math.*, 11, 1846, 381–444.

given polynomial equation, a construction requiring great skill and having no clear methodology.

Like Lagrange, Galois makes use of the notion of substitutions or permutations of the roots. Thus if x_1, x_2, x_3, and x_4 are the four roots of a fourth degree equation, the interchange of x_1 and x_2 in any expression involving the x_i is a substitution. This particular substitution is indicated by

$$\begin{pmatrix} x_1 & x_2 & x_3 & x_4 \\ x_2 & x_1 & x_3 & x_4 \end{pmatrix}.$$

A second substitution is indicated by

$$\begin{pmatrix} x_1 & x_2 & x_3 & x_4 \\ x_3 & x_4 & x_1 & x_2 \end{pmatrix}.$$

To perform the first substitution and then the second one is equivalent to performing the third substitution

$$\begin{pmatrix} x_1 & x_2 & x_3 & x_4 \\ x_4 & x_3 & x_1 & x_2 \end{pmatrix}$$

because, for example, by the first substitution x_1 is replaced by x_2; by the second substitution x_2 is replaced by x_4; and by the third substitution x_1 goes directly into x_4. One says that the *product* of the first two substitutions taken in the order just indicated is the third substitution. There are in all 4! possible substitutions. The set of substitutions is said to form a group because the product of any two substitutions is a member of the set. This notion, which is not of course a formal definition of an abstract group, is due to Galois.

To secure some grasp of Galois's ideas let us consider the equation[10]

$$x^4 + px^2 + q = 0,$$

where p and q are independent. Let R be the field formed by rational expressions in p and q and with coefficients in the field of rational numbers, a typical expression being $(3p^2 - 4q)/(q^2 - 7p)$. One says with Galois that R is the field obtained by adjoining the letters or indeterminates p and q to the rational numbers. This field R is the field or domain of rationality of the coefficients of the given equation and the equation is said to belong to the field R. Like Abel, Galois did not use the terms field or domain of rationality but he did use the concept.

10. Since Galois's own presentation of his ideas was not clear and he introduced so many new notions, we shall utilize an example due to Verriest (see the bibliography at the end of the chapter) to help make Galois's theory clear.

We happen to know that the roots of the fourth degree equation are

$$x_1 = \sqrt{\frac{-p + \sqrt{p^2 - 4q}}{2}}, \qquad x_2 = -\sqrt{\frac{-p + \sqrt{p^2 - 4q}}{2}}$$

$$x_3 = \sqrt{\frac{-p - \sqrt{p^2 - 4q}}{2}}, \qquad x_4 = -\sqrt{\frac{-p - \sqrt{p^2 - 4q}}{2}}.$$

Then it is true that the two relations with coefficients in R

$$x_1 + x_2 = 0 \quad \text{and} \quad x_3 + x_4 = 0$$

hold for the roots. Since our given equation is of the fourth degree there are twenty-four possible substitutions of the roots. The following eight substitutions

$$E = \begin{pmatrix} x_1 & x_2 & x_3 & x_4 \\ x_1 & x_2 & x_3 & x_4 \end{pmatrix}, \qquad E_1 = \begin{pmatrix} x_1 & x_2 & x_3 & x_4 \\ x_2 & x_1 & x_3 & x_4 \end{pmatrix},$$

$$E_2 = \begin{pmatrix} x_1 & x_2 & x_3 & x_4 \\ x_1 & x_2 & x_4 & x_3 \end{pmatrix}, \qquad E_3 = \begin{pmatrix} x_1 & x_2 & x_3 & x_4 \\ x_2 & x_1 & x_4 & x_3 \end{pmatrix},$$

$$E_4 = \begin{pmatrix} x_1 & x_2 & x_3 & x_4 \\ x_3 & x_4 & x_1 & x_2 \end{pmatrix}, \qquad E_5 = \begin{pmatrix} x_1 & x_2 & x_3 & x_4 \\ x_4 & x_3 & x_1 & x_2 \end{pmatrix},$$

$$E_6 = \begin{pmatrix} x_1 & x_2 & x_3 & x_4 \\ x_3 & x_4 & x_2 & x_1 \end{pmatrix}, \qquad E_7 = \begin{pmatrix} x_1 & x_2 & x_3 & x_4 \\ x_4 & x_3 & x_2 & x_1 \end{pmatrix}$$

leave the two relations true in R.[11] One could show that these eight are the only substitutions of the twenty-four which leave invariant *all* relations in R among the roots. These eight are the group of the equation in R. They are a subgroup of the full group. That is, the group of an equation with respect to a field R is the group or subgroup of substitutions on the roots which leave invariant *all* the relations with coefficients in R among the roots of the given equation (whether general or particular). One can say that the number of substitutions that leave all relations in R invariant is a measure of our

11. For the general equation of degree n, that is, with n independent quantities as coefficients, a function of the roots is *invariant* under or unaltered by a substitution on the roots if and only if it remains identical with the original function. If the coefficients are all numerical then a function is unaltered if it remains numerically the same. Thus for $x^3 + x^2 + x + 1 = 0$ the roots are $x_1 = -1$, $x_2 = i$, and $x_3 = -i$. Consider x_2^2. The substitution x_3 for x_2 gives x_3^2. This has the same numerical value as x_2^2. Then x_2^2 is not altered by the substitution. If the coefficients contain some numerical values and some independent quantities then a function of the roots remains unaltered by a substitution on the roots if the function remains numerically the same for all values of the independent quantities (in the domain to which these may be limited) and the numerical values which the roots may assume.

ignorance of the roots because we can not distinguish them under these eight substitutions.

Now consider $x_1^2 - x_3^2$ which equals $\sqrt{p^2 - 4q}$. We adjoin this radical to R, and form the field R', that is, we form the smallest field containing R and $\sqrt{p^2 - 4q}$. Then

(6) $$x_1^2 - x_3^2 = \sqrt{p^2 - 4q}$$

is a relation in R'. Since $x_1 + x_2 = 0$ and $x_3 + x_4 = 0$ we also have

$$x_1^2 = x_2^2 \quad \text{and} \quad x_3^2 = x_4^2.$$

Then in view of these last two facts we can say that the first four of the above eight substitutions leave the relation (6) in R' true, but the last four do not. Then the four substitutions, if they leave *every* true relation in R' among the roots invariant, are the group of the equation in R'. The four are a subgroup of the eight.

Now suppose we adjoin to R' the quantity $\sqrt{(-p - D)/2}$ where $D = \sqrt{p^2 - 4q}$ and form the field R''. Then

$$x_3 - x_4 = 2\sqrt{\frac{-p - D}{2}}$$

is a relation in R''. This relation remains invariant only under the first two substitutions E and E_1 but not under the rest of the eight. Then the group of the equation in R'' consists of these two substitutions, provided every relation in R'' among the roots remains invariant under these two substitutions. The two are a subgroup of the four substitutions.

If we now adjoin to R'' the quantity $\sqrt{(-p + D)/2}$ we get R'''. In R''' we have

$$x_1 - x_2 = 2\sqrt{\frac{-p + D}{2}}.$$

Now the only substitution leaving all relations in R''' true is just E and this is the group of the equation in R'''.

We may see from the above discussion that the group of an equation is a key to its solvability because the group expresses the degree of indistinguishability of the roots. It tells us what we do not know about the roots.

There were many groups, or strictly a group of substitutions and successive subgroups, involved above. Now the *order* of a group (or subgroup) is the number of elements in it. Thus we had groups of the orders 24, 8, 4, 2, and 1. The order of a subgroup always divides the order of the group (sec. 6). The *index* of a subgroup is the order of the group in which it lies divided by the order of the subgroup. Thus the index of the subgroup of order 8 is 3.

The above sketch merely shows the ideas Galois dealt with. His work proceeded as follows: Given an equation general or particular, he showed first how one could find the group G of this equation in the field of the coefficients, that is, the group of substitutions on the roots that leaves invariant every relation among the roots with coefficients in that field. Of course one must find the group of the equation without knowing the roots. In our example above, the group of the quartic equation was of order 8 and the field of the coefficients was R. Having found the group G of the equation, one seeks next the largest subgroup H in G. In our example this was the subgroup of order 4. If there should be two or more largest subgroups we take any one. The determination of H is a matter of pure group theory and can be done. Having found H, one can find by a set of procedures involving only rational operations a function ϕ of the roots whose coefficients belong to R and which does not change value under the substitutions in H but does change under all other substitutions in G. In our example above the function was $x_1^2 - x_3^2$. Actually an infinity of such functions can be obtained. One must of course find such a function without knowing the roots. There is a method of constructing an equation in R one of whose roots is this function ϕ. The degree of this equation is the index of H in G. This equation is called a partial resolvent.[12] In our example the equation is $t^2 - (p^2 - 4q) = 0$ and its degree is 8/4 or 2.

One must now be able to solve the partial resolvent to find the root ϕ. In our example ϕ is $\sqrt{p^2 - 4q}$. One adjoins ϕ to R and obtains a new field R'. Then the group of the original equation with respect to the field R' can be shown to be H.

We now repeat the procedure. We have the group H, of order 4 in our example, and the field R'. We seek next the largest subgroup in H. In our example, it was the subgroup of order 2. Let us call this subgroup K. One can now obtain a function of the roots of the original equation whose coefficients belong to R' and whose value is unchanged by every substitution in K but is changed by other substitutions in H. In our example above this was $x_3 - x_4$. To find this function, ϕ_1 say, without knowing the roots, one constructs an equation in R' having the function ϕ_1 as a root. In our example this equation is $t^2 - 2(-p - \sqrt{p^2 - 4q}) = 0$. The degree of this equation is the index of K with respect to H, that is, 4/2 or 2. This equation is the second partial resolvent.

Now one must be able to solve this resolvent equation and get a root, the function ϕ_1, and adjoin this value to R' thereby forming the field R''. With respect to R'' the group of the equation is K.

We again repeat the process. We find the largest subgroup L in K. In our example this was just the identity substitution E. We seek a function of

12. This use of the word "resolvent" is different from Lagrange's use.

the roots (with coefficients in R'') which retains its value under E but not under the other substitutions of K. In our example such a function was $x_1 - x_2$. To obtain this ϕ_2 without knowing the roots we must construct an equation in R'' having the function ϕ_2 as a root. In our example the equation is $t^2 - 2(-p + \sqrt{p^2 - 4q}) = 0$. The degree of this equation is the index of L in K. In our example this index was $2/1$ or 2. This equation is the third partial resolvent. We must solve this equation to find the value ϕ_2.

By adjoining this root to R'' one gets a field R'''. Let us suppose that we have reached the final stage wherein the group of the original equation in R''' is the identity substitution, E.

Now Galois showed that when the group of an equation with respect to a given field is just E, then the roots of the equation are members of that field. Hence the roots lie in the field R''', and we know the field in which the roots lie because R''' was obtained from the known field R by successive adjunction of known quantities. There is next a straightforward process for finding the roots by rational operations in R'''.

Galois gave a method of finding the group of a given equation, the successive resolvents, and the groups of the equation with respect to the successively enlarged fields of coefficients that result from adjoining the roots of these successive resolvents to the original coefficient field, that is, the successive subgroups of the original group. These processes involve considerable theory but, as Galois pointed out, his work was not intended as a practical method of solving equations.

And now Galois applied the above theory to the problem of solving polynomial equations by rational operations and radicals. Here he introduced another notion of group theory. Suppose H is a subgroup of G. If one multiplies the substitutions of H by any element g of G then one obtains a new collection of substitutions which is denoted by gH, the notation indicating that the substitution g is performed first and then any element of H is applied. If $gH = Hg$ for every g in G then H is called a normal (self-conjugate or invariant) subgroup in G.

We recall that Galois's method of solving an equation called for finding and solving the successive resolvents. Galois showed that when the resolvent that serves to reduce the group of an equation, say from G to H, is a binomial equation $x^p = A$ of prime degree p, then H is a normal subgroup in G (and of index p), and conversely if H is a normal subgroup in G and of prime index p then the corresponding resolvent is a binomial equation of degree p or can be reduced to such. If all of the successive resolvents are binomial equations, then, in view of Gauss's result on binomial equations, we can solve the original equation by radicals. For, we know that we can pass from the initial field to the final field in which the roots lie by successive adjunctions of radicals. Conversely if an equation is solvable by radicals then the set of resolvent equations must exist, and these are binomial equations.

Thus the theory of solvability by radicals is in general outline the same as the theory of solution given earlier except that the series of subgroups

$$G, H, K, L, \cdots, E$$

must each be a maximum normal subgroup (not a subgroup of any larger normal subgroup) of the preceding group. Such a series is called a composition series. The indices of H in G, K in H, and so forth are called the indices of the composition series. If the indices are *prime* numbers the equation is solvable by radicals, and if the indices are not prime the equation is not solvable by radicals. As one proceeds to find the sequence of maximal normal subgroups there may be a choice; that is, there may be more than one maximal normal subgroup of highest order in a given group or subgroup. One may choose any one, though thereafter the subgroups may differ. But the very same set of indices will result though the order in which they appear may differ (see the Jordan-Hölder theorem below). The group G, which contains a composition series of prime indices, is said to be solvable.

How does the Galois theory show that the general nth-degree equation for $n > 4$ is not solvable by radicals whereas for $n \leq 4$ they are? For the general nth-degree equation the group is composed of all the $n!$ substitutions of the n roots. This group is called the symmetric group of degree n. Its order is of course $n!$. It is not difficult to find the composition series for each symmetric group. The maximal normal subgroup, which is called the alternating subgroup, is of order $n!/2$. The only normal subgroup of the alternating group is the identity element. Hence the indices are 2 and $n!/2$. But the number $n!/2$, for $n > 4$, is never prime. Hence the general equation of degree greater than 4 is not solvable by radicals. On the other hand, the quadratic equation can be solved with the aid of a single resolvent equation. The indices of the composition series consist of just the single number 2. The general third degree equation requires for its solution two resolvent equations of the form $y^2 = A$ and $z^3 = B$. These are of course binomial resolvents. The indices of the composition series are 2 and 3. The general fourth degree equation can be solved with four binomial resolvent equations, one of degree 3 and three of degree 2. The indices of the composition series are, then, 2, 3, 2, 2.

For equations with numerical coefficients as opposed to those with literal independent coefficients, Galois gave a theory similar to that described above. However, the process of determining the solvability by radicals is more complicated even though the basic principles are the same.

Galois also proved some special theorems. If one has an irreducible equation of prime degree whose coefficients lie in a field R and whose roots are all rational functions of two of the roots with coefficients in R then the equation is solvable by radicals. He also proved the converse: Every irreducible equation of prime degree which is solvable by radicals has the property that each of its roots is a rational function of two of them with coefficients in R. Such an equation is now called Galoisian. The simplest example of a Galoisian equation is $x^p - A = 0$. The notion is an extension of Abelian equations.

As a postscript Hermite[13] and Kronecker in a letter to Hermite[14] and in a subsequent paper[15] solved the general quintic equation by means of elliptic modular functions. This is analogous to the use of trigonometric functions to solve the irreducible case of the cubic equation.

5. *The Geometric Construction Problems*

The eighteenth-century mathematicians doubted that the famous construction problems could be solved. Galois's work supplied a criterion on constructibility that disposed of some of the famous problems.

Each step of a construction with straightedge and compass calls for finding a point of intersection either of two lines, a line and a circle, or two circles. With the introduction of coordinate geometry it was recognized that in algebraic terms such steps mean the simultaneous solution of either two linear equations, a linear and a quadratic, or two quadratic equations. In any case, the worst that is involved algebraically is a square root. Hence successive quantities found by successive steps or constructions are at worst the result of a chain of *square roots* applied to the given quantities. Accordingly, constructible quantities must lie in fields obtained by adjoining to the field containing the given quantities only square roots of the given or subsequently constructed quantities. We may call such extension fields quadratic extension fields.

In performing the successive constructions a few restrictions must be observed. For example, some of the processes permit the use of an arbitrary line or circle. Thus in bisecting a line segment we can use circles larger than half of the line segment. One must choose this circle in the field or the constructible extension field of the given elements. This can be done.

Also, the extension fields may involve complex elements because, for example, the square root of a negative coordinate may occur. These complex elements are constructible because the real and imaginary parts of the complex quantities that do occur are each roots of a real equation and these roots are constructible.

Given a construction problem, one first sets up an algebraic equation whose solution is the desired quantity. This quantity must belong to some quadratic extension field of the field of the given quantities. In the case of the regular polygon of 17 sides this equation is $x^{17} - 1 = 0$ and the given

13. *Comp. Rend.*, 46, 1858, 508–15 = *Œuvres*, 2, 5–12.
14. *Comp. Rend.*, 46, 1858, 1150–52 = *Werke*, 4, 43–48.
15. *Jour. für Math.*, 59, 1861, 306–10 = *Werke*, 4, 53–62.

quantity can be taken to be the radius of the unit circle. The relevant irreducible equation is $x^{16} + x^{15} + \cdots + 1 = 0$. In terms of Galois theory the necessary and sufficient condition for an equation to be solvable in *square roots* is that the order of the Galois group of the equation be a power of 2. This is the case for the equation $x^{16} + \cdots + 1 = 0$ and the composition series is 2, 2, 2, 2. This means that the resolvents are binomial equations of degree 2 so that only square roots are adjoined to the original rational field determined by the given radius of the unit circle. With this Galois criterion one can prove Gauss's statement that a regular polygon with a prime number p of sides can be constructed with straightedge and compass if and only if the prime number p has the form $2^{2^n} + 1$, that is, if $p = 3, 5, 17, 257, \ldots$ but not for $p = 7, 11, 13, 19, 23, 29, 31, \ldots$. Galois theory can be used to prove also that it is impossible to trisect an arbitrary angle or to double a given cube.

But Galois's criterion does not apply at all to the problem of squaring the circle. Here the given quantity is the radius of the circle. The equation to be solved is $x^2 = \pi r^2$. Though this equation is itself just a quadratic, it is not true that its solutions belong to a quadratic extension field of the field determined by the given quantity because π is not an algebraic irrational (Chap. 41, sec. 2). Galois's work, then, not only answered completely the question of which equations are solvable by algebraic operations but gave a general criterion to determine the constructibility with straightedge and compass of geometrical figures.

So far as the famous construction problems are concerned we should note that before Galois theory was applied to them, Gauss and Wantzel had determined which regular polygons are constructible (sec. 2), and Wantzel in the paper of 1837[16] showed that the general angle could not be trisected nor could a given cube be doubled. He proved that every constructible quantity must satisfy an equation of degree 2^n, and this is not true in the two problems just mentioned.

6. *The Theory of Substitution Groups*

In his work on the solvability of equations (Chap. 25, sec. 2) Lagrange introduced as the key to his analysis functions of the n roots which take the same value under some permutations of the roots. He therefore undertook in the very same papers to study functions for the purpose of determining the different values they can assume among the $n!$ possible values which the $n!$ permutations of the n variables (the roots) might give rise to. Subsequent work by Ruffini, Abel, and Galois lent increased importance to this topic. The fact that a rational function of n letters takes the same value under some

16. *Jour. de Math.*, 2, 1837, 366–72.

collection of permutations or substitutions of the roots means, as we have seen, that this collection is a subgroup of the entire symmetric group. This was observed explicitly by Ruffini in his *Teoria generale delle equazioni* (1799). Hence what Lagrange initiated was one way of studying subgroups of a group of substitutions. The more direct way is, of course, to study the group of substitutions itself and to determine its subgroups. Both methods of studying the structure or composition of substitution groups became an active subject which was pursued as an interest in and for itself though the connection with the solvability of equations was not ignored. The theory of substitution or permutation groups was the first major investigation that ultimately gave rise to the abstract theory of groups. Here we shall note some of the concrete theorems on substitution groups that were obtained during the nineteenth century.

Lagrange himself affirmed one major result which in modern language states that the order of a subgroup divides the order of the group. The proof of this theorem was communicated by Pietro Abbati (1768–1842) to Ruffini in a letter of September 30, 1802, which was published.[17]

Ruffini in his 1799 book introduced, though somewhat vaguely, the notions of transitivity and primitiveness. A permutation group is transitive if each letter of the group is replaced by every other letter under the various permutations of the group. If G is a transitive group and the n symbols or letters may be divided into r distinct subsets σ_i, $i = 1, 2, \ldots, r$, each subset containing s_i symbols such that any permutation of G either permutes the symbols of the σ_i among themselves or replaces these symbols by the symbols of σ_j, this holding for each $i = 1, 2, \ldots, r$, then G is called imprimitive. If no such separation of the n symbols is possible, then the transitive group is called a primitive group. Ruffini also proved that there does not exist a subgroup of order k for all k in a group of order n.

Cauchy, stimulated by the work of Lagrange and Ruffini, wrote a major paper on substitution groups.[18] With the theory of equations in mind, he proved that there is no group in n letters (degree n) whose index relative to the full symmetric group in n letters is less than the largest prime number which does not exceed n unless this index is 2 or 1. Cauchy stated the theorem in the language of function values: The number of different values of a non-symmetric function of n letters cannot be less than the largest prime p smaller than n unless it is 2.

Galois made the largest step in introducing concepts and theorems about substitution groups. His most important concept was the notion of a normal (invariant or self-conjugate) subgroup. Another group concept due to Galois is that of an isomorphism between two groups. This is a one-to-one

17. *Memorie della Società Italiana delle Scienze*, 10, 1803, 385–409.
18. *Jour. de l'Ecole Poly.*, 10, 1815, 1–28 = *Œuvres*, (2), 1, 64–90.

correspondence between the elements of the two groups such that if $a \cdot b = c$ in the first one, then for the corresponding elements in the second, $a' \cdot b' = c'$. He also introduced the notions of simple and composite groups. A group having no invariant subgroup is simple; otherwise it is composite. Apropos of these notions, Galois expressed the conjecture[19] that the smallest simple group whose order is a composite number is the group of order 60.

Unfortunately Galois's work did not become known until Liouville published parts of it in 1846, and even that material was not readily understandable. On the other hand, Lagrange's and Ruffini's work on substitution groups, couched in the language of the values which a function of n letters can take on, became well known. Hence the subject of the solution of equations receded into the background, and when Cauchy returned to the theory of equations he concentrated on substitution groups. During the years 1844 to 1846 he wrote a host of papers. In the major paper[20] he systematized many earlier results and proved a number of special theorems on transitive primitive and imprimitive groups and on intransitive groups. In particular he proved the assertion by Galois that every finite (substitution) group whose order is divisible by a prime p contains at least one subgroup of order p. The major paper was followed by a great number of others published in the *Comptes Rendus* of the Paris Academy for the years 1844–46.[21] Most of this work was concerned with the formal (that is non-numerical) values that functions of n letters can take on under interchange of the letters and with finding functions that take on a given number of values.

After Liouville published some of Galois's work, Serret lectured on it at the Sorbonne and in the third edition of his *Cours* gave a better textual exposition of Galois theory. The work of clarifying Galois's ideas on the solvability of equations and the development of the theory of substitution groups thereafter proceeded hand in hand. In his text Serret gave an improved form of Cauchy's 1815 result. If a function of n letters has less than p values where p is the largest prime smaller than n, then the function cannot have more than two values.

One of the problems Serret stressed in the 1866 text asks for all the groups that one can form with n letters. This problem had already attracted the attention of Ruffini, and he, Cauchy, and Serret himself in a paper of 1850[22] gave a number of partial results, as did Thomas Penyngton Kirkman (1806–95). Despite many efforts and hundreds of limited results the problem is still unsolved.

After Galois, Camille Jordan (1838–1922) was the first to add signifi-

19. *Œuvres*, 1897 ed., 26.
20. *Exercices d'analyse et de physique mathématique*, 3, 1844, 151–252 = *Œuvres*, (2), 13, 171–282.
21. *Œuvres*, (1), Vols. 9 and 10.
22. *Jour. de Math.*, (1), 15, 1850, 45–70.

cantly to Galois theory. In 1869[23] he proved a basic result. Let G_1 be a maximal self-conjugate (normal) subgroup of G_0, G_2 a maximal self-conjugate subgroup of G_1, and so forth until the series terminates in the identity element. This series of subgroups is called a composition series of G_0. If G_{i+1} is any self-conjugate subgroup of order r in G_i whose order is p, then G_i may be decomposed into $\lambda = p/r$ classes. Two elements are in the same class if one is the product of the other and an element of G_{i+1}. If a is any element in one class and b any element in another, the product will be in the same third class. These classes form a group for which G_{i+1} is the identity element and the group is called the quotient group or factor group of G_i by G_{i+1}. It is denoted by G_i/G_{i+1}, a notation introduced by Jordan in 1872. The quotient groups G_0/G_1, G_1/G_2, . . . are called the composition factor groups of G_0 and their orders are known as the composition factors or composition indices. There may be more than one composition series in G_0. Jordan proved that the set of composition factors is invariant except for the order in which they may occur and (Ludwig) Otto Hölder (1859–1937), a professor at the University of Leipzig, showed[24] that the quotient groups themselves were independent of the composition series; that is, the same set of quotient groups would be present for any composition series. The two results are called the Jordan-Hölder theorem.

The knowledge of (finite) substitution groups and their connection with the Galois theory of equations up to 1870 was organized in a masterful book by Jordan, his *Traité des substitutions et des équations algébriques* (1870). In this book Jordan, like almost all of the men preceding him, used as the definition of a substitution group that it is a collection of substitutions such that the product of any two members of the collection belongs to the collection. The other properties that we commonly postulate today in the definition of a group (Chap. 49, sec. 2) were utilized but were brought in as either obvious properties of such groups or as additional conditions but not specified in the definition. The *Traité* presented new results and made explicit for substitution groups the notions of isomorphism (*isomorphisme holoédrique*) and homomorphism (*isomorphisme mériédrique*), the latter being a many-to-one correspondence between two groups such that $a \cdot b = c$ implies $a' \cdot b' = c'$. Jordan added fundamental results on transitive and composite groups. The book also contains Jordan's solution of the problem posed by Abel, to determine the equations of a given degree that are solvable by radicals and to recognize whether a given equation does or does not belong to this class. The groups of the solvable equations are commutative. Jordan called them Abelian and the term Abelian was applied thereafter to commutative groups.

23. *Jour. de Math.*, (2), 14, 1869, 129–46 = *Œuvres*, 1, 241–48.
24. *Math. Ann.*, 34, 1889, 26–56.

Another major theorem on substitution groups was proved shortly after the *Traité* appeared by Ludwig Sylow (1832–1918), a Norwegian professor of mathematics. Cauchy had proved that every group whose order is divisible by a prime number p must contain one or more subgroups of order p. Sylow[25] extended Cauchy's theorem. If the order of a group is divisible by p^α, p being a prime, but not by $p^{\alpha+1}$ then the group contains one and only one system of conjugate subgroups of order p^α.[26] In this same paper Sylow also proved that every group of order p^α is solvable, that is, the indices of a sequence of maximal invariant subgroups are prime.

Quite another approach to substitution groups and ultimately to more general groups was suggested by a purely physical investigation. Auguste Bravais (1811–63), a physicist and mineralogist, studied groups of motions[27] to determine the possible structures of crystals. This study amounts mathematically to the investigation of the group of linear transformations in three variables

$$x'_i = a_{i1}x + a_{i2}y + a_{i3}z, \qquad i = 1, 2, 3,$$

of determinant $+1$ or -1 and it led Bravais to thirty-two classes of symmetric molecular structures that might occur in crystals.

Bravais's work impressed Jordan and he undertook to investigate what he called the analytic representation of groups and what is now called the representation theory for groups. Actually, Serret in the 1866 edition of his *Cours* had considered the representation of substitutions by transformations of the form

$$x' = \frac{ax + b}{cd + d}.$$

But the more useful representation of all types of groups was introduced by Jordan. He sought to represent substitutions by linear transformations of the form

(9)
$$x'_i = \sum_{j=1}^{n} a_{ij}x_j, \qquad i = 1, 2, \cdots, n.$$

Since the substitution groups are finite some restriction had to be placed on the transformations so that the group of transformations would be finite. Galois had considered such transformations[28] and had limited them so that the coefficients and the variables take values in a finite field of prime order.

25. *Math. Ann.*, 5, 1872, 584–94.
26. If H is a subgroup of G and g is any element of G then $g^{-1}Hg$ is a subgroup conjugate to H. H and all its conjugates are said to form a system of conjugate subgroups of G or a complete conjugate set of subgroups.
27. *Jour. de Math.*, 14, 1849, 141–80.
28. *Œuvres*, 1897 ed., 21–23, 27–29.

Jordan in 1878[29] stated that a linear homogeneous substitution (9) of finite period p may be linearly transformed to the canonical form

$$y_i' = \varepsilon_i y_i, \qquad i = 1, 2, \cdots, n$$

where the ε_i's are pth roots of unity. The theorem was proved by a number of men.[30] This was the beginning of a vast amount of research on the determination of the possible linear substitution groups of a given order and of binary and ternary form (two and three variables). Also the determination of the subgroups of given linear substitution groups and the algebraic expressions left invariant by all the members of a group or subgroup stirred up much research.

Directly after noting Bravais's paper Jordan undertook the first major investigation of *infinite* groups. In his paper "Mémoire sur les groupes de mouvements,"[31] Jordan points out that the determination of all the groups of movements (he considered only translations and rotations) is equivalent to the determination of all the possible systems of molecules such that each movement of any one group transforms the corresponding system of molecules into itself. He therefore studied the various types of groups and classified them. The results are not as significant as the fact that his paper initiated the study of geometric transformations under the rubric of groups and the geometers were quick to pick up this line of thought (Chap. 38, sec. 5).

One other development of the middle nineteenth century is both noteworthy and instructive. Arthur Cayley, very much influenced by Cauchy's work, recognized that the notion of substitution group could be generalized. In three papers[32] Cayley introduced the notion of an *abstract* group. He used a general operator symbol θ applied to a system of elements $x, y, z, \ldots$ and spoke of θ so applied producing a function $x', y', z', \ldots$ of $x, y, z, \ldots$. He pointed out that in particular θ may be a substitution. The abstract group contains many operators $\theta, \phi, \ldots$. $\theta\phi$ is a compound (product) of two operations and the compound is associative but not necessarily commutative. His general definition of a group calls for a set of operators $1, \alpha, \beta, \ldots$, all of them different and such that the product of any two of them in either order or the product of any one and itself belongs to the set.[33] He mentions matrices under multiplication and quaternions (under addition) as constituting groups. Unfortunately, Cayley's introduction of the abstract group

29. *Jour. für. Math.*, 84, 1878, 89–215, p. 112 in particular = *Œuvres*, 2, 13–139, p. 36 in particular.
30. See E. H. Moore, *Math. Ann.*, 50, 1898, 215.
31. *Annali di Mat.*, (2), 2, 1868/69, 167–215 and 322–45 = *Œuvres*, 4, 231–302.
32. *Phil. Mag.*, (3), 34, 1849, 527–29 = *Coll. Math. Papers*, 1, 423–24 and (4), 7, 1854, 40–47 and 408–9 = *Papers*, 2, 123–30 and 131–32.
33. *Papers*, 2, 124.

concept attracted no attention at this time, partly because matrices and quaternions were new and not well known and the many other mathematical systems that could be subsumed under the notion of groups were either yet to be developed or were not recognized to be so subsumable. Premature abstraction falls on deaf ears whether they belong to mathematicians or to students.

Bibliography

Abel, N. H.: *Œuvres complètes* (1881), 2 vols., Johnson Reprint Corp., 1964.

Bachmann, P.: "Über Gauss' zahlentheoretische Arbeiten," *Nachrichten König. Ges. der Wiss. zu Gött.*, 1911, 455–518. Also in Gauss's *Werke*, 10, Part 2, 1–69.

Burkhardt, H.: "Endliche discrete Gruppen, *Encyk. der Math. Wiss.*, B. G. Teubner, 1903–15, I, Part 1, 208–26.

———: "Die Anfänge der Gruppentheorie und Paolo Ruffini," *Abhandlungen zur Geschichte der Mathematik*, Heft 6, 1892, 119–59.

Burns, Josephine E.: "The Foundation Period in the History of Group Theory," *Amer. Math. Monthly*, 20, 1913, 141–48.

Dupuy, P.: "La Vie d'Evariste Galois," *Ann. de l'Ecole Norm. Sup.*, (2), 13, 1896, 197–266.

Galois, Evariste: "Œuvres," *Jour. de Math.*, 11, 1846, 381–444.

———: *Œuvres mathématiques*, Gauthier-Villars, 1897.

———: *Ecrits et mémoires mathématiques* (ed. by R. Bourgne and J.-P. Azra), Gauthier-Villars, 1962.

Gauss, C. F.: *Disquisitiones Arithmeticae* (1801), *Werke*, Vol. 1, König. Ges. der Wiss., zu Göttingen, 1870, English translation by Arthur A. Clarke, S.J., Yale University Press, 1966.

Hobson, E. W.: *Squaring the Circle and Other Monographs*, Chelsea (reprint), 1953.

Hölder, Otto: "Galois'sche Theorie mit Anwendungen," *Encyk. der Math. Wiss.*, B. G. Teubner, 1898–1904, I, Part 1, 480–520.

Infeld, Leopold: *Whom the Gods Love: The Story of Evariste Galois*, McGraw-Hill, 1948.

Jordan, Camille: *Œuvres*, 4 vols., Gauthier-Villars, 1961–64.

———: *Traité des substitutions et des équations algébriques* (1870), Gauthier-Villars (reprint), 1957.

Kiernan, B. M.: "The Development of Galois Theory from Lagrange to Artin," *Archive for History of Exact Sciences*, 8, 1971, 40–154.

Lebesgue, Henri: *Notice sur la vie et les travaux de Camille Jordan*, Gauthier-Villars, 1923. Also in Lebesgue's *Notices d'histoire des mathématiques*, pp. 44–65, Institut de Mathématiques, Genève, 1958.

Miller, G. A.: "History of the Theory of Groups to 1900," *Collected Works*, Vol. 1, 427–67, University of Illinois Press, 1935.

Pierpont, James: "Lagrange's Place in the Theory of Substitutions," *Amer. Math. Soc. Bull.*, 1, 1894/95, 196–204.

———: "Early History of Galois's Theory of Equations," *Amer. Math. Soc. Bull.*, 4, 1898, 332–40.

Smith, David Eugene: *A Source Book in Mathematics*, Dover (reprint), 1959, Vol. 1, 232–52, 253–60, 261–66, 278–85.

Verriest, G.: *Œuvres mathématiques d'Evariste Galois* (1897 ed.), 2nd ed., Gauthier-Villars, 1951.

Wiman, A.: "Endliche Gruppen linearer Substitutionen," *Encyk. der Math. Wiss.*, B. G. Teubner, 1898–1904, I, Part 1, 522–54.

Wussing, H. L.: *Die Genesis des abstrakten Gruppenbegriffes*, VEB Deutscher Verlag der Wiss., 1969.

32

Quaternions, Vectors, and Linear Associative Algebras

> Quaternions came from Hamilton after his really good work had been done; and though beautifully ingenious, have been an unmixed evil to those who have touched them in any way. . . . Vector is a useless survival, or offshoot from quaternions, and has never been of the slightest use to any creature.
>
> LORD KELVIN

1. *The Foundation of Algebra on Permanence of Form*

Galois's work on equations solvable by algebraic processes closed a chapter of algebra and, though he introduced ideas such as group and domain of rationality (field) would bear fruit, the fuller exploitation of these ideas had to await other developments. The next major algebraic creation, initiated by William R. Hamilton, opened up totally new domains while shattering age-old convictions as to how "numbers" must behave.

To appreciate the originality of Hamilton's work we must examine the logic of ordinary algebra as it was generally understood in the first half of the nineteenth century. By 1800 the mathematicians were using freely the various types of real numbers and even complex numbers, but the precise definitions of these various types of numbers were not available nor was there any logical justification of the operations with them. Expressions of dissatisfaction with this state of affairs were numerous but were submerged in the mass of new creations in algebra and analysis. The greatest uneasiness seemed to be caused by the fact that letters were manipulated as though they had the properties of the integers; yet the results of these operations were valid when any numbers were substituted for the letters. Since the logic of the various types of numbers was not developed, it was not possible to see that they possessed the same formal properties as the positive integers, and consequently that literal expressions which merely stood for any class of real or complex numbers must possess the same properties—that is, that ordinary algebra is just generalized arithmetic. It seemed as though the algebra of literal expressions possessed a logic of its own, which accounted for its effectiveness and correctness. Hence in the 1830s the mathematicians tackled the problem of justifying the operations with literal or symbolic expressions.

This problem was first considered by George Peacock (1791–1858), professor of mathematics at Cambridge University. To justify the operations with literal expressions that could stand for negative, irrational, and complex numbers he made the distinction between arithmetical algebra and symbolical algebra. The former dealt with symbols representing the positive integers and so was on solid ground. Here only operations leading to positive integers were permissible. Symbolical algebra adopts the rules of arithmetical algebra but removes the restrictions to positive integers. All the results deduced in arithmetical algebra, whose expressions are general in form but particular in value, are results likewise in symbolical algebra where they are general in value as well as in form. Thus $a^m a^n = a^{m+n}$ holds in arithmetical algebra when m and n are positive integers and it therefore holds in symbolical algebra for all m and n. Likewise the series for $(a + b)^n$ when n is a positive integer, if it be exhibited in a general form without reference to a final term, holds for all n. Peacock's argument is known as the principle of the permanence of form.

The explicit formulation of this principle was given in Peacock's "Report on the Recent Progress and Present State of Certain Branches of Analysis,"[1] in which he does not merely report but dogmatically affirms. In symbolical algebra, he says:

1. The symbols are unlimited both in value and in representation.

2. The operations on them, whatever they may be, are possible in all cases.

3. The laws of combination of the symbols are of such a kind as to coincide universally with those in arithmetical algebra when the symbols are arithmetical quantities, and when the operations to which they are subject are called by the same names as in arithmetical algebra.

From these principles he believed he could deduce the principle of permanence of form: "Whatever algebraical forms are equivalent when the symbols are general in form but specific in value [positive integers], will be equivalent likewise when the symbols are general in value as well as in form." Peacock used this principle to justify in particular the operations with complex numbers. He did try to protect his conclusion by the phrase "when the symbols are general in form." Thus one could not state special properties of particular whole numbers in symbolic form and insist that these symbolic statements are general. For example, the decomposition of a composite integer into a product of primes, though expressed symbolically, could not

1. *Brit. Assn. for Adv. of Sci.*, Rept. 3, 1833, 185–352.

be taken over as a statement of symbolical algebra. The principle sanctioned by fiat what was evidently empirically correct but not yet logically established.

This principle Peacock reaffirmed in the second edition of his *Treatise on Algebra*,[2] but here he also introduces a formal science of algebra. In this *Treatise* Peacock states that algebra like geometry is a deductive science. The processes of algebra have to be based on a complete statement of the body of laws that dictate the operations used in the processes. The symbols for the operations have, at least for the deductive science of algebra, no sense other than those given to them by the laws. Thus addition means no more than any process that obeys the laws of addition in algebra. His laws were, for example, the associative and commutative laws of addition and multiplication and the law that if $ac = bc$ and $c \neq 0$ then $a = b$.

Here the principle of permanence of form was derived from the adoption of axioms. This approach paved the way for more abstract thinking in algebra and in particular influenced Boole's thinking on the algebra of logic.

Throughout most of the nineteenth century the view of algebra affirmed by Peacock was accepted. It was supported, for example, by Duncan F. Gregory (1813–44), a great-great-grandson of the seventeenth-century James Gregory. Gregory wrote in a paper "On the Real Nature of Symbolical Algebra":[3]

> The light then in which I would consider symbolical algebra is that it is the science which treats of the combination of operations defined not by their nature, that is, by what they are or what they do, but by the laws of combination to which they are subject. . . . It is true that these laws have been in many cases suggested (as Mr. Peacock has aptly termed it) by the laws of the known operations of number, but the step which is taken from arithmetical to symbolical algebra is that, leaving out of view the nature of the operations which the symbols we use represent, we suppose the existence of classes of unknown operations subject to the same laws. We are thus able to prove certain relations between the different classes of operations, which, when expressed between the symbols, are called algebraical theorems.

In this paper Gregory emphasized the commutative and distributive laws, the terms having been introduced by François-Joseph Servois (1767–1847).[4]

The theory of algebra as the science of symbols and the laws of their combinations was carried further by Augustus De Morgan, who wrote several papers on the structure of algebra.[5] His *Trigonometry and Double Algebra* (1849) also contains his views. The words double algebra meant the algebra of complex numbers, whereas single algebra meant negative numbers. Prior

2. 1842–45; 1st ed., 1830.
3. *Transactions of the Royal Society of Edinburgh*, 14, 1840, 208–16.
4. *Ann. de Math.*, 5, 1814/15, 93–140.
5. *Trans. Camb. Phil. Soc.*, 1841, 1842, 1844, and 1847.

to single algebra is universal arithmetic, which covers the algebra of the positive real numbers. Algebra, De Morgan maintained, is a collection of meaningless symbols and operations with symbols. The symbols are 0, 1, $+$, $-$, $\times$, $\div$, $(\ )^{()}$, and letters. The laws of algebra are the laws which such symbols obey, for example, the commutative law, the distributive law, the laws of exponents, a negative times a positive is negative, $a - a = 0$, $a \div a = 1$, and derived laws. The basic laws are arbitrarily chosen.

By the middle of the nineteenth century the axioms of algebra generally accepted were:

1. Equal quantities added to a third yield equal quantities.
2. $(a + b) + c = a + (b + c)$.
3. $a + b = b + a$.
4. Equals added to equals give equals.
5. Equals added to unequals give unequals.
6. $a(bc) = (ab)c$.
7. $ab = ba$.
8. $a(b + c) = ab + ac$.

The principle of permanence of form rested on these axioms.

It is hard for us to see just what this principle means. It begs the question of why the various types of numbers possess the same properties as the whole numbers. But Peacock, Gregory, and De Morgan sought to make a science out of algebra independent of the properties of real and complex numbers and so regarded algebra as a science of uninterpreted symbols and their laws of combination. In effect it was the justification for assuming that the same fundamental properties hold for all types of numbers. This foundation was not only vague but inelastic. The men insisted on a parallelism between arithmetical and general algebra so rigid that, if maintained, it would destroy the generality of algebra. They do not seem to have realized that a formula that is true with one interpretation of the symbols might not be true with another.

The principle of permanence of form, an arbitrary dictum, could not serve as a solid foundation for algebra. In fact, the developments we shall deal with in this chapter undermined it. The first step, which merely obviated the need for this principle so far as complex numbers were concerned, was made by Hamilton when he founded the logic of complex numbers on the properties of real numbers.

Though by 1830 complex numbers were intuitively well grounded through their representation as points or as directed line segments in the plane, Hamilton, who was concerned with the logic of arithmetic, was not satisfied with just an intuitive foundation. In his paper "Conjugate Functions and on Algebra as the Science of Pure Time,"[6] Hamilton pointed out that a

6. *Trans. Royal Irish Academy*, 17, 1837, 293–422 = *Math. Papers*, 3, 3–96.

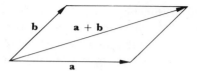

Figure 32.1

complex number $a + bi$ is not a genuine sum in the sense that $2 + 3$ is. The use of the plus sign is a historical accident and bi cannot be added to a. The complex number $a + bi$ is no more than an ordered couple (a, b) of real numbers. The peculiarity which i or $\sqrt{-1}$ introduces in the operations with complex numbers is incorporated by Hamilton in the definitions of the operations with ordered couples. Thus, if $a + bi$ and $c + di$ are two complex numbers then

$$(a, b) \pm (c, d) = (a \pm c, b \pm d)$$
$$(a, b) \cdot (c, d) = (ac - bd, ad + bc)$$

(1)

$$\frac{(a, b)}{(c, d)} = \left(\frac{ac + bd}{c^2 + d^2}, \frac{bc - ad}{c^2 + d^2} \right).$$

The usual associative, commutative, and distributive properties can now be deduced. Under this view of complex numbers, not only are these numbers logically founded on the basis of real numbers, but the hitherto somewhat mysterious $\sqrt{-1}$ is dispensed with entirely. Of course in practice it is still convenient to use the $a + bi$ form and to remember that $\sqrt{-1}\ \sqrt{-1} = -1$. Incidentally, Gauss did say in a letter of 1837 to Wolfgang Bolyai that he had had this notion of ordered couples in 1831. But it was Hamilton's publication that gave the ordered couple concept to the mathematical world.

2. *The Search for a Three-Dimensional "Complex Number"*

The notion of a vector, that is, a directed line segment that might represent the magnitude and direction of a force, a velocity, or an acceleration, entered mathematics quietly. Aristotle knew that forces can be represented as vectors and that the combined action of two forces can be obtained by what is commonly known as the parallelogram law; that is, the diagonal of the parallelogram formed by the two vectors **a** and **b** (Fig. 32.1), gives the magnitude and direction of the resultant force. Simon Stevin employed the parallelogram law in problems of statics, and Galileo stated the law explicitly.

After the geometric representation of complex numbers supplied by Wessel, Argand, and Gauss became somewhat familiar, the mathematicians realized that complex numbers could be used to represent and work with vectors in a plane. For example, if two vectors are represented respectively by say $3 + 2i$ and $2 + 4i$ then the sum of the complex numbers, namely,

$5 + 6i$, represents the sum of the vectors added by means of the parallelogram law. What the complex numbers do for vectors in a plane is to supply an algebra to represent the vectors and operations with vectors. One need not carry out the operations geometrically but can work with them algebraically much as the equation of a curve can be used to represent and work with curves.

This use of complex numbers to represent vectors in a plane became well known by 1830. However, the utility of complex numbers is limited. If several forces act on a body these forces need not lie in one plane. To treat these forces algebraically a three-dimensional analogue of complex numbers is needed. One could use the ordinary Cartesian coordinates (x, y, z) of a point to represent the vector from the origin to the point but there were no operations with the triples of numbers to represent the operations with vectors. These operations, as in the case of complex numbers, would seemingly have to include addition, subtraction, multiplication, and division and moreover obey the usual associative, commutative, and distributive laws so that algebraic operations could be applied freely and effectively. The mathematicians began a search for what was called a three-dimensional complex number and its algebra.

Wessel, Gauss, Servois, Möbius, and others worked on this problem. Gauss[7] wrote a short unpublished note dated 1819 on mutations of space. He thought of the complex numbers as displacements; $a + bi$ was a displacement of a units along one fixed direction followed by a displacement of b units in a perpendicular direction. Hence he tried to build an algebra of a three-component number in which the third component would represent a displacement in a direction perpendicular to the plane of $a + bi$. He arrived at a non-commutative algebra but it was not the effective algebra required by the physicists. Moreover, because he did not publish it, this work had little influence.

The creation of a useful spatial analogue of complex numbers is due to William R. Hamilton (1805–65). Next to Newton, Hamilton is the greatest of the English mathematicians and like Newton he was even greater as a physicist than as a mathematician. At the age of five Hamilton could read Latin, Greek, and Hebrew. At eight he added Italian and French; at ten he could read Arabic and Sanskrit and at fourteen, Persian. A contact with a lightning calculator inspired him to study mathematics. He entered Trinity College in Dublin in 1823 where he was a brilliant student. In 1822, at the age of seventeen he prepared a paper on caustics which was read before the Royal Irish Academy in 1824 but not published. Hamilton was advised to rework and expand it. In 1827 he submitted to the Academy a revision entitled "A Theory of Systems of Rays," which made a science of geometrical

7. *Werke*, 8, 357–62.

optics. Here he introduced what are called the characteristic functions of optics. The paper was published in 1828 in the *Transactions of the Royal Irish Academy*.[8]

In 1827, while still an undergraduate he was appointed Professor of Astronomy at Trinity College, an appointment that carried with it the title of Royal Astronomer of Ireland. His duties as professor were to lecture on science and to manage the astronomical observatory. He did not do much with the latter but he was a fine teacher.

In 1830 and 1832 he published three supplements to "A Theory of Systems of Rays." In the third paper[9] he predicted that a ray of light propagating in special directions in a biaxial crystal would give rise to a cone of refracted rays. This phenomenon was confirmed experimentally by Humphrey Lloyd, a friend and colleague. Hamilton then carried over to dynamics his ideas on optics and in the field of dynamics wrote two very famous papers (Chap. 30) in which he used the characteristic function concept which he had developed for optics. He also gave a system of complete and rigorous integrals for the differential equations of motion of a system of bodies. His major mathematical work was the subject of quaternions, which we shall discuss shortly. His final form of this work he presented in his *Lectures on Quaternions* (1853) and in a two-volume posthumously published *Elements of Quaternions* (1866)

Hamilton could use analogy skillfully to reason from the known to the unknown. Though he had a fine intuition, he did not have great flashes of ideas but worked long and hard on special problems to see what could be general. He was patient and systematic in working out many specific examples and was willing to undertake tremendous calculations to check or prove a point. However, in his publications one finds only the polished compressed general results.

He was deeply religious and this interest was most important to him. Next in order of importance were metaphysics, mathematics, poetry, physics, and general literature. He also wrote poetry. He thought that the geometrical ideas created in his time, the use of infinite elements and imaginary elements in the work of Poncelet and Chasles (Chap. 35), were akin to poetry. Though he was a modest man, he admitted and even emphasized that love of fame moves and cheers great mathematicians.

Hamilton's clarification of the notion of complex numbers enabled him to think more clearly about the problem of introducing a three-dimensional analogue to represent vectors in space. But the immediate effect was to frustrate his efforts. All of the numbers known to mathematicians at this time possessed the commutative property of multiplication and it was natural for

8. *Trans. Royal Irish Academy*, 15, 1828, 69–174 = *Math. Papers*, 1, 1–106.
9. *Trans. Royal Irish Academy*, 17, 1837, 1–144 = *Math. Papers*, 1, 164–293.

Hamilton to believe that the three-dimensional or three-component numbers he sought should possess this same property as well as the other properties that real and complex numbers possess. After some years of effort Hamilton found himself obliged to make two compromises. The first was that his new numbers contained four components and the second, that he had to sacrifice the commutative law of multiplication. Both features were revolutionary for algebra. He called the new numbers quaternions.

With hindsight one can see on geometric grounds that the new "numbers" had to contain four components. The new number regarded as an operator was expected to rotate a given vector about a given axis in space and to stretch or contract the vector. For these purposes, two parameters (angles) are needed to fix the axis of rotation, one parameter must specify the angle of rotation, and the fourth the stretch or contraction of the given vector.

Hamilton himself described his discovery of quaternions:[10]

> Tomorrow will be the fifteenth birthday of the Quaternions. They started into life, or light, full grown, on the 16th of October, 1843, as I was walking with Lady Hamilton to Dublin, and came up to Brougham Bridge. That is to say, I then and there felt the galvanic circuit of thought closed, and the sparks which fell from it were the fundamental equations between I, J, K; *exactly such* as I have used them ever since. I pulled out, on the spot, a pocketbook, which still exists, and made an entry, on which, *at the very moment*, I felt that it might be worth my while to expend the labour of at least ten (or it might be fifteen) years to come. But then it is fair to say that this was because I felt a *problem* to have been at that moment *solved*, an intellectual *want relieved*, which had *haunted* me for at least *fifteen years* before.

He announced the invention of quaternions in 1843 at a meeting of the Royal Irish Academy, spent the rest of his life developing the subject, and wrote many papers on it.

3. *The Nature of Quaternions*

A quaternion is a number of the form

(2)
$$3 + 2\mathbf{i} + 6\mathbf{j} + 7\mathbf{k}$$

wherein the $\mathbf{i}$, $\mathbf{j}$, and $\mathbf{k}$ play somewhat the role that i does in complex numbers. The real part, 3 above, is called the scalar part of the quaternion and the remainder, the vector part. The three coefficients of the vector part are rectangular Cartesian coordinates of a point P while $\mathbf{i}$, $\mathbf{j}$, and $\mathbf{k}$ are called qualitative units that geometrically are directed along the three axes. The criterion of equality of two quaternions is that their scalar parts shall be

10. *North British Review*, 14, 1858, 57.

equal and that the coefficients of their **i**, **j**, and **k** units shall be respectively equal. Two quaternions are added by adding their scalar parts and adding the coefficients of each of the **i**, **j**, and **k** units to form new coefficients for those units. The sum of two quaternions is therefore itself a quaternion.

All the familiar algebraic rules of multiplication are supposed valid in operating with quaternions except that in forming products of the units **i**, **j**, and **k** the following rules, which abandon the commutative law, hold:

$$\mathbf{jk} = \mathbf{i}, \quad \mathbf{kj} = -\mathbf{i}, \quad \mathbf{ki} = \mathbf{j}, \quad \mathbf{ik} = -\mathbf{j}, \quad \mathbf{ij} = \mathbf{k}, \quad \mathbf{ji} = -\mathbf{k}$$
$$(3) \qquad\qquad\qquad \mathbf{i}^2 = \mathbf{j}^2 = \mathbf{k}^2 = -1.$$

Thus if

$$\mathbf{p} = 3 + 2\mathbf{i} + 6\mathbf{j} + 7\mathbf{k} \quad \text{and} \quad \mathbf{q} = 4 + 6\mathbf{i} + 8\mathbf{j} + 9\mathbf{k}$$

then

$$\mathbf{pq} = (3 + 2\mathbf{i} + 6\mathbf{j} + 7\mathbf{k})(4 + 6\mathbf{i} + 8\mathbf{j} + 9\mathbf{k}) = -111 + 24\mathbf{i} + 72\mathbf{j} + 35\mathbf{k}$$

whereas

$$\mathbf{qp} = (4 + 6\mathbf{i} + 8\mathbf{j} + 9\mathbf{k})(3 + 2\mathbf{i} + 6\mathbf{j} + 7\mathbf{k}) = -111 + 28\mathbf{i} + 24\mathbf{j} + 75\mathbf{k}.$$

Hamilton proved that multiplication is associative. This is the first use of that term.[11]

Division of one quaternion by another can also be effected, but the fact that multiplication is not commutative implies that to divide the quaternion **p** by the quaternion **q** can mean to find **r** such that **p** = **qr** or such that **p** = **rq**. The quotient **r** need not be the same in the two cases. The problem of division is best handled by introducing $\mathbf{q}^{-1}$ or $1/\mathbf{q}$. If $\mathbf{q} = a + b\mathbf{i} + c\mathbf{j} + d\mathbf{k}$ one defines $\mathbf{q}'$ to be $a - b\mathbf{i} - c\mathbf{j} - d\mathbf{k}$ and $N(\mathbf{q})$, called the norm of **q**, to be $a^2 + b^2 + c^2 + d^2$. Then $N(\mathbf{q}) = \mathbf{qq}' = \mathbf{q}'\mathbf{q}$. By definition $\mathbf{q}^{-1} = \mathbf{q}'/N(\mathbf{q})$ and $\mathbf{q}^{-1}$ exists if $N(\mathbf{q}) \neq 0$. Also $\mathbf{qq}^{-1} = 1$ and $\mathbf{q}^{-1}\mathbf{q} = 1$. Now to find the **r** such that **p** = **qr** we have $\mathbf{q}^{-1}\mathbf{p} = \mathbf{q}^{-1}\mathbf{qr}$ or $\mathbf{r} = \mathbf{q}^{-1}\mathbf{p}$. To find the **r** such that **p** = **rq** we have $\mathbf{pq}^{-1} = \mathbf{rqq}^{-1}$ or $\mathbf{r} = \mathbf{pq}^{-1}$.

That quaternions can be used to rotate and stretch or contract a given vector into another given vector is readily shown. One must merely show that one can determine a, b, c, and d such that

$$(a + b\mathbf{i} + c\mathbf{j} + d\mathbf{k})(x\mathbf{i} + y\mathbf{j} + z\mathbf{k}) = (x'\mathbf{i} + y'\mathbf{j} + z'\mathbf{k}).$$

By multiplying out the left side as quaternions and equating corresponding coefficients of the left and right sides we get four equations in the unknowns a, b, c, and d. These four equations suffice to determine the unknowns.

Hamilton also introduced an important differential operator. The symbol ∇, which is an inverted Δ—which Hamilton termed "nabla"

11. *Proc. Royal Irish Academy*, 2, 1844, 424–34 = *Math. Papers*, 3, 111–16.

because it looks like an ancient Hebrew musical instrument of that name—stands for the operator

(4) $$\nabla = \mathbf{i}\,\frac{\partial}{\partial x} + \mathbf{j}\,\frac{\partial}{\partial y} + \mathbf{k}\,\frac{\partial}{\partial z}.$$

When applied to a scalar point function $u(x, y, z)$ it produces the vector

(5) $$\nabla u = \frac{\partial u}{\partial x}\,\mathbf{i} + \frac{\partial u}{\partial y}\,\mathbf{j} + \frac{\partial u}{\partial z}\,\mathbf{k}.$$

This vector, which varies from point to point of space, is now called the gradient of u. It represents in magnitude and direction the greatest space rate of increase of u.

Also letting $\mathbf{v} = v_1\mathbf{i} + v_2\mathbf{j} + v_3\mathbf{k}$ denote a continuous vector point function, wherein v_1, v_2, and v_3 are functions of x, y, and z, Hamilton introduced

$$\nabla\mathbf{v} = \left(\mathbf{i}\,\frac{\partial}{\partial x} + \mathbf{j}\,\frac{\partial}{\partial y} + \mathbf{k}\,\frac{\partial}{\partial z}\right)(v_1\mathbf{i} + v_2\mathbf{j} + v_3\mathbf{k})$$

(6) $$= -\left(\frac{\partial v_1}{\partial x} + \frac{\partial v_2}{\partial y} + \frac{\partial v_3}{\partial k}\right) + \left(\frac{\partial v_3}{\partial y} - \frac{\partial v_2}{\partial z}\right)\mathbf{i} + \left(\frac{\partial v_1}{\partial z} - \frac{\partial v_3}{\partial x}\right)\mathbf{j}$$

$$+ \left(\frac{\partial v_2}{\partial x} - \frac{\partial v_1}{\partial y}\right)\mathbf{k}.$$

Thus the result of operating with ∇ on a vector point function $\mathbf{v}$ is to produce a quaternion; the scalar part of this quaternion (except for the minus sign) is what we now call the divergence of $\mathbf{v}$ and the vector part is now called curl $\mathbf{v}$.

Hamilton's enthusiasm for his quaternions was unbounded. He believed that this creation was as important as the calculus and that it would be the key instrument in mathematical physics. He himself made some applications to geometry, optics, and mechanics. His ideas were enthusiastically endorsed by his friend Peter Guthrie Tait (1831–1901), professor of mathematics at Queen's College and later professor of natural history at the University of Edinburgh. In many articles Tait urged physicists to adopt quaternions as the basic tool. He even became involved in long arguments with Cayley, who took a dim view of the usefulness of quaternions. But the physicists ignored quaternions and continued to work with conventional Cartesian coordinates. Nevertheless, as we shall see, Hamilton's work did lead indirectly to an algebra and analysis of vectors that physicists eagerly adopted.

Hamilton's quaternions proved to be of immeasurable importance for algebra. Once mathematicians realized that a meaningful, useful system of "numbers" could be built up which may fail to possess the commutative property of real and complex numbers, they felt freer to consider creations which departed even more from the usual properties of real and complex numbers. This realization was necessary before vector algebra and analysis could be developed because vectors violate more ordinary laws of algebra than do quaternions (sec. 5). More generally, Hamilton's work led to the theory of linear associative algebras (sec. 6). Hamilton himself began work on hypernumbers which contain n components or n-tuples,[12] but it was his work on quaternions which stimulated the new work on linear algebras.

4. Grassmann's Calculus of Extension

While Hamilton was developing his quaternions, another mathematician, Hermann Günther Grassmann (1809–77), who showed no talent for mathematics as a youth and who had no university education in mathematics but later became a teacher of mathematics in the *gymnasium* (high school) at Stettin, Germany, as well as an authority on Sanskrit, was developing an even more audacious generalization of complex numbers. Grassmann had his ideas before Hamilton but did not publish until 1844, one year after Hamilton announced his discovery of quaternions. In that year he published his *Die lineale Ausdehnungslehre* (The Calculus of Extension). Because he shrouded the ideas with mystic doctrines and the exposition was abstract, the more practical-minded mathematicians and physicists found the book vague and unreadable, and as a consequence the work, though highly original, remained little known for years. Grassmann issued a revised edition, entitled *Die Ausdehnungslehre*, in 1862. In it he simplified and amplified the original work but his style and lack of clarity still repelled readers.

Though Grassmann's exposition was almost inextricably bound up with geometrical ideas—he was in fact concerned with n-dimensional geometry—we shall abstract the algebraic notions that proved to be of lasting value. His basic notion, which he called an extensive quantity (*extensive Grösse*), is one type of hypernumber with n components. To study his ideas we shall discuss the case $n = 3$.

Consider two hypernumbers

$$\alpha = \alpha_1 e_1 + \alpha_2 e_2 + \alpha_3 e_3 \quad \text{and} \quad \beta = \beta_1 e_1 + \beta_2 e_2 + \beta_3 e_3,$$

where the α_i and β_i are real numbers and where e_1, e_2, and e_3 are primary or qualitative units represented geometrically by direct line segments of unit length drawn from a common origin so as to determine a right-handed orthogonal system of axes. The $\alpha_i e_i$ are multiples of the primary units and are represented geometrically by lengths α_i along the respective axes, while α is represented by a directed line segment in space whose projections on the axes are the lengths α_i. The same is true for the β_i and β. Grassmann called the directed line segments or line-vectors *Strecke*.

12. *Trans. Royal Irish Academy*, 21, 1848, 199–296 = *Math. Papers*, III, 159–226.

The addition and subtraction of these hypernumbers are defined by

$$(7) \qquad \alpha \pm \beta = (\alpha_1 \pm \beta_1)e_1 + (\alpha_2 \pm \beta_2)e_2 + (\alpha_3 \pm \beta_3)e_3.$$

Grassmann introduced two kinds of multiplications, the inner product and the outer product. For the inner product he postulated that

$$(8) \qquad e_i|e_i = 1, \, e_i|e_j = 0 \qquad \text{for } i \neq j.$$

For the outer product

$$(9) \qquad [e_ie_j] = -[e_je_i], \, [e_ie_i] = 0.$$

These brackets are called units of the second order and are not reduced by Grassmann (whereas Hamilton does) to units of the first order, that is, to the e_i, but are dealt with as though they were equivalent to first order units with $[e_1e_2] = e_3$, and so forth.

From these definitions it follows that the inner product $\alpha|\beta$ of α and β is given by

$$\alpha|\beta = \alpha_1\beta_1 + \alpha_2\beta_2 + \alpha_3\beta_3 \quad \text{and} \quad \alpha|\beta = \beta|\alpha.$$

The numerical value or magnitude a of a hypernumber α is defined as $\sqrt{\alpha|\alpha} = \sqrt{\alpha_1^2 + \alpha_2^2 + \alpha_3^2}$. Thus the magnitude of α is numerically equal to the length of the line-vector which represents it geometrically. If θ denotes the angle between the line-vectors α and β, then

$$\alpha|\beta = ab\left(\frac{\alpha_1\beta_1}{ab} + \frac{\alpha_2\beta_2}{ab} + \frac{\alpha_3\beta_3}{ab}\right) = ab\cos\theta.$$

With the aid of the outer product rule (9) the outer product P of the hypernumbers α and β can be expressed as follows:

$$(10) \quad P = [\alpha\beta] = (\alpha_2\beta_3 - \alpha_3\beta_2)[e_2e_3] + (\alpha_3\beta_1 - \alpha_1\beta_3)[e_3e_1] \\ + (\alpha_1\beta_2 - \alpha_2\beta_1)[e_1e_2].$$

This product is a hypernumber of the second order and is expressed in terms of independent units of the second order. Its magnitude $|P|$ is obtained by means of a definition of the inner product of two hypernumbers of the second order and is

$$|P| = \sqrt{P|P} = \{(\alpha_2\beta_3 - \alpha_3\beta_2)^2 + (\alpha_3\beta_1 - \alpha_1\beta_3)^2 + (\alpha_1\beta_2 - \alpha_2\beta_1)^2\}^{1/2}$$

$$(11) \qquad = ab\left\{1 - \left(\frac{\alpha_1\beta_1}{ab} + \frac{\alpha_2\beta_2}{ab} + \frac{\alpha_3\beta_3}{ab}\right)^2\right\}^{1/2}$$

$$= ab\sin\theta.$$

Hence the magnitude $|P|$ of the outer product $[\alpha\beta]$ is represented geometrically by the area of the parallelogram constructed upon line-vectors which are the geometrical representations of α and β. This area together with a unit line-vector normal to it, whose direction is chosen so that if α is rotated into β about the normal, then the normal will point in the direction of a right-handed screw rotating from α to β, is now called a vectorial area. Grassmann's term was *Plangrösse*.

Grassmann's inner product of two primary hypernumbers for three dimensions is equivalent to the negative of the scalar part of Hamilton's quaternion product of two vectors; and again in the three-dimensional case, Grassmann's outer product, if we replace $[e_2 e_3]$ by e_1 and so forth, is precisely Hamilton's quaternion product of two vectors. However, in the theory of quaternions the vector is a subsidiary part of the quaternion whereas in Grassman's algebra the vector appears as the basic quantity.

Another product was formed by Grassmann by taking the scalar (inner) product of a hypernumber γ with the vector (outer) product $[\alpha\beta]$ of two hypernumbers α and β. This product Q for the three-dimensional case is

$$Q = [\alpha\beta]\gamma = (\alpha_2\beta_3 - \alpha_3\beta_2)\gamma_1 + (\alpha_3\beta_1 - \alpha_1\beta_3)\gamma_2 + (\alpha_1\beta_2 - \alpha_2\beta_1)\gamma_3.$$

In determinant form

$$(12) \qquad\qquad Q = \begin{vmatrix} \alpha_1 & \beta_1 & \gamma_1 \\ \alpha_2 & \beta_2 & \gamma_2 \\ \alpha_3 & \beta_3 & \gamma_3 \end{vmatrix}.$$

Consequently Q can be interpreted geometrically as the volume of a parallelepiped constructed on the line-vectors that represent α, β, and γ. The volume may be positive or negative.

Grassmann considered (for n-component hypernumbers) not only the two types of products described above but also products of higher order. In a paper of 1855,[13] he gave sixteen different types of products for hypernumbers. He also gave the geometrical significance of the products and made applications to mechanics, magnetism, and crystallography.

It might seem as though Grassmann's treatment of hypernumbers of n parts was needlessly general since thus far at least the useful instances of hypernumbers contained at most four parts. Yet Grassmann's thinking helped to lead mathematicians into the theory of tensors (Chap. 48), for tensors, as we shall see, are hypernumbers. Other geometrical ideas and the idea of invariance had yet to make themselves known to mathematicians before the day of tensors was to come. Though thinking about hypernumbers

13. *Jour. für Math.*, 49, 1855, 10–20 and 123–41 = *Ges. Math. und Phys. Werke*, 2, Part I, 199–217.

did lead to various generalizations, no analysis (e.g. calculus) for Grassmann's *n*-dimensional hypernumbers was ever developed. The reason is simply that no applications for such an analysis were found. As we shall see, there is an extensive analysis for tensors but these have their origin in Riemannian geometry.

5. *From Quaternions to Vectors*

Grassmann's work remained neglected for a while whereas, as we have noted, quaternions did receive a great deal of attention almost at once. However, they were not quite what the physicists wanted. They sought a concept that was not divorced from but more closely associated with Cartesian coordinates than quaternions were. The first step in the direction of such a concept was made by James Clerk Maxwell (1831–79), the founder of electromagnetic theory, one of the greatest of mathematical physicists, and professor of physics at Cambridge University.

Maxwell knew Hamilton's work and though he had heard of Grassmann's work he had not seen it. He singled out the scalar and vector parts of Hamilton's quaternion and put the emphasis on these separate notions.[14] However, in his celebrated *A Treatise on Electricity and Magnetism* (1873) he made a greater concession to quaternions and speaks rather of the scalar and vector parts of a quaternion though he does treat these parts as separate entities. A vector, he says (p. 10), requires three quantities (components) for its specification and these can be interpreted as lengths along the three coordinate axes. This vector concept is the vector part of Hamilton's quaternion, and Maxwell states so. Hamilton had introduced a vector function **v** of x, y, and z with components v_1, v_2, and v_3, had applied to it the operator $\nabla = \mathbf{i}(\partial/\partial x) + \mathbf{j}(\partial/\partial y) + \mathbf{k}(\partial/\partial z)$, and obtained the result (6). Thus $\nabla\mathbf{v}$ is a quaternion. But Maxwell separated the scalar part from the vector part and indicated these by $S\nabla\mathbf{v}$ (the scalar part of $\nabla\mathbf{v}$) and $V\nabla\mathbf{v}$ (the vector part of $\nabla\mathbf{v}$). He called $S\nabla\mathbf{v}$ the convergence of **v** because this expression had already appeared many times in fluid dynamics and when **v** is velocity had the meaning of flux or the net quantity per unit volume per unit time which flows through a small area surrounding a point. And he called $V\nabla\mathbf{v}$ the rotation or curl of **v** because this expression too had appeared in fluid dynamics as twice the rate of rotation of the fluid at a point. Clifford later called $-S\nabla\mathbf{v}$ the divergence.

Maxwell then points out that the operator ∇ repeated, gives

$$\nabla^2 = -\frac{\partial^2}{\partial x^2} - \frac{\partial^2}{\partial y^2} - \frac{\partial^2}{\partial z^2}$$

14. *Proc. London Math. Soc.*, 3, 1871, 224–32 = *The Scientific Papers*, Vol. 2, 257–66.

which he calls Laplace's operator. He allows it to act on a scalar function to produce a scalar and on a vector function to produce a vector.[15]

Maxwell noted in his 1871 paper that the curl of a gradient of a scalar function and the divergence of the curl of a vector function are always zero. He also states that the curl of the curl of a vector function $\mathbf{v}$ is the gradient of the divergence of $\mathbf{v}$ minus the Laplacian of $\mathbf{v}$. (This is true in rectangular coordinates only.)

Maxwell often used quaternions as the basic mathematical entity or he at least made frequent reference to quaternions, perhaps to help his readers. Nevertheless his work made clear that vectors were the real tool for physical thinking and not just an abbreviated scheme of writing as some maintained. Thus by Maxwell's time a great deal of vector analysis was created by treating the scalar and vector parts of quaternions separately.

The formal break with quaternions and the inauguration of a new independent subject, three-dimensional vector analysis, was made independently by Josiah Willard Gibbs and Oliver Heaviside in the early 1880s. Gibbs (1839–1903), professor of mathematical physics at Yale College but primarily a physical chemist, had printed (1881 and 1884) for private distribution among his students a small pamphlet on the *Elements of Vector Analysis*.[16] His viewpoint is set forth in an introductory note:

> The fundamental principles of the following analysis are such as are familiar under a slightly different form to students of quaternions. The manner in which the subject is developed is somewhat different from that followed in treatises on quaternions, being simply to give a suitable notation for those relations between vectors, or between vectors and scalars, which seem most important, and which lend themselves most readily to analytical transformations, and to explain some of these transformations. As a precedent for such a departure from quaternionic usage Clifford's *Kinematics* may be cited. In this connection the name Grassmann may be also mentioned, to whose system the following method attaches itself in some respects more closely than to that of Hamilton.

Although printed for private circulation, Gibbs's pamphlet on vector analysis became widely known. The material was finally incorporated in a book

15. Since in vector analysis $\nabla^2 = \nabla \cdot \nabla$, then $\nabla^2 q$ means physically the divergence of the gradient or the divergence of the maximum space rate of change of q. However, the physical meaning is clearer from the fact that the function q which satisfies $\nabla^2 q = 0$ minimizes the Dirichlet integral (see (34) of Chap. 28). The integral is the square of the magnitude of the gradient taken over some volume. Hence $\nabla^2 q = 0$ means that the minimum gradient holds at any point or that the departure from uniformity is a minimum. If $\nabla^2 q$ is not 0 there must be some departure from uniformity and there will be a restoring force. The various equations of mathematical physics, which contain $\nabla^2 q$ in one context or another, state in effect that nature always acts to restore uniformity.

The notation Δ for ∇^2 was introduced by Robert Murphy in 1833.

16. *The Scientific Papers*, 2, 17–90.

written by E. B. Wilson and based on Gibbs's lectures. The book, Gibbs and Wilson: *Vector Analysis*, appeared in 1901.

Oliver Heaviside (1850–1925) was in the early part of his scientific career a telegraph and telephone engineer. He retired to country life in 1874 and devoted himself to writing, principally on the subject of electricity and magnetism. Heaviside had studied quaternions in Hamilton's *Elements* but had been repelled by the many special theorems. He felt that quaternions were too hard for busy engineers to learn and so he built his own vector analysis, which to him was just a shorthand form of ordinary Cartesian coordinates. In papers written during the 1880s in the journal *Electrician* he used this vector analysis freely. Then in his three-volume work *Electromagnetic Theory* (1893, 1899, 1912) he gave a good deal of vector algebra in Volume I. The third chapter of about 175 pages is devoted to vector methods. His development of the subject was essentially in harmony with Gibb's although he did not like Gibbs's notation and adopted his own based on Tait's quaternionic notation.

As formulated by Gibbs and Heaviside, a vector is no more than the vector part of a quaternion but considered independently of any quaternions. Thus a vector $\mathbf{v}$ is

$$\mathbf{v} = a\mathbf{i} + b\mathbf{j} + c\mathbf{k},$$

where $\mathbf{i}$, $\mathbf{j}$, and $\mathbf{k}$ are unit vectors along the x-, y-, and z-axes respectively. The coefficients a, b, and c are real numbers and are called components. Two vectors are equal if the respective components are equal and the sum of two vectors is the vector which has as its components the sum of the respective components of the summands.

Two types of multiplication, both physically useful, were introduced. The first type known as scalar multiplication is defined thus: We multiply $\mathbf{v}$ and $\mathbf{v}' = a'\mathbf{i} + b'\mathbf{j} + c'\mathbf{k}$ as ordinary polynomials and, using the dot as the symbol for multiplication, set

(13a)
$$\mathbf{i}\cdot\mathbf{i} = \mathbf{j}\cdot\mathbf{j} = \mathbf{k}\cdot\mathbf{k} = 1$$
$$\mathbf{i}\cdot\mathbf{j} = \mathbf{j}\cdot\mathbf{i} = \mathbf{i}\cdot\mathbf{k} = \mathbf{k}\cdot\mathbf{i} = \mathbf{j}\cdot\mathbf{k} = \mathbf{k}\cdot\mathbf{j} = 0.$$

Thus $\mathbf{v}\cdot\mathbf{v}' = aa' + bb' + cc'$. This product is no longer a vector but a real number or scalar and is called the scalar product. It possesses a new algebraic feature, for the product of two real numbers or complex numbers or quaternions is always a number of the same kind as we started with. Another surprising property of the scalar product is that it may be zero when neither factor is zero. For example, the product of the vectors $\mathbf{v} = 3\mathbf{i}$ and $\mathbf{v}' = 6\mathbf{j} + 7\mathbf{k}$ is zero.

The scalar product of two vectors is algebraically novel in still another respect—it does not permit an inverse process. That is, we cannot always find a vector or scalar q such that $\mathbf{v}/\mathbf{v}' = q$. Thus if $\mathbf{q}$ were a vector, $\mathbf{q}\cdot\mathbf{v}'$ would

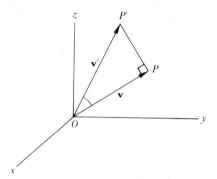

Figure 32.2

be a scalar and not equal to the vector **v**. On the other hand if q were a scalar then, though $q\mathbf{v}'$ is defined as $qa'\mathbf{i} + qb'\mathbf{j} + qc'\mathbf{k}$, it would be rare that $qa' = a$, $qb' = b$ and $qc' = c$ where a, b, and c are the coefficients in **v**. Despite the absence of a quotient the scalar product is useful.

The physical significance of the scalar product is readily shown to be the following: If **v**′ is a force (Fig. 32.2) whose direction and magnitude are represented by the line segment from O to P' then the effectiveness of this force in pushing an object at O, say, in the direction of OP, where OP represents **v**, is the projection of OP' on OP or OP' cos ϕ where ϕ is the angle between OP' and OP. If OP is of unit length, the projection of OP' is precisely the value of the product $\mathbf{v} \cdot \mathbf{v}'$.

The second kind of product of vectors, called the vector product, is defined as follows: We again multiply **v** and **v**′ as polynomials but this time let

$$\mathbf{i} \times \mathbf{i} = \mathbf{j} \times \mathbf{j} = \mathbf{k} \times \mathbf{k} = 0$$

(13b) $\mathbf{i} \times \mathbf{j} = \mathbf{k}, \qquad \mathbf{j} \times \mathbf{i} = -\mathbf{k}, \qquad \mathbf{j} \times \mathbf{k} = \mathbf{i}, \qquad \mathbf{k} \times \mathbf{j} = -\mathbf{i},$

$$\mathbf{k} \times \mathbf{i} = \mathbf{j}, \qquad \mathbf{i} \times \mathbf{k} = -\mathbf{j}.$$

Thus the product, which is indicated by $\mathbf{v} \times \mathbf{v}'$, is

$$\mathbf{v} \times \mathbf{v}' = (bc' - b'c)\mathbf{i} + (ca' - ac')\mathbf{j} + (ab' - b'a)\mathbf{k}.$$

The vector product of two vectors is a vector. Its direction can readily be shown to be perpendicular to that of **v** and **v**′ and pointed in the direction that a right-hand screw moves when it is turned from **v** to **v**′ through the smaller angle. The vector product of two parallel vectors is zero, though neither factor is. Moreover, this product, like the quaternion product, is not commutative. Further, it is not even associative. For example, $\mathbf{i} \times \mathbf{j} \times \mathbf{j}$ can mean $(\mathbf{i} \times \mathbf{j}) \times \mathbf{j} = \mathbf{k} \times \mathbf{j} = -\mathbf{i}$ or $\mathbf{i} \times (\mathbf{j} \times \mathbf{j}) = \mathbf{i} \times 0 = 0$.

There is no inverse to vector multiplication. For the quotient of **v** by **v**′ to be a vector **q** we would have to have

$$\mathbf{v} = \mathbf{v}' \times \mathbf{q}$$

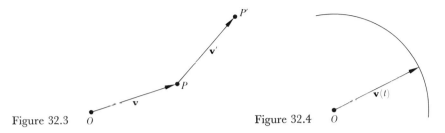

Figure 32.3 O Figure 32.4 O

and this would require, whatever $\mathbf{q}$ be, that $\mathbf{v}'$ be perpendicular to $\mathbf{v}$ which may not be the case to start with. If q were a scalar it would be accidental that $qa'\mathbf{i} + qb'\mathbf{j} + qc'\mathbf{k}$ would equal $\mathbf{v}$.

The vector product, like the scalar product, is suggested by physical situations. Let OP and PP' in Figure 32.3 be the lengths and directions of $\mathbf{v}$ and $\mathbf{v}'$. If $\mathbf{v}'$ is a force whose magnitude and direction are those of PP', the measure of the moment of force exerted by $\mathbf{v}'$ around O is the length of the vector $\mathbf{v} \times \mathbf{v}'$ and is usually taken to have the direction of $\mathbf{v} \times \mathbf{v}'$.

The algebra of vectors is extended to variable vectors and a calculus of vectors. For example, the variable vector $\mathbf{v}(t) = a(t)\mathbf{i} + b(t)\mathbf{j} + c(t)\mathbf{k}$, where $a(t)$, $b(t)$, and $c(t)$ are functions of t, is a vector function. If the various vectors one gets for various values of t are drawn from O as origin (Fig. 32.4) the endpoints of these vectors will trace out a curve. Hence the vector function of a scalar variable t plays a role analogous to the ordinary function

$$y = x^2 + 7,$$

say, and a calculus of vectors has been developed for these vector functions just as there is one for ordinary functions.

The concepts of gradient u,

$$(14) \qquad \nabla u = \frac{\partial u}{\partial x}\mathbf{i} + \frac{\partial u}{\partial y}\mathbf{j} + \frac{\partial u}{\partial z}\mathbf{k},$$

where u is a scalar function of $x, y,$ and z, the divergence of a vector function $\mathbf{v}$,

$$(15) \qquad \nabla \cdot \mathbf{v} = \frac{\partial v_1}{\partial x} + \frac{\partial v_2}{\partial y} + \frac{\partial v_3}{\partial z},$$

where $v_1, v_2,$ and v_3 are the components of $\mathbf{v}$, and the curl of $\mathbf{v}$,

$$(16) \qquad \nabla \times \mathbf{v} = \left(\frac{\partial v_3}{\partial y} - \frac{\partial v_2}{\partial z}\right)\mathbf{i} + \left(\frac{\partial v_1}{\partial z} - \frac{\partial v_3}{\partial x}\right)\mathbf{j} + \left(\frac{\partial v_2}{\partial x} - \frac{\partial v_1}{\partial y}\right)\mathbf{k},$$

were abstracted from quaternions.

Many basic theorems of analysis can be expressed in vector form. Thus in the course of solving the partial differential equation of heat,

Ostrogradsky[17] made use of the following conversion of volume integral to surface integral:

$$\iiint_V \left(\frac{\partial P}{\partial x} + \frac{\partial Q}{\partial y} + \frac{\partial R}{\partial z}\right) dx\, dy\, dz = \iint_S (P \cos \lambda + Q \cos \mu + R \cos \nu)\, dS.$$

Here P, Q, and R are functions of x, y, and z and are components of a vector, and λ, μ, and ν are the direction cosines of the normal to the surface S which bounds the volume V over which the left-hand integral is taken. This theorem, known as the divergence theorem (also Gauss's theorem and Ostrogradsky's theorem) can be expressed in vector form thus: If $\mathbf{F}$ is the vector whose components are P, Q, and R and $\mathbf{n}$ is the normal to S then

(17)
$$\iiint \nabla \cdot \mathbf{F}\, dv = \iint \mathbf{F} \cdot \mathbf{n}\, dS.$$

Likewise Stokes's theorem, which was first stated by him as a question in an examination for the Smith prize at Cambridge in 1854,[18] states in scalar form that

$$\iint_S \left\{ \lambda\left(\frac{\partial R}{\partial y} - \frac{\partial Q}{\partial z}\right) + \mu\left(\frac{\partial P}{\partial z} - \frac{\partial R}{\partial x}\right) + \nu\left(\frac{\partial Q}{\partial x} - \frac{\partial P}{\partial y}\right) \right\} dS$$
$$= \int_C \left(P\frac{\partial x}{\partial s} + Q\frac{\partial y}{\partial s} + R\frac{\partial z}{\partial s}\right) ds,$$

where S is any portion of a surface, C is the curve bounding S, and $x(s)$, $y(s)$, and $z(s)$ are the parametric representation of C. In vector form Stokes's theorem reads

(18)
$$\iint \operatorname{curl} \mathbf{F} \cdot \mathbf{n}\, dS = \int \mathbf{F} \cdot \frac{d\mathbf{r}}{ds}\, ds,$$

where $\mathbf{r}(s)$ is the vector whose components are $x(s)$, $y(s)$, and $z(s)$.

When Maxwell wrote expressions and equations of electromagnetism, especially the equations which now bear his name, he usually wrote out the components of the vectors involved in grad u, div $\mathbf{v}$, and curl $\mathbf{v}$. However, Heaviside wrote Maxwell's equations in vector form (Chap. 28, (52)).

It is true that calculations with vectors and vector functions are often made by resort to Cartesian components but it is highly important to think also in terms of the single entity the vector and in terms of gradient, divergence, and curl. These have a direct physical significance, to say nothing of the fact that complicated technical steps can be performed directly with the

17. *Mém. Acad. Sci. St. Peters.*, (6), 1, 1831, 39–53.
18. The theorem was stated by Lord Kelvin in a letter to Stokes of July 1850.

vectors as when one replaces $\nabla \cdot (\mathbf{u}(x, y, z) \times \mathbf{v}(x, y, z))$ by its equivalent, $\mathbf{v} \cdot \nabla \times \mathbf{u} - \mathbf{u} \cdot \nabla \times \mathbf{v}$. Also integral definitions of gradient, divergence, and curl have been given which make these concepts independent of any coordinate definition. Thus in place of (14) we have, for example,

$$\text{grad } u = \lim_{\Delta\tau \to 0} \frac{1}{\Delta\tau} \int_S u\mathbf{n} \, dS,$$

where S is the boundary of a volume element $\Delta\tau$ and $\mathbf{n}$ is the normal to the surface element dS of S.

While vector analysis was being created and afterward there was much controversy between the proponents of quaternions and the proponents of vectors as to which was more useful. The quaternionists were fanatical about the value of quaternions but the proponents of vector analysis were equally partisan. On one side were aligned the leading supporters of quaternions such as Tait and, on the other, Gibbs and Heaviside. Apropos of the controversy, Heaviside remarked sarcastically that for the treatment of quaternions, quaternions are the best instrument. On the other hand Tait described Heaviside's vector algebra as "a sort of hermaphrodite monster, compounded of the notations of Grassmann and Hamilton." Gibbs's book proved to be of inestimable value in promoting the vector cause.

The issue was finally decided in favor of vectors. Engineers welcomed Gibbs's and Heaviside's vector analysis, though the mathematicians did not. By the beginning of the present century the physicists too were quite convinced that vector analysis was what they wanted. Textbooks on the subject soon appeared in all countries and are now standard. The mathematicians finally followed suit and introduced vector methods in analytic and differential geometry.

The influence of physics in stimulating the creation of such mathematical entities as quaternions, Grassmann's hypernumbers, and vectors should be noted. These creations became part of mathematics. But their significance extends beyond the addition of new subjects. The introduction of these several quantities opened up a new vista—there is not just the one algebra of real and complex numbers but many and diverse algebras.

6. Linear Associative Algebras

From the purely algebraic standpoint quaternions were exciting because they furnished an example of an algebra that had the properties of real numbers and complex numbers except for commutativity of multiplication. During the second half of the nineteenth century many systems of hypernumbers were explored largely to see what varieties could be created and yet retain many properties of the real and complex numbers.

Cayley[19] gave an eight-unit generalization of real quaternions. His units were $1, e_1, e_2, \ldots, e_7$ with

$$e_i^2 = -1, \qquad e_i e_j = -e_j e_i \qquad \text{for } i, j = 1, 2, \cdots, 7 \quad \text{and} \quad i \neq j$$
$$e_1 e_2 = e_3, \qquad e_1 e_4 = e_5, \qquad e_1 e_6 = e_7$$
$$e_2 e_5 = e_7, \qquad e_2 e_4 = -e_6, \qquad e_3 e_4 = e_7, \qquad e_3 e_5 = e_6$$

and the fourteen equations obtained from these last seven by permuting each set of three subscripts cyclically; e.g. $e_2 e_3 = e_1$; $e_3 e_1 = e_2$.

A general number (octonion) x is defined by

$$x = x_0 + x_1 e_1 + \cdots + x_7 e_7,$$

wherein the x_i are real numbers. The norm of x, $N(x)$, is by definition

$$N(x) = x_0^2 + x_1^2 + \cdots + x_7^2.$$

The norm of a product equals the product of the norms. The associative law of multiplication fails in general (as does the commutative law of multiplication). Right- and left-hand division, except by zero, are always possible and unique, a fact overlooked by Cayley and proved by Leonard Eugene Dickson.[20] In later papers Cayley gave other algebras of hypernumbers differing somewhat from the one above.

Hamilton in his *Lectures on Quaternions*[21] also introduced biquaternions, that is, quaternions with complex coefficients. He noted that the product law does not hold for these biquaternions; that is, two non-zero biquaternions may have a product of zero.

William Kingdon Clifford (1845–79), professor of mathematics and mechanics at University College, London, created another type of hyper-number,[22] which he also called biquaternions. If q and Q are real quaternions and if ω is such that $\omega^2 = 1$, and ω commutes with every real quaternion, then $q + \omega Q$ is a biquaternion. Clifford's biquaternions satisfy the product law of multiplication but the multiplication is not associative. In the latter work Clifford introduced the algebras which bear his name. There are Clifford algebras with units, $1, e_1, \ldots, e_{n-1}$ such that the square of each $e_i = -1$ and $e_i e_j = -e_j e_i$ for $i \neq j$. Each product of two or more units is a new unit and so there are 2^n different units. All products are associative. A form is a scalar multiplied by a unit and an algebra is generated by the sum and product of forms.

The flood of new systems of hypernumbers continued to rise and the variety became enormous. In a paper "Linear Associative Algebra" read in

19. *Phil. Mag.*, (3), 26, 1845, 210–13 and 30, 1847, 257–58 = *Coll. Math. Papers*, 1, 127 and 301.

20. *Amer. Math. Soc. Trans.*, 13, 1912, 59–73.

21. 1853, p. 650.

22. *Proc. Lond. Math. Soc.*, 4, 1873, 381–95 = *Coll. Math. Papers*, 181–200 and *Amer. Jour. of Math.*, 1, 1878, 350–58 = *Coll. Math. Papers*, 266–76.

1870 and published in lithographed form in 1871,[23] Benjamin Peirce (1809–80), professor of mathematics at Harvard University, defined and gave a résumé of the linear associative algebras already known by that time. The word linear means that the product of any two primary units is reduced to one of the units just as $\mathbf{i}$ times $\mathbf{j}$ is replaced by $\mathbf{k}$ in quaternions and the word associative means that the multiplication is associative. Addition in these algebras has the usual properties of real and complex numbers. In this paper Peirce introduced the idea of a nilpotent element, that is, an element A such that $A^n = 0$ for some positive integral n, and an idempotent element, that is, $A^n = 1$ for some n. He also showed that an algebra in which at least one element is not nilpotent possesses an idempotent element.

The question of how much freedom there could be in the variety of such algebras had also occurred to mathematicians during the very period in which they were creating specific algebras. Gauss was convinced (*Werke*, 2, 178) that an extension of complex numbers that preserved the basic properties of complex numbers was impossible. It is significant that when Hamilton searched for a three-dimensional algebra to represent vectors in space and settled for quaternions with their lack of commutativity he could not prove that there was no three-dimensional commutative algebra. Nor did Grassman have such a proof.

Later in the century the precise theorems were formulated. In 1878 F. Georg Frobenius (1849–1917)[24] proved that the only linear associative algebras with real coefficients (of the primary units), with a finite number of primary units, a unit element for multiplication, and obeying the product law are those of the real numbers, the complex numbers and real quaternions. The theorem was proved independently by Charles Sanders Peirce (1839–1914) in an appendix to his father's paper.[25] Another key result, which Weierstrass arrived at in 1861, states that the only linear associative algebras with real or complex coefficients (of the primary units), with a finite number of primary units, obeying the product law and with commutative multiplication are those of the real numbers and the complex numbers. Dedekind obtained the same result about 1870. Weierstrass's result was published in 1884[26] and Dedekind's in the next year.[27]

In 1898 Adolf Hurwitz (1859–1919)[20] showed that the real numbers, complex numbers, real quaternions, and Clifford's biquaternions are the only linear associative algebras satisfying the product law.

These theorems are valuable because they tell us what we can expect in

23. *Amer. Jour. of Math.*, 4, 1881, 97–229.
24. *Jour. für Math.*, 84, 1878, 1–63 = *Ges. Abh.*, 1, 343–405.
25. *Amer. Jour. of Math.*, 4, 1881, 225–29.
26. *Nachrichten König. Ges. der Wiss. zu Gött.*, 1884, 395–410 = *Math. Werke*, 2, 311–32.
27. *Nachrichten König. Ges. der Wiss. zu Gött.*, 1885, 141–59 and 1887, 1–7 = *Werke*, 2, 1–27.
28. *Nachrichten König. Ges. der Wiss. zu Gött.*, 1898, 309–16 = *Math. Werke*, 2, 565–71.

extensions of the complex number system if we wish to preserve at least some of its algebraic properties. Had Hamilton known these theorems he would have saved years of labor in his search for a three-dimensional vector algebra.

The study of linear algebras with a finite and even infinite number of generating (primary) units and with or without division continued to be an active subject well into the twentieth century. Such men as Leonard Eugene Dickson and J. H. M. Wedderburn contributed much to the subject.

Bibliography

Clifford, W. K.: *Collected Mathematical Papers* (1882), Chelsea (reprint), 1968.

Collins, Joseph V.: "An Elementary Exposition of Grassmann's *Ausdehnungslehre*," *Amer. Math. Monthly*, 6, 1899, several parts; and 7, 1900, several parts.

Coolidge, Julian L.: *A History of Geometrical Methods*, Dover (reprint), 1963, pp. 252–64.

Crowe, Michael J.: *A History of Vector Analysis*, University of Notre Dame Press, 1967.

Dickson, Leonard E.: *Linear Algebras*, Cambridge University Press, 1914.

Gibbs, Josiah W., and E. B. Wilson: *Vector Analysis* (1901), Dover (reprint), 1960.

Grassmann, H. G.: *Die lineale Ausdehnungslehre* (1844), Chelsea (reprint), 1969.

———: *Gesammelte mathematische und physikalische Werke*, 3 vols., B. G. Teubner, 1894–1911; Vol. 1, Part I, 1–319 contains *Die lineale Ausdehnungslehre*; Vol. 1, Part II, 1–383 contains *Die Ausdehnungslehre*.

Graves, R. P.: *Life of Sir William Rowan Hamilton*, 3 vols., Longmans Green, 1882–89.

Hamilton, Sir Wm. R.: *Elements of Quaternions*, 2 vols., 1866, 2nd ed., 1899–1901, Chelsea (reprint), 1969.

———: *Mathematical Papers*, Cambridge University Press, 1967, Vol. 3.

———: "Papers in Memory of Sir William R. Hamilton," *Scripta Math.*, 1945; also in *Scripta Math.*, 10, 1944, 9–80.

Heaviside, Oliver: *Electromagnetic Theory*, Dover (reprint), 1950, Vol. 1.

Klein, Felix: *Vorlesungen über die Entwicklung der Mathematik im 19. Jahrhundert*, Chelsea (reprint), 1950, Vol. 1, pp. 167–91; Vol. 2, pp. 2–12.

Maxwell, James Clerk: *The Scientific Papers*, 2 vols., Dover (reprint), 1965.

Peacock, George: "Report on the Recent Progress and Present State of Certain Branches of Analysis," *British Assn. for Advancement of Science Report for 1833*, London, 1834.

———: *A Treatise on Algebra*, 2 vols., 2nd ed., Cambridge University Press, 1845; Scripta Mathematica (reprint), 1940.

Shaw, James B.: *Synopsis of Linear Associative Algebra*, Carnegie Institution of Washington, 1907.

Smith, David Eugene: *A Source Book in Mathematics*, Dover (reprint), 1959, Vol. 2, pp. 677–96.

Study, E.: "Theorie der gemeinen und höheren complexen Grössen," *Encyk. der Math. Wiss.*, B. G. Teubner, 1898, I, 147–83.

33
Determinants and Matrices

> Such is the advantage of a well-constructed language that its simplified notation often becomes the source of profound theories.
> P. S. LAPLACE

1. *Introduction*

Though determinants and matrices received a great deal of attention in the nineteenth century and thousands of papers were written on these subjects they do not constitute great innovations in mathematics. The concept of a vector, which from the mathematical standpoint is no more than a collection of ordered triples, nevertheless has a direct physical significance as a force or velocity, and with it one can write down at once mathematically what the physics states. The same applies with all the more cogency to gradient, divergence, and curl. Likewise though mathematically dy/dx is no more than a symbol for a lengthy expression involving the limit of $\Delta y/\Delta x$, the derivative is in itself a powerful concept that enables us to think directly and creatively about physical happenings. Thus though mathematics superficially regarded is no more than a language or a shorthand, its most vital concepts are those that supply keys to new realms of thought. By contrast, determinants and matrices are solely innovations in language. They are shorthand expressions for ideas which already exist in more expanded form. In themselves they say nothing directly that is not already said by equations or transformations albeit in lengthier fashion. Neither determinants nor matrices have influenced deeply the course of mathematics despite their utility as compact expressions and despite the suggestiveness of matrices as concrete groups for the discernment of general theorems of group theory. Nevertheless, both concepts have proved to be highly useful tools and are now part of the apparatus of mathematics.

2. *Some New Uses of Determinants*

Determinants arose in the solution of systems of linear equations (Chap. 25, sec. 3). This problem and elimination theory, transformation of coordinates, change of variables in multiple integrals, solution of systems of differential equations arising in planetary motion, reduction of quadratic forms in three or more variables and of pencils of forms (a pencil is the set $A + \lambda B$, where A and B are specific forms and λ is a parameter) to standard forms all gave rise to various new uses of determinants. This nineteenth-century work followed up directly on the work of Cramer, Bezout, Vandermonde, Lagrange, and Laplace.

The word determinant, used by Gauss for the discriminant of the quadratic form $ax^2 + 2bxy + cy^2$, was applied by Cauchy to the determinants that had already appeared in the eighteenth-century work. The arrangement of the elements in a square array and the double subscript notation are also due to him.[1] Thus a third order determinant is written as (the vertical lines were introduced by Cayley in 1841)

$$(1) \qquad \begin{vmatrix} a_{11} & a_{12} & a_{12} \\ a_{21} & a_{22} & a_{23} \\ a_{31} & a_{32} & a_{33} \end{vmatrix}.$$

In this paper Cauchy gave the first systematic and almost modern treatment of determinants. One of the major results is the multiplication theorem for determinants. Lagrange[2] had already given this theorem for third order determinants but because the rows of his determinant were the coordinates of the vertices of a tetrahedron he was not motivated to generalize. With Cauchy the general theorem, expressed in modern notation, states

$$(2) \qquad |a_{ij}| \cdot |b_{ij}| = |c_{ij}|,$$

where $|a_{ij}|$ and $|b_{ij}|$ stand for nth-order determinants and $c_{ij} = \sum_k a_{ik} b_{kj}$. That is, the term in the ith row and jth column of the product is the sum of the products of the corresponding elements in the ith row of $|a_{ij}|$ and the kth column of $|b_{ij}|$. This theorem had been stated but not satisfactorily proved by Jacques P. M. Binet (1786–1856) in 1812.[3] Cauchy also gave an improved statement and a proof of Laplace's expansion theorem for determinants (Chap. 25, sec. 3).

Heinrich F. Scherk (1798–1885) in his *Mathematische Abhandlungen* (1825) gave several new properties of determinants. He formulated the rules for adding two determinants that have a row or column in common and for multiplying a determinant by a constant. He also stated that the determinant of an array which has as a row a linear combination of two or more other rows is zero, and that the value of a triangular determinant (all the elements

1. *Jour. de l'Ecole Poly.*, 10, 1815, 29–112 = *Œuvres*, (2), 1, 91–169.
2. *Nouv. Mém. de l'Acad. de Berlin*, 1773, 85–128 = *Œuvres*, 3, 577–616.
3. *Jour. de l'Ecole Poly.*, 9, 1813, 280–302.

above or below the main diagonal are zero) is the product of the elements on the main diagonal.

One of the consistent workers in determinant theory over a period of more than fifty years was James Joseph Sylvester (1814–97). After winning the second wranglership in the mathematical tripos he was nevertheless barred from teaching at Cambridge University because he was Jewish. From 1841 to 1845 he was a professor at the University of Virginia. He then returned to London and served from 1845 to 1855 as an actuary and a lawyer. The relatively lowly position of professor in a military academy in Woolwich, England, was offered to him and he served there until 1871. A few years of miscellaneous activity were followed by a professorship at Johns Hopkins University where, starting in 1876, he lectured on invariant theory. He initiated research in pure mathematics in the United States and founded the *American Journal of Mathematics*. In 1884 he returned to England, and at the age of seventy became a professor at Oxford University, a post he held until his death.

Sylvester was a lively, sensitive, stimulating, passionate, and even excitable person. His speeches were brilliant and witty and he presented his ideas enthusiastically and with fire. In his papers he used glowing language. He introduced much new terminology and jokingly likened himself to Adam who gave names to the beasts and the flowers. Though he related many diverse fields, such as mechanics and the theory of invariants, he was not given to systematic and thoroughly worked-out theories. In fact he frequently published guesses, and though many of these were brilliant others were incorrect. He ruefully acknowledged that his friends on the Continent "complimented his powers of divination at the expense of his judgment." His chief contributions were to combinatorial ideas and to abstractions from more concrete developments.

One of Sylvester's major accomplishments was an improved method of eliminating x from an nth degree and an mth degree polynomial. He called it the dialytic method.[4] Thus to eliminate x from the equations

$$a_0 x^3 + a_1 x^2 + a_2 x + a_3 = 0$$
$$b_0 x^2 + b_1 x + b_2 = 0$$

he formed the fifth order determinant

(3)
$$\begin{vmatrix} a_0 & a_1 & a_2 & a_3 & 0 \\ 0 & a_0 & a_1 & a_2 & a_3 \\ b_0 & b_1 & b_2 & 0 & 0 \\ 0 & b_0 & b_1 & b_2 & 0 \\ 0 & 0 & b_0 & b_1 & b_2 \end{vmatrix}.$$

4. *Phil. Mag.*, 16, 1840, 132–35 and 21, 1842, 534–39 = *Coll. Math. Papers*, 1, 54–57 and 86–90.

The vanishing of this determinant is the necessary and sufficient condition that the two equations have a common root. Sylvester gave no proof. The method leads, as Cauchy showed,[5] to the same result as Euler's and Bezout's methods.

The formula for the derivative of a determinant when the elements are functions of t was first given by Jacobi in 1841.[6] If the a_{ij} are functions of t, A_{ij} is the cofactor of a_{ij}, and D is the determinant then

$$\frac{\partial D}{\partial a_{ij}} = A_{ij}$$

and

$$\frac{dD}{dt} = \sum_{i,j} A_{ij} a'_{ij},$$

where the prime denotes differentiation with respect to t.

Determinants were employed in another connection, namely, in the change of variables in a multiple integral. Special results were found first by Jacobi (1832 and 1833). Then Eugène Charles Catalan (1814–94) in 1839[7] gave the result familiar today to students of the calculus. Thus the double integral,

(4)
$$\iint F(x, y)\, dx\, dy$$

under the change of variables given by

(5)
$$x = f(u, v), \qquad y = g(u, v)$$

becomes

(6)
$$\iint G(u, v) \begin{vmatrix} f_u & f_v \\ g_u & g_v \end{vmatrix} du\, dv,$$

where $G(u, v) = F(x(u, v), y(u, v))$. The determinant in (6) is called the Jacobian of x and y with respect to u and v or the functional determinant.

Jacobi devoted a major paper[8] to functional determinants. In this paper Jacobi considers n functions $u_1, \ldots, u_n$ each of which is a function of $x_1, x_2, \ldots, x_n$ and raises the question of when, from such n functions, the x_i can be eliminated so that the u_i's are connected by one equation. If this is not possible the functions u_i are said to be independent. The answer is that

5. *Exercices d'analyse et de physique mathématique*, 1, 1840, 385–422 = *Œuvres*, (2), 11, 466–509.
6. *Jour. für Math.*, 22, 1841, 285–318 = *Werke*, 3, 355–92.
7. *Mémoires couronnés par l'Académie Royale des Sciences et Belles-Lettres de Bruxelles*, 14, 1841.
8. *Jour. für Math.*, 22, 1841, 319–59 = *Werke*, 3, 393–438.

if the Jacobian of the u_i with respect to the x_i vanishes the functions are not independent and conversely. He also gives the product theorem for Jacobians. That is, if the u_i are functions of y_i and the y_i are functions of the x_i, then the Jacobian of the u_i with respect to the x_i is the product of the Jacobian of the u_i with respect to the y_i and the Jacobian of the y_i with respect to the x_i.

3. Determinants and Quadratic Forms

The problem of transforming equations of the conic sections and quadric surfaces to simpler forms by choosing coordinate axes which have the directions of the principal axes had been introduced in the eighteenth century. The classification of quadric surfaces in terms of the signs of the second degree terms when the equation is in standard or canonical form, that is, when the principal axes are the coordinate axes, was given by Cauchy in his *Leçons sur les applications du calcul infinitésimal à la géométrie* (1826).[9] However, it was not clear that the same number of positive and negative terms always results from this reduction to standard form. Sylvester answered this question with his law of inertia of quadratic forms in n variables.[10] It was already known that

$$(7) \qquad \sum_{i,j=1}^{n} a_{ij} x_i x_j$$

can always be reduced to a sum of r squares[11]

$$(8) \qquad y_1^2 + \cdots + y_s^2 - y_{s+1}^2 - \cdots - y_{r-s}^2$$

by a real linear transformation

$$x_i = \sum_j b_{ij} y_j, \qquad i = 1, 2, \cdots, n,$$

with non-vanishing determinant. Sylvester's law states that the number s of positive terms and $r - s$ negative ones is always the same no matter what real transformation is used. Regarding the law as self-evident, Sylvester gave no proof.

The law was rediscovered and proved by Jacobi.[12] If a form is positive or zero for all real values of the variables, it is called positive definite. Then all the signs in (8) are positive, and $r = n$. It is called semidefinite if it can take on positive and negative values (in which case $r < n$), and negative

9. *Œuvres*, (2), 5, 244–85.

10. *Phil. Mag.*, (4), 4, 1852, 138–42 = *Coll. Math. Papers*, 1, 378–81.

11. r is the rank of the form, that is, the rank of the matrix of the coefficients. For the notion of rank see section 4.

12. *Jour. für Math.*, 53, 1857, 265–70 = *Werke*, 3, 583–90; see also 593–98.

definite when the form is always negative or zero (and $r = n$). These terms
were introduced by Gauss in his *Disquisitiones Arithmeticae* (sec. 271).

The further study of the reduction of quadratic forms involves the
notion of the characteristic equation of a quadratic form or of a determinant.
A quadratic form in three variables was usually written in the eighteenth
century and first half of the nineteenth century as

$$Ax^2 + By^2 + Cz^2 + 2Dxy + 2Exz + 2Fyz$$

and in more recent times as

(9) $$a_{11}x_1^2 + a_{22}x_2^2 + a_{33}x_3^2 + 2a_{12}x_1x_2 + 2a_{13}x_1x_3 + 2a_{22}x_2x_3.$$

In the latter notation the form has associated with it the determinant

(10) $$\begin{vmatrix} a_{11} & a_{12} & a_{13} \\ a_{21} & a_{22} & a_{23} \\ a_{31} & a_{32} & a_{33} \end{vmatrix}, \qquad a_{ij} = a_{ji}.$$

The characteristic equation or latent equation of the form or the determinant
is

(11) $$\begin{vmatrix} a_{11} - \lambda & a_{12} & a_{13} \\ a_{21} & a_{22} - \lambda & a_{23} \\ a_{31} & a_{32} & a_{33} - \lambda \end{vmatrix} = 0,$$

and the values of λ which satisfy this equation are called the characteristic
roots or latent roots. From these values of λ the lengths of the principal axes
are readily obtained.[13]

The notion of the characteristic equation appears implicitly in the work
of Euler[14] on the reduction of quadratic forms in three variables to their
principal axes though he failed to prove the reality of the characteristic
roots. The notion of characteristic equation first appears explicitly in
Lagrange's work on systems of linear differential equations,[15] and in
Laplace's work in the same area.[16]

Lagrange, in dealing with the system of differential equations for the
motion of the six planets known in his day, was concerned with the secular
(long-period) perturbations which these exerted on each other. His characteristic equation (also called the secular equation) was that for a sixth order

13. By a linear transformation $x_i' = \sum_j m_{ij}x_j$, $i, j = 1, 2, 3$, the form (9) can be reduced
to $\sum_{j=1}^{3} \lambda_j x_j'^2$ where the λ_j are the characteristic roots of (11). In matrix language the
matrix M of the transformation is orthogonal; that is, the transpose of M equals the
inverse of M.
14. Chapter 5 of the Appendix to his *Introductio* (1748) = *Opera*, (1), 9, 379–92.
15. *Misc. Taur.*, 3, 1762–65 = *Œuvres*, 1, 520–34, and *Mém. de l'Acad. des Sci., Paris,*
1774 = *Œuvres*, 6, 655–66.
16. *Mém. de l'Acad. des Sci., Paris,* 1772, pub. 1775 = *Œuvres*, 8, 325–66.

determinant and the values of λ determined solutions of the system. He was able to decompose the sixth degree equation and obtain information about the roots. Laplace in his *Mécanique céleste* showed that if the planets all move in the same direction then the six characteristic roots are real and distinct. The reality of the characteristic values for the quadratic problem in three variables was established by Hachette, Monge, and Poisson.[17]

Cauchy recognized the common characteristic value problem in the work of Euler, Lagrange, and Laplace. In his *Leçons* of 1826[18] he took up the problem of the reduction of a quadratic form in three variables and showed that the characteristic equation is invariant for any change of rectangular axes. Three years later in his *Exercices de mathématiques*[19] he took up the problem of the secular inequalities of the planetary paths. In the course of this work he showed that two quadratic forms in n variables

$$A = \sum_{i,j=1}^{n} a_{ij}x_ix_j, \qquad B = \sum_{i,j=1}^{n} b_{ij}x_ix_j$$

(Cauchy's B was a sum of squares) could be reduced simultaneously by a linear transformation

$$x_1 = c_{11}x_1' + \cdots + c_{1n}x_n'$$
$$\cdot \quad \cdot \quad \cdot \quad \cdot \quad \cdot \quad \cdot$$
$$x_n = c_{n1}x_1' + \cdots + c_{nn}x_n'$$

to a sum of squares. He also solved the problem of finding the principal axes for forms in any number of variables and in this work he again used the notion of characteristic roots.

His work amounts to the following: If A and B are any two given quadratic forms then one can consider the pencil (family, *Schaar*) of forms $uA + vB$, where u and v are arbitrary parameters. The latent roots of the pencil are the values of the ratio $-u/v$ for which the determinant of the pencil $|ua_{ij} + vb_{ij}|$ is zero. Cauchy proved that the latent roots are all real in the special case when one of the forms is positive definite for all real non-zero values of the variables. Since the determinant of $uA + vB$ is symmetric $(d_{ij} = d_{ji})$ and B could be the identity determinant $(b_{ij} = 0$ for $i \neq j$ and $b_{ii} = 1)$, Cauchy proved that any real symmetric determinant of any order has real characteristic roots. Cauchy's results, duplicated by Jacobi in 1834,[20] excluded equal latent roots. The term characteristic equation is due to Cauchy.[21]

17. Hachette and Monge: *Jour. de l'Ecole Poly.*, 4, 1801–2, 143–69; Poisson and Hachette, *ibid.*, 170–72.
18. *Œuvres*, (2), 5, 244–85.
19. 4, 1829, 140–60 = *Œuvres*, (2), 9, 174–95.
20. *Jour. für Math.*, 12, 1834, 1–69 = *Werke*, 3, 191–268.
21. *Exercices d'analyse et de physique mathématique*, 1, 1840, 53 = *Œuvres*, (2), 11, 76.

The notion of similar determinants also arose from work on transformations. Two determinants A and B are similar if there exists a non-zero determinant P such that $A = P^{-1}BP$. Similar transformations were considered by Cauchy who showed in his *Leçons* of 1826 [22] that they have the same characteristic values. The importance of similar transformations lies in classifying projective transformations (Chap. 38, sec. 5), a problem that for a long time was treated synthetically. If a figure F is related to a figure G by a linear transformation A and if another such transformation B transforms F into F' and G into G' the transformation C which carries F' into G' will have the same properties as A. The transformation $C = BAB^{-1}$ because B^{-1} carries F' into F, A carries F into G, and B carries G into G'.

In 1858 Weierstrass [23] gave a general method of reducing two quadratic forms simultaneously to sums of squares. He also proved that if one of the forms is positive definite the reduction is possible even when some of the latent roots are equal. Weierstrass's interest in this problem arose from the dynamical problem of small oscillations about a position of equilibrium, and he showed by means of his work on quadratic forms that stability is not destroyed by the presence of equal periods in the system, contrary to the suppositions of Lagrange and Laplace.

Sylvester in 1851, [24] working with the contact and the intersection of curves and surfaces of the second degree was led to consider the classification of pencils of such conics and quadric surfaces. In particular he sought the canonical form of any pencil. Writing a pencil in the form $A + \lambda B$ where

$$A = ax^2 + by^2 + cz^2 + 2dyz + 2ezx + 2fxy$$
$$B = Ax^2 + By^2 + Cz^2 + 2Dyz + 2Ezx + 3Fxy$$

he considered the determinant

(12)
$$\begin{vmatrix} a + \lambda A & f + \lambda F & e + \lambda E \\ f + \lambda F & b + \lambda B & d + \lambda D \\ e + \lambda E & d + \lambda D & c + \lambda C \end{vmatrix}.$$

His method of classification introduced the notion of elementary divisors. The elements of the determinant of $A + \lambda B$ are polynomials in λ. Sylvester proved that if all the minors of any one order of $|A + \lambda B|$ have a factor $\lambda + \varepsilon$ in common, this factor will continue to be common to the same system of minors when A and B are simultaneously transformed by a linear transformation of their variables. He also showed that if all the ith order minors have a factor $(\lambda + \varepsilon)^\alpha$, the $(i + h)$-order minors will contain the factor $(\lambda + \varepsilon)^{(h+1)\alpha}$. The various linear factors to the power to which they

22. *Œuvres*, (2), 5, 244–85.
23. *Monatsber. Berliner Akad.*, 1858, 207–20 = *Werke*, 1, 233–46.
24. *Phil. Mag.*, (4), 1, 1851, 119–40 = *Coll. Math. Papers*, 1, 219–40.

occur in the greatest common divisor $D_i(\lambda)$ of the ith order minors for each i are the *elementary divisors* of $|A + \lambda B|$ or of any general determinant. The quotients of $D_i(\lambda)$ by $D_{i-1}(\lambda)$ for each i are called the invariant factors of $|A + \lambda B|$. Sylvester did not prove that the invariant factors constitute a complete set of invariants for the two quadratic forms.

Weierstrass[25] completed the theory of quadratic forms and extended the theory to bilinear forms, a bilinear form being

$$a_{11}x_1y_1 + a_{12}x_1y_2 + \cdots + a_{nn}x_ny_n.$$

Using Sylvester's notion of elementary divisors Weierstrass obtained the canonical form for a pencil $A + \lambda B$, where A and B are not necessarily symmetric but subject to the condition that $|A + \lambda B|$ is not identically zero. He also proved the converse of a theorem due to Sylvester. The converse states that if the determinant of $A + \lambda B$ agrees in its elementary divisors with the determinant of $A' + \lambda B'$ then a pair of linear transformations can be found which will simultaneously transform A into A' and B into B'.

Among the multitude of theorems on determinants are some concerned with the solution of m linear equations in n unknowns. Henry J. S. Smith (1826–83)[26] introduced the terms augmented array and unaugmented array to discuss the existence and number of solutions of, for example, the system

$$a_1x + a_2y = f$$
$$b_1x + b_2y = g$$
$$c_1x + c_2y = h.$$

The augmented and unaugmented arrays are

$$\begin{Vmatrix} a_1 & a_2 & f \\ b_1 & b_2 & g \\ c_1 & c_2 & h \end{Vmatrix} \quad \text{and} \quad \begin{Vmatrix} a_1 & a_2 \\ b_1 & b_2 \\ c_1 & c_2 \end{Vmatrix}.$$

A series of results due to many men including Kronecker and Cayley led to the general results now usually stated in terms of matrices but which in the middle of the nineteenth century were stated in terms of augmented and unaugmented determinants. The general results on m equations in n unknowns with m greater than, equal to, or less than n—the equations can be homogeneous (the constant terms are zero) or nonhomogeneous—are stated, for example, in Charles L. Dodgson's (Lewis Carroll, 1832–1898) *An Elementary Theory of Determinants* (1867). In later texts one finds the present condition: In order that a set of m nonhomogeneous linear equations in n unknowns may be consistent it is necessary and sufficient that the highest order non-vanishing minor be of the same order in the unaugmented and

25. *Monatsber. Berliner Akad.*, 1868, 310–38 = *Werke*, 2, 19–44.
26. *Phil. Trans.*, 151, 1861, 293–326 = *Coll. Math. Papers*, 1, 367–409.

augmented arrays, or in terms of matrix language that the ranks of the two matrices be the same.

New results on determinants were obtained throughout the nineteenth century. As an illustration there is the theorem proved by Hadamard in 1893,[27] though known and proved by many others before and after this date. If the elements of the determinant $D = |a_{ij}|$ satisfy the condition $|a_{ij}| \leq A$ then

$$|D| \leq A^n \cdot n^{n/2}.$$

The above theorems on determinants are but a small sample of the multitude that have been established. Beyond a great variety of other theorems on general determinants there are hundreds of others on determinants of special form such as symmetric determinants $(a_{ij} = a_{ji})$, skew-symmetric determinants $(a_{ij} = -a_{ji})$, orthogonants (determinants of orthogonal coordinate transformations), bordered determinants (determinants extended by the addition of rows and columns), compound determinants (the elements are themselves determinants), and many other special types.

4. Matrices

One could say that the subject of matrices was well developed before it was created. Determinants, as we know, were studied from the middle of the eighteenth century onward. A determinant contains an array of numbers and usually one is concerned with the value of that array, given by the definition of the determinant. However, it was apparent from the immense amount of work on determinants that the array itself could be studied and manipulated for many purposes whether or not the value of the determinant came into question. It remained then to recognize that the array as such should be given an identity independent of the determinant. The array itself is called a matrix. The word matrix was first used by Sylvester[28] when in fact he wished to refer to a rectangular array of numbers and could not use the word determinant, though he was at that time concerned only with the determinants that could be formed from the elements of the rectangular array. Later, as we have already noted in the preceding section, augmented arrays were used freely without any mention of matrices. The basic properties of matrices, as we shall see, were also established in the development of determinants.

It is true, as Arthur Cayley insisted (in the 1855 paper referred to below), that logically the idea of a matrix precedes that of a determinant but historically the order was the reverse and this is why the basic properties of

27. *Bull. des Sci. Math.*, (2), 17, 1893, 240–46 = *Œuvres*, 1, 239–45.
28. *Phil. Mag.*, (3), 37, 1850, 363–70 = *Coll. Math. Papers*, 1, 145–51.

matrices were already clear by the time that matrices were introduced. Thus the general impression among mathematicians that matrices were a highly original and independent creation invented by pure mathematicians when they divined the potential usefulness of the idea is erroneous. Because the uses of matrices were well established it occurred to Cayley to introduce them as distinct entities. He says, "I certainly did not get the notion of a matrix in any way through quaternions; it was either directly from that of a determinant or as a convenient way of expression of the equations:

$$x' = ax + by$$
$$y' = cx + dy."$$

And so he introduced the matrix

$$\begin{pmatrix} a & b \\ c & d \end{pmatrix}$$

which represents the essential information about the transformation. Because Cayley was the first to single out the matrix itself and the first to publish a series of articles on the subject he is generally credited with being the creator of the theory of matrices.

Cayley, born in 1821 of an old and talented English family, showed mathematical ability in school. His teachers persuaded his father to send him to Cambridge instead of putting him in the family business. At Cambridge he was senior wrangler in the mathematical tripos and won the Smith prize. He was elected a fellow of Trinity College in Cambridge and assistant tutor, but left after three years because he would have had to take holy orders. He turned to law and spent the next fifteen years in that profession. During this period he managed to devote considerable time to mathematics and published close to 200 papers. It was during this period, too, that he began his long friendship and collaboration with Sylvester.

In 1863, he was appointed to the newly created Sadlerian professorship of mathematics at Cambridge. Except for the year 1882, spent at Johns Hopkins University at the invitation of Sylvester, he remained at Cambridge until his death in 1895. He was a prolific writer and creator in various subjects, notably the analytic geometry of n dimensions, determinant theory, linear transformations, skew surfaces, and matrix theory. Together with Sylvester, he was the founder of the theory of invariants. For these numerous contributions he received many honors.

Unlike Sylvester, Cayley was a man of even temper, sober judgment, and serenity. He was generous in help and encouragement to others. In addition to his fine work in law and prodigious accomplishments in mathematics, he found time to develop interests in literature, travel, painting, and architecture.

It was in connection with the study of invariants under linear trans-formations (Chap. 39, sec. 2) that Cayley first introduced matrices to simplify the notation involved.[29] Here he gave some basic notions. This was followed by his first major paper on the subject, "A Memoir on the Theory of Matrices."[30]

For brevity we shall state Cayley's definitions for 2 by 2 or 3 by 3 matrices though the definitions apply to n by n matrices and in some cases to rectangular matrices. Two matrices are equal if their corresponding elements are equal. Cayley defines the zero matrix and unit matrix as

$$\begin{pmatrix} 0 & 0 & 0 \\ 0 & 0 & 0 \\ 0 & 0 & 0 \end{pmatrix} \text{ and } \begin{pmatrix} 1 & 0 & 0 \\ 0 & 1 & 0 \\ 0 & 0 & 1 \end{pmatrix}.$$

The sum of two matrices is defined as the matrix whose elements are the sums of the corresponding elements of the two summands. He notes that this definition applies to any two m by n matrices and that addition is associative and convertible (commutative). If m is a scalar and A a matrix then mA is defined as the matrix whose elements are each m times the corresponding element of A.

The definition of multiplication of two matrices Cayley took directly from the representation of the effect of two successive transformations. Thus if the transformation

$$x' = a_{11}x + a_{12}y$$
$$y' = a_{21}x + a_{22}y$$

is followed by the transformation

$$x'' = b_{11}x' + b_{12}y'$$
$$y'' = b_{21}x' + b_{22}y'$$

then the relation between x'', y'' and x, y is given by

$$x'' = (b_{11}a_{11} + b_{12}a_{21})x + (b_{11}a_{12} + b_{12}a_{22})y$$
$$y'' = (b_{21}a_{11} + b_{22}a_{21})x + (b_{21}a_{12} + b_{22}a_{22})y.$$

Hence Cayley defined the product of the matrices to be

$$\begin{pmatrix} b_{11} & b_{12} \\ b_{21} & b_{22} \end{pmatrix} \begin{pmatrix} a_{11} & a_{12} \\ a_{21} & a_{22} \end{pmatrix} = \begin{pmatrix} b_{11}a_{11} + b_{12}a_{21} & b_{11}a_{12} + b_{12}a_{22} \\ b_{21}a_{11} + b_{22}a_{21} & b_{21}a_{12} + b_{22}a_{22} \end{pmatrix}.$$

That is, the element c_{ij} in the product is the sum of the products of the elements in the ith row of the left-hand factor and the corresponding

29. *Jour. für Math.*, 50, 1855, 282–85 = *Coll. Math. Papers*, 2, 185–88.
30. *Phil. Trans.*, 148, 1858, 17–37 = *Coll. Math. Papers*, 2, 475–96.

elements of the jth column of the right-hand factor. Multiplication is associative but not generally commutative. Cayley points out that an m by n matrix can be compounded only with an n by p matrix.

In this same article, he states that the inverse of

$$
\begin{pmatrix} a, & b, & c \\ a', & b', & c' \\ a'', & b'', & c'' \end{pmatrix} \quad \text{is given by} \quad \frac{1}{\nabla} \begin{pmatrix} \partial_a\nabla, & \partial_{a'}\nabla, & \partial_{a''}\nabla \\ \partial_b\nabla, & \partial_{b'}\nabla, & \partial_{b''}\nabla \\ \partial_c\nabla, & \partial_{c'}\nabla, & \partial_{c''}\nabla \end{pmatrix},
$$

where ∇ is the determinant of the matrix and $\partial_x\nabla$ is the co-factor of x in this determinant, that is, the minor of x with the proper sign. The product of a matrix and its inverse is the unit matrix, denoted by I.

When $\nabla = 0$, the matrix is indeterminate (singular, in modern terminology) and has no inverse. Cayley asserted that the product of two matrices may be zero without either being zero, if either one is indeterminate. Actually, Cayley was wrong; both matrices must be indeterminate. For if $AB = 0$, $A \neq 0$, $B \neq 0$, and only A is indeterminate, then the inverse of B, namely B^{-1}, exists, and $ABB^{-1} = O \cdot B^{-1} = 0$. But $BB^{-1} = I$. Therefore, $AI = 0$ or $A = 0$.

The transverse (transposed or conjugate) matrix is defined as the one in which rows and columns are interchanged. The statement is made (without proof) that $(LMN)' = N'M'L'$, where prime denotes transpose. If $M' = M$ then M is called symmetric and if $M' = -M$ then M is skew-symmetric (or alternating). Any matrix can be expressed as the sum of a symmetrical matrix and a skew-symmetrical one.

Another concept carried over from determinant theory is the characteristic equation of a square matrix. For the matrix M it is defined to be

$$|M - xI| = 0,$$

where $|M - xI|$ is the determinant of the matrix $M - xI$ and I is the unit matrix. Thus if

$$M = \begin{pmatrix} a & b \\ c & d \end{pmatrix},$$

the characteristic equation (Cayley does not use the term though it was introduced for determinants by Cauchy [sec. 3]) is

(13) $$x^2 - (a + d)x + (ad - bc) = 0.$$

The roots of this equation are the characteristic roots (eigenvalues) of the matrix.

In the 1858 paper Cayley announced what is now called the Cayley-Hamilton theorem for square matrices of any order. The theorem states that if M is substituted for x in (13) the resulting matrix is the zero matrix.

Cayley states that he has verified the theorem for the 3 by 3 case and that further proof is not necessary. Hamilton's association with this theorem rests on the fact that in introducing in his *Lectures on Quaternions*[31] the notion of a linear vector function $\mathbf{r}'$ of another vector $\mathbf{r}$, a linear transformation from x, y, and z to x', y', and z' is involved. He proved that the matrix of this transformation satisfied the characteristic equation of that matrix, though he did not think formally in terms of matrices.

Other mathematicians found special properties of the characteristic roots of classes of matrices. Hermite[32] showed that if the matrix $M = M^*$, where M^* is the transpose of the matrix formed by replacing each element of M by its complex conjugate (such M's are now called Hermitian), then the characteristic roots are real. In 1861 Clebsch[33] deduced from Hermite's theorem that the non-zero characteristic roots of a real skew-symmetric matrix are pure imaginaries. Then Arthur Buchheim[34] (1859–88) demonstrated that if M is symmetric and the elements are real, the characteristic roots are real, though this result was already established for determinants by Cauchy.[35] Henry Taber (1860–?) in another paper[36] asserted as evident that if

$$x^n - m_1 x^{n-1} + m_2 x^{n-2} - \cdots \pm m_n = 0$$

is the characteristic equation of any square matrix M, then the determinant of M is m_n and if by a principal minor of a matrix we understand the *determinant* of a minor whose diagonal is part of the main diagonal of M, then m_i is the sum of the principal i-rowed minors. In particular, then, m_1, which is also the sum of the characteristic roots, is the sum of the elements along the main diagonal. This sum is called the trace of the matrix. The proofs of Taber's assertions were given by William Henry Metzler (1863–?).[37]

Frobenius[38] raised a question related to the characteristic equation. He sought the minimal polynomial—the polynomial of lowest degree—which the matrix satisfies. He stated that it is formed from the factors of the characteristic polynomial and is unique. It was not until 1904[39] that Kurt Hensel (1861–1941) proved Frobenius's statement of uniqueness. In the same article, Hensel also proved that if $f(x)$ is the minimal polynomial of a matrix M and $g(x)$ is any other polynomial satisfied by M, then $f(x)$ divides $g(x)$.

31. 1853, p. 566.
32. *Comp. Rend.*, 41, 1855, 181–83 = *Œuvres*, 1, 479–81.
33. *Jour. für Math.*, 62, 1863, 232–45.
34. *Messenger of Math.*, (2), 14, 1885, 143–44.
35. *Œuvres*, (2), 9, 174–91.
36. *Amer. Jour. of Math.*, 12, 1890, 337–96.
37. *Amer. Jour. of Math.*, 14, 1891/92, 326–77.
38. *Jour. für Math.*, 84, 1878, 1–63 = *Ges. Ahb.*, 1, 343–405.
39. *Jour. für Math.*, 127, 1904, 116–66.

The notion of the rank of a matrix was introduced by Frobenius[40] in 1879 though in connection with determinants. A matrix A with m rows and n columns (of order m by n) has k-rowed minors of all orders from 1 (the elements of A themselves) to the smaller of the two integers m and n inclusive. A matrix has rank r if and only if it has at least one r-rowed minor whose determinant is not zero while the determinants of all minors of order greater than r are zero.

Two matrices A and B can be related in various ways. They are equivalent if there exist two non-singular matrices U and V such that $A = UBV$. Sylvester had shown in his work on determinants[41] that the greatest common divisor d'_i of the i-rowed minor determinants of B equals the greatest common divisor d_i of the i-rowed minor determinants of A. Then H. J. S. Smith, working with matrices with integral elements, showed[42] that every matrix A of rank ρ is equivalent to a diagonal matrix with elements $h_1, h_2, \ldots, h_\rho$ down the main diagonal and such that h_i divides $h_i + 1$. The quotients $h_1 = d_1, h_2 = d_2/d_1, \ldots$ are called the invariant factors of A. Further if

$$h_i = p_1^{l_{i1}} p_2^{l_{i2}} \cdots p_k^{l_{ik}}$$

(where the p_i are primes) these various powers $p_i^{l_{ij}}$ are the elementary divisors of A. The invariant factors determine the elementary divisors and conversely.

These ideas on invariant factors and elementary divisors, which stem from Sylvester's and Weierstrass's work on determinants (as noted earlier), were carried over to matrices by Frobenius in his 1878 paper. The significance of the invariant factors and elementary divisors for matrices is that the matrix A is equivalent to the matrix B if and only if A and B have the same elementary divisors or invariant factors.

Frobenius did further work with invariant factors in his paper of 1878 and then organized the theory of invariant factors and elementary divisors in logical form.[43] The work in the 1878 paper enabled Frobenius to give the first general proof of the Cayley-Hamilton theorem and to modify the theorem when some of the latent roots (characteristic roots) of the matrix are equal. In this paper he also showed that when $AB^{-1} = B^{-1}A$, in which case there is an unambiguous quotient A/B, then $(A/B)^{-1} = B/A$ and that $(A^{-1})^T = (A^T)^{-1}$, where A^T is the transpose of A.

The subject of orthogonal matrices has received considerable attention. Although the term was used by Hermite in 1854,[44] it was not until 1878 that the formal definition was published by Frobenius (see ref. above). A matrix M is orthogonal if it is equal to the inverse of its transpose, that is, if

40. *Jour. für Math.*, 86, 1879, 146–208 = *Ges. Abh.*, 1, 482–544.
41. *Phil. Mag.*, (4), 1, 1851, 119–40 = *Coll. Math. Papers*, 1, 219–40.
42. *Phil. Trans.*, 151, 1861–62, 293–326 = *Coll. Math. Papers*, 1, 367–409.
43. *Sitzungsber. Akad. Wiss. zu Berlin*, 1894, 31–44 = *Ges. Abh.*, 1, 577–90.
44. *Cambridge and Dublin Math. Jour.*, 9, 1854, 63–67 = *Œuvres*, 1, 290–95.

$M = (M^T)^{-1}$. In addition to the definition, Frobenius proved that if S is a symmetric matrix and T, a skew-symmetric one, an orthogonal matrix can always be written in the form $(S - T)/(S + T)$ or more simply $(I - T)/(I + T)$.

The notion of similar matrices, like many other notions of matrix theory, came from earlier work on determinants as far back as Cauchy's. Two square matrices A and B are similar if there exists a non-singular matrix P such that $B = P^{-1}AP$. The characteristic equations of two similar matrices A and B are the same and the matrices have the same invariant factors and the same elementary divisors. For matrices with complex elements Weierstrass proved this result in his 1868 paper (though he worked with determinants). Since a matrix represents a linear homogeneous transformation similar matrices can be thought of as representing the same transformation but referred to two different coordinate systems.

Using the notion of similar matrices and the characteristic equation, Jordan[45] showed that a matrix can be transformed to a canonical form. If the characteristic equation of a matrix J is

$$f(\lambda) = \lambda^n + b_1\lambda^{n-1} + \cdots + b_n = 0$$

and if

$$f(\lambda) = (\lambda - \lambda_1)^{l_1}\cdots(\lambda - \lambda_k)^{l_k},$$

where the λ_i are distinct, then let

$$J_i = \begin{pmatrix} \lambda_i & 1 & 0 & 0 & \cdots & 0 \\ 0 & \lambda_i & 1 & 0 & \cdots & 0 \\ \cdot & \cdot & \cdot & \cdot & \cdot & \cdot \\ 0 & 0 & 0 & 0 & \cdots & \lambda_i \end{pmatrix}$$

denote an l_ith order matrix. Jordan showed that J can be transformed to a similar matrix having the form

$$\begin{pmatrix} J_1 & 0 & 0 & \cdots & 0 \\ 0 & J_2 & 0 & \cdots & 0 \\ \cdot & \cdot & \cdot & \cdot & \cdot \\ 0 & 0 & 0 & \cdots & J_k \end{pmatrix}.$$

This is the Jordan canonical, or normal, form of a matrix.

The similarity transformation from A to B was also treated by Frobenius under the name of contragredient transformation in his 1878 paper. In the

45. *Traité des substitutions*, 1870, Book II, 88–249.

same discussion he treated the notion of congruent matrices or cogredient transformations. This tells us that if $A = P^T B P$ then A is congruent with B, written $A \overset{c}{=} B$. For example, the transformation of matrix A that results in the simultaneous interchange of the same rows and columns of A is a congruence transformation. Also, a symmetric matrix A of rank r can be reduced by a congruent transformation to a diagonal matrix of the same rank; that is,

$$P^T A P = \begin{bmatrix} d_{11} & 0 & \cdots & 0 & \cdots & 0 \\ 0 & d_{22} & \cdots & 0 & \cdots & 0 \\ \cdot & \cdot & \cdot & \cdot & \cdot & \cdot \\ 0 & 0 & \cdots & d_{rr} & \cdots & 0 \\ \cdot & \cdot & \cdot & \cdot & \cdot & \cdot \\ 0 & 0 & \cdots & 0 & \cdots & 0 \end{bmatrix}.$$

There are many basic theorems on congruent matrices. For example, if S is symmetric and S_1 is congruent to S, then S_1 is symmetric and if S is skew then S_1 is skew.

In his 1892 paper in the *American Journal of Mathematics* Metzler introduced transcendental functions of a matrix, writing each as a power series in a matrix. He established series for e^M, e^{-M}, $\log M$, $\sin M$, and $\sin^{-1} M$. Thus

$$e^M = \sum_{n=0}^{\infty} M^n / n!.$$

The ramifications of the theory of matrices are numerous. Matrices have been used to represent quadratic and bilinear forms. The reduction of such forms to simple canonical forms is the core of the work on the invariants of matrices. They are intimately connected with hypernumbers and Cayley in his 1858 paper developed the idea of treating hypernumbers as matrices.

Both determinants and matrices have been extended to infinite order. Infinite determinants were involved in Fourier's work of determining the coefficients of a Fourier series expansion of a function (Chap. 28, sec. 2) and in Hill's work on the solution of ordinary differential equations (Chap. 29, sec. 7). Isolated papers on infinite determinants were written between these two outstanding nineteenth-century researches but the major activity postdates Hill's.

Infinite matrices were implicitly and explicitly involved in the work of Fourier, Hill, and Poincaré, who completed Hill's work. However, the great impetus to the study of infinite matrices came from the theory of integral equations (Chap. 45). We cannot devote space to the theory of determinants and matrices of infinite order.[46]

46. See the reference to Bernkopf in the bibliography at end of chapter.

In the elementary work on matrices the elements are ordinary real numbers though a great deal of what was done on behalf of the theory of numbers was limited to integral elements. However, they can be complex numbers and indeed most any other quantities. Naturally, the properties the matrices themselves possess depend upon the properties of the elements. Much late nineteenth- and early twentieth-century research has been devoted to the properties of matrices whose elements are members of an abstract field. The importance of matrix theory in the mathematical machinery of modern physics cannot be treated here, but in this connection a prophetic statement made by Tait is of interest. "Cayley is forging the weapons for future generations of physicists."

Bibliography

Bernkopf, Michael: "A History of Infinite Matrices," *Archive for History of Exact Sciences*, 4, 1968, 308–58.

Cayley, Arthur: *The Collected Mathematical Papers*, 13 vols., Cambridge University Press (1889–97), Johnson Reprint Corp., 1963.

Feldman, Richard W., Jr.: (Six articles on matrices with various titles), *The Mathematics Teacher*, 55, 1962, 482–84, 589–90, 657–59; 56, 1963, 37–38, 101–2, 163–64.

Frobenius, F. G.: *Gesammelte Abhandlungen*, 3 vols., Springer-Verlag, 1968.

Jacobi, C. G. J.: *Gesammelte Werke*, Georg Reimer, 1884, Vol. 3.

MacDuffee, C. C.: *The Theory of Matrices*, Chelsea, 1946.

Muir, Thomas: *The Theory of Determinants in the Historical Order of Development* (1906–23), 4 vols., Dover (reprint), 1960.

————: List of writings on the theory of matrices, *Amer. Jour. of Math.*, 20, 1898, 225–28.

Sylvester, James Joseph: *The Collected Mathematical Papers*, 4 vols., Cambridge University Press, 1904–12.

Weierstrass, Karl: *Mathematische Werke*, Mayer und Müller, 1895, Vol. 2.

34
The Theory of Numbers in the Nineteenth Century

It is true that Fourier had the opinion that the principal object of mathematics was public use and the explanation of natural phenomena; but a philosopher like him ought to know that the sole object of the science is the honor of the human spirit and that under this view a problem of [the theory of] numbers is worth as much as a problem on the system of the world.　　　　c. g. j. jacobi

1. *Introduction*

Up to the nineteenth century the theory of numbers was a series of isolated though often brilliant results. A new era began with Gauss's *Disquisitiones Arithmeticae*[1] which he composed at the age of twenty. This great work had been sent to the French Academy in 1800 and was rejected but Gauss published it on his own. In this book he standardized the notation; he systematized the existing theory and extended it; and he classified the problems to be studied and the known methods of attack and introduced new methods. In Gauss's work on the theory of numbers there are three main ideas: the theory of congruences, the introduction of algebraic numbers, and the theory of forms as the leading idea in Diophantine analysis. This work not only began the modern theory of numbers but determined the directions of work in the subject up to the present time. The *Disquisitiones* is difficult to read but Dirichlet expounded it.

Another major nineteenth-century development is analytic number theory, which uses analysis in addition to algebra to treat problems involving the integers. The leaders in this innovation were Dirichlet and Riemann.

2. *The Theory of Congruences*

Though the notion of congruence did not originate with Gauss—it appears in the work of Euler, Lagrange, and Legendre—Gauss introduced the notation

1. Published 1801 = *Werke*, 1.

in the first section of *Disquisitiones* and used it systematically thereafter. The basic idea is simple. The number 27 is congruent to 3 modulo 4,

$$27 \equiv 3 \text{ modulo } 4,$$

because $27 - 3$ is exactly divisible by 4. (The word modulo is often abbreviated to mod.) In general, if a, b, and m are integers

$$a \equiv b \text{ modulo } m$$

if $a - b$ is (exactly) divisible by m or if a and b have the same remainders on division by m. Then b is said to be a residue of a modulo m and a is a residue of b modulo m. As Gauss shows, all the residues of a modulo m, for fixed a and m, are given by $a + km$ where $k = 0, \pm 1, \pm 2, \ldots$.

Congruences with respect to the same modulus can be treated to some extent like equations. Such congruences can be added, subtracted, and multiplied. One can also ask for the solution of congruences involving unknowns. Thus, what values of x satisfy

$$2x \equiv 25 \text{ modulo } 12?$$

This equation has no solutions because $2x$ is even and $2x - 25$ is odd. Hence $2x - 25$ cannot be a multiple of 12. The basic theorem on polynomial congruences, which Gauss re-proves in the second section, had already been established by Lagrange.[2] A congruence of the nth degree

$$Ax^n + Bx^{n-1} + \cdots + Mx + N \equiv 0 \text{ modulo } p$$

whose modulus is a prime number p which does not divide A cannot have more than n noncongruent roots.

In the third section Gauss takes up residues of powers. Here he gives a proof in terms of congruences of Fermat's minor theorem, which, stated in terms of congruences, reads: If p is a prime and a is not a multiple of p then

$$a^{p-1} \equiv 1 \text{ modulo } p.$$

The theorem follows from his study of congruences of higher degree, namely,

$$x^n \equiv a \text{ modulo } m$$

where a and m are relatively prime. This subject was continued by many men after Gauss.

The fourth section of *Disquisitiones* treats quadratic residues. If p is a prime and a is not a multiple of p and if there exists an x such that $x^2 \equiv a$ mod p, then a is a quadratic residue of p; otherwise a is a quadratic nonresidue of p. After proving some subordinate theorems on quadratic residues Gauss gave the first rigorous proof of the law of quadratic reciprocity (Chap.

---

2. *Hist. de l'Acad. de Berlin*, 24, 1768, 192 ff., pub. 1770 = *Œuvres*, 2, 655–726.

25, sec. 4). Euler had given a complete statement much like Gauss's in one paper of his *Opuscula Analytica* of 1783 (Chap. 25, sec. 4). Nevertheless in article 151 of his *Disquisitiones* Gauss says that no one had presented the theorem in as simple a form as he had. He refers to other work of Euler including another paper in the *Opuscula* and to Legendre's work of 1785. Of these papers Gauss says rightly that the proofs were incomplete.

Gauss is supposed to have discovered a proof of the law in 1796 when he was nineteen. He gave another proof in the *Disquisitiones* and later published four others. Among his unpublished papers two more were found. Gauss says that he sought many proofs because he wished to find one that could be used to establish the biquadratic reciprocity theorem (see below). The law of quadratic reciprocity, which Gauss called the gem of arithmetic, is a basic result on congruences. After Gauss gave his proofs, more than fifty others were given by later mathematicians.

Gauss also treated congruences of polynomials. If A and B are two polynomials in x with, say, real coefficients then one knows that one can find unique polynomials Q and R such that

$$A = B \cdot Q + R,$$

where the degree of R is less than the degree of B. One can then say that two polynomials A_1 and A_2 are congruent modulo a third polynomial P if they have the same remainder R on division by P.

Cauchy used this idea[3] to define complex numbers by polynomial congruences. Thus if $f(x)$ is a polynomial with real coefficients then under division by $x^2 + 1$

$$f(x) \equiv a + bx \bmod x^2 + 1$$

because the remainder is of lower degree than the divisor. Here a and b are necessarily real by virtue of the division process. If $g(x)$ is another such polynomial then

$$g(x) \equiv c + dx \bmod x^2 + 1.$$

Cauchy now points out that if A_1, A_2, and B are any polynomials and if

$$A_1 = BQ_1 + R_1 \quad \text{and} \quad A_2 = BQ_2 + R_2,$$

then

$$A_1 + A_2 \equiv R_1 + R_2 \bmod B, \quad \text{and} \quad A_1 A_2 \equiv R_1 R_2 \bmod B.$$

We can now see readily that

$$f(x) + g(x) \equiv (a + c) + (b + d)x \bmod x^2 + 1$$

3. *Exercices d'analyse et de physique mathématique*, 4, 1847, 84 ff. = *Œuvres*, (1), 10, 312–23 and (2), 14, 93–120.

and since $x^2 \equiv -1 \bmod x^2 + 1$ that

$$f(x)g(x) \equiv (ac - bd) + (ad + bc)x \bmod x^2 + 1.$$

Thus the numbers $a + bx$ and $c + dx$ combine like complex numbers; that is, they have the formal properties of complex numbers, x taking the place of i. Cauchy also proved that every polynomial $g(x)$ not congruent to 0 modulo $x^2 + 1$ has an inverse, that is, a polynomial $h(x)$ such that $h(x)g(x)$ is congruent to 1 modulo $x^2 + 1$.

Cauchy did introduce i for x, i being for him a real indeterminate quantity. He then showed that for any

$$f(i) = a_0 + a_1 i + a_2 i^2 + \cdots$$

that

$$f(i) \equiv a_0 - a_2 + a_4 - \cdots + (a_1 - a_3 + a_5 - \cdots)i \text{ modulo } i^2 + 1.$$

Hence any expression involving complex numbers behaves as one of the form $c + di$ and one has all the apparatus needed to work with complex expressions. For Cauchy, then, the polynomials in i, with his understanding about i, take the place of complex numbers and one can put into one class all those polynomials having the same residue modulo $i^2 + 1$. These classes are the complex numbers.

It is interesting that in 1847 Cauchy still had misgivings about $\sqrt{-1}$. He says, "In the theory of algebraic equivalences substituted for the theory of imaginary numbers the letter i ceases to represent the symbolic sign $\sqrt{-1}$, which we repudiate completely and which we can abandon without regret since one does not know what this supposed sign signifies nor what sense to attribute to it. On the contrary we represent by the letter i a real quantity but indeterminate and in substituting the sign $\equiv$ for $=$ we transform what has been called an imaginary equation into an algebraic equivalence relative to the variable i and to the divisor $i^2 + 1$. Since this divisor remains the same in all the formulas one can dispense with writing it."

In the second decade of the century Gauss proceeded to search for reciprocity laws applicable to congruences of higher degree. These laws again involve residues of congruences. Thus for the congruence

$$x^4 \equiv q \bmod p$$

one can define q as a biquadratic residue of p if there is an integral value of x satisfying the equation. He did arrive at a law of biquadratic reciprocity (see below) and a law of cubic reciprocity. Much of this work appeared in papers from 1808 to 1817 and the theorem proper on biquadratic residues was given in papers of 1828 and 1832.[4]

4. *Comm. Soc. Gott.*, 6, 1828, and 7, 1832 = *Werke*, 2, 65–92 and 93–148; also pp. 165–78.

To attain elegance and simplicity in his theory of cubic and biquadratic residues Gauss made use of complex integers, that is, numbers of the form $a + bi$ with a and b integral or 0. In Gauss's work on biquadratic residues it was necessary to consider the case where the modulus p is a prime of the form $4n + 1$ and Gauss needed the complex factors into which prime numbers of the form $4n + 1$ can be decomposed. To obtain these Gauss realized that one must go beyond the domain of the ordinary integers to introduce the complex integers. Though Euler and Lagrange had introduced such integers into the theory of numbers it was Gauss who established their importance.

Whereas in the ordinary theory of integers the units are $+1$ and -1 in Gauss's theory of complex integers the units are ± 1 and $\pm i$. A complex integer is called composite if it is the product of two such integers neither of which is a unit. If such a decomposition is not possible the integer is called a prime. Thus $5 = (1 + 2i)(1 - 2i)$ and so is composite, whereas 3 is a complex prime.

Gauss showed that complex integers have essentially the same properties as ordinary integers. Euclid had proved (Chap. 4, sec. 7) that every integer is uniquely decomposable into a product of primes. Gauss proved that this unique decomposition, which is often referred to as the fundamental theorem of arithmetic, holds also for complex integers provided we do not regard the four unit numbers as different factors. That is, if $a = bc = (ib)(-ic)$, the two decompositions are the same. Gauss also showed that Euclid's process for finding the greatest common divisor of two integers is applicable to the complex integers.

Many theorems for ordinary primes carry over to the complex primes. Thus Fermat's theorem carries over in the form: If p be a complex prime $a + bi$ and k any complex integer not divisible by p then

$$k^{Np-1} \equiv 1 \text{ modulo } p$$

where Np is the norm $(a^2 + b^2)$ of p. There is also a law of quadratic reciprocity for complex integers, which was stated by Gauss in his 1828 paper.

In terms of complex integers Gauss was able to state the law of biquadratic reciprocity rather simply. One defines an uneven integer as one not divisible by $1 + i$. A primary uneven integer is an uneven integer $a + bi$ such that b is even and $a + b - 1$ is even. Thus -7 and $-5 + 2i$ are primary uneven numbers. The law of reciprocity for biquadratic residues states that if α and β are two primary uneven primes and A and B are their norms, then

$$\left(\frac{\alpha}{\beta}\right)_4 = (-1)^{(1/4)(A-1)(1/4)(B-1)}\left(\frac{\beta}{\alpha}\right)_4.$$

The symbol $(\alpha/\beta)_4$ has the following meaning: If p is any complex prime and k is any biquadratic residue not divisible by p, then $(k/p)_4$ is the power i^e of i which satisfies the congruence

$$k^{(Np-1)/4} \equiv 1 \text{ modulo } p$$

wherein Np stands for the norm of p. This law is equivalent to the statement: The biquadratic characters of two primary uneven prime numbers with respect to one another are identical, that is, $(\alpha/\beta)_4 = (\beta/\alpha)_4$, if either of the primes is congruent to 1 modulo 4; but if neither of the primes satisfies the congruence, then the two biquadratic characters are opposite, that is, $(\alpha/\beta)_4 = -(\beta/\alpha)_4$.

Gauss stated this reciprocity theorem but did not publish his proof. This was given by Jacobi in his lectures at Königsberg in 1836–37. Ferdinand Gotthold Eisenstein (1823–52), a pupil of Gauss, published five proofs of the law, of which the first two appeared in 1844.[5]

For cubic reciprocity Gauss found that he could obtain a law by using the "integers" $a + b\rho$ where ρ is a root of $x^2 + x + 1 = 0$ and a and b are ordinary (rational) integers but Gauss did not publish this result. It was found in his papers after his death. The law of cubic reciprocity was first stated by Jacobi[6] and proved by him in his lectures at Königsberg. The first published proof is due to Eisenstein.[7] Upon noting this proof Jacobi claimed[8] that it was precisely the one given in his lectures but Eisenstein indignantly denied any plagiarism.[9] There are also reciprocity laws for congruences of degree greater than four.

3. Algebraic Numbers

The theory of complex integers is a step in the direction of a vast subject, the theory of algebraic numbers. Neither Euler nor Lagrange envisioned the rich possibilities which their work on complex integers opened up. Neither did Gauss.

The theory grew out of the attempts to prove Fermat's assertion about $x^n + y^n = z^n$. The cases $n = 3, 4$, and 5 have already been discussed (Chap. 25, sec. 4). Gauss tried to prove the assertion for $n = 7$ but failed. Perhaps because he was disgusted with his failure, he said in a letter of 1816 to Heinrich W. M. Olbers (1758–1840), "I confess indeed that Fermat's theorem as an isolated proposition has little interest for me, since a multitude of such propositions, which one can neither prove nor refute, can easily be

5. *Jour. für Math.*, 28, 1844, 53–67 and 223–45.
6. *Jour. für Math.*, 2, 1827, 66–69 = *Werke*, 6, 233–37.
7. *Jour. für Math.*, 27, 1844, 289–310.
8. *Jour. für Math.*, 30, 1846, 166–82, p. 172 = *Werke*, 6, 254–74.
9. *Jour. für Math.*, 35, 1847, 135–274 (p. 273).

formulated." This particular case of $n = 7$ was disposed of by Lamé in 1839,[10] and Dirichlet established the assertion for $n = 14$.[11] However, the general proposition was unproven.

It was taken up by Ernst Eduard Kummer (1810–93), who turned from theology to mathematics, became a pupil of Gauss and Dirichlet, and later served as a professor at Breslau and Berlin. Though Kummer's major work was in the theory of numbers, he made beautiful discoveries in geometry which had their origin in optical problems; he also made important contributions to the study of refraction of light by the atmosphere.

Kummer took $x^p + y^p$ where p is prime, and factored it into

$$(x + y)(x + \alpha y) \cdots (x + \alpha^{p-1} y),$$

where α is an imaginary pth root of unity. That is, α is a root of

(1) $$\alpha^{p-1} + \alpha^{p-2} + \cdots + \alpha + 1 = 0.$$

This led him to extend Gauss's theory of complex integers to algebraic numbers insofar as they are introduced by equations such as (1), that is, numbers of the form

$$f(\alpha) = a_0 + a_1\alpha + \cdots + a_{p-2}\alpha^{p-2},$$

where each a_i is an ordinary (rational) integer. (Since α satisfies (1), terms in α^{p-1} can be replaced by terms of lower power.) Kummer called the numbers $f(\alpha)$ complex integers.

By 1843 Kummer made appropriate definitions of integer, prime integer, divisibility, and the like (we shall give the standard definitions in a moment) and then made the mistake of assuming that unique factorization holds in the class of algebraic numbers that he had introduced. He pointed out while transmitting his manuscript to Dirichlet in 1843 that this assumption was necessary to prove Fermat's theorem. Dirichlet informed him that unique factorization holds only for certain primes p. Incidentally, Cauchy and Lamé made the same mistake of assuming unique factorization for algebraic numbers. In 1844 Kummer[12] recognized the correctness of Dirichlet's criticism.

To restore unique factorization Kummer created a theory of ideal numbers in a series of papers starting in 1844.[13] To understand his idea let us consider the domain of $a + b\sqrt{-5}$, where a and b are integers. In this domain

$$6 = 2 \cdot 3 = (1 + \sqrt{-5})(1 - \sqrt{-5})$$

10. *Jour. de Math.*, 5, 1840, 195–211.
11. *Jour. für Math.*, 9, 1832, 390–93 = *Werke*, 1, 189–94.
12. *Jour. de Math.*, 12, 1847, 185–212.
13. *Jour. für Math.*, 35, 1847, 319–26, 327–67.

and all four factors can readily be shown to be prime integers. Then unique decomposition does not hold. Let us introduce, for this domain, the ideal numbers $\alpha = \sqrt{2}$, $\beta_1 = (1 + \sqrt{-5})/\sqrt{2}$, $\beta_2 = (1 - \sqrt{-5})/\sqrt{2}$. We see that $6 = \alpha^2\beta_1\beta_2$. Thus 6 is now uniquely expressed as the product of four factors, all ideal numbers as far as the domain $a + b\sqrt{-5}$ is concerned.[14] In terms of these ideals and other primes factorization in the domain is unique (apart from factors consisting of units). With ideal numbers one can prove some of the results of ordinary number theory in all domains that previously lacked unique factorization.

Kummer's ideal numbers, though ordinary numbers, do not belong to the class of algebraic numbers he had introduced. Moreover, the ideal numbers were not defined in any general way. As far as Fermat's theorem is concerned, with his ideal numbers Kummer did succeed in showing that it was correct for a number of prime numbers. In the first hundred integers only 37, 59, and 67 were not covered by Kummer's demonstration. Then Kummer in a paper of 1857[15] extended his results to these exceptional primes. These results were further extended by Dimitry Mirimanoff (1861–1945), a professor at the University of Geneva, by perfecting Kummer's method.[16] Mirimanoff proved that Fermat's theorem is correct for each n up to 256 if x, y, and z are prime to that exponent n.

Whereas Kummer worked with algebraic numbers formed out of the roots of unity, Richard Dedekind (1831–1916), a pupil of Gauss, who spent fifty years of his life as a teacher at a technical high school in Germany, approached the problem of unique factorization in an entirely new and fresh manner. Dedekind published his results in supplement 10 to the second edition of Dirichlet's *Zahlentheorie* (1871) which Dedekind edited. He extended these results in the supplements to the third and fourth editions of the same book.[17] Therein he created the modern theory of algebraic numbers.

Dedekind's theory of algebraic numbers is a generalization of Gauss's complex integers and Kummer's algebraic numbers but the generalization is somewhat at variance with Gauss's complex integers. A number r that is a root of

$$(2) \qquad a_0x^n + a_1x^{n-1} + \cdots + a_{n-1}x + a_n = 0,$$

where the a_i's are ordinary integers (positive or negative), and that is not a root of such an equation of degree less than n is called an algebraic number of degree n. If the coefficient of the highest power of x in (2) is 1, the solutions are called algebraic integers of degree n. The sum, difference, and product of

---

14. With the introduction of the ideal numbers, 2 and 3 are no longer indecomposable, for $2 = \alpha^2$ and $3 = \beta_1\beta_2$.
15. *Abh. König. Akad. der Wiss. Berlin*, 1858, 41–74.
16. *Jour. für Math.*, 128, 1905, 45–68.
17. 4th ed., 1894 = *Werke*, 3, 2–222.

algebraic integers are algebraic integers, and if an algebraic integer is a rational number it is an ordinary integer.

We should note that under the new definitions an algebraic integer can contain ordinary fractions. Thus $(-13 + \sqrt{-115})/2$ is an algebraic integer of the second degree because it is a root of $x^2 + 13x + 71 = 0$. On the other hand $(1 - \sqrt{-5})/2$ is an algebraic number of degree 2 but not an algebraic integer because it is a root of $2x^2 - 2x + 3 = 0$.

Dedekind introduced next the concept of a number field. This is a collection F of real or complex numbers such that if α and β belong to F then so do $\alpha + \beta$, $\alpha - \beta$, $\alpha\beta$ and, if $\beta \neq 0$, α/β. Every number field contains the rational numbers because if α belongs then so does α/α or 1 and consequently $1 + 1$, $1 + 2$, and so forth. It is not difficult to show that the set of all algebraic numbers forms a field.

If one starts with the rational number field and θ is an algebraic number of degree n then the set formed by combining θ with itself and the rational numbers under the four operations is also a field of degree n. This field may be described alternatively as the smallest field containing the rational numbers and θ. It is also called an extension field of the rational numbers. Such a field does not consist of all algebraic numbers and is a specific algebraic number field. The notation $R(\theta)$ is now common. Though one might expect that the members of $R(\theta)$ are the quotients $f(\theta)/g(\theta)$ where $f(x)$ and $g(x)$ are any polynomials with rational coefficients, one can prove that if θ is of degree n, then any member α of $R(\theta)$ can be expressed in the form

$$\alpha = a_0\theta^{n-1} + a_1\theta^{n-2} + \cdots + a_{n-1},$$

where the a_i are ordinary rational numbers. Moreover, there exist algebraic integers $\theta_1, \theta_2, \ldots, \theta_n$ of this field such that all the algebraic integers of the field are of the form

$$A_1\theta_1 + A_2\theta_2 + \cdots + A_n\theta_n,$$

where the A_i are ordinary positive and negative integers.

A ring, a concept introduced by Dedekind, is essentially any collection of numbers such that if α and β belong, so do $\alpha + \beta$, $\alpha - \beta$, and $\alpha\beta$. The set of all algebraic integers forms a ring as does the set of all algebraic integers of any specific algebraic number field.

The algebraic integer α is said to be divisible by the algebraic integer β if there is an algebraic integer γ such that $\alpha = \beta\gamma$. If j is an algebraic integer which divides every other integer of a field of algebraic numbers then j is called a unit in that field. These units, which include $+1$ and -1, are a generalization of the units $+1$ and -1 of ordinary number theory. The algebraic integer α is a prime if it is not zero or a unit and if any factorization of α into $\beta\gamma$, where β and γ belong to the same algebraic number field, implies that β or γ is a unit in that field.

Now let us see to what extent the fundamental theorem of arithmetic holds. In the ring of *all* algebraic integers there are no primes. Let us consider the ring of integers in a specific algebraic number field $R(\theta)$, say the field $a + b\sqrt{-5}$, where a and b are ordinary rational numbers. In this field unique factorization does not hold. For example,

$$21 = 3 \cdot 7 = (4 + \sqrt{-5})(4 - \sqrt{-5}) = (1 + 2\sqrt{-5})(1 - 2\sqrt{-5}).$$

Each of these last four factors is prime in the sense that it cannot be expressed as a product of the form $(c + d\sqrt{-5})(e + f\sqrt{-5})$ with $c, d, e,$ and f integral.

On the other hand let us consider the field $a + b\sqrt{6}$ where a and b are ordinary rational numbers. If one applies the four algebraic operations to these numbers one gets such numbers. If a and b are restricted to integers one gets the algebraic integers (of degree 2) of this domain. In this domain we can take as an equivalent definition of unit that the algebraic integer M is a unit if $1/M$ is also an algebraic integer. Thus $1, -1, 5 - 2\sqrt{6}$, and $5 + 2\sqrt{6}$ are units. Every integer is divisible by any one of the units. Further, an algebraic integer of the domain is prime if it is divisible only by itself and the units. Now

$$6 = 2 \cdot 3 = \sqrt{6} \cdot \sqrt{6}.$$

It would seem as though there is no unique decomposition into primes. But the factors shown are not primes. In fact

$$6 = 2 \cdot 3 = \sqrt{6} \cdot \sqrt{6} = (2 + \sqrt{6})(-2 + \sqrt{6})(3 + \sqrt{6})(3 - \sqrt{6}).$$

Each of the last four factors is a prime in the domain and unique decomposition does hold in this domain.

In the ring of integers of a specific algebraic number field factorization of the algebraic integers into primes is always possible but *unique* factorization does not generally hold. In fact for domains of the form $a + b\sqrt{-D}$, where D may have any positive integral value not divisible by a square, the unique factorization theorem is valid only when $D = 1, 2, 3, 7, 11, 19, 43, 67,$ and 163, at least for D's up to 10^9.[18] Thus the algebraic numbers themselves do not possess the property of unique factorization.

4. The Ideals of Dedekind

Having generalized the notion of algebraic number, Dedekind now undertook to restore unique factorization in algebraic number fields by a scheme quite different from Kummer's. In place of ideal numbers he introduced

18. H. M. Stark has shown that the above values of D are the only ones possible. See his "On the Problem of Unique Factorization in Complex Quadratic Fields," *Proceedings of Symposia in Pure Mathematics*, XII, 41–56, Amer. Math. Soc., 1969.

classes of algebraic numbers which he called ideals in honor of Kummer's ideal numbers.

Before defining Dedekind's ideals let us note the underlying thought. Consider the ordinary integers. In place of the integer 2, Dedekind considers the class of integers $2m$, where m is any integer. This class consists of all integers divisible by 2. Likewise 3 is replaced by the class $3n$ of all integers divisible by 3. The product 6 becomes the collection of all numbers $6p$, where p is any integer. Then the product $2 \cdot 3 = 6$ is replaced by the statement that the class $2m$ "times" the class $3n$ equals the class $6p$. Moreover, the class $2m$ is a factor of the class $6p$, despite the fact that the former class contains the latter. These classes are examples in the ring of ordinary integers of what Dedekind called ideals. To follow Dedekind's work one must accustom oneself to thinking in terms of classes of numbers.

More generally, Dedekind defined his ideals as follows: Let K be a specific algebraic number field. A set of integers A of K is said to form an ideal if when α and β are any two integers in the set, the integers $\mu\alpha + \nu\beta$, where μ and ν are any other algebraic integers in K, also belong to the set. Alternatively an ideal A is said to be generated by the algebraic integers $\alpha_1, \alpha_2, \ldots, \alpha_n$ of K if A consists of all sums

$$\lambda_1\alpha_1 + \lambda_2\alpha_2 + \cdots + \lambda_n\alpha_n,$$

where the λ_i are any integers of the field K. This ideal is denoted by $(\alpha_1, \alpha_2, \ldots, \alpha_n)$. The zero ideal consists of the number 0 alone and accordingly is denoted by (0). The unit ideal is that generated by the number 1, that is, (1). An ideal A is called principal if it is generated by the single integer α, so that (α) consists of all the algebraic integers divisible by α. In the ring of the ordinary integers every ideal is a principal ideal.

An example of an ideal in the algebraic number field $a + b\sqrt{-5}$, where a and b are ordinary rational numbers, is the ideal generated by the integers 2 and $1 + \sqrt{-5}$. This ideal consists of all integers of the form $2\mu + (1 + \sqrt{-5})\nu$, where μ and ν are arbitrary integers of the field. The ideal also happens to be a principal ideal because it is generated by the number 2 alone in view of the fact that $(1 + \sqrt{-5})2$ must also belong to the ideal generated by 2.

Two ideals $(\alpha_1, \alpha_2, \ldots, \alpha_p)$ and $(\beta_1, \beta_2, \ldots, \beta_q)$ are equal if every member of the former ideal is a member of the latter and conversely. To tackle the problem of factorization we must first consider the product of two ideals. The product of the ideal $A = (\alpha_1, \ldots, \alpha_s)$ and the ideal $B = (\beta_1, \ldots, \beta_t)$ of K is defined to be the ideal

$$AB = (\alpha_1\beta_1, \alpha_1\beta_2, \alpha_2\beta_1, \ldots, \alpha_i\beta_j, \ldots, \alpha_s\beta_t).$$

It is almost evident that this product is commutative and associative. With this definition we may say that A divides B if there exists an ideal C such that

$B = AC$. One writes $A|B$ and A is called a factor of B. As already suggested above by our example of the ordinary integers, the elements of B are *included* in the elements of A and ordinary divisibility is replaced by class inclusion.

The ideals that are the analogues of the ordinary prime numbers are called prime ideals. Such an ideal P is defined to be one which has no factors other than itself and the ideal (1), so that P is not contained in any other ideal of K. For this reason a prime ideal is also called maximal. All of these definitions and theorems lead to the basic theorems for ideals of an algebraic number field K. Any ideal is divisible by only a finite number of ideals and if a prime ideal divides the product AB of two ideals (of the same number class) it divides A or B. Finally the fundamental theorem in the theory of ideals is that every ideal can be factored uniquely into prime ideals.

In our earlier examples of algebraic number fields of the form $a + b\sqrt{D}$, D integral, we found that some permitted unique factorization of the algebraic integers of those fields and others did not. The answer to the question of which do or do not is given by the theorem that the factorization of the integers of an algebraic number field K into primes is unique if and only if all the ideals in K are principal.

From these examples of Dedekind's work we can see that his theory of ideals is indeed a generalization of the ordinary integers. In particular, it furnishes the concepts and properties in the domain of algebraic numbers which enable one to establish unique factorization.

Leopold Kronecker (1823–91), who was Kummer's favorite pupil and who succeeded Kummer as professor at the University of Berlin, also took up the subject of algebraic numbers and developed it along lines similar to Dedekind's. Kronecker's doctoral thesis "On Complex Units," written in 1845 though not published until much later,[19] was his first work in the subject. The thesis deals with the units that can exist in the algebraic number fields created by Gauss.

Kronecker created another theory of fields (domains of rationality).[20] His field concept is more general than Dedekind's because he considered fields of rational functions in any number of variables (indeterminates). Specifically Kronecker introduced (1881) the notion of an indeterminate adjoined to a field, the indeterminate being just a new abstract quantity. This idea of extending a field by adding an indeterminate he made the cornerstone of his theory of algebraic numbers. Here he used the knowledge that had been built up by Liouville, Cantor, and others on the distinction between algebraic and transcendental numbers. In particular he observed that if x is transcendental over a field K (x is an indeterminate) then the field

19. *Jour. für Math.*, 93, 1882, 1–52 = *Werke*, 1, 5–71.
20. "Grundzüge einer arithmetischen Theorie der algebraischen Grössen," *Jour. für Math.*, 92, 1882, 1–122 = *Werke*, 2, 237–387; also published separately by G. Reimer, 1882.

$K(x)$ obtained by adjoining the indeterminate x to K, that is, the smallest field containing K and x, is isomorphic to the field $K[x]$ of rational functions in one variable with coefficients in K.[21] He did stress that the indeterminate was merely an element of an algebra and not a variable in the sense of analysis.[22] He then showed in 1887[23] that to each ordinary prime number p there corresponds within the ring $Q(x)$ of polynomials with rational coefficients a prime polynomial $p(x)$ which is irreducible in the rational field Q. By considering two polynomials to be equal if they are congruent modulo a given prime polynomial $p(x)$, the ring of all polynomials in $Q(x)$ becomes a field of residue classes possessing the same algebraic properties as the algebraic number field $K(\delta)$ arising from the field K by adjoining a root δ of $p(x) = 0$. Here he used the idea Cauchy had already employed to introduce imaginary numbers by using polynomials congruent modulo $x^2 + 1$. In this same work he showed that the theory of algebraic numbers is independent of the fundamental theorem of algebra and of the theory of the complete real number system.

In his theory of fields (in the "Grundzüge") whose elements are formed by starting with a field K and then adjoining indeterminates $x_1, x_2, \ldots, x_n$, Kronecker introduced the notion of a modular system that played the role of ideals in Dedekind's theory. For Kronecker a modular system is a set M of those polynomials in n variables $x_1, x_2, \ldots, x_n$ such that if P_1 and P_2 belong to the set so does $P_1 + P_2$ and if P belongs so does QP where Q is any polynomial in $x_1, x_2, \ldots, x_n$.

A basis of a modular system M is any set of polynomials $B_1, B_2 \ldots$ of M such that every polynomial of M is expressible in the form

$$R_1 B_1 + R_2 B_2 + \cdots,$$

where $R_1, R_2, \ldots$ are constants or polynomials (not necessarily belonging to M). The theory of divisibility in Kronecker's general fields was defined in terms of modular systems much as Dedekind had done with ideals.

The work on algebraic number theory was climaxed in the nineteenth century by Hilbert's famous report on algebraic numbers.[24] This report is primarily an account of what had been done during the century. However, Hilbert reworked all of this earlier theory and gave new, elegant, and powerful methods of securing these results. He had begun to create new ideas in algebraic number theory from about 1892 on and one of the new creations on Galoisian number fields was also incorporated in the report. Subsequently Hilbert and many other men extended algebraic number theory vastly.

21. *Werke*, 2, 253.
22. *Werke*, 2, 339.
23. *Jour. für Math.*, 100, 1887, 490–510 = *Werke*, 3, 211–40.
24. "Die Theorie der algebraischen Zahlkörper" (The Theory of Algebraic Number Fields), *Jahres. der Deut. Math.-Ver.*, 4, 1897, 175–546 = *Ges. Abh.*, 1, 63–363.

However, these later developments, relative Galoisian fields, relative Abelian number fields and class fields, all of which stimulated an immense amount of work in the twentieth century, are of concern primarily to specialists.

Algebraic number theory, originally a scheme for investigating the solutions of problems in the older theory of numbers, has become an end in itself. It has come to occupy a position in between the theory of numbers and abstract algebra, and now number theory and modern higher algebra merge in algebraic number theory. Of course algebraic number theory has also produced new theorems in the ordinary theory of numbers.

5. *The Theory of Forms*

Another class of problems in the theory of numbers is the representation of integers by forms. The expression

$$(3) \qquad\qquad ax^2 + 2bxy + cy^2,$$

wherein a, b, and c are integral, is a binary form because two variables are involved, and it is a quadratic form because it is of the second degree. A number M is said to be represented by the form if for specific integral values of a, b, c, x, and y the above expression equals M. One problem is to find the set of numbers M that are representable by a given form or class of forms. The converse problem, given M and given a, b, and c or some class of a, b, and c, to find the values of x and y that represent M, is equally important. The latter problem belongs to Diophantine analysis and the former may equally well be considered part of the same subject.

Euler had obtained some special results on these problems. However, Lagrange made the key discovery that if a number is representable by one form it is also representable by many other forms, which he called equivalent. The latter could be obtained from the original form by a change of variables

$$(4) \qquad\qquad x = \alpha x' + \beta y', \qquad y = \gamma x' + \delta y'$$

wherein α, β, γ, and δ are integral and $\alpha\delta - \beta\gamma = 1$.[25] In particular, Lagrange showed that for a given discriminant (Gauss used the word determinant) $b^2 - ac$ there is a finite number of forms such that each form with that discriminant is equivalent to one of this finite number. Thus all forms with a given discriminant can be segregated into classes, each class consisting of forms equivalent to one member of that class. This result and some inductively established results by Legendre attracted Gauss's attention. In a bold step Gauss extracted from Lagrange's work the notion of equivalence of forms and concentrated on that. The fifth section of his *Disquisitiones*, by far the largest section, is devoted to this subject.

25. *Nouv. Mém. de l'Acad. de Berlin*, 1773, 263–312; and 1775, 323 ff. = *Œuvres*, 3, 693–795.

Gauss systematized and extended the theory of forms. He first defined equivalence of forms. Let

$$F = ax^2 + 2bxy + cy^2$$

be transformed by means of (4) into the form

$$F' = a'x'^2 + 2b'x'y' + c'y'^2.$$

Then

$$b'^2 - a'c' = (b^2 - ac)(\alpha\delta - \beta\gamma)^2.$$

If now $(\alpha\delta - \beta\gamma)^2 = 1$, the discriminants of the two forms are equal. Then the inverse of the transformation (4) will also contain integral coefficients (by Cramer's rule) and will transform F' into F. F and F' are said to be equivalent. If $\alpha\delta - \beta\gamma = 1$, F and F' are said to be properly equivalent, and if $\alpha\delta - \beta\gamma = -1$, then F and F' are said to be improperly equivalent.

Gauss proved a number of theorems on the equivalence of forms. Thus if F is equivalent to F' and F' to F'' then F is equivalent to F''. If F is equivalent to F', any number M representable by F is representable by F' and in as many ways by one as by the other. He then shows, if F and F' are equivalent, how to find all the transformations from F into F'. He also finds all the representations of a given number M by a form F, provided the values of x and y are relatively prime.

By definition two equivalent forms have the same value for their discriminant $D = b^2 - ac$. However, two forms with the same discriminant are not necessarily equivalent. Gauss shows that all forms with a given D can be segregated into classes; the members of any one class are properly equivalent to each other. Though the number of forms with a given D is infinite, the number of classes for a given D is finite. In each class one form can be taken as representative and Gauss gives criteria for the choice of a simplest representative. The simplest form of all those with determinant D has $a = 1, b = 0, c = -D$. This he calls the principal form and the class to which it belongs is called the principal class.

Gauss then takes up the composition (product) of forms. If the form

$$F = AX^2 + 2BXY + CY^2$$

is transformed into the product of two forms

$$f = ax^2 + 2bxy + cy^2 \quad \text{and} \quad f' = a'x'^2 + 2b'x'y' + c'y'^2$$

by the substitution

$$X = p_1xx' + p_2xy' + p_3x'y + p_4yy'$$
$$Y = q_1xx' + q_2xy' + q_3x'y + q_4yy'$$

then F is said to be transformable into ff'. If further the six numbers

$$p_1q_2 - q_1p_2, p_1q_3 - q_1p_3, p_1q_4 - q_1p_4, p_2q_3 - q_2p_3, p_2q_4 - q_2p_4, p_3q_4 - q_3p_4$$

do not have a common divisor, then F is called a composite of the forms f and f'.

Gauss was then able to prove an essential theorem: If f and g belong to the same class and f' and g' belong to the same class, then the form composed of f and f' will belong to the same class as the form composed of g and g'. Thus one can speak of a class of forms composed of two (or more) given classes. In this composition of classes, the principal class acts as a unit class; that is, if the class K is composed with a principal class, the resulting class will be K.

Gauss now turns to a treatment of ternary quadratic forms

$$Ax^2 + 2Bxy + Cy^2 + 2Dxz + 2Eyz + Fz^2,$$

where the coefficients are integers, and undertakes a study very much like what he has just done for binary forms. The goal as in the case of binary forms is the representation of integers. The theory of ternary forms was not carried far by Gauss.

The objective of the entire work on the theory of forms was, as already noted, to produce theorems in the theory of numbers. In the course of his treatment of forms Gauss shows how the theory can be used to prove any number of theorems about the integers including many that were previously proved by such men as Euler and Lagrange. Thus Gauss proves that any prime number of the form $4n + 1$ can be represented as a sum of squares in one and only one way. Any prime number of the form $8n + 1$ or $8n + 3$ can be represented in the form $x^2 + 2y^2$ (for positive integral x and y) in one and only one way. He shows how to find all the representations of a given number M by the given form $ax^2 + 2bxy + cy^2$ provided the discriminant D is a positive non-square number. Further if F is a primitive form (the values of a, b, and c are relatively prime) with discriminant D and if p is a prime number dividing D, then the numbers not divisible by p which can be represented by F agree in that they are all either quadratic residues of p or non-residues of p.

Among the results Gauss drew from his work on ternary quadratic forms is the first proof of the theorem that every number can be represented as the sum of three triangular numbers. These, we recall, are the numbers

$$1, 3, 6, 10, 15, \cdots, \frac{n^2 + n}{2}, \cdots.$$

He also re-proved the theorem already proved by Lagrange that any positive integer can be expressed as the sum of four squares. Apropos of the former result, it is worth noting that Cauchy read a paper to the Paris Academy in

1815 which established the general result first asserted by Fermat that every integer is the sum of k or fewer k-gonal numbers.[26] (The general k-gonal number is $n + (n^2 - n)(k - 2)/2$.)

The algebraic theory of binary and ternary quadratic forms as presented by Gauss has an interesting geometrical analogue which Gauss himself initiated. In a review which appeared in the *Göttingische Gelehrte Anzeigen* of 1830[27] of a book on ternary quadratic forms by Ludwig August Seeber (1793–1855) Gauss sketched the geometrical representation of his forms and classes of forms.[28] This work is the beginning of a development called the geometrical theory of numbers which first gained prominence when Hermann Minkowski (1864–1909), who served as professor of mathematics at several universities, published his *Geometrie der Zahlen* (1896).

The subject of forms became a major one in the theory of numbers of the nineteenth century. Further work was done by a host of men on binary and ternary quadratic forms and on forms with more variables and of higher degree.[29]

6. *Analytic Number Theory*

One of the major developments in number theory is the introduction of analytic methods and of analytic results to express and prove facts about integers. Actually, Euler had used analysis in number theory (see below) and Jacobi used elliptic functions to obtain results in the theory of congruences and the theory of forms.[30] However, Euler's uses of analysis in number theory were minor and Jacobi's number-theoretic results were almost accidental by-products of his analytic work.

The first deep and deliberate use of analysis to tackle what seemed to be a clear problem of algebra was made by Peter Gustav Lejeune-Dirichlet (1805–59), a student of Gauss and Jacobi, professor at Breslau and Berlin, and then successor to Gauss at Göttingen. Dirichlet's great work, the *Vorlesungen über Zahlentheorie*,[31] expounded Gauss's *Disquisitiones* and gave his own contributions.

The problem that caused Dirichlet to employ analysis was to show that every arithmetic sequence

$$a, a + b, a + 2b, a + 3b, \cdots, a + nb, \cdots,$$

26. *Mém. de l'Acad. des Sci., Paris*, (1), 14, 1813–15, 177–220 = *Œuvres*, (2), 6, 320–53.

27. *Werke*, 2, 188–196.

28. Felix Klein in his *Entwicklung* (see the bibliography at the end of this chapter), pp. 35–39, expands on Gauss's sketch.

29. See works of Smith and Dickson, listed in the bibliography, for further details.

30. *Jour. für Math.*, 37, 1848, 61–94 and 221–54 = *Werke*, 2, 219–88.

31. Published 1863; the second, third, and fourth editions of 1871, 1879, and 1894 were supplemented extensively by Dedekind.

where a and b are relatively prime, contains an infinite number of primes. Euler[32] and Legendre[33] made this conjecture and in 1808 Legendre[34] gave a proof that was faulty. In 1837 Dirichlet[35] gave a correct proof. This result generalizes Euclid's theorem on the infinitude of primes in the sequence $1, 2, 3, \ldots$. Dirichlet's analytical proof was long and complicated. Specifically it used what are now called the Dirichlet series, $\sum_{n=1}^{\infty} a_n n^{-z}$, wherein the a_n and z are complex. Dirichlet also proved that the sum of the reciprocals of the primes in the sequence $\{a + nb\}$ diverges. This extends a result of Euler on the usual primes (see below). In 1841[36] Dirichlet proved a theorem on the primes in progressions of complex numbers $a + bi$.

The chief problem involving the introduction of analysis concerned the function $\pi(x)$ which represents the number of primes not exceeding x. Thus $\pi(8)$ is 4 since 2, 3, 5, and 7 are prime and $\pi(11)$ is 5. As x increases the additional primes become scarcer and the problem was, What is the proper analytical expression for $\pi(x)$? Legendre, who had proved that no rational expression would serve, at one time gave up hope that any expression could be found. Then Euler, Legendre, Gauss, and others surmised that

$$(5) \qquad\qquad \lim_{x \to \infty} \frac{\pi(x)}{x/\log x} = 1.$$

Gauss used tables of primes (he actually studied all the primes up to 3,000,000) to make conjectures about $\pi(x)$ and inferred[37] that $\pi(x)$ differs little from $\int_2^x dt/\log t$. He knew also that

$$\lim_{x \to \infty} \frac{\int_2^x dt/\log t}{x/\log x} = 1.$$

In 1848 Pafnuti L. Tchebycheff (1821–94), a professor at the University of Petrograd, took up the question of the number of prime numbers less than or equal to x and made a big step forward in this old problem. In a key paper, "Sur les nombres premiers"[38] Tchebycheff proved that

$$A_1 < \frac{\pi(x)}{x/\log x} < A_2,$$

where $0.922 < A_1 < 1$ and $1 < A_2 < 1.105$, but did not prove that the function tends to a limit. This inequality was improved by many mathe-

32. *Opuscula Analytica*, 2, 1783.
33. *Mém. de l'Acad. des Sci.*, Paris, 1785, 465–559, pub. 1788.
34. *Théorie des nombres*, 2nd ed., p. 404.
35. *Abh. König. Akad. der Wiss.*, Berlin, 1837, 45–81 and 108–10 = *Werke*, 1, 307–42.
36. *Abh. König. Akad. der Wiss.*, Berlin, 1841, 141–61 = *Werke*, 2, 509–32.
37. *Werke*, 2, 444–47.
38. *Mém. Acad. Sci. St. Peters.*, 7, 1854, 15–33; also *Jour. de Math.*, (1), 17, 1852, 366–90 = *Œuvres*, 1, 51–70.

maticians including Sylvester, who among others doubted in 1881 that the function had a limit. In his work Tchebycheff used what we now call the Riemann zeta function,

$$\zeta(z) = \sum_{n=1}^{\infty} \frac{1}{n^z}$$

though he used it for real values of z. (This series is a special case of Dirichlet's series.) Incidentally he also proved in the same paper that for $n > 3$ there is always at least one prime between n and $2n - 2$.

The zeta function for real z appears in a work of Euler[39] where he introduced

$$\zeta(s) = \sum_{n=1}^{\infty} \frac{1}{n^s} = \prod_{n=1}^{\infty} \left(1 - \frac{1}{p_n^s}\right)^{-1}.$$

Here the p_n's are prime numbers. Euler used the function to prove that the sum of the reciprocals of the prime numbers diverges. For s an even positive integer Euler knew the value of $\zeta(s)$ (Chap. 20, sec. 4). Then in a paper read in 1749[40] Euler asserted that for real s

$$\zeta(1 - s) = 2(2\pi)^{-s} \cos\frac{\pi s}{2} \, \Gamma(s)\zeta(s).$$

He verified the equation to the point where, he said, there was no doubt about it. This relation was established by Riemann in the 1859 paper referred to below. Riemann, using the zeta function for complex z, attempted to prove the prime number theorem, that is, (5) above.[41] He pointed out that to further the investigation one would have to know the complex zeros of $\zeta(z)$. Actually, $\zeta(z)$ fails to converge for $x \leq 1$ when $z = x + iy$, but the values of ζ in the half-plane $x \leq 1$ are defined by analytic continuation. He expressed the hypothesis that all the zeros in the strip $0 \leq x \leq 1$ lie on the line $x = 1/2$. This hypothesis is still unproven.[42]

In 1896 Hadamard,[43] by applying the theory of entire functions (of a complex variable), which he investigated for the purpose of proving the prime number theorem, and by proving the crucial fact that $\zeta(z) \neq 0$ for $x = 1$, was able finally to prove the prime number theorem. Charles-Jean de la Vallée Poussin (1866–1962) obtained the result for the zeta function

39. *Comm. Acad. Sci. Petrop.*, 9, 1737, 160–88, pub. 1744 = *Opera*, (1), 14, 216–44.
40. *Hist. de l'Acad. de Berlin*, 17, 1761, 83–106, pub. 1768 = *Opera*, (1), 15, 70–90.
41. *Monatsber. Berliner Akad.*, 1859, 671–80 = *Werke*, 145–55.
42. In 1914 Godfrey H. Hardy proved (*Comp. Rend.*, 158, 1914, 1012–14 = *Coll. Papers*, 2, 6–9) that an infinity of zeros of $\zeta(z)$ lie on the line $x = \frac{1}{2}$.
43. *Bull. Soc. Math. de France*, 14, 1896, 199–220 = *Œuvres*, 1, 189–210.

and proved the prime number theorem at the same time.[44] This theorem is a central one in analytic number theory.

Bibliography

Bachmann, P.: "Über Gauss' zahlentheoretische Arbeiten," *Nachrichten König. Ges. der Wiss. zu Gött.*, 1911, 455–508; also in Gauss: *Werke*, 10_2, 1–69.

Bell, Eric T.: *The Development of Mathematics*, 2nd ed., McGraw-Hill, 1945, Chaps. 9–10.

Carmichael, Robert D.: "Some Recent Researches in the Theory of Numbers," *Amer. Math. Monthly*, 39, 1932, 139–60.

Dedekind, Richard: *Über die Theorie der ganzen algebraischen Zahlen* (reprint of the eleventh supplement to Dirichlet's *Zahlentheorie*), F. Vieweg und Sohn, 1964.

————: *Gesammelte mathematische Werke*, 3 vols., F. Vieweg und Sohn, 1930–32, Chelsea (reprint), 1968.

————: "Sur la théorie des nombres entiers algébriques," *Bull. des Sci. Math.*, (1), 11, 1876, 278–88; (2), 1, 1877, 17–41, 69–92, 144–64, 207–48 = *Ges. math. Werke*, 3, 263–96.

Dickson, Leonard E.: *History of the Theory of Numbers*, 3 vols., Chelsea (reprint), 1951.

————: *Studies in the Theory of Numbers* (1930), Chelsea (reprint), 1962.

————: "Fermat's Last Theorem and the Origin and Nature of the Theory of Algebraic Numbers," *Annals. of Math.*, (2), 18, 1917, 161–87.

———— et al.: *Algebraic Numbers, Report of Committee on Algebraic Numbers*, National Research Council, 1923 and 1928; Chelsea (reprint), 1967.

Dirichlet, P. G. L.: *Werke* (1889–97); Chelsea (reprint), 1969, 2 vols.

Dirichlet, P. G. L., and R. Dedekind: *Vorlesungen über Zahlentheorie*, 4th ed., 1894 (contains Dedekind's Supplement); Chelsea (reprint), 1968.

Gauss, C. F.: *Disquisitiones Arithmeticae*, trans. A. A. Clarke, Yale University Press, 1965.

Hasse, H.: "Bericht über neuere Untersuchungen und Probleme aus der Theorie der algebraischen Zahlkörper," *Jahres. der Deut. Math.-Verein.*, 35, 1926, 1–55 and 36, 1927, 233–311.

Hilbert, David: "Die Theorie der algebraischen Zahlkörper," *Jahres. der Deut. Math.-Verein.*, 4, 1897, 175–546 = *Gesammelte Abhandlungen*, 1, 63–363.

Klein, Felix: *Vorlesungen über die Entwicklung der Mathematik im 19. Jahrhundert*, Chelsea (reprint), 1950, Vol. 1.

Kronecker, Leopold: *Werke*, 5 vols. (1895–1931), Chelsea (reprint), 1968. See especially, Vol. 2, pp. 1–10 on the law of quadratic reciprocity.

————: *Grundzüge einer arithmetischen Theorie der algebraischen Grössen*, G. Reimer, 1882 = *Jour. für Math.*, 92, 1881/82, 1–122 = *Werke*, 2, 237–388.

Landau, Edmund: *Handbuch der Lehre von der Verteilung der Primzahlen*, B. G. Teubner, 1909, Vol. 1, pp. 1–55.

Mordell, L. J.: "An Introductory Account of the Arithmetical Theory of Algebraic

44. *Ann. Soc. Sci. Bruxelles*, (1), 20 Part II, 1896, 183–256, 281–397.

Numbers and its Recent Development," *Amer. Math. Soc. Bull.*, 29, 1923, 445–63.

Reichardt, Hans, ed.: *C. F. Gauss, Leben und Werk*, Haude und Spenersche Verlags-buchhandlung, 1960, pp. 38–91; also B. G. Teubner, 1957.

Scott, J. F.: *A History of Mathematics*, Taylor and Francis, 1958, Chap. 15.

Smith, David E.: *A Source Book in Mathematics*, Dover (reprint), 1959, Vol. 1, 107–48.

Smith, H. J. S.: Collected Mathematical Papers, 2 vols. (1890–94), Chelsea (reprint), 1965. Vol. 1 contains Smith's *Report on the Theory of Numbers*, which is also published separately by Chelsea, 1965.

Vandiver, H. S.: "Fermat's Last Theorem," *Amer. Math. Monthly*, 53, 1946, 555–78.

35
The Revival of Projective Geometry

> The doctrines of pure geometry often, and in many questions, give a simple and natural way to penetrate to the origin of truths, to lay bare the mysterious chain which unites them, and to make them known individually, luminously and completely.
>
> MICHEL CHASLES

1. *The Renewal of Interest in Geometry*

For over one hundred years after the introduction of analytic geometry by Descartes and Fermat, algebraic and analytic methods dominated geometry almost to the exclusion of synthetic methods. During this period some men, for example the English mathematicians who persisted in trying to found the calculus rigorously on geometry, produced new results synthetically. Geometric methods, elegant and intuitively clear, always captivated some minds. Maclaurin, especially, preferred synthetic geometry to analysis. Pure geometry, then, retained some life even if it was not at the heart of the most vital developments of the seventeenth and eighteenth centuries. In the early nineteenth century several great mathematicians decided that synthetic geometry had been unfairly and unwisely neglected and made a positive effort to revive and extend that approach.

One of the new champions of synthetic methods, Jean-Victor Poncelet, did concede the limitations of the older pure geometry. He says, "While analytic geometry offers by its characteristic method general and uniform means of proceeding to the solution of questions which present themselves . . . while it arrives at results whose generality is without bound, the other [synthetic geometry] proceeds by chance; its way depends completely on the sagacity of those who employ it and its results are almost always limited to the particular figure which one considers." But Poncelet did not believe that synthetic methods were necessarily so limited and he proposed to create new ones which would rival the power of analytic geometry.

Michel Chasles (1793–1880) was another great proponent of geometrical methods. In his *Aperçu historique sur l'origine et le développement des méthodes en géométrie* (1837), a historical study in which Chasles admitted that he ignored the German writers because he did not know the language, he states that the mathematicians of his time and earlier had declared geometry a dead language which in the future would be of no use and influence. Not only does Chasles deny this but he cites Lagrange, who was entirely an analyst, as saying in his sixtieth year,[1] when he had encountered a very difficult problem of celestial mechanics, "Even though analysis may have advantages over the old geometric methods, which one usually, but improperly, calls synthesis, there are nevertheless problems in which the latter appear more advantageous, partly because of their intrinsic clarity and partly because of the elegance and ease of their solutions. There are even some problems for which the algebraic analysis in some measure does not suffice and which, it appears, the synthetic methods alone can master." Lagrange cites as an example the very difficult problem of the attraction of an ellipsoid of revolution exerted on a point (unit mass) on its surface or inside. This problem had been solved purely synthetically by Maclaurin.

Chasles also gives an extract from a letter he received from Lambert Adolphe Quetelet (1796–1874), the Belgian astronomer and statistician. Quetelet says, "It is not proper that most of our young mathematicians value pure geometry so lightly." The young men complain of lack of generality of method, continues Quetelet, but is this the fault of geometry or of those who cultivate geometry, he asks. To counter this lack of generality, Chasles gives two rules to prospective geometers. They should generalize special theorems to obtain the most general result, which should at the same time be simple and natural. Second, they should not be satisfied with the proof of a result if it is not part of a general method or doctrine on which it is naturally dependent. To know when one really has the true basis for a theorem, he says, there is always a principal truth which one will recognize because other theorems will result from a simple transformation or as a ready consequence. The great truths, which are the foundations of knowledge, always have the characteristics of simplicity and intuitiveness.

Other mathematicians attacked analytic methods in harsher language. Carnot wished "to free geometry from the hieroglyphics of analysis." Later in the century Eduard Study (1862–1922) referred to the machine-like process of coordinate geometry as the "clatter of the coordinate mill."

The objections to analytic methods in geometry were based on more than a personal preference or taste. There was, first of all, a genuine question of whether analytic geometry was really geometry since algebra was the essence of the method and results, and the geometric significance of both were hidden. Moreover, as Chasles pointed out, analysis through its formal

---

1. *Nouv. Mém. de l'Acad. de Berlin*, 1773, 121–48, pub. 1775 = *Œuvres*, 3, 617–58.

processes neglects all the small steps which geometry continually makes. The quick and perhaps penetrating steps of analysis do not reveal the sense of what is accomplished. The connection between the starting point and the final result is not clear. Chasles asks, "Is it then sufficient in a philosophic and basic study of a science to know that something is true if one does not know why it is so and what place it should take in the series of truths to which it belongs?" The geometric method, on the other hand, permits simple and intuitively evident proofs and conclusions.

There was another argument which, first voiced by Descartes, still appealed in the nineteenth century. Geometry was regarded as the truth about space and the real world. Algebra and analysis were not in themselves significant truths even about numbers and functions. They were merely methods of arriving at truths, and artificial at that. This view of algebra and analysis was gradually disappearing. Nevertheless the criticism was still vigorous in the early nineteenth century because the methods of analysis were incomplete and even logically unsound. The geometers rightly questioned the validity of the analytic proofs and credited them with merely suggesting results. The analysts could retort only that the geometric proofs were clumsy and inelegant.

The upshot of the controversy is that the pure geometers reasserted their role in mathematics. As if to revenge themselves on Descartes because his creation of analytic geometry had caused the abandonment of pure geometry, the early nineteenth-century geometers made it their objective to beat Descartes at the game of geometry. The rivalry between analysts and geometers grew so bitter that Steiner, who was a pure geometer, threatened to quit writing for Crelle's *Journal für Mathematik* if Crelle continued to publish the analytical papers of Plücker.

The stimulus to revive synthetic geometry came primarily from one man, Gaspard Monge. We have already discussed his valuable contributions to analytic and differential geometry and his inspiring lectures at the Ecole Polytechnique during the years 1795 to 1809. Monge himself did not intend to do more than bring geometry back into the fold of mathematics as a suggestive approach to and an interpretation of analytic results. He sought only to stress both modes of thought. However, his own work in geometry and his enthusiasm for it inspired in his pupils, Charles Dupin, François-Joseph Servois, Charles-Julien Brianchon, Jean-Baptiste Biot (1774–1862), Lazare-Nicholas-Marguerite Carnot, and Jean-Victor Poncelet, the urge to revitalize pure geometry.

Monge's contribution to pure geometry was his *Traité de géométrie descriptive* (1799). This subject shows how to project orthogonally a three-dimensional object on two planes (a horizontal and a vertical one) so that from this representation one can deduce mathematical properties of the object. The scheme is useful in architecture, the design of fortifications, perspective, carpentry, and stonecutting and was the first to treat the projection of a three-dimensional figure into two two-dimensional ones. The ideas and method of descriptive geometry did not prove to be the avenue to subsequent developments in geometry or, for that matter, to any other part of mathematics.

2. Synthetic Euclidean Geometry

Though the geometers who reacted to Monge's inspiration turned to projective geometry, we shall pause to note some new results in synthetic Euclidean geometry. These results, perhaps minor in significance, nevertheless exhibit new themes and the almost inexhaustible richness of this old subject. Actually hundreds of new theorems were produced, of which we can give just a few examples.

Associated with every triangle ABC are nine particular points, the midpoints of the sides, the feet of the three altitudes, and the midpoints of the segments which join the vertices to the point of intersection of the altitudes. All nine points lie on one circle, called the nine-point circle. This theorem was first published by Gergonne and Poncelet.[2] It is often credited to Karl Wilhelm Feuerbach (1800–34), a high-school teacher, who published his proof in *Eigenschaften einiger merkwürdigen Punkte des geradlinigen Dreiecks* (Properties of Some Distinctive Points of the Rectilinear Triangle, 1822). In this book Feuerbach added another fact about the nine-point circle. An excircle (escribed circle) is one which is tangent to one of the sides and to the extensions of the other two sides. (The center of an escribed circle lies on the bisectors of two exterior angles and the remote interior angle.) Feuerbach's theorem states that the nine-point circle is tangent to the inscribed circle and the three excircles.

In a small book published in 1816, *Über einige Eigenschaften des ebenen geradlinigen Dreiecks* (On Some Properties of Plane Rectilinear Triangles), Crelle showed how to determine a point P inside a triangle such that the lines joining P to the vertices of the triangle and the sides of the triangle make equal angles. That is, $\angle 1 = \angle 2 = \angle 3$ in Figure 35.1. There is also a point P' different from P such that $\angle P'AC = \angle P'CB = \angle P'BA$.

The conic sections, we know, were treated definitively by Apollonius as sections of a cone and then introduced as plane loci in the seventeenth century. In 1822 Germinal Dandelin (1794–1847) proved a very interesting theorem about the conic sections in relation to the cone.[3] His theorem states that if two spheres are inscribed in a circular cone so that they are tangent to a given plane cutting the cone in a conic section, the points of contact of the

2. *Ann. de Math.*, 11, 1820/21, 205–20.
3. *Nouv. Mém. de l'Acad. Roy. des Sci.*, Bruxelles, 2, 1822, 169–202.

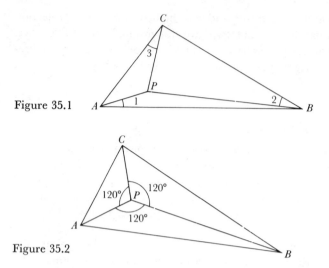

Figure 35.1

Figure 35.2

spheres with the plane are the foci of the conic section, and the intersections of the plane with the planes of the circles along which the spheres touch the cone are the directrices of the conic.

Another interesting theme pursued in the nineteenth century was the solution of maximum and minimum problems by purely geometric methods, that is, without relying upon the calculus of variations. Of the several theorems Jacob Steiner proved by using synthetic methods, the most famous result is the isoperimetric theorem: Of all plane figures with a given perimeter the circle bounds the greatest area. Steiner gave various proofs.[4] Unfortunately, Steiner assumed that there exists a curve that does have maximum area. Dirichlet tried several times to persuade him that his proofs were incomplete on that account but Steiner insisted that this was self-evident. Once, however, he did write (in the first of the 1842 papers):[5] "and the proof is readily made if one assumes that there is a largest figure."

The proof of the existence of a maximizing curve baffled mathematicians for a number of years until Weierstrass in his lectures of the 1870s at Berlin resorted to the calculus of variations.[6] Later Constantin Caratheodory (1873–1950) and Study[7] in a joint paper rigorized Steiner's proofs without employing that calculus. Their proofs (there were two) were direct rather than indirect as in Steiner's method. Hermann Amandus Schwarz, who did great work in partial differential equations and analysis and who

4. *Jour. für Math.*, 18, 1838, 281–96; and 24, 1842, 83–162, 189–250; the 1842 papers are in his *Ges. Werke*, 2, 177–308.
5. *Ges. Werke*, 2, 197.
6. *Werke*, 7, 257–64, 301–2.
7. *Math. Ann.*, 68, 1909, 133–40 = Caratheodory, *Ges. math. Schriften*, 2, 3–11.

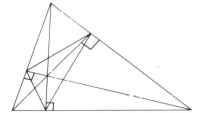

Figure 35.3

served as a professor at several universities including Göttingen and Berlin, gave a rigorous proof for the isoperimetric problem in three dimensions.[8]

Steiner also proved (in the first of the 1842 papers) that of all triangles with a given perimeter, the equilateral has the greatest area. Another of his results[9] states that if A, B, and C are three given points (Fig. 35.2) and if each of the angles of triangle ABC is less than 120°, then the point P for which $PA + PB + PC$ is a minimum is such that each of the angles at P is 120°. If, however, one angle of the triangle, say angle A, is equal to or larger than 120°, then P coincides with A. This result had been proven much earlier by Cavalieri (*Exercitationes Geometricae Sex*, 1647) but it was undoubtedly unknown to Steiner. Steiner also extended the result to n points.

Schwarz solved the following problem: Given an acute-angled triangle, consider all triangles such that each has its vertices on the three sides of the original triangle; the problem is to find the triangle that has least perimeter. Schwarz proved synthetically[10] that the vertices of this triangle of minimum perimeter are the feet of the altitudes of the given triangle (Fig. 35.3).[11]

A theorem novel for Euclidean geometry was discovered in 1899 by Frank Morley, professor of mathematics at Johns Hopkins University, and proofs were subsequently published by many men.[12] The theorem states that if the angle trisectors are drawn at each vertex of a triangle, adjacent trisectors meet at the vertices of an equilateral triangle (Fig. 35.4). The novelty lies in the involvement of angle trisectors. Up to the middle of the nineteenth century no mathematician would have considered such lines because only those elements and figures that are constructible were regarded as having legitimacy in Euclidean geometry. Constructibility guaranteed

8. *Nachrichten König. Ges. der Wiss. zu Gött.*, 1884, 1–13 = *Ges. math. Abh.*, 2, 327–40.

9. *Monatsber. Berliner Akad.*, 1837, 144 = *Ges. Werke*, 2, 93 and 729–31.

10. Paper unpublished, *Ges. Math. Abh.*, 2, 344–45.

11. Schwarz's proof can be found in Richard Courant and Herbert Robbins, *What Is Mathematics?*, Oxford University Press, 1941, pp. 346–49. A proof using the calculus was given by J. F. de' Toschi di Fagnano (1715–97) in the *Acta Eruditorum*, 1775, p. 297. There were less elegant geometrical proofs before Schwarz's.

12. For a proof and references to published proofs see H. S. M. Coxeter, *Introduction to Geometry*, John Wiley and Sons, 1961, pp. 23–25.

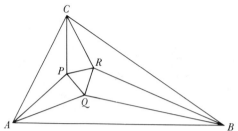

Figure 35.4

existence. However, the conception of what established existence changed as we shall see more clearly when we examine the work on the foundations of Euclidean geometry.

A number of efforts were directed toward reducing the use of straight-edge and compass along the lines initiated by Mohr and Mascheroni (Chap. 12, sec. 2). In his *Traité* of 1822 Poncelet showed that all constructions possible with straightedge and compass (except the construction of circular arcs) are possible with straightedge alone provided that we are given a fixed circle and its center. Steiner re-proved the same result more elegantly in a small book, *Die geometrischen Constructionen ausgeführt mittelst der geraden Linie und eines festen Kreises* (The Geometrical Constructions Executed by Means of the Straight Line and a Fixed Circle).[13] Though Steiner intended the book for pedagogical purposes, he does claim in the preface that he will establish the *conjecture* which a French mathematician had expressed.

The brief sampling above of Euclidean theorems established by synthetic methods should not leave the reader with the impression that analytic geometric methods were not also used. In fact, Gergonne gave analytic proofs of many geometric theorems which he published in the journal he founded, the *Annales de Mathématiques*.

3. *The Revival of Synthetic Projective Geometry*

The major area to which Monge and his pupils turned was projective geometry. This subject had had a somewhat vigorous but short-lived burst of activity in the seventeenth century (Chap. 14) but was submerged by the rise of analytic geometry, the calculus, and analysis. As we have already noted, Desargues's major work of 1639 was lost sight of until 1845 and Pascal's major essay on conics (1639) was never recovered. Only La Hire's books, which used some of Desargues's results, were available. What the nineteenth-century men learned from La Hire's books they often incorrectly credited to him. On the whole, however, these geometers were ignorant of Desargues's and Pascal's work and had to re-create it.

13. Published 1833 = *Werke*, 1, 461–522.

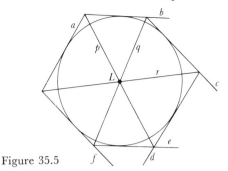

Figure 35.5

The revival of projective geometry was initiated by Lazare N. M. Carnot (1753–1823), a pupil of Monge and father of the distinguished physicist Sadi Carnot. His major work was *Géométrie de position* (1803) and he also contributed the *Essai sur la théorie des transversales* (1806). Monge had espoused the joint use of analysis and pure geometry, but Carnot refused to use analytic methods and started the championship of pure geometry. Many of the ideas we shall shortly discuss more fully are at least suggested in Carnot's work. Thus the principle that Monge called contingent relations and which became known also as the principle of correlativity and more commonly as the principle of continuity is to be found there. To avoid separate figures for various sizes of angles and directions of lines Carnot did not use negative numbers, which he regarded as contradictory, but introduced a complicated scheme called "correspondence of signs."

Among the early nineteenth-century workers in projective geometry we shall just mention François-Joseph Servois and Charles-Julien Brianchon (1785–1864), both of whom made applications of their work to military problems. Though they aided in reconstructing, systematizing, and extending old results, the only new theorem of consequence is Brianchon's famous result,[14] which he proved while still a student at the Ecole. The theorem states that if there are six tangents to a conic (Fig. 35.5), thus forming a circumscribed hexagon, the three lines, each of which joins two opposite vertices, pass through one point. Brianchon derived this theorem by using the pole-polar relationship.

The revival of projective geometry received its main impetus from Poncelet (1788–1867). Poncelet was a pupil of Monge and he also learned much from Carnot. While serving as an officer in Napoleon's campaign against Russia, he was captured and spent the year 1813–14 in a Russian prison at Saratoff. There Poncelet reconstructed without the aid of any books what he had learned from Monge and Carnot and then proceeded to create new results. He later expanded and revised this work and published

14. *Jour de l'Ecole Poly.*, 6, 1806, 297–311.

it as the *Traité des propriétés projectives des figures* (1822). This work was his chief contribution to projective geometry and to the erection of a new discipline. In his later life he was obliged to devote a great deal of time to government service, though he did hold professorships for limited periods.

Poncelet became the most ardent champion of synthetic geometry and even attacked the analysts. He had been friendly with the analyst Joseph-Diez Gergonne (1771–1859) and had published papers in Gergonne's *Annales de Mathématiques*, but his attacks were soon directed to Gergonne also. Poncelet was convinced of the autonomy and importance of pure geometry. Though he admitted the power of analysis he believed that one could give the same power to synthetic geometry. In a paper of 1818, published in Gergonne's *Annales*,[15] he said that the power of analytic methods lay not in the use of algebra but in its generality and this advantage resulted from the fact that the metric properties discovered for a typical figure remain applicable, other than for a possible change of sign, to all related figures which spring from the typical or basic one. This generality could be secured in synthetic geometry by the principle of continuity (which we shall examine shortly).

Poncelet was the first mathematician to appreciate fully that projective geometry was a new branch of mathematics with methods and goals of its own. Whereas the seventeenth-century projective geometers had dealt with specific problems, Poncelet entertained the general problem of seeking all properties of geometrical figures that were common to all sections of any projection of a figure, that is, remain unaltered by projection and section. This is the theme that he and his successors took up. Because distances and angles are altered by projection and section Poncelet selected and developed the theory of involution and of harmonic sets of points but not the concept of cross ratio. Monge had used parallel projection in his work; like Desargues, Pascal, Newton, and Lambert, Poncelet used central projection, that is, projection from a point. This concept Poncelet elevated into a method of approach to geometric problems. Poncelet also considered projective transformation from one space figure to another, of course in purely geometric form. Here he seemed to have lost interest in projective properties and was more concerned with the use of the method in bas-relief and stage design.

His work centers about three ideas. The first is that of homologous figures. Two figures are homologous if one can be derived from the other by one projection and a section, which is called a perspectivity, or by a sequence of projections and sections, that is, a projectivity. In working with homologous figures his plan was to find for a given figure a simpler homologous figure and by studying it find properties which are invariant under projection

---

15. *Ann. de Math.*, 8, 1817/18, 141–55. This paper is reprinted in Poncelet's *Applications d'analyse et de géométrie* (1862–64), 2, 466–76.

and section, and so obtain properties of the more complicated figure. The essence of this method was used by Desargues and Pascal, and Poncelet in his *Traité* praised Desargues's originality in this and other respects.

Poncelet's second leading theme is the principle of continuity. In his *Traité* he phrases it thus: "If one figure is derived from another by a continuous change and the latter is as general as the former, then any property of the first figure can be asserted at once for the second figure." The determination of when both figures are general is not explained. Poncelet's principle also asserts that if a figure should degenerate, as a hexagon does into a pentagon when one side is made to approach zero, any property of the original figure will carry over into an appropriately worded statement for the degenerate figure.

The principle was really not new with Poncelet. In a broad philosophical sense it goes back to Leibniz, who stated in 1687 that when the differences between two cases can be made smaller than any datum in the given, the differences can be made smaller than any given quantity in the result. Since Leibniz's time the principle was recognized and used constantly. Monge began the use of the principle of continuity to establish theorems. He wanted to prove a general theorem but used a special position of the figure to prove it and then maintained that the theorem was true generally, even when some elements in the figures became *imaginary*. Thus to prove a theorem about a line and a surface he would prove it when the line cuts the surface and then maintain that the result holds even when the line no longer cuts the surface and the points of intersection are imaginary. Neither Monge nor Carnot, who also used the principle, gave any justification of it.

Poncelet, who coined the term "principle of continuity," advanced the principle as an absolute truth and applied it boldly in his *Traité*. To "demonstrate" its soundness he takes the theorem on the equality of the products of the segments of intersecting chords of a circle and notes that when the point of intersection moves outside the circle one obtains the equality of the products of the secants and their external segments. Further, when one secant becomes a tangent, the tangent and its external segment become equal and their product continues to equal the product of the other secant and its external segment. All of this was reasonable enough, but Poncelet applied the principle to prove many theorems and, like Monge, extended the principle to make assertions about imaginary figures. (We shall note some examples later.)

The other members of the Paris Academy of Sciences criticized the principle of continuity and regarded it as having only heuristic value. Cauchy, in particular, criticized the principle but unfortunately his criticism was directed at applications made by Poncelet wherein the principle did work. The critics also charged that the confidence which Poncelet and others had in the principle really came from the fact that it could be justified on an algebraic basis. As a matter of fact, the notes Poncelet made in prison show that he did use analysis to test the soundness of the principle. These notes, incidentally, were written up by Poncelet and published by him in two volumes entitled *Applications d'analyse et de géométrie* (1862–64) which is really a revision of his *Traité* of 1822, and in the later work he does use analytic methods. Poncelet admitted that a proof could be based on algebra but he insisted that the principle did not depend on such a proof. However, it is quite certain that Poncelet relied on the algebraic method to see what should be the case and then affirmed the geometric results using the principle as a justification.

Chasles in his *Aperçu* defended Poncelet. Chasles's position was that the algebra is an a posteriori proof of the principle. However, he hedged by pointing out that one must be careful not to carry over from one figure to another any property which depends *essentially* on the elements being real or imaginary. Thus one section of a cone may be a hyperbola, and this has asymptotes. When the section is an ellipse the asymptotes become imaginary. Hence one should not prove a result about the asymptotes alone because these depend upon the particular nature of the section. Also one should not carry the results for a parabola over to the hyperbola because the cutting plane does not have a *general* position in the case of the parabola. Then he discusses the case of two intersecting circles which have a common chord. When the circles no longer intersect, the common chord is imaginary. The fact that the real common chord passes through two *real* points is, he says, an incidental or a contingent property. One must define the chord in some way which does not depend on the fact that it passes through real points when the circles intersect but so that it is a permanent property of the two circles in any position. Thus one can define it as the (real) radical axis, which means that it is a line such that from any point on it the tangents to the two circles are equal, or one can define it by means of the property that any circle drawn about any point of the line as a center cuts the two circles orthogonally.

Chasles, too, insisted that the principle of continuity is suited to treat imaginary elements in geometry. He first explains what one means by the imaginary in geometry. The imaginary elements pertain to a condition or state of a figure in which certain parts are nonexistent, provided these parts are real in another state of the figure. For, he adds, one cannot have any idea of imaginary quantities except by thinking of the related states in which the quantities are real. These latter states are the ones which he called "accidental" and which furnish the key to the imaginary in geometry. To prove results about imaginary elements one has only to take the general condition of the figure in which the elements are real, and then, according to the principle of accidental relations or the principle of continuity, one may conclude that the result holds when the elements are imaginary. "So one sees that the use and the consideration of the imaginary are completely justified." The principle of continuity was accepted during the nineteenth century as intuitively clear and therefore having the status of an axiom. The geometers used it freely and never deemed that it required proof.

Though Poncelet used the principle of continuity to assert results about imaginary points and lines he never gave any general definition of such elements. To introduce some imaginary points he gave a geometrical definition which is complicated and hardly perspicuous. We shall understand these imaginary elements more readily when we discuss them from an algebraic point of view. Despite the lack of clarity in Poncelet's approach he must be credited with introducing the notion of the circular points at infinity, that is, two imaginary points situated on the line at infinity and common to any two circles.[16] He also introduced the imaginary spherical circle which any two spheres have in common. He then proved that two real conics which do not intersect have *two* imaginary common chords, and two conics intersect in four points, real or imaginary.

The third leading idea in Poncelet's work is the notion of pole and polar with respect to a conic. The concept goes back to Apollonius and was used by Desargues (Chap. 14, sec. 3) and others in the seventeenth-century work on projective geometry. Also Euler, Legendre, Monge, Servois, and Brianchon had already used it. But Poncelet gave a general formulation of the transformation from pole to polar and conversely and used it in his *Traité* of 1822 and in his "Mémoire sur la théorie générale des polaires réciproques" presented to the Paris Academy in 1824[17] as a method of establishing many theorems.

One of Poncelet's objectives in studying polar reciprocation with respect to a conic was to establish the principle of duality. The workers in projective geometry had observed that theorems dealing with figures lying in one plane when rephrased by replacing the word "point" by "line" and "line" by "point" not only made sense but proved to be true. The reason for the validity of theorems resulting from such a rephrasing was not clear, and in fact Brianchon questioned the principle. Poncelet thought that the pole and polar relationship was the reason.

However, this relation required the mediation of a conic. Gergonne[18] insisted that the principle was a general one and applied to all statements and theorems except those involving metric properties. Pole and polar were not needed as an intermediary supporting device. He introduced the term "duality" to denote the relationship between the original and new theorem. He also observed that in three-dimensional situations point and plane are dual elements and the line is dual to itself.

To illustrate Gergonne's understanding of the principle of duality let us examine his dualization of Desargues's triangle theorem. We should note

16. *Traité*, 1, 48.
17. *Jour. für Math.*, 4, 1829, 1–71.
18. *Ann. de Math.*, 16, 1825–26, 209–31.

first what the dual of a triangle is. A triangle consists of three points not all on the same line, and the lines joining them. The dual figure consists of three lines not all on the same point, and the points joining them (the points of intersection). The dual figure is again a triangle and so the triangle is called self-dual. Gergonne invented the scheme of writing dual theorems in double columns with the dual alongside the original proposition.

Now let us consider Desargues's theorem, where this time the two triangles and the point O lie in one plane, and let us see what results when we interchange point and line. Gergonne in the 1825–26 paper already referred to wrote this theorem and its dual as follows:

Desargues's Theorem	*Dual of Desargues's Theorem*
If we have two triangles such that lines joining corre-sponding vertices pass through one point O, then corresponding sides intersect in three points on one straight line.	If we have two triangles such that points which are the joins of corre-sponding sides lie on one line O, then corresponding vertices are joined by three lines going through one point.

Here the dual theorem is the converse of the original theorem.

Gergonne's formulation of the general principle of duality was somewhat vague and had deficiencies. Though he was convinced it was a universal principle, he could not justify it and Poncelet rightly objected to the deficiencies. Also he disputed with Gergonne over priority of discovery of the principle (which really belongs to Poncelet) and even accused Gergonne of plagiarism. However, Poncelet did rely upon pole and polar and would not grant that Gergonne had taken a step forward in recognizing the wider application of the principle. Later discussions among Poncelet, Gergonne, Möbius, Chasles, and Plücker clarified the principle for all, and Möbius in his *Der barycentrische Calcul* and Plücker, later, gave a good statement of the relationship of the principle of duality to pole and polar: The notion of duality is independent of conics and quadric forms but agrees with pole and polar when the latter can be used. The logical justification of the general principle of duality was not supplied at this time.

The synthetic development of projective geometry was furthered by Jacob Steiner (1796–1863). He is the first of a German school of geometers who took over French ideas, notably Poncelet's, and favored synthetic methods to the extent of even hating analysis. The son of a Swiss farmer, he himself worked on the farm until he was nineteen years old. Though he was largely self-educated he ultimately became a professor at Berlin. In his younger years he was a teacher in Pestalozzi's school and was impressed by the importance of building up the geometric intuition. It was Pestalozzi's principle to get the student to create the mathematics with the lead of the teacher and the use of the Socratic method. Steiner went to extremes. He

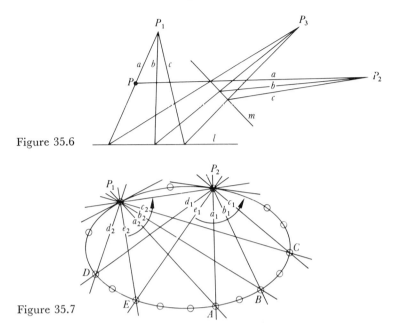

Figure 35.6

Figure 35.7

taught geometry but used no figures, and in training doctoral candidates he darkened the room. In his own later work Steiner took over theorems and proofs published in English and other journals and gave no indication in his own publications that his results had already been established. He had done good original work when younger and sought to maintain his reputation for productivity.

His major work was *Systematische Entwicklung der Abhängigkeit geometrischen Gestalten von einander* (Systematic Development of the Dependence of Geometric Forms on One Another, 1832) and his chief principle was to use projective concepts to build up more complicated structures from simple ones such as points, lines, pencils of lines, planes, and pencils of planes. His results were not especially new but his method was.

To illustrate his principle we shall examine his now standard projective method of defining the conic sections. One starts (Fig. 35.6) with two pencils of lines (families of concurrent lines), say p_1 and p_3, that are perspectively related through a pencil of points on a line l, and the pencils p_3 and p_2 that are perspectively related by means of a pencil of points on another line m. Then the pencils p_1 and p_2 are said to be projectively related. The lines marked a in the pencil with center at P_1 and the pencil with center at P_2 are examples of corresponding lines of the projectivity between the two pencils p_1 and p_2. A conic is now defined as the set of points of intersection of all pairs of corresponding lines of the two projective pencils. Thus P is a point on the conic. Moreover, the conic passes through P_1 and P_2 (Fig. 35.7). In this manner Steiner built up the conics or second degree curves by means of the simpler forms, pencils of lines. However, he did not identify his conics with sections of a cone.

He also built up the ruled quadrics, the hyperboloid of one sheet and the hyperbolic paraboloid, in a similar manner, making projective correspondence the basis of his definitions. Actually his method is not sufficiently general for all of projective geometry.

In his proofs he used cross ratio as a fundamental tool. However, he ignored imaginary elements and referred to them as "the ghosts" or the "shadows of geometry." He also did not use signed quantities though Möbius, whose work we shall soon examine, had already introduced them.

Steiner used the principle of duality from the very outset of his work. Thus he dualized the definition of a conic to obtain a new structure called a line curve. If one starts with two projectively (but not perspectively) related pencils of points then the family of lines (Fig. 35.8) joining the corresponding points of the two pencils is called a line conic. To distinguish such families of lines, which also describe a curve, the usual curve as a locus of points is called a point curve. The tangent lines of a point curve are a line curve and in the case of a conic do constitute the dual curve. Conversely every line conic envelopes a point conic or is the collection of tangents of a point conic.

With Steiner's notion of a dual to a point conic, one can dualize many theorems. Let us take Pascal's theorem and form the dual statement. We shall write the theorem on the left and the new statement on the right.

Pascal's Theorem	*Dual of Pascal's Theorem*
If we take six points, A, B, C, D, E, and F on the point conic, then the lines which join A and B and D and E meet in a point P; the lines which join B and C and E and F meet in a point Q; the lines which join C and D and F and A meet in a point R. The three points P, Q, and R lie on one line l.	If we take six lines a, b, c, d, e, and f on the line conic, then the points which join a and b and d and e are joined by the line p; the points which join b and c and e and f are joined by the line q; the points which join c and d and f and a are joined by the line r. The three lines p, q, and r are on one point L.

Figure 14.12 of Chapter 14 illustrates Pascal's theorem. The dual theorem is the one that Brianchon had discovered by means of the pole and polar relationship (Fig. 35.5 above). Steiner, like Gergonne, did nothing to establish the logical basis for the principle of duality. However, he developed projective geometry systematically by classifying figures and by noting the dual statements as he proceeded. He also studied thoroughly curves and surfaces of the second degree.

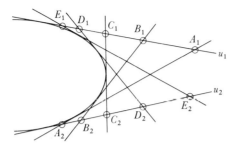

Figure 35.8

Michel Chasles, who devoted his entire life to geometry, followed up on the work of Poncelet and Steiner, though he did not personally know Steiner's work because, as we have noted, Chasles could not read German. Chasles presented his own ideas in his *Traité de géométrie supérieure* (1852) and *Traité des sections coniques* (1865). Since much of Chasles's work either unintentionally duplicated Steiner's or was superseded by more general concepts, we shall note just a few major results that are due to him.

Chasles got the idea of cross ratio from his attempts to understand the lost work of Euclid, *Porisms* (though Steiner and Möbius had already re-introduced it). Desargues, too, had used the concept, but Chasles knew only what La Hire had written about it. Chasles did learn at some time that the idea is also in Pappus because in Note IX of his *Aperçu* (p. 302) he refers to Pappus's use of the idea. One of Chasles's results in this area[19] is that four fixed points of a conic and any fifth point of the conic determine four lines with the same cross ratio.

In 1828 Chasles[20] gave the theorem: Given two sets of collinear points in one-to-one correspondence and such that the cross ratio of any four points on one line equals that of the corresponding points on the other, then the lines joining corresponding points are tangent to a conic that touches the two given lines. This result is the equivalent of Steiner's definition of a line conic, because the cross ratio condition here ensures that the two sets of collinear points are projectively related and the lines joining corresponding points are the lines of Steiner's line conic.

Chasles pointed out that by virtue of the principle of duality lines can be as fundamental as points in the development of plane projective geometry and credits Poncelet and Gergonne with being clear on this point. Chasles also introduced new terminology. The cross ratio he called anharmonic ratio. He introduced the term homography to describe a transformation of a plane into itself or another plane, which carries points into points and lines into lines. This term covers homologous or projectively related figures. He

19. *Correspondance mathématique et physique*, 5, 1829, 6–22.
20. *Correspondance mathématique et physique*, 4, 1828, 363–71.

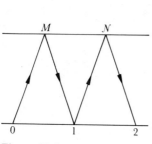

Figure 35.9

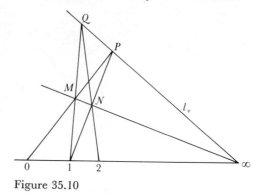

Figure 35.10

added the condition that the transformation must preserve cross ratio but this latter fact can be proved. The transformation that carries points into lines and lines into points he called a correlation.

Though he defended pure geometry Chasles thought analytically but presented his proofs and results geometrically. This approach is called the "mixed method" and was used later by others.

By 1850 the general concepts and goals of projective geometry as distinguished from Euclidean geometry were clear; nevertheless the logical relationship of the two geometries was not clarified. The concept of length was used in projective geometry from Desargues to Chasles. In fact the concept of cross ratio was defined in terms of length. Yet length is not a projective concept because it is not invariant under projective transformation. Karl Georg Christian von Staudt (1798–1867), a professor at Erlangen who was interested in foundations, decided to free projective geometry from dependence on length and congruence. The essence of his plan, presented in his *Geometrie der Lage* (Geometry of Position, 1847), was to introduce an analogue of length on a projective basis. His scheme is called "the algebra of throws." One chooses three arbitrary points on a line and assigns to them the symbols 0, 1, ∞. Then by means of a geometric construction (which in itself comes from Möbius)—a "throw"—a symbol is attached to any arbitrary point P.

To see what the construction amounts to in Euclidean geometry suppose we start with the points labelled 0 and 1 on a line (Fig. 35.9). Through the point M on a parallel line draw $0M$ and then draw $1N$ parallel to $0M$. Now draw $1M$ and draw $N2$ parallel to $1M$. It is immediate that $01 = 12$ because opposite sides of a parallelogram are equal. Thus one carries over the length 01 to 12 by a geometric construction.

Now to treat the projective case we start with three points 0, 1, ∞ (Fig. 35.10). The point ∞ lies on l_∞, the line at infinity, but this is just an ordinary line in projective geometry. Now pick a point M and draw a "parallel" to the line 01 through M. This means that the line through M must meet the line 01 at ∞ and so we draw $M\infty$. Now draw $0M$ and prolong it to l_∞. Next we draw through 1 a "parallel" to $0M$. This means the "parallel" through 1 must meet $0M$ on l_∞. We thereby get the line $1P$ and this determines N. Now draw $1M$ and prolong it until it meets l_∞ at Q. The line through N parallel to $1M$ is QN and where it meets 01 we get a point which is labeled 2.

By this type of construction we can attach "rational coordinates" to points on the line 01∞. To assign irrational numbers to points on the line one must introduce an axiom of continuity (Chap. 41). This notion was not well understood at that time and as a consequence von Staudt's work lacked rigor.

Von Staudt's assignment of coordinates to the points of a line did not involve length. His coordinates, though they were the usual number symbols, served as systematic identification symbols for the points. To add or subtract such "numbers" von Staudt could not then use the laws of arithmetic. Instead he gave geometric constructions that defined the operations with these symbols so that, for example, the sum of the numbers 2 and 3 is the number 5. These operations obeyed all the usual laws of numbers. Thus his symbols or coordinates could be treated as ordinary numbers even though they were built up geometrically.

With these labels attached to his points von Staudt could define the cross ratio of four points. If the coordinates of these points are x_1, x_2, x_3, and x_4, then the cross ratio is defined to be

$$\frac{x_1 - x_3}{x_1 - x_4} \bigg/ \frac{x_2 - x_3}{x_2 - x_4}.$$

Thus von Staudt had the fundamental tools to build up projective geometry without depending on the notions of length and congruence.

A harmonic set of four points is one for which the cross ratio is -1. On the basis of harmonic sets von Staudt gave the fundamental definition that two pencils of points are projectively related when under a one-to-one correspondence of their members a harmonic set corresponds to a harmonic set. Four concurrent lines form a harmonic set if the points in which they meet an arbitrary transversal constitute a harmonic set of points. Thus the projectivity of two pencils of lines can also be defined. With these notions von Staudt defined a collineation of the plane into itself as a one-to-one transformation of point to point and line to line and showed that it carries a set of harmonic elements into a set of harmonic elements.

The principal contribution of von Staudt in his *Geometrie der Lage* was to show that projective geometry is indeed more fundamental than Euclidean. Its concepts are logically antecedent. This book and his *Beiträge zur Geometrie der Lage* (Contributions to the Geometry of Position, 1856, 1857, 1860) revealed projective geometry as a subject independent of distance. However, he did use the parallel axiom of Euclidean geometry, which from a logical standpoint was a blemish because parallelism is not a projective invariant. This blemish was removed by Felix Klein.[21]

4. *Algebraic Projective Geometry*

While the synthetic geometers were developing projective geometry, the algebraic geometers pursued their methods of treating the same subject. The first of the new algebraic ideas was what are now called homogeneous coordinates. One such scheme was created by Augustus Ferdinand Möbius (1790–1868), who like Gauss and Hamilton made his living as an astronomer but devoted considerable time to mathematics. Though Möbius did not take sides in the controversy on synthetic versus algebraic methods his contributions were on the algebraic side.

His scheme for representing the points of a plane by coordinates introduced in his major work, *Der barycentrische Calcul*,[22] was to start with a fixed triangle and to take as the coordinates of any point P in the plane the amount of mass which must be placed at each of the three vertices of the triangle so that P would be the center of gravity of the three masses. When P lies outside the triangle then one or two of the coordinates can be negative. If the three masses are all multiplied by the same constant the point P remains the center of gravity. Hence in Möbius's scheme the coordinates of a point are not unique; only the ratios of the three are determined. The same scheme applied to points in space requires four coordinates. By writing the equations of curves and surfaces in this coordinate system the equations become homogeneous; that is, all terms have the same degree. We shall see examples of the use of homogeneous coordinates shortly.

Möbius distinguished the types of transformations from one plane or space to another. If corresponding figures are equal, the transformation is a congruence and if they are similar, the transformation is a similarity. Next in generality is the transformation that preserves parallelism though not length or shape and this type is called affine (a notion introduced by Euler). The most general transformation that carries lines into lines he called a collineation. Möbius proved in *Der barycentrische Calcul* that every collineation is a projective transformation; that is, it results from a sequence of perspectivities. His proof assumed that the transformation is one-to-one and continuous, but the continuity condition can be replaced by a weaker one. He also gave an analytical representation of the transformation. As Möbius pointed out, one could consider invariant properties of figures under each of the above types of transformations.

21. *Math. Ann.*, 6, 1873, 112–45 = *Ges. Abh.*, 1, 311–43. See also Chap. 38, sec. 3.
22. Published 1827 = *Werke*, 1, 1–388.

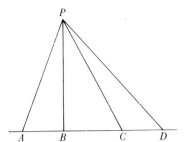

Figure 35.11

Möbius introduced signed elements in geometry not only for the line segments but for areas and volumes. Consequently he was able to give a complete treatment of the notion of the signed cross ratio of four points on a line. He also showed that the cross ratio of four lines of a pencil can be expressed in terms of the sines of the angles at the vertex P (Fig. 35.11) by

$$\frac{\sin APB}{\sin APC} \bigg/ \frac{\sin BPD}{\sin CPD}$$

and this ratio is the same as the cross ratio of the four points A, B, C, and D cut out on any transversal to the lines of the pencil. Hence cross ratio is unaltered by projection and section. Möbius had many other ideas which he developed slowly and did not push very far.

The man who gave the algebraic approach to projective geometry its efficacy and vitality is Julius Plücker (1801–68). After serving as a professor of mathematics at several institutions until 1836, he became a professor of mathematics and physics at Bonn, a position he held for the rest of his life. Plücker was primarily a physicist and in fact an experimental physicist in which activity he made many notable discoveries. From 1863 on he again devoted himself to mathematics.

Plücker, too, introduced homogeneous coordinates but in a manner different from Möbius's. His first notion, trilinear coordinates,[23] was also presented in the second volume of his *Analytisch-geometrische Entwickelungen* (1828 and 1831). He starts with a fixed triangle and takes the coordinates of any point P to be the signed perpendicular distances from P to the sides of the triangle; each distance can be multiplied by the same arbitrary constant. Later in the second volume he introduced a special case which amounts to regarding one line of the triangle as the line at infinity. This is equivalent to replacing the usual rectangular Cartesian coordinates x and y by $x = x_1/x_3$ and $y = x_2/x_3$ so that the equations of curves become homogeneous in x_1, x_2, and x_3. The latter notion is the one which became more widely used.

23. *Jour. für Math.*, 5, 1830, 1–36 = *Wiss. Abh.*, 1, 124–58.

By using homogeneous coordinates and Euler's theorem on homogeneous functions, which states that if $f(tx_1, tx_2, tx_3) = t^n f(x_1, x_2, x_3)$ then

$$x_1 \frac{\partial f}{\partial x_1} + x_2 \frac{\partial f}{\partial x_2} + x_3 \frac{\partial f}{\partial x_3} = nf(x_1, x_2, x_3),$$

Plücker was able to give elegant algebraic representations of geometric ideas. Thus if $f(x_1, x_2, x_3) = 0$ is the equation of a conic with (x_1, x_2, x_3) as the coordinates of a point on the conic then

$$\frac{\partial f}{\partial x_1} x_1' + \frac{\partial f}{\partial x_2} x_2' + \frac{\partial f}{\partial x_3} x_3' = 0$$

can be interpreted, when x_1', x_2', x_3' are the running coordinates, as the equation of the tangent at the point (x_1, x_2, x_3) or, when x_1, x_2, x_3 are the running coordinates, as the equation of the polar line of the arbitrary point (x_1', x_2', x_3') with respect to the conic.

Using homogeneous coordinates Plücker gave the algebraic formulation of the infinitely distant line, the circular points at infinity, and other notions. In the homogeneous coordinate system (x_1, x_2, x_3) the equation of the infinitely distant line is $x_3 = 0$. This line is not exceptional in projective geometry, but in our visualization of the geometric elements each normal point of the Euclidean plane is at a finite position given by $x = x_1/x_3$ and $y = x_2/x_3$ and so we are obliged to think of points on $x_3 = 0$ as infinitely distant.

The equation of a circle

$$(x - a)^2 + (y - b)^2 = r^2$$

becomes on the introduction of homogeneous Cartesian coordinates x_1, x_2, and x_3 through $x = x_1/x_3$ and $y = x_2/x_3$

$$(x_1 - ax_3)^2 + (x_2 - bx_3)^2 = r^2 x_3^2.$$

Since the equation of the infinitely distant line is $x_3 = 0$, the intersection of this line with the circle is given by

$$x_1^2 + x_2^2 = 0 \quad \text{and} \quad x_3 = 0,$$

and this is the equation of the circular points at ∞. These circular points have the coordinates $(1, i, 0)$ and $(1, -i, 0)$ or triples proportional to them. Likewise the equation of the spherical circle (*Kugelkreis*) at infinity is

$$x_1^2 + x_2^2 + x_3^2 = 0 \quad \text{and} \quad x_4 = 0.$$

If we write the equation of a straight line in the homogeneous form (we shall use x, y, and z in place of x_1, x_2, and x_3)

$$Ax + By + Cz = 0$$

and now require that the line pass through the points (x_1, y_1, z_1) and $(1, i, 0)$ then the resulting nonhomogeneous equation of the line is

$$x - x_0 + i(y - y_0) = 0$$

wherein $x_0 = x_1/z_1$, and $y_0 = y_1/z_1$. Likewise the equation of the line through (x_1, y_1, z_1) and $(1, -i, 0)$ is

$$x - x_0 - i(y - y_0) = 0.$$

Each of these lines is perpendicular to itself because the slope equals its negative reciprocal. Sophus Lie called them crazy lines; they are now called isotropic lines.

Plücker's efforts to treat duality algebraically led him to a beautiful idea, line coordinates.[24] If the equation of a line in homogeneous coordinates is

$$ux + vy + wz = 0,$$

and if x, y, and z are fixed quantities, u, v, and w or any three numbers proportional to them are the coordinates of a line in the plane. Then just as an equation $f(x_1, x_2, x_3) = 0$ represents a collection of points, so $f(u, v, w) = 0$ represents a collection of lines or a line curve.

With this notion of line coordinates Plücker was able to give an algebraic formulation and proof of the principle of duality. Given any equation $f(r, s, t) = 0$ if we interpret r, s, and t as the homogeneous coordinates x_1, x_2, and x_3 of a point then we have the equation of a point curve, whereas if we interpret them as u, v, and w we have the dual line curve. Any property proved by algebraic processes for the point curve will, because the algebra is the same under either interpretation of the variables, give rise to the dual property of the line curve.

In this second 1830 paper and in Volume 2 of his *Entwickelungen* Plücker also pointed out that the very same curve regarded as a collection of points can also be regarded as the collection of lines tangent to the curve because the tangents determine the shape as much as the points do. The family of tangents is a line curve and has an equation in line coordinates. The degree of this equation is called the class of the curve, whereas the degree of the equation in point coordinates is called the order.

5. *Higher Plane Curves and Surfaces*

The eighteenth-century men had done some work on curves of degree higher than the second (Chap. 23, sec. 3) but the subject was dormant from 1750 to 1825. Plücker took up third and fourth degree curves and used projective concepts freely in this work.

24. *Jour. für Math.*, 6, 1830, 107–46 = *Wiss. Abh.*, 1, 178–219.

In his *System der analytischen Geometrie* (1834) he used a principle, which though helpful was hardly well grounded, to establish canonical forms of curves. To show that the general curve of order (degree) 4, say, could be put into a particular canonical form he argued that if the number of constants were the same in the two forms then one could be converted into the other. Thus he argued that a ternary (three-variable) form of fourth order could always be put in the form

$$C_4 = pqrs + \mu\Omega^2,$$

where p, q, r, and s are linear and Ω is a quadratic form, because both sides contained 14 constants. In his equations μ was real as were the coefficients.

Plücker also took up the number of intersection points of curves, a topic which had also been pursued in the eighteenth century. He used a scheme for representing the family of all curves which go through the intersections of two nth degree curves C_n' and C_n'' which had been introduced by Lamé in a book of 1818. Any curve C_n passing through these intersections can be expressed as

$$C_n = C_n' + \lambda C_n'' = 0$$

where λ is a parameter.

Using this scheme Plücker gave a clear explanation of Cramer's paradox (Chap. 23, sec. 3). A general curve C_n is determined by $n(n + 3)/2$ points because this is the number of essential coefficients in its equation. On the other hand, since two C_n's intersect in n^2 points, through the $n(n + 3)/2$ points in which two C_n's intersect an infinite number of other C_n's pass. The apparent contradiction was explained by Plücker.[25] Any two nth degree curves do indeed meet in n^2 points. However, only $(n/2)(n + 3) - 1$ points are independent. That is, if we take two nth degree curves through the $(n/2)(n + 3) - 1$ points any other nth degree curve through these points will pass through the remaining $(n - 1)(n - 2)/2$ of the n^2 points. Thus when $n = 4$, 13 points are independent. Any two curves through the 13 points determine 16 points but any other curve through the 13 will pass through the remaining three.

Plücker then took up[26] the theory of intersections of an mth degree curve and an nth degree curve. He regarded the latter as fixed and the curves intersecting it as variable. Using the abridged notation C_n for the expression for the nth degree curve and similar notation for the others he wrote

$$C_m = C_m' + A_{m-n}C_n = 0$$

for the case where $m > n$ so that A_{m-n} is a polynomial of degree $m - n$. From this equation Plücker got the correct method of determining the inter-

25. *Annales de Math.*, 19, 1828, 97–106 = *Wiss. Abh.*, 1, 76–82.
26. *Jour. für Math.*, 16, 1837, 47–54.

sections of a C_n with all curves of degree m. Since according to this equation $m - n + 1$ linearly independent curves (the number of coefficients in A_{m-n}) pass through the intersections of C'_m and C_n, Plücker's conclusion was, given $mn - (n - 1)(n - 2)/2$ arbitrary points on C_n, the remaining $(n - 1)(n - 2)/2$ of the mn intersection points with a C_m are determined. The same result was obtained at about the same time by Jacobi.[27]

In his *System* of 1834 and more explicitly in his *Theorie der algebraischen Curven* (1839) Plücker gave what are now called the Plücker formulas which relate the order n and the class k of a curve and the simple singularities. Let d be the number of double points (singular points at which the two tangents are distinct) and r the number of cusps. To the double points there correspond in the line curve double tangents (a double tangent is actually tangent at two distinct points) the number of which is, say, t. To the cusps there correspond osculating tangents (tangents which cross the curve at inflection points) the number of which is, say, w. Then Plücker was able to establish the following dual formulas:

$$k = n(n - 1) - 2d - 3r \qquad n = k(k - 1) - 2t - 3w$$
$$w = 3n(n - 2) - 6d - 8r \qquad r = 3k(k - 2) - 6t - 8w.$$

The number of any one element includes real and imaginary cases.

In the case when $n = 3$, $d = 0$, and $r = 0$, then w, the number of inflection points, is 9. Up to Plücker's time, De Gua and Maclaurin had proved that a line through two inflection points of the general curve of third order goes through a third one, and the fact that a general C_3 had three real inflection points had been assumed from Clairaut's time on. In his *System* of 1834 Plücker proved that a C_3 has either one or three real inflection points, and in the latter case they lie on one line. He also arrived at the more general result which takes complex elements into account. A general C_3 has nine inflection points of which six are imaginary. To derive this result he used his principle of counting constants to show that

$$C_3 = fgh - l^3,$$

where f, g, h, and l are linear forms, and derived the result of De Gua and Maclaurin. He then showed (with incomplete arguments) that the nine inflection points of C_3 lie three on a line so that there are twelve such lines. Ludwig Otto Hesse (1811–74), who served as a professor at several universities, completed Plücker's proof[28] and showed that the twelve lines can be grouped into four triangles.

As an additional example of the discovery of general properties of curves we shall consider the problem of the inflection points of an nth degree

27. *Jour. für Math.*, 15, 1836, 285–308 = *Werke*, 3, 329–54.
28. *Jour. für Math.*, 28, 1844, 97–107 = *Ges. Abh.*, 123–35.

curve $f(x, y) = 0$. Plücker had expressed the usual calculus condition for an inflection point when $y = f(x)$, namely, $d^2y/dx^2 = 0$, in the form appropriate to $f(x, y) = 0$ and obtained an equation of degree $3n - 4$. Since the original curve and the new curve must have $n(3n - 4)$ intersections, it appeared that the original curve has $n(3n - 4)$ points of inflection. Because this number was too large Plücker suggested that the curve of the equation of degree $3n - 4$ has a tangential contact with each of the n infinite branches of the original curve $f = 0$, so that $2n$ of the common points are not inflection points, and so obtained the correct number, $3n(n - 2)$. Hesse clarified this point [29] by using homogeneous coordinates; that is, he replaced x by x_1/x_3, and y by x_2/x_3, and by using Euler's theorem on homogeneous functions he showed that Plücker's equation for the inflection points can be written as

$$H = \begin{vmatrix} f_{11} & f_{12} & f_{13} \\ f_{21} & f_{22} & f_{23} \\ f_{31} & f_{32} & f_{33} \end{vmatrix} = 0,$$

where the subscripts denote partial derivatives. This equation is of degree $3(n - 2)$ and so meets the curve $f(x_1, x_2, x_3) = 0$ of the nth degree in the correct number of inflection points. The determinant itself is called the Hessian of f, a notion introduced by Hesse.[30]

Plücker among others took up the investigation of quartic curves. He was the first to discover (*Theorie der algebraischen Curven*, 1839) that such curves contain 28 double tangents of which eight at most were real. Jacobi then proved [31] that a curve of nth order has in general $n(n - 2)(n^2 - 9)/2$ double tangents.

The work in algebraic geometry also covered figures in space. Though the representation of straight lines in space had already been introduced by Euler and Cauchy, Plücker in his *System der Geometrie des Raumes* (System of Geometry of Space, 1846) introduced a modified representation

$$x = rz + \rho, \qquad y = sz + \sigma$$

in which the four parameters r, ρ, s, and σ fix the line. Now lines can be used to build up all of space, since, for example, planes are no more than collections of lines, and points are intersections of lines. Plücker then said that if lines are regarded as the fundamental element of space, space is four-dimensional because four parameters are needed to cover all of space with lines. The notion of a four-dimensional space of points he rejected as too

29. *Jour. für Math.*, 41, 1851, 272–84 = *Ges. Abh.*, 263–78.
30. *Jour. für Math.*, 28, 1844, 68–96 = *Ges. Abh.*, 89–122.
31. *Jour. für Math.*, 40, 1850, 237–60 = *Werke*, 3, 517–42.

metaphysical. That the dimension depends on the space-element is the new thought.

The study of figures in space included surfaces of the third and fourth degrees. A ruled surface is generated by a line moving according to some law. The hyperbolic paraboloid (saddle-surface) and hyperboloid of one sheet are examples, as is the helicoid. If a surface of the second degree contains one line it contains an infinity of lines and it is ruled. (It must then be a cone, cylinder, the hyperbolic paraboloid, or the hyperboloid of one sheet.) However, this is not true of cubic surfaces.

As an example of a remarkable property of cubic surfaces there was Cayley's discovery in 1849[32] of the existence of exactly 27 lines on every surface of the third degree. Not all need be real but there are surfaces for which they are all real. Clebsch gave an example in 1871.[33] These lines have special properties. For example, each is cut by ten others. Much further work was devoted to the study of these lines on cubic surfaces.

Among many discoveries concerning surfaces of fourth order one of Kummer's results deserves mention. He had worked with families of lines that represent rays of light, and by considering the associated focal surfaces[34] he was led to introduce a fourth degree surface (and of class four) with 16 double points and 16 double planes as the focal surface of a system of rays of second order. This surface, known as the Kummer surface, embraces as a special case the Fresnel wave surface, which represents the wave front of light propagating in anisotropic media.

The work on synthetic and algebraic projective geometry of the first half of the nineteenth century opened up a brilliant period for geometrical researches of all kinds. The synthetic geometers dominated the period. They attacked each new result to discover some general principle, often not demonstrable geometrically, but from which they nevertheless derived a spate of consequences tied one to the other and to the general principle. Fortunately, algebraic methods were also introduced and, as we shall see, ultimately dominated the field. However, we shall interrupt the history of projective geometry to consider some revolutionary new creations that affected all subsequent work in geometry and, in fact, altered radically the face of mathematics.

Bibliography

Berzolari, Luigi: "Allgemeine Theorie der höheren ebenen algebraischen Kurven," *Encyk. der Math. Wiss.*, B. G. Teubner, 1903–15, III C4, 313–455.

32. *Cambridge and Dublin Math. Jour.*, 4, 1849, 118–32 = *Math. Papers*, 1, 445–56.
33. *Math. Ann.*, 4, 1871, 284–345.
34. *Monatsber. Berliner Akad.*, 1864, 246–60, 495–99.

Boyer, Carl B.: *History of Analytic Geometry*, Scripta Mathematica, 1956, Chaps. 8–9.

Brill, A., and M. Noether: "Die Entwicklung der Theorie der algebraischen Functionen in älterer und neuerer Zeit," *Jahres. der Deut. Math.-Verein.*, 3, 1892/3, 109–566, 287–312 in particular.

Cajori, Florian: *A History of Mathematics*, 2nd ed., Macmillan, 1919, pp. 286–302, 309–14.

Coolidge, Julian L.: *A Treatise on the Circle and the Sphere*, Oxford University Press, 1916.

———: *A History of Geometrical Methods*, Dover (reprint), 1963, Book I, Chap. 5 and Book II, Chap. 2.

———: *A History of the Conic Sections and Quadric Surfaces*, Dover (reprint), 1968.

———: "The Rise and Fall of Projective Geometry," *American Mathematical Monthly*, 41, 1934, 217–28.

Fano, G.: "Gegensatz von synthetischer und analytischer Geometrie in seiner historischen Entwicklung im XIX. Jahrhundert," *Encyk. der Math. Wiss.*, B. G. Teubner 1907–10, III AB4a, 221–88.

Klein, Felix: *Elementary Mathematics from an Advanced Standpoint*, Macmillan, 1939; Dover (reprint), 1945, Geometry, Part 2.

Kötter, Ernst: "Die Entwickelung der synthetischen Geometrie von Monge bis auf Staudt," 1847, *Jahres. der Deut. Math.-Verein.*, Vol. 5, Part II, 1896 (pub. 1901), 1–486.

Möbius, August F.: *Der barycentrische Calcul* (1827), Georg Olms (reprint), 1968. Also in Vol. 1 of *Gesammelte Werke*, pp. 1–388.

———: *Gesammelte Werke*, 4 vols., S. Hirzel, 1885–87; Springer-Verlag (reprint), 1967.

Plücker, Julius: *Gesammelte wissenschaftliche Abhandlungen*, 2 vols., B. G. Teubner, 1895–96.

Schoenflies, A.: "Projektive Geometrie," *Encyk. der Math. Wiss.*, B. G. Teubner, 1907–10, III AB5, 389–480.

Smith, David Eugene: *A Source Book in Mathematics*, Dover (reprint), 1959, Vol. 2, pp. 315–23, 331–45, 670–76.

Steiner, Jacob: *Geometrical Constructions With a Ruler* (a translation of his 1833 book), Scripta Mathematica, 1950.

———: *Gesammelte Werke*, 2 vols., G. Reimer, 1881–82; Chelsea (reprint), 1971.

Zacharias, M.: "Elementargeometrie and elementare nicht-euklidische Geometrie in synthetischer Behandlung," *Encyk. der Math. Wiss.*, B. G. Teubner, 1907–10, III, AB9, 859–1172.

36
Non-Euclidean Geometry

> ...because it seems to be true that many things have, as it were, an epoch in which they are discovered in several places simultaneously, just as the violets appear on all sides in the springtime. WOLFGANG BOLYAI

> The character of necessity ascribed to the truths of mathematics and even the peculiar certainty attributed to them is an illusion. JOHN STUART MILL

1. Introduction

Amidst all the complex technical creations of the nineteenth century the most profound one, non-Euclidean geometry, was technically the simplest. This creation gave rise to important new branches of mathematics but its most significant implication is that it obliged mathematicians to revise radically their understanding of the nature of mathematics and its relation to the physical world. It also gave rise to problems in the foundations of mathematics with which the twentieth century is still struggling. As we shall see, non-Euclidean geometry was the culmination of a long series of efforts in the area of Euclidean geometry. The fruition of this work came in the early nineteenth century during the same decades in which projective geometry was being revived and extended. However, the two domains were not related to each other at this time.

2. The Status of Euclidean Geometry about 1800

Though the Greeks had recognized that abstract or mathematical space is distinct from sensory perceptions of space, and Newton emphasized this point,[1] all mathematicians until about 1800 were convinced that Euclidean geometry was the correct idealization of the properties of physical space and of figures in that space. In fact, as we have already noted, there were many attempts to build arithmetic, algebra, and analysis, whose logical foundation

1. *Principia*, Book I, Def. 8, Scholium.

was obscure, on Euclidean geometry and thereby guarantee the truth of these branches too.

Many men actually voiced their absolute trust in the truth of Euclidean geometry. For example, Isaac Barrow, who built his mathematics including his calculus on geometry, lists eight reasons for the certainty of geometry: the clearness of the concepts, the unambiguous definitions, the intuitive assurance and universal truth of its axioms, the clear possibility and easy imaginability of its postulates, the small number of its axioms, the clear conceivability of the mode by which magnitudes are generated, the easy order of the demonstrations, and the avoidance of things not known.

Barrow did raise the question, How are we sure that the geometric principles do apply to nature? His answer was that these are derived from innate reason. Sensed objects are merely the agents which awaken them. Moreover the principles of geometry had been confirmed by constant experience and would continue to be so because the world, designed by God, is immutable. Geometry is then the perfect and certain science.

It is relevant that the philosophers of the late seventeenth and the eighteenth century also raised the question of how we can be sure that the larger body of knowledge that Newtonian science had produced was true. Almost all, notably Hobbes, Locke, and Leibniz, answered that the mathematical laws, like Euclidean geometry, were inherent in the design of the universe. Leibniz did leave some room for doubt when he distinguished between possible and actual worlds. But the only significant exception was David Hume, who in his *Treatise of Human Nature* (1739) denied the existence of laws or necessary sequences of events in the universe. He contended that these sequences were observed to occur and human beings concluded that they always will occur in the same fashion. Science is purely empirical. In particular the laws of Euclidean geometry are not necessary physical truths.

Hume's influence was negated and indeed superseded by Immanuel Kant's. Kant's answer to the question of how we can be sure that Euclidean geometry applies to the physical world, which he gave in his *Critique of Pure Reason* (1781), is a peculiar one. He maintained that our minds supply certain modes of organization—he called them intuitions—of space and time and that experience is absorbed and organized by our minds in accordance with these modes or intuitions. Our minds are so constructed that they compel us to view the external world in only one way. As a consequence certain principles about space are prior to experience; these principles and their logical consequences, which Kant called a priori synthetic truths, are those of Euclidean geometry. The nature of the external world is known to us only in the manner in which our minds oblige us to interpret it. On the grounds just described Kant affirmed, and his contemporaries accepted, that the physical world must be Euclidean. In any case whether one appealed to experience, relied upon innate truths, or accepted Kant's view, the common conclusion was the uniqueness and necessity of Euclidean geometry.

3. The Research on the Parallel Axiom

Though confidence in Euclidean geometry as the correct idealization of physical space remained unshaken from 300 B.C. to about 1800, one concern did occupy the mathematicians during almost all of that long period. The axioms adopted by Euclid were supposed to be self-evident truths about physical space and about figures in that space. However, the parallel axiom in the form stated by Euclid (Chap. IV, sec. 3) was believed to be somewhat too complicated. No one really doubted its truth and yet it lacked the compelling quality of the other axioms. Apparently even Euclid himself did not like his own version of the parallel axiom because he did not call upon it until he had proved all the theorems he could without it.

A related problem which did not bother as many people but which ultimately came to the fore as equally vital is whether one may assume the existence of infinite straight lines in physical space. Euclid was careful to postulate only that one can produce a (finite) straight line as far as necessary so that even the extended straight line was still finite. Also the peculiar wording of Euclid's parallel axiom, that two lines will meet on that side of the transversal where the sum of the interior angles is less than two right angles, was a way of avoiding the outright assertion that there are pairs of lines that will never meet no matter how far they are extended. Nevertheless Euclid did imply the existence of infinite straight lines for, were they finite, they could not be extended as far as necessary in any given context and he *proved* the existence of parallel lines.

The history of non-Euclidean geometry begins with the efforts to eliminate the doubts about Euclid's parallel axiom. From Greek times to about 1800 two approaches were made. One was to replace the parallel axiom by a more self-evident statement. The other was to try to deduce it from the other nine axioms of Euclid; were this possible it would be a theorem and so be beyond doubt. We shall not give a fully detailed account of this work because this history is readily available.[2]

The first major attempt was made by Ptolemy in his tract on the parallel postulate. He tried to prove the assertion by deducing it from the other nine axioms and Euclid's theorems 1 to 28, which do not depend on the parallel axiom. But Ptolemy assumed unconsciously that two straight lines do not enclose a space and that if *AB* and *CD* are parallel (Fig. 36.1) then whatever holds for the interior angles on one side of *FG* must hold on the other.

The fifth-century commentator Proclus was very explicit about his objection to the parallel axiom. He says, "This [postulate] ought even to be struck out of the postulates altogether; for it is a theorem involving many difficulties, which Ptolemy, in a certain book, set himself to solve, and it requires for the demonstration of it a number of definitions as well as theorems, and the converse of it is actually proved by Euclid himself as a

---

2. See, for example, the reference to Bonola in the bibliography at the end of the chapter.

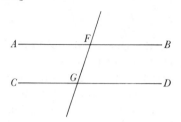

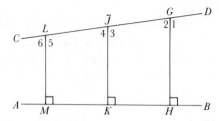

Figure 36.1   Figure 36.2

theorem." Proclus points out that while it is necessary to believe that two
lines will tend toward each other on the side of the transversal where the
sum of the interior angles is less than two right angles, it is not so clear that
these two lines will actually intersect at a finite point. This conclusion is only
probable. For, he continues, there are certain curves that approach each other
more and more but do not actually meet. Thus, a hyperbola approaches but
does not meet its asymptote. Hence might not this be true of the two lines in
Euclid's postulate? He then says that up to a certain sum of the interior
angles on one side of the transversal the two lines might indeed meet; however, for a value slightly greater but still less than two right angles, the lines
might be asymptotic.

Proclus based his own proof of the parallel postulate on an axiom which
Aristotle used to prove the universe is finite. The axiom says, "If from one
point two straight lines forming an angle be produced indefinitely, the
successive distances between the said straight lines [perpendiculars from one
onto the other] will ultimately exceed any finite magnitude." Proclus' proof
was essentially correct but he substituted one questionable axiom for another.

Nasîr-Eddîn (1201–74), the Persian editor of Euclid, likewise gave a
"proof" of Euclid's parallel postulate by assuming that two non-parallel lines
approach each other in one direction and diverge in the other. Specifically,
if AB and CD (Fig. 36.2) are two lines cut by GH, JK, LM, . . ., if these latter
are perpendicular to AB, and if the angles 1, 3, 5, . . . are obtuse while
2, 4, 6, . . . are acute then $GH > JK > LM$ . . . . This fact, Nasîr-Eddîn says,
is clearly seen.

Wallis did some work on the parallel axiom in 1663 which he published
in 1693.[3] First he reproduced Nasîr-Eddîn's work on the parallel axiom which
he had translated for him by a professor of Arabic at Oxford. Incidentally,
this was how Nasîr-Eddîn's work on the parallel axiom became known to
Europe. Wallis then criticized Nasîr-Eddîn's proof and offered his own proof
of Euclid's assertion. His proof rests on the explicit assumption that to each
triangle there is a similar one whose sides have any given ratio to the sides
of the original one. Wallis believed that this axiom was more evident than

3. *Opera*, 2, 669–78.

of arbitrarily small subdivision and arbitrarily large extension. In fact, he said, Euclid's axiom that we can construct a circle with given center and radius presupposes that there is an arbitrarily large radius at our disposal. Hence one can just as well assume the analogue for rectilinear figures such as a triangle.

The simplest of the substitute axioms was suggested by Joseph Fenn in 1769, namely, that two intersecting lines cannot both be parallel to a third straight line. This axiom also appears in Proclus' comments on Proposition 31 of Book I of Euclid's *Elements*. Fenn's statement is entirely equivalent to the axiom given in 1795 by John Playfair (1748–1819): Through a given point P not on a line l, there is only one line in the plane of P and l which does not meet l. This is the axiom used in modern books (which for simplicity usually say there is "one and only one line...").

Legendre worked on the problem of the parallel postulate over a period of about twenty years. His results appeared in books and articles including the many editions of his *Eléments de géométrie*.[4] In one attack on the problem he proved the parallel postulate on the assumption that there exist similar triangles of different sizes; actually his proof was analytical but he assumed that the unit of length does not matter. Then he gave a proof based on the assumption that given any three noncollinear points there exists a circle passing through all three. In still another approach he used all but the parallel postulate and proved that the sum of the angles of a triangle cannot be greater than two right angles. He then observed that under these same assumptions the area is proportional to the defect, that is, two right angles minus the sum of the angles. He therefore tried to construct a triangle twice the size of a given triangle so that the defect of the larger one would be twice that of the given one. Proceeding in this way he hoped to get triangles with larger and larger defects and thus angle sums approaching zero. This result, he thought, would be absurd and so the sum of the angles would have to be 180°. This fact in turn would imply the Euclidean parallel axiom. But Legendre found that the construction reduced to proving that through any given point within a given angle less than 60° one can always draw a straight line which meets both sides of the angle. This he could not prove without Euclid's parallel postulate. In each of the twelve editions (12th ed., 1813) of Legendre's version of Euclid's *Elements* he added appendices which supposedly gave proofs of the parallel postulate but each was deficient because it assumed something implicitly which could not be assumed or assumed an axiom as questionable as Euclid's.

In the course of his researches Legendre,[5] using the Euclidean axioms except for the parallel axiom, proved the following significant theorems: If the sum of the angles of one triangle is two right angles, then it is so in

4. 1st ed., 1794.
5. *Mém. de l'Acad. des Sci., Paris*, 12, 1833, 367–410.

every triangle. Also if the sum is less than two right angles in one triangle it is so in every triangle. Then he gives the proof that if the sum of the angles of any triangle is two right angles, Euclid's parallel postulate holds. This work on the sum of the angles of a triangle was also fruitless because Legendre failed to show (without the aid of the parallel axiom or an equivalent axiom) that the sum of the angles of a triangle cannot be less than two right angles.

The efforts described thus far were mainly attempts to find a more self-evident substitute axiom for Euclid's parallel axiom and many of the proposed axioms did seem intuitively more self-evident. Consequently their creators were satisfied that they had accomplished their objective. However, closer examination showed that these substitute axioms were not really more satisfactory. Some of them made assertions about what happens indefinitely far out in space. Thus, to require that there be a circle through any three points not in a straight line requires larger and larger circles as the three points approach collinearity. On the other hand the substitute axioms that did not involve "infinity" directly, for example, the axiom that there exist two similar but unequal triangles, were seen to be rather complex assumptions and by no means preferable to Euclid's own parallel axiom.

The second group of efforts to solve the problem of the parallel axiom sought to deduce Euclid's assertion from the other nine axioms. The deduction could be direct or indirect. Ptolemy had attempted a direct proof. The indirect method assumes some contradictory assertion in place of Euclid's statement and attempts to deduce a contradiction within this new body of ensuing theorems. For example, since Euclid's parallel axiom is equivalent to the axiom that through a point P not on a line l there is one and only one parallel to l, there are two alternatives to this axiom. One is that there are no parallels to l through P and the other is that there is more than one parallel to l through P. If by adopting each of these in place of the "one parallel" axiom one could show that the new set led to a contradiction, then these alternatives would have to be rejected and the "one parallel" assertion would be *proved*.

The most significant effort of this sort was made by Gerolamo Saccheri (1667–1733), a Jesuit priest and professor at the University of Pavia. He studied the work of Nasîr-Eddîn and Wallis carefully and then adopted his own approach. Saccheri started with the quadrilateral (Fig. 36.3) $ABCD$ in which A and B are right angles and $AC = BD$. It is then easy to prove that $\angle C = \angle D$. Now Euclid's parallel axiom is equivalent to the assertion that angles C and D are right angles. Hence Saccheri considered the two possible alternatives:

(1) the hypothesis of the obtuse angle: $\angle C$ and $\angle D$ are obtuse;
(2) the hypothesis of the acute angle: $\angle C$ and $\angle D$ are acute.

On the basis of the first hypothesis (and the other nine axioms of Euclid)

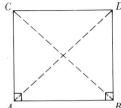

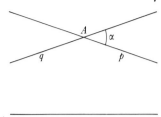

Figure 36.3

Figure 36.4

Saccheri proved that angles C and D must be right angles. Thus with this hypothesis he did deduce a contradiction.

Saccheri next considered the second hypothesis and proved many interesting theorems. He continued until he reached the following theorem: Given any point A and a line b (Fig. 36.4), on the hypothesis of the acute angle there exist in the pencil (family) of lines through A two lines p and q which divide the pencil into two parts. The first of these consists of the lines which intersect b, and the second consists of those lines (lying in angle α) which have a common perpendicular with b somewhere along b. The lines p and q themselves are asymptotic to b. From this result and a lengthy subsequent argument Saccheri deduced that p and b would have a common perpendicular at their common point, which is at infinity. Even though he had not obtained any contradiction, Saccheri found this conclusion and others so repugnant that he decided the hypothesis of the acute angle must be false.

There remained only the hypothesis that angles C and D of Figure 36.3 are right angles. Saccheri had previously proved that when C and D are right angles, the sum of the angles of any triangle equals 180° and that this fact implies Euclid's parallel postulate. He therefore felt justified in concluding that Euclid was upheld and so published his *Euclides ab Omni Naevo Vindicatus* (Euclid Vindicated from All Faults, 1733). However, since Saccheri did not obtain a contradiction on the basis of the acute angle hypothesis the problem of the parallel axiom was still open.

The efforts to find an acceptable substitute for the Euclidean axiom on parallels or to prove that the Euclidean assertion must be a theorem were so numerous and so futile that in 1759 d'Alembert called the problem of the parallel axiom "the scandal of the elements of geometry."

4. *Foreshadowings of Non-Euclidean Geometry*

In his dissertation of 1763 Georg S. Klügel (1739–1812), professor of mathematics at the University of Helmstädt, who knew Saccheri's book, made the remarkable observation that the certainty with which men accepted the truth of the Euclidean parallel axiom was based on experience. This observation introduced for the first time the thought that experience rather than self-evidence substantiated the axioms. Klügel expressed doubt that the Euclidean assertion could be proved. He realized that Saccheri had not arrived at a contradiction but merely at results that seemed at variance with experience.

Klügel's paper suggested work on the parallel axiom to Lambert. In his book, *Theorie der Parallellinien* written in 1766 and published in 1786,[6] Lambert, somewhat like Saccheri, considered a quadrilateral with three right angles and considered the possibilities of the fourth angle being right, obtuse, and acute. Lambert did discard the obtuse angle hypothesis because it led to a contradiction. However, unlike Saccheri, Lambert did not conclude that he had obtained a contradiction from the acute angle hypothesis.

The consequences Lambert deduced from both the obtuse and acute angle hypothesis, respectively, even though the former did lead to a contradiction, are noteworthy. His most remarkable result is that under either hypothesis the area of a polygon of n sides is proportional to the difference between the sum of the angles and $2n - 4$ right angles. (Saccheri had this result for a triangle.) He also noted that the obtuse angle hypothesis gave rise to theorems just like those which hold for figures on the surface of a sphere. And he conjectured that the theorems that followed from the acute angle hypothesis would apply to figures on a sphere of imaginary radius. This led him to write a paper[7] on the trigonometric functions of imaginary angles, that is, iA, where A is a real angle and $i = \sqrt{-1}$, which in effect introduced the hyperbolic functions (Chap. 19, sec. 2). We shall see a little more clearly later just what Lambert's observations mean.

Lambert's views on geometry were quite advanced. He realized that any body of hypotheses which did not lead to contradictions offered a possible geometry. Such a geometry would be a valid logical structure even though it might have little to do with real figures. The latter might be suggestive of a particular geometry but do not control the variety of logically developable geometries. Lambert did not reach the more radical conclusion that was introduced somewhat later by Gauss.

Still another forward step was made by Ferdinand Karl Schweikart (1780–1859), a professor of jurisprudence, who devoted spare time to mathematics. He worked on non-Euclidean geometry during the period in which Gauss devoted some thought to the subject but Schweikart arrived at his conclusions independently. He was, however, influenced by Saccheri's and Lambert's work. In a memorandum of 1816 which he sent to Gauss in 1818 for approval, Schweikart actually distinguished two geometries. There is the geometry of Euclid and a geometry based on the assumption that the sum of the angles of a triangle is not two right angles. This latter geometry he

6. *Magazin für reine und angewandte Mathematik*, 1786, 137–64, 325–58.
7. *Hist. de l'Acad. de Berlin*, 24, 1768, 327–54, pub. 1770 = *Opera Mathematica*, 2, 245–69.

called astral geometry, because it might hold in the space of the stars, and its theorems were those which Saccheri and Lambert had established on the basis of the acute angle hypothesis.

Franz Adolf Taurinus (1794–1874), a nephew of Schweikart, took up his uncle's suggestion to study astral geometry. Though he established in his *Geometriae Prima Elementa* (1826) some new results, notably some analytic geometry, he concluded that only Euclid's geometry could be true of physical space but that the astral geometry was *logically consistent*. Taurinus also showed that the formulas which would hold on a sphere of imaginary radius are precisely those that hold in his astral geometry.

The work of Lambert, Schweikart, and Taurinus constitutes advances that warrant recapitulation. All three and other men such as Klügel and Abraham G. Kästner (1719–1800), a professor at Göttingen, were convinced that Euclid's parallel axiom could not be proven, that is, it is independent of Euclid's other axioms. Further, Lambert, Schweikart, and Taurinus were convinced that it is possible to adopt an alternative axiom contradicting Euclid's and build a logically consistent geometry. Lambert made no assertions about the applicability of such a geometry; Taurinus thought it is not applicable to physical space; but Schweikart believed it might apply to the region of the stars. These three men also noted that the geometry on a real sphere has the properties of the geometry based on the obtuse angle hypothesis (if one leaves aside the contradiction which results from the latter) and the geometry on a sphere of imaginary radius has the properties of the geometry based on the acute angle hypothesis. Thus all three recognized the existence of a non-Euclidean geometry but they missed one fundamental point, namely, that Euclidean geometry is not the only geometry that describes the properties of physical space to within the accuracy for which experience can vouch.

5. *The Creation of Non-Euclidean Geometry*

No major branch of mathematics or even a major specific result is the work of one man. At best, some decisive step or proof may be credited to an individual. This cumulative development of mathematics applies especially to non-Euclidean geometry. If one means by the creation of non-Euclidean geometry the recognition that there can be geometries alternative to Euclid's then Klügel and Lambert deserve the credit. If non-Euclidean geometry means the technical development of the consequences of a system of axioms containing an alternative to Euclid's parallel axiom then most credit must be accorded to Saccheri and even he benefited by the work of many men who tried to find a more acceptable substitute axiom for Euclid's. However, the most significant fact about non-Euclidean geometry is that it can be used to describe the properties of physical space as accurately as Euclidean geometry does. The latter is not the necessary geometry of physical space; its physical truth cannot be guaranteed on any a priori grounds. This realization, which did not call for any technical mathematical development because this had already been done, was first achieved by Gauss.

Carl Friedrich Gauss (1777–1855) was the son of a mason in the German city of Brunswick and seemed destined for manual work. But the director of the school at which he received his elementary education was struck by Gauss's intelligence and called him to the attention of Duke Karl Wilhelm. The Duke sent Gauss to a higher school and then in 1795 to the University of Göttingen. Gauss now began to work hard on his ideas. At eighteen he invented the method of least squares and at nineteen he showed that the 17-sided regular polygon is constructible. These successes convinced him that he should turn from philology to mathematics. In 1798 he transferred to the University of Helmstädt and there he was noticed by Johann Friedrich Pfaff who became his teacher and friend. After finishing his doctor's degree Gauss returned to Brunswick where he wrote some of his most famous papers. This work earned for him in 1807 the appointment as professor of astronomy and director of the observatory at Göttingen. Except for one visit to Berlin to attend a scientific meeting he remained at Göttingen for the rest of his life. He did not like to teach and said so. However, he did enjoy social life, was married twice, and raised a family.

Gauss's first major work was his doctoral thesis in which he proved the fundamental theorem of algebra. In 1801 he published the classic *Disquisitiones Arithmeticae*. His mathematical work in differential geometry, the "Disquisitiones Generales circa Superficies Curvas" (General Investigations of Curved Surfaces, 1827), which, incidentally, was the outcome of his interest in surveying, geodesy, and map-making, is a mathematical landmark (Chap. 37, sec. 2). He made many other contributions to algebra, complex functions, and potential theory. In unpublished papers he recorded his innovative work in two major fields: the elliptic functions and non-Euclidean geometry.

His interests in physics were equally broad and he devoted most of his energy to them. When Giuseppe Piazzi (1746–1826) discovered the planet Ceres in 1801 Gauss undertook to determine its path. This was the beginning of his work in astronomy, the activity that absorbed him most and to which he devoted about twenty years. One of his great publications in this area is his *Theoria Motus Corporum Coelestium* (Theory of Motion of the Heavenly Bodies, 1809). Gauss also earned great distinction in his physical research on theoretical and experimental magnetism. Maxwell says in his *Electricity and Magnetism* that Gauss's studies of magnetism reconstructed the whole science, the instruments used, the methods of observation, and the calculation of results. Gauss's papers on terrestrial magnetism are models of physical research and supplied the best method of measuring the earth's magnetic field. His work on astronomy and magnetism opened up a new and brilliant period of alliance between mathematics and physics.

Though Gauss and Wilhelm Weber (1804–91) did not invent the idea of telegraphy, in 1833 they improved on earlier techniques with a practical device which made a needle rotate right and left depending upon the direction of current sent over a wire. Gauss also worked in optics, which had been neglected since Euler's days, and his investigations of 1838–41 gave a totally new basis for the handling of optical problems.

The universality of Gauss's activities is all the more remarkable because his contemporaries had begun to confine themselves to specialized investigations. Despite the fact that he is acknowledged to be the greatest mathematician at least since Newton, Gauss was not so much an innovator as a transitional figure from the eighteenth to the nineteenth century. Although he achieved some new views which did engage other mathematicians he was oriented more to the past than to the future. Felix Klein describes Gauss's position in these words: We could have a tableau of the development of mathematics if we would imagine a chain of high mountains representing the men of the eighteenth century terminated by an imposing summit— Gauss—then a large and rich region filled with new elements of life. Gauss's contemporaries appreciated his genius and by the time of his death in 1855 he was widely venerated and called the "prince of mathematicians."

Gauss published relatively little of his work because he polished whatever he did partly to achieve elegance and partly to achieve for his demonstrations the maximum of conciseness without sacrificing rigor, at least the rigor of his time. In the case of non-Euclidean geometry he published no definitive work. He said in a letter to Bessel of January 27, 1829, that he probably would never publish his findings in this subject because he feared ridicule, or, as he put it, he feared the clamor of the Boeotians, a figurative reference to a dull-witted Greek tribe. Gauss may have been overly cautious, but one must remember that though some mathematicians had been gradually reaching the climax of the work in non-Euclidean geometry the intellectual world at large was still dominated by Kant's teachings. What we do know about Gauss's work in non-Euclidean geometry is gleaned from his letters to friends, two short reviews in the *Göttingische Gelehrte Anzeigen* of 1816 and 1822 and some notes of 1831 found among his papers after his death.[8]

Gauss was fully aware of the vain efforts to establish Euclid's parallel postulate because this was common knowledge in Göttingen and the whole history of these efforts was thoroughly familiar to Gauss's teacher Kästner. Gauss told his friend Schumacher that as far back as 1792 (Gauss was then fifteen) he had already grasped the idea that there could be a logical geometry in which Euclid's parallel postulate did not hold. By 1794 Gauss had found

8. *Werke*, 8, 157–268, contains all of the above and the letter discussed below.

that in his concept of non-Euclidean geometry the area of a quadrangle must be proportional to the difference between 360° and the sum of the angles. However, at this later date and even up to 1799 Gauss was still trying to deduce Euclid's parallel postulate from other more plausible assumptions and he still believed Euclidean geometry to be the geometry of physical space even though he could conceive of other logical non-Euclidean geometries. However, on December 17, 1799, Gauss wrote to his friend the Hungarian mathematician Wolfgang Farkas Bolyai (1775–1856),

> As for me I have already made some progress in my work. However, the path I have chosen does not lead at all to the goal which we seek [deduction of the parallel axiom], and which you assure me you have reached. It seems rather to compel me to doubt the truth of geometry itself. It is true that I have come upon much which by most people would be held to constitute a proof; but in my eyes it proves as good as nothing. For example, if we could show that a rectilinear triangle whose area would be greater than any given area is possible, then I would be ready to prove the whole of [Euclidean] geometry absolutely rigorously.
>
> Most people would certainly let this stand as an axiom; but I, no! It would, indeed, be possible that the area might always remain below a certain limit, however far apart the three angular points of the triangle were taken.

This passage shows that by 1799 Gauss was rather convinced that the parallel axiom cannot be deduced from the remaining Euclidean axioms and began to take more seriously the development of a new and possibly applicable geometry.

From about 1813 on Gauss developed his new geometry which he first called anti-Euclidean geometry, then astral geometry, and finally non-Euclidean geometry. He became convinced that it was logically consistent and rather sure that it might be applicable. In reviews of 1816 and 1822 and in his letter to Bessel of 1829 Gauss reaffirmed that the parallel postulate could not be proved on the basis of the other axioms in Euclid. His letter to Olbers written in 1817[9] is a landmark. In it Gauss says, "I am becoming more and more convinced that the [physical] necessity of our [Euclidean] geometry cannot be proved, at least not by human reason nor for human reason. Perhaps in another life we will be able to obtain insight into the nature of space, which is now unattainable. Until then we must place geometry not in the same class with arithmetic, which is purely a priori, but with mechanics."

To test the applicability of Euclidean geometry and his non-Euclidean geometry Gauss actually measured the sum of the angles of the triangle formed by three mountain peaks, Brocken, Hohehagen, and Inselsberg. The

9. *Werke*, 8, 177.

sides of this triangle were 69, 85, and 197 km. He found[10] that the sum exceeded 180° by 14".85. The experiment proved nothing because the experimental error was much larger than the excess and so the correct sum could have been 180° or even less. As Gauss realized, the triangle was a small one and since in the non-Euclidean geometry the defect is proportional to the area only a large triangle could possibly reveal any significant departure from an angle sum of 180°.

We shall not discuss the specific non-Euclidean theorems that are due to Gauss. He did not write up a full deductive presentation and the theorems he did prove are much like those we shall encounter in the work of Lobatchevsky and Bolyai. These two men are generally credited with the creation of non-Euclidean geometry. Just what is to their credit will be discussed later but they did publish organized presentations of a non-Euclidean geometry on a deductive synthetic basis with the full understanding that this new geometry was logically as legitimate as Euclid's.

Nikolai Ivanovich Lobatchevsky (1793–1856), a Russian, studied at the University of Kazan and from 1827 to 1846 was professor and rector at that university. He presented his views on the foundations of geometry in a paper before the department of mathematics and physics of the University in 1826. However, the paper was never printed and was lost. He gave his approach to non-Euclidean geometry in a series of papers, the first two of which were published in Kazan journals and the third in the *Journal für Mathematik*.[11] The first was entitled "On the Foundations of Geometry" and appeared in 1829–30. The second, entitled "New Foundations of Geometry with a Complete Theory of Parallels" (1835–37), is a better presentation of Lobatchevsky's ideas. He called his new geometry imaginary geometry for reasons which are perhaps already apparent and will be clearer later. In 1840 he published a book in German, *Geometrische Untersuchungen zur Theorie der Parallellinien* (Geometrical Researches on the Theory of Parallels[12]). In this book he laments the slight interest shown in his writings. Though he became blind he dictated a completely new exposition of his geometry and published it in 1855 under the title *Pangéométrie*.

John (János) Bolyai (1802–60), son of Wolfgang Bolyai, was a Hungarian army officer. On non-Euclidean geometry, which he called absolute geometry, he wrote a twenty-six-page paper "The Science of Absolute Space."[13] This was published as an appendix to his father's book *Tentamen Juventutem Studiosam in Elementa Matheseos* (Essay on the Elements of Mathematics for

---

10. *Werke*, 4, 258.
11. *Jour. für. Math.*, 17, 1837, 295–320.
12. The English translation appears in Bonola. See the bibliography at the end of this chapter.
13. The English translation appears in Bonola. See the bibliography at the end of this chapter.

Studious Youths). Though the two-volume book appeared in 1832–33 and therefore after a publication by Lobatchevsky, Bolyai seems to have worked out his ideas on non-Euclidean geometry by 1825 and was convinced by that time that the new geometry was not self-contradictory. In a letter to his father dated November 23, 1823, John wrote, "I have made such wonderful discoveries that I am myself lost in astonishment." Bolyai's work was so much like Lobatchevsky's that when Bolyai first saw the latter's work in 1835 he thought it was copied from his own 1832–33 publication. On the other hand, Gauss read John Bolyai's article in 1832 and wrote to Wolfgang[14] that he was unable to praise it for to do so would be to praise his own work.

6. *The Technical Content of Non-Euclidean Geometry*

Gauss, Lobatchevsky, and Bolyai had realized that the Euclidean parallel axiom could not be proved on the basis of the other nine axioms and that some such additional axiom was needed to found Euclidean geometry. Since the parallel axiom was an independent fact it was then at least logically possible to adopt a contradictory statement and develop the consequences of the new set of axioms.

To study the technical content of what these men created, it is just as well to take Lobatchevsky's work because all three did about the same thing. Lobatchevsky gave, as we know, several versions which differ only in details. We shall use his 1835–37 paper as the basis for the account here.

Since, as in Euclid's *Elements*, many theorems can be proved which do not depend at all upon the parallel axiom, such theorems are valid in the new geometry. Lobatchevsky devotes the first six chapters of his paper to the proof of these basic theorems. He assumes at the outset that space is infinite, and he is then able to prove that two straight lines cannot intersect in more than one point and that two perpendiculars to the same line cannot intersect.

In his seventh chapter Lobatchevsky boldly rejects the Euclidean parallel axiom and makes the following assumption: Given a line AB and a point C (Fig. 36.5) then all lines through C fall into two classes with respect to AB, namely, the class of lines which meet AB and the class of lines which do not. To the latter belong two lines p and q which form the boundary between the two classes. These two boundary lines are called parallel lines. More precisely, if C is a point at a perpendicular distance a from the line AB, then there exists an angle[15] $\pi(a)$ such that all lines through C which make with the perpendicular CD an angle less than $\pi(a)$ will intersect AB; all other lines through C do not intersect AB.[16] The two lines which make the

14. *Werke*, 8, 220–21.
15. The symbol $\pi(a)$ is standard and so is used here. Actually the π in $\pi(a)$ has nothing to do with the number π.
16. The idea that a specific angle can be associated with a length is due to Lambert.

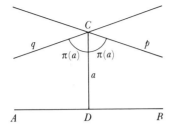

Figure 36.5

angle $\pi(a)$ with AB are the parallels and $\pi(a)$ is called the angle of parallelism. Lines through C other than the parallels and which do not meet AB are called non-intersecting lines, though in Euclid's sense they are parallel to AB and so in this sense Lobatchevsky's geometry contains an infinite number of parallels through C.

If $\pi(a) = \pi/2$ then the Euclidean axiom results. If not, then it follows that $\pi(a)$ increases and approaches $\pi/2$ as a decreases to zero, and $\pi(a)$ decreases and approaches zero as a becomes infinite. The sum of the angles of a triangle is always less than π, decreases as the area of the triangle increases, and approaches π as the area approaches zero. If two triangles are similar then they are congruent.

Now Lobatchevsky turns to the trigonometric part of his geometry. The first step is the determination of $\pi(a)$. The result, if a complete central angle is 2π, is[17]

(1)
$$\tan \frac{\pi(x)}{2} = e^{-x},$$

from which it follows that $\pi(0) = \pi/2$ and $\pi(+\infty) = 0$. The relation (1) is significant in that with each length x it associates a definite angle $\pi(x)$. When $x = 1$, $\tan [\pi(1)/2] = e^{-1}$ so that $\pi(1) = 40°24'$. Thus the unit of length is that length whose angle of parallelism is $40°24'$. This unit of length does not have direct physical significance. It can be physically one inch or one mile. One would choose the physical interpretation which would make the geometry physically applicable.[18]

Then Lobatchevsky deduces formulas connecting sides and angles of the plane triangles of his geometry. In a paper of 1834 he had defined $\cos x$

17. This is a special formulation. In his 1840 work Lobatchevsky gives what amounts to the form usually given in modern texts and which Gauss also has, namely,

(a)
$$\tan \frac{1}{2}\pi(x) = e^{-x/k}$$

where k is a constant, called the space constant. For theoretical purposes the value of k is immaterial. Bolyai also gives the form (a).

18. In the case of the relation $\tan [\pi(x)/2] = e^{-x/k}$, the choice of the value for x which should correspond to, say, $40°24'$ would determine k.

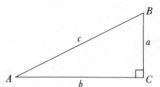

Figure 36.6

and sin x for real x as the real and imaginary parts of e^{ix}. Lobatchevsky's point was to give a purely analytical foundation for trigonometry and so make it independent of Euclidean geometry. The main trigonometric formulas of his geometry are (Fig. 36.6)

$$\cot \pi(a) = \cot \pi(c) \sin A$$
$$\sin A = \cos B \sin \pi(b)$$
$$\sin \pi(c) = \sin \pi(a) \sin \pi(b).$$

These formulas hold in ordinary spherical trigonometry provided that the sides have imaginary lengths. That is, if one replaces a, b, and c in the usual formulas of spherical trigonometry by ia, ib, and ic one obtains Lobatchevsky's formulas. Since the trigonometric functions of imaginary angles are replaceable by hyperbolic functions one might expect to see these latter functions in Lobatchevsky's formulas. They can be introduced by using the relation $\tan(\pi(x)/2) = e^{-x/k}$. Thus the first of the formulas above can be converted into

$$\sinh \frac{a}{k} = \sinh \frac{c}{k} \sin A.$$

Also whereas in the usual spherical geometry the area of a triangle with angles A, B, and C is $r^2(A + B + C - \pi)$, in the non-Euclidean geometry it is $r^2[\pi - (A + B + C)]$ which amounts to replacing r by ir in the usual formula.

By working with an infinitesimal triangle Lobatchevsky derived in his first paper (1829–30) the formula

$$ds = \sqrt{(dy)^2 + \frac{(dx)^2}{\sin^2 \pi(x)}}$$

for the element of arc on a curve $y = f(x)$ at the point (x, y). Then the entire circumference of a circle of radius r can be calculated. It is

$$C = \pi(e^r - e^{-r}).$$

The expression for the area of a circle proves to be

$$A = \pi(e^{r/2} - e^{-r/2})^2.$$

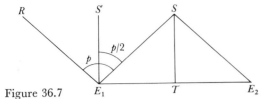

Figure 36.7

He also gives theorems on the area of plane and curved regions and volumes of solids.

The formulas of Euclidean geometry result from the non-Euclidean formulas when the magnitudes are small. Thus if we use the fact that

$$e^r = 1 + r + \frac{r^2}{2!} + \cdots$$

and neglect for small r all but the first two terms then, for example,

$$C = \pi(e^r - e^{-r}) \sim \pi\{1 + r - (1 - r)\} = 2\pi r.$$

In the first paper (1829–30) Lobatchevsky also considered the applicability of his geometry to physical space. The essence of his argument rests on the parallax of stars. Suppose E_1 and E_2 (Fig. 36.7) are the positions of the earth six months apart and S is a star. The parallax p of S is the difference in the directions of E_1S and E_2S measured, say, from the perpendicular E_1S'. If E_1R is the Euclidean parallel to E_2S, then since E_1SE_2 is an isosceles triangle, $\pi/2 - \sphericalangle SE_1E_2$ is half of the change in direction of the star, that is, $p/2$. This angle is $1\rlap{.}''24$ for the star Sirius (Lobatchevsky's value). As long as such an angle is not zero, the line from E_1 to the star cannot be the parallel to TS because the line cuts TS. If, however, there were a lower bound to the various parallaxes of all the stars then any line from E_1 making a smaller angle with E_1S' than this lower bound could be taken to be a parallel to TS through E_1 and this geometry would be equally useful so far as stellar measurements are concerned. But then Lobatchevsky showed that the unit of length in his geometry would have to be, physically, more than a half a million times the radius of the earth's path. In other words, Lobatchevsky's geometry could be applicable only in an enormously large triangle.

7. The Claims of Lobatchevsky and Bolyai to Priority

The creation of non-Euclidean geometry is often used as an example of how an idea occurs independently to different people at about the same time. Sometimes this is regarded as pure coincidence and sometimes as evidence of the spirit of the time working its influence in widely separated quarters. The creation of non-Euclidean geometry by Gauss, Lobatchevsky, and Bolyai is not an example of a simultaneous creation nor is the great credit given to Lobatchevsky and Bolyai justified. It is true, as already noted, that they were the first to publish an avowed non-Euclidean geometry and in this act showed more courage than Gauss did. However, the creation of non-Euclidean geometry is hardly their contribution. We have already pointed out that even Gauss was preceded by Lambert, that Schweikart and Taurinus were independent creators, and that Lambert and Taurinus published their work. Moreover the realization that the new geometry may be applicable to physical space is due to Gauss.

Both Lobatchevsky and Bolyai owe much to Gauss. Lobatchevsky's teacher in Kazan was Johann Martin Bartels (1769–1836), a good friend of Gauss. In fact they spent the years 1805 to 1807 together in Brunswick. Subsequently Gauss and Bartels kept in communication with each other. It is extremely unlikely that Bartels did not pass on to Lobatchevsky, who remained at Kazan as a colleague, Gauss's progress in non-Euclidean geometry. In particular Bartels must have known Gauss's doubts as to the truth of Euclidean geometry.

As for John Bolyai, his father Wolfgang was also a close friend of Gauss and a fellow student in Göttingen from 1796 to 1798. Wolfgang and Gauss not only continued to communicate with each other but discussed the specific subject of the parallel axiom, as one of the quotations above indicates. Wolfgang continued to work hard on the problem of the parallel axiom and sent a purported proof to Gauss in 1804. Gauss showed him that the proof was fallacious. By 1817 Gauss was certain not only that the axiom could not be proved but that a logically consistent and physically applicable non-Euclidean geometry could be constructed. Beyond his communication of 1799 to this effect, Gauss transmitted his later thoughts freely to Wolfgang. Wolfgang continued to work on the problem until he published his *Tentamen* of 1832–33. Since he recommended to his son that he take up the problem of the parallel axiom he almost certainly retailed what he knew.

There are contrary views. The mathematician Friedrich Engel (1861–1941) believed that though Lobatchevsky's teacher Bartels was Gauss's friend, Lobatchevsky could hardly have learned more through this connection than that Gauss doubted the physical correctness of the parallel axiom. But this fact in itself was crucial. However, Engel doubted that Lobatchevsky learned even this much from Gauss, for Lobatchevsky had tried from 1816 on to prove the Euclidean parallel axiom; then recognizing the hopelessness of such efforts finally in 1826 created the new geometry. John Bolyai also tried to prove the Euclidean parallel axiom until about 1820 and then turned to the construction of a new geometry. But these continuing efforts to prove the parallel axiom do not imply ignorance of Gauss's thoughts.[19] Since no one, not even Gauss, had shown that Euclid's

19. However, see George Bruce Halsted, *Amer. Math. Monthly*, 6, 1899, 166–72; and 7, 1900, 247–52.

parallel axiom could not be deduced from the other nine axioms both Lobatchevsky and Bolyai may have decided to try their hand at the problem. Having failed, they could appreciate all the more readily the wisdom of Gauss's views on the subject.

As for the technical content contributed by Lobatchevsky and John Bolyai, though they may have created this independently of their predecessors and of each other, Saccheri's and Lambert's work, to say nothing of Schweikart's and Taurinus's, was widely known in Göttingen and was certainly known to Bartels and Wolfgang Bolyai. And when Lobatchevsky refers in his 1835–37 paper to the futility of the efforts over two thousand years to settle the question of the parallel axiom, by inference he admits to the knowledge of the earlier work.

8. *The Implications of Non-Euclidean Geometry*

We have already stated that the creation of non-Euclidean geometry was the most consequential and revolutionary step in mathematics since Greek times. We shall not treat all the implications of the subject at this time. Instead we shall follow the historical course of events. The impact of the creation and the full realization of its significance were delayed because Gauss did not publish his work on this subject and Lobatchevsky's and Bolyai's work was ignored for about thirty years. Though these men were aware of the importance of their work, the mathematicians generally exhibited their usual reluctance to entertain radical ideas. Also, the key subject in the geometry of the 1830s and 1840s was projective geometry and for this reason too the work on non-Euclidean geometry did not attract the English, French, and German mathematicians. When Gauss's notes and correspondence on non-Euclidean geometry were published after his death in 1855 attention was drawn to the subject. His name gave weight to the ideas and soon thereafter Lobatchevsky's and Bolyai's work was noted by Richard Baltzer (1818–87) in a book of 1866–67. Subsequent developments finally brought mathematicians to the realization of the full import of non-Euclidean geometry.

Gauss did see the most revolutionary implication. The first step in the creation of non-Euclidean geometry was the realization that the parallel axiom could not be proved on the basis of the other nine axioms. It was an independent assertion and so it was possible to adopt a contradictory axiom and develop an entirely new geometry. This Gauss and others did. But Gauss, having realized that Euclidean geometry is not necessarily the geometry of physical space, that is, is not necessarily true, put geometry in the same class with mechanics, and asserted that the quality of truth must be restricted to arithmetic (and its development in analysis). This confidence in arithmetic is in itself curious. Arithmetic at this time had no logical foundation at all. The assurance that arithmetic, algebra, and analysis offered truths about the physical world stemmed entirely from reliance upon experience.

The history of non-Euclidean geometry reveals in a striking manner how much mathematicians are influenced not by the reasoning they perform but by the spirit of the times. Saccheri had rejected the strange theorems of non-Euclidean geometry and concluded that Euclid was vindicated. But one hundred years later Gauss, Lobatchevsky, and Bolyai confidently accepted the new geometry. They believed that their new geometry was logically consistent and hence that this geometry was as valid as Euclid's. But they had no proof of this consistency. Though they proved many theorems and obtained no evident contradictions, the possibility remained open that a contradiction might still be derived. Were this to happen, then the assumption of their parallel axiom would be invalid and, as Saccheri had believed, Euclid's parallel axiom would be a consequence of his other axioms.

Actually Bolyai and Lobatchevsky considered this question of consistency and were partly convinced of it because their trigonometry was the same as for a sphere of imaginary radius and the sphere is part of Euclidean geometry. But Bolyai was not satisfied with this evidence because trigonometry in itself is not a complete mathematical system. Thus despite the absence of any proof of consistency, or of the applicability of the new geometry, which might at least have served as a convincing argument, Gauss, Bolyai, and Lobatchevsky accepted what their predecessors had regarded as absurd. This acceptance was an act of faith. The question of the consistency of non-Euclidean geometry remained open for another forty years.

One more point about the creation of non-Euclidean geometry warrants attention and emphasis. There is a common belief that Gauss, Bolyai, and Lobatchevsky went off into a corner, played with changing the axioms of Euclidean geometry just to satisfy their intellectual curiosity and so created the new geometry. And since this creation has proved to be enormously important for science—a form of non-Euclidean geometry which we have yet to examine has been used in the theory of relativity—many mathematicians have contended that pure intellectual curiosity is sufficient justification for the exploration of any mathematical idea and that the values for science will almost surely ensue as purportedly happened in the case of non-Euclidean geometry. But the history of non-Euclidean geometry does not support this thesis. We have seen that non-Euclidean geometry came about after centuries of work on the parallel axiom. The concern about this axiom stemmed from the fact that it should be, as an axiom, a self-evident truth. Since the axioms of geometry are our basic facts about physical space and vast branches of mathematics and of physical science use the properties of Euclidean geometry the mathematicians wished to be sure that they were relying upon truths. In other words, the problem of the parallel axiom was not only a genuine physical problem but as fundamental a physical problem as there can be.

Bibliography

Bonola, Roberto: *Non-Euclidean Geometry*, Dover (reprint), 1955.

Dunnington, G. W.: *Carl Friedrich Gauss*, Stechert-Hafner, 1960.

Engel, F., and P. Staeckel: *Die Theorie der Parallellinien von Euklid bis auf Gauss*, 2 vols., B. G. Teubner, 1895.

————: *Urkunden zur Geschichte der nichteuklidischen Geometrie*, B. G. Teubner, 1899–1913, 2 vols. The first volume contains the translation from Russian into German of Lobatchevsky's 1829–30 and 1835–37 papers. The second is on the work of the two Bolyais.

Enriques, F.: "Prinzipien der Geometrie," *Encyk. der Math. Wiss.*, B. G. Teubner, 1907–10, III AB1, 1–129.

Gauss, Carl F.: *Werke*, B. G. Teubner, 1900 and 1903, Vol. 8, 157–268; Vol. 9, 297–458.

Heath, Thomas L.: *Euclid's Elements*, Dover (reprint), 1956, Vol. 1, pp. 202–20.

Kagan, V.: *Lobatchevsky and his Contribution to Science*, Foreign Language Pub. House, Moscow, 1957.

Lambert, J. H.: *Opera Mathematica*, 2 vols., Orell Fussli, 1946–48.

Pasch, Moritz, and Max Dehn: *Vorlesungen über neuere Geometrie*, 2nd ed., Julius Springer, 1926, pp. 185–238.

Saccheri, Gerolamo: *Euclides ab Omni Naevo Vindicatus*, English trans. by G. B. Halsted in *Amer. Math. Monthly*, Vols. 1–5, 1894–98; also *Open Court Pub. Co.*, 1920, and Chelsea (reprint), 1970.

Schmidt, Franz, and Paul Staeckel: *Briefwechsel zwischen Carl Friedrich Gauss und Wolfgang Bolyai*, B. G. Teubner, 1899; Georg Olms (reprint), 1970.

Smith, David E.: *A Source Book in Mathematics*, Dover (reprint), 1959, Vol. 2, pp. 351–88.

Sommerville, D. M. Y.: *The Elements of Non-Euclidean Geometry*, Dover (reprint), 1958.

Staeckel, P.: "Gauss als Geometer," *Nachrichten König. Ges. der Wiss. zu Gött.*, 1917, Beiheft, pp. 25–142. Also in Gauss: *Werke*, X_2.

von Walterhausen, W. Sartorius: *Carl Friedrich Gauss*, S. Hirzel, 1856; Springer-Verlag (reprint), 1965.

Zacharias, M.: "Elementargeometrie und elementare nicht-euklidische Geometrie in synthetischer Behandlung," *Encyk. der Math. Wiss.*, B. G. Teubner, 1914–31, III AB9, 859–1172.

37
The Differential Geometry of Gauss and Riemann

> Thou, nature, art my goddess; to thy laws my services are bound . . . .
>
> CARL F. GAUSS

1. *Introduction*

We shall now pick up the threads of the development of differential geometry, particularly the theory of surfaces as founded by Euler and extended by Monge. The next great step in this subject was made by Gauss.

Gauss had devoted an immense amount of work to geodesy and map-making starting in 1816. His participation in actual physical surveys, on which he published many papers, stimulated his interest in differential geometry and led to his definitive 1827 paper "Disquisitiones Generales circa Superficies Curvas" (General Investigations of Curved Surfaces).[1] However, beyond contributing this definitive treatment of the differential geometry of surfaces lying in three-dimensional space Gauss advanced the totally new concept that a surface is a space in itself. It was this concept that Riemann generalized, thereby opening up new vistas in non-Euclidean geometry.

2. *Gauss's Differential Geometry*

Euler had already introduced the idea (Chap. 23, sec. 7) that the coordinates (x, y, z) of any point on a surface can be represented in terms of two parameters u and v; that is, the equations of a surface are given by

$$(1) \qquad x = x(u, v), \qquad y = y(u, v), \qquad z = z(u, v).$$

Gauss's point of departure was to use this parametric representation for the systematic study of surfaces. From these parametric equations we have

$$(2) \qquad dx = a\,du + a'\,dv, \qquad dy = b\,du + b'\,dv, \qquad dz = c\,du + c'\,dv$$

1. *Comm. Soc. Gott.*, 6, 1828, 99–146 = *Werke*, 4, 217–58.

wherein $a = x_u$, $a' = x_v$, and so forth. For convenience Gauss introduces the determinants

$$A = \begin{vmatrix} b & c \\ b' & c' \end{vmatrix}, \qquad B = \begin{vmatrix} c & a \\ c' & a' \end{vmatrix}, \qquad C = \begin{vmatrix} a & b \\ a' & b' \end{vmatrix}$$

and the quantity

$$\Delta = \sqrt{A^2 + B^2 + C^2}$$

which he supposes is not identically 0.

The fundamental quantity on any surface is the element of arc length which in (x, y, z) coordinates is

(3) $$ds^2 = dx^2 + dy^2 + dz^2.$$

Gauss now uses the equations (2) to write (3) as

(4) $$ds^2 = E(u, v)\, du^2 + 2F(u, v)\, du\, dv + G(u, v)\, dv^2,$$

where

$$E = a^2 + b^2 + c^2, \qquad F = aa' + bb' + cc', \qquad G = a'^2 + b'^2 + c'^2.$$

The angle between two curves on a surface is another fundamental quantity. A curve on the surface is determined by a relation between u and v, for then x, y, and z become functions of one parameter, u or v, and the equations (1) become the parametric representation of a curve. One says in the language of differentials that at a point (u, v) a curve or the direction of a curve emanating from the point is given by the ratio $du : dv$. If then we have two curves or two directions emanating from (u, v), one given by $du : dv$ and the other by $du' : dv'$, and if θ is the angle between these directions, Gauss shows that

(5) $$\cos \theta = \frac{E\, du\, du' + F(du\, dv' + du'\, dv) + G\, dv\, dv'}{\sqrt{E\, du^2 + 2F\, du\, dv + G\, dv^2}\,\sqrt{E\, du'^2 + 2F\, du'\, dv' + G\, dv'^2}}.$$

Gauss undertakes next to study the curvature of a surface. His definition of curvature is a generalization to surfaces of the indicatrix used for space curves by Euler and used for surfaces by Olinde Rodrigues.[2] At each point (x, y, z) on a surface there is a normal with a direction attached. Gauss considers a unit sphere and chooses a radius having the direction of the directed normal on the surface. The choice of radius determines a point (X, Y, Z) on the sphere. If we next consider on the surface any small region surrounding (x, y, z) then there is a corresponding region on the sphere surrounding (X, Y, Z). The curvature of the surface at (x, y, z) is defined as the limit of the ratio of the area of the region on the sphere to the area of the corresponding region on the surface as the two areas shrink to their respective

2. *Corresp. sur l'Ecole Poly.*, 3, 1814–16, 162–82.

points. Gauss evaluates this ratio by noting first that the tangent plane at (X, Y, Z) on the sphere is parallel to the one at (x, y, z) on the surface. Hence the ratio of the two areas is the ratio of their projections on the respective tangent planes. To find this latter ratio Gauss performs an amazing number of differentiations and obtains a result which is still basic, namely, that the (total) curvature K of the surface is

(6)
$$K = \frac{LN - M^2}{EG - F^2}$$

wherein

(7)

$$
L = \begin{vmatrix} x_{uu} & y_{uu} & z_{uu} \\ x_u & y_u & z_u \\ x_v & y_v & z_v \end{vmatrix}, \qquad
M = \begin{vmatrix} x_{uv} & y_{uv} & z_{uv} \\ x_u & y_u & z_u \\ x_v & y_v & z_v \end{vmatrix}, \qquad
N = \begin{vmatrix} x_{vv} & y_{vv} & z_{vv} \\ x_u & y_u & z_u \\ x_v & y_v & z_v \end{vmatrix}.
$$

Then Gauss shows that his K is the product of the two principal curvatures at (x, y, z), which had been introduced by Euler. The notion of mean curvature, the average of the two principal curvatures, was introduced by Sophie Germain in 1831.[3]

Now Gauss makes an extremely important observation. When the surface is given by the parametric equations (1), the properties of the surface seem to depend on the functions x, y, z. By fixing u, say $u = 3$, and letting v vary, one obtains a curve on the surface. For the various possible fixed values of u one obtains a family of curves. Likewise by fixing v one obtains a family of curves. These two families are the parametric curves on the surface so that each point is given by a pair of numbers (c, d), say, where $u = c$ and $v = d$ are the parametric curves which pass through the point. These coordinates do not necessarily denote distances any more than latitude and longitude do. Let us think of a surface on which the parametric curves have been determined in some way. Then, Gauss affirms, the geo-metrical properties of the surface are determined solely by the E, F, and G in the expression (4) for ds^2. These functions of u and v are all that matter.

It is certainly the case, as is evident from (4) and (5), that distances and angles on the surface are determined by the E, F, and G. But Gauss's funda-mental expression for curvature, (6) above, depends upon the additional quantities L, M, and N. Gauss now proves that

(8) $\quad K = \dfrac{1}{2H} \left\{ \dfrac{\partial}{\partial u} \left[\dfrac{F}{EH} \dfrac{\partial E}{\partial v} - \dfrac{1}{H} \dfrac{\partial G}{\partial u} \right] + \dfrac{\partial}{\partial v} \left[\dfrac{2}{H} \dfrac{\partial F}{\partial u} - \dfrac{1}{H} \dfrac{\partial E}{\partial v} - \dfrac{F}{EH} \dfrac{\partial E}{\partial u} \right] \right\}$

wherein $H = \sqrt{EG - F^2}$ and is equal to Gauss's Δ defined above. Equation (8), called the Gauss characteristic equation, shows that the curvature K and, in particular in view of (6), the quantity $LN - M^2$ depend only upon

3. *Jour. für Math.*, 7, 1831, 1–29.

E, *F*, and *G*. Since *E*, *F*, and *G* are functions only of the parametric coordinates on the surface, the curvature too is a function only of the parameters and does not depend at all on whether or how the surface lies in three-space.

Gauss had made the observation that the properties of a surface depend only on *E*, *F*, and *G*. However, many properties other than curvature involve the quantities *L*, *M*, and *N* and not in the combination $LN - M^2$ which appears in equation (6). The analytical substantiation of Gauss's point was made by Gaspare Mainardi (1800–79),[4] and independently by Delfino Codazzi (1824–75),[5] both of whom gave two additional relations in the form of differential equations which together with Gauss's characteristic equation, *K* having the value in (6), determine *L*, *M*, and *N* in terms of *E*, *F*, and *G*.

Then Ossian Bonnet (1819–92) proved in 1867[6] the theorem that when six functions satisfy the Gauss characteristic equation and the two Mainardi-Codazzi equations, they determine a surface uniquely except as to position and orientation in space. Specifically, if *E*, *F*, and *G* and *L*, *M*, and *N* are given as functions of *u* and *v* which satisfy the Gauss characteristic equation and the Mainardi-Codazzi equations and if $EG - F^2 \neq 0$, then there exists a surface given by three functions *x*, *y*, and *z* as functions of *u* and *v* which has the first fundamental form

$$E \, du^2 + 2F \, du \, dv + G \, dv^2,$$

and *L*, *M*, and *N* are related to *E*, *F*, and *G* through (7). This surface is uniquely determined except for position in space. (For real surfaces with real coordinates (u, v) we must have $EG - F^2 > 0, E > 0$, and $G \geq 0$). Bonnet's theorem is the analogue of the corresponding theorem on curves (Chap. 23, sec. 6).

The fact that the properties of a surface depend only on the *E*, *F*, and *G* has many implications some of which were drawn by Gauss in his 1827 paper. For example, if a surface is bent without stretching or contracting, the coordinate lines *u* = const. and *v* = const. will remain the same and so *ds* will remain the same. Hence all properties of the surface, the curvature in particular, will remain the same. Moreover, if two surfaces can be put into one-to-one correspondence with each other, that is, if $u' = \phi(u, v)$, $v' = \psi(u, v)$, where *u'* and *v'* are the coordinates of points on the second surface, and if the element of distance at corresponding points is the same on the two surfaces, that is, if

$$E \, du^2 + 2F \, du \, dv + G \, dv^2 = E' \, du'^2 + 2F' \, du' \, dv' + G' \, dv'^2$$

4. *Giornale dell' Istituto Lombardo*, 9, 1856, 385–98.
5. *Annali di Mat.*, (3), 2, 1868–69, 101–19.
6. *Jour. de l'Ecole Poly.*, 25, 1867, 31–151.

wherein E, F, and G are functions of u and v and E', F', G' are functions of u' and v', then the two surfaces, which are said to be *isometric*, must have the same geometry. In particular, as Gauss pointed out, they must have the same total curvature at corresponding points. This result Gauss called a "theorema egregium," a most excellent theorem.

As a corollary, it follows that to move a part of a surface over to another part (which means preserving distance) a necessary condition is that the surface have constant curvature. Thus, a part of a sphere can be moved without distortion to another but this cannot be done on an ellipsoid. (However, bending can take place in fitting a surface or part of a surface onto another under an isometric mapping.) Though the curvatures at corresponding points are equal, if two surfaces do not have constant curvature they need not necessarily be isometrically related. In 1839[7] Ferdinand Minding (1806–85) proved that if two surfaces do have constant and equal curvature then one can be mapped isometrically onto the other.

Another topic of great importance that Gauss took up in his 1827 paper is that of finding geodesics on surfaces. (The term geodesic was introduced in 1850 by Liouville and was taken from geodesy.) This problem calls for the calculus of variations which Gauss uses. He approaches this subject by working with the x, y, z representation and proves a theorem stated by John Bernoulli that the principal normal of a geodesic curve is normal to the surface. (Thus the principal normal at a point of a latitude circle on a sphere lies in the plane of the circle and is not normal to the sphere whereas the principal normal at a point on a longitude circle is normal to the sphere.) Any relation between u and v determines a curve on the surface and the relation which gives a geodesic is determined by a differential equation. This equation, which Gauss merely says is a second order equation in u and v but does not give explicitly, can be written in many forms. One is

$$(9) \qquad \frac{d^2v}{du^2} = n\left(\frac{dv}{du}\right)^3 + (2m - \nu)\left(\frac{dv}{du}\right)^2 + (l - 2\mu)\frac{dv}{du} - \lambda,$$

where n, m, μ, ν, l, and λ are functions of E, F, and G.

One must be careful about assuming the existence of a unique geodesic between two points on a surface. Two nearby points on a sphere have a unique geodesic joining them, but two diametrically opposite points are joined by an infinity of geodesics. Similarly, two points on the same generator of a circular cylinder are connected by a geodesic along the generator but also by an infinite number of geodesic helices. If there is but one geodesic arc between two points in a region, that arc gives the shortest path between them in that region. The problem of actually determining the geodesics on particular surfaces was taken up by many men.

---

7. *Jour. für Math.*, 19, 1839, 370–87.

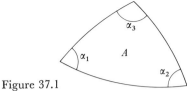

Figure 37.1

In the 1827 article Gauss proved a famous theorem on curvature for a triangle formed by geodesics (Fig. 37.1). Let K be the variable curvature of a surface. $\iint_A K \, dA$ is then the integral of this curvature over the area A. Gauss's theorem applied to the triangle states that

$$\iint_A K \, dA = \alpha_1 + \alpha_2 + \alpha_3 - \pi;$$

that is, the integral of the curvature over a geodesic triangle is equal to the excess of the sum of the angles over 180° or, where the angle sum is less than π, to the defect from 180°. This theorem, Gauss says, ought to be counted as a most elegant theorem. The result generalizes the theorem of Lambert (Chap. 36, sec. 4), which states that the area of a spherical triangle equals the product of its spherical excess and the square of the radius, for in a spherical triangle K is constant and equals $1/R^2$.

One more important piece of work in Gauss's differential geometry must be noted. Lagrange (Chap. 23, sec. 8) had treated the conformal mapping of a surface of revolution into the plane. In 1822 Gauss won a prize offered by the Danish Royal Society of Sciences for a paper on the problem of finding the analytic condition for transforming any surface conformally onto any other surface.[8] His condition, which holds in the neighborhood of corresponding points on the two surfaces, amounts to the fact that a function $P + iQ$, which we shall not specify further, of the parameters T and U by means of which one surface is represented is a function f of $p + iq$ which is the corresponding function of the parameters t and u by which the other surface is represented and $P - iQ$ is $f'(p - iq)$, where f' is f or obtained from f by replacing i by $-i$. The function f depends on the correspondence between the two surfaces, the correspondence being specified by $T = T(t, u)$ and $U(t, u)$. Gauss did not answer the question of whether and in what way a *finite* portion of the surface can be mapped conformally onto the other surface. This problem was taken up by Riemann in his work on complex functions (Chap. 27, sec. 10).

Gauss's work in differential geometry is a landmark in itself. But its implications were far deeper than he himself appreciated. Until this work,

8. *Werke*, 4, 189–216.

surfaces had been studied as figures in three-dimensional Euclidean space. But Gauss showed that the geometry of a surface could be studied by concentrating on the surface itself. If one introduces the u and v coordinates which come from the parametric representation

$$x = x(u, v), \qquad y = y(u, v), \qquad z = z(u, v)$$

of the surface in three-dimensional space and uses the E, F, and G determined thereby then one obtains the Euclidean properties of that surface. However, given these u and v coordinates on the surface and the expression for ds^2 in terms of E, F, and G as functions of u and v, all the properties of the surface follow from this expression. This suggests two vital thoughts. The first is that the surface can be considered as a space in itself because all its properties are determined by the ds^2. One can forget about the fact that the surface lies in a three-dimensional space. What kind of geometry does the surface possess if it is regarded as a space in itself? If one takes the "straight lines" on that surface to be the geodesics, then the geometry is non-Euclidean.

Thus if the surface of the sphere is studied as a space in itself, it has its own geometry and even if the familiar latitude and longitude are used as the coordinates of points, the geometry of that surface is not Euclidean because the "straight lines" or geodesics are arcs of the great circles on the surface. However, the geometry of the spherical surface is Euclidean if it is regarded as a surface in three-dimensional space. The shortest distance between two points on the surface is then the line segment of three-dimensional Euclidean geometry (though it does not lie on the surface). What Gauss's work implied is that there are non-Euclidean geometries at least on surfaces regarded as spaces in themselves. Whether Gauss saw this non-Euclidean interpretation of his geometry of surfaces is not clear.

One can go further. One might think that the proper E, F, and G for a surface is determined by the parametric equations (1). But one could start with the surface, introduce the two families of parametric curves and then pick functions E, F, and G of u and v almost arbitrarily. Then the surface has a geometry determined by these E, F, and G. This geometry is intrinsic to the surface and has no connection with the surrounding space. Consequently the same surface can have *different* geometries depending on the choice of the functions E, F, and G.

The implications are deeper. If one can pick different sets of E, F, and G and thereby determine different geometries on the same surface, why can't one pick different distance functions in our ordinary three-dimensional space? The common distance function in rectangular coordinates is, of course, $ds^2 = dx^2 + dy^2 + dz^2$ and this is obligatory if one starts with Euclidean geometry because it is just the analytic statement of the Pythagorean theorem. However, given the same rectangular Cartesian coordinates for the points of space, one might pick a different expression for ds^2 and obtain a quite different geometry for that space, a non-Euclidean geometry. This extension to any space of the ideas Gauss first obtained by studying surfaces was taken up and developed by Riemann.

3. Riemann's Approach to Geometry

The doubts about what we may believe about the geometry of physical space, raised by the work of Gauss, Lobatchevsky, and Bolyai, stimulated one of the major creations of the nineteenth century, Riemannian geometry. The creator was Georg Bernhard Riemann, the deepest philosopher of geometry. Though the details of Lobatchevsky's and Bolyai's work were unknown to Riemann, they were known to Gauss, and Riemann certainly knew Gauss's doubts as to the truth and necessary applicability of Euclidean geometry. Thus in the field of geometry Riemann followed Gauss whereas in function theory he followed Cauchy and Abel. His investigation of geometry was influenced also by the teachings of the psychologist Johann Friedrich Herbart (1776–1841).

Gauss assigned to Riemann the subject of the foundations of geometry as the one on which he should deliver his qualifying lecture, the *Habilitationsvortrag*, for the title of *Privatdozent*. The lecture was delivered in 1854 to the faculty at Göttingen with Gauss present, and was published in 1868 under the title "Über die Hypothesen, welche der Geometrie zu Grunde liegen" (On the Hypotheses Which Lie at the Foundation of Geometry).[9]

In a paper on the conduction of heat, which Riemann wrote in 1861 to compete for a prize offered by the Paris Academy of Sciences and which is often referred to as his *Pariserarbeit*, Riemann found the need to consider further his ideas on geometry and here he gave some technical elaborations of his 1854 paper. This 1861 paper, which did not win the prize, was published posthumously in 1876 in his *Collected Works*.[10] In the second edition of the *Werke* Heinrich Weber in a note explains Riemann's highly compressed material.

The geometry of space offered by Riemann was not just an extension of Gauss's differential geometry. It reconsidered the whole approach to the study of space. Riemann took up the question of just what we may be certain of about physical space. What conditions or facts are presupposed in the very experience of space before we determine by experience the particular axioms that hold in physical space? One of Riemann's objectives was to show that Euclid's particular axioms were empirical rather than, as had been believed, self-evident truths. He adopted the analytical approach because in

9. *Abh. der Ges. der Wiss. zu Gött.*, 13, 1868, 1–20 = *Werke*, 2nd ed., 272–87. An English translation can be found in W. K. Clifford's *Collected Mathematical Papers*. Also in *Nature*, 8, 1873, 14–36, and in D. E. Smith, *A Source Book in Mathematics*, 411–25.
10. *Werke*, 2nd ed., 1892, 391–404.

geometrical proofs we may be misled by our perceptions to assume facts not explicitly recognized. Thus Riemann's idea was that by relying upon analysis we might start with what is surely a priori about space and deduce the necessary consequences. Any other properties of space would then be known to be empirical. Gauss had concerned himself with this very same problem but of this investigation only the essay on curved surfaces was published. Riemann's search for what is a priori led him to study the local behavior of space or, in other words, the differential geometric approach as opposed to the consideration of space as a whole as one finds it in Euclid or in the non-Euclidean geometry of Gauss, Bolyai, and Lobatchevsky. Before examining the details we should be forewarned that Riemann's ideas as expressed in the lecture and in the manuscript of 1854 are vague. One reason is that Riemann adapted it to his audience, the entire faculty at Göttingen. Part of the vagueness stems from the philosophical considerations with which Riemann began his paper.

Guided to a large extent by Gauss's intrinsic geometry of surfaces in Euclidean space, Riemann developed an intrinsic geometry for any space. He preferred to treat n-dimensional geometry even though the three-dimensional case was clearly the important one and he speaks of n-dimensional space as a manifold. A point in a manifold of n dimensions is represented by assigning special values to n variable parameters, $x_1, x_2, \ldots, x_n$, and the aggregate of all such possible points constitutes the n-dimensional manifold itself, just as the aggregate of the points on a surface constitutes the surface itself. The n variable parameters are called coordinates of the manifold. When the x_i's vary continuously, the points range over the manifold.

Because Riemann believed that we know space only locally he started by defining the distance between two generic points whose corresponding coordinates differ only by infinitesimal amounts. He assumes that the square of this distance is

$$(10) \qquad ds^2 = \sum_{i=1}^{n} \sum_{j=1}^{n} g_{ij}\, dx_i\, dx_j,$$

wherein the g_{ij} are functions of the coordinates $x_1, x_2, \ldots, x_n$, $g_{ij} = g_{ji}$, and the right side of (10) is always positive for all possible values of the dx_i. This expression for ds^2 is a generalization of the Euclidean distance formula

$$ds^2 = dx_1^2 + dx_2^2 + \cdots + dx_n^2.$$

He mentions the possibility of assuming for ds the fourth root of a homogeneous function of the fourth degree in the differentials $dx_1, dx_2, \ldots, dx_n$ but did not pursue this possibility. By allowing the g_{ij} to be functions of the coordinates Riemann provided for the possibility that the nature of the space may vary from point to point.

Though Riemann in his paper of 1854 did not set forth explicitly the following definitions he undoubtedly had them in mind because they parallel what Gauss did for surfaces. A curve on a Riemannian manifold is given by the set of n functions

(11) $$x_1 = x_1(t), x_2 = x_2(t), \cdots, x_n = x_n(t).$$

Then the length of a curve between $t = \alpha$ and $t = \beta$ is defined by

(12) $$l = \int_\alpha^\beta ds = \int_\alpha^\beta \frac{ds}{dt} dt = \int_\alpha^\beta \sqrt{\sum_{i,j=1}^n g_{ij} \frac{dx_i}{dt} \frac{dx_j}{dt}} \, dt.$$

The shortest curve between two given points, $t = \alpha$ and $t = \beta$, the geodesic, is then determinable by the method of the calculus of variations. In the notation of that subject it is the curve for which $\delta \int_\alpha^\beta ds = 0$. One must then determine the particular functions of the form (11) which furnish this shortest path between the two points. In terms of the parameter arc length s, the equations of the geodesics prove to be

$$\frac{d^2x_i}{ds^2} + \sum_{\lambda,\mu} \{^{\lambda\ \mu}_{\ i}\} \frac{dx_\lambda}{ds} \frac{dx_\mu}{ds} = 0, \qquad i, \lambda, \mu = 1, 2, \cdots, n.$$

This is a system of n second order ordinary differential equations.[11]

The angle θ between two curves meeting at a point $(x_1, x_2, \ldots, x_n)$, one curve determined by the directions dx_i/ds, $i = 1, 2, \ldots, n$, and the other by dx_i'/ds', $i = 1, 2, \ldots, n$, where the primes indicate values belonging to the second direction, is defined by the formula

(13) $$\cos \theta = \sum_{i,i'=1}^n g_{ii'} \frac{dx_i}{ds} \frac{dx_i'}{ds'}.$$

By following the procedures which Gauss used for surfaces, a metrical n-dimensional geometry can be developed with the above definitions as a basis. All the metrical properties are determined by the coefficients g_{ij} in the expression for ds^2.

The second major concept in Riemann's 1854 paper is the notion of curvature of a manifold. Through it Riemann sought to characterize Euclidean space and more generally spaces on which figures may be moved about without change in shape or magnitude. Riemann's notion of curvature for any n-dimensional manifold is a generalization of Gauss's notion of total curvature for surfaces. Like Gauss's notion, the curvature of the manifold is defined in terms of quantities determinable on the manifold itself and there

---

11. For the meaning of the brace symbol see (19) below. Riemann did not give these equations explicitly.

is no need to think of the manifold as lying in some higher-dimensional manifold.

Given a point P in the n-dimensional manifold, Riemann considers a two-dimensional manifold at the point and in the n-dimensional manifold. Such a two-dimensional manifold is formed by a singly infinite set of geodesics through the point and tangent to a plane section of the manifold through the point P. Now a geodesic can be described by the point P and a direction at that point. Let $dx_1', dx_2', \ldots, dx_n'$ be the direction of one geodesic and $dx_1'', dx_2'', \ldots, dx_n''$ be the direction of another. Then the ith direction of any one of the singly infinite set of geodesics at P is given by

$$dx_i = \lambda' \, dx_i' + \lambda'' \, dx_i''$$

(subject to the condition $\lambda'^2 + \lambda''^2 + 2\lambda'\lambda'' \cos \theta = 1$ which arises from the condition $\sum g_{ij}(dx_i/ds)(dx_j/ds) = 1$). This set of geodesics forms a two-dimensional manifold which has a Gaussian curvature. Because there is an infinity of such two-dimensional manifolds through P we obtain an infinite number of curvatures at that point in the n-dimensional manifold. But from $n(n-1)/2$ of these measures of curvature, the rest can be deduced. An explicit expression for the measure of curvature can now be derived. This was done by Riemann in his 1861 paper and will be given below. For a manifold which is a surface, Riemann's curvature is exactly Gauss's total curvature. Strictly speaking, Riemann's curvature, like Gauss's, is a property of the metric imposed on the manifold rather than of the manifold itself.

After Riemann had completed his general investigation of n-dimensional geometry and showed how curvature is introduced, he considered more restricted manifolds on which finite spatial forms must be capable of movement without change of size or shape and must be capable of rotation in any direction. This led him to spaces of constant curvature.

When all the measures of curvature at a point are the same and equal to all the measures at any other point, we get what Riemann calls a manifold of constant curvature. On such a manifold it is possible to treat congruent figures. In the 1854 paper Riemann gave the following results but no details: If α be the measure of curvature the formula for the infinitesimal element of distance on a manifold of *constant* curvature becomes (in a suitable coordinate system)

$$(14) \qquad\qquad ds^2 = \frac{\sum\limits_{i=1}^{n} dx_i^2}{(1 + (\alpha/4) \sum x_i^2)^2}.$$

Riemann thought that the curvature α must be positive or zero, so that when $\alpha > 0$ we get a spherical space and when $\alpha = 0$ we get a Euclidean space and conversely. He also believed that if a space is infinitely extended the curvature must be zero. He did, however, suggest that there might be a real surface of constant negative curvature.[12]

To elaborate on Riemann, for $\alpha = a^2 > 0$ and $n = 3$ we get a three-dimensional spherical geometry though we cannot visualize it. The space is finite in extent but boundless. All geodesics in it are of constant length, namely, $2\pi/a$, and return upon themselves. The volume of the space is $2\pi^2/a^3$. For $a^2 > 0$ and $n = 2$ we get the space of the ordinary spherical surface. The geodesics are of course the great circles and are finite. Moreover, any two intersect in two points. Actually it is not clear whether Riemann regarded the geodesics of a surface of constant positive curvature as cutting in one or two points. He probably intended the latter. Felix Klein pointed out later (see the next chapter) that there were two distinct geometries involved.

Riemann also points out a distinction, of which more was made later, between boundlessness [as is the case for the surface of a sphere] and infiniteness of space. Unboundedness, he says, has a greater empirical credibility than any other empirically derived fact such as infinite extent.

Toward the end of his paper Riemann notes that since physical space is a special kind of manifold the geometry of that space cannot be derived only from general notions about manifolds. The properties that distinguish physical space from other triply extended manifolds are to be obtained only from experience. He adds, "It remains to resolve the question of knowing in what measure and up to what point these hypotheses about manifolds are confirmed by experience." In particular the axioms of Euclidean geometry may be only approximately true of physical space. Like Lobatchevsky, Riemann believed that astronomy will decide which geometry fits space. He ends his paper with the prophetic remark: "Either therefore the reality which underlies space must form a discrete manifold or we must seek the ground of its metric relations outside it, in the binding forces which act on it. . . . This leads us into the domain of another science, that of physics, into which the object of our work does not allow us to go today."

This point was developed by William Kingdon Clifford.[13]

> I hold in fact: (1) That small portions of space are of a nature analogous to little hills on a surface which is on the average flat. (2) That this property of being curved or distorted is continually passed on from one portion of space to another after the manner of a wave. (3) That this variation of the curvature of space is really what happens in that phenomenon which we call the motion of matter whether ponderable or ethereal. (4) That in this physical world nothing else takes place but this variation, subject, possibly, to the law of continuity.

12. Such surfaces were already known to Ferdinand Minding (*Jour. für Math.*, 19, 1839, 370–87, pp. 378–80 in particular), including the very one later called the pseudosphere (see Chap. 38, sec. 2). See also Gauss, *Werke*, 8, 265.
13. *Proc. Camb. Phil. Soc.*, 2, 1870, 157–58 = *Math. Papers*, 20–22.

The ordinary laws of Euclidean geometry are not valid for a space whose curvature changes not only from place to place but because of the motion of matter, from time to time. He added that a more exact investigation of physical laws would not be able to ignore these "hills" in space. Thus Riemann and Clifford, unlike most other geometers, felt the need to associate matter with space in order to determine what is true of physical space. This line of thought leads, of course, to the theory of relativity.

  In his *Pariserarbeit* (1861) Riemann returned to the question of when a given Riemannian space whose metric is

$$(15) \qquad ds^2 = \sum_{i,j=1}^{n} g_{ij}\, dx_i\, dx_j$$

might be a space of constant curvature or even a Euclidean space. However, he formulated the more general question of when a metric such as (15) can be transformed by the equations

$$(16) \qquad x_i = x_i(y_1, y_2, \ldots, y_n), \qquad i = 1, 2, \ldots, n$$

into a given metric

$$(17) \qquad ds'^2 = \sum_{i,j=1}^{n} h_{ij}\, dy_i\, dy_j$$

with the understanding of course that ds would equal ds' so that the geometries of the two spaces would be the same except for the choice of coordinates. The transformation (16) is not always possible because, as Riemann points out, there are $n(n+1)/2$ independent functions in (15), whereas the transformation introduces only n functions which might be used to convert the g_{ij} into the h_{ij}.

  To treat the general question Riemann introduced special quantities p_{ijk} which we shall replace by the more familiar Christoffel symbols with the understanding that

$$p_{ijk} = [\begin{smallmatrix} j\,k \\ i \end{smallmatrix}].$$

The Christoffel symbols, denoted in various ways, are

$$(18) \qquad \Gamma_{\alpha\beta,\lambda} = [\begin{smallmatrix} \alpha\ \beta \\ \lambda \end{smallmatrix}] = [\alpha\beta, \lambda] = \frac{1}{2}\left(\frac{\partial g_{\alpha\lambda}}{\partial x_\beta} + \frac{\partial g_{\beta\lambda}}{\partial x_\alpha} - \frac{\partial g_{\alpha\beta}}{\partial x_\lambda}\right)$$

$$(19) \qquad \Gamma_{\alpha\beta}^{\lambda} = \{\begin{smallmatrix} \alpha\ \beta \\ \lambda \end{smallmatrix}\} = \{\alpha\beta, \lambda\} = \sum_i g^{i\lambda}[\begin{smallmatrix} \alpha\ \beta \\ i \end{smallmatrix}]$$

where $g^{i\lambda}$ is the cofactor divided by g of $g_{i\lambda}$ in the determinant of g. Riemann also introduced what is now known as the Riemann four index symbol

$$(20) \quad (\mu\lambda, jk) = R_{\lambda\mu, jk} = \frac{\partial \Gamma_{\lambda j, \mu}}{\partial x_k} - \frac{\partial \Gamma_{\lambda k, \mu}}{\partial x_j} + \sum_{i,\alpha} g^{i\alpha}(\Gamma_{\lambda k,\alpha}\Gamma_{\mu j, i} - \Gamma_{\lambda j,\alpha}\Gamma_{\mu k, i}).$$

Then Riemann shows that a necessary condition that ds^2 be transformable to ds'^2 is

(21)
$$(\alpha\delta, \beta\gamma)' = \sum_{r,k,\iota,h} (rk, ih) \frac{\partial x_r}{\partial y_\alpha} \frac{\partial x_i}{\partial y_\beta} \frac{\partial x_h}{\partial y_\gamma} \frac{\partial x_k}{\partial y_\delta}$$

where the left-hand symbol refers to quantities formed for the ds' metric and (21) holds for all values of α, β, γ, δ, each of which ranges from 1 to n.

And now Riemann turns to the specific question of when a given ds^2 can be transformed to one with constant coefficients. He first derives an explicit ex₁ ession for the curvature of a manifold. The general definition already giv n in the 1854 paper makes use of the geodesic lines issuing from a point O cf the space. Let d and δ determine two vectors or directions of geodesics emanating from O. (Each direction is specified by the components of the tangent to the geodesic.) Then consider the pencil of geodesic vectors emanating from O and given by $\kappa d + \lambda\delta$ where κ and λ are parameters. If one thinks of d and δ as operating on the $x_i = f_i(t)$ which describe any one curve, then there is a meaning for the second differential $(\kappa d + \lambda\delta)^2 = \kappa^2 d^2 + 2\kappa\lambda\, d\delta + \lambda^2\delta^2$. Riemann then forms

(22)
$$\Omega = \delta\delta \sum g_{ij}\, dx_i\, dx_j - 2\, d\delta \sum g_{ij}\, dx_i\, dx_j + dd \sum g_{ij}\, dx_i\, dx_j.$$

Here one understands that the d and δ operate formally on the expressions following them (and d and δ commute) so that

(23)
$$\delta\delta \sum g_{ij}\, dx_i\, dx_j = \delta\left[\sum (\delta g_{ij})\, dx_i\, dx_j + \sum g_{ij}((\delta dx_i)\, dx_j + dx_i\, \delta dx_j)\right]$$

and $\delta g_{ij} = \sum_r (\partial g_{ij}/\partial x_r)\, \delta x_r$. If one calculates Ω one finds that all terms involving third differentials of a function vanish. Only terms involving δx_i, dx_i, $\delta^2 x_i$, δdx_i, and $d^2 x_i$ remain. By calculating these terms and by using the notation

$$p_{ik} = dx_i\, \delta x_k - dx_k\, \delta x_i$$

Riemann obtains

(24)
$$[\Omega] = \sum_{i,k,r,s} (ik, rs) p_{ik} p_{rs}.$$

Now let

$$4\Delta^2 = \sum g_{ij}\, dx_i\, dx_j \cdot \sum g_{ij}\, \delta x_i\, \delta x_j - \left(\sum g_{ij}\, dx_i\, \delta x_j\right)^2.$$

Then the curvature K of a Riemannian manifold is

(25)
$$K = -\frac{[\Omega]}{8\Delta^2}.$$

The overall conclusion is that the necessary and sufficient condition that a given ds^2 can be brought to the form (for $n = 3$)

$$(26) \qquad\qquad ds'^2 = c_1\, dx_1^2 + c_2\, dx_2^2 + c_3\, dx_3^2,$$

where the c_i's are constants, is that all the symbols $(\alpha\beta, \gamma\delta)$ be zero. In case the c_i's are all positive the ds' can be reduced to $dy_1^2 + dy_2^2 + dy_3^2$, that is, the space is Euclidean. As we can see from the value of $[\Omega]$, when K is zero, the space is essentially Euclidean.

It is worth noting that Riemann's curvature for an n-dimensional manifold reduces to Gauss's total curvature of a surface. In fact when

$$ds^2 = g_{11}\, dx_1^2 + 2g_{12}\, dx_1\, dx_2 + g_{22}\, dx_2^2,$$

of the 16 symbols $(\alpha\beta, \gamma\delta)$, 12 are zero and for the remaining four we have $(12, 12) = -(12, 21) = -(21, 12) = (21, 21)$. Then Riemann's K reduces to

$$k = \frac{(12, 12)}{g}.$$

By using (20) this expression can be shown to be equal to Gauss's expression for the total curvature of a surface.

4. *The Successors of Riemann*

When Riemann's essay of 1854 was published in 1868, two years after his death, it created intense interest, and many mathematicians hastened to fill in the ideas he sketched and to extend them. The immediate successors of Riemann were Beltrami, Christoffel, and Lipschitz.

Eugenio Beltrami (1835–1900), professor of mathematics at Bologna and other Italian universities, who knew Riemann's 1854 paper but apparently did not know his 1861 paper, took up the matter of proving that the general expression for ds^2 reduces to the form (14) given by Riemann for a space of constant curvature.[14] Beyond this result and proving a few other assertions by Riemann, Beltrami took up the subject of differential invariants, which we shall consider in the next section.

Elwin Bruno Christoffel (1829–1900), who was a professor of mathematics at Zurich and later at Strasbourg, advanced the ideas in both of Riemann's papers. In two key papers[15] Christoffel's major concern was to reconsider and amplify the theme already treated somewhat sketchily by Riemann in his 1861 paper, namely, when one form

$$F = \sum_{i,j} g_{ij}\, dx_i\, dx_j$$

14. *Annali di Mat.*, (2), 2, 1868–69, 232–55 = *Opere Mat.*, 1, 406–29.
15. *Jour. für Math.*, 70, 1869, 46–70 and 241–45 = *Ges. Math. Abh.*, 1, 352 ff., 378 ff.

can be transformed into another

$$F' = \sum_{i,j} g'_{ij}\, dy_i\, dy_j.$$

Christoffel sought necessary and sufficient conditions. It was in this paper, incidentally, that he introduced the Christoffel symbols.

Let us consider first the two-dimensional case where

$$F = a\, dx^2 + 2b\, dx\, dy + c\, dy^2$$

and

$$F' = A\, dX^2 + 2B\, dX\, dY + C\, dY^2$$

and suppose that x and y may be expressed as functions of X and Y so that F becomes F' under the transformation. Of course $dx = (\partial x/\partial X)\, dX + (\partial x/\partial Y)\, dY$. Now when x, y, dx, and dy are replaced in F by their values in X and Y and when one equates coefficients in this new form of F with those of F' one obtains

$$a\left(\frac{\partial x}{\partial X}\right)^2 + 2b\, \frac{\partial x}{\partial X} \frac{\partial y}{\partial X} + c\left(\frac{\partial y}{\partial X}\right)^2 = A$$

$$a\, \frac{\partial x}{\partial X} \frac{\partial x}{\partial Y} + b\left(\frac{\partial x}{\partial X} \frac{\partial y}{\partial Y} + \frac{\partial x}{\partial Y} \frac{\partial y}{\partial X}\right) + c\, \frac{\partial y}{\partial X} \frac{\partial y}{\partial Y} = B$$

$$a\left(\frac{\partial x}{\partial Y}\right)^2 + 2b\left(\frac{\partial x}{\partial Y} \frac{\partial y}{\partial Y}\right) + c\left(\frac{\partial y}{\partial Y}\right)^2 = C.$$

These are three differential equations for x and y as functions of X and Y. If they can be solved then we know how to transform from F to F'. However, there are only two functions involved. There must then be some relations between a, b, and c on the one hand and A, B, and C on the other. By differentiating the three equations above and further algebraic steps the relation proves to be $K = K'$.

For the n-variable case, Christoffel uses the same technique. He starts with

$$F = \sum g_{rs}\, dx_r\, dx_s$$

and

$$F' = \sum g'_{rs}\, dy_r\, dy_s.$$

The transformation is

$$x_i = x_i(y_1, y_2, \cdots, y_n), \qquad i = 1, 2, \cdots, n.$$

He lets $g = |g_{rs}|$. Then if Δ_{rs} is the cofactor of g_{rs} in the determinant let $g^{pq} = \Delta_{pq}/g$. He, like Riemann, introduces independently the four index symbol (without the comma)

$$(gkhi) = \frac{\partial}{\partial x_i} [gh, k] - \frac{\partial}{\partial x_h} [gi, k] + \sum_p (\{gi, p\}[hk, h] - \{gh, p\}[ik, p]).$$

He then deduces $n(n + 1)/2$ partial differential equations for the x_i as functions of the y_i. A typical one is

$$\sum_{r,s} g_{rs} \frac{\partial x_r}{\partial y_\alpha} \frac{\partial x_s}{\partial y_\beta} = g'_{\alpha\beta}.$$

These equations are the necessary and sufficient conditions that a transformation exist for which $F = F'$.

Partly to treat the integrability of this set of equations and partly because Christoffel wishes to consider forms of degree higher than two in the dx_i, he performs a number of differentiations and algebraic steps which show that

$$(27) \qquad (\alpha\delta\beta\gamma)' = \sum_{g,h,i,k} (gkhi) \frac{\partial x_g}{\partial y_\alpha} \frac{\partial x_h}{\partial y_\beta} \frac{\partial x_i}{\partial y_\gamma} \frac{\partial x_k}{\partial y_\delta},$$

where α, β, γ, and δ take all values from 1 to n. There are $n^2(n^2 - 1)/12$ equations of this form. These equations are the necessary and sufficient conditions for the equivalence of two differential forms of fourth order. Indeed let $d^{(1)}x$, $d^{(2)}x$, $d^{(3)}x$, $d^{(4)}x$ be four sets of differentials of x and likewise for the y's. Then if we have the quadrilinear form

$$G_4 = \sum_{g,k,h,i} (gkhi) d^{(1)}x_g d^{(2)}x_k d^{(3)}x_h d^{(4)}x_i$$

the relations (27) are necessary and sufficient that $G_4 = G'_4$, where G'_4 is the analogue of G_4 in the y variables.

This theory can be generalized to μ-ply differential forms. In fact Christoffel introduces

$$(28) \qquad G_\mu = \sum_{i_1,\cdots,i_\mu} (i_1 i_2 \cdots i_\mu) \underset{1}{\partial x_{i_1}} \underset{2}{\partial x_{i_2}} \cdots \underset{\mu}{\partial x_{i_\mu}}$$

where the term in parentheses is defined in terms of the g_{rs} much as the four index symbol is and the symbol $\underset{i}{\partial}$ is used to distinguish the differentials of the set of x_i from the set obtained by applying $\underset{j}{\partial}$. He then shows that

$$(29) \qquad (\alpha_1 \alpha_2 \cdots \alpha_\mu)' = \sum (i_1 \cdots i_\mu) \frac{\partial x_{i_1}}{\partial y_{\alpha_1}} \cdots \frac{\partial x_{i_\mu}}{\partial y_{\alpha_\mu}}$$

and obtains necessary and sufficient conditions that G_μ be transformable into G'_μ.

He gives next a general procedure whereby from a μ-ply form G_μ a $(\mu + 1)$-ply form $G_{\mu+1}$ can be derived. The key step is to introduce

$$(30) \quad (i_1 i_2 \cdots i_\mu i) = \frac{\partial}{\partial x_i}(i_1 i_2 \cdots i_\mu)$$

$$- \sum_\lambda [\{ii_1, \lambda\}(\lambda i_2 \cdots i_\mu) + \{ii_2, \lambda\}(i_1 \lambda i_3 \cdots i_\mu) + \cdots].$$

These $(\mu + 1)$-index symbols are the coefficients of the $G_{\mu+1}$ form. The procedure Christoffel uses here is what Ricci and Levi-Civita later called covariant differentiation (Chap. 48).

Whereas Christoffel wrote only one key paper on Riemannian geometry, Rudolph Lipschitz, professor of mathematics at Bonn University, wrote a great number appearing in the *Journal für Mathematik* from 1869 on. Though there are some generalizations of the work of Beltrami and Christoffel, the essential subject matter and results are the same as those of the latter two men. He did produce some new results on subspaces of Riemannian and Euclidean n-dimensional spaces.

The ideas projected by Riemann and developed by his three immediate successors suggested hosts of new problems in both Euclidean and Riemannian differential geometry. In particular the results already obtained in the Euclidean case for three dimensions were generalized to curves, surfaces, and higher-dimensional forms in n dimensions. Of many results we shall cite just one.

In 1886 Friedrich Schur (1856–1932) proved the theorem named after him.[16] In accordance with Riemann's approach to the notion of curvature, Schur speaks of the curvature of an orientation of space. Such an orientation is determined by a pencil of geodesics $\mu\alpha + \lambda\beta$ where α and β are the directions of two geodesics issuing from a point. This pencil forms a surface and has a Gauss curvature which Schur calls the Riemannian curvature of that orientation. His theorem then states that if at each point the Riemannian curvature of a space is independent of the orientation then the Riemannian curvature is constant throughout the space. The manifold is then a space of constant curvature.

5. *Invariants of Differential Forms*

It was clear from the study of the question of when a given expression for ds^2 can be transformed by a transformation of the form

$$(31) \quad x_i = x_i(x'_1, x'_2, \cdots, x'_n), \qquad i = 1, 2, \cdots, n$$

16. *Math. Ann.*, 27, 1886, 167–72 and 537–67.

to another such expression with preservation of the value of ds^2 that different coordinate representations can be obtained for the very same manifold. However, the geometrical properties of the manifold must be independent of the particular coordinate system used to represent and study it. Analytically these geometrical properties would be represented by invariants, that is, expressions which retain their form under the change of coordinates and which will consequently have the same value at a given point. The invariants of interest in Riemannian geometry involve not only the fundamental quadratic form, which contains the differentials dx_i and dx_j, but may also contain derivatives of the coefficients and of other functions. They are therefore called differential invariants.

To use the two-dimensional case as an example, if

(32) $$ds^2 = E \, du^2 + 2F \, du \, dv + G \, dv^2$$

is the element of distance for a surface then the Gaussian curvature K is given by formula (8) above. If now the coordinates are changed to

(33) $$u' = f(u, v), \qquad v' = g(u, v)$$

then there is the theorem that if $E \, du^2 + 2F \, du \, dv + G \, dv^2$ transforms into $E' \, du'^2 + 2F' \, du' \, dv' + G' \, dv'^2$ then $K = K'$, where K' is the same expression as in (8) but in the accented variables. Hence the Gaussian curvature of a surface is a scalar invariant. The invariant K is said to be an invariant attached to the form (32) and involves only E, F, and G and their derivatives.

The study of differential invariants was actually initiated in a more limited context by Lamé. He was interested in invariants under transformations from one orthogonal curvilinear coordinate system in three dimensions to another. For rectangular Cartesian coordinates he showed[17] that

(34) $$\Delta_1 \phi = \left(\frac{\partial \phi}{\partial x}\right)^2 + \left(\frac{\partial \phi}{\partial y}\right)^2 + \left(\frac{\partial \phi}{\partial z}\right)^2$$

(35) $$\Delta_2 \phi = \frac{\partial^2 \phi}{\partial x^2} + \frac{\partial^2 \phi}{\partial y^2} + \frac{\partial^2 \phi}{\partial z^2}$$

are differential invariants (he called them differential parameters). Thus if ϕ is transformed into $\phi'(x', y', z')$ under an orthogonal transformation (rotation of axes) then

$$\left(\frac{\partial \phi}{\partial x}\right)^2 + \left(\frac{\partial \phi}{\partial y}\right)^2 + \left(\frac{\partial \phi}{\partial z}\right)^2 = \left(\frac{\partial \phi'}{\partial x'}\right)^2 + \left(\frac{\partial \phi'}{\partial y'}\right)^2 + \left(\frac{\partial \phi'}{\partial z'}\right)^2$$

at the same point whose coordinates are (x, y, z) in the original system and (x', y', z') in the new coordinate system. The analogous equation holds for $\Delta_2 \phi$.

17. *Jour. de l'Ecole Poly.*, 14, 1834, 191–288.

For orthogonal curvilinear coordinates in Euclidean space where ds^2 has the form

(36) $$ds^2 = g_{11}\, du_1^2 + g_{22}\, du_2^2 + g_{33}\, du_3^2$$

Lamé showed (*Leçons sur les coordonnées curvilignes*, 1859, cf. above Chapter 28, sec. 5) that the divergence of the gradient of ϕ, which in rectangular coordinates is given by $\Delta_2\phi$ above, has the invariant form

$$\Delta_2\phi = \frac{1}{\sqrt{g_{11}g_{22}g_{33}}}\left[\frac{\partial}{\partial u_1}\left(\sqrt{\frac{g_{22}g_{33}}{g_{11}}}\frac{\partial\phi}{\partial u_1}\right) + \frac{\partial}{\partial u_2}\left(\sqrt{\frac{g_{33}g_{11}}{g_{22}}}\frac{\partial\phi}{\partial u_2}\right)\right.$$
$$\left. + \frac{\partial}{\partial u_3}\left(\sqrt{\frac{g_{11}g_{22}}{g_{33}}}\frac{\partial\phi}{\partial u_3}\right)\right].$$

Incidentally in this same work Lamé gave conditions on when the ds^2 given by (36) determines a curvilinear coordinate system in Euclidean space and, if it does, how to change to rectangular coordinates.

The investigation of invariants for the theory of surfaces was first made by Beltrami.[18] He gave the two differential invariants

$$\Delta_1\phi = \frac{1}{EG - F^2}\left\{E\left(\frac{\partial\phi}{\partial v}\right)^2 - 2F\frac{\partial\phi}{\partial u}\frac{\partial\phi}{\partial v} + G\left(\frac{\partial\phi}{\partial v}\right)^2\right\}$$

and

$$\Delta_2\phi = \frac{1}{\sqrt{EG - F^2}}\left\{\frac{\partial}{\partial u}\left(\frac{G\phi_u - F\phi_v}{\sqrt{EG - F^2}}\right) + \frac{\partial}{\partial v}\left(\frac{-F\phi_u + E\phi_v}{\sqrt{EG - F^2}}\right)\right\}.$$

These have geometrical meaning. For example in the case of $\Delta_1\phi$, if $\Delta_1\phi = 1$, the curves $\phi(u, v) = $ const. are the orthogonal trajectories of a family of geodesics on the surface.

The search for differential invariants was carried over to quadratic differential forms in n variables. The reason again was that these invariants are independent of particular choices of coordinates; they represent intrinsic properties of the manifold itself. Thus the Riemann curvature is a scalar invariant.

Beltrami, using a method given by Jacobi,[19] succeeded in carrying over to n-dimensional Riemannian spaces the Lamé invariants.[20] Let g as usual be the determinant of the g_{ij} and let g^{ij} be the cofactor divided by g of g_{ij} in g. Then Beltrami showed that Lamé's first invariant becomes

$$\Delta_1(\phi) = \sum_{i,j} g^{ij}\frac{\partial\phi}{\partial x_i}\frac{\partial\phi}{\partial x_j}.$$

18. *Gior. di Mat.*, 2, 1864, 267–82, and succeeding papers in Vols. 2 and 3 = *Opere Mat.*, 1, 107–98.
19. *Jour. für Math.*, 36, 1848, 113–34 = *Werke*, 2, 193–216.
20. *Memorie dell' Accademia delle Scienze dell' Istituto di Bologna*, (2), 8, 1868, 551–90 = *Opere Mat.*, 2, 74–118.

This is the general form for the square of the gradient of ϕ. For the second of Lamé's invariants Beltrami obtained

$$\Delta_2(\phi) = \frac{1}{\sqrt{g}} \sum_i \frac{\partial}{\partial x_i} \left(\sqrt{g} \sum_j g^{ij} \frac{\partial \phi}{\partial x_j} \right).$$

He also introduced the mixed differential invariant

$$\Delta_1(\phi\psi) = \sum_{i,j} g^{ij} \frac{\partial \phi}{\partial x_i} \frac{\partial \psi}{\partial x_j}.$$

This is the general form of the scalar product of the gradients of ϕ and ψ.

Of course the form ds^2 is itself an invariant under a change of coordinates. From this, as we found in the previous section, Christoffel derived higher order differential forms, his G_4 and G_μ, which are also invariants. Moreover, he showed how from G_μ one can derive $G_{\mu+1}$, which is also an invariant. The construction of such invariants was also pursued by Lipschitz. The number and variety are extensive. As we shall see this theory of differential invariants was the inspiration for tensor analysis.

Bibliography

Beltrami, Eugenio: *Opere matematiche*, 4 vols., Ulrico Hoepli, 1902–20.

Clifford, William K.: *Mathematical Papers*, Macmillan, 1882; Chelsea (reprint), 1968.

Coolidge, Julian L.: *A History of Geometrical Methods*, Dover (reprint), 1963, pp. 355–87.

Encyklopädie der Mathematischen Wissenschaften, III, Teil 3, various articles, B. G. Teubner, 1902–7.

Gauss, Carl F.: *Werke*, 4, 192–216, 217–58, Königliche Gesellschaft der Wissenschaften zu Göttingen, 1880. A translation, "General Investigations of Curved Surfaces," has been reprinted by Raven Press, 1965.

Helmholtz, Hermann von: "Über die tatsächlichen Grundlagen der Geometrie," *Wissenschaftliche Abhandlungen*, 2, 610–17.

———: "Über die Tatsachen, die der Geometrie zum Grunde liegen," *Nachrichten König. Ges. der Wiss. zu Gött.*, 15, 1868, 193–221; *Wiss. Abh.*, 2, 618–39.

———: "Über den Ursprung Sinn und Bedeutung der geometrischen Sätze"; English translation, "On the Origin and Significance of Geometrical Axioms," in Helmholtz: *Popular Scientific Lectures*, Dover (reprint), 1962, 223–49. Also in James R. Newman: *The World of Mathematics*, Simon and Schuster, 1956, Vol. 1, 647–68.

Jammer, Max: *Concepts of Space*, Harvard University Press, 1954.

Killing, W.: *Die nicht-euklidischen Raumformen in analytischer Behandlung*, B. G. Teubner, 1885.

Klein, F.: *Vorlesungen über die Entwicklung der Mathematik im 19. Jahrhundert*, Chelsea (reprint), 1950, Vol. 1, 6–62; Vol. 2, 147–206.

Pierpont, James: "Some Modern Views of Space," *Amer. Math. Soc. Bull.*, 32, 1926, 225–58.

Riemann, Bernhard: *Gesammelte mathematische Werke*, 2nd ed., Dover (reprint), 1953, pp. 272–87 and 391–404.

Russell, Bertrand: *An Essay on the Foundations of Geometry* (1897), Dover (reprint), 1956.

Smith, David E.: *A Source Book in Mathematics*, Dover (reprint), 1959, Vol. 2, 411–25, 463–75. This contains translations of Riemann's 1854 paper and Gauss's 1822 paper.

Staeckel, P.: "Gauss als Geometer," *Nachrichten König. Ges. der Wiss. zu Gött.*, 1917, Beiheft, 25–140; also in *Werke*, 10_2.

Weatherburn, C. E.: "The Development of Multidimensional Differential Geometry," *Australian and New Zealand Ass'n for the Advancement of Science*, 21, 1933, 12–28.

38

Projective and Metric Geometry

> But it should always be required that a mathematical subject
> not be considered exhausted until it has become intuitively
> evident . . .   FELIX KLEIN

1. *Introduction*

Prior to and during the work on non-Euclidean geometry, the study of
projective properties was the major geometric activity. Moreover, it was
evident from the work of von Staudt (Chap. 35, sec. 3) that projective
geometry is logically prior to Euclidean geometry because it deals with
qualitative and descriptive properties that enter into the very formation of
geometrical figures and does not use the measures of line segments and
angles. This fact suggested that Euclidean geometry might be some specialization of projective geometry. With the non-Euclidean geometries now at
hand the possibility arose that these, too, at least the ones dealing with spaces
of constant curvature, might be specializations of projective geometry. Hence
the relationship of projective geometry to the non-Euclidean geometries,
which are metric geometries because distance is employed as a fundamental
concept, became a subject of research. The clarification of the relationship
of projective geometry to Euclidean and the non-Euclidean geometries is the
great achievement of the work we are about to examine. Equally vital was the
establishment of the consistency of the basic non-Euclidean geometries.

2. *Surfaces as Models of Non-Euclidean Geometry*

The non-Euclidean geometries that seemed to be most significant after
Riemann's work were those of spaces of constant curvature. Riemann himself had suggested in his 1854 paper that a space of constant positive curvature
in two dimensions could be realized on a surface of a sphere provided the
geodesic on the sphere was taken to be the "straight line." This non-
Euclidean geometry is now referred to as double elliptic geometry for reasons
which will be clearer later. Prior to Riemann's work, the non-Euclidean
geometry of Gauss, Lobatchevsky, and Bolyai, which Klein later called

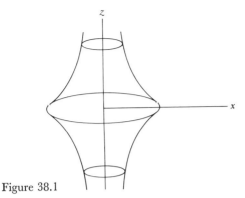

Figure 38.1

hyperbolic geometry, had been introduced as the geometry in a plane in which ordinary (and necessarily infinite) straight lines are the geodesics. The relationship of this geometry to Riemann's varieties was not clear. Riemann and Minding[1] had thought about surfaces of constant negative curvature but neither man related these to hyperbolic geometry.

Independently of Riemann, Beltrami recognized[2] that surfaces of constant curvature are non-Euclidean spaces. He gave a limited representation of hyperbolic geometry on a surface,[3] which showed that the geometry of a restricted portion of the hyperbolic plane holds on a surface of constant negative curvature if the geodesics on this surface are taken to be the straight lines. The lengths and angles on the surface are the lengths and angles of the ordinary Euclidean geometry on the surface. One such surface is known as the pseudosphere (Fig. 38.1). It is generated by revolving a curve called the tractrix about its asymptote. The equation of the tractrix is

$$z = k \log \frac{k + \sqrt{k^2 - x^2}}{x} - \sqrt{k^2 - x^2}$$

and the equation of the surface is

$$z = k \log \frac{\sqrt{k^2 - x^2 - y^2}}{\sqrt{x^2 + y^2}} - \sqrt{k^2 - x^2 - y^2}.$$

The curvature of the surface is $-1/k^2$. Thus the pseudosphere is a model for a limited portion of the plane of Gauss, Lobatchevsky, and Bolyai. On the pseudosphere a figure may be shifted about and just bending will make it conform to the surface, as a plane figure by bending can be fitted to the surface of a circular cylinder.

1. *Jour. für Math.*, 19, 1839, 370–87.
2. *Annali di Mat.*, 7, 1866, 185–204 = *Opere Mat.*, 1, 262–80.
3. *Gior. di Mat.*, 6, 1868, 248–312 = *Opere Mat.*, 1, 374–405.

Beltrami had shown that on a surface of negative constant curvature one can realize a *piece* of the Lobatchevskian plane. However, there is no regular analytic surface of negative constant curvature on which the geometry of the *entire* Lobatchevskian plane is valid. All such surfaces have a singular curve— the tangent plane is not continuous across it—so that a continuation of the surface across this curve will not continue the figures that represent those of Lobatchevsky's geometry. This result is due to Hilbert.[4]

In this connection it is worth noting that Heinrich Liebmann (1874– 1939)[5] proved that the sphere is the only closed analytic surface (free of singularities) of constant positive curvature and so the only one that can be used as a Euclidean model for double elliptic geometry.

The development of these models helped the mathematicians to understand and see meaning in the basic non-Euclidean geometries. One must keep in mind that these geometries, in the two-dimensional case, are fundamentally geometries of the plane in which the lines and angles are the usual lines and angles of Euclidean geometry. While hyperbolic geometry had been developed in this fashion, the conclusions still seemed strange to the mathematicians and had been only grudgingly admitted into mathematics. The double elliptic geometry, suggested by Riemann's differential geometric approach, did not even have an axiomatic development as a geometry of the plane. Hence the only meaning mathematicians could see for it was that provided by the geometry on the sphere. A far better understanding of the nature of these geometries was secured through another development that sought to relate Euclidean and projective geometry.

3. Projective and Metric Geometry

Though Poncelet had introduced the distinction between projective and metric properties of figures and had stated in his *Traité* of 1822 that the projective properties were logically more fundamental, it was von Staudt who began to build up projective geometry on a basis independent of length and angle size (Chap. 35, sec. 3). In 1853 Edmond Laguerre (1834–86), a professor at the Collège de France, though primarily interested in what happens to angles under a projective transformation, actually advanced the goal of establishing metric properties of Euclidean geometry on the basis of

4. *Amer. Math. Soc.*, *Trans.*, 2, 1901, 86–99 = *Ges. Abh.*, 2, 437–48. The proof and further historical details can be found in Appendix V of David Hilbert's *Grundlagen der Geometrie*, 7th ed., B. G. Teubner, 1930. The theorem presupposes that the lines of hyperbolic geometry would be the geodesics of the surface and the lengths and angles would be the Euclidean lengths and angles on the surface.
5. *Nachrichten König. Ges. der Wiss. zu Gött.*, 1899, 44–55; *Math. Ann.*, 53, 1900, 81–112; and 54, 1901, 505–17.

projective concepts by supplying a projective basis for the measure of an angle.[6]

To obtain a measure of the angle between two given intersecting lines one can consider two lines through the origin and parallel respectively to the two given lines. Let the two lines through the origin have the equations (in nonhomogeneous coordinates) $y = x \tan \theta$ and $y = x \tan \theta'$. Let $y - ix$ and $y = -ix$ be two (imaginary) lines from the origin to the circular points at infinity, that is, the points $(1, i, 0)$ and $(1, -i, 0)$. Call these four lines u, u', w, and w' respectively. Let ϕ be the angle between u and u'. Then Laguerre's result is that

$$(1) \qquad \phi = \theta' - \theta = \frac{i}{2} \log (uu', ww'),$$

where (uu', ww') is the cross ratio of the four lines.[7] What is significant about the expression (1) is that it may be taken as the definition of the size of an angle in terms of the projective concept of cross ratio. The logarithm function is, of course, purely quantitative and may be introduced in any geometry.

Independently of Laguerre, Cayley made the next step. He approached geometry from the standpoint of algebra, and in fact was interested in the geometric interpretation of quantics (homogeneous polynomial forms), a subject we shall consider in Chapter 39. Seeking to show that metrical notions can be formulated in projective terms he concentrated on the relation of Euclidean to projective geometry. The work we are about to describe is in his "Sixth Memoir upon Quantics."[8]

Cayley's work proved to be a generalization of Laguerre's idea. The latter had used the circular points at infinity to define angle in the plane. The circular points are really a degenerate conic. In two dimensions Cayley introduced any conic in place of the circular points and in three dimensions he introduced any quadric surface. These figures he called the absolutes. Cayley asserted that all metric properties of figures are none other than projective properties augmented by the absolute or in relation to the absolute. He then showed how this principle led to a new expression for angle and an expression for distance between two points.

He starts with the fact that the points of a plane are represented by homogeneous coordinates. These coordinates are not to be regarded as distances or ratios of distances but as an assumed fundamental notion not

6. *Nouvelles Annales de Mathématiques*, 12, 1853, 57–66 = *Œuvres*, 2, 6–15.

7. The cross ratio is itself a complex number. The coefficient $i/2$ ensures that a right angle has size $\pi/2$. The computation of such cross ratios can be found in texts on projective geometry. See, for example, William C. Graustein: *Introduction to Higher Geometry*, Macmillan, 1933, Chap. 8.

8. *Phil. Trans.*, 149, 1859, 61–91 = *Coll. Math. Papers*, 2, 561–606.

requiring or admitting of explanation. To define distance and size of angle he introduces the quadratic form

$$F(x, x) = \sum_{i,j=1}^{3} a_{ij}x_ix_j, \qquad a_{ij} = a_{ji}$$

and the bilinear form

$$F(x, y) = \sum_{i,j=1}^{3} a_{ij}x_iy_j.$$

The equation $F(x, x) = 0$ defines a conic which is Cayley's absolute. The equation of the absolute in line coordinates is

$$G(u, u) = \sum_{i,j=1}^{3} A^{ij}u_iu_j = 0,$$

where A^{ij} is the cofactor of a_{ij} in the determinant $|a|$ of the coefficients of F.

Cayley now defines the distance δ between two points x and y, where $x = (x_1, x_2, x_3)$ and $y = (y_1, y_2, y_3)$, by the formula

(2) $$\delta = \text{arc cos } \frac{F(x, y)}{[F(x, x)F(y, y)]^{1/2}}.$$

The angle ϕ between two lines whose line coordinates are $u = (u_1, u_2, u_3)$ and $v = (v_1, v_2, v_3)$ is defined by

(3) $$\cos \phi = \frac{G(u, v)}{[G(u, u)G(v, v)]^{1/2}}.$$

These general formulas become simple if we take for the absolute the particular conic $x_1^2 + x_2^2 + x_3^2 = 0$. Then if (a_1, a_2, a_3) and (b_1, b_2, b_3) are the homogeneous coordinates of two points, the distance between them is given by

(4) $$\text{arc cos } \frac{a_1b_1 + a_2b_2 + a_3b_3}{\sqrt{a_1^2 + a_2^2 + a_3^2}\sqrt{b_1^2 + b_2^2 + b_3^2}}$$

and the angle ϕ between two lines whose homogeneous line coordinates are (u_1, u_2, u_3) and (v_1, v_2, v_3) is given by

(5) $$\cos \phi = \frac{u_1v_1 + u_2v_2 + u_3v_3}{\sqrt{u_1^2 + u_2^2 + u_3^2}\sqrt{v_1^2 + v_2^2 + v_3^2}}.$$

With respect to the expression for distance if we use the shorthand that $xy = x_1y_1 + x_2y_2 + x_3y_3$ then, if $a = (a_1, a_2, a_3)$, b and c are three points on a line,

$$\text{arc cos } \frac{ab}{\sqrt{aa}\sqrt{bb}} + \text{arc cos } \frac{bc}{\sqrt{bb}\sqrt{cc}} = \text{arc cos } \frac{ac}{\sqrt{aa}\sqrt{cc}}.$$

That is, the distances add as they should. By taking the absolute conic to be the circular points at infinity, $(1, i, 0)$ and $(1, -i, 0)$, Cayley showed that his formulas for distance and angle reduce to the usual Euclidean formulas.

It will be noted that the expressions for length and angle involve the algebraic expression for the absolute. Generally the analytic expression of any Euclidean metrical property involves the relation of that property to the absolute. Metrical properties are not properties of the figure per se but of the figure in relation to the absolute. This is Cayley's idea of the general projective determination of metrics. The place of the metric concept in projective geometry and the greater generality of the latter were described by Cayley as, "Metrical geometry is part of projective geometry."

Cayley's idea was taken over by Felix Klein (1849–1925) and generalized so as to include the non-Euclidean geometries. Klein, a professor at Göttingen, was one of the leading mathematicians in Germany during the last part of the nineteenth and first part of the twentieth century. During the years 1869–70 he learned the work of Lobatchevsky, Bolyai, von Staudt, and Cayley; however, even in 1871 he did not know Laguerre's result. It seemed to him to be possible to subsume the non-Euclidean geometries, hyperbolic and double elliptic geometry, under projective geometry by exploiting Cayley's idea. He gave a sketch of his thoughts in a paper of 1871 [9] and then developed them in two papers.[10] Klein was the first to recognize that we do not need surfaces to obtain models of non-Euclidean geometries.

To start with, Klein noted that Cayley did not make clear just what he had in mind for the meaning of his coordinates. They were either simply variables with no geometrical interpretation or they were Euclidean distances. But to derive the metric geometries from projective geometry it was necessary to build up the coordinates on a projective basis. Von Staudt had shown (Chap. 35, sec. 3) that it was possible to assign numbers to points by his algebra of throws. But he used the Euclidean parallel axiom. It seemed clear to Klein that this axiom could be dispensed with and in the 1873 paper he shows that this can be done. Hence coordinates and cross ratio of four points, four lines, or four planes can be defined on a purely projective basis.

Klein's major idea was that by specializing the nature of Cayley's absolute quadric surface (if one considers three-dimensional geometry) one could show that the metric, which according to Cayley depended on the nature of the absolute, would yield hyperbolic and double elliptic geometry. When the second degree surface is a real ellipsoid, real elliptic paraboloid, or real hyperboloid of two sheets one gets Lobatchevsky's metric geometry, and when the second degree surface is imaginary one gets Riemann's non-Euclidean geometry (of constant positive curvature). If the absolute is

9. *Nachrichten König. Ges. der Wiss. zu Gött.*, 1871, 419–33 = *Ges. Math. Abh.*, 1, 244–53.
10. *Math. Ann.*, 4, 1871, 573–625; and 6, 1873, 112–45 = *Ges. Math. Abh.*, 1, 254–305, 311–43.

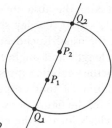

Figure 38.2

taken to be the sphere-circle, whose equation in homogeneous coordinates is $x^2 + y^2 + z^2 = 0$, $t = 0$, then the usual Euclidean metric geometry obtains. Thus the metric geometries become special cases of projective geometry.

To appreciate Klein's ideas let us consider two-dimensional geometry. One chooses a conic in the projective plane; this conic will be the absolute. Its equation is

(6)
$$F = \sum_{i,j=1}^{3} a_{ij}x_i x_j = 0$$

in point coordinates and

(7)
$$G = \sum_{i,j=1}^{3} A^{ij}u_i u_j = 0$$

in line coordinates. To derive Lobatchevsky's geometry the conic must be real, e.g. in plane homogeneous coordinates $x_1^2 + x_2^2 - x_3^2 = 0$; for Riemann's geometry on a surface of constant positive curvature it is imaginary, for example, $x_1^2 + x_2^2 + x_3^2 = 0$; and for Euclidean geometry, the conic degenerates into two coincident lines represented in homogeneous coordinates by $x_3 = 0$ and on this locus one chooses two imaginary points whose equation is $x_1^2 + x_2^2 = 0$, that is, the circular points at infinity whose homogeneous coordinates are $(1, i, 0)$ and $(1, -i, 0)$. In every case the conic has a real equation.

To be specific let us suppose the conic is the one shown in Figure 38.2. If P_1 and P_2 are two points of a line, this line meets the absolute in two points (real or imaginary). Then the distance is taken to be

(8)
$$d = c \log (P_1 P_2, Q_1 Q_2),$$

where the quantity in parentheses denotes the cross ratio of the four points and c is a constant. This cross ratio can be expressed in terms of the co-

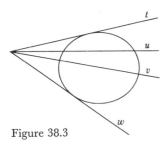

Figure 38.3

ordinates of the points. Moreover, if there are three points P_1, P_2, P_3 on the line then it can readily be shown that

$$(P_1P_2, Q_1Q_2) \cdot (P_2P_3, Q_1Q_2) = (P_1P_3, Q_1Q_2)$$

so that $P_1P_2 + P_2P_3 = P_1P_3$.

Likewise if u and v are two lines (Fig. 38.3) one considers the tangents t and w from their point of intersection to the absolute (the tangents may be imaginary lines); then the angle between u and v is defined to be

$$\phi = c' \log (uv, tw),$$

where again c' is a constant and the quantity in parentheses denotes the cross ratio of the four lines.

To express the values of d and ϕ analytically and to show their dependence upon the choice of the absolute, let the equation of the absolute be given by F and G above. By definition

$$F_{xy} = \sum_{i,j=1}^{3} a_{ij}x_iy_j.$$

One can now show that if $x = (x_1, x_2, x_3)$ and $y = (y_1, y_2, y_3)$ are the coordinates of P_1 and P_2 then

$$d = c \log \frac{F_{xy} + \sqrt{F_{xy}^2 - F_{xx}F_{yy}}}{F_{xy} - \sqrt{F_{xy}^2 - F_{xx}F_{yy}}}.$$

Likewise, if (u_1, u_2, u_3) and (v_1, v_2, v_3) are the coordinates of the two lines, then one can show, using G, that

$$\phi = c' \log \frac{G_{uv} + \sqrt{G_{uv}^2 - G_{uu}G_{vv}}}{G_{uv} - \sqrt{G_{uv}^2 - G_{uu}G_{vv}}}.$$

The constant c' is generally taken to be $i/2$ so as to make ϕ real and a complete central angle 2π.

Klein used the logarithmic expressions above for angle and distance and showed how the metric geometries can be derived from projective geometry.

Thus if one starts with projective geometry then by his choice of the absolute and by using the above expressions for distance and angle one can get the Euclidean, hyperbolic, and elliptic geometries as special cases. The nature of the metric geometry is fixed by the choice of the absolute. Incidentally, Klein's expressions for distance and angle can be shown to be equal to Cayley's.

If one makes a projective (i.e. linear) transformation of the projective plane into itself which transforms the absolute into itself (though points on the absolute go into other points) then because cross ratio is unaltered by a linear transformation, distance and angle will be unaltered. These particular linear transformations which leave the absolute fixed are the rigid motions or congruence transformations of the particular metric geometry determined by the absolute. A general projective transformation will not leave the absolute invariant. Thus projective geometry proper is more general in the transformations it allows.

Another contribution of Klein to non-Euclidean geometry was the observation, which he says he first made in 1871 [11] but published in 1874,[12] that there are two kinds of elliptic geometry. In the double elliptic geometry two points do not always determine a unique straight line. This is evident from the spherical model when the two points are diametrically opposite. In the second elliptic geometry, called single elliptic, two points always determine a unique straight line. When looked at from the standpoint of differential geometry, the differential form ds^2 of a surface of constant positive curvature is (in homogeneous coordinates)

$$ds^2 = \frac{dx_1^2 + dx_2^2 + dx_3^2}{\{1 + (a^2/4)(x_1^2 + x_2^2 + x_3^2)\}^2}.$$

In both cases $a^2 > 0$. However, in the first type the geodesics are curves of finite length $2\pi/a$ or if R is the radius, $2\pi R$, and are closed (return upon themselves). In the second the geodesics are of length π/a or πR and are still closed.

A model of a surface which has the properties of single elliptic geometry and which is due to Klein,[13] is provided by a hemisphere including the boundary. However, one must identify any two points on the boundary which are diametrically opposite. The great circular arcs on the hemisphere are the "straight lines" or geodesics of this geometry and the ordinary angles on the surface are the angles of the geometry. Single elliptic geometry (sometimes called elliptic in which case double elliptic geometry is called spherical) is then also realized on a space of constant positive curvature.

11. *Math. Ann.*, 4, 1871, 604. See also, *Math. Ann.*, 6, 1873, 125; and *Math. Ann.*, 37, 1890, 554–57.
12. *Math. Ann.*, 7, 1874, 549–57; 9, 1876, 476–82 = *Ges. math. Abh.*, 2, 63–77.
13. See the references in footnote 11.

One cannot actually unite the pairs of points which are identified in this model at least in three-dimensional space. The surface would have to cross itself and points that coincide on the intersection would have to be regarded as distinct.

We can now see why Klein introduced the terminology hyperbolic for Lobatchevsky's geometry, elliptic for the case of Riemann's geometry on a surface of constant positive curvature, and parabolic for Euclidean geometry. This terminology was suggested by the fact that the ordinary hyperbola meets the line at infinity in two points and correspondingly in hyperbolic geometry each line meets the absolute in two real points. The ordinary ellipse has no real points in common with the line at infinity and in elliptic geometry, likewise, each line has no real points in common with the absolute. The ordinary parabola has only one real point in common with the line at infinity and in Euclidean geometry (as extended in projective geometry) each line has one real point in common with the absolute.

The import which gradually emerged from Klein's contributions was that projective geometry is really logically independent of Euclidean geometry. Moreover, the non-Euclidean and Euclidean geometries were also seen to be special cases or subgeometries of projective geometry. Actually the strictly logical or rigorous work on the axiomatic foundations of projective geometry and its relations to the subgeometries remained to be done (Chap. 42). But by making apparent the basic role of projective geometry Klein paved the way for an axiomatic development which could start with projective geometry and derive the several metric geometries from it.

4. Models and the Consistency Problem

By the early 1870s several basic non-Euclidean geometries, the hyperbolic and the two elliptic geometries, had been introduced and intensively studied. The fundamental question which had yet to be answered in order to make these geometries legitimate branches of mathematics was whether they were consistent. All of the work done by Gauss, Lobatchevsky, Bolyai, Riemann, Cayley, and Klein might still have proved to be nonsense if contradictions were inherent in these geometries.

Actually the proof of the consistency of two-dimensional double elliptic geometry was at hand, and possibly Riemann appreciated this fact though he made no explicit statement. Beltrami[14] had pointed out that Riemann's two-dimensional geometry of constant positive curvature is realized on a sphere. This model makes possible the proof of the consistency of two-dimensional double elliptic geometry. The axioms (which were not explicit at this time) and the theorems of this geometry are all applicable to

14. *Annali di Mat.*, (2), 2, 1868–69, 232–55 = *Opere Matematiche*, 1, 406–29.

the geometry of the surface of the sphere provided that line in the double elliptic geometry is interpreted as great circle on the surface of the sphere. If there should be contradictory theorems in this double elliptic geometry then there would be contradictory theorems about the geometry of the surface of the sphere. Now the sphere is part of Euclidean geometry. Hence if Euclidean geometry is consistent, then double elliptic geometry must also be so. To the mathematicians of the 1870s the consistency of Euclidean geometry was hardly open to question because, apart from the views of a few men such as Gauss, Bolyai, Lobatchevsky, and Riemann, Euclidean geometry was still the necessary geometry of the physical world and it was inconceivable that there could be contradictory properties in the geometry of the physical world. However, it is important, especially in the light of later developments, to realize that this proof of the consistency of double elliptic geometry depends upon the consistency of Euclidean geometry.

The method of proving the consistency of double elliptic geometry could not be used for single elliptic geometry or for hyperbolic geometry. The hemispherical model of single elliptic geometry cannot be realized in three-dimensional Euclidean geometry though it can be in four-dimensional Euclidean geometry. If one were willing to believe in the consistency of the latter, then one might accept the consistency of single elliptic geometry. However, though n-dimensional geometry had already been considered by Grassmann, Riemann, and others, it is doubtful that any mathematician of the 1870s would have been willing to affirm the consistency of four-dimensional Euclidean geometry.

The case for the consistency of hyperbolic geometry could not be made on any such grounds. Beltrami had given the pseudospherical interpretation, which is a surface in Euclidean space, but this serves as a model for only a limited region of hyperbolic geometry and so could not be used to establish the consistency of the entire geometry. Lobatchevsky and Bolyai had considered this problem (Chap. 36, sec. 8) but had not been able to settle it. As a matter of fact, though Bolyai proudly published his non-Euclidean geometry, there is evidence that he doubted its consistency because in papers found after his death he continued to try to prove the Euclidean parallel axiom.

The consistency of the hyperbolic and single elliptic geometries was established by new models. The model for hyperbolic geometry is due to Beltrami.[15] However, the distance function used in this model is due to Klein and the model is often ascribed to him. Let us consider the two-dimensional case.

Within the Euclidean plane (which is part of the projective plane) one selects a real conic which one may as well take to be a circle (Fig. 38.4).

15. *Annali di Mat.*, 7, 1866, 185–204 = *Opere Mat.*, 1, 262–80; *Gior. di Mat.*, 6, 1868, 284–312 = *Opere Mat.*, 1, 374–405.

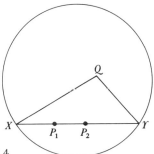

Figure 38.4

According to this representation of hyperbolic geometry the points of the geometry are the points interior to this circle. A line of this geometry is a chord of the circle, say the chord XY (but not including X and Y). If we take any point Q not on XY then we can find any number of lines through Q which do not meet XY. Two of these lines, namely, QX and QY, separate the lines through Q into two classes, those lines which cut XY and those which do not. In other words, the parallel axiom of hyperbolic geometry is satisfied by the points and lines (chords) interior to the circle. Further, let the size of the angle formed by two lines a and b be

$$\sphericalangle(a, b) = \frac{1}{2i} \log (ab, mn),$$

where m and n are the conjugate imaginary tangents from the vertex of the angle to the circle and (ab, mn) is the cross ratio of the four lines a, b, m, and n. The constant $1/2i$ ensures that a right angle has the measure $\pi/2$. The definition of distance between two points is given by formula (8), that is, $d = c \log (PP', XY)$ with c usually taken to be $k/2$. According to this formula as P or P' approaches X or Y, the distance PP' becomes infinite. Hence in terms of this distance, a chord is an infinite line of hyperbolic geometry.

Thus with the projective definitions of distance and angle size, the points, chords, angles, and other figures interior to the circle satisfy the axioms of hyperbolic geometry. Then the theorems of hyperbolic geometry also apply to these figures inside the circle. In this model the axioms and theorems of hyperbolic geometry are really assertions about special figures and concepts (e.g. distance defined in the manner of hyperbolic geometry) of *Euclidean* geometry. Since the axioms and theorems in question apply to these figures and concepts, regarded as belonging to Euclidean geometry, all of the assertions of hyperbolic geometry are theorems of Euclidean geometry. If, then, there were a contradiction in hyperbolic geometry, this contradiction would be a contradiction within Euclidean geometry. But *if Euclidean geometry is consistent*, then hyperbolic geometry must also be. Thus the

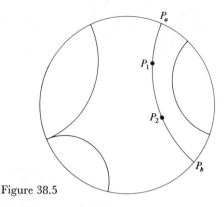

Figure 38.5

consistency of hyperbolic geometry is reduced to the consistency of Euclidean geometry.

The fact that hyperbolic geometry is consistent implies that the Euclidean parallel axiom is independent of the other Euclidean axioms. If this were not the case, that is, if the Euclidean parallel axiom were derivable from the other axioms, it would also be a theorem of hyperbolic geometry for, aside from the parallel axiom, the other axioms of Euclidean geometry are the same as those of hyperbolic geometry. But this theorem would contradict the parallel axiom of hyperbolic geometry and hyperbolic geometry would be inconsistent. The consistency of two-dimensional single elliptic geometry can be shown in the same manner as for hyperbolic geometry because this elliptic geometry is also realized within the projective plane and with the projective definition of distance.

Independently and in connection with his work on automorphic functions Poincaré[16] gave another model which also establishes the consistency of hyperbolic geometry. One form in which this Poincaré model for hyperbolic plane geometry[17] can be expressed takes the absolute to be a circle (Fig. 38.5). Within the absolute the straight lines of the geometry are arcs of circles which cut the absolute orthogonally and straight lines through the center of the absolute. The length of any segment P_1P_2 is given by $\log (P_1P_2, P_aP_b)$, where $(P_1P_2, P_aP_b) = (P_1P_b/P_2P_b)/(P_1P_a/P_2P_a)$, P_a and P_b are the points in which the arc through P_1 and P_2 cuts the absolute, and the lengths P_1P_b, P_2P_b, etc. are the chords. The angle between two intersecting "lines" of this model is the normal Euclidean angle between the two arcs. Two circular arcs which are tangent at a point on the absolute are parallel

16. *Acta Math.*, 1, 1882, 1–62 = *Œuvres*, 2, 108–68; see p. 8 and p. 52 of the paper.

17. This form, attributed to Poincaré, is close to one he gave in the *Bull. Soc. Math. de France*, 15, 1887, 203–16 = *Œuvres*, 11, 79–91. The model described here seems to have been given first by Joseph Wellstein (1869–1919) in H. Weber and J. Wellstein, *Enzyklopädie der Elementar-Mathematik*, 2, 1905, 39–81.

"lines." Since in this model, too, the axioms and theorems of hyperbolic geometry are special theorems of Euclidean geometry, the argument given above apropos of the Beltrami model may be applied here to establish the consistency of hyperbolic geometry. The higher-dimensional analogues of the above models are also valid.

5. Geometry from the Transformation Viewpoint

Klein's success in subsuming the various metric geometries under projective geometry led him to seek to characterize the various geometries not just on the basis of nonmetric and metric properties and the distinctions among the metrics but from the broader standpoint of what these geometries and other geometries which had already appeared on the scene sought to accomplish. He gave this characterization in a speech of 1872, "Vergleichende Betrachtungen über neuere geometrische Forschungen" (A Comparative Review of Recent Researches in Geometry),[18] on the occasion of his admission to the faculty of the University of Erlangen, and the views expressed in it are known as the Erlanger Programm.

Klein's basic idea is that each geometry can be characterized by a group of transformations and that a geometry is really concerned with invariants under this group of transformations. Moreover a subgeometry of a geometry is the collection of invariants under a subgroup of transformations of the original group. Under this definition all theorems of a geometry corresponding to a given group continue to be theorems in the geometry of the subgroup.

Though Klein in his paper does not give the analytical formulations of the groups of transformations he discusses we shall give some for the sake of explicitness. According to his notion of a geometry projective geometry, say in two dimensions, is the study of invariants under the group of transformations from the points of one plane to those of another or to points of the same plane (collineations). Each transformation is of the form

(9)
$$x_1' = a_{11}x_1 + a_{12}x_2 + a_{13}x_3$$
$$x_2' = a_{21}x_1 + a_{22}x_2 + a_{23}x_3$$
$$x_3' = a_{31}x_1 + a_{32}x_2 + a_{33}x_3$$

wherein homogeneous coordinates are presupposed, the a_{ij} are real numbers, and the determinant of the coefficients must not be zero. In nonhomogeneous coordinates the transformations are represented by

(10)
$$x' = \frac{a_{11}x + a_{12}y + a_{13}}{a_{31}x + a_{32}y + a_{33}}$$

$$y' = \frac{a_{21}x + a_{22}y + a_{23}}{a_{31}x + a_{32}y + a_{33}}$$

18. *Math. Ann.*, 43, 1893, 63–100 = *Ges. Math. Abh.*, 1, 460–97. An English translation can be found in the *N.Y. Math. Soc. Bull.*, 2, 1893, 215–49.

and again the determinant of the a_{ij} must not be zero. The invariants under the projective group are, for example, linearity, collinearity, cross ratio, harmonic sets, and the property of being a conic section.

One subgroup of the projective group is the collection of affine transformations.[19] This subgroup is defined as follows: Let any line l_∞ in the projective plane be fixed. The points of l_∞ are called ideal points or points at infinity and l_∞ is called the line at infinity. Other points and lines of the projective plane are called ordinary points and these are the usual points of the Euclidean plane. The affine group of collineations is that subgroup of the projective group which leaves l_∞ invariant (though not necessarily pointwise) and affine geometry is the set of properties and relations invariant under the affine group. Algebraically, in two dimensions and in homogeneous coordinates, the affine transformations are represented by equations (9) above but in which $a_{31} = a_{32} = 0$ and with the same determinant condition. In nonhomogeneous coordinates affine transformations are given by

$$x' = a_{11}x + a_{12}y + a_{13} \qquad \begin{vmatrix} a_{11} & a_{12} \\ a_{21} & a_{22} \end{vmatrix} \neq 0.$$
$$y' = a_{21}x + a_{22}y + a_{23}$$

Under an affine transformation, straight lines go into straight lines, and parallel straight lines into parallel lines. However, lengths and angle sizes are altered. Affine geometry was first noted by Euler and then by Möbius in his *Der barycentrische Calcul*. It is useful in the study of the mechanics of deformations.

The group of any metric geometry is the same as the affine group except that the determinant above must have the value $+1$ or -1. The first of the metric geometries is Euclidean geometry. To define the group of this geometry one starts with l_∞ and supposes there is a fixed involution on l_∞. One requires that this involution has no real double points but has the circular points at ∞ as (imaginary) double points. We now consider all projective transformations which not only leave l_∞ fixed but carry any point of the involution into its corresponding point of the involution, which implies that each circular point goes into itself. Algebraically these transformations of the Euclidean group are represented in nonhomogeneous (two-dimensional) coordinates by

$$x' = \rho(x \cos \theta - y \sin \theta + \alpha) \qquad \rho = \pm 1.$$
$$y' = \rho(x \sin \theta + y \cos \theta + \beta)$$

The invariants are length, size of angle, and size and shape of any figure.

Euclidean geometry as the term is used in this classification is the set of invariants under this class of transformations. The transformations are rotations, translations, and reflections. To obtain the invariants associated with similar figures, the subgroup of the affine group known as the parabolic

19. Klein did not single out this subgroup.

metric group is introduced. This group is defined as the class of projective transformations which leaves the involution on l_∞ invariant, and this means that each pair of corresponding points goes into some pair of corresponding points. In nonhomogeneous coordinates the transformations of the parabolic metric group are of the form

$$x' = ax - by + c$$
$$y' = bex + aey + d$$

wherein $a^2 + b^2 \neq 0$, and $e^2 = 1$. These transformations preserve angle size.

To characterize hyperbolic metric geometry we return to projective geometry and consider an arbitrary, real, nondegenerate conic (the absolute) in the projective plane. The subgroup of the projective group which leaves this conic invariant (though not necessarily pointwise) is called the hyperbolic metric group and the corresponding geometry is hyperbolic metric geometry. The invariants are those associated with congruence.

Single elliptic geometry is the geometry corresponding to the subgroup of projective transformations which leaves a definite imaginary ellipse (the absolute) of the projective plane invariant. The plane of elliptic geometry is the real projective plane and the invariants are those associated with congruence.

Even double elliptic geometry can be encompassed in this transformation viewpoint, but one must start with three-dimensional projective transformations to characterize the two-dimensional metric geometry. The subgroup of transformations consists of those three-dimensional projective transformations which transform a definite sphere (surface) S of the finite portion of space into itself. The spherical surface S is the "plane" of double elliptic geometry. Again the invariants are associated with congruence.

In four metric geometries, that is, Euclidean, hyperbolic, and the two elliptic geometries, the transformations which are permitted in the corresponding subgroup are what are usually called rigid motions and these are the only geometries which permit rigid motions.

Klein introduced a number of intermediate classifications which we shall not repeat here. The scheme below shows the relationships of the principal geometries.

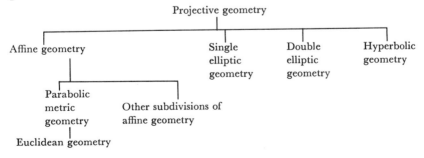

Klein also considered more general geometries than projective. At this time (1872) algebraic geometry was gradually being distinguished as a separate discipline and he characterized this geometry by introducing the transformations which in three dimensions and in nonhomogeneous coordinates read

$$x' = \phi(x, y, z), \qquad y' = \psi(x, y, z), \qquad z' = \chi(x, y, z).$$

He required that the functions ϕ, ψ, and χ be rational and single-valued and that it must be possible to solve for x, y, and z in terms of single-valued rational functions of x', y', z'. These transformations are called Cremona transformations and the invariants under them are the subject matter of algebraic geometry (Chap. 39).

Klein also projected the study of invariants under one-to-one continuous transformations with continuous inverses. This is the class now called homeomorphisms and the study of the invariants under such transformations is the subject matter of topology (Chap. 50). Though Riemann had also considered what are now recognized to be topological problems in his work with Riemann surfaces, the projection of topology as a major geometry was a bold step in 1872.

It has been possible since Klein's days to make further additions and specializations of Klein's classification. But not all of geometry can be incorporated in Klein's scheme. Algebraic geometry today and differential geometry do not come under this scheme.[20] Though Klein's view of geometry did not prove to be all-embracing, it did afford a systematic method of classifying and studying a good portion of geometry and suggested numerous research problems. His "definition" of geometry guided geometrical thinking for about fifty years. Moreover, his emphasis on invariants under transformations carried beyond mathematics to mechanics and mathematical physics generally. The physical problem of invariance under transformation or the problem of expressing physical laws in a manner independent of the coordinate system became important in physical thinking after the invariance of Maxwell's equations under Lorentz transformations (a four-dimensional subgroup of affine geometry) was noted. This line of thinking led to the special theory of relativity.

We shall merely mention here further studies on the classification of geometries which at least in their time attracted considerable attention. Helmholtz and Sophus Lie (1842–99) sought to characterize geometries in which rigid motions are possible. Helmholtz's basic paper, "Über die Thatsachen, die der Geometrie zum Grunde liegen" (On the Facts Which

---

20. Klein does speak, in the case of differential geometry, of the group of transformations which leave the expression for ds^2 invariant. This leads to differential invariants (*Ges. math. Abh.*, 1, 487).

Underlie Geometry),[21] showed that if the motions of rigid bodies are to be possible in a space then Riemann's expression for ds in a space of constant curvature is the only one possible. Lie attacked the same problem by what is called the theory of continuous transformation groups, a theory he had already introduced in the study of ordinary differential equations, and he characterized the spaces in which rigid motions are possible by means of the kinds of groups of transformations which these spaces permit.[22]

6. *The Reality of Non-Euclidean Geometry*

The interest in the classical synthetic non-Euclidean geometries and in projective geometry declined after the work of Klein and Lie partly because the essence of these structures was so clearly exposed by the transformation viewpoint. The feeling of mathematicians so far as the discovery of additional theorems is concerned was that the mine had been exhausted. The rigorization of the foundations remained to be accomplished and this was an active area for quite a few years after 1880 (Chap. 42).

Another reason for the loss of interest in the non-Euclidean geometries was their seeming lack of relevance to the physical world. It is curious that the first workers in the field, Gauss, Lobatchevsky, and Bolyai did think that non-Euclidean geometry might prove applicable when further work in astronomy had been done. But none of the mathematicians who worked in the later period believed that these basic non-Euclidean geometries would be physically significant. Cayley, Klein, and Poincaré, though they considered this matter, affirmed that we would not ever need to improve on or abandon Euclidean geometry. Beltrami's pseudosphere model had made non-Euclidean geometry real in a mathematical sense (though not physically) because it gave a readily visualizable interpretation of Lobatchevsky's geometry but at the expense of changing the line from the ruler's edge to geodesics on the pseudosphere. Similarly the Beltrami-Klein and Poincaré models made sense of non-Euclidean geometry by changing the concepts either of line, distance, or angle-measure, or of all three, and by picturing them in Euclidean space. But the thought that physical space under the usual interpretation of straight line or even under some other interpretation could be non-Euclidean was dismissed. In fact, most mathematicians regarded non-Euclidean geometry as a logical curiosity.

Cayley was a staunch supporter of Euclidean space and accepted the non-Euclidean geometries only so far as they could be realized in Euclidean space by the use of new distance formulas. In 1883 in his presidential address to the British Association for the Advancement of Science[23] he said that

---

21. *Nachrichten König. Ges. der Wiss. zu Gött.*, 15, 1868, 193–221 = *Wiss. Abh.*, 2, 618–39.
22. *Theorie der Transformationsgruppen*, 3, 437–543, 1893.
23. *Collected Math. Papers*, 11, 429–59.

non-Euclidean spaces were a priori a mistaken idea, but non-Euclidean geometries were acceptable because they resulted merely from a change in the
distance function in Euclidean space. He did not grant independent existence
to the non-Euclidean geometries but regarded them as a class of special
Euclidean structures or as a way of representing projective relations in
Euclidean geometry. It was his view that

> Euclid's twelfth [tenth] axiom in Playfair's form of it does not need
> demonstration but is part of our notion of space, of the physical space of
> our own experience—the space, that is, which we become acquainted
> with by experience, but which is the representation lying at the foundation
> of all external experience.
>
> Riemann's view may be said to be that, having *in intellectu* a more
> general notion of space (in fact a notion of non-Euclidean space), we
> learn by experience that space (the physical space of experience) is, if not
> exactly, at least to the highest degree of approximation, Euclidean space.

Klein regarded Euclidean space as the necessary fundamental space. The
other geometries were merely Euclidean with new distance functions. The
non-Euclidean geometries were in effect subordinated to Euclidean geometry.

Poincaré's judgment was more liberal. Science should always try to use
Euclidean geometry and vary the laws of physics where necessary. Euclidean
geometry may not be true but it is most convenient. One geometry cannot
be more true than another; it can only be more convenient. Man creates
geometry and then adapts the physical laws to it to make the geometry and
laws fit the world. Poincaré insisted[24] that even if the angle sum of a triangle
should prove to be greater than 180° it would be better to assume that
Euclidean geometry describes physical space and that light travels along
curves because Euclidean geometry is simpler. Of course events proved he
was wrong. It is not the simplicity of the geometry alone that counts for
science but the simplicity of the entire scientific theory. Clearly the mathematicians of the nineteenth century were still tied to tradition in their
notions about what makes physical sense. The advent of the theory of
relativity forced a drastic change in the attitude toward non-Euclidean
geometry.

The delusion of mathematicians that what they are working on at the
moment is the most important conceivable subject is illustrated again by
their attitude toward projective geometry. The work we have examined in
this chapter does indeed show that projective geometry is fundamental to
many geometries. However, it does not embrace the evidently vital Riemannian geometry and the growing body of algebraic geometry. Neverthe-

24. *Bull. Soc. Math. de France*, 15, 1887, 203–16 = *Œuvres*, 11, 79–91. He expressed this
view again in an article "Les Géométries non-euclidiennes" in the *Revue Générale des
Sciences*, 2, 1891, #23. An English translation is in *Nature*, 45, 1892, 404–7. See also his
Science and Hypothesis, Chapter 3, in *The Foundations of Science*, The Science Press, 1946.

less, Cayley affirmed in his 1859 paper (sec. 3) that, "Projective geometry is all geometry and reciprocally."[25] Bertrand Russell in his *Essay on the Foundations of Geometry* (1897) also believed that projective geometry was necessarily the a priori form of any geometry of physical space. Hermann Hankel, despite the attention he gave to history,[26] did not hesitate to say in 1869 that projective geometry is the royal road to all mathematics. Our examination of the developments already recorded shows clearly that mathematicians can readily be carried away by their enthusiasms.

Bibliography

Beltrami, Eugenio: *Opere matematiche*, Ulrico Hoepli, 1902, Vol. 1.

Bonola, Roberto: *Non-Euclidean Geometry*, Dover (reprint), 1955, pp. 129–264.

Coolidge, Julian L.: *A History of Geometrical Methods*, Dover (reprint), 1963, pp. 68–87.

Klein, Felix: *Gesammelte mathematische Abhandlungen*, Julius Springer, 1921–23, Vols. 1 and 2.

Pasch, Moritz, and Max Dehn: *Vorlesungen über neuere Geometrie*, 2nd ed., Julius Springer, 1926, pp. 185–239.

Pierpont, James: "Non-Euclidean Geometry. A Retrospect," *Amer. Math. Soc. Bull.*, 36, 1930, 66–76.

Russell, Bertrand: *An Essay on the Foundations of Geometry* (1897), Dover (reprint), 1956.

25. Cayley used the term "descriptive geometry" for projective geometry.

26. *Die Entwicklung der Mathematik in den letzten Jahrhunderten* (The Development of Mathematics in the Last Few Centuries), 1869; 2nd ed., 1884.

39
Algebraic Geometry

> In these days the angel of topology and the devil of abstract
> algebra fight for the soul of each individual mathematical
> domain.
>
> HERMANN WEYL

1. Background

While non-Euclidean and Riemannian geometry were being created, the projective geometers were pursuing their theme. As we have seen, the two areas were linked by the work of Cayley and Klein. After the algebraic method became widely used in projective geometry the problem of recognizing what properties of geometrical figures are independent of the coordinate representation commanded attention and this prompted the study of algebraic invariants.

The projective properties of geometrical figures are those that are invariant under linear transformations of the figures. While working on these properties the mathematicians occasionally allowed themselves to consider higher-degree transformations and to seek those properties of curves and surfaces that are invariant under these latter transformations. The class of transformations, which soon superseded linear transformations as the favorite interest, is called birational because these are expressed algebraically as rational functions of the coordinates and the inverse transformations are also rational functions of their coordinates. The concentration on birational transformations undoubtedly resulted from the fact that Riemann had used them in his work on Abelian integrals and functions, and in fact, as we shall see, the first big steps in the study of the birational transformation of curves were guided by what Riemann had done. These two subjects formed the content of algebraic geometry in the latter part of the nineteenth century.

The term algebraic geometry is an unfortunate one because originally it referred to all the work from the time of Fermat and Descartes in which algebra had been applied to geometry; in the latter part of the nineteenth century it was applied to the study of algebraic invariants and birational transformations. In the twentieth century it refers to the last-mentioned field.

2. *The Theory of Algebraic Invariants*

As we have already noted, the determination of the geometric properties of figures that are represented and studied through coordinate representation calls for the discernment of those algebraic expressions which remain invariant under change of coordinates. Alternatively viewed, the projective transformation of one figure into another by means of a linear transformation preserves some properties of the figure. The algebraic invariants represent these invariant geometrical properties.

The subject of algebraic invariants had previously arisen in number theory (Chap. 34, sec. 5) and particularly in the study of how binary quadratic forms

$$(1) \qquad f = ax^2 + 2bxy + cy^2$$

transform when x and y are transformed by the linear transformation T, namely,

$$(2) \qquad x = \alpha x' + \beta y', \qquad y = \gamma x' + \delta y',$$

where $\alpha\delta - \beta\gamma = r$. Application of T to f produces

$$(3) \qquad f' = a'x'^2 + 2b'x'y' + c'y'^2.$$

In number theory the quantities a, b, c, α, β, γ, and δ are integers and $r = 1$. However, it is true generally that the discriminant D of f satisfies the relation

$$(4) \qquad D' = r^2 D.$$

The linear transformations of projective geometry are more general because the coefficients of the forms and the transformations are not restricted to integers. The term algebraic invariants is used to distinguish those arising under these more general linear transformations from the modular invariants of number theory and, for that matter, from the differential invariants of Riemannian geometry.

To discuss the history of algebraic invariant theory we need some definitions. The nth degree form in one variable

$$f(x) = a_0 x^n + a_1 x^{n-1} + \cdots + a_n$$

becomes in homogeneous coordinates the binary form

$$(5) \qquad f(x_1, x_2) = a_0 x_1^n + a_1 x_1^{n-1} x_2 + \cdots + a_n x_2^n.$$

In three variables the forms are called ternary; in four variables, quaternary; etc. The definitions below apply to forms in n variables.

Suppose we subject the binary form to a transformation T of the form (2). Under T the form $f(x_1, x_2)$ is transformed into the form

$$F(X_1, X_2) = A_0 X_1^n + A_1 X_1^{n-1} X_2 + \cdots + A_n X_2^n.$$

The coefficients of F will differ from those of f and the roots of $F = 0$ will differ from the roots of $f = 0$. Any function I of the coefficients of f which satisfies the relationship

$$I(A_0, A_1, \cdots, A_n) = r^w I(a_0, a_1, \cdots, a_n)$$

is called an invariant of f. If $w = 0$, the invariant is called an absolute invariant of f. The degree of the invariant is the degree in the coefficients and the weight is w. The discriminant of a binary form is an invariant, as (4) illustrates. In this case the degree is 2 and the weight is 2. The significance of the discriminant of any polynomial equation $f(x) = 0$ is that its vanishing is the condition that $f(x) = 0$ have equal roots or, geometrically, that the locus of $f(x) = 0$, which is a series of points, has two coincident points. This property is clearly independent of the coordinate system.

If two (or more) binary forms

$$f_1 = a_0 x_1^m + \cdots + a_m x_2^m$$
$$f_2 = b_0 x_1^n + \cdots + b_n x_2^n$$

are transformed by T into

$$F_1 = A_0 X_1^m + \cdots + A_m X_2^m$$
$$F_2 = B_0 X_1^n + \cdots + B_n X_2^n$$

then any function I of the coefficients which satisfies the relationship

(6) $\qquad I(A_0, \cdots, A_m, B_0, \cdots, B_n) = r^w I(a_0, \cdots, a_m, b_0, \cdots, b_n)$

is said to be a joint or simultaneous invariant of the two forms. Thus the linear forms $a_1 x_1 + b_1 x_2$ and $a_2 x_1 + b_2 x_2$ have as a simultaneous invariant the resultant $a_1 b_2 - a_2 b_1$ of the two forms. Geometrically the vanishing of the resultant means that the two forms represent the same point (in homogeneous coordinates). Two quadratic forms $f_1 = a_1 x_1^2 + 2b_1 x_1 x_2 + c_1 x_2^2$ and $f_2 = a_2 x_1^2 + 2b_2 x_1 x_2 + c_2 x_2^2$ possess a simultaneous invariant

$$D_{12} = a_1 c_2 - 2b_1 b_2 + a_2 c_1$$

whose vanishing expresses the fact that f_1 and f_2 represent harmonic pairs of points.

Beyond invariants of a form or system of forms there are covariants. Any function C of the coefficients *and* variables of f which is an invariant under T except for a power of the modulus (determinant) of T is called a covariant of f. Thus, for binary forms, a covariant satisfies the relation

$$C(A_0, A_1, \cdots, A_n, X_1, X_2) = r^w C(a_0, a_1, \cdots, a_n, x_1, x_2).$$

The definitions of absolute and simultaneous covariants are analogous to those for invariants. The degree of a covariant in the coefficients is called its degree and the degree in its variables is called its order. Invariants are thus covariants of order zero. However, sometimes the word invariant is used to mean an invariant in the narrower sense or a covariant.

A covariant of f represents some figure which is not only related to f but projectively related. Thus the Jacobian of two quadratic binary forms $f(x_1, x_2)$ and $\phi(x_1, x_2)$, namely,

$$
\begin{vmatrix}
\dfrac{\partial f}{\partial x_1} & \dfrac{\partial f}{\partial x_2} \\[2ex]
\dfrac{\partial \phi}{\partial x_1} & \dfrac{\partial \phi}{\partial x_2}
\end{vmatrix}
$$

is a simultaneous covariant of weight 1 of the two forms. Geometrically, the Jacobian set equal to zero represents a pair of points which is harmonic to each of the original pairs represented by f and ϕ and the harmonic property is projective.

The Hessian of a binary form introduced by Hesse,[1]

$$
\begin{vmatrix}
\dfrac{\partial^2 f}{\partial x_1^2} & \dfrac{\partial^2 f}{\partial x_1\, \partial x_2} \\[2ex]
\dfrac{\partial^2 f}{\partial x_1\, \partial x_2} & \dfrac{\partial^2 f}{\partial x_2^2}
\end{vmatrix}
$$

is a covariant of weight 2. Its geometric meaning is too involved to warrant space here (Cf. Chap. 35, sec. 5). The concept of the Hessian and its covariance applies to any form in n variables.

The work on algebraic invariants was started in 1841 by George Boole (1815–64) whose results[2] were limited. What is more relevant is that Cayley was attracted to the subject by Boole's work and he interested Sylvester in the subject. They were joined by George Salmon (1819–1904), who was a professor of mathematics at Trinity College in Dublin from 1840 to 1866 and then became a professor of divinity at that institution. These three men did so much work on invariants that in one of his letters Hermite dubbed them the invariant trinity.

In 1841 Cayley began to publish mathematical articles on the algebraic side of projective geometry. The 1841 paper of Boole suggested to Cayley the computation of invariants of nth degree homogeneous functions. He called the invariants derivatives and then hyperdeterminants; the term invariant is due to Sylvester.[3] Cayley, employing ideas of Hesse and Eisenstein on determinants, developed a technique for generating his "derivatives." Then he published ten papers on quantics in the *Philosophical Transactions* from 1854

1. *Jour. für Math.*, 28, 1844, 68–96 = *Ges. Abh.*, 89–122.
2. *Cambridge Mathematical Journal*, 3, 1841, 1–20; and 3, 1842, 106–19.
3. *Coll. Math. Papers*, I, 273.

to 1878.[4] Quantics was the term he adopted for homogeneous polynomials in 2, 3, or more variables. Cayley became so much interested in invariants that he investigated them for their own sake. He also invented a symbolic method of treating invariants.

In the particular case of the binary quartic form

$$f = ax_1^4 + 4bx_1^3x_2 + 6cx_1^2x_2^2 + 4dx_1x_2^3 + ex_2^4$$

Cayley showed that the Hessian H and the Jacobian of f and H are covariants and that

$$g_2 = ae - 4bd + 3c^2$$

and

$$g_3 = \begin{vmatrix} a & b & c \\ b & c & d \\ c & d & e \end{vmatrix}$$

are invariants. To these results Sylvester and Salmon added many more.

Another contributor, Ferdinand Eisenstein, who was more concerned with the theory of numbers, had already found for the binary cubic form[5]

$$f = ax_1^3 + 3bx_1^2x_2 + 3cx_1x_2^2 + dx_2^3$$

that the simplest covariant of the second degree is its Hessian H and the simplest invariant is

$$3b^2c^2 + 6abcd - 4b^3d - 4ac^3 - a^2d^2,$$

which is the determinant of the quadratic Hessian as well as the discriminant of f. Also the Jacobian of f and H is another covariant of order three. Then Siegfried Heinrich Aronhold (1819–84), who began work on invariants in 1849, contributed invariants for ternary cubic forms.[6]

The first major problem that confronted the founders of invariant theory was the discovery of particular invariants. This was the direction of the work from about 1840 to 1870. As we can see, many such functions can be constructed because some invariants such as the Jacobian and the Hessian are themselves forms that have invariants and because some invariants taken together with the original form give a new system of forms that have simultaneous invariants. Dozens of major mathematicians including the few we have already mentioned computed particular invariants.

The continued calculation of invariants led to the major problem of invariant theory, which was raised after many special or particular invariants were found; this was to find a complete system of invariants. What

4. *Coll. Math. Papers*, 2, 4, 6, 7, 10.
5. *Jour. für Math.*, 27, 1844, 89–106, 319–21.
6. *Jour. für Math.*, 55, 1858, 97–191; and 62, 1863, 281–345.

this means is to find for a form of a given number of variables and degree the smallest possible number of rational integral invariants and covariants such that any other rational integral invariant or covariant could be expressed as a rational integral function with numerical coefficients of this complete set. Cayley showed that the invariants and covariants found by Eisenstein for the binary cubic form and the ones he obtained for the binary quartic form are a complete system for the respective cases.[7] This left open the question of a complete system for other forms.

The existence of a finite complete system or basis for binary forms of any given degree was first established by Paul Gordan (1837–1912), who devoted most of his life to the subject. His result[8] is that to each binary form $f(x_1, x_2)$ there belongs a finite complete system of rational integral invariants and covariants. Gordan had the aid of theorems due to Clebsch and the result is known as the Clebsch-Gordan theorem. The proof is long and difficult. Gordan also proved[9] that any finite *system* of binary forms has a finite complete system of invariants and covariants. Gordan's proofs showed how to compute the complete systems.

Various limited extensions of Gordan's results were obtained during the next twenty years. Gordan himself gave the complete system for the ternary quadratic form,[10] for the ternary cubic form,[11] and for a system of two and three ternary quadratics.[12] For the special ternary quartic $x_1^3 x_2 + x_2^3 x_3 + x_3^3 x_1$ Gordan gave a complete system of 54 ground forms.[13]

In 1886 Franz Mertens (1840–1927)[14] re-proved Gordan's theorem for binary systems by an inductive method. He assumed the theorem to be true for any given set of binary forms and then proved it must still be true when the degree of one of the forms is increased by one. He did not exhibit explicitly the finite set of independent invariants and covariants but he proved that it existed. The simplest case, a linear form, was the starting point of the induction and such a form has only powers of itself as covariants.

Hilbert, after writing a doctoral thesis in 1885 on invariants,[15] in 1888[16] also re-proved Gordan's theorem that any given system of binary forms has a finite complete system of invariants and covariants. His proof was a modification of Mertens's. Both proofs were far simpler than Gordan's. But Hilbert's proof also did not present a process for finding the complete system.

7. *Phil. Trans.*, 146, 1856, 101–26 = *Coll. Math. Papers*, 2, 250–75.
8. *Jour. für Math.*, 69, 1868, 323–54.
9. *Math. Ann.*, 2, 1870, 227–80.
10. R. Clebsch and F. Lindemann, *Vorlesungen über Geometrie*, I, 1876, p. 291.
11. *Math. Ann.*, 1, 1869, 56–89, 90–128.
12. Clebsch-Lindemann, p. 288.
13. *Math. Ann.*, 17, 1880, 217–33.
14. *Jour. für Math.*, 100, 1887, 223–30.
15. *Math. Ann.*, 30, 1887, 15–29 = *Ges. Abh.*, 2, 102–16.
16. *Math. Ann.*, 33, 1889, 223–26 = *Ges. Abh.*, 2, 162–64.

In 1888 Hilbert astonished the mathematical community by announcing a totally new approach to the problem of showing that any form of given degree and given number of variables, and any given system of forms in any given number of variables, have a finite complete system of independent rational integral invariants and covariants.[17] The basic idea of this new approach was to forget about invariants for the moment and consider the question: If an infinite system of rational integral expressions in a finite number of variables be given, under what conditions does a finite number of these expressions, a basis, exist in terms of which all the others are expressible as linear combinations with rational integral functions of the same variables as coefficients? The answer is, Always. More specifically, Hilbert's basis theorem, which precedes the result on invariants, goes as follows: By an algebraic form we understand a rational integral homogeneous function in n variables with coefficients in some definite domain of rationality (field). Given a collection of infinitely many forms of any degrees in the n variables, then there is a finite number (a basis) $F_1, F_2, \ldots, F_m$ such that any form F of the collection can be written as

$$F = A_1F_1 + A_2F_2 + \cdots + A_mF_m$$

where $A_1, A_2, \ldots, A_m$ are suitable forms in the n variables (not necessarily in the infinite system) with coefficients in the same domain as the coefficients of the infinite system.

In the application of this theorem to invariants and covariants, Hilbert's result states that for any form or system of forms there is a finite number of rational integral invariants and covariants by means of which every other rational integral invariant or covariant can be expressed as a linear combination of the ones in the finite set. This finite collection of invariants and covariants is the complete invariant system.

Hilbert's existence proof was so much simpler than Gordan's laborious calculation of a basis that Gordan could not help exclaiming, "This is not mathematics; it is theology." However, he reconsidered the matter and said later, "I have convinced myself that theology also has its advantages." In fact he himself simplified Hilbert's existence proof.[18]

In the 1880s and '90s the theory of invariants was seen to have unified many areas of mathematics. This theory was the "modern algebra" of the period. Sylvester said in 1864:[19] "As all roads lead to Rome so I find in my own case at least that all algebraic inquiries, sooner or later, end at the Capitol of modern algebra over whose shining portal is inscribed the Theory of Invariants." Soon the theory became an end in itself, independent of its

17. *Math. Ann.*, 36, 1890, 473–534 = *Ges. Abh.*, 2, 199–257, and in succeeding papers until 1893.
18. *Nachrichten König. Ges. der Wiss. zu Gött.*, 1899, 240–42.
19. *Phil. Trans.*, 154, 1864, 579–666 = *Coll. Math. Papers*, 2, 376–479, p. 380.

origins in number theory and projective geometry. The workers in algebraic invariants persisted in proving every kind of algebraic identity whether or not it had geometrical significance. Maxwell, when a student at Cambridge, said that some of the men there saw the whole universe in terms of quintics and quantics.

On the other hand the physicists of the late nineteenth century took no notice of the subject. Indeed Tait once remarked of Cayley, "Is it not a shame that such an outstanding man puts his abilities to such entirely useless questions?" Nevertheless the subject did make its impact on physics, indirectly and directly, largely through the work in differential invariants.

Despite the enormous enthusiasm for invariant theory in the second half of the nineteenth century, the subject as conceived and pursued during that period lost its attraction. Mathematicians say Hilbert killed invariant theory because he had disposed of all the problems. Hilbert did write to Minkowski in 1893 that he would no longer work in the subject, and said in a paper of 1893 that the most important general goals of the theory were attained. However, this was far from the case. Hilbert's theorem did not show how to compute invariants for any given form or system of forms and so could not provide a single significant invariant. The search for specific invariants having geometrical or physical significance was still important. Even the calculation of a basis for forms of a given degree and number of variables might prove valuable.

What "killed" invariant theory in the nineteenth-century sense of the subject is the usual collection of factors that killed many other activities that were over-enthusiastically pursued. Mathematicians follow leaders. Hilbert's pronouncement, and the fact that he himself abandoned the subject, exerted great influence on others. Also, the calculation of significant specific invariants had become more difficult after the more readily attainable results were achieved.

The computation of algebraic invariants did not end with Hilbert's work. Emmy Noether (1882–1935), a student of Gordan, did a doctoral thesis in 1907 "On Complete Systems of Invariants for Ternary Biquadratic Forms."[20] She also gave a complete system of covariant forms for a ternary quartic, 331 in all. In 1910 she extended Gordan's result to n variables.[21]

The subsequent history of algebraic invariant theory belongs to modern abstract algebra. The methodology of Hilbert brought to the fore the abstract theory of modules, rings, and fields. In this language Hilbert proved that every modular system (an ideal in the class of polynomials in n variables) has a basis consisting of a finite number of polynomials, or every ideal in a polynomial domain of n variables possesses a finite basis provided that in the domain of the coefficients of the polynomials every ideal has a finite basis.

20. *Jour. für Math.*, 134, 1908, 23–90.
21. *Jour. für Math.*, 139, 1911, 118–54.

From 1911 to 1919 Emmy Noether produced many papers on finite bases for various cases using Hilbert's technique and her own. In the subsequent twentieth-century development the abstract algebraic viewpoint dominated. As Eduard Study complained in his text on invariant theory, there was lack of concern for specific problems and only abstract methods were pursued.

3. *The Concept of Birational Transformations*

We saw in Chapter 35 that, especially during the third and fourth decades of the nineteenth century, the work in projective geometry turned to higher-degree curves. However, before this work had gone very far there was a change in the nature of the study. The projective viewpoint means linear transformations in homogeneous coordinates. Gradually transformations of the second and higher degrees came into play and the emphasis turned to birational transformations. Such a transformation, for the case of two non-homogeneous coordinates, is of the form

$$x' = \phi(x, y), \qquad y' = \psi(x, y)$$

where ϕ and ψ are rational functions in x and y and moreover x and y can be expressed as rational functions of x' and y'. In homogeneous coordinates x_1, x_2, and x_3 the transformations are of the form

$$x'_i = F_i(x_1, x_2, x_3), \qquad i = 1, 2, 3$$

and the inverse is

$$x_i = G_i(x'_1, x'_2, x'_3), \qquad i = 1, 2, 3,$$

where F_i and G_i are homogeneous polynomials of degree n in their respective variables. The correspondence is one-to-one except that each of a finite number of points may correspond to a curve.

As an illustration of a birational transformation we have inversion with respect to a circle. Geometrically this transformation (Fig. 39.1) is from M to M' or M' to M by means of the defining equation

$$OM \cdot OM' = r^2,$$

where r is the radius of the circle. Algebraically if we set up a coordinate system at O the Pythagorean theorem leads to

$$(7) \qquad\qquad x' = \frac{r^2 x}{x^2 + y^2}, \qquad y' = \frac{r^2 y}{x^2 + y^2},$$

where M is (x, y) and M' is (x', y'). Under this transformation circles transform into circles or straight lines and conversely. Inversion is a transformation that carries the entire plane into itself and such birational transformations

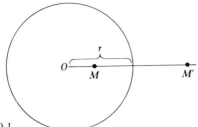

Figure 39.1

are called Cremona transformations. Another example of a Cremona transformation in three (homogeneous) variables is the quadratic transformation

(8) $$x_1' = x_2 x_3, \qquad x_2' = x_3 x_1, \qquad x_3' = x_1 x_2,$$

whose inverse is

$$x_1 = x_2' x_3', \qquad x_2 = x_3' x_1', \qquad x_3 = x_1' x_2'.$$

The term birational transformation is also used in a more general sense, namely, wherein the transformation from the points of one curve into those of another is birational but the transformation need not be birational in the entire plane. Thus (in nonhomogeneous coordinates) the transformation

(9) $$X = x^2, \qquad Y = y$$

is not one-to-one in the entire plane but does take any curve C to the right of the y-axis into another in a one-to-one correspondence.

The inversion transformation was the first of the birational transformations to appear. It was used in limited situations by Poncelet in his *Traité* of 1822 (¶370) and then by Plücker, Steiner, Quetelet, and Ludwig Immanuel Magnus (1790–1861). It was studied extensively by Möbius[22] and its use in physics was recognized by Lord Kelvin,[23] and by Liouville,[24] who called it the transformation by reciprocal radii.

In 1854 Luigi Cremona (1830–1903), who served as a professor of mathematics at several Italian universities, introduced the general birational transformation (of the entire plane into itself) and wrote important papers on it.[25] Max Noether (1844–1921), the father of Emmy Noether, then proved the fundamental result[26] that a plane Cremona transformation can be built

22. *Theorie der Kreisverwandschaft* (Theory of Inversion), *Abh. König. Säch. Ges. der Wiss.*, 2 1855, 529–65 = *Werke*, 2, 243–345.
23. *Jour. de Math.*, 10, 1845, 364–67.
24. *Jour. de Math.*, 12, 1847, 265–90.
25. *Gior. di Mat.*, 1, 1863, 305–11 = *Opere*, 1, 54–61; and 3, 1865, 269–80, 363–76 = *Opere*, 2, 193–218.
26. *Math. Ann.*, 3, 1871, 165–227, p. 167 in particular.

up from a sequence of quadratic and linear transformations. Jacob Rosanes (1842–1922) found this result independently[27] and also proved that all one-to-one algebraic transformations of the plane must be Cremona transformations. The proofs of Noether and Rosanes were completed by Guido Castelnuovo (1865–1952).[28]

4. The Function-Theoretic Approach to Algebraic Geometry

Though the nature of the birational transformation was clear, the development of the subject of algebraic geometry as the study of invariants under such transformations was, at least in the nineteenth century, unsatisfactory. Several approaches were used; the results were disconnected and fragmentary; most proofs were incomplete; and very few major theorems were obtained. The variety of approaches has resulted in marked differences in the languages used. The goals of the subject were also vague. Though invariance under birational transformations has been the leading theme, the subject covers the search for properties of curves, surfaces, and higher-dimensional structures. In view of these factors there are not many central results. We shall give a few samples of what was done.

The first of the approaches was made by Clebsch. (Rudolf Friedrich) Alfred Clebsch (1833–72) studied under Hesse in Königsberg from 1850 to 1854. In his early work he was interested in mathematical physics and from 1858 to 1863 was professor of theoretical mechanics at Karlsruhe and then professor of mathematics at Giessin and Göttingen. He worked on problems left by Jacobi in the calculus of variations and the theory of differential equations. In 1862 he published the *Lehrbuch der Elasticität*. However, his chief work was in algebraic invariants and algebraic geometry.

Clebsch had worked on the projective properties of curves and surfaces of third and fourth degrees up to about 1860. He met Paul Gordan in 1863 and learned about Riemann's work in complex function theory. Clebsch then brought this theory to bear on the theory of curves.[29] This approach is called transcendental. Though Clebsch made the connection between complex functions and algebraic curves, he admitted in a letter to Gustav Roch that he could not understand Riemann's work on Abelian functions nor Roch's contributions in his dissertation.

Clebsch reinterpreted the complex function theory in the following manner: The function $f(w, z) = 0$, wherein z and w are complex variables, calls geometrically for a Riemann surface for z and a plane or a portion of a plane for w or, if one prefers, for a Riemann surface to each point of which a pair of values of z and w is attached. By considering only the real parts of z

27. *Jour. für Math.*, 73, 1871, 97–110.
28. *Atti Accad. Torino*, 36, 1901, 861–74.
29. *Jour. für Math.*, 63, 1864, 189–243.

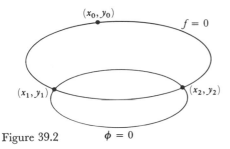

Figure 39.2 $\phi = 0$

and w the equation $f(w, z) = 0$ represents a curve in the real Cartesian plane. z and w may still have complex values satisfying $f(w, z) = 0$ but these are not plotted. This view of real curves with complex points was already familiar from the work in projective geometry. To the theory of birational transformations of the surface corresponds a theory of birational transformations of the plane curve. Under the reinterpretation just described the branch points of the Riemann surface correspond to those points of the curve where a line $x = $ const. meets the curve in two or more consecutive points, that is, is either tangent to the curve or passes through a cusp. A double point of the curve corresponds to a point on the surface where two sheets just touch without any further connection. Higher multiple points on curves also correspond to other peculiarities of Riemann surfaces.

In the subsequent account we shall utilize the following definitions (Cf. Chap. 23, sec. 3): A multiple point (singular point) P of order $k > 1$ of an nth degree plane curve is a point such that a generic line through P cuts the curve in $n - k$ points. The multiple point is ordinary if the k tangents at P are distinct. In counting intersections of an nth degree and an mth degree curve one must take into account the multiplicity of the multiple points on each curve. If it is h on the curve C^n and k on C^m and if the tangents at P of C^n are distinct from those of C^m, then the point of intersection has multiplicity hk. A curve C' is said to be adjoint to a curve C when the multiple points of C are ordinary or cusps and if C' has a point of multiplicity of order $k - 1$ at every multiple point of C of order k.

Clebsch [30] first restated Abel's theorem (Chap. 27, sec. 7) on integrals of the first kind in terms of curves. Abel considered a fixed rational function $R(x, y)$ where x and y are related by any algebraic curve $f(x, y) = 0$, so that y is a function of x. Suppose (Fig. 39.2) $f = 0$ is cut by another algebraic curve

$$\phi(x, y, a_1, a_2, \cdots, a_k) = 0,$$

where the a_i's are the coefficients in $\phi = 0$. Let the intersections of $\phi = 0$ with $f = 0$ be $(x_1, y_1), (x_2, y_2), \cdots, (x_m, y_m)$. (The number m of these is the

30. *Jour. für Math.*, 63, 1864, 189–243.

product of the degrees of f and ϕ.) Given a point (x_0, y_0) on $f = 0$, where y_0 belongs to one branch of $f = 0$, then we can consider the sum

$$I = \sum_{i=1}^{m} \int_{x_0, y_0}^{x_i, y_i} R(x, y)\, dx.$$

The upper limits x_i, y_i all lie on $\phi = 0$ and the integral I is a function of the upper limits. Then there is a characteristic number p of these limits which have to be algebraic functions of the others. This number p depends only on f. Moreover, I can be expressed as the sum of these p integrals and rational and logarithmic functions of the x_i, y_i, $i = 1, 2, \ldots, m$. Further, if the curve $\phi = 0$ is varied by varying the parameters a_1 to a_k then the x_i will also vary and I becomes a function of the a_i through the x_i. The function I of the a_i will be rational in the a_i or at worst involve logarithmic functions of the a_i.

Clebsch also carried over to curves Riemann's concept of Abelian integrals on Riemann surfaces, that is, integrals of the form $\int g(x, y)\, dx$, where g is a rational function and $f(x, y) = 0$. To illustrate the integrals of the first kind consider a plane fourth degree curve C_4 without double points. Here $p = 3$ and there are the three everywhere finite integrals

$$u_1 = \int \frac{x\, dx}{f_y}, \qquad u_2 = \int \frac{y\, dx}{f_y}, \qquad u_3 = \int \frac{dx}{f_y}.$$

What applies to the C_4 carries over to arbitrary algebraic curves $f(x, y) = 0$ of the nth order. In place of the three everywhere finite integrals there are now p such integrals, (where p is the genus of $f = 0$). Each has $2p$ periodicity modules (Chap. 27, sec. 8). The integrals are of the form

$$\int \frac{\phi(x, y)}{\partial f/\partial y}\, dx$$

where ϕ is a polynomial (an adjoint) of precisely the degree $n - 3$ which vanishes at the double points and cusps of $f = 0$.

Clebsch's next contribution [31] was to introduce the notion of genus as a concept for classifying curves. If the curve has d double points then the genus $p = (1/2)(n - 1)(n - 2) - d$. Previously there was the notion of the deficiency of a curve (Chap. 23, sec. 3), that is, the maximum possible number of double points a curve of degree n could possess, namely, $(n - 1)(n - 2)/2$, minus the number it actually does possess. Clebsch showed [32] that for curves with only ordinary multiple points (the tangents are all distinct) the genus is the same as the deficiency and the genus is an invariant under birational transformation of the entire plane into itself.[33]

31. *Jour. für Math.*, 64, 1865, 43–65.
32. *Jour. für Math.*, 64, 1865, 98–100.
33. If the multiple (singular) points are of order r_i then the genus p of a curve C is $(n - 1)(n - 2)/2 - (1/2) \sum r_i(r_i - 1)$, where the summation extends over all multiple points. The genus is a more refined concept.

Clebsch's notion of genus is related to Riemann's connectivity of a Riemann surface. The Riemann surface corresponding to a curve of genus p has connectivity $2p + 1$.

The notion of genus can be used to establish significant theorems about curves. Jacob Lüroth (1844–1910) showed[34] that a curve of genus 0 can be birationally transformed into a straight line. When the genus is 1, Clebsch showed that a curve can be birationally transformed into a third degree curve.

In addition to classifying curves by genus, Clebsch, following Riemann, introduced classes within each genus. Riemann had considered[35] the birational transformation of his surfaces. Thus if $f(w, z) = 0$ is the equation of the surface and if

$$w_1 = R_1(w, z), \qquad z_1 = R_2(w, z)$$

are rational functions and if the inverse transformation is rational then $f(w, z)$ can be transformed to $F(w_1, z_1) = 0$. Two algebraic equations $F(w, z) = 0$ (or their surfaces) can be transformed birationally into one another only if both have the same p value. (The number of sheets need not be preserved.) For Riemann no further proof was needed. It was guaranteed by the intuition.

Riemann (in the 1857 paper) regarded all equations (or the surfaces) which are birationally transformable into each other as belonging to the same class. They have the same genus p. However, there are different classes with the same p value (because the branch-points may differ). The most general class of genus p is characterized by $3p - 3$ (complex) constants (coefficients in the equation) when $p > 1$, by one constant when $p = 1$ and by zero constants when $p = 0$. In the case of elliptic functions $p = 1$ and there is one constant. The trigonometric functions, for which $p = 0$, do not have any arbitrary constant. The number of constants was called by Riemann the class modulus. The constants are invariant under birational transformation. Clebsch likewise put all those curves which are derivable from a given one by a one-to-one birational transformation into one class. Those of one class necessarily have the same genus but there may be different classes with the same genus.

5. The Uniformization Problem

Clebsch then turned his attention to what is called the uniformization problem for curves. Let us note first just what this problem amounts to. Given the equation

$$(10) \qquad\qquad w^2 + z^2 = 1$$

34. *Math. Ann.*, 9, 1876, 163–65.
35. *Jour. für Math.*, 54, 1857, 115–55 = *Werke*, 2nd ed., 88–142.

we can represent it in the parametric form

(11) $z = \sin t, \qquad w = \cos t$

or in the parametric form

(12) $z = \dfrac{2t}{1 + t^2}, \qquad w = \dfrac{1 - t^2}{1 + t^2}.$

Thus even though (10) defines w as a multiple-valued function of z, we can represent z and w as single-valued or uniform functions of t. The parametric equations (11) or (12) are said to uniformize the algebraic equation (10).

For an equation $f(w, z) = 0$ of genus 0 Clebsch[36] showed that each of the variables can be expressed as a rational function of a single parameter. These rational functions are uniformizing functions. When $f = 0$ is interpreted as a curve it is then called unicursal. Conversely if the variables w and z of $f = 0$ are rationally expressible in terms of an arbitrary parameter then $f = 0$ is of genus 0.

When $p = 1$ then Clebsch showed in the same year[37] that w and z can be expressed as rational functions of the parameters ξ and η where η^2 is a polynomial of either the third or fourth degree in ξ. Then $f(w, z) = 0$ or the corresponding curve is called bicursal, a term introduced by Cayley.[38] It is also called elliptic because the equation $(dw/dz)^2 = \eta^2$ leads to elliptic integrals. We can as well say that w and z are expressible as single-valued doubly periodic functions of a single parameter α or as rational functions of $\wp(\alpha)$ where $\wp(\alpha)$ is Weierstrass's function. Clebsch's result on the uniformization of curves of genus 1 by means of elliptic functions of a parameter made it possible to establish for such curves remarkable properties about points of inflection, osculatory conics, tangents from a point to a curve, and other results, many of which had been demonstrated earlier but with great difficulty.

For equations $f(w, z) = 0$ of genus 2, Alexander von Brill (1842–1935) showed[39] that the variables w and z are expressible as rational functions of ξ and η where η^2 is now a polynomial of the fifth or sixth degree in ξ.

Thus functions of genus 0, 1, and 2 can be uniformized. For function $f(w, z) = 0$ of genus greater than 2 the thought was to employ more general functions, namely automorphic functions. In 1882 Klein[40] gave a general uniformization theorem but the proof was not complete. In 1883 Poincaré announced[41] his general uniformization theorem but he too had no complete

36. *Jour. für Math.*, 64, 1865, 43–65.
37. *Jour. für Math.*, 64, 1865, 210–70.
38. *Proc. Lon. Math. Soc.*, 4, 1871–73, 347–52 = *Coll. Math. Papers*, 8, 181–87.
39. *Jour. für Math.*, 65, 1866, 269–83.
40. *Math. Ann.*, 21, 1883, 141–218 = *Ges. math. Abh.*, 3, 630–710.
41. *Bull. Soc. Math. de France*, 11, 1883, 112–25 = *Œuvres*, 4, 57–69.

proof. Both Klein and Poincaré continued to work hard to prove this theorem but no decisive result was obtained for twenty-five years. In 1907 Poincaré[42] and Paul Koebe (1882–1945)[43] independently gave a proof of this uniformization theorem. Koebe then extended the result in many directions. With the theorem on uniformization now rigorously established an improved treatment of algebraic functions and their integrals has become possible.

6. *The Algebraic-Geometric Approach*

A new direction of work in algebraic geometry begins with the collaboration of Clebsch and Gordan during the years 1865–70. Clebsch was not satisfied merely to show the significance of Riemann's work for curves. He sought now to establish the theory of Abelian integrals on the basis of the algebraic theory of curves. In 1865 he and Gordan joined forces in this work and produced their *Theorie der Abelschen Funktionen* (1866). One must appreciate that at this time Weierstrass's more rigorous theory of Abelian integrals was not known and Riemann's foundation—his proof of existence based on Dirichlet's principle—was not only strange but not well established. Also at this time there was considerable enthusiasm for the theory of invariants of algebraic forms (or curves) and for projective methods as the first stage, so to speak, of the treatment of birational transformations.

Although the work of Clebsch and Gordan was a contribution to algebraic geometry, it did not establish a purely algebraic theory of Riemann's theory of Abelian integrals. They did use algebraic and geometric methods as opposed to Riemann's function-theoretic methods but they also used basic results of function theory and the function-theoretic methods of Weierstrass. In addition they took some results about rational functions and the intersection point theorem as given. Their contribution amounted to starting from some function-theoretic results and, using algebraic methods, obtaining new results previously established by function-theoretic methods. Rational transformations were the essence of the algebraic method.

They gave the first *algebraic* proof for the invariance of the genus p of an algebraic curve under rational transformations, using as a definition of p the degree and number of singularities of $f = 0$. Then, using the fact that p is the number of linearly independent integrals of the first kind on $f(x_1, x_2, x_3) = 0$ and that these integrals are everywhere finite, they showed that the transformation

$$\rho x_i = \psi_i(y_1, y_2, y_3), \qquad i = 1, 2, 3,$$

42. *Acta Math.*, 31, 1908, 1–63 = *Œuvres*, 4, 70–139.
43. *Math. Ann.*, 67, 1909, 145–224.

transforms an integral of the first kind into an integral of the first kind so that p is invariant. They also gave new proofs of Abel's theorem (by using function-theoretic ideas and methods).

Their work was not rigorous. In particular they, too, in the Plücker tradition counted arbitrary constants to determine the number of inter-section points of a C_m with a C_n. Special kinds of double points were not investigated. The significance of the Clebsch-Gordan work for the theory of algebraic functions was to express clearly in algebraic form such results as Abel's theorem and to use it in the study of Abelian integrals. They put the algebraic part of the theory of Abelian integrals and functions more into the foreground and in particular established the theory of transformations on its own foundations.

Clebsch and Gordan had raised many problems and left many gaps. The problems lay in the direction of new algebraic investigations for a purely algebraic theory of algebraic functions. The work on the algebraic approach was continued by Alexander von Brill and Max Noether from 1871 on; their key paper was published in 1874.[44] Brill and Noether based their theory on a celebrated residual theorem (*Restsatz*) which in their hands took the place of Abel's theorem. They also gave an algebraic proof of the Riemann-Roch theorem on the number of constants which appear in algebraic functions $F(w, z)$ which become infinite nowhere except in m prescribed points of a C_n. According to this theorem the most general algebraic function which fulfills this condition has the form

$$F = C_1 F_1 + C_2 F_2 + \cdots + C_\mu F_\mu + C_{\mu+1}$$

where

$$\mu = m - p + \tau,$$

τ is the number of linearly independent functions ϕ (of degree $n - 3$) which vanish in the m prescribed points, and p is the genus of the C_n. Thus if the C_n is a C_4 without double points, then $p = 3$ and the ϕ's are straight lines. For this case when

$$m = 1, \text{ then } \tau = 2 \quad \text{and } \mu = 1 - 3 + 2 = 0;$$
$$m = 2, \text{ then } \tau = 1, \quad \text{and } \mu = 2 - 3 + 1 = 0;$$
$$m = 3, \text{ then } \tau = 1 \text{ or } 0 \text{ and } \mu = 1 \text{ or } 0.$$

When $\mu = 0$ there is no algebraic function which becomes infinite in the given points. When $m = 3$, there is one and only one such function provided the three given points be on a straight line. If the three points do lie on a line $v = 0$, this line cuts the C_4 in a fourth point. We choose a line $u = 0$ through this point and then $F_1 = u/v$.

44. "Über die algebraischen Funktionen und ihre Anwendung in der Geometrie," *Math. Ann.*, 7, 1874, 269–310.

This work replaces Riemann's determination of the most general algebraic function having given points at which it becomes infinite. Also the Brill-Noether result transcends the projective viewpoint in that it deals with the geometry of points on the curve C_n given by $f = 0$, whose mutual relations are not altered by a one-to-one birational transformation. Thus for the first time the theorems on points of intersection of curves were established algebraically. The counting of constants as a method was dispensed with.

The work in algebraic geometry continued with the detailed investigation of algebraic space curves by Noether[45] and Halphen.[46] Any space curve C can be projected birationally into a plane curve C_1. All such C_1 coming from C have the same genus. The genus of C is therefore defined to be that of any such C_1 and the genus of C is invariant under birational transformation of the space.

The topic which has received the greatest attention over the years is the study of singularities of plane algebraic curves. Up to 1871, the theory of algebraic functions considered from the algebraic viewpoint had limited itself to curves which had distinct or separated double points and at worst only cusps (*Rückkehrpunkte*). Curves with more complicated singularities were believed to be treatable as limiting cases of curves with double points. But the actual limiting procedure was vague and lacked rigor and unity. The culmination of the work on singularities is two famous transformation theorems. The first states that every plane irreducible algebraic curve can be transformed by a Cremona transformation to one having no singular points other than multiple points with distinct tangents. The second asserts that by a transformation birational only on the curve every plane irreducible algebraic curve can be transformed into another having only double points with distinct tangents. The reduction of curves to these simpler forms facilitates the application of many of the methodologies of algebraic geometry.

However, the numerous proofs of these theorems, especially the second one, have been incomplete or at least criticized by mathematicians (other than the author). There really are two cases of the second theorem, real curves in the projective plane and curves in the complex function theory sense where x and y each run over a complex plane. Noether[47] in 1871 used a sequence of quadratic transformations which are one-to-one in the entire plane to prove the first theorem. He is generally credited with the proof but actually he merely indicated a proof which was perfected and modified by many writers.[48] Kronecker, using analysis and algebra, developed a method for proving the second theorem. He communicated this method verbally to

45. *Jour. für Math.*, 93, 1882, 271–318.
46. *Jour. de l'Ecole Poly.*, Cahier 52, 1882, 1–200 = *Œuvres*, 3, 261–455.
47. *Nachrichten König. Ges. der Wiss. zu Gött.*, 1871, 267–78.
48. See also Noether, *Math. Ann.*, 9, 1876, 166–82; and 23, 1884, 311–58.

Riemann and Weierstrass in 1858, lectured on it from 1870 on and published it in 1881.[49] The method used rational transformations, which with the aid of the equation of the given plane curve are one-to-one, and transformed the singular case into the "regular one;" that is, the singular points become just double points with distinct tangents. The result, however, was not stated by Kronecker and is only implicit in his work.

This second theorem to the effect that all multiple points can be reduced to double points by birational transformations on the curve was first explicitly stated and proved by Halphen in 1884.[50] Many other proofs have been given but none is universally accepted.

7. *The Arithmetic Approach*

In addition to the transcendental approach and the algebraic-geometric approach there is what is called the arithmetic approach to algebraic curves, which is, however, in concept at least, purely algebraic. This approach is really a group of theories which differ greatly in detail but which have in common the construction and analysis of the integrands of the three kinds of Abelian integrals. This approach was developed by Kronecker in his lectures,[51] by Weierstrass in his lectures of 1875–76, and by Dedekind and Heinrich Weber in a joint paper.[52] The approach is fully presented in the text by Kurt Hensel and Georg Landsberg: *Theorie der algebraischen Funktionen einer Variabeln* (1902).

The central idea of this approach comes from the work on algebraic numbers by Kronecker and Dedekind and utilizes an analogy between the algebraic integers of an algebraic number field and the algebraic functions on the Riemann surface of a complex function. In the theory of algebraic numbers one starts with an irreducible polynomial equation $f(x) = 0$ with integral coefficients. The analogue for algebraic geometry is an irreducible polynomial equation $f(\zeta, z) = 0$ whose coefficients of the powers of ζ are polynomials in z (with, say, real coefficients). In number theory one then considers the field $R(x)$ generated by the coefficients of $f(x) = 0$ and one of its roots. In the geometry one considers the field of all $R(\zeta, z)$ which are algebraic and one-valued on the Riemann surface. One then considers in the number theory the integral algebraic numbers. To these there correspond the algebraic functions $G(\zeta, z)$ which are entire, that is, become infinite only at $z = \infty$. The decomposition of the algebraic integers into real prime factors

49. *Jour. für Math.*, 91, 1881, 301–34 = *Werke*, 2, 193–236.
50. Reproduced in an appendix to a French edition (1884) of G. Salmon's *Higher Plane Curves* and in E. Picard's *Traité d'analyse*, 2, 1893, 364 ff. = *Œuvres*, 4, 1–93.
51. *Jour. für Math.*, 91, 1881, 301–34 and 92, 1882, 1–122 = *Werke*, 2, 193–387.
52. "Theorie der algebraischen Funktionen einer Veränderlichen," *Jour. für Math.*, 92, 1882, 181–290 = Dedekind's *Werke*, I, 238–350.

and units respectively corresponds to the decomposition of the $G(\zeta, z)$ into factors such that each vanishes at one point only of the Riemann surface and factors that vanish nowhere, respectively. Where Dedekind introduced ideals in the number theory to discuss divisibility, in the geometric analogue one replaces a factor of a $G(\zeta, z)$ which vanishes at one point of the Riemann surface by the collection of all functions of the field of $R(\zeta, z)$ which vanish at that point. Dedekind and Weber used this arithmetic method to treat the field of algebraic functions and they obtained the classic results.

Hilbert[53] continued what is essentially the algebraic or arithmetic approach to algebraic geometry of Dedekind and Kronecker. One principal theorem, Hilbert's *Nullstellensatz*, states that every algebraic structure (figure) of arbitrary extent in a space of arbitrarily many homogeneous variables $x_1, \ldots, x_n$ can always be represented by a finite number of homogeneous equations

$$F_1 = 0, F_2 = 0, \cdots, F_\mu = 0$$

so that the equation of any other structure containing the original one can be represented by

$$M_1 F_1 + \cdots + M_\mu F_\mu = 0,$$

where the M's are arbitrary homogeneous integral forms whose degree must be so chosen that the left side of the equation is itself homogeneous.

Hilbert following Dedekind called the collection of the $M_i F_i$ a module (the term is now ideal and module now is something more general). One can state Hilbert's result thus: Every algebraic structure of R_n determines the vanishing of a finite module.

8. The Algebraic Geometry of Surfaces

Almost from the beginning of work in the algebraic geometry of curves, the theory of surfaces was also investigated. Here too the direction of the work turned to invariants under linear and birational transformations. Like the equation $f(x, y) = 0$, the polynomial equation $f(x, y, z) = 0$ has a double interpretation. If $x, y,$ and z take on real values then the equation represents a two-dimensional surface in three-dimensional space. If, however, these variables take on complex values, then the equation represents a four-dimensional manifold in a six-dimensional space.

The approach to the algebraic geometry of surfaces paralleled that for curves. Clebsch employed function-theoretic methods and introduced[54]

53. "Über die Theorie der algebraischen Formen," *Math. Ann.*, 36, 1890, 473–534 = *Ges. Abh.*, 2, 199–257.
54. *Comp. Rend.*, 67, 1868, 1238–39.

double integrals which play the rôle of Abelian integrals in the theory of curves. Clebsch noted that for an algebraic surface $f(x, y, z) = 0$ of degree m with isolated multiple points and ordinary multiple lines, certain surfaces of degree $m - 4$ ought to play the role which the adjoint curves of degree $m - 3$ play with respect to a curve of degree m. Given a rational function $R(x, y, z)$ where x, y, and z are related by $f(x, y, z) = 0$, if one seeks the double integrals

$$\int \int R(x, y, z) \, dx \, dy,$$

which always remain finite when the integrals extend over a two-dimensional region of the four-dimensional surface, one finds that they are of the form

$$\int \int \frac{Q(x, y, z)}{f_z} \, dx \, dy$$

where Q is a polynomial of degree $m - 4$. $Q = 0$ is an adjoint surface which passes through the multiple lines of $f = 0$ and has a multiple line of order $k - 1$ at least in every multiple line of f of order k and has a multiple point of order $q - 2$ at least in every isolated multiple point of f of order q. Such an integral is called a double integral of the first kind. The number of linearly independent integrals of this class, which is the number of essential constants in $Q(x, y, z)$, is called the geometrical genus p_g of $f = 0$. If the surface has no multiple lines of points

$$p_g = \frac{(m - 1)(m - 2)(m - 3)}{6}.$$

Max Noether[55] and Hieronymus G. Zeuthen (1839–1920)[56] proved that p_g is an invariant under birational transformations of the surface (not of the whole space).

Up to this point the analogy with the theory of curves is good. The double integrals of the first kind are analogous to the Abelian integrals of the first kind. But now a first difference becomes manifest. It is necessary to calculate the number of essential constants in the polynomials Q of degree $m - 4$ which behave at multiple points of the surface in such a manner that the integral remains finite. But one can find by a precise formula the number of conditions thus involved only for a polynomial of sufficiently large degree N. If one puts into this formula $N = m - 4$, one might find a number different from p_g. Cayley[57] called this new number the numerical (arithmetic) genus p_n of the surface. The most general case is where $p_n = p_g$. When

55. *Math. Ann.*, 2, 1870, 293–316.
56. *Math. Ann.*, 4, 1871, 21–49.
57. *Phil. Trans.*, 159, 1869, 201–29 = *Coll. Math. Papers*, 6, 329–58; and *Math. Ann.*, 3, 1871, 526–29 = *Coll. Math. Papers*, 8, 394–97.

the equality does not hold one has $p_n < p_g$ and the surface is called irregular; otherwise it is called regular. Then Zeuthen[58] and Noether[59] established the invariance of the number p_n when it is not equal to p_g.

Picard[60] developed a theory of double integrals of the second kind. These are the integrals which become infinite in the manner of

$$(13) \qquad \int \int \left(\frac{\partial U}{\partial x} + \frac{\partial V}{\partial y} \right) dx \, dy$$

where U and V are rational functions of x, y, and z and $f(x, y, z) = 0$. The number of different integrals of the second kind, different in the sense that no linear combination of these integrals reduces to the form (13), is finite; this is a birational invariant of the surface $f = 0$. But it is not true here, as in the case of curves, that the number of distinct Abelian integrals of the second kind is $2p$. This new invariant of algebraic surfaces does not appear to be tied to the numerical or the geometrical genus.

Far less has been accomplished for the theory of surfaces than for curves. One reason is that the possible singularities of surfaces are much more complicated. There is the theorem of Picard and Georges Simart proven by Beppo Levi (1875–1928)[61] that any (real) algebraic surface can be birationally transformed into a surface free of singularities which must, however, be in a space of five dimensions. But this theorem does not prove to be too helpful.

In the case of curves the single invariant number, the genus p, is capable of definition in terms of the characteristics of the curve or the connectivity of the Riemann surface. In the case of $f(x, y, z) = 0$ the number of characterizing arithmetical birational invariants is unknown.[62] We shall not attempt to describe further the few limited results for the algebraic geometry of surfaces.

The subject of algebraic geometry now embraces the study of higher-dimensional figures (manifolds or varieties) defined by one or more algebraic equations. Beyond generalization in this direction, another type, namely, the use of more general coefficients in the defining equations, has also been undertaken. These coefficients can be members of an abstract ring or field and the methods of abstract algebra are applied. The several methods of pursuing algebraic geometry as well as the abstract algebraic formulation introduced in the twentieth century have led to sharp differences in language and methods of approach so that one class of workers finds it very difficult to understand another. The emphasis in this century has been on the abstract algebraic approach. It does seem to offer sharp formulations of theorems and

58. *Math. Ann.*, 4, 1871, 21–49.
59. *Math. Ann.*, 8, 1875, 495–533.
60. *Jour. de Math.*, (5), 5, 1899, 5–54, and later papers.
61. *Annali di Mat.*, (2), 26, 1897, 219–53.
62. The manifolds involved cannot be characterized even topologically.

proofs thereby settling much controversy about the meaning and correctness of the older results. However, much of the work seems to have far more bearing on algebra than on geometry.

Bibliography

Baker, H. F.: "On Some Recent Advances in the Theory of Algebraic Surfaces," *Proc. Lon. Math. Soc.*, (2), 12, 1912–13, 1–40.

Berzolari, L.: "Allgemeine Theorie der höheren ebenen algebraischen Kurven," *Encyk. der Math. Wiss.*, B. G. Teubner, 1903–15, III C4, 313–455.

————: "Algebraische Transformationen und Korrespondenzen," *Encyk. der Math. Wiss.*, B. G. Teubner, 1903–15, III, 2, 2nd half B, 1781–2218. Useful for results on higher-dimensional figures.

Bliss, G. A.: "The Reduction of Singularities of Plane Curves by Birational Transformations," *Amer. Math. Soc. Bull.*, 29, 1923, 161–83.

Brill, A., and M. Noether: "Die Entwicklung der Theorie der algebraischen Funktionen," *Jahres. der Deut. Math.-Verein.*, 3, 1892–93, 107–565.

Castelnuovo, G., and F. Enriques: "Die algebraischen Flächen vom Gesichtspunkte der birationalen Transformationen," *Encyk. der Math. Wiss.*, B. G. Teubner, 1903–15, III C6b, 674–768.

————: "Sur quelques récents résultats dans la théorie des surfaces algébriques," *Math. Ann.*, 48, 1897, 241–316.

Cayley, A.: *Collected Mathematical Papers*, Johnson Reprint Corp., 1963, Vols. 2, 4, 6, 7, 10, 1891–96.

Clebsch, R. F. A.: "Versuch einer Darlegung und Würdigung seiner Wissenschaftlichen Leistungen," *Math. Ann.*, 7, 1874, 1–55. An article by friends of Clebsch.

Coolidge, Julian L.: *A History of Geometrical Methods*, Dover (reprint), 1963, pp. 195–230, 278–92.

Cremona, Luigi: *Opere mathematiche*, 3 vols., Ulrico Hoepli, 1914–17.

Hensel, Kurt, and Georg Landsberg: *Theorie der algebraischen Funktionen einer Variabeln* (1902), Chelsea (reprint), 1965, pp. 694–702 in particular.

Hilbert, David: *Gesammelte Abhandlungen*, Julius Springer, 1933, Vol. 2.

Klein, Felix: *Vorlesungen über die Entwicklung der Mathematik im 19 Jahrhundert*, 1, 155–66, 295–319; 2, 2–26, Chelsea (reprint), 1950.

Meyer, Franz W.: "Bericht über den gegenwärtigen Stand der Invariantentheorie," *Jahres. der Deut. Math.-Verein.*, 1, 1890–91, 79–292.

National Research Council: *Selected Topics in Algebraic Geometry*, Chelsea (reprint), 1970.

Noether, Emmy: "Die arithmetische Theorie der algebraischen Funktionen einer Veränderlichen in ihrer Beziehung zu den übrigen Theorien and zu der Zahlentheorie," *Jahres. der Deut. Math.-Verein.*, 28, 1919, 182–203.

40
The Instillation of Rigor in Analysis

> But it would be a serious error to think that one can find
> certainty only in geometrical demonstrations or in the
> testimony of the senses.   A. L. CAUCHY

1. Introduction

By about 1800 the mathematicians began to be concerned about the looseness
in the concepts and proofs of the vast branches of analysis. The very concept
of a function was not clear; the use of series without regard to convergence
and divergence had produced paradoxes and disagreements; the controversy
about the representations of functions by trigonometric series had introduced
further confusion; and, of course, the fundamental notions of derivative and
integral had never been properly defined. All these difficulties finally brought
on dissatisfaction with the logical status of analysis.

Abel, in a letter of 1826 to Professor Christoffer Hansteen,[1] complained
about "the tremendous obscurity which one unquestionably finds in analysis.
It lacks so completely all plan and system that it is peculiar that so many men
could have studied it. The worst of it is, it has never been treated stringently.
There are very few theorems in advanced analysis which have been demonstrated in a logically tenable manner. Everywhere one finds this miserable
way of concluding from the special to the general and it is extremely peculiar
that such a procedure has led to so few of the so-called paradoxes."

Several mathematicians resolved to bring order out of chaos. The leaders
of what is often called the critical movement decided to rebuild analysis
solely on the basis of arithmetical concepts. The beginnings of the movement
coincide with the creation of non-Euclidean geometry. An entirely different
group, except for Gauss, was involved in the latter activity and it is therefore
difficult to trace any direct connection between it and the decision to found
analysis on arithmetic. Perhaps the decision was reached because the hope of

1. Œuvres, 2, 263–65.

grounding analysis on geometry, which many seventeenth-century men often asserted could be done, was blasted by the increasing complexity of the eighteenth-century developments in analysis. However, Gauss had already expressed his doubts as to the truth of Euclidean geometry as early as 1799, and in 1817 he decided that truth resided only in arithmetic. Moreover, during even the early work by Gauss and others on non-Euclidean geometry, flaws in Euclid's development had already been noted. Hence it is very likely that both factors caused distrust of geometry and prompted the decision to found analysis on arithmetical concepts. This certainly was what the leaders of the critical movement undertook to do.

Rigorous analysis begins with the work of Bolzano, Cauchy, Abel, and Dirichlet and was furthered by Weierstrass. Cauchy and Weierstrass are best known in this connection. Cauchy's basic works on the foundations of analysis are his *Cours d'analyse algébrique*,[2] *Résumé des leçons sur le calcul infinitésimal*,[3] and *Leçons sur le calcul différentiel*.[4] Actually Cauchy's rigor in these works is loose by modern standards. He used phrases such as "approach indefinitely," "as little as one wishes," "last ratios of infinitely small increments," and "a variable approaches its limit." However, if one compares Lagrange's *Théorie des fonctions analytiques*[5] and his *Leçons sur le calcul des fonctions*[6] and the influential book by Lacroix, *Traité du calcul différentiel et du calcul intégral*[7] with the *Cours d'analyse algébrique* of Cauchy one begins to see the striking difference between the mathematics of the eighteenth century and that of the nineteenth. Lagrange, in particular, was purely formal. He operated with symbolic expressions. The underlying concepts of limit, continuity, and so on are not there.

Cauchy is very explicit in his introduction to the 1821 work that he seeks to give rigor to analysis. He points out that the free use for all functions of the properties that hold for algebraic functions, and the use of divergent series are not justified. Though Cauchy's work was but one step in the direction of rigor, he himself believed and states in his *Résumé* that he had brought the ultimate in rigor into analysis. He did give the beginnings of precise proofs of theorems and properly limited assertions at least for the elementary functions. Abel in his paper of 1826 on the binomial series (sec. 5) praised this achievement of Cauchy: "The distinguished work [the *Cours d'analyse*] should be read by everyone who loves rigor in mathematical investigations." Cauchy abandoned the explicit representations of Euler and the power series of Lagrange and introduced new concepts to treat functions.

2. 1821, *Œuvres*, (2) III.
3. 1823, *Œuvres*, (2), IV, 1–261.
4. 1829, *Œuvres*, (2), IV, 265–572.
5. 1797; 2nd ed., 1813 = *Œuvres*, 9.
6. 1801; 2nd ed., 1806 = *Œuvres*, 10.
7. 3 vols., 1st ed., 1797–1800; 2nd ed., 1810–19.

2. *Functions and Their Properties*

The eighteenth century mathematicians had on the whole believed that a function must have the same analytic expression throughout. During the latter part of the century, largely as a consequence of the controversy over the vibrating-string problem, Euler and Lagrange allowed functions that have different expressions in different domains and used the word continuous where the same expression held and discontinuous at points where the expression changed form (though in the modern sense the entire function could be continuous). While Euler, d'Alembert, and Lagrange had to reconsider the concept of function, they did not arrive at any widely accepted definition nor did they resolve the problem of what functions could be represented by trigonometrical series. However, the gradual expansion in the variety and use of functions forced mathematicians to accept a broader concept.

Gauss in his earlier work meant by a function a closed (finite analytical) expression and when he spoke of the hypergeometric series $F(\alpha, \beta, \gamma, x)$ as a function of α, β, γ, and x he qualified it by the remark, "insofar as one can regard it as a function." Lagrange had already used a broader concept in regarding power series as functions. In the second edition of his *Mécanique analytique* (1811–15) he used the word function for almost any kind of dependence on one or more variables. Even Lacroix in his *Traité* of 1797 had already introduced a broader notion. In the introduction he says, "Every quantity whose value depends on one or several others is called a function of the latter, whether one knows or one does not know by what operations it is necessary to go from the latter to the first quantity." Lacroix gives as an example a root of an equation of the fifth degree as a function of its coefficients.

Fourier's work opened up even more widely the question of what a function is. On the one hand, he insisted that functions need not be representable by any analytic expressions. In his *The Analytical Theory of Heat*[8] he says, "In general the function $f(x)$ represents a succession of values or ordinates each of which is arbitrary. . . . We do not suppose these ordinates to be subject to a common law; they succeed each other in any manner whatever. . . ." Actually he himself treated only functions with a finite number of discontinuities in any finite interval. On the other hand, to a certain extent Fourier was supporting the contention that a function must be representable by an analytic expression, though this expression was a Fourier series. In any case Fourier's work shook the eighteenth-century belief that all functions were at worst extensions of algebraic functions. The algebraic functions and even the elementary transcendental functions were no longer the prototype of functions. Since the properties of algebraic functions could no longer be carried over to all functions, the question then arose as to what one really

8. English translation, p. 430, Dover (reprint), 1955.

means by a function, by continuity, differentiability, integrability, and other properties.

In the positive reconstructions of analysis which many men undertook the real number system was taken for granted. No attempt was made to analyze this structure or to build it up logically. Apparently the mathematicians felt they were on sure ground as far as this area was concerned.

Cauchy begins his 1821 work with the definition of a variable. "One calls a quantity which one considers as having to successively assume many values different from one another a variable." As for the concept of function, "When variable quantities are so joined between themselves that, the value of one of these being given, one may determine the values of all the others, one ordinarily conceives these diverse quantities expressed by means of the one among them, which then takes the name independent variable; and the other quantities expressed by means of the independent variable are those which one calls functions of this variable." Cauchy is also explicit that an infinite series is one way of specifying a function. However, an analytical expression for a function is not required.

In a paper on Fourier series which we shall return to later, "Über die Darstellung ganz willkürlicher Functionen durch Sinus-und Cosinusreihen" (On the Representation of Completely Arbitrary Functions by Sine and Cosine Series),[9] Dirichlet gave the definition of a (single-valued) function which is now most often employed, namely, that y is a function of x when to each value of x in a given interval there corresponds a unique value of y. He added that it does not matter whether throughout this interval y depends upon x according to one law or more or whether the dependence of y on x can be expressed by mathematical operations. In fact in 1829[10] he gave the example of a function of x which has the value c for all rational values of x and the value d for all irrational values of x.

Hankel points out that the best textbooks of at least the first half of the century were at a loss as to what to do about the function concept. Some defined a function essentially in Euler's sense; others required that y vary with x according to some law but did not explain what law meant; some used Dirichlet's definition; and still others gave no definition. But all deduced consequences from their definitions which were not logically implied by the definitions.

The proper distinction between continuity and discontinuity gradually emerged. The careful study of the properties of functions was initiated by Bernhard Bolzano (1781–1848), a priest, philosopher, and mathematician of Bohemia. Bolzano was led to this work by trying to give a purely arithmetical proof of the fundamental theorem of algebra in place of Gauss's first proof (1799) which used geometric ideas. Bolzano had the correct

9. *Repertorium der Physik*, 1, 1837, 152–74 = *Werke*, 1, 135–60.
10. *Jour. für Math.*, 4, 1829, 157–69 = *Werke*, 1, 117–32.

concepts for the establishment of the calculus (except for a theory of real numbers), but his work went unnoticed for half a century. He denied the existence of infinitely small numbers (infinitesimals) and infinitely large numbers, both of which had been used by the eighteenth-century writers. In a book of 1817 whose long title starts with *Rein analytischer Beweis* (see the bibliography) Bolzano gave the proper definition of continuity, namely, $f(x)$ is continuous in an interval if at any x in the interval the difference $f(x + \omega) - f(x)$ can be made as small as one wishes by taking ω sufficiently small. He proves that polynomials are continuous.

Cauchy, too, tackled the notions of limit and continuity. As with Bolzano the limit concept was based on purely arithmetical considerations. In the *Cours* (1821) Cauchy says, "When the successive values attributed to a variable approach indefinitely a fixed value so as to end by differing from it by as little as one wishes, this last is called the limit of all the others. Thus, for example, an irrational number is the limit of diverse fractions which furnish closer and closer approximate values of it." This example was a bit unfortunate because many took it to be a definition of irrational numbers in terms of limit whereas the limit could have no meaning if irrationals were not already present. Cauchy omitted it in his 1823 and 1829 works.

In the preface to his 1821 work Cauchy says that to speak of the continuity of functions he must make known the principal properties of infinitely small quantities. "One says [*Cours*, p. 5] that a variable quantity becomes infinitely small when its numerical value decreases indefinitely in such a way as to converge to the limit 0." Such variables he calls infinitesimals. Thus Cauchy clarifies Leibniz's notion of infinitesimal and frees it of metaphysical ties. Cauchy continues, "One says that a variable quantity becomes infinitely large when its numerical value increases indefinitely in such a manner as to converge to the limit ∞." However, ∞ means not a fixed quantity but something indefinitely large.

Cauchy is now prepared to define continuity of a function. In the *Cours* (pp. 34–35) he says. "Let $f(x)$ be a function of the variable x, and suppose that, for each value of x intermediate between two given limits [bounds], this function constantly assumes a finite and unique value. If, beginning with a value of x contained between these limits, one assigns to the variable x an infinitely small increment α, the function itself will take on as an increment the difference $f(x + \alpha) - f(x)$ which will depend at the same time on the new variable α and on the value of x. This granted, the function $f(x)$ will be, between the two limits assigned to the variable x, a *continuous* function of the variable if, for each value of x intermediate between these two limits, the numerical value of the difference $f(x + \alpha) - f(x)$ decreases indefinitely with that of α. In other words, *the function* f(x) *will remain continuous with respect to* x *between the given limits, if, between these limits, an infinitely small increment of the variable always produces an infinitely small increment of the function itself.*

"We also say that the function $f(x)$ is a continuous function of x in the neighborhood of a particular value assigned to the variable x, as long as it [the function] is continuous between those two limits of x, no matter how close together, which enclose the value in question." He then says that $f(x)$ is discontinuous at x_0 if it is not continuous in every interval around x_0.

In his *Cours* (p. 37) Cauchy asserted that if a function of several variables is continuous in each one separately it is a continuous function of all the variables. This is not correct.

Throughout the nineteenth century the notion of continuity was explored and mathematicians learned more about it, sometimes producing results that astonished them. Darboux gave an example of a function which took on all intermediate values between two given values in passing from $x = a$ to $x = b$ but was not continuous. Thus a basic property of continuous functions is not sufficient to insure continuity.[11]

Weierstrass's work on the rigorization of analysis improved on Bolzano, Abel, and Cauchy. He, too, sought to avoid intuition and to build on arithmetical concepts. Though he did this work during the years 1841–56 when he was a high-school teacher much of it did not become known until 1859 when he began to lecture at the University of Berlin.

Weierstrass attacked the phrase "a variable approaches a limit," which unfortunately suggests time and motion. He interprets a variable simply as a letter standing for any one of a set of values which the letter may be given. Thus motion is eliminated. A continuous variable is one such that if x_0 is any value of the set of values of the variable and δ any positive number there are other values of the variable in the interval $(x_0 - \delta, x_0 + \delta)$.

To remove the vagueness in the phrase "becomes and remains less than any given quantity," which Bolzano and Cauchy used in their definitions of continuity and limit of a function, Weierstrass gave the now accepted definition that $f(x)$ is continuous at $x = x_0$ if given any positive number ε, there exists a δ such that for all x in the interval $|x - x_0| < \delta$, $|f(x) - f(x_0)| < \varepsilon$. A function $f(x)$ has a limit L at $x = x_0$ if the same statement holds but with L replacing $f(x_0)$. A function $f(x)$ is continuous in an interval of x values if it is continuous at each x in the interval.

During the years in which the notion of continuity itself was being refined, the efforts to establish analysis rigorously called for the proof of many theorems about continuous functions which had been accepted intuitively. Bolzano in his 1817 publication sought to prove that if $f(x)$ is negative for $x = a$ and positive for $x = b$, then $f(x)$ has a zero between a and b. He considered the sequence of functions (for fixed x)

(1) $$F_1(x), F_2(x), F_3(x), \cdots, F_n(x), \cdots$$

11. Consider $y = \sin(1/x)$ for $x \neq 0$ and $y = 0$ for $x = 0$. This function goes through all values from one assumed at a negative value of x to one assumed at a positive value of x. However, it is not continuous at $x = 0$.

and introduced the theorem that if for n large enough we can make the difference $F_{n+r} - F_n$ less than any given quantity, no matter how large r is, then there exists a fixed magnitude X such that the sequence comes closer and closer to X, and indeed as close as one wishes. His determination of the quantity X was somewhat obscure because he did not have a clear theory of the real number system and of irrational numbers in particular on which to build. However, he had the idea of what we now call the Cauchy condition for the convergence of a sequence (see below).

In the course of the proof Bolzano established the existence of a least upper bound for a bounded set of real numbers. His precise statement is: If a property M does not apply to all values of a variable quantity x, but to all those that are smaller than a certain u, there is always a quantity U which is the largest of those of which it can be asserted that all smaller x possess the property M. The essence of Bolzano's proof of this lemma was to divide the bounded interval into two parts and select a particular part containing an infinite number of members of the set. He then repeats the process until he closes down on the number which is the least upper bound of the given set of real numbers. This method was used by Weierstrass in the 1860s, with due credit to Bolzano, to prove what is now called the Weierstrass-Bolzano theorem. It establishes for any bounded infinite set of points the existence of a point such that in every neighborhood of it there are points of the set.

Cauchy had used without proof (in one of his proofs of the existence of roots of a polynomial) the existence of a minimum of a continuous function defined over a closed interval. Weierstrass in his Berlin lectures proved for any continuous function of one or more variables defined over a closed bounded domain the existence of a minimum value and a maximum value of the function.

In work inspired by the ideas of Georg Cantor and Weierstrass, Heine defined uniform continuity for functions of one or several variables[12] and then proved[13] that a function which is continuous on a closed bounded interval of the real numbers is uniformly continuous. Heine's method introduced and used the following theorem: Let a closed interval $[a, b]$ and a countably infinite set Δ of intervals, all in $[a, b]$, be given such that every point x of $a \leq x \leq b$ is an interior point of at least one of the intervals of Δ. (The endpoints a and b are regarded as interior points when a is the left-hand end of an interval and b the right-hand end of another interval.) Then a set consisting of a *finite* number of the intervals of Δ has the same property, namely, every point of the closed interval $[a, b]$ is an interior point of at least one of this finite set of intervals (a and b can be endpoints).

Emile Borel (1871–1956), one of the leading French mathematicians of this century, recognized the importance of being able to select a finite number of covering intervals and first stated it as an independent theorem

12. *Jour. für Math.*, 71, 1870, 353–65.
13. *Jour. für Math.*, 74, 1872, 172–88.

for the case when the original set of intervals Δ is countable.[14] Though many German and French mathematicians refer to this theorem as Borel's, since Heine used the property in his proof of uniform continuity the theorem is also known as the Heine-Borel theorem. The merit of the theorem, as Lebesgue pointed out, is not in the proof of it, which is not difficult, but in perceiving its importance and enunciating it as a distinct theorem. The theorem applies to closed sets in any number of dimensions and is now basic in set theory.

The extension of the Heine-Borel theorem to the case where a finite set of covering intervals can be selected from an uncountably infinite set is usually credited to Lebesgue who claimed to have known the theorem in 1898 and published it in his *Leçons sur l'intégration* (1904). However, it was first published by Pierre Cousin (1867–1933) in 1895.[15]

3. *The Derivative*

D'Alembert was the first to see that Newton had essentially the correct notion of the derivative. D'Alembert says explicitly in the *Encyclopédie* that the derivative must be based on the limit of the ratio of the differences of dependent and independent variables. This version is a reformulation of Newton's prime and ultimate ratio. D'Alembert did not go further because his thoughts were still tied to geometric intuition. His successors of the next fifty years still failed to give a clear definition of the derivative. Even Poisson believed that there are positive numbers that are not zero, but which are smaller than any given number however small.

Bolzano was the first (1817) to define the derivative of $f(x)$ as the quantity $f'(x)$ which the ratio $[f(x + \Delta x) - f(x)]/\Delta x$ approaches indefinitely closely as Δx approaches 0 through positive and negative values. Bolzano emphasized that $f'(x)$ was not a quotient of zeros or a ratio of evanescent quantities but a number which the ratio above approached.

In his *Résumé des leçons*[16] Cauchy defined the derivative in the same manner as Bolzano. He then unified this notion and the Leibnizian differentials by defining dx to be any finite quantity and dy to be $f'(x)\ dx$.[17] In other words, one introduces two quantities dx and dy whose ratio, by definition, is $f'(x)$. Differentials have meaning in terms of the derivative and are merely an auxiliary notion that could be dispensed with logically but are convenient as a way of thinking or writing. Cauchy also pointed out what the differential expressions used throughout the eighteenth century meant in terms of derivatives.

He then clarified the relation between $\Delta y/\Delta x$ and $f'(x)$ through the mean

14. *Ann. de l'Ecole Norm. Sup.*, (3), 12, 1895, 9–55.
15. *Acta Math.*, 19, 1895, 1–61.
16. 1823, *Œuvres*, (2), 4, 22.
17. Lacroix in the first edition of his *Traité* had already defined dy in this manner.

value theorem, that is, $\Delta y = f'(x + \theta\,\Delta x)\,\Delta x$, where $0 < \theta < 1$. The theorem itself was known to Lagrange (Chap. 20, sec. 7). Cauchy's proof of the mean value theorem used the continuity of $f'(x)$ in the interval Δx.

Though Bolzano and Cauchy had rigorized (somewhat) the notions of continuity and the derivative, Cauchy and nearly all mathematicians of his era believed and many texts "proved" for the next fifty years that a continuous function must be differentiable (except of course at isolated points such as $x = 0$ for $y = 1/x$). Bolzano did understand the distinction between continuity and differentiability. In his *Funktionenlehre*, which he wrote in 1834 but did not complete and publish,[18] he gave an example of a continuous function which has no finite derivative at any point. Bolzano's example, like his other works, was not noticed.[19] Even if it had been published in 1834 it probably would have made no impression because the curve did not have an analytic representation, and for mathematicians of that period functions were still entities given by analytical expressions.

The example that ultimately drove home the distinction between continuity and differentiability was given by Riemann in the *Habilitationsschrift*, the paper of 1854 he wrote to qualify as a *Privatdozent* at Göttingen, "Über die Darstellbarkeit einer Function durch eine trigonometrische Reihe," (On the Representability of a Function by a Trigonometric Series).[20] (The paper on the foundations of geometry (Chap. 37, sec. 3) was given as a qualifying *lecture*.) Riemann defined the following function. Let (x) denote the difference between x and the nearest integer and let $(x) = 0$ if it is halfway between two integers. Then $-1/2 < (x) < 1/2$. Now $f(x)$ is defined as

$$f(x) = \frac{(x)}{1} + \frac{(2x)}{4} + \frac{(3x)}{9} + \cdots.$$

This series converges for all values of x. However, for $x = p/2n$ where p is an integer prime to $2n$, $f(x)$ is discontinuous and has a jump whose value is $\pi^2/8n^2$. At all other values of x, $f(x)$ is continuous. Moreover, $f(x)$ is discontinuous an infinite number of times in every arbitrarily small interval. Nevertheless, $f(x)$ is integrable (sec. 4). Moreover, $F(x) = \int f(x)\,dx$ is continuous for all x but fails to have a derivative where $f(x)$ is discontinuous. This pathological function did not attract much attention until it was published in 1868.

An even more striking distinction between continuity and differentiability was demonstrated by the Swiss mathematician Charles Cellérier

18. *Schriften*, 1, Prague, 1930. It was edited and published by K. Rychlik, Prague, 1930.
19. In 1922 Rychlik proved that the function was nowhere differentiable. See Gerhard Kowalewski, "Über Bolzanos nichtdifferenzierbare stetige Funktion," *Acta Math.*, 44, 1923, 315–19. This article contains a description of Bolzano's function.
20. *Abh. der Ges. der Wiss. zu Gött.*, 13, 1868, 87–132 = *Werke*, 227–64.

(1818–89). In 1860 he gave an example of a function which is continuous but nowhere differentiable, namely,

$$f(x) = \sum_{n=1}^{\infty} a^{-n} \sin a^n x$$

in which a is a large positive integer. This was not published, however, until 1890.[21] The example that attracted the most attention is due to Weierstrass. As far back as 1861 he had affirmed in his lectures that any attempt to prove that differentiability follows from continuity must fail. He then gave the classic example of a continuous nowhere differentiable function in a lecture to the Berlin Academy on July 18, 1872.[22] Weierstrass communicated his example in a letter of 1874 to Du Bois-Reymond and the example was first published by the latter.[23] Weierstrass's function is

$$f(x) = \sum_{n=0}^{\infty} b^n \cos (a^n \pi x)$$

wherein a is an odd integer and b a positive constant less than 1 such that $ab > 1 + (3\pi/2)$. The series is uniformly convergent and so defines a continuous function. The example given by Weierstrass prompted the creation of many more functions that are continuous in an interval or everywhere but fail to be differentiable either on a dense set of points or at any point.[24]

The historical significance of the discovery that continuity does not imply differentiability and that functions can have all sorts of abnormal behavior was great. It made mathematicians all the more fearful of trusting intuition or geometrical thinking.

4. *The Integral*

Newton's work showed how areas could be found by reversing differentiation. This is of course still the essential method. Leibniz's idea of area and volume as a "sum" of elements such as rectangles or cylinders [the definite integral] was neglected. When the latter concept was employed at all in the eighteenth century it was loosely used.

Cauchy stressed defining the integral as the limit of a sum instead of the inverse of differentiation. There was at least one major reason for the change.

21. *Bull. des Sci. Math.*, (2), 14, 1890, 142–60.
22. *Werke*, 2, 71–74.
23. *Jour. für Math.*, 79, 1875, 21–37.
24. Other examples and references can be found in E. J. Townsend, *Functions of Real Variables*, Henry Holt, 1928, and in E. W. Hobson, *The Theory of Functions of a Real Variable*, 2, Chap. 6, Dover (reprint), 1957.

Fourier, as we know, dealt with discontinuous functions, and the formula for the coefficients of a Fourier series, namely,

$$a_n = \frac{1}{\pi} \int_0^{2\pi} f(x) \cos nx \, dx, \qquad b_n = \frac{1}{\pi} \int_0^{2\pi} f(x) \sin nx \, dx$$

calls for the integrals of such functions. Fourier regarded the integral as a sum (the Leibnizian view) and so had no difficulty in handling even discontinuous $f(x)$. The problem of the analytical meaning of the integral when $f(x)$ is discontinuous had, however, to be considered.

Cauchy's most systematic attack on the definite integral was made in his *Résumé* (1823) wherein he also points out that it is necessary to establish the existence of the definite integral and indirectly of the antiderivative or primitive function before one can use them. He starts with continuous functions.

For continuous $f(x)$ he gives[25] the precise definition of the integral as the limit of a sum. If the interval $[x_0, X]$ is subdivided by the x-values, $x_1, x_2, \ldots, x_{n-1}$, with $x_n = X$, then the integral is

$$\lim_{n \to \infty} \sum_{i=1}^{n} f(\xi_i)(x_i - x_{i-1}),$$

where ξ_i is any value of x in $[x_{i-1}, x_i]$. The definition presupposes that $f(x)$ is continuous over $[x_0, X]$ and that the length of the largest subinterval approaches zero. The definition is arithmetical. Cauchy shows the integral exists no matter how the x_i and ξ_i are chosen. However, his proof was not rigorous because he did not have the notion of uniform continuity. He denotes the limit by the notation proposed by Fourier $\int_{x_0}^{X} f(x) \, dx$ in place of

$$\int f(x) \, dx \begin{bmatrix} x = b \\ x = a \end{bmatrix}$$

often employed by Euler for antidifferentiation.

Cauchy then defines

$$F(x) - \int_{x_0}^{x} f(x) \, dx$$

and shows that $F(x)$ is continuous in $[x_0, X]$. By forming

$$\frac{F(x + h) - F(x)}{h} = \frac{1}{h} \int_x^{x+h} f(x) \, dx$$

and using the mean value theorem for integrals Cauchy proves that

$$F'(x) = f(x).$$

25. *Résumé*, 81–84 = *Œuvres*, (2), 4, 122–27.

This is the fundamental theorem of the calculus, and Cauchy's presentation is the first demonstration of it. Then, after showing that all primitives of a given $f(x)$ differ by a constant he defines the indefinite integral as

$$\int f(x)\ dx = \int_a^x f(x)\ dx + C.$$

He points out that

$$\int_a^b f'(x) = f(b) - f(a)$$

presupposes $f'(x)$ continuous. Cauchy then treats the singular (improper) integrals where $f(x)$ becomes infinite at some value of x in the interval of integration or where the interval of integration extends to ∞. For the case where $f(x)$ has a discontinuity at $x = c$ at which value $f(x)$ may be bounded or not Cauchy defines

$$\int_a^b f(x)\ dx = \lim_{\varepsilon_1 \to 0} \int_a^{c-\varepsilon_1} f(x)\ dx + \lim_{\varepsilon_2 \to 0} \int_{c+\varepsilon_2}^b f(x)\ dx$$

when these limits exist. When $\varepsilon_1 = \varepsilon_2$ we get what Cauchy called the principal value.

The notions of area bounded by a curve, length of a curve, volume bounded by surfaces and areas of surfaces had been accepted as intuitively understood, and it had been considered one of the great achievements of the calculus that these quantities could be calculated by means of integrals. But Cauchy, in keeping with his goal of arithmetizing analysis, *defined* these geometric quantities by means of the integrals which had been formulated to calculate them. Cauchy unwittingly imposed a limitation on the concepts he defined because the calculus formulas impose restrictions on the quantities involved. Thus the formula for the length of arc of a curve given by $y = f(x)$ is

$$s = \int_a^b \sqrt{1 + (y')^2}\ dx$$

and this formula presupposes the differentiability of $f(x)$. The question of what are the most general definitions of areas, lengths of curves, and volumes was to be raised later (Chap. 42, sec. 5).

Cauchy had proved the existence of an integral for any continuous integrand. He had also defined the integral when the integrand has jump discontinuities and infinities. But with the growth of analysis the need to consider integrals of more irregularly behaving functions became manifest. The subject of integrability was taken up by Riemann in his paper of 1854 on trigonometric series. He says that it is important at least for mathematics, though not for physical applications, to consider the broader conditions under which the integral formula for the Fourier coefficients holds.

Riemann generalized the integral to cover functions $f(x)$ defined and bounded over an interval $[a, b]$. He breaks up this interval into subintervals[26] $\Delta x_1, \Delta x_2, \ldots, \Delta x_n$ and defines the oscillation of $f(x)$ in Δx_i as the difference between the greatest and least value of $f(x)$ in Δx_i. Then he proves that a necessary and sufficient condition that the sums

$$S = \sum_{i=1}^{n} f(x_i)\, \Delta x_i,$$

where x_i is any value of x in Δx_i, approach a unique limit (that the integral exists) as the maximum Δx_i approaches zero is that the sum of the intervals Δx_i in which the oscillation of $f(x)$ is greater than any given number λ must approach zero with the size of the intervals.

Riemann then points out that this condition on the oscillations allows him to replace continuous functions by functions with isolated discontinuities and also by functions having an everywhere dense set of points of discontinuity. In fact the example he gave of an integrable function with an infinite number of discontinuities in every arbitrarily small interval (sec. 3) was offered to illustrate the generality of his integral concept. Thus Riemann dispensed with continuity and piecewise continuity in the definition of the integral.

In his 1854 paper Riemann with no further remarks gives another necessary and sufficient condition that a bounded function $f(x)$ be integrable on $[a, b]$. It amounts to first setting up what are now called the upper and lower sums

$$S = M_1\,\Delta x_1 + \cdots + M_n\,\Delta x_n$$
$$s = m_1\,\Delta x_1 + \cdots + m_n\,\Delta x_n$$

where m_i and M_i are the least and greatest values of $f(x)$ in Δx_i. Then letting $D_i = M_i - m_i$, Riemann states that the integral of $f(x)$ over $[a, b]$ exists if and only if

$$\lim_{\max \Delta x \to 0} \{D_1\,\Delta x_1 + D_2\,\Delta x_2 + \cdots + D_n\,\Delta x_n\} = 0$$

for all choices of Δx_i filling out the interval $[a, b]$. Darboux completed this formulation and proved that the condition is necessary and sufficient.[27] There are many values of S each corresponding to a partition of $[a, b]$ into Δx_i. Likewise there are many values of s. Each S is called an upper sum and each s a lower sum. Let the greatest lower bound of the S be J and the least upper bound of the s be I. It follows that $I \leq J$. Darboux's theorem then states that the sums S and s tend respectively to J and I when the number of

26. For brevity we use Δx_i for the subintervals and their lengths.
27. *Ann. de l'Ecole Norm. Sup.*, (2), 4, 1875, 57–112.

Δx_i is increased indefinitely in such a way that the maximum subinterval approaches zero. A bounded function is said to be integrable on $[a, b]$ if $J = I$.

Darboux then shows that a bounded function will be integrable on $[a, b]$ if and only if the discontinuities in $f(x)$ constitute a set of measure zero. By the latter he meant that the points of discontinuity can be enclosed in a finite set of intervals whose total length is arbitrarily small. This very formulation of the integrability condition was given by a number of other men in the same year (1875). The terms upper integral and the notation $\overline{\int_a^b} f(x)\, dx$ for the greatest lower bound J of the S's, and lower integral and the notation $\underline{\int_a^b} f(x)\, dx$ for the least upper bound I of the s's were introduced by Volterra.[28]

Darboux also showed in the 1875 paper that the fundamental theorem of the calculus holds for functions integrable in the extended sense. Bonnet had given a proof of the mean value theorem of the differential calculus which did not use the continuity of $f'(x)$.[29] Darboux using this proof, which is now standard, showed that

$$\int_a^b f'(x)\, dx = f(b) - f(a)$$

when f' is merely integrable in the Riemann-Darboux sense. Darboux's argument was that

$$f(b) - f(a) = \sum_{i=1}^n f(x_i) - f(x_{i-1}),$$

where $a = x_0 < x_1 < x_2 < \cdots < x_n = b$. By the mean value theorem

$$\sum f(x_i) - f(x_{i-1}) = \sum f'(t_i)(x_i - x_{i-1}),$$

where t_i is some value in (x_{i-1}, x_i). Now if the maximum Δx_i, or $x_i - x_{i-1}$, approaches zero then the right side of this last equation approaches $\int_a^b f'(x)\, dx$ and the left side is $f(b) - f(a)$.

One of the favorite activities of the 1870s and the 1880s was to construct functions with various infinite sets of discontinuities that would still be integrable in Riemann's sense. In this connection H. J. S. Smith[30] gave the first example of a function nonintegrable in Riemann's sense but for which the points of discontinuity were "rare." Dirichlet's function (sec. 2) is also nonintegrable in this sense, but it is discontinuous everywhere.

The notion of integration was then extended to unbounded functions and

28. *Gior. di Mat.*, 19, 1881, 333–72.
29. Published in Serret's *Cours de calcul différentiel et intégral*, 1, 1868, 17–19.
30. *Proc. Lon. Math. Soc.*, 6, 1875, 140–53 = *Coll. Papers*, 2, 86–100.

to various improper integrals. The most significant extension was made in the next century by Lebesgue (Chap. 44). However, as far as the elementary calculus was concerned the notion of integral was by 1875 sufficiently broad and rigorously founded.

The theory of double integrals was also tackled. The simpler cases had been treated in the eighteenth century (Chap. 19, sec. 6). In his paper of 1814 (Chap. 27, sec. 4) Cauchy showed that the order of integration in which one evaluates a double integral $\int \int f(x, y) \, dx \, dy$ does matter if the integrand is discontinuous in the domain of integration. Specifically Cauchy pointed out [31] that the repeated integrals

$$\int_0^1 dy \left(\int_0^1 f(x, y) \, dx \right), \; \int_0^1 dx \left(\int_0^1 f(x, y) \, dy \right)$$

need not be equal when f is unbounded.

Karl J. Thomae (1840–1921) extended Riemann's theory of integration to functions of two variables.[32] Then Thomae in 1878[33] gave a simple example of a bounded function for which the second repeated integral exists but the first is meaningless.

In the examples of Cauchy and Thomae the double integral does not exist. But in 1883[34] Du Bois-Reymond showed that even when the double integral exists the two repeated integrals need not. In the case of double integrals too the most significant generalization was made by Lebesgue.

5. Infinite Series

The eighteenth-century mathematicians used series indiscriminately. By the end of the century some doubtful or plainly absurd results from work with infinite series stimulated inquiries into the validity of operations with them. Around 1810 Fourier, Gauss, and Bolzano began the exact handling of infinite series. Bolzano stressed that one must consider convergence and criticized in particular the loose proof of the binomial theorem. Abel was the most outspoken critic of the older uses of series.

In his 1811 paper and his *Analytical Theory of Heat* Fourier gave a satisfactory definition of convergence of an infinite series, though in general he worked freely with divergent series. In the book (p. 196 of the English edition) he describes convergence to mean that as n increases the sum of n terms approaches a fixed value more and more closely and should differ from it only by a quantity which becomes less than any given magnitude. Moreover, he recognized that convergence of a series of functions may obtain only

31. *Mémoire* of 1814; see, in particular, p. 394 of *Œuvres*, (1), 1.
32. *Zeit. für Math. und Phys.*, 21, 1876, 224–27.
33. *Zeit. für Math. und Phys.*, 23, 1878, 67–68.
34. *Jour. für Math.*, 94, 1883, 273–90.

in an interval of x values. He also stressed that a necessary condition for convergence is that the terms approach zero in value. However, the series $1 - 1 + 1 + \cdots$ still fooled him; he took its sum to be $1/2$.

The first important and strictly rigorous investigation of convergence was made by Gauss in his 1812 paper "Disquisitiones Generales Circa Seriem Infinitam" (General Investigations of Infinite Series)[35] wherein he studied the hypergeometric series $F(\alpha, \beta, \gamma, x)$. In most of his work he called a series convergent if the terms from a certain one on decrease to zero. But in his 1812 paper he noted that this is not the correct concept. Because the hypergeometric series can represent many functions for different choices of α, β, and γ it seemed desirable to him to develop an exact criterion for convergence for this series. The criterion is laboriously arrived at but it does settle the question of convergence for the cases it was designed to cover. He showed that the hypergeometric series converges for real and complex x if $|x| < 1$ and diverges if $|x| > 1$. For $x = 1$, the series converges if and only if $\alpha + \beta < \gamma$ and for $x = -1$ the series converges if and only if $\alpha + \beta < \gamma + 1$. The unusual rigor discouraged interest in the paper by mathematicians of the time. Moreover, Gauss was concerned with particular series and did not take up general principles of the convergence of series.

Though Gauss is often mentioned as one of the first to recognize the need to restrict the use of series to their domains of convergence he avoided any decisive position. He was so much concerned to solve concrete problems by numerical calculation that he used Stirling's divergent development of the gamma function. When he did investigate the convergence of the hypergeometric series in 1812 he remarked[36] that he did so to please those who favored the rigor of the ancient geometers, but he did not state his own stand on the subject. In the course of his paper[37] he used the development of $\log(2 - 2\cos x)$ in cosines of multiples of x even though there was no proof of the convergence of this series and there could have been no proof with the techniques available at the time. In his astronomical and geodetic work Gauss, like the eighteenth-century men, followed the practice of using a finite number of terms of an infinite series and neglecting the rest. He stopped including terms when he saw that the succeeding terms were numerically small and of course did not estimate the error.

Poisson too took a peculiar position. He rejected divergent series[38] and even gave examples of how reckoning with divergent series can lead to false results. But he nevertheless made extensive use of divergent series in his representation of arbitrary functions by series of trigonometric and spherical functions.

35. *Comm. Soc. Gott.*, 2, 1813, = *Werke*, 3, 125–62 and 207–29.
36. *Werke*, 3, 129.
37. *Werke*, 3, 156.
38. *Jour. de l'Ecole Poly.*, 19, 1823, 404–509.

Bolzano in his 1817 publication had the correct notion of the condition for the convergence of a sequence, the condition now ascribed to Cauchy. Bolzano also had clear and correct notions about the convergence of series. But, as we have already noted, his work did not become widely known.

Cauchy's work on the convergence of series is the first extensive significant treatment of the subject. In his *Cours d'analyse* Cauchy says, "Let

$$s_n = u_0 + u_1 + u_2 + \cdots + u_{n-1}$$

be the sum of the first n terms [of the infinite series which one considers], n designating a natural number. If, for constantly increasing values of n, the sum s_n approaches indefinitely a certain limit s, the series will be called *convergent,* and the limit in question will be called the *sum* of the series.[39] On the contrary, if while n increases indefinitely, the sum s_n does not approach a fixed limit, the series will be called *divergent* and will have no sum."

After defining convergence and divergence Cauchy states (*Cours*, p. 125) the Cauchy convergence criterion, namely, a sequence $\{S_n\}$ converges to a limit S if and only if $S_{n+r} - S_n$ can be made less in absolute value than any assignable quantity for all r and sufficiently large n. Cauchy proves this condition is necessary but merely remarks that if the condition is fulfilled, the convergence of the sequence is assured. He lacked the knowledge of properties of real numbers to make the proof.

Cauchy then states and proves specific tests for the convergence of series with positive terms. He points out that u_n must approach zero. Another test (*Cours*, 132–35) requires that one find the limit or limits toward which the expression $(u_n)^{1/n}$ tends as n becomes infinite, and designate the greatest of these limits by k. Then the series will be convergent if $k < 1$ and divergent if $k > 1$. He also gives the ratio test which uses $\lim_{n \to \infty} u_{n+1}/u_n$. If this limit is less than 1 the series converges and if greater than 1, the series diverges. Special tests are given if the ratio is 1. There follow comparison tests and a logarithmic test. He proves that the sum $u_n + v_n$ of two convergent series converges to the sum of the separate sums and the analogous result for product. Series with some negative terms, Cauchy shows, converge when the series of absolute values of the terms converge, and he then deduces Leibniz's test for alternating series.

Cauchy also considered the sum of a series

$$\sum u_n(x) = u_1(x) + u_2(x) + u_3(x) + \cdots$$

in which all the terms are continuous, single-valued real functions. The theorems on the convergence of series of constant terms apply here to determine an interval of convergence. He also considers series with complex functions as terms.

39. The correct notion of the limit of a sequence was given by Wallis in 1655 (*Opera*, 1695, 1, 382) but was not taken up.

Lagrange was the first to state Taylor's theorem with a remainder but Cauchy in his 1823 and 1829 texts made the important point that the infinite Taylor series converges to the function from which it is derived if the remainder approaches zero. He gives the example $e^{-x^2} + e^{-1/x^2}$ of a function whose Taylor series does not converge to the function. In his 1823 text he gives the example e^{-1/x^2} of a function which has all derivatives at $x = 0$ but has no Taylor expansion around $x = 0$. Here he contradicts by an example Lagrange's assertion in his *Théorie des fonctions* (Ch. V., Art. 30) that if $f(x)$ has at x_0 all derivatives then it can be expressed as a Taylor series which converges to $f(x)$ for x near x_0. Cauchy also gave[40] an alternative form for the remainder in Taylor's formula.

Cauchy here made some additional missteps with respect to rigor. In his *Cours d'analyse* (pp. 131–32) he states that $F(x)$ is continuous if when $F(x) = \sum_1^\infty u_n(x)$ the series is convergent and the $u_n(x)$ are continuous. In his *Résumé des leçons*[41] he says that if the $u_n(x)$ are continuous and the series converges then one may integrate the series term by term; that is,

$$\int_a^b F \, dx = \sum_1^\infty \int_a^b u_n \, dx.$$

He overlooked the need for uniform convergence. He also asserts for continuous functions[42] that

$$\frac{\partial}{\partial u} \int_a^b f(x, u) \, dx = \int_a^b \frac{\partial f}{\partial u} \, dx.$$

Cauchy's work inspired Abel. Writing to his former teacher Holmboë from Paris in 1826 Abel said[43] Cauchy "is at present the one who knows how mathematics should be treated." In that year[44] Abel investigated the domain of convergence of the binomial series

$$1 + mx + \frac{m(m-1)}{2} x^2 + \frac{m(m-1)(m-2)}{3!} x^3 + \cdots$$

with m and x complex. He expressed astonishment that no one had previously investigated the convergence of this most important series. He proves first that if the series

$$f(\alpha) = v_0 + v_1\alpha + v_2\alpha^2 + \cdots,$$

wherein the v_i's are constants and α is real, converges for a value δ of α then it will converge for every smaller value of α, and $f(\alpha - \beta)$ for β approaching

40. *Exercices de mathématiques*, 1, 1826, 5 = *Œuvres*, (2), 6, 38–42.
41. 1823, *Œuvres*, (2), 4, p. 237.
42. *Exercices de mathématiques*, 2, 1827 = *Œuvres*, (2), 7, 160.
43. *Œuvres*, 2, 259.
44. *Jour. für Math.*, 1, 1826, 311–39 = *Œuvres*, 1, 219–50.

0 will approach $f(\alpha)$ when α is equal to or smaller than δ. The last part says that a convergent *power* series is a continuous function of its argument up to and including δ, for α can be δ.

In this same 1826 paper[45] Abel corrected Cauchy's error on the continuity of the sum of a convergent series of continuous functions. He gave the example of

(2)
$$\sin x - \frac{\sin 2x}{2} + \frac{\sin 3x}{3} \cdots$$

which is discontinuous when $x = (2n + 1)\pi$ and n is integral, though the individual terms are continuous.[46] Then by using the *idea* of uniform convergence he gave a correct proof that the sum of a uniformly convergent series of continuous functions is continuous in the interior of the interval of convergence. Abel did not isolate the property of uniform convergence of a series.

The notion of uniform convergence of a series $\sum_1^\infty u_n(x)$ requires that given any ε, there exists an N such that for all $n > N$, $|S(x) - \sum_1^n u_n(x)| < \varepsilon$ for all x in some interval. $S(x)$ is of course the sum of the series. This notion was recognized in and for itself by Stokes, a leading mathematical physicist,[47] and independently by Philipp L. Seidel (1821–96).[48] Neither man gave the precise formulation. Rather both showed that if a sum of a series of continuous functions is discontinuous at x_0 then there are values of x near x_0 for which the series converges arbitrarily slowly. Also neither related the need for uniform convergence to the justification of integrating a series term by term. In fact, Stokes accepted[49] Cauchy's use of term-by-term integration. Cauchy ultimately recognized the need for uniform convergence[50] in order to assert the continuity of the sum of a series of continuous functions but even he at that time did not see the error in his use of term-by-term integration of series.

Actually Weierstrass[51] had the notion of uniform convergence as early as 1842. In a theorem that duplicates unknowingly Cauchy's theorem on the existence of power series solutions of a system of first order ordinary differential equations, he affirms that the series converge uniformly and so constitute analytic functions of the complex variable. At about the same time

45. *Œuvres*, 1, 224.
46. The series (2) is the Fourier expansion of $x/2$ in the interval $-\pi < x < \pi$. Hence the series represents the periodic function which is $x/2$ in each 2π interval. Then the series converges to $\pi/2$ when x approaches $(2n + 1)\pi$ from the left and the series converges to $-\pi/2$ when x approaches $(2n + 1)\pi$ from the right.
47. *Trans. Camb. Phil. Soc.*, 8_5, 1848, 533–83 = *Math. and Phys. Papers*, 1, 236–313.
48. *Abh. der Bayer. Akad. der Wiss.*, 1847/49, 379–94.
49. *Papers*, 1, 242, 255, 268, and 283.
50. *Comp. Rend.*, 36, 1853, 454–59 = *Œuvres*, (1), 12, 30–36.
51. *Werke*, 1, 67–85.

Weierstrass used the notion of uniform convergence to give conditions for the integration of a series term by term and conditions for differentiation under the integral sign.

Through Weierstrass's circle of students the importance of uniform convergence was made known. Heine emphasized the notion in a paper on trigonometric series.[52] Heine may have learned of the idea through Georg Cantor who had studied at Berlin and then came to Halle in 1867 where Heine was a professor of mathematics.

During his years as a high-school teacher Weierstrass also discovered that any continuous function over a closed interval of the real axis can be expressed in that interval as an absolutely and uniformly convergent series of polynomials. Weierstrass included also functions of several variables. This result[53] aroused considerable interest and many extensions of this result to the representation of complex functions by a series of polynomials or a series of rational functions were established in the last quarter of the nineteenth century.

It had been assumed that the terms of a series can be rearranged at will. In a paper of 1837[54] Dirichlet proved that in an absolutely convergent series one may group or rearrange terms and not change the sum. He also gave examples to show that the terms of any conditionally convergent series can be rearranged so that the sum is altered. Riemann in a paper written in 1854 (see below) proved that by suitable rearrangement of the terms the sum could be any given number. Many more criteria for the convergence of infinite series were developed by leading mathematicians from the 1830s on throughout the rest of the century.

6. *Fourier Series*

As we know Fourier's work showed that a wide class of functions can be represented by trigonometric series. The problem of finding precise conditions on the functions which would possess a convergent Fourier series remained open. Efforts by Cauchy and Poisson were fruitless.

Dirichlet took an interest in Fourier series after meeting Fourier in Paris during the years 1822–25. In a basic paper "Sur la convergence des séries trigonométriques"[55] Dirichlet gave the first set of *sufficient* conditions that the Fourier series representing a given $f(x)$ converge and converge to $f(x)$. The proof given by Dirichlet is a refinement of that sketched by Fourier in the concluding sections of his *Analytical Theory of Heat*. Consider $f(x)$ either

52. *Jour. für Math.*, 71, 1870, 353–65.
53. *Sitzungsber. Akad. Wiss. zu Berlin*, 1885, 633–39, 789–905 = *Werke*, 3, 1–37.
54. *Abh. König. Akad. der Wiss., Berlin*, 1837, 45–81 = *Werke*, 1, 313–342 = *Jour. de Math.*, 4, 1839, 393–422.
55. *Jour. für Math.*, 4, 1829, 157–69 = *Werke*, 1, 117–32.

given periodic with period 2π or given in the interval $[-\pi, \pi]$ and defined to be periodic in each interval of length 2π to the left and right of $[-\pi, \pi]$. Dirichlet's conditions are:

(a) $f(x)$ is single-valued and bounded.
(b) $f(x)$ is piecewise continuous; that is, it has only a finite number of discontinuities in the (closed) period.
(c) $f(x)$ is piecewise monotone; that is, it has only a finite number of maxima and minima in one period.

The $f(x)$ can have different analytic representations in different parts of the fundamental period.

Dirichlet's method of proof was to make a direct summation of n terms and to investigate what happens as n becomes infinite. He proved that for any given value of x the sum of the series is $f(x)$ provided $f(x)$ is continuous at that value of x and is $(1/2)[f(x-0) + f(x+0)]$ if $f(x)$ is discontinuous at that value of x.

In his proof Dirichlet had to give a careful discussion of the limiting values of the integrals

$$\int_0^a f(x) \frac{\sin \mu x}{\sin x} \, dx, \qquad a > 0$$

$$\int_a^b f(x) \frac{\sin \mu x}{\sin x} \, dx, \qquad b > a > 0$$

as μ increases indefinitely. These are still called the Dirichlet integrals.

It was in connection with this work that he gave the function which is c for rational values of x and d for irrational values of x (sec. 2). He had hoped to generalize the notion of integral so that a larger class of functions could still be representable by Fourier series converging to these functions, but the particular function just noted was intended as an example of one which could not be included in a broader notion of integral.

Riemann studied for a while under Dirichlet in Berlin and acquired an interest in Fourier series. In 1854 he took up the subject in his *Habilitations-schrift* (probationary essay) at Göttingen,[56] "Über die Darstellbarkeit einer Function durch eine trigonometrische Reihe," which aimed to find necessary and sufficient conditions that a function must satisfy so that at a point x in the interval $[-\pi, \pi]$ the Fourier series for $f(x)$ should converge to $f(x)$.

Riemann did prove the fundamental theorem that if $f(x)$ is bounded and integrable in $[-\pi, \pi]$ then the Fourier coefficients

$$(3) \qquad a_n = \frac{1}{\pi} \int_{-\pi}^{\pi} f(x) \cos nx \, dx, \qquad b_n = \frac{1}{\pi} \int_{-\pi}^{\pi} f(x) \sin nx \, dx$$

56. *Abh. der Ges. der Wiss. zu Gött.*, 13, 1868, 87–132 = *Werke*, 227–64.

approach zero as n tends to infinity. The theorem showed too that for bounded and integrable $f(x)$ the convergence of its Fourier series at a point in $[-\pi, \pi]$ depends only on the behavior of $f(x)$ in the neighborhood of that point. However, the problem of finding necessary *and* sufficient conditions on $f(x)$ so that its Fourier series converges to $f(x)$ was not and has not been solved.

Riemann opened up another line of investigation. He considered *trigonometric* series but did not require that the coefficients be determined by the formula (3) for the Fourier coefficients. He starts with the series

$$(4) \qquad \sum_{1}^{\infty} a_n \sin nx + \frac{b_0}{2} + \sum_{1}^{\infty} b_n \cos nx$$

and defines

$$A_0 = \frac{1}{2} b_0, \qquad A_n(x) = a_n \sin nx + b_n \cos nx.$$

Then the series (4) is equal to

$$f(x) = \sum_{n=0}^{\infty} A_n(x).$$

Of course $f(x)$ has a value only for those values of x for which the series converges. Let us refer to the series itself by Ω. Now the terms of Ω may approach zero for all x or for some x. These two cases are treated separately by Riemann.

If a_n and b_n approach zero, the terms of Ω approach zero for all x. Let $F(x)$ be the function

$$F(x) = C + C'x + A_0 \frac{x^2}{2} - A_1 - \frac{A_2}{4} - \cdots - \frac{A_n}{n^2} \cdots$$

which is obtained by two successive integrations of Ω. Riemann shows that $F(x)$ converges for all x and is continuous in x. Then $F(x)$ can itself be integrated. Riemann now proves a number of theorems about $F(x)$ which lead in turn to necessary and sufficient conditions for a series of the form (4) to converge to a given function $f(x)$ of period 2π. He then gives a necessary and sufficient condition that the trigonometric series (4) converge at a particular value of x, with a_n and b_n still approaching 0 as n approaches ∞.

Next he considers the alternate case where $\lim_{n \to \infty} A_n$ depends on the value of x and gives conditions which hold when the series Ω is convergent for particular values of x and a criterion for convergence at particular values of x.

He also shows that a given $f(x)$ may be integrable and yet not have a Fourier series representation. Further there are nonintegrable functions to which the series Ω converges for an infinite number of values of x taken between arbitrarily close limits. Finally a trigonometric series can converge for an infinite number of values of x in an arbitrarily small interval even though a_n and b_n do not approach zero for all x.

The nature of the convergence of Fourier series received further attention after the introduction of the concept of uniform convergence by Stokes and Seidel. It had been known since Dirichlet's time that the series were, in general, only conditionally convergent, if at all, and that their convergence depended upon the presence of positive and negative terms. Heine noted in a paper of 1870[57] that the usual proof that a bounded $f(x)$ is uniquely represented between $-\pi$ and π by a Fourier series is incomplete because the series may not be uniformly convergent and so cannot be integrated term by term. This suggested that there may nevertheless exist nonuniformly converging trigonometric series which do represent a function. Moreover, a continuous function might be representable by a Fourier series and yet the series might not be uniformly convergent. These problems gave rise to a new series of investigations seeking to establish the uniqueness of the representation of a function by a trigonometric series and whether the coefficients are necessarily the Fourier coefficients. Heine in the above-mentioned paper proved that a Fourier series which represents a bounded function satisfying the Dirichlet conditions is uniformly convergent in the portions of the interval $[-\pi, \pi]$ which remain when arbitrarily small neighborhoods of the points of discontinuity of the function are removed from the interval. In these neighborhoods the convergence is necessarily nonuniform. Heine then proved that if the uniform convergence just specified holds for a trigonometric series which represents a function, then the series is unique.

The second result, on uniqueness, is equivalent to the statement that if a trigonometric series of the form

(5)
$$\frac{a_0}{2} + \sum_{n=1}^{\infty} (a_n \cos nx + b_n \sin nx)$$

is uniformly convergent and represents zero where convergent, that is, except on a finite set P of points, then the coefficients are all zero and of course then the series represents zero throughout $[-\pi, \pi]$.

The problems associated with the uniqueness of trigonometric and Fourier series attracted Georg Cantor, who studied Heine's work. Cantor began his investigations by seeking uniqueness criteria for trigonometric series representations of functions. He proved[58] that when $f(x)$ is represented by a trigonometric series convergent for all x, there does not exist a different series of the same form which likewise converges for every x and

57. *Jour. für Math.*, 71, 1870, 353–65.
58. *Jour. für Math.*, 72, 1870, 139–42 = *Ges. Abh.*, 80–83.

represents the same $f(x)$. Another paper[59] gave a better proof for this last result.

The uniqueness theorem he proved can be restated thus: If, for all x, there is a convergent representation of zero by a trigonometric series, then the coefficients a_n and b_n are zero. Then Cantor demonstrates in the 1871 paper that the conclusion still holds even if the convergence is renounced for a finite number of x values. This paper was the first of a sequence of papers in which Cantor treats the sets of exceptional values of x. He extended[60] the uniqueness result to the case where an infinite set of exceptional values is permitted. To describe this set he first defined a point p to be a limit of a set of points S if every interval containing p contains infinitely many points of S. Then he introduced the notion of the derived set of a set of points. This derived set consists of the limit points of the original set. There is, then, a second derived set, that is, the derived set of the derived set, and so forth. If the nth derived set of a given set is a finite set of points then the given set is said to be of the nth kind or nth order (or of the first species). Cantor's final answer to the question of whether a function can have two different trigonometric series representations in the interval $[-\pi, \pi]$ or whether zero can have a non-zero Fourier representation is that if in the interval a trigonometric series adds up to zero for all x except those of a point set of the nth kind (at which one knows nothing more about the series) then all the coefficients of the series must be zero. In this 1872 paper Cantor laid the foundation of the theory of point sets which we shall consider in a later chapter. The problem of uniqueness was pursued by many other men in the last part of the nineteenth century and the early part of the twentieth.[61]

For about fifty years after Dirichlet's work it was believed that the Fourier series of any function continuous in $[-\pi, \pi]$ converged to the function. But Du Bois-Reymond[62] gave an example of a function continuous in $(-\pi, \pi)$ whose Fourier series did not converge at a particular point. He also constructed another continuous function whose Fourier series fails to converge at the points of an everywhere dense set. Then in 1875[63] he proved that if a trigonometric series of the form

$$a_0 + \sum_1^\infty (a_n \cos nx + b_n \sin nx)$$

converged to $f(x)$ in $[-\pi, \pi]$ and if $f(x)$ is integrable (in a sense even more general than Riemann's in that $f(x)$ can be unbounded on a set of the first

59. *Jour. für Math.*, 73, 1871, 294–6 = *Ges. Abh.*, 84–86.

60. *Math. Ann.*, 5, 1872, 123–32 = *Ges. Abh.*, 92–102.

61. Details can be found in E. W. Hobson, *The Theory of Functions of a Real Variable*, Vol. 2, 656–98.

62. *Nachrichten König. Ges. der Wiss. zu Gött.*, 1873, 571–82.

63. *Abh. der Bayer. Akad. der Wiss.*, 12, 1876, 117–66.

species) then the series must be the Fourier series for $f(x)$. He also showed[64] that any Fourier series of a function that is Riemann integrable can be integrated term by term even though the series is not uniformly convergent.

Many men then took up the problem already answered in one way by Dirichlet, namely, to give sufficient conditions that a function $f(x)$ have a Fourier series which converges to $f(x)$. Several results are classical. Jordan gave a sufficient condition in terms of the concept of a function of bounded variation, which he introduced.[65] Let $f(x)$ be bounded in $[a, b]$ and let $a = x_0, x_1, \ldots, x_{n-1}, x_n = b$ be a mode of division (partition) of this interval. Let $y_0, y_1, \ldots, y_{n-1}, y_n$ be the values of $f(x)$ at these points. Then, for every partition

$$\sum_{0}^{n-1} (y_{r+1} - y_r) = f(b) - f(a)$$

let t denote

$$\sum_{0}^{n-1} |y_{r+1} - y_r|.$$

To every mode of subdividing $[a, b]$ there is a t. When corresponding to all possible modes of division of $[a, b]$, the sums t have a least upper bound then f is defined to be of bounded variation in $[a, b]$.

Jordan's sufficient condition states that the Fourier series for the integrable function $f(x)$ converges to

$$\frac{1}{2} [f(x + 0) + f(x - 0)]$$

at every point for which there is a neighborhood in which $f(x)$ is of bounded variation.[66]

During the 1860s and 1870s the properties of the Fourier coefficients were also examined and among the important results obtained were what is called Parseval's theorem (who stated it under more restricted conditions, Chap. 29, sec. 3) according to which if $f(x)$ and $[f(x)]^2$ are Riemann integrable in $[-\pi, \pi]$ then

$$\frac{1}{\pi} \int_{-\pi}^{\pi} [f(x)]^2 \, dx = 2a_0^2 + \sum_{1}^{\infty} (a_n^2 + b_n^2),$$

64. *Math. Ann.*, 22, 1883, 260–68.
65. *Comp. Rend.*, 92, 1881, 228–30 = *Œuvres*, 4, 393–95 and *Cours d'analyse*, 2, 1st ed., 1882, Ch. V.
66. *Cours d'analyse*, 2nd ed., 1893, 1, 67–72.

and if $f(x)$ and $g(x)$ and their squares are Riemann integrable then

$$\frac{1}{\pi}\int_{-\pi}^{\pi} f(x)\,g(x)\,dx = 2a_0\alpha_0 + \sum_{1}^{\infty}(a_n\alpha_n + b_n\beta_n)$$

where a_n, b_n, α_n, and β_n are the Fourier coefficients of $f(x)$ and $g(x)$ respectively.

7. The Status of Analysis

The work of Bolzano, Cauchy, Weierstrass, and others supplied the rigor in analysis. This work freed the calculus and its extensions from all dependence upon geometrical notions, motion, and intuitive understandings. From the outset these researches caused a considerable stir. After a scientific meeting at which Cauchy presented his theory on the convergence of series Laplace hastened home and remained there in seclusion until he had examined the series in his *Mécanique céleste*. Luckily every one was found to be convergent. When Weierstrass's work became known through his lectures, the effect was even more noticeable. The improvements in rigor can be seen by comparing the first edition of Jordan's *Cours d'analyse* (1882–87) with the second (1893–96) and the third edition (3 vols., 1909–15). Many other treatises incorporated the new rigor.

The rigorization of analysis did not prove to be the end of the investigation into the foundations. For one thing, practically all of the work presupposed the real number system but this subject remained unorganized. Except for Weierstrass who, as we shall see, considered the problem of the irrational number during the 1840s, all the others did not believe it necessary to investigate the logical foundations of the number system. It would appear that even the greatest mathematicians must develop their capacities to appreciate the need for rigor in stages. The work on the logical foundations of the real number system was to follow shortly (Chap. 41).

The discovery that continuous functions need not have derivatives, that discontinuous functions can be integrated, the new light on discontinuous functions shed by Dirichlet's and Riemann's work on Fourier series, and the study of the variety and the extent of the discontinuities of functions made the mathematicians realize that the rigorous study of functions extends beyond those used in the calculus and the usual branches of analysis where the requirement of differentiability usually restricts the class of functions. The study of functions was continued in the twentieth century and resulted in the development of a new branch of mathematics known as the theory of functions of a real variable (Chap. 44).

Like all new movements in mathematics, the rigorization of analysis did not go unopposed. There was much controversy as to whether the refinements in analysis should be pursued. The peculiar functions that were introduced were attacked as curiosities, nonsensical functions, funny functions, and as mathematical toys perhaps more intricate but of no more consequence than magic squares. They were also regarded as diseases or part of the morbid pathology of functions and as having no bearing on the important problems of pure and applied mathematics. These new functions, violating laws deemed perfect, were looked upon as signs of anarchy and chaos which mocked the order and harmony previous generations had sought. The many hypotheses which now had to be made in order to state a precise theorem were regarded as pedantic and destructive of the elegance of the eighteenth-century classical analysis, "as it was in paradise," to use Du Bois-Reymond's phrasing. The new details were resented as obscuring the main ideas.

Poincaré, in particular, distrusted this new research. He said:[67]

> Logic sometimes makes monsters. For half a century we have seen a mass of bizarre functions which appear to be forced to resemble as little as possible honest functions which serve some purpose. More of continuity, or less of continuity, more derivatives, and so forth. Indeed from the point of view of logic, these strange functions are the most general; on the other hand those which one meets without searching for them, and which follow simple laws appear as a particular case which does not amount to more than a small corner.
>
> In former times when one invented a new function it was for a practical purpose; today one invents them purposely to show up defects in the reasoning of our fathers and one will deduce from them only that.

Charles Hermite said in a letter to Stieltjes, "I turn away with fright and horror from this lamentable evil of functions which do not have derivatives."

Another kind of objection was voiced by Du Bois-Reymond.[68] His concern was that the arithmetization of analysis separated analysis from geometry and consequently from intuition and physical thinking. It reduced analysis "to a simple game of symbols where the written signs take on the arbitrary significance of the pieces in a chess or card game."

The issue that provoked the most controversy was the banishment of divergent series notably by Abel and Cauchy. In a letter to Holmboë, written in 1826, Abel says,[69]

> The divergent series are the invention of the devil, and it is a shame to base on them any demonstration whatsoever. By using them, one may draw any conclusion he pleases and that is why these series have produced so many fallacies and so many paradoxes. . . . I have become prodigiously attentive to all this, for with the exception of the geometrical series, there does not exist in all of mathematics a single infinite series the

---

67. *L'Enseignement mathématique*, 1, 1899, 157–62 = *Œuvres*, 2, 129–34.
68. *Théorie générale des fonctions*, 1887, 61.
69. *Œuvres*, 2, 256.

sum of which has been determined rigorously. In other words, the things which are most important in mathematics are also those which have the least foundation.

However, Abel showed some concern about whether a good idea had been overlooked because he continues his letter thus: "That most of these things are correct in spite of that is extraordinarily surprising. I am trying to find a reason for this; it is an exceedingly interesting question." Abel died young and so never did pursue the matter.

Cauchy, too, had some qualms about ostracizing divergent series. He says in the introduction of his *Cours* (1821), "I have been forced to admit diverse propositions which appear somewhat deplorable, for example, that a divergent series cannot be summed." Despite this conclusion Cauchy continued to use divergent series as in notes appended to the publication in 1827[70] of a prize paper written in 1815 on water waves. He decided to look into the question of why divergent series proved so useful and as a matter of fact he ultimately did come close to recognizing the reason (Chap. 47).

The French mathematicians accepted Cauchy's banishment of divergent series. But the English and the Germans did not. In England the Cambridge school defended the use of divergent series by appealing to the principle of permanence of form (Chap. 32, sec. 1). In connection with divergent series the principle was first used by Robert Woodhouse (1773–1827). In *The Principles of Analytic Calculation* (1803, p. 3) he points out that in the equation

$$(6) \qquad \frac{1}{1-r} = 1 + r + r^2 + \cdots$$

the equality sign has "a more extended signification" than just numerical equality. Hence the equation holds whether the series diverges or not.

Peacock, too, applied the principle of permanence of forms to operations with divergent series.[71] On page 267 he says, "Thus since for $r < 1$, (6) above holds, then for $r = 1$ we do get $\infty = 1 + 1 + 1 + \cdots$. For $r > 1$ we get a negative number on the left and, because the terms on the right continually increase, a quantity more than ∞ on the right." This Peacock accepts. The point he tries to make is that the series can represent $1/(1 - r)$ for all r. He says,

> If the operations of algebra be considered as general, and the symbols which are subject to them as unlimited in value, it will be impossible to avoid the formation of divergent as well as convergent series; and if such series be considered as the results of operations which are definable, apart from the series themselves, then it will not be very important to

70. *Mém. des sav. étrangers*, 1, 1827, 3–312; see *Œuvres*, (1), 1, 238, 277, 286.
71. *Report on the Recent Progress and Present State of Certain Branches of Analysis*, Brit. Assn. for Adv. of Science, 3, 1833, 185–352.

enter into such an examination of the relation of the arithmetical values of the successive terms as may be necessary to ascertain their convergence or divergence; for under such circumstances, they must be considered as equivalent forms representing their generating function, and as possessing for the purposes of such operations, equivalent properties. . . . The attempt to exclude the use of divergent series in symbolical operations would necessarily impose a limit upon the universality of algebraic formulas and operations which is altogether contrary to the spirit of science. . . . It would necessarily lead to a great and embarrassing multiplication of cases: it would deprive almost all algebraical operations of much of their certainty and simplicity.

Augustus De Morgan, though much sharper and more aware than Peacock of the difficulties in divergent series, was nevertheless under the influence of the English school and also impressed by the results obtained by the use of divergent series despite the difficulties in them. In 1844 he began an acute and yet confused paper on "Divergent Series"[72] with these words, "I believe it will be generally admitted that the heading of this paper describes the only subject yet remaining, of an elementary character, on which a serious schism exists among mathematicians as to absolute correctness or incorrectness of results." The position De Morgan took he had already declared in his *Differential and Integral Calculus*:[73] "The history of algebra shows us that nothing is more unsound than the rejection of any method which naturally arises, on account of one or more apparently valid cases in which such a method leads to erroneous results. Such cases should indeed teach caution, but not rejection; if the latter had been preferred to the former, negative quantities, and still more their square roots, would have been an effectual bar to the progress of algebra . . . and those immense fields of analysis over which even the rejectors of divergent series now range without fear, would have been not so much as discovered, much less cultivated and settled. . . . The motto which I should adopt against a course which seems to me calculated to stop the progress of discovery would be contained in a word and a symbol—remember $\sqrt{-1}$." He distinguishes between the arithmetic and algebraic significance of a series. The algebraic significance holds in all cases. To account for some of the false conclusions obtained with divergent series he says in the 1844 paper (p. 187) that integration is an arithmetic and not an algebraic operation and so could not be applied without further thought to divergent series. But the derivation of

$$\frac{1}{1-r} = 1 + r + r^2 + \cdots$$

72. *Trans. Camb. Philo. Soc.*, 8, Part II, 1844, 182–203, pub. 1849.
73. London, 1842, p. 566.

by starting with $y = 1 + ry$, replacing y on the right by $1 + ry$, and continuing thus, he accepts because it is algebraic. Likewise from $z = 1 + 2z$ one gets $z = 1 + 2 + 4 + \cdots$. Hence $-1 = 1 + 2 + 4 + \cdots$ and this is right. He accepts the entire theory (as of that date) of trigonometric series but would be willing to reject it if one could give one instance where $1 - 1 + 1 - 1 + \cdots$ does not equal $1/2$ (see Chap. 20).

Many other prominent English mathematicians gave other kinds of justification for the acceptance of divergent series, some going back to an argument of Nicholas Bernoulli (Chap. 20, sec. 7) that the series (6) contains a remainder r^∞ or $r^\infty/(1 - r^\infty)$. This must be taken into account (though they did not indicate how). Others said the sum of a divergent series is algebraically true but arithmetically false.

Some German mathematicians used the same arguments as Peacock though they used different words, such as syntactical operations as opposed to arithmetical operations or literal as opposed to numerical. Martin Ohm [74] said, "An infinite series (leaving aside any question of convergence or divergence) is completely suited to represent a given expression if one can be sure of having the correct law of development of the series. Of the *value* of an infinite series one can speak only if it converges." Arguments in Germany in favor of the legitimacy of divergent series were advanced for several more decades.

The defense of the use of divergent series was not nearly so foolish as it might seem, though many of the arguments given in behalf of the series were, perhaps, far-fetched. For one thing in the whole of eighteenth-century analysis the attention to rigor or proof was minimal and this was acceptable because the results obtained were almost always correct. Hence mathematicians became accustomed to loose procedures and arguments. But, even more to the point, many of the concepts and operations which had caused perplexities, such as complex numbers, were shown to be correct after they were fully understood. Hence mathematicians thought that the difficulties with divergent series would also be cleared up when a better understanding was obtained and that divergent series would prove to be legitimate. Further the operations with divergent series were often bound up with other little understood operations of analysis such as the interchange of the order of limits, integration over discontinuities of an integrand and integration over an infinite interval, so that the defenders of divergent series could maintain that the false conclusions attributed to the use of divergent series came from other sources of trouble.

One argument which might have been brought up is that when an analytic function is expressed in some domain by a power series, what Weierstrass called an element, this series does indeed carry with it the

74. *Aufsätze aus dem Gebiet der höheren Mathematik* (Essays in the Domain of Higher Mathematics, 1823).

"algebraic" or "syntactical" properties of the function and these properties are carried beyond the domain of convergence of the element. The process of analytic continuation uses this fact. Actually there was sound mathematical substance in the concept of divergent series which accounted for their usefulness. But the recognition of this substance and the final acceptance of divergent series had to await a new theory of infinite series (Chap. 47).

Bibliography

Abel, N. H.: *Œuvres complètes*, 2 vols., 1881, Johnson Reprint Corp., 1964.

————: *Mémorial publié à l'occasion du centenaire de sa naissance*, Jacob Dybwad, 1902. Letters to and from Abel.

Bolzano, B.: *Rein analytischer Beweis des Lehrsatzes, dass zwischen je zwei Werthen, die ein entgegengesetztes Resultat gewähren, wenigstens eine reele Wurzel der Gleichung liege*, Gottlieb Hass, Prague, 1817 = *Abh. Königl. Böhm. Ges. der Wiss.*, (3), 5, 1814–17, pub. 1818 = *Ostwald's Klassiker der exakten Wissenschaften* #153, 1905, 3–43. Not contained in Bolzano's *Schriften*.

————: *Paradoxes of the Infinite*, Routledge and Kegan Paul, 1950. Contains a historical survey of Bolzano's work.

————: *Schriften*, 5 vols., Königlichen Böhmischen Gesellschaft der Wissenschaften, 1930–48.

Boyer, Carl B.: *The Concepts of the Calculus*, Dover (reprint), 1949, Chap. 7.

Burkhardt, H.: "Über den Gebrauch divergenter Reihen in der Zeit von 1750–1860," *Math. Ann.*, 70, 1911, 169–206.

————: "Trigonometrische Reihe und Integrale," *Encyk. der Math. Wiss.*, II A12, 819–1354, B. G. Teubner, 1904–16.

Cantor, Georg: *Gesammelte Abhandlungen* (1932), Georg Olms (reprint), 1962.

Cauchy, A. L.: *Œuvres*, (2), Gauthier-Villars, 1897–99, Vols. 3 and 4.

Dauben, J. W.: "The Trigonometric Background to Georg Cantor's Theory of Sets," *Archive for History of Exact Sciences*, 7, 1971, 181–216.

Dirichlet, P. G. L.: *Werke*, 2 vols. Georg Reimer, 1889–97, Chelsea (reprint), 1969.

Du Bois-Reymond, Paul: *Zwei Abhandlungen über unendliche und trigonometrische Reihen* (1871 and 1874), Ostwald's Klassiker #185; Wilhelm Engelmann, 1913.

Freudenthal, H.: "Did Cauchy Plagiarize Bolzano?," *Archive for History of Exact Sciences*, 7, 1971, 375–92.

Gibson, G. A.: "On the History of Fourier Series," *Proc. Edinburgh Math. Soc.*, 11, 1892/93, 137–66.

Grattan-Guinness, I.: "Bolzano, Cauchy and the 'New Analysis' of the Nineteenth Century," *Archive for History of Exact Sciences*, 6, 1970, 372–400.

————: *The Development of the Foundations of Mathematical Analysis from Euler to Riemann*, Massachusetts Institute of Technology Press, 1970.

Hawkins, Thomas W., Jr.: *Lebesgue's Theory of Integration: Its Origins and Development*, University of Wisconsin Press, 1970, Chaps. 1–3.

Manheim, Jerome H.: *The Genesis of Point Set Topology*, Macmillan, 1964, Chaps. 1–4.

Pesin, Ivan N.: *Classical and Modern Integration Theories*, Academic Press, 1970, Chap. 1.

Pringsheim, A.: "Irrationalzahlen und Konvergenz unendlichen Prozesse," *Encyk. der Math. Wiss.*, IA3, 47–147, B. G. Teubner, 1898–1904.

Reiff, R.: *Geschichte der unendlichen Reihen*, H. Lauppsche Buchhandlung, 1889; Martin Sändig (reprint), 1969.

Riemann, Bernhard: *Gesammelte mathematische Werke*, 2nd. ed. (1902), Dover (reprint), 1953.

Schlesinger, L.: "Über Gauss' Arbeiten zur Funktionenlehre," *Nachrichten König. Ges. der Wiss. zu Gött.*, 1912, Beiheft, 1–43. Also in Gauss's *Werke*, 10_2, 77 ff.

Schoenflies, Arthur M.: "Die Entwicklung der Lehre von den Punktmannig-faltigkeiten," *Jahres. der Deut. Math.-Verein.*, 8_2, 1899, 1–250.

Singh, A. N.: "The Theory and Construction of Non-Differentiable Functions," in E. W. Hobson: *Squaring the Circle and Other Monographs*, Chelsea (reprint), 1953.

Smith, David E.: *A Source Book in Mathematics*, Dover (reprint), 1959, Vol. 1, 286–91, Vol. 2, 635–37.

Stolz, O.: "B. Bolzanos Bedeutung in der Geschichte der Infinitesimalrechnung," *Math. Ann.*, 18, 1881, 255–79.

Weierstrass, Karl: *Mathematische Werke*, 7 vols., Mayer und Müller, 1894–1927.

Young, Grace C.: "On Infinite Derivatives," *Quart. Jour. of Math.*, 47, 1916, 127–75.

41

The Foundations of the Real and Transfinite Numbers

> God made the integers; all else is the work of man.
>
> LEOPOLD KRONECKER

1. *Introduction*

One of the most surprising facts in the history of mathematics is that the logical foundation of the real number system was not erected until the late nineteenth century. Up to that time not even the simplest properties of positive and negative rational numbers and irrational numbers were logically established, nor were these numbers defined. Even the logical foundation of complex numbers had not been long in existence (Chap. 32, sec. 1), and that foundation presupposed the real number system. In view of the extensive development of algebra and analysis, all of which utilized the real numbers, the failure to consider the precise structure and properties of the real numbers shows how illogically mathematics progresses. The intuitive understanding of these numbers seemed adequate and mathematicians were content to operate on this basis.

The rigorization of analysis forced the realization that the lack of clarity in the number system itself had to be remedied. For example, Bolzano's proof (Chap. 40, sec. 2) that a continuous function that is negative for $x = a$ and positive for $x = b$ is zero for some value of x between a and b floundered at a critical point because he lacked an adequate understanding of the structure of the real number system. The closer study of limits also showed the need to understand the real number system, for rational numbers can have an irrational limit and conversely. Cauchy's inability to prove the sufficiency of his criterion for the convergence of a sequence likewise resulted from his lack of understanding of the structure of the number system. The study of discontinuities of functions representable by Fourier series revealed the same deficiency. It was Weierstrass who first pointed out that to establish carefully the properties of continuous functions he needed the theory of the arithmetic continuum.

979

Another motivation to erect the foundations of the number system was the desire to secure the truth of mathematics. One consequence of the creation of non-Euclidean geometry was that geometry had lost its status as truth (Chap. 36, sec. 8), but it still seemed that the mathematics built on the ordinary arithmetic must be unquestionable reality in some philosophical sense. As far back as 1817, in his letter to Olbers, Gauss[1] had distinguished arithmetic from geometry in that only the former was purely a priori. In his letter to Bessel of April 9, 1830,[2] he repeats the assertion that only the laws of arithmetic are necessary and true. However, the foundation of the number system that would dispel any doubts about the truth of arithmetic and of the algebra and analysis built on that base was lacking.

It is very much worth noting that before the mathematicians appreciated that the number system itself must be analyzed, the problem that had seemed most pertinent was to build the foundations of *algebra*, and in particular to account for the fact that one could use letters to represent real and complex numbers and yet operate with letters by means of properties accepted as true for the positive integers. To Peacock, De Morgan, and Duncan Gregory, algebra in the early nineteenth century was an ingenious but also an ingenuous complex of manipulatory schemes with some rhyme but very little reason; it seemed to them that the crux of the current confusion lay in the inadequate foundation for algebra. We have already seen how these men resolved this problem (Chap. 32, sec. 1). The late nineteenth-century men realized that they must probe deeper on behalf of analysis and clarify the structure of the entire real number system. As a by-product they would also secure the logical structure of algebra, for it was already intuitively clear that the various types of numbers possessed the same formal properties. Hence if they could establish these properties on a sound foundation, they could apply them to letters that stood for any of these numbers.

2. *Algebraic and Transcendental Numbers*

A step in the direction of an improved understanding of irrational numbers was the mid-nineteenth-century work on algebraic and transcendental irrationals. The distinction between algebraic and transcendental irrationals had been made in the eighteenth century (Chap. 25, sec. 1). The interest in this distinction was heightened by the nineteenth-century work on the solution of equations, because this work revealed that not all algebraic irrationals could be obtained by algebraic operations on rational numbers. Moreover, the problem of determining whether e and π were algebraic or transcendental continued to attract mathematicians.

1. *Werke*, 8, 177.
2. *Werke*, 8, 201.

Up to 1844, the question of whether there were any transcendental irrationals was open. In that year Liouville[3] showed that any number of the form

$$\frac{a_1}{10} + \frac{a_2}{10^{2!}} + \frac{a_3}{10^{3!}} + \cdots,$$

where the a_i are arbitrary integers from 0 to 9, is transcendental.

To prove this, Liouville first proved some theorems on the approximation of algebraic irrationals by rational numbers. By definition (Chap. 25, sec. 1) an algebraic number is any number, real or complex, that satisfies an algebraic equation

$$a_0 x^n + a_1 x^{n-1} + \cdots + a_n = 0$$

where the a_i are integers. A root is said to be an algebraic number of degree n if it satisfies an equation of the nth degree but of no lower degree. Some algebraic numbers are rational; these are of degree one. Liouville proved that if p/q is any approximation to an algebraic irrational x of degree n, with p and q integral, then there exists a positive number M such that

$$\left| x - \frac{p}{q} \right| > \frac{M}{q^n}.$$

This means that any rational approximation to an algebraic irrational of degree n by any p/q must be less accurate than M/q^n. Alternatively we may say that if x is an algebraic irrational of degree n, there is a positive number M such that the inequality

$$\left| x - \frac{p}{q} \right| < \frac{M}{q^\mu}$$

has no solutions in integers p and q for $\mu = n$ and hence for $\mu \leq n$. Then x is transcendental if for a fixed M and for every positive integer μ the inequality has a solution p/q. By showing that his irrationals satisfy this last criterion, Liouville proved they were transcendental.

The next big step in the recognition of specific transcendental numbers was Hermite's proof in 1873[4] that e is transcendental. After obtaining this result, Hermite wrote to Carl Wilhelm Borchardt (1817–80), "I do not dare to attempt to show the transcendence of π. If others undertake it, no one will be happier than I about their success, but believe me, my dear friend, this cannot fail to cost them some effort."

That π is transcendental had already been suspected by Legendre (Chap. 25, sec. 1). Ferdinand Lindemann (1852–1939) proved this in

3. *Comp. Rend.*, 18, 1844, 910–11, and *Jour. de Math.*, (1), 16, 1851, 133–42.
4. *Comp. Rend.*, 77, 1873, 18–24, 74–79, 226–33, 285–93 = *Œuvres*, 2, 150–81.

1882[5] by a method that does not differ essentially from Hermite's. Linde-mann established that if $x_1, x_2, \ldots, x_n$ are distinct algebraic numbers, real or complex, and $p_1, p_2, \ldots, p_n$ are algebraic numbers and not all zero, then the sum

$$p_1 e^{x_1} + p_2 e^{x_2} + \cdots + p_n e^{x_n}$$

cannot be 0. If we take $n = 2$, $p_1 = 1$, and $x_2 = 0$, we see that e^{x_1} cannot be algebraic for an x_1 that is algebraic and nonzero. Since x_1 can be 1, e is transcendental. Now it was known that $e^{i\pi} + 1 = 0$; hence the number $i\pi$ cannot be algebraic. Then π is not, because i is, and the product of two algebraic numbers is algebraic. The proof that π is transcendental disposed of the last point in the famous construction problems of geometry, for all constructible numbers are algebraic.

One mystery about a fundamental constant remains. Euler's constant (Chap. 20, sec. 4)

$$C = \lim_{n \to \infty} \left(1 + \frac{1}{2} + \cdots + \frac{1}{n} - \log n\right),$$

which is approximately 0.577216 and which plays an important role in analysis, notably in the study of the gamma and zeta functions, is not known to be rational or irrational.

3. The Theory of Irrational Numbers

By the latter part of the nineteenth century the question of the logical structure of the real number system was faced squarely. The irrational numbers were considered to be the main difficulty. However, the development of the meaning and properties of irrational numbers presupposes the establishment of the rational number system. The various contributors to the theory of irrational numbers either assumed that the rational numbers were so assuredly known that no foundation for them was needed or gave some hastily improvised scheme.

Curiously enough, the erection of a theory of irrationals required not much more than a new point of view. Euclid in Book V of the *Elements* had treated incommensurable ratios of magnitudes and had defined the equality and inequality of such ratios. His definition of equality (Chap. 4, sec. 5) amounted to dividing the rational numbers m/n into two classes, those for which m/n is less than the incommensurable ratio a/b of the magnitudes a and b and those for which m/n is greater. It is true that Euclid's logic was deficient because he never defined an incommensurable ratio. Moreover, Euclid's development of the theory of proportion, the equality of two incom-

5. *Math. Ann.*, 20, 1882, 213–25.

mensurable ratios, was applicable only to geometry. Nevertheless, he did have the essential idea that could have been used sooner to define irrational numbers. Actually Dedekind did make use of Euclid's work and acknowledged this debt;[6] Weierstrass too may have been guided by Euclid's theory. However, hindsight is easier than foresight. The long delay in taking advantage of some reformulation of Euclid's ideas is readily accounted for. Negative numbers had to be fully accepted so that the complete rational number system would be available. Moreover, the need for a theory of irrationals had to be felt, and this happened only when the arithmetization of analysis had gotten well under way.

In two papers, read before the Royal Irish Academy in 1833 and 1835 and published as "Algebra as the Science of Pure Time," William R. Hamilton offered the first treatment of irrational numbers.[7] He based his notion of all the numbers, rationals and irrationals, on time, an unsatisfactory basis for mathematics (though regarded by many, following Kant, as a basic intuition). After presenting a theory of rational numbers he introduced the idea of partitioning the rationals into two classes (the idea will be described more fully in connection with Dedekind's work) and defined an irrational number as such a partition. He did not complete the work.

Apart from this unfinished work all pre-Weierstrassian introductions of irrationals used the notion that an irrational is the limit of an infinite sequence of rationals. But the limit, if irrational, does not exist logically until irrationals are defined. Cantor[8] points out that this logical error escaped notice for some time because the error did not lead to subsequent difficulties. Weierstrass, in lectures at Berlin beginning in 1859, recognized the need for and gave a theory of irrational numbers. A publication by H. Kossak, *Die Elemente der Arithmetik* (1872), claimed to present this theory but Weierstrass disowned it.

In 1869 Charles Méray (1835–1911), an apostle of the arithmetization of mathematics and the French counterpart of Weierstrass, gave a definition of the irrationals based on the rationals.[9] Georg Cantor also gave a theory, which he needed to clarify the ideas on point sets that he used in his 1871 work on Fourier series. This was followed one year later by the theory of (Heinrich) Eduard Heine, which appeared in the *Journal für Mathematik*,[10] and one by Dedekind, which he published in *Stetigkeit und irrationale Zahlen*.[11]

The various theories of irrational numbers are in essence very much alike; we shall therefore confine ourselves to giving some indication of the

6. *Essays*, p. 40.
7. *Trans. Royal Irish Academy*, 17, 1837, 293–422 = *Math. Papers*, 3, 3–96.
8. *Math. Ann.*, 21, 1883, p. 566.
9. *Revue des Sociétés Savants*, 4, 1869, 280–89.
10. *Jour. für Math.*, 74, 1872, 172–88.
11. Continuity and Irrational Numbers, 1872 = *Werke*, 3, 314–34.

theories of Cantor and Dedekind. Cantor [12] starts with the rational numbers. In his 1883 paper,[13] wherein he gives more details about his theory of irrational numbers, he says (p. 565) that it is not necessary to enter into the rational numbers because this had been done by Hermann Grassmann in his *Lehrbuch der Arithmetik* (1861) and J. H. T. Muller (1797–1862) in his *Lehrbuch der allgemeinen Arithmetik* (1855). Actually these presentations did not prove definitive. Cantor introduced a new class of numbers, the real numbers, which contain rational real and irrational real numbers. He builds the real numbers on the rationals by starting with any sequence of rationals that obeys the condition that for any given ε, all the members except a finite number differ from each other by less than ε, or that

$$\lim_{n \to \infty} (a_{n+m} - a_n) = 0$$

for arbitrary m. Such a sequence he calls a fundamental sequence. Each such sequence is, by definition, a real number that we can denote by b. Two such sequences (a_ν) and (b_ν) are the same real number if and only if $|a_\nu - b_\nu|$ approaches zero as ν becomes infinite.

For such sequences three possibilities present themselves. Given any arbitrary rational number, the members of the sequence for sufficiently large ν are all smaller in absolute value than the given number; or, from a given ν on, the members are all larger than some definite positive rational member ρ; or, from a given ν on, the members are all smaller than some definite negative rational number $-\rho$. In the first case $b = 0$; in the second $b > 0$; in the third $b < 0$.

If (a_ν) and (a'_ν) are two fundamental sequences, denoted by b and b', then one can show that $(a_\nu \pm a'_\nu)$ and $(a_\nu \cdot a'_\nu)$ are fundamental sequences. These define $b \pm b'$ and $b \cdot b'$. Moreover, if $b \neq 0$, then (a'_ν/a_ν) is also a fundamental sequence that defines b'/b.

The rational real numbers are included in the above definition of real numbers, because, for example, any sequence (a_ν) with a_ν equal to the rational number a for each ν defines the rational real number a.

Now one can define the equality and inequality of any two real numbers. Indeed $b = b'$, $b > b'$, or $b < b'$, according as $b - b'$ equals 0, is greater than 0 or is less than 0.

The next theorem is crucial. Cantor proves that if (b_ν) is any sequence of real numbers (rational or irrational), and if $\lim_{\nu \to \infty} (b_{\nu + \mu} - b_\nu) = 0$ for arbitrary μ, then there is a unique real number b, determined by a fundamental sequence (a_ν) of rational a_ν such that

$$\lim_{\nu \to \infty} b_\nu = b.$$

12. *Math. Ann.*, 5, 1872, 123–32 = *Ges. Abh.*, 92–102.
13. *Math. Ann.*, 21, 1883, 545–91 = *Ges. Abh.*, 165–204.

That is, the formation of fundamental sequences of real numbers does not create the need for still newer types of numbers that can serve as limits of these fundamental sequences, because the already existing real numbers suffice to provide the limits. In other words, from the standpoint of providing limits for fundamental sequences (or what amounts to the same thing, sequences which satisfy Cauchy's criterion of convergence), the real numbers are a complete system.

Dedekind's theory of irrational numbers, presented in his book of 1872 mentioned above, stems from ideas he had in 1858. At that time he had to give lectures on the calculus and realized that the real number system had no logical foundation. To prove that a monotonically increasing quantity that is bounded approaches a limit he, like the other authors, had to resort to geometrical evidence. (He says that this is still the way to do it in the first treatment of the calculus, especially if one does not wish to lose much time.) Moreover, many basic arithmetic theorems were not proven. He gives as an example the fact that $\sqrt{2} \cdot \sqrt{3} = \sqrt{6}$ had not yet been strictly demonstrated.

He then states that he presupposes the development of the rational numbers, which he discusses briefly. To approach the irrational numbers he asks first what is meant by geometrical continuity. Contemporary and earlier thinkers—for example, Bolzano—believed that continuity meant the existence of at least one other number between any two, the property now known as denseness. But the rational numbers in themselves form a dense set. Hence denseness is not continuity.

Dedekind obtained the suggestion for the definition of an irrational number by noting that in every division of the line into two classes of points such that every point in one class is to the left of each point in the second, there is *one and only one point* that produces the division. This fact makes the line continuous. For the line it is an axiom. He carried this idea over to the number system. Let us consider, Dedekind says, any division of the rational numbers into two classes such that any number in the first class is less than any number in the second. Such a division of the rational numbers he calls a cut. If the classes are denoted by A_1 and A_2, then the cut is denoted by (A_1, A_2). From some cuts, namely, those determined by a rational number, there is either a largest number in A_1 or a smallest number in A_2. Conversely every cut in the rationals in which there is a largest number in the first class or a smallest in the second is determined by a rational number.

But there are cuts that are not determined by rational numbers. If we put into the first class all negative rational numbers and all positive ones whose squares are less than 2, and put into the second class all the other rationals, then this cut is not determined by a rational number. To each such cut "we create a new irrational member α which is fully defined by this cut; we will say that the number α corresponds to this cut or that it brings about this cut." Hence there corresponds to each cut one and only one either rational or irrational number.

Dedekind's language in introducing irrational numbers leaves a little to be desired. He introduces the irrational α as corresponding to the cut and defined by the cut. But he is not too clear about where α comes from. He should say that the irrational number α is no more than the cut. In fact Heinrich Weber told Dedekind this, and in a letter of 1888 Dedekind replied that the irrational number α is not the cut itself but is something distinct, which corresponds to the cut and which brings about the cut. Likewise, while the rational numbers generate cuts, they are *not* the same as the cuts. He says we have the mental power to create such concepts.

He then defines when one cut (A_1, A_2) is less than or greater than another cut (B_1, B_2). After having defined inequality, he points out that the real numbers possess three provable properties: (1) If $\alpha > \beta$ and $\beta > \gamma$, then $\alpha > \gamma$. (2) If α and γ are two different real numbers, then there is an infinite number of different numbers which lie between α and γ. (3) If α is any real number, then the real numbers are divided into two classes A_1 and A_2, each of which contains an infinite number of members, and each member of A_1 is less than α and each member of A_2 is greater than α. The number α itself can be assigned to either class. The class of real numbers now possesses *continuity*, which he expresses thus: If the set of all real numbers is divided into two classes A_1 and A_2 such that each member of A_1 is less than all members of A_2, then there exists one and only one number α which brings about this division.

He defines next the operations with real numbers. Addition of the cuts (A_1, A_2) and (B_1, B_2) is defined thus: If c is any rational number, then we put it in the class C_1 if there is a number a_1 in A_1 and a number b_1 in B_1 such that $a_1 + b_1 \geq c$. All other rational numbers are put in the class C_2. This pair of classes C_1 and C_2 forms a cut (C_1, C_2) because every member of C_1 is less than every member of C_2. The cut (C_1, C_2) is the sum of (A_1, A_2) and (B_1, B_2). The other operations, he says, are defined analogously. He can now establish properties such as the associative and commutative properties of addition and multiplication. Though Dedekind's theory of irrational numbers, with minor modifications such as the one indicated above, is logically satisfactory, Cantor criticized it because cuts do not appear naturally in analysis.

There are other approaches to the theory of irrational numbers beyond those already mentioned or described. For example, Wallis in 1696 had identified rational numbers and periodic decimal numbers. Otto Stolz (1842–1905), in his *Vorlesungen über allgemeine Arithmetik*,[14] showed that every irrational number can be represented as a nonperiodic decimal and this fact can be used as a defining property.

14. 1886, 1, 109–19.

It is apparent from these various approaches that the logical definition of the irrational number is rather sophisticated. Logically an irrational number is not just a single symbol or a pair of symbols, such as a ratio of two integers, but an infinite collection, such as Cantor's fundamental sequence or Dedekind's cut. The irrational number, logically defined, is an intellectual monster, and we can see why the Greeks and so many later generations of mathematicians found such numbers difficult to grasp.

Advances in mathematics are not greeted with universal approbation. Hermann Hankel, himself the creator of a logical theory of rational numbers, objected to the theories of irrational numbers.[15] "Every attempt to treat the irrational numbers formally and without the concept of [geometric] magnitude must lead to the most abstruse and troublesome artificialities, which, even if they can be carried through with complete rigor, as we have every right to doubt, do not have a higher scientific value."

4. *The Theory of Rational Numbers*

The next step in the erection of foundations for the number system was the definition and deduction of the properties of the rational numbers. As already noted, one or two efforts in this direction preceded the work on irrational numbers. Most of the workers on the rational numbers assumed that the nature and properties of the ordinary integers were known and that the problem was to establish logically the negative numbers and fractions.

The first such effort was made by Martin Ohm (1792–1872), a professor in Berlin and brother of the physicist, in his *Versuch eines vollkommen consequenten Systems der Mathematik* (Study of a Complete Consistent System of Mathematics, 1822). Then Weierstrass, in lectures given during the 1860s, derived the rational numbers from the natural numbers by introducing the positive rationals as couples of natural numbers, the negative integers as another type of couple of natural numbers, and the negative rationals as couples of negative integers. This idea was utilized independently by Peano, and so we shall present it in more detail later in connection with his work. Weierstrass did not feel the need to clarify the logic of the integers. Actually his theory of rational numbers was not free of difficulties. However, in his lectures from 1859 on, he did affirm correctly that once the whole numbers were admitted there was no need for further axioms to build up the real numbers.

The key problem in building up the rational number system was the founding of the ordinary integers by some process and the establishment of the properties of the integers. Among those who worked on the theory of the

15. *Theorie der complexen Zahlensystem*, 1867, p. 46–47.

integers, a few believed that the whole numbers were so fundamental that no logical analysis of them could be made. This position was taken by Kronecker, who was motivated by philosophical considerations into which we shall enter more deeply later. Kronecker too wished to arithmetize analysis, that is, to found analysis on the integers; but he thought one could not go beyond recognition of knowledge of the integers. Of these man has a fundamental intuition. "God made the integers," he said, "all else is the work of man."

Dedekind gave a theory of the integers in his *Was sind und was sollen die Zahlen*.[16] Though published in 1888, the work dates from 1872 to 1878. He used set-theoretic ideas, which by this time Cantor had already advanced, and which were to assume great importance. Nevertheless, Dedekind's approach was so complicated that it was not accorded much attention.

The approach to the integers that best suited the axiomatizing proclivities of the late nineteenth century was to introduce them entirely by a set of axioms. Using results obtained by Dedekind in the above-mentioned work, Giuseppe Peano (1858–1932) first accomplished this in his *Arithmetices Principia Nova Methodo Exposita* (1889).[17] Since Peano's approach is very widely used we shall review it.

Peano used a great deal of symbolism because he wished to sharpen the reasoning. Thus ε means to belong to; $\supset$ means implies; N_0 means the class of natural numbers; and $a+$ denotes the next natural number after a. Peano used this symbolism in his presentation of all of mathematics, notably in his *Formulario mathematico* (5 vols., 1895–1908). He used it also in his lectures, and the students rebelled. He tried to satisfy them by passing all of them, but that did not work, and he was obliged to resign his professorship at the University of Turin.

Though Peano's work influenced the further development of symbolic logic and the later movement of Frege and Russell to build mathematics on logic, his work must be distinguished from that of Frege and Russell. Peano did *not* wish to build mathematics on logic. To him, logic was the servant of mathematics.

Peano started with the undefined concepts (cf. Chap. 42, sec. 2) of "set," "natural numbers," "successor," and "belong to." His five axioms for the natural numbers are:

(1) 1 is a natural number.
(2) 1 is not the successor of any other natural number.
(3) Each natural number a has a successor.
(4) If the successors of a and b are equal then so are a and b.

16. The Nature and Meaning of Numbers = *Werke*, 3, 335–91.
17. *Opere scelte*, 2, 20–55, and *Rivista di Matematica*, 1, 1891, 87–102, 256–57 = *Opere scelte*, 3, 80–109.

(5) If a set S of natural numbers contains 1 and if when S contains any number a it also contains the successor of a, then S contains all the natural numbers.

This last axiom is the axiom of mathematical induction.

Peano also adopted the reflexive, symmetric, and transitive axioms for equality. That is, $a = a$; if $a = b$, then $b = a$; and if $a = b$ and $b = c$, then $a = c$. He defined addition by the statements that for each pair of natural numbers a and b there is a unique sum such that

$$a + 1 = a+$$
$$a + (b+) = (a+b) +.$$

Likewise, multiplication was defined by the statement that to each pair of natural numbers a and b there is a unique product such that

$$a \cdot 1 = a$$
$$a \cdot b+ = (a \cdot b) + a.$$

He then established all the familiar properties of natural numbers.

From the natural numbers and their properties it is straightforward to define and establish the properties of the negative whole numbers and the rational numbers. One can first define the positive and negative integers as a new class of numbers, each an ordered pair of natural numbers. Thus (a, b) where a and b are natural numbers is an integer. The intuitive meaning of (a, b) is $a - b$. Hence when $a > b$, the couple represents the usual positive integer, and when $a < b$, the couple represents the usual negative integer. Suitable definitions of the operations of addition and multiplication lead to the usual properties of the positive and negative integers.

Given the integers, one introduces the rational numbers as ordered couples of integers. Thus if A and B are integers, the ordered couple (A, B) is a rational number. Intuitively (A, B) is A/B. Again suitable definitions of the operations of addition and multiplication of the couples lead to the usual properties of the rational numbers.

Thus, once the logical approach to the natural numbers was attained, the problem of building up the foundations of the real number system was completed. As we have already noted, generally the men who worked on the theory of irrationals assumed that the rational numbers were so thoroughly understood that they could be taken for granted, or made only a minor gesture toward clarifying them. After Hamilton had grounded the complex numbers on the real numbers, and after the irrationals were defined in terms of the rational numbers, the logic of this last class was finally created. The historical order was essentially the reverse of the logical order required to build up the complex number system.

5. *Other Approaches to the Real Number System*

The essence of the approaches to the logical foundations of the real number system thus far described is to obtain the integers and their properties in some manner and, with these in hand, then to derive the negative numbers, the fractions, and finally the irrational numbers. The logical base of this approach is some series of assertions concerning the natural numbers only, for example, Peano's axioms. All the other numbers are constructed. Hilbert called the above approach the genetic method (he may not have known Peano's axioms at this time but he knew other approaches to the natural numbers). He grants that the genetic method may have pedagogical or heuristic value but, he says, it is logically more secure to apply the axiomatic method to the entire real number system. Before we state his reasons, let us look at his axioms.[18]

He introduces the undefined term, number, denoted by $a, b, c, \ldots$, and then gives the following axioms:

I. *Axioms of Connection*

I_1. From the number a and the number b there arises through addition a definite number c; in symbols

$$a + b = c \quad \text{or} \quad c = a + b.$$

I_2. If a and b are given numbers, then there exists one and only one number x and also one and only one number y so that

$$a + x = b \quad \text{and} \quad y + a = b.$$

I_3. There is a definite number, denoted by 0, so that for each a

$$a + 0 = a \quad \text{and} \quad 0 + a = a.$$

I_4. From the number a and the number b there arises by another method, by multiplication, a definite number c; in symbols

$$ab = c \quad \text{or} \quad c = ab.$$

I_5. If a and b are arbitrary given numbers and a not 0, then there exists one and only one number x, and also one and only one number y such that

$$ax = b \quad \text{and} \quad ya = b.$$

I_6. There exists a definite number, denoted by 1, such that for each a we have

$$a \cdot 1 = a \quad \text{and} \quad 1 \cdot a = a.$$

18. *Jahres. der Deut. Math.-Verein.*, 8, 1899, 180–84; this article is not in Hilbert's *Gesammelte Abhandlungen*. It is in his *Grundlagen der Geometrie*, 7th ed., Appendix 6.

II. *Axioms of Calculation*

II_1. $a + (b + c) = (a + b) + c.$

II_2. $a + b = b + a.$

II_3. $a(bc) = (ab)c.$

II_4. $a(b + c) = ab + ac.$

II_5. $(a + b)c = ac + bc.$

II_6. $ab = ba.$

III. *Axioms of Order*

III_1. If a and b are any two different numbers, then one of these is always greater than the other; the latter is said to be smaller; in symbols

$$a > b \quad \text{and} \quad b < a.$$

III_2. If $a > b$ and $b > c$ then $a > c.$

III_3. If $a > b$ then it is always true that

$$a + c > b + c \quad \text{and} \quad c + a > c + b.$$

III_4. If $a > b$ and $c > 0$ then $ac > bc$ and $ca > cb.$

IV. *Axioms of Continuity*

IV_1. (Axiom of Archimedes) If $a > 0$ and $b > 0$ are two arbitrary numbers then it is always possible to add a to itself a sufficient number of times so that

$$a + a + \cdots + a > b.$$

IV_2. (Axiom of Completeness) It is not possible to adjoin to the system of numbers any collection of things so that in the combined collection the preceding axioms are satisfied; that is, briefly put, the numbers form a system of objects which cannot be enlarged with the preceding axioms continuing to hold.

Hilbert points out that these axioms are not independent; some can be deduced from the others. He then affirms that the objections against the existence of infinite sets (sec. 6) are not valid for the above conception of the real numbers. For, he says, we do not have to think about the collection of all possible laws in accordance with which the elements of a fundamental sequence (Cantor's sequences of rational numbers) can be formed. We have but to consider a closed system of axioms and conclusions that can be deduced from them by a finite number of logical steps. He does point out that it is necessary to prove the consistency of this set of axioms, but when this is done the objects defined by it, the real numbers, exist in the mathematical sense. Hilbert was not aware at this time of the difficulty of proving the consistency of axioms for real numbers.

To Hilbert's claim that his axiomatic method is superior to the genetic method, Bertrand Russell replied that the former has the advantage of theft over honest toil. It assumes at once what can be built up from a much smaller set of axioms by deductive arguments.

As in the case of almost every significant advance in mathematics, the creation of the theory of real numbers met with opposition. Du Bois-Reymond, whom we have already cited as opposing the arithmetization of analysis, said in his *Théorie générale des fonctions* of 1887, [19]

> No doubt with help from so-called axioms, from conventions, from philosophic propositions contributed *ad hoc*, from unintelligible extensions of originally clear concepts, a system of arithmetic can be constructed which resembles in every way the one obtained from the concept of magnitude, in order thus to isolate the computational mathematics, as it were, by a cordon of dogmas and defensive definitions....But in this way one can also invent other arithmetic systems. Ordinary arithmetic is just the one which corresponds to the concept of linear magnitude.

Despite attacks such as this, the completion of the work on the real numbers seemed to mathematicians to resolve all the logical problems the subject had faced. Arithmetic, algebra, and analysis were by far the most extensive part of mathematics and this part was now securely grounded.

6. *The Concept of an Infinite Set*

The rigorization of analysis had revealed the need to understand the structure of sets of real numbers. To treat this problem Cantor had already introduced (Chap. 40, sec. 6) some notions about infinite sets of points, especially the sets of the first species. Cantor decided that the study of infinite sets was so important that he undertook to study infinite sets as such. This study, he expected, would enable him to distinguish clearly the different infinite sets of discontinuities.

The central difficulty in the theory of sets is the very concept of an infinite set. Such sets had naturally come to the attention of mathematicians and philosophers from Greek times onward, and their very nature and seemingly contradictory properties had thwarted any progress in understanding them. Zeno's paradoxes are perhaps the first indication of the difficulties. Neither the infinite divisibility of the straight line nor the line as an infinite set of discrete points seemed to permit rational conclusions about motion. Aristotle considered infinite sets, such as the set of whole numbers, and denied the existence of an infinite set of objects as a fixed entity. For him, sets could be only potentially infinite (Chap. 3, sec. 10).

19. Page 62, French ed. of *Die allgemeine Funktionentheorie*, 1882.

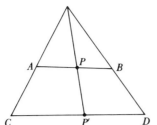

Figure 41.1

Proclus, the commentator on Euclid, noted that since a diameter of a circle divides it into halves and since there is an infinite number of diameters, there must be twice that number of halves. This seems to be a contradiction to many, Proclus says, but he resolves it by saying that one cannot speak of an actual infinity of diameters or parts of a circle. One can speak only of a larger and larger number of diameters or parts of a circle. In other words, Proclus accepted Aristotle's concept of a potential infinity but not an actual infinity. This avoids the problem of a double infinity equaling an infinity.

Throughout the Middle Ages philosophers took one side or the other on the question of whether there can be an actual infinite collection of objects. It was noted that the points of two concentric circles could be put into one-to-one correspondence with each other by associating points on a common radius. Yet one circumference was longer than the other.

Galileo struggled with infinite sets and rejected them because they were not amenable to reason. In his *Two New Sciences* (pp. 18–40 of the English translation), he notes that the points of two unequal lengths AB and CD (Fig. 41.1) can be put into one-to-one correspondence with each other and so presumably contain the same number of points. He also notes that the whole numbers can be put into one-to-one correspondence with their squares merely by assigning each number to its square. But this leads to different "amounts" of infinities, which Galileo says cannot be. All infinite quantities are the same and cannot be compared.

Gauss, in his letter to Schumacher of July 12, 1831,[20] says, "I protest against the use of an infinite quantity as an actual entity; this is never allowed in mathematics. The infinite is only a manner of speaking, in which one properly speaks of limits to which certain ratios can come as near as desired, while others are permitted to increase without bound." Cauchy, like others before him, denied the existence of infinite sets because the fact that a part can be put into one-to-one correspondence with the whole seemed contradictory to him.

The polemics on the various problems involving sets were endless and

20. *Werke*, 8, 216.

involved metaphysical and even theological arguments. The attitude of most mathematicians toward this problem was to ignore what they could not solve. On the whole they also avoided the explicit recognition of actually infinite sets, though they used infinite series, for example, and the real number system. They would speak of points of a line and yet avoid saying that the line is composed of an infinite number of points. This avoidance of troublesome problems was hypocritical, but it did suffice to build classical analysis. However, when the nineteenth century faced the problem of instituting rigor in analysis, it could no longer side-step many questions about infinite sets.

7. The Foundation of the Theory of Sets

Bolzano, in his *Paradoxes of the Infinite* (1851), which was published three years after his death, was the first to take positive steps toward a definitive theory of sets. He defended the existence of actually infinite sets and stressed the notion of equivalence of two sets, by which he meant what was later called the one-to-one correspondence between the elements of the two sets. This notion of equivalence applied to infinite sets as well as finite sets. He noted that in the case of infinite sets a part or subset could be equivalent to the whole and insisted that this must be accepted. Thus the real numbers between 0 and 5 can be put into one-to-one correspondence with the real numbers between 0 and 12 through the formula $y = 12x/5$, despite the fact that the second set of numbers contains the first set. Numbers could be assigned to infinite sets and there would be different transfinite numbers for different infinite sets, though Bolzano's assignment of transfinite numbers was incorrect according to the later theory of Cantor.

Bolzano's work on the infinite was more philosophical than mathematical and did not make sufficiently clear the notion of what was called later the power of a set or the cardinal number of a set. He, too, encountered properties that appeared paradoxical to him, and these he cites in his book. He decided that transfinite numbers were not needed to found the calculus and so did not pursue them farther.

The creator of the theory of sets is Georg Cantor (1845–1918) who was born in Russia of Danish-Jewish parentage but moved to Germany with his parents. His father urged him to study engineering and Cantor entered the University of Berlin in 1863 with that intention. There he came under the influence of Weierstrass and turned to pure mathematics. He became *Privatdozent* at Halle in 1869 and professor in 1879. When he was twenty-nine he published his first revolutionary paper on the theory of infinite sets in the *Journal für Mathematik*. Although some of its propositions were deemed faulty by the older mathematicians, its overall originality and brilliance attracted attention. He continued to publish papers on the theory of sets and on transfinite numbers until 1897.

Cantor's work which resolved age-old problems and reversed much previous thought, could hardly be expected to receive immediate acceptance. His ideas on transfinite ordinal and cardinal numbers aroused the hostility of the powerful Leopold Kronecker, who attacked Cantor's ideas savagely over more than a decade. At one time Cantor suffered a nervous breakdown, but resumed work in 1887. Even though Kronecker died in 1891, his attacks left mathematicians suspicious of Cantor's work.

Cantor's theory of sets is spread over many papers and so we shall not attempt to indicate the specific papers in which each of his notions and theorems appear. These papers are in the *Mathematische Annalen* and the *Journal für Mathematik* from 1874 on.[21] By a set Cantor meant a collection of definite and separate objects which can be entertained by the mind and to which we can decide whether or not a given object belongs. He says that those who argue for only potentially infinite sets are wrong, and he refutes the earlier arguments of mathematicians and philosophers against actually infinite sets. For Cantor a set is infinite if it can be put into one-to-one correspondence with part of itself. Some of his set-theoretic notions, such as limit point of a set, derived set, and set of the first species, were defined and used in a paper on trigonometric series;[22] these we have already described in the preceding chapter (sec. 6). A set is closed if it contains all its limit points. It is open if every point is an interior point, that is, if each point may be enclosed in an interval that contains only points of the set. A set is perfect if each point is a limit point and the set is closed. He also defined the union and intersection of sets. Though Cantor was primarily concerned with sets of points on a line or sets of real numbers, he did extend these notions of set theory to sets of points in n-dimensional Euclidean space.

He sought next to distinguish infinite sets as to "size" and, like Bolzano, decided that one-to-one correspondence should be the basic principle. Two sets that can be put into one-to-one correspondence are equivalent or have the same power. (Later the term "power" became "cardinal number.") Two sets may be unequal in power. If of two sets of objects M and N, N can be put into one-to-one correspondence with a subset of M but M cannot be put into one-to-one correspondence with a subset of N, the power of M is larger than that of N.

Sets of numbers were of course the most important, and so Cantor illustrates his notion of equivalence or power with such sets. He introduces the term "enumerable" for any set that can be put into one-to-one correspondence with the positive integers. This is the smallest infinite set. Then

21. *Ges. Abh.*, 115–356.
22. *Math. Ann.*, 5, 1872, 122–32 = *Ges. Abh.*, 92–102.

Cantor proved that the set of rational numbers is enumerable. He gave one proof in 1874.[23] However, his second proof[24] is the one now most widely used and we shall describe it.

The rational numbers are arranged thus:

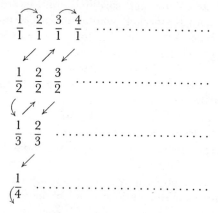

It will be noted that all those in any one diagonal have the same sum of numerator and denominator. Now one starts with $1/1$ and follows the arrows assigning the number 1 to $1/1$, 2 to $2/1$, 3 to $1/2$, 4 to $1/3$, and so on. Every rational number will be reached at some stage and to each one a finite integer will be assigned. Hence the above set of rational numbers (in which some appear many times) is in one-to-one correspondence with the integers. Then if duplicates are eliminated the set of rational numbers will still be infinite and necessarily enumerable since this is the smallest infinite set.

Still more surprising is Cantor's proof in the 1874 paper just referred to that the set of all algebraic numbers, that is, the set of all numbers that are solutions of all algebraic equations

$$a_0 x^n + a_1 x^{n-1} + \cdots + a_n = 0,$$

where the a_i are integers, is also enumerable.

To prove this, he assigns to any algebraic equation of degree n the height N defined by

$$N = n - 1 + |a_0| + |a_1| + \cdots + |a_n|,$$

where the a_i are the coefficients of the equation. The height N is an integer. To each N there corresponds only a finite number of algebraic equations and hence only a finite number of algebraic numbers, say $\phi(N)$. Thus $\phi(1) = 1$; $\phi(2) = 2$; $\phi(3) = 4$. He starts with $N = 1$ and labels the corresponding algebraic numbers from 1 to n_1; the algebraic numbers that have

23. *Jour. für Math.*, 77, 1874, 258–62 = *Ges. Abh.*, 115–18.
24. *Math. Ann.*, 46, 1895, 481–512 = *Ges. Abh.*, 283–356, pp. 294–95 in particular.

height 2 are labeled from $n_1 + 1$ to n_2; and so forth. Because each algebraic number will be reached and be assigned to one and only one integer, the set of algebraic numbers is enumerable.

In his correspondence with Dedekind in 1873, Cantor posed the question of whether the set of real numbers can be put into one-to-one correspondence with the integers, and some weeks later he answered in the negative. He gave two proofs. The first (in the 1874 article just referred to) is more complicated than the second,[25] which is the one most often used today. It also has the advantage, as Cantor pointed out, of being independent of technical considerations about irrational numbers.

Cantor's second proof that the real numbers are uncountable (non-enumerable) begins by assuming that the real numbers between 0 and 1 are countable (enumerable). Let us write each as a decimal and let us agree that a number such as $1/2$ will be written as $.4999\ldots$. If these real numbers are countable, then we can assign each one to an integer n, thus:

$$1 \leftrightarrow 0.\ a_{11}\ a_{12}\ a_{13}\ \cdots$$
$$2 \leftrightarrow 0.\ a_{21}\ a_{22}\ a_{23}\ \cdots$$
$$3 \leftrightarrow 0.\ a_{31}\ a_{32}\ a_{33}\ \cdots$$
$$\cdot\quad\cdot\quad\cdot\quad\cdot\quad\cdot\quad\cdot\quad\cdot$$

Now let us define a real number between 0 and 1 thus: Let $b = 0.b_1 b_2 b_3 \ldots$, where $b_k = 9$ if $a_{kk} = 1$ and $b_k = 1$ if $a_{kk} \neq 1$. This real number differs from any of those in the above correspondence. However, this was supposed to contain all the real numbers between 0 and 1. Hence there is a contradiction.

Since the real numbers are uncountable and the algebraic numbers are countable, there must be transcendental irrationals. This is Cantor's nonconstructive existence proof, which should be compared with Liouville's actual construction of transcendental irrationals (sec. 2).

In 1874 Cantor occupied himself with the equivalence of the points of a line and the points of R^n (n-dimensional space) and sought to prove that a one-to-one correspondence between these two sets was impossible. Three years later he proved that there is such a correspondence. He wrote to Dedekind,[26] "I see it but I do not believe it."

The idea[27] used to set up this one-to-one correspondence can be exhibited readily if we set up such a correspondence between the points of the unit square and the points of the segment $(0, 1)$. Let (x, y) be a point of the unit square and z a point of the unit interval. Let x and y be represented by infinite decimals so that in a finite decimal terminating in zero, we replace the 0 by an infinite sequence of 9's. We now break up x and y

25. *Jahres. der Deut. Math.-Verein.*, 1, 1890/91, 75–78 = *Ges. Abh.*, 278–81.
26. *Briefwechsel Cantor-Dedekind*, p. 34.
27. *Jour. für Math.*, 84, 1878, 242–58 = *Ges. Abh.*, 119–33.

into groups of decimals, each group ending with the first nonzero digit encountered. Thus

$$x = .3\ 002\ 03\ 04\ 6\ \cdots$$
$$y = .01\ 6\ 07\ 8\ 09\ \cdots.$$

Form

$$z = .3\ 01\ 002\ 6\ 03\ 07\ 04\ 8\ 6\ 09\ \cdots$$

by choosing as the groups in z the first group from x, then the first from y, and so forth. If two x's or y's differ in some digit then the corresponding z's will differ. Hence to each (x, y) there is a unique z. Given a z, one breaks up its decimal representation into the groups just described and forms the x and y by reversing the above process. Again two different z's will yield two different pairs (x, y) so that to each z there is a unique (x, y). The one-to-one correspondence just described is not continuous; roughly stated, this means that neighboring z-points do not necessarily go into neighboring (x, y)-points, nor conversely.

Du Bois-Reymond objected to this proof.[28] "It appears repugnant to common sense. The fact is that this is simply the conclusion of a type of reasoning which allows the intervention of idealistic fictions, wherein one lets them play the role of genuine quantities even though they are not even limits of representations of quantities. This is where the paradox resides."

8. *Transfinite Cardinals and Ordinals*

Having demonstrated the existence of sets with the same power and different powers, Cantor pursued this concept of the power of a set and introduced a theory of cardinal and ordinal numbers in which the transfinite cardinals and ordinals are the striking elements. Cantor developed this work in a series of papers in the *Mathematische Annalen* from 1879 to 1884, all under the title "Über unendliche lineare Punktmannichfaltigkeiten" (On Infinite Linear Aggregates of Points). Then he wrote two definitive papers in 1895 and 1897 in the same journal.[29]

In the fifth paper on linear aggregates[30] Cantor opens with the observation,

> The description of my investigations in the theory of aggregates has reached a stage where their continuation has become dependent on a generalization of real positive integers beyond the present limits; a

28. Page 167, French ed. (1887) of his *Die allgemeine Funktionentheorie* (1882).
29. *Math. Ann.*, 46, 1895, 481–512, and 49, 1897, 207–46 = *Ges. Abh.*, 282–351; an English translation of these two papers may be found in Georg Cantor, *Contributions to the Founding of a Theory of Transfinite Numbers*, Dover (reprint), no date.
30. 1883, *Ges. Abh.*, 165.

generalization which takes a direction in which, as far as I know, nobody has yet looked.

I depend on this generalization of the number concept to such an extent that without it I could not freely take even small steps forward in the theory of sets. I hope that this situation justifies or, if necessary, excuses the introduction of seemingly strange ideas into my arguments. In fact the purpose is to generalize or extend the series of real integers beyond infinity. Daring as this might appear, I express not only the hope but also the firm conviction that in due course this generalization will be acknowledged as a quite simple, appropriate, and natural step. Still I am well aware that by adopting such a procedure I am putting myself in opposition to widespread views regarding infinity in mathematics and to current opinions on the nature of number.

He points out that his theory of infinite or transfinite numbers is distinct from the concept of infinity wherein one speaks of a variable becoming infinitely small or infinitely large. Two sets which are in one-to-one correspondence have the same power or cardinal number. For finite sets the cardinal number is the usual number of objects in the set. For infinite sets new cardinal numbers are introduced. The cardinal number of the set of whole numbers he denoted by $\aleph_0$. Since the real numbers cannot be put into one-to-one correspondence with the whole numbers, the set of real numbers must have another cardinal number which is denoted by c, the first letter of continuum. Just as in the case of the concept of power, if two sets M and N are such that N can be put into one-to-one correspondence with a subset of M but M cannot be put into one-to-one correspondence with a subset of N, the cardinal number of M is greater than that of N. Then $c > \aleph_0$.

To obtain a cardinal number larger than a given one,[31] one considers any set M which the second cardinal number represents and considers the set N of all the subsets of M. Among the subsets of M are the individual elements of M, all pairs of elements of M, and so on. Now it is certainly possible to set up a one-to-one correspondence between M and a subset of N because one subset of N consists of all the individual members of N (each regarded as a set and as a member of N), and these members are the very members of M. It is not possible to set up a one-to-one correspondence between M and all the members of N. For suppose such a one-to-one correspondence were set up. Let m be any member of M and let us consider all the m's such that the subsets (members) of N that are associated with m's of M do *not* contain the m to which they correspond under this supposed one-to-one correspondence. Let η be the set of all these m's. η is of course a member of N. Cantor affirms that η is not included in the supposed one-to-one correspondence. For if η corresponded to some m of M, and η contained m,

31. Cantor, *Ges. Abh.*, 278–80.

that would be a contradiction of the very definition of η. On the other hand if η did not contain m, it should because η was by definition the set of all m's which were not contained in the corresponding subsets of N. Hence the assumption that there is a one-to-one correspondence between the elements m of M and of N, which consists of all the subsets of M, leads to a contradiction. Then the cardinal number of the set consisting of all the subsets of a given set is larger than that of the given set.

Cantor defined the sum of two cardinal numbers as the cardinal number of the set that is the union of the (disjoined) sets represented by the summands. Cantor also defined the product of any two cardinal numbers. Given two cardinal numbers α and β, one takes a set M represented by α and a set N represented by β and forms the pairs of elements (m, n) where m is any element of M and n of N. Then the cardinal number of the set of all possible pairs is the product of α and β.

Powers of cardinal numbers are also defined. If we have a set M of m objects and a set N of n objects, Cantor defines the set m^n, and this amounts to the set of permutations of m objects n at a time with repetitions of the m objects permitted. Thus if $m = 3$ and $n = 2$ and we have m_1, m_2, m_3, the permutations are

$$m_1 m_1 \qquad m_2 m_2 \qquad m_3 m_1$$

$$m_1 m_2 \qquad m_2 m_1 \qquad m_3 m_2$$

$$m_1 m_3 \qquad m_2 m_3 \qquad m_3 m_3.$$

Cantor defines the cardinal number of this number of permutations to be a^β where α is the cardinal number of M and β that of N. He then proves that $2^{\aleph_0} = c$.

Cantor calls attention to the fact that his theory of cardinal numbers applies in particular to finite cardinals, and so he has given "the most natural, shortest, and most rigorous foundation for the theory of finite numbers."

The next concept is that of ordinal number. He has already found the need for this concept in his introduction of successive derived sets of a given point set. He now introduces it abstractly. A set is simply ordered if any two elements have a definite order so that given m_1 and m_2 either m_1 precedes m_2 or m_2 precedes m_1; the notation is $m_1 < m_2$ or $m_2 < m_1$. Further if $m_1 < m_2$ and $m_2 < m_3$, then simple order also implies $m_1 < m_3$; that is, the order relationship is transitive. An ordinal number of an ordered set M is the order type of the order in the set. Two ordered sets are similar if there is a one-to-one correspondence between them and if, when m_1 corresponds to n_1 and m_2 to n_2 and $m_1 < m_2$, then $n_1 < n_2$. Two similar sets have the same type or ordinal number. As examples of ordered sets, we may use any finite set of numbers in any given order. For a finite set, no matter what the order is, the ordinal number is the same and the symbol for it can be taken to be the cardinal number of the set of numbers in the set. The ordinal number of the set of positive integers in their natural order is denoted by ω. On the other hand, the set of positive integers in decreasing order, that is

$$\ldots, 4, 3, 2, 1$$

is denoted by $*\omega$. The set of positive and negative integers and zero in the usual order has the ordinal number $*\omega + \omega$.

Then Cantor defines the addition and multiplication of ordinal numbers. The sum of two ordinal numbers is the ordinal number of the first ordered set plus the ordinal number of the second ordered set taken in that specific order. Thus the set of positive integers followed by the first five integers, that is,

$$1, 2, 3, \cdots, 1, 2, 3, 4, 5,$$

has the ordinal number $\omega + 5$. Also the equality and inequality of ordinal numbers is defined in a rather obvious way.

And now he introduces the full set of transfinite ordinals, partly for their own value and partly to define precisely higher transfinite cardinal numbers. To introduce these new ordinals he restricts the simply ordered sets to well-ordered sets.[32] A set is well-ordered if it has a first element in the ordering and if every subset has a first element. There is a hierarchy of ordinal numbers and cardinal numbers. In the first class, denoted by Z_1, are the finite ordinals

$$1, 2, 3, \cdots.$$

In the second class, denoted by Z_2, are the ordinals

$$\omega, \omega + 1, \omega + 2, \cdots, 2\omega, 2\omega + 1, \cdots, 3\omega, 3\omega + 1, \cdots, \omega^2, \omega^3, \cdots, \omega^\omega.$$

Each of these ordinals is the ordinal of a set whose cardinal number is $\aleph_0$.

The set of ordinals in Z_2 has a cardinal number. The set is not enumerable and so Cantor introduces a new cardinal number $\aleph_1$ as the cardinal number of the set Z_2. $\aleph_1$ is then shown to be the next cardinal after $\aleph_0$.

The ordinals of the third class, denoted by Z_3, are

$$\Omega, \Omega + 1, \Omega + 2, \cdots, \Omega + \Omega, \cdots.$$

These are the ordinal numbers of well-ordered sets, each of which has $\aleph_1$ elements. However, the set of ordinals Z_3 has more than $\aleph_1$ elements, and Cantor denotes the cardinal number of the set Z_3 by $\aleph_2$. This hierarchy of ordinals and cardinals can be continued indefinitely.

32. *Math. Ann.*, 21, 1883, 545–86 = *Ges. Abh.*, 165–204.

Now Cantor had shown that given any set, it is always possible to create a new set, the set of subsets of the given set, whose cardinal number is larger than that of the given set. If $\aleph_0$ is the given set, then the cardinal number of the set of subsets is $2^{\aleph_0}$. Cantor had proved that $2^{\aleph_0} = c$, where c is the cardinal number of the continuum. On the other hand he introduced $\aleph_1$ through the ordinal numbers and proved that $\aleph_1$ is the next cardinal after $\aleph_0$. Hence $\aleph_1 \leq c$, but the question as to whether $\aleph_1 = c$, known as the continuum hypothesis, Cantor, despite arduous efforts, could not answer. In a list of outstanding problems presented at the International Congress of Mathematicians in 1900, Hilbert included this (Chap. 43, sec. 5; see also Chap. 51, sec. 8).

For general sets M and N it is possible that M cannot be put into one-to-one correspondence with any subset of N and N cannot be put into one-to-one correspondence with a subset of M. In this case, though M and N have cardinal numbers α and β, say, it is not possible to say that $\beta = \alpha$, $\alpha < \beta$, or $\alpha > \beta$. That is, the two cardinal numbers are not comparable. For well-ordered sets Cantor was able to prove that this situation cannot arise. It seemed paradoxical that there should be non-well-ordered sets whose cardinal numbers cannot be compared. But this problem, too, Cantor could not solve.

Ernst Zermelo (1871–1953) took up the problem of what to do about the comparison of the cardinal numbers of sets that are not well-ordered. In 1904[33] he proved, and in 1908[34] gave a second proof, that every set can be well-ordered (in some rearrangement). To make the proof he had to use what is now known as the axiom of choice (Zermelo's axiom), which states that given any collection of nonempty, disjoined sets, it is possible to choose just one member from each set and so make up a new set. The axiom of choice, the well-ordering theorem, and the fact that any two sets may be compared as to size (that is, if their cardinal numbers are α and β, either $\alpha = \beta$, $\alpha < \beta$, or $\alpha > \beta$) are equivalent principles.

9. The Status of Set Theory by 1900

Cantor's theory of sets was a bold step in a domain that, as already noted, had been considered intermittently since Greek times. It demanded strict application of purely rational arguments, and it affirmed the existence of infinite sets of higher and higher power, which are entirely beyond the grasp of human intuition. It would be singular if these ideas, far more revolutionary than most others previously introduced, had not met with opposition. The doubts as to the soundness of this development were rein-

33. *Math. Ann.*, 59, 1904, 514–16.
34. *Math. Ann.*, 65, 1908, 107–28.

forced by questions raised by Cantor himself and by others. In letters to Dedekind of July 28 and August 28, 1899,[35] Cantor asked whether the set of all cardinal numbers is itself a set, because if it is it would have a cardinal number larger than any other cardinal. He thought he answered this in the negative by distinguishing between consistent and inconsistent sets. However, in 1897 Cesare Burali-Forti (1861–1931) pointed out that the sequence of *all* ordinal numbers, which is well-ordered, should have the greatest of all ordinal numbers as its ordinal number.[36] Then this ordinal number is greater than *all* the ordinal numbers. (Cantor had already noted this difficulty in 1895.) These and other unresolved problems, called paradoxes, were beginning to be noted by the end of the nineteenth century.

The opposition did make itself heard. Kronecker, as we have already observed, opposed Cantor's ideas almost from the start. Felix Klein was by no means in sympathy with them. Poincaré[37] remarked critically, "But it has happened that we have encountered certain paradoxes, certain apparent contradictions which would have pleased Zeno of Elea and the school of Megara. . . . I think for my part, and I am not the only one, that the important point is never to introduce objects that one cannot define completely in a finite number of words." He refers to set theory as an interesting "pathological case." He also predicted (in the same article) that "Later generations will regard [Cantor's] *Mengenlehre* as a disease from which one has recovered." Hermann Weyl spoke of Cantor's hierarchy of alephs as a fog on a fog.

However, many prominent mathematicians were impressed by the uses to which the new theory had already been put. At the first International Congress of Mathematicians in Zurich (1897), Adolf Hurwitz and Hadamard indicated important applications of the theory of transfinite numbers to analysis. Additional applications were soon made in the theory of measure (Chap. 44) and topology (Chap. 50). Hilbert spread Cantor's ideas in Germany, and in 1926[38] said, "No one shall expel us from the paradise which Cantor created for us." He praised Cantor's transfinite arithmetic as "the most astonishing product of mathematical thought, one of the most beautiful realizations of human activity in the domain of the purely intelligible."[39] Bertrand Russell described Cantor's work as "probably the greatest of which the age can boast."

35. *Ges. Abh.*, 445–48.
36. *Rendiconti del Circolo Matematico di Palermo*, 11, 1897, 154–64 and 260.
37. *Proceedings of the Fourth Internat. Cong. of Mathematicians*, Rome, 1908, 167–82; *Bull. des Sci. Math.*, (2), 32, 1908, 168–90 = extract in *Œuvres*, 5, 19–23.
38. *Math. Ann.*, 95, 1926, 170 = *Grundlagen der Geometrie*, 7th ed., 1930, 274.
39. *Math. Ann.*, 95, 1926, 167 = *Grundlagen der Geometrie*, 7th ed., 1930, 270. The article "Über das Unendliche," from which the above quotations were taken, appears also in French in *Acta Math.*, 48, 1926, 91–122. It is not included in Hilbert's *Gesammelte Abhandlungen*.

Bibliography

Becker, Oskar: *Grundlagen der Mathematik in geschichtlicher Entwicklung*, Verlag Karl Alber, 1954, pp. 217–316.

Boyer, Carl B.: *A History of Mathematics*, John Wiley and Sons, 1968, Chap. 25.

Cantor, Georg: *Gesammelte Abhandlungen*, 1932, Georg Olms (reprint), 1962.

————: *Contributions to the Founding of the Theory of Transfinite Numbers*, Dover (reprint), no date. This contains an English translation of Cantor's two key papers of 1895 and 1897 and a very helpful introduction by P. E. B. Jourdain.

Cavaillès, Jean: *Philosophie mathématique*, Hermann, 1962. Also contains the Cantor-Dedekind correspondence translated into French.

Dedekind, R.: *Essays on the Theory of Numbers*, Dover (reprint), 1963. Contains an English translation of Dedekind's "Stetigkeit und irrationale Zahlen" and "Was sind und was sollen die Zahlen." Both essays are in Dedekind's *Werke*, 3, 314–34 and 335–91.

Fraenkel, Abraham A.: "Georg Cantor," *Jahres. der Deut. Math.-Verein.*, 39, 1930, 189–266. A historical account of Cantor's work.

Helmholtz, Hermann von: *Counting and Measuring*, D. Van Nostrand, 1930. English translation of Helmholtz's *Zählen und Messen*, *Wissenschaftliche Abhandlungen*, 3, 356–91.

Manheim, Jerome H.: *The Genesis of Point Set Topology*, Macmillan, 1964, pp. 76–110.

Meschkowski, Herbert: *Ways of Thought of Great Mathematicians*, Holden-Day, 1964, pp. 91–104.

————: *Evolution of Mathematical Thought*, Holden-Day, 1965, Chaps. 4–5.

————: *Probleme des Unendlichen: Werk und Leben Georg Cantors*, F. Vieweg und Sohn, 1967.

Noether, E. and J. Cavaillès: *Briefwechsel Cantor-Dedekind*, Hermann, 1937.

Peano G.: *Opere scelte*, 3 vols., Edizioni Cremonese, 1957–59.

Schoenflies, Arthur M.: *Die Entwickelung der Mengenlehre und ihre Anwendungen*, two parts, B. G. Teubner, 1908, 1913.

Smith, David Eugene: *A Source Book in Mathematics*, Dover (reprint), 1959, Vol. 1, pp. 35–45, 99–106.

Stammler, Gerhard: *Der Zahlbegriff seit Gauss*, Georg Olms, 1965.

42
The Foundations of Geometry

> Geometry is nothing if it be not rigorous . . . . The methods of Euclid are, by almost universal consent, unexceptional in point of rigor.
>
> H. J. S. SMITH (1873)

> It has been customary when Euclid, considered as a textbook, is attacked for his verbosity or his obscurity or his pedantry, to defend him on the ground that his logical excellence is transcendent, and affords an invaluable training to the youthful powers of reasoning. This claim, however, vanishes on a close inspection. His definitions do not always define, his axioms are not always indemonstrable, his demonstrations require many axioms of which he is quite unconscious. A valid proof retains its demonstrative force when no figure is drawn, but very many of Euclid's earlier proofs fail before this test. . . . The value of his work as a masterpiece of logic has been very grossly exaggerated.
>
> BERTRAND RUSSELL (1902)

1. The Defects in Euclid

Criticism of Euclid's definitions and axioms (Chap. 4, sec. 10) dates back to the earliest known commentators, Pappus and Proclus. When the Europeans were first introduced to Euclid during the Renaissance, they too noted flaws. Jacques Peletier (1517–82), in his *In Euclidis Elementa Geometrica Demonstrationum* (1557), criticized Euclid's use of superposition to prove theorems on congruence. Even the philosopher Arthur Schopenhauer said in 1844 that he was surprised that mathematicians attacked Euclid's parallel postulate rather than the axiom that figures which coincide are equal. He argued that coincident figures are automatically identical or equal and hence no axiom is needed, or coincidence is something entirely empirical, which belongs not to pure intuition (*Anschauung*) but to external sensuous experience. Moreover, the axiom presupposes the mobility of figures; but that which is movable in space is matter, and hence outside geometry. In the nineteenth century it was generally recognized that the method of superposition either

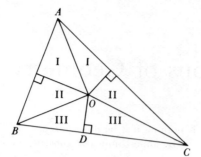

Figure 42.1

rested on unstated axioms or should be replaced by another approach to congruence.

Some critics did not like as an axiom the statement that all right angles are equal and sought to prove it, of course on the basis of the other axioms. Christophorus Clavius (1537–1612), an editor of Euclid's work, noted the absence of an axiom guaranteeing the existence of a fourth proportional to three given magnitudes (Chap. 4, sec. 5). Rightly Leibniz commented that Euclid relied upon intuition when he asserted (Book 1, Proposition 1) that two circles, each of which passes through the center of the other, have a point in common. Euclid assumed, in other words, that a circle is some kind of continuous structure and so must have a point where it is cut by another circle.

The shortcomings in Euclid's presentation of geometry were also noted by Gauss. In a letter to Wolfgang Bolyai of March 6, 1832,[1] Gauss pointed out that to speak of a part of a plane inside a triangle calls for the proper foundation. He also says, "In a complete development such words as 'between' must be founded on clear concepts, which can be done, but which I have not found anywhere." Gauss made additional criticisms of the definition of the straight line[2] and of the definition of the plane as a surface in which a line joining any two points of the plane must lie.[3]

It is well known that many "proofs" of false results can be made because Euclid's axioms do not dictate where certain points must lie in relation to others. There is, for example, the "proof" that every triangle is isoceles. One constructs the angle bisector at A of triangle ABC and the perpendicular bisector of side BC (Fig. 42.1). If these two lines are parallel, the angle bisector is perpendicular to BC and the triangle is isosceles. We suppose, then, that the lines meet at O, say, and we shall still "show" that the triangle is isosceles. We now draw the perpendiculars OF to AB and OE to AC.

1. *Werke*, 8, 222.
2. *Werke*, 8, 196.
3. *Werke*, 8, 193–95 and 200.

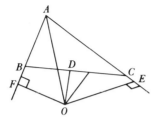

Figure 42.2

Then the triangles marked I are congruent, and $OF = OE$. The triangles marked III are also congruent, and $OB = OC$. Consequently the triangles marked II are congruent, and $FB = EC$. From the triangles marked I we have $AF = AE$. Then $AB = AC$ and the triangle is isosceles.

One might question the position of the point O, and indeed one can show that it must lie outside the triangle on the circumscribed circle. If, however, one draws Figure 42.2, it can again be "proved" that triangle ABC is isosceles. The flaw is that of the two points E and F one must lie inside and the other outside of the respective sides of the triangle. But this means that we must be able to determine the correct position of F with respect to A and B and E with respect to A and C before starting the proof. Of course one should not rely upon drawing a correct figure to determine the locations of E and F, but this was precisely what Euclid and the mathematicians up to 1800 did. Euclidean geometry was supposed to have offered accurate proofs of theorems suggested intuitively by figures, but actually it offered intuitive proofs of accurately drawn figures.

Though criticisms of the logical structure of Euclid's *Elements* were launched almost from the time it was written, they were not widely known or the defects were regarded as minor. The *Elements* was generally taken to be the model of rigor. However, the work on non-Euclidean geometry made mathematicians aware of the full extent of the deficiencies in Euclid's structure, for in carrying out proofs they had to be especially critical of what they were accepting. The recognition of the many deficiencies finally obliged the mathematicians to undertake the reconstruction of the foundations of Euclidean geometry and of other geometries that contained the same weaknesses. This activity became extensive in the last third of the nineteenth century.

2. Contributions to the Foundations of Projective Geometry

In the 1870s the work on projective geometry in relation to the metric geometries revealed that projective geometry is the fundamental one (Chap. 38). Perhaps for this reason, the foundational work began with projective geometry. However, nearly all of the writers were equally concerned to build up the metric geometries, either on the basis of projective geometry or independently. Hence the books and papers of the late nineteenth and early twentieth centuries dealing with the foundations of geometry cannot be segregated under distinct topics.

The work on non-Euclidean geometry had brought the realization that geometries are man-made constructions bearing upon physical space but not necessarily exact idealizations of it. This fact implied that several major changes had to be incorporated in any axiomatic approach to geometry. These were recognized and stressed by Moritz Pasch (1843–1930), who was the first to make major contributions to the foundations of geometry. His *Vorlesungen über neuere Geometrie* (1st ed., 1882; 2nd ed., revised by Max Dehn, 1926) is a groundbreaking work.

Pasch noted that Euclid's common notions, such as point and line, were really not defined. Defining a point as that which has no parts means little, for what is the meaning of parts? In fact, Pasch pointed out, as had Aristotle and a few later mathematicians such as Peacock and Boole, that some concepts must be undefined or else either the process of definition would be unending or mathematics would rest on physical concepts. Once some undefined concepts are selected, others can be defined in terms of these. Thus in geometry point, line, plane (Pasch also used, in his first edition, congruence of line segments) might be chosen as the undefined terms. The choice is not unique. Since there are undefined terms, the question arises as to what properties of these concepts can be used to make proofs about them. The answer given by Pasch is that the axioms make assertions about the undefined terms and these are the only assertions about them that we may use. As Gergonne put it as far back as 1818,[4] the undefined concepts are defined implicitly by the axioms.

As for the axioms, Pasch continues, though some may be suggested by experience, once the set is selected it must be possible to execute all proofs without further reference to experience or to the physical meaning of the concepts. Moreover, the axioms are by no means self-evident truths but just assumptions designed to yield the theorems of any particular geometry. He says in his *Vorlesungen* (2nd ed., p. 90),

> . . . if geometry is to become a genuine deductive science, it is essential that the way in which inferences are made should be altogether independent of the *meaning* of the geometrical concepts, and also of the diagrams; all that need be considered are the relationships between the geometrical concepts asserted by the propositions and definitions. In the course of deduction it is both advisable and useful to bear in mind the meaning of the geometrical concepts used, but this is *in no way essential*; in fact it is

4. *Ann. de Math.*, 9, 1818/19, 1.

precisely when this becomes necessary that a gap occurs in the deduction and (when it is not possible to supply the deficiency by modifying the reasoning) we are forced to admit the inadequacy of the propositions invoked as the means of proof.

Pasch did believe that the concepts and axioms should bear on experience, but logically this was irrelevant.

In his *Vorlesungen* Pasch gave axioms for projective geometry, but many of the axioms or their analogues were equally important for the axiomatization of Euclidean and the non-Euclidean geometries when built up as independent subjects. Thus he was first to give a set of axioms for the order of points on a line (or the concept of betweenness). Such axioms must also be incorporated in a complete set for any one of the metric geometries. We shall see below what the order axioms amount to.

His method of building projective geometry was to add the point, line, and plane at infinity to the proper points, lines, and planes. Then he introduced coordinates (on a geometric basis), using the throw construction of von Staudt and Klein (Chap. 35, sec. 3), and finally the algebraic representation of projective transformations. The non-Euclidean and Euclidean geometries were introduced as special cases on a geometric basis by distinguishing the proper and improper lines and points à la Felix Klein.

A more satisfactory approach to projective geometry was given by Peano.[5] This was followed by the work of Mario Pieri (1860–1904), "I Principii della geometria di posizione";[6] Federigo Enriques (1871–1946, *Lezioni di geometria proiettiva*, 1898); Eliakim Hastings Moore (1862–1932);[7] Friedrich H. Schur (1856–1932);[8] Alfred North Whitehead (1861–1947, *The Axioms of Projective Geometry*);[9] and Oswald Veblen (1880–1960) and John W. Young (1879–1932).[10] The last two men gave a completely independent set. The classic text, Veblen and Young's *Projective Geometry* (2 vols., 1910 and 1918), carries out Klein's organization of geometry by starting with projective geometry on a strict axiomatic basis and then specializing this geometry by choosing different absolute quadrics (Chap. 38, sec. 3) to obtain Euclidean and the several non-Euclidean geometries. Their axioms are general enough to include geometries with a finite number of points, geometries with only rational points, and geometries with complex points.

One more point about many of the axiomatic systems for projective

5. *Rivista di Matematica = Revue de Mathématiques*, 4, 1894, 51–90 = *Opere scelte*, 3, 115–57.

6. *Memorie della Reale Accademia delle Scienze di Torino*, (2), 48, 1899, 1–62.

7. *Amer. Math. Soc. Trans.*, 3, 1902, 142–58.

8. *Math. Ann.*, 55, 1902, 265–92.

9. Cambridge University Press, 1906.

10. *Amer. Jour. of Math.*, 30, 1908, 347–78.

geometry and those we shall look at in a moment for Euclidean geometry is worth noting. Some of Euclid's axioms are existence axioms (Chap. 4, sec. 3). To guarantee the logical existence of figures, the Greeks used construction with line and circle. The nineteenth-century foundational work revised the notion of existence, partly to supply deficiencies in Euclid's handling of this topic, and partly to broaden the notion of existence so that Euclidean geometry could include points, lines, and angles not necessarily constructible with line and circle. We shall see what the new kind of existence axioms amount to in the systems we are about to examine.

3. *The Foundations of Euclidean Geometry*

In his *I Principii di geometria* (1889), Giuseppe Peano gave a set of axioms for Euclidean geometry. He, too, stressed that the basic elements are undefined. He laid down the principle that there should be as few undefined concepts as possible and he used point, segment, and motion. The inclusion of motion seems somewhat surprising in view of the criticism of Euclid's use of superposition; however, the basic objection is not to the concept of motion but to the lack of a proper axiomatic basis if it is to be used. A similar set was given by Peano's pupil Pieri[11] who adopted point and motion as undefined concepts. Another set, using line, segment, and congruence of segments as undefined elements, was given by Giuseppe Veronese (1854–1917) in his *Fondamenti di geometria* (1891).

The system of axioms for Euclidean geometry that seems simplest in its concepts and statements, hews closest to Euclid's, and has gained most favor is due to Hilbert, who did not know the work of the Italians. He gave the first version in his *Grundlagen der Geometrie* (1899) but revised the set many times. The following account is taken from the seventh (1930) edition of this book. In his use of undefined concepts and the fact that their properties are specified solely by the axioms, Hilbert follows Pasch. No explicit meaning need be assigned to the undefined concepts. These elements, point, line, plane, and others, could be replaced, as Hilbert put it, by tables, chairs, beer mugs, or other objects. Of course, if geometry deals with "things," the axioms are certainly not self-evident truths but must be regarded as arbitrary even though, in fact, they are suggested by experience.

Hilbert first lists his undefined concepts. They are point, line, plane, lie on (a relation between point and line), lie on (a relation between point and plane), betweenness, congruence of pairs of points, and congruence of angles. The axiom system treats plane and solid Euclidean geometry in one set and the axioms are broken up into groups. The first group contains axioms on existence:

---

11. *Memorie della Reale Accademia delle Scienze di Torino*, (2), 49, 1899, 173–222.

I. *Axioms of Connection*

I_1. To each two points A and B there is one line a which lies on A and B.

I_2. To each two points A and B there is not more than one line which lies on A and B.

I_3. On each line there are at least two points. There exist at least three points which do not lie on one line.

I_4. To any three points A, B, and C which do not lie on one line there is a plane α which lies on [contains] these three points. On each plane there is [at least] one point.

I_5. To any three points A, B, and C not on one line there is not more than one plane containing these three points.

I_6. If two points of a line lie on a plane α then every point of the line lies on α.

I_7. If two planes α and β have a point A in common, then they have at least one more point B in common.

I_8. There are at least four points not lying on the same plane.

The second group of axioms supplies the most serious omission in Euclid's set, namely, axioms about the relative order of points and lines:

II. *Axioms of Betweenness*

II_1. If a point B lies between points A and C then A, B, and C are three different points on one line and B also lies between C and A.

II_2. To any two points A and C there is at least one point B on the line AC such that C lies between A and B.

II_3. Among any three points on a line there is not more than one which lies between the other two.

Axioms II_2 and II_3 amount to making the line infinite.

Definition. Let A and B be two points of a line a. The pair of points A, B or B, A is called the segment AB. The points between A and B are called points of the segment or points interior to the segment. A and B are called endpoints of the segment. All other points of line a are said to be outside the segment.

II_4. (Pasch's Axiom) Let A, B, and C be the three points not on one line and let a be any line in the plane of A, B, and C but which does not go through (lie on) A, B, or C. If a goes through a point of the segment

AB then it must also go through a point of the segment AC or one of the segment BC.

III. *Axioms of Congruence*

III$_1$. If A, B are two points of a line a and A' is a point of a or another line a', then on a given side (previously defined) of A' on the line a' one can find a point B' such that the segment AB is congruent to $A'B'$. In symbols $AB \equiv A'B'$.

III$_2$. If $A'B'$ and $A''B''$ are congruent to AB then $A'B' \equiv A''B''$.

This axiom limits Euclid's "Things equal to the same thing are equal to each other" to line segments.

III$_3$. Let AB and BC be segments without common interior points on a line a and let $A'B'$ and $B'C'$ be segments without common interior points on a line a'. If $AB \equiv A'B'$ and $BC \equiv B'C'$, then $AC \equiv A'C'$.

This amounts to Euclid's "Equals added to equals gives equals" applied to line segments.

III$_4$. Let $\sphericalangle(h, k)$ lie in a plane α and let a line a' lie in a plane α' and a definite side of a' in α' be given. Let h' be a ray of a' which emanates from the point O'. Then in α' there is one and only one ray k' such that $\sphericalangle(h, k)$ is congruent to $\sphericalangle(h', k')$ and all inner points of $\sphericalangle(h', k')$ lie on the given side of a'. Each angle is congruent to itself.

III$_5$. If for two triangles ABC and $A'B'C'$ we have $AB \equiv A'B'$, $AC \equiv A'C'$, and $\sphericalangle BAC \equiv \sphericalangle B'A'C'$ then $\sphericalangle ABC \equiv \sphericalangle A'B'C'$.

This last axiom can be used to prove that $\sphericalangle ACB = \sphericalangle A'C'B'$. One considers the same two triangles and the same hypotheses. However, by taking first $AC \equiv A'C'$ and then $AB \equiv A'B'$, we are entitled to conclude that $\sphericalangle ACB \equiv \sphericalangle A'C'B'$ because the wording of the axiom applied to the new order of the hypotheses yields this new conclusion.

IV. *The Axiom on Parallels*

Let a be a line and A a point not on a. Then in the plane of a and A there exists at most one line through A which does not meet a.

The existence of at least one line through A which does not meet a can be proved and hence is not needed in this axiom.

V. *Axioms of Continuity*

V$_1$. (Axiom of Archimedes) If AB and CD are any two segments, then there exists on the line AB a number of points $A_1, A_2, \ldots, A_n$ such that the segments $AA_1, A_1A_2, A_2A_3, \ldots, A_{n-1}A_n$ are congruent to CD and such that B lies between A and A_n.

V₂. (Axiom of Linear Completeness) The points of a line form a collection of points which, satisfying axioms I_1, I_2, II, III, and V_1, cannot be extended to a larger collection which continues to satisfy these axioms.

This axiom amounts to requiring enough points on the line so that the points can be put into one-to-one correspondence with the real numbers. Though this fact had been used consciously and unconsciously since the days of coordinate geometry, the logical basis for it had not previously been stated.

With these axioms Hilbert proved some of the basic theorems of Euclidean geometry. Others completed the task of showing that all of Euclidean geometry does follow from the axioms.

The arbitrary character of the axioms of Euclidean geometry, that is, their independence of physical reality, brought to the fore another problem, the consistency of this geometry. As long as Euclidean geometry was regarded as the truth about physical space, any doubt about its consistency seemed pointless. But the new understanding of the undefined concepts and axioms required that the consistency be established. The problem was all the more vital because the consistency of the non-Euclidean geometries had been reduced to that of Euclidean geometry (Chap. 38, sec. 4). Poincaré brought this matter up in 1898[12] and said that we could believe in the consistency of an axiomatically grounded structure if we could give it an arithmetic interpretation. Hilbert proceeded to show that Euclidean geometry is consistent by supplying such an interpretation.

He identifies (in the case of plane geometry) point with the ordered pair of real numbers[13] (a, b) and line with the ratio $(u:v:w)$ in which u and v are not both 0. A point lies on a line if

$$ua + vb + w = 0.$$

Congruence is interpreted algebraically by means of the expressions for translation and rotation of analytic geometry; that is, two figures are congruent if one can be transformed into the other by translation, reflection in the x-axis, and rotation.

After every concept has been interpreted arithmetically and it is clear that the axioms are satisfied by the interpretation, Hilbert's argument is that the theorems must also apply to the interpretation because they are logical consequences of the axioms. If there were a contradiction in Euclidean geometry, then the same would hold of the algebraic formulation of geometry,

12. *Monist*, 9, 1898, p. 38.
13. Strictly he uses a more limited set of real numbers.

which is an extension of arithmetic. Hence *if arithmetic is consistent,* so is Euclidean geometry. The consistency of arithmetic remained open at this time (see Chap. 51).

It is desirable to show that no one of the axioms can be deduced from some or all of the others of a given set, for if it can be deduced there is no need to include it as an axiom. This notion of independence was brought up and discussed by Peano in the 1894 paper already referred to, and even earlier in his *Arithmetices Principia* (1889). Hilbert considered the independence of his axioms. However, it is not possible in his system to show that each axiom is independent of all the others because the meaning of some of them depends upon preceding ones. What Hilbert did succeed in showing was that all the axioms of any one group cannot be deduced from the axioms of the other four groups. His method was to give consistent interpretations or models that satisfy the axioms of four groups but do not satisfy all the axioms of the fifth group.

The proofs of independence have a special bearing on non-Euclidean geometry. To establish the independence of the parallel axiom, Hilbert gave a model that satisfies the other four groups of axioms but does not satisfy the Euclidean parallel axiom. His model uses the points interior to a Euclidean sphere and special transformations which take the boundary of the sphere into itself. Hence the parallel axiom cannot be a consequence of the other four groups because if it were, the model as a part of Euclidean geometry would possess contradictory properties on parallelism. This same proof shows that non-Euclidean geometry is possible because if the Euclidean parallel axiom is independent of the other axioms, a denial of this axiom must also be independent; for if it were a consequence, the full set of Euclidean axioms would contain a contradiction.

Hilbert's system of axioms for Euclidean geometry, which first appeared in 1899, attracted a great deal of attention to the foundations of Euclidean geometry and many men gave alternative versions using different sets of undefined elements or variations in the axioms. Hilbert himself, as we have already noted, kept changing his system until he gave the 1930 version. Among the numerous alternative systems we shall mention just one. Veblen [14] gave a set of axioms based on the undefined concepts of point and order. He showed that each of his axioms is independent of the others, and he also established another property, namely, categoricalness. This notion was first clearly stated and used by Edward V. Huntington (1874–1952) in a paper devoted to the real number system.[15] (He called the notion sufficiency.) A set of axioms $P_1, P_2, \ldots, P_n$ connecting a set of undefined symbols $S_1, S_2, \ldots, S_m$ is said to be categorical if between the elements of any two assemblages, each of which contains undefined symbols and satisfies the

14. *Amer. Math. Soc. Trans.,* 5, 1904, 343–84.
15. *Amer. Math. Soc. Trans.,* 3, 1902, 264–79.

axioms, it is possible to set up a one-to-one correspondence of the undefined concepts which is preserved by the relationships asserted by the axioms; that is, the two systems are isomorphic. In effect categoricalness means that all interpretations of the axiom system differ only in language. This property would not hold if, for example, the parallel axiom were omitted, because then Euclidean and hyperbolic non-Euclidean geometry would be nonisomorphic interpretations of the reduced set of axioms.

Categoricalness implies another property which Veblen called disjunctive and which is now called completeness. A set of axioms is called complete if it is impossible to add another axiom that is independent of the given set and consistent with the given set (without introducing new primitive concepts). Categoricalness implies completeness, for if a set A of axioms were categorical but not complete it would be possible to introduce an axiom S such that S and not-S are consistent with the set A. Then, since the original set A is categorical, there would be isomorphic interpretations of A together with S and A together with not-S. But this would be impossible because the corresponding propositions in the two interpretations must hold, whereas S would apply to one interpretation and not-S to the other.

4. Some Related Foundational Work

The clear delineation of the axioms for Euclidean geometry suggested the corresponding investigations for the several non-Euclidean geometries. One of the nice features of Hilbert's axioms is that the axioms for hyperbolic non-Euclidean geometry are obtained at once by replacing the Euclidean parallel axiom by the Lobatchevsky-Bolyai axiom. All the other axioms in Hilbert's system remain the same.

To obtain axioms for either single or double elliptic geometry, one must not only abandon the Euclidean parallel axiom in favor of an axiom to the effect that any two lines have one point in common (single elliptic) or at least one in common (double elliptic), but one must also change other axioms. The straight line of these geometries is not infinite but has the properties of a circle. Hence one must replace the order axioms of Euclidean geometry by order axioms that describe the order relations of points on a circle. Several such axiom systems have been given. George B. Halsted (1853–1922), in his *Rational Geometry*,[16] and John R. Kline (1891–1955)[17] have given axiomatic bases for double elliptic geometry; and Gerhard Hessenberg (1874–1925)[18] gave a system of axioms for single elliptic geometry.

16. 1904, pp. 212–47.
17. *Annals of Math.*, (2), 18, 1916/17, 31–44.
18. *Math. Ann.*, 61, 1905, 173–84.

Another class of investigations in the foundations of geometry is to consider the consequences of denying or just omitting one or more of a set of axioms. Hilbert in his independence proofs had done this himself, for the essence of such a proof is to construct a model or interpretation that satisfies all of the axioms except the one whose independence is to be established. The most significant example of an axiom that was denied is, of course, the parallel axiom. Interesting results have come from dropping the axiom of Archimedes, which can be stated as in Hilbert's V_1. The resulting geometry is called non-Archimedean; in it there are segments such that the multiple of one by any whole number, however large, need not exceed another. In *Fondamenti di geometria*, Giuseppe Veronese constructed such a geometry. He also showed that the theorems of this geometry approximate as closely as one wishes those of Euclidean geometry.

Max Dehn (1878–1952) also obtained[19] many interesting theorems by omitting the Archimedean axiom. For example, there is a geometry in which the angle sum is two right angles, similar but noncongruent triangles exist, and an infinity of straight lines that are parallel to a given line may be drawn through a given point.

Hilbert pointed out that the axiom of continuity, axiom V_2, need not be used in constructing the theory of areas in the plane. But for space Max Dehn proved[20] the existence of polyhedra having the same volume though not decomposable into mutually congruent parts (even after the addition of congruent polyhedra). Hence in three dimensions the axiom of continuity is needed.

The foundation of Euclidean geometry was approached in an entirely different manner by some mathematicians. Geometry, as we know, had fallen into disfavor because mathematicians found that they had unconsciously accepted facts on an intuitive basis, and their supposed proofs were consequently incomplete. The danger that this would continually recur made them believe that the only sound basis for geometry would be arithmetic. The way to erect such a basis was clear. In fact, Hilbert had given an arithmetic interpretation of Euclidean geometry. What had to be done now, for plane geometry say, was not to interpret point as the pair of numbers (x, y) but to *define* point to *be* the pair of numbers, to define a line as a ratio of three numbers $(u:v:w)$, to define the point (x, y) as being on the line (u, v, w) if and only if $ux + vy + w = 0$, to define circle as the set of all (x, y) satisfying the equation $(x - a)^2 + (y - b)^2 = r^2$, and so on. In other words, one would use the analytic geometry equivalents of purely geometric notions as the definitions of the geometrical concepts and algebraic methods to prove theorems. Since analytic geometry contains in algebraic form the

19. *Math. Ann.*, 53, 1900, 404–39.
20. *Math. Ann.*, 55, 1902, 465–78.

complete counterpart of all that exists in Euclidean geometry, there was no question that the arithmetic foundations could be obtained. Actually the technical work involved had really been done, even for n-dimensional Euclidean geometry, for example by Grassmann in his *Calculus of Extension*; and Grassmann himself proposed that this work serve as the foundation for Euclidean geometry.

5. *Some Open Questions*

The critical investigation of geometry extended beyond the reconstruction of the foundations. Curves had of course been used freely. The simpler ones, such as the ellipse, had secure geometrical and analytical definitions. But many curves were introduced only through equations and functions. The rigorization of analysis had included not only a broadening of the concept of function but the construction of very peculiar functions, such as continuous functions without derivatives. That the unusual functions were troublesome from the geometric standpoint is readily seen. Thus the curve representing Weierstrass's example of a function which is continuous everywhere but is differentiable nowhere certainly did not fit the usual concept, because the lack of a derivative means that such a curve cannot have a tangent anywhere. The question that arose was, Are the geometrical representations of such functions curves? More generally, what is a curve?

Jordan gave a definition of a curve.[21] It is the set of points represented by the continuous functions $x = f(t)$, $y = g(t)$, for $t_0 \le t \le t_1$. For some purposes Jordan wished to restrict his curves so that they did not possess multiple points. He therefore required that $f(t) \ne f(t')$ and $g(t) \ne g(t')$ for t and t' in (t_0, t_1) or that to each (x, y) there is one t. Such a curve is now called a Jordan curve.

It was in this work that he added the notion of closed curve,[22] which requires merely that $f(t_0) = f(t_1)$ and $g(t_0) = g(t_1)$, and stated the theorem that a closed curve divides the plane into two parts, an inside and an outside. Two points of the same region can always be joined by a polygonal path that does not cut the curve. Two points not in the same region cannot be joined by any polygonal line or continuous curve that does not cut the simple closed curve. The theorem is more powerful than it seems at first sight because a simple closed curve can be quite crinkly in shape. In fact, since the functions $f(t)$ and $g(t)$ need be only continuous, the full variety of complicated continuous functions is involved. Jordan himself and many distinguished mathematicians gave incorrect proofs of the theorem. The first rigorous proof is due to Veblen.[23]

21. *Cours d'analyse*, 1st ed. Vol. 3, 1887, 593; 2nd ed., Vol. 1, 1893, p. 90.
22. First ed., p. 593; 2nd ed., p. 98.
23. *Amer. Math. Soc. Trans.*, 6, 1905, 83–98, and 14, 1913, 65–72.

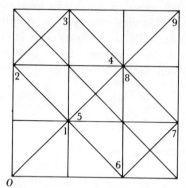

Figure 42.3   0

However, Jordan's definition of a curve, though satisfactory for many applications, was too broad. In 1890 Peano[24] discovered that a curve meeting Jordan's definition can run through all the points of a square at least once. Peano gave a detailed arithmetic description of the correspondence between the points of the interval [0, 1] and the points of the square which, in effect, specifies two functions $x = f(t)$, $y = g(t)$ that are single-valued and continuous for $0 \le t \le 1$ and such that x and y take on values belonging to each point of the unit square. However, the correspondence of (x, y) to t is not single-valued, nor is it continuous. A one-to-one continuous correspondence from the t-values to the (x, y)-values is impossible; that is, both $f(t)$ and $g(t)$ cannot be continuous. This was proved by Eugen E. Netto (1846–1919).[25]

The geometrical interpretation of Peano's curve was given by Arthur M. Schoenflies (1853–1928)[26] and E. H. Moore.[27] One maps the line segment [0, 1] into the nine segments shown in Figure 42.3, and then within each subsquare breaks up the segment contained there into the same pattern but making the transition from one subsquare to the next one a continuous one. The process is repeated *ad infinitum,* and the limiting point set covers the original square. Ernesto Cesàro (1859–1906)[28] gave the analytical form of Peano's f and g.

Hilbert[29] gave another example of the continuous mapping of a unit segment onto the square. Divide the unit segment (Fig. 42.4) and the square into four equal parts, thus:

24. *Math. Ann.,* 36, 1890, 157–60 = *Opere scelte,* 1, 110–15.
25. *Jour. für Math.,* 86, 1879, 263–68.
26. *Jahres. der Deut. Math.-Verein.,* 8₂, 1900, 121–25.
27. *Amer. Math. Soc. Trans.,* 1, 1900, 72–90.
28. *Bull. des Sci. Math.,* (2), 21, 1897, 257–66.
29. *Math. Ann.,* 38, 1891, 459–60 = *Ges. Abh.,* 3, 1–2.

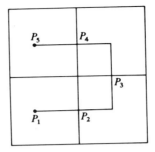

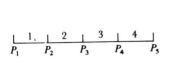

Figure 42.4

Travel through each subsquare so that the path shown corresponds to the unit segment. Now divide the unit square into 16 subsquares numbered as shown in Figure 42.5 and join the centers of the 16 subsquares as shown.

We continue the process of dividing each subsquare into four parts, numbering them so that we can traverse the entire set by a continuous path. The desired curve is the limit of the successive polygonal curves formed at each stage. Since the subsquares and the parts of the unit segment both contract to a point as the subdivision continues, we can see intuitively that each point on the unit segment maps into one point on the square. In fact, if we fix on one point in the unit segment, say $t = 2/3$, then the image of this point is the limit of the successive images of $t = 2/3$ which appear in the successive polygons.

These examples show that the definition of a curve Jordan suggested is not satisfactory because a curve, according to this definition, can fill out a square. The question of what is meant by a curve remained open. Felix Klein remarked in 1898[30] that nothing was more obscure than the notion of a curve. This question was taken up by the topologists (Chap. 50, sec. 2).

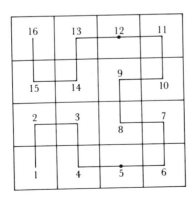

Figure 42.5

30. *Math. Ann.*, 50, 1898, 586.

Beyond the problem of what is meant by a curve, the extension of analysis to functions without derivatives also raised the question of what is meant by the length of a curve. The usual calculus formula

$$L = \int_a^b (1 + y'^2)^{1/2} \, dx,$$

where $y = f(x)$, calls, at the very least, for the existence of the derivative. Hence, the concept no longer applies to the non-differentiable functions. Various efforts to generalize the concept of length of curve were made by Du Bois-Reymond, Peano, Ludwig Scheeffer (1859–85), and Jordan, using either generalized integral definitions or geometric concepts. The most general definition was formulated in terms of the notion of measure, which we shall examine in Chapter 44.

A similar difficulty was noted for the concept of area of a surface. The concept favored in the texts of the nineteenth century was to inscribe in the surface a polyhedron with triangular faces. The limit of the sum of the areas of these triangles when the sides approach 0 was taken to be the area of the surface. Analytically, if the surface is represented by

$$x = \phi(u, v), \qquad y = \psi(u, v), \qquad z = \chi(u, v)$$

then the formula for surface area becomes

$$\int_D \sqrt{A^2 + B^2 + C^2} \, du \, dv,$$

where A, B, and C are the Jacobians of y and z, x and z, and x and y, respectively. Again the question arose of what the definition should be if x, y, and z do not possess derivatives. To complicate the situation, in a letter to Hermite, H. A. Schwarz gave an example[31] in which the choice of the triangles leads to an infinite surface area even for any ordinary cylinder.[32] The theory of surface area was also reconsidered in terms of the notion of measure.

By 1900 no one had proved that every closed plane curve, as defined by Jordan and Peano, encloses an area. Helge von Koch (1870–1924)[33] complicated the area problem by giving an example of a continuous but non-differentiable curve with infinite perimeter which bounds a finite area. One starts with the equilateral triangle ABC (Fig. 42.6) of side $3s$. On the middle third of each side construct an equilateral triangle of side s and delete the base of each triangle. There will be three such triangles. Then on each *outside* segment of length s of the new figure construct on each middle

31. *Ges. Math. Abh.*, 2, pp. 309–11.
32. This example can be found in James Pierpont, *The Theory of Functions of Real Variables*, Dover (reprint), 1959, Vol. 2, p. 26.
33. *Acta Math.*, 30, 1906, 145–76.

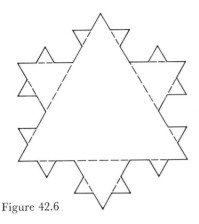

Figure 42.6

third an equilateral triangle of side $s/3$ and delete the base of each triangle. There will be twelve such triangles. Then on the outside segments of the resulting figure, construct a new equilateral triangle of side $s/9$. There will be 48 of these triangles. The perimeters of the successive figures are $9s$, $12s$, $16s$, . . . and these perimeters become infinite. However, the area of the limiting figure is finite. For, by the well-known formula for the area of an equilateral triangle in terms of its side, namely, if the side is b, the area is $(b^2/4)\sqrt{3}$, then the area of the original triangle is $[(3s)^2/4]\sqrt{3}$. The area of the first three triangles added is $3 \cdot (s^2/4)\sqrt{3}$. Since the side of the next triangles added is $s/3$ and there are 12, these areas are $12(s/3)^2\sqrt{3}/4 = (s^2/3)\sqrt{3}$. The sum of the areas is

$$S = \frac{9s^2}{4}\sqrt{3} + \frac{3}{4}s^2\sqrt{3} + \frac{s^2}{3}\sqrt{3} + \frac{4s^2}{27}\sqrt{3} + \cdots .$$

This is an infinite geometric progression (aside from the first term) with common ratio $4/9$. Then

$$S = \frac{9s^2}{4}\sqrt{3} + \frac{(3/4)s^2\sqrt{3}}{1 - 4/9} = \frac{18}{5}s^2\sqrt{3}.$$

The Peano and Hilbert curves also raised the question of what we mean by dimension. The square is two-dimensional in itself, but, as the continuous image of a curve, should be one-dimensional. Moreover, Cantor had shown that the points of a line segment can be put into one-to-one correspondence with the points of a square (Chap. 41, sec. 7). Though this correspondence is not continuous from the line segment to the square or the other way, it did show that dimension is not a matter of multiplicity of points. Nor is it a matter of the number of coordinates needed to fix the position of a point, as Riemann and Helmholtz had thought, because the Peano curve assigns a unique (x, y) to each value of t.

In the light of these difficulties we see that the rigorization of geometry certainly did not answer all the questions that were raised. Many were settled by the topologists and analysts of the next century. The very fact that questions about fundamental concepts continued to arise illustrates once more that mathematics does not grow as a logical structure. Advances into new fields and even the perfection of old ones reveal new and unsuspected defects. Beyond the resolution of the problems involving curves and surfaces we have yet to see whether the ultimate stage in rigor was reached by the foundational work in analysis, the real number system, and basic geometry.

Bibliography

Becker, Oskar: *Grundlagen der Mathematik in geschichtlicher Entwicklung*, Karl Alber, 1954, 199–212.

Enriques, Federigo: "Prinzipien der Geometrie," *Encyk. der Math. Wiss.*, B. G. Teubner, 1907–10, III AB1, 1–129.

Hilbert, David: *Grundlagen der Geometrie*, 7th ed., B. G. Teubner, 1930.

Pasch, M. and M. Dehn: *Vorlesungen über neuere Geometrie*, 2nd ed., Julius Springer, 1926, 185–271.

Peano, Giuseppe: *Opere scelte*, 3 vols., Edizioni Cremonese, 1957–59.

Reichardt, Hans: *C. F. Gauss, Leben und Werke*, Haude und Spenersche, 1960, 111–50.

Schmidt, Arnold: "Zu Hilberts Grundlegung der Geometrie," in Hilbert's *Gesammelte Abhandlungen*, 2, 404–14.

Singh, A. N.: "The Theory and Construction of Non-Differentiable Functions," in E. W. Hobson: *Squaring the Circle and Other Monographs*, Chelsea (reprint), 1953.

43
Mathematics as of 1900

I have not hesitated in 1900, at the Congress of Mathematicians in Paris, to call the nineteenth century the century of the theory of functions.   VITO VOLTERRA

1. *The Chief Features of the Nineteenth-Century Developments*

It was true in the nineteenth as in the two preceding centuries that the progress in mathematics brought with it larger changes barely perceptible in the year-to-year developments but vital in themselves and in their effect on future developments. The vast expansion in subject matter and the opening of new fields as well as the extension of older ones are of course apparent. Algebra received a totally new impetus with Galois; geometry again became active and was radically altered with the introduction of non-Euclidean geometry and the revival of projective geometry; number theory flowed over into analytic number theory; and analysis was immeasurably broadened by the introduction of complex function theory and the expansion in ordinary and partial differential equations. From the standpoint of technical development, complex function theory was the most significant of the new creations. But from the standpoint of intellectual importance and ultimate effect on the nature of mathematics, the most consequential development was non-Euclidean geometry. As we shall see, its effects were far more revolutionary than we have thus far pointed out. The circle within which mathematical studies appeared to be enclosed at the beginning of the century was broken at all points, and mathematics exploded into a hundred branches. The flood of new results contradicted sharply the leading opinion at the end of the eighteenth century that the mine of mathematics was exhausted.

Mathematical activity during the nineteenth century expanded in other respects. The number of mathematicians increased enormously as a consequence of the democratization of learning. Though Germany, France, and Britain were the major centers, Italy reappears in the arena, and the United States, with Benjamin Peirce, G. W. Hill, and Josiah Willard Gibbs, enters for the first time. In 1863 the United States founded the National Academy of Sciences. However, unlike the Royal Society of London, the

Academy of Sciences of Paris, and the Academy of Sciences of Berlin, the National Academy has not been a scientific meeting place at which papers have been presented and reviewed. It does publish a journal, the *Proceedings of the Academy*. More mathematical societies were organized (Chap. 26, sec. 6) for the meeting of research men, the presentation of papers, and the sponsorship of journals. By the end of the century the number of journals devoted partly or entirely to mathematical research increased to about 950. In 1897 the practice of holding an international congress every four years was begun.

Accompanying the explosion of mathematical activity was a less healthy development. The many disciplines became autonomous, each featuring its own terminology and methodology. The pursuit of any one imposed the assumption of more specialized and more difficult problems, requiring more and more ingenious ideas, rich inspirations, and less perspicuous proofs. To make progress, mathematicians had to acquire a great deal of background in theory and technical facility. Specialization became apparent in the work of Abel, Jacobi, Galois, Poncelet, and others. Though some stress was laid on interrelationships among the many branches through such notions as groups, linear transformations, and invariance, the overall effect was a segregation into numerous distinct and unrelated divisions. It did seem to Felix Klein in 1893 that the specialization and divergence of the various branches could be overcome by means of the concepts just mentioned, but the hope was vain. Cauchy and Gauss were the last men to know the subject as a whole, though Poincaré and Hilbert were almost universal men.

From the nineteenth century on one finds mathematicians who work only in small corners of mathematics; quite naturally, each rates the importance of his area above all others. His publications are no longer for a large public but for particular colleagues. Most articles no longer contain any indication of their connection with the larger problems of mathematics, are hardly accessible to many mathematicians, and are certainly not palatable to a large circle.

Beyond its achievements in subject matter, the nineteenth century reintroduced rigorous proof. No matter what individual mathematicians may have thought about the soundness of their results, the fact is that from about 200 B.C. to about 1870 almost all of mathematics rested on an empirical and pragmatic basis. The concept of deductive proof from explicit axioms had been lost sight of. It is one of the astonishing revelations of the history of mathematics that this ideal of the subject was, in effect, ignored during the two thousand years in which its content expanded so extensively. Though some earlier efforts to rigorize analysis were made, notably by Lagrange (Chap. 19, sec. 7), the more characteristic note was sounded by Lacroix (Chap. 26, sec. 3). Fourier's work makes a modern analyst's hair stand on end; and as far as Poisson was concerned, the derivative and integral were just shorthand for the difference quotient and a finite sum. The movement to shore up the foundations, initiated by Bolzano and Cauchy, undoubtedly arose from the concern for the rapidly increasing mass of mathematics that rested on the loose foundations of the calculus. The movement was accelerated by Hamilton's discovery of the non-commutative quaternions, which surely challenged the uncritically accepted principles of number. But even more disturbing was the creation of non-Euclidean geometry. Not only did this destroy the very notion of the self-evidency of axioms and their too-superficial acceptance, but the work revealed inadequacies in proofs that had been regarded as the soundest in all of mathematics. The mathematicians realized that they had been gullible and had relied upon intuition.

By 1900 the goal of establishing mathematics rigorously seemed to have been achieved, and the mathematicians were almost smug about this accomplishment. In his address before the Second International Congress in Paris,[1] Poincaré boasted, "Have we at last attained absolute rigor? At each stage of its evolution our forerunners believed they too had attained it. If they were deceived are we not deceived like them? . . . Now in analysis today, if we care to take pains to be rigorous, there are only syllogisms or appeals to the intuition of pure number that can not possibly deceive us. One may say today that absolute rigor has been attained." When one considers the key results in the foundations of the number system and geometry and the erection of analysis on the basis of the number system, one can see reason for gloating. Mathematics now had the foundation that almost all men were happy to accept.

The precise formulation of basic concepts such as the irrational number, continuity, integral, and derivative was not greeted enthusiastically by all mathematicians. Many did not understand the new ε–δ language and regarded the precise definitions as fads, unnecessary for the comprehension of mathematics or even for rigorous proof. These men felt that intuition was good enough, despite the surprises of continuous functions without derivatives, space-filling curves, and curves without length. Emile Picard said, apropos of the rigor in partial differential equations, ". . . true rigor is productive, being distinguished in this from another rigor which is purely formal and tiresome, casting a shadow over the problems it touches."[2]

Despite the fact that geometry too had been rigorized, one consequence of the rigorization movement was that number and analysis took precedence over geometry. The recognition by mathematicians, during and after the creation of non-Euclidean geometry, that they had unconsciously relied upon intuition in accepting the proofs of Euclidean geometry made them fearful that they would continue to do so in all geometric reasoning. Hence they preferred a mathematics built upon number. Many favored going further

1. *Comp. Rendu du Deuxième Congrès Internat. des Math.*, 1900, pub. 1902, pp. 121–22.
2. *Amer. Math. Soc. Bull.*, 11, 1904/05, 417; see also Chap. 40, sec. 7, and Chap. 41, secs. 5 and 9.

and building up all of geometry on number, which could be done through analytic geometry in the manner already described. Thus most mathematicians spoke of the arithmetization of mathematics, though it would have been more accurate to speak of the arithmetization of analysis. Where Plato could say that "God eternally geometrizes," Jacobi, even by the middle of the century, said, "God ever arithmetizes." At the Second International Congress Poincaré asserted, "Today there remain in analysis only integers and finite and infinite systems of integers, interrelated by a net of relations of equality or inequality. Mathematics, as we say, has been arithmetized." Pascal had said, "*Tout ce qui passe la Géométrie nous passe.*"[3] In 1900 the mathematicians preferred to say, "*Tout ce qui passe l'Arithmétique nous passe.*"

The erection of the logical foundations of mathematics, quite apart from whether arithmetic or geometry was the preferred basis, completed another step in the break from metaphysics. The vagueness of the foundations and justifications of mathematical steps had been evaded in the eighteenth and early nineteenth centuries by allusions to metaphysical arguments which, though never made explicit, were mentioned as grounds for accepting the mathematics. The axiomatization of the real numbers and geometry gave mathematics a clear-cut, independent, and self-sufficient basis. The recourse to metaphysics was no longer needed. As Lord Kelvin put it, "Mathematics is the only good metaphysics."

The rigorization of mathematics may have filled a nineteenth-century need, but it also teaches us something about the development of the subject. The newly founded logical structure presumably guaranteed the soundness of mathematics; but the guarantee was somewhat of a sham. Not a theorem of arithmetic, algebra, or Euclidean geometry was changed as a consequence, and the theorems of analysis had only to be more carefully formulated. In fact, all that the new axiomatic structures and rigor did was substantiate what mathematicians knew had to be the case. Indeed the axioms had to yield the existing theorems rather than determine them. All of which means that mathematics rests not on logic but on sound intuitions. Rigor, as Jacques Hadamard pointed out, merely sanctions the conquests of the intuition; or, as Hermann Weyl stated, logic is the hygiene the mathematician practices to keep his ideas healthy and strong.

2. *The Axiomatic Movement*

The rigorization of mathematics was achieved by axiomatizing the various branches. The essence of an axiomatic development, in accordance with the pattern we have examined in Chapters 41 and 42, is to start with undefined terms whose properties are specified by the axioms; the goal of the work is to

3. "All that transcends geometry transcends our comprehension."

derive the consequences of the axioms. In addition the independence, consistency, and categoricalness of the axioms (notions we have already examined in the two preceding chapters) are to be established for each system.

By the early part of the twentieth century the axiomatic method not only permitted the establishment of the logical foundations of many old and newer branches of mathematics, but also revealed precisely what assumptions underlie each branch and made possible the comparison and clarification of the relationships of various branches. Hilbert was enthusiastic about the values of this method. In discussing the perfect state mathematics had presumably attained by founding each of its branches on sound axiomatic bases, Hilbert remarked,[4]

> Indeed the axiomatic method is and remains the one suitable and indispensable aid to the spirit of every exact investigation no matter in what domain; it is logically unassailable and at the same time fruitful; it guarantees thereby complete freedom of investigation. To proceed axiomatically means in this sense nothing else than to think with knowledge of what one is about. While earlier without the axiomatic method one proceeded naively in that one believed in certain relationships as dogma, the axiomatic approach removes this naiveté and yet permits the advantages of belief.

Again, in the last part of his "Axiomatisches Denken,"[5] he praised the method:

> Everything that can be the object of mathematical thinking, as soon as the erection of a theory is ripe, falls into the axiomatic method and thereby directly into mathematics. By pressing to ever deeper layers of axioms . . . we can obtain deeper insights into scientific thinking and learn the unity of our knowledge. Especially by virtue of the axiomatic method mathematics appears called upon to play a leading role in all knowledge.

The opportunity to explore new problems by omitting, negating, or varying in some other manner the axioms of established systems enticed many mathematicians. This activity and the erection of axiomatic bases for the various branches of mathematics are known as the axiomatic movement. It continues to be a favorite activity. In part its great attraction is explained by the fact that after sound axiomatic bases for the major branches have been erected, variations of the kind just described are relatively easy to introduce and explore. However, any new development in mathematics has always attracted a number of men who seek fields wide-open to exploration or are sincerely convinced that the future of mathematics lies in that particular area.

4. *Abh. Math. Seminar der Hamburger Univ.*, 1, 1922, 157–77 = *Ges. Abh.*, 3, 157–77.
5. *Math. Ann.*, 78, 1918, 405–15 = *Ges. Abh.*, 3, 145–56.

3. *Mathematics as Man's Creation*

From the standpoint of the future development of mathematics, the most significant happening of the century was the acquisition of the correct view of the relationship of mathematics to nature. Though we have not treated the views on mathematics of many of the men whose work we have described, we have said that the Greeks, Descartes, Newton, Euler, and many others believed mathematics to be the accurate description of real phenomena and that they regarded their work as the uncovering of the mathematical design of the universe. Mathematics did deal with abstractions, but these were no more than the ideal forms of physical objects or happenings. Even such concepts as functions and derivatives were demanded by real phenomena and served to describe them.

Beyond what we have already reported that supports this view of mathematics, the position of mathematicians on the number of dimensions that can be considered in geometry shows clearly how closely mathematics had been tied to reality. Thus, in the first book of the *Heaven*, Aristotle says, "The line has magnitude in one way, the plane in two ways, and the solid in three ways, and beyond these there is no other magnitude because the three are all. . . . There is no transfer into another kind, like the transfer from length to area and from area to solid." In another passage he says, ". . . no magnitude can transcend three because there are no more than three dimensions," and adds, "for three is the perfect number." In his *Algebra*, John Wallis regarded a higher-dimensional space as a "monster in nature, less possible than a chimera or a centaure." He says, "Length, Breadth and Thickness take up the whole of space. Nor can Fansie imagine how there should be a Fourth Local Dimension beyond these Three." Cardan, Descartes, Pascal, and Leibniz had also considered the possibility of a fourth dimension and rejected it as absurd. As long as algebra was tied to geometry, the product of more than three quantities was also rejected. Jacques Ozanam pointed out that a product of more than three letters will be a magnitude of "as many dimensions as there are letters, but it would only be imaginary because in nature we do not know of any quantity which has more than three dimensions."

The idea of a mathematical geometry of more than three dimensions was rejected even in the early nineteenth century. Möbius, in his *Der barycentrische Calcul* (1827), pointed out that geometrical figures that could not be superposed in three dimensions because they are mirror images of each other could be superposed in four dimensions. But then he says,[6] "Since, however, such a space cannot be thought about, the superposition is impossible." Kummer in the 1860s mocked the idea of a four-dimensional geometry. The objections that all of these men made to a higher-dimensional

6. *Ges. Werke*, 1, 172.

geometry were sound as long as geometry was identified with the study of physical space.

But gradually and unwittingly mathematicians began to introduce concepts that had little or no direct physical meaning. Of these, negative and complex numbers were the most troublesome. It was because these two types of numbers had no "reality" in nature that they were still suspect at the beginning of the nineteenth century, even though freely utilized by then. The geometrical representation of negative numbers as distances in one direction on a line and of complex numbers as points or vectors in the complex plane, which, as Gauss remarked of the latter, gave them intuitive meaning and so made them admissible, may have delayed the realization that mathematics deals with man-made concepts. But then the introduction of quaternions, non-Euclidean geometry, complex elements in geometry, n-dimensional geometry, bizarre functions, and transfinite numbers forced the recognition of the artificiality of mathematics.

In this connection the impact of non-Euclidean geometry has already been noted (Chap. 36, sec. 8) and the impact of n-dimensional geometry must now be observed. The concept appears innocuously in the analytical work of d'Alembert, Euler, and Lagrange. D'Alembert suggested thinking of time as a fourth dimension in his article "Dimension" in the *Encyclopédie*. Lagrange, in studying the reduction of quadratic forms to standard forms, casually introduces forms in n variables. He, too, used time as a fourth dimension in his *Mécanique analytique* (1788) and in his *Théorie des fonctions analytiques* (1797). He says in the latter work, "Thus we may regard mechanics as a geometry of four dimensions and analytic mechanics as an extension of analytic geometry." Lagrange's work put the three spatial coordinates and the fourth one representing time on the same footing. Further, George Green in his paper of 1828 on potential theory did not hesitate to consider potential problems in n dimensions; he says of the theory, "It is no longer confined, as it was, to the three dimensions of space."

These early involvements in n dimensions were not intended as a study of geometry proper. They were natural generalizations of analytical work that was no longer tied to geometry. In part, this introduction of n-dimensional language was intended only as a convenience and aid to analytical thinking. It was helpful to think of $(x_1, x_2, \ldots, x_n)$ as a point and of an equation in n variables as a hypersurface in n-dimensional space, for by thinking in terms of what these mean in three-dimensional geometry one might secure some insight into the analytical work. In fact Cauchy[7] actually emphasized that the concept of n-dimensional space is useful in many analytical investigations, especially those of number theory.

However, the serious study of n-dimensional geometry, though not

7. *Comp. Rend.*, 24, 1847, 885–87 = *Œuvres*, (1), 10, 292–95.

implying a physical space of n dimensions, was also undertaken in the nineteenth century; the founder of this abstract geometry is Grassmann, in his *Ausdehnungslehre* of 1844. There one finds the concept of n-dimensional geometry in full generality. Grassmann said in a note published in 1845,

> My Calculus of Extension builds the abstract foundation of the theory of space; that is, it is free of all spatial intuition and is a pure mathematical science; only the special application to [physical] space constitutes geometry.
>
> However the theorems of the Calculus of Extension are not merely translations of geometrical results into an abstract language; they have a much more general significance, for while the ordinary geometry remains bound to the three dimensions of [physical] space, the abstract science is free of this limitation.

Grassmann adds that geometry in the usual sense is improperly regarded as a branch of pure mathematics, but that it is really a branch of applied mathematics since it deals with a subject not created by the intellect but given to it. It deals with matter. But, he says, it should be possible to create a purely intellectual subject that would deal with extension as a concept, rather than with the space perceived by the sensations. Thus Grassmann's work is representative of the development asserting that pure thought can build arbitrary constructions that may or may not be physically applicable.

Cayley, independently of Grassmann, also undertook to treat n-dimensional geometry analytically, and, as he says, "without recourse to any metaphysical notions." In the *Cambridge Mathematical Journal* of 1845,[8] Cayley published "Chapters in the Analytical Geometry of N-Dimensions." This work gives analytical results in n variables, which for $n = 3$ state known theorems about surfaces. Though he did nothing especially novel in n-dimensional geometry, the concept is fully grasped there.

By the time that Riemann gave his *Habilitationsvortrag* of 1854, "Über die Hypothesen welche die Geometrie zu Grunde liegen," he had no hesitation in dealing with n-dimensional manifolds, though he was primarily concerned with the geometry of three-dimensional physical space. Those who followed up on this basic paper—Helmholtz, Lie, Christoffel, Beltrami, Lipschitz, Darboux, and others—continued to work in n-dimensional space.

The notion of n-dimensional geometry encountered stiff-necked resistance from some mathematicians even long after it was introduced. Here, as in the case of negative and complex numbers, mathematics was progressing beyond concepts suggested by experience, and mathematicians had yet to grasp that their subject could consider concepts created by the mind and was no longer, if it ever had been, a reading of nature.

8, 4. 119–27 = *Collected Math. Papers*, 1, 55–62,

However after about 1850, the view that mathematics can introduce and deal with rather arbitrary concepts and theories that do not have immediate physical interpretation but may nevertheless be useful, as in the case of quaternions, or satisfy a desire for generality, as in the case of n-dimensional geometry, gained acceptance. Hankel in his *Theorie der complexen Zahlensysteme* (1867, p. 10) defended mathematics as "purely intellectual, a pure theory of forms, which has for its object, not the combination of quantities or of their images, the numbers, but things of thought to which there could correspond effective objects or relations even though such a correspondence is not necessary."

In defense of his creation of transfinite numbers as existing, real definite quantities, Cantor claimed that mathematics is distinguished from other fields by its freedom to create its own concepts without regard to transient reality. He said[9] in 1883, "Mathematics is entirely free in its development and its concepts are restricted only by the necessity of being noncontradictory and coordinated to concepts previously introduced by precise definitions. . . . The essence of mathematics lies in its freedom." He preferred the term "free mathematics" over the usual form, "pure mathematics."

The new view of mathematics extended to the older, physically grounded branches. In his *Universal Algebra* (1898), Alfred North Whitehead says (p. 11),

> . . . Algebra does not depend on Arithmetic for the validity of its laws of transformation. If there were such a dependence it is obvious that as soon as algebraic expressions are arithmetically unintelligible all laws respecting them must lose their validity. But the laws of Algebra, though suggested by Arithmetic, do not depend on it. They depend entirely on the conventions by which it is stated that certain modes of grouping the symbols are to be considered as identical. This assigns certain properties to the marks which form the symbols of Algebra.

Algebra is a logical development independent of meaning. "It is obvious that we can take any marks we like and manipulate them according to any rule we choose to assign" (p. 4). Whitehead does point out that such arbitrary manipulations of symbols can be frivolous and only constructions to which some meaning can be attached or which have some use are significant.

Geometry, too, cut its bonds to physical reality. As Hilbert pointed out in his *Grundlagen* of 1899, geometry speaks of things whose properties are specified in the axioms. Though Hilbert referred only to the strategy by which we must approach mathematics for the purpose of examining its logical structure, he nevertheless supported and encouraged the view that mathematics is quite distinct from the concepts and laws of nature.

9. *Math. Ann.*, 21, 1883, 563–64 = *Ges. Abh.*, 182.

4. *The Loss of Truth*

The introduction and gradual acceptance of concepts that have no immediate counterparts in the real world certainly forced the recognition that mathematics is a human, somewhat arbitrary creation, rather than an idealization of the realities in nature, derived solely from nature. But accompanying this recognition and indeed propelling its acceptance was a more profound discovery—mathematics is not a body of truths about nature. The development that raised the issue of truth was non-Euclidean geometry, though its impact was delayed by the characteristic conservatism and closed-mindedness of all but a few mathematicians. The philosopher David Hume (1711–76) had already pointed out that nature did not conform to fixed patterns and necessary laws; but the dominant view, expressed by Kant, was that the properties of physical space were Euclidean. Even Legendre in his *Eléments de géométrie* of 1794 still believed that the axioms of Euclid were self-evident truths.

With respect to geometry, at least, the view that seems correct today was first expressed by Gauss. Early in the nineteenth century he was convinced that geometry was an empirical science and must be ranked with mechanics, whereas arithmetic and analysis were a priori truths. Gauss wrote to Bessel in 1830,[10]

> According to my deepest conviction the theory of space has an entirely different place in our a priori knowledge than that occupied by pure arithmetic. There is lacking throughout our knowledge of the former the complete conviction of necessity (also of absolute truth) which is characteristic of the latter; we must add in humility, that if number is merely the product of our mind, space has a reality outside our mind whose laws we cannot a priori completely prescribe.

However Gauss seems to have had conflicting views, because he also expressed the opinion that all of mathematics is man-made. In a letter to Bessel of November 21, 1811, in which he speaks of functions of a complex variable he says,[11] "One should never forget that the functions, like all mathematical constructions, are only our own creations, and that when the definition with which one begins ceases to make sense, one should not ask: What is it, but what is it convenient to assume in order that it remain significant?"

Despite Gauss's views on geometry, most mathematicians thought there were basic truths in it. Bolyai thought that the absolute truths in geometry were those axioms and theorems common to Euclidean and hyperbolic geometry. He did not know elliptic geometry, and so in his time still did not

10. *Werke*, 8, 201.
11. *Werke*, 10, 363.

appreciate that many of these common axioms were not common to all geometries.

In his 1854 paper "On the Hypotheses Which Underlie Geometry," Riemann still believed that there were some propositions about space that were a priori, though these did not include the assertion that physical space is truly Euclidean. It was, however, locally Euclidean.

Cayley and Klein remained attached to the reality of Euclidean geometry (see also Chap. 38, sec. 6). In his presidential address to the British Association for the Advancement of Science,[12] Cayley said, ". . . not that the propositions of geometry are only approximately true, but that they remain absolutely true in regard to that Euclidean space which has so long been regarded as being the physical space of our experience." Though they themselves had worked in non-Euclidean geometries, they regarded the latter as novelties that result when new distance functions are introduced in Euclidean geometry. They failed to see that non-Euclidean is as basic and as applicable as Euclidean geometry.

In the 1890s Bertrand Russell took up the question of what properties of space are necessary to and are presupposed by experience. That is, experience would be meaningless if any of these a priori properties were denied. In his *Essay on the Foundations of Geometry* (1897), he agrees that Euclidean geometry is not a priori knowledge. He concludes, rather, that projective geometry is a priori for all geometry, an understandable conclusion in view of the importance of that subject around 1900. He then adds to projective geometry as a priori the axioms common to Euclidean and all the non-Euclidean geometries. The latter facts, the homogeneity of space, finite dimensionality, and a concept of distance make measurement possible. The facts that space is three-dimensional and that our actual space *is* Euclidean he considers to be empirical.

That the metrical geometries can be derived from projective geometry by the introduction of a metric Russell regards as a technical achievement having no philosophical significance. Metrical geometry is a logically subsequent and separate branch of mathematics and is not a priori. With respect to Euclidean and the several basic non-Euclidean geometries, he departs from Cayley and Klein and regards all these geometries as being on an equal footing. Since the only metric spaces that possess the above properties are the Euclidean, hyperbolic, and single and double elliptic, Russell concludes that these are the only possible metrical geometries, and of course Euclidean is the only physically applicable one. The others are of philosophical importance in showing that there can be other geometries. With hindsight it is now possible to see that Russell had replaced the Euclidean bias by a projective bias.

---

12. *Report of the Brit. Assn. for the Adv. of Sci.*, 1883, 3–37 = *Coll. Math. Papers*, 11, 429–59.

Though mathematicians were slow to recognize the fact, clearly seen by Gauss, that there is no assurance at all to the physical truth of Euclidean geometry, they gradually came around to that view and also to the related conviction of Gauss that the truth of mathematics resides in arithmetic and therefore also in analysis. For example, Kronecker in his essay "Über den Zahlbegriff" (On the Number Concept)[13] maintained the truth of the arithmetical disciplines but denied it to geometry. Gottlob Frege, about whose work we shall say more later, also insisted on the truth of arithmetic.

However, even arithmetic and the analysis built on it soon became suspect. The creation of non-commutative algebras, notably quaternions and matrices, certainly raised the question of how one can be sure that ordinary numbers possess the privileged property of truth about the real world. The attack on the truth of arithmetic came first from Helmholtz. After he had insisted, in a famous essay,[14] that our knowledge of physical space comes only from experience and depends on the existence of rigid bodies to serve, among other purposes, as measuring rods, in his *Zählen und Messen* (Counting and Measuring, 1887) he attacked the truths of arithmetic. He regards as the main problem in arithmetic the meaning or the validity of the objective application of quantity and equality to experience. Arithmetic itself may be just a consistent account of the consequences of the arithmetical operations. It deals with symbols and can be regarded as a game. But these symbols are applied to real objects and to relations among them and give results about real workings of nature. How is this possible? Under what conditions are the numbers and operations applicable to real objects? In particular, what is the objective meaning of the equality of two objects and what character must physical addition have to be treated as arithmetical addition?

Helmholtz points out that the applicability of numbers is neither an accident nor proof of the truths of the laws of numbers. *Some* kinds of experience suggest them and to these they are applicable. To apply numbers to real objects, Helmholtz says, objects must not disappear, or merge with one another, or divide in two. One raindrop added physically to another does not produce two raindrops. Only experience can tell us whether the objects of a physical collection retain their identity so that the collection has a definite number of objects in it. Likewise, knowing when equality between physical quantities can be applied also depends on experience. Any assertion of quantitative equality must satisfy two conditions. If the objects are exchanged they must remain equal. Also, if object a equals c and object b equals c, object a must equal object b. Thus we may speak of the equality of weights and intervals of time, because for these objects equality can be determined. But two pitches may, as far as the ear is concerned, equal an

13. *Jour. für Math.*, 101, 1887, 337–55 = *Werke*, 3, 249–74.
14. *Nachrichten König. Ges. der Wiss. zu Gött.*, 15, 1868, 193–221 = *Wiss. Abh.*, 2, 618–39.

intermediate one and yet the ear might distinguish the original two. Here things equal to the same thing are not equal to each other. One cannot add the values of electrical resistance connected in parallel to obtain the total resistance, nor can one combine in any way the indices of refraction of different media.

By the end of the nineteenth century, the view that all the axioms of mathematics are arbitrary prevailed. Axioms were merely to be the basis for the deduction of consequences. Since the axioms were no longer truths about the concepts involved in them, the physical meaning of these concepts no longer mattered. This meaning could, at best, be a heuristic guide when the axioms bore some relation to reality. Thus even the concepts were severed from the physical world. By 1900 mathematics had broken away from reality; it had clearly and irretrievably lost its claim to the truth about nature, and had become the pursuit of necessary consequences of arbitrary axioms about meaningless things.

The loss of truth and the seeming arbitrariness, the subjective nature of mathematical ideas and results, deeply disturbed many men who considered this a denigration of mathematics. Some therefore adopted a mystical view that sought to grant some reality and objectivity to mathematics. These mathematicians subscribed to the idea that mathematics is a reality in itself, an independent body of truths, and that the objects of mathematics are given to us as are the objects of the real world. Mathematicians merely discover the concepts and their properties. Hermite, in a letter to Stieltjes,[15] said, "I believe that the numbers and functions of analysis are not the arbitrary product of our minds; I believe that they exist outside of us with the same character of necessity as the objects of objective reality; and we find or discover them and study them as do the physicists, chemists and zoologists."

Hilbert said at the International Congress in Bologna in 1928,[16] "How would it be above all with the truth of our knowledge and with the existence and progress of science if there were no truth in mathematics? Indeed in professional writings and public lectures there often appears today a skepticism and despondency about knowledge; this is a certain kind of occultism which I regard as damaging."

Godfrey H. Hardy (1877–1947), an outstanding analyst of the twentieth century, said in 1928,[17] "Mathematical theorems are true or false; their truth or falsity is absolutely independent of our knowledge of them. In *some* sense, mathematical truth is part of objective reality." He expressed the same view in his book *A Mathematician's Apology* (1967 ed., p. 123): "I believe that mathematical reality lies outside us, that our function is to discover or observe

15. *C. Hermite-T. Stieltjes Correspondance*, Gauthier-Villars, 1905, 2, p. 398.
16. *Atti del Congresso*, 1, 1929, 141 = *Grundlagen der Geometrie*, 7th ed., p. 323.
17. *Mind*, 38, 1929, 1–25.

it and that the theorems which we describe grandiloquently as our 'creations,' are simply the notes of our observations."

5. Mathematics as the Study of Arbitrary Structures

The nineteenth-century mathematicians were primarily concerned with the study of nature, and physics certainly was the major inspiration for the mathematical work. The greatest men—Gauss, Riemann, Fourier, Hamilton, Jacobi, and Poincaré—and the less well-known men—Christoffel, Lipschitz, Du Bois-Reymond, Beltrami, and hundreds of others—worked directly on physical problems and on mathematical problems arising out of physical investigations. Even the men commonly regarded as pure mathematicians, Weierstrass, for example, worked on physical problems. In fact, more than in any earlier century, physical problems supplied the suggestions and directions for mathematical investigations, and highly complex mathematics was created to master them. Fresnel had remarked that "Nature is not embarrassed by difficulties of analysis" but mathematicians were not deterred and overcame them. The only major branch that had been pursued for intrinsic aesthetic satisfaction, at least since Diophantus' work, was the theory of numbers.

However, in the nineteenth century for the first time, mathematicians not only carried their work far beyond the needs of science and technology but raised and answered questions that had no bearing on real problems. The *raison d'être* of this development might be described as follows. The two-thousand-year-old conviction that mathematics was the truth about nature was shattered. But the mathematical theories now recognized to be arbitrary had nevertheless proved useful in the study of nature. Though the existing theories historically owed much to suggestions from nature, perhaps new theories constructed solely by the mind might also prove useful in the representation of nature. Mathematicians then should feel free to create arbitrary structures. This idea was seized upon to justify a new freedom in mathematical research. However, since a few structures already in evidence by 1900, and many of those created since, seemed so artificial and so far removed from even potential application, their sponsors began to defend them as desirable in and for themselves.

The gradual rise and acceptance of the view that mathematics should embrace arbitrary structures that need have no bearing, immediate or ultimate, on the study of nature led to a schism that is described today as pure versus applied mathematics. Such a break from tradition could not but generate controversy. We can take space to cite only a few of the arguments on either side.

Fourier had written, in the preface to his *Analytical Theory of Heat*, "The profound study of nature is the most fertile source of mathematical discoveries. This study offers not only the advantage of a well-determined goal but the advantage of excluding vague questions and useless calculations. It is a means of building analysis itself and of discovering the ideas which matter most and which science must always preserve. The fundamental ideas are those which represent the natural happenings." He also stressed the application of mathematics to socially useful problems.

Though Jacobi had done first-class work in mechanics and astronomy, he took issue with Fourier. He wrote to Legendre on July 2, 1830,[18] "It is true that Fourier is of the opinion that the principal object of mathematics is the public utility and the explanation of natural phenomena; but a scientist like him ought to know that the unique object of science is the honor of the human spirit and on this basis a question of [the theory of] numbers is worth as much as a question about the planetary system. . . ."

Throughout the century, as more men became disturbed by the drift to pure mathematics, voices were raised in protest. Kronecker wrote to Helmholtz, "The wealth of your practical experience with sane and interesting problems will give to mathematics a new direction and a new impetus. . . . One-sided and introspective mathematical speculation leads into sterile fields."

Felix Klein, in his *Mathematical Theory of the Top*, (1897, pp. 1–2) stated, "It is the great need of the present in mathematical science that the pure science and those departments of physical science in which it finds its most important applications should again be brought into the intimate association which proved so fruitful in the works of Lagrange and Gauss." And Emile Picard, speaking in the early part of this century (*La Science moderne et son état actuel*, 1908), warned against the tendency to abstractions and pointless problems.

Somewhat later, Felix Klein spoke out again.[19] Fearing that the freedom to create arbitrary structures was being abused, he emphasized that arbitrary structures are "the death of all science. The axioms of geometry are . . . not arbitrary but sensible statements which are, in general, induced by space perception and are determined as to their precise content by expediency." To justify the non-Euclidean axioms Klein pointed out that visualization can verify the Euclidean parallel axiom only within certain limits. On another occasion he pointed out that "whoever has the privilege of freedom should also bear responsibility." By "responsibility" Klein meant service in the investigation of nature.

Despite the warnings, the trend to abstractions, to generalization of existing results for the sake of generalization, and the pursuit of arbitrarily chosen problems continued. The reasonable need to study an entire class of

18. *Ges. Werke*, 1, 454–55.
19. *Elementary Mathematics from an Advanced Standpoint*, Macmillan, 1939; Dover (reprint), 1945, Vol. 2, p. 187.

problems in order to learn more about concrete cases and to abstract in order to get at the essence of a problem became excuses to tackle generalities and abstractions in and for themselves.

Partly to counter the trend to generalization, Hilbert not only stressed that concrete problems are the lifeblood of mathematics, but took the trouble in 1900 to publish a list of twenty-three outstanding ones (see the bibliography) and to cite them in a talk he gave at the Second International Congress of Mathematicians in Paris. Hilbert's prestige did cause many men to tackle these problems. No honor could be more avidly sought than solving a problem posed by so great a man. But the trend to free creations, abstractions, and generalizations was not stemmed. Mathematics broke away from nature and science to pursue its own course.

6. *The Problem of Consistency*

Mathematics, from a logical standpoint, was by the end of the nineteenth century a collection of structures each built on its own system of axioms. As we have already noted, one of the necessary properties of any such structure is the consistency of its axioms. As long as mathematics was regarded as the truth about nature, the possibility that contradictory theorems could arise did not occur; and indeed the thought would have been regarded as absurd. When the non-Euclidean geometries were created, their seeming variance with reality did raise the question of their consistency. As we have seen, this question was answered by making the consistency of the non-Euclidean geometries depend upon that of Euclidean geometry.

By the 1880s the realization that neither arithmetic nor Euclidean geometry is true made the investigation of the consistency of these branches imperative. Peano and his school began in the 1890s to consider this problem. He believed that clear tests could be devised that would settle it. However, events proved that he was mistaken. Hilbert did succeed in establishing the consistency of Euclidean geometry (Chap. 42, sec. 3) on the assumption that arithmetic is consistent. But the consistency of the latter had not been established, and Hilbert posed this problem as the second in the list he presented at the Second International Congress in 1900; in his "Axiomatisches Denken"[20] he stressed it as the basic problem in the foundations of mathematics. Many other men became aware of the importance of the problem. In 1904 Alfred Pringsheim (1850–1941)[21] emphasized that the truth mathematics seeks is neither more nor less than consistency. We shall examine in Chapter 51 the work on this problem.

20. *Math. Ann.*, 78, 1918, 405–15 = *Ges. Abh.*, 145–56.
21. *Jahres. der Deut. Math.-Verein*, 13, 1904, 381.

7. A Glance Ahead

The pace of mathematical creation has expanded steadily since 1600, and this is certainly true of the twentieth century. Most of the fields pursued in the nineteenth century were further developed in the twentieth. However, the details of the newer work in these fields would be of interest only to specialists. We shall therefore limit our account of twentieth-century work to those fields that first became prominent in this period. Moreover, we shall consider only the beginnings of those fields. Developments of the second and third quarters of this century are too recent to be properly evaluated. We have noted many areas pursued vigorously and enthusiastically in the past, which were taken by their advocates to be the essence of mathematics, but which proved to be passing fancies or to have little consequential impact on the course of mathematics. However confident mathematicians of the last half-century may be that their work is of the utmost importance, the place of their contributions in the history of mathematics cannot be decided at the present time.

Bibliography

Fang, J.: *Hilbert*, Paideia Press, 1970. Sketches of Hilbert's mathematical work.

Hardy, G. H.: *A Mathematician's Apology*, Cambridge University Press, 1940 and 1967.

Helmholtz, H. von: *Counting and Measuring*, D. Van Nostrand, 1930. Translation of *Zählen und Messen = Wissenschaftliche Abhandlungen*, 3, 356–91.

——: "Über den Ursprung Sinn und Bedeutung der geometrischen Sätze"; English translation: "On the Origin and Significance of Geometrical Axioms," in Helmholtz: *Popular Scientific Lectures*, Dover (reprint), 1962, pp. 223–49. Also in James R. Newman: *The World of Mathematics*, Simon and Schuster, 1956, Vol. 1, pp. 647–68. See also Helmholtz's *Wissenschaftliche Abhandlungen*, 2, 640–60.

Hilbert, David: "Sur les problèmes futurs des mathématiques," *Comptes Rendus du Deuxième Congrès International des Mathématiciens*, Gauthier-Villars, 1902, 58–114. Also in German, in *Nachrichten König. Ges. der Wiss. zu Gött.*, 1900, 253–97, and in Hilbert's *Gesammelte Abhandlungen*, 3, 290–329. English translation in *Amer. Math. Soc. Bull.*, 8, 1901/2, 437–79.

Klein, Felix: "Über Arithmetisirung der Mathematik," *Ges. Math. Abh.*, 2, 232–40. English translation in *Amer. Math. Soc. Bull.*, 2, 1895/6, 241–49.

Pierpont, James: "On the Arithmetization of Mathematics," *Amer. Math. Soc. Bull.*, 5, 1898/9, 394–406.

Poincaré, Henri: *The Foundations of Science*, Science Press, 1913. See especially pp. 43–91.

Reid, Constance: *Hilbert*, Springer-Verlag, 1970. A biography.

44
The Theory of Functions of
Real Variables

> If Newton and Leibniz had thought that continuous functions
> do not necessarily have a derivative—and this is the general
> case—the differential calculus would never have been
> created.
> EMILE PICARD

1. *The Origins*

The theory of functions of one or more real variables grew out of the attempt
to understand and clarify a number of strange discoveries that had been made
in the nineteenth century. Continuous but non-differentiable functions,
series of continuous functions whose sum is discontinuous, continuous
functions that are not piecewise monotonic, functions possessing bounded
derivatives that are not Riemann integrable, curves that are rectifiable but
not according to the calculus definition of arc length, and nonintegrable
functions that are limits of sequences of integrable functions—all seemed to
contradict the expected behavior of functions, derivatives, and integrals.
Another motivation for the further study of the behavior of functions came
from the research on Fourier series. This theory, as built up by Dirichlet,
Riemann, Cantor, Ulisse Dini (1845–1918), Jordan, and other mathematicians of the nineteenth century, was a quite satisfactory instrument for
applied mathematics. But the properties of the series, as thus far developed,
failed to give a theory that could satisfy the pure mathematicians. Unity,
symmetry, and completeness of relation between function and series were
still wanting.

The research in the theory of functions emphasized the theory of the
integral because it seemed that most of the incongruities could be resolved
by broadening that notion. Hence to a large extent this work may be
regarded as a direct continuation of the work of Riemann, Darboux, Du
Bois-Reymond, Cantor, and others (Chap. 40, sec. 4).

2. *The Stieltjes Integral*

Actually the first extension of the notion of integral came from a totally different class of problems than those just described. In 1894 Thomas Jan Stieltjes (1856–94) published his "Recherches sur les fractions continues,"[1] a most original paper in which he started from a very particular question and solved it with rare elegance. This work suggested problems of a completely novel nature in the theory of analytic functions and in the theory of functions of a real variable. In particular, in order to represent the limit of a sequence of analytic functions Stieltjes was obliged to introduce a new integral that generalized the Riemann-Darboux concept.

Stieltjes considers a positive distribution of mass along a line, a generalization of the point concept of density which, of course, had already been used. He remarks that such a distribution of mass is given by an increasing function $\phi(x)$ which specifies the total mass in the interval $[0, x]$ for $x > 0$, the discontinuities of ϕ corresponding to masses concentrated at a point. For such a distribution of mass in an interval $[a, b]$ he formulates the Riemann sums

$$\sum_{i=0}^{n} f(\xi_i)(\phi(x_{i+1}) - \phi(x_i))$$

wherein the $x_0, x_1, \ldots, x_n$ are a partition of $[a, b]$ and ξ_i is within $[x_i, x_{i+1}]$. He then showed that when f is continuous in $[a, b]$ and the maximum subinterval of the partitions approaches 0, the sums approach a limit which he denoted by $\int_a^b f(x)\, d\phi(x)$. Though he used this integral in his own work, Stieltjes did not push further the integral notion itself, except that for the interval $(0, \infty)$ he defined

$$\int_0^\infty f(x)\, d\phi(x) = \lim_{b \to \infty} \int_0^b f(x)\, d\phi(x).$$

His integral concept was not taken up by mathematicians until much later, when it did find many applications (see Chap. 47, sec. 4).

3. *Early Work on Content and Measure*

Quite another line of thought led to a different generalization of the notion of integral, the Lebesgue integral. The study of the sets of discontinuities of functions suggested the question of how to measure the extent or "length" of the set of discontinuities, because the extent of these discontinuities determines the integrability of the function. The theory of content and later the theory of

1. *Ann. Fac. Sci. de Toulouse*, 8, 1894, J.1–122, and 9, 1895, A.1–47 = *Œuvres complètes*, 2, 402–559.

measure were introduced to extend the notion of length to sets of points that are not full intervals of the usual straight line.

The notion of content is based on the following idea: Consider a set E of points distributed in some manner over an interval $[a, b]$. To be loose for the moment, suppose that it is possible to enclose or cover these points by small subintervals of $[a, b]$ so that the points of E are either interior to one of the subintervals or at worst an endpoint. We reduce the lengths of these subintervals more and more and add others if necessary to continue to enclose the points of E, while reducing the sum of the lengths of the sub-intervals. The greatest lower bound of the sum of those subintervals that cover points of E is called the (outer) content of E. This loose formulation is not the definitive notion that was finally adopted, but it may serve to indicate what the men were trying to do.

A notion of (outer) content was given by Du Bois-Reymond in his *Die allgemeine Funktionentheorie* (1882), Axel Harnack (1851–88) in his *Die Elemente der Differential- und Integralrechnung* (1881), Otto Stolz,[2] and Cantor.[3] Stolz and Cantor also extended the notion of content to two and higher-dimensional sets using rectangles, cubes, and so forth in place of intervals.

The use of this notion of content, which was unfortunately not satis-factory in all respects, nevertheless revealed that there were nowhere dense sets (that is, the set lies in an interval but is not dense in any subinterval of that interval) of positive content and that functions with discontinuities on such sets were not integrable in Riemann's sense. Also there were functions with bounded nonintegrable derivatives. But mathematicians of this time, the 1880s, thought that Riemann's notion of the integral could not be extended.

To overcome limitations in the above theory of content and to rigorize the notion of area of a region, Peano (*Applicazioni geometriche del calcolo infinitesimale*, 1887) introduced a fuller and much improved notion of content. He introduced an inner and outer content for regions. Let us consider two dimensions. The inner content is the least upper bound of all polygonal regions contained within the region R and the outer content is the greatest lower bound of all polygonal regions containing the region R. If the inner and outer content are equal, this common value is the area. For a one-dimensional set the idea is similar but uses intervals instead of polygons. Peano pointed out that if $f(x)$ is non-negative in $[a, b]$ then

$$\underline{\int_a^b} f \, dx = C_i(R) \quad \text{and} \quad \overline{\int_a^b} = C_e(R),$$

where the first integral is the least upper bound of the lower Riemann sums of f on $[a, b]$ and the second is the greatest lower bound of the upper Riemann

2. *Math. Ann.*, 23, 1884, 152–56.
3. *Math. Ann.*, 23, 1884, 453–88 = *Ges. Abh.*, 210–46.

sums and $C_i(R)$ and $C_e(R)$ are the inner and outer content of the region R bounded above by the graph of f. Thus f is integrable if and only if R has content in the sense that $C_i(R) = C_e(R)$.

Jordan made the most advanced step in the nineteenth-century theory of content (*étendue*). He too introduced an inner and outer content,[4] but formulated the concept somewhat more effectively. His definition for a set of points E contained in $[a, b]$ starts with the outer content. One covers E by a finite set of subintervals of $[a, b]$ such that each point of E is interior to or an endpoint of one of these subintervals. The greatest lower bound of the sum of all such sets of subintervals that contain at least one point of E is the outer content of E. The inner content of E is defined to be the least upper bound of the sum of the subintervals that enclose only points of E in $[a, b]$. If the inner and outer content of E are equal, then E has content. The same notion was applied by Jordan to sets in n-dimensional space except that rectangles and the higher-dimensional analogues replace the subintervals. Jordan could now prove what is called the additivity property: The content of the sum of a *finite* number of disjoined sets with content is the sum of the contents of the separate sets. This was not true for the earlier theories of content, except Peano's.

Jordan's interest in content derived from the attempt to clarify the theory of double integrals taken over some plane region E. The definition usually adopted was to divide the plane into squares R_{ij} by lines parallel to the coordinate axes. This partition of the plane induces a partition of E into E_{ij}'s. Then, by definition,

$$\int_E f(x, y)\, dE = \lim_{\Delta x, \Delta y \to 0} \sum_{i,j} f(x_i, y_j) a(R_{ij}),$$

where $a(R_{ij})$ denotes the area of R_{ij} and the sum is over all R_{ij} interior to E and all R_{ij} that contain any points of E but may also contain points outside of E. For the integral to exist it is necessary to show that the R_{ij} that are not entirely interior to E can be neglected or that the sum of the areas of the R_{ij} that contain boundary points of E approaches 0 with the dimensions of the R_{ij}. It had been generally assumed that this was the case and Jordan himself did so in the first edition of his *Cours d'analyse* (Vol. 2, 1883). However the discovery of such peculiar curves as Peano's square-filling curve made the mathematicians more cautious. If E has two-dimensional Jordan content, then one can neglect the R_{ij} that contain the boundary points of E. Jordan was also able to obtain results on the evaluation of double integrals by repeated integration.

The second edition of Jordan's *Cours d'analyse* (Vol. 1, 1893) contains Jordan's treatment of content and its application to integration. Though

4. *Jour. de Math.*, (4), 8, 1892, 69–99 = *Œuvres*, 4, 427–57.

superior to that of his predecessors, Jordan's definition of content was not quite satisfactory. According to it an open bounded set does not necessarily have content and the set of rational points contained in a bounded interval does not have content.

The next step in the theory of content was made by Borel. Borel was led to study the theory, which he called measure, while working on the sets of points on which series representing complex functions converge. His *Leçons sur la théorie des fonctions* (1898) contains his first major work on the subject. Borel discerned the defects in the earlier theories of content and remedied them.

Cantor had shown that every open set U on the line is the union of a *denumerable* family of open intervals, no two having a point in common. In place of approaching U by enclosing it in a finite set of intervals Borel, using Cantor's result, proposed as the measure of a bounded open set U the sum of the lengths of the component intervals. He then defined the measure of the sum of a *countable* number of disjoined measurable sets as the sum of the individual measures and the measure of the set $A - B$, if A and B are measurable and B is contained in A, as the difference of the measures. With these definitions he could attach a measure to sets formed by adding any countable number of disjoined measurable sets and to the difference of any two measurable sets A and B if A contains B. He then considered sets of measure 0 and showed that a set of measure greater than 0 is non-denumerable.

Borel's theory of measure was an improvement over Peano's and Jordan's notions of content, but it was not the final word, nor did he consider its application to integration.

4. *The Lebesgue Integral*

The generalization of measure and the integral that is now considered definitive was made by Henri Lebesgue (1875–1941), a pupil of Borel and a professor at the Collège de France. Guided by Borel's ideas and also by those of Jordan and Peano, he first presented his ideas on measure and the integral in his thesis, "Intégrale, longueur, aire." [5] His work superseded the nineteenth-century creations and, in particular, improved on Borel's theory of measure.

Lebesgue's theory of integration is based on his notion of measure of sets of points and both ideas apply to sets in n-dimensional space. For illustrative purposes we shall consider the one-dimensional case. Let E be a set of points in $a \leq x \leq b$. The points of E can be enclosed as interior points in a finite or *countably infinite* set of intervals $d_1, d_2, \ldots$ lying in $[a, b]$. (The endpoints of $[a, b]$ can be endpoints of a d_i.) It can be shown that the set of intervals $\{d_i\}$

5. *Annali di Mat.*, (3), 7, 1902, 231–59.

can be replaced by a set of non-overlapping intervals δ_1, δ_2, ... such that every point of E is an interior point of one of the intervals or the common endpoint of two adjacent intervals. Let $\sum \delta_n$ denote the sum of the lengths δ_i. The lower bound of $\sum \delta_n$ for all possible sets $\{\delta_i\}$ is called the exterior measure of E and denoted by $m_e(E)$. The interior measure $m_i(E)$ of E is defined to be the exterior measure of the set $C(E)$, that is, the complement of E in $[a, b]$ or the points of $a \leq x \leq b$ not in E.

Now one can prove a number of subsidiary results, including the fact that $m_i(E) \leq m_e(E)$. The set E is defined to be measurable if $m_i(E) = m_e(E)$ and the measure $m(E)$ is taken to be this common value. Lebesgue showed that a union of a countable number of measurable sets that are pairwise disjoined has as its measure the sum of the measures of the component sets. Also all Jordan measurable sets are Lebesgue measurable and the measure is the same. Lebesgue's notion of measure differs from Borel's by the adjunction of a part of a set of measure 0 in the sense of Borel. Lebesgue also called attention to the existence of nonmeasurable sets.

His next significant notion is that of a measurable function. Let E be a bounded measurable set on the x-axis. The function $f(x)$, defined on all points of E, is said to be measurable in E if the set of points of E for which $f(x) > A$ is measurable for every constant A.

Finally we arrive at Lebesgue's notion of an integral. Let $f(x)$ be a bounded and measurable function on the measurable set E contained in $[a, b]$. Let A and B be the greatest lower and least upper bounds of $f(x)$ on E. Divide the interval $[A, B]$ (on the y-axis) into n partial intervals

$$[A, l_1], [l_1, l_2], \cdots, [l_{n-1}, B],$$

with $A = l_0$ and $B = l_n$. Let e_r be the set of points of E for which $l_{r-1} \leq f(x) \leq l_r, r = 1, 2, \ldots, n$. Then $e_1, e_2, \ldots, e_n$ are measurable sets. Form the sums S and s where

$$S = \sum_1^n l_r m(e_r), \qquad s = \sum_1^n l_{r-1} m(e_r).$$

The sums S and s have a greatest lower bound J and a least upper bound I, respectively. Lebesgue showed that for bounded measurable functions $I = J$, and this common value is the Lebesgue integral of $f(x)$ on E. The notation is

$$I = \int_E f(x)\, dx.$$

If E consists of the entire interval $a \leq x \leq b$, then we use the notation $\int_a^b f(x)\, dx$, but the integral is understood in the Lebesgue sense. If $f(x)$ is Lebesgue integrable and the value of the integral is finite, then $f(x)$ is said to be summable, a term introduced by Lebesgue. An $f(x)$ that is Riemann integrable on $[a, b]$ is Lebesgue integrable but not necessarily conversely. If $f(x)$ is integrable in both senses, the values of the two integrals are the same.

The generality of the Lebesgue integral derives from the fact that a Lebesgue integrable function need not be continuous almost everywhere (that is, except on a set of measure 0). Thus the Dirichlet function, which is 1 for rational values of x and 0 for irrational values of x in $[a, b]$, is totally discontinuous and though not Riemann integrable is Lebesgue integrable. In this case $\int_a^b f(x)\, dx = 0$.

The notion of the Lebesgue integral can be extended to more general functions, for example unbounded functions. If $f(x)$ is Lebesgue integrable but not bounded in the interval of integration, the integral converges absolutely. Unbounded functions may be Lebesgue integrable but not Riemann integrable and conversely.

For practical purposes the Riemann integral suffices. In fact Lebesgue showed (*Leçons sur l'intégration et la recherche des fonctions primitives*, 1904) that a bounded function is Riemann integrable if and only if the points of discontinuity form a set of measure 0. But for theoretical work the Lebesgue integral affords simplifications. The new theorems rest on the countable additivity of Lebesgue measure as contrasted with the finite additivity of Jordan content.

To illustrate the simplicity of theorems using Lebesgue integration, we have the result given by Lebesgue himself in his thesis. Suppose $u_1(x)$, $u_2(x), \ldots$ are summable functions on a measurable set E and $\sum_1^\infty u_n(x)$ converges to $f(x)$. Then $f(x)$ is measurable. If in addition $s_n(x) = \sum_1^n u_n(x)$ is uniformly bounded ($|s_n(x)| \leq B$ for all x in E and all n), then it is a theorem that $f(x)$ is Lebesgue integrable on $[a, b]$ and

$$\int_a^b f(x)\, dx = \lim_{n \to \infty} \int_a^b s_n(x)\, dx.$$

If we were working with Riemann integrals we would need the additional hypothesis that the sum of the series is integrable; this case for the Riemann integral is a theorem due to Cesare Arzelà (1847–1912).[6] Lebesgue made his theorem the cornerstone in the exposition of his theory in his *Leçons sur l'intégration*.

The Lebesgue integral is especially useful in the theory of Fourier series and most important contributions were made in this connection by Lebesgue himself.[7] According to Riemann, the Fourier coefficients a_n and b_n of a bounded and integrable function tend to 0 as n becomes infinite. Lebesgue's generalization states that

$$\lim_{n \to \infty} \int_a^b f(x) \begin{cases} \sin nx \\ \cos nx \end{cases} dx = 0,$$

6. *Atti della Accad. dei Lincei, Rendiconti*, (4), 1, 1885, 321–26, 532–37, 566–69.
7. E.g., *Ann. de l'Ecole Norm. Sup.*, (3), 20, 1903, 453–85.

where $f(x)$ is any function, bounded or not, that is Lebesgue integrable. This fact is now referred to as the Riemann-Lebesgue lemma.

In this same paper of 1903 Lebesgue showed that if f is a bounded function represented by a trigonometric series, that is,

$$f(x) = \frac{a_0}{2} + \sum_{1}^{\infty} (a_n \cos nx + b_n \sin nx),$$

then the a_n and b_n are Fourier coefficients. In 1905[8] Lebesgue gave a new sufficient condition for the convergence of the Fourier series to a function $f(x)$ that included all previously known conditions.

Lebesgue also showed (*Leçons sur les séries trigonométriques*, 1906, p. 102) that term-by-term integration of a Fourier series does not depend on the uniform convergence of the series to $f(x)$ itself. What does hold is that

$$\int_{-\pi}^{x} f(x)\, dx = a_0(x + \pi) + \sum_{1}^{\infty} \frac{1}{n} (a_n \sin nx + b_n (\cos n\pi - \cos nx)),$$

where x is any point in $[-\pi, \pi]$, for any $f(x)$ that is Lebesgue integrable, whether or not the original series for $f(x)$ converges. And the new series converges uniformly to the left side of the equation in the interval $[-\pi, \pi]$.

Further, Parseval's theorem that

$$\frac{1}{\pi} \int_{-\pi}^{\pi} [f(x)]^2\, dx = 2a_0^2 + \sum_{1}^{\infty} (a_n^2 + b_n^2)$$

holds for any $f(x)$ whose square is Lebesgue integrable in $[-\pi, \pi]$ (*Leçons*, 1906, p. 100). Then Pierre Fatou (1878–1929) proved[9] that

$$\frac{1}{\pi} \int_{-\pi}^{\pi} f(x)g(x)\, dx = 2a_0\alpha_0 + \sum_{1}^{\infty} (a_n\alpha_n + b_n\beta_n),$$

where a_n and b_n, and α_n and β_n are the Fourier coefficients for $f(x)$ and $g(x)$ whose squares are Lebesgue integrable in $[-\pi, \pi]$. Despite these advances in the theory of Fourier series, there is no known property of an $f(x)$ Lebesgue integrable in $[-\pi, \pi]$ that is necessary and sufficient for the convergence of its Fourier series.

Lebesgue devoted most of his efforts to the connection between the notions of integral and of primitive functions (indefinite integrals). When Riemann introduced his generalization of the integral, the question was posed whether the correspondence between definite integral and primitive function, valid for continuous functions, held in the more general case. But it is possible to give examples of functions f integrable in Riemann's sense and such that $\int_a^x f(t)\,dt$ does not have a derivative (not even a right or left derivative) at certain points. Conversely Volterra showed in 1881 [10] that a function $F(x)$ can have a bounded derivative in an interval I that is not integrable in Riemann's sense over the interval. By a subtle analysis Lebesgue showed that if f is integrable in his sense in $[a, b]$, then $F(x) = \int_a^x f(t)\,dt$ has a derivative equal to $f(x)$ almost everywhere, that is, except on a set of measure zero (*Leçons sur l'intégration*). Conversely, if a function g is differentiable in $[a, b]$ and if its derivative $g' = f$ is bounded, then f is Lebesgue integrable and the formula $g(x) - g(a) = \int_a^x f(t)\,dt$ holds. However, as Lebesgue established, the situation is much more complex if g' is not bounded. In this case g' is not necessarily integrable, and the first problem is to characterize the functions g for which g' exists almost everywhere and is integrable. Limiting himself to the case where one of the derived numbers [11] of g is finite everywhere, Lebesgue showed that g is necessarily a function of bounded variation (Chap. 40, sec. 6). Finally Lebesgue established (in the 1904 book) the reciprocal result. A function g of bounded variation admits a derivative almost everywhere and g' is integrable. But one does not necessarily have

$$(1) \qquad\qquad g(x) - g(a) = \int_a^x g'(t)\,dt;$$

the difference between the two members of this equation is a nonconstant function of bounded variation with derivative zero almost everywhere. As for the functions of bounded variation g for which (1) does hold, these have the following property: The total variation of g in an open set U (that is, the sum of the total variations of g in each of the connected components of U) tends to 0 with the measure of U. These functions were called absolutely continuous by Giuseppe Vitali (1875–1932), who studied them in detail.

Lebesgue's work also advanced the theory of multiple integrals. Under his definition of the double integral, the domain of functions for which the double integral can be evaluated by repeated integration is extended. Lebesgue gave a result in his thesis of 1902, but the better result was given by Guido Fubini (1879–1943): [12] If $f(x, y)$ is summable over the measurable set G, then

(a) $f(x, y)$ as a function of x and as a function of y is summable for almost all y and x, respectively;

10. *Gior. di Mat.*, 19, 1881, 333–72 = *Opere Mat.*, 1, 16–48.
11. The two right derived numbers of g are the two limits,

$$\limsup_{h \to 0, h > 0} \frac{g(x + h) - g(x)}{h}, \qquad \liminf_{h \to 0, h > 0} \frac{g(x + h) - g(x)}{h}.$$

The two left derived numbers are defined similarly.
12. *Atti della Accad. dei Lincei, Rendiconti*, (5), 16, 1907, 608–14.

(b) the set of points (x_0, y_0) for which either $f(x, y_0)$ or $f(x_0, y)$ is not summable has measure 0;

(c) $$\iint_G f(x, y) \, dG = \int dy \left(\int f(x, y) \, dx \right) = \int dx \left(\int f(x, y) \, dy \right)$$

where the outer integrals are taken over the set of points y (respectively x) for which $f(x, y)$ as a function of x (as a function of y, respectively) are summable.

Finally, in 1910[13] Lebesgue arrived at results for multiple integrals that extended those for the derivatives of single integrals. He associated with a function f integrable in every compact region of R^n, the set function (as opposed to functions of numerical variables) $F(E) = \int_E f(x) \, dx$ (x represents n coordinates) defined for each integrable domain E of R^n. This concept generalizes the indefinite integral. He observed that the function F possesses two properties:

(1) It is completely additive; that is, $F(\sum_1^\infty E_n) = \sum_1^\infty F(E_n)$ where the E_n are pairwise disjoined measurable sets.
(2) It is absolutely continuous in the sense that $F(E)$ tends to 0 with the measure of E.

The essential part of this paper of Lebesgue was to show the converse of this proposition, that is, to define a derivative of $F(E)$ at a point P of n-dimensional space. Lebesgue arrived at the following theorem: If $F(E)$ is absolutely continuous and additive, then it possesses a finite derivative almost everywhere, and F is the indefinite integral of that summable function which is the derivative of F where it exists and is finite but is otherwise arbitrary at the remaining points.

The principal tool in the proof is a covering theorem due to Vitali,[14] which remains fundamental in this area of integration. But Lebesgue did not stop there. He indicated the possibility of generalizing the notion of functions of bounded variation by considering functions $F(E)$ where E is a measurable set, the functions being completely additive and such that $\sum_n |F(E_n)|$ remains bounded for every denumerable partition of E into measurable subsets E_n. It would be possible to cite many other theorems of the calculus that have been generalized by Lebesgue's notion of the integral.

Lebesgue's work, one of the great contributions of this century, did win approval but, as usual, not without some resistance. We have already noted (Chap. 40, sec. 7) Hermite's objections to functions without derivatives. He tried to prevent Lebesgue from publishing a "Note on Non-Ruled Surfaces

13. *Ann. de l'Ecole Norm. Sup.*, (3), 27, 1910, 361–450.
14. *Atti Accad. Torino*, 43, 1908, 229–46.

Applicable to the Plane,"[15] in which Lebesgue treated non-differentiable surfaces. Lebesgue said many years later in his *Notice* (p. 14, see bibliography),

> Darboux had devoted his *Mémoire* of 1875 to integration and to functions without derivatives; he therefore did not experience the same horror as Hermite. Nevertheless I doubt whether he ever entirely forgave my "Note on Applicable Surfaces." He must have thought that those who make themselves dull in this study are wasting their time instead of devoting it to useful research.

Lebesgue also said (*Notice*, p. 13),

> To many mathematicians I became the man of the functions without derivatives, although I never at any time gave myself completely to the study or consideration of such functions. And since the fear and horror which Hermite showed was felt by almost everybody, whenever I tried to take part in a mathematical discussion there would always be an analyst who would say, "This won't interest you; we are discussing functions having derivatives." Or a geometer would say it in his language: "We're discussing surfaces that have tangent planes."

5. *Generalizations*

We have already indicated the advantages of Lebesgue integration in generalizing older results and in formulating neat theorems on series. In subsequent chapters we shall meet additional applications of Lebesgue's ideas. The immediate developments in the theory of functions were many extensions of the notion of integral. Of these we shall merely mention one by Johann Radon (1887–1956), which embraces both Stieltjes's and Lebesgue's integral and is in fact known as the Lebesgue-Stieltjes integral.[16] The generalizations cover not only broader or different notions of integrals on point sets of n-dimensional Euclidean space but on domains of more general spaces such as spaces of functions. The applications of these more general concepts are now found in the theory of probability, spectral theory, ergodic theory, and harmonic analysis (generalized Fourier analysis).

Bibliography

Borel, Emile: *Notice sur les travaux scientifiques de M. Emile Borel*, 2nd ed., Gauthier-Villars, 1921.
Bourbaki, Nicolas: *Eléments d'histoire de mathématiques*, Hermann, 1960, pp. 246–59.
Collingwood, E. F.: "Emile Borel," *Jour. Lon. Math. Soc.*, 34, 1959, 488–512.

15. *Comp. Rend.*, 128, 1899, 1502–05.
16. *Sitzungsber. der Akad. der Wiss. Wien*, 122, Abt. IIa, 1913, 1295–1438.

Fréchet, M.: "La Vie et l'œuvre d'Emile Borel," *L'Enseignement Mathématique*, (2), 11, 1965, 1–94.

Hawkins, T. W., Jr.: *Lebesgue's Theory of Integration: Its Origins and Development*, University of Wisconsin Press, 1970, Chaps. 4–6.

Hildebrandt, T. H.: "On Integrals Related to and Extensions of the Lebesgue Integral," *Amer. Math. Soc. Bull.*, 24, 1918, 113–77.

Jordan, Camille: *Œuvres*, 4 vols., Gauthier-Villars, 1961–64.

Lebesgue, Henri: *Measure and the Integral*, Holden-Day, 1966, pp. 176–94. Translation of the French *La Mesure des grandeurs*.

————: *Notice sur les travaux scientifiques de M. Henri Lebesgue*, Edouard Privat, 1922.

————: *Leçons sur l'intégration et la recherche des fonctions primitives*, Gauthier-Villars, 1904, 2nd ed., 1928.

McShane, E. J.: "Integrals Devised for Special Purposes," *Amer. Math. Soc. Bull.*, 69, 1963, 597–627.

Pesin, Ivan M.: *Classical and Modern Integration Theories*, Academic Press, 1970.

Plancherel, Michel: "Le Développement de la théorie des séries trigonométriques dans le dernier quart de siècle," *L'Enseignement Mathématique*, 24, 1924/25, 19–58.

Riesz, F.: "L'Evolution de la notion d'intégrale depuis Lebesgue," *Annales de l'Institut Fourier*, 1, 1949, 29–42.

45
Integral Equations

> Nature is not embarrassed by difficulties of analysis.
>
> AUGUSTIN FRESNEL

1. *Introduction*

An integral equation is an equation in which an unknown function appears under an integral sign and the problem of solving the equation is to determine that function. As we shall soon see, some problems of mathematical physics lead directly to integral equations, and other problems, which lead first to ordinary or partial differential equations, can be handled more expeditiously by converting them to integral equations. At first, solving integral equations was described as inverting integrals. The term integral equations is due to Du Bois-Reymond.[1]

As in other branches of mathematics, isolated problems involving integral equations occurred long before the subject acquired a distinct status and methodology. Thus Laplace in 1782[2] considered the integral equation for $g(t)$ given by

$$(1) \qquad f(x) = \int_{-\infty}^{\infty} e^{-xt} g(t) \, dt.$$

As equation (1) now stands, it is called the Laplace transform of $g(t)$. Poisson[3] discovered the expression for $g(t)$, namely,

$$g(t) = \frac{1}{2\pi i} \int_{a-i\infty}^{a+i\infty} e^{xt} f(x) \, dx$$

for large enough a. Another of the noteworthy results that really belong to the history of integral equations stems from Fourier's famous 1811 paper on the theory of heat (Chap. 28, sec. 3). Here one finds

$$f(x) = \int_{0}^{\infty} \cos(xt) u(t) \, dt$$

1. *Jour. für Math.*, 103, 1888, 228.
2. *Mém. de l'Acad. des Sci.*, Paris, 1782, 1–88, pub. 1785, and 1783, 423–67, pub. 1786 = *Œuvres*, 10, 209–91, p. 236 in particular.
3. *Jour. de l'Ecole Poly.*, 12, 1823, 1–144, 249–403.

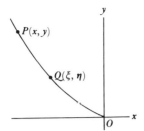

Figure 45.1

and the inversion formula

$$u(t) = \frac{2}{\pi} \int_0^\infty \cos(xt) f(x) \, dx.$$

The first conscious direct use and solution of an integral equation go back to Abel. In two of his earliest published papers, the first published in an obscure journal in 1823[4] and the second published in the *Journal für Mathematik*,[5] Abel considered the following mechanics problem: A particle starting at P slides down a smooth curve (Fig. 45.1) to the point O. The curve lies in a vertical plane. The velocity acquired at O is independent of the shape of the curve but the time required to slide from P to O is not. If (ξ, η) are the coordinates of any point Q between P and O and s is the arc OQ, then the velocity of the particle at Q is given by

$$\frac{ds}{dt} = -\sqrt{2g(x - \xi)}$$

where g is the gravitational constant. Hence

$$t = \frac{-1}{\sqrt{2g}} \int_P^Q \frac{ds}{\sqrt{x - \xi}}.$$

Now s can be expressed in terms of ξ. Suppose s is $v(\xi)$. Then the whole time of descent T from P to O is given by

$$T(x) = \frac{1}{\sqrt{2g}} \int_0^x \frac{v'(\xi) \, d\xi}{\sqrt{x - \xi}}.$$

The time T clearly depends upon x for any curve. The problem Abel set was, given T as a function of x, to find $v(\xi)$. If we introduce

$$f(x) = \sqrt{2g} \, T(x)$$

4. *Magazin for Naturwidenskaberne*, 1, 1823 = *Œuvres*, 1, 11–27.
5. *Jour. für Math*, 1, 1826, 153–57 = *Œuvres*, 1, 97–101.

the problem becomes to determine v from the equation

$$f(x) = \int_0^x \frac{v'(\xi)}{\sqrt{x - \xi}}\, d\xi.$$

Abel obtained the solution

$$v(\xi) = \int_0^\xi \frac{f(x)\, dx}{(\xi - x)^{1/2}}.$$

His methods—he gave two—were special and not worth noting.

Actually Abel undertook to solve the more general problem

(2) $$f(x) = \int_a^x \frac{u(\xi)\, d\xi}{(x - \xi)^\lambda}, \qquad 0 < \lambda < 1$$

and obtained

$$u(z) = \frac{\sin \lambda \pi}{\pi} \frac{d}{dz} \int_a^z \frac{f(x)\, dx}{(z - x)^{1-\lambda}}.$$

Liouville, who worked independently of Abel, solved special integral equations from 1832 on.[6] A more significant step by Liouville[7] was to show how the solution of certain differential equations can be obtained by solving integral equations. The differential equation to be solved is

(3) $$y'' + [\rho^2 - \sigma(x)]y = 0$$

over the interval $a \leq x \leq b$; ρ is a parameter. Let $u(x)$ be the particular solution that satisfies the initial conditions

(4) $$u(a) = 1, \qquad u'(a) = 0.$$

This function will also be a solution of the nonhomogeneous equation

$$y'' + \rho^2 y = \sigma(x)u(x).$$

Then by a basic result on ordinary differential equations,

(5) $$u(x) = \cos \rho(x - a) + \frac{1}{\rho}\int_a^x \sigma(\xi) \sin \rho(x - \xi)u(\xi)\, d\xi.$$

Thus if we can solve this integral equation we shall have obtained that solution of the differential equation (3) that satisfies the initial conditions (4).

Liouville obtained the solution by a method of successive substitutions attributed to Carl G. Neumann, whose work *Untersuchungen über das logarithmische und Newton'sche Potential* (1877) came thirty years later. We shall not describe Liouville's method because it is practically identical with the one given by Volterra, which is to be described shortly.

6. *Jour. de l'Ecole Poly.*, 13, 1832, 1–69.
7. *Jour. de Math.*, 2, 1837, 16–35.

The integral equations treated by Abel and Liouville are of basic types. Abel's is of the form

(6)
$$f(x) = \int_a^x K(x, \xi)u(\xi) \, d\xi,$$

and Liouville's of the form

(7)
$$u(x) = f(x) + \int_a^x K(x, \xi)u(\xi) \, d\xi.$$

In both of these $f(x)$ and $K(x, \xi)$ are known, and $u(\xi)$ is the function to be determined. The terminology used today, introduced by Hilbert, refers to these equations as the first and second kind, respectively, and $K(x, \xi)$ is called the kernel. As stated, they are also referred to as Volterra's equations, whereas when the upper limit is a fixed number b, they are called Fredholm's equations. Actually the Volterra equations are special cases, respectively, of Fredholm's because one can always take $K(x, \xi) = 0$ for $\xi > x$ and then regard the Volterra equations as Fredholm equations. The special case of the equation of the second kind in which $f(x) \equiv 0$ is called the homogeneous equation.

By the middle of the nineteenth century the chief interest in integral equations centered around the solution of the boundary-value problem associated with the potential equation

(8)
$$\Delta u = u_{xx} + u_{yy} = 0.$$

The equation holds in a given plane area that is bounded by some curve C. If the boundary value of u is some function $f(s)$ given as a function of arc length s along C, then a solution of this potential problem can be represented by

$$u(x, y) = \frac{1}{2\pi} \int_C \rho(s) \log \frac{1}{r(s; x, y)} \, ds,$$

wherein $r(s; x, y)$ is the distance from a point s to any point (x, y) in the interior or boundary and $\rho(s)$ is an unknown function satisfying for $s = (x, y)$ on C

(9)
$$f(s) = \frac{1}{2\pi} \int_C \rho(t) \log \frac{1}{r(t; x, y)} \, dt.$$

This is an integral equation of the first kind for $\rho(t)$. Alternatively, if one takes as a solution of (8) with the same boundary condition

$$v(x, y) = \frac{1}{2\pi} \int_C \phi(s) \frac{\partial}{\partial n} \left(\log \frac{1}{r(s; x, y)} \right) ds,$$

where $\partial/\partial n$ denotes the normal derivative to the boundary, then $\phi(s)$ must satisfy the integral equation

$$(10) \qquad f(s) = \frac{1}{2}\phi(s) + \frac{1}{2\pi}\int_C \phi(t)\frac{\partial}{\partial n}\left(\log\frac{1}{r(t;x,y)}\right) dt,$$

an integral equation of the second kind. These equations were solved by Neumann for convex areas in his *Untersuchungen* and later publications.

Another problem of partial differential equations was tackled through integral equations. The equation

$$(11) \qquad \Delta u + \lambda u = f(x,y)$$

arises in the study of wave motion when the time dependence of the corresponding hyperbolic equation

$$\Delta u - \frac{1}{c^2}u_{tt} = f(x,y),$$

usually taken to be $e^{-i\omega t}$, is eliminated. It was known (Chap. 28, sec. 8) that the homogeneous case of (11) subject to boundary conditions has nontrivial solutions only for a discrete set of λ-values, called eigenvalues or characteristic values. Poincaré in 1894[8] considered the inhomogeneous case (11) with complex λ. He was able to produce a function, meromorphic in λ, which represented a unique solution of (11) for any λ which is not an eigenvalue, and whose residues produce eigenfunctions for the homogeneous case, that is, when $f = 0$.

On the basis of these results, Poincaré in 1896[9] considered the equation

$$u(x) + \lambda\int_a^b K(x,y)u(y)\,dy = f(x),$$

which he derived from (11), and affirmed that the solution is a meromorphic function of λ. This result was established by Fredholm in a paper we shall consider shortly.

The conversion of differential equations to integral equations, which is illustrated by the above examples, became a major technique for solving initial- and boundary-value problems of ordinary and partial differential equations and was the strongest impetus for the study of integral equations.

2. The Beginning of a General Theory

Vito Volterra (1860–1940), who succeeded Beltrami as professor of mathematical physics at Rome, is the first of the founders of a general theory of

8. *Rendiconti del Circolo Matematico di Palermo*, 8, 1894, 57–155 = *Œuvres*, 9, 123–96.
9. *Acta Math.*, 20, 1896–97, 59–142 = *Œuvres*, 9, 202–72. See also the Hellinger and Toeplitz reference in the bibliography, p. 1354.

integral equations. He wrote papers on the subject from 1884 on and principal ones in 1896 and 1897.[10] Volterra contributed a method of solving
integral equations of the second kind,

$$(12) \qquad f(s) = \phi(s) + \int_a^b K(s, t)\phi(t) \, dt$$

wherein $\phi(s)$ is unknown and $K(s, t) = 0$ for $t > s$. Volterra wrote this
equation as

$$f(s) = \phi(s) + \int_a^s K(s, t)\phi(t) \, dt.$$

His method of solution was to let

$$f_1(s) = -\int_a^b K(s, t)f(t) \, dt$$

$$(13) \qquad \cdots\cdots\cdots\cdots\cdots\cdots\cdots\cdots$$

$$f_n(s) = -\int_a^b K(s, t)f_{n-1}(t) \, dt$$

$$\cdots\cdots\cdots\cdots\cdots\cdots\cdots\cdots$$

and take $\phi(s)$ to be

$$(14) \qquad \phi(s) = f(s) + \sum_{p=1}^{\infty} f_p(s).$$

For his kernel $K(s, t)$ Volterra was able to prove the convergence of (14),
and if one substitutes (14) in (12), one can show that it is a solution. This
substitution gives

$$\phi(s) = f(s) - \int_a^b K(s, t)f(t) \, dt + \int_a^b \int_a^b K(s, r)K(r, t)f(t) \, dr \, dt + \cdots,$$

which can be written in the form

$$(15) \qquad \phi(s) = f(s) + \int_a^b \overline{K}(s, t)f(t) \, dt,$$

where the kernel $\overline{K}$ (later called the solving kernel or resolvent by Hilbert) is

$$\overline{K}(s, t) = -K(s, t) + \int_a^b K(s, r)K(r, t) \, dr$$

$$- \int_a^b \int_a^b K(s, r)K(r, w)K(w, t) \, dr \, dw + \cdots.$$

10. *Atti della Accad. dei Lincei, Rendiconti*, (5), 5, 1896, 177–85, 289–300; *Atti Accad. Torino*, 31,
1896, 311–23, 400–8, 557–67, 693–708; *Annali di Mat.*, (2), 25, 1897, 139–78; all are in his
Opere matematiche, 2, 216–313.

Equation (15) is the representation obtained earlier for a particular integral equation by Liouville and credited to Neumann. Volterra also solved integral equations of the first kind $f(s) = \int_a^s K(x, s)\phi(x)\,dx$ by reducing them to equations of the second kind.

In 1896 Volterra observed that an integral equation of the first kind is a limiting form of a system of n linear algebraic equations in n unknowns as n becomes infinite. Erik Ivar Fredholm (1866–1927), professor of mathematics at Stockholm, concerned with solving the Dirichlet problem, took up this idea in 1900[11] and used it to solve integral equations of the second kind, that is, equations of the form (12), without, however, the restriction on $K(s, t)$.

We shall write the equation Fredholm tackled in the form

$$(16) \qquad u(x) = f(x) + \lambda \int_a^b K(x, \xi)u(\xi)\,d\xi,$$

though the parameter λ was not explicit in his work. However, what he did is more intelligible in the light of later work if we exhibit it. To be faithful to Fredholm's formulas, one should set $\lambda = 1$ or regard it as implicitly involved in K.

Fredholm divided the x-interval $[a, b]$ into n equal parts by the points

$$a, x_1 = a + \delta, x_2 = a + 2\delta, \cdots, x_n = a + n\delta = b.$$

He then replaced the definite integral in (16) by the sum

$$(17) \qquad u_n(x) = f(x) + \sum_{j=1}^{n} \lambda K(x, x_j)u_n(x_j)\,\delta.$$

Now equation (17) is supposed to hold for all values of x in $[a, b]$. Hence it should hold for $x = x_1, x_2, \ldots, x_n$. This gives the system of n equations

$$(18) \qquad -\sum_{j=1}^{n} \lambda K(x_i, x_j)u_n(x_j)\,\delta + u_n(x_i) = f(x_i), \qquad i = 1, 2, \cdots, n.$$

This system is a set of n nonhomogeneous linear equations for determining the n unknowns $u_n(x_1), u_n(x_2), \ldots, u_n(x_n)$.

In the theory of linear equations the following result was known: If the matrix

$$S_n = \begin{Vmatrix} 1 + a_{11} & a_{12} & a_{13} & \cdots & a_{1n} \\ a_{21} & 1 + a_{22} & a_{23} & \cdots & a_{2n} \\ \cdots\cdots\cdots\cdots\cdots\cdots\cdots\cdots\cdots\cdots\cdots\cdots \\ a_{n1} & a_{n2} & a_{n3} & \cdots & 1 + a_{nn} \end{Vmatrix}$$

11. *Acta Math.*, 27, 1903, 365–90.

then the determinant $D(n)$ of S_n has the following expansion:[12]

$$D(n) = 1 + \frac{1}{1!} \sum_{r_1} a_{r_1 r_1} + \frac{1}{2!} \sum_{r_1, r_2} \begin{vmatrix} a_{r_1 r_1} & a_{r_1 r_2} \\ a_{r_2 r_1} & a_{r_2 r_2} \end{vmatrix} + \cdots$$

$$+ \frac{1}{n!} \sum_{r_1, \ldots, r_n} \begin{vmatrix} a_{r_1 r_1} & a_{r_1 r_2} & \cdots & a_{r_1 r_n} \\ a_{r_2 r_1} & a_{r_2 r_2} & \cdots & a_{r_2 r_n} \\ \cdots & \cdots & \cdots & \cdots \\ a_{r_n r_1} & a_{r_n r_2} & \cdots & a_{r_n r_n} \end{vmatrix}$$

where $r_1, r_2, \ldots, r_n$ run independently over all the values from 1 to n. By expanding the determinant of the coefficients in (18), and then letting n become infinite, Fredholm obtained the determinant

$$(19) \quad D(\lambda) = 1 - \lambda \int_a^b K(\xi_1, \xi_1)\, d\xi_1$$

$$+ \frac{\lambda^2}{2!} \int_a^b \int_a^b \begin{vmatrix} K(\xi_1, \xi_1) & K(\xi_1, \xi_2) \\ K(\xi_2, \xi_1) & K(\xi_2, \xi_2) \end{vmatrix} d\xi_1\, d\xi_2 - \cdots.$$

This he called the determinant of (16) or of the kernel K. Likewise, by considering the cofactor of the element in the νth row and μth column of the determinant of the coefficients in (18) and letting n become infinite, Fredholm obtained the function

$$(20) \quad D(x, y, \lambda) = \lambda K(x, y) - \lambda^2 \int_a^b \begin{vmatrix} K(x, y) & K(x, \xi_1) \\ K(\xi_1, y) & K(\xi_1, \xi_1) \end{vmatrix} d\xi_1$$

$$+ \frac{\lambda^3}{2} \int_a^b \int_a^b \begin{vmatrix} K(x, y) & K(x, \xi_1) & K(x, \xi_2) \\ K(\xi_1, y) & K(\xi_1, \xi_1) & K(\xi_1, \xi_2) \\ K(\xi_2, y) & K(\xi_2, \xi_1) & K(\xi_2, \xi_2) \end{vmatrix} d\xi_1\, d\xi_2 - \cdots.$$

Fredholm called $D(x, y, \lambda)$ the first minor of the kernel K because it plays the role analogous to first minors in the case of n linear equations in n unknowns. He also called the zeros of the integral analytic function $D(\lambda)$ the roots of $K(x, y)$. By applying Cramer's rule to the system of linear equations (18) and by letting n become infinite, Fredholm inferred the form of the solution of (16). He then proved that it was correct by direct substitution and could assert the following results: If λ is not one of the roots of K, that is, if $D(\lambda) \neq 0$, (16) has one and only one (continuous) solution, namely,

$$(21) \quad u(x, \lambda) = f(x) + \int_a^b \frac{D(x, y, \lambda)}{D(\lambda)} f(y)\, dy.$$

Further, if λ is a root of $K(x, y)$, then (16) has either no continuous solution or an infinite number of them.

12. A fine exposition can be found in Gerhard Kowalewski, *Integralgleichungen*, Walter de Gruyter, 1930, pp. 101–34.

Fredholm obtained further results involving the relation between the homogeneous equation

$$(22) \qquad u(x) = \lambda \int_a^b K(x, \xi)u(\xi) \, d\xi$$

and the inhomogeneous equation (16). It is almost evident from (21) that when λ is not a root of K the only continuous solution of (22) is $u \equiv 0$. Hence he considered the case when λ is a root of K. Let $\lambda = \lambda_1$ be such a root. Then (22) has the infinite number of solutions

$$c_1 u_1(x) + c_2 u_2(x) + \cdots + c_n u_n(x),$$

where the c_i's are arbitrary constants; the $u_1, u_2, \ldots, u_n$, called principal solutions, are linearly independent; and n depends upon λ_1. The quantity n is called the index of λ_1 [which is not the multiplicity of λ_1 as a zero of $D(\lambda)$]. Fredholm was able to determine the index of any root λ_i and to show that the index can never exceed the multiplicity (which is always finite). The roots of $D(\lambda) = 0$ are called the characteristic values of $K(x, y)$ and the set of roots is called the spectrum. The solutions of (22) corresponding to the characteristic values are called eigenfunctions or characteristic functions.

And now Fredholm was able to establish what has since been called the Fredholm alternative theorem. In the case where λ is a characteristic value of K not only does the integral equation (22) have n independent solutions but the associated or adjoint equation which has the transposed kernel, namely,

$$u(x) = \lambda \int_a^b K(\xi, x)u(\xi) \, d\xi,$$

also has n solutions $\psi_1(x), \ldots, \psi_n(x)$ for the same characteristic value and then the nonhomogeneous equation (16) is solvable if and only if

$$(23) \qquad \int_a^b f(x)\psi_i(x) \, dx = 0, \qquad i = 1, 2, \cdots, n.$$

These last few results parallel very closely the theory of a system of linear algebraic equations, homogeneous and nonhomogeneous.

3. The Work of Hilbert

A lecture by Erik Holmgren (b. 1872) in 1901 on Fredholm's work on integral equations, which had already been published in Sweden, aroused Hilbert's interest in the subject. David Hilbert (1862–1943), the leading mathematician of this century, who had already done superb work on algebraic numbers, algebraic invariants, and the foundations of geometry, now turned his attention to integral equations. He says that an investigation of the subject showed him that it was important for the theory of definite integrals, for the development of arbitrary functions in series (of special functions or trigonometric functions), for the theory of linear differential equations, for potential theory, and for the calculus of variations. He wrote a series of six papers from 1904 to 1910 in the *Nachrichten von der Königlichen Gesellschaft der Wissenschaften zu Göttingen* and reproduced these in his book *Grundzüge einer allgemeinen Theorie der linearen Integralgleichungen* (1912). During the latter part of this work he applied integral equations to problems of mathematical physics.

Fredholm had used the analogy between integral equations and linear algebraic equations, but instead of carrying out the limiting processes for the infinite number of algebraic equations, he boldly wrote down the resulting determinants and showed that they solved the integral equations. Hilbert's first work was to carry out a rigorous passage to the limit on the finite system of linear equations.

He started with the integral equation

$$(24) \qquad f(s) = \phi(s) - \lambda \int_0^1 K(s, t)\phi(t) \, dt,$$

wherein $K(s, t)$ is continuous. The parameter λ is explicit and plays a significant role in the subsequent theory. Like Fredholm, Hilbert divided up the interval $[0, 1]$ into n parts so that p/n or q/n ($p, q = 1, 2, \ldots, n$) denotes a coordinate in the interval $[0, 1]$. Let

$$K_{pq} = K\left(\frac{p}{n}, \frac{q}{n}\right), \qquad f_p = f\left(\frac{p}{n}\right), \qquad \phi_q = \phi\left(\frac{q}{n}\right).$$

Then from (24) we obtain the system of n equations in n unknowns $\phi_1, \ldots, \phi_n$, namely,

$$f_p = \phi_p - \lambda \sum_{q=1}^n K_{pq}\phi_q, \qquad p = 1, 2, \cdots, n.$$

After reviewing the theory of solution of a finite system of n linear equations in n unknowns, Hilbert considers equation (24). For the kernel K of (24) the eigenvalues are defined to be zeros of the power series

$$\delta(\lambda) = 1 + \sum_{n=1}^\infty (-1)^n d_n \lambda^n,$$

where the coefficients d_n are given by

$$d_n = \frac{1}{n!} \int_0^1 \cdots \int_0^1 |\{K(s_i, s_j)\}| \, ds_1 \cdots ds_n.$$

Here $|\{K(s_i, s_j)\}|$ is the determinant of the n by n matrix $\{K(s_i, s_j)\}$, $i, j = 1, 2, \ldots, n$, and the s_i are values of t in the interval $[0, 1]$. To indicate Hilbert's major result we need the intermediate quantities

$$\Delta_p(x, y) = \frac{1}{p!} \int_0^1 \cdots \int_0^1 \begin{vmatrix} 0 & x(s_1) & \cdots & x(s_p) \\ y(s_1) & K(s_1, s_2) & \cdots & K(s_1, s_p) \\ \vdots & \vdots & \vdots & \vdots \\ y(s_p) & K(s_p, s_1) & \cdots & K(s_p, s_p) \end{vmatrix} ds_1 \cdots ds_p,$$

wherein $x(r)$ and $y(r)$ are arbitrary continuous functions of r on $[0, 1]$, and

$$\Delta(\lambda; x, y) = \sum_{p=1}^{\infty} (-1)^p \Delta_p(x, y) \lambda^{p-1}.$$

Hilbert next defines

$$\Delta^*(\lambda; s, t) = \lambda \Delta(\lambda; x, y) - \delta(\lambda)$$

wherein now $x(r) = K(s, r)$ and $y(r) = K(r, t)$. He then proves that if $\overline{K}$ is defined by

$$\overline{K}(s, t) = \frac{\Delta^*(\lambda; s, t)}{-\delta(\lambda)}$$

for values of λ for which $\delta(\lambda) \neq 0$, then

$$K(s, t) = \overline{K}(s, t) - \lambda \int_0^1 \overline{K}(s, r) K(r, t) \, dr$$

$$= \overline{K}(s, t) - \lambda \int_0^1 K(s, r) \overline{K}(r, t) \, dr.$$

Finally if ϕ is taken to be

(25) $$\phi(r) = f(r) + \lambda \int_0^1 \overline{K}(r, t) f(t) \, dt,$$

then ϕ is a solution of (24). The proofs of various steps in this theory involve a number of limit considerations on expressions which occur in Hilbert's treatment of the finite system of linear equations.

Thus far Hilbert showed that for any continuous (not necessarily symmetric) kernel $K(s, t)$ and for any value of λ such that $\delta(\lambda) \neq 0$, there exists the solving function (resolvent) $\overline{K}(s, t)$, which has the property that (25) solves equation (24).

Now Hilbert assumes $K(s, t)$ to be symmetric, which enables him to use facts about symmetric matrices in the finite case, and shows that the zeros of $\delta(\lambda)$, that is, the eigenvalues of the symmetric kernel, are real. Then the zeros of $\delta(\lambda)$ are ordered according to increasing absolute values (for equal absolute values the positive zero is taken first and any multiplicities are to be counted). The eigenfunctions of (24) are now defined by

$$\phi^k(s) = \left(\frac{\lambda_k}{\Delta^*(\lambda_k; s^*, s^*)}\right)^{1/2} \Delta^*(\lambda_k; s, s^*),$$

where s^* is chosen so that $\Lambda^*(\lambda_k; s^*, s^*) \neq 0$ and λ_k is any eigenvalue of $K(s, t)$.

The eigenfunctions associated with the separate eigenvalues can be chosen to be an orthonormal (orthogonal and normalized[13]) set and for each eigenvalue λ_k and for each eigenfunction belonging to λ_k

$$\phi^k(s) = \lambda_k \int_0^1 K(s, t)\phi^k(t) \, dt.$$

With these results Hilbert is able to prove what is called the generalized principal axis theorem for symmetric quadratic forms. First, let

(26)
$$\sum_{p=1}^n \sum_{q=1}^n k_{pq}x_p x_q$$

be an n-dimensional quadratic form in the n variables $x_1, x_2, \ldots, x_n$. This can be written as (Kx, x) where K is the matrix of the k_{pq}, x stands for the vector $(x_1, x_2, \ldots, x_n)$ and (Kx, x) is the inner product (scalar product) of the two vectors Kx and x. Suppose K has the n distinct eigenvalues $\lambda_1, \lambda_2, \ldots, \lambda_n$. Then for any fixed λ_k the equations

(27)
$$0 = \phi_p - \lambda_p \sum_{q=1}^n k_{pq}\phi_q, \qquad p = 1, 2, \cdots, n$$

have the solution

$$\phi^k = (\phi_1^k, \phi_2^k, \cdots, \phi_n^k),$$

which is a unique solution up to a constant multiple. It is then possible, as Hilbert showed, to write

(28)
$$K(x, x) = \sum_{k=1}^n \frac{1}{\lambda_k} \frac{(\phi^k, x)^2}{(\phi^k, \phi^k)}$$

wherein the parentheses again denote an inner product of vectors.

Hilbert's generalized principal axis theorem reads as follows: Let $K(s, t)$ be a continuous symmetric function of s and t. Let $\phi^p(s)$ be the normalized eigenfunction belonging to the eigenvalue λ_p of the integral equation

---

13. Normalization means that $\phi^k(s)$ is modified so that $\int_0^1 (\phi^k)^2 \, ds = 1$.

(24). Then for arbitrary continuous $x(s)$ and $y(s)$, the following relation holds:

(29) $$\int_a^b \int_a^b K(s, t)x(s)y(t) \, ds \, dt = \sum_{p=1}^{\alpha} \frac{1}{\lambda_p} \left(\int_a^b \phi^p(s)x(s) \, ds \right) \left(\int_a^b \phi^p(s)y(s) \, ds \right),$$

where $\alpha = n$ or ∞ depending on the number of eigenvalues and in the latter case the sum converges uniformly and absolutely for all $x(s)$ and $y(s)$ which satisfy

$$\int_a^b x^2(s) \, ds < \infty \quad \text{and} \quad \int_a^b y^2(s) \, ds < \infty.$$

The generalization of (28) to (29) becomes apparent if we first define $\int_a^b u(s)v(s) \, ds$ as the inner product of any two functions $u(s)$ and $v(s)$ and denote it by (u, v). Now replace $y(s)$ in (29) by $x(s)$ and replace the left side of (28) by integration instead of summation.

Hilbert proved next a famous result, later called the Hilbert-Schmidt theorem. If $f(s)$ is such that for some continuous $g(s)$

(30) $$f(s) = \int_a^b K(s, t)g(t) \, dt$$

then

(31) $$f(s) = \sum_{p=1}^{\infty} c_p \phi^p,$$

where the ϕ^p are the orthonormal eigenfunctions of K and

(32) $$c_p = \int_a^b \phi^p(s)f(s) \, ds.$$

Thus an "arbitrary" function $f(s)$ can be expressed as an infinite series in the eigenfunctions of K with coefficients c_p that are the "Fourier" coefficients of the expansion.

Hilbert, in the preceding work, had carried out limiting processes that generalized results on finite systems of linear equations and finite quadratic forms to integrals and integral equations. He decided that a treatment of infinite quadratic forms themselves, that is, quadratic forms with infinitely many variables, would "form an essential completion of the well-known theory of quadratic forms with finitely many variables." He therefore took up what may be called purely algebraic questions. He considers the infinite bilinear form

$$K(x, y) = \sum_{p,q=1}^{\infty} k_{pq} x_p y_q,$$

and by passing to the limit of results for bilinear and quadratic forms in $2n$ and n variables respectively, obtains basic results. The details of the work are considerable and we shall only note some of the results. Hilbert first obtains an expression for a resolvent form $\overline{K}(\lambda; x, x)$, which has the peculiar feature that it is the sum of expressions, one for each of a discrete set of values of λ, and of an integral over a set of λ belonging to a continuous range. The discrete set of λ values belongs to the point spectrum of K, and the continuous set to the continuous or band spectrum. This is the first significant use of continuous spectra, which had been observed for partial differential equations by Wilhelm Wirtinger (b. 1865) in 1896.[14]

To get at the key result for quadratic forms, Hilbert introduces the notion of a bounded form. The notation (x, x) denotes the inner (scalar) product of the vector $(x_1, x_2, \ldots, x_n, \ldots)$ with itself and (x, y) has the analogous meaning. Then the form $K(x, y)$ is said to be bounded if $|K(x, y)| \leq M$ for all x and y such that $(x, x) \leq 1$ and $(y, y) \leq 1$. Boundedness implies continuity, which Hilbert defines for a function of infinitely many variables.

Hilbert's key result is the generalization to quadratic forms in infinitely many variables of the more familiar principal axis theorem of analytic geometry. He proves that there exists an orthogonal transformation T such that in the new variables x_i', where $x' = Tx$, K can be reduced to a "sum of squares." That is, every bounded quadratic form $K(x, x) = \sum_{p,q=1}^{\infty} k_{pq} x_p x_q$ can be transformed by a unique orthogonal transformation into the form

$$(33) \qquad K(x, x) = \sum_{i=1}^{\infty} k_i x_i^2 + \int_{(s)} \frac{do(\mu, \xi)}{\mu},$$

where the k_i are the reciprocal eigenvalues of K. The integral, which we shall not describe further, is over a continuous range of eigenvalues or a continuous spectrum.

To eliminate the continuous spectrum, Hilbert introduces the concept of complete continuity. A function $F(x_1, x_2, \ldots)$ of infinitely many variables is said to be completely continuous at $a = (a_1, a_2, \ldots)$ if

$$\lim_{\substack{\varepsilon_1 \to 0, \\ \varepsilon_2 \to 0, \\ \cdots}} F(a_1 + \varepsilon_1, a_2 + \varepsilon_2, \cdots) = F(a_1, a_2, \cdots)$$

whenever $\varepsilon_1, \varepsilon_2, \ldots$ are allowed to run through any value system $\varepsilon_1^{(h)}, \varepsilon_2^{(h)}, \ldots$ having the limit

$$\lim_{h \to \infty} \varepsilon_1^{(h)} = 0, \lim_{h \to \infty} \varepsilon_2^{(h)} = 0, \cdots.$$

This is a stronger requirement than continuity as previously introduced by Hilbert.

14. *Math. Ann.*, 48, 1897, 365–89.

For a quadratic form $K(x, x)$ to be completely continuous it is sufficient that $\sum_{p,q=1}^{\infty} k_{pq}^2 < \infty$. With the requirement of complete continuity, Hilbert is able to prove that if K is a completely continuous bounded form, then by an orthogonal transformation it can be brought into the form

$$(34) \qquad K(x, x) = \sum_j k_j x_j^2,$$

where the k_j are reciprocal eigenvalues and the $(x_1, x_2, \ldots)$ satisfy the condition that $\sum_1^{\infty} x_i^2$ is finite.

Now Hilbert applies his theory of quadratic forms in infinitely many variables to integral equations. The results in many instances are not new but are obtained by clearer and simpler methods. Hilbert starts this new work on integral equations by defining the important concept of a complete orthogonal system of functions $\{\phi_p(s)\}$. This is a sequence of functions all defined and continuous on the interval $[a, b]$ with the following properties:

(a) orthogonality:

$$\int_a^b \phi_p(s)\phi_q(s) = \delta_{pq}, \qquad p, q = 1, 2, \cdots.$$

(b) completeness: for every pair of functions u and v defined on $[a, b]$

$$\int_a^b u(s)v(s)\, ds = \sum_{p=1}^{\infty} \int_a^b \phi_p(s)u(s)\, ds \int_a^b \phi_p(s)v(s)\, ds.$$

The value

$$u_p^* = \int_a^b \phi_p(s)u(s)\, ds$$

is called the Fourier coefficient of $u(s)$ with respect to the system $\{\phi_p\}$.

Hilbert shows that a complete orthonormal system can be defined for any finite interval $[a, b]$, for example, by the use of polynomials. Then a generalized Bessel's inequality is proved, and finally the condition

$$\int_a^b u^2(s)\, ds = \sum_{p=1}^{\infty} u_p^{*2}$$

is shown to be equivalent to completeness.

Hilbert now turns to the integral equation

$$(35) \qquad f(s) = \phi(s) + \int_a^b K(s, t)\phi(t)\, dt.$$

The kernel $K(s, t)$, not necessarily symmetric, is developed in a double "Fourier" series by means of the coefficients

$$a_{pq} = \int_a^b \int_a^b K(s, t)\phi_p(s)\phi_q(t) \, ds \, dt.$$

It follows that

$$\sum_{p,q=1}^{\infty} a_{pq}^2 \leq \int_a^b \int_a^b K^2(s, t) \, ds \, dt.$$

Also, if

$$a_p = \int_a^b \phi_p(s)f(s) \, ds,$$

that is, if the a_p are the "Fourier" coefficients of $f(s)$, then $\sum_{p=1}^{\infty} a_p^2 < \infty$.

Hilbert next converts the above integral equation into a system of infinitely many linear equations in infinitely many unknowns. The idea is to look at solving the integral equation for $\phi(s)$ as a problem of finding the "Fourier" coefficients of $\phi(s)$. Denoting the as yet unknown coefficients by $x_1, x_2, \ldots$, he gets the following linear equations:

$$\text{(36)} \qquad x_p + \sum_{q=1}^{\infty} a_{pq}x_q = a_p, \qquad p = 1, 2, \cdots.$$

He proves that if this system has a unique solution, then the integral equation has a unique continuous solution, and when the linear homogeneous system associated with (36) has n linearly independent solutions, then the homogeneous integral equation associated with (35),

$$\text{(37)} \qquad 0 = \phi(s) + \int_a^b K(s, t)\phi(t) \, dt,$$

has n linearly independent solutions. In this case the original nonhomogeneous integral equation has a solution if and only if $\psi^{(h)}(s)$, $h = 1, 2, \ldots, n$, which are the n linearly independent solutions of the transposed homogeneous equation

$$\phi(s) + \int_a^b K(t, s)\phi(t) \, dt = 0$$

and which also exist when (37) has n solutions, satisfy the conditions

$$\text{(38)} \qquad \int_a^b \psi^{(h)}(s)f(s) \, ds = 0, \qquad h = 1, 2, \cdots, n.$$

Thus the Fredholm alternative theorem is obtained: Either the equation

$$\text{(39)} \qquad f(s) = \phi(s) + \int_a^b K(s, t)\phi(t) \, dt$$

has a unique solution for all f or the associated homogeneous equation has n linearly independent solutions. In the latter case (39) has a solution if and only if the orthogonality conditions (38) are satisfied.

Hilbert turns next to the eigenvalue problem

$$(40) \qquad f(s) = \phi(s) - \lambda \int_a^b K(s, t)\phi(t) \, dt$$

where K is now symmetric. The symmetry of K implies that its "Fourier" coefficients determine a quadratic form $K(x, x)$ which is completely continuous. He shows that there exists an orthogonal transformation T whose matrix is $\{l_{pq}\}$ such that

$$K(x', x') = \sum_{p=1}^{\infty} \mu_p x_p'^2,$$

where the μ_p are the reciprocal eigenvalues of the quadratic form $K(x, x)$. The eigenfunctions $\{\phi_p(s)\}$ for the kernel $K(s, t)$ are now defined by

$$L_p(K(s)) = \sum_{q=1}^{\infty} l_{pq} \int_0^b K(s, t)\Phi_q(t) \, dt = \mu_p \phi_p(s)$$

where the $\Phi_q(t)$ are a given complete orthonormal set. The $\phi_p(s)$ [as distinct from the $\Phi_q(t)$] are shown to form an orthonormal set and to satisfy

$$\phi_p(s) = \lambda_p \int_a^b K(s, t)\phi_p(t) \, dt$$

where $\lambda_p = 1/\mu_p$. Thus Hilbert shows anew the existence of eigenfunctions for the homogeneous case of (40) and for every finite eigenvalue of the quadratic form $K(x, x)$ associated with the kernel $K(s, t)$ of (40).

Now Hilbert establishes again (Hilbert-Schmidt theorem) that if $f(s)$ is any continuous function for which there is a g so that

$$\int_a^b K(s, t)g(t) \, dt = f(s),$$

then f is representable as a series in the eigenfunctions of K which is uniformly and absolutely convergent (cf. [31]). Hilbert uses this result to show that the homogeneous equation associated with (40) has no nontrivial solutions except at the eigenvalues λ_p. Then the Fredholm alternative theorem takes the form: For $\lambda \neq \lambda_p$ equation (40) has a unique solution. For $\lambda = \lambda_p$, equation (40) has a solution if and only if the n_p conditions

$$\int_a^b \phi_{p+j}(s)f(s) = 0, \qquad j = 1, 2, \cdots, n_p$$

are satisfied where the $\phi_{p+j}(s)$ are the n_p eigenfunctions associated with λ_p. Finally, he proves anew the extension of the principal axis theorem:

$$\int_a^b \int_a^b K(s,\,t)u(s)u(t)\ ds\ dt = \sum_{p=1}^{\infty} \frac{1}{\lambda_p}\left\{ \int_a^b u(t)\phi_p(t)\ dt \right\}^2,$$

where $u(s)$ is an arbitrary (continuous) function and wherein all ϕ_p associated with any λ_p are included in the summation.

This later work (1906) of Hilbert dispensed with Fredholm's infinite determinants. In it he showed directly the relation between integral equations and the theory of complete orthogonal systems and the expansion of functions in such systems.

Hilbert applied his results on integral equations to a variety of problems in geometry and physics. In particular, in the third of the six papers he solved Riemann's problem of constructing a function holomorphic in a domain bounded by a smooth curve when the real or the imaginary part of the boundary value is given or both are related by a given linear equation.

One of the most noteworthy achievements in Hilbert's work, which appears in the 1904 and 1905 papers, is the formulation of Sturm-Liouville boundary-value problems of differential equations as integral equations. Hilbert's result states that the eigenvalues and eigenfunctions of the differential equation

$$(41) \qquad\qquad \frac{d}{dx}\left(p(x)\,\frac{du}{dx} \right) + q(x)u + \lambda u = 0$$

subject to the boundary conditions

$$u(a) = 0, \qquad u(b) = 0$$

(and even more general boundary conditions) are the eigenvalues and eigenfunctions of

$$(42) \qquad\qquad \phi(x) - \lambda \int_a^b G(x,\,\xi)\phi(\xi)\ d\xi = 0,$$

where $G(x,\,\xi)$ is the Green's function for (41), that is, a particular solution of

$$\frac{d}{dx}\left(p\,\frac{du}{dx} \right) + q(x)u = 0$$

which satisfies certain differentiability conditions and whose first partial derivative $\partial G/\partial x$ has a jump singularity at $x = \xi$ equal to $-1/p(\xi)$. Similar results hold for partial differential equations. Thus integral equations are a way of solving ordinary and partial differential equations.

To recapitulate Hilbert's major results, first of all he established the general spectral theory for symmetric kernels K. Only twenty years earlier, it had required great mathematical efforts (Chap. 28, sec. 8) to prove the existence of the lowest oscillating frequency for a membrane. With integral equations, constructive proof of the existence of the whole series of frequencies and of the actual eigenfunctions was obtained under very general conditions on the oscillating medium. These results were first derived, using Fredholm's theory, by Emile Picard.[15] Another noteworthy result due to Hilbert is that the development of a function in the eigenfunctions belonging to an integral equation of the second kind depends on the solvability of the corresponding integral equation of the first kind. In particular, Hilbert discovered that the success of Fredholm's method rested on the notion of complete continuity, which he carried over to bilinear forms and studied intensively. Here he inaugurated the spectral theory of bilinear symmetric forms.

After Hilbert showed how to convert problems of differential equations to integral equations, this approach was used with increasing frequency to solve physical problems. Here the use of a Green's function to convert has been a major tool. Also, Hilbert himself showed,[16] in problems of gas dynamics, that one can go directly to integral equations. This direct recourse to integral equations is possible because the concept of summation proves as fundamental in some physical problems as the concept of rate of change which leads to differential equations is in other problems. Hilbert also emphasized that not ordinary or partial differential equations but integral equations are the necessary and natural starting point for the theory of expansion of functions in series, and that the expansions obtained through differential equations are just special cases of the general theorem in the theory of integral equations.

4. The Immediate Successors of Hilbert

Hilbert's work on integral equations was simplified by Erhard Schmidt (1876–1959), professor at several German universities, who used methods originated by H. A. Schwarz in potential theory. Schmidt's most significant contribution was his generalization in 1907 of the concept of eigenfunction to integral equations with non-symmetric kernels.[17]

Friedrich Riesz (1880–1956), a Hungarian professor of mathematics, in 1907 also took up Hilbert's work.[18] Hilbert had treated integral equations of the form

$$f(s) = \phi(s) + \int_a^b K(s, t)\phi(t)\, dt,$$

15. *Rendiconti del Circolo Matematico di Palermo*, 22, 1906, 241–59.
16. *Math. Ann.*, 72, 1912, 562–77 = *Grundzüge*, Chap. 22.
17. *Math. Ann.*, 63, 1907, 433–76 and 64, 1907, 161–74.
18. *Comp. Rend.*, 144, 1907, 615–19, 734–36, 1409–11.

where f and K are continuous. Riesz sought to extend Hilbert's ideas to more general functions $f(s)$. Toward this end it was necessary to be sure that the "Fourier" coefficients of f could be determined with respect to a given orthonormal sequence of functions $\{\phi_p\}$. He was also interested in discovering under what circumstances a given sequence of numbers $\{a_p\}$ could be the Fourier coefficients of some function f relative to a given orthonormal sequence $\{\phi_p\}$.

Riesz introduced functions whose squares are Lebesgue integrable and obtained the following theorem: Let $\{\phi_p\}$ be an orthonormal sequence of Lebesgue square integrable functions all defined on the interval $[a, b]$. If $\{a_p\}$ is a sequence of real numbers, then the convergence of $\sum_{p=1}^{\infty} a_p^2$ is a necessary and sufficient condition for there to exist a function f such that

$$\int_a^b f(x)\phi_p(x)\, dx = a_p$$

for each ϕ_p and a_p. The function f proves to be Lebesgue square integrable. This theorem establishes a one-to-one correspondence between the set of Lebesgue square integrable functions and the set of square summable sequences through the mediation of any orthonormal sequence of Lebesgue square integrable functions.

With the introduction of Lebesgue integrable functions, Riesz was also able to show that the integral equation of the second kind

$$f(s) = \phi(s) + \int_a^b K(s, t)\phi(t)\, dt$$

can be solved under the relaxed conditions that $f(s)$ and $K(s, t)$ are Lebesgue square integrable. Solutions are unique up to a function whose Lebesgue integral on $[a, b]$ is 0.

In the same year that Riesz published his first papers, Ernst Fischer (1875–1959), a professor at the University of Cologne, introduced the concept of convergence in the mean.[19] A sequence of functions $\{f_n\}$ defined on an interval $[a, b]$ is said to converge in the mean if

$$\lim_{m,n \to \infty} \int_a^b (f_n(x) - f_m(x))^2\, dx = 0,$$

and $\{f_n\}$ is said to converge in the mean to f if

$$\lim_{n \to \infty} \int_a^b (f - f_n)^2\, dx = 0.$$

The integrals are Lebesgue integrals. The function f is uniquely determined to within a function defined over a set of measure 0, that is, a function $g(x) \neq 0$, called a null function, which satisfies the condition $\int_a^b g^2(x)\, dx = 0$.

19. *Comp. Rend.*, 144, 1907, 1022–24.

The set of functions which are Lebesgue square integrable on an interval $[a, b]$ was denoted later by $L^2(a, b)$ or simply by L^2. Fischer's main result is that $L^2(a, b)$ is complete in the mean; that is, if the functions f_n belong to $L^2(a, b)$ and if $\{f_n\}$ converges in the mean, then there is a function f in $L^2(a, b)$ such that $\{f_n\}$ converges in the mean to f. This completeness property is the chief advantage of square summable functions. Fischer then deduced Riesz's above theorem as a corollary, and this result is known as the Riesz-Fischer theorem. Fischer emphasized in a subsequent note [20] that the use of Lebesgue square integrable functions was essential. No smaller set of functions would suffice.

The determination of a function $f(x)$ that belongs to a given set of Fourier coefficients $\{a_n\}$ with respect to a given sequence of orthonormal functions $\{g_n\}$, or the determination of an f satisfying

$$\int_a^b g_n(x) f(x)\, dx = a_n, \qquad n = 1, 2, \cdots,$$

which had arisen in Riesz's 1907 works, is called the moment problem (Lebesgue integration is understood). In 1910 [21] Riesz sought to generalize this problem. Because in this new work Riesz used the Hölder inequalities

$$\sum_{i=1}^n |a_i b_i| \leq \left(\sum_{i=1}^n |a_i|^p\right)^{1/p} \left(\sum_{i=1}^n |b_i|^q\right)^{1/q}$$

and

$$\left|\int_M f(x) g(x)\, dx\right| \leq \left(\int_M |f|^p\, dx\right)^{1/p} \left(\int_M |g|^q\, dx\right)^{1/q},$$

where $1/p + 1/q = 1$, and other inequalities, he was obliged to introduce the set L^p of functions f measurable on a set M and for which $|f|^p$ is Lebesgue integrable on M. His first important theorem is that if a function $h(x)$ is such that the product $f(x)h(x)$ is integrable for every f in L^p, then h is in L^q and, conversely, the product of an L^p function and an L^q function is always (Lebesgue) integrable. It is understood that $p > 1$ and $1/p + 1/q = 1$.

Riesz also introduced the concepts of strong and weak convergence. The sequence of functions $\{f_n\}$ is said to converge strongly to f (in the mean of order p) if

$$\lim_{n \to \infty} \int_a^b |f_n(x) - f(x)|^p\, dx = 0.$$

20. *Comp. Rend.*, 144, 1907, 1148–50.
21. *Math. Ann.*, 69, 1910, 449–97.

The sequence $\{f_n\}$ is said to converge weakly to f if

$$\int_a^b |f_n(x)|^p \, dx < M$$

for M independent of n, and if for every x in $[a, b]$

$$\lim_{n \to \infty} \int_a^x (f_n(t) - f(t)) \, dt = 0.$$

Strong convergence implies weak convergence. (The modern definition of weak convergence, if $\{f_n\}$ belongs to L^p and if f belongs to L^p and if

$$\lim_{n \to \infty} \int_a^b (f(x) - f_n(x))g(x) \, dx = 0$$

holds for every g in L^q, then $\{f_n(x)\}$ converges weakly to f, is equivalent to Riesz's.)

In the same 1910 paper Riesz extended the theory of the integral equation

$$\phi(x) - \lambda \int_a^b K(x, t)\phi(t) \, dt = f(x)$$

to the case where the given f and the unknown ϕ are functions in L^p. The results on solution of the eigenvalue problem for this integral equation are analogous to Hilbert's results. What is more striking is that, to carry out his work, Riesz introduced the abstract concept of an operator, formulated for it the Hilbertian concept of complete continuity, and treated abstract operator theory. We shall say more about this abstract approach in the next chapter. Among other results, Riesz proved that the continuous spectrum of a real completely continuous operator in L^2 is empty.

5. *Extensions of the Theory*

The importance Hilbert attached to integral equations made the subject a world-wide fad for a considerable length of time, and an enormous literature, most of it of ephemeral value, was produced. However, some extensions of the subject have proved valuable. We can merely name them.

The theory of integral equations presented above deals with linear integral equations; that is, the unknown function $u(x)$ enters linearly. The theory has been extended to nonlinear integral equations, in which the unknown function enters to the second or higher degree or in some more complicated fashion.

Moreover, our brief sketch has said little about the conditions on the given functions $f(x)$ and $K(x, \xi)$ which lead to the many conclusions. If these functions are not continuous and the discontinuities are not limited, or if the interval $[a, b]$ is replaced by an infinite interval, many of the results are altered or at least new proofs are needed. Thus even the Fourier transform

$$f(x) = \sqrt{\frac{2}{\pi}} \int_0^\infty \cos{(x\xi)}u(\xi) \, d\xi,$$

which can be regarded as an integral equation of the first kind and has as its solution the inverse transform

$$u(x) = \sqrt{\frac{2}{\pi}} \int_0^\infty \cos{(x\xi)}f(\xi) \, d\xi,$$

has just two eigenvalues ± 1 and each has an infinite number of eigenfunctions. These cases are now studied under the heading of singular integral equations. Such equations cannot be solved by the methods applicable to the Volterra and Fredholm equations. Moreover, they exhibit a curious property, namely, there are continuous intervals of λ-values or band spectra for which there are solutions. The first significant paper on this subject is due to Hermann Weyl (1885–1955).[22]

The subject of existence theorems for integral equations has also been given a great deal of attention. This work has been devoted to linear and nonlinear integral equations. For example, existence theorems for

$$y(x) = f(x) + \int_a^x K(x, s, y(s)) \, ds,$$

which includes as a special case the Volterra equation of the second kind

$$y(x) = f(x) + \int_a^x K(x, s)y(s) \, ds,$$

have been given by many mathematicians.

Historically, the next major development was an outgrowth of the work on integral equations. Hilbert regarded a function as given by its Fourier coefficients. These satisfy the condition that $\sum_1^\infty a_p^2$ is finite. He had also introduced sequences of real numbers $\{x_n\}$ such that $\sum_{n=1}^\infty x_n^2$ is finite. Then Riesz and Fischer showed that there is a one-to-one correspondence between Lebesgue square summable functions and square summable sequences of their Fourier coefficients. The square summable sequences can be regarded as the coordinates of points in an infinite-dimensional space, which is a generalization of n-dimensional Euclidean space. Thus functions can be regarded as points of a space, now called Hilbert space, and the integral $\int_a^b K(x, y)u(x) \, dx$ can be regarded as an operator transforming $u(x)$ into itself or another function. These ideas suggested for the study of integral equations an abstract approach that fitted into an incipient abstract approach to the calculus of variations. This new approach is now known as functional analysis and we shall consider it next.

22. *Math. Ann.*, 66, 1908, 273–324 = *Ges. Abh.*, 1, 1–86.

Bibliography

Bernkopf, M.: "The Development of Function Spaces with Particular Reference to their Origins in Integral Equation Theory," *Archive for History of Exact Sciences*, 3, 1966, 1–96.

Bliss, G. A.: "The Scientific Work of E. H. Moore," *Amer. Math. Soc. Bull.*, 40, 1934, 501–14.

Bocher, M.: *An Introduction to the Study of Integral Equations*, 2nd ed., Cambridge University Press, 1913.

Bourbaki, N.: *Eléments d'histoire des mathématiques*, Hermann, 1960, pp. 230–45.

Davis, Harold T.: *The Present State of Integral Equations*, Indiana University Press, 1926.

Hahn, H.: "Bericht über die Theorie der linearen Integralgleichungen," *Jahres. der Deut. Math.-Verein.*, 20, 1911, 69–117.

Hellinger, E.: *Hilberts Arbeiten über Integralgleichungen und unendliche Gleichungssysteme*, in Hilbert's *Gesam. Abh.*, 3, 94–145, Julius Springer, 1935.

———: "Begründung der Theorie quadratischer Formen von unendlichvielen Veränderlichen," *Jour. für Math.*, 136, 1909, 210–71.

Hellinger, E., and O. Toeplitz: "Integralgleichungen und Gleichungen mit unendlichvielen Unbekannten," *Encyk. der Math. Wiss.*, B. G. Teubner, 1923–27, Vol. 2, Part 3, 2nd half, 1335–1597.

Hilbert, D.: *Grundzüge einer allgemeinen Theorie der linearen Integralgleichungen*, 1912, Chelsea (reprint), 1953.

Reid, Constance: *Hilbert*, Springer-Verlag, 1970.

Volterra, Vito: *Opere matematiche*, 5 vols., Accademia Nazionale dei Lincei, 1954–62.

Weyl, Hermann: *Gesammelte Abhandlungen*, 4 vols, Springer-Verlag, 1968.

46

Functional Analysis

1. *The Nature of Functional Analysis*

By the late nineteenth century it was apparent that many domains of
mathematics dealt with transformations or operators acting on functions.
Thus even the ordinary differentiation operation and its inverse antidifferentiation act on a function to produce a new function. In problems of the
calculus of variations wherein one deals with integrals such as

$$J = \int_a^b F(x, y, y') \, dx,$$

the integral can be regarded as operating on a class of functions $y(x)$ of
which that one is sought which happens to maximize or minimize the integral.
The area of differential equations offers another class of operators. Thus the
differential operator

$$L = \frac{d^2}{dx^2} + p(x) \frac{d}{dx} + q(x),$$

acting on a class $y(x)$ of functions, converts these to other functions. Of
course to solve the differential equation, one seeks a particular $y(x)$ such
that L acting on that $y(x)$ yields 0 and perhaps satisfies initial or boundary
conditions. As a final example of operators we have integral equations. The
right-hand side of

$$f(x) = \int_a^b K(x, y)u(x) \, dx$$

can be regarded as an operator acting on various $u(x)$ to produce new
functions, though again, as in the case of differential equations, the $u(x)$
that solves the equation is transformed into $f(x)$.

The idea that motivated the creation of functional analysis is that all of these operators could be considered under one abstract formulation of an operator acting on a class of functions. Moreover, these functions could be regarded as elements or points of a space. Then the operator transforms points into points and in this sense is a generalization of ordinary transformations such as rotations. Some of the above operators carry functions into real numbers, rather than functions. Those operators that do yield real or complex numbers are today called functionals and the term operator is more commonly reserved for the transformations that carry functions into functions. Thus the term functional analysis, which was introduced by Paul P. Lévy (1886–) when functionals were the key interest, is no longer appropriate. The search for generality and unification is one of the distinctive features of twentieth century mathematics, and functional analysis seeks to achieve these goals.

2. The Theory of Functionals

The abstract theory of functionals was initiated by Volterra in work concerned with the calculus of variations. His work on functions of lines (curves), as he called the subject, covers a number of papers.[1] A function of lines was, for Volterra, a real-valued function F whose values depend on all the values of functions $y(x)$ defined on some interval $[a, b]$. The functions themselves were regarded as points of a space for which a neighborhood of a point and the limit of a sequence of points can be defined. For the functionals $F[y(x)]$ Volterra offered definitions of continuity, derivative and differential. However, these definitions were not adequate for the abstract theory of the calculus of variations and were superseded. His definitions were in fact criticized by Hadamard.[2]

Even before Volterra commenced his work, the notion that a collection of functions $y(x)$ all defined on some common interval be regarded as points of a space had already been suggested. Riemann, in his thesis,[3] spoke of a collection of functions forming a connected closed domain (of points of a space). Giulio Ascoli (1843–1896)[4] and Cesare Arzelà[5] sought to extend to sets of functions Cantor's theory of sets of points, and so regarded functions as points of a space. Arzelà also spoke of functions of lines. Hadamard suggested at the First International Congress of Mathematicians in 1897[6]

1. *Atti della Reale Accademia dei Lincei*, (4), 3, 1887, 97–105, 141–46, 153–58 = *Opere matematiche*, 1, 294–314, and others of the same and later years.
2. *Bull. Soc. Math. de France*, 30, 1902, 40–43 = *Œuvres*, 1, 401–4.
3. *Werke*, p. 30.
4. *Memorie della Reale Accademia dei Lincei*, (3), 18, 1883, 521–86.
5. *Atti della Accad. dei Lincei, Rendiconti*, (4), 5, 1889, 342–48.
6. *Verhandlungen des ersten internationalen Mathematiker-Kongresses*, Teubner, 1898, 201–2.

that curves be considered as points of a set. He was thinking of the family of all continuous functions defined over [0, 1], a family that arose in his work on partial differential equations. Emile Borel made the same suggestion[7] for a different purpose, namely, the study of arbitrary functions by means of series.

Hadamard, too, undertook the study of functionals[8] on behalf of the calculus of variations. The term functional is due to him. According to Hadamard, a functional $U[y(t)]$ is linear if when $y(t) = \lambda_1 y_1(t) + \lambda_2 y_2(t)$, wherein λ_1 and λ_2 are constants, then $U[y(t)] = \lambda_1 U[y_1(t)] + \lambda_2 U[y_2(t)]$.

The first major effort to build up an abstract theory of function spaces and functionals was made by Maurice Fréchet (1878–), a leading French professor of mathematics, in his doctoral thesis of 1906.[9] In what he called the functional calculus he sought to unify in abstract terms the ideas in the work of Cantor, Volterra, Arzelà, Hadamard, and others.

To gain the largest degree of generality for his function spaces, Fréchet took over the collection of basic notions about sets developed by Cantor, though for Fréchet the points of the sets are functions. He also formulated more generally the notion of a limit of a set of points. This notion was not defined explicitly but was characterized by properties general enough to embrace the kinds of limit found in the concrete theories Fréchet sought to unify. He introduced a class L of spaces, the L denoting that a limit concept exists for each space. Thus if A is a space of class L and the elements are $A_1, A_2, \ldots$ chosen at random, it must be possible to determine whether or not there exists a unique element A, called the limit of the sequence $\{A_n\}$ when it exists, and such that

(a) If $A_i = A$ for every i, then $\lim \{A_i\} = A$.
(b) If A is the limit of $\{A_n\}$ then A is the limit of every infinite subsequence of $\{A_n\}$.

Then Fréchet introduced a number of concepts for any space of class L. Thus the derived set E' of a set E consists of the set of points of E which are limits of sequences belonging to E. E is closed if E' is contained in E. E is perfect if $E' = E$. A point A of E is an interior point of E (in the narrow sense) if A is not the limit of any sequence not in E. A set E is compact if either it has finitely many elements or if every infinite subset of E has at least one limit element. If E is closed and compact it is called extremal. (Fréchet's "compact" is the modern relatively sequentially compact, and his "extremal" is today's sequentially compact.) Fréchet's first important theorem is a generalization of the closed nested interval theorem: If $\{E_n\}$ is a

7. *Verhandlungen*, 204–5.
8. *Comp. Rend.*, 136, 1903, 351–54 = *Œuvres*, 1, 405–8.
9. *Rendiconti del Circolo Matematico di Palermo*, 22, 1906, 1–74.

decreasing sequence, that is, E_{n+1} is contained in E_n, of closed subsets of an extremal set, then the intersection of all the E_n is not empty.

Fréchet now considers functionals (he calls them functional operations). These are real-valued functions defined on a set E. He defines the continuity of a functional thus: A functional U is said to be continuous at an element A of E if $\lim_{n \to \infty} U(A_n) = U(A)$ for all sequences $\{A_n\}$ contained in E and converging to A. He also introduced semicontinuity of functionals, a notion introduced for ordinary functions by René Baire (1874–1932) in 1899.[10] U is upper semicontinuous in E if $U(A) \geq \lim \sup U(A_n)$ for the $\{A_n\}$ just described. It is lower semicontinuous if $U(A) \leq \lim \inf U(A_n)$.[11]

With these definitions Fréchet was able to prove a number of theorems on functionals. Thus every functional continuous on an extremal set E is bounded and attains its maximum and minimum on E. Every upper semicontinuous functional defined on an extremal set E is bounded above and attains its maximum on E.

Fréchet introduced next concepts for sequences and sets of functionals such as uniform convergence, quasi-uniform convergence, compactness, and equicontinuity. For example, the sequence of functionals $\{U_n\}$ converges uniformly to U if given any positive quantity ε, $|U_n(A) - U(A)| < \varepsilon$ for n sufficiently large but independent of A in E. He was then able to prove generalizations to functionals of theorems previously obtained for real functions.

Having treated the class of spaces L, Fréchet defines more specialized spaces such as neighborhood spaces, redefines the concepts introduced for spaces with limit points, and proves theorems analogous to those already described but often with better results because the spaces have more properties.

Finally he introduces metric spaces. In such a space a function playing the role of distance (*écart*) and denoted by (A, B) is defined for each two points A and B of the space and this function satisfies the conditions:

(a) $(A, B) = (B, A) \geq 0$;
(b) $(A, B) = 0$ if and only if $A = B$;
(c) $(A, B) + (B, C) \geq (A, C)$.

Condition C is called the triangle inequality. Such a space, he says, is of class $\mathscr{E}$. For such spaces, too, Fréchet is able to prove a number of theorems about the spaces and the functionals defined on them much like those previously proven for the more general spaces.

Fréchet gave some examples of function spaces. Thus the set of all real functions of a real variable, all continuous on the same interval I with the

---

10. *Annali di Mat.*, (3), 13, 1899, 1–122.
11. The inferior limit is the smallest limit point of the sequence $U(A_n)$.

écart of any two functions f and g defined to be $\max_{x \in I} |f(x) - g(x)|$, is a space
of class $\mathcal{E}$. Today this *écart* is called the maximum norm.

Another example offered by Fréchet is the set of all sequences of real
numbers. If $x = (x_1, x_2, \ldots)$ and $y = (y_1, y_2, \ldots)$ are any two sequences,
the *écart* of x and y is defined to be

$$(x, y) = \sum_{p=1}^{\infty} \frac{1}{p!} \frac{|x_p - y_p|}{1 + |x_p - y_p|}.$$

This, as Fréchet remarks, is a space of countably infinite dimensions.

Using his spaces $\mathcal{E}$, Fréchet[12] succeeded in giving a definition of
continuity, differential, and differentiability of a functional. Though these
were not entirely adequate for the calculus of variations, his definition of
differential is worth noting because it is the core of what did prove to be
satisfactory. He assumes that there exists a linear functional $L(\eta(x))$ such
that

$$F[y(x) + \eta(x)] = F(y) + L(\eta) + \varepsilon M(\eta),$$

where $\eta(x)$ is a variation on $y(x)$, $M(\eta)$ is the maximum absolute value
of $\eta(x)$ on $[a, b]$, and ε approaches 0 with M. Then $L(\eta)$ is the differential of
$F(y)$. He also presupposes the continuity of $F(y)$, which is more than can be
satisfied in problems of the calculus of variations.

Charles Albert Fischer (1884–1922)[13] then improved Volterra's definition of the derivative of a functional so that it did cover the functionals
used in the calculus of variations; the differential of a functional could then
be defined in terms of the derivative.

So far as the basic definitions of properties of functionals needed for the
calculus of variations are concerned, the final formulations were given by
Elizabeth Le Stourgeon (1881–1971).[14] The key concept, the differential
of a functional, is a modification of Fréchet's. The functional $F(y)$ is said
to have a differential at $y_0(x)$ if there exists a linear functional $L(\eta)$ such that
for all arcs $y_0 + \eta$ in the neighborhood of y_0 the relation

$$F(y_0 + \eta) = F(y_0) + L(\eta) + M(\eta)\varepsilon(\eta)$$

holds where $M(\eta)$ is the maximum absolute value of η and η' on the interval
$[a, b]$ and $\varepsilon(\eta)$ is a quantity vanishing with $M(\eta)$. She also defined second
differentials.

Both Le Stourgeon and Fischer deduced from their definitions of
differentials necessary conditions for the existence of a minimum of a
functional that are of a type applicable to the problems of the calculus of
variations. For example, a necessary condition that a functional $F(y)$ have

12. *Amer. Math. Soc. Trans.*, 15, 1914, 135–61.
13. *Amer. Jour. of Math.*, 35, 1913, 369–94.
14. *Amer. Math. Soc. Trans.*, 21, 1920, 357–83.

a minimum for $y = y_0$ is that $L(\eta)$ vanish for every $\eta(x)$ that is continuous and has continuous first derivatives on $[a, b]$ and such that $\eta(a) = \eta(b) = 0$. It is possible to deduce the Euler equation from the condition that the first differential vanish, and by using a definition of the second differential of a functional (which several writers already named had given), it is possible to deduce the necessity of the Jacobi condition of the calculus of variations.

The definitive work, at least up to 1925, on the theory of functionals required for the calculus of variations was done by Leonida Tonelli (1885–1946), a professor at the universities of Bologna and Pisa. After writing many papers on the subject from 1911 on, he published his *Fondamenti di calcolo delle variazioni* (2 vols., 1922, 1924), in which he considers the subject from the standpoint of functionals. The classical theory had relied a great deal on the theory of differential equations. Tonelli's aim was to replace existence theorems for differential equations by existence theorems for minimizing curves of integrals. Throughout his work the concept of the lower semicontinuity of a functional is a fundamental one because the functionals are not continuous.

Tonelli first treats sets of curves and gives theorems that insure the existence of a limiting curve in a class of curves. Ensuing theorems insure that the usual integral, but in the parametric form

$$\int_{t_1}^{t_2} F(t, x(t), y(t), x', y') \, dt,$$

will be lower semicontinuous as a function of $x(t)$ and $y(t)$. (Later he considers the more fundamental nonparametric integrals.) He derives the four classical necessary conditions of the calculus of variations for the standard types of problems. The main emphasis in the second volume is on existence theorems for a great variety of problems deduced on the basis of the notion of semicontinuity. That is, given an integral of the above form, by imposing conditions on it as a functional and on the class of curves to be considered, he proves there is a curve in the class which minimizes the integral. His theorems deal with absolute and relative minima and maxima.

To some extent Tonelli's work pays dividends in differential equations, for his existence theorems imply the existence of the solutions of the differential equations which in the classical approach supply the minimizing curves as solutions. However, his work was limited to the basic types of problems of the calculus of variations. Though the abstract approach was pursued subsequently by many men, the progress made in the application of the theory of functionals to the calculus of variations has not been great.

3. Linear Functional Analysis

The major work in functional analysis sought to provide an abstract theory for integral equations as opposed to the calculus of variations. The properties of functionals needed for the latter field are rather special and do not hold for functionals in general. In addition, the nonlinearity of these functionals created difficulties that were irrelevant to the functionals and operators that embraced integral equations. While concrete extensions of the theory of solution of integral equations were being made by Schmidt, Fischer, and Riesz, these men and others also began work on the corresponding abstract theory.

The first attempt at an abstract theory of linear functionals and operators was made by the American mathematician E. H. Moore starting in 1906.[15] Moore realized that there were features common to the theory of linear equations in a finite number of unknowns, the theory of infinitely many equations in an infinite number of unknowns, and the theory of linear integral equations. He therefore undertook to build an abstract theory, called General Analysis, which would include the above more concrete theories as special cases. His approach was axiomatic. We shall not present the details because Moore's influence was not extensive, nor did he achieve effective methodology. Moreover, his symbolic language was strange and difficult for others to follow.

The first influential step toward an abstract theory of linear functionals and operators was taken in 1907 by Erhard Schmidt[16] and Fréchet.[17] Hilbert, in his work on integral equations, had regarded a function as given by its Fourier coefficients in an expansion with respect to an orthonormal sequence of functions. These coefficients and the values he attached to the x_i in his theory of quadratic forms in infinitely many variables are sequences $\{x_n\}$ such that $\sum_1^\infty x_n^2$ is finite. However, Hilbert did not regard these sequences as coordinates of a point in space, nor did he use geometrical language. This step was taken by Schmidt and Fréchet. By regarding each sequence $\{x_n\}$ as a point, functions were represented as points of an infinite-dimensional space. Schmidt also introduced complex as well as real numbers in the sequences $\{x_n\}$. Such a space has since been called a Hilbert space. Our account follows Schmidt's work.

The elements of Schmidt's function spaces are infinite sequences of complex numbers, $z = \{z_p\}$, with

$$\sum_{p=1}^{\infty} |z_p|^2 < \infty.$$

Schmidt introduced the notation $\|z\|$ for $\{\sum_{p=1}^\infty z_p \bar{z}_p\}^{1/2}$; $\|z\|$ was later called the norm of z. Following Hilbert, Schmidt used the notation (z, w) for

15. See, for example, *Atti del IV Congresso Internazionale dei Matematici* (1908), 2, *Reale Accademia dei Lincei*, 1909, 98–114, and *Amer. Math. Soc. Bull.*, 18, 1911/12, 334–62.
16. *Rendiconti del Circolo Matematico di Palermo*, 25, 1908, 53–77.
17. *Nouvelles Annales de Mathématiques*, 8, 1908, 97–116, 289–317.

$\sum_{p=1}^{\infty} z_p w_p$, so that $\|z\| = \sqrt{(z, \bar{z})}$. (The modern practice is to define (z, w) as $\sum z_p \bar{w}_p$.) Two elements z and w of the space are called orthogonal if and only if $(z, \bar{w}) = 0$. Schmidt then proves a generalized Pythagorean theorem, namely, if $z_1, z_2, \ldots, z_n$ are n mutually orthogonal elements of the space, then

$$w = \sum_{p=1}^{n} z_p$$

implies

$$\|w\|^2 = \sum_{p=1}^{n} \|z_p\|^2.$$

It follows that the n mutually orthogonal elements are linearly independent. Schmidt also obtained Bessel's inequality in this general space; that is, if $\{z_n\}$ is an infinite sequence of orthonormal elements, so that $(z_p, \bar{z}_q) = \delta_{pq}$, and if w is any element, then

$$\sum_{p=1}^{\infty} |(w, \bar{z}_p)|^2 \le \|w\|^2.$$

Also Schwarz's inequality and the triangle inequality are proved for the norm.

A sequence of elements $\{z_n\}$ is said to converge strongly to z if $\|z_n - z\|$ approaches 0 and every strong Cauchy sequence, i.e. every sequence for which $\|z_p - z_q\|$ approaches 0 as p and q approach ∞, is shown to converge to an element z so that the space of sequences is complete. This is a vital property.

Schmidt next introduces the notion of a (strongly) closed subspace. A subset A of his space H is said to be a closed subspace if it is a closed subset in the sense of the convergence just defined and if it is algebraically closed; that is, if w_1 and w_2 are elements of A, then $a_1 w_1 + a_2 w_2$, where a_1 and a_2 are any complex numbers, is also an element of A. Such closed subspaces are shown to exist. One merely takes any sequence $\{z_n\}$ of linearly independent elements and takes all finite linear combinations of elements of $\{z_n\}$. The closure of this collection of elements is an algebraically closed subspace.

Now let A be any fixed closed subspace. Schmidt first proves that if z is any element of the space, there exist unique elements w_1 and w_2 such that $z = w_1 + w_2$ where w_1 is in A and w_2 is orthogonal to A, which means that w_2 is orthogonal to every element in A. (This result is today called the projection theorem; w_1 is the projection of z in A.) Further, $\|w_2\| = \min \|y - z\|$ where y is any element in A and this minimum is assumed only for $y = w_1$. $\|w_2\|$ is called the distance between z and A.

In 1907 both Schmidt and Fréchet remarked that the space of square summable (Lebesgue integrable) functions has a geometry completely analogous to the Hilbert space of sequences. This analogy was elucidated when, some months later, Riesz, making use of the Riesz-Fischer theorem (Chap. 45, sec. 4), which establishes a one-to-one correspondence between Lebesgue measurable square integrable functions and square summable sequences of real numbers, pointed out that in the set L^2 of square summable functions a distance can be defined and one can use this to build a geometry of this space of functions. This notion of distance between any two square summable functions of the space L^2, all defined on an interval $[a, b]$, was in fact also defined by Fréchet[18] as

(1)
$$\sqrt{\int_a^b [f(x) - g(x)]^2 \, dx},$$

where the integral is understood in the Lebesgue sense; and it is also understood that two functions that differ only on a set of measure 0 are considered equal. The square of the distance is also called the mean square deviation of the functions. The scalar product of f and g is defined to be $(f, g) = \int_a^b f(x)g(x) \, dx$. Functions for which $(f, g) = 0$ are called orthogonal. The Schwarz inequality,

$$\int_a^b f(x)\, g(x) \, dx \leq \sqrt{\int_a^b f^2 \, dx} \sqrt{\int_a^b g^2 \, dx},$$

and other properties that hold in the space of square summable sequences, apply to the function space; in particular this class of square summable functions forms a complete space. Thus the space of square summable functions and the space of square summable sequences that are the Fourier coefficients of these functions with respect to a fixed complete orthonormal system of functions can be identified.

So far as abstract function spaces are concerned, we should recall (Chap. 45, sec. 4) the introduction of L^p spaces, $1 < p < \infty$, by Riesz. These spaces are also complete in the metric $d(f_1, f_2) = \left(\int_a^b |f_1 - f_2|^p \, dx \right)^{1/p}$.

Though we shall soon look into additional creations in the area of abstract spaces, the next developments concern functionals and operators. In the 1907 paper just referred to, in which he introduced the metric or *écart* for functions in L^2 space, and in another paper of that year,[19] Fréchet proved that for every continuous linear functional $U(f)$ defined on L^2 there is a unique $u(x)$ in L^2 such that for every f in L^2

$$U(f) = \int_a^b f(x)u(x) \, dx.$$

18. *Comp. Rend.*, 444, 1907, 1414–16.
19. *Amer. Math. Soc. Trans.*, 8, 1907, 433–46.

This result generalizes one obtained by Hadamard in 1903.[20] In 1909 Riesz[21] generalized this result by expressing $U(f)$ as a Stieltjes integral; that is,

$$U(f) = \int_a^b f(x) \, du(x).$$

Ricsz himself generalized this result to linear functionals A that satisfy the condition that for all f in L^p

$$A(f) \leq M \left[\int_a^b |f(x)|^p \, dx \right]^{1/p},$$

where M depends only on A. Then there is a function $a(x)$ in L^q, unique up to the addition of a function with zero integral, such that for all f in L^p

(2) $$U(f) = \int_a^b a(x) f(x) \, dx.$$

This result is called the Riesz representation theorem.

The central part of functional analysis deals with the abstract theory of the operators that occur in differential and integral equations. This theory unites the eigenvalue theory of differential and integral equations and linear transformations operating in n-dimensional space. Such an operator, for example,

$$g(x) = \int_a^b k(x, y) f(y) \, dy$$

for given k, assigns f to g and satisfies some additional conditions. In the notation A for the abstract operator and the notation $g = Af$, linearity means

(3) $$A(\lambda_1 f_1 + \lambda_2 f_2) = \lambda_1 A f_1 + \lambda_2 A f_2,$$

where λ_i is any real or complex constant. The indefinite integral $g(x) = \int_a^x f(t) \, dt$ and the derivative $f'(x) = Df(x)$ are linear operators for the proper classes of functions. Continuity of the operator A means that if the sequence of functions f_n approaches f in the sense of the limit in the space of functions, then $A f_n$ must approach Af.

The abstract analogue of what one finds in integral equations when the kernel $k(x, y)$ is symmetric is the self-adjointness property of the operator A. If for any two functions

$$(Af_1, f_2) = (f_1, Af_2),$$

20. *Comp. Rend.*, 136, 1903, 351–54 = *Œuvres*, 1, 405–8.
21. *Comp. Rend.*, 149, 1909, 974–77, and *Ann. de l'Ecole Norm. Sup.*, 28, 1911, 33 ff.

where (Af_1, f_2) denotes the inner or scalar product of two functions of the space, then A is said to be self-adjoint. In the case of integral equations, if

$$Af_1 = \int_a^b k(x, y) f(y) \, dy,$$

then

$$(Af_1, f_2) = \int_a^b \int_a^b k(x, y) f_1(y) f_2(x) \, dy \, dx$$

$$(f_1, Af_2) = \int_a^b \int_a^b k(x, y) f_2(y) f_1(x) \, dy \, dx,$$

and $(Af_1, f_2) = (f_1, Af_2)$ if the kernel is symmetric. For arbitrary self-adjoint operators the eigenvalues are real, and the eigenfunctions corresponding to distinct eigenvalues are orthogonal to each other.

A healthy start on the abstract operator theory, which is the core of functional analysis, was made by Riesz in his 1910 *Mathematische Annalen* paper, in which he introduced L^p spaces (Chap. 45, sec. 4). He set out in that paper to generalize the solutions of the integral equation

$$\phi(x) - \lambda \int_a^b K(x, t)\phi(t) \, dt = f(x)$$

to functions in L^p spaces. Riesz thought of the expression

$$\int_a^b K(s, t)\phi(t) \, dt$$

as a transformation acting on the function $\phi(t)$. He called it a functional transformation and denoted it by $T(\phi(t))$. Moreover, since the $\phi(t)$ with which Riesz was concerned were in L^p space, the transformation takes functions into the same or another space. In particular a transformation or operator that takes functions of L^p into functions of L^p is called linear in L^p if it satisfies (3) and if T is bounded; that is, there is a constant M such that for all f in L^p that satisfy

$$\int_a^b |f(x)|^p \, dx \le 1$$

then

$$\int_a^b |T(f(x))|^p \, dx \le M^p.$$

Later the least upper bound of such M was called the norm of T and denoted by $\|T\|$.

Riesz also introduced the notion of the adjoint or transposed operator for T. For any g in L^q and T operating in L^p,

$$(4) \qquad \int_a^b T(f(x)) g(x) \, dx$$

defines for fixed g and varying f in L^p a functional on L^p. Hence by the Riesz representation theorem there is a function $\psi(x)$ in L^q, unique up to a function whose integral is 0, such that

$$(5) \qquad \int_a^b T(f(x)) g(x) \, dx = \int_a^b f(x) \psi(x) \, dx.$$

The adjoint or transpose of T, denoted by T^*, is now defined to be that operator in L^q, depending for fixed T only on g, which assigns to g the ψ of equation (5); that is, $T^*(g) = \psi$. (In modern notation T^* satisfies $(Tf, g) = (f, T^*g)$.) T^* is a linear transformation in L^q and $\|T^*\| = \|T\|$.

Riesz now considers the solution of

$$(6) \qquad T(\phi(x)) = f(x),$$

where T is a linear transformation in L^p, f is known, and ϕ is unknown. He proves that (6) has a solution if and only if

$$\left| \int_a^b f(x) g(x) \, dx \right| \leq M \left(\int_a^b |T^*(g(x))|^q \, dx \right)^{1/q}$$

for all g in L^q. He is thereby led to the notion of the inverse transformation or operator T^{-1} and in the very same thought to T^{*-1}. With the aid of the adjoint operator he proves the existence of the inverses.

In his 1910 paper Riesz introduced the notation

$$(7) \qquad \phi(x) - \lambda K(\phi(x)) = f(x),$$

where K now stands for $\int_a^b K(x, t) * dt$ and where $*$ stands for the function on which K operates. His additional results are limited to L^2 wherein $K = K^*$. To treat the eigenvalue problem of integral equations he introduces Hilbert's notion of complete continuity, but now formulated for abstract operators. An operator K in L^2 is said to be completely continuous if K maps every weakly convergent sequence (Chap. 45, sec. 4) of functions into a strongly convergent one, that is, $\{f_n\}$ converging weakly implies $\{K(f_n)\}$ converging strongly. He does prove that the spectrum of (7) is discrete (that is, there is no continuous spectrum of a symmetric K) and that the eigenfunctions associated with the eigenvalues are orthogonal.

Another approach to abstract spaces, using the notion of a norm, was also initiated by Riesz.[22] However, the general definition of normed spaces

22. *Acta Math.*, 41, 1918, 71–98.

was given during the years 1920 to 1922 by Stefan Banach (1892–1945), Hans Hahn (1879–1934), Eduard Helly (1884–1943), and Norbert Wiener (1894–1964). Though there is much overlap in the work of these men and the question of priority is difficult to settle, it is Banach's work that had the greatest influence. His motivation was the generalization of integral equations.

The essential feature of all this work, Banach's in particular,[23] was to set up a space with a norm but one which is no longer defined in terms of an inner product. Whereas in L^2, $\|f\| = (f,f)^{1/2}$ it is not possible to define the norm of a Banach space in this way because an inner product is no longer available.

Banach starts with a space E of elements denoted by $x, y, z, \ldots$, whereas $a, b, c, \ldots$ denote real numbers. The axioms for his space are divided into three groups. The first group contains thirteen axioms which specify that E is a commutative group under addition, that E is closed under multiplication by a real scalar, and that the familiar associative and distributive relations are to hold among various operations on the real numbers and the elements.

The second group of axioms characterizes a norm on the elements (vectors) of E. The norm is a real-valued function defined on E and denoted by $\|x\|$. For any real number a and any x in E the norm has the following properties:

(a) $\|x\| \geq 0$;
(b) $\|x\| = 0$ if and only if $x = 0$;
(c) $\|ax\| = |a| \cdot \|x\|$;
(d) $\|x + y\| \leq \|x\| + \|y\|$.

The third group contains just a completeness axiom, which states that if $\{x_n\}$ is a Cauchy sequence in the norm, i.e. if $\lim_{n,p \to \infty} \|x_n - x_p\| = 0$, then there is an element x in E such that $\lim_{n \to \infty} \|x_n - x\| = 0$.

A space satisfying these three groups of axioms is called a Banach space or a complete normed vector space. Though a Banach space is more general than a Hilbert space, because the inner product of two elements is not presupposed to define the norm, as a consequence the key notion of two elements being orthogonal is lost in a Banach space that is not also a Hilbert space. The first and third sets of conditions hold also for Hilbert space, but the second set is weaker than conditions on the norm in Hilbert space. Banach spaces include L^p spaces, the space of continuous functions, the space of bounded measurable functions, and other spaces, provided the appropriate norm is used.

With the concept of the norm Banach is able to prove a number of

23. *Fundamenta Mathematicae*, 3, 1922, 133–81.

familiar facts for his spaces. One of the key theorems states: Let $\{x_n\}$ be a set of elements of E such that

$$\sum_{p=1}^{\infty} \|x_p\| < \infty.$$

Then $\sum_{p=1}^{\infty} x_p$ converges in the norm to an element x of E.

After proving a number of theorems, Banach considers operators defined on the space but whose range is in another space E_1, which is also a Banach space. An operator F is said to be continuous at x_0 relative to a set A if $F(x)$ is defined for all x of A; x_0 belongs to A and to the derived set of A; and whenever $\{x_n\}$ is a sequence of A with limit x_0, then $F(x_n)$ approaches $F(x_0)$. He also defines uniform continuity of F relative to a set A and then turns his attention to sequences of operators. The sequence $\{F_n\}$ of operators is said to converge in the norm to F on a set A if for every x in A, $\lim_{n \to \infty} F_n(x) = F(x)$.

An important class of operators introduced by Banach is the set of continuous additive ones. An operator F is additive if for all x and y, $F(x + y) = F(x) + F(y)$. An additive continuous operator proves to have the property that $F(ax) = aF(x)$ for any real number a. If F is additive and continuous at one element (point) of E, it follows that it is continuous everywhere and it is bounded; that is, there is a constant M, depending only on F, so that $\|F(x)\| \le M\|x\|$ for every x of E. Another theorem asserts that if $\{F_n\}$ is a sequence of additive continuous operators and if F is an additive operator such that for each x, $\lim_{n \to \infty} F_n(x) = F(x)$, then F is continuous and there exists an M such that for all n, $\|F_n(x)\| \le M\|x\|$.

In this paper Banach proves theorems on the solution of the abstract formulation of integral equations. If F is a continuous operator with domain and range in the space E and if there exists a number M, $0 < M < 1$, so that for all x' and x'' in E, $\|F(x') - F(x'')\| \le M\|x' - x''\|$, then there exists a unique element x in E that satisfies $F(x) = x$. More important is the theorem: Consider the equation

(8) $$x + hF(x) = y$$

where y is a known function in E, F is an additive continuous operator with domain and range in E, and h is a real number. Let M be the least upper bound of all those numbers M' satisfying $\|F(x)\| \le M'\|x\|$ for all x. Then for every y and every value of h satisfying $|hM| < 1$ there exists a function x satisfying (8) and

$$x = y + \sum_{n=1}^{\infty} (-1)^n h^n F^{(n)}(y),$$

where $F^{(n)}(y) = F(F^{(n-1)}(y))$. This result is one form of the spectral radius theorem, and it is a generalization of Volterra's method of solving integral equations.

Banach in 1929[24] introduced another important notion in the subject of functional analysis, the notion of the dual or adjoint space of a Banach space. The idea was also introduced independently by Hahn,[25] but Banach's work was more thorough. This dual space is the space of all continuous bounded linear functionals on the given space. The norm for the space of functionals is taken to be the bounds of the functionals, and the space proves to be a complete normed linear space, that is, a Banach space. Actually Banach's work here generalizes Riesz's work on L^p and L^q spaces, where $q = p/(p - 1)$, because the L^q space is equivalent to the dual in Banach's sense of the L^p space. The connection with Banach's work is made evident by the Riesz representation theorem (2). In other words, Banach's dual space has the same relationship to a given Banach space E that L^q has to L^p.

Banach starts with the definition of a continuous linear functional, that is, a continuous real-valued function whose domain is the space E, and proves that each functional is bounded. A key theorem, which generalizes one in Hahn's work, is known today as the Hahn-Banach theorem. Let p be a real-valued functional defined on a complete normed linear space R and let p satisfy for all x and y in R

(a) $p(x + y) \le p(x) + p(y);$

(b) $p(\lambda x) = \lambda p(x)$ for $\lambda \ge 0.$

Then there exists an additive functional f on R that satisfies

$$-p(-x) \le f(x) \le p(x)$$

for all x in R. There are a number of other theorems on the set of continuous functionals defined on R.

The work on functionals leads to the notion of adjoint operator. Let R and S be two Banach spaces and let U be a continuous, linear operator with domain in R and range in S. Denote by R^* and S^* the set of bounded linear functionals defined on R and S respectively. Then U induces a mapping from S^* into R^* as follows: If g is an element of S^*, then $g(U(x))$ is well-defined for all x in R. But by the linearity of U and g this is also a linear functional defined on R, that is, $g(U(x))$ is a member of R^*. Put otherwise, if $U(x) = y$, then $g(y) = f(x)$ where f is a functional of R^*. The induced mapping U^* from S^* to R^* is called the adjoint of U.

With this notion Banach proves that if U^* has a continuous inverse, then $y = U(x)$ is solvable for any y of S. Also if $f = U^*(g)$ is solvable for every f of R^*, then U^{-1} exists and is continuous on the range of U, where the range of U in S is the set of all y for which there is a g of S^* with the property that $g(y) = 0$ whenever $U^*(g) = 0$. (This last statement is a generalized form of the Fredholm alternative theorem.)

Banach applied his theory of adjoint operators to Riesz operators, which Riesz introduced in his 1918 paper. These are operators U of the form $U = I - \lambda V$ where I is the identity operator and V is a completely continuous operator. The abstract theory can then be applied to the space L^2 of functions defined on $[0, 1]$ and to the operators

$$U_\lambda(x) = x(s) - \lambda \int_0^1 K(s, t) x(t)\, dt,$$

where $\int_0^1 \int_0^1 K^2(s, t)\, ds\, dt < \infty$. The general theory, applied to

(9)
$$y(s) = x(s) - \lambda \int_0^1 K(s, t) x(t)\, dt$$

and its associated transposed equation

(10)
$$f(s) = g(s) - \lambda \int_0^1 K(t, s) g(t)\, dt,$$

tells us that if λ_0 is an eigenvalue of (9), then λ_0 is an eigenvalue of (10) and conversely. Further, (9) has a finite number of linearly independent eigenfunctions associated with λ_0 and the same is true of (10). Also (9) does not have a solution for all y when $\lambda = \lambda_0$. In fact a necessary and sufficient condition that (9) have a solution is that the n conditions

$$\int_0^1 y(t) g_0^{(p)}(t)\, dt = 0, \qquad p = 1, 2, \cdots, n$$

be satisfied, where $g_0^{(1)}, \ldots, g_0^{(n)}$ is a set of linearly independent solutions of

$$0 = g(s) - \lambda_0 \int_0^1 K(t, s) g(t)\, dt.$$

4. The Axiomatization of Hilbert Space

The theory of function spaces and operators seemed in the 1920s to be heading only toward abstraction for abstraction's sake. Even Banach made no use of his work. This state of affairs led Hermann Weyl to remark, "It was not merit but a favor of fortune when, beginning in 1923 . . . the spectral theory of Hilbert space was discovered to be the adequate mathematical instrument of quantum mechanics." Quantum mechanical research showed that the observables of a physical system can be represented by linear symmetric operators in a Hilbert space and the eigenvalues and eigenvectors (eigenfunctions) of the particular operator that represents energy are the energy levels of an electron in an atom and corresponding stationary quantum states of the system. The differences in two eigenvalues give the frequencies of the emitted quantum of light and thus define the radiation spectrum of the substance. In 1926 Erwin Schrödinger presented his quantum theory based on differential equations. He also showed its identity with Werner Heisenberg's theory of infinite matrices (1925), which the latter had applied to quantum theory. But a general theory unifying Hilbert's work and the eigenfunction theory for differential equations was missing.

The use of operators in quantum theory stimulated work on an abstract theory of Hilbert space and operators; this was first undertaken by John von Neumann (1903–57) in 1927. His approach treated both the space of square summable sequences and the space of L^2 functions defined on some common interval.

In two papers[26] von Neumann presented an *axiomatic* approach to Hilbert space and to operators in Hilbert space. Though origins of von Neumann's axioms can be seen in the work of Norbert Wiener, Weyl, and Banach, von Neumann's work was more complete and influential. His major objective was the formulation of a general eigenvalue theory for a large class of operators called Hermitian.

He introduced an L^2 space of complex-valued, measurable, and square integrable functions defined on any measurable set E of the complex plane. He also introduced the analogue, the complex sequence space, that is, the set of all sequences of complex numbers $a_1, a_2, \ldots$ with the property that $\sum_{p=1}^{\infty} |a_p|^2 < \infty$. The Riesz-Fischer theorem had shown that there exists a one-to-one correspondence between the functions of the function space and the sequences of the sequence space, as follows: In the function space choose a complete orthonormal sequence of functions $\{\phi_n\}$. Then if f is an element of the function space, the Fourier coefficients of f in an expansion with respect to the $\{\phi_n\}$ are a sequence of the sequence space; and conversely, if one starts with such a sequence there is a unique (up to a function whose integral is 0) function of the L^2 function space which has this sequence as its Fourier coefficients with respect to the $\{\phi_n\}$.

Further, if one defines the inner product (f, g) in the function space by

$$(f, g) = \int_E f(z)\overline{g(z)}\, dz,$$

where $\overline{g(z)}$ is the complex conjugate of $g(z)$, and in the sequence space for two sequences a and b by

$$(a, b) = \sum_{p=1}^{\infty} a_p \overline{b}_p,$$

then, if f corresponds to a and g to b, $(f, g) = (a, b)$.

26. *Math. Ann.*, 102, 1929/30, 49–131, and 370–427 = *Coll. Works*, 2, 3–143.

A Hermitian operator R in these spaces is defined to be a linear operator which has the property that for all f and g in its domain $(Rf, g) = (f, Rg)$. Likewise, in the sequence space, $(Ra, b) = (a, Rb)$.

Von Neumann's theory prescribes an axiomatic approach to both the function space and the sequence space. He proposed the following axiomatic foundation:

(A) H is a linear vector space. That is, there is an addition and scalar multiplication defined on H so that if f_1 and f_2 are elements of H and a_1 and a_2 are any complex numbers, then $a_1 f_1 + a_2 f_2$ is also an element of H.

(B) There exists on H an inner product or a complex-valued function of any two vectors f and g, denoted by (f, g), with the properties: (a) $(af, g) = a(f, g)$; (b) $(f_1 + f_2, g) = (f_1, g) + (f_2, g)$; (c) $(f, g) = \overline{(g, f)}$; (d) $(f, f) \geq 0$; (e) $(f, f) = 0$ if and only if $f = 0$.

Two elements f and g are called orthogonal if $(f, g) = 0$. The norm of f, denoted by $\|f\|$, is $\sqrt{(f, f)}$. The quantity $\|f - g\|$ defines a metric on the space.

(C) In the metric just defined H is separable, that is, there exists in H a countable set dense in H relative to the metric $\|f - g\|$.

(D) For every positive integral n there exists a set of n linearly independent elements of H.

(E) H is complete. That is, if $\{f_n\}$ is such that $\|f_n - f_m\|$ approaches 0 as m and n approach ∞, then there is an f in H such that $\|f - f_n\|$ approaches 0 as n approaches ∞. (This convergence is equivalent to strong convergence.)

From these axioms a number of simple properties follow: the Schwarz inequality $\|(f, g)\| \leq \|f\| \cdot \|g\|$, the fact that any complete orthonormal set of elements of H must be countable, and Parseval's inequality, that is, $\sum_{p=1}^{\infty} \|(f, \phi_p)\|^2 \leq \|f\|^2$.

Von Neumann then treats linear subspaces of H and projection operators. If M and N are closed subspaces of H, then $M - N$ is defined to be the set of all elements of M that are orthogonal to every element of N. The projection theorem follows: Let M be a closed subspace of H. Then any f of H can be written in one and only one way as $f = g + h$ where g is in M and h is in $H - M$. The projection operator P_M is defined by $P_M(f) = g$; that is, it is the operator defined on all of H which projects the elements f into their components in M.

In his second paper von Neumann introduces two topologies in H, strong and weak. The strong topology is the metric topology defined through the norm. The weak topology, which we shall not make precise, provides a system of neighborhoods associated with weak convergence.

Von Neumann presents a number of results on operators in Hilbert space. The linearity of the operators is presupposed, and integration is generally understood in the sense of Lebesgue-Stieltjes. A linear bounded transformation or operator is one that transforms elements of one Hilbert space into another, satisfies the linearity condition (3), and is bounded; that is, there exists a number M such that for all f of the space on which the operator R acts

$$\|R(f)\| \leq M\|f\|.$$

The smallest possible value of M is called the modulus of R. This last condition is equivalent to the continuity of the operator. Also continuity at a single point together with linearity is sufficient to guarantee continuity at all points and hence boundedness.

There is an adjoint operator R^*; for Hermitian operators $R = R^*$, and R is said to be self-adjoint. If $RR^* = R^*R$ for any operator, then R is said to be normal. If $RR^* = R^*R = I$, where I is the identity operator, then R is the analogue of an orthogonal transformation and is called unitary. For a unitary R, $\|R(f)\| = \|f\|$.

Another of von Neumann's results states that if R is Hermitian on all of the Hilbert space and satisfies the weaker closed condition, namely that if f_n approaches f and $R(f_n)$ approaches g so that $R(f) = g$, then R is bounded. Further, if R is Hermitian, the operator $I - \lambda R$ has an inverse for all complex and real λ exterior to an interval (m, M) of the real axis in which the values of $(R(f),f)$ lie when $\|f\| = 1$. A more fundamental result is that to every linear bounded Hermitian operator R there correspond two operators E_- and E_+ (*Einzeltransformationen*) having the following properties:

(a) $E_-E_- = E_-, E_+E_+ = E_+, I = E_- + E_+$.
(b) E_- and E_+ are commutative and commutative with any operator commutative with R.
(c) RE_- and RE_+ are respectively negative and positive (RE_+ is positive if $(RE_+(f),f) \geq 0$).
(d) For all f such that $Rf = 0$, $E_-f = 0$ and $E_+f = f$.

Von Neumann also established a connection between Hermitian and unitary operators, namely, if U is unitary and R is Hermitian, then $U = e^{iR}$.

Von Neumann subsequently extended his theory to unbounded operators; though his contributions and those of others are of major importance, an account of these developments would take us too far into recent developments.

Applications of functional analysis to the generalized moment problem, statistical mechanics, existence and uniqueness theorems in partial differential equations, and fixed-point theorems have been and are being made. Functional analysis now plays a role in the calculus of variations and in the theory of representations of continuous compact groups. It is also involved in algebra, approximate computations, topology, and real variable theory. Despite this variety of applications, there has been a deplorable absence of new applications to the large problems of classical analysis. This failure has disappointed the founders of functional analysis.

Bibliography

Bernkopf, M.: "The Development of Function Spaces with Particular Reference to their Origins in Integral Equation Theory," *Archive for History of Exact Sciences*, 3, 1966, 1–96.

———: "A History of Infinite Matrices," *Archive for History of Exact Sciences*, 4, 1968, 308–58.

Bourbaki, N.: *Eléments d'histoire des mathématiques*, Hermann, 1960, 230–45.

Dresden, Arnold: "Some Recent Work in the Calculus of Variations," *Amer. Math. Soc. Bull.*, 32, 1926, 475–521.

Fréchet, M.: *Notice sur les travaux scientifiques de M. Maurice Fréchet*, Hermann, 1933.

Hellinger, E. and O. Toeplitz: "Integralgleichungen und Gleichungen mit unendlichvielen Unbekannten," *Encyk. der Math. Wiss.*, B. G. Teubner, 1923–27, Vol. 2, Part III, 2nd half, 1335–1597.

Hildebrandt, T. H.: "Linear Functional Transformations in General Spaces," *Amer. Math. Soc. Bull.*, 37, 1931, 185–212.

Lévy, Paul: "Jacques Hadamard, sa vie et son œuvre," *L'Enseignement Mathématique*, (2), 13, 1967, 1–24.

McShane, E. J.: "Recent Developments in the Calculus of Variations," *Amer. Math. Soc. Semicentennial Publications*, 2, 1938, 69–97.

Neumann, John von: *Collected Works*, Pergamon Press, 1961, Vol. 2.

Sanger, Ralph G.: "Functions of Lines and the Calculus of Variations," *University of Chicago Contributions to the Calculus of Variations for 1931–32*, University of Chicago Press, 1933, 193–293.

Tonelli, L.: "The Calculus of Variations," *Amer. Math. Soc. Bull.*, 31, 1925, 163–72.

Volterra, Vito: *Opere matematiche*, 5 vols., Accademia Nazionale dei Lincei, 1954–62.

47
Divergent Series

It is indeed a strange vicissitude of our science that those series which early in the century were supposed to be banished once and for all from rigorous mathematics should, at its close, be knocking at the door for readmission.

JAMES PIERPONT

The series is divergent; therefore we may be able to do something with it.

OLIVER HEAVISIDE

1. Introduction

The serious consideration from the late nineteenth century onward of a subject such as divergent series indicates how radically mathematicians have revised their own conception of the nature of mathematics. Whereas in the first part of the nineteenth century they accepted the ban on divergent series on the ground that mathematics was restricted by some inner requirement or the dictates of nature to a fixed class of correct concepts, by the end of the century they recognized their freedom to entertain any ideas that seemed to offer any utility.

We may recall that divergent series were used throughout the eighteenth century, with more or less conscious recognition of their divergence, because they did give useful approximations to functions when only a few terms were used. After the dawn of rigorous mathematics with Cauchy, most mathematicians followed his dictates and rejected divergent series as unsound. However, a few mathematicians (Chap. 40, sec. 7) continued to defend divergent series because they were impressed by their usefulness, either for the computation of functions or as analytical equivalents of the functions from which they were derived. Still others defended them because they were a method of discovery. Thus De Morgan[1] says, "We must admit that many series are such as we cannot at present safely use, except as a means of discovery, the results of which are to be subsequently verified and the most

---

1. *Trans. Camb. Phil. Soc.*, 8, Part II, 1844, 182–203, pub. 1849.

determined rejector of all divergent series doubtless makes this use of them in his closet. . . ."

Astronomers continued to use divergent series even after they were banished because the exigencies of their science required them for purposes of computation. Since the first few terms of such series offered useful numerical approximations, the astronomers ignored the fact that the series as a whole were divergent, whereas the mathematicians, concerned with the behavior not of the first ten or twenty terms but with the character of the entire series, could not base a case for such series on the sole ground of utility.

However, as we have already noted (Chap. 40, sec. 7), both Abel and Cauchy were not without concern that in banishing divergent series they were discarding something useful. Cauchy not only continued to use them (see below) but wrote a paper with the title "Sur l'emploi légitime des séries divergentes,"[2] in which, speaking of the Stirling series for $\log \Gamma(x)$ or $\log m!$ (Chap. 20, sec. 4), Cauchy points out that the series, though divergent for all values of x, can be used in computing $\log \Gamma(x)$ when x is large and positive. In fact he showed that having fixed the number n of terms taken, the absolute error committed by stopping the summation at the nth term is less than the absolute value of the next succeeding term, and the error becomes smaller as x increases. Cauchy tried to understand why the approximation furnished by the series was so good, but failed.

The usefulness of divergent series ultimately convinced mathematicians that there must be some feature which, if culled out, would reveal why they furnished good approximations. As Oliver Heaviside put it in the second volume of *Electromagnetic Theory* (1899), "I must say a few words on the subject of generalized differentiation and divergent series. . . . It is not easy to get up any enthusiasm after it has been artificially cooled by the wet blanket of the rigorists. . . . There will have to be a theory of divergent series, or say a larger theory of functions than the present, including convergent and divergent series in one harmonious whole." When Heaviside made this remark, he was unaware that some steps had already been taken.

The willingness of mathematicians to move on the subject of divergent series was undoubtedly strengthened by another influence that had gradually penetrated the mathematical atmosphere: non-Euclidean geometry and the new algebras. The mathematicians slowly began to appreciate that mathematics is man-made and that Cauchy's definition of convergence could no longer be regarded as a higher necessity imposed by some superhuman power. In the last part of the nineteenth century they succeeded in isolating the essential property of those divergent series that furnished useful approximations to functions. These series were called asymptotic by Poincaré, though during the century they were called semiconvergent, a term introduced by

2. *Comp. Rend.*, 17, 1843, 370–76 = *Œuvres*, (1), 8, 18-25.

Legendre in his *Essai de la théorie des nombres* (1798, p. 13) and used also for oscillating series.

The theory of divergent series has two major themes. The first is the one already briefly described, namely, that some of these series may, for a fixed number of terms, approximate a function better and better as the variable increases. In fact, Legendre, in his *Traité des fonctions elliptiques* (1825–28), had already characterized such series by the property that the error committed by stopping at any one term is of the order of the first term omitted. The second theme in the theory of divergent series is the concept of summability. It is possible to define the sum of a series in entirely new ways that give finite sums to series that are divergent in Cauchy's sense.

2. The Informal Uses of Divergent Series

We have had occasion to describe eighteenth-century work in which both convergent and divergent series were employed. In the nineteenth century, both before and after Cauchy banned divergent series, some mathematicians and physicists continued to use them. One new application was the evaluation of integrals in series. Of course the authors were not aware at the time that they were finding either complete asymptotic series expansions or first terms of asymptotic series expansions of the integrals.

The asymptotic evaluation of integrals goes back at least to Laplace. In his *Théorie analytique des probabilités* (1812) [3] Laplace obtained by integration by parts the expansion for the error function

$$Erfc(T) = \int_T^\infty e^{-t^2}\, dt = \frac{e^{-T^2}}{2T}\left\{1 - \frac{1}{2T^2} + \frac{1\cdot 3}{(2T^2)^2} - \frac{1\cdot 3\cdot 5}{(2T^2)^3} + \cdots\right\}.$$

He remarked that the series is divergent, but he used it to compute $Erfc(T)$ for large values of T.

Laplace also pointed out in the same book [4] that

$$\int \phi(x)\{u(x)\}^s\, dx$$

when s is large depends on the values of $u(x)$ near its stationary points, that is, the values of x for which $u'(x) = 0$. Laplace used this observation to prove that

$$s! \sim s^{s+1/2}e^{-s}\sqrt{2\pi}\left(1 + \frac{1}{12s} + \frac{1}{288s^2} + \cdots\right),$$

a result that can be obtained as well from Stirling's approximation to $\log s!$ (Chap. 20, sec. 4).

3. Third ed., 1820, 88–109 = *Œuvres*, 7, 89–110.
4. *Œuvres*, 7, 128–31.

Laplace had occasion, in his *Théorie analytique des probabilités*,[5] to consider integrals of the form

(1)
$$f(x) = \int_a^b g(t)e^{xh(t)} dt,$$

where g may be complex, h and t are real, and x is large and positive; he remarked that the main contribution to the integral comes from the immediate neighborhood of those points in the range of integration in which $h(t)$ attains its absolute maximum. The contribution of which Laplace spoke is the first term of what we would call today an asymptotic series evaluation of the integral. If $h(t)$ has just one maximum at $t = a$, then Laplace's result is

$$f(x) \sim g(a)e^{xh(a)} \sqrt{\frac{-\pi}{2xh''(a)}}$$

as x approaches ∞.

If in place of (1) the integral to be evaluated is

(2)
$$f(x) = \int_a^b g(t)e^{ixh(t)} dt$$

when t and x are real and x is large, then $|e^{ixh(t)}|$ is a constant and Laplace's method does not apply. In this case a method adumbrated by Cauchy in his major paper on the propagation of waves,[6] now called the principle of stationary phase, is applicable. The principle states that the most relevant contribution to the integral comes from the immediate neighborhoods of the stationary points of $h(t)$, that is, the points at which $h'(t) = 0$. This principle is intuitively reasonable because the integrand can be thought of as an oscillating current or wave with amplitude $|g(t)|$. If t is time, then the velocity of the wave is proportional to $xh'(t)$, and if $h'(t) \neq 0$, the velocity of the oscillations increases indefinitely as x becomes infinite. The oscillations are then so rapid that during a full period $g(t)$ is approximately constant and $xh(t)$ is approximately linear, so that the integral over a full period vanishes. This reasoning fails at a value of t where $h'(t) = 0$. Thus the stationary points of $h(t)$ are likely to furnish the main contribution to the asymptotic value of $f(x)$. If τ is a value of t at which $h'(t) = 0$ and $h''(\tau) > 0$, then

$$f(x) \sim \sqrt{\frac{2\pi}{xh''(\tau)}} \, g(\tau)e^{ixh(\tau) + i\pi/4}$$

as x becomes infinite.

This principle was used by Stokes in evaluating Airy's integral (see

5. Third ed., 1820, Vol. 1, Part 2, Chap. 1 = *Œuvres*, 7, 89–110.
6. *Mém. de l'Acad. des Sci., Inst. France*, 1, 1827, Note 16 = *Œuvres*, (1), 1, 230.

below) in a paper of 1856[7] and was formulated explicitly by Lord Kelvin.[8] The first satisfactory proof of the principle was given by George N. Watson (1886–1965).[9]

In the first few decades of the nineteenth century, Cauchy and Poisson evaluated many integrals containing a parameter in series of powers of the parameter. In the case of Poisson the integrals arose in geophysical heat conduction and elastic vibration problems, whereas Cauchy was concerned with water waves, optics, and astronomy. Thus Cauchy, working on the diffraction of light,[10] gave divergent series expressions for the Fresnel integrals:

$$\int_0^m \cos\left(\frac{\pi}{2} z^2\right) dz = \frac{1}{2} - N \cos\frac{\pi}{2} m^2 + M \sin\frac{\pi}{2} m^2$$

$$\int_0^m \sin\left(\frac{\pi}{2} z^2\right) dz = \frac{1}{2} - M \cos\frac{\pi}{2} m^2 - N \sin\frac{\pi}{2} m^2,$$

where

$$M = \frac{1}{m\pi} - \frac{1\cdot 3}{m^5\pi^3} + \frac{1\cdot 3\cdot 5\cdot 7}{m^9\pi^5} - \cdots, \qquad N = \frac{1}{m^3\pi^2} - \frac{1\cdot 3\cdot 5}{m^7\pi^4} + \cdots.$$

During the nineteenth century several other methods, such as the method of steepest descent, were invented to evaluate integrals. The full theory of all these methods and the proper understanding of what the approximations amounted to, whether single terms or full series, had to await the creation of the theory of asymptotic series.

Many of the integrals that were expanded by the above-described methods arose first as solutions of differential equations. Another use of divergent series was to solve differential equations directly. This use can be traced back at least as far as Euler's work[11] wherein, concerned with solving the non-uniform vibrating-string problem (Chap. 22, sec. 3), he gave an asymptotic series solution of an ordinary differential equation that is essentially Bessel's equation of order $r/2$, where r is integral.

Jacobi[12] gave the asymptotic form of $J_n(x)$ for large x:

$$J_n(x) \sim \left(\frac{2}{\pi x}\right)^{1/2}\left[\cos\left(x - \frac{n\pi}{2} - \frac{\pi}{4}\right)\left\{1 - \frac{(4n^2 - 1^2)(4n^2 - 3^2)}{2!\,(8x)^2} + \cdots\right\} - \right.$$
$$\left. \sin\left(x - \frac{n\pi}{2} - \frac{\pi}{4}\right)\left\{\frac{4n^2 - 1^2}{1!\,8x} - \frac{(4n^2 - 1^2)(4n^2 - 3^2)(4n^2 - 5^2)}{3!\,(8x)^3} + \cdots\right\}\right].$$

7. *Trans. Camb. Phil. Soc.*, 9, 1856, 166–87 = *Math. and Phys. Papers*, 2, 329–57.
8. *Phil. Mag.*, (5), 23, 1887, 252–55 = *Math. and Phys. Papers*, 4, 303–6.
9. *Proc. Camb. Phil. Soc.*, 19, 1918, 49–55.
10. *Comp. Rend.*, 15, 1842, 554–56 and 573–78 = *Œuvres*, (1), 7, 149–57.
11. *Novi Comm. Acad. Sci. Petrop.*, 9, 1762/3, 246–304, pub. 1764 = *Opera*, (2), 10, 293–343.
12. *Astronom. Nach.*, 28, 1849, 65–94 = *Werke*, 7, 145–74.

A somewhat different use of divergent series in the solution of differential equations was introduced by Liouville. He sought[13] approximate solutions of the differential equation

(3)
$$\frac{d}{dx}\left(p\frac{dy}{dx}\right) + (\lambda^2 q_0 + q_1)y = 0,$$

where p, q_0, and q_1 are positive functions of x and λ is a parameter; the solution is to be obtained in $a \le x \le b$. Here, as opposed to boundary-value problems where discrete values of λ are sought, he was interested in obtaining some approximate form of y for large values of λ. To do this he introduced the variables

(4)
$$t = \int_{x_0}^{x}\left(\frac{q_0}{p}\right)^{1/2} dx, \qquad w = (q_0 p)^{1/4}y$$

and obtained

(5)
$$\frac{d^2w}{dt^2} + \lambda^2 w = rw,$$

where

$$r = (q_0 p)^{-1/4}\frac{d^2}{dt^2}(q_0 p)^{1/4} - \frac{q_1}{q_0}.$$

He then used a process that amounts in modern language to the solution by successive approximations of an integral equation of the Volterra type, namely,

$$w(t) = c_1 \cos \lambda t + c_2 \sin \lambda t + \int_{t_0}^{t}\frac{\sin \lambda(t-s)}{\lambda}r(s)w(s)\,ds.$$

Liouville now argued that for sufficiently large values of λ the first approximation to the solutions of (5) should be

(6)
$$w \sim c_1 \cos \lambda t + c_2 \sin \lambda t.$$

If the values of w and t given by (4) are now used to obtain the approximate solution of (3), we have from (6) that

(7)
$$y \sim c_1 \frac{1}{(q_0 p)^{1/4}} \cos\left\{\lambda \int_{x_0}^{x}\left(\frac{q_0}{p}\right)^{1/2} dx\right\} + c_2 \frac{1}{(q_0 p)^{1/4}} \sin\left\{\lambda \int_{x_0}^{x}\left(\frac{q_0}{p}\right)^{1/2} dx\right\}.$$

Though Liouville was not aware of it, he had obtained the first term of an asymptotic series solution of (3) for large λ.

The same method was used by Green[14] in the study of the propagation

13. *Jour. de Math.*, 2, 1837, 16–35.
14. *Trans. Camb. Phil. Soc.*, 6, 1837, 457–62 = *Math. Papers*, 225–30.

of waves in a channel. The method has been generalized slightly to equations of the form

$$(8) \qquad\qquad y'' + \lambda^2 q(x, \lambda)y = 0,$$

in which λ is a large positive parameter and x may be real or complex. The solutions are now usually expressed as

$$(9) \qquad\qquad y \sim q^{-1/4} \exp\left(\pm i\lambda \int_0^x q^{1/2}\, dx\right)\left[1 + O\left(\frac{1}{\lambda}\right)\right].$$

The error term $O(1/\lambda)$ implies that the exact solution would contain a term $F(x, \lambda)/\lambda$ where $|F(x, \lambda)|$ is bounded for all x in the domain under consideration and for $\lambda > \lambda_0$. The form of the error term is valid in a restricted domain of the complex x-plane. Liouville and Green did not supply the error term or conditions under which their solutions were valid. The more general and precise approximation (9) is explicit in papers by Gregor Wentzel (1898–),[15] Hendrik A. Kramers (1894–1952),[16] Léon Brillouin (1889–1969),[17] and Harold Jeffreys (1891–),[18] and is familiarly known as the WKBJ solution. These men were working with Schrödinger's equation in quantum theory.

In a paper read in 1850,[19] Stokes considered the value of Airy's integral

$$(10) \qquad\qquad W = \int_0^\infty \cos\frac{\pi}{2}\,(w^3 - mw)\, dw$$

for large $|m|$. This integral represents the strength of diffracted light near a caustic. Airy had given a series for W in powers of m which, though convergent for all m, was not useful for calculation when $|m|$ is large. Stokes's method consisted in forming a differential equation of which the integral is a particular solution and solving the differential equation in terms of divergent series that might be useful for calculation (he called such series semiconvergent).

After showing that $U = (\pi/2)^{1/3}\, W$ satisfies the Airy differential equation

$$(11) \qquad\qquad \frac{d^2 U}{dn^2} + \frac{n}{3}\, U = 0, \qquad n = \left(\frac{\pi}{2}\right)^{2/3} m,$$

Stokes proved that for positive n

$$(12) \qquad U = An^{-1/4}\left(R\cos\frac{2}{3}\sqrt{\frac{n^3}{3}} + S\sin\frac{2}{3}\sqrt{\frac{n^3}{3}}\right)$$

$$+ Bn^{-1/4}\left(R\sin\frac{2}{3}\sqrt{\frac{n^3}{3}} - S\cos\frac{2}{3}\sqrt{\frac{n^3}{3}}\right),$$

15. *Zeit. für Physik*, 38, 1926, 518–29.
16. *Zeit. für Physik*, 39, 1926, 828–40.
17. *Comp. Rend.*, 183, 1926, 24–26).
18. *Proc. London Math. Soc.*, (2), 23, 1923, 428–36.
19. *Trans. Camb. Phil. Soc.*, 9, 1856, 166–87 = *Math. and Phys. Papers*, 2, 329–57.

where

(13) $$R = 1 - \frac{1 \cdot 5 \cdot 7 \cdot 11}{1 \cdot 2 \cdot 16^2 \cdot 3n^3} + \frac{1 \cdot 5 \cdot 7 \cdot 11 \cdot 13 \cdot 17 \cdot 19 \cdot 23}{1 \cdot 2 \cdot 3 \cdot 4 \cdot 16^4 \cdot 3^2 n^6} \cdots ,$$

(14) $$S = \frac{1 \cdot 5}{1 \cdot 16(3n^3)^{1/2}} - \frac{1 \cdot 5 \cdot 7 \cdot 11 \cdot 13 \cdot 17}{1 \cdot 2 \cdot 3 \cdot 16^3 (3n^3)^{3/2}} + \cdots .$$

The quantities A and B that make U yield the integral were determined by a special argument (wherein Stokes uses the principle of stationary phase). He also gave an analogous result for n negative.

The series (12) and the one for negative n behave like convergent series for some number of terms but are actually divergent. Stokes observed that they can be used for calculation. Given a value of n, one uses the terms from the first up to the one that becomes smallest for the value of n in question. He gave a qualitative argument to show why the series are useful for numerical work.

Stokes encountered a special difficulty with the solutions of (11) for positive and negative n. He was not able to pass from the series for which n is positive to the series for which n is negative by letting n vary through 0 because the series have no meaning for $n = 0$. He therefore tried to pass from positive to negative n through complex values of n, but this did not yield the correct series and constant multipliers.

What Stokes did discover, after some struggle,[20] was that if for a certain range of the amplitude of n a general solution was represented by a certain linear combination of two asymptotic series each of which is a solution, then in a neighboring range of the amplitude of n it was by no means necessary for the same linear combination of the two fundamental asymptotic expansions to represent the same general solution. He found that the constants of the linear combination changed abruptly as certain lines given by amp. $n = $ const. were crossed. These are now called Stokes lines.

Though Stokes had been primarily concerned with the evaluation of integrals, it was clear to him that divergent series could be used generally to solve differential equations. Whereas Euler, Poisson, and others had solved individual equations in such terms, their results appeared to be tricks that produced answers for specific physical problems. Stokes actually gave several examples in the 1856 and 1857 papers.

The above work on the evaluation of integrals and the solution of differential equations by divergent series is a sample of what was done by many mathematicians and physicists.

3. The Formal Theory of Asymptotic Series

The full recognition of the nature of those divergent series that are useful in the representation and calculation of functions and a formal definition

20. *Trans. Camb. Phil. Soc.*, 10, 1857, 106–28 = *Math. and Phys. Papers*, 4, 77–109.

of these series were achieved by Poincaré and Stieltjes independently in 1886. Poincaré called these series asymptotic while Stieltjes continued to use the term semiconvergent. Poincaré[21] took up the subject in order to further the solution of linear differential equations. Impressed by the useful-ness of divergent series in astronomy, he sought to determine which were useful and why. He succeeded in isolating and formulating the essential property. A series of the form

$$(15) \qquad\qquad a_0 + \frac{a_1}{x} + \frac{a_2}{x^2} + \cdots,$$

where the a_i are independent of x, is said to represent the function $f(x)$ asymptotically for large values of x whenever

$$(16) \qquad\qquad \lim_{x \to \infty} x^n \left[f(x) - \left(a_0 + \frac{a_1}{x} + \cdots + \frac{a_n}{x^n} \right) \right] = 0$$

for $n = 0, 1, 2, 3, \ldots$. The series is generally divergent but may in special cases be convergent. The relationship of the series to $f(x)$ is denoted by

$$f(x) \sim a_0 + \frac{a_1}{x} + \frac{a_2}{x^2} + \cdots.$$

Such series are expansions of functions in the neighborhood of $x = \infty$. Poincaré in his 1886 paper considered real x-values. However, the definition holds also for complex x if $x \to \infty$ is replaced by $|x| \to \infty$, though the validity of the representation may then be confined to a sector of the complex plane with vertex at the origin.

The series (15) is asymptotic to $f(x)$ in the neighborhood of $x = \infty$. However, the definition has been generalized, and one speaks of the series

$$a_0 + a_1 x + a_2 x^2 + \cdots$$

as asymptotic to $f(x)$ at $x = 0$ if

$$\lim_{x \to 0} \frac{1}{x^n} \left[f(x) - \sum_{0}^{n-1} a_i x^i \right] = a_n.$$

Though in the case of some asymptotic series one knows what error is committed by stopping at a definite term, no such information about the numerical error is known for general asymptotic series. However, asymptotic series can be used to give rather accurate numerical results for large x by employing only those terms for which the magnitude of the terms decreases as one takes more and more terms. The order of the magnitude of the error at any stage is equal to the magnitude of the first term omitted.

21. *Acta Math.*, 8, 1886, 295–344 = *Œuvres*, 1, 290–332.

Poincaré proved that the sum, difference, product, and quotient of two functions are represented asymptotically by the sum, difference, product, and quotient of their separate asymptotic series, provided that the constant term in the divisor series is not zero. Also, if

$$f(x) \sim a_0 + \frac{a_1}{x} + \frac{a_2}{x^2} + \cdots,$$

then

$$\int_{x_0}^{x} f(z)\, dz \sim C + a_0 x + a_1 \log x - \frac{a_2}{x} - \frac{1}{2}\frac{a_3}{x^2} - \cdots.$$

The use of integration involves a slight generalization of the original definition, namely,

$$\phi(x) \sim f(x) + g(x)\left(a_0 + \frac{a_1}{x} + \frac{a_2}{x^2} + \cdots\right)$$

if

$$\frac{\phi(x) - f(x)}{g(x)} \sim a_0 + \frac{a_1}{x} + \frac{a_2}{x^2} + \cdots$$

even when $f(x)$ and $g(x)$ do not themselves have asymptotic series representations. As for differentiation, if $f'(x)$ is known to have an asymptotic series expansion, then it can be obtained by differentiating the asymptotic series for $f(x)$.

If a given function has an asymptotic series expansion it is unique, but the converse is not true because, for example, $(1 + x)^{-1}$ and $(1 + e^{-x}) \cdot (1 + x)^{-1}$ have the same asymptotic expansion.

Poincaré applied his theory of asymptotic series to differential equations, and there are many such uses in his treatise on celestial mechanics, *Les Méthodes nouvelles de la mécanique céleste*.[22] The class of equations treated in his 1886 paper is

(17) $$P_n(x)y^{(n)} + P_{n-1}(x)y^{(n-1)} + \cdots + P_0(x)y = 0$$

where the $P_i(x)$ are polynomials in x. Actually Poincaré treated only the second order case but the method applies to (17).

The only singular points of equation (17) are the zeros of $P_n(x)$ and $x = \infty$. For a regular singular point (*Stelle der Bestimmtheit*) there are convergent expressions for the integrals given by Fuchs (Chap. 29, sec. 5). Consider then an irregular singular point. By a linear transformation this point can be removed to ∞, while the equation keeps its form. If P_n is of the pth degree, the condition that $x = \infty$ shall be a regular singular point

22. Vol. 2, Chap. 8, 1893.

is that the degrees of $P_{n-1}, P_{n-2}, \ldots, P_0$ be at most $p - 1, p - 2, \ldots, p - n$ respectively. For an irregular singular point one or more of these degrees must be greater. Poincaré showed that for a differential equation of the form (17), wherein the degrees of the P_i do not exceed the degree of P_n, there exist n series of the form

$$e^{ax}x^{\alpha}\left(A_0 + \frac{A_1}{x} + \cdots\right)$$

and the series satisfy the differential equation formally. He also showed that to each such series there corresponds an exact solution in the form of an integral to which the series is asymptotic.

Poincaré's results are included in the following theorem due to Jakob Horn (1867–1946).[23] He treats the equation

(18) $y^{(n)} + a_1(x)y^{(n-1)} + \cdots + a_n(x)y^n = 0,$

where the coefficients are rational functions of x and are assumed to be developable for large positive x in convergent or asymptotic series of the form

$$a_r(x) = x^{rk}\left[a_{r,0} + \frac{a_{r,1}}{x} + \frac{a_{r,2}}{x^2} + \cdots\right], \qquad r = 1, 2, \cdots, n,$$

k being some positive integer or 0. If for the above equation (18) the roots $m_1, m_2, \ldots, m_n$ of the characteristic equation, that is, the algebraic equation

$$m^n + a_{1,0}m^{n-1} + \cdots + a_{n,0} = 0,$$

are distinct, equation (18) possesses n linearly independent solutions $y_1, y_2, \ldots, y_n$ which are developable asymptotically for large positive values of x in the form

$$y_r \sim e^{f_r(x)}x^{\rho_r} \sum_{j=0}^{\infty} \frac{A_{r,j}}{x^j}, \qquad r = 1, 2, \cdots, n,$$

where $f_r(x)$ is a polynomial of degree $k + 1$ in x, the coefficient of whose highest power in x is $m_r/(k + 1)$, while ρ_r and $A_{r,j}$ are constants with $A_{r,0} = 1$. The results of Poincaré and Horn have been extended to various other types of differential equations and generalized by inclusion of the cases where the roots of the characteristic equation are not necessarily distinct.

The existence, form, and range of the asymptotic series solution when the independent variable in (18) is allowed to take on complex values was first taken up by Horn.[24] A general result was given by George David Birkhoff (1884–1944), one of the first great American mathematicians.[25]

23. *Acta Math.*, 24, 1901, 289–308.
24. *Math. Ann.*, 50, 1898, 525–56.
25. *Amer. Math. Soc. Trans.*, 10, 1909, 436–70 = *Coll. Math. Papers*, 1, 201–35.

Birkhoff in this paper did not consider equation (18) but the more general system

$$(19) \qquad \frac{dy_i}{dx} = \sum_{j=1}^{n} a_{ij}(x) y_j, \qquad i = 1, 2, \ldots, n,$$

in which for $|x| > R$ we have for each a_{ij}

$$a_{ij}(x) \sim a_{ij} x^q + a_{ij}^{(1)} x^{q-1} + \cdots + a_{ij}^{(q)} + a_{ij}^{(q+1)} \frac{1}{x} + \cdots$$

and for which the characteristic equation in α

$$|a_{ij} - \delta_{ij}\alpha| = 0$$

has distinct roots. He gave asymptotic series solutions for the y_i which hold in various sectors of the complex plane with vertices at $x = 0$.

Whereas Poincaré's and the other expansions above are in powers of the independent variable, further work on asymptotic series solutions of differential equations turned to the problem first considered by Liouville (sec. 2) wherein a parameter is involved. A general result on this problem was given by Birkhoff.[26] He considered

$$(20) \qquad \frac{d^n z}{dx^n} + \rho a_{n-1}(x, \rho) \frac{d^{n-1} z}{dx^{n-1}} + \cdots + \rho^n a_0(x, \rho) z = 0$$

for large $|\rho|$ and for x in the interval $[a, b]$. The functions $a_i(x, \rho)$ are supposed analytic in the complex parameter ρ at $\rho = \infty$ and have derivatives of all orders in the real variable x. The assumptions on $a_i(x, \rho)$ imply that

$$a_i(x, \rho) = \sum_{j=0}^{\infty} a_{ij}(x) \rho^{-j}$$

and that the roots of $w_1(x), w_2(x), \ldots, w_n(x)$ of the characteristic equation

$$w^n + a_{n-1,0}(x) w^{n-1} + \cdots + a_{00}(x) = 0$$

are distinct for each x. He proves that there are n independent solutions

$$z_1(x, \rho), \cdots, z_n(x, \rho)$$

of (20) that are analytic in ρ in a region S of the ρ plane (determined by the argument of ρ), such that for any integer m and large $|\rho|$

$$(21) \qquad z_i(x, \rho) = u_i(x, \rho) + \exp\left[\rho \int_a^x w_i(t)\, dt\right] E_0 \rho^{-m},$$

26. *Amer. Math. Soc. Trans.*, 9, 1908, 219–31 and 380–82 = *Coll. Math. Papers*, 1, 1–36.

where

$$(22) \qquad u_i(x, \rho) = \exp\left[\rho \int_a^x w_i(t)\, dt \sum_{j=0}^{m-1} u_{ij}(x)\rho^{-j}\right]$$

and E_0 is a function of x, ρ, and m bounded for all x in $[a, b]$ and ρ in S. The $u_{ij}(x)$ are themselves determinable. The result (21), in view of (22), states that z_i is given by a series in $1/\rho$ up to $1/\rho^{m-1}$ plus a remainder term, namely, the second term on the right, which contains $1/\rho^m$. Moreover, since m is arbitrary, one can take as many terms in $1/\rho$ as one pleases into the expression for $u_i(x, \rho)$. Since E_0 is bounded, the remainder term is of higher order in $1/\rho$ than $u_i(x, \rho)$. Then the infinite series, which one can obtain by letting m become infinite, is asymptotic to $z_i(x, \rho)/\exp\left[\rho \int_a^x w_i(t)\, dt\right]$ in Poincaré's sense.

In Birkhoff's theorem the asymptotic series for complex ρ is valid only in a sector S of the complex ρ plane. The Stokes phenomenon enters. That is, the analytic continuation of $z_i(x, \rho)$ across a Stokes line is not given by the analytic continuation of the asymptotic series for $z_i(x, \rho)$.

The use of asymptotic series or of the WKBJ approximation to solutions of differential equations has raised another problem. Suppose we consider the equation

$$(23) \qquad\qquad y'' + \lambda^2 q(x)y = 0$$

where x ranges over $[a, b]$. The WKBJ approximation for large λ, in view of (7), gives two solutions for $x > 0$ and two for $x < 0$. It breaks down at the value of x where $q = 0$. Such a point is called a transition point, turning point, or Stokes point. The exact solutions of (23) are, however, finite at such a point. The problem is to relate the WKBJ solutions on each side of the transition point so that they represent the same exact solution over the interval $[a, b]$ in which the differential equation is being solved. To put the problem more specifically, consider the above equation in which $q(x)$ is real for real x and such that $q(x) = 0$, $q'(x) \neq 0$ for $x = 0$. Suppose also that $q(x)$ is negative for x positive (or the reverse). Given a linear combination of the two WKBJ solutions for $x < 0$ and another for $x > 0$, the question arises as to which of the solutions holding for $x > 0$ should be joined to that for $x < 0$. Connection formulas provide the answer.

The scheme for crossing a zero of $q(x)$ was first given by Lord Rayleigh[27] and extended by Richard Gans (1880–1954),[28] who was familiar with Rayleigh's work. Both of these men were working on the propagation of light in a varying medium.

27. *Proc. Roy. Soc.*, A 86, 1912, 207–26 = *Sci. Papers*, 6, 71–90.
28. *Annalen der Phys.*, (4), 47, 1915, 709–36.

The first systematic treatment of connection formulas was given by Harold Jeffreys [29] independently of Gans's work. Jeffreys considered the equation

$$(24) \qquad \frac{d^2y}{dx^2} + \lambda^2 X(x)y = 0,$$

where x is real, λ is large and real, and $X(x)$ has a simple zero at, say, $x = 0$. He derived formulas connecting the asymptotic series solutions of (24) for $x > 0$ and $x < 0$ by using an approximating equation for (24), namely,

$$(25) \qquad \frac{d^2y}{dx^2} + \lambda^2 xy = 0$$

which replaces $X(x)$ in (24) by a linear function in x. The solutions of (25) are

$$(26) \qquad y_{\pm}(x) = x^{1/2} J_{\pm 1/3}(\xi)$$

where $\xi = (2/3)\lambda x^{3/2}$. The asymptotic expansions for large x of these solutions can be used to join the asymptotic solutions of (24) on each side of $x = 0$. There are many details on the ranges of x and λ that must be considered in the joining process but we shall not enter into them. Extension of the work on connection formulas for the cases where $X(x)$ in (24) has multiple zeros or several distinct zeros and for more complicated second order and higher-order equations and for complex x and λ has appeared in numerous papers.

The theory of asymptotic series, whether used for the evaluation of integrals or the approximate solution of differential equations, has been extended vastly in recent years. What is especially worth noting is that the mathematical development shows that the eighteenth- and nineteenth-century men, notably Euler, who perceived the great utility of divergent series and maintained that these series could be used as analytical equivalents of the functions they represented, that is, that operations on the series corresponded to operations on the functions, were on the right track. Even though these men failed to isolate the essential rigorous notion, they saw intuitively and on the basis of results that divergent series were intimately related to the functions they represented.

4. Summability

The work on divergent series described thus far has dealt with finding asymptotic series to represent functions either known explicitly or existing implicitly as solutions of ordinary differential equations. Another problem that mathematicians tackled from about 1880 on is essentially the converse of finding asymptotic series. Given a series divergent in Cauchy's sense,

29. *Proc. London Math. Soc.*, (2), 23, 1922–24, 428–36.

can a "sum" be assigned to the series? If the series consists of variable terms this "sum" would be a function for which the divergent series might or might not be an asymptotic expansion. Nevertheless the function might be taken to be the "sum" of the series, and this "sum" might serve some useful purposes, even though the series will certainly not converge to it or may not be usable to calculate approximate values of the function.

To some extent the problem of summing divergent series was actually undertaken before Cauchy introduced his definitions of convergence and divergence. The mathematicians encountered divergent series and sought sums for them much as they did for convergent series, because the distinction between the two types was not sharply drawn and the only question was, What is the appropriate sum? Thus Euler's principle (Chap. 20, sec. 7) that a power series expansion of a function has as its sum the value of the function from which the series is derived gave a sum to the series even for values of x for which the series diverges in Cauchy's sense. Likewise, in his transformation of series (Chap. 20, sec. 4), he converted divergent series to convergent ones without doubting that there should be a sum for practically all series. However, after Cauchy did make the distinction between convergence and divergence, the problem of summing divergent series was broached on a different level. The relatively naive assignments in the eighteenth century of sums to all series were no longer acceptable. The new definitions prescribe what is now called *summability*, to distinguish the notion from convergence in Cauchy's sense.

With hindsight one can see that the notion of summability was really what the eighteenth- and early nineteenth-century men were advancing. This is what Euler's methods of summing just described amount to. In fact in his letter to Goldbach of August 7, 1745, in which he asserted that the sum of a power series is the value of the function from which the series is derived, Euler also asserted that every series must have a sum, but since the word sum implies the usual process of adding, and this process does not lead to the sum in the case of a divergent series such as $1 - 1! + 2! - 3! + \cdots$, we should use the word value for the "sum" of a divergent series.

Poisson too introduced what is now recognized to be a summability notion. Implied in Euler's definition of sum as the value of the function from which the series comes is the idea that

$$(27) \qquad \sum_{n=0}^{\infty} a_n = \lim_{x \to 1-} \sum_{n=0}^{\infty} a_n x^n.$$

(The notation $1-$ means that x approaches 1 from below.) According to (27) the sum of $1 - 1 + 1 - 1 + \cdots$ is

$$\lim_{x \to 1-} (1 - x + x^2 - x^3 + \cdots) = \lim_{x \to 1-} (1 + x)^{-1} = \frac{1}{2}.$$

Poisson[30] was concerned with the series

$$\sin \theta + \sin 2\theta + \sin 3\theta + \cdots$$

which diverges except when θ is a multiple of π. His concept, expressed for the full Fourier series

$$(28) \qquad \frac{a_0}{2} + \sum_n (a_n \cos n\theta + b_n \sin n\theta),$$

is that one should consider the associated power series

$$(29) \qquad \frac{a_0}{2} + \sum_n (a_n \cos n\theta + b_n \sin n\theta)r^n$$

and define the sum of (28) to be the limit of the series (29) as r approaches 1 from below. Of course Poisson did not appreciate that he was suggesting a definition of a sum of a divergent series because, as already noted, the distinction between convergence and divergence was in his time not a critical one.

The definition used by Poisson is now called Abel summability because it was also suggested by a theorem due to Abel,[31] which states that if the power series

$$f(x) = \sum_{n=0}^{\infty} a_n x^n$$

has a radius of convergence r and converges for $x = r$, then

$$(30) \qquad \lim_{x \to r^-} f(x) = \sum_{n=0}^{\infty} a_n r^n.$$

Then the function $f(x)$ defined by the series in $-r < x \leq r$ is continuous on the left at $x = r$. However, if $\sum a_n$ does not converge and the limit (30) does exist for $r = 1$, then one has a definition of sum for the divergent series. This formal definition of summability for series divergent in Cauchy's sense was not introduced until the end of the nineteenth century in a connection soon to be described.

One of the motivations for reconsidering the summation of divergent series, beyond their continued usefulness in astronomical work, was what has been called the boundary-value (*Grenzwert*) problem in the theory of analytic functions. A power series $\sum a_n x^n$ may represent an analytic function in a circle of radius r but not for values of x on the circle. The problem was whether one could find a concept of sum such that the power series might have a sum for $|x| = r$ and such that this sum might even be the value of

30. *Jour. de l'Ecole Poly.*, 11, 1820, 417–89.
31. *Jour. für Math.*, 1, 1826, 311–39 = *Œuvres*, 1, 219–50.

$f(x)$ as $|x|$ approaches r. It was this attempt to extend the range of the power series representation of an analytic function that motivated Frobenius, Hölder, and Ernesto Cesàro. Frobenius[32] showed that if the power series $\sum a_n x^n$ has the interval of convergence $-1 < x < 1$, and if

(31) $$s_n = a_0 + a_1 + \cdots + a_n,$$

then

$$\lim_{x \to 1-} \sum_{n=0}^{\infty} a_n x^n = \lim_{n \to \infty} \frac{s_0 + s_1 + \cdots + s_n}{n + 1}$$

when the right-hand limit exists. Thus the power series normally divergent for $x = 1$ can have a sum. Moreover if $f(x)$ is the function represented by the power series, Frobenius's definition of the value of the series at $x = 1$ agrees with $\lim_{x \to 1-} f(x)$.

Divorced from its connection with power series, Frobenius's work suggested a summability definition for divergent series. If $\sum a_n$ is divergent and s_n has the meaning in (31), then one can take as the sum

$$\sum_{n=0}^{\infty} a_n = \lim_{n \to \infty} S_n = \lim_{n \to \infty} \frac{s_0 + s_1 + \cdots + s_n}{n + 1}$$

if this limit exists. Thus for the series $1 - 1 + 1 - 1 + \cdots$, the S_n have the values $1, 1/2, 2/3, 2/4, 3/5, 1/2, 4/7, 1/2, \ldots$ so that $\lim_{n \to \infty} S_n = 1/2$. If $\sum a_n$ converges, then Frobenius's "sum" gives the usual sum. This idea of averaging the partial sums of a series can be found in the older literature. It was used for special types of series by Daniel Bernoulli[33] and Joseph L. Raabe (1801–59).[34]

Shortly after Frobenius published his paper, Hölder[35] produced a generalization. Given the series $\sum a_n$, let

$$s_n^{(0)} = s_n$$

$$s_n^{(1)} = \frac{1}{n + 1} \left(s_0^{(0)} + s_1^{(0)} + \cdots + s_n^{(0)} \right)$$

$$s_n^{(2)} = \frac{1}{n + 1} \left(s_0^{(1)} + s_1^{(1)} + \cdots + s_n^{(1)} \right)$$

$$\cdots \cdots \cdots \cdots \cdots \cdots$$

$$s_n^{(r)} = \frac{1}{n + 1} \left(s_0^{(r-1)} + s_1^{(r-1)} + \cdots + s_n^{(r-1)} \right).$$

32. *Jour. für Math.*, 89, 1880, 262–64 = *Ges. Abh.*, 2, 8–10.
33. *Comm. Acad. Sci. Petrop.*, 16, 1771, 71–90.
34. *Jour. für Math.*, 15, 1836, 355–64.
35. *Math. Ann.*, 20, 1882, 535–49.

Then the sum s is given by

(32)
$$s = \lim_{n \to \infty} s_n^{(r)}$$

if this limit exists for some r. Hölder's definition is now known as summability (H, r).

Hölder gave an example. Consider the series

$$-\frac{1}{(1 + x)^2} = -1 + 2x - 3x^2 + 4x^3 - \cdots.$$

This series diverges when $x = 1$. However for $x = 1$,

$$s_0 = -1, s_1 = 1, s_2 = -2, s_3 = 2, s_4 = -3, \cdots.$$

Then

$$s_0^{(1)} = -1, s_1^{(1)} = 0, s_2^{(1)} = -\frac{2}{3}, s_3^{(1)} = 0, s_4^{(1)} = -\frac{3}{5}, \cdots;$$

$$s_0^{(2)} = -1, s_1^{(2)} = -\frac{1}{2}, s_2^{(2)} = -\frac{5}{9}, s_3^{(2)} = -\frac{5}{12}, s_4^{(2)} = -\frac{34}{75}, \cdots.$$

It is almost apparent that $\lim_{n \to \infty} s_n^{(2)} = -1/4$, and this is the Hölder sum $(H, 2)$. It is also the value that Euler assigned to the series on the basis of his principle that the sum is the value of the function from which the series is derived.

Another of the now-standard definitions of summability was given by Cesàro, a professor at the University of Naples.[36] Let the series be $\sum_{i=0}^{\infty} a_i$ and let s_n be $\sum_{i=0}^{n} a_i$. Then the Cesàro sum is

(33)
$$s = \lim_{n \to \infty} \frac{S_n^{(r)}}{D_n^{(r)}}, \qquad r \text{ integral and } \geq 0,$$

where

$$S_n^{(r)} = s_n + r s_{n-1} + \frac{r(r + 1)}{2!} s_{n-2} + \cdots + \frac{r(r + 1) \cdots (r + n - 1)}{n!} s_0$$

and

$$D_n^{(r)} = \frac{(r + 1)(r + 2) \cdots (r + n)}{n!}.$$

The case $r = 1$ includes Frobenius's definition. Cesàro's definition is now referred to as summability (C, r). The methods of Hölder and Cesàro give the same results. That Hölder summability implies Cesàro's was proved by Konrad Knopp (1882–1957) in an unpublished dissertation of 1907; the converse was proved by Walter Schnee (b. 1885).[37]

36. *Bull. des Sci. Math.*, (2), 14, 1890, 114–20.
37. *Math. Ann.*, 67, 1909, 110–25.

An interesting feature of some of the definitions of summability when applied to power series with radius of convergence 1 is that they not only give a sum that agrees with $\lim_{x \to 1-} f(x)$, where $f(x)$ is the function whose power series representation is involved, but have the further property that they preserve a meaning in regions where $|x| > 1$ and in these regions furnish the analytic continuation of the original power series.

Further progress in finding a "sum" for divergent series received its motivation from a totally different direction, the work of Stieltjes on continued fractions. The fact that continued fractions can be converted into divergent or convergent series and conversely was utilized by Euler.[38] Euler sought (Chap. 20, secs. 4 and 6) a sum for the divergent series

$$(34) \qquad\qquad 1 - 2! + 3! - 4! + 5! - \cdots.$$

In his article on divergent series [39] and in correspondence with Nicholas Bernoulli (1687–1759),[40] Euler first proved that

$$(35) \qquad\qquad x - (1!)x^2 + (2!)x^3 - (3!)x^4 + \cdots$$

formally satisfies the differential equation

$$x^2 \frac{dy}{dx} + y = x,$$

for which he obtained the integral solution

$$(36) \qquad\qquad y = \int_0^\infty \frac{xe^{-t}}{1 + xt} \, dt.$$

Then by using rules he derived for converting convergent series into continued fractions Euler transformed (35) into

$$(37) \qquad\qquad \frac{x}{1+} \frac{x}{1+} \frac{x}{1+} \frac{2x}{1+} \frac{2x}{1+} \frac{3x}{1+} \frac{3x}{1+} \cdots.$$

This work contains two features. On the one hand, Euler obtained an integral that can be taken to be the "sum" of the divergent series (35); the latter is in fact asymptotic to the integral. On the other hand he showed how to convert divergent series into continued fractions. In fact he used the continued fraction for $x = 1$ to calculate a value for the series (34).

There was incidental work of this nature during the latter part of the eighteenth century and a good deal of the nineteenth, of which the most noteworthy is due to Laguerre.[41] He proved, first of all, that the integral

38. *Novi Comm. Acad. Sci. Petrop.*, 5, 1754/5, 205–37, pub. 1760 = *Opera*, (1), 14, 585–617, and *Nova Acta Acad. Sci. Petrop.*, 2, 1784, 36–45, pub. 1788 = *Opera*, (1), 16, 34–43.
39. *Novi Comm. Acad. Sci. Petrop.*, 5, 1754/5, 205–37, pub. 1760 = *Opera*, (1), 14, 585–617.
40. Euler's *Opera Posthuma*, 1, 545–49.
41. *Bull. Soc. Math. de France*, 7, 1879, 72–81 = *Œuvres*, 1, 428–37.

(36) could be expanded into the continued fraction (37). He also treated the divergent series

(38) $$1 + x + 2! \, x^2 + 3! \, x^3 + \cdots.$$

Since

$$m! = \Gamma(m + 1) = \int_0^\infty e^{-z} z^m \, dz,$$

the series can be written as

$$\int_0^\infty e^{-z} \, dz + x \int_0^\infty e^{-z} z \, dz + x^2 \int_0^\infty e^{-z} z^2 \, dz + \cdots.$$

If we formally interchange summation and integration we obtain

$$\int_0^\infty e^{-z}(1 + xz + x^2 z^2 + \cdots) \, dz$$

or

(39) $$f(x) = \int_0^\infty e^{-z} \frac{1}{1 - zx} \, dz.$$

The $f(x)$ thus derived is analytic for all complex x except real and positive values and can be taken to be the "sum" of the series (38).

In his thesis of 1886 Stieltjes took up the study of divergent series.[42] Here Stieltjes introduced the very same definition of a series asymptotic to a function that Poincaré had introduced, but otherwise confined himself to the computational aspects of some special series.

Stieltjes did go on to study continued fraction expansions of divergent series and wrote two celebrated papers of 1894–95 on this subject.[43] This work, which is the beginning of an analytic theory of continued fractions, considered questions of convergence and the connection with definite integrals and divergent series. It was in these papers that he introduced the integral bearing his name.

Stieltjes starts with the continued fraction

(40) $$\frac{1}{a_1 z +} \frac{1}{a_2 +} \frac{1}{a_3 z +} \frac{1}{a_4 +} \frac{1}{a_5 z +} \cdots \frac{1}{a_{2n} +} \frac{1}{a_{2n+1} z +} \cdots,$$

where the a_n are positive real numbers and z is complex. He then shows that when the series $\sum_{n=1}^\infty a_n$ diverges, the continued fraction (40) converges

42. "Recherches sur quelques séries semi-convergentes," *Ann. de l'Ecole Norm. Sup.*, (3), 3, 1886, 201–58 = *Œuvres complètes*, 2, 2–58.
43. *Ann. Fac. Sci. de Toulouse*, 8, 1894, J. 1–122, and 9, 1895, A. 1–47 = *Œuvres complètes*, 2, 402–559.

to a function $F(z)$ which is analytic in the complex plane except along the negative real axis and at the origin, and

$$(41) \qquad\qquad F(z) = \int_0^\infty \frac{d\phi(u)}{u + z}.$$

When $\sum a_n$ converges, the even and odd partial sums of (40) converge to distinct limits $F_1(z)$ and $F_2(z)$ where

$$F_1(z) = \int_0^\infty \frac{dg_1(u)}{z + u}, \qquad F_2(z) = \int_0^\infty \frac{dg_2(u)}{z + u}.$$

Now it was known that the continued fraction (40) could be formally developed into a series

$$(42) \qquad\qquad \frac{C_0}{z} - \frac{C_1}{z^2} + \frac{C_2}{z^3} - \frac{C_3}{z^4} + \cdots$$

with positive C_i. The correspondence (with some restrictions) is also reciprocal. To every series (42) there corresponds a continued fraction (40) with positive a_n. Stieltjes showed how to determine the C_i from the a_n and in the case where $\sum a_n$ is divergent he showed that the ratio C_n/C_{n-1} increases. If it has a finite limit λ, the series converges for $|z| > \lambda$, but if the ratio increases without limit then the series diverges for all z.

The relation between the series (42) and the continued fraction (40) is more detailed. Although the continued fraction converges if the series does, the converse is not true. When the series (42) diverges one must distinguish two cases, according as $\sum a_n$ is divergent or convergent. In the former case, as we have noted, the continued fraction gives one and only one functional equivalent, which can be taken to be the sum of the divergent series (42). When $\sum a_n$ is convergent two different functions are obtained from the continued fraction, one from the even convergents and the other from the odd convergents. But to the series (42) (now divergent) there corresponds an infinite number of functions each of which has the series as its asymptotic development.

Stieltjes's results have also this significance: they indicate a division of divergent series into at least two classes, those series for which there is properly a single functional equivalent whose expansion is the series and those for which there are at least two functional equivalents whose expansions are the series. The continued fraction is only the intermediary between the series and the integral; that is, given the series one obtains the integral through the continued fraction. Thus, a divergent series belongs to one or more functions, which functions can be taken to be the sum of the series in a new sense of sum.

Stieltjes also posed and solved an inverse problem. To simplify the statement a bit, let us suppose that $\phi(u)$ is differentiable so that the integral (41) can be written as

$$\int_0^\infty \frac{f(u)}{z + u}\, du.$$

To the divergent series (42) and in the case where $\sum a_n$ is divergent there corresponds an integral of this form. The problem is, given the series to find $f(u)$. A formal expansion of the integral shows that

(43) $C_n = \int_0^\infty f(u)u^n\, du, \qquad n = 0, 1, 2, \cdots.$

Hence knowing the C_n one must determine $f(u)$ satisfying the infinite set of equations (43). This is what Stieltjes called the "problem of moments." It does not admit a unique solution, for Stieltjes himself gave a function

$$f(u) = e^{-\sqrt[4]{u}} \sin \sqrt[4]{u}$$

which makes $C_n = 0$ for all n. If the supplementary condition is imposed that $f(u)$ shall be positive between the limits of integration, then only a single $f(u)$ is possible.

The systematic development of the theory of summable series begins with the work of Borel from 1895 on. He first gave definitions that generalize Cesàro's. Then, taking off from Stieltjes's work, he gave an integral definition.[44] If the process used by Laguerre is applied to any series of the form

(44) $a_0 + a_1 x + a_2 x^2 + \cdots$

having a finite radius of convergence (including 0), we are led to the integral

(45) $\int_0^\infty e^{-z} F(zx)\, dz$

where

$$F(u) = 1 + a_1 u + \frac{a_2}{2!} u^2 + \cdots + \frac{a_n}{n!} u^n + \cdots.$$

This integral is the expression on which Borel built his theory of divergent series. It was taken by Borel to be the sum of the series (44). The series $F(u)$ is called the associated series of the original series.

If the original series (44) has a radius of convergence R greater than 0, the associated series represents an entire function. Then the integral $\int_0^\infty e^{-z} F(zx)\, dz$ has a sense if x lies within the circle of convergence, and the values of the integral and series are identical. But the integral may have a sense for values of x outside the circle of convergence, and in this case the

44. *Ann. de l'Ecole Norm. Sup.*, (3), 16, 1899, 9–136.

integral furnishes an analytic continuation of the original series. The series is said by Borel to be *summable* (in the sense just explained) at a point x where the integral has a meaning.

If the original series (44) is divergent ($R = 0$), the associated series may be convergent or divergent. If it is convergent over only a portion of the plane $u = zx$ we understand by $F(u)$ the value not merely of the associated series but of its analytic continuation. Then the integral $\int_0^\infty e^{-z} F(zx) \, dz$ may have a meaning and is, as we see, obtained from the original divergent series. The determination of the region of x-values in which the original series is summable was undertaken by Borel, both when the original series is convergent ($R > 0$) and divergent ($R = 0$).

Borel also introduced the notion of absolute summability. The original series is absolutely summable at a value of x when

$$\int_0^\infty e^{-z} F(zx) \, dz$$

is absolutely convergent and the successive integrals

$$\int_0^\infty e^{-z} \left| \frac{d^\lambda F(zx)}{dz^\lambda} \right| dz, \qquad \lambda = 1, 2, \cdots$$

have a sense. Borel then shows that a divergent series, if absolutely summable, can be manipulated precisely as a convergent series. The series, in other words, represents a function and can be manipulated in place of the function. Thus the sum, difference, and product of two absolutely summable series is absolutely summable and represents the sum, difference, and product respectively of the two functions represented by the separate series. The analogous fact holds for the derivative of an absolutely summable series. Moreover the sum in the above sense agrees with the usual sum in the case of a convergent series, and subtraction of the first k terms reduces the "sum" of the entire series by the sum of these k terms. Borel emphasized that any satisfactory definition of summability must possess these properties, though not all do. He did not require that any two definitions necessarily yield the same sum.

These properties made possible the immediate application of Borel's theory to differential equations. If, in fact,

$$P(x, y, y', \ldots, y^{(n)}) = 0$$

is a differential equation which is holomorphic in x at the origin and is algebraic in y and its derivatives, any absolutely summable series

$$a_0 + a_1 x + a_2 x^2 + \cdots$$

which satisfies the differential equation formally defines an analytic function that is a solution of the equation. For example, the Laguerre series

$$1 + x + 2! \, x^2 + 3! \, x^3 + \cdots$$

satisfies formally

$$x^2 \frac{d^2y}{dx^2} + (x - 1)y = -1,$$

and so the function (cf. (39))

$$f(x) = \int_0^\infty \frac{e^{-z} \, dz}{1 - zx}$$

must be a solution of the equation.

Once the notion of summability gained some acceptance, dozens of mathematicians introduced a variety of new definitions that met some or all of the requirements imposed by Borel and others. Many of the definitions of summability have been extended to double series. Also, a variety of problems has been formulated and many solved that involve the notion of summability. For example, suppose a series is summable by some method. What additional conditions can be imposed on the series so that, granted its summability, it will also be convergent in Cauchy's sense? Such theorems are called Tauberian after Alfred Tauber (b. 1866). Thus Tauber proved[45] that if $\sum a_n$ is Abelian summable to s and na_n approaches 0 as n becomes infinite, then $\sum a_n$ converges to s.

The concept of summability does then allow us to give a value or sum to a great variety of divergent series. The question of what is accomplished thereby necessarily arises. If a given infinite series were to arise directly in a physical situation, the appropriateness of any definition of sum would depend entirely on whether the sum is physically significant, just as the physical utility of any geometry depends on whether the geometry describes physical space. Cauchy's definition of sum is the one that usually fits because it says basically that the sum is what one gets by continually adding more and more terms in the ordinary sense. But there is no logical reason to prefer this concept of sum to the others that have been introduced. Indeed the representation of functions by series is greatly extended by employing the newer concepts. Thus Leopold Fejér (1880–1959), a student of H. A. Schwarz, showed the value of summability in the theory of Fourier series. In 1904[46] Fejér proved that if in the interval $[-\pi, \pi]$, $f(x)$ is bounded and (Riemann) integrable, or if unbounded the integral $\int_{-\pi}^{\pi} f(x) \, dx$ is absolutely convergent, then at every point in the interval at which $f(x + 0)$ and $f(x - 0)$ exist, the Frobenius sum of the Fourier series

$$\frac{a_0}{2} + \sum_{n=1}^{\infty} (a_n \cos nx + b_n \sin nx)$$

45. *Monatshefte für Mathematik und Physik*, 8, 1897, 273–77.
46. *Math. Ann.*, 58, 1904, 51–69.

is $[f(x + 0) + f(x - 0)]/2$. The conditions on $f(x)$ in this theorem are weaker than in previous theorems on the convergence of Fourier series to $f(x)$ (cf. Chap. 40, sec. 6).

Fejér's fundamental result was the beginning of an extensive series of fruitful investigations on the summability of series. We have had numerous occasions to view the need to represent functions by infinite series. Thus in meeting the initial condition in the solution of initial- and boundary-value problems of partial differential equations, it is usually necessary to represent the given initial $f(x)$ in terms of the eigenfunctions that are obtained from the application of the boundary conditions to the ordinary differential equations that result from the method of separation of variables. These eigenfunctions can be Bessel functions, Legendre functions, or any one of a number of types of special functions. Whereas the convergence in the Cauchy sense of such a series of eigenfunctions to the given $f(x)$ may not obtain, the series may indeed be summable to $f(x)$ in one or another of the senses of summability and the initial condition is thereby satisfied. These applications of summability represent a great success for the concept.

The construction and acceptance of the theory of divergent series is another striking example of the way in which mathematics has grown. It shows, first of all, that when a concept or technique proves to be useful even though the logic of it is confused or even nonexistent, persistent research will uncover a logical justification, which is truly an afterthought. It also demonstrates how far mathematicians have come to recognize that mathematics is man-made. The definitions of summability are not the natural notion of continually adding more and more terms, the notion which Cauchy merely rigorized; they are artificial. But they serve mathematical purposes, including even the mathematical solution of physical problems; and these are now sufficient grounds for admitting them into the domain of legitimate mathematics.

Bibliography

Borel, Emile: *Notice sur les travaux scientifiques de M. Emile Borel*, 2nd ed., Gauthier-Villars, 1921.

——: *Leçons sur les séries divergentes*, Gauthier-Villars, 1901.

Burkhardt, H.: "Trigonometrische Reihe und Integrale," *Encyk. der Math. Wiss.*, B. G. Teubner, 1904–16, II A12, 819–1354.

——: "Über den Gebrauch divergenter Reihen in der Zeit von 1750–1860," *Math. Ann.*, 70, 1911, 169–206.

Carmichael, Robert D.: "General Aspects of the Theory of Summable Series," *Amer. Math. Soc. Bull.*, 25, 1918/19, 97–131.

Collingwood E. F.: "Emile Borel," *Jour. Lon. Math. Soc.*, 34, 1959, 488–512.

Ford, W. B.: "A Conspectus of the Modern Theories of Divergent Series," *Amer. Math. Soc. Bull.*, 25, 1918/19, 1–15.

Hardy, G. H.: *Divergent Series*, Oxford University Press, 1949. See the historical notes at the ends of the chapters.

Hurwitz, W. A.: "A Report on Topics in the Theory of Divergent Series," *Amer. Math. Soc. Bull.*, 28, 1922, 17–36.

Knopp, K.: "Neuere Untersuchungen in der Theorie der divergenten Reihen," *Jahres. der Deut. Math.-Verein.*, 32, 1923, 43–67.

Langer, Rudolf E.: "The Asymptotic Solution of Ordinary Linear Differential Equations of the Second Order," *Amer. Math. Soc. Bull.*, 40, 1934, 545–82.

McHugh, J. A. M.: "An Historical Survey of Ordinary Linear Differential Equations with a Large Parameter and Turning Points," *Archive for History of Exact Sciences*, 7, 1971, 277–324.

Moore, C. N.: "Applications of the Theory of Summability to Developments in Orthogonal Functions," *Amer. Math. Soc. Bull.*, 25, 1918/19, 258–76.

Plancherel, Michel: "Le Développement de la théorie des séries trigonométriques dans le dernier quart de siècle," *L'Enseignement Mathématique*, 24, 1924/25, 19–58.

Pringsheim, A.: "Irrationalzahlen und Konvergenz unendlicher Prozesse," *Encyk. der Math. Wiss.*, B. G. Teubner, 1898–1904, IA3, 47–146.

Reiff, R.: *Geschichte der unendlichen Reihen*, H. Lauppsche Buchhandlung, 1889, Martin Sandig (reprint), 1969.

Smail, L. L.: *History and Synopsis of the Theory of Summable Infinite Processes*, University of Oregon Press, 1925.

Van Vleck, E. B.: "Selected Topics in the Theory of Divergent Series and Continued Fractions," *The Boston Colloquium of the Amer. Math. Soc.*, 1903, Macmillan, 1905, 75–187.

48

Tensor Analysis and Differential Geometry

> Either therefore the reality which underlies space must form
> a discrete manifold or we must seek the ground of its metric
> in relations outside it, in the binding forces which act on it.
> This leads us into the domain of another science, that of
> physics, into which the object of our work does not allow us
> to go today.   BERNHARD RIEMANN

1. *The Origins of Tensor Analysis*

Tensor analysis is often described as a totally new branch of mathematics,
created *ab initio* either to meet some specific objective or just to delight
mathematicians. It is actually no more than a variation on an old theme,
namely, the study of differential invariants associated primarily with a
Riemannian geometry. These invariants, we may recall (Chap. 37, sec. 5),
are expressions that retain their form and value under any change in the
coordinate system because they represent geometrical or physical properties.

The study of differential invariants had been launched by Riemann,
Beltrami, Christoffel, and Lipschitz. The new approach was initiated by
Gregorio Ricci-Curbastro (1853–1925), a professor of mathematics at the
University of Palermo. He was influenced by Luigi Bianchi, whose work
had followed Christoffel's. Ricci sought to expedite the search for geometrical properties and the expression of physical laws in a form invariant under
change of coordinates. He did his major work on the subject during the years
1887–96, though he and an Italian school continued to work on it for twenty
or more years after 1896. In the main period Ricci worked out his approach
and a comprehensive notation for the subject, which he called the absolute
differential calculus. In an article published in 1892[1] Ricci gave the first
systematic account of his method and applied it to some problems in differential geometry and physics.

1. *Bull. des Sci. Math.*, (2), 16, 1892, 167–89 = *Opere*, 1, 288–310.

Nine years later Ricci and his famous pupil Tullio Levi-Civita (1873–1941) collaborated on a comprehensive paper, "Methods of the Absolute Differential Calculus and Their Applications."[2] The work of Ricci and Levi-Civita gave a more definitive formulation of this calculus. The subject became known as tensor analysis after Einstein gave it this name in 1916. In view of the many changes in notation made by Ricci and later by Levi-Civita and Ricci, we shall use the notation which has by now become rather standard.

2. *The Notion of a Tensor*

To get at the notion of a tensor as introduced by Ricci, let us consider the function $A(x^1, x^2, \ldots, x^n)$. By A_i we shall mean $\partial A/\partial x^i$. Then the expression

$$(1) \qquad \sum A_j \, dx^j$$

is a differential invariant under transformations of the form

$$(2) \qquad x^i = f_i(y^1, y^2, \ldots, y^n).$$

It is assumed that the functions f_i possess all necessary derivatives and that the transformation is reversible, so that

$$(3) \qquad y^i = g_i(x^1, x^2, \ldots, x^n).$$

Under the transformation (2) the expression (1) becomes

$$(4) \qquad \sum \bar{A}_j(y^1, y^2, \cdots, y^n) \, dy^j.$$

However, $\bar{A}_j$ does not equal A_j. Rather

$$(5) \qquad \bar{A}_j = \frac{\partial \bar{A}}{\partial y^j} = A_1 \frac{\partial x^1}{\partial y^j} + A_2 \frac{\partial x^2}{\partial y^j} + \cdots + A_n \frac{\partial x^n}{\partial y^j}$$

wherein it is understood that the x^i in the A_i are replaced by their values in terms of the y^i. Thus the $\bar{A}_j$ can be related to the A_j by the specific law of transformation (5) wherein the first derivatives of the transformation are involved.

Ricci's idea was that instead of concentrating on the invariant differential form (1), it would be sufficient and more expeditious to treat the set

$$A_1, A_2, \cdots, A_n$$

and to call this set of components a tensor provided that, under a change of coordinates, the new set of components

$$\bar{A}_1, \bar{A}_2, \cdots, \bar{A}_n$$

2. *Math. Ann.*, 54, 1901, 125–201 = Ricci, *Opere*, 2, 185–271.

are related to the original set by the law of transformation (5). It is this explicit emphasis on the set or system of functions and the law of transformation that marks Ricci's approach to the subject of differential invariants. The set of A_j, which happen to be the components of the gradient of the scalar function A, is an example of a covariant tensor of rank 1. The notion of a set or system of functions characterizing an invariant quantity was in itself not new because vectors were already well known in Ricci's time; these are represented by their components in a coordinate system and are also subject to a law of transformation if the vector is to remain, as it should be, invariant under a change of coordinates. However, the new systems that Ricci introduced were far more general and the emphasis on the law of transformation was also new.

As another example of the point of view introduced by Ricci, let us consider the expression for the element of distance. This is given by

$$(6) \qquad ds^2 = \sum_{i,j=1}^{n} g_{ij}\, dx^i\, dx^j.$$

Under change of coordinates the value of the distance ds must remain the same on geometrical grounds. However, if we perform the transformation (2) and write the new expression in the form

$$(7) \qquad \overline{ds^2} = \sum_{i,j=1}^{n} G_{ij}\, dy^i\, dy^j,$$

then $g_{ij}(x^1, \ldots, x^n)$ will not equal $G_{ij}(y^i, \ldots, y^n)$ (when the values of the y^i represent the same point as the x^i do). What does hold is that

$$(8) \qquad G_{kl} = \sum_{i,j=1}^{n} g_{ij} \frac{\partial x^i}{\partial y^k} \frac{\partial x^j}{\partial y^l},$$

when the x^i in the g_{ij} are replaced by their values in terms of the y^i. To see that (8) holds we have but to replace dx^i in (6) by

$$dx^i = \sum_{k=1}^{n} \frac{\partial x^i}{\partial y^k} dy^k$$

and dx^j in (6) by

$$dx^j = \sum_{l=1}^{n} \frac{\partial x^j}{\partial y^l} dy^l$$

and extract the coefficient of $dy^k\, dy^l$. Thus, though G_{kl} is not at all g_{kl}, we do know how to obtain G_{kl} from the g_{kl}. The set of n^2 coefficients g_{ik} of the fundamental quadratic form are a tensor, a covariant tensor of rank 2, and the law of transformation is that given by (8).

Ricci also introduced contravariant tensors. Let us consider the inverse of the transformation hitherto considered. If this inverse is

(9) $$y^j = g_j(x^1, x^2, \cdots, x^n),$$

then

(10) $$dy^j = \sum_{k=1}^{n} \frac{\partial y^j}{\partial x^k} \, dx^k.$$

If we now regard the dx^k as a set of quantities constituting a tensor, then we see that, of course, $dy^j \neq dx^j$, but we can obtain the dy^j from the dx^j by the law of transformation illustrated in (10). The set of elements dx^k is called a contravariant tensor of rank 1, the term contravariant pointing to the presence of the $\partial y^j/\partial x^k$ in the transformation, as opposed to the derivatives $\partial x^i/\partial y^j$ which appear in (8) and (5). Thus the very differentials of the transformation variables form a contravariant tensor of rank 1.

Correspondingly, we can have a tensor of rank 2 that transforms contravariantly in both indices. If the set of functions $A^{kl}(x^1, x^2, \ldots, x^n)$, $k, l = 1, 2, \ldots, n$, transform under (9) so that

(11) $$\bar{A}^{ij} = \sum_{k,l=1}^{n} \frac{\partial y^i}{\partial x^k} \frac{\partial y^j}{\partial x^l} A^{kl},$$

then this set is a contravariant tensor of rank 2. Moreover, we can have what are called mixed tensors, which transform covariantly in some indices and contravariantly in others. For example, the set A^k_{ij}, $i, j, k = 1, 2, \ldots, n$, denotes a mixed tensor wherein, following Ricci, the lower indices are the ones in which it transforms covariantly and the upper index is the one in which it transforms contravariantly. The tensor whose elements are A^k_{ij} is called a tensor of rank 3. We can have covariant, contravariant, and mixed tensors of rank r. An n-dimensional tensor of rank r will have n^r components. Equation (21) in Chapter 37 shows that Riemann's four-index symbol (rk, ih) is a covariant tensor of rank 4. A covariant tensor of rank 1 is a vector. To it Levi-Civita associated the contravariant vector which is defined as follows: If the set of λ_i are the components of a covariant vector, then the set

$$\lambda^i = \sum_{k=1}^{n} g^{ik} \lambda_k$$

is the associated contravariant vector; g^{ik} is a quotient whose numerator is the cofactor of g_{ik} in the determinant of the g_{ik} and whose denominator is g, the value of the determinant.

There are operations on tensors. Thus if we have two tensors of the same kind, that is, having the same number of covariant and the same number of contravariant indices, we may add them by adding the components with the identical indices. Thus

$$A_i^j + B_i^j = C_i^j.$$

One must and can show that the C_i^j constitute a tensor covariant in the index i and contravariant in the index j.

One may multiply any two tensors whose indices run from 1 to n. An example may suffice to show the idea. Thus

$$A_i^h B_j^k = C_{ij}^{hk},$$

and one can show that the tensor with the n^4 components C_{ij}^{hk} is covariant in the lower indices and contravariant in the upper ones. There is no operation of division of tensors.

The operation of contraction is illustrated by the following example: Given the tensor A_{ir}^{hs}, we define the quantity

$$B_i^h = \sum_{r=1}^{n} A_{ir}^{hr},$$

wherein on the right side we add the components. We could show that the set of quantities B_i^h is a tensor of rank 2, covariant in the index i and contravariant in the index h.

To sum up, a tensor is a set of functions (components), fixed relative to one frame of reference or coordinate system, that transform under change of coordinates in accordance with certain laws. Each component in one coordinate system is a linear homogeneous function of the components in another coordinate system. If the components of one tensor equal those of another when both are expressed in one coordinate system, they will be equal in all coordinate systems. In particular, if the components vanish in one system they vanish in all. Equality of tensors is then an invariant with respect to change of reference system. The physical, geometrical, or even purely mathematical significance which a tensor possesses in one coordinate system is preserved by the transformation so that it obtains again in the second coordinate system. This property is vital in the theory of relativity, wherein each observer has his own coordinate system. Since the true physical laws are those that hold for all observers, to reflect this independence of coordinate system these laws are expressed as tensors.

With the tensor concept in hand, one can re-express many of the concepts of Riemannian geometry in tensor form. Perhaps the most important is the curvature of the space. Riemann's concept of curvature (Chap. 37, sec. 3) can be formulated as a tensor in many ways. The modern expressions use the summation convention introduced by Einstein, namely, that if an index is repeated in a product of two symbols, then summation is understood. Thus

$$g^{ij}\lambda_j = \sum_{j-1}^{n} g^{ij}\lambda_j.$$

In this notation the curvature tensor is (cf. [20] of Chap. 37)

$$R_{\lambda\mu\rho\sigma} = \frac{\partial}{\partial x^\rho}[\mu\sigma, \lambda] - \frac{\partial}{\partial x^\sigma}[\mu\rho, \lambda] + \{\mu\rho, \varepsilon\}[\lambda\sigma, \varepsilon] - \{\mu\sigma, \varepsilon\}[\lambda\rho, \varepsilon]$$

or equivalently

$$R^i_{jlk} = \frac{\partial}{\partial x^l}\{jk, i\} - \frac{\partial}{\partial x^k}\{jl, i\} - [\{sk, i\}\{jl, s\} - \{sl, i\}\{jk, s\}]$$

wherein the brackets denote Christoffel symbols of the first kind and the braces, symbols of the second kind. Either form is now called the Riemann-Christoffel curvature tensor. By reason of certain relationships (which we shall not describe) among the components, the number of distinct components of this tensor is $n^2(n^2 - 1)/12$. For $n = 4$, which is the case in the general theory of relativity, the number of distinct components is 20. In a two-dimensional Riemannian space there is just one distinct component, which can be taken to be R_{1212}. In this case Gauss's total curvature K proves to be

$$K = \frac{R_{1212}}{g},$$

where g is the determinant of the g_{ij} or $g_{11}g_{22} - g_{12}^2$. If all components vanish the space is Euclidean.

From the Riemann-Christoffel tensor Ricci obtained by contraction what is now called the Ricci tensor or the Einstein tensor. The components R_{jl} are $\sum_{k=1}^{n} R^k_{jlk}$. This tensor for $n = 4$ was used by Einstein[3] to express the curvature of his space-time Riemannian geometry.

3. Covariant Differentiation

Ricci also introduced into tensor analysis[4] an operation that he and Levi-Civita later called covariant differentiation. This operation had already appeared in Christoffel's and Lipschitz's work.[5] Christoffel had given a method (Chap. 37, sec. 4) whereby from differential invariants involving

3. *Zeit. für Math. und Phys.*, 62, 1914, 225–61.
4. *Atti della Accad. dei Lincei, Rendiconti*, (4), 3, 1887, 15–18 = *Opere*, 1, 199–203.
5. *Jour. für Math.*, 70, 1869, 46–70 and 241–45, and 71–102.

the derivatives of the fundamental form for ds^2 and of functions $\phi(x_1, x_2, \ldots, x_n)$ one could derive invariants involving higher derivatives. Ricci appreciated the importance of this method for his tensor analysis and adopted it.

Whereas Christoffel and Lipschitz treated covariant differentiation of the entire form, Ricci, in accordance with his emphasis on the components of a tensor, worked with these. Thus if $A_i(x^1, x^2, \ldots, x^n)$ is a covariant component of a vector or tensor of rank 1, the covariant derivative of A_i is not simply the derivative with respect to x^l, but the tensor of second rank

$$(12) \qquad A_{i,l} = \frac{\partial A_i}{\partial x^l} - \sum_{j=1}^{n} \{il, j\} A_j,$$

where the braces indicate the Christoffel symbol of the second kind. Likewise, if A_{ik} is a component of a covariant tensor of rank 2, then its covariant derivative with respect to x^l is given by

$$(13) \qquad A_{ik,l} = \frac{\partial A_{ik}}{\partial x^l} - \sum_{j=1}^{n} \{il, j\} A_{jk} - \sum_{j=1}^{n} \{kl, j\} A_{ij}.$$

For a contravariant tensor with components A^i of rank 1 the covariant derivative $A^i_{,l}$ is given by

$$A^i_{,l} = \frac{\partial A^i}{\partial x^l} + \sum_{j=1}^{n} A^j \{jl, i\},$$

and this is a mixed tensor of the second order. For the mixed tensor with components A^h_i the covariant derivative is

$$A^h_{i,l} = \frac{\partial A^h_i}{\partial x^l} - \sum_{j=1}^{n} A^h_j \{il, j\} + \sum_{j=1}^{n} A^j_i \{jl, h\}.$$

The covariant derivative of a scalar invariant ϕ is the covariant vector whose components are given by $\phi_{,i} = \partial \phi / \partial x_i$. This vector is called the gradient of the scalar invariant.

From the purely mathematical standpoint, the covariant derivative of a tensor is a tensor of one rank higher in the covariant indices. This fact is important because it enables one to treat such derivatives within the framework of tensor analysis. It also has geometrical meaning. Suppose we have a constant vector field in the plane, that is, a set of vectors one at each point but all having the same magnitude and direction. Then the components of any vector, expressed with respect to a rectangular coordinate system, are also constant. However, the components of these vectors with respect to the polar coordinate system, one component along the radius vector and the other perpendicular to the radius vector, change from point to point, because the directions in which the components are taken change from point to point.

If one takes the derivatives with respect to the coordinates, r and θ say, of these components, then the rate of change expressed by these derivatives reflects the change in the components caused by the coordinate system and not by any change in the vectors themselves. The coordinate systems used in Riemannian geometry are curvilinear. The effect of the curvilinearity of the coordinates is given by the Christoffel symbol of the second kind (here denoted by braces). The full covariant derivative of a tensor gives the actual rate of change of the underlying physical or geometrical quantity represented by the original tensor as well as the change due to the underlying coordinate system.

In Euclidean spaces, where the ds^2 can always be reduced to a sum of squares with constant coefficients, the covariant derivative reduces to the ordinary derivative because the Christoffel symbols are 0. Also the covariant derivative of each g_{ij} in a Riemannian metric is 0. This last fact was proved by Ricci[6] and is called Ricci's lemma.

The concept of covariant differentiation enables us to express readily for tensors generalizations of notions already known in vector analysis but now treatable in Riemannian geometry. Thus if the $A_i(x^1, x^2, \ldots, x^n)$ are the components of an n-dimensional vector A, then

$$(14) \qquad \theta = \sum_{i,l=1}^{n} g^{il} A_{i,l},$$

where g^{il} has been defined above, is a differential invariant. When the fundamental metric is a rectangular Cartesian coordinate system (in Euclidean space), the constants $g^{il} = 0$ except when $i = l$, and in this case the covariant and ordinary derivatives are identical. Then (14) becomes

$$\theta = \sum_{i=1}^{n} \frac{\partial A_i}{\partial x^i},$$

and this is the n-dimensional analogue in Euclidean space of what is called the divergence in three dimensions. Hence (14) is also called the divergence of the tensor with components A_i. One can also show, by using (14), that if A is a scalar point function then the divergence of the gradient of A is given by

$$(15) \qquad \Delta_2 A = \frac{1}{\sqrt{g}} \sum_{l=1}^{n} \frac{\partial}{\partial x^i} (\sqrt{g}\, A^l),$$

where

$$A^l = \sum_{i=1}^{n} g^{il} \frac{\partial A}{\partial x^i} = \sum_{i=1}^{n} g^{il} A_i.$$

6. *Atti della Accad. dei Lincei, Rendiconti*, (4), 5, 1889, 112–18 = *Opere*, 1, 268–75.

This expression for $\Delta_2 A$ is also Beltrami's expression for $\Delta_2 A$ in a Riemannian geometry (Chap. 37, sec. 5).

Though Ricci and Levi-Civita devoted much of their 1901 paper to developing the technique of tensor analysis, they were primarily concerned to find differential invariants. They posed the following general problem: Given a positive differential quadratic form ϕ and an arbitrary number of associated functions S, to determine all the absolute differential invariants that one can form from the coefficients of ϕ, the functions S, and the derivatives of the coefficients and functions up to a definite order m. They gave a complete solution. It is sufficient to find the *algebraic* invariants of the system consisting of the fundamental differential quadratic form ϕ, the covariant derivatives of any associated functions S up to order m, and, for $m > 1$, a certain quadrilinear form G_4 whose coefficients are the Riemann expressions (ih, jk) and its covariant derivatives up to order $m - 2$.

They conclude their paper by showing how some partial differential equations and physical laws can be expressed in tensor form so as to render them independent of the coordinate system. This was Ricci's avowed goal. Thus tensor analysis was used to express the mathematical invariance of physical laws many years before Einstein used it for this purpose.

4. Parallel Displacement

From 1901 to 1915, research on tensor analysis was limited to a very small group of mathematicians. However, Einstein's work changed the picture. Albert Einstein (1879–1955), while engaged as an engineer in the Swiss patent office, greatly stirred the scientific world by the announcement of his restricted or special theory of relativity.[7] In 1914 Einstein accepted a call to the Prussian Academy of Science in Berlin, as successor to the celebrated physical chemist Jacobus Van't Hoff (1852–1911). Two years later he announced his general theory of relativity.[8]

Einstein's revolutionary views on the relativity of physical phenomena aroused intense interest among physicists, philosophers, and mathematicians throughout the world. Mathematicians especially were excited by the nature of the geometry Einstein found it expedient to use in the creation of his theories.

The exposition of the restricted theory, involving the properties of four-dimensional pseudo-Euclidean manifolds (space-time), is best made with the aid of vectors and tensors, but the exposition of the general theory, involving properties of four-dimensional Riemannian manifolds (space-time), demands the use of the special calculus of tensors associated with such manifolds.

7. *Annalen der Phys.*, 17, 1905, 891–921; an English translation can be found in the Dover edition of A. Einstein, *The Principle of Relativity* (1951).
8. *Annalen der Physik*, 49, 1916, 769–822.

Fortunately, the calculus had already been developed but had not as yet attracted particular notice from physicists.

Actually Einstein's work on the restricted theory did not use Riemannian geometry or tensor analysis.[9] But the restricted theory did not involve the action of gravitation. Einstein then began to work on the problem of dispensing with the force of gravitation and accounting for its effect by imposing a structure on his space-time geometry so that objects would automatically move along the same paths as those derived by assuming the action of the gravitational force. In 1911 he made public a theory that accounted in this manner for a gravitational force that had a constant direction throughout space, knowing of course that this theory was unrealistic. Up to this time Einstein had used only the simplest mathematical tools and had even been suspicious of the need for "higher mathematics," which he thought was often introduced to dumbfound the reader. However, to make progress on his problem he discussed it in Prague with a colleague, the mathematician Georg Pick, who called his attention to the mathematical theory of Ricci and Levi-Civita. In Zurich Einstein found a friend, Marcel Grossmann (1878–1936), who helped him learn the theory; and with this as a basis, he succeeded in formulating the general theory of relativity.

To represent his four-dimensional world of three space coordinates and a fourth coordinate denoting time, Einstein used the Riemannian metric

$$(16) \qquad ds^2 = \sum_{i,j=1}^{4} g_{ij}\, dx_i\, dx_j,$$

wherein x_4 represents time. The g_{ij} were chosen so as to reflect the presence of matter in various regions. Moreover, since the theory is concerned with the determination of lengths, time, mass, and other physical quantities by different observers who are moving in arbitrary fashion with respect to each other, the "points" of space-time are represented by different coordinate systems, one attached to each observer. The relation of one coordinate system to another is given by a transformation

$$x_i = \phi_i(y_1, y_2, \ldots, y_4), \qquad i = 1, \ldots, 4.$$

The laws of nature are those relationships or expressions that are the same for all observers. Hence they are invariants in the mathematical sense.

From the standpoint of mathematics, the importance of Einstein's work was, as already noted, the enlargement of interest in tensor analysis and Riemannian geometry. The first innovation in tensor analysis following upon the theory of relativity is due to Levi-Civita. In 1917, improving on an idea

9. The metric is $ds^2 = dx^2 + dy^2 + dz^2 - c^2\, dt^2$. This is a space of constant curvature. Any section by a plane $t = $ const. is Euclidean.

of Ricci, he introduced[10] the notion of parallel displacement or parallel transfer of a vector. The idea was also created independently by Gerhard Hessenberg in the same year.[11] In 1906 Brouwer had introduced it for surfaces of constant curvature. The objective in the notion of parallel displacement is to define what is meant by parallel vectors in a Riemannian space. The difficulty in doing so may be seen by considering the surface of a sphere which, considered as a space in itself and with distance on the surface given by arcs of great circles, is a Riemannian space. If a vector, say one starting on a circle of latitude and pointing north (the vector will be tangent to the sphere), is moved by having the initial point follow the circle and is kept parallel to itself in the Euclidean three-space, then when it is halfway around the circle it is no longer tangent to the sphere and so is not in that space. To obtain a notion of parallelism of vectors suitable for a Riemannian space, the Euclidean notion is generalized, but some of the familiar properties are lost in the process.

The geometrical idea used by Levi-Civita to define parallel transfer or displacement is readily understood for a surface. Consider a curve C on the surface, and let a vector with one endpoint on C be moved parallel to itself in the following sense: At each point of C there is a tangent plane. The envelope of this family of planes is a developable surface, and when this surface is flattened out onto a plane, the vectors parallel along C are truly parallel in the Euclidean plane.

Levi-Civita generalized this idea to fit n-dimensional Riemannian spaces. It is true in the Euclidean plane that when a vector is carried parallel to itself and its initial point follows a straight line—a geodesic in the plane— the vector always makes the same angle with the line. Accordingly, parallelism in a Riemannian space is defined thus: When a vector in the space moves so that its initial point follows a geodesic, then the vector must continue to make the same angle with the geodesic (tangent to the geodesic). In particular, a tangent to a geodesic remains parallel to itself as it moves along a geodesic. By definition the vector continues to have the same magnitude. It is understood that the vector remains in the Riemannian space, even if that space is imbedded in a Euclidean space. The definition of parallel transfer also requires that the angle between two vectors be preserved as each is moved parallel to itself along the same curve C. In the general case of parallel transfer around an arbitrary closed curve C, the initial vector and the final vector will usually not have the same (Euclidean) direction. The deviation in direction will depend on the path C. Thus consider a vector starting at a point P on a circle of latitude on a sphere and tangent to the sphere along a meridian. When carried by *parallel transfer* around the circle it will end up at

10. *Rendiconti del Circolo Matematico di Palermo*, 42, 1917, 173–205.
11. *Math. Ann.*, 78, 1918, 187–217.

P tangent to the surface; but if ϕ is the co-latitude of P, it will make an angle $2\pi - 2\pi \cos \phi$ with the original vector.

If one uses the general definition of parallel displacement along a curve of a Riemannian space, he obtains an analytic condition. The differential equation satisfied by the components X^α of a contravariant vector under parallel transfer along a curve is (summation understood)

$$(17) \qquad \frac{dX^\alpha}{dt} + \{\beta\gamma, \alpha\}X^\beta \frac{du^\gamma}{dt} = 0, \qquad \alpha = 1, 2, \cdots, n,$$

wherein it is presupposed that the $u^i(t)$, $i = 1, 2, \ldots, n$, define a curve. For a covariant vector X_α the condition is

$$(18) \qquad \frac{dX_\alpha}{dt} - \{\alpha l, j\}X_j \frac{du^l}{dt} = 0, \qquad \alpha = 1, 2, \cdots, n.$$

These equations can be used to define parallel transfer along any curve C. The solution determined uniquely by the values of the components at a definite P is a vector having values at each point of C and by definition parallel to the initial vector at P. Equation (18) states that the covariant derivative of X_α is 0.

Once the notion of parallel displacement is introduced, one can describe the curvature of a space in terms of it, specifically in terms of the change in an infinitesimal vector upon parallel displacement by infinitesimal steps. Parallelism is at the basis of the notion of curvature even in Euclidean space, because the curvature of an infinitesimal arc depends on the change in direction of the tangent vector over the arc.

5. *Generalizations of Riemannian Geometry*

The successful use of Riemannian geometry in the theory of relativity revived interest in that subject. However, Einstein's work raised an even broader question. He had incorporated the gravitational effect of mass in space by using the proper functions for the g_{ij}. As a consequence the geodesics of his space-time proved to be precisely the paths of objects moving freely as, for example, the earth does around the sun. Unlike the situation in Newtonian mechanics, no gravitational force was required to account for the path. The elimination of gravity suggested another problem, namely, to account for attraction and repulsion of electric charges in terms of the metric of space. Such an accomplishment would supply a unified theory of gravitation and electromagnetism. This work led to generalizations of Riemannian geometry known collectively as non-Riemannian geometries.

In Riemannian geometry the ds^2 ties together the points of the space. It specifies how points are related to each other by prescribing the distance between points. In the non-Riemannian geometries the connection between points is specified in ways that do not necessarily rely upon a metric. The variety of these geometries is great, and each has a development as extensive as Riemannian geometry itself. Hence we shall give only some examples of the basic ideas in these geometries.

This area of work was initiated primarily by Hermann Weyl[12] and the class of geometries he introduced is known as the geometry of affinely connected spaces. In Riemannian geometry the proof that the covariant derivative of a tensor is itself a tensor depends only on relations of the form

$$(19) \qquad \overline{\{ik, h\}} = \{ab, j\} \frac{\partial x^a}{\partial y^i} \frac{\partial x^b}{\partial y^k} \frac{\partial y^h}{\partial x^j} + \frac{\partial^2 x^j}{\partial y^i \partial y^k} \frac{\partial y^h}{\partial x^j},$$

wherein the left side is the transform of $\{ab, j\}$ under a change of coordinates from the x^i to the y^i. These relations are satisfied by the Christoffel symbols, and the symbols themselves are defined in terms of the coefficients of the fundamental form. Consider instead functions L^i_{jk} and $\bar{L}^i_{jk}$ of the x^i and y^i, respectively, which satisfy the same relations (19) but are specified only as given functions unrelated to the fundamental quadratic form. A set of functions L^i_{jk} associated with a space V_n and having the transformation property (19) is said to constitute an affine connection. The functions are called the coefficients of affine connection and the space V_n is said to be affinely connected or to be an affine space. Riemannian geometry is the special case in which the coefficients of affine connection are the Christoffel symbols of the second kind and are derived from the fundamental tensor of the space. Given the L-functions, concepts such as covariant differentiation, curvature, and other notions analogous to those in Riemannian geometry can be introduced. However, one cannot speak of the magnitude of a vector in this new geometry.

In an affinely connected space a curve such that its tangents are parallel with respect to the curve (in the sense of parallel displacement for that space) is called a path of the space. Paths are thus a generalization of the geodesics of a Riemannian space. All affinely connected spaces that have the same L's have the same paths. Thus the geometry of affinely connected spaces dispenses with the Riemannian metric. Weyl did derive Maxwell's equations from the properties of the space, but the theory as a whole did not accord with other established facts.

Another non-Riemannian geometry, due to Luther P. Eisenhart (1876–1965) and Veblen,[13] and called the geometry of paths, proceeds somewhat differently. It starts with n^3 given functions $\Gamma^i_{\lambda\mu}$ of $x^1, \ldots, x^n$. The system of n differential equations

$$(20) \qquad \frac{d^2 x^i}{ds^2} + \sum_{\lambda\mu} \Gamma^i_{\lambda\mu} \frac{dx^\lambda}{ds} \frac{dx^\mu}{ds} = 0, \qquad i = 1, 2, \cdots, n,$$

12. *Mathematische Zeitschrift*, 2, 1918, 384–411 = *Ges. Abh.*, 2, 1–28.
13. *Proceedings of the National Academy of Sciences*, 8, 1922, 19–23.

with $\Gamma^i_{\lambda\mu} = \Gamma^i_{\mu\lambda}$, defines a family of curves called paths. These are the geodesics of the geometry. (In Riemannian geometry equations (20) are precisely those for geodesics.) Given the geodesics in the sense just described, one can build up the geometry of paths in a manner quite analogous to that pursued in Riemannian geometry.

A different generalization of Riemannian geometry is due to Paul Finsler (1894–) in his (unpublished) thesis of 1918 at Göttingen. The Riemannian ds^2 is replaced by a more general function $F(x, dx)$ of the coordinates and differentials. There are restrictions on F to insure the possibility of minimizing the integral $\int F(x, (dx/dt))\, dt$ and obtaining thereby the geodesics.

The attempts to generalize the concept of Riemannian geometry so as to incorporate electromagnetic as well as gravitational phenomena have thus far failed. However, mathematicians have continued to work on the abstract geometries.

Bibliography

Cartan, E.: "Les récentes généralisations de la notion d'espace," *Bull. des Sci. Math.*, 48, 1924, 294–320.

Pierpont, James: "Some Modern Views of Space," *Amer. Math. Soc. Bull.*, 32, 1926, 225–58.

Ricci-Curbastro, G.: *Opere*, 2 vols., Edizioni Cremonese, 1956–57.

Ricci-Curbastro, G., and T. Levi-Civita: "Méthodes de calcul différentiel absolu et leurs applications," *Math. Ann.*, 54, 1901, 125–201.

Thomas, T. Y.: "Recent Trends in Geometry," *Amer. Math. Soc. Semicentennial Publications* II, 1938, 98–135.

Weatherburn, C. E.: *The Development of Multidimensional Differential Geometry*, Australian and New Zealand Association for the Advancement of Science, 21, 1933, 12–28.

Weitzenbock, R.: "Neuere Arbeiten der algebraischen Invariantentheorie. Differentialinvarianten," *Encyk. der Math. Wiss.*, B. G. Teubner, 1902–27, III, Part III, E1, 1–71.

Weyl, H.: *Mathematische Analyse des Raumproblems* (1923), Chelsea (reprint), 1964.

49
The Emergence of Abstract Algebra

Perhaps I may without immodesty lay claim to the appella-
tion of Mathematical Adam, as I believe that I have given
more names (passed into general circulation) of the creatures
of the mathematical reason than all the other mathematicians
of the age combined.   J. J. SYLVESTER

1. *The Nineteenth-Century Background*

In abstract algebra, as in the case of most twentieth-century developments,
the basic concepts and goals were fixed in the nineteenth century. The fact
that algebra can deal with collections of objects that are not necessarily real
or complex numbers was demonstrated in a dozen nineteenth-century creations. Vectors, quaternions, matrices, forms such as $ax^2 + bxy + cy^2$, hypernumbers of various sorts, transformations, and substitutions or permutations
are examples of objects that were combined under operations and laws of
operation peculiar to the respective collections. Even the work on algebraic
numbers, though it dealt with classes of complex numbers, brought to the
fore the variety of algebras because it demonstrated that only some properties
are applicable to these classes as opposed to the entire complex number
system.

These various classes of objects were distinguished in accordance with
the properties that the operations on them possessed; and we have seen that
such notions as group, ring, ideal, and field, and subordinate notions such as
subgroup, invariant subgroup, and extension field were introduced to identify
the sets of properties. However, nearly all of the nineteenth-century work
on these various types of algebras dealt with the concrete systems mentioned
above. It was only in the last decades of the nineteenth century that the
mathematicians appreciated that they could move up to a new level of
efficiency by integrating many separate algebras through abstraction of their
common content. Thus permutation groups, the groups of classes of forms
treated by Gauss, hypernumbers under addition, and transformation groups
could all be treated in one swoop by speaking of a set of elements or things
subject to an operation whose nature is specified only by certain abstract

properties, the foremost of these being that the operation applied to two elements of the set produces a third element of the set. The same advantages could be achieved for the various collections that formed rings and fields. Though the idea of working with abstract collections preceded the axiomatics of Pasch, Peano, and Hilbert, the latter development undoubtedly accelerated the acceptance of the abstract approach to algebras.

Thus arose abstract algebra as the conscious study of entire classes of algebras, which individually were not only concrete but which served purposes in specific areas as substitution groups did in the theory of equations. The advantage of obtaining results that might be useful in many specific areas by considering abstract versions was soon lost sight of, and the study of abstract structures and the derivation of their properties became an end in itself.

Abstract algebra has been one of the favored fields of the twentieth century and is now a vast area. We shall present only the beginnings of the subject and indicate the almost unlimited opportunities for continued research. The great difficulty in discussing what has been done in this field is terminology. Aside from the usual difficulties, that different authors use different terms and that terms change meaning from one period to another, abstract algebra is marked and marred by the introduction of hundreds of new terms. Every minor variation in concept is distinguished by a new and often imposing sounding term. A complete dictionary of the terms used would fill a large book.

2. Abstract Group Theory

The first abstract structure to be introduced and treated was the group. A great many of the basic ideas of abstract group theory can be found implicitly and explicitly at least as far back as 1800. It is a favorite activity of historians, now that the abstract theory is in existence, to trace how many of the abstract ideas were foreshadowed by the concrete work of Gauss, Abel, Galois, Cauchy, and dozens of other men. We shall not devote space to this reexamination of the past. The only significant point that bears mentioning is that, once the abstract notion was acquired, it was relatively easy for the founders of abstract group theory to obtain ideas and theorems by rephrasing this past work.

Before examining the development of the abstract group concept, it may be well to know where the men were heading. The abstract definition of a group usually used today calls for a collection of elements, finite or infinite in number, and an operation which, when applied to any two elements in the collection, results in an element of the collection (the closure property). The operation is associative; there is an element, e, say, such that for any element a of the group, $ae = ea = a$; and to each element a there exists an inverse element a' such that $aa' = a'a = e$. When the operation is commutative the group is called commutative or Abelian and the operation is called addition and denoted by $+$. The element e is then denoted by 0 and called the zero element. If the operation is not commutative it is called multiplication and the element e is denoted by 1 and called the identity.

The abstract group concept and the properties that should be attached to it were slow in coming to light. We may recall (Chap. 31, sec. 6) that Cayley had proposed the abstract group in 1849, but the merit of the notion was not recognized at the time. In 1858 Dedekind,[1] far in advance of his time, gave an abstract definition of finite groups which he derived from permutation groups. Again in 1877[2] he observed that his modules of algebraic numbers, which call for $\alpha + \beta$ and $\alpha - \beta$ belonging when α and β do, can be generalized so that the elements are no longer algebraic numbers and the operation can be general but must have an inverse and be commutative. Thus he suggested an abstract finite commutative group. Dedekind's understanding of the value of abstraction is noteworthy. He saw clearly in his work on algebraic number theory the value of structures such as ideals and fields. He is the effective founder of abstract algebra.

Kronecker,[3] taking off from work on Kummer's ideal numbers, also gave what amounts to the abstract definition of a finite Abelian group similar to Cayley's 1849 concept. He specifies abstract elements, an abstract operation, the closure property, the associative and commutative properties, and the existence of a unique inverse element of each element. He then proves some theorems. Among the various powers of any element θ, there is one which equals the unit element, 1. If ν is the smallest exponent for which θ^ν equals the unit element, then to each divisor μ of ν there is an element ϕ for which $\phi^\mu = 1$. If θ^ρ and ϕ^σ both equal 1 and ρ and σ are the smallest numbers for which this holds and are relatively prime, then $(\theta\phi)^{\rho\sigma} = 1$. Kronecker also gave the first proof of what is now called a basis theorem. There exists a fundamental finite system of elements θ_1, θ_2, θ_3, ... such that the products

$$\theta_1^{h_1}\theta_2^{h_2}\theta_3^{h_3}\cdots, \qquad h_i = 1, 2, 3, \cdots, n_i,$$

represent all elements of the group just once. The lowest possible values $n_1, n_2, n_3, \ldots$ that correspond to $\theta_1, \theta_2, \theta_3, \ldots$ (that is, for which $\theta_i^{n_i} = 1$) are such that each is divisible by the following one and the product $n_1 n_2 n_3 \cdots$ equals the number n of elements in the group. Moreover all the prime factors of n are in n_1.

In 1878 Cayley wrote four more papers on finite abstract groups.[4] In

1. *Werke*, 3, 439–46.
2. *Bull. des Sci. Math.*, (2), 1, 1877, 17–41, p. 41 in particular $=$ *Werke*, 3, 262–96.
3. *Monatsber. Berliner Akad.*, 1870, 881–89 $=$ *Werke*, 1, 271–82.
4. *Math. Ann.*, 13, 1878, 561–65; *Proc. London Math. Soc.*, 9, 1878, 126–33; *Amer. Jour. of Math.*, 1, 1878, 50–52 and 174–76; all in Vol. 10 of his *Collected Math. Papers*.

these, as in his 1849 and 1854 papers, he stressed that a group can be considered as a general concept and need not be limited to substitution groups, though, he points out, every (finite) group can be represented as a substitution group. These papers of Cayley had more influence than his earlier ones because the time was ripe for an abstraction that embraced more than substitution groups.

In a joint paper, Frobenius and Ludwig Stickelberger (1850–1936)[5] made the advance of recognizing that the abstract group concept embraces congruences and Gauss's composition of forms as well as the substitution groups of Galois. They mention groups of infinite order.

Though Eugen Netto, in his book *Substitutionentheorie und ihre Anwendung auf die Algebra* (1882), confined his treatment to substitution groups, his wording of his concepts and theorems recognized the abstractness of the concepts. Beyond putting together results established by his predecessors, Netto treated the concepts of isomorphism and homomorphism. The former means a one-to-one correspondence between two groups such that if $a \cdot b = c$, where a, b, and c are elements of the first group, then $a' \cdot b' = c'$, where a', b', and c' are the corresponding elements of the second group. A homomorphism is a many-to-one correspondence in which again $a \cdot b = c$ implies $a' \cdot b' = c'$.

By 1880 new ideas on groups came into the picture. Klein, influenced by Jordan's work on permutation groups, had shown in his Erlanger Programm (Chap. 38, sec. 5) that infinite transformation groups, that is, groups with infinitely many elements, could be used to classify geometries. These groups, moreover, are continuous in the sense that arbitrarily small transformations are included in any group or, alternatively stated, the parameters in the transformations can take on all real values. Thus in the transformations that express rotation of axes the angle θ can take on all real values. Klein and Poincaré in their work on automorphic functions had utilized another kind of infinite group, the discrete or noncontinuous group (Chap. 29, sec. 6).

Sophus Lie, who had worked with Klein around 1870, took up the notion of continuous transformation groups, but for other purposes than the classification of geometries. He had observed that most of the ordinary differential equations that had been integrated by older methods were invariant under classes of continuous transformation groups, and he thought he could throw light on the solution of differential equations and classify them.

In 1874 Lie introduced his general theory of transformation groups.[6] Such a group is represented by

$$(1) \qquad x_i' = f_i(x_1, x_2, \cdots, x_n, a_1, \cdots, a_n), \qquad i = 1, 2, \cdots, n,$$

5. *Jour. für. Math.*, 86, 1879, 217–62 = Frobenius, *Ges. Abh.*, 1, 545–90.
6. *Nachrichten König. Ges. der Wiss. zu Gött.*, 1874, 529–42 = *Ges. Abh.*, 5, 1–8.

where the f_i are analytic in the x_i and a_i. The a_i are parameters as opposed to the x_i, which are variables, and $(x_1, x_2, \ldots, x_n)$ stands for a point in n-dimensional space. Both the parameters and the variables may take on real or complex values. Thus in one dimension the class of transformations

$$x' = \frac{ax + b}{cx + d},$$

where a, b, c, and d take on all real values, is a continuous group. The groups represented by (1) are called finite, the word finite referring to the number of parameters. The number of transformations is, of course, infinite. The one-dimensional case is a three-parameter group because only the ratios of a, b, and c to d matter. In the general case the product of two transformations

$$x_i' = f_i(x_i, \cdots, x_n, a_1, \cdots, a_n)$$
$$x_i'' = f_i(x_i', \cdots, x_n', b_1, \cdots, b_n)$$

is

$$x_i''' = f_i(x_i, \cdots, x_n, c_1, \cdots, c_n),$$

where the c_i are functions of the a_i and b_i. In the case of one variable Lie spoke of the group as a simply extended manifold; for n variables he spoke of an arbitrarily extended manifold.

   In a paper of 1883 on continuous groups, published in an obscure Norwegian journal,[7] Lie also introduced infinite continuous transformation groups. These are not defined by equations such as (1) but by means of differential equations. The resulting transformations do not depend upon a finite number of continuous parameters but on arbitrary functions. There is no abstract group concept corresponding to these infinite continuous groups, and though much work has been done on them, we shall not pursue it here.

   It is perhaps of interest that Klein and Lie at the beginning of their work defined a group of transformations as one possessing only the closure property. The other properties, such as the existence of an inverse to each transformation, were established by using the properties of the transformations, or, as in the case of the associative law, were used as obvious properties of transformations. Lie recognized during the course of his work that one should postulate as part of the definition of a group the existence of an inverse to each element.

   By 1880 four main types of groups were known. These are the discontinuous groups of finite order, exemplified by substitution groups; the infinite discontinuous (or discrete) groups, such as occur in the theory of automorphic functions; the finite continuous groups of Lie exemplified by the transformation groups of Klein and the more general analytic transformations of Lie; and the infinite continuous groups of Lie defined by differential equations.

7. *Ges. Abh.*, 5, 314–60.

With the work of Walther von Dyck (1856–1934), the three main roots of group theory—the theory of equations, number theory, and infinite transformation groups—were all subsumed under the abstract group concept. Dyck was influenced by Cayley and was a student of Felix Klein. In 1882 and 1883[8] he published papers on abstract groups that included discrete and continuous groups. His definition of a group calls for a set of elements and an operation that satisfy the closure property, the associative but not the commutative property, and the existence of an inverse element of each element.

Dyck worked more explicitly with the notion of the generators of a group, which is implicit in Kronecker's basis theorem and explicit in Netto's work on substitution groups. The generators are a fixed subset of independent elements of a group such that every member of the group can be expressed as the product of powers of the generators and their inverses. When there are no restrictions on the generators the group is called a free group. If A_1, $A_2, \ldots$ are the generators, then an expression of the form

$$A_1^{\mu_1} A_2^{\mu_2} \cdots,$$

where the μ_i are positive or negative integers, is called a word. There may be relations among the generators, and these would be of the form

$$F_i(A_i) = 1;$$

that is, a word or combination of words equals the identity element of the group. Dyck then shows that the presence of relations implies an invariant subgroup and a factor group $\bar{G}$ of the free group G. In his 1883 paper he applied the abstract group theory to permutation groups, finite rotation groups (symmetries of polyhedra), number-theoretic groups, and transformation groups.

Sets of independent postulates for an abstract group were given by Huntington,[9] E. H. Moore,[10] and Leonard E. Dickson (1874–1954).[11] These as well as other postulate systems are minor variations of each other.

Having arrived at the abstract notion of a group, the mathematicians turned to proving theorems about abstract groups that were suggested by known results for concrete cases. Thus Frobenius[12] proved Sylow's theorem (Chap. 31, sec. 6) for finite abstract groups. Every finite group whose order, that is, the number of elements, is divisible by the νth power of a prime p always contains a subgroup of order p^ν.

Beyond searching concrete groups for properties that may hold for abstract groups, many men introduced concepts directly for abstract groups.

8. *Math. Ann.*, 20, 1882, 1–44 and 22, 1883, 70–118.
9. *Amer. Math. Soc. Bull.*, 8, 1902, 296–300 and 388–91, and *Amer. Math. Soc. Trans.*, 6, 1905, 181–97.
10. *Amer. Math. Soc. Trans.*, 3, 1902, 485–92, and 6, 1905, 179–80.
11. *Amer. Math. Soc. Trans.*, 6, 1905, 198–204.
12. *Jour. für Math.*, 100, 1887, 179–81 = *Ges. Abh.*, 2, 301–3.

Dedekind[13] and George A. Miller (1863–1951)[14] treated non-Abelian groups in which every subgroup is a normal (invariant) subgroup. Dedekind in his 1897 paper and Miller[15] introduced the notions of commutator and commutator subgroup. If s and t are any two elements of a group G, the element $s^{-1}t^{-1}st$ is called the commutator of s and t. Both Dedekind and Miller used this notion to prove theorems. For example, the set of all commutators of all (ordered) pairs of elements of a group G generates an invariant subgroup of G. The automorphisms of a group, that is, the one-to-one transformations of the members of a group into themselves under which if $a \cdot b = c$ then $a' \cdot b' = c'$, were studied on an abstract basis by Hölder[16] and E. H. Moore.[17]

The further development of abstract group theory has pursued many directions. One of these taken over from substitution groups by Hölder in the 1893 paper is to find all the groups of a given order, a problem Cayley had also mentioned in his 1878 papers.[18] The general problem has defied solution. Hence particular orders have been investigated, such as p^2q^2 where p and q are prime. A related problem has been the enumeration of intransitive and primitive and imprimitive groups of various degrees (the number of letters in a substitution group).

Another direction of research has been the determination of composite or solvable groups and simple groups, that is, those which have no invariant subgroups (other than the identity). This problem of course derives from Galois theory. Hölder, after introducing the abstract notion of a factor group,[19] treated simple groups[20] and composite groups.[21] Among results are the fact that a cyclic group of prime order is simple and so is the alternating group of all even permutations on n letters for $n \geq 5$. Many other finite simple groups have been found.

As for solvable groups, Frobenius devoted several papers to the problem. He found, for example,[22] that all groups whose order is not divisible by the square of a prime are solvable.[23] The problem of investigating which groups

13. *Math. Ann.*, 48, 1897, 548–61 = *Werke*, 2, 87–102.
14. *Amer. Math. Soc. Bull.*, 4, 1898, 510–15 = *Coll. Works*, 1, 266–69.
15. *Amer. Math. Soc. Bull.*, 4, 1898, 135–39 = *Coll. Works*, 1, 254–57.
16. *Math. Ann.*, 43, 1893, 301–412.
17. *Amer. Math. Soc. Bull.*, 1, 1895, 61–66, and 2, 1896, 33–43.
18. *Coll. Math. Papers*, 10, 403.
19. *Math. Ann.*, 34, 1889, 26–56.
20. *Math. Ann.*, 40, 1892, 55–88, and 43, 1893, 301–412.
21. *Math. Ann.*, 46, 1895, 321–422.
22. *Sitzungsber. Akad. Wiss. zu Berlin*, 1893, 337–45, and 1895, 1027–44 = *Ges. Abh.*, 2, 565–73, 677–94.
23. A recent result of major importance was obtained by Walter Feit (1930–) and John G. Thompson (*Pacific Jour. of Math.*, 13, Part 2, 1963, 775–1029). All finite groups of odd order are solvable. The suggestion that this might be the case was made by Burnside in 1906.

are solvable is part of the broader problem of determining the structure of a given group.

Dyck in his papers of 1882 and 1883 had introduced the abstract idea of a group defined by generators and relations among the generators. Given a group defined in terms of a finite number of generators and relations, the identity or word problem, as formulated by Max Dehn,[24] is the task of determining whether any "word" or product of elements is equal to the unit element. Any set of relations may be given because at worst the trivial group consisting only of the identity satisfies them. To decide whether a group given by generators and relations is trivial is not trivial. In fact there is no effective procedure. For one defining relation Wilhelm Magnus (1907–) showed[25] that the word problem is solvable. But the general problem is not.[26]

Another famous unsolved problem of ordinary group theory is Burnside's problem. Any finite group has the properties that it is finitely generated and every element has finite order. In 1902[27] William Burnside (1852–1927) asked whether the converse was true; that is, if a group G is finitely generated and if every element has finite order, is G finite? This problem has attracted a great deal of attention and only specializations of it have received solutions. Still another problem, the isomorphism problem, is to determine when two groups, each defined by generators and relations, are isomorphic.

One of the surprising turns in group theory is that shortly after the abstract theory had been launched, the mathematicians turned to representations by more concrete algebras in order to obtain results for the abstract groups. Cayley had pointed out in his 1854 paper that any finite abstract group can be represented by a group of permutations. We have also noted (Chap. 31, sec. 6) that Jordan in 1878 introduced the representation of substitution groups by linear transformations. These transformations or their matrices have proved to be the most effective representation of abstract groups and are called linear representations.

A matrix representation of a group G is a homomorphic correspondence of the elements g of G to a set of non-singular square matrices $A(g)$ of fixed order and with complex elements. The homomorphism implies that

$$A(g_i g_j) = A(g_i)A(g_j)$$

for all g_i and g_j of G. There are many matrix representations of a group G because the order (number of rows or columns) can be altered, and even for a given order the correspondence can be altered. Also one can add two

24. *Math. Ann.*, 71, 1911, 116–44.
25. *Math. Ann.*, 106, 1932, 295–307.
26. This was proved in 1955 by P. S. Novikov. See *American Math. Soc. Translations* (2), 9, 1958, 1–122.
27. *Quart. Jour. of Math.*, 33, 230–38.

representations. If for each element g of G, A_g is the corresponding matrix of one representation of order m and B_g is the corresponding matrix of order n, then

$$\begin{pmatrix} A_g & 0 \\ 0 & B_g \end{pmatrix}$$

is another representation, which is called the sum of the separate representations. Likewise if

$$E_g = \begin{pmatrix} B_g & C_g \\ 0 & D_g \end{pmatrix}$$

is another representation, when B_g and D_g are non-singular matrices of orders m and n respectively, the B_g and D_g are also representations, and of lower order than E_g. E_g is called a graduated representation; it and any representation equivalent to it ($F_g^{-1} E_g F_g$ is an equivalent representation if F is non-singular and of the order of E_g) is called reducible. A representation not equivalent to a graduated one is called irreducible. The basic idea of an irreducible representation consisting of a set of linear transformations in n variables is that it is a homomorphic or isomorphic representation in which it is impossible to choose $m < n$ linear functions of the variables which are transformed among themselves by every operation of the group they represent. A representation that is equivalent to the sum of irreducible representations is called completely reducible.

Every finite group has a particular representation called regular. Suppose the elements are labeled $g_1, g_2, \ldots, g_n$. Let a be any one of the g's. We consider an n by n matrix. Suppose $ag_i = g_j$. Then we place a 1 in the (i, j) place of the matrix. This is done for all g_i and the fixed a. All the other elements of the matrix are taken to be 0. The matrix so obtained corresponds to a. There is such a matrix for each g of the group and this set of matrices is a left regular representation. Likewise by forming the products $g_i a$ we get a right regular representation. By reordering the g's of the group we can get other regular representations. The notion of a regular representation was introduced by Charles S. Peirce in 1879.[28]

The representation of substitution groups by linear transformations of the form

$$x_i' = \sum_j a_{ij} x_j, \qquad i = 1, 2, \cdots, n,$$

initiated by Jordan was broadened to the study of representations of all finite abstract groups by Frobenius, Burnside, Theodor Molien (1861–1941), and Issai Schur (1875–1941) in the latter part of the nineteenth century and

28. *Amer. Jour. of Math.*, 4, 1881, 221–25.

the first part of the twentieth. Frobenius[29] introduced for finite groups the notions of reducible and completely reducible representations and showed that a regular representation contains all the irreducible representations. In other papers published from 1897 to 1910, some in conjunction with Schur, he proved many other results, including the fact that there are only a few irreducible representations, out of which all the others are composed.

Burnside[30] gave another major result, a necessary and sufficient condition on the coefficients of a group of linear transformations in n variables in order that the group be reducible. The fact that any finite group of linear transformations is completely reducible was first proved by Heinrich Maschke (1853–1908).[31] Representation theory for finite groups has led to important theorems for abstract groups. In the second quarter of this century, representation theory was extended to continuous groups, but this development will not be pursued here.

Aiding in the study of group representations is the notion of group character. This notion, which can be traced back to the work of Gauss, Dirichlet, and Heinrich Weber (see note 35), was formulated abstractly by Dedekind for Abelian groups in the third edition of Dirichlet's *Vorlesungen über Zahlentheorie* (1879). A character of a group is a function $x(s)$ defined on all the elements s such that it is not zero for any s and $x(ss') = x(s)x(s')$. Two characters are distinct if $x(s) \neq x'(s)$ for at least one s of the group.

This notion was generalized to all finite groups by Frobenius. After having formulated a rather complex definition,[32] he gave a simpler definition,[33] which is now standard. The character function is the trace (sum of the main diagonal elements) of the matrices of an irreducible representation of the group. The same concept was applied later by Frobenius and others to infinite groups.

Group characters furnish in particular a determination of the minimum number of variables in terms of which a given finite group can be represented as a linear transformation group. For commutative groups they permit a determination of all subgroups.

Displaying the usual exuberance for the current vogue, many mathematicians of the late nineteenth and early twentieth centuries thought that all mathematics worth remembering would ultimately be comprised in the theory of groups. Klein in particular, though he did not like the formalism of *abstract* group theory, favored the group concept because he thought it would unify mathematics. Poincaré was equally enthusiastic. He said,[34] ". . . the

29. *Sitzungsber. Akad. Wiss. zu Berlin*, 1897, 994–1015 = *Ges. Abh.*, 3, 82–103.
30. *Proc. London Math. Soc.*, (2), 3, 1905, 430–34.
31. *Math. Ann.*, 52, 1899, 363–68.
32. *Sitzungsber. Akad. Wiss. zu Berlin*, 1896, 985–1021 = *Ges. Abh.*, 3, 1–37.
33. *Sitzungsber. Akad. Wiss. zu Berlin*, 1897, 994–1015 = *Ges. Abh.*, 3, 82–103.
34. *Acta Math.*, 38, 1921, 145.

theory of groups is, so to say, the whole of mathematics divested of its matter and reduced to pure form."

3. *The Abstract Theory of Fields*

The concept of a field R generated by n quantities $a_1, a_2, \ldots, a_n$, that is, the set of all quantities formed by adding, subtracting, multiplying, and dividing these quantities (except division by 0) and the concept of an extension field formed by adjoining a new element λ not in R, are in Galois's work. His fields were the domains of rationality of the coefficients of an equation and extensions were made by adjunction of a root. The concept also has a quite different origin in Dedekind's and Kronecker's work on algebraic numbers (Chap. 34, sec. 3), and in fact the word "field" (*Körper*) is due to Dedekind.

The abstract theory of fields was initiated by Heinrich Weber, who had already espoused the abstract viewpoint for groups. In 1893[35] he gave an abstract formulation of Galois theory wherein he introduced (commutative) fields as extensions of groups. A field, as Weber specifies, is a collection of elements subject to two operations, called addition and multiplication, which satisfy the closure condition, the associative and commutative laws, and the distributive law. Moreover each element must have a unique inverse under each operation, except for division by 0. Weber stressed group and field as the two major concepts of algebra. Somewhat later, Dickson[36] and Huntington[37] gave independent postulates for a field.

To the fields that were known in the nineteenth century, the rational, real and complex numbers, algebraic number fields, and fields of rational functions in one or several variables, Kurt Hensel added another type, p-adic fields, which initiated new work in algebraic numbers (*Theorie der algebraischen Zahlen*, 1908). Hensel observed, first of all, that any ordinary integer D can be expressed in one and only one way as a sum of powers of a prime p. That is,

$$D = d_0 + d_1 p + \cdots + d_k p^k,$$

in which d_i is some integer from 0 to $p - 1$. For example,

$$14 = 2 + 3 + 3^2$$
$$216 = 2 \cdot 3^3 + 2 \cdot 3^4.$$

Similarly any rational number r (not 0) can be written in the form

$$r = \frac{a}{b} p^n,$$

35. *Math. Ann.*, 43, 1893, 521–49. For groups see *Math. Ann.*, 20, 1882, 301–29.
36. *Amer. Math. Soc. Trans.*, 4, 1903, 13–20, and 6, 1905, 198–204.
37. *Amer. Math. Soc. Trans.*, 4, 1903, 31–37, and 6, 1905, 181–97.

where a and b are integers not divisible by p and n is 0 or a positive or negative integer. Hensel generalized on these observations and introduced p-adic numbers. These are expressions of the form

$$(2) \qquad \sum_{i=-\rho}^{\infty} c_i p^i,$$

where p is a prime and the coefficients, the c_i's, are ordinary rational numbers reduced to their lowest form whose denominator is not divisible by p. Such expressions need not in general have values as ordinary numbers. However, by definition they are mathematical entities.

Hensel defined the four basic operations with these numbers and showed that they are a field. A subset of the p-adic numbers can be put into one-to-one correspondence with the ordinary rational numbers, and in fact this subset is isomorphic to the rational numbers in the full sense of an isomorphism between two fields. In the field of p-adic numbers, Hensel defined units, integral p-adic numbers, and other notions analogous to those of the ordinary rational numbers.

By introducing polynomials whose coefficients are p-adic numbers, Hensel was able to speak of p-adic roots of polynomial equations and extend to these roots all of the concepts of algebraic number fields. Thus there are p-adic integral algebraic numbers and more general p-adic algebraic numbers, and one can form fields of p-adic algebraic numbers that are extensions of the "rational" p-adic numbers defined by (2). In fact, all of the ordinary theory of algebraic numbers is carried over to p-adic numbers. Surprisingly perhaps, the theory of p-adic algebraic numbers leads to results on ordinary algebraic numbers. It has also been useful in treating quadratic forms and has led to the notion of valuation fields.

The growing variety of fields motivated Ernst Steinitz (1871–1928), who was very much influenced by Weber's work, to undertake a comprehensive study of abstract fields; this he did in his fundamental paper, *Algebraischen Theorie der Körper*.[38] All fields, according to Steinitz, can be divided into two principal types. Let K be a field and consider all subfields of K (for example, the rational numbers are a subfield of the real numbers). The elements common to all the subfields are also a subfield, called the prime field P of K. Two types of prime fields are possible. The unit element e is contained in P and, therefore, so are

$$e, 2e, \cdots, ne, \cdots.$$

These elements are either all different or there exists an ordinary integer p such that $pe = 0$. In the first case P must contain all fractions ne/me, and

38. *Jour. für Math.*, 137, 1910, 167–309.

since these elements form a field, P must be isomorphic to the field of rational numbers and K is said to have characteristic 0.

If, on the other hand, $pe = 0$, it is readily shown that the smallest such p must be a prime and the field must be isomorphic to the field of integral residues modulo p, that is, $0, 1, \ldots, p - 1$. Then K is said to be a field of characteristic p. Any subfield of K has the same characteristic. Then $pa = pea = 0$; that is, all expressions in K can be reduced modulo p.

From the prime field P in either of the types just described, the original field K can be obtained by the process of adjunction. The method is to take an element a in K but not in P and to form all rational functions $R(a)$ of a with coefficients in P and then, if necessary, to take b not in $R(a)$ and do the same with b, and to continue the process as long as necessary.

If one starts with an arbitrary field K one can make various types of adjunctions. A simple adjunction is obtained by adjoining a single element x. The enlarged field must contain all expressions of the form

$$(3) \qquad a_0 + a_1 x + \cdots + a_n x^n,$$

where the a_i are elements of K. If these expressions are all different, then the extended field is the field $K(x)$ of all rational functions of x with coefficients in K. Such an adjunction is called a transcendental adjunction and $K(x)$ is called a transcendental extension. If some of the expressions (3) are equal, one can show that there must exist a relation (using α for x)

$$f(\alpha) = \alpha^m + b_1 \alpha^{m-1} + \cdots + b_m = 0$$

with the b_i in K and with $f(x)$ irreducible in K. Then the expressions

$$C_1 \alpha^{m-1} + \cdots + C_m$$

with the C_i in K constitute a field $K(\alpha)$ formed by the adjunction of α to K. This field is called a simple algebraic extension of K. In $K(\alpha)$, $f(x)$ has a root, and conversely, if an arbitrary irreducible $f(x)$ in K is chosen, then one can construct a $K(\alpha)$ in which $f(x)$ has a root.

A fundamental result due to Steinitz is that every field can be obtained from its prime field by first making a series of (possibly infinite) transcendental adjunctions and then a series of algebraic adjunctions to the transcendental field. A field K' is said to be an algebraic extension of K if it can be obtained by successive simple algebraic adjunctions. If the number of adjunctions is finite, K' is said to be of finite rank.

Not every field can be enlarged by algebraic adjunctions. For example, the complex numbers cannot be because every $f(x)$ is reducible in this field. Such a field is algebraically complete. Steinitz also proved that for every field K there is a unique algebraically complete field K' which is algebraic over K in the sense that all other algebraically complete fields over K (containing K) contain a subfield equivalent to K'.

Steinitz considered also the problem of determining the fields in which the Galois theory of equations holds. To say that Galois theory holds in a field means the following: A Galois field $\overline{K}$ over a given field K is an algebraic field in which every irreducible $f(x)$ in K either remains irreducible or decomposes into a product of linear factors. To every Galois field $\overline{K}$ there exists a set of automorphisms, each of which transforms the elements of $\overline{K}$ into other elements of $\overline{K}$ and such that $\alpha \pm \beta$ and $\alpha\beta$ correspond to $\alpha' \pm \beta'$ and $\alpha'\beta'$ while all elements of K remain invariant (correspond to themselves). The set of automorphisms forms a group G, the Galois group of $\overline{K}$ with respect to K. The main theorem of Galois theory asserts that there is a unique correspondence between the subgroups of G and the subfields of $\overline{K}$ such that to any subgroup G' of G there corresponds a subfield K' consisting of all elements left invariant by G' and conversely. Galois theory is said to hold for those fields in which this theorem holds. Steinitz's result is essentially that Galois theory holds in those fields of finite rank that can be obtained from a given field by a series of adjunctions of roots of irreducible $f(x)$ which have no equal roots. Fields in which all irreducible $f(x)$ have no equal roots are called separable. (Steinitz said *vollkommen* or complete.)

The theory of fields also includes, as Steinitz's classification indicates, finite fields of characteristic p. A simple example of the latter is the set of all residues (remainders) modulo a prime p. The concept of a finite field is due to Galois. In 1830 he published a definitive paper, "Sur la théorie des nombres."[39] Galois wished to solve congruences

$$F(x) \equiv 0 \bmod p,$$

where p is a prime and $F(x)$ is a polynomial of degree n. He took $F(x)$ to be irreducible (modulo p), so that the congruence did not have integral or irrational roots. This obliged him to consider other solutions, which were suggested by the imaginary numbers. Galois denoted one of the roots of $F(x)$ by i (which is not $\sqrt{-1}$). He then considered the expression

$$(4) \qquad a_0 + a_1 i + a_2 i^2 + \cdots + a_{n-1} i^{n-1},$$

where the a_i are whole numbers. When these coefficients assume separately all the least positive residues, modulo p, this expression can take on only p^ν values. Let α be one of these non-zero values, of which there are $p^\nu - 1$. The powers of α also have the form (4). Hence these powers cannot be all different. There must then be at least one power $\alpha^m = 1$, where m is the smallest of such values. Then there will be m different values

$$(5) \qquad 1, \alpha, \alpha^2, \ldots, \alpha^{m-1}.$$

If we multiply these m quantities by an expression β of the same form, we obtain a new group of quantities different from (5) and from each other.

39. *Bulletin des Sciences Mathématiques de Férussac*, 13, 1830, 428–35 = *Œuvres*, 1897, 15–23.

Multiplying the set (5) by γ will produce more such quantities, until we obtain all of the form (4). Hence m must divide $p^\nu - 1$ or $\alpha^{p^\nu - 1} = 1$ and so $\alpha^{p^\nu} = \alpha$. The p^ν values of the form (4) constitute a finite field. Galois had shown in this concrete situation that the number of elements in a Galois field of characteristic p is a power of p.

E. H. Moore[40] showed that any finite abstract field is isomorphic to a Galois field of order p^n, p a prime. There is such a field for every prime p and positive integer n. The characteristic of each field is p. Joseph H. M. Wedderburn (1882–1948), a professor at Princeton University,[41] and Dickson proved simultaneously that any finite field is necessarily commutative (in the multiplication operation). A great deal of work has been done to determine the structure of the additive groups contained in Galois fields and the structure of the fields themselves.

4. Rings

Though the structures rings and ideals were well known and utilized in Dedekind's and Kronecker's work on algebraic numbers, the abstract theory is entirely a product of the twentieth century. The word ideal had already been adopted (Chap. 34, sec. 4). Kronecker used the word "order" for ring; the latter term was introduced by Hilbert.

Before discussing the history, it may be well to be clear about the modern meanings of the concepts. An abstract ring is a collection of elements that form a commutative group with respect to one operation and are subject to a second operation applicable to any two elements. The second operation is closed and associative but may or may not be commutative. There may or may not be a unit element. Moreover the distributive law $a(b + c) = ab + ac$ and $(b + c)a = ba + ca$ holds.

An ideal in a ring R is a sub-ring M such that if a belongs to M and r is any element of R, then ar and ra belong to M. If only ar belongs to M, then M is called a right ideal. If only ra belongs to M, then M is a left ideal. If an ideal is both right and left, it is called a two-sided ideal. The unit ideal is the entire ring. The ideal (a) generated by one element a consists of all elements of the form

$$ra + na,$$

where r belongs to R and n is any whole number. If R has a unit element, then $ra + na = ra + nea = (r + ne)a = r'a$, where r' is now any element of R. The ideal generated by one element is called a principal ideal. Any ideal other than 0 and R is a proper ideal. Similarly if $a_1, a_2, \ldots, a_m$ are m given elements of a ring R with unit element the set of all sums $r_1 a_1 + \cdots + r_m a_m$

40. N.Y. Math. Soc. Bull., 3, 1893, 73–78.
41. Amer. Math. Soc. Trans., 6, 1905, 349–52.

with coefficients r in R is an ideal of R denoted by $(a_1, a_2, \ldots, a_m)$. It is the smallest ideal containing $a_1, a_2, \ldots, a_m$. A commutative ring R is called Noetherian if every ideal has this form.

An ideal M in a ring, since it is a subgroup of the additive group of the ring, divides the ring into residue classes. Two elements of R, a and b, are congruent relative to M if $a - b$ belongs to M or $a \equiv b \pmod{M}$. Under a homomorphism T from a ring R to a ring R', which calls for $T(ab) = (Ta) \cdot (Tb)$, $T(a + b) = Ta + Tb$ and $T1 = 1'$, the elements of R which correspond to the zero element of R' constitute an ideal called the kernel of R, and R' is isomorphic to the ring of residue classes of R modulo the kernel. Conversely, given an ideal L in R one may form a ring R modulo L and a homomorphism of R into R modulo L which has L as its kernel. R modulo L or R/L is called a quotient ring.

The definition of a ring does not call for the existence of an inverse to each element under multiplication. If an inverse (except for 0) and the unit element both exist, the ring is called a division ring (division algebra) and it is in effect a non-commutative (or skew) field. Wedderburn's result already noted (1905) showed that a finite division ring is a commutative field. Up to 1905 the only division algebras known were commutative fields and quaternions. Then Dickson created a number of new ones, commutative and non-commutative. In 1914 he[42] and Wedderburn[43] gave the first examples of non-commutative fields with centers (the set of all elements which commute with each other) of rank n^2.[44]

In the late nineteenth century a great variety of concrete linear associative algebras were created (Chap. 32, sec. 6). These algebras, abstractly considered, are rings, and when the theory of abstract rings was formulated it absorbed and generalized the work on these concrete algebras. This theory of linear associative algebras and the whole subject of abstract algebra received a new impulse when Wedderburn, in his paper "On Hypercomplex Numbers,"[45] took up results of Elie Cartan (1869–1951)[46] and generalized them. The hypercomplex numbers are, we may recall, numbers of the form

$$(6) \qquad x = x_1 e_1 + x_2 e_2 + \cdots + x_n e_n,$$

where the e_i are qualitative units and the x_i are real or complex numbers. Wedderburn replaced the x_i by members of an arbitrary field F. He called

42. *Amer. Math. Soc. Trans.*, 15, 1914, 31–46.

43. *Amer. Math. Soc. Trans.*, 15, 1914, 162–66.

44. In 1958 Michel Kervaire (1927–) in *Proceedings of the National Academy of Sciences*, 44, 1958, 280–83 and John Milnor (1931–) in *Annals of Math.* (2), 68, 1958, 444–49, both using a result of Raoul Bott (1923–), proved that the only possible division algebras with real coefficients, if one does not assume the associative and commutative laws of multiplication, are the real and complex numbers, quaternions, and the Cayley numbers.

45. *Proc. London Math. Soc.*, (2), 6, 1907, 77–118.

46. *Ann. Fac. Sci. de Toulouse*, 12B, 1898, 1–99 = *Œuvres*, Part II, Vol. 1, 7–105.

these generalized linear associative algebras simply algebras. To treat these generalized algebras he had to abandon the methods of his predecessors because an arbitrary field F is not algebraically closed. He also adopted and perfected Benjamin Peirce's technique of idempotents.

In Wedderburn's work, then, an algebra consists of all linear combinations of the form (6) with coefficients now in a field F. The number of e_i, called basal units, is finite and is the order of the algebra. The sum of two such elements is given by

$$\sum x_i e_i + \sum y_i e_i = \sum (x_i + y_i) e_i.$$

The scalar product of an element x of the algebra and an element a of the field F is defined as

$$a \sum x_i e_i = \sum a x_i e_i$$

and the product of two elements of the algebra is defined by

$$\left(\sum x_i e_i\right)\left(\sum y_i e_i\right) = \sum_{i,j} x_i y_i e_i e_j,$$

which is completed by a table expressing all products $e_i e_j$ as some linear combination of the e_i with coefficients in F. The product is required to satisfy the associative law. One can always add a unit element (modulus) 1 such that $x \cdot 1 = 1 \cdot x = x$ for every x and then the elements of the algebra include the field F of the coefficients.

Given an algebra A, a subset B of elements which itself forms an algebra is called a sub-algebra. If x belongs to A and y to B and yx and xy both belong to B, then B is called an invariant sub-algebra. If an algebra A is the sum of two invariant sub-algebras with no common element, then A is called reducible. It is also said to be the direct sum of the sub-algebras.

A simple algebra is one having no invariant sub-algebra. Wedderburn also used and modified Cartan's notion of a semi-simple algebra. To define this notion, Wedderburn made use of the notion of nilpotent elements. An element x is nilpotent if $x^n = 0$ for some integer n. An element x is said to be properly nilpotent if xy and also yx are nilpotent for every y in an algebra A. It can be shown that the set of properly nilpotent elements of an algebra A form an invariant sub-algebra. Then a semi-simple algebra A is one having no nilpotent invariant sub-algebra.

Wedderburn proved that every semi-simple algebra can be expressed as a direct sum of irreducible algebras and each irreducible algebra is equivalent to the direct product of a matrix algebra and a division algebra (primitive algebra in Wedderburn's terminology). This means that each element of the irreducible algebra can be taken to be a matrix whose elements are members of the division algebra. Since semi-simple algebras can be reduced to the direct sum of several simple algebras, this theorem amounts to the determination of all semi-simple algebras. Still another result uses the notion of a total matrix algebra, which is just the algebra of all n by n matrices. An algebra for which the coefficient field F is the complex numbers and which contains no properly nilpotent elements is equivalent to a direct sum of total matrix algebras. This sample of results obtained by Wedderburn may give some indication of the work done on the generalized linear associative algebras.

The theory of rings and ideals was put on a more systematic and axiomatic basis by Emmy Noether, one of the few great women mathematicians, who in 1922 became a lecturer at Göttingen. Many results on rings and ideals were already known when she began her work, but by properly formulating the abstract notions she was able to subsume these results under the abstract theory. Thus she reexpressed Hilbert's basis theorem (Chap. 39, sec. 2) as follows: A ring of polynomials in any number of variables over a ring of coefficients that has an identity element and a finite basis, itself has a finite basis. In this reformulation she made the theory of invariants a part of abstract algebra.

A theory of ideals for polynomial domains had been developed by Emanuel Lasker (1868–1941)[47] in which he sought to give a method of deciding whether a given polynomial belongs to an ideal generated by r polynomials. In 1921[48] Emmy Noether showed that this ideal theory for polynomials can be deduced from Hilbert's basis theorem. Thereby a common foundation was created for the ideal theory of integral algebraic numbers and integral algebraic functions (polynomials). Noether and others penetrated much farther into the abstract theory of rings and ideals and applied it to rings of differential operators and other algebras. However, an account of this work would take us too far into special developments.

5. Non-associative Algebras

Modern ring theory, or, more properly, an extension of ring theory, also includes non-associative algebras. The product operation is non-associative and non-commutative. The other properties of linear associative algebras are applicable. There are today several important non-associative algebras. Historically the most important is the type called a Lie algebra. It is customary in such algebras to denote the product of two elements a and b by $[a, b]$. In place of the associative law the product operation satisfies two conditions,

$$[a, b] = -[b, a] \quad \text{and} \quad [a, [b, c]] + [b, [c, a]] + [c, [a, b]] = 0.$$

47. *Math. Ann.*, 60, 1905, 20–116.
48. *Math. Ann.*, 83, 1921, 24–66.

The second property is called the Jacobi identity. Incidentally, the vector product of two vectors satisfies the two conditions.

An ideal in a Lie algebra L is a sub-algebra L_1 such that the product of any element of L and any element of L_1 is in L_1. A simple Lie algebra is one that has no nontrivial ideals. It is semi-simple if it has no Abelian ideals.

Lie algebras arose out of Lie's efforts to study the structure of his continuous transformation groups. To do this Lie introduced the notion of infinitesimal transformations.[49] Roughly speaking, an infinitesimal transformation is one that moves points an infinitesimal distance. Symbolically it is represented by Lie as

$$(7) \qquad x_i' = x_i + \delta t X_i(x_1, x_2, \cdots, x_n),$$

where δt is an infinitesimally small quantity, or

$$(8) \qquad \delta x_i = \delta t X_i(x_1, x_2, \cdots, x_n).$$

The δt is a consequence of a small change in the parameters of the group. Thus suppose a group of transformations is given by

$$x_1 = \phi(x, y, a) \quad \text{and} \quad y_1 = \psi(x, y, a).$$

Let a_0 be the value of the parameter for which ϕ and ψ are the identity transformation so that

$$x = \phi(x, y, a_0), \qquad y = \psi(x, y, a_0).$$

If a_0 is changed to $a_0 + \delta a$, then by Taylor's theorem

$$x_1 = \phi(x, y, a_0) + \frac{\partial \phi}{\partial a} \delta a + \cdots$$

$$y_1 = \psi(x, y, a_0) + \frac{\partial \psi}{\partial a} \delta a + \cdots$$

so that neglecting higher powers of δa gives

$$\delta x = x_1 - x = \frac{\partial \phi}{\partial a} \delta a, \qquad \delta y = y_1 - y = \frac{\partial \psi}{\partial a} \delta a.$$

For the fixed a_0, $\partial \phi / \partial a$ and $\partial \psi / \partial a$ are functions of x and y so that

$$\frac{\partial \phi}{\partial a} = \xi(x, y), \qquad \frac{\partial \psi}{\partial a} = \eta(x, y)$$

and

$$(9) \qquad \delta x = \xi(x, y)\, \delta a, \qquad \delta y = \eta(x, y)\, \delta a.$$

If δa is δt we get the form (7) or (8). The equations (9) represent an infinitesimal transformation of the group.

49. *Archiv for Mathematik Naturvidenskab*, 1, 1876, 152–93 = *Ges. Abh.*, 5, 42–75.

If $f(x, y)$ is an analytic function of x and y, the effect of an infinitesimal transformation on it is to replace $f(x, y)$ by $f(x + \xi\, \delta a, y + \eta\, \delta a)$, and by applying Taylor's theorem one finds that to first order

$$\delta f = \left(\xi \frac{\partial f}{\partial x} + \eta \frac{\partial f}{\partial y}\right) \delta a.$$

The operator

$$\xi \frac{\partial}{\partial x} + \eta \frac{\partial}{\partial y}$$

is another way of *representing* the infinitesimal transformation (9) because knowledge of one gives the other. Such operators can be added and multiplied in the usual sense of differential operators.

The number of independent infinitesimal transformations or the number of corresponding independent operators is the number of parameters in the original group of transformations. The infinitesimal transformations or the corresponding operators, now denoted by $X_1, X_2, \ldots, X_n$, do determine the Lie group of transformations. But, equally important, they are themselves generators of a group. Though the product $X_i X_j$ is not a linear operator, the expression called the alternant of X_i and X_j,

$$X_i X_j - X_j X_i,$$

is a linear operator and is denoted by $[X_i, X_j]$. With this product operation, the group of operators becomes a Lie algebra.

Lie had begun the work on finding the structure of his simple finite (continuous) groups with r parameters. He found four main classes of algebras. Wilhelm K. J. Killing (1847–1923)[50] found that these classes are correct for all simple algebras but that in addition there are five exceptional cases of 14, 52, 78, 133, and 248 parameters. Killing's work contained gaps and Elie Cartan undertook to fill them.

In his thesis, *Sur la structure des groupes de transformations finis et continus*,[51] Cartan gave the complete classification of all simple Lie algebras over the field of complex values for the variables and parameters. Like Killing, Cartan found that they fall into four general classes and the five exceptional algebras. Cartan constructed the exceptional algebras explicitly. In 1914[52] Cartan determined all the simple algebras with real values for the parameters and variables. These results are still basic.

The use of representations to study Lie algebras has been pursued much as in the case of abstract groups. Cartan, in his thesis and in a paper of 1913,[53]

50. *Math. Ann.*, 31, 1888, 252–90, and in Vols. 33, 34, and 36.
51. 1894; 2nd ed., Vuibert, Paris, 1933 = *Œuvres*, Part I, Vol. 1, 137–286.
52. *Ann. de l'Ecole Norm. Sup.*, 31, 1914, 263–355 = *Œuvres*, Part I, Vol. 1, 399–491.
53. *Bull. Soc. Math. de France*, 41, 1913, 53–96 = *Œuvres*, Part I, Vol. 1, 355–98.

found irreducible representations of the simple Lie algebras. A key result
was obtained by Hermann Weyl.[54] Any representation of a semi-simple Lie
algebra (over an algebraically closed field of characteristic 0) is completely
reducible.

6. *The Range of Abstract Algebra*

Our few indications of the accomplishments in the field of abstract algebra
certainly do not give the full picture of what was created even in the first
quarter of this century. It may, however, be helpful to indicate the vast area
that was opened up by the conscious turn to abstraction.

Up to about 1900 the various algebraic subjects that had been studied,
whether matrices, the algebras of forms in two, three, or n variables, hypernumbers, congruences, or the theory of solution of polynomial equations, had
been based on the real and complex number systems. However, the abstract
algebraic movement introduced abstract groups, rings, ideals, division
algebras, and fields. Beyond investigating the properties of such abstract
structures and relationships as isomorphism and homomorphism, mathematicians now found it possible to take almost any algebraic subject and
raise questions about it by replacing the real and complex numbers with any
one of the abstract structures. Thus in place of matrices with complex
elements, one can study matrices with elements that belong to a ring or field.
Similarly one can take problems of the theory of numbers and, replacing the
positive and negative integers and 0 by a ring, reconsider every question that
has previously been investigated for the integers. One can even consider
functions and power series with coefficients in an arbitrary field.

Such generalizations have indeed been made. We have noted that
Wedderburn in his 1907 work generalized previous work on linear associative
algebras (hypernumbers) by replacing the real or complex coefficients by
any field. One can replace the field by a ring and investigate the theorems
that hold under this change. Even the theory of equations with coefficients
in arbitrary or finite fields has been studied.

As another example of the modern tendency to generalize, consider
quadratic forms. These were important with integer coefficients in the study
of the representation of integers as sums of squares and with real coefficients
as the representation of conic and quadric surfaces. In the twentieth century
quadratic forms are studied with any and every field as coefficients. As more
abstract structures are introduced, these can be used as the base or coefficient
field of older algebraic theories, and the process of generalization goes on
indefinitely. This use of abstract concepts calls for the use of abstract algebraic techniques; thus many formerly distinct subjects were absorbed in

---

54. *Mathematische Zeitschrift*, 23, 1925, 271–309, and 24, 1926, 328–95 = *Ges. Abh.*, 2,
543–647.

abstract algebra. This is the case with large parts of number theory, including algebraic numbers.

However, abstract algebra has subverted its own role in mathematics. Its concepts were formulated to unify various seemingly diverse and dissimilar mathematical domains as, for example, group theory did. Having formulated the abstract theories, mathematicians turned away from the original concrete fields and concentrated on the abstract structures. Through the introduction of hundreds of subordinate concepts, the subject has mushroomed into a welter of smaller developments that have little relation to each other or to the original concrete fields. Unification has been succeeded by diversification and specialization. Indeed, most workers in the domain of abstract algebra are no longer aware of the origins of the abstract structures, nor are they concerned with the application of their results to the concrete fields.

Bibliography

Artin, Emil: "The Influence of J. H. M. Wedderburn on the Development of Modern Algebra," *Amer. Math. Soc. Bull.*, 56, 1950, 65–72.

Bell, Eric T.: "Fifty Years of Algebra in America, 1888–1938," *Amer. Math. Soc. Semicentennial Publications*, II, 1938, 1–34.

Bourbaki, N.: *Eléments d'histoire des mathématiques*, Hermann, 1960, pp. 110–28.

Cartan, Elie: "Notice sur les travaux scientifiques," *Œuvres complètes*, Gauthier-Villars, 1952–55, Part I, Vol. 1, pp. 1–98.

Dicke, Auguste: *Emmy Noether, 1882–1935*, Birkhäuser Verlag, 1970.

Dickson, L. E.: "An Elementary Exposition of Frobenius's Theory of Group-Characters and Group-Determinants," *Annals of Math.*, 4, 1902, 25–49.

————: *Linear Algebras*, Cambridge University Press, 1914.

————: *Algebras and Their Arithmetics* (1923), G. E. Stechert (reprint), 1938.

Frobenius, F. G.: *Gesammelte Abhandlungen*, 3 vols., Springer-Verlag, 1968.

Hawkins, Thomas: "The Origins of the Theory of Group Characters," *Archive for History of Exact Sciences*, 7, 1971, 142–70.

MacLane, Saunders: "Some Recent Advances in Algebra," *Amer. Math. Monthly*, 46, 1939, 3–19. Also in Albert, A. A., ed.: *Studies in Modern Algebra*, The Math. Assn. of Amer., 1963, pp. 9–34.

————: "Some Additional Advances in Algebra," in Albert, A. A., ed.: *Studies in Modern Algebra*, The Math. Assn. of Amer., 1963, pp. 35–58.

Ore, Oystein: "Some Recent Developments in Abstract Algebra," *Amer. Math. Soc. Bull.*, 37, 1931, 537–48.

————: "Abstract Ideal Theory," *Amer. Math. Soc. Bull.*, 39, 1933, 728–45.

Steinitz, Ernst: *Algebraische Theorie der Körper*, W. de Gruyter, 1910; 2nd rev. ed., 1930; Chelsea (reprint), 1950. The first edition is the same as the article in *Jour. für Math.*, 137, 1910, 167–309.

Wiman, A.: "Endliche Gruppen linearer Substitutionen," *Encyk. der Math. Wiss.*, B. G. Teubner, 1898–1904, I, Part 1, 522–54.

Wussing, H. L.: *Die Genesis des abstrakten Gruppenbegriffes*, VEB Deutscher Verlag der Wissenschaften, 1969.

50
The Beginnings of Topology

> I believe that we lack another analysis properly geometric or
> linear which expresses location directly as algebra expresses
> magnitude.
>
> G. W. LEIBNIZ

1. *The Nature of Topology*

A number of developments of the nineteenth century crystallized in a new
branch of geometry, now called topology but long known as analysis situs.
To put it loosely for the moment, topology is concerned with those properties
of geometric figures that remain invariant when the figures are bent, stretched,
shrunk, or deformed in any way that does not create new points or fuse
existing points. The transformation presupposes, in other words, that there
is a one-to-one correspondence between the points of the original figure
and the points of the transformed figure, and that the transformation carries
nearby points into nearby points. This latter property is called continuity,
and the requirement is that the transformation and its inverse both be
continuous. Such a transformation is called a homeomorphism. Topology
is often loosely described as rubber-sheet geometry, because if the figures
were made of rubber, it would be possible to deform many figures into
homeomorphic figures. Thus a rubber band can be deformed into and is
topologically the same as a circle or a square, but it is not topologically the
same as a figure eight, because this would require the fusion of two points
of the band.

Figures are often thought of as being in a surrounding space. For the
purposes of topology two figures can be homeomorphic even though it is
not possible to transform topologically the entire space in which one figure
lies into the space containing the second figure. Thus if one takes a long
rectangular strip of paper and joins the two short ends, he obtains a cylindrical
band. If instead one end is twisted through 360° and then the short ends are
joined, the new figure is topologically equivalent to the old one. But it is
not possible to transform the three-dimensional space into itself topologically
and carry the first figure into the second one.

Topology, as it is understood in this century, breaks down into two somewhat separate divisions: point set topology, which is concerned with geometrical figures regarded as collections of points with the entire collection often regarded as a space; and combinatorial or algebraic topology, which treats geometrical figures as aggregates of smaller building blocks just as a wall is a collection of bricks. Of course notions of point set topology are used in combinatorial topology, especially for very general geometric structures.

Topology has had numerous and varied origins. As with most branches of mathematics, many steps were made which only later were recognized as belonging to or capable of being subsumed under one new subject. In the present case the possibility of a distinct significant study was at least outlined by Klein in his Erlanger Programm (Chap. 38, sec. 5). Klein was generalizing the types of transformation studied in projective and algebraic geometry and he was already aware through Riemann's work of the importance of homeomorphisms.

2. Point Set Topology

The theory of point sets as initiated by Cantor (Chap. 41, sec. 7) and extended by Jordan, Borel, and Lebesgue (Chap. 44, secs. 3 and 4) is not *eo ipso* concerned with transformations and topological properties. On the other hand, a set of points regarded as a space is of interest in topology. What distinguishes a space as opposed to a mere set of points is some concept that binds the points together. Thus in Euclidean space the distance between points tells us how close points are to each other and in particular enables us to define limit points of a set of points.

The origins of point set topology have already been related (Chap. 46, sec. 2). Fréchet in 1906, stimulated by the desire to unify Cantor's theory of point sets and the treatment of functions as points of a space, which had become common in the calculus of variations, launched the study of abstract spaces. The rise of functional analysis with the introduction of Hilbert and Banach spaces gave additional importance to the study of point sets as spaces. The properties that proved to be relevant for functional analysis are topological largely because limits of sequences are important. Further, the operators of functional analysis are transformations that carry one space into another.[1]

As Fréchet pointed out, the binding property need not be the Euclidean distance function. He introduced (Chap. 46, sec. 2) several different concepts that can be used to specify when a point is a limit point of a sequence of points. In particular he generalized the notion of distance by introducing

---

1. The definitions of the basic properties of point sets, such as compactness and separability, have had different meanings for different authors and are still not standardized. We shall use the present commonly understood meanings.

the class of metric spaces. In a metric space, which can be two-dimensional
Euclidean space, one speaks of the neighborhood of a point and means all
those points whose distance from the point is less than some quantity ε,
say. Such neighborhoods are circular. One could use square neighborhoods
as well. However, it is also possible to suppose that the neighborhoods,
certain subsets of a given set of points, are specified in some way, even
without the introduction of a metric. Such spaces are said to have a neighborhood topology. This notion is a generalization of a metric space. Felix
Hausdorff (1868–1942), in his *Grundzüge der Mengenlehre* (Essentials of Set
Theory, 1914), used the notion of a neighborhood (which Hilbert had
already used in 1902 in a special axiomatic approach to Euclidean plane
geometry) and built up a definitive theory of abstract spaces on this notion.

Hausdorff defines a topological space as a set of elements x together
with a family of subsets U_x associated with each x. These subsets are called
neighborhoods and must satisfy the following conditions:

(a) To each point x there is at least one neighborhood U_x which
contains the point x.

(b) The intersection of two neighborhoods of x contains a neighborhood
of x.

(c) If y is a point in U_x there exists a U_y such that $U_y \subseteq U_x$.

(d) If $x \neq y$, there exist U_x and U_y such that $U_x \cdot U_y = 0$.

Hausdorff also introduced countability axioms:

(a) For each point x, the set of U_x is at most countable.

(b) The set of all distinct neighborhoods is countable.

The groundwork in point set topology consists in defining several basic
notions. Thus a limit point of a set of points in a neighborhood space is
one such that every neighborhood of the point contains other points of the
set. A set is open if every point of the set can be enclosed in a neighborhood
that contains only points of the set. If a set contains all its limit points, then
it is closed. A space or a subset of a space is called compact if every infinite
subset has a limit point. Thus the points on the usual Euclidean line are
not a compact set because the infinite set of points corresponding to the
positive integers has no limit point. A set is connected if, no matter how it
is divided into disjoined sets, at least one of these contains limit points of
the other. The curve of $y = \tan x$ is not connected but the curve of $y = \sin 1/x$
plus the interval $(-1, 1)$ of the Y-axis is connected. Separability, introduced
by Fréchet in his 1906 thesis, is another basic concept. A space is called
separable if it has a denumerable subset whose closure, the set plus its limit
points, is the space itself.

The notions of continuous transformations and homeomorphism can
now also be introduced. A continuous transformation usually presupposes

that to each point of one space there is associated a unique point of the second or image space and that given any neighborhood of an image point there is a neighborhood of the original point (or each original point if there are many) whose points map into the neighborhood in the image space. This concept is no more than a generalization of the $\varepsilon - \delta$ definition of a continuous function, the ε specifying the neighborhood of a point in the image space, and the δ a neighborhood of the original point. A homeomorphism between two spaces S and T is a one-to-one correspondence that is continuous both ways; that is, the transformations from S to T and from T to S are continuous. The basic task of point set topology is to discover properties that are invariant under continuous transformations and homeomorphisms. All of the properties mentioned above are topological invariants.

Hausdorff added many results to the theory of metric spaces. In particular he added to the notion of completeness, which Fréchet had introduced in his 1906 thesis. A space is complete if every sequence $\{a_n\}$ that satisfies the condition that given ε, there exists an N, such that $|a_n - a_m| < \varepsilon$ for all m and n greater than N, has a limit point. Hausdorff proved that every metric space can be extended to a complete metric space in one and only one way.

The introduction of abstract spaces raised several questions that prompted much research. For example, if a space is defined by neighborhoods, is it necessarily metrizable; that is, is it possible to introduce a metric that preserves the structure of the space so that limit points remain limit points? This question was raised by Fréchet. One result, due to Paul S. Urysohn (1898–1924), states that every normal topological space can be metrized.[2] A normal space is one in which two disjoined closed sets can each be enclosed in an open set and the two open sets are disjoined. A related result of some importance is also due to Urysohn. He proved[3] that every separable metric space, that is, every metric space in which a countable subset is dense in the space, is homeomorphic to a subset of the Hilbert cube; the cube consists of the space of all infinite sequences $\{x_i\}$ such that $0 \leq x_i \leq 1/i$ and in which distance is defined by $d = \sqrt{\sum_{i=1}^{\infty} (x_i - y_i)^2}$.

The question of dimension, as already noted, was raised by Cantor's demonstration of a one-to-one correspondence of line and plane (Chap. 41, sec. 7) and by Peano's curve, which fills out a square (Chap. 42, sec. 5). Fréchet (already working with abstract spaces) and Poincaré saw the need for a definition of dimension that would apply to abstract spaces and yet grant to line and plane the dimensions usually assumed for them. The definition that had been tacitly accepted was the number of coordinates needed to fix the points of a space. This definition was not applicable to general spaces.

2. *Math. Ann.*, 94, 1925, 262–95.
3. *Math. Ann.*, 94, 1925, 309–15.

In 1912[4] Poincaré gave a recursive definition. A continuum (a closed connected set) has dimension n if it can be separated into two parts whose common boundary consists of continua of dimension $n - 1$. Luitzen E. J. Brouwer (1881–1967) pointed out that the definition does not apply to the cone with two nappes, because the nappes are separated by a point. Poincaré's definition was improved by Brouwer,[5] Urysohn,[6] and Karl Menger (1902–).[7]

The Menger and Urysohn definitions are similar and both are credited with the now generally accepted definition. Their concept assigns a local dimension. Menger's formulation is this: The empty set is defined to be of dimension -1. A set M is said to be n-dimensional at a point P if n is the smallest number for which there are arbitrarily small neighborhoods of P whose boundaries in M have dimension less than n. The set M is called n-dimensional if its dimension in all of its points is less than or equal to n but is n in at least one point.

Another widely accepted definition is due to Lebesgue.[8] A space is n-dimensional if n is the least number for which coverings by closed sets of arbitrarily small diameter contain points common to $n + 1$ of the covering sets. Euclidean spaces have the proper dimension under any of the latter definitions and the dimension of any space is a topological invariant.

A key result in the theory of dimension is a theorem due to Menger (*Dimensionstheorie*, 1928, p. 295) and A. Georg Nöbeling (1907–).[9] It asserts that every n-dimensional compact metric space is homeomorphic to some subset of the $(2n + 1)$-dimensional Euclidean space.

Another problem raised by the work of Jordan and Peano was the very definition of a curve (Chap. 42, sec. 5). The answer was made possible by the work on dimension theory. Menger[10] and Urysohn[11] defined a curve as a one-dimensional continuum, a continuum being a closed connected set of points. (The definition requires that an open curve such as the parabola be closed by a point at infinity.) This definition excludes space-filling curves and renders the property of being a curve invariant under homeomorphisms.

The subject of point set topology has continued to be enormously active. It is relatively easy to introduce variations, specializations, and generalizations of the axiomatic bases for the various types of spaces. Hundreds of concepts have been introduced and theorems established, though the ultimate value of these concepts is dubious in most cases. As in other

4. *Revue de Métaphysique et de Morale*, 20, 1912, 483–504.

5. *Jour. für Math.*, 142, 1913, 146–52.

6. *Fundamenta Mathematicae*, 7, 1925, 30–137 and 8, 1926, 225–359.

7. *Monatshefte für Mathematik und Physik*, 33, 1923, 148–60 and 34, 1926, 137–61.

8. *Fundamenta Mathematicae*, 2, 1921, 256–85.

9. *Math. Ann.*, 104, 1930, 71–80.

10. *Monatshefte für Mathematik und Physik*, 33, 1923, 148–60 and *Math. Ann.*, 95, 1926, 277–306.

11. *Fundamenta Mathematicae*, 7, 1925, 30–137, p. 93 in part.

fields, mathematicians have not hesitated to plunge freely and broadly into point set topology.

3. The Beginnings of Combinatorial Topology

As far back as 1679 Leibniz, in his *Characteristica Geometrica*, tried to formulate basic geometric properties of geometrical figures, to use special symbols to represent them, and to combine these properties under operations so as to produce others. He called this study analysis situs or *geometria situs*. He explained in a letter to Huygens of 1679[12] that he was not satisfied with the coordinate geometry treatment of geometric figures because, beyond the fact that the method was not direct or pretty, it was concerned with magnitude, whereas "I believe we lack another analysis properly geometric or linear which expresses location [*situs*] directly as algebra expresses magnitude." Leibniz's few examples of what he proposed to build still involved metric properties even though he aimed at geometric algorithms that would furnish solutions of purely geometric problems. Perhaps because Leibniz was vague about the kind of geometry he sought, Huygens was not enthusiastic about his idea and his symbolism. To the extent that he was at all clear, Leibniz envisioned what we now call combinatorial topology.

A combinatorial property of geometric figures is associated with Euler, though it was known to Descartes in 1639 and, through the latter's unpublished manuscript, to Leibniz in 1675. If one counts the number of vertices, edges, and faces of any closed convex polyhedron—for example, a cube—then $V - E + F = 2$. This fact was published by Euler in 1750.[13] In 1751 he submitted a proof.[14] Euler's interest in this relation was to use it to classify polyhedra. Though he had discovered a property of all closed convex polyhedra, Euler did not think of invariance under continuous transformation. Nor did he define the class of polyhedra for which the relation held.

In 1811 Cauchy[15] gave another proof. He removed the interior of a face and stretched the remaining figure out on a plane. This gives a polygon for which $V - E + F$ should be one. He established the latter by triangulating the figure and then counting the changes as the triangles are removed one by one. This proof, inadequate because it supposes that any closed convex polyhedron is homeomorphic with a sphere, was accepted by nineteenth-century mathematicians.

Another well-known problem, which was a curiosity at the time but

12. Leibniz, *Math. Schriften*, 1 Abt., Vol. 2, 1850, 19–20 = Gerhardt, *Der Briefwechsel von Leibniz mit Mathematikern*, 1, 1899, 568 = Chr. Huygens, *Œuv. Comp.*, 8, No. 2192.
13. *Novi Comm. Acad. Sci. Petrop.*, 4, 1752–53, 109–40, pub. 1758 = *Opera*, (1), 26, 71–93.
14. *Novi Comm. Acad. Sci. Petrop.*, 4, 1752–53, 140–60, pub. 1758 = *Opera*, (1), 26, 94–108.
15. *Jour. de l'Ecol. Poly.*, 9, 1813, 68–86 and 87–98 = *Œuvres*, (2), 1, 7–38.

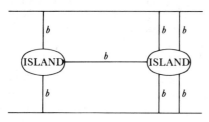

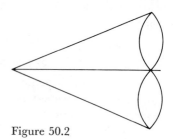

Figure 50.1   Figure 50.2

whose topological nature was later appreciated, is the Koenigsberg bridge problem. In the Pregelarme, a river flowing through Koenigsberg, there are two islands and seven bridges (marked b in Fig. 50.1). The villagers amused themselves by trying to cross all seven bridges in one continuous walk without recrossing any one. Euler, then at St. Petersburg, heard of the problem and found the solution in 1735.[16] He simplified the representation of the problem by replacing the land by points and the bridges by line segments or arcs and obtained Figure 50.2. The question Euler then framed was whether it was possible to describe this figure in one continuous motion of the pencil without recrossing any arcs. He proved it was not possible in the above case and gave a criterion as to when such paths are or are not possible for given sets of points and arcs.

Gauss gave frequent utterance[17] to the need for the study of basic geometric properties of figures but made no outstanding contribution. In 1848 Johann B. Listing (1806–82), a student of Gauss in 1834 and later professor of physics at Göttingen, published *Vorstudien zur Topologie*, in which he discussed what he preferred to call the geometry of position but, since this term was used for projective geometry by von Staudt, he used the term topology. In 1858 he began a new series of topological investigations that were published under the title *Der Census raümlicher Complexe* (Survey of Spatial Complexes).[18] Listing sought qualitative laws for geometrical figures. Thus he attempted to generalize the Euler relation $V - E + F = 2$.

The man who first formulated properly the nature of topological investigations was Möbius, who was an assistant to Gauss in 1813. He had classified the various geometrical properties, projective, affine, similarity, and congruence and then in 1863 in his "Theorie der elementaren Verwandschaft" (Theory of Elementary Relationships)[19] he proposed studying the

16. *Comm. Acad. Sci. Petrop.*, 8, 1736, 128–40, pub. 1741. An English translation of this paper can be found in James R. Newman: *The World of Mathematics*, Simon and Schuster, 1956, Vol. 1, 573–80.

17. *Werke*, 8, 270–86.

18. *Abh. der Ges. der Wiss. zu Gött.*, 10, 1861, 97–180, and as a book in 1862.

19. *Königlich Sächsischen Ges. der Wiss. zu Leipzig*, 15, 1863, 18–57 = *Werke*, 2, 433–71.

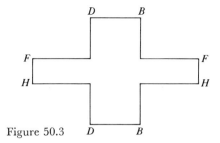

Figure 50.3

relationship between two figures whose points are in one-to-one correspondence and such that neighboring points correspond to neighboring points.
He began by studying the *geometria situs* of polyhedra. He stressed that a
polyhedron can be considered as a collection of two-dimensional polygons,
which, since each piece can be triangulated, would make a polyhedron a
collection of triangles. This idea proved to be basic. He also showed[20] that
some surfaces could be cut up and laid out as polygons with proper identification of sides. Thus a double ring could be represented as a polygon (Fig.
50.3), provided that the edges that are lettered alike are identified.

In 1858 he and Listing independently discovered one-sided surfaces,
of which the Möbius band is best known (Fig. 50.4). This figure is formed
by taking a rectangular strip of paper, twisting it at one short edge through
180°, then joining this edge to the opposite edge. Listing published it in
Der Census; the figure is also described in a publication by Möbius.[21] As far
as the band is concerned, its one-sidedness may be characterized by the fact
that it can be painted by a continuous sweep of the brush so that the entire
surface is covered. If an untwisted band is painted on one side then the brush
must be moved over an edge to get onto the other face. One-sidedness
may also be defined by means of a perpendicular to the surface. Let it
have a definite direction. If it can be moved arbitrarily over the surface and
have the same direction when it returns to its original position, the surface
is said to be two-sided. If the direction is reversed, the surface is one-sided.

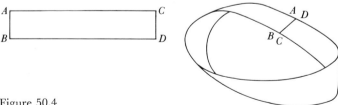

Figure 50.4

20. *Werke*, 2, 518–59.
21. *Königlich Sächsischen Ges. der Wiss. zu Leipzig*, 17, 1865, 31–68 = *Werke*, 2, 473–512;
see also p. 519.

On the Möbius band the perpendicular will return to the point on the "opposite side" with reverse direction.

Still another problem that was later seen to be topological in nature is called the map problem. The problem is to show that four colors suffice to color all maps so that countries with at least an arc as a common boundary are differently colored. The conjecture that four colors will always suffice was made in 1852 by Francis Guthrie (d. 1899), a little-known professor of mathematics, at which time his brother Frederick communicated it to De Morgan. The first article devoted to it was Cayley's;[22] in it he says that he could not obtain a proof. The proof was attempted by a number of mathematicians, and though some published proofs were accepted for a time, they have been shown to be fallacious and the problem is still open.

The greatest impetus to topological investigations came from Riemann's work in complex function theory. In his thesis of 1851 on complex functions and in his study of Abelian functions,[23] he stressed that to work with functions some theorems of analysis situs were indispensable. In these investigations he found it necessary to introduce the connectivity of Riemann surfaces. Riemann defined connectivity in the following manner: "If upon the surface F [with boundaries] there can be drawn n closed curves $a_1, a_2, \ldots, a_n$ which neither individually nor in combination completely bound a part of this surface F, but with whose aid every other closed curve forms the complete boundary of a part of F, the surface is said to be $(n + 1)$-fold connected." To reduce the connectivity of a surface (with boundaries) Riemann states,

> By means of a crosscut [*Querschnitt*], that is, a line lying in the interior of the surface and going from a boundary point to a boundary point, an $(n + 1)$-fold connected surface can be changed into an n-fold connected one, F'. The parts of the boundary arising from the cutting play the role of boundary even during the further cutting so that a crosscut can pass through no point more than once but can end in one of its earlier points. ... To apply these considerations to a surface without boundary, a closed surface, we must change it into a bounded one by the specialization of an arbitrary point, so that the first division is made by means of this point and a crosscut beginning and ending in it, hence by a closed curve.

Riemann gives the example of an anchor ring or torus (Fig. 50.5) which is three-fold connected [genus 1 or one-dimensional Betti number 2] and can be changed into a simply connected surface by means of a closed curve abc and a crosscut $ab'c'$.

Riemann had thus classified surfaces according to their connectivity and, as he himself realized, had introduced a topological property. In

22. *Proceedings of the Royal Geographical Society*, 1, 1879, 259–61 = *Coll. Math. Papers*, 11, 7–8.

23. *Jour. für Math.*, 54, 1857, 105–10 = *Werke*, 91–96; see also *Werke*, 479–82.

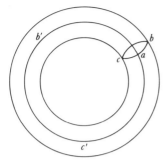

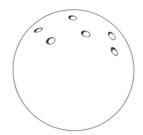

Figure 50.5   Figure 50.6. Sphere with p holes

terms of genus, the term used by the algebraic geometers of the latter part of the nineteenth century, Riemann had classified closed surfaces by means of their genus p, $2p$ being the number of closed curves [loop cuts or *Rück-kehrschnitte*] needed to make the surface simply connected and $2p + 1$ to cut the surface into two distinct pieces. He regarded as intuitively evident that if two closed (orientable) Riemann surfaces are topologically equivalent they have the same genus. He also observed that all closed (algebraic) surfaces of genus zero, that is, simply connected, are topologically (and conformally and birationally) equivalent. Each can be mapped topologically on a sphere.

Since the structure of Riemann surfaces is complicated and a topologically equivalent figure has the same genus, some mathematicians sought simpler structures. William K. Clifford showed[24] that the Riemann surface of an n-valued function with w branch-points can be transformed into a sphere with p holes where $p = (w/2) - n + 1$ (Fig. 50.6). There is the likelihood that Riemann knew and used this model. Klein suggested another topological model, a sphere with p handles (Fig. 50.7).[25]

The study of the topological equivalence of closed surfaces was made by many men. To state the chief result it is necessary to note the concept of orientable surfaces. An orientable surface is one that can be triangulated and each (curvilinear) triangle can be oriented so that any side common to two triangles has opposite orientations induced on it. Thus the sphere is orientable but the projective plane (see below) is non-orientable. This fact was discovered by Klein.[26] The chief result, clarified by Klein in this paper, is that two orientable closed surfaces are homeomorphic if and only if they have the same genus. For orientable surfaces with boundaries, as Klein also pointed

24. *Proc. Lon. Math. Soc.*, 8, 1877, 292–304 = *Math. Papers*, 241–54.
25. *Über Riemanns Theorie der algebraischen Funktionen und ihrer Integrale*, B. G. Teubner, 1882; Dover reprint in English, 1963. Also in Klein's *Ges. Math. Abh.*, 3, 499–573.
26. *Math. Ann.*, 7, 1874, 549–57 = *Ges. Math. Abh.*, 2, 63–77.

Figure 50.7. Sphere with p handles

out, the equality of the number of boundary curves must be added to the above condition. The theorem had been proved by Jordan.[27]

The complexity of even two-dimensional closed figures was emphasized by Klein's introduction in 1882 (see sec. 23 of the reference in note 25) of the surface now called the Klein bottle (Fig. 50.8). The neck enters the bottle without intersecting it and ends smoothly joined to the base along C. Along D the surface is uninterrupted and yet the tube enters the surface. It has no edge, no inside, and no outside; it is one-sided and has a one-dimensional connectivity number of 3 or a genus of 1. It cannot be constructed in three dimensions.

The projective plane is another example of a rather complex closed surface. Topologically the plane can be represented by a circle with diametrically opposite points identified (Fig. 50.9). The infinitely distant line is represented by the semicircumference CAD. The surface is closed and its connectivity number is 1 or its genus is 0. It can also be formed by pasting the edge of a circle along the edge of a Möbius band (which has just a single edge), though again the figure cannot be constructed in three dimensions without having points coincide that should be distinct.

Still another impetus to topological research came from algebraic geometry. We have already related (Chap. 39, sec. 8) that the geometers

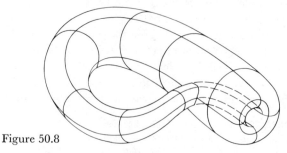

Figure 50.8

27. *Jour. de Math.*, (2), 11, 1866, 105–9 = *Œuvres*, 4, 85–89.

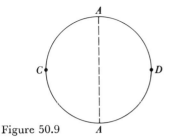

Figure 50.9

had turned to the study of the four-dimensional "surfaces" that represent the domain of algebraic functions of two complex variables and had introduced integrals on these surfaces in the manner analogous to the theory of algebraic functions and integrals on the two-dimensional Riemann surfaces. To study these four-dimensional figures, their connectivity was investigated and the fact learned that such figures cannot be characterized by a single number as the genus characterizes Riemann surfaces. Some investigations by Emile Picard around 1890 revealed that at least a one-dimensional and a two-dimensional connectivity number would be needed to characterize such surfaces.

The need to study the connectivity of higher-dimensional figures was appreciated by Enrico Betti (1823–92), a professor of mathematics at the University of Pisa. He decided that it was just as useful to take the step to n dimensions. He had met Riemann in Italy, where the latter had gone for several winters to improve his health, and from him Betti learned about Riemann's own work and the work of Clebsch. Betti[28] introduced connectivity numbers for each dimension from 1 to $n - 1$. The one-dimensional connectivity number is the number of closed curves that can be drawn in the geometrical structure that do not divide the surface into disjoined regions. (Riemann's connectivity number was 1 greater.) The two-dimensional connectivity number is the number of closed surfaces in the figure that collectively do not *bound* any three-dimensional region of the figure. The higher-dimensional connectivity numbers are similarly defined. The closed curves, surfaces, and higher-dimensional figures involved in these definitions are called cycles. (If a surface has edges, the curves must be crosscuts; that is, a curve that goes from a point on one edge to a point on another edge. Thus the one-dimensional connectivity number of a finite hollow tube (without ends) is 1, because a crosscut from one edge to the other can be drawn without disconnecting the surface.) For the four-dimensional structures used to represent complex algebraic functions $f(x, y, z) = 0$, Betti showed that the one-dimensional connectivity number equals the three-dimensional connectivity number.

28. *Annali di Mat.*, (2), 4, 1870–71, 140–58.

4. The Combinatorial Work of Poincaré

Toward the end of the century, the only domain of combinatorial topology that had been rather fully covered was the theory of closed surfaces. Betti's work was just the beginning of a more general theory. The man who made the first systematic and general attack on the combinatorial theory of geometrical figures, and who is regarded as the founder of combinatorial topology, is Henri Poincaré (1854–1912). Poincaré, professor of mathematics at the University of Paris, is acknowledged to be the leading mathematician of the last quarter of the nineteenth and the first part of this century, and the last man to have had a universal knowledge of mathematics and its applications. He wrote a vast number of research articles, texts, and popular articles, which covered almost all the basic areas of mathematics and major areas of theoretical physics, electromagnetic theory, dynamics, fluid mechanics, and astronomy. His greatest work is *Les Méthodes nouvelles de la mécanique céleste* (3 vols., 1892–99). Scientific problems were the motivation for his mathematical research.

Before he undertook the combinatorial theory we are about to describe, Poincaré contributed to another area of topology, the qualitative theory of differential equations (Chap. 29, sec. 8). That work is basically topological because it is concerned with the form of the integral curves and the nature of the singular points. The contribution to combinatorial topology was stimulated by the problem of determining the structure of the four-dimensional "surfaces" used to represent algebraic functions $f(x, y, z) = 0$ wherein x, y, and z are complex. He decided that a systematic study of the analysis situs of general or n-dimensional figures was necessary. After some notes in the *Comptes Rendus* of 1892 and 1893, he published a basic paper in 1895,[29] followed by five lengthy supplements running until 1904 in various journals. He regarded his work on combinatorial topology as a systematic way of studying n-dimensional geometry, rather than as a study of topological invariants.

In his 1895 paper Poincaré tried to approach the theory of n-dimensional figures by using their analytical representations. He did not make much progress this way, and he turned to a purely geometric theory of manifolds, which are generalizations of Riemann surfaces. A figure is a closed n-dimensional manifold if each point possesses neighborhoods that are homeomorphic to the n-dimensional interior of an $n - 1$ sphere. Thus the circle (and any homeomorphic figure) is a one-dimensional manifold. A spherical surface or a torus is a two-dimensional manifold. In addition to closed manifolds there are manifolds with boundaries. A cube or a solid torus is a three-manifold with boundary. At a boundary point a neighborhood is only part of the interior of a two-sphere.

29. *Jour. de l'Ecole Poly.*, (2), 1, 1895, 1–121 = *Œuvres*, 6, 193–288.

The method Poincaré finally adopted appears in his first supplement.[30] Though he used curved cells or pieces of his figures and treated manifolds, we shall formulate his ideas in terms of complexes and simplexes, which were introduced later by Brouwer. A simplex is merely the n-dimensional triangle. That is, a zero-dimensional simplex is a point; a one-dimensional simplex is a line segment; a two-dimensional simplex is a triangle; a three-dimensional one is a tetrahedron; and the n-dimensional simplex is the generalized tetrahedron with $n + 1$ vertices. The lower-dimensional faces of a simplex are themselves simplexes. A complex is any finite set of simplexes such that any two simplexes meet, if at all, in a common face, and every face of every simplex of the complex is a simplex of the complex. The simplexes are also called cells.

For the purposes of combinatorial topology each simplex or cell of every dimension is given an orientation. Thus a two-simplex E^2 (a triangle) with vertices a_0, a_1, and a_2 can be oriented by choosing an order, say $a_0a_1a_2$. Then any order of the a_i which is obtainable from this by an even number of permutations of the a_i is said to have the same orientation. Thus E^2 is given by $(a_0a_1a_2)$ or $(a_2a_0a_1)$ or $(a_1a_2a_0)$. Any order derivable from the basic one by an odd number of permutations represents the oppositely oriented simplex. Thus $-E^2$ is given by $(a_0a_2a_1)$ or $(a_1a_0a_2)$ or $(a_2a_1a_0)$.

The boundary of a simplex consists of the next lower-dimensional simplexes contained in the given one. Thus the boundary of a two-simplex consists of three one-simplexes. However, the boundary must be taken with the proper orientation. To obtain the oriented boundary one adopts the following rule: The simplex E^k

$$a_0a_1a_2\cdots a_k$$

induces the orientation

(1) $$(-1)^i(a_0a_1\cdots a_{i-1}a_{i+1}\cdots a_k)$$

on each of the $(k-1)$-dimensional simplexes on its boundary. The E_i^{k-1} may have the orientation given by (1), in which case the incidence number, which represents the orientation of E_i^{k-1} relative to E^k, is 1, or it may have the opposite orientation, in which case its incidence number is -1. Whether the incidence number is 1 or -1, the basic fact is that the boundary of the boundary of E^k is 0.

Given any complex one can form a linear combination of its k-dimensional, oriented simplexes. Thus if E_i^k is an oriented k-dimensional simplex,

(2) $$C^k = c_1E_1^k + c_2E_2^k + \cdots + c_lE_l^k,$$

where the c_i are positive or negative integers, is such a combination and is called a chain. The numbers c_i merely tell us how many times a given simplex

30. *Rendiconti del Circolo Matematico di Palermo*, 13, 1899, 285–343 = *Œuvres*, 6, 290–337.

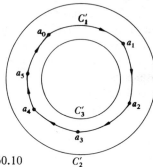

Figure 50.10

is to be counted, a negative number implying also a change in the orientation of the simplex. Thus if our figure is a tetrahedron determined by four points $(a_0a_1a_2a_3)$, we could form the chain $C^3 = 5E_1^3$. The boundary of any chain is the sum of all the next lower-dimensional simplexes on all simplexes of the chain, each taken with the proper incidence number and with the multiplicity appearing in (2). Since the boundary of a chain is the sum of the boundaries of each of the simplexes appearing in the chain, the boundary of the boundary of the chain is zero.

A chain whose boundary is zero is called a cycle. That is, some chains are cycles. Among cycles some are boundaries of other chains. Thus the boundary of the simplex E_1^3 is a cycle that bounds E_1^3. However, if our original figure were not the three-dimensional simplex but merely the surface, we would still have the same cycle but it would not bound in the figure under consideration, namely, the surface. As another example, the chain (Fig. 50.10)

$$C_1^1 = (a_0a_1) + (a_1a_2) + (a_2a_3) + \cdots + (a_5a_0)$$

is a cycle because the boundary of a_1a_2, for example, is $a_2 - a_1$ and the boundary of the entire chain is zero. But C_1^1 does not bound any two-dimensional chain. This is intuitively obvious because the complex in question is the circular ring and the inner hole is not part of the figure.

It is possible for two separate cycles not to bound but for their sum or difference to bound a region. Thus the sum (or difference) of C_2^1 and C_3^1 (Fig. 50.10) bounds the area in the entire ring. Two such cycles are said to be dependent. In general the cycles $C_1^k, \ldots, C_r^k$ are said to be dependent if

$$\sum_{i=1}^{r} c_i C_i^k$$

bounds and not all the c_i are zero.

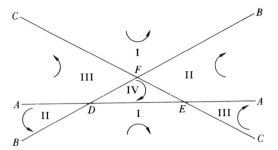

Figure 50.11

Poincaré introduced next the important quantities he called the Betti numbers (in honor of Enrico Betti). For each dimension of possible simplexes in a complex, the number of independent cycles of that dimension is called the Betti number of that dimension (Poincaré actually used a number that is 1 more than Betti's connectivity number.) Thus in the case of the ring, the zero-dimensional Betti number is 1, because any point is a cycle but two points bound the sequence of line segments joining them. The one-dimensional Betti number is 1 because there are one-cycles that do not bound but any two such one-cycles (their sum or difference) do bound. The two-dimensional Betti number of the ring is zero because no chain of two simplexes is a cycle. To appreciate what these numbers mean, one can compare the ring with the circle itself (including its interior). In the latter figure every one-dimensional cycle bounds so that the one-dimensional Betti number is zero.

In this 1899 paper, Poincaré also introduced what he called torsion coefficients. It is possible in more complicated structures—for example, the projective plane—to have a cycle that does not bound but two times this cycle does bound. Thus if the simplexes are oriented as shown in Figure 50.11, the boundary of the four triangles is twice the line BB. (We must remember that AB and BA are the same line segment.) The number 2 is called a torsion coefficient and the corresponding cycle BB a torsion cycle. There may be a finite number of such independent torsion cycles.

It is clear from even these very simple examples that the Betti numbers and torsion coefficients of a geometric figure do somehow distinguish one figure from another, as the circular ring is distinguished from the circle.

In his first supplement (1899) and in the second,[31] Poincaré introduced a method for computing the Betti numbers of a complex. Each simplex E_q of dimension q has simplexes of dimension $q - 1$ on its boundary. These have incidence numbers of $+1$ or -1. A $(q - 1)$-dimensional simplex that does not lie on the boundary of E_q is given the incidence number zero. It

31. *Proc. Lon. Math. Soc.*, 32, 1900, 277–308 = *Œuvres*, 6, 338–70.

is now possible to set up a rectangular matrix that shows the incidence numbers ε_{ij}^q of the jth $(q - 1)$-dimensional simplex with respect to the ith q-dimensional simplex. There is such a matrix T_q for each dimension q other than zero. T_q will have as many rows as there are q-dimensional simplexes in the complex and as many columns as there are $(q - 1)$-dimensional simplexes. Thus T_1 gives the incidence relations of the vertices to the edges. T_2 gives the incidence relations of the one-dimensional simplexes to the two-dimensional ones; and so forth. By elementary operations on the matrices it is possible to make all the elements not on the main diagonal zero and those on this diagonal positive integers or zero. Suppose that γ_q of these diagonal elements are 1. Then Poincaré shows that the q-dimensional Betti number p_q (which is one more than Betti's connectivity number) is

$$p_q = \alpha_q - \gamma_{q+1} - \gamma_q + 1$$

where α_q is the number of q-dimensional simplexes.

Poincaré distinguished complexes with torsion from those without. In the latter case all the numbers in the main diagonal for all q are 0 or 1. Larger values indicate the presence of torsion.

He also introduced the characteristic $N(K^n)$ of an n-dimensional complex K^n. If this has α_k k-dimensional simplexes, then by definition

$$N(K^n) = \sum_{k=0}^{n} (-1)^k \alpha_k.$$

This quantity is a generalization of the Euler number $V - E + F$. Poincaré's result on this characteristic is that if p_k is the kth Betti number of K^n then [32]

$$N(K^n) = \sum_{k=0}^{n} (-1)^k p_k.$$

This result is called the Euler-Poincaré formula.

In his 1895 paper Poincaré introduced a basic theorem, known as the duality theorem. It concerns the Betti numbers of a closed manifold. An n-dimensional closed manifold, as already noted, is a complex for which each point has a neighborhood homeomorphic to a region of n-dimensional Euclidean space. The theorem states that in a closed orientable n-dimensional manifold the Betti number of dimension p equals the Betti number of dimension $n - p$. His proof, however, was not complete.

In his efforts to distinguish complexes, Poincaré introduced (1895) one other concept that now plays a considerable role in topology, the fundamental group of a complex, also known as the Poincaré group or the first

32. These p_k are 1 less than Poincaré's. We use here the statement that is familiar today.

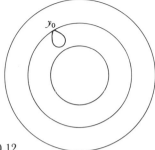

Figure 50.12

homotopy group. The idea arises from considering the distinction between simply and multiply connected plane regions. In the interior of a circle all closed curves can be shrunk to a point. However, in a circular ring some closed curves, those that surround the inner circular boundary, cannot be shrunk to a point, whereas those closed curves that do not surround the inner boundary can be shrunk to a point.

The more precise notion is best approached by considering the closed curves that start and end at a point y_0 of the complex. Then those curves that can be deformed by continuous motion in the space of the complex onto each other are homotopic to each other and are regarded as one class. Thus the closed curves starting and ending at y_0 (Fig. 50.12) in the circular ring and not enclosing the inner boundary are one class. Those which, starting and ending at y_0, do enclose the inner boundary are another class. Those which, starting and ending at y_0 enclose the inner boundary n times, are another class.

It is now possible to define an operation of one class on another, which geometrically amounts to starting at y_0 and traversing any curve of one class, then traversing any curve of the second class. The order in which the curves are chosen and the direction in which a curve is traversed are distinguished. The classes then form a group, called the fundamental group of the complex with respect to the base point y_0. This non-commutative group is now denoted by $\pi_1(K, y_0)$ where K is the complex. In reasonable complexes the group does not actually depend upon the point. That is, the groups at y_0 and y_1, say, are isomorphic. For the circular ring the fundamental group is infinite cyclic. The simply connected domain usually referred to in analysis, e.g. the circle and its interior, has only the identity element for its fundamental group. Just as the circle and the circular ring differ in their homotopy groups, so it is possible to describe higher-dimensional complexes that differ markedly in this respect.

Poincaré bequeathed some major conjectures. In his second supplement he asserted that any two closed manifolds that have the same Betti numbers and torsion coefficients are homeomorphic. But in the fifth supplement[33] he gave an example of a three-dimensional manifold that has the Betti numbers and torsion coefficients of the three-dimensional sphere (the surface of a four-dimensional solid sphere) but is not simply connected. Hence he added simple connectedness as a condition. He then showed that there are three-dimensional manifolds with the same Betti numbers and torsion coefficients but which have different fundamental groups and so are not homeomorphic. However, James W. Alexander (1888–1971), a professor of mathematics at Princeton University and later at the Institute for Advanced Study, showed[34] that two three-dimensional manifolds may have the same Betti numbers, torsion coefficients, and fundamental group and yet not be homeomorphic.

In his fifth supplement (1904), Poincaré made a somewhat more restricted conjecture, namely, that every simply connected, closed, orientable three-dimensional manifold is homeomorphic to the sphere of that dimension. This famous conjecture has been generalized to read: Every simply connected, closed, n-dimensional manifold that has the Betti numbers and torsion coefficients of the n-dimensional sphere is homeomorphic to it. Neither Poincaré's conjecture nor the generalized one has been proven.[35]

Another famous conjecture, called the *Hauptvermutung* (most important conjecture) of Poincaré, asserts that if T_1 and T_2 are simplicial (not necessarily straightedged) subdivisions of the same three-manifold, then T_1 and T_2 have subdivisions that are isomorphic.[36]

5. Combinatorial Invariants

The problem of establishing the invariance of combinatorial properties is to show that any complex homeomorphic to a given complex in the *point set sense* has the same combinatorial properties as the given complex. The proof that the Betti numbers and torsion coefficients are combinatorial invariants was first given by Alexander.[37] His result is that if K and K_1 are any simplicial subdivisions (not necessarily rectilinear) of any two homeomorphic (as point sets) polyhedra P and P_1, then the Betti numbers and

33. *Rendiconti del Circolo Matematico di Palermo*, 18, 1904, 45–110 = *Œuvres*, 6, 435–98.
34. *Amer. Math. Soc. Trans.*, 20, 1919, 339–42.
35. The generalized conjecture has been proved for $n \geq 5$ by Stephen Smale (*Amer. Math. Soc. Bull.*, 66, 1960, 373–75), John R. Stallings (*ibid.*, 485–88) and E. C. Zeeman (*ibid.*, 67, 1961, 270).
36. The *Hauptvermutung* has been shown to be true for finite simplicial complexes (which are more general than manifolds) of dimension less than three and to be false for such complexes of dimension greater than five. It is correct for manifolds of dimension less than or equal to three and an open problem for manifolds of dimension four or greater. See John Milnor, *Annals of Math.*, (2), 74, 1961, 575–90.
37. *Amer. Math. Soc. Trans.*, 16, 1915, 148–54.

torsion coefficients of P and P_1 are equal. The converse is not true, so that the equality of the Betti numbers and torsion coefficients of two complexes does not guarantee the homeomorphism of the two complexes.

Another invariant of importance was contributed by L. E. J. Brouwer. Brouwer became interested in topology through problems in function theory. He sought to prove that there are $3g - 3$ classes of conformally equivalent Riemann surfaces of genus $g > 1$ and was led to take up the related topological problems. He proved[38] the invariance of the dimension of a complex in the following sense: If K is an n-dimensional, simplicial subdivision of a polyhedron P, then every simplicial subdivision of P and every such division of any polyhedron homeomorphic to P is also an n-dimensional complex.

The proofs of this theorem, as well as Alexander's theorem, are made by using a method due to Brouwer,[39] namely, simplicial approximations of continuous transformations. Simplicial transformations (simplexes into simplexes) are themselves no more than the higher-dimensional analogues of continuous transformations, while the simplicial approximation of continuous transformations is analogous to linear approximation applied to continuous functions. If the domain in which the approximation is made is small, then the approximation serves to represent the continuous transformation for the purposes of the invariance proofs.

6. Fixed Point Theorems

Beyond serving to distinguish complexes from one another, combinatorial methods have produced fixed point theorems that are of geometric significance and also have applications to analysis. By introducing notions (which we shall not take up) such as the class of a mapping of one complex into another[40] and the degree of a mapping,[41] Brouwer was able to treat first what are called singular points of vector fields on a manifold. Consider the circle S^1, the spherical surface S^2, and the n-sphere $\sum_{i=1}^{n+1} x_i^2 = 1$ in Euclidean $(n + 1)$-dimensional space. On S^1 it is possible to have a tangent vector at each point and such that the lengths and directions of these vectors vary continuously around the circle while no vector has length zero. There is, one says, a continuous tangent vector field without singularities on S^1. However, no such field can exist on S^2. Brouwer proved[42] that what happens on S^2 must happen on every even-dimensional sphere; that is, no continuous vector field can exist on even-dimensional spheres, so that there must be at least one singular point.

38. *Math. Ann.*, 70, 1910/11, 161–65, and 71, 1911/12, 305–13.
39. *Math. Ann.*, 71, 1911/12, 97–115.
40. *Proceedings Koninklijke Akademie von Wettenschappen te Amsterdam*, 12, 1910, 785–94.
41. *Math. Ann.*, 71, 1911/12, 97–115.
42. *Math. Ann.*, 71, 1911/12, 97–115.

Closely related to the theory of singular points is the theory of continuous transformations of complexes into themselves. Of special interest in such transformations are the fixed points. If we denote by $f(x)$ the transform of a point x under such a transformation, then a fixed point is one for which $f(x) = x$. We may for any point x introduce a vector from x to $f(x)$. In the case of a fixed point the vector is indeterminate and the point is a singular point. The basic theorem on fixed points is due to Brouwer.[43] It applies to the n-dimensional simplex (or a homeomorph of it) and states that every continuous transformation of the n-simplex into itself possesses at least one fixed point. Thus a continuous transformation of a circular disc into itself must possess at least one fixed point. In the same paper Brouwer proved that every one-to-one continuous transformation of a sphere of even dimension into itself that can be deformed into the identity transformation possesses at least one fixed point.

Shortly before his death in 1912, Poincaré showed[44] that periodic orbits would exist in a restricted three-body problem if a certain topological theorem holds. The theorem asserts the existence of at least two fixed points in an annular region between two circles under a topological transformation of the region that carries each circle into itself, moving one in one direction and the other in the opposite direction while preserving the area. This last "theorem" of Poincaré was proved by George D. Birkhoff.[45]

Fixed point theorems were generalized to infinite-dimensional function spaces by Birkhoff and Oliver D. Kellogg in a joint paper[46] and applied to prove the existence of solutions of differential equations by Jules P. Schauder (1899–1940),[47] and by Schauder and Jean Leray (1906–) in a joint paper.[48] A key theorem used for such applications states that if T is a continuous mapping of a closed, convex, compact set of a Banach space into itself, then T has a fixed point.

The use of fixed point theorems to establish the existence of solutions of differential equations is best understood from a rather simple example. Consider the differential equation

$$\frac{dy}{dx} = F(x, y)$$

in the interval $0 \leq x \leq 1$ and the initial condition $y = 0$ at $x = 0$. The solution $\phi(x)$ satisfies the equation

$$\phi(x) = \int_0^x F(x, \phi(x)) \, dx.$$

43. *Math. Ann.*, 71, 1911/12, 97–115.
44. *Rendiconti del Circolo Matematico di Palermo*, 33, 1912, 375–407 = *Œuvres*, 6, 499–538.
45. *Amer. Math. Soc. Trans.*, 14, 1913, 14–22 = *Coll. Math. Papers*, 1, 673–81.
46. *Amer. Math. Soc. Trans.*, 23, 1922, 96–115 = Birkhoff, *Coll. Math. Papers*, 3, 255–74.
47. *Studia Mathematica*, 2, 1930, 170–79.
48. *Ann. de l'Ecole Norm. Sup.*, 51, 1934, 45–78.

We introduce the general transformation

$$g(x) = \int_0^x F(x, f(x)) \, dx$$

where $f(x)$ is an arbitrary function. This transformation associates f with g and can be shown to be continuous on the space of continuous functions $f(x)$ defined on $[0, 1]$. The solution ϕ that we seek is a fixed point of that function space. If one can show that the function space satisfies the conditions that validate a fixed point theorem, then the existence of ϕ is established. The fixed point theorems applicable to function spaces do precisely that. The method illustrated by this simple example enables us to establish the existence of solutions of nonlinear partial differential equations that are common in the calculus of variations and hydrodynamics.

7. *Generalizations and Extensions*

The ideas of Poincaré and Brouwer were seized upon by a number of men who have extended the scope of topology so much that it is now one of the most active fields of mathematics. Brouwer himself extended the Jordan curve theorem.[49] This theorem (Chap. 42, sec. 5) can be stated as follows: Let S^2 be a two-sphere (surface) and J a closed curve (topologically an S^1) on S^2. Then the zero-dimensional Betti number of $S^2 - J$ is 2. Since this Betti number is the number of components, J separates S^2 into two regions. Brouwer's generalization states that an $(n-1)$-dimensional manifold separates the n-dimensional Euclidean space R_n into two regions. Alexander[50] generalized Poincaré's duality theorem and indirectly the Jordan curve theorem. Alexander's theorem states that the r-dimensional Betti number of a complex K in the n-dimensional sphere S^n is equal to the $(n-r-1)$-dimensional Betti number of the complementary domain $S^n - K$, $r \neq 0$ and $r \neq n - 1$. For $r = 0$ the Betti number of K equals 1 plus the $(n-1)$-dimensional Betti number of $S^n - K$ and for $r = n - 1$ the Betti number of K equals the zero-dimensional Betti number of $S^n - K$ minus 1.[51] This theorem generalizes the Jordan curve theorem, for if we take K to be the $(n-1)$-dimensional sphere S^{n-1} then the theorem states that the $(n-1)$-dimensional Betti number of S^{n-1}, which is 1, equals the zero-dimensional Betti number of $S^n - S^{n-1}$ minus 1 so that the zero-dimensional Betti number of $S^n - S^{n-1}$ is 2 and S^{n-1} divides S^n into two regions.

The definitions of the Betti numbers have been modified and generalized

49. *Math. Ann.*, 71, 1911/12, 314–19.
50. *Amer. Math. Soc. Trans.*, 23, 1922, 333–49.
51. Alexander stated his theorem under the condition that the coefficients of his chains were integers modulo 2 (see next paragraph). Our statement is for the usual integral coefficients.

in various ways. Veblen and Alexander[52] introduced chains and cycles modulo 2; that is, in place of oriented simplexes one uses unoriented ones but the integral coefficients are taken modulo 2. The boundaries of chains are also counted in the same way. Alexander then introduced[53] coefficients modulo m for chains and cycles. Solomon Lefschetz (1884–) suggested rational numbers as coefficients.[54] A still broader generalization was made by Lev S. Pontrjagin (1908–1960),[55] in taking the coefficients of chains to be elements of an Abelian group. This concept embraces the previously mentioned classes of coefficients as well as another class also utilized, namely, the real numbers modulo 1. All these generalizations, though they do lead to more general theorems, have not enhanced the power of Betti numbers and torsion coefficients to distinguish complexes.

Another change in the formulation of basic combinatorial properties, made during the years 1925 to 1930 by a number of men and possibly suggested by Emmy Noether, was to recast the theory of chains, cycles, and bounding cycles into the language of group theory. Chains of the same dimension can be added to each other in the obvious way, that is, by adding coefficients of the same simplex; and since cycles are chains they too can be added and their sum is a cycle. Thus chains and cycles form groups. Given a complex K, to each k-dimensional chain there is a $(k-1)$-dimensional boundary chain and the sum of two chains has as its boundary the sum of the separate boundary chains. Hence the relation of chain to boundary establishes a homomorphism of the group $C^k(K)$ of k-dimensional chains into a subgroup $H^{k-1}(K)$ of the group of $(k-1)$-dimensional chains. The set of all k-cycles $(k > 0)$ is a subgroup $Z^k(K)$ of $C^k(K)$ and goes under this homomorphism into the identity element or 0 of $C^{k-1}(K)$. Since every boundary chain is a cycle, $H^{k-1}(K)$ is a subgroup of $Z^{k-1}(K)$.

With these facts we may make the following definition: For any $k \geq 0$, the factor group of $Z^k(K)$, that is, the k-dimensional cycles, modulo the subgroup $H^k(K)$ of the bounding cycles, is called the kth homology group of K and denoted by $B^k(K)$. The number of linearly independent generators of this factor group is called the kth Betti number of the complex and denoted by $p^k(K)$. The kth homology group may also contain finite cyclic groups and these correspond to the torsion cycles. In fact the orders of these finite groups are the torsion coefficients. With this group-theoretic formulation of the homology groups of a complex, many older results can be reformulated in group language.

The most significant generalization of the early part of this century was to introduce homology theory for general spaces, such as compact

52. *Annals of Math.*, (2), 14, 1913, 163–78.
53. *Amer. Math. Soc. Trans.*, 28, 1926, 301–29.
54. *Annals of Math.*, (2), 29, 1928, 232–54.
55. *Annals of Math.*, (2), 35, 1934, 904–14.

metric spaces as opposed to figures that are complexes to start with. The basic schemes were introduced by Paul S. Alexandroff (1896–),[56] Leopold Vietoris (1891–),[57] and Eduard Čech (1893–1960).[58] The details will not be given, since they involve a totally new approach to homology theory. However, we should note that this work marks a step toward the fusion of point set and combinatorial topology.

Bibliography

Bouligand, Georges: *Les Définitions modernes de la dimension*, Hermann, 1935.

Dehn, M., and P. Heegard: "Analysis Situs," *Encyk. der Math. Wiss.*, B. G. Teubner, 1907–10, III AB3, 153–220.

Franklin, Philip: "The Four Color Problem," *Scripta Mathematica*, 6, 1939, 149–56, 197–210.

Hadamard, J.: "L'Œuvre mathématique de Poincaré, *Acta Math.*, 38, 1921, 203–87.

Manheim, J. H.: *The Genesis of Point Set Topology*, Pergamon Press, 1964.

Osgood, William F.: "Topics in the Theory of Functions of Several Complex Variables," *Madison Colloquium*, American Mathematical Society, 1914, pp. 111–230.

Poincaré, Henri: *Œuvres*, Gauthier-Villars, 1916–56, Vol. 6.

Smith, David E.: *A Source Book in Mathematics*, Dover (reprint), 1959, Vol. 2, 404–10.

Tietze, H., and L. Vietoris: "Beziehungen zwischen den verschiedenen Zweigen der Topologie," *Encyk. der Math. Wiss.*, B. G. Teubner, 1914–31, III AB13, 141–237.

Zoretti, L., and A. Rosenthal: "Die Punktmengen," *Encyk. der Math. Wiss.*, B. G. Teubner, 1923–27, II C9A, 855–1030.

56. *Annals of Math.*, (2), 30, 1928/29, 101–87.
57. *Math. Ann.*, 97, 1927, 454–72.
58. *Fundamenta Mathematicae*, 19, 1932, 149–83.

51
The Foundations of Mathematics

> Logic is invincible because in order to combat logic it is necessary to use logic.   PIERRE BOUTROUX

> We know that mathematicians care no more for logic than logicians for mathematics. The two eyes of exact science are mathematics and logic: the mathematical sect puts out the logical eye, the logical sect puts out the mathematical eye, each believing that it can see better with one eye than with two.   AUGUSTUS DE MORGAN

1. *Introduction*

By far the most profound activity of twentieth-century mathematics has been the research on the foundations. The problems thrust upon the mathematicians, and others that they voluntarily assumed, concern not only the nature of mathematics but the validity of deductive mathematics.

Several activities converged to bring foundational problems to a head in the first part of the century. The first was the discovery of contradictions, euphemistically called paradoxes, notably in set theory. One such contradiction, the Burali-Forti paradox, has already been noted (Chap. 41, sec. 9). A number of others were discovered in the first few years of this century. Clearly the discovery of contradictions disturbed the mathematicians deeply. Another problem that had gradually been recognized and that emerged into the open early in this century was the consistency of mathematics (Chap. 43, sec. 6). In view of the paradoxes of set theory consistency had to be established especially in this area.

During the latter part of the nineteenth century a number of men had begun to reconsider the foundations of mathematics and, in particular, the relationship of mathematics to logic. Research in this area (about which we shall say more later) suggested to some mathematicians that mathematics could be founded on logic. Others questioned the universal application of logical principles, the meaningfulness of some existence proofs, and even the reliance upon logical proof as the substantiation of mathematical results.

Controversies that had been smoldering before 1900 broke out into open fire when the paradoxes and the consistency problem added fuel. Thereupon the question of the proper foundation for all mathematics became vital and of widespread concern.

2. *The Paradoxes of Set Theory*

Following hard upon the discovery by Cantor and Burali-Forti of the paradox involving ordinal numbers came a number of other paradoxes or antinomies. Actually the word paradox is ambiguous, for it can refer to a seeming contradiction. But what the mathematicians actually encountered were unquestionably contradictions. Let us see first what they were.

One paradox was put in popular form by Bertrand Russell (1872–1970) in 1918, as the "barber" paradox. A village barber, boasting that he has no competition, advertises that of course he does not shave those people who shave themselves, but does shave all those who do not shave themselves. One day it occurs to him to ask whether he should shave himself. If he should shave himself, then by the first half of his assertion, he should not shave himself; but if he does not shave himself, then in accordance with his boast, he must shave himself. The barber is in a logical predicament.

Still another paradox was formulated by Jules Richard (b. 1862).[1] A simplified version of it was given by G. G. Berry and Russell and published by the latter.[2] The simplified paradox, also known as Richard's, reads as follows: Every integer can be described in words requiring a certain number of letters. For example, the number 36 can be described as thirty-six or as four times nine. The first description uses nine letters and the second thirteen. There is no one way to describe any given number, but this is not essential. Now let us divide all the positive whole numbers into two groups, the first to include all those that can be described (in at least one way) in 100 letters or fewer and the second group to include all those numbers that require a minimum of 101 letters, no matter how described. Only a finite number of numbers can have a description in 100 or fewer letters, for there are at most 27^{100} expressions with 100 or fewer letters (and some of these are meaningless). There is then a smallest integer in the second group. It can be described by the phrase, "the least integer not describable in one hundred or fewer letters." But this phrase requires fewer than 100 letters. Hence, the least integer not describable in 100 or fewer letters can be described in fewer than 100 letters.

Let us consider another form of this paradox, first stated by Kurt Grelling (1886–1941) and Leonard Nelson (1882–1927) in 1908 and

---

1. *Revue Générale des Sciences*, 16, 1905, 541.
2. *Proc. Lon. Math. Soc.*, (2), 4, 1906, 29–53.

published in an obscure journal.[3] Some words are descriptive of themselves. For example, the word "polysyllabic" is polysyllabic. On the other hand, the word "monosyllabic" is not monosyllabic. We shall call those words that are not descriptive of themselves heterological. In other words the word X is heterological if X is not itself X. Now let us replace X by the word "heterological." Then the word "heterological" is heterological if heterological is not heterological.

Cantor pointed out, in a letter to Dedekind of 1899, that one could not speak of the set of all sets without ending in a contradiction (Chap. 41, sec. 9). This is essentially what is involved in Russell's paradox (*The Principles of Mathematics*, 1903, p. 101). The class of all men is not a man. But the class of all ideas is an idea; the class of all libraries is a library; and the class of all sets with cardinal number greater than 1 is such a set. Hence, some classes are not members of themselves and some are. This description of classes includes all and the two types are mutually exclusive. Let us now denote by M the class of all classes that are members of themselves and by N the class of all classes that are not members of themselves. Now N is itself a class and we ask whether it belongs to M or to N. If N belongs to N then N is a member of itself and so must belong to M. On the other hand, if N is a member of M, since M and N are mutually exclusive classes, N does not belong to N. Hence N is not a member of itself and should, by virtue of the definition of N, belong to N.

The cause of all these paradoxes, as Russell and Whitehead point out, is that an object is defined in terms of a class of objects that contains the object being defined. Such definitions are also called impredicative and occur particularly in set theory. This type of definition is also used, as Zermelo noted in 1908, to define the lower bound of a set of numbers and to define other concepts of analysis. Hence classical analysis contains paradoxes.

Cantor's proof of the nondenumerability of the set of real numbers (Chap. 41, sec. 7) also uses such an impredicative set. A one-to-one correspondence is assumed to hold between the set of *all* positive integers and the set M of all real numbers. Then to each integer k, there corresponds the set $f(k)$. Now $f(k)$ does or does not contain k. Let N be the set of all k such that k does not belong to $f(k)$. This set N (taken in some order) is a real number. Hence there should be, by the initial one-to-one correspondence, an integer n such that n corresponds to N. Now if n belongs to N, it should not by the definition of N. If n does not belong to N, then by the definition of N it should belong to N. The definition of the set N is impredicative because k belongs to N if and only if there exists a set K in M such that $K = f(k)$ and k does not belong to K. Thus in defining N we make use of the totality M of sets which contains N as a member. That is, to define N, N must already be in the set M.

3. *Abhandlungen der Friesschen Schule*, 2, 1908, 301–24.

It is rather easy to fall unwittingly into the trap of introducing impredicative definitions. Thus if one defines the class of all classes that contain more than five members, he has defined a class that contains itself. Likewise the statement, the set S of all sets definable in twenty-five or fewer words, define S impredicatively.

These paradoxes jolted the mathematicians, while they compromised not only set theory but large portions of classical analysis. Mathematics as a logical structure was in a sad state, and mathematicians looked back longingly to the happier days before the paradoxes were recognized.

3. The Axiomatization of Set Theory

It is perhaps not surprising that the mathematicians' first recourse was to axiomatize Cantor's rather freely formulated and, as some are wont to say today, naive set theory. The axiomatization of geometry and the number system had resolved logical problems in those areas, and it seemed likely that axiomatization would clarify the difficulties in set theory. The task was first undertaken by the German mathematician Ernst Zermelo who believed that the paradoxes arose because Cantor had not restricted the concept of a set. Cantor in 1895[4] had defined a set as a collection of distinct objects of our intuition or thought. This was rather vague, and Zermelo therefore hoped that clear and explicit axioms would clarify what is meant by a set and what properties sets should have. Cantor himself was not unaware that his concept of a set was troublesome. In a letter to Dedekind of 1899[5] he had distinguished between consistent and inconsistent sets. Zermelo thought he could restrict his sets to Cantor's consistent ones and these would suffice for mathematics. His axiom system[6] contained fundamental concepts and relations that are defined only by the statements in the axioms themselves. Among such concepts was the notion of a set itself and the relation of belonging to a set. No properties of sets were to be used unless granted by the axioms. The existence of an infinite set and such operations as the union of sets and the formation of subsets were also provided for in the axioms. Notably Zermelo included the axiom of choice (Chap. 41, sec. 8).

Zermelo's plan was to admit into set theory only those classes that seemed least likely to generate contradictions. Thus the null class, any finite class, and the class of natural numbers seemed safe. Given a safe class, certain classes formed from it, such as any subclass, the union of safe classes, and the class of all subsets of a safe class, should be safe classes. However, he avoided complementation for, while x might be a safe class,

4. *Math. Ann.*, 46, 1895, 481–512 = *Ges. Abh.*, 282–356.
5. *Ges. Abh.*, 443–48.
6. *Math. Ann.*, 65, 1908, 261–81.

the complement of x, that is, all non-x, in some large universe of objects might not be safe.

Zermelo's development of set theory was improved by Abraham A. Fraenkel (1891–1965).[7] Additional changes were made by von Neumann.[8] The hope of avoiding the paradoxes rests in the case of the Zermelo-Fraenkel system on restricting the types of sets that are admitted while admitting enough to serve the foundations of analysis. Von Neumann's idea was a little more daring. He makes the distinction between classes and sets. Classes are sets so large that they are not contained in other sets or classes, whereas sets are more restricted classes and may be members of a class. Thus sets are the safe classes. As von Neumann pointed out, it was not the admission of the classes that led to contradictions but their being treated as members of other classes.

Zermelo's formal set theory, as modified by Fraenkel, von Neumann, and others, is adequate for developing the set theory required for practically all of classical analysis and avoids the paradoxes to the extent that, as yet, no one has discovered any within the theory. However, the consistency of the axiomatized set theory has not been demonstrated. Apropos of the open question of consistency Poincaré remarked, "We have put a fence around the herd to protect it from the wolves but we do not know whether some wolves were not already within the fence."

Beyond the problem of consistency, the axiomatization of set theory used the axiom of choice, which is needed to establish parts of standard analysis, topology, and abstract algebra. This axiom was considered objectionable by a number of mathematicians, among them Hadamard, Lebesgue. Borel, and Baire, and in 1904, when Zermelo used it to prove the well-ordering theorem (Chap. 41, sec. 8), a host of objections flooded the journals.[9] The questions of whether this axiom was essential and whether it was independent of the others were raised and remained unanswered for some time (see sec. 8).

The axiomatization of set theory, despite the fact that it left open such questions as consistency and the role of the axiom of choice, might have put mathematicians at ease with respect to the paradoxes and have led to a decline in the interest in foundations. But by this time several schools of thought on the foundations of mathematics, no doubt stirred into life by the paradoxes and the problem of consistency, had become active and contentious. To the proponents of these philosophies the axiomatic method as practiced by Zermelo and others was not satisfactory. To some it was objec-

7. *Math. Ann.*, 86, 1921/22, 230–37, and many later papers.
8. *Jour. für Math.*, 154, 1925, 219–40, and later papers.
9. The views of these men are expressed in a famous exchange of letters. See the *Bull. Soc. Math. de France*, 33, 1905, 261–73. Also in E. Borel: *Leçons sur la théorie des fonctions*, Gauthier-Villars, 4th ed., 1950, 150–58.

tionable because it presupposed the logic it used, whereas by this time logic itself and its relation to mathematics was under investigation. Others, more radical, objected to the reliance upon any kind of logic, particularly as it was applied to infinite sets. To understand the arguments that various schools of thought propounded, we must go back in time somewhat.

4. The Rise of Mathematical Logic

One development that caused new controversies as well as dissatisfaction with the axiomatization of set theory concerns the role of logic in mathematics; it arose from the nineteenth-century mathematization of logic. This development has its own history.

The power of algebra to symbolize and even mechanize geometrical arguments had impressed both Descartes and Leibniz, among others (Chap. 13, sec. 8), and both envisioned a broader science than the algebra of numbers. They contemplated a general or abstract science of reasoning that would operate somewhat like ordinary algebra but be applicable to reasoning in all fields. As Leibniz put it in one of his papers, "The universal mathematics is, so to speak, the logic of the imagination," and ought to treat "all that which in the domain of the imagination is susceptible of exact determination." With such a logic one might build any edifice of thought from its simple elements to more and more complicated structures. The universal algebra would be part of logic but an algebraicized logic. Descartes began modestly by attempting to construct an algebra of logic; an incomplete sketch of this work is extant.

In pursuit of the same broad goal as Descartes's, Leibniz launched a more ambitious program. He had paid attention to logic throughout his life and rather early became entranced with the scheme of the medieval theologian Raymond Lull (1235–1315) whose book *Ars Magna et Ultima* offered a naive mechanical method for producing new ideas by combining existing ones but who did have the concept of a universal science of logic that would be applicable to all reasoning. Leibniz broke from scholastic logic and from Lull but became impressed with the possibility of a broad calculus that would enable man to reason in all fields mechanically and effortlessly. Leibniz says of his plan for a universal symbolic logic that such a science, of which ordinary algebra is but a small part, would be limited only by the necessity of obeying the laws of formal logic. One could name it, he said, "algebraico-logical synthesis."

This general science was to provide, first of all, a rational, universal language that would be adapted to thinking. The concepts, having been resolved into primitive distinct and nonoverlapping ones, could be combined in an almost mechanical way. He also thought that symbolism would be necessary in order to keep the mind from getting lost. Here the influence of algebraic symbolism on his thinking is clear. He wanted a symbolic language capable of expressing human thoughts unambiguously and aiding deduction. This symbolic language was his "universal characteristic."

In 1666 Leibniz wrote his *De Arte Combinatoria*,[10] which contains among other matters his early plans for his universal system of reasoning. He then wrote numerous fragments, which were never published but which are available in the edition of his philosophical writings (see note 10). In his first attempt he associated with each primitive concept a prime number; any concept composed of several primitive ones was represented by the product of the corresponding primes. Thus if 3 represents "man" and 7 "rational," 21 would represent "rational man." He then sought to translate the usual rules of the syllogism into this scheme but was not successful. He also tried, at another time, to use special symbols in place of prime numbers, where again complex ideas would be represented by combinations of symbols. Actually Leibniz thought that the number of primitive ideas would be few, but this proved to be erroneous. Also one basic operation, conjunction, for compounding primitive ideas did not suffice.

He also commenced work on an algebra of logic proper. Directly and indirectly Leibniz had in his algebra concepts we now describe as logical addition, multiplication, identity, negation, and the null class. He also called attention to the desirability of studying abstract relations such as inclusion, one-to-one correspondence, many-to-one correspondences, and equivalence relations. Some of these, he recognized, have the properties of symmetry and transitivity. Leibniz did not complete this work; he did not get beyond the syllogistic rules, which he himself recognized do not encompass all the logic mathematics uses. Leibniz described his ideas to l'Hospital and others, but they paid no attention. His logical works remained unedited until the beginning of the twentieth century and so had little direct influence. During the eighteenth and early nineteenth centuries a number of men sketched attempts similar to Leibniz's but got no further than he had.

A more effective if less ambitious step was taken by Augustus De Morgan. De Morgan published *Formal Logic* (1847) and many papers, some of which appeared in the *Transactions of the Cambridge Philosophical Society*. He sought to correct defects of and improve on Aristotelian logic. In his *Formal Logic* he added a new principle to Aristotelian logic. In the latter the premises "Some M's are A's" and "Some M's are B's" permit no conclusion; and, in fact, this logic says that the middle term M must be used universally, that is, "All M's" must occur. But De Morgan pointed out that from "Most M's are A's" and "Most M's are B's," it follows of necessity that "Some A's are B's." De Morgan put this fact in quantitative form. If there are m

---

10. Pub. 1690 = G. W. Leibniz: *Die philosophischen Schriften*, ed. by C. I. Gerhardt, 1875–90, Vol. 4, 27–102.

of the M's, and a of the M's are in A and b of the M's are in B, then there are at least $(a + b - m)$ A's that are B's. The point of De Morgan's observation is that the terms may be quantified. He was able as a consequence to introduce many more valid forms of the syllogism. Quantification also eliminated a defect in Aristotelian logic. The conclusion "Some A's are B's" which in Aristotelian logic can be drawn from "All A's are B's" implies the existence of A's but they need not exist.

De Morgan also initiated the study of the logic of relations. Aristotelian logic is devoted primarily to the relationship "to be" and either asserts or denies this relationship. As De Morgan pointed out, this logic could not prove that if a horse is an animal, then a horse's tail is an animal's tail. It certainly could not handle a relation such as x loves y. De Morgan introduced symbolism to handle relations but did not carry this subject very far.

In the area of symbolic logic De Morgan is widely known for what are now called De Morgan's laws. As he stated them,[11] the contrary of an aggregate is the compound of the contraries of the aggregates; the contrary of a compound is the aggregate of the contraries of the components. In logical notation these laws read

$$1 - (x + y) = (1 - x)(1 - y).$$
$$1 - xy = (1 - x) + (1 - y).$$

The contribution of symbolism to an algebra of logic was the major step made by George Boole (1815–1864), who was largely self-taught and became professor of mathematics at Queens College in Cork. Boole was convinced that the symbolization of language would rigorize logic. His *Mathematical Analysis of Logic*, which appeared on the same day as De Morgan's *Formal Logic*, and his *An Investigation of the Laws of Thought* (1854) contain his major ideas.

Boole's approach was to emphasize extensional logic, that is, a logic of classes, wherein sets or classes were denoted by $x, y, z, \ldots$ whereas the symbols $X, Y, Z, \ldots$ represented individual members. The universal class was denoted by 1 and the empty or null class by 0. He used xy to denote the intersection of two sets (he called the operation election), that is, the set of elements common to both x and y, and $x + y$ to denote the set consisting of all the elements in x and in y. (Strictly for Boole addition or union applied only to disjoined sets; W. S. Jevons [1835–82] generalized the concept.) The complement of x is denoted by $1 - x$. More generally $x - y$ is the class of x's that are not y's. The relation of inclusion, that is, x is contained in y, he wrote as $xy = x$. The equal sign denoted the identity of the two classes.

Boole believed that the mind grants to us at once certain elementary processes of reasoning that are the axioms of logic. For example, the law

11. *Trans. Camb. Phil. Soc.*, 10, 1858, 173–230.

of contradiction, that A cannot be both B and not B, is axiomatic. This is expressed by

$$x(1 - x) = 0.$$

It is also obvious to the mind that

$$xy = yx$$

and so this commutative property of intersection is another axiom. Equally obvious is the property

$$xx = x.$$

This axiom is a departure from ordinary algebra. Boole also accepted as axiomatic that

$$x + y = y + x$$

and

$$x(u + v) = xu + xv.$$

With these axioms the law of excluded middle could be stated in the form

$$x + (1 - x) = 1;$$

that is, everything is x or not x. Every X is Y becomes $x(1 - y) = 0$. No X is Y reads as $xy = 0$. Some X are Y is denoted by $xy \neq 0$ and some X are not Y by $x(1 - y) \neq 0$.

From the axioms Boole planned to deduce the laws of reasoning by applying the processes permitted by the axioms. As trivial conclusions he had that $1 \cdot x = x$ and $0 \cdot x = 0$. A slightly more involved argument is illustrated by the following. From

$$x + (1 - x) = 1$$

it follows that

$$z[x + (1 - x)] = z \cdot 1$$

and then that

$$zx + z(1 - x) = z.$$

Thus the class of objects z consists of those that are in x and those that are in $1 - x$.

Boole observed that the calculus of classes could be interpreted as a calculus of propositions. Thus if x and y are propositions instead of classes, then xy is the joint assertion of x and y and $x + y$ is the assertion of x or y or both. The statement $x = 1$ would mean that the proposition x is true and $x = 0$ that x is false. $1 - x$ would mean the denial of x. However Boole did not get very far with his calculus of propositions.

Both De Morgan and Boole can be regarded as the reformers of Aristotelian logic and the initiators of an algebra of logic. The effect of their work was to build a science of logic which thenceforth was detached from philosophy and attached to mathematics.

The calculus of propositions was advanced by Charles S. Peirce. Peirce distinguished between a proposition and a propositional function. A proposition, John is a man, contains only constants. A propositional function, x is a man, contains variables. Whereas a proposition is true or false, a propositional function is true for some values of the variable and false for others. Peirce also introduced propositional functions of two variables, for example, x knows y.

The men who had built symbolic logic thus far were interested in logic and in mathematizing that subject. With the work of Gottlob Frege (1848–1925), professor of mathematics at Jena, mathematical logic takes a new direction, the one that is pertinent to our account of the foundations of mathematics. Frege wrote several major works, *Begriffsschrift* (Calculus of Concepts, 1879), *Die Grundlagen der Arithmetik* (The Foundation of Arithmetic, 1884), and *Grundgesetze der Arithmetik* (The Fundamental Laws of Arithmetic; Vol. 1, 1893; Vol. 2, 1903). His works are characterized by precision and thoroughness of detail.

In the area of logic proper, Frege expanded on the use of variables, quantifiers, and propositional functions; most of this work was done independently of his predecessors, including Peirce. In his *Begriffsschrift* Frege gave an axiomatic foundation to logic. He introduced many distinctions that acquired great importance later, for example the distinction between the statement of a proposition and the assertion that it is true. The assertion is denoted by placing the symbol ⊢ in front of the proposition. He also distinguished between an object x and the set $\{x\}$ containing just x, and between an object belonging to a set and the inclusion of one set in another. Like Peirce, he used variables and propositional functions, and he indicated the quantification of his propositional functions, that is, the domain of the variable or variables for which they are true. He also introduced (1879) the concept of material implication: A implies B means either that A is true and B is true or A is false and B is true or A is false and B is false. This interpretation of implication is more convenient for mathematical logic. The logic of relations was also taken up by Frege; thus the relation of order involved in stating, for example, that a is greater than b was important in his work.

Having built up logic on explicit axioms, he proceeded in his *Grundlagen* to his real goal, to *build mathematics as an extension of logic*. He expressed the concepts of arithmetic in terms of the logical concepts. Thus the definitions and laws of number were derived from logical premises. We shall examine this construction in connection with the work of Russell and Whitehead.

Unfortunately Frege's symbolism was quite complex and strange to mathematicians. His work was in fact not well known until it was discovered by Russell. Rather ironic too is the fact that, just as the second volume of *Grundgesetze* was about to go to press, he received a letter from Russell, who informed him of the paradoxes of set theory. At the close of Volume 2 (p. 253), Frege remarks, "A scientist can hardly meet with anything more undesirable than to have the foundation give way just as the work is finished. I was put in this position by a letter from Mr. Bertrand Russell when the work was nearly through the press."

5. *The Logistic School*

We left the account of the work on the foundations of mathematics at the point where the axiomatization of set theory had provided a foundation that avoided the known paradoxes and yet served as a logical basis for the existing mathematics. We did point out that this approach was not satisfactory to many mathematicians. That the consistency of the real number system and the theory of sets remained to be proved was acknowledged by all; consistency was no longer a minor matter. The use of the axiom of choice was controversial. But beyond these problems was the overall question of what the proper foundation for mathematics was. The axiomatic movement of the late nineteenth century and the axiomatization of set theory had proceeded on the basis that the logic employed by mathematics could be taken for granted. But by the early nineteen-hundreds there were several schools of thought that were no longer content with this presupposition. The school led by Frege sought to rebuild logic and to build mathematics on logic. This plan, as we have already noted, was set back by the appearance of the paradoxes, but it was not abandoned. It was in fact independently conceived and pursued by Bertrand Russell and Alfred North Whitehead. Hilbert, already impressed by the need to establish consistency, began to formulate his own systematic foundation for mathematics. Still another group of mathematicians, known as intuitionists, was dissatisfied with the concepts and proofs introduced in late nineteenth-century analysis. These men were adherents of a philosophical position that not only could not be reconciled with some of the methodology of analysis but also challenged the role of logic. The development of these several philosophies was the major undertaking in the foundations of mathematics; its outcome was to open up the entire question of the nature of mathematics. We shall examine each of these three major schools of thought.

The first of these is known as the logistic school and its philosophy is called logicism. The founders are Russell and Whitehead. Independently of Frege, they had the idea that mathematics is derivable from logic and therefore is an extension of logic. The basic ideas were sketched by Russell in his *Principles of Mathematics* (1903); they were developed in the detailed work by Whitehead and Russell, the *Principia Mathematica* (3 vols., 1910–13). Since the *Principia* is the definitive version, we shall base our account on it.

This school starts with the development of logic itself, from which mathematics follows without any axioms of mathematics proper. The development of logic consists in stating some axioms of logic, from which theorems are deduced that may be used in subsequent reasoning. Thus the laws of logic receive a formal derivation from axioms. The *Principia* also has undefined ideas, as any axiomatic theory must have since it is not possible to define all the terms without involving an infinite regress of definitions. Some of these undefined ideas are the notion of an elementary proposition, the notion of a propositional function, the assertion of the truth of an elementary proposition, the negation of a proposition, and the disjunction of two propositions.

Russell and Whitehead explain these notions, though, as they point out, this explanation is not part of the logical development. By a proposition they mean simply any sentence stating a fact or a relationship: for example, John is a man; apples are red; and so forth. A propositional function contains a variable, so that substitution of a value for that variable gives a proposition. Thus "X is an integer" is a propositional function. The negation of a proposition is intended to mean, "It is not true that the proposition holds," so that if p is the proposition that John is a man, the negation of p, denoted by $\sim p$, means, "It is not true that John is a man," or "John is not a man." The disjunction of two propositions p and q, denoted by $p \vee q$, means p or q. The meaning of "or" here is that intended in the sentence, "Men or women may apply." That is, men may apply; women may apply; and both may apply. In the sentence, "That person is a man or a woman," "or" has the more common meaning of either one or the other but not both. Mathematics uses "or" in the first sense, though sometimes the second sense is the only one possible. For example, "the triangle is isosceles or the quadrilateral is a parallelogram" illustrates the first sense. We also say that every number is positive or negative. Here additional facts about positive and negative numbers say that both cannot be true. Thus the assertion $p \vee q$ means p and q, $\sim p$ and q, and p and $\sim q$.

A most important relationship between propositions is implication, that is, the truth of one proposition compelling the truth of another. In the *Principia* implication, $p \supset q$, is defined by $\sim p \vee q$, which in turn means $\sim p$ and q, p and q, or $\sim p$ and $\sim q$. As an illustration consider the implication, If X is a man, then X is mortal. Here the state of affairs could be

X is not a man and X is mortal;
X is a man and X is mortal;
X is not a man and X is not mortal.

Any one of these possibilities is allowable. What the implication forbids is

X is a man and X is not mortal.

Some of the postulates of the *Principia* are:

(a) Anything implied by a true elementary proposition is true.
(b) $(p \lor p) \supset p$.
(c) $q \supset (p \lor q)$.
(d) $(p \lor q) \supset (q \lor p)$.
(e) $[p \lor (q \lor r)] \supset [q \lor (p \lor r)]$.
(f) The assertion of p and the assertion $p \supset q$ permits the assertion of q.

The independence of these postulates and their consistency cannot be proved because the usual methods do not apply. From these postulates the authors proceed to deduce theorems of logic and ultimately arithmetic and analysis. The usual syllogistic rules of Aristotle occur as theorems.

To illustrate how even logic itself has been formalized and made deductive, let us note a few theorems of the early part of *Principia Mathematica*:

2.01. $(p \supset \sim p) \supset \sim p$.

This is the principle of *reductio ad absurdum*. In words, if the assumption of p implies that p is false, then p is false.

2.05. $[q \supset r] \supset [(p \supset q) \supset (p \supset r)]$.

This is one form of the syllogism. In words, if q implies r, then if p implies q, p implies r.

2.11. $p \lor \sim p$.

This is the principle of excluded middle: p is true or p is false.

2.12. $p \supset \sim(\sim p)$.

In words, p implies that not-p is false.

2.16. $(p \supset q) \supset (\sim q \supset \sim p)$.

If p implies q, then not-q implies not-p.

Propositions are a step to propositional functions that treat sets by means of properties rather than by naming the objects in a set. Thus the propositional function "x is red" denotes the set of all red objects.

If the members of a set are individual objects, then the propositional functions applying to such members are said to be of type 0. If the members of a set are themselves propositional functions, any propositional function applying to such members is said to be of type 1. And generally propositional functions whose variables are of types less than and equal to n are of type $n + 1$.

The theory of types seeks to avoid the paradoxes, which arise because a collection of objects contains a member that itself can be defined only in terms of the collection. The resolution by Russell and Whitehead of this difficulty was to require that "whatever involves all members of a collection must not itself be a member of the collection." To carry out this restriction in the *Principia*, they specify that a (logical) function cannot have as one of its arguments anything defined in terms of the function itself. They then discuss the paradoxes and show that the theory of types avoids them.

However, the theory of types leads to classes of statements that must be carefully distinguished by type. If one attempts to build mathematics in accordance with the theory of types, the development becomes exceedingly complex. For example, in the *Principia* two objects a and b are equal if for every property $P(x)$, $P(a)$ and $P(b)$ are equivalent propositions (each implies the other). According to the theory of types, P may be of different types because it may contain variables of various orders as well as the individual objects a or b, and so the definition of equality must apply for all types of P; in other words, there is an infinity of relations of equality, one for each type of property. Likewise, an irrational number defined by the Dedekind cut proves to be of higher type than a rational number, which in turn is of higher type than a natural number, and so the continuum consists of numbers of different types. To escape this complexity, Russell and Whitehead introduced the axiom of reducibility, which affirms the existence, for each propositional function of whatever type, of an equivalent propositional function of type zero.

Having treated propositional functions, the authors take up the theory of classes. A class, loosely stated, is the set of objects satisfying some propositional function. Relations are then expressed as classes of couples satisfying propositional functions of two variables. Thus "x judges y" expresses a relation. On this basis the authors are prepared to introduce the notion of cardinal number.

The definition of a cardinal number is of considerable interest. It depends upon the previously introduced relation of one-to-one correspondence between classes. If two classes are in one-to-one correspondence, they are called similar. The relationship of similarity is proven to be reflexive, symmetric, and transitive. All similar classes possess a common property, and this is their number. However, similar classes may have more than one common property. Russell and Whitehead get around this, as had Frege, by defining the number of a class as the class of all classes that are similar to the given class. Thus the number 3 is the class of all three-membered classes and the denotation of all three-membered classes is $\{x, y, z\}$ with $x \neq y \neq z$. Since the definition of number presupposes the concept of one-to-one correspondence, it would seem as though the definition is circular. The authors point out, however, that a relation is one-to-one if, when x

and x' have the relation to y, then x and x' are identical, and when x has the relation to y and y', then y and y' are identical. Hence the concept of one-to-one correspondence does not involve the number 1.

Given the cardinal or natural numbers, it is possible to build up the real and complex number systems, functions, and in fact all of analysis. Geometry can be introduced through numbers. Though the details in the *Principia* differ somewhat, our own examination of the foundations of the number system and of geometry (Chaps. 41 and 42) shows that such constructions are logically possible without additional axioms.

This, then, is the grand program of the logistic school. What it does with logic itself is quite a story, which we are skimming over here briefly. What it does for mathematics, and this we must emphasize, is to found mathematics on logic. No axioms of mathematics are needed; mathematics becomes no more than a natural extension of the laws and subject matter of logic. But the postulates of logic and all their consequences are arbitrary and, moreover, formal. That is, they have no content; they have merely form. As a consequence, mathematics too has no content, but merely form. The physical meanings we attach to numbers or to geometric concepts are not part of mathematics. It was with this in mind that Russell said that mathematics is the subject in which we never know what we are talking about nor whether what we are saying is true. Actually, when Russell started this program in the early part of the century, he (and Frege) thought the axioms of logic were truths. But he abandoned this view in the 1937 edition of the *Principles of Mathematics*.

The logistic approach has received much criticism. The axiom of reducibility aroused opposition, for it is quite arbitrary. It has been called a happy accident and not a logical necessity; it has been said that the axiom has no place in mathematics, and that what cannot be proved without it cannot be regarded as proved at all. Others called the axiom a sacrifice of the intellect. Moreover, the system of Russell and Whitehead was never completed and is obscure in numerous details. Many efforts were made later to simplify and clarify it.

Another serious philosophical criticism of the entire logistic position is that if the logistic view is correct, then all of mathematics is a purely formal, logico-deductive science whose theorems follow from the laws of thought. Just how such a deductive elaboration of the laws of thought can represent wide varieties of natural phenomena such as acoustics, electromagnetics, and mechanics seems unexplained. Further, in the creation of mathematics perceptual or imaginative intuition must supply new concepts, whether or not derived from experience. Otherwise, how could new knowledge arise? But in the *Principia* all concepts reduce to logical ones.

The formalization of the logistic program apparently does not represent mathematics in any real sense. It presents us with the husk, not the kernel.

Poincaré said, snidely (*Foundations of Science*, p. 483), "The logistic theory is not sterile; it engenders contradictions." This is not true if one accepts the theory of types, but this theory, as noted, is artificial. Weyl also attacked logicism; he said that this complex structure "taxes the strength of our faith hardly less than the doctrines of the early Fathers of the Church or of the Scholastic philosophers of the Middle Ages."

Despite the criticisms, the logistic philosophy is accepted by many mathematicians. The Russell-Whitehead construction also made a contribution in another direction. It carried out a thorough axiomatization of logic in entirely symbolic form and so advanced enormously the subject of mathematical logic.

6. *The Intuitionist School*

A radically different approach to mathematics has been undertaken by a group of mathematicians called intuitionists. As in the case of logicism, the intuitionist philosophy was inaugurated during the late nineteenth century when the rigorization of the number system and geometry was a major activity. The discovery of the paradoxes stimulated its further development.

The first intuitionist was Kronecker, who expressed his views in the 1870s and 80s. To Kronecker, Weierstrass's rigor involved unacceptable concepts, and Cantor's work on transfinite numbers and set theory was not mathematics but mysticism. Kronecker was willing to accept the whole numbers because these are clear to the intuition. These "were the work of God." All else was the work of man and suspect. In his essay of 1887, "Über den Zahlbegriff" (On the Number Concept),[12] he showed how some types of numbers, fractions for example, could be defined in terms of the whole numbers. Fractional numbers as such were acceptable as a convenience of notation. The theory of irrational numbers and of continuous functions he wished to strip away. His ideal was that every theorem of analysis should be interpretable as giving relations among the integers only.

Another objection Kronecker made to many parts of mathematics was that they did not give constructive methods or criteria for determining in a finite number of steps the objects with which they dealt. Definitions should contain the means of calculating the object defined in a finite number of steps, and existence proofs should permit the calculation to any required degree of accuracy of the quantity whose existence is being established. Algebraists were content to say that a polynomial $f(x)$ may have a rational factor, in which case $f(x)$ is reducible. In the contrary case it is irreducible. In his *Festschrift* "Grundzüge einer arithmetischen Theorie der algebraischen Grössen" (Elements of an Arithmetic Theory of Algebraic Quantities)[13]

12. *Jour. für Math.*, 101, 1887, 337–55 = *Werke*, 3, 251–74.
13. *Jour. für Math.*, 92, 1882, 1–122 = *Werke*, 2, 237–387.

Kronecker said, "The definition of reducibility is devoid of a sure foundation until a *method* is given by means of which it can be decided whether a given function is irreducible or not."

Again, though the several theories of the irrational numbers give definitions as to when two real numbers a and b are equal or when $a > b$ or $b > a$, they do not give criteria to determine which alternative holds in a given case. Hence Kronecker objected to such definitions. They are definitions only in appearance. The entire theory of irrationals was unsatisfactory to him and one day he said to Lindemann, who had proved that π is a transcendental irrational, "Of what use is your beautiful investigation regarding π? Why study such problems, since irrational numbers are non-existent?"

Kronecker himself did little to develop the intuitionist philosophy except to criticize the absence of constructive procedures for determining quantities whose existence was merely established. He tried to rebuild algebra but made no efforts to reconstruct analysis. Kronecker produced fine work in arithmetic and algebra which did not conform to his own requirements because, as Poincaré remarked,[14] he temporarily forgot his own philosophy.

Kronecker had no supporters of his philosophy in his day and for almost twenty-five years no one pursued his ideas. However, after the paradoxes were discovered, intuitionism was revived and became a widespread and serious movement. The next strong advocate was Poincaré. His opposition to set theory because it gave rise to paradoxes has already been noted. Nor would he accept the logistic program for rescuing mathematics. He ridiculed attempts to base mathematics on logic on the ground that mathematics would reduce to an immense tautology. He also mocked the (to him) highly artificial derivation of number. Thus in the *Principia* 1 is defined as $\hat{\alpha}\{\exists x \cdot \alpha = i'x\}$. Poincaré said sarcastically that this was an admirable definition to give to people who never heard of the number 1.

In *Science and Method* (*Foundations of Science*, p. 480) he stated,

> Logistic has to be made over, and one is none too sure of what can be saved. It is unnecessary to add that only Cantorism and Logistic are meant; true mathematics, that which serves some useful purpose, may continue to develop according to its own principles without paying any attention to the tempests raging without, and it will pursue step by step its accustomed conquests which are definitive and which it will never need to abandon.

Poincaré objected to concepts that cannot be defined in a finite number of words. Thus a set chosen in accordance with the axiom of choice is not really defined when a choice has to be made from each of a transfinite num-

14. *Acta Math.*, 22, 1899, 17.

ber of sets. He also contended that arithmetic cannot be justified by an axiomatic foundation. Our intuition precedes such a structure. In particular, mathematical induction is a fundamental intuition and not just an axiom that happens to be useful in some system of axioms. Like Kronecker, he insisted that all definitions and proof should be constructive.

He agreed with Russell that the source of the paradoxes was the definition of collections or sets that included the object defined. Thus the set A of all sets contains A. But A cannot be defined until each member of A is defined, and if A is one member the definition is circular. Another example of an impredicative definition is the definition of the maximum value of a continuous function defined over a closed interval as the largest value that the function takes on in this interval. Such definitions were common in analysis and especially in the theory of sets.

Further criticisms of the current logical state of mathematics were developed and discussed in an exchange of letters among Borel, Baire, Hadamard, and Lebesgue.[15] Borel supported Poincaré's assertion that the integers cannot be founded axiomatically. He too criticized the axiom of choice because it calls for a nondenumerable infinity of choices, which is inconceivable to the intuition. Hadamard and Lebesgue went further and said that even a denumerable infinity of arbitrary successive choices is not more intuitive because it calls for an infinity of operations, which it is impossible to conceive as being effectively realized. For Lebesgue the difficulties all reduced to knowing what one means when one says that a mathematical object exists. In the case of the axiom of choice he argued that if one merely "thinks" of a way of choosing, may one then not change his choices in the course of his reasoning? Even the choice of a single object in one set, Lebesgue maintained, raises the same difficulties. One must know the object "exists" which means that one must name the choice explicitly. Thus Lebesgue rejected Cantor's proof of the existence of transcendental numbers. Hadamard pointed out that Lebesgue's objections led to a denial of the existence of the set of all real numbers, and Borel drew exactly the same conclusion.

All of the above objections by intuitionists were sporadic and fragmented. The systematic founder of modern intuitionism is Brouwer. Like Kronecker, much of his mathematical work, notably in topology, was not in accord with his philosophy, but there is no question as to the seriousness of his position. Commencing with his doctoral dissertation, *On the Foundations of Mathematics* (1907), Brouwer began to build up the intuitionist philosophy. From 1918 on he expanded and expounded his views in papers in various journals, including the *Mathematische Annalen* of 1925 and 1926.

Brouwer's intuitionist position stems from a broader philosophy. The fundamental intuition, according to Brouwer, is the occurrence of perceptions in a time sequence. "Mathematics arises when the subject of twoness,

15. See note 9.

which results from the passage of time, is abstracted from all special occurrences. The remaining empty form [the relation of n to $n + 1$] of the common content of all these twonesses becomes the original intuition of mathematics and repeated unlimitedly creates new mathematical subjects." Thus by unlimited repetition the mind forms the concept of the successive natural numbers. This idea that the whole numbers derive from the intuition of time had been maintained by Kant, William R. Hamilton in his "Algebra as a Science of Time," and the philosopher Arthur Schopenhauer.

Brouwer conceives of mathematical thinking as a process of construction that builds its own universe, independent of the universe of our experience and somewhat as a free design, restricted only in so far as it is based upon the fundamental mathematical intuition. This fundamental intuitive concept must not be thought of as an undefined idea, such as occurs in postulational theories, but rather as something *in terms of which* all undefined ideas that occur in the various mathematical systems are to be intuitively conceived, if they are indeed to serve in mathematical thinking.

Brouwer holds that "in this constructive process, bound by the obligation to notice with care which theses are acceptable to the intuition and which are not, lies the only possible foundation for mathematics." Mathematical ideas are imbedded in the human mind *prior to language, logic, and experience*. The intuition, not experience or logic, determines the soundness and acceptability of ideas. It must of course be remembered that these statements concerning the role of experience are to be taken in the philosophical sense, not the historical sense.

The mathematical objects are for Brouwer acquired by intellectual construction, wherein the basic numbers $1, 2, 3, \ldots$ furnish the prototype of such a construction. The possibility of the unlimited repetition of the empty form, the step from n to $n + 1$, leads to infinite sets. However, Brouwer's infinite is the potential infinity of Aristotle, whereas modern mathematics, as founded for example by Cantor, makes extensive use of actually infinite sets whose elements are all present "at once."

In connection with the intuitionist notion of infinite sets, Weyl, who belonged to the intuitionist school, says that

> ... the sequence of numbers which grows beyond any stage already reached ... is a manifold of possibilities opening to infinity; it remains forever in the status of creation, but is not a closed realm of things existing in themselves. That we blindly converted one into the other is the true source of our difficulties, including the antinomies—a source of more fundamental nature than Russell's vicious circle principle indicated. Brouwer opened our eyes and made us see how far classical mathematics, nourished by a belief in the absolute that transcends all human possibilities of realization, goes beyond such statements as can claim real meaning and truth founded on evidence.

The world of mathematical intuition is opposed to the world of causal perceptions. In this causal world, not in mathematics, belongs language, which serves there for the understanding of common dealings. Words or verbal connections are used to communicate truths. Language serves to evoke copies of ideas in men's minds by symbols and sounds. But thoughts can never be completely symbolized. These remarks apply also to mathematical language, including symbolic language. Mathematical ideas are independent of the dress of language and in fact far richer.

Logic belongs to language. It offers a system of rules that permit the deduction of further verbal connections and also are intended to communicate truths. However, these latter truths are not such before they are experienced, nor is it guaranteed that they can be experienced. Logic is not a reliable instrument to uncover truths and can deduce no truths that are not obtainable just as well in some other way. Logical principles are the regularity observed a posteriori in the language. They are a device for manipulating language, or they are the theory of representation of language. The most important advances in mathematics are not obtained by perfecting the logical form but by modifying the basic theory itself. Logic rests on mathematics, not mathematics on logic.

Since Brouwer does not recognize any a priori obligatory logical principles, he does not recognize the mathematical task of deducing conclusions from axioms. Mathematics is not bound to respect the rules of logic, and for this reason the paradoxes are unimportant even if we were to accept the mathematical concepts and constructions the paradoxes involve. Of course, as we shall see, the intuitionists do not accept all these concepts and proofs.

Weyl[16] expands on the role of logic:

> According to his [Brouwer's] view and reading of history, classical logic was abstracted from the mathematics of finite sets and their subsets. . . . Forgetful of this limited origin, one afterwards mistook that logic for something above and prior to all mathematics, and finally applied it, without justification, to the mathematics of infinite sets. This is the Fall and original sin of set theory, for which it is justly punished by the antinomics. It is not that such contradictions showed up that is surprising, but that they showed up at such a late stage of the game.

In the realm of logic there are some clear, *intuitively* acceptable logical principles or procedures that can be used to assert new theorems from old ones. These principles are part of the fundamental mathematical intuition. However, not all logical principles are acceptable to the basic intuition and one must be critical of what has been sanctioned since the days of Aristotle. Because mathematicians have applied freely these Aristotelian laws, they

16. *Amer. Math. Monthly*, 53, 1946, 2–13 = *Ges. Abh.*, 4, 268–79.

have produced antinomies. The intuitionists therefore proceed to analyze which logical principles are allowable in order that the usual logic conform to and properly express the correct intuitions.

As a specific example of a logical principle that is applied too freely, Brouwer cites the law of excluded middle. This principle, which asserts that every meaningful statement is true or false, is basic to the indirect method of proof. It arose historically in the application of reasoning to subsets of finite sets and was abstracted therefrom. It was then accepted as an independent a priori principle and was unjustifiably applied to infinite sets. Whereas for finite sets one can decide whether all elements possess a certain property P by testing each one, this procedure is no longer possible for infinite sets. One may happen to know that an element of the infinite set does not possess the property or it may be that by the very construction of the set we know or can prove that every element has the property. In any case, one cannot use the law of excluded middle to prove the property holds.

Hence if one proves that not all elements of an infinite set possess a property, then the conclusion that there exists at least one element which does not have the property is rejected by Brouwer. Thus from the denial that $a^b = b^a$ holds for all numbers the intuitionists do not conclude that there exists an a and b for which $a^b \neq b^a$. Consequently many existence proofs are not accepted by the intuitionists. The law of excluded middle can be used in cases where the conclusion can be reached in a finite number of steps, for example, to decide the question of whether a book contains misprints. In other cases the intuitionists deny the possibility of a decision.

The denial of the law of excluded middle gives rise to a new possibility, undecidable propositions. The intuitionists maintain, with respect to *infinite* sets, that there is a third state of affairs, namely, there may be propositions which are neither provable nor unprovable. As an example of such a proposition, let us define the kth position in the decimal expansion of π to be the position of the first zero which is followed by the integers $1 \ldots 9$. Aristotelian logic says that k either exists or does not exist and mathematicians following Aristotle may then proceed to argue on the basis of these two possibilities. Brouwer would reject all such arguments, for we do not know whether we shall ever be able to prove that it does or does not exist. Hence all reasoning about the number k is rejected by the intuitionists. Thus there are sensible mathematical questions which may never be settled on the basis of the statements contained in the axioms of mathematics. The questions seem to us to be decidable but actually our basis for expecting that they must be decidable is really nothing more than that they involve mathematical concepts.

With respect to the concepts they will accept as legitimate for mathematical discussion, the intuitionists insist on constructive definitions. For Brouwer, as for all intuitionists, the infinite exists in the sense that one can always find a finite set larger than the given one. To discuss any other type of infinite, the intuitionists demand that one give a method of constructing or defining this infinite in a finite number of steps. Thus Brouwer rejects the aggregates of set theory.

The requirement of constructibility is another ground for excluding any concept whose existence is established by indirect reasoning, that is, by the argument that the nonexistence leads to a contradiction. Aside from the fact that the existence proof may use the objectionable law of excluded middle, to the intuitionists this proof is not satisfactory because they want a constructive definition of the object whose existence is being established. The constructive definition must permit determination to any desired accuracy in a finite number of steps. Euclid's proof of the existence of an infinite number of primes (Chap. 4, sec. 7) is nonconstructive; it does not afford the determination of the nth prime. Hence it is not acceptable. Further, if one proved merely the existence of integers x, y, z, and n satisfying $x^n + y^n = z^n$, the intuitionist would not accept the proof. On the other hand, the definition of a prime number is constructive, for it can be applied to determine in a finite number of steps whether a number is prime. The insistence on a constructive definition applies especially to infinite sets. A set constructed by the axiom of choice applied to infinitely many sets would not be acceptable.

Weyl said of nonconstructive existence proofs (*Philosophy of Mathematics and Natural Science*, p. 51) that they inform the world that a treasure exists without disclosing its location. Proof through postulation cannot replace construction without loss of significance and value. He also pointed out that adherence to the intuitionist philosophy means the abandonment of the existence theorems of classical analysis—for example, the Weierstrass-Bolzano theorem. A bounded monotonic set of real numbers does not necessarily have a limit. For the intuitionists, if a function of a real variable exists in their sense then it is *ipso facto* continuous. Transfinite induction and its applications to analysis and most of the theory of Cantor are condemned outright. Analysis, Weyl says, is built on sand.

Brouwer and his school have not limited themselves to criticism but have sought to build up a new mathematics on the basis of constructions they accept. They have succeeded in saving the calculus with its limit processes, but their construction is very complicated. They also reconstructed elementary portions of algebra and geometry. Unlike Kronecker, Weyl and Brouwer do allow some kinds of irrational numbers. Clearly the mathematics of the intuitionists differs radically from what mathematicians had almost universally accepted before 1900.

7. The Formalist School

The third of the principal philosophies of mathematics is known as the formalist school and its leader was Hilbert. He began work on this philosophy in 1904. His motives at that time were to provide a basis for the number system without using the theory of sets and to establish the consistency of arithmetic. Since his own proof of the consistency of geometry reduced to the consistency of arithmetic, the consistency of the latter was a vital open question. He also sought to combat Kronecker's contention that the irrationals must be thrown out. Hilbert accepted the actual infinite and praised Cantor's work (Chap. 41, sec. 9). He wished to keep the infinite, the pure existence proofs, and concepts such as the least upper bound whose definition appeared to be circular.

Hilbert presented one paper on his views at the International Congress of 1904.[17] He did no more on this subject for fifteen years; then, moved by the desire to answer the intuitionists' criticisms of classical analysis, he took up problems of the foundations and continued to work on them for the rest of his scientific career. He published several key papers during the nineteen-twenties. Gradually a number of men took up his views.

Their mature philosophy contains many doctrines. In keeping with the new trend that any foundation for mathematics must take cognizance of the role of logic, the formalists maintain that logic must be treated simultaneously with mathematics. Mathematics consists of several branches and each branch is to have its own axiomatic foundation. This must consist of logical and mathematical concepts and principles. Logic is a sign language that puts mathematical statements into formulas and expresses reasoning by formal processes. The axioms merely express the rules by which formulas follow from one another. All signs and symbols of operation are freed from their significance with respect to content. Thus all meaning is eliminated from the mathematical symbols. In his 1926 paper[18] Hilbert says the objects of mathematical thought are the symbols themselves. The symbols are the essence; they no longer stand for idealized physical objects. The formulas may imply intuitively meaningful statements, but these implications are not part of mathematics.

Hilbert retained the law of excluded middle because analysis depends upon it. He said,[19] "Forbidding a mathematician to make use of the principle of excluded middle is like forbidding an astronomer his telescope or a boxer the use of his fists." Because mathematics deals only with symbolic expressions, all the rules of Aristotelian logic can be applied to these formal expressions. In this new sense the mathematics of infinite sets is possible. Also, by avoiding the explicit use of the word "all," Hilbert hoped to avoid the paradoxes.

To formulate the logical axioms Hilbert introduced symbolism for

17. *Proc. Third Internat. Congress of Math.*, Heidelberg, 1904, 174–85 = *Grundlagen der Geom.*, 7th ed., 247–61; English trans. in *Monist*, 15, 1905, 338–52.
18. *Math. Ann.*, 95, 1926, 161–90 = *Grundlagen der Geometrie*, 7th ed., 262–88. See note 20.
19. Weyl, *Amer. Math. Soc. Bull.*, 50, 1944, 637 = *Ges. Abh.*, 4, 157.

concepts and relations such as "and," "or," "negation," "there exists," and the like. Luckily the logical calculus (symbolic logic) had already been developed (for other purposes) and so, Hilbert says, he has at hand what he needs. All the above symbols are the building blocks for the ideal expressions —the formulas.

To handle the infinite, Hilbert uses, aside from ordinary noncontroversial axioms, the transfinite axiom

$$A(\tau A) \to A(a).$$

This, he says, means: If a predicate A applies to the fiducial object τA, it applies to all objects a. Thus suppose A stands for being corruptible. If Aristides the Just is fiducial and corruptible, then everybody is corruptible.

Mathematical proof will consist of this process: the assertion of some formula; the assertion that this formula implies another; the assertion of the second formula. A sequence of such steps in which the asserted formulas or the implications are preceding axioms or conclusions will constitute the proof of a theorem. Also, substitution of one symbol for another or a group of symbols is a permissible operation. Thus formulas are derived by applying the rules for manipulating the symbols of previously established formulas.

A proposition is true if and only if it can be obtained as the last of a sequence of propositions such that every proposition of the sequence is either an axiom in the formal system or is itself derived by one of the rules of deduction. Everyone can check as to whether a given proposition has been obtained by a proper sequence of propositions. Thus under the formalist view truth and rigor are well defined and objective.

To the formalist, then, mathematics proper is a collection of formal systems, each building its own logic along with its mathematics, each having its own concepts, its own axioms, its own rules for deducing theorems such as rules about equality or substitution, and its own theorems. The development of each of these deductive systems is the task of mathematics. Mathematics becomes not a subject about something, but a collection of formal systems, in each of which formal expressions are obtained from others by formal transformations. So much for the part of Hilbert's program that deals with mathematics proper.

However, we must now ask whether the deductions are free of contradictions. This cannot necessarily be observed intuitively. But to show noncontradiction, all we need to show is that one can never arrive at the formal statement $1 = 2$. (Since by a theorem of logic any other false proposition implies this proposition, we may confine ourselves to this one.)

Hilbert and his students Wilhelm Ackermann (1896–1962), Paul Bernays (1888–), and von Neumann gradually evolved, during the years 1920 to 1930, what is known as Hilbert's *Beweistheorie* [proof theory] or metamathematics, a method of establishing the consistency of any formal system. In metamathematics Hilbert proposed to use a special logic that was to be basic and free of all objections. It employs concrete and finite reasoning of a kind universally admitted and very close to the intuitionist principles. Controversial principles such as proof of existence by contradiction, trans- finite induction, and the axiom of choice are not used. Existence proofs must be constructive. Since a formal system can be unending, metamathematics must entertain concepts and questions involving at least potentially infinite systems. However, only finitary methods of proof should be used. There should be no reference either to an infinite number of structural properties of formulas or to an infinite number of manipulations of formulas.

Now the consistency of a major part of classical mathematics can be reduced to that of the arithmetic of the natural numbers (number theory) much as this theory is embodied in the Peano axioms, or to a theory of sets sufficiently rich to yield Peano's axioms. Hence the consistency of the arithmetic of the natural numbers became the center of attention.

Hilbert and his school did demonstrate the consistency of simple formal systems and they believed they were about to realize the goal of proving the consistency of arithmetic and of the theory of sets. In his article "Über das Unendliche"[20] he says,

> In geometry and physical theory the proof of consistency is accomplished by reducing it to the consistency of arithmetic. This method obviously fails in the proof for arithmetic itself. Since our proof theory . . . makes this last step possible, it constitutes the necessary keystone in the structure of mathematics. And in particular what we have twice experienced, first in the paradoxes of the calculus and then in the paradoxes of set theory, cannot happen again in the domain of mathematics.

But then Kurt Gödel (1906–) entered the picture. Gödel's first major paper was "Über formal unentscheidbare Sätze der *Principia Mathematica* und verwandter Systeme I."[21] Here Gödel showed that the consistency of a system embracing the usual logic and number theory cannot be established if one limits himself to such concepts and methods as can formally be represented in the system of number theory. What this means in effect is that the consistency of number theory cannot be established by the narrow logic permissible in metamathematics. Apropos of this result, Weyl said that God exists since mathematics is consistent and the devil exists since we cannot prove the consistency.

The above result of Gödel's is a corollary of his more startling result.

20. *Math. Ann.*, 95, 1926, 161–90 = *Grundlagen der Geometrie*, 7th ed., 262–88. An English translation can be found in Paul Benacerraf and Hilary Putnam: *Philosophy of Mathematics*, 134–181, Prentice-Hall, 1964.
21. *Monatshefte für Mathematik und Physik*, 38, 1931, 173–98; see the bibliography.

The major result (Gödel's incompleteness theorem) states that if any formal theory T adequate to embrace number theory is consistent and if the axioms of the formal system of arithmetic are axioms or theorems of T, then T is incomplete. That is, there is a statement S of number theory such that neither S nor not-S is a theorem of the theory. Now either S or not-S is true; there is, then, a true statement of number theory which is not provable. This result applies to the Russell-Whitehead system, the Zermelo-Fraenkel system, and Hilbert's axiomatization of number theory. It is somewhat ironic that Hilbert in his address at the International Congress in Bologna of 1928 (see note 22) had criticized the older proofs of completeness through categoricalness (Chap. 42, sec. 3) but was very confident that his own system was complete. Actually the older proofs involving systems containing the natural numbers were accepted as valid only because set theory had not been axiomatized but was used on a naive basis.

Incompleteness is a blemish in that the formal system is not adequate to prove all the assertions frameable in the system. To add insult to injury, there are assertions that are undecidable but are intuitively true in the system. Incompleteness cannot be remedied by adjoining S or $\sim S$ as an axiom, for Gödel proved that *any* system embracing number theory must contain an undecidable proposition. Thus while Brouwer made clear that what is intuitively certain falls short of what is mathematically proved, Gödel showed that the intuitively certain goes beyond mathematical proof.

One of the implications of Gödel's theorem is that no system of axioms is adequate to encompass, not only all of mathematics, but even any one significant branch of mathematics, because any such axiom system is incomplete. There exist statements whose concepts belong to the system, which cannot be proved within the system but can nevertheless be shown to be true by nonformal arguments, in fact by the logic of metamathematics. This implication, that there are limitations on what can be achieved by axiomatization, contrasts sharply with the late nineteenth-century view that mathematics is coextensive with the collection of axiomatized branches. Gödel's result dealt a death blow to comprehensive axiomatization. This inadequacy of the axiomatic method is not in itself a contradiction, but it was surprising, because mathematicians had expected that any true statement could certainly be established within the framework of some axiomatic system. Of course the above arguments do not exclude the possibility of new methods of proof that would go beyond what Hilbert's metamathematics permit.

Hilbert was not convinced that these blows destroyed his program. He argued that even though one might have to use concepts outside a formal system, they might still be finite and intuitively concrete and so acceptable. Hilbert was an optimist. He had unbounded confidence in the power of man's reasoning and understanding. At the talk he gave at the 1928

International Congress[22] he had asserted, ". . . to the mathematical understanding there are no bounds . . . in mathematics there is no Ignorabimus
[we shall not know]; rather we can always answer meaningful questions
. . . our reason does not possess any secret art but proceeds by quite definite
and statable rules which are the guarantee of the absolute objectivity of its
judgment." Every mathematician, he said, shares the conviction that each
definite mathematical problem must be capable of being solved. This
optimism gave him courage and strength, but it barred him from understanding that there could be undecidable mathematical problems.

The formalist program, successful or not, was unacceptable to the
intuitionists. In 1925 Brouwer blasted away at the formalists.[23] Of course, he
said, axiomatic, formalistic treatments will avoid contradictions, but nothing
of mathematical value will be obtained in this way. A false theory is none the
less false even if not halted by a contradiction, just as a criminal act is
criminal whether or not forbidden by a court. Sarcastically he also remarked,
"To the question, where shall mathematical rigor be found, the two parties
give different answers. The intuitionist says, in the human intellect; the
formalist says, on paper." Weyl too attacked Hilbert's program. "Hilbert's
mathematics may be a pretty game with formulas, more amusing even than
chess; but what bearing does it have on cognition, since its formulas admittedly have no material meaning by virtue of which they could express
intuitive truths." In defense of the formalist philosophy, one must point out
that it is only for the purposes of proving consistency, completeness, and
other properties that mathematics is reduced to meaningless formulas. As for
mathematics as a whole, even the formalists reject the idea that it is simply a
game; they regard it as an objective science.

Hilbert in turn charged Brouwer and Weyl with trying to throw overboard everything that did not suit them and dictatorially promulgating an
embargo.[24] He called intuitionism a treason to science. (Yet in his metamathematics he limited himself to intuitively clear logical principles.)

8. Some Recent Developments

None of the proposed solutions of the basic problems of the foundations—the
axiomatization of set theory, logicism, intuitionism, or formalism—achieved
the objective of providing a universally acceptable approach to mathematics.
Developments since Gödel's work of 1931 have not essentially altered the
picture. However, a few movements and results are worth noting. A number

---

22. *Atti Del Congresso Internazionale Dei Matematici*, I, 135–41 = *Grundlagen der Geometrie*, 7th
ed., 313–23.
23. *Jour. für Math.*, 154, 1925, 1.
24. *Abh. Math. Seminar der Hamburger Univ.*, 1, 1922, 157–77 = *Ges. Abh.*, 3, 157–77.

of men have erected compromise approaches to mathematics that utilize features of two basic schools. Others, notably Gerhard Gentzen (1909–45), a member of Hilbert's school, have loosened the restrictions on the methods of proof allowed in Hilbert's metamathematics and, for example, by using transfinite induction (induction over the transfinite numbers), have thereby managed to establish the consistency of number theory and restricted portions of analysis.[25]

Among other significant results, two are especially worth noting. In *The Consistency of the Axiom of Choice and of the Generalized Continuum Hypothesis with the Axioms of Set Theory* (1940, rev. ed., 1951), Gödel proved that if the Zermelo-Fraenkel system of axioms without the axiom of choice is consistent, then the system obtained by adjoining this axiom is consistent; that is, the axiom cannot be disproved. Likewise the continuum hypothesis that there is no cardinal number between $\aleph_0$ and $2^{\aleph_0}$, is consistent with the Zermelo-Fraenkel system (without the axiom of choice). In 1963 Paul J. Cohen (1934–), a professor of mathematics at Stanford University, proved[26] that the latter two axioms are independent of the Zermelo-Fraenkel system; that is, they cannot be proved on the basis of that system. Moreover, even if one retained the axiom of choice in the Zermelo-Fraenkel system, the continuum hypothesis could not be proved. These results imply that we are free to construct new systems of mathematics in which either or both of the two controversial axioms are denied.

All of the developments since 1930 leave open two major problems: to prove the consistency of unrestricted classical analysis and set theory, and to build mathematics on a strictly intuitionistic basis or to determine the limits of this approach. The source of the difficulties in both of these problems is infinity as used in infinite sets and infinite processes. This concept, which created problems even for the Greeks in connection with irrational numbers and which they evaded in the method of exhaustion, has been a subject of contention ever since and prompted Weyl to remark that mathematics is the science of infinity.

The question as to the proper logical basis for mathematics and the rise particularly of intuitionism suggest that, in a larger sense, mathematics has come full circle. The subject started on an intuitive and empirical basis. Rigor became a goal with the Greeks, and though more honored in the breach until the nineteenth century, it seemed for a moment to be achieved. But the efforts to pursue rigor to the utmost have led to an impasse in which there is no longer any agreement on what it really means. Mathematics remains alive and vital, but only on a pragmatic basis.

There are some who see hope for resolution of the present impasse. The French group of mathematicians who write under the pseudonym of Nicolas

25. *Math. Ann.*, 112, 1936, 493–565.
26. *Proceedings of the National Academy of Sciences*, 50, 1963, 1143–48; 51, 1964, 105–10.

Bourbaki, offer this encouragement:[27] "There are now twenty-five centuries during which the mathematicians have had the practice of correcting their errors and thereby seeing their science enriched, not impoverished; this gives them the right to view the future with serenity."

Whether or not the optimism is warranted, the present state of mathematics has been aptly described by Weyl:[28] "The question of the ultimate foundations and the ultimate meaning of mathematics remains open; we do not know in what direction it will find its final solution or even whether a final objective answer can be expected at all. 'Mathematizing' may well be a creative activity of man, like language or music, of primary originality, whose historical decisions defy complete objective rationalization."

Bibliography

Becker, Oskar: *Grundlagen der Mathematik in geschichtlicher Entwicklung*, Verlag Karl Alber, 1956, 317–401.

Beth, E. W.: *Mathematical Thought: An Introduction to the Philosophy of Mathematics*, Gordon and Breach, 1965.

Bochenski, I. M.: *A History of Formal Logic*, University of Notre Dame Press, 1962; Chelsea (reprint), 1970.

Boole, George: *An Investigation of the Laws of Thought* (1854), Dover (reprint), 1951.

————: *The Mathematical Analysis of Logic* (1847), Basil Blackwell (reprint), 1948.

————: *Collected Logical Works*, Open Court, 1952.

Bourbaki, N.: *Eléments d'histoire des mathématiques*, 2nd ed., Hermann, 1969, 11–64.

Brouwer, L. E. J.: "Intuitionism and Formalism," *Amer. Math. Soc. Bull.*, 20, 1913/14, 81–96. An English translation of Brouwer's inaugural address as professor of mathematics at Amsterdam.

Church, Alonzo: "The Richard Paradox," *Amer. Math. Monthly*, 41, 1934, 356–61.

Cohen, Paul J., and Reuben Hersh: "Non-Cantorian Set Theory," *Scientific American*, Dec. 1967, 104–16.

Couturat, L.: *La Logique de Leibniz d'après des documents inédits*, Alcan, 1901.

De Morgan, Augustus: *On the Syllogism and Other Logical Writings*, Yale University Press, 1966. A collection of his papers edited by Peter Heath.

Dresden, Arnold: "Brouwer's Contribution to the Foundations of Mathematics," *Amer. Math. Soc. Bull.*, 30, 1924, 31–40.

Enriques, Federigo: *The Historic Development of Logic*, Henry Holt, 1929.

Fraenkel, A. A.: "The Recent Controversies About the Foundations of Mathematics," *Scripta Mathematica*, 13, 1947, 17–36.

Fraenkel, A. A., and Y. Bar-Hillel: *Foundations of Set Theory*, North-Holland, 1958.

Frege, Gottlob: *The Foundations of Arithmetic*, Blackwell, 1953, English and German; also English translation only, Harper and Bros., 1960.

————: *The Basic Laws of Arithmetic*, University of California Press, 1965.

27. *Journal of Symbolic Logic*, 14, 1949, 2–8.
28. *Obituary Notices of Fellows of the Royal Soc.*, 4, 1944, 547–53 = *Ges. Abh.*, 4, 121–29, p. 126 in part.

Gerhardt, C. I., ed.: *Die philosophischen Schriften von G. W. Leibniz*, 1875–80, Vol. 7.

Gödel, Kurt: *On Formally Undecidable Propositions of* Principia Mathematica *and Related Systems*, Basic Books, 1965.

————: "What Is Cantor's Continuum Problem?," *Amer. Math. Monthly*, 54, 1947, 515–25.

————: *The Consistency of the Axiom of Choice and of the Generalized Continuum Hypothesis with the Axioms of Set Theory*, Princeton University Press, 1940; rev. ed., 1951.

Kneale, William and Martha: *The Development of Logic*, Oxford University Press, 1962.

Kneebone, G. T.: *Mathematical Logic and the Foundations of Mathematics*, D. Van Nostrand, 1963. See the Appendix especially on developments since 1939.

Leibniz, G. W.: *Logical Papers*, edited and translated by G. A. R. Parkinson, Oxford University Press, 1966.

Lewis, C. I.: *A Survey of Symbolic Logic*, Dover (reprint), 1960, pp. 1–117.

Meschkowski, Herbert: *Probleme des Unendlichen, Werk und Leben Georg Cantors*, F. Vieweg und Sohn, 1967.

Mostowski, Andrzej: *Thirty Years of Foundational Studies*, Barnes and Noble, 1966.

Nagel, E., and J. R. Newman: *Gödel's Proof*, New York University Press, 1958.

Poincaré, Henri: *The Foundations of Science*, Science Press, 1946, 448–85. This is a reprint in one volume of *Science and Hypothesis, The Value of Science*, and *Science and Method*.

Rosser, J. Barkley: "An Informal Exposition of Proofs of Gödel's Theorems and Church's Theorem," *Journal of Symbolic Logic*, 4, 1939, 53–60.

Russell, Bertrand: *The Principles of Mathematics*, George Allen and Unwin, 1903; 2nd ed., 1937.

Scholz, Heinrich: *Concise History of Logic*, Philosophical Library, 1961.

Styazhkin, N. I.: *History of Mathematical Logic from Leibniz to Peano*, Massachusetts Institute of Technology Press, 1969.

Van Heijenoort, Jean: *From Frege to Gödel*, Harvard University Press, 1967. Translations of key papers on logic and the foundations of mathematics.

Weyl, Hermann: "Mathematics and Logic," *Amer. Math. Monthly*, 53, 1946, 2–13 = *Ges. Abh.*, 4, 268–79.

————: *Philosophy of Mathematics and Natural Science*, Princeton University Press, 1949.

Wilder, R. L.: "The Role of the Axiomatic Method," *Amer. Math. Monthly*, 74, 1967, 115–27.

Abbreviations

Journals whose titles have been written out in full in the text are not listed here.

Abh. der Bayer. Akad. der Wiss. Abhandlungen der Königlich Bayerischen Akademie der Wissenschaften (München)

Abh. der Ges. der Wiss. zu Gött. Abhandlungen der Königlichen Gesellschaft der Wissenschaften zu Göttingen

Abh. König. Akad. der Wiss., Berlin Abhandlungen der Königlich Preussischen Akademie der Wissenschaften zu Berlin

Abh. Königlich Böhm. Ges. der Wiss. Abhandlungen der Königlichen Böhmischen Gesellschaft der Wissenschaften

Abh. Math. Seminar der Hamburger Univ. Abhandlungen aus dem Mathematischen Seminar Hamburgischen Universität

Acta Acad. Sci. Petrop. Acta Academiae Scientiarum Petropolitanae

Acta Erud. Acta Eruditorum

Acta Math. Acta Mathematica

Acta Soc. Fennicae Acta Societatis Scientiarum Fennicae

Amer. Jour. of Math. American Journal of Mathematics

Amer. Math. Monthly American Mathematical Monthly

Amer. Math. Soc. Bull. American Mathematical Society, Bulletin

Amer. Math. Soc. Trans. American Mathematical Society, Transactions

Ann. de l'Ecole Norm. Sup. Annales Scientifiques de l'Ecole Normale Supérieure

Ann. de Math. Annales de Mathématiques Pures et Appliquées

Ann. Fac. Sci. de Toulouse Annales de la Faculté des Sciences de Toulouse

Ann. Soc. Sci. Bruxelles Annales de la Société Scientifique de Bruxelles

Annali di Mat. Annali di Matematica Pura ed Applicata

Annals of Math. Annals of Mathematics

Astronom. Nach. Astronomische Nachrichten

Atti Accad. Torino Atti della Reale Accademia delle Scienze di Torino

Atti della Accad. dei Lincei, Rendiconti Atti della Reale Accademia dei Lincei, Rendiconti

Brit. Assn. for Adv. of Sci. British Association for the Advancement of Science

Bull. des Sci. Math. Bulletin des Sciences Mathématiques

Bull. Soc. Math. de France Bulletin de la Société Mathématique de France

Cambridge and Dublin Math. Jour. Cambridge and Dublin Mathematical Journal

Comm. Acad. Sci. Petrop. Commentarii Academiae Scientiarum Petropolitanae

Comm. Soc. Gott. Commentationes Societatis Regiae Scientiarum Gottingensis Recentiores

Comp. Rend. Comptes Rendus

Corresp. sur l'Ecole Poly. Correspondance sur l'Ecole Polytechnique

Encyk. der Math. Wiss. Encyklopädie der Mathematischen Wissenschaften

Gior. di Mat. Giornale di Matematiche

Hist. de l'Acad. de Berlin Histoire de l'Académie Royale des Sciences et des Belles-Lettres de Berlin

Hist. de l'Acad. des Sci., Paris Histoire de l'Académie Royale des Sciences avec les Mémoires de Mathématique et de Physique

Jahres. der Deut. Math.-Verein. Jahresbericht der Deutschen Mathematiker-Vereinigung

Jour. de l'Ecole Poly. Journal de l'Ecole Polytechnique

Jour. de Math. Journal de Mathématiques Pures et Appliquées

Jour. des Sçavans Journal des Sçavans

Jour. für Math. Journal für die Reine und Angewandte Mathematik

Jour. Lon. Math. Soc. Journal of the London Mathematical Society

Königlich Sächsischen Ges. der Wiss. zu Leipzig Berichte über die Verhandlungen der Königlich Sächsischen Gesellschaft der Wissenschaften zu Leipzig

Math. Ann. Mathematische Annalen

Mém. de l'Acad. de Berlin See *Hist. de l'Acad. de Berlin*

Mém. de l'Acad. des Sci., Paris See *Hist. de l'Acad. des Sci., Paris*; after 1795, Mémoires de l'Académie des Sciences de l'Institut de France

Mém. de l'Acad. Sci. de St. Peters. Mémoires de l'Académie Impériale des Sciences de Saint-Petersbourg

Mém. des sav. étrangers Mémoires de Mathématique et de Physique Présentés à l'Académie Royal des Sciences, par Divers Sçavans, et Lus dans ses Assemblées

Mém. divers Savans See *Mém. des sav. étrangers*

Misc. Berolin. Miscellanea Berolinensia; also as *Hist. de l'Acad. de Berlin (q.v.)*

Misc. Taur. Miscellanea Philosophica-Mathematica Societatis Privatae Taurinensis (published by Accademia delle Scienze di Torino)

Monatsber. Berliner Akad. Monatsberichte der Königlich Preussischen Akademie der Wissenschaften zu Berlin

N.Y. Math. Soc. Bull. New York Mathematical Society, Bulletin

Nachrichten König. Ges. der Wiss. zu Gött. Nachrichten von der Königlichen Gesellschaft der Wissenschaften zu Göttingen

Nou. Mém. de l'Acad. Roy. des Sci., Bruxelles Nouveaux Mémoires de l'Académie Royale des Sciences, des Lettres, et des Beaux-Arts de Belgique

Nouv. Bull. de la Soc. Philo. Nouveau Bulletin de la Société Philomatique de Paris

Nouv. Mém. de l'Acad. de Berlin Nouveaux Mémoires de l'Académie Royale des Sciences et des Belles-Lettres de Berlin

Nova Acta Acad. Sci. Petrop. Nova Acta Academiae Scientiarum Petropolitanae

Nova Acta Erud. Nova Acta Eruditorum

Novi Comm. Acad. Sci. Petrop. Novi Commentarii Academiae Scientiarum Petropolitanae

Phil. Mag. The Philosophical Magazine

Philo. Trans. Philosophical Transactions of the Royal Society of London

Proc. Camb. Phil. Soc. Cambridge Philosophical Society, Proceedings

Proc. Edinburgh Math. Soc. Edinburgh Mathematical Society, Proceedings

Proc. London Math. Soc. Proceedings of the London Mathematical Society

Proc. Roy. Soc. Proceedings of the Royal Society of London

Proc. Royal Irish Academy Proceedings of the Royal Irish Academy

Quart. Jour. of Math. Quarterly Journal of Mathematics

Scripta Math. Scripta Mathematica

Sitzungsber. Akad. Wiss zu Berlin Sitzungsberichte der Königlich Preussischen Akademie der Wissenschaften zu Berlin

Sitzungsber. der Akad. der Wiss., Wien Sitzungsberichte der Kaiserlichen Akademie der Wissenschaften zu Wien. Mathematisch-Naturwissenschaftlichen Klasse

Trans. Camb. Phil. Soc. Cambridge Philosophical Society, Transactions

Trans. Royal Irish Academy Transactions of the Royal Irish Academy

Zeit. für Math. und Phys. Zeitschrift für Mathematik und Physik

Zeit. für Physik Zeitschrift für Physik

Indexes

Name Index

Subject Index